Pearson

Get the Most Out of
MyMathLab®

MyMathLab, Pearson's online homework, tutorial, and assessment program, creates personalized experiences for students and provides powerful tools for instructors. With a wealth of tested and proven resources, each course can be tailored to fit your specific needs.

Learning in any Environment

- Because classroom formats and student needs continually change and evolve, MyMathLab has built-in flexibility to accommodate various course designs and formats.

- With an updated and streamlined design, students and instructors can access MyMathLab from most mobile devices.

Personalized Learning

Not every student learns the same way or at the same rate. Thanks to our advances in adaptive learning technology capabilities, you no longer have to teach as if they do.

- MyMathLab's **adaptive Study Plan** acts as a personal tutor, updating in real-time based on student performance throughout the course to provide personalized recommendations for practice. You can now assign the Study Plan as a prerequisite to a test or quiz with **Companion Study Plan Assignments**.

- MyMathLab can **personalize homework assignments** for students based on their performance on a test or quiz. This way, students can focus on just the topics they have not yet mastered.

- **MyMathLab with Integrated Review courses**—available with selected titles for Developmental Mathematics through Calculus—can be used for just-in-time prerequisite review or for co-requisite courses. These courses provide videos on review topics, along with pre-made, assignable skill-review quizzes and personalized homework assignments integrated throughout your regular MyMathLab course content.

Visit www.mymathlab.com and click Get Trained to make sure you're getting the most out of MyMathLab.

Get the Most Out of Your Developmental Math MyMathLab Courses

Offer Step-by-Step Exercise Support

New Workspace Assignments allow students to naturally write out their work, step-by-step, and receive instant feedback at each step. Specific, relevant hints and videos are available to guide students through the steps. Each student's work is captured in the MyMathLab gradebook so you can easily pinpoint exactly where you need to focus your instruction.

Get Students Engaged

Learning Catalytics™—a student response tool that uses students' smartphones, tablets, or laptops to engage them in more interactive tasks and thinking— is available through any MyMathLab course to foster student engagement and peer-to-peer learning. You can generate class discussion, guide your lecture, and promote peer-to-peer learning with real-time analytics.

Provide Adaptive Just-in-Time Practice

When students struggle on an exercise, Skill Builder Assignments provide just-in-time, targeted support to help them build up the requisite skills needed to complete their homework assignment. As students progress, these assignments adapt to provide support exercises that are personalized to each student's activity and performance throughout the course.

Use Data to Tailor Your Course

A comprehensive gradebook with enhanced reporting functionality helps you efficiently manage your course. The new Reporting Dashboard presents student performance data at the class, section, and program levels in an accessible, visual manner. Item Analysis allows you to track class-wide understanding at the exercise level, so that you can refine your lectures or assignments to address just-in-time student needs.

Intermediate Algebra

Fourth Edition

Michael Sullivan, III
Joliet Junior College

Katherine R. Struve
Columbus State Community College

with contributions by

Jessica Bernards
Portland Community College

Wendy Fresh
Portland Community College

 Pearson

VP, Courseware Portfolio Management: Chris Hoag
Director, Courseware Portfolio Management: Michael Hirsch
Courseware Portfolio Manager: Mary Beckwith
Courseware Portfolio Management Assistant: Alison Oehman
Content Producer: Tamela Ambush
Managing Producer: Karen Wernholm
Producer: Audra Walsh
Manager, Courseware QA: Mary Durnwald
Manager, Content Development: Rebecca Williams
Product Marketing Manager: Alicia Frankel
Field Marketing Managers: Jennifer Crum and Lauren Schur
Product Marketing Assistant: Hanna Lafferty
Senior Author Support/Technology Specialist: Joe Vetere
Manager, Rights and Permissions: Gina Cheselka
Manufacturing Buyer: Carol Melville, LSC Communications
Associate Director of Design: Blair Brown
Text Design, Cover Design, and Composition: Cenveo Publisher Services
Cover Images: (grass) Val Lawless/Shutterstock; (cap) Angelo Gilardelli/Shutterstock;
    (duck tape) Mexrix/Shutterstock

Library of Congress Cataloging-in-Publication Data

Names: Sullivan, Michael, III, 1967– | Struve, Katherine R.
Title: Intermediate algebra / Michael Sullivan, III, Joliet Junior College,
    Katherine R. Struve, Columbus State Community College.
Description: 4th edition. | Boston : Pearson, [2018]
Identifiers: LCCN 2016020915| ISBN 9780134555805 (hardcover) | ISBN
    0134555805 (hardcover) | ISBN 9780134556789 (hardcover) | ISBN 013455678X
    (hardcover)
Subjects: LCSH: Algebra—Textbooks.
Classification: LCC QA154.3 .S785 2018 | DDC 512.9—dc23
LC record available at https://lccn.loc.gov/2016020915

2   17

ISBN 13: 978-0-13-455580-5
ISBN 10: 0-13-455580-5

To my children, Michael, Kevin, and Marissa.
—Michael Sullivan

To my Family: Judy, Ruth, Leon, David, Patti, Terry, Madison, Mackenizie, Morgan, and Tegan
—Katherine R. Struve

About the Authors

With training in mathematics, statistics, and economics, Michael Sullivan, III has a varied teaching background that includes 27 years of instruction in both high school and college-level mathematics. He is currently a full-time professor of mathematics at Joliet Junior College. Michael has numerous textbooks in publication, including an Introductory Statistics series and a Precalculus series, which he writes with his father, Michael Sullivan.

Michael believes that his experiences writing texts for college-level math and statistics courses give him a unique perspective as to where students are headed once they leave the developmental mathematics tract. This experience is reflected in the philosophy and presentation of his developmental text series. When not in the classroom or writing, Michael enjoys spending time with his three children, Michael, Kevin, and Marissa, and playing golf. Now that his two sons are getting older, he has the opportunity to do both at the same time!

Kathy Struve has been a classroom teacher for nearly 35 years, first at the high school level and, for the past 27 years, at Columbus State Community College. Kathy embraces classroom diversity: diversity of students' age, learning styles, and previous learning success. She is aware of the challenges of teaching mathematics at a large, urban community college, where students have varied mathematics backgrounds and may enter college with a high level of mathematics anxiety.

Kathy served as Lead Instructor of the Developmental Algebra sequence at Columbus State, where she developed curriculum, conducted workshops, and provided leadership to adjunct faculty in the mathematics department. She embraces the use of technology in instruction, and has taught web and hybrid classes in addition to traditional face-to-face emporium-style classes. She is always looking for ways to more fully involve students in the learning process. In her spare time Kathy enjoys spending time with her two adult daughters, her four granddaughters, and biking, hiking, and traveling with her husband.

Contents

Preface

We would like to thank the reviewers, class testers, and users of the third edition of *Intermediate Algebra* who helped to make the book an overwhelming success. Their thoughtful comments and suggestions provided strong guidance for improvements to the fourth edition that we believe will enhance this solid, student-friendly text.

Intermediate Algebra is a gateway course to other college-level mathematics courses. The goal of the course is to provide students with the mathematical skills that are prerequisites for courses such as College Algebra, Elementary Statistics, Liberal-Arts Math, and Mathematics for Teachers. In addition, Intermediate Algebra must expose students to a variety of mathematical concepts that build on one another and that range from the basics such as linear equations to sophisticated concepts such as exponential functions.

Of particular importance in this course are rigor and mathematical thinking. It is imperative that the coverage be sufficiently rigorous to teach students how to study math successfully. At the same time, the course must develop students' ability to think mathematically. The rigor in the course exists both in the material presented (such as a more thorough development of functions) and in the array of problems and examples. As a result, it should be clear to students that this course is not simply a rehash of Elementary Algebra.

Most students have seen at least some of the content of this course at some point in their high school careers or in other college coursework. For some students, success at studying and facility with math concepts did not develop during their previous contact with the material, and they need a fresh start. In addition, the number of nontraditional students who have lost some of their math skills over the course of time continues to grow, especially at community colleges. Nontraditional students often are highly motivated to succeed because of their life experiences, yet they may be rusty at the business of "going to school." For nontraditional students, this course refreshes and reinforces their study skills as well as their mathematical skills.

To address the many needs and the diversity of today's Intermediate Algebra students, we have been guided by the following ideas as broad goals for this text:

- Provide the student with a strong conceptual foundation in mathematics through a clear, comprehensive presentation of topics with a special emphasis on functions.

- Present a variety of innovative pedagogical features, tools, study tips, and easy-to use aids to help students see the value of the text as an important resource and guide that will increase their success in the course.

- Provide comprehensive exercise sets with paired exercises that build problem-solving skills, show a variety of applications of mathematics, and reinforce mathematical concepts for students.

- Streamline the Intermediate Algebra course through the strategic placement of topics that will provide instructors with the flexibility to review material, as needed, instead of reteaching it.

New to the Fourth Edition

The revision of this text takes advantage of MyMathLab as a tool for learning. To address the needs of students who are exposed to the material almost exclusively through MyMathLab, we have introduced the following new features based on some of the hallmark features of the text to the program.

- Discovery activities using **applets** have been developed. These explorations are carefully crafted to allow students to develop understanding of mathematical concepts through experiential learning. The applets and guided exercises written that utilize the applets may be found in MyMathLab. The applets may also be accessed using the QR code at the beginning of the section.

- **Guided Exercises** are now available in MyMathLab based on the popular Showcase Examples. Showcase Examples from the text are easy to recognize with the words "How To" in the example title and provide step-by-step solutions to examples. This example structure was written into 55 *new* MyMathLab exercises that require students to respond to questions as the steps to solving problems are developed, similar to the "Help Me Solve This" feature of MyMathLab. This keeps the student completely engaged in the learning process and develops their conceptual understanding of the content. These exercises are easy to identify in the Assignment Builder as they are designated "How-To-#.# Ex#-<title abbreviation>. For example, "How-To-1.5 Ex3 How To Graph an Equation."

- **Quick Response (QR) codes** now appear at each section opener, at section-level exercises, and as part of the Chapter Tests. Students can simply use a QR scanner from their smart phone for easy access to the popular Author in Action lecture videos, select end of section exercise videos, the discovery Applets, and the Chapter Test Prep videos.

- The authors developed a **Premade Author Created MyMathLab** course that utilizes all the new MyMathLab features. Each section has two MyMathLab assignments.

  - The first assignment is a multimedia assignment that incorporates the Author in Action lecture videos, the new applet discovery exercises, the new "How To" guided exercises, and the Quick Check exercises from the text. Recall, the Quick Check exercises follow many of the examples in the text. To assist students in utilizing the text, the Textbook learning aid for each Quick Check exercise will link *directly* to the corresponding example in the text. All learning aids, with the exception of "View an Example", will be available for this portion of the homework. Our experience as instructors has been that too many students rely on this learning aid while doing homework, thereby reducing the effect of homework as students simply mimic the View an Example content.

  - The second assignment is based on the Skill Building and Mixed Practice exercises from the text. Skill building exercises are tied to objectives within the text, so the Textbook learning aid will link directly to the objective within the section. The idea is to reduce the amount of guidance provided to the student (compared with Quick Check exercises) so they are more responsible for identifying the problem type. The Mixed Practice exercises are based on multiple concepts learned within the section or text, so the Textbook learning aid is linked to the section. The student must determine the problem type based on Quick Check and Skill Building exercise experience. The "View an Example" learning aid is disabled for this exercise set as well. Because this text has Skill Builder available in MyMathLab, you may consider reducing the number of exercises in the second assignment. By checking the Skill Builder box, the assignments will adapt to provide support exercises personalized to each student's needs.

Build a Strong Foundation Through a Functions Approach

The approach that we take in Intermediate Algebra is that the function is the overriding theme of the text. The reason for this stress on functions is twofold. First, Intermediate Algebra is not a terminal course but rather a gateway to the future, and functions form the basis for much study in mathematics. The introduction of functions helps make the "jump" from Intermediate Algebra to College Algebra less severe because students feel more comfortable with functions and function notation. Second, today's students like to learn in context so that they can see the relevancy of the material. The function provides a great way to present the usefulness of the material we are teaching.

Develop an Effective Text for Use In and Out of the Classroom

Given the hectic lives led by most students, coupled with the anxiety and trepidation with which they approach this course, an outstanding developmental mathematics text must provide pedagogical support that makes the text valuable to students as they study and do assignments. Pedagogy must be presented within a framework that teaches students how to study math; pedagogical devices must also address what students see as the "mystery" of mathematics—and solve that mystery.

To encourage students and to clarify the material, we developed a set of pedagogical features that help students develop good study skills, garner an understanding of the connections between topics, and work smarter in the process. The pedagogy used is based upon the more than 70 years of classroom teaching experience that the authors bring to this text.

Examples are often the determining factor in how valuable a textbook is to a student. Students look to examples to provide them with guidance and instruction when they need it most—the times when they are away from the instructor and the classroom. We have developed several example formats in an attempt to provide superior guidance and instruction for students. The formats include:

Innovative Sullivan/Struve Examples

The innovative *Sullivan/Struve Example* has a two-column format in which annotations are provided to the **left** of the algebra, rather than the right, as is the practice in most texts. Because we read from **left to right,** placing the annotation on the left will make more sense to the student. It becomes clear that the annotation describes what we are about to do instead of what was just done. The annotations may be thought of as the teacher's voice offering clarification immediately before writing the next step in the solution on the board. Consider the following:

EXAMPLE 6 **Solving Linear Inequalities**

Solve the inequality: $x - 4 \geq 5x + 12$

Solution
$$x - 4 \geq 5x + 12$$

Add 4 to both sides: $x - 4 + 4 \geq 5x + 12 + 4$

$$x \geq 5x + 16$$

Subtract 5x from both sides: $x - 5x \geq 5x + 16 - 5x$

$$-4x \geq 16$$

Divide both sides by -4. Don't forget to change the direction of the inequality:
$$\frac{-4x}{-4} \leq \frac{16}{-4}$$

$$x \leq -4$$

The solution using set-builder notation is $\{x \mid x \leq -4\}$. The solution using interval notation is $(-\infty, -4]$. See Figure 20 for the graph of the solution set. ●

Figure 20

> **Quick ✓**
> *In Problems 20–22, solve each linear inequality. Write the solution using set-builder notation and interval notation. Graph the solution set.*
> **20.** $3x + 1 > x - 5$ **21.** $-2x + 1 \leq 3x + 11$
>
> **22.** $-5x + 12 < x - 3$

Showcase Examples

Showcase Examples are used strategically to introduce key topics or important problem solving techniques. These examples provide "how-to" instruction by offering a guided, step-by-step approach to solving a problem. Students can then immediately see how each of the steps is employed. We remind students that the *Showcase Example* is meant to provide "how-to" instruction by including the words "how to" in the example title. The *Showcase Example* has a three-column format in which the left column describes a step, the middle column provides a brief annotation, as needed, to explain the step, and the right column presents the algebra. With this format, students can see each step in the problem-solving process in context so that the steps make more sense. This approach is more effective than simply stating each step in the text.

(**EXAMPLE 9**)   **How to Solve an Inequality Involving** $>$

Solve the inequality $|2x - 5| > 3$. Express the solution set in set-builder notation and interval notation. Graph the solution set.

Step-by-Step Solution

Step 1: The inequality is in the form $|u| > a$, where $u = 2x - 5$ and $a = 3$. Rewrite the inequality as a compound inequality that does not involve absolute value.

$$|2x - 5| > 3$$
$|u| > a$ means $u < -a$ or $u > a$: $2x - 5 < -3$ or $2x - 5 > 3$

Step 2: Solve each inequality separately.

	$2x - 5 < -3$		$2x - 5 > 3$
Add 5 to both sides:	$2x < 2$	Add 5 to both sides:	$2x > 8$
Divide both sides by 2:	$x < 1$	Divide both sides by 2:	$x > 4$

Step 3: Find the union of the solution sets of each inequality.

The solution set is $\{x \mid x < 1 \text{ or } x > 4\}$ or, using interval notation, $(-\infty, 1) \cup (4, \infty)$. See Figure 47 for the graph of the solution set.

Figure 47

●

Work Smart

$$|u| > a$$
CANNOT be written as
$$-a > u > a$$

Quick ✓

34. $|x - 9| > 6$ is equivalent to $x - 9 > 6$ or $x - 9 \_\_\_ -6$.

In Problems 35–40, solve each inequality. Express the solution set in set-builder notation and interval notation. Graph the solution set.

35. $|x + 3| > 4$ **36.** $|4x - 3| \geq 5$

37. $|-3x + 2| > 7$ **38.** $|2x + 5| - 2 > -2$

39. $|6x - 5| \geq 0$ **40.** $|2x + 1| > -3$

Quick Check Exercises

Placed at the conclusion of most examples, the *Quick Check* exercises provide students with an opportunity for immediate reinforcement. By working the problems that mirror the example just presented, students get instant feedback and gain confidence in their understanding of the concept. All *Quick Check* exercises answers are provided in the back of the text. The *Quick Check* exercises should be assigned as homework to encourage students to read, consult, and use the text regularly.

Superior Exercise Sets: Paired with Purpose

Students learn algebra by doing algebra. The superior end-of-section exercise sets in this text provide students with ample practice of both procedures and concepts. The exercises are paired and present problem types with every possible derivative. The exercises also present a gradual increase in difficulty level. The early, basic exercises keep the student's focus on as few "levels of understanding" as possible. The later or higher-numbered exercises are "multi-task" (or Mixed Practice) exercises where students are required to utilize multiple skills, concepts, or problem-solving techniques.

Throughout the textbook, the exercise sets are grouped into nine categories—some of which appear only as needed:

1. **Are You Prepared For This Section?** problems are located at the opening of the section. They are problems that deal with prerequisite material for the section along with page references so students may remediate, if necessary. Answers to the Prepared? . . . problems appear as a footnote on the page.

2. **Quick Check** exercises, which provide the impetus to get students into the text, follow most examples and are numbered sequentially as the first problems in each section exercise set. By doing these problems as homework and the first exercises attempted, the student is directed into the material in the section. If a student gets stuck, he or she will learn that the example immediately preceding the Quick Check exercise illustrates the concepts needed to solve the problem.

3. **Building Skills** exercises are drill problems that develop the student's understanding of the procedures and skills in working with the methods presented in the section. These exercises can be linked back to a single objective in the section. Notice that the Building Skills problems begin the numbering scheme where the Quick Checks leave off. For example, if the last Quick Check exercise is Problem 20, then we begin the Building Skills exercises with Problem 21. This serves as a reminder that Quick Check exercises should be assigned as homework.

4. **Mixed Practice** exercises are also drill problems, but they offer a comprehensive assessment of the skills learned in the section by asking problems that relate to more than one concept or objective. In addition, we may present problems from previous sections so students must first recognize the type of problem and then employ the appropriate technique to solve the problem.

5. **Applying the Concepts** exercises are problems that allow students to see the relevance of the material learned within the section. Problems in this category either are situational problems that use material learned in the section to solve "real-world" problems or are problems that ask a series of questions to enhance a student's conceptual understanding of the mathematics presented in the section.

6. **Extending the Concepts** exercises can be thought of as problems that go beyond the basics. Within this block of exercises an instructor will find a variety of problems to sharpen students' critical-thinking skills.

7. **Explaining the Concepts** problems require students to think about the big picture concepts of the section and express these ideas in their own words. It is our belief that students need to improve their ability to communicate complicated ideas both orally and in writing. When they are able to explain mathematical methods or concepts to another individual, they have truly mastered the ideas. These problems can serve as a basis for classroom discussion or can be used as writing assignments.

8. Starting with Chapter 5, we provide **Synthesis Review** exercises to help students grasp the "big picture" of algebra—once they have a sufficient conceptual foundation to build upon from their work in Chapters R through 4. Synthesis Review exercises ask students to perform a single operation (adding, solving, and so on) on several objects (polynomials, rational expressions, and so on). The student is then asked to discuss the similarities and differences in performing the same operation on the different objects.

9. Finally, we include **technology exercises.** Instructors' philosophies about the use of graphing technology, such as graphing calculators, GeoGebra, or Desmos to solve problems vary considerably. Because instructors disagree about the value of these tools, we have made an effort to make graphing technology entirely optional. When appropriate, technology exercises are included at the close of a section's exercise set. Also included in the technology exercises are the new applet explorations. The applets may be found in MyMathLab or using the Quick Response (QR) code located in the section opener ribbon.

Problem Icons In addition to the carefully structured categories of exercises, selected problems are flagged with icons to denote that:

- Problems whose number is green have complete worked-out solutions found in MyMathLab.
- △ These problems focus on geometry concepts.

Hallmark Features of Sullivan/Struve

Author in Action Videos

The Author in Action videos are videos of the authors presenting the content. Most of the videos are from the authors' actual classroom lectures. This makes the videos authentic and gives the viewer the sense of participating in the lecture. The videos are tied to the objectives and under 12 minutes in length. For those objectives that require more than 12 minutes, we have multiple videos. Students are alerted to the availability of a video with the ⊳ icon. The videos are available in MyMathLab, the Multimedia Textbook (in MyMathLab), or through a Quick Response (QR) code located in the section opener ribbon. The videos are captioned in English and Spanish.

Video Notebook

A video notebook is available, which is ideal for online, emporium/redesign courses, or inverted (flipped) classrooms. This video notebook assists students in taking thorough, organized, and understandable notes as they watch the Author in Action videos by asking students to complete definitions, procedures, and examples based on the content of the videos.

Streamlining Intermediate Algebra: *Getting Ready for Chapter . . .* Review Sections

To maintain the pace of the course, we created several **_Getting Ready_** sections that review material taught in Elementary Algebra courses. The *Getting Ready* sections are designed to allow students to brush up on topics and skills as needed before beginning the chapters in the Intermediate Algebra text where the skills will be used or further developed. These optional, yet integrated, sections provide the student with timely review. They also streamline the Intermediate Algebra course by providing the instructors with the flexibility to decide if the *Getting Ready* sections should be covered in their entirety, briefly reviewed, or skipped, depending upon the needs of their students. *Getting Ready* review sections have been placed before Chapters 4, 5, and 6 in the text.

Quick Check Exercises: Encourage Study Skills that Lead to Independent Learning

What is one of the overarching goals of an education? We believe it is to learn to solve problems independently. In particular, we would like to see students develop the ability to pick up a text or manual and teach themselves the skills they need. In our mathematics classes, however, we are often frustrated because students rarely read the text and often struggle to understand the concepts independently.

To encourage students to use the text more effectively and to help them achieve greater success in the course, we have structured the exercises in the fourth edition of our text differently from other mathematics textbooks. The aim of this structure is to get students "into the text" in order to increase their ability and confidence to work any math problem—particularly when they are away from the classroom and an instructor who can help.

Each section's exercise set begins with *Quick Check* exercises. The *Quick Checks* are consecutively numbered. The end-of-section exercises begin their numbering scheme based on where the *Quick Checks* end. For example:

- Section 1.1: *Quick Checks* end at Problem 30, so the end-of-section exercise set starts with Problem 31 (see page 57).
- Section 1.2: *Quick Checks* end at Problem 30, so the end-of-section exercise set starts with Problem 31 (see page 70).

The *Quick Checks* follow most examples and provide the platform for students to get "into the text." By integrating these exercises into the exercise set, we direct students to the instructional material in that section. Our hope is that students will then become more aware of the instructional value of the text and will be more likely to succeed when studying away from the classroom and the instructor.

Answer annotations to Quick Checks and exercises have been placed directly next to each problem in the Annotated Instructor's Edition to make it easier for instructors to create assignments.

We have used the same background color for the Quick Checks and the exercise sets to reinforce the connection between them visually. The colored background will also make the Quick Checks easier to find on the page.

Answers to Selected Exercises at the back of the text integrate the answers to *every* Quick Check exercise with the answers to *every odd* problem from the section exercise sets.

Study Skills and Student Success

We have included study skills and student success as regular themes throughout this text starting with *Section R.1, Success in Mathematics*. In addition to this dedicated section that covers many of the basics that are essential to success in any math course, we have included several recurring study aids that appear in the margin. These features were designed to anticipate the student's needs and to provide immediate help—as if the teacher were looking over his or her shoulder. These margin features include *In Other Words; Work Smart;* and *Work Smart: Study Skills*.

Section R.1: *Success in Mathematics* focuses the student on basic study skills, including what to do during the first week of the term; what to do before, during, and after class; how to use the text effectively; and how to prepare for an exam.

In Other Words helps to address the difficulty that students have in reading mathematically precise definitions and theorems by explaining them in easier to understand language.

Work Smart provides "tricks of the trade" hints, tips, reminders, and alerts. It also identifies some common errors to avoid and helps students work more efficiently.

Work Smart: Study Skills reminds students of study skills that will help them to succeed at various points in the course. Attention to these practices will help them to become better, more proficient learners.

Test Preparation and Student Success

The Chapter Tests in this text and the companion Chapter Test Prep Videos have been designed to help students make the most of their valuable study time.

Chapter Test In preparation for their classroom test, students should take the practice test to make sure they understand the key topics in the chapter. The exercises in the Chapter Tests have been crafted to reflect the level and types of exercises a student is likely to see on a classroom test.

Chapter Test Prep Video   The Chapter Test Prep Videos provide students with help at the critical juncture when they are studying for a test. The videos present step-by-step solutions to the exact exercises found in each of the book's Chapter Tests. Easy video navigation allows students instant access to the worked-out solutions to the exercises they want to study or review. These videos are available in MyMathLab or may be accessed using the QR code in the Chapter Test ribbon.

How It All Fits Together: The Big Picture

Another important role of the pedagogy in this text is to help students see and understand the connection among the mathematical topics being presented. Several section-opening and margin features help to reinforce connections:

The Big Picture: Putting It Together **(Chapter Opener)**   This feature is based on how we start each chapter in the classroom—with a quick sketch of what we plan to cover. Before tackling a chapter, we tie concepts and techniques together by summarizing material covered previously and then relate these ideas to material we are about to discuss. It is important for students to understand that content truly builds from one chapter to the next. We find that students need to be reminded that the familiar operations of addition, subtraction, multiplication, and division are being applied to different or more complex objects.

Are You Prepared for This Section?   As part of this building process, we think it is important to remind students of specific concepts or skills that they will need from earlier in the course to be successful within a given section. The *Are You Prepared?* feature that begins each section not only provides a list of prerequisite skills that a student should understand before tackling the content of a new section but also offers a short set of problems to test students' preparedness. Answers to the problems are provided in a footnote on the same page, and a cross-reference to the material in the text is provided so that the student can remediate when necessary.

Putting the Concepts Together **(Mid-Chapter Review)**   Each chapter has a group of exercises at the appropriate point in the chapter, entitled *Putting the Concepts Together*. These exercises serve as a review—synthesizing material introduced up to that point in the chapter. The exercises in these mid-chapter reviews are carefully chosen to assist students in seeing the "big picture."

Synthesis Review Exercises   Starting with Chapter 5, we provide Synthesis Review exercises to help students grasp the "big picture" of algebra—once they have a sufficient conceptual foundation to build upon from their work in Chapters R through 4.

Cumulative Review   Learning algebra is a building process, and building involves considerable reinforcement. The Cumulative Review exercises at the end of each odd-numbered chapter, starting with Chapter 1, help students reinforce and solidify their knowledge by revisiting concepts and using them in context. This way, studying for the final exam should be fairly easy. Cumulative Reviews for each even-numbered chapter can be found on the Instructor's Resource Center. Answers to all cumulative review problems appear in the back of the text.

In Closing

When we started writing this textbook, we discussed improvements we could make in coverage; in staples such as examples and problems; and in any pedagogical features that we found truly useful. After writing and rewriting, and reading many thoughtful reviews from instructors, we focused on the following features of the text to set it apart.

- **Functions**   are introduced early and revisited often throughout the course. This integration helps prepare students for the quantitative courses that they will take after Intermediate Algebra.

- The innovative ***Sullivan/Struve Examples*** and ***Showcase Examples*** provide students with superior guidance and instruction when they need it most—when they are away from the instructor and the classroom. Each of the margin features

In Other Words, Work Smart, and *Work Smart: Study Skills* are designed to improve study skills, make the textbook easier to navigate, and increase student success.

- **Exercise Sets: Paired with Purpose**—The exercise sets are structured to assess student understanding of vocabulary, concepts, drill, problem solving, and applications. The exercise sets are graded in difficulty level to build confidence and to enhance students' mathematical thinking. The *Quick Check* exercises provide students with immediate reinforcement and instant feedback to determine their understanding of the concepts presented in the examples. *Putting the Concepts Together* and *Synthesis Review* help students see the big picture and provide a structure for learning each new concept and skill in the course.

- The text is written to streamline Intermediate Algebra (and distinguish it from Elementary Algebra) through a single-chapter presentation of linear equations and inequalities along with the strategic placement of *Getting Ready* review sections that provide instructors with the flexibility to review material instead of reteaching it.

Resources for Success
MyMathLab Online Course for Sullivan/Struve, *Intermediate Algebra*, 4th edition

To give students a consistent tone, voice, and teaching method, this text's approach is tightly integrated throughout its accompanying MyMathLab course, making learning the material as seamless as possible. This course contains all of MyMathLab's powerful features, in addition to specific Sullivan/Struve tools.

Premade Author Created MyMathLab Course

A premade course developed by the authors with a guided learning path for students allows instructors the ease of quick start-up, and encourages students to learn and retain the concepts in order to be more successful on their homework. The learning path guides students to first take advantage of the learning resources at their disposal, including videos and new applets, before directing them to their assignments, which are premade. The MyMathLab course is set up to help instructors get the most out of their course, but all assignments are able to be tailored to instructors' needs.

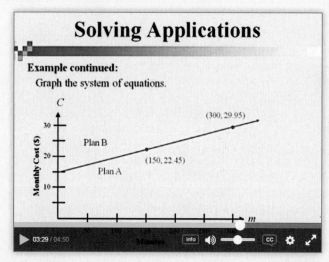

Robust Video Program

The wealth of video resources in the MyMathLab course give students just-in-time help at home, in the lab, or on the go. Video resources include:
– Author-in-Action videos featuring author Mike Sullivan's actual classroom lecture
– Example-level solution clips
– Chapter Test Prep videos

New QR codes located throughout the textbook give students instant, easy access to all the videos at their fingertips.

New Applets

New applets developed by the authors let students interact with the math in a visual, tangible way. These animations allow students to explore and manipulate the mathematical concepts, leading to more durable understanding, and corresponding exercises in MyMathLab make them truly assignable.

Guided Exercises

In addition to MyMathLab's hallmark interactive exercises, Guided Exercises walk students through each step of the problem-solving process, giving them a truly guided, step-by-step learning experience. These are based on the "How To" exercises from the text and were written by the authors.

Resources for Success

Instructor Resources

Annotated Instructor's Edition

ISBN 10: 013455678X **ISBN 13:** 9780134556789
The AIE provides annotations for instructors, including answers and teaching tips.

The following resources can be downloaded from www.pearsonhighered.com* or in MyMathLab.

Instructor Solutions Manual

This manual provides worked-out solutions to all exercises in the text.

Instructor's Resource Manual

This manual includes resources designed to help both new and experienced instructors with course preparation and classroom management. This includes mini-lectures for each section of the text, chapter by chapter teaching tips, sample syllabi, and more.

PowerPoints

These slides present key concepts and definitions from the text.

TestGen

TestGen® (www.pearsoned.com/testgen) enables instructors to build, edit, print, and administer tests using a computerized bank of questions developed to cover all the objectives of the text.

Student Resources

Author in Action videos

Available in MyMathLab, these videos feature each objective presented by the authors with detailed explanations and examples.

Student Solutions Manual

ISBN 10: 0134556763 **ISBN 13:** 9780134556765
This manual contains complete worked solutions to the odd-numbered problems in the end-of-section exercise sets and all of the Quick Checks and end-of-chapter exercises.

Video Notebook

ISBN 10: 0134592069 **ISBN 13:** 9780134592060
The Video Notebook is an unbound, three-hole-punched workbook/note-taking guide that students use in conjunction with the Sullivan/Struve/Mazzarella "Author in Action" videos. The notebook helps them develop organized notes as they work along with the videos.

- A Video Guide for each section is organized by learning objective. Typically, there is one Author in Action video per objective, and students are asked to write down important definitions and procedures and work through key examples as they watch the video.

- The clean layout and ample space let students write out full definitions and show all work for the examples.

- The unbound, loose-leaf format allows students to insert additional notes from class and/or homework—so they can build a course notebook and good study skills for future classes!

Do The Math Workbook

ISBN 10: 0134556674 **ISBN 13:** 9780134556673
This workbook offers a collection of 5-Minute Warm-Up exercises, Guided Practice exercises, and Do the Math exercises for each section in the text. These worksheets can be used as in-class assignments, as an in-lab study assignment, or for homework.

www.mymathlab.com

Acknowledgments

Textbooks are written by authors but evolve through the efforts of many people. We would like to extend our thanks to the following individuals for their important contributions to the project. From Pearson: Mary Beckwith, whose editorial expertise has been an invaluable asset; Michelle Renda for her innovative marketing ideas; Chris Hoag for her support and encouragement; Heather Scott and the design team for the attractive and functional design; Rose Kernan, Patty Bergin and Lauren Morse for keeping a watchful eye and managing countless production details and finally, the Pearson Arts & Sciences sales team for their confidence and support of our books.

We would also like to thank Brad Davis and John Bialas for their attention to details and consistency in accuracy checking the text and answer sections. A huge thanks goes out to Val Villegas, who created the outstanding classroom video notebooks that accompany the text. We offer many thanks to all the instructors from across the country who participated in reviewer conferences and focus groups, reviewed or class-tested some aspect of the manuscript, and taught from the previous editions. Their insights and ideas form the backbone of this text. Hundreds of instructors contributed their time, energy, and ideas to help us shape this text. We will attempt to thank them all here. We apologize for any omissions.

The following individuals, many of whom reviewed or class-tested the first edition, provided direction and guidance in shaping the fourth edition.

Marwan Abu–Sawwa, *Florida Community College—Jacksonville*
MaryAnne Anthony, *Santa Ana College*
Darla Aguilar, *Pima State University*
Grant Alexander, *Joliet Junior College*
Philip Anderson, *South Plains College*
Mary Lou Baker, *Columbia State Community College*
Bill Bales, *Rogers State*
Tony Barcellos, *American River College*
John Beachy, *Northern Illinois University*
Donna Beatty, *Ventura College*
David Bell, *Florida Community College—Jacksonville*
Sandy Berry, *Hinds Community College*
John Bialas, *Joliet Junior College*
Linda Blanco, *Joliet Junior College*
Kevin Bodden, *Lewis and Clark College*
Rebecca Bonk, *Joliet Junior College*
Cherie Bowers, *Santa Ana College*
Becky Bradshaw, *Lake Superior College*
Lori Braselton, *Georgia Southern University*
Tim Britt, *Jackson State Community College*
Beverly Broomell, *Suffolk Community College*
Joanne Brunner, *Joliet Junior College*
Hien Bui, *Hillsborough Community College—Dale Mabry*
Connie Buller, *Metropolitan Community College*
Annette Burden, *Youngstown State University*
James Butterbach, *Joliet Junior College*
Marc Campbell, *Daytona Beach Community College*
Elena Catoiu, *Joliet Junior College*
Nancy Chell, *Anne Arundel Community College*

John F. Close, *Salt Lake Community College*
Bobbi Cook, *Indian River Community College*
Carlos Corona, *San Antonio College*
Faye Dang, *Joliet Junior College*
Shirley Davis, *South Plains College*
Vivian Dennis-Monzingo, *Eastfield College*
Alvio Dominguez, *Miami Dade Community College—Wolfson*
Karen Driskell, *South Plains College*
Thomas Drucker, *University of Wisconsin—Whitewater*
Brenda Dugas, *McNeese State University*
Doug Dunbar, *Okaloosa-Walton Junior College*
Laura Dyer, *Southwestern Illinois State University*
Bill Echols, *Houston Community College—Northwest*
Laura Egner, *Joliet Junior College*
Jason Eltrevoog, *Joliet Junior College*
Nancy Eschen, *Florida Community College—Jacksonville*
Mike Everett, *Santa Ana College*
Phil Everett, *Ohio State University*
Scott Fallstrom, *Shoreline Community College*
Betsy Farber, *Bucks County Community College*
Fitzroy Farqharson, *Valencia Community College—West*
Stephanie Fluke, *Austin Community College*
Jacqueline Fowler, *South Plains College*
Dorothy French, *Community College of Philadelphia*
Randy Gallaher, *Lewis and Clark College*
Sanford Geraci, *Broward Community College*

Donna Gerken, *Miami Dade Community College—Kendall*
Adrienne Goldstein, *Miami Dade Community College—Kendall*
Marion Graziano, *Montgomery County Community College*
Susan Grody, *Broward Community College*
Tom Grogan, *Cincinnati State University*
Barbara Grover, *Salt Lake Community College*
Shawna Haider, *Salt Lake Community College*
Margaret Harris, *Milwaukee Area Technical College*
Teresa Hasenauer, *Indian River Community College*
Mary Henderson, *Okaloosa-Walton Junior College*
Celeste Hernandez, *Richland College*
Paul Hernandez, *Palo Alto College*
Pete Herrera, *Southwestern College*
Bob Hervey, *Hillsborough Community College—Dale Mabry*
Teresa Hodge, *Broward Community College*
Sandee House, *Georgia Perimeter College*
Becky Hubiak, *Tidewater Community College—Virginia Beach*
Sally Jackman, *Richland College*
John Jarvis, *Utah Valley State College*
Nancy Johnson, *Broward Community College*
Steven Kahn, *Anne Arundel Community College*
Linda Kass, *Bergen Community College*
Donna Katula, *Joliet Junior College*
Mohammed Kazemi, *University of North Carolina—Charlotte*
Doreen Kelly, *Mesa Community College*

Mike Kirby, *Tidewater Community College—Virginia Beach*
Keith Kuchar, *College of Dupage*
Carla Kulinsky, *Salt Lake Community College*
Julie Labbiento, *Leigh Carbon Community College*
Kathy Lavelle, *Westchester Community College*
Deanna Li, *North Seattle Community College*
Brian Macon, *Valencia Community College—West*
Iris Magee, *Los Angeles City College*
Lynn Marecek, *Santa Ana College*
Jim Matovina, *Community College of Southern Nevada*
Jean McArthur, *Joliet Junior College*
Michael McComas, *Marshall University*
Mikal McDowell, *Cedar Valley College*
Lee McEwen, *Ohio State University*
David McGuire, *Joliet Junior College*
Angela McNulty, *Joliet Junior College*
Debbie McQueen, *Fullerton College*
Judy Meckley, *Joliet Junior College*
Lynette Meslinsky, *Erie Community College—City Campus*
Kausha Miller, *Lexington Community College*
Chris Mizell, *Okaloosa Walton Junior College*
Jim Moore, *Madison Area Technical College*
Ronald Moore, *Florida Community College—Jacksonville*
Elizabeth Morrison, *Valencia Community College—West*
Roya Namavar, *Rogers State*
Hossein Navid-Tabrizi, *Houston Community College*
Carol Nessmith, *Georgia Southern University*
Kim Neuburger, *Portland Community College*

Larry Newberry, *Glendale Community College*
Elsie Newman, *Owens Community College*
Charlotte Newsome, *Tidewater Community College*
Charles Odion, *Houston Community College*
Viann Olson, *Rochester Community and Technical College*
Linda Padilla, *Joliet Junior College*
Carol Perry, *Marshall Community and Technical College*
Faith Peters, *Miami Dade Community College—Wolfson*
Dr. Eugenia Peterson, *Richard J. Daley College*
Jean Pierre-Victor, *Richard J. Daley College*
Philip Pina, *Florida Atlantic University*
Carol Poos, *Southwestern Illinois University*
Elise Price, *Tarrant County Community College*
R.B. Pruitt, *South Plains College*
William Radulovich, *Florida Community College—Jacksonville*
Pavlov Rameau, *Miami Dade Community College—Wolfson*
David Ray, *University of Tennessee—Martin*
Nancy Ressler, *Oakton Community College*
Michael Reynolds, *Valencia Community College—West*
George Rhys, *College of the Canyons*
Jorge Romero, *Hillsborough Community College—Dale Mabry*
David Ruffato, *Joliet Junior College*
Carol Rychly, *Augusta State University*
David Santos, *Community College of Philadelphia*
Togba Sapolucia, *Houston Community College*

Doug Smith, *Tarrant Community College*
Catherine J.W. Snyder, *Alfred State College*
Gisela Spieler-Persad, *Rio Hondo College*
Raju Sriram, *Okaloosa-Walton Junior College*
Patrick Stevens, *Joliet Junior College*
Bryan Stewart, *Tarrant Community College*
Jennifer Strehler, *Oakton Community College*
Elizabeth Suco, *Miami Dade Community College—Wolfson*
Katalin Szucs, *East Carolina University*
KD Taylor, *Utah Valley State College*
Mary Ann Teel, *University of North Texas*
Suzanne Topp, *Salt Lake Community College*
Suzanne Trabucco, *Nassau Community College*
Jo Tucker, *Tarrant Community College*
Bob Tuskey, *Joliet Junior College*
Mary Vachon, *San Joaquin Delta College*
Carol Walker, *Hinds Community College*
Kim Ward, *Eastern Connecticut State University*
Richard Watkins, *Tidewater Community College*
Natalie Weaver, *Daytona Beach Community College*
Darren Wiberg, *Utah Valley State College*
Rachel Wieland, *Bergen Community College*
Christine Wilson, *Western Virginia University*
Brad Wind, *Miami Dade Community College—North*
Roberta Yellott, *McNeese State University*
Steve Zuro, *Joliet Junior College*
Jon Anderson, *Utah Valley University*
Darren Wiberg, *Utah Valley University*
Rick Woodmansee, *Sacramento City College*

Additional Acknowledgments

We also would like to extend thanks to our colleagues at Joliet Junior College and Columbus State Community College, who provided encouragement, support, and the teaching environment where the ideas and teaching philosophies in this text were developed.

Michael Sullivan, III
Katherine R. Struve

R Review

The image to the left of a sun carrier shell fossil (*Stellaria solaris*) demonstrates how a specific sequence of numbers, called the Fibonacci sequence, occurs frequently in nature. Problem 142 in Section R.3 explores the Fibonacci sequence further.

The Big Picture: Putting It Together

As the "R" in the title implies, this chapter is a review. The purpose of this chapter is to help you recall mathematical concepts that you learned in earlier courses. The topics in this chapter are important building blocks that will help you succeed in this course.

Your instructor may or may not decide to cover this chapter, depending on the course syllabus. Regardless, as you proceed through the book, references will be made to Chapter R so that you can use it as a "just-in-time" review.

Outline

R.1 Success in Mathematics
R.2 Sets and Classification of Numbers
R.3 Operations on Signed Numbers; Properties of Real Numbers
R.4 Order of Operations
R.5 Algebraic Expressions

R.1 Success in Mathematics

Let's start by having a discussion about the big picture goals of the course and how this text can help you be successful at mathematics. Our first "big picture" goal is to develop algebraic skills and gain an appreciation for the power of algebra and mathematics. But there is also a second "big picture" goal. By studying mathematics, we develop a sense of logic and exercise the part of our brains that deals with logical thinking. The examples and problems in this text are like the crunches we do in a gym to exercise our bodies. The goal of running or walking is to get from point A to point B, so doing fifty crunches on a mat does not accomplish that goal, but crunches do make our upper bodies, backs, and hearts stronger for when we need to run or walk.

Logical thinking can assist us in solving difficult everyday problems, and solving algebra problems "builds the muscles" in the part of our brain that performs logical thinking. So, when you are studying algebra and getting frustrated with the amount of work that needs to be done, and you say, "My brain hurts," remember that just like an athlete, you must practice to be successful. But as is also true of an athlete, practice needs to be on a regular basis, not just before "the big game."

Another phrase to keep in mind is "Success breeds success." Mathematics is everywhere. You already are successful at doing some everyday mathematics. With practice, you can take your initial successes and become even more successful. Have you ever done any of the following everyday activities?

- Double a cookie recipe.
- Leave a tip at a restaurant.
- Figure out how many calories your bowl of breakfast cereal provides.
- Compare the distances between cities as you plan a vacation.
- Order the appropriate number of gallons of paint to cover the walls of a room.
- Buy a car and take out a car loan with interest.
- Compare the price per ounce of different sizes of jars of peanut butter or jam.
- Exchange American dollars for Canadian dollars.
- Find the final price on a t-shirt after applying a 20%-off coupon.

You may do five or ten mathematical activities in a single day! The everyday mathematics that you already know is the foundation for your success in this course.

1 What to Do the First Week of the Semester

The first week of the semester gives you the opportunity to prepare for a successful course. Here are the things you should do:

1. **Pick a good seat.** Choose a seat that gives you a good view of the room. Sit close enough to the front so you can easily see the board and hear the professor.

2. **Read the syllabus to learn about your instructor and the course.** Take note of your instructor's name, office location, e-mail address, telephone number, and office hours. Pay attention to any additional help such as tutoring centers, videos, software, and online tutorials. Be sure you fully understand all of the instructor's policies for the class, including those on absences, missed exams or quizzes, and homework. Know important dates and put them in your planner, tablet, computer, or phone. Ask questions.

3. **Learn the names of some of your classmates and exchange contact information.** One of the best ways to learn math is through group study sessions. Try to create time each week to study with your classmates. Knowing how to get in contact with classmates is also useful, because you can obtain the assignment for the day if you ever miss class.

4. **Budget your time.** Most students have a tendency to "bite off more than they can chew." To help with time management, consider the following general rule: Plan on studying *at least* two hours outside of class for each hour in class. So, if you enrolled in a four-hour math class, you should set aside at least eight hours each week to study for the course. You will also need to set aside time for other courses. Consider your work schedule and personal life when creating your time budget. A blank time chart is provided in the exercises for you to use to manage your time.

5. **Get a notebook or binder that is dedicated to math alone.** Don't use this notebook for any other classes, just math, so you stay organized.

❷ What to Do Before, During, and After Class

Now that the semester is under way, we present the following ideas for what to do before, during, and after each class meeting. These suggestions may sound overwhelming, but by following them, you will increase your chances to be successful in mathematics (and other courses). Also, you will find that following this plan will make studying for exams much easier.

Before Class

1. Read the section or sections that will be covered in the upcoming class meeting. Watch the video lectures that accompany the text. There is a QR code ▦ at the beginning of each section that will take you to the video lectures.

2. Based on your reading and watching, write down a list of questions. Your questions will probably be answered through the lecture, but if not, you can then ask any questions that are not answered completely. Also, write down any important new vocabulary and formulas in your math notebook.

3. Arrive early and make sure you are mentally prepared for class. Your mind should be alert and ready to concentrate for the entire class. A good night's sleep and healthy meals high in protein can help.

During Class

1. Stay alert. Do not doze off or daydream during class. Understanding the classroom discussion will be very difficult when you "return to class."

2. Avoid distractions such as cell phones and websites that do not pertain to the class.

3. Take thorough notes. It is normal not to understand certain topics the first time you hear them in a lecture. However, this does not mean you should throw your hands up in despair. Rather, continue to take class notes.

4. Do not be afraid to ask questions. In fact, instructors love questions, for two reasons. First, if one student has a question, other students are likely to have the same question. Second, by asking questions, you teach the teacher what topics cause difficulty.

After Class

1. Reread (and possibly rewrite) your class notes. You may be amazed at how often your confusion during class disappears after studying your in-class notes after class.

2. Reread the section. This is an especially important step. Once you have heard the lecture, the section will make more sense and you will understand much more.

3. Do your homework as soon as possible. **Homework is not optional.** There is an old Chinese proverb that says,

> I hear … and I forget
> I see … and I remember
> I do … and I understand

This proverb applies to any situation in life in which you want to succeed. Would a pianist expect to be the best if she did not practice? The only way you are going to learn algebra is by doing algebra.

4. When you get a problem wrong, try to figure out *why* you got the problem wrong. Once you figure out why, work a similar problem. If you can not discover your error, be sure to ask for help. If possible, connect with others in your class to ask and answer questions about the homework.

5. If you have questions, visit your professor during office hours. You can also ask someone in your study group or go to the tutoring center on campus, if available.

❸ How to Use the Text Effectively

This text was developed so that there is more than one way to learn the material. All of the features in the text are here to help you succeed. These features are based on techniques we use in class. The paragraphs that follow outline the features that appear, an explanation of the purpose of each feature, and how each can be used to help you succeed in this course.

Are You Prepared for This Section? Warming Up

Beginning with Section 1.1, each section starts with a short set of review problems. These problems ask questions about material that was presented earlier in the course and is needed for the upcoming section. Complete the problems to be sure you understand the material that the new section is based on. Answers appear as footnotes on the page where the problems appear. Check your answers. If you get a problem wrong, or if you do not know how to do a problem, go back to the section listed and review the material.

Objectives: A "Road Map" Through the Course

To the left of the "Are you Prepared for This Section" problems is a list of objectives to be covered in the section. If you follow the objectives, you will get a good idea of the section's "big picture"—the important concepts, techniques, and procedures that it introduces.

The objectives are numbered. (See the numbered headline at the beginning of this section.) When we begin discussing a particular objective within the section, the objective number appears along with the stated objective.

Examples: Where to Look for Information

Examples are meant to provide you with guidance and instruction when you are away from the instructor and the classroom. With this in mind, we have developed two special example formats.

Step-by-Step Examples have a three-column format. The left column describes a step, the middle column briefly explains the step, and the right column presents the algebra. Thus, the left and middle columns can be thought of as your instructor's voice during a lecture. *Step-by-Step Examples* introduce key topics or important problem-solving strategies. They provide easy-to-understand, practical instructions, and can be identified by the words "How to ..." in the example's title.

Annotated Examples have a two-column format with explanations to the left of the algebra. The explanation clearly describes what operations we are about to perform and the order in which we will do them. Again, annotations are like your instructor's voice as he or she writes each step of the solution on the board.

Authors in Action: Lecture Videos to Help You Learn

In each section there are QR codes ▦ that take you to the accompanying video lectures. Every objective has one or more classroom lecture videos, marked by a ▶ icon, of the authors teaching their students. These "live" classroom lectures can be used to supplement your instructor's presentations and your reading of the text. They can also be found in the Multimedia Library of MyMathLab.

In Other Words: Math in Everyday Language

Have you ever been given a math definition in class and said, "What in the world does that mean?" We have heard that from our students. So we added the "In Other Words" feature, which restates mathematical definitions in everyday language. This margin feature will help you understand the language of mathematics better.

Work Smart

These "tricks of the trade" that appear in the margin can help you solve problems. They also show alternative problem-solving approaches. There is often more than one way to solve a math problem!

Work Smart: Study Skills

These margin notes highlight the study skills required for success in this and other mathematics courses.

Exercises: A Unique Numbering Scheme

As teachers, we know that students typically jump right to the exercises after attending class. This means they may skip all of the examples and explanations of concepts in the section. To help you use the book most effectively to learn the math, we have structured the exercises differently from other texts you have used. Our structure is designed to encourage the reading of the book while boosting your confidence and ability to work any mathematical problem. Thus the exercises in each section are broken into as many as eight parts. Each exercise set will have some, or all, of the following exercise types.

1. Quick Checks
2. Building Skills
3. Mixed Practice
4. Applying the Concepts
5. Extending the Concepts
6. Explaining the Concepts
7. Synthesis Review
8. Technology Exercises

1. **Quick Checks: Learning to Ride a Bicycle with Training Wheels** Do you remember when you were first learning to ride a bicycle? Training wheels were placed on the bicycle to assist you in learning balance. The Quick Checks are like exercises with training wheels. These exercises appear right after the example(s) that illustrate the concept being taught. So if you get stuck on a Quick Check problem, you can simply consult the example(s) immediately preceding it, rather than searching through the text. These Quick Checks are intended for you to complete within one day of the class. If you do this, your retention will increase, making the rest of your work much easier! The answers to all Quick Checks are in the back of the book. For an example, see Quick ✓ on page 19 in Section R.3.

2. **Building Skills: Learning to Ride a Bicycle with Assistance** Once you felt ready to ride without training wheels, you probably had an adult follow closely behind you, holding the bicycle for balance and building your confidence. The Building Skills problems serve a similar purpose. They are linked to the objectives within the section, so the directions for a problem indicate which objective is being developed. As a result, you know exactly which objective (but not

exactly which example) to consult if you get stuck. For an example, see page 31 in Section R.3.

3. **Mixed Practice: Now You Are Ready to Ride!** After mastering training wheels and learning to balance with assistance, you are ready to ride alone. This stage corresponds to the Mixed Practice exercises. These exercises include problems that develop your ability to see the big picture of mathematics. They are not linked to a particular objective, and they require you to determine the appropriate approach to solving a problem on your own. For an example, see page 31 in Section R.3.

4. **Applying the Concepts: Where Will I Ever Use This Stuff?** The Applying the Concepts exercises not only illustrate the application of mathematics in your life, but also provide problems that test your conceptual understanding of the mathematics. For an example, see page 32 in Section R.3.

5. **Extending the Concepts: Stretching Your Mind** Sometimes we need to be challenged further. These exercises extend your skills to a new level and provide further insight into where mathematics can be used. For an example, see page 32 in Section R.3.

6. **Explaining the Concepts: Verbalize Your Understanding** These problems require you to express the section's big-picture concepts in your own words. Students need to improve their ability to communicate complicated ideas (both oral and written). If you truly understand the material in the section, you should be able to articulate the concepts clearly. For an example, see page 32 in Section R.3.

7. **Synthesis Review: Seeing the Forest for the Trees** The Synthesis Review exercises, which begin in Chapter 5, can help you grasp the "big picture" of algebra. These exercises ask you to perform a single operation (adding, solving, and so on) on several objects (polynomials, rational expressions, and so on). For an example, see page 409 in Section 5.1.

8. **Technology Exercises** Technology can be a great way to verify answers and to help visualize results. These exercises illustrate how technology, such as graphing calculators or Geogebra, can be incorporated into the material of the section. Sections that include Geogebra exercises will include a URL that will lead you to the Geogebra applet associated with those questions. For an example, see page 120 in Section 1.6.

Chapter Review

The chapter review is arranged by section. For each section, we state key concepts, key terms, and objectives. We also list the examples, the page numbers from the text that illustrate each objective, and the review exercises that assess your understanding of each objective. If you get a problem wrong, use this feature to determine where to look in the book to help you work the problem.

Chapter Test

Once you think you are prepared for the exam, take the chapter test at the end of the chapter. If you do well on the chapter test, chances are you will do well on your in-class exam. Be sure to take the test under the same conditions that you will face in class. For example, try setting a timer to mimic the time constraints felt in class during a test, or take the test at a desk in a quiet room. If you are unsure how to solve a problem in the chapter test, watch the Chapter Test Prep Videos, available in MyMathLab or on YouTube, which show an instructor solving each chapter test problem. These videos can also be accessed by using the QR code found at the beginning of the chapter test.

Cumulative Review: Reinforcing Your Knowledge

The building process of learning algebra involves a lot of reinforcement. For this reason, we provide cumulative reviews at the end of every odd-numbered chapter starting with Chapter 1. Do these cumulative reviews after each chapter test, so that you are always refreshing your memory. This way, studying for the final exam should be easier.

④ How to Prepare for an Exam

The following steps are time-tested suggestions to help you prepare for an exam.

Step 1: Revisit your homework and the chapter review problems. About one week before your exam, start to redo your homework assignments. If you do not understand a topic, seek out help. Work the problems in the chapter review as well. These problems are linked to the section objectives. If you get a problem wrong, identify the objective and the examples that illustrate the objective. Then review this material and try the problem in the chapter review again. If you still get the problem wrong, seek out help.

Step 2: Test yourself. A day or two before the exam, take the chapter test under test conditions. For example, this means if your test is timed, to time yourself; if you are only allowed certain resources for the exam, only use those resources. Be sure to check your answers in the back of the book or the Chapter Test Prep Videos. If you get any problems wrong, determine why and remedy the situation. Do not gamble on what might be on the exam; master everything.

Step 3: View the Chapter Test Prep Videos. These videos show step-by-step solutions to the problems found in each of the book's chapter tests. Follow the worked-out solutions to any of the exercises on the chapter test that you want to study or review.

Work Smart: Study Skills

Do not "cram" for an exam by pulling an "all-nighter."

Step 4: Arrive early and well rested to the exam. Be sure to arrive early at the location of the exam. Prepare your mind for the exam. Be sure you are well rested. Do not try to pull "all-nighters." It does not work since your brain can only process a certain amount of material at a time. You should be reviewing material you already know the night before an exam, not learning new material.

R.1 Exercises MyMathLab®

1. Why do you want to be successful in mathematics? Are your goals positive or negative? If you stated your goal negatively ("Just get me out of this course!"), can you restate it positively?

2. Name three activities in your daily life that involve the use of math (for instance, playing cards, operating your computer, or reading a credit card bill).

3. What is your instructor's name?

4. What are your instructor's office hours? Where is your instructor's office?

5. What is your instructor's e-mail address?

6. Does your class have a website? Do you know how to access it? What information is located on the website?

7. Are there tutors available for this course? If so, where are they located? When are they available?

8. Name two other students in your class. What is their contact information? When can you meet with them to study?

9. List at least two things that you should do before class begins.

10. List at least two things you should do during class.

11. List at least two things you should do after class.

12. What is the point of the Chinese proverb on page 3?

13. What are the "Are you Prepared for This Section" problems? How should they be used?

14. Name two features that appear in the margins. What is the purpose of each of them?

15. Name the categories of exercises that appear in this book.

16. How should the chapter review material be used?

17. How should the chapter test be used?

18. What are the Chapter Test Prep Videos? How can you access them?

19. List the four steps that should be followed when preparing for an exam. Can you think of other methods of preparing for an exam that have worked for you?

20. How is mathematics like doing crunches at the gym?

21. Use the chart to help manage your time. Be sure to fill in the time allocated to various activities in your life, including each of your classes, studying, work, and leisure. Remember, for every hour you are in class, you should spend two hours outside of class studying the material.

	Monday	Tuesday	Wednesday	Thursday	Friday	Saturday	Sunday
7 am							
8 am							
9 am							
10 am							
11 am							
Noon							
1 pm							
2 pm							
3 pm							
4 pm							
5 pm							
6 pm							
7 pm							
8 pm							
9 pm							

R.2 Sets and Classification of Numbers

Objectives

1. Use Set Notation
2. Classify Numbers
3. Approximate Decimals by Rounding or Truncating
4. Plot Points on the Real Number Line
5. Use Inequalities to Order Real Numbers

1 Use Set Notation

A **set** is a well-defined collection of objects. "Well-defined" means that there is a rule for determining whether a given object is in the set. For example, the students enrolled in Intermediate Algebra at your college is a set. The collection of numbers 0, 1, 2, 3, 4, 5, 6, 7, 8, and 9 may also be identified as a set. If D represents this set of numbers, then

$$D = \{0, 1, 2, 3, 4, 5, 6, 7, 8, 9\}$$

This notation uses braces { } to enclose the objects, or **elements,** in the set. This method of representing a set is called the **roster method.**

EXAMPLE 1 | **Using the Roster Method**

Write the set that represents the vowels.

Solution
The vowels are $a, e, i, o,$ and u, so write

$$V = \{a, e, i, o, u\}$$

Another way to denote a set is to use **set-builder notation.** The numbers in the set $D = \{0, 1, 2, 3, 4, 5, 6, 7, 8, 9\}$ are called digits. Using set-builder notation, the set D of digits can be written as

$$D = \{x \mid x \text{ is a digit}\}$$

In algebra, letters such as x, y, a, b, and c are used to represent numbers. When the letter can be any number in a set of numbers, it is called a **variable.** In the set D, the letter x can represent any digit, so x is a variable that can take on the value 0, 1, 2, 3, 4, 5, 6, 7, 8, or 9.

> **In Other Words**
> The symbol "|" means "such that", so read the set to the right as "D is the set of all x such that x is a digit."

EXAMPLE 2 **Using Set-Builder Notation**

Use set-builder notation to represent the following sets.

   (a) The set of all even digits
   (b) The set of all odd digits

Solution

   (a) Let E represent the set of all even digits, so that

$$E = \{x \mid x \text{ is an even digit}\}$$

   (b) Let O represent the set of all odd digits, so that

$$O = \{x \mid x \text{ is an odd digit}\}$$ ●

> **Quick ✓**
> **1.** A ___ is a well-defined collection of objects.
> **2.** The objects in a set are called _____.
>
> *In Problems 3 and 4, use set-builder notation and the roster method to represent each set.*
>
> **3.** The set of all digits less than 5
> **4.** The set of all digits greater than or equal to 6

When the elements in a set are listed, an element is never listed more than once. For example, do not write $\{1, 2, 3, 2\}$; write $\{1, 2, 3\}$. Also, the order in which the elements are listed does not matter. For example, $\{2, 3\}$ and $\{3, 2\}$ represent the same set.

More notation for describing sets will now be introduced.

Set Notation

• If two sets A and B have the same elements, then say that A **equals** B and write $A = B$.

• If every element of a set A is also an element of a set B, then say that A is a **subset** of B and write $A \subseteq B$.

• If $A \subseteq B$ and $A \neq B$, then say that A is a **proper subset** of B and write $A \subset B$. Put another way, A is a proper subset of B if all elements in A are also in B and there are elements in B that are not in A.

• If a set A has no elements, it is called the **empty set,** or **null set,** and is denoted by the symbol $\varnothing$ or $\{\ \}$. The empty set is a subset of every set; that is $\varnothing \subseteq A$ for any set A.

Work Smart

The empty set is represented as either $\varnothing$ or $\{\ \}$. Never write $\{\varnothing\}$ to represent the empty set because this notation means a set that contains the set called the empty set.

When working with sets, we usually designate a **universal set,** which is the set of all elements of interest to us. For instance, in Example 2, we were interested in the set of all digits, so the universal set is the set of all digits.

It is often helpful to draw pictures of sets because the pictures help us visualize relations among sets. Pictures of sets are called **Venn diagrams,** in honor of John Venn (1834 – 1923). In Venn diagrams, sets are represented as circles enclosed in a rectangle. The rectangle represents the universal set. For example, if $A = \{1, 2\}$, $B = \{1, 2, 3, 4, 5\}$, and the universal set is $U = \{0, 1, 2, 3, 4, 5, 6, 7, 8, 9\}$, then $A \subset B$. Figure 1 illustrates the relation between sets A and B in a Venn diagram.

Figure 1
Venn diagram with $A \subset B$.

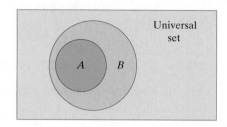

EXAMPLE 3 **Using Set Notation**

Let $A = \{0, 1, 2, 3, 4, 5\}$, $B = \{3, 4, 5\}$, $C = \{5, 4, 3\}$, and $D = \{3, 4, 5, 6\}$.
Write True or False for each statement.

(a) $B \subseteq A$ **(b)** $D \subseteq A$ **(c)** $B = C$
(d) $C = D$ **(e)** $B \subset C$ **(f)** $\varnothing \subseteq C$

Solution

(a) The statement $B \subseteq A$ is True because all the elements in B are also elements in A.

(b) The statement $D \subseteq A$ is False because there is an element in D, 6, that is not in A.

(c) The statement $B = C$ is True because sets B and C have the same elements.

(d) The statement $C = D$ is False because there is an element in D, 6, that is not in C.

(e) In order for B to be a proper subset of C, it must be the case that all elements in B are also elements in C. In addition, there must be at least one element in C that is not in B. Because $B = C$, the statement $B \subset C$ is False. Note, however, that $B \subseteq C$ is True and $B \subset D$ is also True.

(f) The statement $\varnothing \subseteq C$ is True because the empty set is a subset of every set.   ●

Quick ✓

5. *True or False* The order in which elements are listed in a set does not matter.

6. If every element of a set A is also an element of a set B, then say that A is a _____ of B and write $A \subseteq B$.

7. *True or False* If a set has no elements, it is called the empty set and is denoted $\{\varnothing\}$.

In Problems 8–11, let $A = \{a, b, c, d, e, f, g\}$, $B = \{a, b, c\}$, $C = \{c, d\}$, and $D = \{c, b, a\}$. Write True or False for each statement. Be sure to justify your answer.

8. $B \subseteq A$ 9. $B = C$ 10. $B \subset D$ 11. $\varnothing \subseteq A$

The symbol $\in$ (which is read "is an element of") is used to indicate that a particular element is in a set. For example, $7 \in \{1, 3, 5, 7, 9\}$ means "7 is an element of the set $\{1, 3, 5, 7, 9\}$." If an element is not in a set, use the symbol $\notin$ (which is read "is not an element of"). For example, write "b is not a vowel" as $b \notin \{a, e, i, o, u\}$.

EXAMPLE 4   **Using Set Notation**

Write True or False for each statement.

    (a) $3 \in \{ x \mid x \text{ is a digit} \}$

    (b) $h \notin \{ a, e, i, o, u \}$

    (c) $\dfrac{1}{2} \in \left\{ x \mid x = \dfrac{p}{q}, \text{ where } p \text{ and } q \text{ are digits}, q \neq 0 \right\}$

Solution

    (a) The statement $3 \in \{ x \mid x \text{ is a digit} \}$ is True because 3 is a digit.

    (b) The statement $h \notin \{ a, e, i, o, u \}$ is True because h is not an element of the set $\{ a, e, i, o, u \}$.

    (c) The statement $\dfrac{1}{2} \in \left\{ x \mid x = \dfrac{p}{q}, \text{ where } p \text{ and } q \text{ are digits}, q \neq 0 \right\}$ is True because $\dfrac{1}{2}$ is of the form $\dfrac{p}{q}$, where $p = 1$ and $q = 2$.   ●

Work Smart

Using correct notation is important:

$3 \subset \{ 1, 2, 3 \}$ is incorrect.

$3 \in \{ 1, 2, 3 \}$ is correct.

Quick ✓

In Problems 12–14, answer True or False for each statement.

12. $5 \in \{ 0, 1, 2, 3, 4, 5 \}$

13. Michigan $\notin$ { Illinois, Indiana, Michigan, Wisconsin }

14. $\dfrac{8}{3} \in \left\{ x \mid x = \dfrac{p}{q}, \text{ where } p \text{ and } q \text{ are digits}, q \neq 0 \right\}$

Summary Set Notation

$A = B$	means	set A and set B have the same elements.
$A \subseteq B$	means	all the elements in set A are also elements in set B.
$A \subset B$	means	all the elements in set A are also elements in set B, but there is at least one element in set B that is not in set A.
$\varnothing$ or $\{\ \}$	means	the empty set. The set has no elements.
$5 \in A$	means	5 is an element in set A.
$5 \notin A$	means	5 is not an element in set A.

▶ ❷ **Classify Numbers**

Sets are discussed because it is helpful to classify the various kinds of numbers that we deal with as sets.

Definition

The **natural numbers,** or **counting numbers,** are the numbers in the set $\mathbb{N} = \{ 1, 2, 3, \ldots \}$.

The three dots in the definition above, called an *ellipsis,* indicate that the pattern continues indefinitely. The counting numbers are often used to count things. For example, we can count the cars that arrive at a McDonald's drive-through, or we can count the letters in the alphabet.

    Suppose you had $7 and purchased a drink and a hot dog at a baseball game for $7. Can the counting numbers be used to describe the amount of money you have left?

No, because zero is not part of the natural numbers. A new number system is needed to describe the remaining amount. We need the *whole number system*.

> **Definition**
>
> The **whole numbers** are the numbers in the set $W = \{0, 1, 2, 3, \ldots\}$.

The whole numbers consist of the set of counting numbers together with the number 0, so $\mathbb{N} \subset W$.

Now suppose you have a balance of $100 in your checking account and you write a check for $120. Can the whole numbers be used to describe your new balance? No, because the set of whole numbers does not contain negative numbers. You need the *integers*.

> **Definition**
>
> The **integers** are the numbers in the set $\mathbb{Z} = \{\ldots, -3, -2, -1, 0, 1, 2, 3, \ldots\}$.

> **In Other Words**
>
> The set of natural numbers is a subset of the set of whole numbers. The set of whole numbers is a subset of the set of integers.

Notice that the set of natural numbers is a proper subset of the set of whole numbers, and the set of whole numbers is a proper subset of the set of integers. That is, $\mathbb{N} \subset W$ and $W \subset \mathbb{Z}$. As the number system is expanded, new and usually more complicated problems can be discussed. For example, the whole numbers allow the discussion of the absence of something because they include zero, but the counting numbers do not. The integers allow us to deal with problems involving both negative and positive quantities, such as profit (positive counting numbers) and loss (negative counting numbers).

To represent a portion of a dollar or a portion of a whole pie, our number system is enlarged to include *rational numbers*.

> **In Other Words**
>
> A rational number is a number that can be expressed as a fraction, where the numerator is any integer and the denominator is a nonzero integer.

> **Definition**
>
> A **rational number** is a number that can be expressed as a quotient, $\dfrac{p}{q}$, of two integers. The integer p is the **numerator,** and the integer q, which cannot be 0, is the **denominator.** The set of rational numbers are the numbers
>
> $$\mathbb{Q} = \left\{ x \,\middle|\, x = \frac{p}{q}, \text{ where } p \text{ and } q \text{ are integers}, q \neq 0 \right\}.$$

Examples of rational numbers are $\dfrac{3}{4}, \dfrac{4}{3}, \dfrac{0}{6}, -\dfrac{4}{5}$, and $\dfrac{33}{5}$. Because $\dfrac{p}{1} = p$ for any integer p, it follows that the set of integers is a proper subset of the set of rational numbers ($\mathbb{Z} \subset \mathbb{Q}$). For example, 5 is a rational number because it can be written as $\dfrac{5}{1}$, but more specifically, it is an integer. More specifically than that, it is a counting number.

Rational numbers can also be represented as *decimals*. The **decimal** representation of a rational number is found by carrying out the division indicated. For example,

Work Smart

To write $\dfrac{2}{11}$ as a decimal, write $11)\overline{2}$ and carry out the division as follows:

$$
\begin{array}{r}
0.181 \\
11)\overline{2.000} \\
\underline{11} \\
90 \\
\underline{88} \\
20 \\
\underline{11} \\
9
\end{array}
$$

and so on to get $0.1818\ldots = 0.\overline{18}$.

$$\frac{4}{5} = 0.8 \qquad \frac{7}{2} = 3.5 \qquad -\frac{2}{3} = -0.6666\ldots = -0.\overline{6} \qquad \frac{2}{11} = 0.181818\ldots = 0.\overline{18}$$

Notice the line above the 6 in $-0.\overline{6}$. This **repeat bar** represents the fact that the pattern continues. Similarly, $0.\overline{18}$ means the block of numbers 18 will continue indefinitely to the right of the decimal point.

Every rational number may be represented by a decimal that either **terminates** $\Big($ as in the case of $\dfrac{4}{5} = 0.8$ and $\dfrac{7}{2} = 3.5 \Big)$ or is **nonterminating** with a repeating block of decimals $\Big($ as in the case of $-\dfrac{2}{3} = -0.\overline{6}$ and $\dfrac{2}{11} = 0.\overline{18} \Big)$.

Work Smart

Although it is true that $\sqrt{2}$ is an irrational number, not all numbers involving the $\sqrt{\ }$ symbol are irrational. For example, $\sqrt{4} = 2$, is a rational number.

What if a decimal neither terminates nor has a block of digits that repeat? These decimals represent a set of numbers called *irrational numbers*. Every **irrational number** may be represented by a decimal that neither terminates nor repeats. This means that irrational numbers cannot be written in the form $\dfrac{p}{q}$, where p and q are integers and $q \neq 0$. An example of an irrational number is π, whose value is approximately 3.14159265359. Another example of an irrational number is $\sqrt{2}$. This is the number whose square is 2. Its value is approximately 1.414213562.

> **Definition**
>
> Together, the set of rational numbers and the set of irrational numbers form the set of **real numbers.** The set of real numbers is denoted using the symbol $\mathbb{R}$.

Figure 2 shows the relationships among the various types of numbers. The universal set in Figure 2 is the set of all real numbers. It is extremely important to be able to distinguish among the various number systems—in particular, it is important to know the difference between a rational number and an irrational number. Notice that if a number is rational, it cannot be irrational, and vice versa.

Figure 2

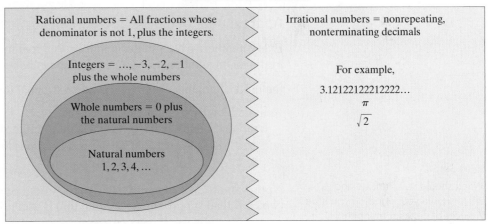

The Set of Real Numbers

EXAMPLE 5   **Classifying Numbers in a Set**

List the numbers in the set

$$\left\{ 3, -\frac{3}{5}, -1, 0, \sqrt{2}, 5.\overline{94}, 4.2122122212222\ldots, -\frac{8}{2}, \frac{\pi}{2} \right\}$$

that are

(a) Natural numbers   (b) Whole numbers   (c) Integers

(d) Rational numbers   (e) Irrational numbers   (f) Real numbers

Solution

(a) 3 is the only natural number.

(b) 0 and 3 are the whole numbers.

(c) $3, -1, 0,$ and $-\dfrac{8}{2}$ are the integers $\left(-\dfrac{8}{2}$ is an integer because it simplifies to $-4 \right)$.

(d) $3, -\dfrac{3}{5}, -1, 0, 5.\overline{94},$ and $-\dfrac{8}{2}$ are the rational numbers.

(e) $\sqrt{2}, 4.2122122212222\ldots,$ and $\dfrac{\pi}{2}$ are the irrational numbers. Note that $4.2122122212222\ldots$ is irrational because it has infinitely many nonrepeating digits.

(f) All of the numbers are real numbers.   ●

Quick ✓

15. The _____ numbers are numbers in the set $\{0, 1, 2, 3, 4, \dots\}$.

16. The numbers in the set $\left\{ x \,\middle|\, x = \dfrac{p}{q}, \text{ where } p \text{ and } q \text{ are integers}, q \neq 0 \right\}$ are called _____ numbers.

17. The set of _____ numbers have decimal representations that neither terminate nor repeat.

18. *True or False* The rational numbers are a subset of the set of irrational numbers.

19. *True or False* If a number is expressed as a decimal, then it is rational.

20. *True or False* Every rational number is a real number.

21. *True or False* If a number is rational, it cannot be irrational.

In Problems 22–27, list the numbers in the set
$$\left\{ \frac{7}{3}, -9, 10, 4.\overline{56}, 5.7377377737777 \dots, \frac{0}{3}, \pi, -\frac{4}{7}, \frac{12}{4}, \sqrt{11} \right\} \text{ that are}$$

22. Natural numbers   **23.** Whole numbers

24. Integers   **25.** Rational numbers

26. Irrational numbers   **27.** Real numbers

▶ ❸ Approximate Decimals by Rounding or Truncating

Every number written in decimal form is a real number that may either be rational or irrational. In addition, every real number can be represented by a decimal. For example, the rational number $\dfrac{3}{4}$ can be written in decimal form as 0.75. The rational number $\dfrac{2}{3}$ is equivalent to 0.666 ... or $0.\overline{6}$ as a decimal.

Irrational numbers have decimals that neither terminate nor repeat. The irrational numbers $\sqrt{2}$ and π have decimal representations that begin as follows:

$$\sqrt{2} = 1.414213\dots \qquad \pi = 3.14159\dots$$

Approximations are used to write irrational numbers as decimals. The symbol $\approx$ (which is read "approximately equal to") is used to do so. For example,

$$\sqrt{2} \approx 1.4142 \qquad \pi \approx 3.1416$$

In approximating decimals, either *round* or *truncate* to a given number of decimal places. The number of decimal places determines the location of the *final digit* in the decimal approximation.

Work Smart

When should you truncate and when should you round? If you have 7 pieces of pizza for four people, how many whole pieces should each person get? If you divide, you get 1.75 pieces. Rounding to the nearest whole digit gives 2 pieces and truncating gives 1 piece. Better truncate to make sure you have enough for everyone!

> **Definition**
>
> **Truncation:** Drop all the digits that follow the specified final digit in the decimal.
> **Rounding:** Identify the specified final digit in the decimal. If the next digit is 5 or more, add 1 to the final digit; if the next digit is 4 or less, leave the final digit as it is. Then truncate all digits to the right of the final digit.

EXAMPLE 6   **Approximating a Decimal by Truncating and by Rounding**

Approximate 13.9463 to two decimal places by

  (a) Truncating   **(b)** Rounding

Solution

We want to approximate the decimal to two decimal places, so the final digit is 4:

$$13.9463$$

  (a) To truncate, remove all digits to the right of the final digit, 4, to get 13.94.

  (b) To round to two decimal places, find the digit to the right of the final digit. It is 6. Because 6 is 5 or more, add 1 to the final digit 4 to get 13.95.

Quick ✓

In Problems 28 and 29, approximate each number by (a) truncating and (b) rounding to the indicated number of decimal places.

28. 5.694392; three decimal places   **29.** −4.9369102; two decimal places

---

The Graphing Calculator: Does Your Calculator Truncate or Round?

Calculators are capable of displaying only a certain number of decimal places. Most scientific calculators display eight digits, and most graphing calculators display ten to twelve digits. When a number has more digits than the calculator can display, the calculator will either round or truncate.

To see whether your calculator rounds or truncates, divide 2 by 3. How many digits do you see? Is the last digit a 6 or a 7? If it is a 6, your calculator truncates; if it is a 7, your calculator rounds. Figure 3 shows the result on a graphing calculator. Does the calculator shown in Figure 3 round or truncate? Because the last digit displayed is a 7, it rounds.

Figure 3

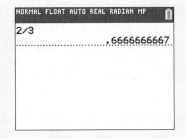

▶ **4 Plot Points on the Real Number Line**

Figure 4
The real number line.

The real numbers can be represented by points on a line called the **real number line.** Every real number corresponds to a point on the line, and each point on the line has a unique real number associated with it.

To construct a real number line, pick a point on a line somewhere in the center, and label it O. This point, called the **origin,** corresponds to the real number 0. See Figure 4. The point 1 unit to the right of O corresponds to the real number 1. The distance between 0 and 1 determines the **scale** of the number line. For example, the point associated with the number 2 is twice as far from O as 1 is. Notice that an arrowhead on the right end of the line indicates the direction in which the numbers increase. Points to the left of the origin correspond to the real numbers −1, −2, and so on.

> **In Other Words**
> Think of the real number line as the graph of the set of all real numbers.

Definition

The real number associated with a point P is called the **coordinate** of P. The **real number line** is the set of all points that have been assigned coordinates.

EXAMPLE 7   **Plotting Points on the Real Number Line**

On the real number line, plot the points with coordinates $0, 5, -1, 1.5, -\dfrac{1}{2}$.

Solution
Draw a real number line with a scale of 1 and then plot the points. See Figure 5.

Figure 5

Quick ✓

30. On the real number line, plot the points with coordinates $0, 3, -2, \dfrac{1}{2}$, and 3.5.

The real number line consists of three classes (or categories) of real numbers, as shown in Figure 6.

Figure 6

- The **negative real numbers** are the coordinates of points to the left of the origin O.
- The real number **zero** is the coordinate of the origin O.
- The **positive real numbers** are the coordinates of points to the right of the origin O.

Figure 7

(a) $a < b$

(b) $a = b$

(c) $a > b$

⑤ Use Inequalities to Order Real Numbers

Given two numbers (points) a and b, a must be to the left of b, the same as b, or to the right of b. See Figure 7.

If a is to the left of b, say "a is less than b" and write $a < b$. If a is at the same location as b, then say that "a is equal to b" and write $a = b$. If a is to the right of b, say "a is greater than b" and write $a > b$. If a is either less than or equal to b, write $a \le b$. Similarly, $a \ge b$ means that a is either greater than or equal to b. Collectively, the symbols $<$, $>$, $\le$, and $\ge$ are called **inequality symbols.**

Note that $a < b$ and $b > a$ mean the same thing, so we can write $2 < 3$ (2 is to the left of 3) or $3 > 2$ (3 is to the right of 2).

EXAMPLE 8 Using Inequality Symbols

(a) $2 < 5$ because the coordinate 2 lies to the left of the coordinate 5 on the real number line.

(b) $-1 > -3$ because -1 lies to the right of -3 on the real number line.

(c) $3.5 \le \dfrac{7}{2}$ because $3.5 = \dfrac{7}{2}$.

(d) $\dfrac{5}{6} > \dfrac{4}{5}$ because $\dfrac{5}{6} = 0.8\overline{3}$ and $\dfrac{4}{5} = 0.8$, so $\dfrac{5}{6}$ lies to the right of $\dfrac{4}{5}$ on the real number line. ●

Quick ✓

In Problems 31–36, replace the question mark by $<$, $>$, or $=$, whichever is correct.

31. $3 \ ? \ 6$

32. $-3 \ ? \ -2$

33. $\dfrac{2}{3} \ ? \ \dfrac{1}{2}$

34. $\dfrac{5}{7} \ ? \ 0.7$

35. $\dfrac{2}{3} \ ? \ \dfrac{10}{15}$

36. $\pi \ ? \ 3.14$

Notice in Example 8 that the inequality symbol always points to the smaller number. Inequalities of the form $a < b$ or $b > a$ are **strict inequalities,** whereas inequalities of the form $a \leq b$ or $b \geq a$ are **nonstrict** (or **weak**) **inequalities**.

Based upon the discussion so far, we conclude that

Work Smart: Study Skills

Selected problems in exercise sets are identified by a green color. For extra help, you can find worked video solutions to these problems in MyMathLab or by using the QR code in each section.

$$a > 0 \quad \text{is equivalent to} \quad a \text{ is positive}$$
$$a < 0 \quad \text{is equivalent to} \quad a \text{ is negative}$$

Read $a > 0$ as "a is positive" or "a is greater than 0." If $a \geq 0$, then either $a > 0$ or $a = 0$, and read this as "a is nonnegative" or "a is greater than or equal to 0."

R.2 Exercises MyMathLab®

Exercise numbers in green have complete video solutions in MyMathLab or may be accessed using the QR code to the right.

*Problems **1–36** are the Quick ✓ s that follow the EXAMPLES.*

Building Skills

In Problems 37–42, write each set using the roster method. See Objective 1.

37. $\{x \mid x \text{ is a whole number less than } 6\}$

38. $\{x \mid x \text{ is a natural number less than } 4\}$

39. $\{x \mid x \text{ is an integer between } -3 \text{ and } 5\}$

40. $\{x \mid x \text{ is an integer between } -4 \text{ and } 6\}$

41. $\{x \mid x \text{ is a natural number less than } 1\}$

42. $\{x \mid x \text{ is a whole number less than } 0\}$

In Problems 43–50, let $A = \{1, 3, 5, 7, 9\}$, $B = \{2, 4, 6, 8\}$, $C = \{1, 2, 3, 4, 5, 6, 7, 8, 9\}$, and $D = \{8, 6, 4, 2\}$. Write True or False for each statement. Then justify your answer. See Objective 1.

43. $B \subseteq C$ 44. $A \subseteq C$ 45. $B \subset D$ 46. $A \subset C$

47. $B = D$ 48. $B \subseteq D$ 49. $\varnothing \subset C$ 50. $\varnothing \subset B$

In Problems 51–54, fill in the blank with the appropriate symbol, $\in$ or $\notin$. See Objective 1.

51. $\dfrac{1}{2}$ ___ $\{x \mid x \text{ is an integer}\}$

52. $4.\overline{5}$ ___ $\{x \mid x \text{ is a rational number}\}$

53. π ___ $\{x \mid x \text{ is a real number}\}$

54. 0 ___ $\{x \mid x \text{ is a natural number}\}$

In Problems 55–58, list the numbers in each set that are (a) Natural numbers, (b) Integers, (c) Rational numbers, (d) Irrational numbers, (e) Real numbers. See Objective 2.

55. $A = \left\{ -5, 4, \dfrac{4}{3}, -\dfrac{7}{5}, 5.\overline{1}, \pi \right\}$

56. $B = \left\{ 13, 0, -4.5656\ldots, 2.43, \sqrt{2} \right\}$

57. $C = \left\{ 100, -5.423, \dfrac{8}{7}, \sqrt{2} + 4, -64 \right\}$

58. $D = \left\{ 15, -\dfrac{6}{1}, 7.3, \sqrt{2} + \pi \right\}$

In Problems 59–62, approximate each number (a) by truncating and (b) by rounding to the indicated number of decimal places. See Objective 3.

59. 19.93483; 4 decimal places

60. -93.432101; 2 decimal places

61. 0.06345; 1 decimal place

62. 9.9999; 2 decimal places

In Problems 63 and 64, plot the points whose coordinates are given on the real number line. See Objective 4.

63. $2, 0, -3, 1.5, -\dfrac{3}{2}$ 64. $4, -5, 2.5, \dfrac{5}{3}, -\dfrac{1}{2}$

In Problems 65–70, replace the question mark by $<$, $>$, or $=$, whichever is correct. See Objective 5.

65. $-5 \; ? \; -3$ 66. $4 \; ? \; 2$ 67. $\dfrac{3}{2} \; ? \; 1.5$

68. $\dfrac{2}{3} \; ? \; \dfrac{2}{5}$ 69. $\dfrac{1}{3} \; ? \; 0.3$ 70. $-\dfrac{8}{3} \; ? \; -\dfrac{8}{5}$

Applying the Concepts

71. **Death Valley** Death Valley in California is the lowest point in the United States, with an elevation that is 282 feet below sea level. Express this elevation as an integer. (SOURCE: *Information Please Almanac*)

72. **Dead Sea** The Dead Sea, Israel–Jordan, is the lowest point in the world, with an elevation that is 1349 feet below sea level. Express this elevation as an integer. (SOURCE: *Information Please Almanac*)

73. **Best Buy** Best Buy lost $9.87 per share of common stock in 2015. Express this loss as a rational number. (SOURCE: *Yahoo!Finance*)

74. **Cisco Systems** Cisco Systems stock lost $0.37 in a recent trading day. Express this loss as a rational number. (SOURCE: *Yahoo!Finance*)

75. **Golf** In the game of golf, your score is often given in relation to par. For example, if par is 72 and a player shoots a 66, then he is 6 under par. Express this score as an integer.

76. **It's a Little Chilly!** The normal high temperature in Las Vegas, Nevada, on January 10 is 55°F. On January 1, 2016, the temperature was 7°F below normal. Express the departure from normal as an integer. (SOURCE: *USA Today*)

Extending the Concepts

77. Research the history of the set of irrational numbers. Your research should concentrate on the Greek cult called the Pythagoreans. Write a report on your findings.

78. Research the origins of the number 0. Is there any single person who can claim its discovery?

79. The first known computation of the decimal approximation of the number π is attributed to Archimedes around 200 B.C. Research Archimedes and find out his approximation.

80. The irrational number e is attributed to Leonhard Euler. Research Euler and find out the decimal approximation of e.

Explaining the Concepts

81. Are there any real numbers that are both rational and irrational? Are there any real numbers that are neither? Explain your reasoning.

82. Explain why the sum of a rational number and an irrational number must be irrational.

83. Explain what a set is. Give an example of a set.

84. Explain why it is impossible to list the set of rational numbers using the roster method.

85. Explain the difference between a subset and a proper subset.

86. Describe the difference between 0.45 and $0.\overline{45}$. Are both rational?

87. Explain the circumstances under which rounding and truncating will both result in the same decimal approximation.

88. Is there a positive real number "closest" to 0?

Technology Exercises

89. Use technology to express $\dfrac{8}{7}$ rounded to three decimal places. Express $\dfrac{8}{7}$ truncated to three decimal places.

90. Use technology to express $\dfrac{19}{7}$ rounded to four decimal places. Express $\dfrac{19}{7}$ truncated to four decimal places.

R.3 Operations on Signed Numbers; Properties of Real Numbers

Objectives

1. Compute the Absolute Value of a Real Number
2. Add and Subtract Signed Numbers
3. Multiply and Divide Signed Numbers
4. Perform Operations on Fractions
5. State the Associative and Distributive Properties

The symbols used in algebra for the operations of addition, subtraction, multiplication, and division and the words used to describe the results of these operations are shown in Table 1 on the following page.

In algebra, avoid using the multiplication sign $\times$ (to avoid confusion with the often-used x) and the division sign $\div$ (to avoid confusion with $+$). Also, two expressions placed next to each other without an operation symbol, as in ab or $(a)(b)$, are understood to be **factors** that are to be multiplied.

It is also preferable not to use mixed numbers in algebra. When mixed numbers are used, addition is understood—for example, $2\dfrac{3}{4}$ means $2 + \dfrac{3}{4}$. But in algebra, the absence

Table 1

Operation	Symbol	Result of Operation
Addition	$a + b$	**Sum:** a plus b
Subtraction	$a - b$	**Difference:** a minus b b is subtracted from a
Multiplication	$a \cdot b, (a) \cdot b, a \cdot (b), (a) \cdot (b),$ $ab, (a)b, a(b), (a)(b)$	**Product:** a times b
Division	a/b or $\dfrac{a}{b}$	**Quotient:** a divided by b

of an operation symbol between two terms is taken to mean multiplication. To avoid any confusion, $2\dfrac{3}{4}$ will be written as 2.75 in decimal format or as an improper fraction, $\dfrac{11}{4}$.

The symbol $=$, which is called an **equal sign** and read as "equals" or "is," is used to express the idea that the expression on the left side of the equal sign is equivalent to the expression on the right.

 1 **Compute the Absolute Value of a Real Number**

The real number line is used to describe the concept of *absolute value*.

Work Smart

The absolute value of a number can never be negative because it represents a distance.

Definition

The **absolute value** of a number a, written $|a|$, is the distance from 0 to a on the real number line.

Figure 8

For example, because the distance from 0 to 3 on the real number line is 3, the absolute value of 3, $|3|$, is 3. Because the distance from 0 to -4 on the real number line is 4, $|-4| = 4$. See Figure 8.

EXAMPLE 1   **Computing Absolute Value**

(a) $|8| = 8$ because the distance from 0 to 8 on the real number line is 8.

(b) $|-5| = 5$ because the distance from 0 to -5 on the real number line is 5.

(c) $|0| = 0$ because the distance from 0 to 0 on the real number line is 0.   ●

Quick ✓

1. In the expression $a \cdot b$, the expressions a and b are called _____.

2. *True or False* The absolute value of a number is always positive.

In Problems 3 and 4, evaluate each expression.

3. $|6|$

4. $|-10|$

▶ ❷ Add and Subtract Signed Numbers

The following rules are used to add two real numbers.

In Other Words

To add two real numbers that have the same sign, add their absolute values and then keep the sign of the original numbers. To add two real numbers with different signs, subtract the smaller absolute value from the larger absolute value. The sign of the sum will be the sign of the number whose absolute value is larger.

Adding Two Nonzero Real Numbers

The approach to adding two real numbers depends on the signs of the two numbers.

1. *Both Positive:* Add the numbers.
2. *Both Negative:* Add the absolute values of the numbers. The sum will be negative.
3. *One Positive, One Negative:* Determine the absolute value of each number. Subtract the smaller absolute value from the larger absolute value.

   - If the larger absolute value was originally the positive number, then the sum is positive.
   - If the larger absolute value was originally the negative number, then the sum is negative.
   - If the two absolute values are equal, then the sum is 0.

EXAMPLE 2 **Adding Two Real Numbers**

Perform the indicated operation.

(a) $-2 + (-3)$ (b) $9.3 + (-6.4)$

Solution

(a) Both numbers are negative, so first find their absolute values: $|-2| = 2$ and $|-3| = 3$. Now add the absolute values: $2 + 3 = 5$. Because both numbers to be added are negative, the sum must be negative. So

$$-2 + (-3) = -5$$

(b) One number is positive and the other is negative. Find the absolute value of each number: $|9.3| = 9.3$ and $|-6.4| = 6.4$. Subtract the smaller absolute value from the larger absolute value and get $9.3 - 6.4 = 2.9$. The larger absolute value was originally positive, so the sum is positive:

$$9.3 + (-6.4) = 2.9$$

Work Smart

Another way to add two real numbers is to use a number line. In Example 2(a), place a point at -2 on the real number line. Then move three places to the left (since -3 is being added). See Figure 9. End up at the point whose coordinate is -5, so $-2 + (-3) = -5$.

Figure 9

Quick ✓

In Problems 5–9, perform the indicated operation.

5. $18 + (-6)$ **6.** $-21 + 10$ **7.** $-5.4 + (-1.2)$

8. $-6.5 + 4.3$ **9.** $-9 + 9$

What do you notice about the sum in Problem 9 from the Quick Check above? The result of Problem 9 is true in general.

Work Smart

The additive inverse of a, $-a$, is sometimes called the *negative* of a or the *opposite* of a. Be careful when using these terms because they suggest that the additive inverse is a negative number, which may not be true! For example, the additive inverse of -3 is 3, a positive number.

Additive Inverse Property

For any real number a, there is a real number $-a$, called the **additive inverse,** or **opposite,** of a, having the following property:

$$a + (-a) = -a + a = 0$$

For example, the opposite of 3 is -3, so $-3 + 3 = 0$. -3 and 3 are additive inverses, or opposites. The opposite of -42 is 42, and the opposite of $-\dfrac{5}{8}$ is $\dfrac{5}{8}$. This gives the following property:

Double Negative Property

For any real number a,

$$-(-a) = a$$

$\boxed{\text{EXAMPLE 3}}$ **Finding an Additive Inverse**

(a) The additive inverse of 6 is -6 because $6 + (-6) = 0$.

(b) The additive inverse of -10 is $-(-10) = 10$ because $-10 + 10 = 0$.   ●

Quick ✓

10. For any real number a, there is a real number $-a$, called the _____ _____, or _____, of a such that $a + (-a) = -a + a = \_.$

11. For any real number a, $-(-a) = \_.$

In Problems 12–16, determine the additive inverse of the given real number.

12. 5   **13.** $\dfrac{4}{5}$ **14.** -12 **15.** $-\dfrac{5}{3}$ **16.** 0

The real number 0 is the only number that can be added to any real number a and result in the same number a. This is called the *Identity Property of Addition.*

Identity Property of Addition

For any real number a,

$$0 + a = a + 0 = a$$

That is, the sum of any number and 0 is that number. We call 0 the **additive identity**.

The Identity Property of Addition is used throughout the entire course (and future courses) to create a new expression that is equivalent to a previous expression. For example, $a = a + 0 = a + 6 + (-6)$ because $6 + (-6) = 0$.

Addition is also *commutative*. That is to say, the same result happens whether $a + b$ or $b + a$ is computed.

Commutative Property of Addition

If a and b are real numbers, then

$$a + b = b + a$$

In Other Words

The same results occur whether $a + b$ or $b + a$ is computed.

Now a more formal definition of absolute value is presented.

Definition

The **absolute value** of a real number a, denoted by the symbol $|a|$, is defined by the rules

$$|a| = a \quad \text{if } a \geq 0 \qquad \text{and} \qquad |a| = -a \quad \text{if } a < 0$$

This definition of absolute value is sometimes called the algebraic definition. For example, $|12| = 12$ and $|-13| = -(-13) = 13$.

The additive inverse can be used to define subtraction between two real numbers.

Definition

If a and b are real numbers, then the **difference** $a - b$, read "a minus b" or "a less b," is defined as

$$a - b = a + (-b)$$

Based on this definition, note that subtracting b from a is really just adding the additive inverse of b to a.

EXAMPLE 4   **Working with Differences**

Evaluate each expression.

(a) $10 - 4$ (b) $-7.3 - (-4.2)$

Solution

(a) $10 - 4 = 10 + (-4) = 6$

(b) Notice that a negative number is being subtracted. Subtracting -4.2 is the same as adding 4.2, so

$$-7.3 - (-4.2) = -7.3 + 4.2$$
$$= -3.1$$

●

Quick ✓

In Problems 17–22, evaluate each expression.

17. $6 - 2$ **18.** $4 - 13$ **19.** $-3 - 8$

20. $12.5 - 3.4$ **21.** $-8.5 - (-3.4)$ **22.** $-6.9 - 9.2$

❸ **Multiply and Divide Signed Numbers**

Multiplication is repeated addition. For example, $3 \cdot 5$ is equivalent to adding 5 three times. That is,

$$3 \cdot 5 = \underbrace{5 + 5 + 5}_{\text{Add 5 three times}} = 15$$

Also, because

$$5 \cdot 3 = 3 + 3 + 3 + 3 + 3 = 15$$

multiplication of two real numbers a and b is commutative, just like addition.

Commutative Property of Multiplication

If a and b are real numbers, then

$$a \cdot b = b \cdot a$$

When multiplying real numbers, follow these rules for determining the sign of the product:

> ### Rules of Signs for Multiplying Two Real Numbers
> **1.** The product of two positive real numbers is positive.
> **2.** The product of one positive real number and one negative real number is negative.
> **3.** The product of two negative real numbers is positive.

(**EXAMPLE 5**) **Multiplying Signed Numbers**

(a) $3(-5) = -(3 \cdot 5) = -15$ **(b)** $-7 \cdot 3 = -(7 \cdot 3) = -21$

(c) $(-9)(-4) = 36$ **(d)** $-1.5(2.6) = -3.9$ ●

Work Smart

In Example 5(a), $3(-5)$ means to add -5 three times. That is, $3(-5)$ means

$$-5 + (-5) + (-5)$$

which is -15. In Example 5(b),

$$-7 \cdot 3 = 3 \cdot (-7)$$
$$= -7 + (-7) + (-7)$$
$$= -21$$

Do you see why a positive times a negative is negative?

Quick ✓

23. *True or False* The product of two negative real numbers is positive.

24. *True or False* Addition and multiplication are commutative.

In Problems 25–29, multiply.

25. $-6(8)$ **26.** $12(-5)$ **27.** $4 \cdot 14$

28. $-7(-15)$ **29.** $-1.9(-2.7)$

The real number 1 has an interesting property. Recall that the expression $3 \cdot 5$ is equivalent to $5 + 5 + 5$. Therefore, $1 \cdot 5$ means to add 5 one time, so, $1 \cdot 5 = 5$. This result is true in general.

> ### Identity Property of Multiplication
> For any real number a,
> $$a \cdot 1 = 1 \cdot a = a$$
> 1 is called the **multiplicative identity.**

In Other Words

The product of any number and 1 is that number.

The multiplicative identity lets us create equivalent expressions. For example, the expressions $\frac{4}{5}$ and $\frac{4}{5} \cdot \frac{3}{3}$ are equivalent because $\frac{3}{3} = 1$.

For each *nonzero* real number a, there is a real number $\frac{1}{a}$, called the *multiplicative inverse* of a, having the following property:

In Other Words

When reciprocals are multiplied, their product is 1.

> ### Multiplicative Inverse Property
> For each *nonzero* real number a, there is a real number $\frac{1}{a}$, called the **multiplicative inverse** or **reciprocal** of a, having the following property:
> $$a \cdot \frac{1}{a} = \frac{1}{a} \cdot a = 1 \qquad a \neq 0$$

EXAMPLE 6 **Finding the Multiplicative Inverse or Reciprocal**

(a) The multiplicative inverse, or reciprocal, of 5 is $\frac{1}{5}$.

(b) The multiplicative inverse, or reciprocal, of -4 is $-\frac{1}{4}$.

(c) The multiplicative inverse of $\frac{2}{3}$ is $\frac{3}{2}$ because $\frac{2}{3} \cdot \frac{3}{2} = 1$. ●

Work Smart

Do not confuse the multiplicative inverse or reciprocal with the additive inverse or opposite. The multiplicative inverse, or reciprocal, of a number does not change signs. It only reverses the numbers in the numerator and denominator.

Quick ✓

30. The additive inverse of a, $-a$, is also called the _____ of a. The multiplicative inverse of a, $\frac{1}{a}$, is also called the _____ of a.

In Problems 31–34, find the multiplicative inverse or reciprocal of the given real number.

31. 10   **32.** -8 **33.** $\frac{2}{5}$ **34.** $-\frac{1}{5}$

The idea behind the multiplicative inverse is used to define division of real numbers.

Definition

If a is a real number and b is a nonzero real number, the **quotient** $\frac{a}{b}$, read as "a divided by b" or "the ratio of a to b," is defined as

$$\frac{a}{b} = a \cdot \frac{1}{b} \quad \text{if } b \neq 0$$

For example, $\frac{5}{8} = 5 \cdot \frac{1}{8}$. Because division of real numbers can be represented as multiplication, the same rules of signs that apply to multiplication also apply to division.

Rules of Signs for Dividing Two Real Numbers

1. The quotient of two positive numbers is positive.
2. The quotient of one positive real number and one negative real number is negative.
3. The quotient of two negative real numbers is positive.

Put another way, if a and b are real numbers and $b \neq 0$, then

$$-\frac{a}{b} = \frac{-a}{b} = \frac{a}{-b} \quad \text{and} \quad \frac{-a}{-b} = \frac{a}{b}$$

Additional properties of the numbers 0 and 1 are now introduced on the next page.

Multiplication by Zero

For any real number a, the product of a and 0 is always 0; that is,

$$a \cdot 0 = 0 \cdot a = 0$$

Division Properties

For any nonzero real number a,

$$\frac{0}{a} = 0 \qquad \frac{a}{a} = 1 \qquad \frac{a}{1} = a \qquad \frac{a}{0} \text{ is undefined}$$

Perhaps you are wondering what $\frac{0}{0}$ equals. The answer, which may surprise you, is that the value of $\frac{0}{0}$ cannot be determined! Why? If $\frac{0}{0} = n$, then $0 \cdot n = 0$. But $0 \cdot n = 0$ for any real number n. For this reason, we say that $\frac{0}{0}$ is **indeterminate.**

4 Perform Operations on Fractions

The following property is used to write a fraction in *lowest terms*. A fraction is said to be in **lowest terms** if the numerator and the denominator share no common factor other than 1.

Reduction Property

If $a, b,$ and c are real numbers, then

$$\frac{ac}{bc} = \frac{a}{b} \text{ if } b \neq 0, c \neq 0$$

EXAMPLE 7   **Using the Reduction Property**

(a)
$$\frac{5 \cdot 3}{5 \cdot 4} = \frac{\cancel{5} \cdot 3}{\cancel{5} \cdot 4}$$

Divide out the 5's: $= \dfrac{3}{4}$

(b) Factor the numerator and denominator and divide out any common factors.

$$\frac{18}{12} = \frac{6 \cdot 3}{6 \cdot 2}$$

Divide out the 6's: $= \dfrac{3}{2}$

●

Quick ✓

35. A fraction is written in _____ _____ if the numerator and denominator share no common factor other than 1.

In Problems 36–39, use the Reduction Property to simplify each expression.

36. $\dfrac{2 \cdot 6}{2 \cdot 5}$ **37.** $\dfrac{33}{24}$ **38.** $-\dfrac{24}{20}$ **39.** $\dfrac{4}{9}$

We now have all the tools needed to perform arithmetic operations on rational numbers.

Arithmetic of Rational Numbers

$$\frac{a}{b} \cdot \frac{c}{d} = \frac{ac}{bd} \qquad\qquad \text{if } b \neq 0, d \neq 0$$

$$\frac{a}{b} \div \frac{c}{d} = \frac{\dfrac{a}{b}}{\dfrac{c}{d}} = \frac{a}{b} \cdot \frac{d}{c} = \frac{ad}{bc} \qquad \text{if } b \neq 0, c \neq 0, d \neq 0$$

$$\frac{a}{c} + \frac{b}{c} = \frac{a+b}{c} \qquad\qquad \text{if } c \neq 0$$

$$\frac{a}{b} + \frac{c}{d} = \frac{ad}{bd} + \frac{bc}{bd} = \frac{ad+bc}{bd} \qquad \text{if } b \neq 0, d \neq 0$$

EXAMPLE 8   **Multiplying and Dividing Rational Numbers**

Perform the indicated operation. Express your answer in lowest terms.

▶ **(a)** $\dfrac{8}{3} \cdot \dfrac{15}{4}$ ▶ **(b)** $-\dfrac{3}{5} \div \dfrac{6}{7}$

Solution

(a) Multiply the numerators and then multiply the denominators using $\dfrac{a}{b} \cdot \dfrac{c}{d} = \dfrac{ac}{bd}$.

$$\frac{8}{3} \cdot \frac{15}{4} = \frac{8 \cdot 15}{3 \cdot 4}$$

Factor the numerator: $\qquad = \dfrac{4 \cdot 2 \cdot 5 \cdot 3}{3 \cdot 4}$

Divide out like factors: $\qquad = \dfrac{4 \cdot 2 \cdot 5 \cdot 3}{3 \cdot 4}$

$$= \frac{2 \cdot 5}{1}$$

$$= \frac{10}{1}$$

$$= 10$$

Work Smart

Notice that when all the factors in the denominator of Example 8(a) divide out, the denominator is 1.

(b) Rewrite the division problem as an equivalent multiplication problem using $\dfrac{a}{b} \div \dfrac{c}{d} = \dfrac{a}{b} \cdot \dfrac{d}{c}$.

$$-\frac{3}{5} \div \frac{6}{7} = -\frac{3}{5} \cdot \frac{7}{6}$$

$$= -\frac{3 \cdot 7}{5 \cdot 3 \cdot 2}$$

$$= -\frac{7}{10}$$

Quick ✓

In Problems 40–43, perform the indicated operation. Express each answer in lowest terms.

40. $\dfrac{5}{3} \cdot \dfrac{12}{25}$

41. $\dfrac{2}{3}\left(-\dfrac{5}{4}\right)$

42. $\dfrac{4}{3} \div \dfrac{8}{3}$

43. $\dfrac{-\dfrac{10}{3}}{-\dfrac{5}{12}}$

EXAMPLE 9 **Adding and Subtracting Rational Numbers with the Same Denominator**

Perform the indicated operation. Express your answer in lowest terms.

(a) $\dfrac{5}{12} + \dfrac{11}{12}$

(b) $\dfrac{4}{15} - \dfrac{7}{15}$

Solution

(a) Because the denominators are the same, add the numerators and write the sum over the common denominator.

$$\frac{5}{12} + \frac{11}{12} = \frac{5 + 11}{12}$$

$$= \frac{16}{12}$$

Factor the numerator and denominator. Divide out like factors: $\quad = \dfrac{\cancel{4} \cdot 4}{\cancel{4} \cdot 3}$

$$= \frac{4}{3}$$

(b) Because the denominators are the same, subtract the numerators and write the difference over the common denominator.

$$\frac{4}{15} - \frac{7}{15} = \frac{4 - 7}{15}$$

$$= \frac{-3}{15}$$

$$= \frac{-1 \cdot \cancel{3}}{\cancel{3} \cdot 5}$$

$$= -\frac{1}{5}$$

Quick ✓

In Problems 44–47, perform the indicated operation. Express your answer in lowest terms.

44. $\dfrac{3}{11} + \dfrac{2}{11}$

45. $\dfrac{8}{15} - \dfrac{13}{15}$

46. $-\dfrac{3}{7} - \dfrac{11}{7}$

47. $\dfrac{8}{5} - \dfrac{3}{5}$

To add two rational numbers with different denominators, first find the **least common denominator (LCD)**—the smallest number that all the denominators have as a common multiple. The next example illustrates the idea.

EXAMPLE 10 **How to Find the Least Common Denominator**

Find the least common denominator of the rational numbers $\frac{8}{15}$ and $\frac{5}{12}$. Then rewrite each rational number with the least common denominator.

Step-by-Step Solution

Step 1: Factor each denominator as the product of prime factors.	$15 = 5 \cdot 3 \qquad \begin{aligned} 12 &= 4 \cdot 3 \\ &= 2 \cdot 2 \cdot 3 \end{aligned}$
Step 2: Write the factor(s) that both denominators have in common. Then write the remaining factors.	The common factor is 3. The remaining factors are 5, 2, and 2.
Step 3: Multiply the factors listed in Step 2. The product is the least common denominator (LCD).	$\begin{aligned} \text{LCD} &= 3 \cdot 5 \cdot 2 \cdot 2 \\ &= 60 \end{aligned}$
Step 4: Rewrite $\frac{8}{15}$ and $\frac{5}{12}$ with a denominator of 60. Multiply $\frac{8}{15}$ by $\frac{4}{4}$. Multiply $\frac{5}{12}$ by $\frac{5}{5}$.	$\frac{8}{15} = \frac{8}{15} \cdot \frac{4}{4} = \frac{8 \cdot 4}{15 \cdot 4} = \frac{32}{60}$ $\frac{5}{12} = \frac{5}{12} \cdot \frac{5}{5} = \frac{5 \cdot 5}{12 \cdot 5} = \frac{25}{60}$

Quick ✓

48. The _____ _____ _____ is the smallest number that each denominator has as a common multiple.

In Problems 49 and 50, find the least common denominator (LCD) of each pair of rational numbers. Then rewrite each rational number with the LCD.

49. $\frac{3}{20}$ and $\frac{2}{15}$

50. $\frac{5}{18}$ and $-\frac{1}{45}$

EXAMPLE 11 **Adding Rational Numbers Using the Least Common Denominator**

Perform the indicated operation.

(a) $\frac{8}{15} + \frac{5}{12}$ **(b)** $-\frac{3}{10} - \frac{2}{15}$

Solution

(a) From Example 10, we know that the LCD of $\frac{8}{15}$ and $\frac{5}{12}$ is 60. Rewrite each fraction with a denominator of 60:

$$\frac{8}{15} + \frac{5}{12} = \frac{8}{15} \cdot \frac{4}{4} + \frac{5}{12} \cdot \frac{5}{5}$$

$$= \frac{32}{60} + \frac{25}{60}$$

$$\frac{a}{c} + \frac{b}{c} = \frac{a+b}{c}: \quad = \frac{32+25}{60}$$

$$= \frac{57}{60}$$

$$= \frac{19}{20}$$

(b) First, write the subtraction problem as an equivalent addition problem using $a - b = a + (-b)$:

$$-\frac{3}{10} - \frac{2}{15} = \frac{-3}{10} + \left(\frac{-2}{15}\right)$$

$$\text{LCD} = 30: \quad = \frac{-3}{10} \cdot \frac{3}{3} + \left(\frac{-2}{15} \cdot \frac{2}{2}\right)$$

$$= \frac{-9}{30} + \left(\frac{-4}{30}\right)$$

$$= \frac{-9 + (-4)}{30}$$

$$= \frac{-13}{30}$$

$$\frac{-a}{b} = -\frac{a}{b}: \quad = -\frac{13}{30}$$

Quick ✓

In Problems 51–54, perform the indicated operation. Express your answer in lowest terms.

51. $\dfrac{3}{20} + \dfrac{2}{15}$

52. $\dfrac{5}{14} - \dfrac{11}{21}$

53. $-\dfrac{4}{25} - \dfrac{7}{30}$

54. $-\dfrac{5}{18} - \dfrac{1}{45}$

⑤ State the Associative and Distributive Properties

The order in which three real numbers are added or multiplied does not affect the final result. This property is called the *Associative Property*.

▶ **EXAMPLE 12**   **Illustrating the Associative Property**

(a) $17 + (3 + 5) = 17 + 8 = 25$
$(17 + 3) + 5 = 20 + 5 = 25$

Therefore,
$17 + (3 + 5) = (17 + 3) + 5$

(b) $13(2 \cdot 5) = 13 \cdot 10 = 130$
$(13 \cdot 2) \cdot 5 = 26 \cdot 5 = 130$

Therefore,
$13(2 \cdot 5) = (13 \cdot 2) \cdot 5$

Associative Properties of Addition and Multiplication

If a, b, and c are real numbers, then

$$a + (b + c) = (a + b) + c = a + b + c$$
$$a(bc) = (ab) \cdot c = abc$$

▶ The next property will be used throughout the course and in future courses.

The Distributive Property

If a, b, and c are real numbers, then

$$a(b + c) = ab + ac$$
$$(a + b) \cdot c = ac + bc$$

EXAMPLE 13　**Using the Distributive Property**

Use the Distributive Property to remove the parentheses.

(a) $2(x + 3)$　(b) $-3(2y + 1)$　(c) $(z - 4) \cdot 3$　(d) $\dfrac{1}{2}(4x - 10)$

Solution

(a) $2(x + 3) = 2 \cdot x + 2 \cdot 3 = 2x + 6$

(b) $-3(2y + 1) = -3 \cdot 2y + (-3)1 = -6y - 3$

(c) $(z - 4) \cdot 3 = z \cdot 3 - 4 \cdot 3 = 3z - 12$

(d) $\dfrac{1}{2}(4x - 10) = \dfrac{1}{2} \cdot 4x - \dfrac{1}{2} \cdot 10 = \dfrac{4x}{2} - \dfrac{10}{2} = 2x - 5$　●

When parentheses are preceded by a minus sign, such as in $-(3x + 7)$, the Distributive Property can be used as shown below.

$$-(3x + 7) = -1(3x + 7) = -1 \cdot 3x + -1 \cdot 7 = -3x - 7$$

We can generalize this property.

> **In Other Words**
> The opposite of a sum is the sum of the opposites.

The Opposite of a Sum

For any real numbers a and b,

$$-(a + b) = -a + (-b) = -a - b$$

Quick ✓

55. The Distributive Property states that $a(b + c) = $ __ + __.

56. $-(a + b) = $ _____

In Problems 57–62, use the Distributive Property to remove the parentheses.

57. $5(x + 3)$　　　　　　　　**58.** $-6(x + 1)$

59. $-4(z - 8)$　　　　　　　**60.** $\dfrac{1}{3}(6x + 9)$

61. $-(11p + 8)$　　　　　　　**62.** $-(-9 - 4t)$

Summary　**Properties of the Real Number System**

If $a, b,$ and c are real numbers,

- Identity Properties　　　　　$a + 0 = 0 + a = a; \quad 1 \cdot a = a \cdot 1 = a$
- Inverse Properties　　　　　$a + (-a) = (-a) + a = 0; \quad a \cdot \dfrac{1}{a} = \dfrac{1}{a} \cdot a = 1 \, (a \neq 0)$
- Double Negative Property　　$-(-a) = a$
- Commutative Properties　　　$a + b = b + a; \quad a \cdot b = b \cdot a$
- Multiplication Property of 0　$a \cdot 0 = 0 \cdot a = 0$
- Division Properties　　　　　$\dfrac{0}{a} = 0, a \neq 0; \quad \dfrac{a}{a} = 1, a \neq 0;$

  　　　　　　　　　　　　　$\dfrac{a}{0}$ is undefined, $a \neq 0$
- Reduction Property　　　　　$\dfrac{ac}{bc} = \dfrac{a}{b}, b \neq 0, c \neq 0$
- Associative Properties　　　　$(a + b) + c = a + (b + c); \quad (ab)c = a(bc)$
- Distributive Property　　　　$a(b + c) = ab + ac$
- Opposite of a Sum　　　　　$-(a + b) = -a - b$

R.3 Exercises MyMathLab®

*Problems **1–62** are the Quick ✔ s that follow the **EXAMPLES.***

Building Skills

In Problems 63–66, evaluate each expression. See Objective 1.

63. $\left|\dfrac{2}{3}\right|$ **64.** $\left|\dfrac{5}{6}\right|$ **65.** $\left|-\dfrac{8}{3}\right|$ **66.** $\left|-\dfrac{7}{2}\right|$

In Problems 67–72, perform the indicated operation. See Objective 2.

67. $-13 + 4$

68. $-6 + 10$

69. $12 - 5$

70. $9 + (-3)$

71. $4.3 - 6.8$

72. $-8.2 - 4.5$

In Problems 73–78, perform the indicated operation. See Objective 3.

73. $4(-8)$

74. $-5(-15)$

75. $-6(-14)$

76. $7(-15)$

77. $-4.3(8.5)$

78. $-10.4(-0.6)$

In Problems 79–82, use the Reduction Property to simplify each expression. See Objective 4.

79. $\dfrac{28}{12}$ **80.** $\dfrac{25}{40}$ **81.** $\dfrac{30}{25}$ **82.** $\dfrac{40}{16}$

In Problems 83–94, perform the indicated operation. See Objective 4.

83. $\dfrac{3}{4} \cdot \dfrac{20}{9}$

84. $\dfrac{2}{5} \cdot \dfrac{15}{8}$

85. $\dfrac{2}{3}\left(-\dfrac{9}{14}\right)$

86. $-\dfrac{7}{3}\left(-\dfrac{12}{35}\right)$

87. $\dfrac{7}{4} \div \dfrac{21}{8}$

88. $-\dfrac{10}{3} \div \dfrac{15}{21}$

89. $\dfrac{\frac{2}{5}}{\frac{8}{25}}$

90. $\dfrac{\frac{18}{7}}{\frac{3}{14}}$

91. $-\dfrac{5}{12} + \dfrac{7}{12}$

92. $\dfrac{8}{5} - \dfrac{18}{5}$

93. $\dfrac{7}{15} + \dfrac{9}{20}$

94. $\dfrac{3}{8} - \dfrac{7}{18}$

In Problems 95–104, use the Distributive Property to remove the parentheses. See Objective 5.

95. $2(x + 4)$

96. $3(y - 5)$

97. $-3(z - 2)$

98. $-5(x - 4)$

99. $(x - 10) \cdot 3$

100. $(3x + y) \cdot 2$

101. $\dfrac{3}{4}(8x - 12)$

102. $-\dfrac{2}{3}(3x + 15)$

103. $-(5z + 17)$

104. $-(11 - 8k)$

Mixed Practice

In Problems 105–126, perform the indicated operation. Express all rational numbers in lowest terms.

105. $|13 - 16|$

106. $|-4| + 12$

107. $\dfrac{51}{42} - \left(\dfrac{-8}{35}\right)$

108. $-\dfrac{18}{45} - \dfrac{23}{24}$

109. $5 \div \dfrac{15}{4}$

110. $\dfrac{2}{3} \div 8$

111. $|6.2 - 9.5|$

112. $|-5.4 + 10.5|$

113. $-|-8 \cdot (4)|$

114. $-|-5 \cdot 9|$

115. $\left|-\dfrac{1}{2} - \dfrac{4}{5}\right|$

116. $\left|\dfrac{4}{3} - \dfrac{8}{7}\right|$

117. $\left|-\dfrac{5}{6} - \dfrac{3}{10}\right|$

118. $\left|\dfrac{4}{15} - \dfrac{1}{6}\right|$

119. $\dfrac{21}{32} + (-5)$

120. $-\dfrac{10}{21} + 6$

121. $\dfrac{\frac{2}{3}}{6}$

122. $\dfrac{20}{\frac{5}{4}}$

123. $\dfrac{18}{0}$

124. $-\dfrac{7}{0}$

125. $\dfrac{0}{20}$

126. $\dfrac{0}{5}$

In Problems 127–134, state the property that is being illustrated.

127. $5 \cdot 3 = 3 \cdot 5$

128. $9 + (-9) = 0$

129. $5 \cdot \dfrac{1}{5} = 1$

130. $\dfrac{a}{a} = 1, a \neq 0$

131. $\dfrac{42}{10} = \dfrac{21}{5}$

132. $3(x - 4) = 3x - 12$

133. $3 + (4 + 5) = (3 + 4) + 5$

134. $\dfrac{0}{6} = 0$

Applying the Concepts

135. Age of Presidents The youngest president at the time of inauguration was Theodore Roosevelt (42 years of age). The oldest president at the time of inauguration was Ronald Reagan (69 years of age). What is the difference in age at the time of inauguration between the oldest and youngest presidents?

136. Life Expectancy In South Korea, the life expectancy for females born in 1950 was 49 years of age. The life expectancy for females born in 2016 was 83 years of age. Compute the difference in life expectancy between 1950 and 2016.

137. Football The Chicago Bears obtained the following yardages for each of the first three plays of a game: 4, −3, 8. How many total yards did they gain for the first three plays? If 10 yards are required for a first down, did the Bears obtain a first down?

138. Balancing a Checkbook At the beginning of the month, Paul had $400 in his checking account. During the month he wrote four checks for $20, $45, $60, and $105. He also made a deposit in the amount of $150. What is Paul's balance at the end of the month?

139. Peaks and Valleys In the United States, the highest elevation is Mount Denali in Alaska (20,320 feet above sea level); the lowest elevation is Death Valley in California (282 feet below sea level). What is the difference between the highest and lowest elevations?

140. More Peaks and Valleys In Louisiana, the highest elevation is Driskill Mountain (535 feet above sea level); the lowest elevation is New Orleans (8 feet below sea level). What is the difference between the highest and lowest elevations?

Extending the Concepts

141. Illustrate why the product of two positive numbers is positive. Illustrate why the product of a positive number and a negative number is negative.

142. The Fibonacci Sequence The Fibonacci sequence is a famous sequence of numbers that were discovered by Leonardo Fibonacci of Pisa. The numbers in the sequence are 1, 1, 2, 3, 5, 8, 13, 21, 34, 55, . . . , where each term after the second term is the sum of the two preceding terms.

(a) Compute the ratio of consecutive terms in the sequence. That is, compute $\dfrac{1}{1}, \dfrac{2}{1}, \dfrac{3}{2}, \dfrac{5}{3}$, and so on.

(b) What number does the ratio approach? This number is called the **golden ratio** and has application in many different areas.

(c) Research Fibonacci numbers and cite three different applications.

Problems 143–148 use the following definition.

If P and Q are two points on a real number line with coordinates a and b, respectively, the **distance between P and Q,** denoted by $d(P, Q)$, is

$$d(P, Q) = |b - a|$$

Since $|b - a| = |a - b|$, it follows that $d(P, Q) = d(Q, P)$.

143. Plot the points $P = -4$ and $Q = 10$ on the real number line, and then find $d(P, Q)$.

144. Plot the points $P = -2$ and $Q = 6$ on the real number line, and then find $d(P, Q)$.

145. Plot the points $P = -3.2$ and $Q = 7.2$ on the real number line, and then find $d(P, Q)$.

146. Plot the points $P = -9.3$ and $Q = 1.6$ on the real number line, and then find $d(P, Q)$.

147. Plot the points $P = -\dfrac{10}{3}$ and $Q = \dfrac{6}{5}$ on the real number line, and then find $d(P, Q)$.

148. Plot the points $P = -\dfrac{7}{5}$ and $Q = 6$ on the real number line, and then find $d(P, Q)$.

149. Why Is the Product of Two Negatives Positive? In this problem, we use the Distributive Property to illustrate why the product of two negative real numbers is positive.

(a) Express the product of any real number a and 0.

(b) Use the Additive Inverse Property to write 0 from part (a) as $b + (-b)$.

(c) Use the Distributive Property to distribute the a into the expression in part (b).

(d) Suppose that $a < 0$ and $b > 0$. What can be said about the product ab? Now, what must be true regarding the product $a(-b)$ in order for the sum to be zero?

150. Math for the Future Find the set of all ratios $\dfrac{p}{q}$ such that $p \in A$ and $q \in B$.

$$A = \{-9, -3, -1, 1, 3, 9\}, B = \{-4, -2, -1, 1, 2, 4\}$$

Explaining the Concepts

151. Explain why 0 does not have a multiplicative inverse.

152. Why does $2(4 \cdot 5)$ not equal $(2 \cdot 4) \cdot (2 \cdot 5)$?

153. Explain the flaw in the following reasoning:

$$\frac{4 + 3}{2 + 5} = \frac{2 \cdot 2 + 3}{2 + 5} = \frac{2 \cdot 2 + 3}{2 + 5} = \frac{2 + 3}{1 + 5} = \frac{5}{6}$$

154. Is subtraction commutative? Support your conclusion with an example.

155. Is subtraction associative? Support your conclusion with an example.

156. Is division commutative? Support your conclusion with an example.

157. Is division associative? Support your conclusion with an example.

158. Explain why $\frac{1}{4} + \frac{1}{8} = \frac{3}{8}$.

Technology Exercises

Many calculators have the ability to compute absolute value. Figure 10 shows the results of Example 1 using a graphing calculator. Calculators have the ability to add, subtract, multiply, and divide rational numbers and write the solution in lowest terms! Figure 11 shows the results of Examples 8(b) and 11(b).

Figure 10

Figure 11

In Problems 159–168, use a calculator to perform the indicated operation. Express all rational numbers in lowest terms.

159. $5.4 - 9.2$

160. $-2.9 + (-6.3)$

161. $-3(6.4)$

162. $-5.4(-4.8)$

163. $\frac{3}{10} - \frac{4}{15}$

164. $-\frac{3}{8} + \frac{1}{10}$

165. $|4.5(-3.2)|$

166. $|-3.65|(5.4)$

167. $\frac{6}{5} \cdot \frac{45}{12}$

168. $-\frac{3}{4} \div \frac{9}{20}$

R.4 Order of Operations

Objectives

1 Evaluate Real Numbers with Exponents

2 Use the Order of Operations to Evaluate Expressions

▶ 1 Evaluate Real Numbers with Exponents

Integer exponents indicate repeated multiplication. For example,

$$3^4 = 3 \cdot 3 \cdot 3 \cdot 3 = 81$$

In the expression 3^4, 3 is the **base** and 4 is the **exponent** or **power.** The power tells us the number of times to use the base as a factor.

Definition

If a is a real number and n is a positive integer, the expression a^n represents the product of n factors of a. That is,

$$a^n = \underbrace{a \cdot a \cdot \,\cdots\, \cdot a}_{n \text{ factors}}$$

Thus, we have

$$a^1 = a$$
$$a^2 = a \cdot a$$
$$a^3 = a \cdot a \cdot a$$

and so on.

a^n is read as "a raised to the power n" or as "a to the nth power." a^2 is usually read as "a squared" and a^3 as "a cubed."

> **EXAMPLE 1**   **Evaluating Expressions Containing Exponents**
>
> Evaluate each expression.
>
> **(a)** 2^3 **(b)** $\left(\frac{1}{2}\right)^2$ **(c)** 5^4

(continued)

Solution

(a) The expression 2^3 means that the base, 2, is used as a factor 3 times, so
$$2^3 = 2 \cdot 2 \cdot 2 = 8$$

(b) $\left(\dfrac{1}{2}\right)^2 = \dfrac{1}{2} \cdot \dfrac{1}{2} = \dfrac{1}{4}$ (c) $5^4 = 5 \cdot 5 \cdot 5 \cdot 5 = 625$ ●

EXAMPLE 2 **Evaluating Expressions Containing Exponents**

Evaluate each expression.

(a) $(-3)^4$ (b) $(-3)^5$ (c) -3^4 (d) $-(-3)^4$

Solution

(a) $(-3)^4 = (-3)(-3)(-3)(-3) = 81$

(b) $(-3)^5 = (-3)(-3)(-3)(-3)(-3) = -243$

(c) The expression -3^4 means the additive inverse, or opposite, of 3^4. That is,
$$-3^4 = -(3)(3)(3)(3) = -81.$$

(d) $-(-3)^4 = -(-3)(-3)(-3)(-3) = -81$ ●

Work Smart

The expressions $(-3)^4$ and -3^4 are different. $(-3)^4$ means to use (-3) as a factor four times, while -3^4 means to determine the additive inverse of 3^4. That is,
$$(-3)^4 = (-3)(-3)(-3)(-3) = 81$$
and
$$-3^4 = -(3)(3)(3)(3) = -81$$
Another way to look at -3^4 is $-1 \cdot 3^4$.

The results of Examples 2(a) and 2(b) suggest that raising a negative number to an even integer exponent yields a positive number, while raising a negative number to an odd integer exponent yields a negative number. This conclusion is correct in general.

Quick ✓

1. In the expression 5^6, the number 5 is the ＿＿ and 6 is the ＿＿＿＿ or ＿＿.

2. *True or False* $-7^4 = (-7)(-7)(-7)(-7)$

In Problems 3–8, evaluate each expression.

3. 4^3 4. $(-7)^2$ 5. $(-10)^3$ 6. $\left(\dfrac{2}{3}\right)^3$ 7. -8^2 8. $-(-5)^3$

---

❷ **Use the Order of Operations to Evaluate Expressions**

To evaluate an expression containing both multiplication and addition, such as $2 + 3 \cdot 7$, order of operation dictates the following:

> If the two operations of addition and multiplication separate three numbers, always multiply first, then add.

Therefore, $2 + 3 \cdot 7 = 2 + 21 = 23$.

▶ **EXAMPLE 3** **Finding the Value of an Expression**

Evaluate each expression.

(a) $3 + 4 \cdot 5$ (b) $8 \cdot 2 + 1$

Solution

(a) Don't forget to use the order of operations! Multiply before adding.
$$3 + 4 \cdot 5 = 3 + 20$$
$$= 23$$

(b) $8 \cdot 2 + 1 = 16 + 1$
$$= 17$$ ●

Quick ✓

In Problems 9 and 10, evaluate each expression.

9. $5 \cdot 2 + 6$ **10.** $3 \cdot 2 + 5 \cdot 6$

When addition is to take place before multiplication, parentheses will be used. For example, to evaluate $(2 + 3) \cdot 7$, add 2 and 3 first, then multiply the sum by 7.

▶ **EXAMPLE 4** | **Finding the Value of an Expression**

(a) $(6 + 2) \cdot 4 = 8 \cdot 4$
$= 32$

(b) $(8 - 3)(7 + 3) = 5 \cdot 10$
$= 50$ ●

Quick ✓

In Problems 11–14, evaluate each expression.

11. $4(5 + 3)$ **12.** $8(9 - 3)$

13. $(12 - 4)(18 - 13)$ **14.** $(4 + 9)(6 - 4)$

Work Smart

Do not use the Reduction Property across addition.
This is incorrect:

$$\frac{4 + 5}{6 + 12} = \frac{\cancel{2} \cdot 2 + 5}{\cancel{2} \cdot 3 + 12}$$

$$= \frac{2 + 5}{3 + 12} = \frac{7}{15}$$

This is correct:

$$\frac{4 + 5}{6 + 12} = \frac{9}{18}$$

$$= \frac{1}{2}$$

In expressions like $\dfrac{4 + 5}{6 + 12}$, it is understood that the division bar acts like parentheses. This means evaluate the expressions in the numerator and denominator FIRST, and then divide. That is,

$$\frac{4 + 5}{6 + 12} = \frac{(4 + 5)}{(6 + 12)} = \frac{9}{18} = \frac{\cancel{9} \cdot 1}{\cancel{9} \cdot 2} = \frac{1}{2}$$

This rule will be extremely important throughout the course.

In addition to parentheses, grouping symbols include brackets [ ], braces { }, and absolute value symbols | |.

If one set of grouping symbols is embedded within a second set of grouping symbols, as in $2 + [3 + 2(9 - 4)]$, simplify within the innermost grouping symbols first.

EXAMPLE 5 | **Finding the Value of an Expression**

Evaluate each expression.

(a) $\dfrac{3 + 7}{6 + 4 \cdot 6}$ **(b)** $9[8 + 4(10 - 7)]$

Solution

(a) $\dfrac{3 + 7}{6 + 4 \cdot 6} = \dfrac{3 + 7}{6 + 24}$

$$= \frac{10}{30}$$

$$= \frac{1}{3}$$

Work Smart

When one set of parentheses is embedded within another set of parentheses, simplify the expression in the innermost parentheses first.

(b) Evaluate the expression in the innermost grouping symbols first.

$$9[8 + 4(10 - 7)] = 9[8 + 4(3)]$$
$$= 9[8 + 12]$$
$$= 9 \cdot 20$$
$$= 180 \quad ●$$

Quick ✓

In Problems 15–18, evaluate each expression.

15. $\dfrac{3 + 7}{4 + 9}$

16. $1 - 4 + 8 \cdot 2 + 5$

17. $25[2(8 - 3) - 8]$

18. $\dfrac{3 + 5}{2(9 - 4)}$

▶ When are exponents evaluated in the order of operations? In the expression $3 \cdot 2^4$, do we multiply 3 and 2 first and then raise this product to the 4th power to get 1296, or do we first raise 2 to the 4th power to get 16 and then multiply this by 3 to obtain 48? Because $2^4 = 2 \cdot 2 \cdot 2 \cdot 2$, we have $3 \cdot 2^4 = 3 \cdot 2 \cdot 2 \cdot 2 \cdot 2 = 48$, which indicates that exponents should be evaluated before multiplication.

Order of Operations

1. Evaluate expressions within **parentheses** first. When an expression has more than one set of grouping symbols, begin with the innermost grouping symbols and work outward.
2. Evaluate expressions containing **exponents,** working from left to right.
3. Perform **multiplication** and **division,** working from left to right.
4. Perform **addition** and **subtraction,** working from left to right.

Quick ✓

19. The order of operations is (1) _____ , (2) _____ ,
(3) _____ and _____ , (4) _____ and _____ .

In Problems 20–24, evaluate each expression.

20. $6 + 5 \cdot 2$

21. $(3 + 9) \cdot 4$

22. $\dfrac{7 + 5}{4 + 10}$

23. $4 + [(8 - 3) \cdot 2]$

24. $4(6 - 2) - 9$

EXAMPLE 6 **Evaluating Expressions Using the Order of Operations**

Evaluate each expression.

(a) $6 + 4 \cdot 3^2$

(b) $\dfrac{1}{4} \cdot 2^3 + 9 \cdot 4 - 2 \cdot 5^2$

Solution

(a) Evaluate the exponent first:

$$6 + 4 \cdot 3^2 = 6 + 4 \cdot 9$$
$$\text{Multiply: } = 6 + 36$$
$$= 42$$

(b) $\dfrac{1}{4} \cdot 2^3 + 9 \cdot 4 - 2 \cdot 5^2 = \dfrac{1}{4} \cdot 8 + 9 \cdot 4 - 2 \cdot 25$

$$\text{Multiply: } = 2 + 36 - 50$$
$$\text{Add from left to right: } = 38 - 50$$
$$= -12$$

EXAMPLE 7   **Evaluating Expressions Using the Order of Operations**

Evaluate each expression.

(a) $\dfrac{2 + 3 \cdot 4^2}{5(6 - 2)}$

(b) $-2\left|-(-2)^3 - 4(10 - 7)^2\right|$

Solution

(a)

$$\dfrac{2 + 3 \cdot 4^2}{5(6 - 2)} = \dfrac{2 + 3 \cdot 16}{5 \cdot 4}$$

$$= \dfrac{2 + 48}{20}$$

$$= \dfrac{50}{20}$$

$$= \dfrac{5 \cdot 10}{2 \cdot 10}$$

Divide out the common factor, 10: $= \dfrac{5}{2}$

(b) The absolute value bars form a grouping symbol, so simplify within the absolute value bars first.

Evaluate inner parentheses first:
↓

$$-2\left|-(-2)^3 - 4(10 - 7)^2\right| = -2\left|-(-2)^3 - 4(3)^2\right|$$

Evaluate exponents: $= -2\left|-(-8) - 4 \cdot 9\right|$

$$= -2\left|8 - 36\right|$$

$$= -2\left|-28\right|$$

$$= -2 \cdot 28$$

$$= -56$$

Quick ✓

In Problems 25–30, evaluate each expression.

25. $-8 + 2 \cdot 5^2$

26. $5 \cdot 3 - 3 \cdot 2^3$

27. $5(10 - 8)^2$

28. $3(-2)^2 + 6 \cdot 3 - 3 \cdot 4^2$

29. $\dfrac{4 + 6^2}{2 \cdot 3 + 2}$

30. $-3\left|7^2 - 2(8 - 5)^3\right|$

R.4 Exercises MyMathLab® Exercise numbers in green have complete video solutions in MyMathLab or may be accessed using the QR code to the right.

*Problems **1–30** are the Quick ✓ s that follow the* EXAMPLES.

Building Skills

In Problems 31–38, evaluate each expression. See Objective 1.

31. $(-3)^4$

32. $(-4)^3$

33. -5^4

34. -2^4

35. $-(-2)^3$

36. $-(-7)^3$

37. $-\left(\dfrac{2}{3}\right)^2$

38. $-\left(\dfrac{3}{4}\right)^4$

Mixed Practice

In Problems 39–74, evaluate each expression.

39. $4^2 - 3^2$

40. $(-3)^2 + (-2)^2$

41. $3 \cdot 2 + 9$

42. $-5 \cdot 3 + 12$

43. $4 + 2(6 - 2)$

44. $2 + 5(8 - 5)$

45. $-2[10 - (3 - 7)]$

46. $3[15 - (7 - 3)]$

47. $\dfrac{5 - (-7)}{4}$

48. $\dfrac{12 - 4}{-2}$

49. $\left|3 \cdot 2 - 4 \cdot 5\right|$

50. $\left|6 \cdot 2 - 5 \cdot 3\right|$

51. $12 \cdot \dfrac{2}{3} - 5 \cdot 2$ **52.** $15 \cdot \dfrac{3}{5} + 4 \cdot 3$

53. $3[2 + 3(1 + 5)]$ **54.** $2[25 - 2(10 - 4)]$

55. $(3^2 - 3)(3 - (-3)^3)$ **56.** $-2(5 - 2) - (-5)^2$

57. $|3(6 - 3^2)|$ **58.** $-2(4 + |2 \cdot 3 - 5^2|)$

59. $|4[2 \cdot 5 + (-3) \cdot 4]|$ **60.** $|6(3 \cdot 2 - 10)|$

61. $\dfrac{2 \cdot 5 + 15}{2^2 + 3 \cdot 2}$ **62.** $\dfrac{2 \cdot 3^2}{4^2 - 4}$

63. $\dfrac{2(4 + 8)}{3 + 3^2}$ **64.** $\dfrac{3(5 + 2^2)}{2 \cdot 3^3}$

65. $\dfrac{6[12 - 3(5 - 2)]}{5[21 - 2(4 + 5)]}$ **66.** $\dfrac{4[3 + 2(8 - 6)]}{5[14 - 2(2 + 3)]}$

67. $\left(\dfrac{2}{3}\right)^2 \left(\dfrac{1 + 2^3}{2^3 - 2}\right)$ **68.** $\left(\dfrac{3^2}{29 - 3 \cdot 2^3}\right) \cdot \dfrac{5}{4 + 5}$

69. $\dfrac{3^3 - 2^4 \cdot 3}{4(3^2 - 2 \cdot 3)}$ **70.** $2^5 + 4(-5) + 4^2$

71. $\dfrac{2 \cdot 4 - 5}{4^2 + (-2)^3} + \dfrac{3^2}{2^3}$ **72.** $\dfrac{5(37 - 6^2)}{6 \cdot 2 - 3^2} + \dfrac{7 \cdot 2 - 4^2}{5 + 4}$

73. $\dfrac{\dfrac{2 \cdot 3 + 3^2}{2 \cdot 5 - 8} + \dfrac{4}{3}}{\dfrac{7(5 - 3)}{14 - 2^3}}$ **74.** $\dfrac{\dfrac{4}{4^2 - 1} - \dfrac{3}{5(7 - 5)}}{\dfrac{-(-2)^2}{4 \cdot 7 + 2}}$

Applying the Concepts

In Problems 75–78, insert parentheses in order to make each statement true.

75. $3 \cdot 7 - 2 = 15$ **76.** $-2 \cdot 3 - 5 = 4$

77. $3 + 5 \cdot 6 - 3 = 18$ **78.** $3 + 5 \cdot 6 - 3 = 24$

△ **79. Geometry** The surface area of a closed right circular cylinder whose radius is 5 inches and height is 12 inches is given approximately by $2 \cdot 3.1416 \cdot 5^2 + 2 \cdot 3.1416 \cdot 5 \cdot 12$. Evaluate this expression, rounding your answer to two decimal places.

△ **80. Geometry** The surface area of a sphere whose radius is 3 centimeters is given approximately by $4 \cdot 3.1416 \cdot 3^2$. Evaluate this expression, rounding your answer to two decimal places.

81. Hitting a Golf Ball The height (in feet) of a golf ball hit with an initial speed of 100 feet per second after 3 seconds is given by $-16 \cdot 3^2 + 50 \cdot 3$. Evaluate this expression in order to determine the height of the golf ball after 3 seconds.

82. Horsepower The horsepower rating of an engine is $\dfrac{10^2 \cdot 8}{2.5}$. Evaluate this expression in order to determine the horsepower rating of the engine.

Math for the Future *Problems 83 and 84 show some computations required in statistics. Completely simplify each expression.*

83. $\dfrac{105 + 80 + 115 + 95 + 105}{5}$

84. $\dfrac{(105 - 100)^2 + (80 - 100)^2 + (115 - 100)^2 + (95 - 100)^2 + (105 - 100)^2}{4}$

Explaining the Concepts

85. Explain why $\dfrac{2 + 7}{2 + 9} \neq \dfrac{7}{9}$.

86. Develop an example that illustrates why we perform multiplication before addition when there are no grouping symbols.

87. Develop an example that illustrates why we evaluate exponents before multiplication when there are no grouping symbols.

88. Explain the difference between -4^3 and $(-4)^3$.

Technology Exercises

Calculators can be used to evaluate exponential expressions. Figures 12(a) and (b) show the results of Examples 2(a), (b), and (c) using a graphing calculator.

Figure 12

(a)   (b)

Graphing calculators know the order of operations. Figure 13(a) on the following page shows the result of Example 6(b), and Figure 13(b) the result of Example 7(a), using a graphing calculator. Be careful with the placement of parentheses when using the calculator!

Figure 13

(a)   (b)

In Problems 89–96, use technology to evaluate each expression. When necessary, express answers rounded to two decimal places.

89. $\dfrac{4}{5} - \left(\dfrac{2}{3}\right)^2$

90. $3 - \left(\dfrac{6}{5}\right)^3$

91. $\dfrac{4^2 + 1}{13}$

92. $\dfrac{3^2 - 2^3}{1 + 3 \cdot 2}$

93. $\dfrac{3.5^3}{1.3^2} - 6.2^3$

94. $2.3^4 \cdot \dfrac{4}{11} - (3.7)^2 \cdot \dfrac{8}{3}$

95. $4.3[9.3^2 - 4(34.2 + 18.5)]$

96. $6.3^2 + 4.2^2$

R.5 Algebraic Expressions

Objectives

❶ Translate English Expressions into Mathematical Language

❷ Evaluate Algebraic Expressions

❸ Simplify Algebraic Expressions by Combining Like Terms

❹ Determine the Domain of a Variable

In Other Words

In algebra, letters of the alphabet are used to represent numbers.

Remember that if a letter is used to represent *any* number from a given set of numbers, it is called a **variable**. A **constant** is either a fixed number, such as 5 or $\sqrt{2}$, or a letter that represents a fixed (possibly unspecified) number. For example, in Einstein's Theory of Relativity, $E = mc^2$, E and m are variables that represent total energy and mass, respectively, while c is a constant that represents the speed of light (299,792,458 meters per second).

An **algebraic expression** is any combination of variables, constants, grouping symbols such as parentheses ( ) and brackets [ ], and mathematical operations such as addition, subtraction, multiplication, division, and exponents. The following are examples of algebraic expressions.

$$3x + 4 \qquad 2y^2 - 5y - 12z \qquad \frac{5v^4 - v}{4 - v}$$

Work Smart

An algebraic expression is not the same as an algebraic equation. An expression might be $x + 3$ while an equation might be $x + 3 = 7$. Do you see the difference?

▶ ❶ **Translate English Expressions into Mathematical Language**

One of the neat features of mathematics is that various English phrases can be translated into a few math symbols. Table 2 lists some English words and phrases and their corresponding math symbols.

Table 2 Math Symbols and the Words They Represent

Add (+)	Subtract (−)	Multiply (·)	Divide (/)
sum	difference	product	quotient
plus	minus	times	divided by
more than	subtracted from	of	per
exceeds by	less	twice	ratio
in excess of	less than		
added to	decreased by		
increased by			

EXAMPLE 1 **Writing English Phrases Using Math Symbols**

Express each English phrase as an algebraic expression.

(a) The sum of 3 and 8

(b) The quotient of 50 and some number y

(c) 9 less than 15

(d) Twice the sum of a number x and 5

Solution

(a) A sum implies the $+$ symbol, so "the sum of 3 and 8" is represented mathematically as $3 + 8$.

(b) "The quotient of 50 and some number y" is represented mathematically as $\dfrac{50}{y}$.

(c) "9 less than 15" is represented mathematically as $15 - 9$.

(d) "Twice the sum of a number x and 5" is represented as $2(x + 5)$. Notice that the example said "twice the sum." (The English phrase represented as $2x + 5$ might be "the sum of twice a number and 5." Do you see the difference?) ●

Quick ✓

1. A _____ is a letter used to represent any number from a given set of numbers.

2. A _____ is either a fixed number or a letter used to represent a fixed (possibly unspecified) number.

In Problems 3–8, express each English phrase using an algebraic expression.

3. The sum of 3 and 11

4. The product of 6 and 7

5. The quotient of y and 4

6. The difference of 3 and z

7. Twice the difference of x and 3

8. The difference of twice a number x and 3

▶ ❷ Evaluate Algebraic Expressions

To **evaluate an algebraic expression,** substitute a numerical value for each variable in the expression, and simplify the result.

EXAMPLE 2 **Evaluating an Algebraic Expression**

Evaluate each expression for the given value of the variable.

(a) $4x + 3$ for $x = 5$ (b) $\dfrac{-z^2 + 4z}{z + 1}$ for $z = 9$ (c) $|10y - 8|$ for $y = \dfrac{1}{2}$

Solution

(a) Substitute 5 for x in the expression $4x + 3$.
$$4(5) + 3 = 20 + 3 = 23$$

(b) Substitute 9 for z in the expression $\dfrac{-z^2 + 4z}{z + 1}$.
$$\frac{-(9)^2 + 4 \cdot 9}{9 + 1} = \frac{-81 + 36}{10} = \frac{-45}{10} = \frac{-9 \cdot \cancel{5}}{\cancel{5} \cdot 2} = \frac{-9}{2} = -\frac{9}{2}$$

(c) Substitute $\dfrac{1}{2}$ for y in the expression $|10y - 8|$.
$$\left|10 \cdot \frac{1}{2} - 8\right| = |5 - 8| = |-3| = 3$$
●

Work Smart

Be sure to use parentheses when you substitute a number in for a variable raised to an exponent.

Quick ✔

Quick ✔

9. To _____ an algebraic expression, substitute a numerical value for each variable in the expression, and simplify the result.

In Problems 10–13, evaluate each expression for the given value of the variable.

10. $-5x + 3$ for $x = 2$

11. $y^2 - 6y + 1$ for $y = -4$

12. $\dfrac{w + 8}{3w}$ for $w = 4$

13. $|4x - 5|$ for $x = \dfrac{1}{2}$

EXAMPLE 3

Figure 14

$x + 10$

x

Evaluating an Algebraic Expression

The algebraic expression $2x + 2(x + 10)$ represents the perimeter of a rectangular field whose length is 10 yards more than its width. See Figure 14. Evaluate the algebraic expression for $x = 4, 8,$ and 10.

Solution
Evaluate the algebraic expression $2x + 2(x + 10)$ for each value of x.

$x = 4$: $2(4) + 2(4 + 10) = 8 + 2(14) = 8 + 28 = 36$ yards

$x = 8$: $2(8) + 2(8 + 10) = 16 + 2(18) = 16 + 36 = 52$ yards

$x = 10$: $2(10) + 2(10 + 10) = 20 + 2(20) = 20 + 40 = 60$ yards ●

Quick ✔

14. The algebraic expression $102x$ approximates the number of Japanese yen that you could purchase for x dollars. Evaluate the algebraic expression for $x = 100, 1000,$ and 10,000 dollars. (Source: *Yahoo! Finance*)

15. The algebraic expression $\dfrac{5}{9}(x - 32)$ represents the temperature in degrees Celsius that is equivalent to x degrees Fahrenheit. Determine the equivalent temperature for $x = 32, 86,$ and 212 degrees Fahrenheit.

❸ Simplify Algebraic Expressions by Combining Like Terms

One way to simplify algebraic expressions is to combine *like terms*. A **term** is a number or the product of a number and one or more variables raised to a power. In algebraic expressions, terms are separated by addition signs. See Table 3, which lists the terms in various algebraic expressions.

Table 3	
Algebraic Expression	**Terms**
$5x + 4$	$5x, 4$
$7x^2 - 8x + 3 = 7x^2 + (-8x) + 3$	$7x^2, -8x, 3$
$3x^2 + 7y^2$	$3x^2, 7y^2$

Notice that in the algebraic expression $7x^2 - 8x + 3$, we first rewrite it with only addition signs as $7x^2 + (-8x) + 3$ to identify the terms.

Terms that have the same variable(s) and the same exponent(s) on the variable(s) are **like terms.** For example, $3x$ and $8x$ are like terms because both terms have the variable x raised to the 1st power. Also, $-4x^3y$ and $10x^3y$ are like terms because both have the variable x raised to the 3rd power, along with y raised to the 1st power. The numerical factor of the term is the **coefficient.** For example, the coefficient of $3x$ is 3; the coefficient of $-4x^3y$ is -4. Terms that have no number as a factor, such as xy, have the coefficient one (1) because $xy = 1 \cdot xy$. The coefficient of the term $-z$ is negative one (-1) because

$-z = -1 \cdot z$. If a term consists of just a constant, the coefficient is the number itself. For example, the coefficient of the term 19 is 19.

To combine like terms use the Distributive Property "in reverse."

▶ **EXAMPLE 4** **Combining Like Terms**

Simplify each algebraic expression by combining like terms.

(a) $5x + 3x$ (b) $z + 7z - 5$

Solution

(a) $5x + 3x = (5 + 3)x = 8x$

(b) Since $z = 1 \cdot z$ because of the multiplicative identity, it follows that

$$z + 7z - 5 = 1z + 7z - 5$$

Use the Distributive Property "in reverse": $= (1 + 7)z - 5$

Combine like terms: $= 8z - 5$ ●

> **Quick ✓**
>
> **16.** Terms that have the same variable(s) and the same exponent(s) on the variables are called _____.
>
> **17.** The coefficient of the term $-mn$ is __.
>
> *In Problems 18–21, simplify each expression by combining like terms.*
>
> **18.** $4x - 9x$ **19.** $-2x^2 + 13x^2$
>
> **20.** $-5x - 3x + 6 - 3$ **21.** $6x - 10x - 4y + 12y$

Sometimes terms must be rearranged using the Commutative Property of Addition to combine like terms.

EXAMPLE 5 **Combining Like Terms Using the Commutative and Distributive Properties**

Simplify each algebraic expression.

(a) $13y^2 + 8 - 4y^2 + 3$ (b) $12z + 5 + 5z - 8z - 2$

Solution

(a) Use the Commutative Property to rearrange the terms.

$$13y^2 + 8 - 4y^2 + 3 = 13y^2 - 4y^2 + 8 + 3$$

Use the Distributive Property "in reverse": $= (13 - 4)y^2 + 8 + 3$

Combine like terms: $= 9y^2 + 11$

(b) Rearrange terms before using the Distributive Property.

$$12z + 5 + 5z - 8z - 2 = 12z + 5z - 8z + 5 - 2$$

Use the Distributive Property "in reverse": $= (12 + 5 - 8)z + 5 - 2$

Combine like terms: $= 9z + 3$ ●

> **Quick ✓**
>
> *In Problems 22–24, simplify each expression by combining like terms.*
>
> **22.** $10y - 3 + 5y + 2$ **23.** $0.5x^2 + 1.3 + 1.8x^2 - 0.4$
>
> **24.** $4z + 6 - 8z - 3 - 2z$

It may also be necessary to remove parentheses using the Distributive Property before collecting like terms.

▶ **EXAMPLE 6**   **Combining Like Terms Using the Distributive Property**

Simplify each algebraic expression.

(a) $5(x - 3) - 2x$ **(b)** $x + 4 - 2(x + 3)$

Solution

(a) First, use the Distributive Property to remove the parentheses.

$$5(x - 3) - 2x = 5x - 15 - 2x$$

Rearrange terms: $= 5x - 2x - 15$

Combine like terms: $= 3x - 15$

(b) Distribute -2 to remove the parentheses.

$$x + 4 - 2(x + 3) = x + 4 - 2x - 6$$

Rearrange terms: $= x - 2x + 4 - 6$

Combine like terms: $= -x - 2$ ●

EXAMPLE 7   **Combining Like Terms Using the Distributive Property**

Simplify each algebraic expression.

(a) $5(x - 3) - (7x - 4)$ **(b)** $\dfrac{1}{2}(4x + 3) + \dfrac{8x - 3}{5}$

Solution

(a) When grouping symbols are preceded by a minus sign, multiply the terms in the grouping symbol by -1.

$$5(x - 3) - (7x - 4) = 5(x - 3) - 1(7x - 4)$$

Use the Distributive Property: $= 5x - 15 - 7x + 4$

Rearrange terms: $= 5x - 7x - 15 + 4$

Combine like terms: $= -2x - 11$

(b) Use the fact that $\dfrac{a}{b} = \dfrac{1}{b} \cdot a$, so $\dfrac{8x - 3}{5} = \dfrac{1}{5}(8x - 3)$.

$$\frac{1}{2}(4x + 3) + \frac{8x - 3}{5} = \frac{1}{2}(4x + 3) + \frac{1}{5}(8x - 3)$$

Use the Distributive Property: $= \dfrac{1}{2} \cdot 4x + \dfrac{1}{2} \cdot 3 + \dfrac{1}{5} \cdot 8x + \dfrac{1}{5} \cdot (-3)$

Simplify: $= 2x + \dfrac{3}{2} + \dfrac{8}{5}x - \dfrac{3}{5}$

Rearrange terms; rewrite fractions with least common denominator: $= \dfrac{10}{5}x + \dfrac{8}{5}x + \dfrac{15}{10} - \dfrac{6}{10}$

Combine like terms: $= \dfrac{18}{5}x + \dfrac{9}{10}$ ●

Quick ✓

In Problems 25–30, simplify each expression by combining like terms.

25. $3(x - 2) + x$

26. $5(y + 3) - 10y - 4$

27. $3(z + 4) - 2(3z + 1)$

28. $-4(x - 2) - (2x + 4)$

29. $\dfrac{1}{2}(6x + 4) - \dfrac{15x + 5}{5}$

30. $\dfrac{5x - 1}{3} + \dfrac{5x + 9}{2}$

▶ ④ Determine the Domain of a Variable

In some algebraic expressions, the variable may be allowed to take on values from only a certain set of numbers. For example, in the expression $\dfrac{1}{x}$, the variable x cannot take on the value 0, because this would cause division by 0, which is not defined.

> **Definition**
>
> The set of values that a variable may assume is called the **domain of the variable.**

EXAMPLE 8 **Determining the Domain of a Variable**

Determine which of the following numbers are in the domain of the variable x for the expression $\dfrac{4}{x + 3}$.

(a) $x = 3$ (b) $x = 0$ (c) $x = -3$

Solution

Determine whether the value of the variable causes division by 0. That is, determine whether the value of the variable causes $x + 3$ to equal 0. If it does, exclude it from the domain.

(a) When $x = 3$, the denominator is $x + 3 = 3 + 3 = 6$, so 3 is in the domain of the variable.

(b) When $x = 0$, the denominator is $x + 3 = 0 + 3 = 3$, so 0 is in the domain.

(c) When $x = -3$, the denominator $x + 3 = -3 + 3 = 0$, so -3 is NOT in the domain.   ●

Quick ✓

31. The set of values that a variable may assume is called the _____ of the variable.

In Problems 32–34, determine which of the following numbers are in the domain of the variable.

(a) $x = 2$ (b) $x = 0$ (c) $x = 4$ (d) $x = -3$

32. $\dfrac{2}{x - 4}$

33. $\dfrac{x}{x + 3}$

34. $\dfrac{x + 3}{x^2 + x - 6}$

R.5 Exercises MyMathLab® Exercise numbers in green have complete video solutions in MyMathLab or may be accessed using the QR code to the right.

*Problems **1–34** are the Quick ✓s that follow the* **EXAMPLES.**

Building Skills

In Problems 35–46, express each English phrase using an algebraic expression. See Objective 1.

35. The sum of 5 and a number x

36. The difference of 10 and a number y

37. The product of 4 and a number z

38. The ratio of a number x and 5

39. A number y decreased by 7

40. A number z increased by 30

41. Twice the sum of a number t and 4

42. The sum of a number x and 5 divided by 10

43. Three less than five times a number x

44. Three times a number z increased by the quotient of z and 8

45. The quotient of some number y and 3 increased by the product of 6 and some number x

46. Twice some number x decreased by the ratio of a number y and 3

In Problems 47–62, evaluate each expression for the given value of the variable. See Objective 2.

47. $4x + 3$ for $x = 2$

48. $-5x + 1$ for $x = -3$

49. $x^2 + 5x - 3$ for $x = -2$

50. $y^2 - 4y + 5$ for $y = 3$

51. $4 - z^2$ for $z = -5$

52. $3 + z - 2z^2$ for $z = -4$

53. $\dfrac{2w}{w^2 + 2w + 1}$ for $w = 3$

54. $\dfrac{4z + 3}{z^2 - 4}$ for $z = 3$

55. $\dfrac{v^2 + 2v + 1}{v^2 + 3v + 2}$ for $v = 5$

56. $\dfrac{2x^2 + 5x + 2}{x^2 + 5x + 6}$ for $x = 3$

57. $|5x - 4|$ for $x = -5$

58. $|x^2 - 6x + 1|$ for $x = 2$

59. $(x + 2y)^2$ for $x = 3, y = -4$

60. $(a - 5b)^2$ for $a = 1, b = 3$

61. $\dfrac{(x + 2)^2}{|4x - 10|}$ for $x = 1$

62. $\dfrac{|3 - 5z|}{(z - 4)^2}$ for $z = 4$

In Problems 63–92, simplify each expression by combining like terms. See Objective 3.

63. $3x - 2x$

64. $5y + 2y$

65. $-4z - 2z + 3$

66. $8x - 9x + 1$

67. $13z + 2 - 14z - 7$

68. $-10x + 6 + 4x - x + 1$

69. $\dfrac{3}{4}x + \dfrac{1}{6}x$

70. $\dfrac{3}{10}y + \dfrac{4}{15}y$

71. $2x + 3x^2 - 5x + x^2$

72. $-x - 3x^2 + 4x - x^2$

73. $-1.3x - 3.4 + 2.9x + 3.4$

74. $2.5y - 1.8 - 1.4y + 0.4$

75. $3x - 2 - x + 3 - 5x$

76. $10y + 3 - 2y + 6 + y$

77. $-2(5x - 4) - (4x + 1)$

78. $3(2y + 5) - 6(y + 2)$

79. $5(z + 2) - 6z$

80. $\dfrac{1}{2}(20x - 14) + \dfrac{1}{3}(6x + 9)$

81. $\dfrac{2}{5}(5x - 10) + \dfrac{1}{4}(8x + 4)$

82. $4(w + 2) + 3(4w + 3)$

83. $2(v - 3) + 5(2v - 1)$

84. $-4(w - 3) - (2w + 1)$

85. $\dfrac{3}{5}(10x + 4) - \dfrac{8x + 3}{2}$

86. $\dfrac{4}{3}(5y + 1) - \dfrac{2}{5}(3y - 4)$

87. $\dfrac{5}{6}\left(\dfrac{3}{10}x - \dfrac{2}{5}\right) + \dfrac{2}{3}\left(\dfrac{1}{6}x + \dfrac{1}{2}\right)$

88. $\dfrac{1}{4}\left(\dfrac{2}{3}x - \dfrac{1}{2}\right) + \dfrac{1}{10}\left(\dfrac{5}{2}x - \dfrac{15}{4}\right)$

89. $4.3\,(1.2x - 2.3) + 9.3x - 5.6$

90. $0.4\,(2.9x - 1.6) - 2.7\,(0.3x + 6.2)$

91. $6.2\,(x - 1.4) - 5.4\,(3.2x - 0.6)$

92. $9.3\,(0.2x - 0.8) + 3.8\,(1.3x + 6.3)$

In Problems 93–98, determine which of the following numbers are in the domain of the variable. See Objective 4.

(a) $x = 5$ **(b)** $x = -1$ **(c)** $x = -4$ **(d)** $x = 0$

93. $\dfrac{3}{x - 5}$

94. $\dfrac{7}{x + 1}$

95. $\dfrac{x + 1}{x + 5}$

96. $\dfrac{x + 4}{x - 1}$

97. $\dfrac{x}{x^2 - 5x}$

98. $\dfrac{x + 1}{x^2 + 5x + 4}$

Applying the Concepts

△ **99. Volume of a Cube** The algebraic expression s^3 represents the volume of a cube whose sides are length s. See the figure. Evaluate the algebraic expression for $s = 1, 2, 3,$ and 4 inches.

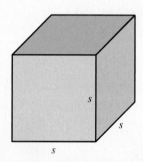

△ **100. Area of a Triangle** The algebraic expression $\dfrac{1}{2}\,h(h + 2)$ represents the area of a triangle whose base is 2 centimeters longer than its height h. See the figure. Evaluate the algebraic expression for $h = 2, 5,$ and 10 centimeters.

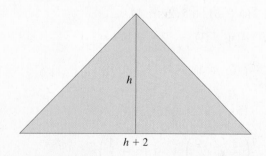

101. Projectile Motion The algebraic expression $-16t^2 + 75t$ represents the height (in feet) of a golf ball hit at an angle of 30° to the horizontal after t seconds.

  (a) Evaluate the algebraic expression for $t = 0, 1, 2, 3,$ and 4 seconds.

  (b) Use the results of part (a) to describe what happens to the golf ball as time passes.

102. Cost of Production The algebraic expression $30x + 1000$ represents the cost of manufacturing x watches in a day. Evaluate the algebraic expression for $x = 20, 30,$ and 40 watches.

103. How Old Is Tony? Suppose that Bob is x years of age. Write a mathematical expression for the following: "Tony is 5 years older than Bob." Evaluate the expression if Bob is $x = 13$ years of age.

104. Getting a Discount Suppose the regular price of a computer is p dollars. Write a mathematical expression for the following: "I'll give you $50 off regular price." Evaluate the expression if the regular price of the computer is $890.

105. Getting a Big Discount Suppose the regular price of a computer is p dollars. Write a mathematical expression for the following: "I'll give you half off regular price." Evaluate the expression if the regular price of the computer is $900.

106. How Old Is Marissa's Mother? Suppose that Marissa is x years of age. Write a mathematical expression for the following: The age of "Marissa's mother is twice the sum of Marissa's age and 3." How old is Marissa's mother if Marissa is $x = 18$ years of age?

Math for the Future *For Problems 107 and 108, evaluate the expression* $\dfrac{X - \mu}{\sigma}$ *for the given values.*

107. $X = 120, \mu = 100, \sigma = 15$

108. $X = 40, \mu = 50, \sigma = 10$

Extending the Concepts

In Problems 109–114, write an English phrase that would translate into the given mathematical expression.

109. $2z - 5$ **110.** $5x + 3$ **111.** $2\,(z - 5)$

112. $5\,(x + 3)$ **113.** $\dfrac{z + 3}{2}$ **114.** $\dfrac{t}{3} - 2t$

Explaining the Concepts

115. Explain the difference between a variable and a constant.

116. Explain the difference between a term and a factor.

117. Explain what like terms are. Explain how the Distributive Property is used to combine like terms.

118. Explain the difference between the direction "evaluate the algebraic expression" and "simplify the algebraic expression."

Technology Exercises

Technology can be used to evaluate an algebraic expression. Figure 15 shows the results of Example 2 using a graphing calculator.

Figure 15

(a)   (b)

(c)

In Problems 119–126, use technology to evaluate each algebraic expression for the given values of the variable.

119. $-4x + 3$ for **(a)** $x = 0$ **(b)** $x = -3$

120. $-5x + 9$ for **(a)** $x = 4$ **(b)** $x = -3$

121. $4x^2 - 8x + 3$ for **(a)** $x = 5$ **(b)** $x = -2$

122. $-9x^2 + x - 5$ for **(a)** $x = 3$ **(b)** $x = -4$

123. $\dfrac{3z - 1}{z^2 + 1}$ for **(a)** $z = -2$ **(b)** $z = 8$

124. $\dfrac{2y^2 + 5}{3y - 1}$ for **(a)** $y = 3$ **(b)** $y = -8$

125. $|-9x + 5|$ for **(a)** $x = 8$ **(b)** $x = -3$

126. $|-3x^2 + 5x - 2|$ for **(a)** $x = 6$ **(b)** $x = -4$

127. Use technology to evaluate $\dfrac{2x + 1}{x - 5}$ when $x = 5$. What result is displayed? Why?

128. Use technology to evaluate $\dfrac{x + 2}{(x + 2)(x - 2)}$ when $x = -2$. What result is displayed? Why?

1 Linear Equations and Inequalities

Did you know that the amount of income tax an individual or a couple pays to the federal government can be found by solving a linear equation? See Problems 103 and 104 in Section 1.1.

The Big Picture: Putting It Together

Chapter R reviewed skills learned in earlier courses. These skills will be used throughout the text.

Our discussion of *algebra* will now begin. The word "algebra" is derived from the Arabic word *al-jabr*. The word is part of the title of a ninth-century work, "Hisâb al-jabr w'al-muqâbalah," written during the golden age of Islamic science and mathematics by Muhammad ibn Mûqâ al-Khowârizî.

The word *al-jabr* means "restoration," a reference to the fact that if a number is added to one side of an equation, then it must also be added to the other side in order to "restore" the equality. The title of the work "Hisâb al-jabr w'al-muqâbalah" means "the science of restoring and canceling." Today, algebra means a great deal more.

Chapter 1 consists of Part I, a review of linear equations and inequalities in one variable, and Part II, a review of linear equations and inequalities in two variables.

1.1 Linear Equations in One Variable

Objectives

❶ Determine Whether a Number Is a Solution to an Equation
❷ Solve Linear Equations
❸ Determine Whether an Equation Is a Conditional Equation, an Identity, or a Contradiction

Are You Prepared for This Section?

Before getting started, complete the following problems. If you get a problem wrong, go back to the section cited and review the material.

P1. Determine the additive inverse of 5. [Section R.3, p. 21]

P2. Determine the multiplicative inverse of -3. [Section R.3, pp. 23–24]

P3. Simplify: $\dfrac{1}{5} \cdot 5x$. [Section R.3, p. 25]

P4. Find the Least Common Denominator of $\dfrac{3}{8}$ and $\dfrac{5}{12}$. [Section R.3, pp. 27–28]

P5. Use the Distributive Property to remove the parentheses: $6(z - 2)$. [Section R.3, pp. 29–30]

P6. What is the coefficient of $-4x$? [Section R.5, pp. 41–42]

P7. Simplify by combining like terms: $4(y - 2) - y + 5$. [Section R.5, pp. 41–44]

P8. Evaluate the expression $-5(x + 3) - 8$ when $x = -2$. [Section R.5, pp. 40–41]

P9. Is $x = 3$ in the domain of $\dfrac{2}{x + 3}$? Is $x = -3$ in the domain? [Section R.5, p. 44]

▶ ❶ Determine Whether a Number Is a Solution to an Equation

An **equation in one variable** is a statement made up of two expressions that are equal, where at least one of the expressions contains the variable. The expressions are called the **sides** of the equation. Examples of equations in one variable are

$$2y + 5 = 0 \qquad 4x + 5 = -2x + 10 \qquad \frac{3}{z + 2} = 9$$

This section focuses on solving *linear equations in one variable*.

> ### Definition
>
> A **linear equation in one variable** is an equation with one unknown that is written to the first power. Linear equations in one variable can be written in the form
>
> $$ax + b = 0$$
>
> where a and b are real numbers and $a \neq 0$.

The following are all examples of linear equations in one variable because they can be written in the form $ax + b = 0$ with a little algebraic manipulation.

$$4x - 3 = 12 \qquad \frac{2}{3}y + \frac{1}{5} = \frac{2}{15} \qquad -0.73p + 1.23 = 1.34p + 8.05$$

Because an equation is a statement, it can be either true or false, depending on the value of the variable. Any value of the variable that results in a true statement is called a **solution** of the equation. If a value of the variable results in a true statement, we say that the value **satisfies** the equation. To determine whether a number satisfies an equation, we replace the variable with the number. If the left side of the equation equals the right side, then the statement is true and the number is a solution.

In Other Words

In the equation $2y + 5 = 0$, the expression $2y + 5$ is the *left side* of the equation and 0 is the *right side*. In this equation, only the left side contains the variable, y. In the equation $4x + 5 = -2x + 10$, the expression $4x + 5$ is the *left side* of the equation and $-2x + 10$ is the *right side*. In this equation, both sides have expressions that contain the variable, x.

Work Smart

Although there is no exponent written on x in the equation $ax + b = 0$, it is important to remember that $x = x^1$.

Prepared? ... Answers **P1.** -5
P2. $-\dfrac{1}{3}$ **P3.** x **P4.** 24 **P5.** $6z - 12$
P6. -4 **P7.** $3y - 3$ **P8.** -13
P9. Yes; No

EXAMPLE 1 **Determining Whether a Number Is a Solution to a Linear Equation**

Determine whether the following numbers are solutions to the equation:

$$3(x - 1) = -2x + 12$$

(a) $x = 5$ **(b)** $x = 3$

Solution

(a) Let $x = 5$ in the equation and simplify.

$$3(x - 1) = -2x + 12$$

Let $x = 5$: $3(5 - 1) \stackrel{?}{=} -2(5) + 12$

Simplify: $3(4) \stackrel{?}{=} -10 + 12$

$$12 \neq 2 \quad \text{False}$$

In Other Words
The symbol $\stackrel{?}{=}$ is used to indicate that we are unsure whether the left side of the equation equals the right side of the equation.

Because the left side of the equation does not equal the right side of the equation, we do not have a true statement. Therefore, $x = 5$ is not a solution.

(b) $3(x - 1) = -2x + 12$

Let $x = 3$: $3(3 - 1) \stackrel{?}{=} -2(3) + 12$

Simplify: $3(2) \stackrel{?}{=} -6 + 12$

$$6 = 6 \quad \text{True}$$

Because the left and right sides of the equation are equal, we have a true statement. Therefore, $x = 3$ is a solution to the equation.   ●

Quick ✓

1. The equation $3x + 5 = 2x - 3$ is a _____ equation in one variable. The expressions $3x + 5$ and $2x - 3$ are called _____ of the equation.

2. The _____ of a linear equation is the value or values of the variable that satisfy the equation.

In Problems 3–5, determine which of the given numbers are solutions to the equation.

3. $-5x + 3 = -2; x = -2, x = 1, x = 3$

4. $3x + 2 = 2x - 5; x = 0, x = 6, x = -7$

5. $-3(z + 2) = 4z + 1; z = -3, z = -1, z = 2$

Work Smart

The directions "solve," "simplify," and "evaluate" are different! We *solve* equations, where **equa**tions have an **equal** sign. We *simplify* algebraic expressions to form equivalent algebraic expressions. We *evalu*ate algebraic expressions to find the **valu**e of the expression for a specific value of the variable.

② **Solve Linear Equations**

To **solve an equation** means to find ALL the solutions of the equation. The set of all solutions to the equation is called the **solution set** of the equation.

One method for solving equations uses a series of *equivalent equations* beginning with the original equation and ending with a solution.

Definition

Two or more equations with the same solutions are **equivalent equations.**

But how are equivalent equations found? One way is to use the *Addition Property of Equality*.

In Other Words

The Addition Property says that whatever you add to one side of an equation, you must also add to the other side.

Addition Property of Equality

The **Addition Property of Equality** states that for real numbers a, b, and c,

$$\text{if } a = b, \quad \text{then} \quad a + c = b + c$$

For example, if $x = 3$, then $x + 2 = 3 + 2$ (2 was added to both sides of the equation). Because $a - b$ is equivalent to $a + (-b)$, the Addition Property can be used to add a real number to each side of an equation or to subtract a real number from each side of an equation. This property will be used a great deal in algebra.

An equivalent equation can also be found by using the *Multiplication Property of Equality*.

In Other Words

The Multiplication Property says that whenever you multiply one side of an equation by a nonzero expression, you must also multiply the other side by the same nonzero expression.

Multiplication Property of Equality

The **Multiplication Property of Equality** states that for real numbers a, b, and c where $c \neq 0$,

$$\text{if } a = b, \quad \text{then} \quad ac = bc$$

For example, if $5x = 30$, then $\frac{1}{5} \cdot 5x = \frac{1}{5} \cdot 30$. Also, because $\frac{a}{b} = a \cdot \frac{1}{b}$, the Multiplicative Property of Equality says that we may divide both sides of an equation by the same nonzero number and the result will be an equivalent equation. Thus if $5x = 30$, then it is also true that $\frac{5x}{5} = \frac{30}{5}$. So the Multiplication Property can be used to multiply or divide each side of an equation by some nonzero number.

EXAMPLE 2

Using the Addition and Multiplication Properties to Solve a Linear Equation

Solve the linear equation: $\frac{1}{3}x - 2 = 4$

Solution

The goal in solving any linear equation is to get the variable by itself with a coefficient of 1—that is, to **isolate the variable.**

Work Smart

The number in front of the variable expression is the coefficient. For example, the coefficient in the expression $2x$ is 2.

$$\frac{1}{3}x - 2 = 4$$

Addition Property of Equality; add 2 to both sides:
$$\left(\frac{1}{3}x - 2\right) + 2 = 4 + 2$$

Simplify:
$$\frac{1}{3}x = 6$$

Multiplication Property of Equality; multiply both sides by 3:
$$3\left(\frac{1}{3}x\right) = 3 \cdot 6$$

Simplify:
$$x = 18$$

Check
$$\frac{1}{3}x - 2 = 4$$

Let $x = 18$ in the original equation:
$$\frac{1}{3}(18) - 2 \stackrel{?}{=} 4$$
$$6 - 2 \stackrel{?}{=} 4$$
$$4 = 4 \quad \text{True}$$

Because $x = 18$ satisfies the equation, the solution of line equation is 18, or the solution set is $\{18\}$.

6. *True or False* The Addition Property of Equality says that whatever you add to one side of an equation you must also add to the other side.

7. To _____ the variable means to get the variable by itself with a coefficient of 1.

In Problems 8–10, solve each equation and verify your solution.

8. $3x + 8 = 17$ 9. $-4a - 7 = 1$ 10. $5y + 1 = 2$

Often, it is necessary to combine like terms or use the Distributive Property to eliminate parentheses before the Addition or Multiplication Properties can be used. As usual, the goal is to isolate the variable.

▶ **EXAMPLE 3**   **Solving a Linear Equation by Combining Like Terms**

Solve the linear equation: $3y - 2 + 5y = 2y + 5 + 4y + 3$

Solution

$$3y - 2 + 5y = 2y + 5 + 4y + 3$$

Combine like terms: $$8y - 2 = 6y + 8$$

Subtract 6y from both sides: $$8y - 2 - 6y = 6y + 8 - 6y$$

$$2y - 2 = 8$$

Add 2 to both sides: $$2y - 2 + 2 = 8 + 2$$

$$2y = 10$$

Divide both sides by 2: $$\frac{2y}{2} = \frac{10}{2}$$

$$y = 5$$

Check $$3y - 2 + 5y = 2y + 5 + 4y + 3$$

Let $y = 5$ in the original equation: $$3(5) - 2 + 5(5) \stackrel{?}{=} 2(5) + 5 + 4(5) + 3$$

$$15 - 2 + 25 \stackrel{?}{=} 10 + 5 + 20 + 3$$

$$38 = 38 \quad \text{True}$$

Because $y = 5$ satisfies the equation, the solution of the equation is 5, or the solution set is $\{5\}$.   ●

In Problems 11–13, solve each linear equation and verify your solution.

11. $2x + 3 + 5x + 1 = 4x + 10$

12. $4b + 3 - b - 8 - 5b = 2b - 1 - b - 1$

13. $2w + 8 - 7w + 1 = 3w - 1 + 2w - 5$

▶ **EXAMPLE 4**   **Solving a Linear Equation Using the Distributive Property**

Solve the linear equation: $4(x + 3) = x - 3(x - 2)$

Solution

Use the Distributive Property $$4(x + 3) = x - 3(x - 2)$$

to remove parentheses: $$4x + 12 = x - 3x + 6$$

Combine like terms: $$4x + 12 = -2x + 6$$

$$\text{Add } 2x \text{ to both sides: } 4x + 12 + 2x = -2x + 6 + 2x$$

$$6x + 12 = 6$$

$$\text{Subtract 12 from both sides: } 6x + 12 - 12 = 6 - 12$$

$$6x = -6$$

$$\text{Divide both sides by 6: } \frac{6x}{6} = \frac{-6}{6}$$

$$x = -1$$

Check

$$4(x + 3) = x - 3(x - 2)$$

Let $x = -1$ in the original equation: $\quad 4(-1 + 3) \overset{?}{=} -1 - 3(-1 - 2)$

$$4(2) \overset{?}{=} -1 - 3(-3)$$

$$8 \overset{?}{=} -1 + 9$$

$$8 = 8 \quad \text{True}$$

Because $x = -1$ satisfies the equation, the solution of the equation is -1, or the solution set is $\{-1\}$.

Quick ✓

In Problems 14–17, solve each linear equation. Be sure to verify your solution.

14. $4(x - 1) = 12$

15. $-2(x - 4) - 6 = 3(x + 6) + 4$

16. $4(x + 3) - 8x = 3(x + 2) + x$

17. $5(x - 3) + 3(x + 3) = 2x - 3$

The steps for solving a linear equation are summarized below. Not every step is always necessary.

Solving a Linear Equation

Step 1: Remove any parentheses using the Distributive Property.

Step 2: Combine like terms on each side of the equation.

Step 3: Use the Addition Property of Equality to get the terms with the variable on one side of the equation and the constants on the other side.

Step 4: Use the Multiplication Property of Equality to get the coefficient of the variable to equal 1.

Step 5: Check the solution to verify that it satisfies the original equation.

▶ **Linear Equations with Fractions or Decimals**

To rewrite a linear equation that contains fractions as an equivalent equation without fractions, multiply both sides of the equation by the Least Common Denominator (LCD) of all the fractions in the equation.

EXAMPLE 5 **How to Solve a Linear Equation That Contains Fractions**

Solve the linear equation: $\dfrac{y + 1}{4} + \dfrac{y - 2}{10} = \dfrac{y + 7}{20}$

Step-by-Step Solution

Before following the steps, rewrite the equation without fractions by multiplying both sides of the equation by the LCD, which is 20.

$$20\left(\frac{y + 1}{4} + \frac{y - 2}{10}\right) = 20\left(\frac{y + 7}{20}\right)$$

Now follow Steps 1–5 for solving a linear equation.

Step 1: Remove all parentheses using the Distributive Property.

$$20\left(\frac{y + 1}{4} + \frac{y - 2}{10}\right) = 20\left(\frac{y + 7}{20}\right)$$

Use the Distributive Property: $20 \cdot \dfrac{y + 1}{4} + 20 \cdot \dfrac{y - 2}{10} = 20 \cdot \dfrac{y + 7}{20}$

Divide out common factors: $5(y + 1) + 2(y - 2) = y + 7$

Use the Distributive Property: $5y + 5 + 2y - 4 = y + 7$

Step 2: Combine like terms on each side of the equation.

$$7y + 1 = y + 7$$

Step 3: Use the Addition Property of Equality to get the terms with the variable on one side of the equation and the constants on the other side.

Subtract y from both sides: $7y + 1 - y = y + 7 - y$

$6y + 1 = 7$

Subtract 1 from both sides: $6y + 1 - 1 = 7 - 1$

$6y = 6$

Step 4: Use the Multiplication Property of Equality to get the coefficient on the variable to equal 1.

Divide both sides by 6: $\dfrac{6y}{6} = \dfrac{6}{6}$

$y = 1$

Step 5: Check the solution to verify that it satisfies the original equation.

$$\frac{y + 1}{4} + \frac{y - 2}{10} = \frac{y + 7}{20}$$

Let $y = 1$ in the original equation: $\dfrac{1 + 1}{4} + \dfrac{1 - 2}{10} \overset{?}{=} \dfrac{1 + 7}{20}$

$\dfrac{2}{4} + \dfrac{-1}{10} \overset{?}{=} \dfrac{8}{20}$

Rewrite each rational number with LCD = 20: $\dfrac{2}{4} \cdot \dfrac{5}{5} + \dfrac{-1}{10} \cdot \dfrac{2}{2} \overset{?}{=} \dfrac{8}{20}$

$\dfrac{10}{20} + \dfrac{-2}{20} \overset{?}{=} \dfrac{8}{20}$

$\dfrac{8}{20} = \dfrac{8}{20}$ True

Because $y = 1$ satisfies the equation, the solution of the equation is 1, or the solution set is $\{1\}$.

●

Quick ✓

18. To solve a linear equation containing fractions, we can multiply each side of the equation by the _____ _____ _____ to clear the fractions.

In Problems 19–22, solve each linear equation and verify your solution.

19. $\dfrac{3y}{2} + \dfrac{y}{6} = \dfrac{10}{3}$

20. $\dfrac{3x}{4} - \dfrac{5}{12} = \dfrac{5x}{6}$

21. $\dfrac{x + 2}{6} + 2 = \dfrac{5}{3}$

22. $\dfrac{4x + 3}{9} - \dfrac{2x + 1}{2} = \dfrac{1}{6}$

To rewrite a linear equation that contains decimals as an equivalent equation without decimals, multiply both sides of the equation by a power of 10 that removes the decimals.

For example, multiplying 0.7 $\left(\text{or }\dfrac{7}{10}\right)$ by 10 "eliminates" the decimal, and multiplying 0.03 $\left(\text{or }\dfrac{3}{100}\right)$ by 100 "eliminates" the decimal.

EXAMPLE 6 | **Solving a Linear Equation That Contains Decimals**

Solve the linear equation: $0.5x - 0.4 = 0.3x + 0.2$

Solution

Multiplying both sides of the equation by 10 will "eliminate" the decimal because each of the decimals is written to the tenths position.

$$10(0.5x - 0.4) = 10(0.3x + 0.2)$$

Use the Distributive Property: $\quad 10(0.5x) - 10(0.4) = 10(0.3x) + 10(0.2)$

$$5x - 4 = 3x + 2$$

Subtract 3x from both sides: $\quad 5x - 4 - 3x = 3x + 2 - 3x$

$$2x - 4 = 2$$

Add 4 to both sides: $\quad 2x - 4 + 4 = 2 + 4$

$$2x = 6$$

Divide both sides by 2: $\quad \dfrac{2x}{2} = \dfrac{6}{2}$

$$x = 3$$

$$0.5x - 0.4 = 0.3x + 0.2$$

Check

Let x = 3 in the original equation: $\quad 0.5(3) - 0.4 \overset{?}{=} 0.3(3) + 0.2$

$$1.5 - 0.4 \overset{?}{=} 0.9 + 0.2$$

$$1.1 = 1.1 \quad \text{True}$$

Because $x = 3$ satisfies the equation, the solution of the equation is 3, or the solution set is $\{3\}$. ●

Quick ✓

In Problems 23 and 24, solve each linear equation. Be sure to verify your solution.

23. $0.07x - 1.3 = 0.05x - 1.1$ **24.** $0.4(y + 3) = 0.5(y - 4)$

▶ ❸ Determine Whether an Equation Is a Conditional Equation, an Identity, or a Contradiction

Each equation studied so far has had one solution. Equations may have one solution, no solution, or infinitely many solutions. Equations can be classified based on the number of their solutions.

The equations that have been solved so far are called *conditional equations*.

> **Definition**
>
> A **conditional equation** is an equation that is true for some values of the variable and false for other values of the variable.

For example, the equation $x + 7 = 10$ is a conditional equation because it is true when $x = 3$ and false for every other real number x.

> **Definition**
>
> An equation that is false for every value of the variable is a **contradiction.**

For example, the equation $3x + 8 = 3x + 6$ is a contradiction because it is false for any value of x. Contradictions are recognized by creating equivalent equations. For example, if $3x$ is subtracted from both sides of $3x + 8 = 3x + 6$, the result is $8 = 6$, which is clearly false. Contradictions have no solution, so the solution set is the empty set and is written as $\varnothing$ or $\{\ \}$.

> **Definition**
>
> An equation that is satisfied for every value of the variable for which both sides of the equation are defined is called an **identity**.

In Other Words

Conditional equations are true for some values of the variable and false for others. Contradictions are false for all values of the variable. Identities are true for all allowed values of the variable.

For example, $2x + 3 + x + 8 = 3x + 11$ is an identity because any real number x satisfies the equation. Identities are recognized by creating equivalent equations. For example, if like terms are combined in the equation $2x + 3 + x + 8 = 3x + 11$, the result is $3x + 11 = 3x + 11$, which is true no matter what value of x is chosen. Therefore, the solution set of equations that are identities, where each side of the equation is a linear expression, is the set of all real numbers, which can be expressed as either $\{x \mid x \text{ is any real number}\}$ or $\mathbb{R}$.

EXAMPLE 7 **Classifying an Equation**

Solve the equation $3(x + 3) - 6x = 5(x + 1) - 8x$. Then state whether it is an identity, a contradiction, or a conditional equation.

Solution

As with any equation, the goal is to isolate the variable.

$$3(x + 3) - 6x = 5(x + 1) - 8x$$

Use the Distributive Property: $\quad 3x + 9 - 6x = 5x + 5 - 8x$

Combine like terms: $\quad -3x + 9 = -3x + 5$

Add 3x to both sides: $\quad -3x + 9 + 3x = -3x + 5 + 3x$

$$9 = 5$$

Work Smart

In the solution to Example 7, we obtained the equation

$$-3x + 9 = -3x + 5$$

You may recognize at this point that the equation is a contradiction and express the solution set as $\varnothing$ or $\{\ \}$.

Because the last statement, $9 = 5$, is false, the equation is a contradiction. The original equation is a contradiction and has no solution. The solution set is $\varnothing$ or $\{\ \}$. ●

EXAMPLE 8 **Classifying an Equation**

Solve the equation $-4(x - 2) + 3(4x + 2) = 2(4x + 7)$. Then state whether the equation is an identity, a contradiction, or a conditional equation.

Solution

$$-4(x - 2) + 3(4x + 2) = 2(4x + 7)$$

Use the Distributive Property: $\quad -4x + 8 + 12x + 6 = 8x + 14$

Combine like terms: $\quad 8x + 14 = 8x + 14$

The equation $8x + 14 = 8x + 14$ is true for all real numbers x. So the original equation is an identity, and its solution set is $\{x \mid x \text{ is any real number}\}$ or $\mathbb{R}$. ●

What would happen if we continued to solve the equation in Example 8? The result may be the following:

$$8x + 14 - 8x = 8x + 14 - 8x$$

$$14 = 14$$

Because the statement $14 = 14$ is true for all real numbers x, the solution set of the original equation is all real numbers.

Quick ✔

25. A(n) _____ _____ is an equation that is true for some values of the variable and false for others.

26. A(n) _____ is an equation that is false for every value of the variable. A(n) _____ is an equation that is satisfied by every allowed choice of the variable.

In Problems 27–30, solve the equation and state whether it is an identity, a contradiction, or a conditional equation.

27. $4(x + 2) = 4x + 2$

28. $3(x - 2) = 2x - 6 + x$

29. $-4x + 2 + x + 1 = -4(x + 2) + 11$

30. $-3(z + 1) + 2(z - 3) = z + 6 - 2z - 15$

1.1 Exercises MyMathLab® Exercise numbers in green have complete video solutions in MyMathLab or may be accessed using the QR code to the right.

*Problems **1–30** are the Quick ✔s that follow the **EXAMPLES**.*

Building Skills

In Problems 31–36, determine which of the numbers are solutions to the given equation. See Objective 1.

31. $8x - 10 = 6; x = -2, x = 1, x = 2$

32. $-4x - 3 = -15; x = -2, x = 1, x = 3$

33. $5m - 3 = -3m + 5; m = -2, m = 1, m = 3$

34. $6x + 1 = -2x + 9; x = -2, x = 1, x = 4$

35. $4(x - 1) = 3x + 1; x = -1, x = 2, x = 5$

36. $3(t + 1) - t = 4t + 9; t = -3, t = -1, t = 2$

In Problems 37–56, solve each linear equation, and be sure to verify your solution. See Objective 2.

37. $3x + 1 = 7$

38. $8x - 6 = 18$

39. $4z + 3 = 2$

40. $-6x - 5 = 13$

41. $-3w + 2w + 5 = -4$

42. $-7t - 3 + 5t = 11$

43. $5x + 2 - 2x + 3 = 7x + 2 - x + 5$

44. $-6x + 2 + 2x + 9 + x = 5x + 10 - 6x + 11$

45. $3(x + 2) = -6$

46. $4(z - 2) = 12$

47. $\dfrac{4y}{5} - \dfrac{14}{15} = \dfrac{y}{3}$

48. $\dfrac{3x}{2} + \dfrac{x}{6} = -\dfrac{5}{3}$

49. $\dfrac{4x + 3}{9} - \dfrac{2x + 1}{2} = \dfrac{1}{6}$

50. $\dfrac{2x + 1}{3} - \dfrac{6x - 1}{4} = -\dfrac{5}{12}$

51. $\dfrac{y}{10} + 6 = \dfrac{y}{4} + 12$

52. $\dfrac{n}{5} + 3 = \dfrac{n}{2} + 6$

53. $0.5x - 3.2 = -1.7$

54. $0.3z + 0.8 = -0.1$

55. $0.14x + 2.23 = 0.09x + 1.98$

56. $0.12y - 5.26 = 0.05y + 1.25$

In Problems 57–72, solve the equation. Identify each equation as an identity, a contradiction, or a conditional equation. Be sure to verify your solution. See Objective 3.

57. $4(x + 1) = 4x$

58. $5(s + 3) = 3s + 2s$

59. $4m + 1 - 6m = 2(m + 3) - 4m$

60. $10(x - 1) - 4x = 2x - 1 + 4(x + 1)$

61. $2(y + 1) - 3(y - 2) = 5y + 8 - 6y$

62. $8(w + 2) - 3w = 7(w + 2) + 2(1 - w)$

63. $\dfrac{x}{4} + \dfrac{3x}{10} = -\dfrac{33}{20}$

64. $\dfrac{z - 2}{4} + \dfrac{2z - 3}{6} = 7$

65. $3p - \dfrac{p}{4} = \dfrac{11p}{4} + 1$

66. $\dfrac{r}{2} + 2(r - 1) = \dfrac{5r}{2} + 4$

67. $\dfrac{2x + 1}{2} - \dfrac{x + 1}{5} = \dfrac{23}{10}$

68. $\dfrac{3x + 1}{4} - \dfrac{7x - 4}{2} = \dfrac{26}{3}$

69. $0.4(z + 1) - 0.7z = -0.1z + 0.7 - 0.2z - 0.3$

70. $0.9(z - 3) - 0.2(z - 5) = 0.4(z + 1) + 0.3z - 2.1$

71. $\dfrac{1}{3}(2x - 3) + 2 = \dfrac{5}{6}(x + 3) - \dfrac{11}{12}$

72. $\dfrac{4}{5}(y - 4) + 3 = \dfrac{2}{3}(y + 1) + \dfrac{4}{15}$

Mixed Practice

In Problems 73–90, solve each equation and verify the solution. State whether the equation is an identity, a contradiction, or a conditional equation.

73. $7y - 8 = -7$

74. $4y + 5 = 7$

75. $4a + 3 - 2a + 4 = 5a - 7 + a$

76. $-5x + 5 + 3x + 7 = 5x - 6 + x + 12$

77. $4(p + 3) = 3(p - 2) + p + 18$

78. $7(x + 2) = 5(x - 2) + 2(x + 12)$

79. $4b - 3(b + 1) - b = 5(b - 1) - 5b$

80. $13z - 8(z + 1) = 2(z - 3) + 3z$

81. $8(4x + 6) = 11 - (x + 7)$

82. $5(2x + 3) = 9 - (x + 5)$

83. $\dfrac{m + 1}{4} + \dfrac{5}{6} = \dfrac{2m - 1}{12}$

84. $\dfrac{z - 4}{6} - \dfrac{2z + 1}{9} = \dfrac{1}{3}$

85. $0.3x - 1.3 = 0.5x - 0.7$

86. $-0.8y + 0.3 = 0.2y - 3.7$

87. $-0.8(x + 1) = 0.2(x + 4)$

88. $0.5(x + 3) = 0.2(x - 6)$

89. $\dfrac{1}{4}(x - 4) + 3 = \dfrac{1}{3}(2x + 6) - \dfrac{5}{6}$

90. $\dfrac{1}{5}(2a - 5) - 4 = \dfrac{1}{2}(a + 4) - \dfrac{7}{10}$

Applying the Concepts

91. Find a such that the solution set of $ax + 3 = 15$ is $\{-3\}$.

92. Find a such that the solution set of $ax + 6 = 20$ is $\{7\}$.

93. Find a such that the solution set of $a(x - 1) = 3(x - 1)$ is the set of all real numbers.

94. Find a such that the solution set of $ax + 3 = 2x + 3(x + 1)$ is the set of all real numbers.

In Section R.5, we introduced the domain of a variable. The domain of a variable is the set of all values that a variable may assume. Recall that division by zero is not defined, so any value of the variable that results in division by zero must be excluded from the domain. In Problems 95–100, determine which values of the variable must be excluded from the domain.

95. $\dfrac{5}{2x + 1}$

96. $\dfrac{-3}{5x + 8}$

97. $\dfrac{3x + 7}{4x - 3}$

98. $\dfrac{2x}{3x + 1}$

99. $\dfrac{6x - 2}{3(x + 1) - 6}$

100. $\dfrac{-2x + 7}{4(x - 3) + 2}$

101. Interest Suppose you have a credit card debt of $2000. Last month, the bank charged you $25 interest on the debt. The solution to the equation $25 = \dfrac{2000}{12} \cdot r$ represents the annual interest rate r on the credit card. Find the annual interest rate on the credit card.

102. How Much Do I Make? Last week, before taxes, you earned $539 after working 26 hours at your regular hourly rate and 6 hours at time-and-a-half. The solution to the equation $26x + 9x = 539$ represents your regular hourly rate, x. Determine your regular hourly rate. What does 9 represent in the equation?

103. Paying Your Taxes You are single and have just determined that you paid $4412.50 in federal income tax in 2015. The solution to the equation $4412.5 = 0.15(I - 9225) + 922.50$ represents your annual adjusted income I for 2015. Determine, to the nearest dollar, your annual adjusted income for 2015. (SOURCE: *Internal Revenue Service*)

104. Paying Your Taxes You are married and have just determined that you paid $11,037.50 in federal income tax in 2015. The solution to the equation $11,037.5 = 0.25(I - 74,900) + 10,312.50$ represents your annual adjusted income I for 2015. Determine, to the nearest dollar, your annual adjusted income for 2015. (SOURCE: *Internal Revenue Service*)

Explaining the Concepts

105. Explain the difference between $4(x + 1) - 2$ and $4(x + 1) = 2$. In general, what is the difference between an algebraic expression and an equation?

106. Explain the difference between the directions "solve" and "simplify."

107. Make up a linear equation that has one solution. Make up an equation that has no solution. Make up an equation that is an identity. Comment on the differences and similarities in making up these equations.

1.2 An Introduction to Problem Solving

Objectives

1. Translate English Sentences into Mathematical Statements
2. Model and Solve Direct Translation Problems
3. Model and Solve Mixture Problems
4. Model and Solve Uniform Motion Problems

Are You Prepared for This Section?

Before getting started, complete the following problems. If you get a problem wrong, go back to the section cited and review the material.

In Problems P1–P6, express each English phrase as an algebraic expression. [Section R.5, pp. 39–40]

P1. The sum of 12 and a number z

P2. The product of 4 and x

P3. A number y decreased by 87

P4. The quotient of z and 12

P5. Four times the sum of x and 7

P6. The sum of four times a number x and 7

Work Smart

We learned in the last section that an equation is a statement in which two algebraic expressions, separated by an equal sign, are equal.

▶ 1 Translate English Sentences into Mathematical Statements

In English, a complete sentence must contain a subject and a verb, so expressions or "phrases" are not complete sentences. For example, the expression "5 more than a number x" does not contain a verb and therefore is not a complete sentence. The statement "5 more than a number x is 18" is a complete sentence because it contains a subject (5 more than "a number x") and a verb ("is"), so we can translate it into a mathematical statement. Mathematical statements are represented symbolically through equations. In English, statements can be true or false. Mathematical statements can be true or false as well.

Look back at page 39 in Section R.5 for a review of key words that translate into mathematical symbols. Table 1 provides a summary of words that typically translate into an equal sign.

Table 1 Words That Translate into an Equal Sign				
is	yields	are	equals	is equivalent to
was	gives	results in	is equal to	

EXAMPLE 1 **Translating English Sentences into Mathematical Statements**

Translate each sentence into a mathematical statement. Do not solve the equation.

(a) Five more than a number x is 20.

(b) Four times the sum of a number z and 3 is 15.

(c) The difference of x and 5 equals the quotient of x and 2.

Solution

Work Smart

The English statement "The sum of four times a number z and 3 is 15" would be expressed mathematically as $4z + 3 = 15$. Do you see the difference between this statement and the one in Example 1(b)?

(a) 5 more than a number x ‿ is ‿ 20
$$x + 5 \quad\quad = \quad 20$$

(b) The expression "Four times the sum of" tells us to find the sum first and then multiply this result by 4.

Four times the sum of a number z and 3 ‿ is ‿ 15
$$4(z + 3) \quad\quad = \quad 15$$

(c) The difference of x and 5 ‿ equals ‿ the quotient of x and 2
$$x - 5 \quad\quad = \quad \frac{x}{2}$$

Prepared?...Answers **P1.** $12 + z$ **P2.** $4x$

P3. $y - 87$ **P4.** $\dfrac{z}{12}$ **P5.** $4(x + 7)$

P6. $4x + 7$

Quick ✔

1. Mathematical statements are represented symbolically as _____.

In Problems 2–6, translate each English statement into a mathematical statement. Do not solve the equation.

2. The sum of *x* and 7 results in 12.

3. The product of 3 and *y* is equal to 21.

4. Two times the sum of *n* and 3 is 5.

5. The difference of *x* and 10 equals the quotient of *x* and 2.

6. The sum of two times *n* and 3 is 5.

▶ An Introduction to Problem Solving and Mathematical Models

Every day, everyone encounters various types of problems that must be solved. **Problem solving** is the ability to use information, tools, and our own skills to achieve a goal. For example, suppose Kevin wants a glass of water, but he is too short to reach the sink. Kevin has a problem. To solve the problem, he finds a step stool and pulls it over to the sink. He uses the step stool to climb on the counter, opens the kitchen cabinet, and pulls out a cup. He then crawls along the counter top, turns on the faucet, and proceeds to fill his cup with water. Problem solved!

Of course, Kevin could solve the problem in other ways. Just as there are various ways to solve life's everyday problems, there are many ways to solve problems using mathematics. However, regardless of the approach, there are always some common aspects in solving any problem. For example, regardless of how Kevin ultimately ends up with his cup of water, someone must get a cup from the cabinet and someone must turn on the faucet.

One of the purposes of learning algebra is to be able to solve certain types of problems. To solve these problems, translate the verbal description of the problem into equations that can be solved. The process of turning a verbal description into a mathematical equation is **mathematical modeling.** The equation that is developed is called the **mathematical model.**

Not all models are mathematical. In general, a **model** is a way of using graphs, pictures, equations, or even verbal descriptions to represent a real-life situation. Because the world is a complex place, it is helpful to simplify information to develop a model. For example, a map is a model of our road system. Maps don't show all the details of the system (such as trees, buildings, or potholes), but they do a good job of describing how to get from point A to point B. Mathematical models are similar in that we often make assumptions regarding our world in order to make the mathematics less complicated.

Every problem is unique in some way. However, because many problems are similar, they can be categorized. In this text, five categories of problems will be solved.

Five Categories of Problems

1. Direct Translation—problems in which English is directly translated into an equation by using key words in the verbal description.

2. Mixture—problems where two or more quantities are combined in some fashion.

3. Geometry—problems where the unknown quantities are related through geometric formulas.

4. Uniform Motion—problems where an object travels at a constant speed.

5. Work—problems where two or more entities join forces to complete a job.

This section concentrates on direct translation problems, mixture problems, and uniform motion problems. The guidelines on the next page will help solve any category of problem.

Work Smart

When you solve a problem by making a mathematical model, check your work to make sure you have the right answer. Typically, errors can happen in two ways:

- One type of error occurs if you correctly translate the problem into a model but then make an error solving the equation. This type of error is usually easy to find.

- However, if you misinterpret the problem and develop an incorrect model, then the solution you obtain may still satisfy your model, but it probably will not be the correct solution to the original problem. We can check for this type of error by determining whether the solution is reasonable. Does your answer make sense? Always be sure that you are answering the question that is being asked.

Solving Problems With Mathematical Models

Step 1: Identify What You Are Looking For   Read the problem very carefully, perhaps two or three times. Identify the type of problem and the information you wish to learn. It is fairly typical for the last sentence in the problem to indicate what you need to solve for.

Step 2: Give Names to the Unknowns   Assign variables to the unknown quantities. Choose a variable that is representative of the unknown quantity. For example, use t for time, making sure to include units, such as hours in this case.

Step 3: Translate the Problem into Mathematics   Read the problem again. This time, after each sentence is read, determine whether the sentence can be translated into a mathematical statement or expression in terms of the variables identified in Step 2. It is often helpful to create a table, chart, or figure. Combine the mathematical statements or expressions into an equation that can be solved.

Step 4: Solve the Equation(s) Found in Step 3   Solve the equation for the variable.

Step 5: Check the Reasonableness of Your Answer   Check your answer with the wording of the original problem to be sure that it makes sense. If it does not, go back and try again.

Step 6: Answer the Question   Write your answer in a complete sentence.

---

❷ Model and Solve Direct Translation Problems

Let's solve some **direct translation** problems, which can be set up by reading the problem and translating the verbal description into an equation.

▶ (EXAMPLE 2)   **Finding Consecutive Integers**

The sum of three consecutive odd integers results in 45. Find the integers.

Solution

Step 1: Identify   This is a direct translation problem. Look for three odd integers. The odd integers are 1, 3, 5, and so on.

Step 2: Name   If x represents the first odd integer, then $x + 2$ is the next odd integer, and $x + 4$ is the third odd integer.

Step 3: Translate   It is given that their sum is 45, so

$$\underset{\text{First Integer}}{x} \; + \; \underset{\text{Second Integer}}{x + 2} \; + \; \underset{\text{Third Integer}}{x + 4} \; = 45 \qquad \text{The Model}$$

Step 4: Solve   Now solve the equation.

$$x + x + 2 + x + 4 = 45$$

$$\text{Combine like terms:} \qquad 3x + 6 = 45$$

$$\text{Subtract 6 from both sides:} \qquad 3x = 39$$

$$\text{Divide both sides by 3:} \qquad x = 13$$

Step 5: Check   Because x represents the first odd integer, the remaining two odd integers are $13 + 2 = 15$ and $13 + 4 = 17$. It is always a good idea to make sure your answer is reasonable. Because $13 + 15 + 17 = 45$, this is the right answer!

Step 6: Answer the Question   The three consecutive odd integers are 13, 15, and 17.

Quick ✓

7. *True or False* If *n* represents the first of two consecutive odd integers, $n + 1$ represents the next consecutive odd integer.

In Problems 8 and 9, translate each English statement into an equation and solve the equation.

8. The sum of three consecutive even integers is 60. Find the integers.

9. The sum of three consecutive integers is 78. Find the integers.

▶ **EXAMPLE 3** **Finding Hourly Wage**

Before taxes, Marissa earned $725 one week after working 52 hours. Her employer pays time-and-a-half for all hours worked in excess of 40 hours. What is Marissa's hourly wage?

Solution

Step 1: Identify In this direct translation problem, find Marissa's hourly wage.

Step 2: Name Let *w* represent Marissa's hourly wage in dollars.

Step 3: Translate It is known that Marissa earned $725 by working 40 hours at her regular wage and 12 hours at 1.5 times her regular wage. If her regular wage is *w* dollars, then her overtime wage is 1.5*w* dollars. Therefore, her total salary is

$$\underbrace{40w}_{\text{Regular Earnings}} + \underbrace{12(1.5w)}_{\text{Overtime Earnings}} = \underbrace{725}_{\text{Total Earnings}} \qquad \text{The Model}$$

Step 4: Solve

$$40w + 12(1.5w) = 725$$

Simplify: $\qquad 40w + 18w = 725$

Combine like terms: $\qquad 58w = 725$

Divide both sides by 58: $\qquad \dfrac{58w}{58} = \dfrac{725}{58}$

$$w = 12.50$$

Step 5: Check If Marissa's hourly wage is $12.50, then for the first 40 hours she earned $12.50(40) = \$500$ and for the next 12 hours she earned $1.5(\$12.50)(12) = \$225$. Her total salary was $\$500 + \$225 = \$725$. This checks with the information presented in the problem.

Step 6: Answer the Question Marissa's hourly wage is $12.50. ●

Quick ✓

10. Before taxes, Melody earned $735 one week after working 46 hours. Her employer pays time-and-a-half for all hours worked in excess of 40 hours. What is Melody's hourly wage?

11. An inheritance of $800,000 is to be divided among Avery, Connor, and Olivia in the following manner: Olivia is to receive $\dfrac{3}{4}$ of what Connor gets, while Avery gets $\dfrac{1}{4}$ of what Connor gets. How much does each receive?

EXAMPLE 4   **Choosing Take-Out Pizza**

Domino's Pizza charges $13.25 plus $1.95 per topping for their large 14-inch pizza. Extreme Pizza charges $14.45 plus $1.70 per topping for the same size pizza. For how many toppings will the two orders cost the same?

(SOURCE: *Domino's and Extreme Pizza*)

Solution

Step 1: Identify   This is a direct translation problem. Find the number of toppings for which the two orders cost the same.

Step 2: Name   Let t represent the number of toppings on a large 14-inch pizza.

Step 3: Translate   The cost of a 14-inch pizza at Domino's Pizza is $13.25 plus $1.95 for each topping. So if one topping is picked, the cost is $13.25 + 1.95(1)$ = $15.20. For 2 toppings, the cost is $13.25 + 1.95(2)$ = $17.15. In general, for t toppings, the cost is $13.25 + 1.95t$ dollars. Similar logic reveals that the cost of a 14-inch pizza at Extreme Pizza is $14.45 + 1.70t$ dollars. To find the number of toppings for which the cost for the two pizzas will be the same, solve:

$$\text{Cost for Domino's pizza} = \text{Cost for Extreme Pizza pizza}$$
$$13.25 + 1.95t = 14.45 + 1.70t \quad \text{The Model}$$

Step 4: Solve

	$13.25 + 1.95t = 14.45 + 1.70t$
Subtract 13.25 from both sides:	$1.95t = 1.20 + 1.70t$
Subtract 1.70t from both sides:	$0.25t = 1.20$
Divide both sides by 0.25:	$t = 4.8$

Step 5: Check   For $t = 4.8$ toppings, the cost of a Domino's pizza will be $13.25 + 1.95(4.8)$ = $22.61, and the cost of an Extreme Pizza pizza will be $14.45 + 1.70(4.8)$ = $22.61. They are the same! Of course, because you cannot order a partial topping, round 4.8 topping to 5 toppings.

Step 6: Answer the Question   The two orders will cost roughly the same when 5 toppings are ordered.   ●

Quick ✔

12. You need to rent a moving truck. EZ-Rental charges $35 per day plus $0.15 per mile. Do It Yourself Rental charges $20 per day plus $0.25 per mile. For how many miles will the cost of renting be the same?

13. You are comparing cell phone plans. Verizon offers two plans you are considering. Plan A charges $79.99 per month for 900 minutes of talk and text, with a $0.40 per-minute rate after the allowance. Plan B charges $89.99 per month for unlimited minutes of talk and text. For how many minutes will the monthly cost be the same? (SOURCE: *Verizon*)

There are many types of direct translation problems. One type of direct translation problem is a "percent problem." These problems often involve discounts or mark-ups that businesses use in determining their prices.

Percent means *divided by 100* or *per hundred*. The symbol % denotes percent, so 45% means 45 out of 100 or $\dfrac{45}{100}$ or 0.45. In applications involving percents, we often see the word "of," as in 20% of 60. Remember that the word "of" often translates into multiplication in mathematics, so 20% of 60 means

$$20\% \cdot 60 \quad \text{or} \quad 0.20 \cdot 60$$

When dealing with percents and the price of goods, it is helpful to remember the following:

$$\text{Original Price} - \text{Discount} = \text{Sale Price}$$
$$\text{Wholesale Price} + \text{Markup} = \text{Selling Price}$$

▶ **EXAMPLE 5**   **Finding the Original Price**

Everything in your favorite shoe store is marked at a discount of 40% off. If the sale price of a pair of boots is $144, what was the original price?

Solution

Step 1: Identify This is a direct translation problem involving percents. Find the original price of the boots.

Step 2: Name Let p represent the original price in dollars.

Step 3: Translate The original price minus the discount is the sale price, $144, so

$$p - \text{discount} = 144$$

The discount was 40% off of the original price, so the discount is $0.40p$. Substitute this expression into the equation $p - \text{discount} = 144$:

$$p - 0.40p = 144 \quad \text{The Model}$$

Work Smart

You can eliminate the decimals by multiplying both sides of the equation by 10.

Step 4: Solve
$$p - 0.40p = 144$$
Combine like terms: $\quad 0.60p = 144$

Divide both sides by 0.60: $\quad \dfrac{0.60p}{0.60} = \dfrac{144}{0.60}$

Simplify: $\quad p = 240$

Step 5: Check If p, the original price of the boots, was $240, then the discount would be $0.4(\$240) = \$96$. Subtracting $96 from the original price gives $144, the sale price. This agrees with the information in the problem.

Step 6: Answer the Question The original price of the boots was $240.   ●

Quick ✓

14. Percent means "divided by _____."

15. What is 40% of 100?

16. 8 is 5% of what number?

17. 15 is what percent of 20?

18. Suppose you have just entered your favorite clothing store and find that everything in the store is marked at a discount of 30% off. If the sale price of a shirt is $21, what was its original price?

19. A Milex Tune-Up automotive facility marks up its parts 35%. Suppose that Milex charges its customers $1.62 for each spark plug it installs. What is Milex's cost for each spark plug?

In Other Words

"Per annum" is another way of indicating an annual or yearly rate.

In many direct translation problems, a known formula can be used to help solve. One of those formulas is the **Simple Interest Formula**. **Interest** is money paid for the use of money. The total amount borrowed is called the **principal.** The principal can be in the form of a loan (an individual borrows from the bank) or a deposit (the bank borrows from the individual). The **rate of interest,** expressed as a percent, is the amount charged for the use of the principal for a given period of time, usually for a year (that is, on a *per annum* basis).

> **Simple Interest Formula**
>
> If a principal of P dollars is borrowed for a period of t years at an annual interest rate r, expressed as a decimal, then the interest I charged is
>
> $$I = Prt$$
>
> Interest charged according to this formula is called **simple interest.**

▶ **EXAMPLE 6** **Computing Credit Card Interest**

Yolanda has a credit card balance of $2800. Each month, the credit card company charges 14% annual simple interest on any outstanding balances. How much interest will Yolanda be charged on this loan after one month? What is her credit card balance after one month? Round answers to the nearest penny.

Solution

Find the interest I charged on the loan. Because the interest rate is given as an annual rate, the length of time the money is borrowed must be expressed in years. One month is $\frac{1}{12}$ of a year, so $t = \frac{1}{12}$. The outstanding balance, or principal, is $P = \$2800$. The annual interest rate is $r = 14\% = 0.14$. Substituting into $I = Prt$ yields

$$I = (\$2800)(0.14)\left(\frac{1}{12}\right) = \$32.67$$

Yolanda will owe the amount borrowed, $2800, plus accrued interest, $32.67, for a total of $2800 + $32.67 = $2832.67. ●

> **Quick ✓**
>
> **20.** _____ is money paid for the use of money. The total amount borrowed is called the _____.
>
> **21.** Suppose that Dave has a car loan of $6500. The bank charges 6% annual simple interest. What is the interest charge on Dave's car loan after 1 month?
>
> **22.** Suppose that you have $1400 in a savings account. The bank pays 1.5% annual simple interest. What would be the interest paid after 6 months? What is the balance in the account?

❸ **Model and Solve Mixture Problems**

In **mixture problems,** two or more items are combined to form a third item. The different types of mixture problems all rely on the same basic idea:

Portion from Item A + Portion from Item B = Whole or Total

One type of mixture problem is an interest problem.

▶ **EXAMPLE 7** **Financial Planning**

Sergio has $15,000 to invest. His goal is to earn 5%, or $750, annually. His financial advisor recommends investing some of the money in corporate bonds that pay 6% and the rest in government-backed Treasury bonds paying 3%. How much should Sergio place in each investment in order to achieve his goal?

Solution

Step 1: Identify This is a mixture problem involving simple interest. Sergio needs to know how much to invest in corporate bonds and how much to invest in Treasury bonds in order to earn $750 in interest.

(continued)

Work Smart

Whenever you have two unknowns that have a given relationship, label one of the parts as the variable and the other part as the total minus the variable.

Step 2: Name Let b represent the amount invested in corporate bonds; then $\$15{,}000 - b$ represents the amount invested in Treasury bonds, both in dollars.

Step 3: Translate Organize the given information in Table 2.

Table 2

	Principal $	Rate %	Time Yr	Interest $
Corporate Bonds	b	0.06	1	$0.06b$
Treasury Bonds	$15{,}000 - b$	0.03	1	$0.03(15{,}000 - b)$
Total	15,000	0.05	1	$0.05(15{,}000) = \$750$

Sergio wants to earn 5% each year on his principal of $15,000. The total interest is the sum of the interest from the corporate bonds and the Treasury bonds.

Interest from corporate bonds + Interest from Treasury bonds $= \$750$

$$0.06b + 0.03(15{,}000 - b) = 750 \quad \text{The Model}$$

Step 4: Solve
$$0.06b + 0.03(15{,}000 - b) = 750$$
Use the Distributive Property: $\quad 0.06b + 450 - 0.03b = 750$
Combine like terms: $\quad 0.03b + 450 = 750$
Subtract 450 from both sides: $\quad 0.03b = 300$
Divide both sides by 0.03: $\quad b = 10{,}000$

Step 5: Check If Sergio invests $10,000 in corporate bonds and $15{,}000 - \$10{,}000 = \$5000$ in Treasury bonds, the simple interest earned each year on the corporate bonds is $(\$10{,}000)(0.06)(1) = \$600$, and that on the Treasury bonds is $(\$5000)(0.03)(1) = \$150$. The total interest earned is $\$600 + \$150 = \$750$, which is the amount he wanted to earn.

Step 6: Answer the Question Sergio should invest $10,000 in corporate bonds and $5000 in Treasury bonds. ●

Quick ✓

23. Sophia has recently retired and requires an extra $5400 per year in income. She has $90,000 to invest and can invest in either an Aaa-rated bond that pays 5% per annum or a B-rated bond paying 9% per annum. How much should Sophia place in each investment in order to achieve her goal?

24. Steve has $25,000 to invest and wishes to earn an overall annual rate of return of 5%. His financial advisor recommends that he invest some of the money in a CD paying 2% per annum and the rest in a corporate bond paying 6% per annum. How much should Steve place in each investment in order to achieve his goal?

Mixing two quantities together creates a new blend. For example, a chef might mix buckwheat flour with wheat flour to make buckwheat pancakes. Or a coffee shop might mix two different types of coffee to create a new coffee blend.

EXAMPLE 8 **Blending Coffees**

A coffee shop manager wishes to form a new blend of coffee. She wants to mix distinctive-tasting Sumatra beans that sell for $12 per pound with milder Brazilian beans that sell for $8 per pound to get 50 pounds of the new blend. The new blend will sell for $9 per pound, and the revenue from selling the new blend and that from selling the beans separately will be the same. See Figure 1. How many pounds of the Sumatra beans and how many pounds of Brazilian beans will be required?

Figure 1

Sumatra	Brazilian	Blend
$12 per pound	$8 per pound	$9 per pound

(The Blend bag is labeled "50 pounds.")

Solution

Step 1: Identify This is a blend problem. Find the number of pounds of Sumatra beans and the number of pounds of Brazilian beans that are required in the new blend.

Step 2: Name Let s represent the required number of pounds of Sumatra beans. Thus $50 - s$ will be the required number of pounds of Brazilian beans.

Step 3: Translate The revenue from selling the new blend and that from selling the beans separately will be the same. This means that if the blend contains one pound of Sumatra and one pound of Brazilian, $\$12(1) + \$8(1) = \$20$ will be collected, which is what would be collected if the beans were sold separately.

We set up Table 3.

Table 3

	Price $/pound	Number of Pounds	Revenue
Sumatra	12	s	$12s$
Brazilian	8	$50 - s$	$8(50 - s)$
Blend	9	50	$9(50) = 450$

In general, if the blend contains s pounds of Sumatra beans, collect $\$12s$. If it contains $50 - s$ pounds of Brazilian beans, collect $\$8(50 - s)$. All 50 pounds of the blend, selling at $9 per pound, will sell for $\$9(50) = \$450$.

$$\left(\begin{array}{c}\text{Price per pound}\\\text{of Sumatra}\end{array}\right)\left(\begin{array}{c}\text{Pounds of}\\\text{Sumatra}\end{array}\right) + \left(\begin{array}{c}\text{Price per pound}\\\text{of Brazilian}\end{array}\right)\left(\begin{array}{c}\text{Pounds of}\\\text{Brazilian}\end{array}\right) = \left(\begin{array}{c}\text{Price per pound}\\\text{of blend}\end{array}\right)\left(\begin{array}{c}\text{Pounds of}\\\text{blend}\end{array}\right)$$

$$\underbrace{12 \cdot s}_{\text{Revenue from Sumatra}} + \underbrace{8 \cdot (50 - s)}_{\text{Revenue from Brazilian}} = \underbrace{\$9 \cdot 50}_{\text{Revenue from Mixture}}$$

$$12s + 8(50 - s) = 450 \quad \text{The Model}$$

Step 4: Solve Solve the equation for s.

	$12s + 8(50 - s) = 450$
Use the Distributive Property to remove the parentheses:	$12s + 400 - 8s = 450$
Combine like terms:	$4s + 400 = 450$
Subtract 400 from both sides:	$4s = 50$
Divide both sides by 4:	$s = 12.5$

Step 5: Check If the manager mixes 12.5 pounds of Sumatra beans with $50 - 12.5 = 37.5$ pounds of Brazilian beans, the total revenue will be $\$12(12.5) + \$8(37.5) = \$150 + \$300 = \$450$. This equals the revenue from the blend and checks with the given information.

Step 6: Answer the Question The manager should mix 12.5 pounds of Sumatra beans with 37.5 pounds of Brazilian beans.

Quick ✔

25. Suppose you want to blend two teas to obtain 10 pounds of a new blend selling for $3.50 per pound. Tea A sells for $4.00 per pound, and Tea B sells for $2.75 per pound. Assuming that the revenue from selling the new blend and that from selling the tea separately will be the same, determine the number of pounds of each tea required in the blend.

26. "We're Nuts!" sells cashews for $6.00 per pound and peanuts for $1.50 per pound. The manager has decided to make a trail mix. She wants the mix to sell for $3.00 per pound, and there should be no loss in revenue from selling the trail mix rather than selling the nuts alone. How many pounds of cashews and peanuts are required to create 30 pounds of trail mix?

EXAMPLE 9 **Juice Mixture**

Sierra originally has 8 ounces of a drink that contains only 10% juice. She wants to make her drink healthier by mixing it with 4 ounces of pure orange juice (100% juice). See Figure 2. What percent of the new drink mixture is juice? How many ounces of pure juice will the new drink contain?

Figure 2

| 8 ounce drink with | + | 4 ounce drink with | = | 12 ounce drink |
| 10% juice | | pure juice | | with ___% juice |

Solution

Step 1: Identify This is a mixture problem. Find the percent of the new drink mixture that is pure juice and determine how many ounces of juice that will be in the new drink.

Step 2: Name Let p represent the percent of juice in the new mixture.

Step 3: Translate Organize the given information in Table 4.

Table 4

	Ounces	% Juice	Total Ounces Juice
original drink	8	10% or 0.10	$8(0.10) = 0.8$
pure orange juice	4	100% or 1.00	$4(1.00) = 4$
new drink	$8 + 4 = 12$	p	$12p$

Sierra has 8 ounces of the original drink that contains 10% juice and 4 ounces of the pure orange juice (100% juice). The total ounces of juice in the new drink mixture is the sum of the total number of ounces of juice in the original drink and total number of ounces of juice in the pure orange juice.

Ounces of juice in original drink $+$ Ounces of juice in pure orange juice $=$ Total Ounces of juice in the new drink

$$0.8 + 4 = 12p \quad \text{The Model}$$

Step 4: Solve
$$0.8 + 4 = 12p$$
Combine like terms: $\qquad 4.8 = 12p$
Divide both sides by 12: $\qquad p = 0.4$

Step 5: Check For $p = 0.4$, Sierra will end up with $12(0.4) = 4.8$ ounces of juice in her 12-ounce drink. In looking at the third column of Table 4, when the total ounces of juice in the original drink is added to the total ounces of juice in the pure orange juice you also get $0.8 + 4 = 4.8$ ounces.

Step 6: Answer the Question After mixing 8 ounces of a drink that contains 10% juice with 4 ounces of pure orange juice that is 100% juice, Sierra will end up with a new drink mixture that is 40% juice, and contains 4.8 ounces of juice.   ●

▶ ❹ Model and Solve Uniform Motion Problems

Objects that move at a constant velocity (speed) are in **uniform motion.** Interpret an object's average speed as its constant velocity. For example, a car traveling at an average velocity of 40 miles per hour is in uniform motion.

Uniform Motion Formula

If an object moves at an average speed r, the distance d covered in time t is given by the formula

$$d = rt$$

Note that **rate** is another term for speed.

EXAMPLE 10) **Uniform Motion**

Roger and Bill decide to have a 10-mile race. Roger can run at an average speed of 12 miles per hour, and Bill can run at an average speed of 10 miles per hour. To "even things up," Roger agrees to give Bill a head start of 9 minutes. When will Roger catch up to Bill?

Solution

Step 1: Identify This is a uniform motion problem. Find the number of hours it will take for Roger to catch up to Bill if each man runs at his average speed.

Step 2: Name Let t represent the number of hours it takes for Roger to catch up to Bill. Since Bill has a head start of 9 minutes, or $\dfrac{9}{60} = 0.15$ of an hour, he will have been running for $t + 0.15$ hours.

Step 3: Translate Figure 3 illustrates the situation. We will set up Table 5.

Figure 3

Table 5

	Rate, mph	Time, hours	Distance, miles
Roger	12	t	$12t$
Bill	10	$t + 0.15$	$10(t + 0.15)$

(continued)

Roger will travel $12t$ miles since his average speed is 12 miles per hour and he runs for t hours. Bill will travel $10(t + 0.15)$ miles, since his average speed is 10 miles per hour and he runs for $t + 0.15$ hours. Because each travels the same distance, we set up the following model:

$$\text{Distance Roger runs} = \text{Distance Bill runs}$$
$$12t = 10(t + 0.15) \quad \text{The Model}$$

Step 4: Solve
$$12t = 10(t + 0.15)$$

Use the Distributive Property: $\quad 12t = 10t + 1.5$

Subtract $10t$ from both sides: $\quad 2t = 1.5$

Divide both sides by 2: $\quad t = \dfrac{1.5}{2} = 0.75$

Step 5: Check It should take Roger 0.75 hour to catch up to Bill. After 0.75 hour, Roger will have traveled $12(0.75) = 9$ miles, and Bill will have traveled $10(0.75 + 0.15) = 9$ miles. It checks!

Step 6: Answer the Question It will take Roger 0.75 hour, or 45 minutes, to catch up to Bill.

Quick ✓

27. Objects that move at a constant velocity are said to be in _____ _____.

28. Another term for speed is ____.

29. A Chevrolet Cruze left Omaha, Nebraska, traveling at an average velocity of 40 miles per hour. Two hours later, a Dodge Charger left Omaha, Nebraska, traveling on the same road at an average velocity of 60 miles per hour. When will the Charger catch up to the Cruze? How far will each car have traveled?

30. A train leaves Union Station traveling at an average velocity of 50 miles per hour. Four hours later, a helicopter follows along the train tracks traveling at an average velocity of 90 miles per hour. When will the helicopter catch up to the train? How far will each have traveled?

1.2 Exercises MyMathLab®

*Problems **1–30** are the Quick ✓ s that follow the EXAMPLES.*

Building Skills

31. What is 25% of 40?

32. What is 150% of 70?

33. 12 is 30% of what?

34. 50 is 90% of what?

35. 30 is what percent of 80?

36. 90 is what percent of 120?

In Problems 37–46, translate each English statement into a mathematical statement. Then solve the equation. See Objective 1.

37. The sum of a number x and 12 is 20.

38. The difference between 10 and a number z is 6.

39. Twice the sum of y and 3 is 16.

40. The sum of two times y and 3 is 16.

41. The difference between w and 22 equals three times w.

42. The sum of x and 4 results in twice x.

43. Four times a number x is equivalent to the sum of two times x and 14.

44. Five times a number x is equivalent to the difference of three times x and 10.

45. 80% of a number is equivalent to the sum of the number and 5.

46. 40% of a number equals the difference between the number and 10.

Applying the Concepts

47. Number Sense Grant is thinking of two numbers. He says that one of the numbers is twice the other number and the sum of the numbers is 39. What are the numbers?

48. Number Sense Pattie is thinking of two numbers. She says that one of the numbers is 8 more than the other number and the sum of the numbers is 56. What are the numbers?

49. Consecutive Integers The sum of three consecutive integers is 75. Find the integers.

50. Consecutive Integers The sum of four consecutive odd integers is 104. Find the integers.

51. Computing Grades Going into the final exam, which counts as two grades, Kendra has test scores of 84, 78, 64, and 88. What score does Kendra need on the final exam in order to have an average of 80?

52. Computing Grades Going into the final exam, which counts as three grades, Mark has test scores of 65, 79, 83, and 68. What score does Mark need on the final exam in order to have an average of 70?

53. Comparing Printers Jacob is trying to decide between two laser printers, one manufactured by Hewlett-Packard, the other by Brother. Both have similar features and warranties, so price is the determining factor. The Hewlett-Packard costs $180 and printing costs are approximately $0.03 per page. The Brother costs $230 and printing costs are approximately $0.01 per page. How many pages need to be printed for the two printers to cost the same?

54. Comparing Job Offers Maria has just been offered two sales jobs. The first job offer is a base monthly salary of $2500 plus a commission of 3% of total sales. The second job offer is a base monthly salary of $1500 plus a commission of 3.5% of total sales. For what level of monthly sales are the salaries offered by these two jobs equivalent?

55. Hourly Wage Before taxes, Jim earned $564 one week after working 30 hours at his regular wage, 6 hours at time-and-a-half on Saturday, and 4 hours at double-time on Sunday. What is Jim's hourly wage?

56. Sharing the Cost of a Pizza Judy and Linda agree to share the cost of a $21 pizza based on how much each ate. If Judy ate $\frac{3}{4}$ of the amount that Linda ate, how much should each pay?

57. Sales Tax In the state of Connecticut there is a state sales tax of 6.35% on all goods purchased. If Jan buys a television for $599, what will be the final bill, including state sales tax?

58. Sales Tax In Austin, Texas, there is a sales tax of 8.25% on all goods purchased. If Megan buys a sofa for $450, what will be the final bill, including the sales tax?

59. Markups A new 2016 Honda Accord has a list price of $25,965. Suppose that the dealer mark-up on this car is 9.3%. What is the dealer's cost?

60. Markups Suppose that the price of a new Intermediate Algebra text is $95. The bookstore has a policy of marking texts up 30%. What is the cost of the text to the bookstore?

61. Discount Pricing Suppose that you just received an e-mail alert indicating that 32-gigabyte flash drives have been discounted by 58%. If the sale price of the flash drive is now $9.99, what was the original price?

62. Discount Pricing Suppose that you just received an e-mail alert from Kohls indicating that the fall line of clothing has just been discounted by 30% and knit polo shirts are now $28. What was the original price of a polo shirt?

63. Cars A Mazda 6s weighs 136 pounds more than a Nissan Altima. A Honda Accord EX weighs 119 pounds more than a Nissan Altima. The total weight of all three cars is 9834 pounds. How much does each car weigh? (SOURCE: *Road and Track magazine*)

64. Cars The Honda Accord EX has 1.9 cubic feet less cargo space than a Mazda 6s. Together the cars have 31.3 cubic feet of cargo space. How much cargo space does each car have? (SOURCE: *Road and Track magazine*)

65. Finance A total of $20,000 is going to be split between Adam and Krissy with Adam receiving $3000 less than Krissy. How much will each get?

66. Finance A total of $40,000 is going to be invested in stocks and bonds. A financial advisor recommends that $6000 more be invested in stocks than in bonds. How much is invested in stocks? How much is invested in bonds?

67. Investments Suppose that your long-lost Aunt Sara has left you an unexpected inheritance of $24,000. You have decided to invest the money rather than spending it on frivolous purchases. Your financial advisor has recommended that you diversify by placing some of the money in stocks and some in bonds. Based upon current market conditions, she has recommended that the amount in bonds equal three-fifths of the amount invested in stocks. How much should be invested in stocks? How much should be invested in bonds?

68. Investments Jack and Diane have $60,000 to invest. Their financial advisor has recommended that they diversify by placing some of the money in stocks and some in bonds. Based upon current market conditions, he has recommended that the amount in bonds equal two-thirds of the amount invested in stocks. How much should be invested in stocks? How much should be invested in bonds?

69. Simple Interest Elena has a credit card balance of $2500. The credit card company charges 14% per annum simple interest. What is the interest charge on Elena's credit card after 1 month?

70. Simple Interest Faye has a home equity loan of $70,000. The bank charges Faye 6% per annum simple interest. What is the interest charge on Faye's loan after 1 month?

71. Banking A bank has loaned out $500,000, part of it at 6% per annum and the rest of it at 11% per annum. If the bank receives $43,750 in interest each year, how much was loaned at 6%?

72. Banking Patrick is a loan officer at a bank. He has $2,000,000 to lend out and has two loan programs. His home equity loan is currently priced at 6% per annum, and his unsecured personal loan is priced at 14%. The bank president wants Patrick to earn a rate of return of 12% on the $2,000,000 available. How much should Patrick lend out at 6%?

73. Investments Pedro wants to invest his $25,000 bonus check. His investment advisor has recommended that he put some of the money in a bond fund that yields 5% per annum and the rest in a stock fund that yields 9% per annum. If Pedro wants to earn $1875 each year from his investments, how much should he place in each investment?

74. Investments Johnny is a shrewd 8-year-old. For Christmas, his grandparents gave him $10,000. Johnny decides to invest some of the money in a savings account that pays 2% per annum and the rest in a stock fund paying 10% per annum. Johnny wants his investments to yield 7% per annum. How much should Johnny put into each account?

75. Candy "Sweet Tooth!" candy store sells chocolate-covered almonds for $6.50 per pound and chocolate-covered peanuts for $4.00 per pound. The manager decides to make a bridge mix that combines the almonds with the peanuts. She wants the bridge mix to sell for $5.00 per pound, and there should be no loss in revenue from selling the bridge mix rather than the almonds and peanuts alone. How many pounds of chocolate-covered almonds and how many pounds of chocolate-covered peanuts are required to create 50 pounds of bridge mix?

76. Coins Diana has been saving nickels and dimes. She opened up her piggy bank and found that it contained 48 coins worth $4.50. Determine how many nickels and dimes were in the piggy bank.

77. More Coins Bobby has been saving quarters and dimes. He opened up his piggy bank and found that it contained 47 coins worth $9.50. Determine how many dimes and quarters were in the piggy bank.

78. Gold The purity of gold is measured in karats, with pure gold being 24 karats. Other purities of gold are expressed as proportional parts of pure gold. For example, 18-karat gold is $\frac{18}{24}$, or 75%, pure gold; 12-karat gold is $\frac{12}{24}$, or 50%, pure gold; and so on. How much pure gold should be mixed with 12-karat gold to obtain 72 grams of 18-karat gold?

79. Nancy needs 8 ounces of whole milk for a recipe, but only has cream and low fat milk. In order to get 8 ounces of whole milk she needs to create a mix of cream and low fat milk to make a miture that contains 3.25% butterfat. If cream contains 25% butterfat and low fat milk contains 1% butterfat, how many ounces of each does she need to combine to create the 8 ounces of whole milk?

80. Antifreeze The cooling system of a car has a capacity of 15 liters. The system is currently filled with a mixture that is 40% antifreeze. How much of this mixture should be drained and replaced with pure antifreeze so that the system is filled with a solution that is 50% antifreeze?

81. A Biathlon Suppose that you have entered a 62-mile biathlon that consists of a run and a bicycle race. During your run, your average velocity is 8 miles per hour, and during your bicycle race, your average velocity is 20 miles per hour. You finish the race in 4 hours. What is the distance of the run? What is the distance of the bicycle race?

82. A Biathlon Suppose that you have entered a 15-mile biathlon that consists of a run and swim. During your run, your average velocity is 7 miles per hour, and during your swim, your average velocity is 2 miles per hour. You finish the race in 2.5 hours. What is the length of the swim? What is the distance of the run?

83. Collision Course Two cars that are traveling toward each other are 455 miles apart. One car is traveling 10 miles per hour faster than the other car. The cars pass after 3.5 hours. How fast is each car traveling?

84. Collision Course Two planes that are traveling toward each other are 720 miles apart. One plane is traveling 40 miles per hour faster than the other. The planes pass after 0.75 hour. How fast is each plane traveling?

85. Boats Two boats leave a port at the same time, one traveling north and the other traveling south. The northbound boat travels at 12 miles per hour (mph) faster than the southbound boat. If the southbound boat is traveling at 25 mph, how long will it be before they are 155 miles apart?

86. Cyclists Two cyclists leave a city at the same time, one going east and the other going west. The westbound cyclist bikes at 3 mph faster than the eastbound cyclist. If after 6 hours they are 162 miles apart, how fast is each cyclist riding?

87. Walking At 10:00 A.M. two people leave their homes that are 15 miles apart and begin walking toward each other. If one person walks at a rate that is 2 mph faster than the other and they meet after 1.5 hours, how fast was each person walking?

88. Trains At 9:00 A.M., two trains are 715 miles apart traveling toward each other on parallel tracks. If one train is traveling at a rate that is 10 miles per hour (mph) faster than the other train and they meet after 5.5 hours, how fast is each train traveling?

Extending the Concepts

89. Computing Average Speed On a recent trip to Florida, we averaged 50 miles per hour. On the return trip we averaged 60 miles per hour. What do you think the average speed of the trip to Florida and back was? Defend your position. Algebraically, determine the average speed of the trip to Florida and back.

90. Discount Pricing Suppose that you are the manager of a clothing store and have just purchased 100 shirts for $15 each. After 1 month of selling the shirts at the regular price, you plan to have a sale giving 30% off the original selling price. However, you want to make a profit of $6 on each shirt at the sale price. What should you price the shirts at initially to ensure this?

91. Critical Thinking Make up an applied problem that would result in the equation $x - 0.05x = 60$.

92. Critical Thinking Make up an applied problem that would result in the equation $10 + 0.14x = 50$.

93. Uniform Motion Suppose that you are walking along the side of train tracks at 2 miles per hour. A train that is traveling 20 miles per hour in the same direction requires 1 minute to pass you. How long is the train?

Explaining the Concepts

94. Consider the phrase "... the revenue from selling the new blend and that from selling the beans separately will be the same," which appeared in Example 8. Explain what this means.

95. How is mathematical modeling related to problem solving?

96. Why do we make assumptions when creating mathematical models?

97. Name the different categories of problems presented in this section. Name two different kinds of mixture problems.

98. Think of two models that you use in your everyday life. Are any of them mathematical models? Describe your models to members of your class.

1.3 Using Formulas to Solve Problems

Objectives

1 Solve for a Variable in a Formula

2 Use Formulas to Solve Problems

Are You Prepared for This Section?

Before getting started, complete the following problems. If you get a problem wrong, go back to the section cited and review the material.

In Problems P1 and P2, (a) round and (b) truncate each decimal to the indicated number of places. [Section R.2, pp. 14–15]

P1. 3.00343; three decimal places

P2. 14.957; two decimal places

Prepared?...Answers P1. (a) 3.003
(b) 3.003 **P2. (a)** 14.96 **(b)** 14.95

A known relation that exists between two or more variables can be used to solve certain types of problems. A **formula** is an equation that describes how two or more variables are related.

EXAMPLE 1 | **Answering Questions with Formulas**

Work Smart

Think of a formula as another kind of mathematical model.

The information in Table 6 gives descriptions, in words, of known relations and the corresponding formulas.

Table 6

Verbal Description	Formula
Geometry: The area A of a rectangle is the product of its length l and width w.	$A = lw$
Physics: The energy of an object in motion, kinetic energy K, is one-half the product of the mass m and the square of the velocity v.	$K = \dfrac{1}{2}mv^2$
Grading: The final grade, G, in your Intermediate Algebra class is the average of your four exam grades, A, B, C, and D.	$G = \dfrac{A + B + C + D}{4}$
Finance: The future value of money, A, is the product of the present value P and the sum of 1 and the annual interest rate r.	$A = P(1 + r)$

Quick ✓

1. A _____ is an equation that describes how two or more variables are related.

In Problems 2–5, translate the verbal description into a mathematical formula.

2. The area A of a circle is the product of the number π and the square of its radius r.

3. The volume V of a right circular cylinder is the product of the number π, the square of its radius r, and its height h.

4. The daily cost C of manufacturing computers is \$175 times the number of computers manufactured x, plus \$7000.

5. The distance s that an object free-falls is one-half the product of acceleration due to gravity g and the square of time t.

▶ ❶ **Solve for a Variable in a Formula**

To "solve for a variable" means to isolate the variable on one side of the equation with all the other variables and constants, if any, on the other side, by forming equivalent equations. For example, the formula for the area of a rectangle, $A = lw$, is solved for A because A is by itself on one side of the equation. The steps to follow when solving a formula for a certain variable are identical to those to follow when solving an equation.

EXAMPLE 2 | **Solving for a Variable in a Formula**

Figure 4

The volume V of a cone is given by the formula $V = \dfrac{1}{3}\pi r^2 h$, where r is the radius and h is the height of the cone. See Figure 4.

    (a) Solve the formula for h.

    (b) Use the result from part (a) to find the height of a cone if its volume is 50π cubic feet and its radius is 5 feet.

Solution

    (a) To solve the formula for h, isolate h on one side of the equation.

$$V = \frac{1}{3}\pi r^2 h$$

Multiply both sides by 3: $\quad 3 \cdot V = 3 \cdot \frac{1}{3}\pi r^2 h$

Divide out common factors: $\quad 3V = \pi r^2 h$

Divide both sides by πr^2: $\quad \dfrac{3V}{\pi r^2} = \dfrac{\pi r^2 h}{\pi r^2}$

Divide out common factors: $\quad \dfrac{3V}{\pi r^2} = h$

If $a = b$, then $b = a$: $\quad h = \dfrac{3V}{\pi r^2}$

(b) Substituting $V = 50\pi$ ft^3 and $r = 5$ ft into $h = \dfrac{3V}{\pi r^2}$, we obtain

$$h = \frac{3\left(50\pi \text{ ft}^3\right)}{\pi\left(5 \text{ ft}\right)^2}$$

$$h = \frac{150\pi \text{ ft}^3}{25\pi \text{ ft}^2}$$

$$h = 6 \text{ feet}$$

Work Smart

Solving for a variable is just like solving an equation with one unknown. When solving for a variable, treat all the other variables as constants.

Geometry Formulas

A review of a few common terms from geometry follows:

Definitions

The **perimeter** is the sum of the lengths of all the sides of a figure.

The **area** is the amount of space enclosed by a two-dimensional figure, measured in square units.

The **surface area** of a solid is the sum of the areas of the surfaces of a three-dimensional figure.

The **volume** is the amount of space occupied by a three-dimensional figure, measured in cubic units.

The **radius** r of a circle is the line segment that extends from the center of the circle to any point on the circle.

The **diameter** of a circle is any line segment that extends from one point on the circle through the center to a second point on the circle. The length of a diameter is two times the length of the radius, $d = 2r$.

In circles, we use the term **circumference** to mean the perimeter.

Work Smart

When working with formulas, keep track of the units. Verify that the units in your answer are reasonable.

Formulas from geometry are useful in solving many types of problems. Some of these formulas are provided in Table 7.

Table 7

Plane Figures	Formulas	Plane Figures	Formulas
Square s	**Area:** $A = s^2$ **Perimeter:** $P = 4s$	Trapezoid 	**Area:** $A = \frac{1}{2}h(B + b)$ **Perimeter:** $P = a + b + c + B$
Rectangle 	**Area:** $A = lw$ **Perimeter:** $P = 2l + 2w$	Parallelogram 	**Area:** $A = bh$ **Perimeter:** $P = 2a + 2b$

(continued)

Table 7 (*continued*)

Plane Figures	Formulas	Plane Figures	Formulas
Triangle	**Area:** $A = \dfrac{1}{2}bh$ **Perimeter:** $P = a + b + c$	Circle	**Area:** $A = \pi r^2$ **Circumference:** $C = 2\pi r$ $\qquad\qquad\quad = \pi d$

Solids	Formulas	Solids	Formulas
Cube	**Volume:** $V = s^3$ **Surface Area:** $S = 6s^2$	Right Circular Cylinder	**Volume:** $V = \pi r^2 h$ **Surface Area:** $S = 2\pi r^2 + 2\pi rh$
Rectangular Solid	**Volume:** $V = lwh$ **Surface Area:** $S = 2lw + 2lh + 2wh$	Cone	**Volume:** $V = \dfrac{1}{3}\pi r^2 h$
Sphere	**Volume:** $V = \dfrac{4}{3}\pi r^3$ **Surface Area:** $S = 4\pi r^2$		

Quick ✔

6. The _____ is the amount of space enclosed by a two-dimensional figure and is measured in square units, whereas _____ is the amount of space occupied by a three-dimensional figure and is measured in cubic units.

7. *True or False* The perimeter of a rectangle 6 feet long and 4 feet wide is 24 square feet.

8. The area A of a triangle is given by the formula $A = \dfrac{1}{2}bh$, where b is the base of the triangle and h is the height.

   (a) Solve the formula for h.

   (b) Find the height of the triangle whose area is 10 square inches and whose base is 4 inches.

9. The perimeter P of a parallelogram is given by the formula $P = 2a + 2b$, where a is the length of one side of the parallelogram and b is the length of the adjacent side.

   (a) Solve the formula for b.

   (b) Find the length of one side of a parallelogram whose perimeter is 60 cm and length of the other side is 20 cm.

EXAMPLE 3   **Solving for a Variable in a Formula**

The formula $Y = C + bY + I + G + N$ is a model used in economics to describe the total income of an economy. In the model, Y is income, C is consumption, I is investment in capital, G is government spending, N is net exports, and b is a constant. Solve the formula for Y.

Solution

Get all terms with Y on the same side of the equal sign.

$$Y = C + bY + I + G + N$$

Subtract bY from both sides: $$Y - bY = C + bY + I + G + N - bY$$

Combine like terms: $$Y - bY = C + I + G + N$$

Use the Distributive Property
"in reverse" to isolate Y: $$Y(1 - b) = C + I + G + N$$

Divide both sides by $1 - b$: $$\frac{Y(1 - b)}{1 - b} = \frac{C + I + G + N}{1 - b}$$

Simplify: $$Y = \frac{C + I + G + N}{1 - b}$$

Quick ✔

In Problems 10–13, solve for the indicated variable.

10. $I = Prt$ for P

11. $Ax + By = C$ for y

12. $2xh - 4x = 3h - 3$ for h

13. $S = na + (n - 1)d$ for n

❷ **Use Formulas to Solve Problems**

Formulas are often used to solve certain types of word problems. Use the problem-solving steps from page 61.

(**EXAMPLE 4**)   **Finding the Dimensions of a Window**

The perimeter of a rectangular picture window is 466 inches. The length of the window is 55 inches more than the width. See Figure 5. Find the dimensions of the window.

Solution

Step 1: Identify This is a geometry problem that requires the formula for the perimeter of a rectangle. Determine the dimensions of the rectangular window. That is, find the window's length and width.

Step 2: Name Let w represent the width in inches. The length is 55 inches more than the width, so $l = w + 55$.

Step 3: Translate The perimeter of a rectangle is $P = 2l + 2w$. Substitute the known values into this formula.

$$P = 2l + 2w$$

$P = 466; l = w + 55$: $466 = 2(w + 55) + 2w$ The Model

Figure 5

w

$w + 55$

Step 4: Solve $466 = 2(w + 55) + 2w$

Distribute the 2: $466 = 2w + 110 + 2w$

Combine like terms: $466 = 4w + 110$

Subtract 110 from both sides: $356 = 4w$

Divide both sides by 4: $89 = w$

Step 5: Check If the width of the window is 89 inches, then the length is $89 + 55 = 144$ inches. The perimeter of the window is $2(144) + 2(89) = 466$ inches. It checks!

Step 6: Answer the Question The window's width is 89 inches and its length is 144 inches.

There is an interesting side note to the result of Example 4. The ratio of the window's length to its width is $\frac{144}{89} \approx 1.618$. Rectangles whose dimensions form this ratio are called **golden rectangles.** Golden rectangles are said to have dimensions that are "pleasing to the eye." The golden rectangle was first constructed by the Greek philosopher Pythagoras in the sixth century B.C. These rectangles are used regularly in architecture (the Parthenon) and in art (the Mona Lisa). See Figure 6.

Figure 6

Dagli Orti (A)/Picture Desk, Inc./ Kobal Collection

Quick ✓

14. The perimeter of a rectangular pool is 180 feet. If the length of the pool is 10 feet more than the width, find the dimensions of the pool.

15. The opening of a rectangular bookcase has a perimeter of 224 inches. If the height of the bookcase is 32 inches more than the width, determine the dimensions of the opening of the bookcase.

⊙ (**EXAMPLE 5**)   **Constructing a Soup Can**

Figure 7

$r = 1.375''$

A can of Campbell's soup has a surface area of 46.5 square inches. See Figure 7. The surface area S of a right circular cylinder is $S = 2\pi r^2 + 2\pi rh$, where r is the radius of the can and h is its height. Find the height of a can of Campbell's soup if its radius is 1.375 inches. Round your answer to two decimal places.

Solution

Step 1: Identify This is a geometry problem that requires the formula for the surface area of a cylinder. Find the height of the can of soup.

Step 2: Name Let h represent the height, in inches, of the can of soup.

Step 3: Translate It is given that the surface area S of the can is 46.5 square inches. The radius is 1.375 inches. Substitute these values into the formula for the surface area of the can.

$$S = 2\pi r^2 + 2\pi rh$$
$S = 46.5; r = 1.375:$ $46.5 = 2\pi(1.375)^2 + 2\pi(1.375)h$ The Model

Step 4: Solve Solve for h. To avoid round-off error, do not compute any of the values until the last calculation.

Subtract $2\pi(1.375)^2$ from $46.5 = 2\pi(1.375)^2 + 2\pi(1.375)h$

both sides: $46.5 - 2\pi(1.375)^2 = 2\pi(1.375)h$

Divide both sides by $\dfrac{46.5 - 2\pi(1.375)^2}{2\pi(1.375)} = h$

$2\pi(1.375):$

Work Smart

Round-off error occurs when decimals are continually rounded during the course of solving a problem. The more times we round, the more inaccurate the results may be. So do not do any rounding until the last step.

Figure 8

Use a calculator to evaluate this expression. Figure 8 shows the output from a graphing calculator. Thus $h = 4.01$ inches, rounded to two decimal places.

Step 5: Check The surface area S of the can with height 4.01 inches is

$$S = 2\pi r^2 + 2\pi r h$$
$$= 2\pi (1.375)^2 + 2\pi (1.375)(4.01)$$
$$\approx 46.5 \text{ square inches}$$

Step 6: Answer the Question The height of the can is 4.01 inches, rounded to two decimal places.

Work Smart

Make sure to put parentheses around the numerator and parentheses around the denominator to keep them in order in your calculator. See Figure 8.

Quick ✓

16. A can of peaches has a surface area of 51.8 square inches. The surface area S of a right circular cylinder is $S = 2\pi r^2 + 2\pi r h$, where r is the radius of the can and h is the height of the can. Find the height of a can of peaches if its radius is 1.5 inches. Round your answer to two decimal places.

1.3 Exercises MyMathLab® Exercise numbers in green have complete video solutions in MyMathLab or may be accessed using the QR code to the right.

Problems 1–16 are the Quick ✓s that follow the EXAMPLES.

Building Skills

In Problems 17–20, translate the verbal description into a mathematical formula.

17. Force F equals the product of mass m and acceleration a.

18. The area A of a triangle is one-half the product of its base b and its height h.

19. The volume V of a sphere is four-thirds the product of the number π and the cube of its radius r.

20. The revenue R of selling computers is $800 times the number of computers sold x.

In Problems 21–32, solve the formula for the indicated variable. See Objective 1.

21. **Uniform Motion** Solve $d = rt$ for r.

22. **Direct Variation** Solve $y = kx$ for k.

23. **Algebra** Solve $y - y_1 = m(x - x_1)$ for m.

24. **Algebra** Solve $y = mx + b$ for m.

25. **Statistics** Solve $Z = \dfrac{x - \mu}{\sigma}$ for x.

26. **Statistics** Solve $E = \dfrac{Z \cdot \sigma}{\sqrt{n}}$ for $\sqrt{n}$.

27. **Newton's Law of Gravitation** Solve $F = G\dfrac{m_1 m_2}{r^2}$ for m_1.

28. **Sequences** Solve $S - rS = a - ar^5$ for S.

29. **Finance** Solve $A = P + Prt$ for P.

30. **Bernoulli's Equation** Solve $p + \dfrac{1}{2}\rho v^2 + \rho gy = a$ for ρ.

31. **Temperature Conversion** Solve $C = \dfrac{5}{9}(F - 32)$ for F.

32. **Trapezoid** Solve $A = \dfrac{1}{2}h(B + b)$ for b.

In Problems 33–40, solve for y.

33. $2x + y = 13$

34. $-4x + y = 12$

35. $9x - 3y = 15$

36. $4x + 2y = 20$

37. $4x + 3y = 13$

38. $5x - 6y = 18$

39. $\dfrac{1}{2}x + \dfrac{1}{6}y = 2$

40. $\dfrac{2}{3}x - \dfrac{5}{2}y = 5$

Applying the Concepts

41. Cylinders The volume V of a right circular cylinder is given by the formula $V = \pi r^2 h$, where r is the radius and h is the height.

(a) Solve the formula for h.

(b) Find the height of a right circular cylinder whose volume is 32π cubic inches and whose radius is 2 inches.

42. Cylinders The surface area S of a right circular cylinder is given by the formula $S = 2\pi rh + 2\pi r^2$, where r is the radius and h is the height.

(a) Solve the formula for h.

(b) Determine the height of a right circular cylinder whose surface area is 72π square centimeters and whose radius is 4 centimeters.

43. Maximum Heart Rate The model $M = -0.711A + 206.3$ was developed by Londeree and Moeschberger to determine the maximum heart rate M of an individual who is age A. (SOURCE: Londeree and Moeschberger, "Effect of Age and Other Factors on HR max," *Research Quarterly for Exercise and Sport,* 53(4), 297–304)

(a) Solve the model for A.

(b) According to this model, what is the age of an individual whose maximum heart rate is 160?

44. Maximum Heart Rate The model $M = -0.85A + 217$ was developed by Miller to determine the maximum heart rate M of an individual who is age A. (SOURCE: Miller et al., "Predicting Max HR," *Medicine and Science in Sports and Exercise,* 25(9), 1077–1081)

(a) Solve the model for A.

(b) According to this model, what is the age of an individual whose maximum heart rate is 160?

45. Finance The formula $A = P(1 + r)^t$ can be used to relate the future value A of a deposit of P dollars in an account that earns an annual interest rate r (expressed as a decimal) after t years.

(a) Solve the formula for P.

(b) How much would you have to deposit today in order to have $5000 in 5 years in a bank account that pays 4% annual interest?

46. Federal Income Taxes For a single filer with an annual adjusted income in 2015 over $37,450 but less than $90,751, the federal income tax T for an annual adjusted income I is found using the formula $T = 0.25(I - 37,450) + 5156.25$.

(a) Solve the formula for I.

(b) Determine the adjusted income of a single filer whose tax bill is $14,780.

47. Supplementary Angles Two angles are **supplementary** if the sum of the measures of the angles is 180°. If one angle is 30° more than its supplement, find the measures of the two angles.

48. Supplementary Angles See Problem 47. If one angle is twice the measure of its supplement, find the measures of the two angles.

49. Complementary Angles Two angles are **complementary** if the sum of the measures of the angles is 90°. If one angle is 10° more than 3 times its complement, find the measures of the two angles.

50. Complementary Angles See Problem 49. If one angle is 30° less than twice its complement, find the measures of the two angles.

51. Dimensions of a Window The perimeter of a rectangular window is 26 feet. The width of the window is 3 feet more than the length. What are the dimensions of the window?

52. Dimensions of a Window The perimeter of a rectangular window is 120 inches. The length of the window is twice its width. What are the dimensions of the window?

53. Art An artist wants to place a piece of round stained glass into a square that is made of copper wire. See the figure. If the perimeter of the square is 40 inches, what is the area of the largest circular piece of stained glass that can fit into the copper square?

54. Art An artist wants to place two circular pieces of stained glass into a rectangle that is made of copper wire. See the figure. The perimeter of the rectangle is 36 centimeters, and the length is twice the width. Find the area of each circle, assuming they are to be as large as possible and still fit inside the rectangle.

55. Angles in a Triangle The sum of the measures of the interior angles in a triangle is 180°. The measure of the second angle is 15° more than the measure of the first angle. The measure of the third angle is 45° more than the measure of the first angle. Find the measures of the interior angles in the triangle.

$$x° + y° + z° = 180°$$

56. Angles in a Triangle See Problem 55. The measure of the second angle is 3 times the measure of the first angle. The measure of the third angle is 20° more than the measure of the first angle. Find the measures of the interior angles in the triangle.

57. Designing a Patio Suppose that you wish to build a rectangular cement patio that is to have a perimeter of 80 feet. The length is to be 5 feet more than the width.

(a) Find the dimensions of the patio.

(b) If the building code requires that the cement be 4 inches deep, how much cement do you have to purchase?

58. Designing a Foundation You have just purchased a circular gazebo whose diameter is 12 feet. Before you have it delivered, you must lay a cement foundation for the gazebo.

(a) Find the area of the base of the gazebo.

(b) If the building code requires that the cement be 4 inches deep, how much cement do you have to purchase?

Extending the Concepts

59. Critical Thinking Suppose that you have just purchased a swimming pool whose diameter is 25 feet. You decide to build around the pool a deck that is 3 feet wide.

(a) What is the exact area of the deck?

(b) The building code requires that the pool must be enclosed in a fence. How much fence is needed to encircle the pool and deck?

(c) If the fence costs $25 per linear foot to install, how much will the fence cost?

60. Critical Thinking Suppose that you wish to install a window with the dimensions given in the figure.

(a) What is the area of the opening of the window?

(b) What is the perimeter of the window?

(c) If glass costs $8.25 per square foot, what is the cost of the glass for the window?

61. If the radius of a circle is doubled, does the area double? Explain. If the length of the sides of a cube is doubled, what happens to the volume?

1.4 Linear Inequalities in One Variable

Objectives

1 Represent Inequalities Using the Real Number Line and Interval Notation

2 Understand the Properties of Inequalities

3 Solve Linear Inequalities

4 Solve Problems Involving Linear Inequalities

Are You Prepared for This Section?

Before getting started, complete the following problems. If you get a problem wrong, go back to the section cited and review the material.

In Problems P1–P4, replace the question mark by $<$, $>$, or $=$ to make the statement true. [Section R.2, pp. 16–17]

P1. $3\ ?\ 6$ **P2.** $-3\ ?\ -6$ **P3.** $\dfrac{1}{2}\ ?\ 0.5$ **P4.** $\dfrac{2}{3}\ ?\ \dfrac{3}{5}$

P5. *True or False* The inequality $\geq$ is called a strict inequality. [Section R.2, p. 17]

P6. Use set-builder notation to represent the set of all digits that are divisible by 3. [Section R.2, p. 9]

An **inequality in one variable** is a statement involving two expressions, at least one containing the variable, separated by the inequality symbol $<$, $\leq$, $>$, or $\geq$. To **solve an inequality** means to find all values of the variable for which the statement is true. These values are called **solutions** of the inequality. The set of all solutions is called the **solution set.**

> ### Definition
>
> A **linear inequality in one variable** is an inequality that can be written in the form
> $$ax + b < c \quad \text{or} \quad ax + b \leq c \quad \text{or} \quad ax + b > c \quad \text{or} \quad ax + b \geq c$$
> where a, b, and c are real numbers and $a \neq 0$.

For example, the following are all linear inequalities involving one variable:

$$x - 4 > 9 \qquad 5x - 1 \leq 14 \qquad 8z < 0 \qquad 5x - 1 \geq 3x + 8$$

Before methods for solving linear inequalities are discussed, two new ways to represent the solution set need to be introduced: *interval notation* and graphing the solution set on a real number line.

1 Represent Inequalities Using the Real Number Line and Interval Notation

Suppose that a and b are two real numbers and $a < b$. The notation

$$a < x < b$$

means that x is a number *between* a and b. So the expression $a < x < b$ is equivalent to the two inequalities $a < x$ and $x < b$. Similarly, the expression $a \leq x \leq b$ is equivalent to the two inequalities $a \leq x$ and $x \leq b$. Define $a \leq x < b$ and $a < x \leq b$ similarly. Expressions such as $-2 < x < 5$ and $x \geq 5$ are in **inequality notation.**

Even though the expression $3 \geq x \geq 2$ is technically correct, it is customary to order the numbers in an inequality from smaller values to larger values. So express $3 \geq x \geq 2$ as $2 \leq x \leq 3$.

A statement such as $3 \leq x \leq 1$ is false because there is no number x for which $3 \leq x$ and $x \leq 1$. Never mix the inequalities' directions, as in $2 \leq x \geq 3$.

Interval notation is used to represent the solution set of an inequality.

Work Smart

Remember that the inequalities $<$ and $>$ are called strict inequalities, while $\leq$ and $\geq$ are called nonstrict inequalities.

Prepared?...Answers **P1.** $<$ **P2.** $>$
P3. $=$ **P4.** $>$ **P5.** False
P6. $\{x \mid x \text{ is a digit that is divisible by 3}\}$

Definitions

Let a and b represent two real numbers with $a < b$.

A **closed interval,** denoted by $[a, b]$, consists of all real numbers x for which $a \leq x \leq b$.

An **open interval,** denoted by (a, b), consists of all real numbers x for which $a < x < b$.

The **half-open,** or **half-closed, intervals** are $(a, b]$, consisting of all real numbers x for which $a < x \leq b$, and $[a, b)$, consisting of all real numbers x for which $a \leq x < b$.

In each of these definitions, a is the **left endpoint** and b is the **right endpoint** of the interval.

The symbol ∞ (which is read "infinity") is not a real number but a notational device that indicates unboundedness in the positive direction. In other words, the symbol ∞ means the inequality has no right endpoint. The symbol $-\infty$ (which is read "negative infinity") is also not a number but a notational device that indicates unboundedness in the negative direction. The symbol $-\infty$ means the inequality has no left endpoint. The symbols ∞ and $-\infty$ allow us to define five other kinds of intervals.

Work Smart

The symbols ∞ and $-\infty$ are never included as endpoints because they are not real numbers. Therefore, always use parentheses when $-\infty$ or ∞ is an endpoint.

Interval Notation

$[a, \infty)$	consists of all real numbers x where $x \geq a$
(a, ∞)	consists of all real numbers x where $x > a$
$(-\infty, a]$	consists of all real numbers x where $x \leq a$
$(-\infty, a)$	consists of all real numbers x where $x < a$
$(-\infty, \infty)$	consists of all real numbers x (that is, $-\infty < x < \infty$)

Another way to represent inequalities is to draw a graph on the real number line. The inequality $x > 3$ or the interval $(3, \infty)$ consists of all numbers x that lie to the right of 3 on the real number line. Graph these values by shading the real number line to the right of 3. Use a *parenthesis* on the endpoint to indicate that 3 is not included in the set. See Figure 9.

To graph the inequality $x \geq 3$ or the interval $[3, \infty)$, shade to the right of 3 but use a *bracket* on the endpoint to indicate that 3 is included in the set. See Figure 10.

Table 8 summarizes interval notation, inequality notation, and their graphs.

Figure 9

$x > 3$

Figure 10

$x \geq 3$

Table 8

Interval Notation	Inequality Notation	Graph
The open interval (a, b)	$\{x \mid a < x < b\}$	
The closed interval $[a, b]$	$\{x \mid a \leq x \leq b\}$	
The half-open interval $[a, b)$	$\{x \mid a \leq x < b\}$	
The half-open interval $(a, b]$	$\{x \mid a < x \leq b\}$	
The interval $[a, \infty)$	$\{x \mid x \geq a\}$	
The interval (a, ∞)	$\{x \mid x > a\}$	
The interval $(-\infty, a]$	$\{x \mid x \leq a\}$	
The interval $(-\infty, a)$	$\{x \mid x < a\}$	
The interval $(-\infty, \infty)$	$\{x \mid x \text{ is a real number}\}$	

EXAMPLE 1   **Using Interval Notation and Graphing Inequalities**

Write each inequality using interval notation. Graph the inequality.

(a) $-2 \le x \le 4$ (b) $1 < x \le 5$

Solution

(a) $-2 \le x \le 4$ describes all numbers x between -2 and 4, inclusive. In interval notation, write $[-2, 4]$. To graph $-2 \le x \le 4$, place brackets at -2 and 4 and shade in between. See Figure 11.

Figure 11

(b) $1 < x \le 5$ describes all numbers x greater than 1 and less than or equal to 5. In interval notation, write $(1, 5]$. To graph $1 < x \le 5$, place a parenthesis at 1 and a bracket at 5 and shade in between. See Figure 12.

Figure 12

EXAMPLE 2   **Using Interval Notation and Graphing Inequalities**

Write each inequality using interval notation. Graph the inequality.

(a) $x < 2$ (b) $x \ge -3$

Solution

(a) $x < 2$ describes all numbers x less than 2. In interval notation, write $(-\infty, 2)$. To graph $x < 2$, place a parenthesis at 2, and then shade to the left. See Figure 13.

Figure 13

(b) $x \ge -3$ describes all numbers x greater than or equal to -3. In interval notation, write $[-3, \infty)$. To graph $x \ge -3$, place a bracket at -3 and then shade to the right. See Figure 14.

Figure 14

Quick ✓

1. A(n) _____ _____, denoted $[a, b]$, consists of all real numbers x for which $a \le x \le b$.
2. In the interval (a, b), a is called the _____ _____ and b is called the _____ _____ of the interval.
3. *True or False* The inequality $x \le -11$ is written $[-11, -\infty)$ in interval notation.

In Problems 4–7, write each inequality in interval notation. Graph the inequality.

4. $-3 \le x \le 2$ 5. $3 \le x < 6$

6. $x \le 3$ 7. $\dfrac{1}{2} < x < \dfrac{7}{2}$

EXAMPLE 3 **Using Inequality Notation and Graphing Inequalities**

Write each interval in inequality notation involving x. Graph the inequality.

(a) $[-2, 4)$ **(b)** $(1, 5)$

Solution

(a) The interval $[-2, 4)$ consists of all numbers x for which $-2 \le x < 4$. See Figure 15 for the graph.

(b) The interval $(1, 5)$ consists of all numbers x for which $1 < x < 5$. See Figure 16 for the graph.

Figure 15

$$\begin{array}{c} \longmapsto\!\longrightarrow \\ -3 \ -2 \ -1 \ \ 0 \ \ 1 \ \ 2 \ \ 3 \ \ 4 \ \ 5 \end{array}$$

Figure 16

$$\begin{array}{c} \longmapsto\!\longrightarrow \\ -1 \ \ 0 \ \ 1 \ \ 2 \ \ 3 \ \ 4 \ \ 5 \ \ 6 \ \ 7 \end{array}$$

●

EXAMPLE 4 **Using Inequality Notation and Graphing Inequalities**

Write each interval in inequality notation involving x. Graph the inequality.

(a) $\left[\dfrac{3}{2}, \infty\right)$ **(b)** $(-\infty, 1)$

Solution

(a) The interval $\left[\dfrac{3}{2}, \infty\right)$ consists of all numbers x for which $x \ge \dfrac{3}{2}$. See Figure 17 for the graph.

(b) The interval $(-\infty, 1)$ consists of all numbers x for which $x < 1$. See Figure 18 for the graph.

Figure 17

$$\begin{array}{c} \longmapsto\!\longrightarrow \\ -1 \ \ 0 \ \ 1 \ \tfrac{3}{2} 2 \ \ 3 \ \ 4 \ \ 5 \ \ 6 \ \ 7 \end{array}$$

Figure 18

$$\begin{array}{c} \longleftarrow\!\longmapsto \\ -4 \ -3 \ -2 \ -1 \ \ 0 \ \ 1 \ \ 2 \ \ 3 \ \ 4 \end{array}$$

●

Quick ✓

In Problems 8–11, write each interval as an inequality. Graph the inequality.

8. $(0, 5]$ **9.** $(-6, 0)$

10. $(5, \infty)$ **11.** $\left(-\infty, \dfrac{8}{3}\right]$

❷ **Understand the Properties of Inequalities**

 Consider the inequality $2 < 5$. Because $2 < 5$, if 4 is added to both sides of the inequality, then $2 + 4 < 5 + 4$ or $6 < 9$. Also, since $3 > -1$, then $3 + (-2) > -1 + (-2)$ or $1 > -3$ so, adding the same quantity to both sides of an inequality does not change the direction of the inequality.

In Other Words

The Addition Property states that the direction of the inequality does not change when the same quantity is added to each side of the inequality.

Addition Property of Inequalities

For real numbers a, b, and c,

$$\text{If} \quad a < b, \quad \text{then} \quad a + c < b + c$$
$$\text{If} \quad a > b, \quad \text{then} \quad a + c > b + c$$

Because $a - b$ is equivalent to $a + (-b)$, the Addition Property of Inequalities can also be used to subtract a real number from each side of an inequality without changing the direction of the inequality.

What happens when both sides are multiplied by a nonzero constant? Because $3 < 5$, multiply both sides of $3 < 5$ by 2. The left side becomes 6 and the right side becomes 10. Certainly $6 < 10$, so the direction of the inequality did not change.

Now multiply both sides of the inequality $3 < 5$ by -2. The left side becomes -6 and the right side becomes -10. Because $-6 > -10$, the direction of the inequality is reversed.

These results are true in general and lead us to the next properties.

> **In Other Words**
>
> The Multiplication Property states that if both sides of an inequality are multiplied by a positive real number, the direction of the inequality is unchanged. If both sides of an inequality are multiplied by a negative real number, the direction of the inequality is reversed.

Multiplication Properties of Inequalities

Let a, b, and c be real numbers.

$$\text{If } a < b \text{ and } c > 0, \text{ then } ac < bc$$
$$\text{If } a > b \text{ and } c > 0, \text{ then } ac > bc$$
$$\text{If } a < b \text{ and } c < 0, \text{ then } ac > bc$$
$$\text{If } a > b \text{ and } c < 0, \text{ then } ac < bc$$

Because the quotient $\dfrac{a}{b}$ is equivalent to $a \cdot \dfrac{1}{b}$, the Multiplication Properties of Inequality can also be used to divide both sides of an inequality by a nonzero real number.

Quick ✓

In Problems 12–16, write the inequality that results from following the directions. Then state the inequality property you have illustrated.

12. Add 5 to each side of the inequality $4 < 7$.

13. Subtract 3 from each side of the inequality $x + 3 > -6$.

14. Multiply both sides of the inequality $2 < 8$ by $\dfrac{1}{2}$.

15. Divide both sides of the inequality $-6 < 9$ by -3.

16. Divide both sides of the inequality $5x < 30$ by 5.

▶ ❸ Solve Linear Inequalities

Two inequalities that have exactly the same solution set are called **equivalent inequalities.** To solve linear inequalities, use a process similar to the one used to solve linear equations: Create a series of equivalent inequalities to arrive at a solution.

Although not essential, it is easier to read an inequality when the variable is on the left side and the constant is on the right. Rewrite inequalities using the fact that

> **In Other Words**
>
> If the sides of an inequality are interchanged, the direction of the inequality is reversed.

$$a < x \qquad \text{is equivalent to} \qquad x > a$$
$$\text{and}$$
$$a > x \qquad \text{is equivalent to} \qquad x < a$$

EXAMPLE 5 **How to Solve a Linear Inequality**

Solve the inequality $3x - 2 > 13$. Graph the solution set.

Step-by-Step Solution

The goal in solving any linear inequality is to get the variable by itself with a coefficient of 1.

Step 1: Isolate the term containing the variable.

Add 2 to both sides
(Addition Property):
$$3x - 2 > 13$$
$$3x - 2 + 2 > 13 + 2$$
$$3x > 15$$

Step 2: Get a coefficient of 1 on the variable.

Divide both sides by 3 (Multiplication Property): $\dfrac{3x}{3} > \dfrac{15}{3}$

$$x > 5$$

The solution using set-builder notation is $\{x \mid x > 5\}$. The solution using interval notation is $(5, \infty)$. Figure 19 shows the graph of the solution set.

Figure 19

Quick ✓

In Problems 17–19, solve each linear inequality. Write the solution using set-builder notation and interval notation. Graph the solution set.

17. $x + 3 > 5$

18. $\dfrac{1}{3}x \le 2$

19. $-2x + 1 \le 13$

(**EXAMPLE 6**) **Solving Linear Inequalities**

Solve the inequality: $x - 4 \ge 5x + 12$

Solution

$$x - 4 \ge 5x + 12$$

Add 4 to both sides: $\quad x - 4 + 4 \ge 5x + 12 + 4$

$$x \ge 5x + 16$$

Subtract 5x from both sides: $\quad x - 5x \ge 5x + 16 - 5x$

$$-4x \ge 16$$

Divide both sides by -4. Don't forget to change the direction of the inequality: $\quad \dfrac{-4x}{-4} \le \dfrac{16}{-4}$

$$x \le -4$$

The solution using set-builder notation is $\{x \mid x \le -4\}$. The solution using interval notation is $(-\infty, -4\,]$. See Figure 20 for the graph of the solution set.

Figure 20

Quick ✓

In Problems 20–22, solve each linear inequality. Write the solution using set-builder notation and interval notation. Graph the solution set.

20. $3x + 1 > x - 5$

21. $-2x + 1 \le 3x + 11$

22. $-5x + 12 < x - 3$

▶ (**EXAMPLE 7**) **Solving Linear Inequalities**

Solve the inequality: $3(x - 1) + 2x < 6x + 3$

Solution

$$3(x - 1) + 2x < 6x + 3$$

Distribute the 3: $\quad 3x - 3 + 2x < 6x + 3$

Combine like terms: $\quad 5x - 3 < 6x + 3$

Add 3 to both sides: $\quad 5x - 3 + 3 < 6x + 3 + 3$

$$5x < 6x + 6$$

Subtract 6x from both sides: $\quad 5x - 6x < 6x + 6 - 6x$

$$-x < 6$$

Multiply both sides of the inequality by -1: $\quad (-1)(-x) > (-1)6$

$$x > -6$$

Figure 21

The solution using set-builder notation is $\{x \mid x > -6\}$. The solution using interval notation is $(-6, \infty)$. See Figure 21 for the graph of the solution set.

Quick ✓

In Problems 23–25, solve each linear inequality. Write the solution using set-builder notation and interval notation. Graph the solution set.

23. $4(x - 2) < 3x - 4$

24. $-2(x + 1) \geq 4(x + 3)$

25. $7 - 2(x + 1) \leq 3(x - 5)$

EXAMPLE 8

Solving Linear Inequalities Involving Fractions

Solve the inequality: $\dfrac{2x + 1}{3} > \dfrac{x - 2}{2}$

Solution

Clear the linear inequality of fractions by multiplying both sides of the inequality by 6, the least common denominator.

$$6 \cdot \left(\frac{2x + 1}{3}\right) > 6 \cdot \left(\frac{x - 2}{2}\right)$$

$$2(2x + 1) > 3(x - 2)$$

Distribute: $\qquad 4x + 2 > 3x - 6$

Subtract 2 from both sides: $\qquad 4x + 2 - 2 > 3x - 6 - 2$

$$4x > 3x - 8$$

Subtract 3x from both sides: $\qquad 4x - 3x > 3x - 8 - 3x$

$$x > -8$$

The solution using set-builder notation is $\{x \mid x > -8\}$. The solution using interval notation is $(-8, \infty)$. See Figure 22 for the graph of the solution set. ●

Figure 22

Quick ✓

In Problems 26–28, solve each linear inequality. Write the solution using set-builder notation and interval notation. Graph the solution set.

26. $\dfrac{3x + 1}{5} \geq 2$

27. $\dfrac{2}{5}x + \dfrac{3}{10} < \dfrac{1}{2}$

28. $\dfrac{1}{2}(x + 3) > \dfrac{1}{3}(x - 4)$

▶ ❹ **Solve Problems Involving Linear Inequalities**

One of the first things to do to solve a word problem is look for key words. Certain phrases can give a clue about the type of inequality problem given. Some of these phrases are shown in Table 9.

Table 9

Phrase	Inequality	Phrase	Inequality
At least	$\geq$	No more than	$\leq$
No less than	$\geq$	At most	$\leq$
More than	$>$	Fewer than	$<$
Greater than	$>$	Less than	$<$

To solve applications involving linear inequalities, use the problem-solving steps from Section 1.2 on page 61.

EXAMPLE 9 **Comparing Credit Cards**

US Bank offers several credit card options. You qualify for the SKYPASS Visa Signature card, which charges an annual fee of $80 plus 12.99% annual interest on outstanding balances. You also qualify for the SKYPASS Visa Classic card, which has an annual fee of $50 plus 14.24% annual interest on outstanding balances. What annual balances will result in the Signature card costing less than the Classic card? (SOURCE: *US Bank*)

Solution

Step 1: Identify Find the credit card balance for which the Signature credit card costs less than the Classic card. The phrase "costs less" implies that this is an inequality problem.

Step 2: Name Let b represent the credit card balance in dollars on each card.

Step 3: Translate Because each card charges an annual fee plus simple interest, the cost for each card will be "annual fee $+$ interest."

Recall from Section 1.2 the simple interest formula is

$$I = Prt$$

where I is the interest charged, P is the principal, r is the annual interest rate, and t is time.

In this problem, the principal is the credit card balance, b. The annual interest rate r is 0.1299 (for the Signature card) or 0.1424 (for the Classic card). Because annual cost is being discussed, let $t = 1$ for one year.

To find what balance results in the Signature card costing less than the Classic card, use the following inequality:

Signature Card Cost $<$ Classic Card Cost

Annual Fee for Signature		Interest Charged for Signature		Annual Fee for Classic		Interest Charged for Classic	
80	$+$	$0.1299\,b$	$<$	50	$+$	$0.1424\,b$	The Model

Step 4: Solve Solve the inequality for b. $\quad 80 + 0.1299b < 50 + 0.1424b$

Subtract 80 from both sides: $\qquad\qquad 0.1299b < -30 + 0.1424b$

Subtract $0.1424b$ from both sides: $\qquad -0.0125b < -30$

Divide both sides by -0.0125 and
reverse the inequality symbol: $\qquad\qquad b > 2400$

Step 5: Check If the balance is $2400, then the annual cost for the Signature card is $80 + 0.1299(2400) = \$391.76$, and the annual cost for the Classic card is $50 + 0.1424(2400) = \$391.76$. What about balances greater than $2400, say $2500? The annual cost for the Signature card is $80 + 0.1299(2500) = \$404.75$. The annual cost for the Classic card is $50 + 0.1424(2500) = \$406.00$.

Step 6: Answer the Question The Signature card costs less than the Classic card if the annual balance is greater than $2400. ●

> **Quick ✓**
>
> **29.** You are choosing between two credit cards. Bank A's card has an annual fee of $24 and charges 9.95% simple interest. Bank B's card has no annual fee but charges 14.95% simple interest. For what annual balance will the Bank A card cost less than the Bank B card?
>
> **30.** Suppose the daily revenue from selling x boxes of candy is given by the equation $R = 12x$. The daily cost of operating the store and making the candy is given by the equation $C = 8x + 96$. For how many boxes of candy will revenue exceed costs? That is, solve $R > C$.

1.4 Exercises MyMathLab® Exercise numbers in green have complete video solutions in MyMathLab or may be accessed using the QR code to the right.

*Problems **1–30** are the Quick ✔ s that follow the EXAMPLES.*

Building Skills

In Problems 31–38, write each inequality using interval notation. Graph the inequality. See Objective 1.

31. $2 \leq x \leq 10$ **32.** $1 < x < 7$

33. $-4 \leq x < 0$ **34.** $-8 < x \leq 1$

35. $x \geq 6$ **36.** $x < 0$

37. $x < \dfrac{3}{2}$ **38.** $x \geq -\dfrac{5}{2}$

In Problems 39–46, write each interval as an inequality involving x. Graph each inequality. See Objective 1.

39. $(1, 8)$ **40.** $[-2, 3]$

41. $(-5, 1]$ **42.** $[1, 4)$

43. $(-\infty, 5)$ **44.** $(2, \infty)$

45. $[3, \infty)$ **46.** $(-\infty, 8]$

In Problems 47–54, fill in the blank with the correct inequality symbol. State which property of inequalities is being utilized. See Objective 2.

47. If $x - 3 < 7$, then x __ 10.

48. If $2x - 5 > 6$, then $2x$ __ 11.

49. If $\dfrac{1}{3}x > 5$, then x __ 15.

50. If $4x > 36$, then x __ 9.

51. If $2x + 5 \leq 9$, then $2x$ __ 4.

52. If $\dfrac{2}{5}x + 6 \leq 8$, then $\dfrac{2}{5}x$ __ 2.

53. If $-2x \geq 10$, then x __ -5.

54. If $-6x < 30$, then x __ -5.

In Problems 55–92, solve each linear inequality. Write the solution using set-builder notation and interval notation. Graph the solution set. See Objective 3.

55. $x - 4 \leq 2$ **56.** $x + 6 < 9$

57. $6x < 24$ **58.** $4x \geq 20$

59. $-7x < 21$ **60.** $-8x > 32$

61. $\dfrac{4}{15}x > \dfrac{8}{5}$ **62.** $\dfrac{3}{8}x < \dfrac{9}{16}$

63. $3x + 2 > 11$ **64.** $5x - 4 \leq 16$

65. $-3x + 1 > 13$ **66.** $-6x - 5 < 13$

67. $6x + 5 \leq 3x + 2$ **68.** $8x + 3 \geq 5x - 9$

69. $-3x + 1 < 2x + 11$ **70.** $3x + 4 \geq 5x - 8$

71. $3(x - 3) < 2(x + 4)$

72. $3(x - 2) + 5 > 4(x + 1) + x$

73. $4(x + 1) - 2x \geq 5(x - 2) + 2$

74. $-3(x + 4) + 5x < 4(x + 3) - 14$

75. $0.5x + 4 \leq 0.2x - 5$

76. $2.3x - 1.2 > 1.8x + 0.4$

77. $\dfrac{3x + 1}{4} < \dfrac{1}{2}$

78. $\dfrac{2x - 3}{3} > \dfrac{4}{3}$

79. $\dfrac{1}{2}(x - 4) > \dfrac{3}{4}(2x + 1)$

80. $\dfrac{1}{3}(3x + 5) < \dfrac{1}{6}(x + 4)$

81. $\dfrac{3}{5} - x > \dfrac{5}{3}$

82. $\dfrac{2}{3} - \dfrac{5}{6}x > 2$

83. $-5(x - 3) \geq 3[4 - (x + 4)]$

84. $-3(2x + 1) \leq 2[3x - 2(x - 5)]$

85. $4(3x - 1) - 5(x + 4) \geq 3[2 - (x + 3)] - 6x$

86. $7(x + 2) - 4(2x + 3) < -2[5x - 2(x + 3)] + 7x$

87. $\dfrac{2}{3}(4x - 1) - \dfrac{4}{9}(x - 4) > \dfrac{5}{12}(2x + 3)$

88. $\dfrac{5}{6}(3x - 2) - \dfrac{2}{3}(4x - 1) < -\dfrac{2}{9}(2x + 5)$

89. $\dfrac{4x - 3}{3} < 3$

90. $\dfrac{2}{5}x + \dfrac{3}{10} < \dfrac{1}{2}$

91. $\dfrac{2}{3}x < \dfrac{1}{4}(2x + 3)$

92. $\dfrac{x}{12} \geq \dfrac{x}{2} - \dfrac{2x + 1}{4}$

Mixed Practice

In Problems 93–106, solve each linear inequality. Write the solution using set-builder notation and interval notation. Graph the solution set.

93. $y + 8 > -7$

94. $y - 5 \geq 7$

95. $-3a \leq -21$

96. $-5x < 30$

97. $13x - 5 > 10x - 6$

98. $4x + 3 \geq -6x - 2$

99. $4(x + 2) \leq 3(x - 2)$

100. $5(y + 7) < 6(y + 4)$

101. $3(4 - 3x) > 6 - 5x$

102. $2(5 - x) - 3 \leq 4 - 5x$

103. $2[4 - 3(x + 1)] \leq -4x + 8$

104. $3[1 + 2(x - 4)] \geq 3x + 3$

105. $\dfrac{x}{2} + \dfrac{3}{4} \geq \dfrac{3}{8}$

106. $\dfrac{b}{3} + \dfrac{5}{6} < \dfrac{11}{12}$

Applying the Concepts

107. Find the set of all x such that the sum of twice x and 5 is at least 13.

108. Find the set of all x such that the difference between 3 times x and 2 is less than 7.

109. Find the set of all z such that the product of 4 and z minus 3 is no more than 9.

110. Find the set of all y such that the sum of twice y and 3 is greater than 13.

111. **Computing Grades** In order to earn an A in Mr. Ruffatto's Intermediate Algebra course, Jackie must earn at least 540 points. Thus far, Jackie has earned 90, 83, 95, and 90 points on her four exams. The final exam, which counts as 200 points, is rapidly approaching. How many points does Jackie need to earn on the final to earn an A in Mr. Ruffatto's class?

112. **Computing Grades** In order to earn an A in Mrs. Padilla's Intermediate Algebra course, Mark must obtain an average score of at least 90. On his first four exams, Mark scored 94, 83, 88, and 92. The final exam counts as two test scores. What score does Mark need on the final to earn an A in Mrs. Padilla's class?

113. **McDonald's** Suppose that you have ordered one medium order of French fries and one 16-ounce triple-thick chocolate shake from McDonald's. The fries have 19 grams of fat, and the shake has 21 grams of fat. Each McDonald's hamburger has 9 grams of fat. How many hamburgers can you order and still keep the total fat content of the meal to no more than 67 grams? (SOURCE: *McDonald's Corporation*)

114. **Burger King** Suppose that you have ordered one medium onion ring and one 16-ounce chocolate shake from Burger King. The onion rings have 21 grams of fat and the shake has 21 grams of fat. Each cheeseburger has 13 grams of fat. How many cheeseburgers can you order and still keep the total fat content of the meal to no more than 81 grams? (SOURCE: *Burger King*)

115. **Payload Restrictions** An Airbus A320 has a maximum payload of 45,686 pounds. Suppose that on a flight from Chicago to New Orleans, United Airlines sold out the flight at 179 seats. Assuming that the average passenger weighs 150 pounds, determine the weight of the luggage and other cargo that the plane can carry. (SOURCE: *Airbus*)

116. **Moving Trucks** A 15-foot moving truck from Budget costs $39.95 per day plus $0.65 per mile. If your budget allows for you to spend at most $125.75, what is the number of miles you can drive? (SOURCE: *Budget*)

117. **Health Benefits** The average monthly benefit B, in dollars, for individuals on disability is given by the equation $B = 19.25t + 585.72$, where t is the number of years since 1990. In what years did the monthly benefit B exceed $1000? That is, solve $B > 1000$.

118. **Health Expenditures** Total private health expenditures H, in billions of dollars, are given by the equation $H = 26t + 411$, where t is the number of years since 1990. In what years will total private health expenditures exceed $1.25 trillion ($1250 billion)? That is, solve $26t + 411 > 1250$.

119. **Commissions** Susan sells computer systems. Her annual base salary is $34,000. She also earns a commission of 1.2% on the sale price of all computer systems that she sells. For what value of the computer systems sold will Susan's annual salary be at least $100,000?

120. **Commissions** Al sells used cars for a Chevy dealer. His annual base salary is $24,300. He also earns a commission of 3% on the sale price of the cars that he sells. For what value of the cars sold will Al's annual salary be more than $60,000? If used cars at this particular dealership sell for an average of $15,000, how many cars does Al have to sell to meet his salary goal?

121. Supply and Demand The quantity demanded of custom monogrammed shirts is given by the equation $D = 1000 - 20p$, where p is the price of a shirt. The quantity supplied of custom monogrammed shirts is given by the equation $S = -200 + 10p$. For what prices will quantity supplied exceed quantity demanded, thereby resulting in a surplus of shirts? That is, solve $S > D$.

122. Supply and Demand The quantity supplied of digital cameras is given by the equation $S = -2800 + 13p$, where p is the price of a camera. The quantity demanded of digital cameras is given by the equation $D = 1800 - 12p$. For what prices will quantity demanded exceed quantity supplied, thereby resulting in a shortage of cameras? That is, solve $D > S$.

Extending the Concepts

123. Solve the linear inequality
$$3(x + 2) + 2x > 5(x + 1).$$

124. Solve the linear inequality
$$-3(x - 2) + 7x > 2(2x + 5).$$

Explaining the Concepts

125. Write a brief paragraph that explains the circumstances under which the direction, or sense, of an inequality changes.

126. Explain why the inequality $5 < x < 1$ is false.

127. Explain why we never mix inequalities as in $4 < x > 7$.

Putting the Concepts Together (Sections 1.1–1.4)

We designed these problems so that you can review Sections 1.1 through 1.4 and show your mastery of the concepts. Take time to work these problems before proceeding with the next section. The answers are at the back of the text on page AN-4.

1. Determine which, if any, of the following are solutions to
$$5(2x - 3) + 1 = 2x - 6$$
  (a) $x = -3$ **(b)** $x = 1$

In Problems 2 and 3, solve the equation.

2. $3(2x - 1) + 6 = 5x - 2$

3. $\dfrac{7}{3}x + \dfrac{4}{5} = \dfrac{5x + 12}{15}$

Determine whether the equation is an identity, a contradiction, or a conditional equation.

4. $5 - 2(x + 1) + 4x = 6(x + 1) - (3 + 4x)$

In Problems 5 and 6, translate the English statement into a mathematical statement. Do not solve the equation.

5. The difference of a number x and 3 is two more than half the number.

6. The quotient of a number x and 2 is less than the number increased by 5.

7. Mixture Two acid solutions are available to a chemist. One is a 20% nitric acid solution, and the other is a 40% nitric acid solution. How much of each type of solution should be mixed together to form 16 liters of a 35% nitric acid solution?

8. Travel Two cars leave from the same location and travel in opposite directions along a straight road. One car travels 30 miles per hour, while the other travels at 45 miles per hour. How long will it take the two cars to be 255 miles apart?

9. Solve $3x - 2y = 4$ for y.

10. Solve the formula $A = P + Prt$ for r.

11. The volume of a right circular cylinder is given by the formula $V = \pi r^2 h$, where r is the radius of the cylinder and h is the height of the cylinder.

  (a) Solve the formula for h.

  (b) Use the result from part (a) to find the height of a right circular cylinder with volume $V = 294\pi$ inches^3 and radius $r = 7$ inches.

12. Write the following inequalities in interval notation. Graph the inequality.
    (a) $x > -3$
    (b) $2 < x \leq 5$

13. Write the interval in inequality notation involving x. Graph the inequality.
    (a) $(-\infty, -1.5]$
    (b) $(-3, 1]$

In Problems 14–16, solve the inequality and graph the solution set on a real number line.

14. $2x + 3 \leq 4x - 9$

15. $-3 > 3x - (x + 5)$

16. $x - 9 \leq x + 3(2 - x)$

17. **Birthday Party** A recreational center offers a children's birthday party for $75 plus $7 for each child. How many children can Logan invite to his birthday party if the budget for the party is no more than $250?

Part II: Linear Equations and Inequalities in Two Variables

1.5 Rectangular Coordinates and Graphs of Equations

Objectives

1. Plot Points in the Rectangular Coordinate System
2. Determine Whether an Ordered Pair Is a Point on the Graph of an Equation
3. Graph an Equation Using the Point-Plotting Method
4. Identify the Intercepts from the Graph of an Equation
5. Interpret Graphs

Are You Prepared for This Section?

Before getting started, complete the following problems. If you get a problem wrong, go back to the section cited and review the material.

P1. Plot the following points on the real number line: $-2, 4, 0, \dfrac{1}{2}$.   [Section R.2, pp. 15–16]

P2. Determine which of the following are solutions to the equation $3x - 5(x + 2) = 4$.
   (a) $x = 0$ (b) $x = -3$ (c) $x = -7$ [Section 1.1, pp. 49–50]

P3. Evaluate the expression $2x^2 - 3x + 1$ for the given values of the variable.
   (a) $x = 0$ (b) $x = 2$ (c) $x = -3$ [Section R.5, pp. 40–41]

P4. Solve the equation $3x + 2y = 8$ for y.   [Section 1.3, pp. 73–74]

P5. Evaluate $|-4|$.   [Section R.3, p. 19]

▶ ❶ Plot Points in the Rectangular Coordinate System

In Section R.2, when a point was located on the real number line, the point was in one dimension. Now we will work in two dimensions.

Begin by drawing two real number lines—one horizontal and one vertical—that intersect at right (90°) angles. The horizontal line is called the **x-axis** and the vertical line is called the **y-axis.** The point where the **axes** (plural of "axis") intersect is called the **origin, O.** See Figure 23.

Figure 23

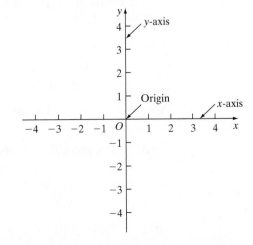

The origin O has a value of 0 on each axis. On the x-axis, points to the right of O are positive real numbers, and points to the left are negative. On the y-axis, points above O are positive real numbers, and points below are negative. In Figure 23 we label the x-axis "x" and the y-axis "y". An arrow at the end of each axis denotes the positive direction.

The coordinate system in Figure 23 is called a **rectangular** or **Cartesian coordinate system,** named after René Descartes (1596–1650), a French mathematician, philosopher, and theologian. The plane formed by the x-axis and the y-axis is often called the **xy-plane,** and the x-axis and y-axis are called the **coordinate axes.**

Any point P in the rectangular coordinate system can be represented by using an **ordered pair** (x, y) of real numbers. If $x > 0$, then P is x units to the right of the y-axis; if $x < 0$, then P is $|x|$ units to the left of the y-axis. If $y > 0$, then P is y units above the x-axis; if $y < 0$, then P is $|y|$ units below the x axis. The ordered pair (x, y) is also called the **coordinates** of P.

The origin O has coordinates $(0, 0)$. Any point on the x-axis has coordinates of the form $(x, 0)$, and any point on the y-axis has coordinates of the form $(0, y)$.

If point P has coordinates (x, y), then x is the **x-coordinate,** or **abscissa,** of P, and y is the **y-coordinate,** or **ordinate,** of P.

Look back at Figure 23. Notice that the x- and y-axes divide the plane into four separate regions or **quadrants.** In quadrant I, both the x- and y-coordinate are positive; in quadrant II, x is negative and y is positive; in quadrant III, both x and y are negative; and in quadrant IV, x is positive and y is negative. Points on the coordinate axes do not belong to a quadrant. See Figure 24.

Figure 24

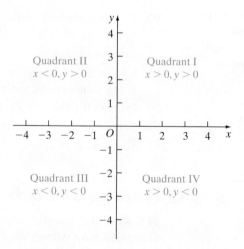

EXAMPLE 1

Plotting Points in the Rectangular Coordinate System and Determining the Quadrant in Which the Point Lies

Plot the points in the xy-plane. State which quadrant each point is in or state that the point lies on the x- or y- axis.

(a) $A(3, 2)$ (b) $B(-2, 4)$ (c) $C(-1, -3)$

(d) $D(3, -4)$ (e) $E(-2, 0)$

Solution

First, draw a rectangular, or Cartesian, coordinate system. See Figure 25(a). Now plot the points.

(a) To plot point $A(3, 2)$, begin at the origin O, travel 3 units to the right, and then travel 2 units up. Label the point A. See Figure 25(b). Point A is in quadrant I because both x and y are positive.

Figure 25

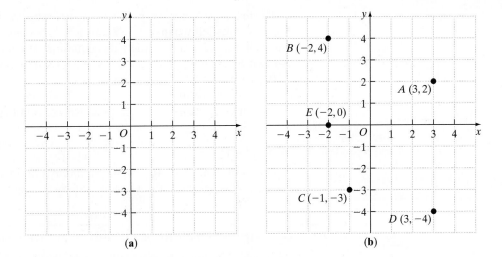

(a)   (b)

(b) To plot point $B(-2, 4)$, begin at the origin O, and travel 2 units to the left and then 4 units up. Label the point B. See Figure 25(b). Point B is in quadrant II.

(c) See Figure 25(b). Point C is in quadrant III.

(d) See Figure 25(b). Point D is in quadrant IV.

(e) See Figure 25(b). Point E is not in a quadrant because it lies on the x-axis.   ●

Quick ✓

1. The point where the x-axis and the y-axis intersect in the Cartesian coordinate system is called the _____.

2. *True or False* If a point lies in quadrant III of the Cartesian coordinate system, then both x and y are negative.

In Problems 3 and 4, plot each point in the xy-plane. Tell in which quadrant or on which coordinate axis each point lies.

3. **(a)** $A(5, 2)$ 4. **(a)** $A(-3, 2)$
     (b) $B(4, -2)$ **(b)** $B(-4, 0)$
     (c) $C(0, -3)$ **(c)** $C(3, -2)$
     (d) $D(-4, -3)$ **(d)** $D(6, 1)$

▶ ❷ Determine Whether an Ordered Pair Is a Point on the Graph of an Equation

Recall from Section 1.1 that the solution of an equation must be a single value of the variable, the empty set, or all real numbers. Now a method for representing the solution to an *equation in two variables* will be presented.

Definition

An **equation in two variables,** say x and y, is a statement in which the algebraic expressions involving x and y are equal. The expressions are called **sides** of the equation.

Since an equation is a statement, it may be true or false, depending on the values of the variables. Any values of the variable that make the equation true **satisfy** the equation.

For example, the following are all equations in two variables.

$$x^2 = y + 2 \qquad 3x + 2y = 6 \qquad y = -4x + 5$$

The first equation, $x^2 = y + 2$, is satisfied when $x = 3$ and $y = 7$ because $3^2 = 7 + 2$. It is also satisfied when $x = -2$ and $y = 2$ because $(-2)^2 = 2 + 2$. In fact, infinitely many choices of x and y satisfy this equation. However, some choices of x and y do not satisfy it. For example $x = 3$ and $y = 4$ does not satisfy the equation because $3^2 \neq 4 + 2$.

If two values of x and y, respectively, satisfy an equation, then the point whose coordinates are (x, y) is on the *graph of the equation*. For the remainder of the course, say "the point (x, y)" rather than "the point whose coordinates are (x, y)" for the sake of brevity.

> **In Other Words**
> The graph of an equation is a geometric way of representing the set of all points that make the equation a true statement.

Definition

The **graph of an equation in two variables** x and y is the set of all points whose coordinates, (x, y), in the xy-plane satisfy the equation.

EXAMPLE 2 | **Determining Whether a Point is on the Graph of an Equation**

Determine whether the following points are on the graph of $3x - y = 6$.

(a) $(2, 0)$ **(b)** $(1, -2)$ **(c)** $\left(\dfrac{1}{2}, -\dfrac{9}{2}\right)$

Solution

(a) For $(2, 0)$, check to see whether $x = 2$, $y = 0$ satisfies the equation $3x - y = 6$.

$$3x - y = 6$$
$$\text{Let } x = 2, y = 0: \quad 3(2) - 0 \stackrel{?}{=} 6$$
$$6 = 6 \quad \text{True}$$

The statement is true, so $(2, 0)$ is on the graph of the equation.

(b) For $(1, -2)$, we have
$$3x - y = 6$$
$$\text{Let } x = 1, y = -2: \quad 3(1) - (-2) \stackrel{?}{=} 6$$
$$3 + 2 \stackrel{?}{=} 6$$
$$5 = 6 \quad \text{False}$$

The statement $5 = 6$ is false, so the point $(1, -2)$ is not on the graph.

(c) For $\left(\dfrac{1}{2}, -\dfrac{9}{2}\right)$, we have
$$3x - y = 6$$
$$\text{Let } x = \frac{1}{2}, y = -\frac{9}{2}: \quad 3\left(\frac{1}{2}\right) - \left(-\frac{9}{2}\right) \stackrel{?}{=} 6$$
$$\frac{3}{2} + \frac{9}{2} \stackrel{?}{=} 6$$
$$\frac{12}{2} \stackrel{?}{=} 6$$
$$6 = 6 \quad \text{True}$$

The statement is true, so the point $\left(\dfrac{1}{2}, -\dfrac{9}{2}\right)$ is on the graph. ●

Quick ✓

5. *True or False* The graph of an equation in two variables x and y is the set of all points whose coordinates, (x, y), in the xy-plane satisfy the equation.

6. Determine whether the following points are on the graph of $2x - 4y = 12$.

     (a) $(2, -3)$ **(b)** $(2, -2)$ **(c)** $\left(\dfrac{3}{2}, -\dfrac{9}{4}\right)$

7. Determine whether the following points are on the graph of $y = x^2 + 3$.

     (a) $(1, 4)$ **(b)** $(-2, -1)$ **(c)** $(-3, 12)$

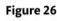 ❸ Graph an Equation Using the Point-Plotting Method

One way to graph an equation is the **point-plotting method.** With this method, choose values for one of the variables and use the equation to find the corresponding values of the other variable. It does not matter whether we choose values of x and use the equation to find the corresponding values of y or choose y and find x. Let convenience determine which variable to begin with when using the point-plotting method.

EXAMPLE 3   **How to Graph an Equation by Plotting Points**

Graph the equation $y = -2x + 4$ by plotting points.

Step-by-Step Solution

Step 1: Find points (x, y) that satisfy the equation. To determine these points, choose values of x (do you see why?) and use the equation to determine the corresponding values of y. See Table 10.

Table 10

x	$y = -2x + 4$	(x, y)
-3	$y = -2(-3) + 4 = 10$	$(-3, 10)$
-2	$y = -2(-2) + 4 = 8$	$(-2, 8)$
-1	$y = -2(-1) + 4 = 6$	$(-1, 6)$
0	$y = -2(0) + 4 = 4$	$(0, 4)$
1	$y = -2(1) + 4 = 2$	$(1, 2)$
2	$y = -2(2) + 4 = 0$	$(2, 0)$
3	$y = -2(3) + 4 = -2$	$(3, -2)$

Step 2: Plot the ordered pairs listed in the third column of Table 10 as shown in Figure 26(a). Now connect the points to obtain the graph of the equation (*a line*), as shown in Figure 26(b).

Figure 26

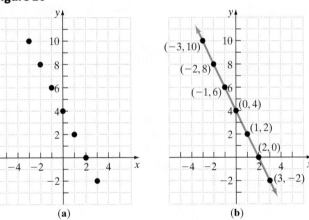

(a)   (b)

The graph of the equation in Figure 26(b) does not show all of the equation's points such as the point $(8, -12)$. Use arrows on the ends of the graph to indicate that the pattern continues. It is important to show enough of the graph so that the rest of the graph is obvious. Such a graph is called a **complete graph.**

EXAMPLE 4   **Graphing an Equation by Plotting Points**

Graph the equation $y = x^2$ by plotting points.

Solution

Table 11 on the following page shows several points on the graph.

Table 11

x	$y = x^2$	(x, y)
-4	$y = (-4)^2 = 16$	$(-4, 16)$
-3	$y = (-3)^2 = 9$	$(-3, 9)$
-2	$y = (-2)^2 = 4$	$(-2, 4)$
-1	$y = (-1)^2 = 1$	$(-1, 1)$
0	$y = (0)^2 = 0$	$(0, 0)$
1	$y = (1)^2 = 1$	$(1, 1)$
2	$y = (2)^2 = 4$	$(2, 4)$
3	$y = (3)^2 = 9$	$(3, 9)$
4	$y = (4)^2 = 16$	$(4, 16)$

In Figure 27(a), plot the ordered pairs listed in Table 11. In Figure 27(b), connect the points in a smooth curve and put arrows on either end to show the graph will continue indefinitely in that direction.

Figure 27

(a)

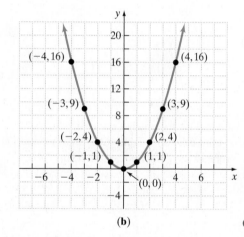

(b)

Work Smart

Note that in Figure 27, different scales are used on the x- and y-axes.

Two questions that you might be asking yourself right now are "How do I know how many points are sufficient?" and "How do I know which x-values (or y-values) I should choose in order to obtain points on the graph?" Often, the type of equation to graph indicates the number of points that are necessary. For example, as shown in the next section, if the equation is of the form $y = mx + b$, then its graph is a line, and only two points are required to obtain the graph (as in Example 3). Other times, more points are required. At this stage in your math career, plot quite a few points to obtain a complete graph. However, as your experience and knowledge grow, you will become more efficient in obtaining complete graphs.

Work Smart

Experience will play a huge role in determining which x-values to choose in creating a table of values. For the time being, start by choosing values of x around $x = 0$, as in Table 11.

Quick ✓

In Problems 8–10, graph each equation using the point-plotting method.

8. $y = 3x + 1$ **9.** $2x + 3y = 8$ **10.** $y = x^2 + 3$

EXAMPLE 5   **Graphing the Equation $x = y^2$**

Graph the equation $x = y^2$ by plotting points.

Solution

Because the equation is solved for x, choose values of y and use the equation to find the corresponding values of x. See Table 12. Plot the ordered pairs listed in Table 12 and connect the points in a smooth curve. See Figure 28.

Table 12

y	$x = y^2$	(x, y)
-3	$x = (-3)^2 = 9$	$(9, -3)$
-2	$x = (-2)^2 = 4$	$(4, -2)$
-1	$x = (-1)^2 = 1$	$(1, -1)$
0	$x = 0^2 = 0$	$(0, 0)$
1	$x = 1^2 = 1$	$(1, 1)$
2	$x = 2^2 = 4$	$(4, 2)$
3	$x = 3^2 = 9$	$(9, 3)$

Figure 28

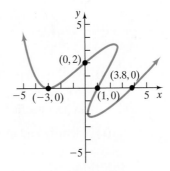

Quick ✓

In Problems 11 and 12, graph each equation using the point-plotting method.

11. $x = y^2 + 2$ **12.** $x = (y - 1)^2$

Figure 29

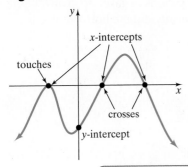

▶ ❹ Identify the Intercepts from the Graph of an Equation

A complete graph needs to include the *intercepts* of the graph.

Definition

The **intercepts** are the points, if any, where a graph crosses or touches the coordinate axes. An **x-intercept** is the point where the graph crosses or touches the x-axis. A **y-intercept** is the point where the graph crosses or touches the y-axis.

See Figure 29 for an illustration. Notice that an x-intercept exists when $y = 0$ and a y-intercept exists when $x = 0$.

(**EXAMPLE 6**) **Finding Intercepts from a Graph**

Find the intercepts of the graph shown in Figure 30. What are the x-intercepts? What are the y-intercepts?

Figure 30

Solution

The intercepts of the graph are the points $(-3, 0)$, $(0, 2)$, $(1, 0)$, and $(3.8, 0)$. The x-intercepts are $(-3, 0)$, $(1, 0)$, and $(3.8, 0)$; the y-intercept is $(0, 2)$.

Quick ✓

13. The points, if any, at which a graph crosses or touches a coordinate axis are called _____.

14. *True or False* An *x*-intercept exists when $x = 0$.

15. Find the intercepts of the graph shown in the figure. What are the *x*-intercepts? What are the *y*-intercepts?

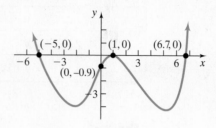

⑤ Interpret Graphs

Graphs help us to visualize a relationship between two variables or quantities. A graph is a "picture" that illustrates the relationship between two variables. Visualizing allows important information to be seen and conclusions drawn about the relationship between the two variables.

EXAMPLE 7 **Interpreting a Graph**

The graph in Figure 31 shows the profit P for selling x gallons of gasoline in an hour at a gas station. The vertical axis represents the profit, and the horizontal axis represents the number of gallons of gasoline sold.

Figure 31

(a) What is the profit if 150 gallons of gasoline are sold?

(b) How many gallons of gasoline are sold when profit is highest? What is the highest profit?

(c) Identify and interpret the intercepts.

Solution

(a) Locate the point on the graph with *x*-coordinate 150. The *y*-coordinate at this point is 200. The profit from selling 150 gallons of gasoline is $200.

(b) The profit is highest when 375 gallons of gasoline are sold. The highest profit is $565.

(c) The intercepts are $(0, -200)$, $(100, 0)$, and $(750, 0)$. For $(0, -200)$: If 0 gallons of gasoline are sold, the profit is $-\$200$. The negative profit is due to the fact that the company has $0 in revenue but has hourly expenses of $200. For $(100, 0)$: When 100 gallons of gasoline are sold, there is $0 profit. The company sells just enough gas to pay its bills. For $(750, 0)$: If 750 gallons of gasoline are sold, the profit is $0. The 750 gallons sold represents the maximum number of gallons that the station can pump and still break even.

Quick ✓

16. The graph shown represents the cost C (in thousands of dollars) of refining x gallons (in thousands) of gasoline per hour. The vertical axis represents the cost, and the horizontal axis represents the number of gallons of gasoline refined.

(a) What is the cost of refining 250 thousand gallons of gasoline per hour?

(b) What is the cost of refining 400 thousand gallons of gasoline per hour?

(c) In the context of the problem, explain the meaning of the graph ending at 700 thousand gallons of gasoline.

(d) Identify and interpret the intercept.

1.5 Exercises MyMathLab® Exercise numbers in green have complete video solutions in MyMathLab or may be accessed using the QR code to the right.

*Problems **1–16** are the Quick ✓s that follow the EXAMPLES.*

Building Skills

17. Determine the coordinates of each of the points plotted. Tell in which quadrant or on what coordinate axis each point lies. See Objective 1.

18. Determine the coordinates of each of the points plotted. Tell in which quadrant or on what coordinate axis each point lies. See Objective 1.

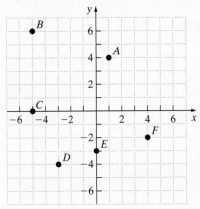

In Problems 19 and 20, plot each point in the xy-plane. Tell in which quadrant or on what coordinate axis each point lies. See Objective 1.

19. $A(3, 5)$
    $B(-2, -6)$
    $C(5, 0)$
    $D(1, -6)$
    $E(0, 3)$
    $F(-4, 1)$

20. $A(-3, 1)$
    $B(-6, 0)$
    $C(2, -5)$
    $D(-6, -2)$
    $E(1, 2)$
    $F(0, -5)$

In Problems 21–26, determine whether the given points are on the graph of the equation. See Objective 2.

21. $2x + 5y = 12$
    (a) $(1, 2)$
    (b) $(-2, 3)$
    (c) $(-4, 4)$
    (d) $\left(-\dfrac{3}{2}, 3\right)$

22. $-4x + 3y = 18$
    (a) $(1, 7)$
    (b) $(0, 6)$
    (c) $(-3, 10)$
    (d) $\left(\dfrac{3}{2}, 4\right)$

23. $y = -2x^2 + 3x - 1$
    (a) $(-2, -15)$
    (b) $(3, 10)$
    (c) $(0, 1)$
    (d) $(2, -3)$

24. $y = x^3 - 3x$
    (a) $(2, 2)$
    (b) $(3, 8)$
    (c) $(-3, -18)$
    (d) $(0, 0)$

25. $y = |x - 3|$
    (a) $(1, 4)$
    (b) $(4, 1)$
    (c) $(-6, 9)$
    (d) $(0, 3)$

26. $x^2 + y^2 = 1$
    (a) $(0, 1)$
    (b) $(1, 1)$
    (c) $\left(\dfrac{1}{2}, \dfrac{1}{2}\right)$
    (d) $\left(\dfrac{\sqrt{3}}{2}, \dfrac{1}{2}\right)$

In Problems 27–54, graph each equation by plotting points. See Objective 3.

27. $y = 4x$

28. $y = 2x$

29. $y = -\dfrac{1}{2}x$

30. $y = -\dfrac{1}{3}x$

31. $y = x + 3$

32. $y = x - 2$

33. $y = -3x + 1$

34. $y = -4x + 2$

35. $y = \dfrac{1}{2}x - 4$

36. $y = -\dfrac{1}{2}x + 2$

37. $2x + y = 7$

38. $3x + y = 9$

39. $y = -x^2$

40. $y = x^2 - 2$

41. $y = 2x^2 - 8$

42. $y = -2x^2 + 8$

43. $y = |x|$

44. $y = |x| - 2$

45. $y = |x - 1|$

46. $y = -|x|$

47. $y = x^3$

48. $y = -x^3$

49. $y = x^3 + 1$

50. $y = x^3 - 2$

51. $x^2 - y = 4$

52. $x^2 + y = 5$

53. $x = y^2 - 1$

54. $x = y^2 + 2$

In Problems 55–58, the graph of an equation is given. List the intercepts of the graph. See Objective 4.

55.

56.

57.

58.

Applying the Concepts

59. If $(a, 4)$ is a point on the graph of $y = 4x - 3$, what is a?

60. If $(a, -2)$ is a point on the graph of $y = -3x + 5$, what is a?

61. If $(3, b)$ is a point on the graph of $y = x^2 - 2x + 1$, what is b?

62. If $(-2, b)$ is a point on the graph of $y = -2x^2 + 3x + 1$, what is b?

63. Area of a Window The Property Brothers wish to put a new window in a home. They want the perimeter of the window to be 100 feet. The graph below shows the relation between the width, x, of the opening and the area of the opening.

(a) What is the area of the opening if the width is 10 feet?

(b) What is the width of the opening in order for area to be a maximum? What is the maximum area of the opening?

(c) Identify and interpret the intercepts.

64. Projectile Motion The graph to the right shows the height, in feet, of a ball thrown straight up with an initial speed of 80 feet per second from an initial height of 96 feet after t seconds.

(a) What is the height of the object after 1.5 seconds?

(b) At what time is the height a maximum? What is the maximum height?

(c) Identify and interpret the intercepts.

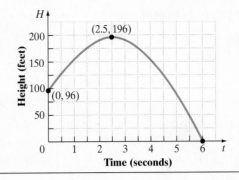

65. Cell Phones We all struggle with selecting a cell phone plan. The graph below shows the relation between the monthly cost of a cell phone and the number of minutes of talk used, m. (SOURCE: *Verizon.com*)

(a) What is the cost of talking for 200 minutes in a month? 400 minutes?

(b) What is the cost of talking 8000 minutes in a month?

(c) Identify and interpret the intercept.

66. Wind Chill It is 10° Celsius outside. The wind is calm but then gusts up to 20 meters per second. You feel the chill go right through your bones. The following graph shows the relation between the wind chill temperature (in degrees Celsius) and wind speed (in meters per second).

(a) What is the wind chill if the wind is blowing 4 meters per second?

(b) What is the wind chill if the wind is blowing 20 meters per second?

(c) Identify and interpret the intercepts.

67. Plot the points $(4, 0)$, $(4, 2)$, $(4, -3)$, and $(4, -6)$. Describe the set of all points of the form $(4, y)$ where y is a real number.

68. Plot the points $(4, 2)$, $(1, 2)$, $(0, 2)$, and $(-3, 2)$. Describe the set of all points of the form $(x, 2)$ where x is a real number.

Extending the Concepts

69. Draw a graph of an equation that contains two x-intercepts, $(-2, 0)$ and $(3, 0)$. At the x-intercept $(-2, 0)$ the graph crosses the x-axis; at the x-intercept $(3, 0)$, the graph touches the x-axis. Compare your graph with those of your classmates. How are they similar? How are they different?

70. Draw a graph that contains the points $(-3, -1)$, $(-1, 1)$, $(0, 3)$, and $(1, 5)$. Compare your graph with those of your classmates. How many of the graphs are straight lines? How many are "curved"?

71. Make up an equation that is satisfied by the points $(2, 0)$, $(4, 0)$, and $(1, 0)$. Compare your equation with those of your classmates. How are they similar? How are they different?

72. Make up an equation that is satisfied by the points $(0, 3)$, $(1, 3)$, and $(-4, 3)$. Compare your equation with those of your classmates. How many are the same?

Explaining the Concepts

73. Explain what is meant by a complete graph.

74. Explain what the graph of an equation represents.

75. What is the point-plotting method for graphing an equation?

76. What is the y-coordinate of a point that is an x-intercept? What is the x-coordinate of a point that is a y-intercept?

Technology Exercises

Just as we have graphed equations using point plotting, the graphing calculator also graphs equations by plotting points. Figure 32 shows the graph of $y = x^2$, and Table 13 shows points on the graph of $y = x^2$, using a graphing calculator.

Figure 32

Table 13

X	Y1
-5	25
-4	16
-3	9
-2	4
-1	1
0	0
1	1
2	4
3	9
4	16
5	25

X= -5

In Problems 77–84, use technology to draw a complete graph of each equation. Remember, a complete graph shows all the interesting features of a graph, including intercepts.

77. $y = 3x - 9$

78. $y = -5x + 8$

79. $y = -x^2 + 8$

80. $y = 2x^2 - 4$

81. $y + 2x^2 = 13$

82. $y - x^2 = -15$

83. $y = x^3 - 6x + 1$

84. $y = -x^3 + 3x$

1.6 Linear Equations in Two Variables

Objectives

1. Graph Linear Equations Using Point Plotting
2. Graph Linear Equations Using Intercepts
3. Graph Vertical and Horizontal Lines
4. Find the Slope of a Line Given Two Points
5. Interpret Slope as an Average Rate of Change
6. Graph a Line Given a Point and Its Slope
7. Use the Point-Slope Form of a Line
8. Identify the Slope and y-Intercept of a Line from Its Equation
9. Find the Equation of a Line Given Two Points

Are You Prepared for This Section?

Before getting started, complete the following problems. If you get a problem wrong, go back to the section cited and review the material.

P1. Solve: $3x + 12 = 0$ [Section 1.1, pp. 50–52]

P2. Solve: $2x + 5 = 13$ [Section 1.1, pp. 50–52]

P3. Solve for y: $3x - 2y = 10$ [Section 1.3, pp. 74–75]

P4. Evaluate: $\dfrac{5 - 2}{-2 - 4}$ [Section R.4, pp. 35–36]

P5. Evaluate: $\dfrac{-7 - 2}{-5 - (-2)}$ [Section R.4, pp. 35–36]

P6. Distribute: $-2(x + 3)$ [Section R.3, pp. 29–30]

▶ ❶ Graph Linear Equations Using Point Plotting

In Section 1.5, how to graph any equation using the point-plotting method was discussed. Remember, the graph of an equation is the set of all ordered pairs (x, y) that satisfy the equation.

Now methods for graphing a specific type of equation called a *linear equation in two variables* are going to be discussed.

Definition

A **linear equation in two variables** is an equation of the form

$$Ax + By = C$$

where A, B, and C are real numbers. A and B cannot both be 0.

Prepared?...Answers   **P1.** $\{-4\}$

P2. $\{4\}$ **P3.** $y = \dfrac{3}{2}x - 5$ **P4.** $-\dfrac{1}{2}$

P5. 3   **P6.** $-2x - 6$

A linear equation written in the form $Ax + By = C$ is in **standard form.***
Some examples of linear equations in standard form are

$$3x - 4y = 9 \qquad \frac{1}{2}x + \frac{2}{3}y = 4 \qquad 3x = 9 \qquad -2y = 5$$

The graph of a linear equation is a **line.**

EXAMPLE 1 **Graphing a Linear Equation Using the Point-Plotting Method**

Graph the linear equation: $4x + 2y = 6$

Solution
To graph the linear equation, choose various values for x and then use the equation to find the corresponding values of y. For this equation, let $x = -2, -1, 0,$ and 1.

$x = -2$:	$4(-2) + 2y = 6$		$x = -1$:	$4(-1) + 2y = 6$
	$-8 + 2y = 6$			$-4 + 2y = 6$
Add 8 to both sides:	$2y = 14$		Add 4 to both sides:	$2y = 10$
Divide both sides by 2:	$y = 7$		Divide both sides by 2:	$y = 5$

$(-2, 7)$ is on the graph. $\qquad\qquad (-1, 5)$ is on the graph.

$x = 0$:	$4(0) + 2y = 6$		$x = 1$:	$4(1) + 2y = 6$
	$0 + 2y = 6$			$4 + 2y = 6$
	$2y = 6$		Subtract 4 from both sides:	$2y = 2$
Divide both sides by 2:	$y = 3$		Divide both sides by 2:	$y = 1$

$(0, 3)$ is on the graph. $\qquad\qquad (1, 1)$ is on the graph.

Table 14 shows the points that are on the graph of $4x + 2y = 6$. Plot these ordered pairs in the Cartesian plane and connect the points in a straight line. See Figure 33.

Table 14

x	y	(x, y)
−2	7	$(-2, 7)$
−1	5	$(-1, 5)$
0	3	$(0, 3)$
1	1	$(1, 1)$

Figure 33

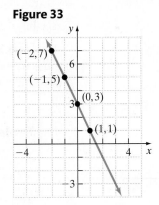

Work Smart

In Example 1, we chose to pick x-values and use the equation to find the corresponding y-values; however, we could also have chosen y-values and used the equation to find the corresponding x-values.

Work Smart

We recommend that you find at least three points when graphing linear equations to be sure your graph is correct. Also, remember that a complete graph is a graph that shows all the interesting features of the graph, such as its intercepts.

Although only two points are needed to obtain a complete graph of a linear equation, to guard against error, plot at least three points.

Quick ✓

1. A(n) _____ _____ is an equation of the form $Ax + By = C$, where $A, B,$ and C are real numbers. A and B cannot both be 0.

2. The graph of a linear equation is a ___.

In Problems 3–5, graph each linear equation using the point-plotting method.

3. $y = 2x - 3$ $\qquad$ **4.** $\dfrac{1}{2}x + y = 2$ $\qquad$ **5.** $-6x + 3y = 12$

---

*Some texts refer to this as the general form.

▶ ❷ Graph Linear Equations Using Intercepts

Figure 34

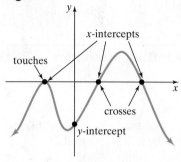

As indicated in Section 1.5, a complete graph should display any intercepts. The intercepts of the graph of an equation are the points, if any, where the graph crosses or touches the coordinate axes. See Figure 34. To find an equation's intercepts algebraically, use the fact that at an *x*-intercept, the value of *y* is 0, and at a *y*-intercept, the value of *x* is 0.

Finding Intercepts

- To find the *x*-intercept(s), if any, of the graph of an equation, let $y = 0$ in the equation and solve for *x*.
- To find the *y*-intercept(s), if any, of the graph of an equation, let $x = 0$ in the equation and solve for *y*.

The procedure given can be used to find the intercepts of any type of equation. Use this procedure to find the intercepts of a linear equation.

EXAMPLE 2

Graphing a Linear Equation by Finding Its Intercepts

Graph the linear equation $3x + 2y = 12$ by finding its intercepts.

Solution

To find the *x*-intercept, let $y = 0$ and solve the equation $3x + 2y = 12$ for *x*.

$$\text{Let } y = 0: \quad 3x + 2(0) = 12$$
$$3x + 0 = 12$$
$$3x = 12$$
$$\text{Divide both sides by 3:} \quad x = 4$$

The *x*-intercept is $(4, 0)$.

To find the *y*-intercept, let $x = 0$ and solve the equation $3x + 2y = 12$ for *y*.

$$\text{Let } x = 0: \quad 3(0) + 2y = 12$$
$$0 + 2y = 12$$
$$2y = 12$$
$$\text{Divide both sides by 2:} \quad y = 6$$

The *y*-intercept is $(0, 6)$.

Find one additional point on the graph by letting $x = 2$ (or any other value of *x* besides 0 or 4), and find *y* to be 3. Table 15 shows points on the graph. Plot the points and connect them in a line, obtaining the graph in Figure 35.

Table 15

x	y	(x, y)
0	6	(0, 6)
4	0	(4, 0)
2	3	(2, 3)

Figure 35

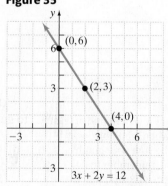

Work Smart

Equations in one variable have no solution, one solution, or infinitely many solutions. Because the procedure for finding intercepts of linear equations in two variables results in a linear equation in one variable, linear equations can have no *x*-intercepts, one *x*-intercept, or infinitely many *x*-intercepts. The same applies to *y*-intercepts.

Quick ✓

6. *True or False* To find the *x*-intercept(s), if any, of the graph of an equation, let $y = 0$ in the equation and solve for *x*.

In Problems 7 and 8, graph each linear equation by finding its intercepts.

7. $x + y = 4$

8. $4x - 5y = 20$

EXAMPLE 3 **Graphing a Linear Equation by Finding Its Intercepts**

Graph the linear equation $x + 3y = 0$ by finding its intercepts.

Solution

x-intercept:

Let $y = 0$: $\quad x + 3(0) = 0$
$$x + 0 = 0$$
$$x = 0$$

The x-intercept is $(0, 0)$.

y-intercept:

Let $x = 0$: $\quad 0 + 3y = 0$
$$3y = 0$$
Divide both sides by 3: $\quad y = 0$

The y-intercept is $(0, 0)$.

Work Smart

We chose $x = -3$ and $x = 3$ to avoid fractions. This makes plotting the points easier.

Because both the x- and y-intercepts are $(0, 0)$, find *two* additional points on the graph. When $x = 3$, for example, $y = -1$. And when $x = -3$, $y = 1$. Plot the points $(0, 0)$, $(-3, 1)$, and $(3, -1)$, connect them using a line, and obtain the graph in Figure 36.

Figure 36

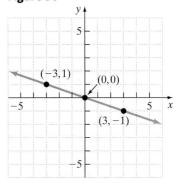

Example 3 shows that any equation of the form $Ax + By = 0$, where $A \neq 0$ and $B \neq 0$, has only one intercept at $(0, 0)$. Therefore, to graph equations of this form, find two additional points on the graph.

Quick ✔

In Problem 9, graph the equation by finding its intercepts.

9. $3x - 2y = 0$

⊳ ❸ **Graph Vertical and Horizontal Lines**

In the equation of a line, $Ax + By = C$, A and B cannot both be zero. But what if $A = 0$ or $B = 0$?

EXAMPLE 4 **Graphing a Vertical Line**

Graph the equation $x = 3$ using the point-plotting method.

Solution

Because the equation $x = 3$ can be written as $1x + 0y = 3$, the graph is a line. For the equation $x = 3$, no matter what value of y is chosen, the corresponding value of x is always going to be 3. Therefore, points such as $(3, -2)$, $(3, -1)$, $(3, 0)$, $(3, 1)$, and $(3, 2)$ are all on the line. See Figure 37.

Figure 37

$x = 3$

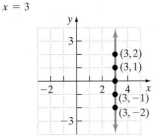

Based on the results of Example 4, the equation of a vertical line can be written as:

Equation of a Vertical Line

A **vertical line** is given by an equation of the form

$$x = a$$

where $(a, 0)$ is the x-intercept.

Now let's look at equations that lead to graphs that are horizontal lines.

EXAMPLE 5 **Graphing a Horizontal Line**

Graph the equation $y = -2$ using the point-plotting method.

Solution

Because the equation $y = -2$ can be written as $0x + 1y = -2$, the graph is a line. For the equation $y = -2$, no matter what value of x is chosen, the corresponding value of y is always going to be -2. Therefore, points such as $(-2, -2), (-1, -2), (0, -2), (1, -2)$, and $(2, -2)$ are all on the line. See Figure 38. ●

Figure 38
$y = -2$

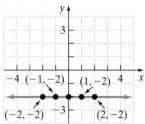

Based on the results of Example 5, the equation of a horizontal line can be written as:

Equation of a Horizontal Line

A **horizontal line** is given by an equation of the form

$$y = b$$

where $(0, b)$ is the y-intercept.

Quick ✓

10. The equation of a vertical line is of the form _ $= a$, where $(a, 0)$ is the x-intercept. The equation of a horizontal line is the form _ $= b$, where $(0, b)$ is the y-intercept.

In Problems 11–13, graph each equation.

11. $x = 5$ 12. $y = -4$ 13. $-3x + 4 = 1$

▶ ❹ Find the Slope of a Line Given Two Points

Consider the staircase shown in Figure 39(a). If a line is drawn at the edge of each step (in blue), it can be seen that each step has exactly the same horizontal **run** and the same vertical **rise.** The ratio of the rise to the run is called the *slope* of the line. Thus slope is a numerical measure of the steepness of the line. For example, the staircase in Figure 39(a) has a run of 7 inches and a rise of 6 inches, so the line's slope is $\dfrac{\text{Rise}}{\text{Run}} = \dfrac{6 \text{ inches}}{7 \text{ inches}} = \dfrac{6}{7}$. In Figure 39(b), the slope of the line is $\dfrac{\text{Rise}}{\text{Run}} = \dfrac{6 \text{ inches}}{10 \text{ inches}} = \dfrac{6}{10} = \dfrac{3}{5}$. And in Figure 39(c), the line's slope is $\dfrac{\text{Rise}}{\text{Run}} = \dfrac{6 \text{ inches}}{4 \text{ inches}} = \dfrac{6}{4} = \dfrac{3}{2}$.

Figure 39

(a) Slope $= \dfrac{6}{7}$ (b) Slope $= \dfrac{6}{10} = \dfrac{3}{5}$ (c) Slope $= \dfrac{6}{4} = \dfrac{3}{2}$

The slope of a line using rectangular coordinates is defined as follows:

Definition

Let $P = (x_1, y_1)$ and $Q = (x_2, y_2)$ be two distinct points. If $x_1 \neq x_2$, the **slope** m of the nonvertical line L containing P and Q is defined by the formula

$$m = \frac{y_2 - y_1}{x_2 - x_1}, \quad x_1 \neq x_2$$

If $x_1 = x_2$, then L is a vertical line and the slope m of L is **undefined** (since this results in division by 0).

The accepted symbol for the slope of a line is m. It comes from the French word *monter*, which means "to ascend or climb." Figure 40(a) illustrates the slope of a nonvertical line; Figure 40(b) shows a vertical line.

Figure 40

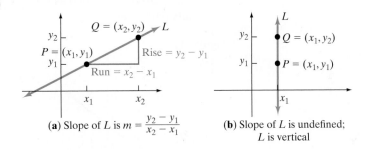

(a) Slope of L is $m = \frac{y_2 - y_1}{x_2 - x_1}$

(b) Slope of L is undefined; L is vertical

Notice that the slope m of a nonvertical line may be viewed as

$$m = \frac{y_2 - y_1}{x_2 - x_1} = \frac{\text{rise}}{\text{run}}$$

The slope m of a nonvertical line can also be written as

$$m = \frac{y_2 - y_1}{x_2 - x_1} = \frac{\text{Change in } y}{\text{Change in } x} = \frac{\Delta y}{\Delta x}$$

In Other Words

Slope is the change in y divided by the change in x.

The symbol Δ is the Greek letter delta, which comes from the Greek word for "difference," *diaphora*. In mathematics, read Δ as "change in." Thus read $\frac{\Delta y}{\Delta x}$ as "change in y divided by change in x." The slope m of a nonvertical line measures the amount that y changes (the vertical change) as x changes from x_1 to x_2 (the horizontal change).

Figure 41

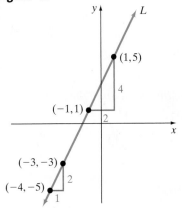

Comments Regarding the Slope of a Nonvertical Line

1. Any two different points on the line can be used to compute the slope of the line shown in Figure 41. The slope m of the line L is given by

$$m = \frac{-5 - (-3)}{-4 - (-3)} = \frac{-2}{-1} = 2 \quad \text{or} \quad m = \frac{5 - 1}{1 - (-1)} = \frac{4}{2} = 2$$

This result is due to the fact that the two triangles in Figure 41 are similar (the measure of the angles is the same in both triangles). Therefore, the ratios of the sides are proportional.

2. The slope of a line may be computed from $P = (x_1, y_1)$ to $Q = (x_2, y_2)$ or from Q to P because

$$m = \frac{y_2 - y_1}{x_2 - x_1} = \frac{-(y_1 - y_2)}{-(x_1 - x_2)} = \frac{y_1 - y_2}{x_1 - x_2}$$

Work Smart

It doesn't matter whether we compute the slope of the line from point P to Q or from point Q to P.

EXAMPLE 6 **Finding and Interpreting the Slope of a Line**

Find and interpret the slope of the line containing the points $(3, 6)$ and $(-2, 2)$.

Solution

Plot the points $P = (x_1, y_1) = (3, 6)$ and $Q = (x_2, y_2) = (-2, 2)$ and draw a line through them, as shown in Figure 42. The slope of the line is

$$m = \frac{y_2 - y_1}{x_2 - x_1} = \frac{\Delta y}{\Delta x} = \frac{2 - 6}{-2 - 3} = \frac{-4}{-5} = \frac{4}{5}$$

The slope could also be computed as

$$m = \frac{y_1 - y_2}{x_1 - x_2} = \frac{6 - 2}{3 - (-2)} = \frac{4}{5}$$

Interpret a slope of $\frac{4}{5}$ as follows: For every 5-unit increase in x (run), y increases by 4 units (rise). Or for every 5-unit decrease in x, y decreases by 4 units. See Figure 42(a). Slope may also be described as follows: For every 4-unit increase in y (rise), x increases by 5 units (run). See Figure 42(b).

Figure 42

(a)

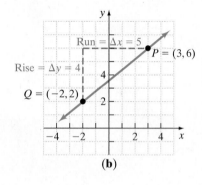

(b)

Quick ✓

14. If a line is vertical, then its slope is _____.

15. On a line, for every 10-foot run there is a 4-foot rise. The slope of the line is _.

16. *True or False* If $P = (x_1, y_1)$ and $Q = (x_2, y_2)$ are two distinct points with $x_1 \neq x_2$, the slope m of the line L containing P and Q is defined by the formula

$$m = \frac{x_2 - x_1}{y_2 - y_1}, \quad y_1 \neq y_2.$$

17. *True or False* If the slope of a line is $\frac{1}{2}$, then when x increases by 2, y increases by 1.

In Problems 18–21, find and interpret the slope of the line containing the points.

18. $(0, 3); (3, 12)$ **19.** $(-1, 3); (3, -4)$

20. $(3, 2); (-3, 2)$ **21.** $(-2, 4); (-2, -1)$

EXAMPLE 7 **Finding Slopes of Different Lines Each of Which Contains (3, 5)**

Find the slope of the lines L_1, L_2, L_3, and L_4 containing the following pairs of points. Graph the lines in the Cartesian plane.

L_1: $P(3, 5), Q_1(5, 8)$ L_2: $P(3, 5), Q_2(6, 5)$

L_3: $P(3, 5), Q_3(5, -2)$ L_4: $P(3, 5), Q_4(3, 0)$

Solution

Let $m_1, m_2, m_3,$ and m_4 denote the slopes of the lines $L_1, L_2, L_3,$ and $L_4,$ respectively. Then

$$m_1 = \frac{8-5}{5-3} = \frac{3}{2} \qquad\qquad m_2 = \frac{5-5}{6-3} = \frac{0}{3} = 0$$

$$m_3 = \frac{-2-5}{5-3} = \frac{-7}{2} = -\frac{7}{2} \qquad m_4 = \frac{0-5}{3-3} = \frac{-5}{0} \quad \text{undefined}$$

The four lines are shown in Figure 43.

Figure 43

(a)   (b)

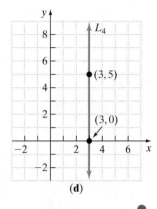

(c)   (d)

Based on Figure 43, we have the following properties of slope:

Properties of Slope

- When the slope of a line is positive, the line slants upward from left to right, as shown by L_1 in Figure 43(a).
- When the slope of a line is zero, the line is horizontal, as shown by L_2 in Figure 43(b).
- When the slope of a line is negative, the line slants downward from left to right, as shown by L_3 in Figure 43(c).
- When the slope of a line is undefined, the line is vertical, as shown by L_4 in Figure 43(d).

Quick ✓

22. *True or False* If the slope of a line is negative, then the line slants downward from left to right.

23. A horizontal line has slope = __.

24. Find the slope of the lines $L_1, L_2, L_3,$ and L_4 containing the following pairs of points. Graph all four lines in the same Cartesian plane.

L_1: $P(1, 3), Q_1(6, 4)$ L_2: $P(1, 3), Q_2(1, 8)$

L_3: $P(1, 3), Q_3(-3, 7)$ L_4: $P(1, 3), Q_4(-4, 3)$

▶ ❺ Interpret Slope as an Average Rate of Change

The slope m of a nonvertical line measures the amount that y changes as x changes from x_1 to x_2. The slope of a line is also called the **average rate of change** of y with respect to x.

In applications, it is often asked how the change in one variable will affect some other variable. For example, if your income increases by \$1000, how much will your spending (on average) change? Or if the speed of your car increases by 10 miles per hour, how much (on average) will your car's gas mileage change?

EXAMPLE 8   **Slope as an Average Rate of Change**

A strain of *E. coli* Beu 397-recA441 is placed into a Petri dish at 30° Celsius and allowed to grow. The data shown in Table 16 are collected. The population is measured in grams and the time in hours. The population growth is shown in Figure 44.

Table 16

Time (hours), x	Population (grams), y
0	0.09
1	0.12
2	0.16
3	0.22
4	0.29
5	0.39

SOURCE: Dr. Polly Lavery, Joliet Junior College

Figure 44

Growth of *E. coli*

[graph: Population (grams) vs. Time (hours), curve rising from about 0.10 at time 0 to about 0.39 at time 5]

(a) Compute and interpret the average rate of change in the population between hours 0 and 1.

(b) Compute and interpret the average rate of change in the population between hours 3 and 4.

(c) Based on your results to parts (a) and (b), do you think that the population grows linearly? Why?

Solution

(a) To find the average rate of change, compute the slope of the line between the points $(0, 0.09)$ and $(1, 0.12)$.

$$m = \text{average rate of change} = \frac{0.12 - 0.09}{1 - 0} = 0.03 \text{ gram per hour}$$

The population of *E. coli* grew at the rate of 0.03 gram per hour between hours 0 and 1.

(b) The slope of the line between the points $(3, 0.22)$ and $(4, 0.29)$ is

$$m = \text{average rate of change} = \frac{0.29 - 0.22}{4 - 3} = 0.07 \text{ gram per hour}$$

The population growth rate was 0.07 gram per hour between hours 3 and 4.

(c) The population is not growing linearly, because the average rate of change (slope) is not the same (or constant), from point to point (or hour to hour). ●

Quick ✓

25. The data to the left represent the total revenue that would be received from selling *x* bicycles at Gibson's Bicycle Shop.

(a) Plot the ordered pairs (x, y) on a graph and connect the points with straight lines.

(b) Compute and interpret the average rate of change in the revenue between 0 and 25 bicycles sold.

(c) Compute and interpret the average rate of change in the revenue between 102 and 150 bicycles sold.

(d) Based on your results to parts (a), (b), and (c), do you think the revenue grows linearly? Why?

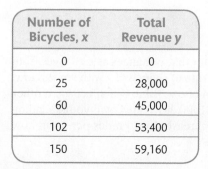

Number of Bicycles, x	Total Revenue y
0	0
25	28,000
60	45,000
102	53,400
150	59,160

⑥ Graph a Line Given a Point and Its Slope

One reason why slope is found is that it can be used to help us graph lines.

EXAMPLE 9 **Graphing a Line Given a Point and Its Slope**

Draw a graph of the line that contains the point $(1, 2)$ and has a slope of

(a) 3

(b) $-\dfrac{3}{2}$

Solution

Figure 45

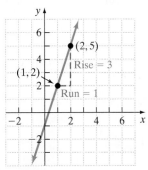

(a) The slope $= \dfrac{\text{Rise}}{\text{Run}} = \dfrac{\Delta y}{\Delta x} = 3 = \dfrac{3}{1}$. Thus as x increases by 1 unit, y increases by 3 units. Start at $(1, 2)$ and move 1 unit to the right and then 3 units up, and arrive at the point $(2, 5)$. The graph is the line through the points $(1, 2)$ and $(2, 5)$. See Figure 45.

(b) The slope $= \dfrac{\text{Rise}}{\text{Run}} = \dfrac{\Delta y}{\Delta x} = -\dfrac{3}{2} = \dfrac{-3}{2}$. Thus as x increases by 2 units, y decreases by 3 units. Start at $(1, 2)$ and move 2 units to the right and then 3 units down, and arrive at the point $(3, -1)$. The graph is the line through the points $(1, 2)$ and $(3, -1)$. See Figure 46.

It is perfectly acceptable to set $\dfrac{\Delta y}{\Delta x} = -\dfrac{3}{2} = \dfrac{3}{-2}$ and move left 2 units from $(1, 2)$ and then up 3 units. We would then end up at $(-1, 5)$, which is also on the graph of the line in Figure 46. ●

Figure 46

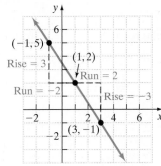

Quick ✓

26. Draw a graph of the line that contains the point $(-1, 3)$ and has a slope of

(a) $\dfrac{1}{3}$

(b) -4

(c) 0

⑦ Use the Point-Slope Form of a Line

Another reason why slope is found is that it can help to find the equation of a line. Suppose L is a nonvertical line that has slope m and contains the point (x_1, y_1). See Figure 47.

Figure 47

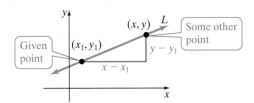

For any other point (x, y) on L, it is known from the formula for the slope of a line that

$$m = \dfrac{y - y_1}{x - x_1}$$

Multiplying both sides by $x - x_1$, rewrite this equation as

$$y - y_1 = m(x - x_1)$$

Point-Slope Form of an Equation of a Line

An equation of a nonvertical line with slope m that contains the point (x_1, y_1) is

Slope
↓

$$y - y_1 = m(x - x_1)$$

↑   ↑

Given Point

EXAMPLE 10   **Using the Point-Slope Form of an Equation of a Line**

Find the equation of a line that has slope 3 and contains the point $(-2, 5)$. Graph the line.

Solution

Because the slope and a point on the line are given, use the point-slope form of a line with $m = 3$ and $(x_1, y_1) = (-2, 5)$.

$$y - y_1 = m(x - x_1)$$

$m = 3, x_1 = -2, y_1 = 5:$ $y - 5 = 3[x - (-2)]$

$$y - 5 = 3(x + 2)$$

Figure 48

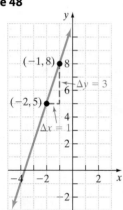

See Figure 48 for a graph of the line.   ●

Quick ✓

27. An equation in the form $y - y_1 = m(x - x_1)$ is said to be in _____ form.

In Problems 28–31, find an equation of the line with the given properties. Graph the line.

28. $m = 2, (x_1, y_1) = (3, 5)$

29. $m = -4, (x_1, y_1) = (-2, 3)$

30. $m = \dfrac{1}{3}, (x_1, y_1) = (3, -4)$

31. $m = 0, (x_1, y_1) = (4, -2)$

▶ ❽ **Identify the Slope and y-Intercept of a Line from Its Equation**

Solve the equation in Example 10 for y:

$$y - 5 = 3(x + 2)$$

Distribute the 3: $y - 5 = 3x + 6$

Add 5 to both sides: $y = 3x + 11$

The coefficient of x, 3, is the slope of the line; the y-intercept of the line is $(0, 11)$ because $y = 3(0) + 11 = 11$. An equation written in the form $y = mx + b$, is in *slope-intercept form*.

Slope-Intercept Form of an Equation of a Line

An equation of a line with slope m and y-intercept $(0, b)$ is

$$y = mx + b$$

EXAMPLE 11   **Finding the Slope and y-Intercept of a Line from Its Equation**

Write the equation $x - 3y = 9$ in slope-intercept form. Find the slope m and y-intercept $(0, b)$ of the line. Graph the line.

Solution
To put the equation in slope-intercept form, $y = mx + b$, solve the equation for y.

$$x - 3y = 9$$

Subtract x from both sides: $\qquad -3y = -x + 9$

Divide both sides by -3: $\qquad y = \dfrac{-x + 9}{-3}$

Divide -3 into both terms in the numerator: $\qquad y = \dfrac{1}{3}x - 3$

Because $y = \dfrac{1}{3}x - 3$ is in the form $y = mx + b$, the coefficient of x, $\dfrac{1}{3}$, is the slope, and $(0, -3)$ is the y-intercept.

Graph the line by plotting a point at $(0, -3)$. Then use the slope to find an additional point by moving right 3 units and up 1 unit from the point $(0, -3)$ to the point $(3, -2)$. Draw a line through the two points and obtain the graph shown in Figure 49.

Figure 49

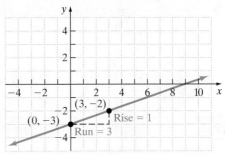

Quick ✓

32. An equation in the form $y = mx + b$ is said to be in _____ form.

In Problems 33–36, find the slope and y-intercept of each line. Graph the line.

33. $3x - y = 2$ **34.** $6x + 2y = 8$

35. $3x - 2y = 7$ **36.** $7x + 3y = 0$

▶ ❾ Find the Equation of a Line Given Two Points

It is known that two points are all that are needed to graph a line. If two points are given, an equation of the line through the points can be found by first finding the slope of the line and then using the point-slope form of a line.

EXAMPLE 12   **How to Find an Equation of a Line from Two Points**

Find the equation of a line through the points $(-1, 4)$ and $(2, -5)$. If possible, write the equation in slope-intercept form. Graph the line.

Step-by-Step Solution

Step 1: Find the slope of the line containing the points.

Let $(x_1, y_1) = (-1, 4)$ and $(x_2, y_2) = (2, -5)$. Substitute these values into the formula for the slope of a line.

$$m = \frac{y_2 - y_1}{x_2 - x_1} = \frac{-5 - 4}{2 - (-1)} = \frac{-9}{3} = -3$$

Step 2: Use the point-slope form of a line to find the equation.

With $m = -3$, $x_1 = -1$, and $y_1 = 4$, we have

$$y - y_1 = m(x - x_1)$$
$$y - 4 = -3[x - (-1)]$$
$$y - 4 = -3(x + 1)$$

Step 3: Solve the equation for y.

Distribute the -3: $\qquad y - 4 = -3x - 3$

Add 4 to both sides: $\qquad y = -3x + 1$

(continued)

Work Smart

In Step 2 of Example 12, we chose to use $x_1 = -1$ and $y_1 = 4$, but we could also have used $x_1 = 2$ and $y_1 = -5$. Choose the values of x and y that make the algebra easiest.

The slope of the line is -3 and the y-intercept is $(0, 1)$. See Figure 50 for the graph.

Figure 50

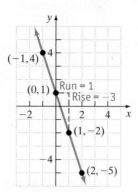

Quick ✓

In Problems 37–39, find the equation of the line containing the given points. If possible, write the equation in slope-intercept form. Graph the line.

37. $(1, 3); (4, 9)$ **38.** $(-2, 4); (2, 2)$

39. $(-4, 6); (3, 6)$

EXAMPLE 13 **Finding an Equation of a Line from Two Points**

Find the equation of a line through the points $(-3, 2)$ and $(-3, -2)$. If possible, write the equation in slope-intercept form. Graph the line.

Figure 51

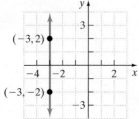

Solution

Let $(x_1, y_1) = (-3, 2)$ and $(x_2, y_2) = (-3, -2)$. Substitute these values into the slope formula.

$$m = \frac{y_2 - y_1}{x_2 - x_1} = \frac{-2 - 2}{-3 - (-3)} = \frac{-4}{0} \quad \text{undefined}$$

The slope is undefined, so the line is vertical. The equation of the line is $x = -3$. See Figure 51 for the graph.

Work Smart

If you plot the points first, it will be clear that the line through the points is vertical.

Quick ✓

40. Find an equation of the line containing the points $(3, 2)$ and $(3, -4)$. If possible, write the answer in slope-intercept form. Graph the line.

Work Smart

To determine which equation of a line to use, ask yourself "What information do I know?"

If you know the slope and a point that isn't the y-intercept, then use the point-slope form. If you know the slope and the y-intercept, use the slope-intercept form. If you know two points, use the slope formula with the point-slope form. If the slope is undefined, the line is vertical.

Summary Equations of Lines

Form of Line	Formula	Comments
Horizontal Line	$y = b$	Graph is a horizontal line (slope is 0) with y-intercept $(0, b)$.
Vertical Line	$x = a$	Graph is a vertical line (undefined slope) with x-intercept $(a, 0)$.
Point-slope	$y - y_1 = m(x - x_1)$	Useful for finding the equation of a line given a point and a slope or two points.
Slope-intercept	$y = mx + b$	Useful for quickly determining the slope and y-intercept of the line.
Standard	$Ax + By = C$	Useful for finding the x- and y-intercepts.

1.6 Exercises MyMathLab® Exercise numbers in green have complete video solutions in MyMathLab or may be accessed using the QR code to the right.

*Problems **1–40** are the Quick ✓ s that follow the **EXAMPLES**.*

Building Skills

In Problems 41–48, graph each linear equation by plotting points. See Objective 1.

41. $x - 2y = 6$

42. $2x - y = -8$

43. $3x + 2y = 12$

44. $-5x + y = 10$

45. $\frac{2}{3}x + y = 6$

46. $2x - \frac{3}{2}y = 10$

47. $5x - 3y = 6$

48. $-7x + 3y = 9$

In Problems 49–58, graph each linear equation by finding its intercepts. See Objective 2.

49. $3x + y = 6$

50. $-2x + y = 4$

51. $5x - 3y = 15$

52. $-4x + 3y = 24$

53. $\frac{1}{3}x - \frac{1}{2}y = 1$

54. $\frac{1}{4}x + \frac{1}{5}y = 2$

55. $2x + y = 0$

56. $4x + 3y = 0$

57. $\frac{2}{3}x - \frac{1}{2}y = 0$

58. $-\frac{3}{2}x + \frac{3}{4}y = 0$

In Problems 59–64, graph each linear equation. See Objective 3.

59. $x = -5$

60. $x = 5$

61. $y = 1$

62. $y = 6$

63. $2y + 8 = -6$

64. $3y + 20 = -10$

In Problems 65–68, (a) find the slope of the line and (b) interpret the slope. See Objective 4.

65.

66.

67.

68.

In Problems 69–80, plot each pair of points and graph the line containing them. Determine the slope of the line. See Objective 4.

69. $(0, 0); (1, 5)$

70. $(0, 0); (-2, 5)$

71. $(-2, 3); (1, -6)$

72. $(3, -1); (-2, 11)$

73. $(-2, 3); (3, 7)$

74. $(1, -4); (-1, 3)$

75. $(-3, 2); (4, 2)$

76. $(-3, 1); (2, 1)$

77. $(10, 2); (10, -3)$

78. $(4, 1); (4, -3)$

79. $\left(\frac{1}{2}, \frac{5}{3}\right); \left(\frac{9}{4}, \frac{11}{6}\right)$

80. $\left(\frac{7}{3}, \frac{5}{2}\right); \left(\frac{13}{9}, \frac{13}{4}\right)$

In Problems 81–90, graph the line that contains the given point and has slope m. Do not find an equation of the line. See Objective 6.

81. $m = 3; (1, 2)$

82. $m = 2; (-1, 4)$

83. $m = -2; (-3, 1)$

84. $m = -4; (-1, 5)$

85. $m = \frac{1}{3}; (-3, 4)$

86. $m = \frac{4}{3}; (-2, -5)$

87. $m = -\frac{3}{2}; (2, 5)$

88. $m = -\frac{1}{2}; (3, 3)$

89. $m = 0; (1, 2)$

90. m is undefined; $(-5, 2)$

In Problems 91 and 92, the slope and a point on a line are given. Use the information to find three additional points on the line. Answers may vary.

91. $m = \frac{5}{2}; (-2, 3)$

92. $m = -\frac{2}{3}; (1, -3)$

In Problems 93–96, find an equation of the line. Express your answer in slope-intercept form. See Objective 7.

93.

94.

95. **96.**

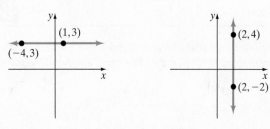

In Problems 97–106, find an equation of the line that has the given slope and contains the given point. Express your answer in slope-intercept form, if possible. See Objective 7.

97. $m = 2; (0, 0)$

98. $m = -1; (0, 0)$

99. $m = -3; (-1, 1)$

100. $m = 4; (2, -1)$

101. $m = \dfrac{4}{3}; (3, 2)$

102. $m = \dfrac{1}{2}; (2, 1)$

103. $m = -\dfrac{5}{4}; (-2, 4)$

104. $m = -\dfrac{4}{3}; (1, -3)$

105. m undefined; $(6, 1)$

106. $m = 0; (3, -2)$

In Problems 107–118, find an equation of the line that contains the given points. Express your answer in slope-intercept form, if possible. See Objective 9.

107. $(0, 0); (5, 7)$

108. $(0, 0); (4, -3)$

109. $(3, 2); (4, 7)$

110. $(1, 3); (3, 7)$

111. $(-2, 1); (5, -2)$

112. $(-3, 1); (1, 6)$

113. $(-1, -3); (-1, 5)$

114. $(-3, -4); (1, -4)$

115. $(1, 3); (-3, -7)$

116. $(-5, 1); (1, -1)$

117. $(2, 4); (-4, 4)$

118. $(3, 1); (3, -4)$

In Problems 119–130, find the slope and y-intercept of each line. Graph the line. See Objective 8.

119. $y = 2x - 1$

120. $y = 3x + 2$

121. $y = -4x$

122. $y = -7x$

123. $2x + y = 3$

124. $-3x + y = 1$

125. $4x + 2y = 8$

126. $3x + 6y = 12$

127. $x - 4y - 2 = 0$

128. $2x - 5y - 10 = 0$

129. $x = 3$

130. $y = -4$

Applying the Concepts

131. Find an equation for the x-axis.

132. Find an equation for the y-axis.

133. **Maximum Heart Rate** The data below represent the maximum number of heartbeats that a healthy individual should have during a 15-second interval of time while exercising for different ages.

Age, x	Maximum Number of Heartbeats, y
20	50
30	47.5
40	45
50	42.5
60	40
70	37.5

SOURCE: American Heart Association

(a) Plot the ordered pairs (x, y) on a graph and connect the points with straight lines.

(b) Compute and interpret the average rate of change in the maximum number of heartbeats between 20 and 30 years of age.

(c) Compute and interpret the average rate of change in the maximum number of heartbeats between 50 and 60 years of age.

(d) Based on your results for parts (a), (b), and (c), do you think that the maximum number of heartbeats is linearly related to age? Why?

134. Raisins The following data represent the weight (in grams) of a box of raisins and the number of raisins in the box.

Weight (in grams), x	Number of Raisins, y
42.3	82
42.5	86
42.6	89
42.7	91
42.8	93

SOURCE: *Jennifer Maxwell, student at Joliet Junior College*

(a) Plot the ordered pairs (x, y) on a graph and connect the points with straight lines.
(b) Compute and interpret the average rate of change in the number of raisins between 42.3 and 42.5 grams.
(c) Compute and interpret the average rate of change in the number of raisins between 42.7 and 42.8 grams.
(d) Based on your results for parts (a), (b), and (c), do you think that the number of raisins is linearly related to weight? Why?

135. Average Income An individual's income varies with age. The following data show the average income of individuals of different ages in the United States for 2010.

Age, x	Average Income, y
20	$10,036
30	$31,914
40	$42,224
50	$44,731
60	$41,296
70	$25,877

SOURCE: *Statistical Abstract, 2012*

(a) Plot the ordered pairs (x, y) on a graph and connect the points with straight lines.
(b) Compute and interpret the average rate of change in average income between 20 and 30 years of age.
(c) Compute and interpret the average rate of change in average income between 50 and 60 years of age.
(d) Based on your results for parts (a), (b), and (c), do you think that average income is linearly related to age? Why?

136. U.S. Population The following data represent the population of the United States between 1930 and 2015.

Year, x	Population, y
1930	123,202,624
1940	132,164,569
1950	151,325,798
1960	179,323,175
1970	203,302,031
1980	226,542,203
1990	248,709,873
2000	281,421,906
2010	308,745,538
2015	321,418,820

SOURCE: *U.S. Census Bureau*

(a) Plot the ordered pairs (x, y) on a graph and connect the points with straight lines.
(b) Compute and interpret the average rate of change in population between 1930 and 1940.
(c) Compute and interpret the average rate of change in population between 2000 and 2010.
(d) Based on your results for parts (a), (b), and (c), do you think that population is linearly related to the year? Why?

137. Measuring Temperature The relationship between Celsius (°C) and Fahrenheit (°F) degrees for measuring temperature is linear. Find an equation relating °C and °F if 0°C corresponds to 32°F and 100°C corresponds to 212°F. Use the equation to find the Celsius measure of 60°F.

138. Building Codes As a result of the Americans with Disabilities Act (ADA, 1990), the building code states that access ramps must have a slope not steeper than $\dfrac{1}{12}$. Interpret what this result means.

139. Which of the following equations might have the graph shown? (More than one answer is possible.)

(a) $y = 3x - 1$
(b) $y = -2x + 3$
(c) $y = 2x + 3$
(d) $3x - 2y = 4$
(e) $-3x + 2y = -4$

140. Which of the following equations might have the graph shown? (More than one answer is possible.)

(a) $y = 2x - 5$

(b) $y = -x + 2$

(c) $y = -\dfrac{2}{3}x - 3$

(d) $4x + 3y = -5$

(e) $-2x + y = -4$

Explaining the Concepts

141. Name the five forms of equations of lines given in this section.

142. What type of line has one x-intercept but no y-intercept?

143. What type of line has one y-intercept but no x-intercept?

144. What type of line has one x-intercept and one y-intercept?

145. Are there any lines that have no intercepts? Explain your answer.

Technology Exercises

For Problems 146 and 147 open the "Slope" Geogebra applet. The applet may be found using the QR code at the beginning of this section or through the Multimedia Library in MyMathLab.

The grade of a mountain comes from the *rise*, or vertical distance between two points on the mountain, and the *run*, or horizontal distance between two points on the mountain. See the figure. The grade of the mountain is the slope of a line connecting two points on the *mountain* and is defined as:

$$m = \frac{\text{rise}}{\text{run}} = \frac{\text{vertical change}}{\text{horizontal change}}$$

Because slope is defined in terms of rise and run, the value of the slope will change as the rise and run change.

146. Exploration: Slope When given the graph of a line, one way to find the slope is to choose two points on the line and find the rise and run between these two points. The ratio of the rise to the run is the slope of the line. If possible, express the slope in simplest form.

(a) Move point A to $(-3, 6)$ and point B to $(3, 7)$. What is the rise? What is the run? What is the slope?

(b) Move point A to $(1, 4)$ and point B to $(3, 1)$. What is the rise? What is the run? What is the slope?

(c) Move point A to $(1, 2)$ and point B to $(7, 2)$. What is the rise? What is the run? What is the slope?

(d) Move point A to $(-4, -3)$ and point B to $(-4, 6)$ What is the rise? What is the run? What is the slope?

(e) Move point A to $(-2, -3)$ and point B to $(4, 3)$. What is the slope?

(f) Move point A to $(-2, 7)$ and point B to $(6, -1)$ What is the slope?

147. Exploration: Slope Move the points A and B to wherever you choose on the line in order to see the position of the skier. Use the words positive, negative, zero, and undefined to fill in the blanks.

(a) When the skier is going "uphill" moving left to right, the slope of the mountain is _____.

(b) When the skier is going "downhill" moving left to right, the slope of the mountain is _____.

(c) When the skier is on flat land, the slope of the mountain is _____.

(d) When the skier is moving vertically the slope of the mountain is _____.

148. Exploration: Slope and y-intercept *Open the "Slope and y-intercept" Geogebra applet. The applet may be found using the QR code at the beginning of this section or through the Multimedia Library in MyMathLab.*

In the upper left corner are two "sliders". The top one, labeled m, controls the slope of the line, and the bottom one, labeled b, controls the y-intercept of the line. The slider changes the slope in increments of $\dfrac{1}{4}$ and the y-intercept in integer increments. If you are using a keyboard, it may be easier to use the arrow keys, instead of the mouse, to move the slider.

(a) Use the sliders to obtain the graph of $y = 2x$. Compare this to the graphs of $y = 2x + 3$ and $y = 2x - 4$. What do all the graphs have in common. What is different about all the graphs? In general, describe the graph of $y = 2x + b$.

(b) Use the sliders to obtain the graph of $y = 2x + 1$. Compare this to the graphs of $y = 4x + 1$ and $y = \dfrac{1}{2}x + 1$. What do all the graphs have in common? What is different about all the graphs? In general, describe the graph of $y = mx + 1$ for $m > 0$.

(c) Use the sliders to obtain the graph of $y = -2x + 1$. Compare this to the graphs of $y = -4x + 1$ and $y = -\dfrac{1}{2}x + 1$. What do all the graphs have in common? What is different about all the graphs? In general, describe the graph of $y = mx + 1$ for $m < 0$.

(d) Consider each of the following slopes: $4, \dfrac{1}{4}, \dfrac{1}{2}$,

and 2. Rank the slopes in terms of "steepness" from least "steep" to most "steep." What conclusion can you draw from this observation?

(e) Consider each of the following slopes:

$-4, -\dfrac{1}{4}, -\dfrac{1}{2}$, and -2. Rank the slopes in terms

of "steepness" from least "steep" to most "steep."

(f) What can you conclude about the appearance of a line when its slope is close to zero?

(g) What can you conclude about the appearance of a line when its slope approaches infinity or negative infinity (as $|m|$ becomes large)?

1.7 Parallel and Perpendicular Lines

Objectives

1. Define Parallel Lines
2. Find Equations of Parallel Lines
3. Define Perpendicular Lines
4. Find Equations of Perpendicular Lines

Are You Prepared for This Section?

Before getting started, complete the following problems. If you get a problem wrong, go back to the section cited and review the material.

P1. Determine the reciprocal of 3.   [Section R.3, pp. 23–24]

P2. Determine the reciprocal of $-\dfrac{3}{5}$.   [Section R.3, pp. 23–24]

▶ 1 Define Parallel Lines

When two lines (in the Cartesian plane) do not intersect (that is, they have no points in common), they are said to be *parallel*.

Work Smart

The words "if and only if" given in the definition mean that there are two statements being made:
(1) If two nonvertical lines are parallel, then their slopes are equal and they have different *y*-intercepts.
(2) If two nonvertical lines have equal slopes and different *y*-intercepts, then they are parallel.

> **Parallel Lines**
>
> Two nonvertical lines are **parallel** if and only if their slopes are equal and they have different *y*-intercepts. Vertical lines are parallel if they have different *x*-intercepts.

Figure 52(a) shows nonvertical parallel lines. Figure 52(b) shows vertical parallel lines.

Figure 52
Parallel lines

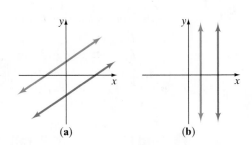

(a)   (b)

EXAMPLE 1

Determining Whether Two Lines Are Parallel

Determine whether the given lines are parallel.

(a) $L_1: 4x + y = 8$

   $L_2: 6x + 2y = 12$

(b) $L_1: -3x + 2y = 6$

   $L_2: 6x - 4y = 8$

Solution

Find the slope and y-intercept of each line by writing their equations in slope-intercept form. If the slopes are the same but the y-intercepts are different, then the lines are parallel.

(a) Solve L_1 for y: $4x + y = 8$

Subtract 4x
from both sides: $y = -4x + 8$

The slope of L_1 is -4 and the y-intercept is $(0, 8)$.

Solve L_2 for y: $6x + 2y = 12$

Subtract 6x from
both sides: $2y = -6x + 12$

Divide both sides
by 2: $y = \dfrac{-6x + 12}{2}$

Divide each term in
the numerator by 2: $y = -3x + 6$

The slope of L_2 is -3 and the y-intercept is $(0, 6)$.

Because the lines have different slopes, they are not parallel.

(b) Solve L_1 for y: $-3x + 2y = 6$

Add 3x to both sides: $2y = 3x + 6$

Divide both sides by 2: $y = \dfrac{3x + 6}{2}$

Divide each term in
the numerator by 2: $y = \dfrac{3}{2}x + 3$

The slope of L_1 is $\dfrac{3}{2}$ and the y-intercept is $(0, 3)$.

Solve L_2 for y: $6x - 4y = 8$

Subtract 6x from
both sides: $-4y = -6x + 8$

Divide both sides
by −4: $y = \dfrac{-6x + 8}{-4}$

Divide each term in
the numerator by −4: $y = \dfrac{3}{2}x - 2$

The slope of L_2 is $\dfrac{3}{2}$ and the y-intercept is $(0, -2)$.

The lines have the same slope but different y-intercepts, so they are parallel. ●

Quick ✓

1. Two lines are parallel if and only if they have the same _____ and different _____.

In Problems 2–4, determine whether the two lines are parallel.

2. $L_1: y = 3x + 1$

   $L_2: y = -3x - 3$

3. $L_1: 6x + 3y = 3$

   $L_2: -8x - 4y = 12$

4. $L_1: -3x + 5y = 10$

   $L_2: 6x + 10y = 10$

▶ ❷ Find Equations of Parallel Lines

Let's now discuss how to find the equation of a line that is parallel to a given line.

(**EXAMPLE 2**)   **How to Find the Equation of a Line Parallel to a Given Line**

Find an equation for the line that is parallel to $4x + 2y = 2$ and contains the point $(-2, 3)$. Graph the lines.

Step-by-Step Solution

Step 1: Find the slope of the given line by putting the equation in slope-intercept form.

$$4x + 2y = 2$$

Subtract $4x$ from both sides: $2y = -4x + 2$

Divide both sides by 2: $y = -2x + 1$

The slope of the line is -2.

Step 2: Use the point-slope form of a line with the given point and the slope found in Step 1 to find the equation of the parallel line.

$$y - y_1 = m(x - x_1)$$

$m = -2, x_1 = -2, y_1 = 3$: $y - 3 = -2[x - (-2)]$

Step 3: Put the equation in slope-intercept form by solving for y.

$$y - 3 = -2(x + 2)$$

Distribute the -2: $y - 3 = -2x - 4$

Add 3 to both sides: $y = -2x - 1$

Figure 53

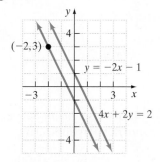

The line parallel to $4x + 2y = 2$ that contains $(-2, 3)$ is $y = -2x - 1$. Notice that the slopes of the two lines are the same, but the y-intercepts are different. Figure 53 shows the graph of the parallel lines. ●

Quick ✓

In Problems 5 and 6, find the equation of the line that contains the given point and is parallel to the given line. Write the line in slope-intercept form. Graph the lines.

5. $(5, 8); y = 3x + 1$ **6.** $(-2, 4); 3x + 2y = 10$

▶ ❸ **Define Perpendicular Lines**

Two lines that intersect at a right angle $(90°)$, are **perpendicular.** See Figure 54.

Figure 54
Perpendicular lines

Use the slopes of the lines to determine whether two lines are perpendicular.

Work Smart

If $m_1 \cdot m_2 = -1$, then $m_1 = \dfrac{-1}{m_2}$.

In addition, remember that the phrase "negative reciprocal" does not mean that the slope of the perpendicular line must be negative. It means that the slopes are opposite in sign.

Perpendicular Lines

Two nonvertical lines are **perpendicular** if and only if the product of their slopes is -1. Alternatively, two nonvertical lines are perpendicular if their slopes are negative reciprocals of each other. Any vertical line is perpendicular to any horizontal line.

EXAMPLE 3   **Finding the Slope of a Line Perpendicular to a Given Line**

Find the slope of any line that is perpendicular to a line whose slope is $\dfrac{5}{4}$.

Solution

The given line's slope is $\dfrac{5}{4}$. Any line whose slope is $-\dfrac{4}{5}$ will be perpendicular to the given line, because $\dfrac{5}{4}\left(-\dfrac{4}{5}\right) = -1$.

●

Quick ✓

7. Two lines are perpendicular if and only if the product of their slopes is ___.
8. Find the slope of any line perpendicular to a line whose slope is -3.

EXAMPLE 4   **Determining Whether Two Lines Are Perpendicular**

Determine whether the given lines are perpendicular.

(a) $L_1: y = 4x + 1$
  $L_2: y = -4x - 3$

(b) $L_1: y = \dfrac{2}{3}x - 5$

  $L_2: y = -\dfrac{3}{2}x + 2$

Solution

(a) The slope of L_1 is $m_1 = 4$. The slope of L_2 is $m_2 = -4$. Because the product of the slopes $m_1 m_2 = 4(-4) = -16 \neq -1$, the lines are not perpendicular.

(b) The slope of L_1 is $m_1 = \dfrac{2}{3}$. The slope of L_2 is $m_2 = -\dfrac{3}{2}$. The product of the slopes is $m_1 \cdot m_2 = \dfrac{2}{3}\left(-\dfrac{3}{2}\right) = -1$, so the lines are perpendicular.

●

Quick ✓

In Problems 9–11, determine whether the given lines are perpendicular.

9. $L_1: y = 5x - 3$

  $L_2: y = -\dfrac{1}{5}x - 4$

10. $L_1: 4x - y = 3$
   $L_2: x - 4y = 2$

11. $L_1: 2y + 4 = 0$
   $L_2: 3x - 6 = 0$

▶ ❹ **Find Equations of Perpendicular Lines**

Now find the equation of a line perpendicular to a second line.

EXAMPLE 5   **How to Find the Equation of a Line Perpendicular to a Given Line**

Find an equation of the line that is perpendicular to the line $2x + 5y = 10$ and contains the point $(4, -1)$. Write the equation in slope-intercept form. Graph the two lines.

Step-by-Step Solution

Step 1: Find the slope of the given line by putting the equation in slope-intercept form.	$2x + 5y = 10$

Subtract $2x$ from both sides: $\quad 5y = -2x + 10$

Divide both sides by 5: $\quad y = -\dfrac{2}{5}x + 2$

The slope of the line is $-\dfrac{2}{5}$.

Step 2: Find the slope of the perpendicular line.

The slope of the perpendicular line is the negative reciprocal of $-\dfrac{2}{5}$, which is $\dfrac{5}{2}$.

Step 3: Use the point-slope form of a line with the given point and the slope found in Step 2 to find the equation of the perpendicular line.

$$y - y_1 = m(x - x_1)$$

$m = \dfrac{5}{2}, x_1 = 4, y_1 = -1: \quad y - (-1) = \dfrac{5}{2}(x - 4)$

Step 4: Put the equation in slope-intercept form by solving for y.

$$y + 1 = \dfrac{5}{2}(x - 4)$$

Distribute the $\dfrac{5}{2}$: $\quad y + 1 = \dfrac{5}{2}x - 10$

Subtract 1 from both sides: $\quad y = \dfrac{5}{2}x - 11$

The equation of the line perpendicular to $2x + 5y = 10$ through $(4, -1)$ is $y = \dfrac{5}{2}x - 11$. See Figure 55.

Figure 55

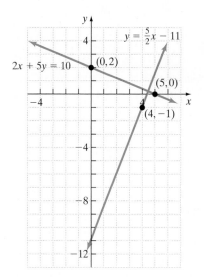

Quick ✓

In Problems 12–14, find the equation of the line that contains the given point and is perpendicular to the given line. Write the line in slope-intercept form. Graph the lines.

12. $(-4, 2); y = 2x + 1$ **13.** $(-3, -4); 3x - 4y = 8$

14. $(3, -2); x = 3$

1.7 Exercises MyMathLab®

Exercise numbers in green have complete video solutions in MyMathLab or may be accessed using the QR code to the right.

*Problems **1–14** are the Quick ✔s that follow the **EXAMPLES**.*

Building Skills

In Problems 15–18, a slope of a line is given. Determine (a) the slope of any line parallel to the line whose slope is given and (b) the slope of any line perpendicular to the line whose slope is given. See Objectives 1 and 3.

15. $m = 5$

16. $m = -\dfrac{8}{5}$

17. $m = -\dfrac{5}{6}$

18. $m = 0$

In Problems 19-26, determine whether the given linear equations are parallel, perpendicular, or neither. See Objectives 1 and 3.

19. $y = 5x + 4$
   $y = 5x - 7$

20. $y = 3x - 1$
   $y = -\dfrac{1}{3}x - 5$

21. $8x + y = 12$
   $2x - 8y = 3$

22. $-3x - y = 3$
   $6x + 2y = 9$

23. $-4x + 2y = 12$
   $x + 2y = 6$

24. $10x - 3y = 5$
   $5x + 6y = 3$

25. $-x + \dfrac{1}{3}y = \dfrac{1}{3}$
   $x - \dfrac{1}{3}y = \dfrac{5}{3}$

26. $\dfrac{1}{2}x - \dfrac{3}{2}y = 3$
   $2x + \dfrac{2}{3}y = 1$

In Problems 27–32, find an equation of the line L. Express your answer in slope-intercept form. See Objectives 2 and 4.

27.

L is parallel to $y = \dfrac{3}{2}x$

28.

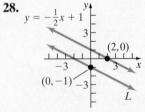

L is parallel to $y = -\dfrac{1}{2}x + 1$

29.

L is perpendicular to $y = -2x - 1$

30.

L is perpendicular to $y = \dfrac{2}{3}x + 1$

31.

L is perpendicular to $x = -2$

32.

L is perpendicular to $y = 3$

In Problems 33–46, find an equation of the line with the given properties. Express your answer in slope-intercept form. Graph the lines. See Objectives 2 and 4.

33. Parallel to $y = 2x + 3$ through the point $(3, 1)$

34. Parallel to $y = -3x + 1$ through the point $(2, 5)$

35. Perpendicular to $y = -2x + 1$ through the point $(2, 3)$

36. Perpendicular to $y = 4x + 3$ through the point $(4, 1)$

37. Parallel to $y = 1$ through the point $(-1, -3)$

38. Parallel to $x = -2$ through the point $(2, 5)$

39. Perpendicular to $x = 1$ through the point $(1, 3)$

40. Perpendicular to $y = 8$ through the point $(2, -4)$

41. Parallel to $3x - y = 2$ through the point $(1, 5)$

42. Parallel to $2x + y = 5$ through the point $(-4, 3)$

43. Perpendicular to $4x + 3y - 1 = 0$ through the point $(-4, 1)$

44. Perpendicular to $-2x + 5y - 3 = 0$ through the point $(2, -3)$

45. Parallel to $5x + 2y = 1$ through the point $(-2, -3)$

46. Perpendicular to $3x + y = 1$ through the point $(3, -1)$

Mixed Practice

In Problems 47–52, two points on L_1 and two points on L_2 are given. Plot the points in the Cartesian plane and draw a line through the points. Compute the slope of the line containing these points and determine whether the lines are parallel, perpendicular, or neither.

47. L_1: $(1, 2)$; $(6, 5)$
$\quad L_2$: $(-2, 3)$; $(1, -2)$

48. L_1: $(1, 1)$; $(4, 3)$
$\quad L_2$: $(-1, 3)$; $(3, -3)$

49. L_1: $(-2, 4)$; $(1, -3)$
$\quad L_2$: $(-1, -6)$; $(-4, 1)$

50. L_1: $(-3, 0)$; $(0, 2)$
$\quad L_2$: $(4, 8)$; $(2, 5)$

51. L_1: $(0, 5)$; $(1, 3)$
$\quad L_2$: $(-4, 6)$; $(0, 14)$

52. L_1: $(1, -3)$; $(5, -4)$
$\quad L_2$: $(0, 4)$; $(8, 2)$

Applying the Concepts

△ **53. Geometry** Given the points $A = (1, 1)$, $B = (4, 3)$, and $C = (2, 6)$,

(a) Plot the points in a Cartesian plane. Connect the points to form a triangle.

(b) Verify that the triangle is a right triangle by showing that the line segment $\overline{AB}$ is perpendicular to the line segment $\overline{BC}$ and therefore forms a right angle.

△ **54. Geometry** Given the points $A = (-2, -2)$, $B = (3, 1)$, and $C = (-5, 3)$,

(a) Plot the points in a Cartesian plane. Connect the points to form a triangle.

(b) Verify that the triangle is a right triangle by showing that the line segment $\overline{AB}$ is perpendicular to the line segment $\overline{AC}$ and therefore forms a right angle.

△ **55. Geometry** In geometry, we learn that a **parallelogram** is a quadrilateral in which both pairs of opposite sides are parallel. Given the points $A = (2, 2)$, $B = (7, 3)$, $C = (8, 6)$, and $D = (3, 5)$,

(a) Plot the points in a Cartesian plane. Connect the points to form a quadrilateral.

(b) Verify that the quadrilateral is a parallelogram by showing that the opposite sides are parallel.

△ **56. Geometry** In geometry, we learn that a parallelogram is a quadrilateral in which both pairs of opposite sides are parallel. Given the points $A = (-2, -1)$, $B = (4, 1)$, $C = (5, 5)$, and $D = (-1, 3)$,

(a) Plot the points in a Cartesian plane. Connect the points to form a quadrilateral.

(b) Verify that the quadrilateral is a parallelogram by showing that the opposite sides are parallel.

Extending the Concepts

57. Find A so that $Ax + 4y = 12$ is perpendicular to $4x + y = 3$.

58. Find B so that $-6x + By = 3$ is perpendicular to $2x - 3y = 8$.

59. The figure shows the graph of two parallel lines. Which of the following pairs of equations might have such a graph?

(a) $y = x + 3$
$\quad y = -x - 1$

(b) $y = 2x + 3$
$\quad y = 2x + 1$

(c) $x - 2y = 4$
$\quad x - 2y = -3$

(d) $-2x + y = 5$
$\quad -2x + y = 2$

(e) $x - y = 3$
$\quad 3x - 3y = 9$

60. The figure shows the graph of two perpendicular lines. Which of the following pairs of equations might have such a graph?

(a) $y = 3x + 4$
$\quad y = -\dfrac{1}{3}x - 2$

(b) $-2x + y = 3$
$\quad x + 2y = 1$

(c) $2x + 3y = -2$
$\quad 3x - 2y = 5$

(d) $3x + 4y = 5$
$\quad -3x + 4y = -2$

(e) $x - 2y = 6$
$\quad 2y + x = 2$

Explaining the Concepts

61. If two nonvertical lines have the same x-intercept but different y-intercepts, can they be parallel? Explain your answer.

62. Why don't we say that a horizontal line is perpendicular to a vertical line if they have slopes that are negative reciprocals of each other?

Technology Exercises

63. Exploration: Parallel Lines *Open the "Parallel and Perpendicular Lines" Geogebra applet. The applet may be found using the QR code at the beginning of this section or through the Multimedia Library in MyMathLab.*

In the upper left corner are two pairs of "sliders". The top one in each pair, labeled slope$_1$ and slope$_2$, controls the slope of the line, and the bottom one, labeled yintercept$_1$ and yintercept$_2$, controls the y-intercept of the line. The slider changes the slope in increments of $\frac{1}{4}$ and the y-intercept in integer increments. If you are using a keyboard, it may be easier to use the arrow keys, instead of the mouse, to move the slider.

(a) Use the sliders for y_1 to create the graph of $y_1 = 2x + 1$. Now, use the sliders for y_2 to find an equation that is parallel to y_1 The applet will show the words "Parallel Lines!" once you have successfully found a pair.

(b) Use the sliders for y_1 to create the graph of $y_1 = \frac{1}{2}x + 3$. Now, use the sliders for y_2 to find an equation that is parallel to y_1. The applet will show the words "Parallel Lines!" once you have successfully found a pair.

(c) Use the sliders for y_1 to create the graph of $y_1 = -2x + 2$. Now, use the sliders for y_2 to find an equation that is parallel to y_1. The applet will show the words "Parallel Lines!" once you have successfully found a pair.

(d) What relationship did you notice exists between a pair of parallel lines? Specifically, what is the relationship between the slopes (if any) and what is the relationship between the y-intercepts (if any)?

64. **Exploration: Perpendicular Lines** *Open the "Parallel and Perpendicular Lines" Geogebra applet. The applet may be found using the QR code at the beginning of this section or through the Multimedia Library in MyMathLab.*

If you are using a keyboard, it is easier to move the sliders with the arrow keys instead of a mouse.

(a) Use the sliders for y_1 to create the graph of $y_1 = 2x + 1$. Now, use the sliders for y_2 to find an equation that is perpendicular to y_1. The applet will show the words "Perpendicular Lines!" once you have successfully found a pair.

(b) Use the sliders for y_1 to create the graph of $y_1 = \frac{1}{2}x + 3$. Now, use the sliders for y_2 to find an equation that is perpendicular to y_1. The applet will show the words "Perpendicular Lines!" once you have successfully found a pair.

(c) Use the sliders for y_1 to create the graph of $y_1 = -2x + 2$. Now, use the sliders for y_2 to find an equation that is perpendicular to y_1. The applet will show the words "Perpendicular Lines!" once you have successfully found a pair.

(d) What relationship did you notice exists between a pair of perpendicular lines? Specifically, what is the relationship between the slopes (if any) and what is the relationship between the y-intercepts (if any)?

1.8 Linear Inequalities in Two Variables

Objectives

1 Determine Whether an Ordered Pair Is a Solution to a Linear Inequality

2 Graph Linear Inequalities

3 Solve Problems Involving Linear Inequalities

Are You Prepared for This Section?

Before getting started, complete the following problems. If you get a problem wrong, go back to the section cited and review the material.

P1. Determine whether $x = 4$ satisfies the inequality $3x + 1 \geq 7$.   [Section 1.4, pp. 86–88]

P2. Solve the inequality: $-4x - 3 > 9$ [Section 1.4, pp. 86–88]

P3. *True or False* The inequality $5 > 3$ is a strict inequality.   [Section R.2, p. 17]

In Section 1.4, inequalities in one variable were solved. In this section, linear inequalities in two variables will be discussed.

▶ 1 **Determine Whether an Ordered Pair Is a Solution to a Linear Inequality**

A linear inequality in two variables is an inequality in one of the forms

$$Ax + By < C \qquad Ax + By > C \qquad Ax + By \leq C \qquad Ax + By \geq C$$

where A, B, and C are real numbers and A and B are not both zero.

If the inequality symbol is replaced with an equal sign, the equation of a line, $Ax + By = C$ is obtained. The line separates the xy-plane into two regions called **half-planes.** See Figure 56.

Prepared?...Answers   **P1.** Satisfies
P2. $\{x \mid x < -3\}$ or $(-\infty, -3)$
P3. True

Figure 56

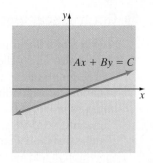

A linear inequality in two variables x and y is **satisfied** by an ordered pair (a, b) if a true statement results when x is replaced by a and y is replaced by b.

EXAMPLE 1 **Determining Whether an Ordered Pair Is a Solution to a Linear Inequality in Two Variables**

Determine which of the following ordered pairs are solutions to the linear inequality $3x + y < 7$.

(a) $(3, 2)$ (b) $(-3, 1)$ (c) $(1, -2)$

Solution

(a) Let $x = 3$ and $y = 2$ in the inequality. If a true statement results, then $(3, 2)$ is a solution to the inequality.

$$3x + y < 7$$
$$x = 3, y = 2: \quad 3(3) + 2 \overset{?}{<} 7$$
$$9 + 2 \overset{?}{<} 7$$
$$11 \overset{?}{<} 7 \quad \text{False}$$

The statement is false, so $(3, 2)$ is not a solution to the inequality.

(b) Let $x = -3$ and $y = 1$ in the inequality. If a true statement results, then $(-3, 1)$ is a solution to the inequality.

$$3x + y < 7$$
$$x = -3, y = 1: \quad 3(-3) + 1 \overset{?}{<} 7$$
$$-9 + 1 \overset{?}{<} 7$$
$$-8 \overset{?}{<} 7 \quad \text{True}$$

The statement is true, so $(-3, 1)$ is a solution to the inequality.

(c) Let $x = 1$ and $y = -2$ in the inequality.

$$3x + y < 7$$
$$x = 1, y = -2: \quad 3(1) + (-2) \overset{?}{<} 7$$
$$3 - 2 \overset{?}{<} 7$$
$$1 \overset{?}{<} 7 \quad \text{True}$$

The statement is true, so $(1, -2)$ is a solution to the inequality.   ●

Quick ✓

1. If we replace the inequality symbol in $Ax + By > C$ with an equal sign, we obtain the equation of a line, $Ax + By = C$. The line separates the xy-plane into two regions called _____.

2. Determine which of the following ordered pairs are solutions to the linear inequality $-2x + 3y \geq 3$.

(a) $(4, 1)$ (b) $(-1, 2)$ (c) $(2, 3)$ (d) $(0, 1)$

▶ ❷ Graph Linear Inequalities

A **graph of a linear inequality in two variables** x and y consists of all points (x, y) whose coordinates satisfy the inequality.

The graph of any linear inequality in two variables may be obtained by graphing the equation corresponding to the inequality, using dashes if the inequality is strict ($<$ or $>$) and a solid line if the inequality is nonstrict ($\leq$ or $\geq$). This graph will separate the xy-plane into two half-planes. In each half-plane, either all points satisfy the inequality or no points satisfy the inequality. So the use of a single test point is all that is required to obtain the graph of a linear inequality in two variables.

| EXAMPLE 2 | **How to Graph a Linear Inequality in Two Variables** |

Graph the linear inequality: $3x + y < 7$

Step-by-Step Solution

Step 1: Replace the inequality symbol with an equal sign and graph the corresponding line. If the inequality is strict ($<$ or $>$), graph the line as a dashed line. If the inequality is nonstrict ($\leq$ or $\geq$), graph the line as a solid line.

Replace $<$ with $=$ to obtain $3x + y = 7$. Graph the line $3x + y = 7$ by first writing it in slope-intercept form: $y = -3x + 7$. The y-intercept of the line is $(0, 7)$ and the slope is -3, so plot the point $(0, 7)$. Use slope $= -\dfrac{3}{1} = \dfrac{\text{rise}}{\text{run}}$ to count right 1 unit and down 3 units and plot the point $(1, 4)$. Graph the line through $(0, 7)$ and $(1, 4)$, using a dashed line because the inequality is strict. See Figure 57(a).

Step 2: Select any test point not on the line—the origin $(0, 0)$ is often a good choice—and determine whether the point satisfies the inequality. If it does, shade the half-plane containing that point; otherwise, shade the other half-plane.

$$3x + y < 7$$
$$\text{Test Point: } (0, 0): \quad 3(0) + 0 \overset{?}{<} 7$$
$$0 \overset{?}{<} 7 \quad \text{True}$$

Because 0 is less than 7, the point $(0, 0)$ satisfies the inequality. Therefore, shade the half-plane containing the point $(0, 0)$. See Figure 57(b). The shaded region represents the solutions to the linear inequality.

Figure 57

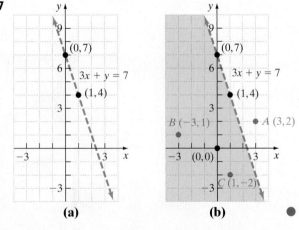

(a)   (b)

Only one test point is needed to obtain the graph of the inequality. Why? Consider the inequality $3x + y < 7$ presented in Examples 1 and 2. Notice that $A(3, 2)$ does not satisfy the inequality, while $B(-3, 1)$ and $C(1, -2)$ satisfy the inequality. Notice that point A is not in the shaded region of Figure 57(b), while points B and C are in the shaded region. So if a point does not satisfy the inequality, then none of the points in the half-plane containing that point satisfy the inequality. If a point does satisfy the inequality, then all the points in the half-plane containing the point satisfy the inequality.

Work Smart

An alternative to using test points is to solve the inequality for y. If the inequality is of the form $y >$ or $y \geq$, shade above the line. If the inequality is of the form $y <$ or $y \leq$, shade below the line.

> **Graphing a Linear Inequality in Two Variables**
>
> **Step 1:** Replace the inequality symbol with an equal sign, and graph the resulting equation. If the inequality is strict ($<$ or $>$), use a dashed line; if the inequality is nonstrict ($\leq$ or $\geq$) use a solid line. The graph separates the xy-plane into two half-planes.
>
> **Step 2:** Select a test point P that is not on the line.
>
> **(a)** If the coordinates of P satisfy the inequality, then shade the half-plane containing P.
>
> **(b)** If the coordinates of P do not satisfy the inequality, then shade the half-plane that does not contain P.

Quick ✓

3. *True or False* The graph of a linear inequality is a line.

4. *True or False* In a graph of a linear inequality in two variables with a strict inequality, the line separating the two half-planes should be dashed.

In Problems 5 and 6, graph each linear inequality.

5. $y < -2x + 3$

6. $6x - 3y \leq 15$

EXAMPLE 3 **Graphing a Linear Inequality in Two Variables**

Graph the linear inequality: $y \geq \dfrac{1}{2}x$

Solution

Graph the corresponding line $y = \dfrac{1}{2}x$. The line is solid because the inequality is nonstrict. See Figure 58(a).

Now, select any test point not on the line. The line contains the origin, so we decide to use $(2, 0)$ as the test point.

Work Smart

When the line is of the form $Ax + By = 0$, do not use $(0, 0)$ as a test point because this line contains the origin and will not help to determine the correct shading.

Work Smart

Notice the inequality is of the form $y \geq$, so we shade above the line.

$$y \geq \frac{1}{2}x$$

$$\text{Test Point: } (2, 0): \quad 0 \stackrel{?}{\geq} \frac{1}{2} \cdot 2$$

$$0 \stackrel{?}{\geq} 1 \qquad \text{False}$$

Because 0 is not greater than 1, the point $(2, 0)$ does not satisfy the inequality. Therefore, shade the half-plane that does not contain $(2, 0)$. The shaded region in Figure 58(b) represents the solutions to the linear inequality.

Figure 58

(a)

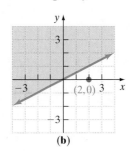

(b)

Quick ✓

7. Graph the linear inequality: $2x + y < 0$

▶ ❸ Solve Problems Involving Linear Inequalities

Linear inequalities involving two variables can be used to solve problems in areas such as manufacturing, sales, or (as in the next example) nutrition.

EXAMPLE 4 | **Saturated Fat Intake**

Randy really enjoys Wendy's Junior Cheeseburgers and Biggie French Fries. However, he knows that his intake of saturated fat during lunch should not exceed 16 grams. Each Junior Cheeseburger contains 6 grams of saturated fat, and each order of Biggie Fries contains 3 grams of saturated fat. (SOURCE: *wendys.com*)

(a) Write a linear inequality that describes all the combinations of Junior Cheeseburgers and Biggie Fries that Randy can order.

(b) Can Randy eat two Junior Cheeseburgers and one order of Biggie Fries during lunch and stay within his allotment of saturated fat?

(c) Can Randy eat three Junior Cheeseburgers and one order of Biggie Fries during lunch and stay within his allotment of saturated fat?

Solution

(a) Use the first three steps in the problem-solving strategy given in Section 1.2 on page 61.

Step 1: Identify Determine the number of Junior Cheeseburgers and Biggie Fries Randy can eat while not exceeding 16 grams of saturated fat.

Step 2: Name the Unknowns Let x represent the number of Junior Cheeseburgers that Randy eats, and let y represent the number of Biggie Fries Randy eats.

Step 3: Translate One Junior Cheeseburger has 6 grams of saturated fat. Two cheeseburgers have $6(2) = 12$ grams. In general, x Junior Cheeseburgers have $6x$ grams of saturated fat. Similar logic for the Biggie Fries tells us that y Biggie Fries have $3y$ grams of saturated fat. The words "should not exceed' imply a $\leq$ inequality. Therefore, a linear inequality that describes the combinations Randy can have is

$$6x + 3y \leq 16 \quad \text{The Model}$$

(b) Letting $x = 2$ and $y = 1$, we obtain

$$6(2) + 3(1) \overset{?}{\leq} 16$$
$$15 \overset{?}{\leq} 16 \quad \text{True}$$

Because the inequality is true, Randy can eat two Junior Cheeseburgers and one order of Biggie Fries and remain within 16-grams of saturated fat.

(c) Letting $x = 3$ and $y = 1$, we obtain

$$6(3) + 3(1) \overset{?}{\leq} 16$$
$$21 \overset{?}{\leq} 16 \quad \text{False}$$

Because the inequality is false, Randy cannot eat three Junior Cheeseburgers and one order of Biggie Fries and remain within his limit. ●

Quick ✓

8. Avery is on a diet that requires that he consume no more than 900 calories for lunch. He really enjoys Wendy's Homestyle Chicken fillet and Frosties. Each chicken fillet has 560 calories, and each Frosty has 300 calories.

(a) Write a linear inequality that describes Avery's options for eating at Wendy's.

(b) Can Avery eat one Homestyle Chicken fillet and one Frosty and stay within his allotment of calories?

(c) Can Avery eat two Homestyle Chicken fillets and one Frosty and stay within his allotment of calories?

1.8 Exercises   MyMathLab®

Exercise numbers in green have complete video solutions in MyMathLab or may be accessed using the QR code to the right.

*Problems **1–8** are the Quick ✓s that follow the **EXAMPLES.***

Building Skills

In Problems 9–12, determine whether the given points are solutions to the linear inequality. See Objective 1.

9. $x + 3y < 6$
  (a) $(0, 1)$
  (b) $(-2, 4)$
  (c) $(8, -1)$

10. $2x + y > -3$
  (a) $(2, -1)$
  (b) $(1, -3)$
  (c) $(-5, 4)$

11. $-3x + 4y \geq 12$
  (a) $(-4, 2)$
  (b) $(0, 2)$
  (c) $(0, 3)$

12. $2x - 5y \leq 2$
  (a) $(1, 0)$
  (b) $(3, 0)$
  (c) $(1, 2)$

In Problems 13–32, graph each inequality. See Objective 2.

13. $y > 3$

14. $y < -2$

15. $x \geq -2$

16. $x < 7$

17. $y < 5x$

18. $y \geq \dfrac{2}{3}x$

19. $y > 2x + 3$

20. $y < -3x + 1$

21. $y \leq \dfrac{1}{2}x - 5$

22. $y \geq -\dfrac{4}{3}x + 5$

23. $3x + y \leq 4$

24. $-4x + y \geq -5$

25. $2x + 5y \leq -10$

26. $3x + 4y \geq 12$

27. $-4x + 6y > 24$

28. $-5x + 3y < 30$

29. $\dfrac{x}{2} + \dfrac{y}{3} < 1$

30. $\dfrac{x}{3} - \dfrac{y}{4} \leq 1$

31. $\dfrac{2}{3}x - \dfrac{3}{2}y \geq 2$

32. $\dfrac{5}{4}x - \dfrac{3}{5}y \leq 2$

Applying the Concepts

33. Nutrition Sammy goes to McDonald's for lunch. He is on a diet that requires that he consume no more than 125 grams of carbohydrates for lunch. Sammy enjoys McDonald's Filet-o-Fish sandwiches and French fries. Each Filet-o-Fish has 39 grams of carbohydrates, and each small order of French fries has 29 grams of carbohydrates. (SOURCE: *McDonald's Corp.*)

  (a) Write a linear inequality that describes Sammy's options for eating at McDonald's.
  (b) Can Sammy eat two Filet-o-Fish and one order of fries and meet his carbohydrate requirement?
  (c) Can Sammy eat three Filet-o-Fish and one order of fries and meet his carbohydrate requirement?

34. Salesperson Juanita sells two different computer models. For each Model A computer sold she makes $45, and for each Model B computer sold she makes $65. Juanita set a monthly goal of earning at least $4000.

  (a) Write a linear inequality that describes Juanita's options for making her sales goal.
  (b) Will Juanita make her sales goal if she sells 50 Model A and 28 Model B computers?
  (c) Will Juanita make her sales goal if she sells 41 Model A and 33 Model B computers?

35. Production Planning Acme Switch Company is a small manufacturing firm that makes two different styles of microwave switches. Switch A requires 2 hours to assemble, and Switch B requires 1.5 hours to assemble. Suppose that there are at most 80 hours of assembly time available each week.

  (a) Write a linear inequality that describes Acme's options for making the microwave switches.
  (b) Can Acme manufacture 24 of Switch A and 41 of Switch B in a week?
  (c) Can Acme manufacture 16 of Switch A and 45 of Switch B in a week?

36. Budget Constraints Johnny can spend no more than $3.00 that he got from his grandparents. He goes to the candy store and wants to buy gummy bears that cost $0.10 each and suckers that cost $0.25 each.

  (a) Write a linear inequality that describes Johnny's options for buying candy.
  (b) Can Johnny buy 18 gummy bears and 5 suckers?
  (c) Can Johnny buy 19 gummy bears and 4 suckers?

Extending the Concepts

In Problems 37–40, determine the linear inequality whose graph is given.

37.

38.

39.

40.

Explaining the Concepts

41. If an ordered pair does not satisfy a linear inequality, then we shade the half-plane opposite the point. Explain why this works.

42. Explain why we cannot use a test point that lies on the line separating the two half-planes when determining the solutions to a linear inequality involving two variables.

Technology Exercises

Technology can be used to graph linear inequalities in two variables. Figure 59 shows the result of Example 2 using Desmos.

Figure 59

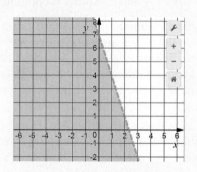

In Problems 43-54, graph the inequalities using technology.

43. $y > 3$

44. $y < -2$

45. $y < 5x$

46. $y \geq \dfrac{2}{3}x$

47. $y > 2x + 3$

48. $y < -3x + 1$

49. $y \leq \dfrac{1}{2}x - 5$

50. $y \geq -\dfrac{4}{3}x + 5$

51. $3x + y \leq 4$

52. $-4x + y \geq -5$

53. $2x + 5y \leq -10$

54. $3x + 4y \geq 12$

Chapter 1 Activity: Pass the Paper

Focus: Solving equations and inequalities

Time: 15 minutes

Group size: 4

Materials needed: One blank piece of notebook paper per group member

To the right are two equations and two inequalities. In this activity you will work together to solve these problems by following the procedure below. Be sure to read through the entire procedure together before beginning the activity so all group members will understand the procedure.

Procedure

1. Write one of the problems shown at the top of your paper. Be sure that each group member chooses a different problem.

2. Two lines below the original problem, write out the first step for solving the problem.

3. Fold the top of the paper down to cover the original problem, leaving your first step visible.

4. Pass your paper to a different group member. You might want to arrange your seats so that the papers can be passed around in a circle.

5. Continue solving the problems one step at a time, covering the step above yours and passing the paper to the next group member, until all problems are solved.

6. As a group, discuss the solutions and decide whether they are correct. If any solutions are incorrect, solve them correctly together.

Problems:

Solve the equation or inequality.

(a) $4 - (x - 3) = -10 + 5(x + 1)$

(b) $3(x - 2) + 8 = 5x - 2(x - 1)$

(c) $3(x - 4) - 5x > 2x + 12$

(d) $4(x + 1) - 2x \geq 5x - 11$

Chapter 1 Review

Section 1.1   Linear Equations in One Variable

KEY CONCEPTS

- **Linear Equation in One Variable**
  An equation equivalent to one of the form $ax + b = 0$, where a and b are real numbers with $a \neq 0$.

- **Addition Property of Equality**
  For real numbers a, b, and c, if $a = b$, then $a + c = b + c$.

- **Multiplication Property of Equality**
  For real numbers a, b, and c where $c \neq 0$, if $a = b$, then $ac = bc$.

KEY TERMS

Equation in one variable
Sides of the equation
Solution
Satisfies
Solve an equation
Solution set
Equivalent equations
Conditional equation
Contradiction
Identity

You Should Be Able To...	EXAMPLE	Review Exercises
① Determine whether a number is a solution to an equation (p. 49)	Example 1	1–4
② Solve linear equations (p. 50)	Examples 2 through 6	5–18
③ Determine whether an equation is a conditional equation, an identity, or a contradiction (p. 55)	Examples 7 and 8	5–14

In Problems 1–4, determine which of the numbers, if any, are solutions to the given equation.

1. $3x - 4 = 6 + x$; $x = 5, x = 6$

2. $-1 - 4x = 2(3 - 2x) - 7$; $x = -2, x = -1$

3. $4y - (1 - y) + 5 = -6 - 2(3y - 5) - 2y$; $y = -2, y = 0$

4. $\dfrac{w - 7}{3} - \dfrac{w}{4} = -\dfrac{7}{6}$; $w = -14, w = 7$

In Problems 5–14, solve the linear equation. State whether the equation is an identity, a contradiction, or a conditional equation.

5. $2w + 9 = 15$

6. $-4 = 8 - 3y$

7. $2x + 5x - 1 = 20$

8. $7x + 5 - 8x = 13$

9. $-2(x - 4) = 8 - 2x$

10. $3(2r + 1) - 5 = 9(r - 1) - 3r$

11. $\dfrac{2y + 3}{4} - \dfrac{y}{2} = 5$

12. $\dfrac{x}{3} + \dfrac{2x}{5} = \dfrac{x - 20}{15}$

13. $0.2(x - 6) + 1.75 = 4.25 + 0.1(3x + 10)$

14. $2.1w - 3(2.4 - 0.2w) = 0.9(3w - 5) - 2.7$

In Problems 15 and 16, determine which values of the variable must be excluded from the domain.

15. $\dfrac{8}{2x + 3}$

16. $\dfrac{6x - 5}{6(x - 1) + 3}$

17. State Income Tax A resident of Oregon completes her state tax return and determines that she paid $2370 in state income tax in 2015. The solution to the equation $2370 = 0.09(x - 8250) + 511.50$ represents her Oregon taxable income x in 2015. Solve the equation to determine her Oregon taxable income. (SOURCE: *Oregon Department of Revenue*)

18. Movie Club The DVD club to which you belong offers unlimited DVDs at $10 off the regular price if you buy 1 at the regular price. You purchase 5 DVDs through this offer and spend $69.75 (not including tax and shipping). The solution to the equation $x + 4(x - 10) = 69.75$ represents the regular club price x for a DVD. Solve the equation to determine the regular club price for a DVD.

Section 1.2   An Introduction to Problem Solving

KEY CONCEPTS

- **Simple Interest Formula**
  $I = Prt$, where I is interest, P is principal, r is the per annum interest rate expressed as a decimal, and t is time in years
- **Uniform Motion Formula**
  $d = rt$, where d is distance, r is average speed, and t is time

KEY TERMS

Problem solving	Principal
Mathematical modeling	Rate of interest
Mathematical model	Simple interest
Direct translation	Mixture problems
Interest	Uniform motion

You Should Be Able To...	EXAMPLE	Review Exercises
❶ Translate English sentences into mathematical statements (p. 59)	Example 1	19–22
❷ Model and solve direct translation problems (p. 61)	Examples 2 through 6	23–28
❸ Model and solve mixture problems (p. 65)	Examples 7 through 9	29–32
❹ Model and solve uniform motion problems (p. 69)	Example 10	33–34

In Problems 19–22, translate each of the following English statements into a mathematical statement. Do not solve the equation.

19. The sum of three times a number x and 7 is 22.

20. The difference of a number x and 3 is equivalent to the quotient of the number x and 2.

21. 20% of a number x equals the difference of the number x and 12.

22. The product of 6 and a number x is the same as 4 less than twice the number x.

For Problems 23 and 24, translate each English statement into a mathematical statement. Then solve the equation.

23. Shawn is 8 years older than Payton and the sum of their ages is 18. What are their ages?

24. The sum of five consecutive odd integers is 125. Find the integers.

25. Computing Grades Logan is in an elementary statistics course and has test scores of 85, 81, 84, and 77. If the final exam counts the same as two tests, what score does Logan need on the final to have an average of 80?

26. Home Equity Loans On February 18, 2016, Bank of America offered a home equity line of credit at a rate of 2.240% *annual* simple interest. If Cherie has such a credit line with a balance of $15,000, how much interest will she accrue at the end of *1 month*?

27. Discounted Price Suppose that REI sells a 0° sleeping bag at the discounted price of $94.50. If this price represents a discount of 30% off the original selling price, find the original price.

28. Minimum Wage On July 24, 2009, the federal minimum wage was increased 10.7% to $7.25. Determine the federal minimum wage prior to July 24, 2009.

29. Making a Mixture CoffeeAM sells chocolate-covered blueberries for $10.95 per pound and chocolate-covered strawberries for $13.95 per pound. The company wants to create a mix of the two that will sell for $12.95 per pound with no loss in revenue. How many pounds of each treat should be used to make 12 pounds of the mix?

30. A Sports Mix The Candy Depot sells baseball gumballs for $3.50 per pound and soccer gumballs for $4.50 per pound. The company wants to sell a "sports mix" that sells for $3.75 per pound with no loss in revenue. How many pounds of each gumball type should be included to make 10 pounds of the mix?

31. Investments Angie received an $8000 bonus and wants to invest the money. She can invest part of the money at 8% simple interest with a moderate risk and the rest at 18% simple interest with a high risk. She wants an overall annual return of 12% but does not want to risk losing any more than necessary. How much should she invest at each rate to reach her goal?

32. Antifreeze A 2012 Chevrolet Malibu has an engine coolant system capacity of 7.5 quarts. If the system is currently filled with a mixture that is 30% antifreeze, how much of this mixture should be drained and replaced with pure antifreeze so that the system is filled with a mixture that is 50% antifreeze?

33. Road Trip On a 300-mile trip to Chicago, Josh drove part of the time at 60 miles per hour and the remainder of the trip at 70 miles per hour. If the total trip took 4.5 hours, for how many miles did Josh drive at a rate of 60 miles per hour?

34. Uniform Motion An F15 Strike Eagle near New York City and an F14 Tomcat near San Diego are about 2200 miles apart and traveling toward each other. The F15 is traveling 200 miles per hour faster than the F14, and the planes pass each other after 50 minutes. How fast is each plane traveling?

Section 1.3   Using Formulas to Solve Problems

KEY CONCEPTS

- **Geometry Formulas** (see pages 75–76)

KEY TERMS

Formula
Golden rectangle
Supplementary angles
Complementary angles

You Should Be Able To...	EXAMPLE	Review Exercises
1 Solve for a variable in a formula (p. 74)	Examples 2 and 3	35–44
2 Use formulas to solve problems (p. 77)	Examples 4 and 5	45–52

In Problems 35–40, solve for the indicated variable.

35. Solve $y = \dfrac{k}{x}$ for x.

36. Solve $F = \dfrac{9}{5}C + 32$ for C.

37. Solve $P = 2L + 2W$ for W.

38. Solve $\rho = m_1 v_1 + m_2 v_2$ for m_2.

39. Solve $PV = nRT$ for T.

40. Solve $S = 2LW + 2LH + 2WH$ for W.

In Problems 41–44, solve for y.

41. $3x + 4y = 2$

42. $-5x + 4y = 10$

43. $48x - 12y = 60$

44. $\dfrac{2}{5}x + \dfrac{1}{3}y = 8$

45. Temperature Conversions To convert temperatures from Fahrenheit to Celsius, we can use the formula $C = \dfrac{5}{9}(F - 32)$. If the melting point for platinum is 3221.6°F, convert this temperature to degrees Celsius.

46. Angles in a Triangle The measure of each congruent angle in an isosceles triangle is 30 degrees larger than the measure of the remaining angle. Determine the measures of all three angles.

47. Window Dimensions The perimeter of a rectangular window is 76 feet. The window is 8 feet longer than it is wide. Find the dimensions of the window.

48. Cell Phone Plan A cell phone company charges a monthly fee of $59.99 and a per-minute charge of $0.40 after the first 900 minutes of talk. The monthly cost is given by $C = 59.99 + 0.40(x - 900)$, where x is the number of minutes used.

(a) Solve the equation for x.

(b) How many full minutes can Debbie use in one month on this plan if she does not want to spend more than $100 on her cell phone bill in one month?

49. Concrete Rick has 80 cubic feet of concrete to pour for his new patio. If the patio is rectangular and Rick wants it to be 12 feet by 18 feet, how thick will the patio be if he uses all the concrete?

50. Right Circular Cones The lateral surface area for a frustum of a right circular cone is given by

$$A = \pi s(R + r)$$

where s is the slant height of the frustum, R is the radius of the base, and r is the radius of the top.

(a) Solve the equation for r.

(b) If the frustum of a right circular cone has a lateral surface area of 10π square feet, a slant height of 2 feet, and a base whose radius is 3 feet, what is the radius of the top of the frustum?

51. Heating Bills The winter energy charge C for Illinois Power residential service is computed using the formula $C = 7.48 + 0.08674x$ for x kilowatt-hours (kwh).

(a) Solve the equation for x.

(b) How many kwh were used if the winter energy charge is $206.03? Round to the nearest whole number.

52. Supplementary and Complementary Angles The supplement of an angle and the complement of the same angle sum to 150°. What is the measure of the angle?

Section 1.4 Linear Inequalities in One Variable

KEY CONCEPTS

- **Linear Inequality in One Variable**
  An inequality of the form $ax + b < c$, $ax + b \leq c$, $ax + b > c$, or $ax + b \geq c$, where a, b, and c are real numbers with $a \neq 0$.

- **Interval Notation versus Inequality Notation**

INTERVAL NOTATION	INEQUALITY NOTATION	GRAPH
The open interval (a, b)	$\{x \mid a < x < b\}$	
The closed interval $[a, b]$	$\{x \mid a \leq x \leq b\}$	
The half-open interval $[a, b)$	$\{x \mid a \leq x < b\}$	
The half-open interval $(a, b]$	$\{x \mid a < x \leq b\}$	
The interval $[a, \infty)$	$\{x \mid x \geq a\}$	
The interval (a, ∞)	$\{x \mid x > a\}$	
The interval $(-\infty, a]$	$\{x \mid x \leq a\}$	
The interval $(-\infty, a)$	$\{x \mid x < a\}$	
The interval $(-\infty, \infty)$	$\{x \mid x \text{ is a real number}\}$	

- **Addition Property of Inequalities**
  For real numbers a, b, and c.

  If $a < b$, then $a + c < b + c$.

  If $a > b$, then $a + c > b + c$.

- **Multiplication Properties of Inequalities**
  For real numbers a, b, and c.

  If $a < b$ and $c > 0$, then $ac < bc$.

  If $a > b$ and $c > 0$, then $ac > bc$.

  If $a < b$ and $c < 0$, then $ac > bc$.

  If $a > b$ and $c < 0$, then $ac < bc$.

KEY TERMS

Solve an inequality
Solutions
Solution set
Interval notation
Closed interval
Open interval
Half-open or half-closed interval
Left endpoint
Right endpoint
Equivalent inequalities

You Should Be Able To...	EXAMPLE	Review Exercises
❶ Represent inequalities using the real number line and interval notation (p. 82)	Examples 1 through 4	53–56
❷ Understand the properties of inequalities (p. 85)		57–58
❸ Solve linear inequalities (p. 86)	Examples 5 through 8	59–68
❹ Solve problems involving linear inequalities (p. 88)	Example 9	69–72

In Problems 53 and 54, write each inequality using interval notation and graph the inequality.

53. $2 < x \leq 7$

54. $x > -2$

In Problems 55 and 56, write each interval as an inequality involving x and graph the inequality.

55. $(-\infty, 4]$

56. $[-1, 3)$

In Problems 57 and 58, use the Addition Property and/or the Multiplication Property to find a and b.

57. If $5 \leq x \leq 9$, then $a \leq 2x - 3 \leq b$.

58. If $-2 < x < 0$, then $a < 3x + 5 < b$.

In Problems 59–68, solve each linear inequality. Express your solution using set-builder notation and interval notation. Graph the solution set.

59. $3x + 12 \leq 0$

60. $2 < 1 - 3x$

61. $-7 \leq 3(h+1) - 8$

62. $-7x - 8 < -22$

63. $3(p-2) + (5-p) > 2 - (p-3)$

64. $2(x+1) + 1 > 2(x-2)$

65. $5(x-1) - 7x > 2(2-x)$

66. $0.03x + 0.10 > 0.52 - 0.07x$

67. $-\dfrac{4}{9}w + \dfrac{7}{12} < \dfrac{5}{36}$

68. $\dfrac{2}{5}y - 20 > \dfrac{2}{3}y + 12$

69. Octoberfest The German Club plans to rent a hall for its annual Octoberfest banquet. The hall costs $150 to rent plus $7.50 for each person who attends. If the club does not want to spend more than $600 for the event, how many people can attend the banquet?

70. Car Rentals A Ford Taurus at Enterprise Rent-a-Car rents for $43.46 per day. You receive 150 free miles per day but are charged $0.25 per mile for any additional miles. How many miles can you drive per day, on average, and not exceed your daily budget of $60.00?

71. Fund Raising A middle school band sells $1 candy bars at a carnival to raise money for new instruments. The band pays $50.00 to rent a booth and must pay the candy company $0.60 for each bar sold. How many bars must the band sell to make a profit?

72. Movie Club A DVD club offers unlimited DVDs for $9.95 if you purchase one for $24.95. How many DVDs can you purchase without spending more than $72.00?

Section 1.5   Rectangular Coordinates and Graphs of Equations

KEY CONCEPTS

- **Graph of an Equation in Two Variables**
  The set of all ordered pairs (x, y) in the xy-plane that satisfy the equation

- **Intercepts**
  The points, if any, where a graph crosses or touches the coordinate axes

KEY TERMS

x-axis	Quadrants
y-axis	Equation in two
Origin	variables
Rectangular or Cartesian	Sides
coordinate system	Satisfy
xy-plane	Graph of an equation
Coordinate axes	in two variables
Ordered pair	Point-plotting method
Coordinates	Complete graph
x-coordinate	Intercept
y-coordinate	x-intercept
Abscissa	y-intercept
Ordinate	

You Should Be Able To...	EXAMPLE	Review Exercises
1 Plot points in the rectangular coordinate system (p. 93)	Example 1	73, 74
2 Determine whether an ordered pair is a point on the graph of an equation (p. 95)	Example 2	75, 76
3 Graph an equation using the point-plotting method (p. 97)	Examples 3 through 5	77–82
4 Identify the intercepts from the graph of an equation (p. 99)	Example 6	83
5 Interpret graphs (p. 100)	Example 7	84

In Problems 73 and 74, plot each point in the same xy-plane. Tell in which quadrant or on what coordinate axis each point lies.

73. $A(2, -4)$
$B(-1, -3)$
$C(0, 4)$
$D(-5, 1)$
$E(1, 0)$
$F(4, 3)$

74. $A(3, 0)$
$B(1, 5)$
$C(-3, -5)$
$D(-1, 4)$
$E(5, -2)$
$F(0, -5)$

In Problems 75 and 76, determine whether the given points are on the graph of the equation.

75. $3x - 2y = 7$
  (a) $(3, 1)$
  (b) $(2, -1)$
  (c) $(4, 0)$
  (d) $\left(\dfrac{1}{3}, -3\right)$

76. $y = 2x^2 - 3x + 2$
  (a) $(-1, 3)$
  (b) $(1, 1)$
  (c) $(-2, 16)$
  (d) $\left(\dfrac{1}{2}, \dfrac{3}{2}\right)$

In Problems 77–82, graph each equation by plotting points.

77. $y = x + 2$

78. $2x + y = 3$

79. $y = -x^2 + 4$

80. $y = |x + 2| - 1$

81. $y = x^3 + 2$

82. $x = y^2 + 1$

In Problem 83, the graph of an equation is given. List the intercepts of the graph.

83.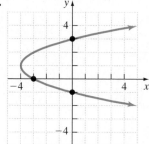

84. Cell Phones A cell phone company offers a plan for $59.99 per month for 900 minutes with additional minutes costing $0.40 per minute. The graph below shows the monthly cost, in dollars, when x minutes are used.

(a) If you talk for 2250 minutes in a month, how much is your monthly bill?

(b) Use the graph to estimate your monthly bill if you talk for 9500 minutes.

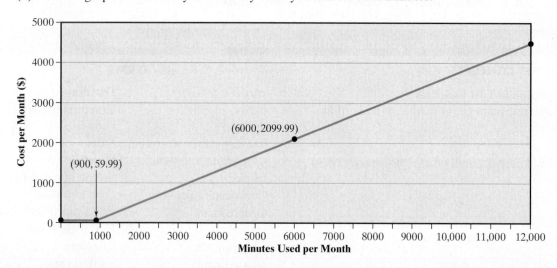

Section 1.6   Linear Equations in Two Variables

KEY CONCEPTS

- **Standard Form of a Line**
  $Ax + By = C$, where A, B, and C are real numbers. A and B are not both 0.

- **Finding Intercepts**
  x-intercept(s): Let $y = 0$ in the equation and solve for x.
  y-intercept(s): Let $x = 0$ in the equation and solve for y.

- **Equation of a Vertical Line**
  $x = a$, where $(a, 0)$ is the x-intercept

- **Equation of a Horizontal Line**
  $y = b$, where $(0, b)$ is the y-intercept

- **Slope of a Line**
  Let $P = (x_1, y_1)$ and $Q = (x_2, y_2)$ be two distinct points. The slope m of the nonvertical line L containing P and Q is defined by the formula

  $$m = \frac{y_2 - y_1}{x_2 - x_1}, \quad x_1 \neq x_2$$

  If $x_1 = x_2$, then L is a vertical line and the slope m of L is undefined (because this results in division by 0).

KEY TERMS

Linear equation
Standard form
Line
Vertical line
Horizontal line
Run
Rise
Slope
Undefined slope
Average rate of change

- When the slope of a line is positive, the line slants upward from left to right.
- When the slope of a line is negative, the line slants downward from left to right.
- When the slope of a line is zero, the line is horizontal.
- When the slope of a line is undefined, the line is vertical.
- **Point-Slope Form of a Line**
  An equation of a nonvertical line of slope m that contains the point (x_1, y_1) is $y - y_1 = m(x - x_1)$.
- **Slope-Intercept Form of a Line**
  An equation of a nonvertical line with slope m and y-intercept $(0, b)$ is $y = mx + b$.

You Should Be Able To...	EXAMPLE	Review Exercises
❶ Graph linear equations using point plotting (p. 104)	Example 1	85–88
❷ Graph linear equations using intercepts (p. 106)	Examples 2 and 3	89–92
❸ Graph vertical and horizontal lines (p. 107)	Examples 4 and 5	93–95
❹ Find the slope of a line given two points (p. 108)	Examples 6 and 7	96–99
❺ Interpret slope as an average rate of change (p. 111)	Example 8	100
❻ Graph a line given a point and its slope (p. 113)	Example 9	101–104
❼ Use the point-slope form of a line (p. 113)	Example 10	105–108
❽ Identify the slope and y-intercept of a line from its equation (p. 114)	Example 11	113–114
❾ Find the equation of a line given two points (p. 115)	Examples 12 and 13	109–112

In Problems 85–88, graph each linear equation by plotting points.

85. $x + y = 7$ **86.** $x - y = -4$

87. $5x - 2y = 6$ **88.** $-3x + 2y = 8$

In Problems 89–92, graph each linear equation by finding its intercepts.

89. $5x + 3y = 30$ **90.** $4x + 3y = 0$

91. $\dfrac{3}{4}x - \dfrac{1}{2}y = 1$ **92.** $4x + y = 8$

In Problems 93–95, graph each linear equation.

93. $x = 4$ **94.** $y = -8$ **95.** $3x + 5 = -1$

In Problems 96 and 97, (a) find the slope of the line and (b) interpret the slope.

96.

97.

In Problems 98 and 99, plot each pair of points and determine the slope of the line containing them. Graph the line.

98. $(-1, 5); (2, -1)$ **99.** $(4, 5); (0, -1)$

100. Illinois's Population The following data represent the population of Illinois between 1940 and 2010.

Year, x	Population, y
1940	7,897,241
1950	8,712,176
1960	10,081,158
1970	11,110,285
1980	11,427,409
1990	11,430,602
2000	12,419,293
2010	12,830,632

Source: *U.S. Census Bureau*

(a) Plot the ordered pairs (x, y) on a graph and connect the points with straight lines.

(b) Compute and interpret the average rate of change in population between 1940 and 1950.

(c) Compute and interpret the average rate of change in population between 1980 and 1990.

(d) Compute and interpret the average rate of change in population between 2000 and 2010.

(e) Based on the results to parts (a), (b), (c), and (d), do you think that population is linearly related to the year? Why?

In Problems 101–104, graph the line that contains the point P and has slope m. Do not find the equation of the line.

101. $m = 4; P(-1, -5)$

102. $m = -\dfrac{2}{3}; P(3, 2)$

103. m is undefined; $P(2, -4)$

104. $m = 0; P(-3, 1)$

In Problems 105 and 106, find an equation of the line. Express your answer in either slope-intercept or standard form, whichever you prefer.

105.

106.

In Problems 107 and 108, find an equation of the line that has the given slope and contains the given point. Express your answer in either slope-intercept or standard form, whichever you prefer.

107. $m = -1; (3, 2)$

108. $m = \dfrac{3}{5}; (-10, -4)$

In Problems 109–112, find an equation of the line containing the given points. Express your answer in either slope-intercept or standard form, whichever you prefer.

109. $(6, 2); (-3, 5)$

110. $(-2, 3); (4, 3)$

111. $(4, -1); (1, -7)$

112. $(-1, 2); (8, -1)$

In Problems 113–114, find the slope and y-intercept of each line. Graph the line.

113. $y = 4x - 6$

114. $2x + 3y = 12$

Section 1.7   Parallel and Perpendicular Lines

KEY CONCEPTS

- **Parallel Lines**
  Two lines are parallel if they have the same slope but different y-intercepts.

- **Slopes of Perpendicular Lines**
  Two lines are perpendicular if the product of their slopes is -1. Alternatively, two lines are perpendicular if their slopes are negative reciprocals of each other.

KEY TERMS

Parallel lines
Perpendicular lines

You Should Be Able To...	EXAMPLE	Review Exercises
❶ Define parallel lines (p. 121)	Example 1	115, 117–120
❷ Find equations of parallel lines (p. 122)	Example 2	121–123
❸ Define perpendicular lines (p. 123)	Examples 3 and 4	116, 117–120
❹ Find equations of perpendicular lines (p. 124)	Example 5	124–126

In Problems 115–116, the slope of a line L is $m = -\dfrac{3}{8}$.

115. Determine the slope of a line that is parallel to *L*.

116. Determine the slope of a line that is perpendicular to *L*.

In Problems 117–120, determine whether the given pairs of linear equations are parallel, perpendicular, or neither.

117. $x - 3y = 9$
$9x + 3y = -3$

118. $6x - 8y = 16$
$3x + 4y = 28$

119. $2x - y = 3$
$-6x + 3y = 0$

120. $x = 2$
$y = 2$

In Problems 121–126, find an equation of the line with the given properties. Express your answer in slope-intercept form, if possible. Graph the lines.

121. Parallel to $y = -2x - 5$ through $(1, 2)$

122. Parallel to $5x - 2y = 8$ through $(4, 3)$

123. Parallel to $x = -3$ through $(1, -4)$

124. Perpendicular to $y = 3x + 7$ through $(6, 2)$

125. Perpendicular to $3x + 4y = 6$ through $(-3, -2)$

126. Perpendicular to $x = 2$ through $(5, -4)$

Section 1.8   Linear Inequalities in Two Variables

KEY CONCEPT

* Linear inequalities in two variables are inequalities in one of the forms

$$Ax + By < C \quad Ax + By > C \quad Ax + By \le C \quad Ax + By \ge C$$

where A, B, and C are real numbers and A and B are not both zero.

KEY TERMS

Half-planes
Satisfied
Graph of a linear inequality in two variables

You Should Be Able To...	EXAMPLE	Review Exercises
❶ Determine whether an ordered pair is a solution to a linear inequality (p. 128)	Example 1	127–128
❷ Graph linear inequalities (p. 130)	Examples 2 and 3	129–134
❸ Solve problems involving linear inequalities (p. 132)	Example 4	135–136

In Problems 127 and 128, determine whether the given points are solutions to the linear inequality.

127. $5x + 3y \le 15$
    (a) $(4, -2)$
    (b) $(-6, 15)$
    (c) $(5, -1)$

128. $x - 2y > -4$
    (a) $(2, 3)$
    (b) $(5, -2)$
    (c) $(-1, 3)$

In Problems 129–134, graph each inequality.

129. $y < 3x - 2$

130. $2x - 4y \le 12$

131. $3x + 4y > 20$

132. $y \ge 5$

133. $2x + 3y < 0$

134. $x > -8$

135. **Entertainment Budget** Ethan's entertainment budget permits him to spend a maximum of $60 per month on movie tickets and MP3 music downloads. Movie tickets cost on average $7.50 each. MP3 music downloads average $2.00 each.

    (a) Write a linear inequality that describes Ethan's options for spending the $60 maximum budget.

    (b) Can Ethan buy 5 movie tickets and 15 music downloads?

    (c) Can Ethan buy 6 movie tickets and 5 music downloads?

136. **Fund Raising** For a fund-raiser, the Math Club agrees to sell candy bars and candles. The club's profit will be 50¢ for each candy bar it sells and $2.00 for each candle it sells. The club needs to earn at least $1000 in order to pay for an upcoming field trip.

    (a) Write a linear inequality that describes the combination of candy bars and candles that must be sold.

    (b) Will selling 500 candy bars and 350 candles earn enough for the trip?

    (c) Will selling 600 candy bars and 400 candles earn enough for the trip?

Chapter 1 Test   *Step-by-step test solutions are found on the Chapter Test Prep Videos available in* MyMathLab®, *on* You Tube®, *or may be accessed using the QR code to the right.*

1. Determine which, if any, of the following are solutions to $3(x - 7) + 5 = x - 4$.
   (a) $x = 6$
   (b) $x = -2$

2. Write the following inequalities in interval notation, and graph them on a real number line.

   (a) $x > -4$
   (b) $3 < x \leq 7$

In Problems 3 and 4, translate the English statement into a mathematical statement. Do not attempt to solve.

3. Three times a number x, decreased by 8, is 4 more than the number x.

4. Two-thirds of a number x, increased by twice the difference of the number x and 5, is more than 7.

In Problems 5 and 6, solve the equation. Determine whether the equation is an identity, a contradiction, or a conditional equation.

5. $5x - (x - 2) = 6 + 2x$

6. $7 + x - 3 = 3(x + 1) - 2x$

In Problems 7–9, solve the inequality and graph the solution set on a real number line.

7. $x + 2 \leq 3x - 4$

8. $4x + 7 > 2x - 3(x - 2)$

9. $-x + 4 \leq x + 3$

10. Solve $7x + 4y = 3$ for y.

11. **Computer Sales** Glen works as a computer salesman and earns $400 weekly plus 8% commission on his weekly sales. If he wants to make at least $750 in a week, how much must his sales be?

12. **Party Costs** A recreational center offers a children's birthday party for $75 plus $5 for each child. How many children were at Payton's birthday party if the total cost for the party was $145?

13. **Sandbox** Rick is building a rectangular sandbox for his daughter. He wants the length of the sandbox to be 2 feet more than the width, and he has 20 feet of lumber to build the frame. Find the dimensions of the sandbox.

14. **Mixture** Two acid solutions are available to a chemist. One is a 10% nitric acid solution, and the other is a 40% nitric acid solution. How much of each type of solution should be mixed together to form 12 liters of a 20% nitric acid solution?

15. **Ironman Race** The last leg of the Ironman competition is a 26.2-mile run. Contestant A runs at a constant rate of 8 miles per hour. If Contestant B starts the run 30 minutes after Contestant A and runs at a constant rate of 10 miles per hour, how long will it take Contestant B to catch up to Contestant A?

16. Plot the following ordered pairs in the same xy-plane. Tell in which quadrant or on what coordinate axis each point lies.

    $A(3, -4), B(0, 2), C(3, 0), D(2, 1), E(-1, -4), F(-3, 5)$

17. Determine whether the ordered pair is a point on the graph of the equation $y = 3x^2 + x - 5$.
    (a) $(-2, 4)$
    (b) $(-1, -3)$
    (c) $(2, 9)$

In Problems 18 and 19, graph the equations by plotting points.

18. $y = 4x - 1$

19. $y = 4x^2$

20. Identify the intercepts from the graph below.

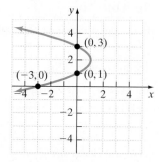

21. The following graph represents the speed of a car over time.

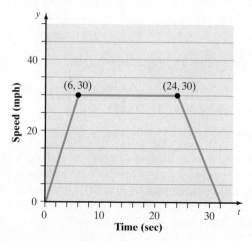

    (a) What is the speed of the car at 6 seconds?

    (b) Identify and interpret the intercepts.

In Problems 22–26, graph each linear equation, using any appropriate method.

22. $x - y = 8$

23. $3x + 5y = 0$

24. $3x + 2y = 12$

25. $\dfrac{3}{2}x - \dfrac{1}{4}y = 1$

26. $x = -7$

27. Find and interpret the slope of the line containing the points $(5, -2)$ and $(-1, 6)$.

28. Draw a graph of the line that contains the point $(2, -4)$ and has a slope of $-\dfrac{3}{5}$. Do not find the equation of the line.

29. Determine whether the graphs of the following pair of linear equations are parallel, perpendicular, or neither.

$$8x - 2y = 1$$
$$x + 4y = -2$$

In Problems 30–33, find the equation of the line with the given properties. Express your answer in either slope-intercept or standard form, whichever you prefer.

30. Through the point $(-3, 1)$ and having a slope of 4

31. Through the points $(6, 1)$ and $(-3, 7)$

32. Parallel to $x - 5y = 15$ and through the point $(5, -1)$

33. Perpendicular to $3x - y = 4$ and through the point $(6, 2)$

34. Determine whether the given points are solutions to the linear inequality $3x - y > 10$.

(a) $(3, -1)$ (b) $(4, 5)$ (c) $(5, 3)$

In Problems 35 and 36, graph each linear inequality.

35. $y \le -2x + 1$ **36.** $5x - 2y < 0$

37. Area of a Circle The following data show the relationship between the diameter of a circle and the area of that circle.

Diameter (feet), x	Area (square feet), y
1	0.79
3	7.07
6	28.27
7	38.48
8	50.27
10	78.54
13	132.73

(a) Plot the ordered pairs (x, y) on a graph and connect the points with straight lines.

(b) Compute and interpret the average rate of change in area between diameter lengths of 1 and 3 feet.

(c) Compute and interpret the average rate of change in area between diameter lengths of 10 and 13 feet.

(d) Based upon the results to parts (a), (b), and (c), do you think that the area of the circle is linearly related to the diameter? Why?

Cumulative Review Chapters R–1

1. Approximate each number by (i) truncating and (ii) rounding to the indicated number of decimal places.

(a) 27.2357; three decimal places.

(b) 1.0729; one decimal place.

2. Plot the points $-4, -\dfrac{5}{2}, 0$, and $\dfrac{7}{2}$ on a real number line.

In Problems 3–8, evaluate the expressions.

3. $-|-14|$

4. $-3 + 4 - 7$

5. $\dfrac{-3(12)}{-6}$

6. $(-3)^4$

7. $5 - 2(1 - 4)^3 + 5 \cdot 3$

8. $\dfrac{2}{3} + \dfrac{1}{2} - \dfrac{1}{4}$

9. Evaluate $3x^2 + 2x - 7$ when $x = 2$.

10. Simplify: $4a^2 - 6a + a^2 - 12 + 2a - 1$

11. Determine whether the given values are in the domain of x for the expression $\dfrac{x + 3}{x^2 + x - 2}$.

(a) $x = -2$ (b) $x = 0$

12. Use the Distributive Property to remove parentheses and then simplify: $3(x + 2) - 4(2x - 1) + 8$

13. Determine whether $x = 3$ is a solution to the equation $x - (2x + 3) = 5x - 1$.

In Problems 14 and 15, solve the equation.

14. $4x - 3 = 2(3x - 2) - 7$

15. $\dfrac{x + 1}{3} = x - 4$

16. Solve $2x - 5y = 6$ for y.

In Problems 17 and 18, solve the inequality and graph the solution set on a real number line.

17. $\dfrac{x + 3}{2} \le \dfrac{3x - 1}{4}$

18. $5(x - 3) \ge 7(x - 4) + 3$

19. Plot the following ordered pairs in the same Cartesian plane.

$$A(-3, 0), \quad B(4, -2), \quad C(1, 5),$$
$$D(0, 3), \quad E(-4, -5), \quad F(-5, 2)$$

In Problems 20 and 21, graph the linear equation using the method you prefer.

20. $y = -\dfrac{1}{2}x + 4$ **21.** $4x - 5y = 15$

In Problems 22 and 23, find the equation of the line with the given properties. Express your answer in either slope-intercept or standard form, whichever you prefer.

22. Through the points $(3, -2)$ and $(-6, 10)$

23. Parallel to $y = -3x + 10$ and through the point $(-5, 7)$

24. Graph $x - 3y > 12$.

25. Computing Grades Shawn really wants an A in his geometry class. His four exam scores are 94, 95, 90, and 97. The final exam is worth two exam scores. To have an A, his average must be at least 93. For what range of scores on the final exam will Shawn be able to earn an A in the course?

26. Body Mass Index The body mass index (BMI) of a person who is 62 inches tall and weighs x pounds is given by $0.2x - 2$. A person with a BMI of 30 or more is considered obese. For what weights would a person 62 inches tall be considered obese?

27. Supplementary Angles Two angles are supplementary. The measure of the larger angle is 15 degrees more than twice the measure of the smaller angle. Find the angle measures.

28. Cylinders Max has 100 square inches of aluminum with which to make a closed cylinder. If the radius of the cylinder must be 2 inches, how tall will the cylinder be? (Round to the nearest hundredth of an inch.)

29. Consecutive Integers Find three consecutive even integers such that the sum of the first two is 22 more than the third.

2 Relations, Functions, and More Inequalities

Your cell phone bill depends on the cost of the data plan you choose. The revenue a Brownie troop earns from selling Girl Scout cookies depends on the number of boxes of cookies they sell. The area of a circle depends on the length of its radius. These three relations are examples of *functions*. Did you know that your weekly earnings are a function of the number of hours you work? See Problem 87 in Section 2.2.

The Big Picture: Putting It Together

In Chapter 1, linear equations in two variables were reviewed. The solution of a linear equation in two variables was graphed using the rectangular coordinate system. This chapter begins with the concept of a function, probably the single most important concept of algebra. Then it discusses compound inequalities, absolute value equations and inequalities, and the graphs of solutions to these equations and inequalities.

Outline

2.1 Relations

Objectives

① Understand Relations
② Find the Domain and the Range of a Relation
③ Graph a Relation Defined by an Equation

Are You Prepared for This Section?

Before getting started, complete the following problems. If you get a problem wrong, go back to the section cited and review the material.

P1. Write the inequality $-4 \le x \le 4$ in interval notation.   [Section 1.4, pp. 82–85]

P2. Write the interval $[2, \infty)$ using an inequality.   [Section 1.4, pp. 82–85]

P3. Plot the ordered pairs $(-2, 4)$, $(3, -1)$, $(0, 5)$, and $(4, 0)$ in the rectangular coordinate system.   [Section 1.5, pp. 93–95]

P4. Graph the equation: $2x + 5y = 10$ [Section 1.6, pp. 104–107]

P5. Graph the equation $y = x^2 - 3$ by plotting points.   [Section 1.5, pp. 97–99]

▶ ① Understand Relations

When the value of one variable is related to the value of a second variable, it is called a *relation*. For example, an individual's level of education is related to annual income. Engine size is related to gas mileage.

> **Definition**
>
> A **relation** exists when the elements in one set are associated with elements in a second set. If x and y are two distinct elements in these sets, and if a relation exists between x and y, then x **corresponds** to y, or y **depends on** x, which is written as $x \rightarrow y$. A relation where y depends on x may also be written as an ordered pair (x, y).

EXAMPLE 1 **Illustrating a Relation**

The data presented in Figure 1 show a correspondence between states and senators in 2016 for randomly selected senators. A possible name for the relation might be "is represented in the U.S. Senate by." Thus the relation reveals that "Indiana is represented by Dan Coats," for example. In Figure 1, **mapping** is used to represent the relation by drawing an arrow from an element in the set "state" to an element in the set "senator."

Figure 1

This relation could also be represented by using ordered pairs in the form (state, senator) as follows:

{ (Indiana, Dan Coats), (Washington, Maria Cantwell),
(North Carolina, Richard Burr), (Oregon, Ron Wyden),
(Oregon, Jeff Merkley), (Maine, Susan Collins) }

Prepared?...Answers **P1.** $[-4, 4]$
P2. $x \ge 2$
P3.

P4.

P5.

Quick ✓

1. If a relation exists between x and y, then say that x _____ to y or that y _____ on x, and write $x \rightarrow y$.

2. Use the map to represent the relation as a set of ordered pairs.

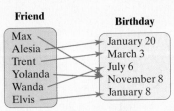

3. Use the set of ordered pairs to represent the relation as a map.

$$\{(1,3),(5,4),(8,4),(10,13)\}$$

❷ Find the Domain and the Range of a Relation

▶ In a relation we say that y depends on x and can write the relation as a set of ordered pairs (x, y). The set of all x can be thought of as the **inputs** of the relation. The set of all y can be thought of as the **outputs** of the relation. This interpretation of a relation can be used to define *domain* and *range*.

> **Definition**
> The **domain** of a relation is the set of all inputs of the relation. The **range** is the set of all outputs of the relation.

(**EXAMPLE 2**)　**Finding the Domain and the Range of a Relation**

Find the domain and the range of the relation presented in Figure 1 from Example 1.

Solution
The domain of the relation is the set of all inputs, therefore, the domain of the relation is

{ Indiana, Washington, North Carolina, Oregon, Maine }

The range of the relation is the set of all outputs. Therefore, the range of the relation is

{ Dan Coats, Maria Cantwell, Richard Burr, Ron Wyden,
Jeff Merkley, Susan Collins }　　　　　●

Work Smart
Never list elements in the domain or range more than once.

　　Notice that Oregon was not listed twice in the domain. The domain and the range are sets, and elements in a set should not be listed more than once. Also, the order in which the elements in the domain or range are listed does not matter.

Quick ✓
4. The _____ of a relation is the set of all inputs of the relation. The _____ is the set of all outputs of the relation.

5. State the domain and the range of the relation.

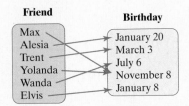

6. State the domain and the range of the relation.

$$\{(1,3),(5,4),(8,4),(10,13)\}$$

Relations can also be represented by plotting a set of ordered pairs. The set of all x-coordinates represents the domain of the relation, and the set of all y-coordinates represents the range of the relation.

EXAMPLE 3 **Finding the Domain and the Range of a Relation**

Figure 2 shows the graph of a relation. Identify the domain and the range of the relation.

Figure 2

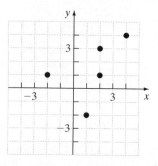

Work Smart

Write the points as ordered pairs to assist in finding the domain and range.

Solution

The ordered pairs in the graph are $(-2, 1)$, $(1, -2)$, $(2, 1)$, $(2, 3)$, and $(4, 4)$, so the domain is the set of all x-coordinates: $\{-2, 1, 2, 4\}$. The range is the set of all y-coordinates: $\{-2, 1, 3, 4\}$.

Quick ✓

7. Identify the domain and the range of the relation shown in the figure.

A third way to define a relation is by a graph. Remember that the graph of an equation is the set of all ordered pairs (x, y) that make the equation a true statement. If a graph exists for some ordered pair (x, y), then the x-coordinate is in the domain and the y-coordinate is in the range.

EXAMPLE 4 **Identifying the Domain and the Range of a Relation from Its Graph**

Figure 3 shows the graph of a relation. Determine the domain and the range of the relation.

Figure 3

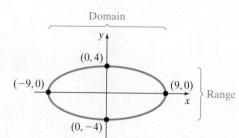

Work Smart

You can think of the domain as the shadow created by the graph on the *x*-axis by vertical beams of light. The range can be thought of as the shadow created by the graph on the *y*-axis by horizontal beams of light.

Solution

The domain of the relation consists of all *x*-coordinates at which the graph exists. The graph exists at all *x*-values between -9 and 9, inclusive. Therefore, the domain is $\{x \mid -9 \le x \le 9\}$ or, using interval notation, $[-9, 9]$.

The range of the relation consists of all *y*-coordinates at which the graph exists. The graph exists at all *y*-values between -4 and 4, inclusive. Therefore, the range is $\{y \mid -4 \le y \le 4\}$ or, using interval notation, $[-4, 4]$. ●

Quick ✓

8. *True or False* If the graph of a relation does not exist at $x = 3$, then 3 is not in the domain of the relation.

9. *True or False* The range of a relation is always the set of all real numbers.

In Problems 10 and 11, identify the domain and range of the relation from its graph.

10.

11.

Work Smart

Relations can be defined by
1. Mapping
2. Sets of ordered pairs
3. Graphs
4. Equations

 ❸ **Graph a Relation Defined by an Equation**

Another way to define a relation (instead of a map, a set of ordered pairs, or a graph) is to use equations such as $x + y = 4$ or $x = y^2$. When relations are defined by equations, graph the relation in order to visualize how the variables are related. As seen in Example 4, the graph of the relation also helps identify its domain and range.

EXAMPLE 5 **Relations Defined by Equations**

Graph the relation $y = -x^2 + 4$. Find its domain and range using the graph.

Solution

The relation says to take the input *x*, square it, multiply this result by -1, and then add 4 to get the output *y*. Use the point-plotting method to graph the relation. Table 1 shows some points on the graph. Figure 4 shows a graph of the relation.

Figure 4

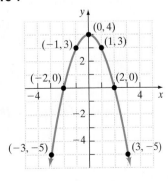

Table 1

x	$y = -x^2 + 4$	(x, y)
-3	$-(-3)^2 + 4 = -5$	$(-3, -5)$
-2	$-(-2)^2 + 4 = 0$	$(-2, 0)$
-1	3	$(-1, 3)$
0	4	$(0, 4)$
1	3	$(1, 3)$
2	0	$(2, 0)$
3	-5	$(3, -5)$

(continued)

The graph extends indefinitely to the left and to the right (that is, the graph exists for all x-values). Therefore, the domain of the relation is the set of all real numbers, or $\{x \mid x$ is any real number$\}$, or, using interval notation, $(-\infty, \infty)$. Notice that there are no y-values greater than 4, but the graph exists everywhere for y-values less than or equal to 4. The range of the relation is $\{y \mid y \leq 4\}$ or, using interval notation, $(-\infty, 4]$. ●

Quick ✓

In Problems 12–14, graph each relation. Use the graph to identify the domain and range.

12. $y = 3x - 8$ **13.** $y = x^2 - 8$ **14.** $x = y^2 + 1$

2.1 Exercises MyMathLab®

Exercise numbers in green have complete video solutions in MyMathLab or may be accessed using the QR code to the right.

Problems 1–14 are the Quick ✓s that follow the EXAMPLES.

Building Skills

In Problems 15–18, write each relation as a set of ordered pairs. Then identify the domain and the range of the relation. See Objectives 1 and 2.

15.

SOURCE: *Information Please Almanac*

16.

SOURCE: *wholesaleinsurance.net*

17.

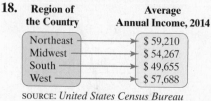

SOURCE: *United States Census Bureau*

18.

SOURCE: *United States Census Bureau*

In Problems 19–24, write each relation as a map. Then identify the domain and the range of the relation. See Objectives 1 and 2.

19. $\{(-3, 4), (-2, 6), (-1, 8), (0, 10), (1, 12)\}$

20. $\{(-2, 6), (-1, 3), (0, 0), (1, -3), (2, 6)\}$

21. $\{(-2, 4), (-1, 2), (0, 0), (1, 2), (2, 4)\}$

22. $\{(-2, -8), (-1, -1), (0, 0), (1, 1), (2, 8)\}$

23. $\{(0, -4), (-1, -1), (-2, 0), (-1, 1), (0, 4)\}$

24. $\{(-3, 0), (0, 3), (3, 0), (0, -3)\}$

In Problems 25–32, identify the domain and the range of the relation from the graph. See Objective 2.

25.

26.

27.

28.

29.

30.

31.

32.

In Problems 33–54, graph the relation to identify the domain and the range of the relation. Note: These relations were graphed in Problems 33–54 in Section 1.5. See Objective 3.

33. $y = -3x + 1$

34. $y = -4x + 2$

35. $y = \dfrac{1}{2}x - 4$

36. $y = -\dfrac{1}{2}x + 2$

37. $2x + y = 7$

38. $3x + y = 9$

39. $y = -x^2$

40. $y = x^2 - 2$

41. $y = 2x^2 - 8$

42. $y = -2x^2 + 8$

43. $y = |x|$

44. $y = |x| - 2$

45. $y = |x - 1|$

46. $y = -|x|$

47. $y = x^3$

48. $y = -x^3$

49. $y = x^3 + 1$

50. $y = x^3 - 2$

51. $x^2 - y = 4$

52. $x^2 + y = 5$

53. $x = y^2 - 1$

54. $x = y^2 + 2$

Applying the Concepts

55. Area of a Window Chip Gaines wishes to put a new window in his home. He wants the perimeter of the window to be 100 feet. The graph in the next column shows the relation between the width, x, of the opening (in feet) and the area of the opening.

(a) Determine the domain and the range of the relation.

(b) Explain why the domain obtained in part (a) makes sense.

56. Projectile Motion The graph below shows the height, in feet, of a ball thrown straight up with an initial speed of 80 feet per second from an initial height of 96 feet after t seconds. Determine the domain and the range of the relation.

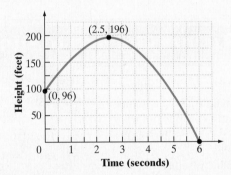

57. Cell Phones The graph below shows the relation between the monthly cost, C, of a cell phone and the number of anytime minutes used, m.

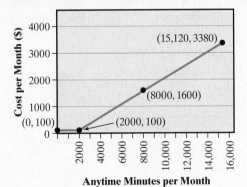

(a) Determine the domain and the range of the relation.

(b) If anytime minutes are from 7:00 A.M. to 7:00 P.M. Monday through Friday, explain why the domain obtained in part (a) makes sense, assuming there are 21 nonweekend days per month.

58. Wind Chill It is 10° Celsius outside. The wind is calm but then gusts up to 20 meters per second. You feel the chill go right through your bones. The graph below shows the relation between the wind chill temperature (in degrees Celsius) and wind speed, v (in meters per second). Determine the domain and the range of the relation.

Extending the Concepts

59. Draw the graph of a relation whose domain is all real numbers, but whose range is a single real number. Compare your graph with those of your classmates. How are they similar?

60. Draw the graph of a relation whose domain is a single real number, but whose range is all real numbers. Compare your graph with those of your classmates. How are they similar?

Explaining the Concepts

61. Explain what a relation is. Be sure to include an explanation of domain and range.

62. State the four methods for describing a relation that are presented in this section. When is using ordered pairs most appropriate? When is using a graph most appropriate? Support your opinion.

2.2 An Introduction to Functions

Objectives

❶ Determine Whether a Relation Expressed as a Map or Ordered Pairs Represents a Function

❷ Determine Whether a Relation Expressed as an Equation Represents a Function

❸ Determine Whether a Relation Expressed as a Graph Represents a Function

❹ Find the Value of a Function

❺ Find the Domain of a Function

❻ Work with Applications of Functions

Are You Prepared for This Section?

Before getting started, complete the following problems. If you get a problem wrong, go back to the section cited and review the material.

P1. Evaluate the expression $2x^2 - 5x$ for

    (a) $x = 1$ **(b)** $x = 4$ **(c)** $x = -3$ [Section R.5, pp. 40–41]

P2. Evaluate $\dfrac{3}{2x + 1}$ for $x = -\dfrac{1}{2}$.   [Section R.5, pp. 40–41]

P3. Express the inequality $x \le 5$ using interval notation.   [Section 1.4, pp. 82–85]

P4. Express the interval $(2, \infty)$ using set-builder notation.   [Section 1.4, pp. 82–85]

▶ ❶ **Determine Whether a Relation Expressed as a Map or Ordered Pairs Represents a Function**

One of the most important concepts in algebra is now presented—the *function*. A function is a special type of relation. To understand functions, let's revisit the relation in Example 1 from Section 2.1, shown again in Figure 5. In this correspondence between states and their senators, we named the relation "is represented in the U.S. Senate by."

Figure 5

In this relation, if you were asked to name the senator who represents Oregon, you could respond, "Ron Wyden" or "Jeff Merkley." In other words, the input "state" does not correspond to a single output "senator."

Let's consider the relation in Figure 6(a) on the next page, a correspondence between states and their populations. If asked for the population that corresponds

Prepared?...Answers P1. (a) -3
(b) 12 **(c)** 33 **P2.** undefined
P3. $(-\infty, 5]$ **P4.** $\{x \mid x > 2\}$

to North Carolina, you could only respond, "10,043 thousand." In other words, each input "state" corresponds to exactly one output "population."

Figure 6(b) is a relation that shows a correspondence between animals and life expectancies. If asked to determine the life expectancy of a dog you could only respond, "11 years." You would have the same answer about the life expectancy of a cat.

Figure 6

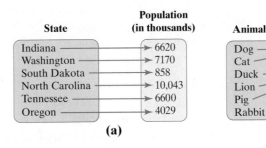

(a) (b)

What do the relations in Figures 6(a) and 6(b) have in common that is missing from the relation in Figure 5? In the relations in Figure 6(a) and 6(b), each input corresponds to only one output. In Figure 5, however, the input Oregon corresponds to two outputs—Ron Wyden and Jeff Merkley. This leads to the definition of a *function*.

> **In Other Words**
> For a relation to be a function, each input may have only one output.

Definition

A **function** is a relation in which each input, or element in the domain of the relation, corresponds to exactly one output, or element in the range of the relation.

The idea behind functions is predictability. If an input is known, a function can be used to determine the output with 100% certainty, as seen in Figures 6(a) and (b). Nonfunctions do not have this predictability (Figure 5).

EXAMPLE 1   **Determining Whether a Relation Represents a Function**

Determine whether the relation is a function. If so, state its domain and range.

(a) See Figure 7(a). The domain represents the length (mm) of the right humerus, and the range represents the length (mm) of the right tibia for each of five rats that had been sent to space.

(b) See Figure 7(b). The domain represents the age of six people, and the range represents their height.

(c) See Figure 7(c). The domain represents the age of five males, and the range represents their HDL (or "good") cholesterol (mg/dL).

Figure 7

(a)

SOURCE: *NASA Life Sciences Data Archive*

(b)

SOURCE: *diamonds.com*

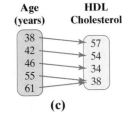

(c)

Solution

(a) The relation in Figure 7(a) is a function because each element in the domain corresponds to exactly one element in the range. The domain is { 24.80, 24.59, 24.29, 23.81, 24.87 }. The range is { 36.05, 35.57, 34.58, 34.20, 34.73 }.

(continued)

(b) The relation in Figure 7(b) is not a function because an element in the domain, 0.86, corresponds to two elements in the range. A single price cannot be determined for the 0.86-carat diamonds.

(c) The relation in Figure 7(c) is a function because each element in the domain corresponds to exactly one element in the range. Notice that it is okay for more than one element in the function's domain to correspond to the same element in the range. The domain of the function is { 38, 42, 46, 55, 61 }. The range of the function is { 57, 54, 34, 38 }.   ●

Quick ✓

1. A _____ is a relation in which each element in the domain of the relation corresponds to exactly one element in the range of the relation.

2. *True or False*   Every relation is a function.

In Problems 3 and 4, determine whether the relation is a function. If so, state its domain and range.

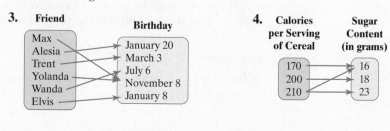

3. Friend / Birthday
   Max, Alesia, Trent, Yolanda, Wanda, Elvis → January 20, March 3, July 6, November 8, January 8

4. Calories per Serving of Cereal / Sugar Content (in grams)
   170, 200, 210 → 16, 18, 23

Work Smart

All functions are relations, but not all relations are functions!

A function can also be thought of as a set of ordered pairs (x, y) in which no ordered pairs have the same x-coordinate and different y-coordinates.

EXAMPLE 2   **Determining Whether a Relation Represents a Function**

Determine whether the relation is a function. If so, state its domain and range.

(a) $\{ (1, 3), (-1, 4), (0, 6), (2, 8) \}$
(b) $\{ (-2, 6), (-1, 3), (0, 2), (1, 3), (2, 6) \}$
(c) $\{ (0, 3), (1, 4), (4, 5), (9, 5), (4, 1) \}$

Solution

(a) This relation is a function because no ordered pairs have the same x-coordinate and different y-coordinates. The domain of the function is the set of all x-coordinates, $\{ -1, 0, 1, 2 \}$. The range of the function is the set of all y-coordinates, $\{ 3, 4, 6, 8 \}$.

(b) This relation is a function because no ordered pairs have the same x-coordinate and different y-coordinates. The domain is, $\{ -2, -1, 0, 1, 2 \}$. The range is, $\{ 2, 3, 6 \}$.

(c) This relation is not a function because two ordered pairs, $(4, 5)$ and $(4, 1)$, have the same x-coordinate and different y-coordinates.   ●

Work Smart

In a function, two different inputs can correspond to the same output, but two different outputs cannot be the result of a single input.

In Example 2(b), notice that inputs -2 and 2 each correspond to output 6. This does not violate the definition of a function – two different x-coordinates can have the same y-coordinate. A violation of the definition occurs when two ordered pairs have the same x-coordinate and different y-coordinates, as in Example 2(c).

Quick ✔

In Problems 5 and 6, determine whether each relation is a function. If so, state its domain and range.

5. $\{(-3, 3), (-2, 2), (-1, 1), (0, 0), (1, 1)\}$

6. $\{(-3, 2), (-2, 5), (-1, 8), (-3, 6)\}$

▶ ❷ Determine Whether a Relation Expressed as an Equation Represents a Function

We now know how to identify when a relation defined by a map or ordered pairs is a function. In Section 2.1, relations were also expressed as equations. We will now address the circumstances under which equations are functions.

To determine whether an equation, where y depends on x, is a function, it is often easiest to solve the equation for y. If each value of x corresponds to exactly one value of y, the equation is a function; otherwise, it is not.

EXAMPLE 3 | **Determining Whether an Equation Is a Function**

Determine whether the equation $y = 3x + 5$ shows y as a function of x.

Solution
The rule for getting from x to y is to multiply x by 3 and then add 5. Since only one output y can result by performing these operations on any input x, the equation is a function. ●

EXAMPLE 4 | **Determining Whether an Equation Is a Function**

Determine whether the equation $y = \pm x^2$ shows y as a function of x.

> **In Other Words**
> The symbol $\pm$ is a shorthand device and is read "plus or minus." For example, ± 4 means "negative four or positive four."

Solution
Notice that for any single value of x (other than 0), two values of y result. For example, if $x = 2$, then $y = \pm 4$ (-4 or $+4$). Since a single x corresponds to more than one y, the equation is not a function. ●

Quick ✔

In Problems 7–9, determine whether each equation shows y as a function of x.

7. $y = -2x + 5$ **8.** $y = \pm 3x$ **9.** $y = x^2 + 5x$

▶ ❸ Determine Whether a Relation Expressed as a Graph Represents a Function

Remember that the graph of an equation is the set of all ordered pairs (x, y) that satisfy the equation. For a relation to be a function, each number x in the domain can correspond to only one number y in the range. This means that a graph is *not* a function if two points with the same x-coordinate have different y-coordinates. This leads to the following test.

Vertical Line Test

A set of points in the xy-plane is the graph of a function if and only if every vertical line intersects the graph in at most one point.

EXAMPLE 5   **Using the Vertical Line Test to Identify Graphs of Functions**

Which of the graphs in Figure 8 are graphs of functions?

Figure 8

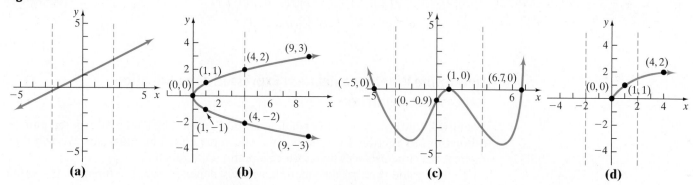

(a)   (b)   (c)   (d)

Solution

The graphs in Figures 8(a), (c), and (d) are functions because every vertical line intersects these graphs in at most one point. The graph in Figure 8(b) is not a function, because a vertical line intersects the graph in more than one point. ●

Does Example 5 show you why the vertical line test works? If a vertical line intersects the graph of an equation in two or more points, then the same x-coordinate corresponds to two or more different y-coordinates, and we have violated the definition of a function.

Quick ✓

10. *True or False* For a graph to be a function, any vertical line can intersect the graph in at most one point.

In Problems 11 and 12, use the vertical line test to determine whether the graph is a function.

11.   12.

④ Find the Value of a Function

▶ Functions are often denoted by letters such as f, F, g, G, and so on. If f is a function, use special notation, called *function notation*, to represent the function. The following table shows some equations where y is a function of x. In each case, denote the function by the letter f.

Equation in Two Variables	Function Notation
$y = \dfrac{1}{2}x + 5$	$f(x) = \dfrac{1}{2}x + 5$
$y = -3x + 1$	$f(x) = -3x + 1$
$y = 2x^2 - 3x$	$f(x) = 2x^2 - 3x$

Work Smart

Be careful with function notation. In the expression $y = f(x)$, y is the dependent variable, x is the independent variable, and f is the name given to a rule that relates the input x to the output y.

Notice that $f(x)$ replaces the variable y. If f is a function, then for each number x in its domain, the corresponding value in the range is denoted by $f(x)$, which is read as "f of x" or as "f at x." Understand that $f(x)$ does not mean "f times x."

We call $f(x)$ the **value of the function f at the number x**; $f(x)$ is the number that results when the function is applied to x. Finding the value of a function parallels finding the value of y in an equation in two variables when y is a function of x. For example, "Given the equation $y = \dfrac{1}{2}x + 5$, find the value of y when $x = 4$" means the same as "Given the function $f(x) = \dfrac{1}{2}x + 5$, evaluate $f(4)$." In both cases, replace the value of x with 4 and evaluate.

EXAMPLE 6   **Finding Values of a Function**

If $f(x) = \dfrac{3}{2}x - 6$ and $g(x) = x^2 + 6x$, evaluate:

(a) $f(10)$ **(b)** $f(-8)$ **(c)** $g(3)$ **(d)** $g(-2)$

Solution

(a) Wherever an x is in the equation defining the function f, substitute 10.

$$f(10) = \frac{3}{2}(10) - 6$$
$$= 15 - 6$$
$$= 9$$

So $f(10) = 9$.

(b) Substitute -8 for x in the expression $\dfrac{3}{2}x - 6$ to get

$$f(-8) = \frac{3}{2}(-8) - 6$$
$$= -12 - 6$$
$$= -18$$

So $f(-8) = -18$.

(c) Substitute 3 for x in the expression $x^2 + bx$ to get.

$$g(3) = (3)^2 + 6(3)$$
$$= 9 + 18$$
$$= 27$$

Work Smart

Notice the use of parentheses when -2 is squared in Example 6(d) Remember, $-2^2 \neq (-2)^2$.

(d) Substitute -2 for x in the expression $x^2 + 6x$ to get

$$g(-2) = (-2)^2 + 6(-2)$$
$$= 4 + (-12)$$
$$= -8$$

Quick ✓

In Problems 13–16, let $f(x) = 3x + 2$ and $g(x) = -2x^2 + x + 3$. Evaluate each function.

13. $f(4)$ **14.** $f(-2)$ **15.** $g(-3)$ **16.** $g(1)$

For a function $y = f(x)$, the variable x is called the **independent variable**, because it can be assigned any of the numbers in the domain. The variable y is called the **dependent variable**, because its value depends on x.

The independent variable is also called the **argument** of the function. For example, if f is the function defined by $f(x) = x^2$, then f tells us to square the argument. Thus, $f(2)$ means to square 2, $f(a)$ means to square a, and $f(x + h)$ means to square the quantity $x + h$.

The notation $f(x)$ plays a dual role—it represents the rule for getting from the input to the output and it represents the output y of the function. In Example 6(c), the rule for getting from the input to the output is given by $g(x) = x^2 + 6x$. In other words, the function says to "take some input x, square it, and add the result to six times the input x." If the input is 3, then $g(3)$ represents the output, 27.

▷ (**EXAMPLE 7**) **Evaluate a Function**

For the function $F(z) = 4z + 7$, evaluate:

(a) $F(z + 3)$ **(b)** $F(z) + F(3)$

Solution

(a) Wherever a z is in the equation defining F, substitute $z + 3$ to get

$$F(z + 3) = 4(z + 3) + 7$$
$$= 4z + 12 + 7$$
$$= 4z + 19$$

(b) $F(z) + F(3) = \underbrace{4z + 7}_{F(z)} + \underbrace{4 \cdot 3 + 7}_{F(3)}$

$$= 4z + 7 + 12 + 7$$
$$= 4z + 26$$

●

Quick ✓

17. In the function $H(q) = 2q^2 - 5q + 1$, H is called the _____ variable, and q is called the _____ variable or _____.

In Problems 18 and 19, let $f(x) = 2x - 5$. Evaluate each function.

18. $f(x - 2)$ **19.** $f(x) - f(2)$

Summary **Important Facts About Functions**

1. Each x in the domain has exactly one corresponding y in the range.
2. Letters such as f are used to denote the function. It represents the rule used to get from an x in the domain to $f(x)$ in the range.
3. If $y = f(x)$, then x is the independent variable or argument of f, and y is the dependent variable or the value of f at x.

▷ **⑤ Find the Domain of a Function**

In working with functions, it is important to know which inputs are possible and produce an output.

In Other Words

The domain of a function is the set of all inputs for which the function gives an output that is a real number.

Definition

When only the equation of a function is given, the **domain of f** is the set of real numbers x for which $f(x)$ is a real number.

When identifying the domain of a function, do not forget that division by zero is undefined. Exclude values of the variable that result in division by zero.

<table>
<tr><td>EXAMPLE 8</td></tr>
</table>

EXAMPLE 8   **Finding the Domain of a Function**

Find the domain of each of the following functions:

(a) $G(x) = x^2 + 1$ **(b)** $g(z) = \dfrac{z - 3}{z + 1}$

Solution

Work Smart

In Example 8(b), it is understood that $\{z \mid z \neq -1\}$ includes all real numbers except -1.

(a) The function G squares a number x and then adds 1 to the result. These operations can be performed on any real number, so the domain of G is the set of all real numbers, which we can express as $\{x \mid x \text{ is a real number}\}$ or, using interval notation, $(-\infty, \infty)$.

(b) The function g involves division. Since division by 0 is not defined, the denominator $z + 1$ cannot be 0. Therefore, z cannot equal -1. The domain of g is $\{z \mid z \neq -1\}$.   ●

Quick ✓

20. When only the equation of a function f is given, the _____ of f is the set of real numbers x for which $f(x)$ is a real number.

In Problems 21 and 22, find the domain of each function.

21. $f(x) = 3x^2 + 2$ **22.** $h(x) = \dfrac{x + 1}{x - 3}$

In an application, a function's domain may be restricted by physical or geometric considerations, in addition to mathematical restrictions. For example, the domain of $f(x) = x^2$ is the set of all real numbers. However, if f is used to find the area of a square given side x, then restrict the domain of f to the positive real numbers, since the length of a side cannot be 0 or negative.

EXAMPLE 9   **Finding the Domain of a Function**

The number N of computers produced at a Dell Computers' manufacturing facility in one day after t hours is given by the function, $N(t) = 336t - 7t^2$. What is the domain of this function?

Solution

The independent variable is t, the number of hours in the day. Therefore the function's domain is $\{t \mid 0 \leq t \leq 24\}$, or the interval $[0, 24]$.   ●

Quick ✓

23. The function $A(r) = \pi r^2$ gives the area of a circle A as a function of the radius r. What is the domain of the function?

▶ ⑥ **Work with Applications of Functions**

In practice, the symbols used for the independent and dependent variables should remind us of what they represent. For example, in economics, C is used for cost and q for quantity, so that $C(q)$ represents the cost of manufacturing q units of a good. Here C is the dependent variable, q is the independent variable, and $C(q)$ is the rule that tells us how to get the output C from the input q.

EXAMPLE 10   **Life-Cycle Hypothesis**

The Life-Cycle Hypothesis from economics was presented by Franco Modigliani in 1954. It states that income is a function of age. The function

$$I(a) = -55a^2 + 5119a - 54{,}448$$

represents the relation between average annual income I and age a.

(a) Identify the dependent and independent variables.

(b) Evaluate $I(20)$ and explain what $I(20)$ represents.

Solution

(a) Because income depends on age, the dependent variable is income, I, and the independent variable is age, a.

(b) Let $a = 20$ in the function.

$$I(20) = -55(20)^2 + 5119(20) - 54{,}448$$
$$= 25{,}932$$

The average annual income of a 20-year-old is $25,932.

Quick ✓

24. In 2010, the *Deepwater Horizon* oil explosion spilled millions of gallons of oil in the Gulf of Mexico. The oil slick takes the shape of a circle. Suppose that the area A (in square miles) of the circle contaminated with oil can be determined using the function $A(t) = 0.25\pi t^2$, where t represents the number of days since the rig exploded.

(a) Identify the dependent and independent variables.

(b) Evaluate $A(30)$ and explain what $A(30)$ means.

2.2 Exercises MyMathLab®

Exercise numbers in green have complete video solutions in MyMathLab or may be accessed using the QR code to the right.

*Problems **1–24** are the Quick ✓s that follow the EXAMPLES.*

Building Skills

In Problems 25–34, determine whether each relation represents a function. State the domain and the range of each relation. See Objective 1.

25.

26.

27.

28.

29. $\{(0, 3), (1, 4), (2, 5), (3, 6)\}$

30. $\{(-1, 4), (0, 1), (1, -2), (2, -5)\}$

31. $\{(-3, 5), (1, 5), (4, 5), (7, 5)\}$

32. $\{(-2, 3), (-2, 1), (-2, -3), (-2, 9)\}$

33. $\{ (-10, 1), (-5, 4), (0, 3), (-5, 2) \}$

34. $\{ (-5, 3), (-2, 1), (5, 1), (7, -3) \}$

In Problems 35–44, determine whether each equation shows y as a function of x. See Objective 2.

35. $y = 2x + 9$

36. $y = -6x + 3$

37. $2x + y = 10$

38. $6x - 3y = 12$

39. $y = \pm 5x$

40. $y = \pm 2x^2$

41. $y = x^2 + 2$

42. $y = x^3 - 3$

43. $x + y^2 = 10$

44. $y^2 = x$

In Problems 45–52, determine whether the graph is that of a function. See Objective 3.

45.

46.

47.

48.

49.

50.

51.

52.

In Problems 53–60, find the indicated value of each function. See Objective 4.

    (a) $f(0)$ **(b)** $f(3)$ **(c)** $f(-2)$

53. $f(x) = 2x + 3$

54. $f(x) = 3x + 1$

55. $f(x) = -5x + 2$

56. $f(x) = -2x - 3$

57. $f(x) = x^2 - 3x$

58. $f(x) = 2x^2 + 5x$

59. $f(x) = -x^2 + x + 3$

60. $f(x) = -x^2 + 2x - 5$

In Problems 61–64, find the indicated value of each function. See Objective 4.

    (a) $f(-x)$ **(b)** $f(x + 2)$ **(c)** $f(2x)$
    (d) $-f(x)$ **(e)** $f(x + h)$

61. $f(x) = 2x - 5$ **62.** $f(x) = 4x + 3$

63. $f(x) = 7 - 5x$ **64.** $f(x) = 8 - 3x$

In Problems 65–72, evaluate each function. See Objective 4.

65. $f(x) = x^2 + 3; f(2)$

66. $f(x) = -2x^2 + x + 1; f(-3)$

67. $s(t) = -t^3 - 4t; s(-2)$

68. $g(h) = -h^2 + 5h - 1; g(4)$

69. $F(x) = |x - 2|; F(-3)$

70. $G(z) = 2|z + 5|; G(-6)$

71. $F(z) = \dfrac{z + 2}{z - 5}; F(4)$

72. $h(q) = \dfrac{3q^2}{q + 2}; h(2)$

In Problems 73–80, find the domain of each function. See Objective 5.

73. $f(x) = 4x + 7$ **74.** $G(x) = -8x + 3$

75. $F(z) = \dfrac{2z + 1}{z - 5}$ **76.** $H(x) = \dfrac{x + 5}{2x + 1}$

77. $f(x) = 3x^4 - 2x^2$ **78.** $s(t) = 2t^2 - 5t + 1$

79. $G(x) = \dfrac{3x - 5}{3x + 1}$ **80.** $H(q) = \dfrac{1}{6q + 5}$

Applying the Concepts

81. If $f(x) = 3x^2 - x + C$ and $f(3) = 18$, what is the value of C?

82. If $f(x) = -2x^2 + 5x + C$ and $f(-2) = -15$, what is the value of C?

83. If $f(x) = \dfrac{2x + 5}{x - A}$ and $f(0) = -1$, what is the value of A?

84. If $f(x) = \dfrac{-x + B}{x - 5}$ and $f(3) = -1$, what is the value of B?

△ **85. Geometry** Express the area A of a circle as a function of its radius, r. Determine the area of a circle whose radius is 4 inches. That is, find $A(4)$.

△ **86. Geometry** Express the area A of a triangle as a function of its height h, assuming that the length of the base is 8 centimeters. Determine the area of this triangle if its height is 5 centimeters. That is, find $A(5)$.

87. Salary Express the gross salary G of Jackie, who earns $15 per hour, as a function of the number of hours worked, h. Then determine the gross salary of Jackie if she works 25 hours. That is, find $G(25)$.

88. Commissions Roberta is a commissioned salesperson. She earns a base weekly salary of $250 per week plus 15% of the sales price of items sold. Express her gross salary G as a function of the price p of items sold. Determine the weekly gross salary of Roberta if the value of items sold is $10,000. That is, find $G(10,000)$.

89. Population as a Function of Age The function $P(a) = -1.02a^2 + 57.36a + 3694.1$ represents the population (in thousands) of U.S. residents in 2014, P, that are a years of age.
SOURCE: United States Census Bureau

  (a) Identify the dependent and independent variables.

  (b) Evaluate $P(20)$. Provide a verbal explanation of the meaning of $P(20)$.

  (c) Evaluate $P(0)$. Provide a verbal explanation of the meaning of $P(0)$.

90. Number of Rooms The function $N(r) = -1.33r^2 + 14.68r - 17.09$ represents the number of housing units (in millions), N, in 2015 that have r rooms, where $1 \leq r \leq 9$.
SOURCE: United States Census Bureau

  (a) Identify the dependent and independent variables.

  (b) Evaluate $N(3)$. Provide a verbal explanation of the meaning of $N(3)$.

  (c) Why is it unreasonable to evaluate $N(0)$?

91. Revenue Function The function $R(p) = -p^2 + 200p$ represents the daily revenue R earned from selling MP3 players at p dollars for $0 \leq p \leq 200$.

  (a) Identify the dependent and independent variables.

  (b) Evaluate $R(50)$. Provide a verbal explanation of the meaning of $R(50)$.

  (c) Evaluate $R(120)$. Provide a verbal explanation of the meaning of $R(120)$.

92. Average Trip Length The function $T(x) = 0.01x^2 - 0.12x + 8.89$ represents the average vehicle trip length T (in miles) x years since 1969.

  (a) Identify the dependent and independent variables.

  (b) Evaluate $T(35)$. Provide a verbal explanation of the meaning of $T(35)$.

  (c) Evaluate $T(0)$. Provide a verbal explanation of the meaning of $T(0)$.

△ **93. Geometry** The volume V of a sphere as a function of its radius r is given by $V(r) = \dfrac{4}{3}\pi r^3$. What is the domain of this function?

△ **94. Geometry** The area A of a triangle as a function of its height h, assuming that the length of the base is 5 centimeters, is $A = \dfrac{5}{2}h$. What is the domain of the function?

95. Salary The gross salary G of Kevin as a function of the number of hours worked, h, is given by $G(h) = 22.5h$. What is the domain of the function if he can work up to 60 hours per week?

96. Commissions Carlos is a commissioned salesperson. He earns a base weekly salary of $350 plus 12% of the sales price of items sold. His gross salary G as a function of the price p of items sold is given by $G(p) = 350 + 0.12p$. What is the domain of the function?

97. Demand for Hot Dogs Suppose the function $D(p) = 1200 - 10p$ represents the demand for hot dogs, whose price is p, at a baseball game. Find the domain of the function.

98. Revenue Function The function $R(p) = -p^2 + 200p$ represents the daily revenue earned from selling MP3 players at p dollars for $0 \leq p \leq 200$. Explain why any p greater than $200 is not in the domain of the function.

Extending the Concepts

99. Math for the Future: College Algebra A **piecewise-defined function** is a function defined by more than one equation. For example, the absolute value function $f(x) = |x|$ is actually defined by two equations: $f(x) = x$ if $x \geq 0$ and $f(x) = -x$ if $x < 0$. We can combine these equations into one expression as

$$f(x) = \begin{cases} -x & x < 0 \\ x & x \geq 0 \end{cases}$$

To evaluate $f(3)$, we recognize that $3 \geq 0$, so we use the rule $f(x) = x$ and obtain $f(3) = 3$. To evaluate $f(-4)$, we recognize that $-4 < 0$, so we use the rule $f(x) = -x$ and obtain $f(-4) = -(-4) = 4$.

(a) $f(x) = \begin{cases} x + 3 & x < 0 \\ -2x + 1 & x \geq 0 \end{cases}$

(i) Find $f(3)$.   **(ii)** Find $f(-2)$.   **(iii)** Find $f(0)$.

(b) $f(x) = \begin{cases} -3x + 1 & x < -2 \\ x^2 & x \geq -2 \end{cases}$

(i) Find $f(-4)$.   **(ii)** Find $f(2)$.   **(iii)** Find $f(-2)$.

100. Math for the Future: Calculus

(a) If $f(x) = 3x + 7$, find $\dfrac{f(x + h) - f(x)}{h}$.

(b) If $f(x) = -2x + 1$, find $\dfrac{f(x + h) - f(x)}{h}$.

Explaining the Concepts

101. Investigate when the use of function notation $y = f(x)$ first appeared. Start by researching Lejeune Dirichlet.

102. Are all relations functions? Are all functions relations? Explain your answers.

103. Explain what a function is. Be sure to include the terms *domain* and *range* in your explanation.

104. Explain why the vertical line test can be used to identify the graph of a function.

105. What are the four forms of a function presented in this section?

106. Explain why the terms *independent variable* for x and *dependent variable* for y make sense in the function $y = f(x)$.

Technology Exercises

Graphing calculators have the ability to evaluate any function you wish. Figure 9 shows the results obtained in Example 6(c) and 6(d) using a graphing calculator.

Figure 9

NORMAL FLOAT AUTO REAL RADIAN MP
Plot1 Plot2 Plot3
∎\Y₁⊟X²+6X
∎\Y₂=
∎\Y₃=
∎\Y₄=
∎\Y₅=
∎\Y₆=
∎\Y₇=
∎\Y₈=

NORMAL FLOAT AUTO REAL RADIAN MP
Y₁(3)
27.
Y₁(⁻2)
⁻8.
∎

In Problems 107–114, use technology to find the value of each function.

107. $f(x) = x^2 + 3; f(2)$

108. $f(x) = -2x^2 + x + 1; f(-3)$

109. $F(x) = |x - 2|; F(-3)$

110. $G(z) = 2|z + 5|; G(-6)$

111. $H(x) = \sqrt{4x - 3}; H(7)$

112. $g(h) = \sqrt{2h + 1}; g(4)$

113. $F(z) = \dfrac{0.1z + 2.2}{0.2z - 4.8}; F(4)$

114. $h(q) = \dfrac{1.2q^2}{q + 2.8}; h(2)$

2.3 Functions and Their Graphs

Objectives

❶ Graph a Function

❷ Obtain Information from the Graph of a Function

❸ Know Properties and Graphs of Basic Functions

❹ Interpret Graphs of Functions

Are You Prepared for This Section?

Before getting started, complete the following problems. If you get a problem wrong, go back to the section cited and review the material.

P1. Solve: $3x - 12 = 0$ [Section 1.1, pp. 50–52]

P2. Graph $y = x^2$ by point-plotting.   [Section 1.5, pp. 97–99]

Prepared?...Answers **P1.** $\{4\}$

P2.

❶ **Graph a Function**

The graph of the linear equation $y = -2x + 4$ (see Figure 10 on the next page) passes the vertical line test, so the equation is a function. This can be written as $f(x) = -2x + 4$. The graph of a function is the same as the graph of the equation that defines the function. The horizontal axis represents the independent variable and the vertical axis represents the dependent variable. When graphing functions, label the horizontal axis x and the vertical axis either y or by the name of the function.

Figure 10

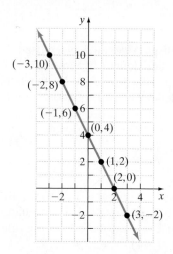

Definition

When a function is defined by an equation in x and y, the **graph of the function** is the set of *all* ordered pairs (x, y) such that $y = f(x)$.

So, if $f(3) = 8$, the point whose ordered pair is $(3, 8)$ is on the graph of $y = f(x)$.

EXAMPLE 1

Graphing a Function

Graph the function $f(x) = |x|$.

Solution

To graph $f(x) = |x|$, first determine some ordered pairs $(x, f(x)) = (x, y)$ such that $y = |x|$. See Table 2. Now plot the ordered pairs (x, y) from Table 2 and connect the points as shown in Figure 11.

Table 2

x	$f(x) = \lvert x\rvert$	$(x, f(x))$
-3	$\lvert -3\rvert = 3$	$(-3, 3)$
-2	$\lvert -2\rvert = 2$	$(-2, 2)$
-1	$\lvert -1\rvert = 1$	$(-1, 1)$
0	$\lvert 0\rvert = 0$	$(0, 0)$
1	$\lvert 1\rvert = 1$	$(1, 1)$
2	$\lvert 2\rvert = 2$	$(2, 2)$
3	$\lvert 3\rvert = 3$	$(3, 3)$

Figure 11

$f(x) = |x|$

Quick ✓

1. When a function is defined by an equation in x and y, the _____ of the _____ is the set of all ordered pairs (x, y) such that $y = f(x)$.

2. If $f(4) = -7$, then the point whose ordered pair is $(\_, \_)$ is on the graph of $y = f(x)$.

In Problems 3–5, graph each function.

3. $f(x) = -2x + 9$ **4.** $f(x) = x^2 + 2$ **5.** $f(x) = |x - 2|$

❷ Obtain Information from the Graph of a Function

▶ Remember, the domain of a function is the set of all inputs, and the range is the set of all outputs of the function. The domain and the range of a function can be found from its graph.

EXAMPLE 2   **Finding the Domain and Range of a Function from Its Graph**

Figure 12 shows the graph of a function.

    (a) Determine the function's domain and range.

    (b) Identify the intercepts.

Figure 12

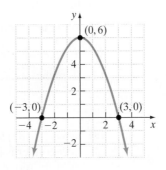

Solution

> **In Other Words**
>
> When the graph of a function is given, its domain may be viewed as the shadow created by the graph on the x-axis by vertical beams of light. Its range can be viewed as the shadow created by the graph on the y-axis by horizontal beams of light.

    (a) The domain of the function consists of all of the graph's x-coordinates. Because the graph exists for all real numbers x, the domain is $\{x \mid x \text{ is any real number}\}$, or the interval, $(-\infty, \infty)$.

        The range of the function consists of all of the graph's y-coordinates. Because the graph exists for all real numbers y less than or equal to 6, the range is $\{y \mid y \le 6\}$, or the interval, $(-\infty, 6]$.

    (b) The x-intercepts are $(-3, 0)$ and $(3, 0)$. The y-intercept is $(0, 6)$.   ●

Quick ✓

6. Use the graph of the function to answer parts (a) and (b).

    (a) Determine the domain and range of the function.

    (b) Identify the intercepts.

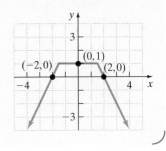

The next example illustrates how to obtain information about a function from its graph. Remember, if $(1, 5)$ is a point on the graph of f, then $f(1) = 5$.

EXAMPLE 3   **Obtaining Information from the Graph of a Function**

The Wonder Wheel is a Ferris wheel in Coney Island. See Figure 13 on the next page. Let f be the distance (in feet) above the ground of a person riding in a car on the Wonder Wheel as a function of time x (in minutes). Use the graph of f in Figure 14 to answer the following questions.

(continued)

Figure 13

Figure 14

(a) Find $f(1.5)$ and $f(3)$. Interpret these values.

(b) What is the domain of f?

(c) What is the range of f?

(d) For what values of x does $f(x) = 85$? That is, solve $f(x) = 85$.

Solution

(a) Since $(1.5, 85)$ is on the graph of f, then $f(1.5) = 85$. After 1.5 minutes, a Wonder Wheel rider is 85 feet above the ground. Similarly, since $(3, 160)$ is on the graph, $f(3) = 160$. After 3 minutes, a Wonder Wheel rider is 160 feet above the ground.

(b) To determine the domain of f, notice that the graph exists for each number x between 0 and 4, inclusive. Therefore, the domain of f is $\{x \mid 0 \le x \le 4\}$, or the interval $[0, 4]$.

(c) The points on the graph have y-coordinates between 10 and 160, inclusive. Therefore, the range of f is $\{y \mid 10 \le y \le 160\}$, or the interval $[10, 160]$.

(d) Since $(0.5, 85)$, $(1.5, 85)$, $(2.5, 85)$, and $(3.5, 85)$ are the only points on the graph for which $y = f(x) = 85$, the solution set to the equation $f(x) = 85$ is $\{0.5, 1.5, 2.5, 3.5\}$.

Quick ✓

7. If the point $(3, 8)$ is on the graph of a function f, then $f(\_) = \_$. If $g(-2) = 4$, then $(\_\_, \_)$ is a point on the graph of g.

8. Use the graph of $y = f(x)$ to answer the following questions.

    (a) Find $f(-3)$ and $f(1)$.

    (b) What is the domain of f?

    (c) What is the range of f?

    (d) Identify the intercepts.

    (e) For what value of x does $f(x) = 15$? That is, solve $f(x) = 15$.

▶ **EXAMPLE 4**   **Obtaining Information about the Graph of a Function**

Consider the function $f(x) = 2x - 5$.

(a) Is the point $(3, -1)$ on the graph of the function?

(b) If $x = 1$, what is $f(x)$? Based on this result, what point is on the graph of the function?

(c) If $f(x) = 3$, what is x? Based on this result, what point is on the graph of f?

Solution

(a) If $x = 3$, then

$$f(x) = 2x - 5$$
$$f(3) = 2(3) - 5 = 6 - 5 = 1$$

Since $f(3) = 1$, the point $(3, 1)$ is on the graph but $(3, -1)$ is not.

(b) If $x = 1$, then

$$f(1) = 2(1) - 5 = 2 - 5 = -3$$

Therefore, the point $(1, -3)$ is on the graph of f.

(c) If $f(x) = 3$, then

$$f(x) = 3$$
$$2x - 5 = 3$$

Add 5 to both sides: $2x = 8$

Divide both sides by 2: $x = 4$

If $f(x) = 3$, then $x = 4$. Therefore, the point $(4, 3)$ is on the graph of f. ●

Work Smart: Study Skills

Do not confuse the directions "Find $f(3)$" with "If $f(x) = 3$, what is x?" Write down and study errors that you commonly make so that you can avoid them.

Quick ✓

9. Consider the function $f(x) = -3x + 7$.

(a) Is the point $(-2, 1)$ on the graph of the function?

(b) If $x = 3$, what is $f(x)$? Based on this result, what point is on the graph of the function?

(c) If $f(x) = -8$, what is x? What point is on the graph of f?

▶ The Zero of a Function

In Other Words

A zero of a function is an input that makes the function equal 0.

If $f(r) = 0$ for some number r, then r is a **zero** of f. For example, if $f(x) = x^2 - 4$, then -2 and 2 are zeros of f because $f(-2) = 0$ and $f(2) = 0$. The zeros of a function can be identified from its graph by identifying the x-intercepts of the graph. Why? If $f(r) = 0$, then the point $(r, 0)$ is on the graph of f, and any point with coordinates $(r, 0)$ is an x-intercept of the graph.

| EXAMPLE 5 | **Finding the Zeros of a Function from Its Graph** |

Find the zeros of the function f whose graph is shown in Figure 15.

Solution
The x-intercepts of the graph are $(-3, 0)$ and $(3, 0)$. Therefore, the zeros of f are -3 and 3. ●

Figure 15

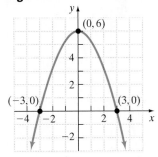

Quick ✓

In Problems 10–12, determine whether the value is a zero of the function.

10. $f(x) = 2x + 6; -3$ 11. $g(x) = x^2 - 2x - 3; 1$

12. $h(z) = -z^3 + 4z; 2$

13. Find the zeros of the function f whose graph is shown.

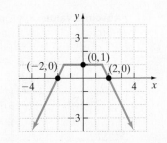

▶ ❸ Know Properties and Graphs of Basic Functions

In Table 3, we list a number of basic functions, their properties, and their graphs.

Table 3

Function	Properties	Graph
Linear Function $f(x) = mx + b$ m and b are real numbers	• Domain and range are all real numbers. • Graph is nonvertical line with slope $= m$ and y-intercept $= (0, b)$.	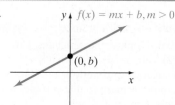
Identity Function (special type of linear function) $f(x) = x$	• Domain and range are all real numbers. • Graph is a line with slope of $m = 1$ and y-intercept $= (0, 0)$. • The line consists of all points for which the x-coordinate equals the y-coordinate.	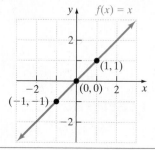
Constant Function (special type of linear function) $f(x) = b$ b is a real number	• Domain is the set of all real numbers, and range is the set consisting of a single number b. • Graph is a horizontal line with slope $m = 0$ and y-intercept of $(0, b)$.	
Square Function $f(x) = x^2$	• Domain is the set of all real numbers, and range is the set of nonnegative real numbers. • The intercept of the graph is $(0, 0)$.	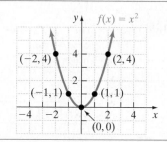
Cube Function $f(x) = x^3$	• Domain and range are the set of all real numbers. • The intercept of the graph is $(0, 0)$.	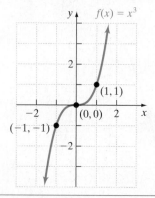
Absolute Value Function $f(x) = |x|$	• Domain is the set of all real numbers, and range is the set of nonnegative real numbers. • The intercept of the graph is $(0, 0)$. • If $x \geq 0$, then $f(x) = x$, and the graph of f is part of the line $y = x$; if $x < 0$, then $f(x) = -x$, and the graph of f is part of the line $y = -x$.	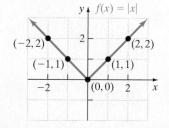

④ **Interpret Graphs of Functions**

The graph of a function can be used to give a visual description of many different scenarios. Consider the following example.

EXAMPLE 6 **Graphing a Verbal Description**

Maria decides to take a walk. She leaves her house and walks 3 blocks in 2 minutes at a constant speed. She realizes that she left her front door unlocked, so she runs home in 1 minute. It takes Maria 1 minute to find her keys and lock the door. She next runs 10 blocks in 3 minutes and then rests for 1 minute. She walks 4 more blocks in 10 minutes, and finally hitches a ride home with her neighbor, who happens to drive by, and gets home in 2 minutes. Draw a graph of Maria's distance from home (in blocks) as a function of time.

Solution

Because distance from home is a function of time, draw a Cartesian plane with the horizontal axis representing the independent variable, time, and the vertical axis representing the dependent variable, distance from home.

The ordered pair $(0, 0)$ corresponds to starting the walk. The ordered pair $(2, 3)$ represents being 3 blocks from home after 2 minutes. Start the graph at $(0, 0)$ and then draw a straight line from $(0, 0)$ to $(2, 3)$. Next, draw a straight line to $(3, 0)$, which represents the return trip home to lock the door. Draw a line segment from $(3, 0)$ to $(4, 0)$ to represent the time it takes to lock the door. Draw a line segment from $(4, 0)$ to $(7, 10)$, which represents the 10 block run in 3 minutes. Now draw a horizontal line from $(7, 10)$ to $(8, 10)$. This represents the resting period. Draw a line from $(8, 10)$ to $(18, 14)$ to represent the 4-block walk in 10 minutes. Finally, draw a line segment from $(18, 14)$ to $(20, 0)$ to represent the ride home. See Figure 16.

Figure 16

Quick ✓

14. Clara decides to take a walk. She leaves her house and walks 5 blocks in 5 minutes at a constant speed. She realizes that she left her front door unlocked, so she runs home in 2 minutes. It takes her 1 minute to find her keys and lock the door. She next jogs 8 blocks in 5 minutes, and then runs 3 blocks in 1 minute. After resting for 2 minutes she walks home in 10 minutes. Draw a graph of Clara's distance from home (in blocks) as a function of time (in minutes).

2.3 Exercises MyMathLab® Exercise numbers in green have complete video solutions in MyMathLab or may be accessed using the QR code to the right.

Problems **1–14** *are the* **Quick ✓** *s that follow the* **EXAMPLES**.

Building Skills

In Problems 15–22, graph each function. See Objective 1.

15. $f(x) = 4x - 6$

16. $g(x) = -3x + 5$

17. $h(x) = x^2 - 2$

18. $F(x) = x^2 + 1$

19. $G(x) = |x - 1|$

20. $H(x) = |x + 1|$

21. $g(x) = x^3$

22. $h(x) = x^3 - 3$

In Problems 23–32, for each graph of a function, find (a) the domain and the range, (b) the intercepts, if any, and (c) the zeros, if any. See Objective 2.

23.

24.

25.

26.

27.

28.

29.

30.

31.

32.

33. Use the graph of the function f shown to answer parts (a) – (l).

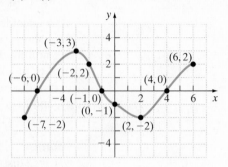

(a) Find $f(-7)$.

(b) Find $f(-3)$.

(c) Find $f(6)$.

(d) Is $f(2)$ positive or negative?

(e) For what numbers x is $f(x) = 0$?

(f) What is the domain of f?

(g) What is the range of f?

(h) What are the x-intercepts?

(i) What is the y-intercept?

(j) For what numbers x is $f(x) = -2$?

(k) For what number x is $f(x) = 3$?

(l) What are the zeros of f?

34. Use the graph of the function g shown to answer parts (a) – (l).

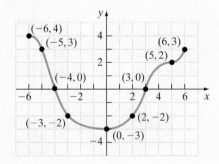

(a) Find $g(-3)$.

(b) Find $g(5)$.

(c) Find $g(6)$.

(d) Is $g(-5)$ positive or negative?

(e) For what numbers x is $g(x) = 0$?

(f) What is the domain of g?

(g) What is the range of g?

(h) What are the x-intercepts?

(i) What is the y-intercept?

(j) For what numbers x is $g(x) = -2$?

(k) For what number x is $g(x) = 3$?

(l) What are the zeros of g?

35. Use the table of values for the function F to answer questions (a) – (e).

x	$F(x)$
−4	0
−2	3
−1	5
0	2
3	−6

(a) What is $F(-2)$?
(b) What is $F(3)$?
(c) For what number(s) x is $F(x) = 5$?
(d) What is the x-intercept of the graph of F?
(e) What is the y-intercept of the graph of F?

36. Use the table of values for the function G to answer questions (a) – (e).

x	$G(x)$
−7	−3
−4	0
0	5
3	8
7	5

(a) What is $G(3)$?   **(b)** What is $G(7)$?
(c) For what number(s) x is $G(x) = 5$?
(d) What is the x-intercept of the graph of G?
(e) What is the y-intercept of the graph of G?

In Problems 37–40, answer the questions about the given function. See Objective 2.

37. $f(x) = 4x - 9$

(a) Is the point $(2, 1)$ on the graph of the function?
(b) If $x = 3$, what is $f(x)$? What point is on the graph of the function?
(c) If $f(x) = 7$, what is x? What point is on the graph of f?
(d) Is 2 a zero of f?

38. $f(x) = 3x + 5$

(a) Is the point $(-2, 1)$ on the graph of the function?
(b) If $x = 4$, what is $f(x)$? What point is on the graph of the function?
(c) If $f(x) = -4$, what is x? What point is on the graph of f?
(d) Is −2 a zero of f?

39. $g(x) = -\dfrac{1}{2}x + 4$

(a) Is the point $(4, 2)$ on the graph of the function?
(b) If $x = 6$, what is $g(x)$? What point is on the graph of the function?
(c) If $g(x) = 10$, what is x? What point is on the graph of g?
(d) Is 8 a zero of g?

40. $H(x) = \dfrac{2}{3}x - 4$

(a) Is the point $(3, -2)$ on the graph of the function?
(b) If $x = 6$, what is $H(x)$? What point is on the graph of the function?
(c) If $H(x) = -4$, what is x? What point is on the graph of H?
(d) Is 6 a zero of H?

In Problems 41–46, match each graph to the function listed whose graph most resembles the one given. See Objective 3.

(a) Constant function   **(b)** Linear function
(c) Square function   **(d)** Cube function
(e) Absolute value function   **(f)** Identity function

41.   **42.**

43.   **44.**

45.   **46.**

In Problems 47–50, sketch the graph of each function. Label at least three points. See Objective 3.

47. $f(x) = x^2$ **48.** $f(x) = x^3$

49. $f(x) = |x|$ **50.** $f(x) = 4$

Applying the Concepts

51. Match each of the following functions with the graph on the following page that best describes the situation.

(a) The distance from ground level of a person who is jumping on a trampoline as a function of time
(b) The cost of a telephone call as a function of time
(c) The height of a human as a function of time
(d) The height of a rocket as a function of time
(e) The book value of a machine that is depreciated by equal amounts each year as a function of the year

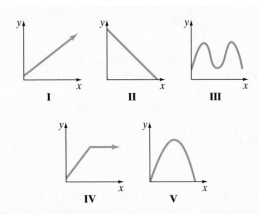

52. Match each of the following functions with the graph that best describes the situation.

(a) The average high temperature each day as a function of the day of the year
(b) The height of a human as a function of their age
(c) The distance that a person rides her bicycle at a constant speed as a function of time
(d) The temperature of a pizza after it is removed from the oven as a function of time
(e) The value of a car as a function of time

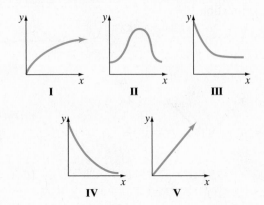

53. Pulse Rate Consider the following scenario: Zach starts jogging on a treadmill. His resting pulse rate is 70 beats per minute (b.p.m.). As he continues to jog on the treadmill, his pulse increases at a constant rate until, after 10 minutes, his pulse is 120 b.p.m. He then starts jogging faster and his pulse increases at a constant rate for 2 minutes, at which time his pulse is up to 150 b.p.m. He then begins a cooling-off period for 7 minutes until his pulse goes down to 110 b.p.m. He then gets off the treadmill and his pulse returns to 70 b.p.m. after 12 minutes. Draw a graph of Zach's pulse as a function of time.

54. Altitude of an Airplane Suppose that a plane is flying from Chicago to New Orleans. The plane leaves the gate and taxis for 5 minutes. The plane takes off and gets up to 10,000 feet after 5 minutes. The plane continues to ascend at a constant rate until it reaches its cruising altitude of 35,000 feet after another 25 minutes. For the next 80 minutes, the plane maintains a constant height of 35,000 feet. The plane then descends at a constant rate until it

lands after 20 minutes. It requires 5 minutes to taxi to the gate. Draw a graph of the height of the plane as a function of time.

55. Height of a Swing An 8-year-old girl gets on a swing and starts swinging for 10 minutes. Draw a graph that represents the height of the child from the ground as a function of time.

56. Temperature of Pizza Marissa is hungry and would like a pizza. Her mother pulls a frozen pizza out of the freezer and puts it in the oven. After 12 minutes the pizza is done, but Mom lets the pizza cool for 5 minutes before serving it to Marissa. Draw a graph that represents the temperature of the pizza as a function of time.

57. The graph below shows the weight in pounds of a person as a function of his age in years. Describe the weight of the individual over the course of his life.

58. The following graph shows the depth of a lake (in feet) as a function of time (in days). Describe the depth of the lake over the course of the 400 days.

Extending the Concepts

59. Draw a graph of a function f with the following characteristics: x-intercepts: $(-4, 0)$, $(-1, 0)$, and $(2, 0)$; y-intercept: $(0, -2)$; $f(-3) = 7$ and $f(3) = 8$.

60. Draw a graph of a function f with the following characteristics: x-intercepts: $(-3, 0)$, $(2, 0)$, and $(5, 0)$; y-intercept: $(0, 3)$; $f(3) = -2$.

Explaining the Concepts

61. Using the definition of a function, explain why the graph of a function can have at most one y-intercept.

62. Explain what the domain of a function is. In your explanation, discuss how domains are determined in applications.

63. Explain what the range of a function is.

64. Explain the relationship between the x-intercepts of a function and its zeros.

65. Why are y-intercepts not considered zeros of a function?

Putting the Concepts Together (Sections 2.1–2.3)

We designed these problems so that you can review Sections 2.1–2.3 and show your mastery of the concepts. Take time to work these problems before proceeding with the next section. The answers are at the back of the text on page AN-11–12.

1. Explain why the following relation is a function. Then express the function as a set of ordered pairs.

2. Determine which of the following relations represent functions.

(a) $y = x^3 - 4x$　　　　(b) $y = \pm 4x + 3$

3. Is the following relation a function? If so, state the domain and range.

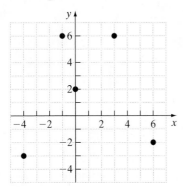

4. Explain why the relation whose graph is below is a function. If the name of the function is f, find $f(5)$.

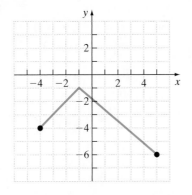

5. What is the zero of the function whose graph is shown?

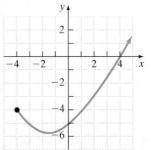

6. Let $f(x) = -5x + 3$ and $g(x) = -2x^2 + 5x - 1$. Find the value of each of the following.

(a) $f(4)$　　　　(b) $g(-3)$
(c) $f(x) - f(4)$　　　　(d) $f(x - 4)$

7. Find the domain of each of the following functions.

(a) $G(h) = h^2 + 4$　　　　(b) $F(w) = \dfrac{w - 4}{3w + 1}$

8. Graph $f(x) = |x| - 2$. Use the graph to state the domain and the range of f.

9. Vertical Motion The graph below shows the height, h, in feet, of a ball thrown straight up with an initial speed of 40 feet per second from an initial height of 80 feet after t seconds.

(a) Find and interpret $h(2.5)$.
(b) What is the domain of h?
(c) What is the range of h?
(d) For what value of t does $h(t) = 105$?

10. Consider the function $f(x) = 5x - 2$.

(a) Is the point $(3, 12)$ on the graph of the function?
(b) If $x = -2$, what is $f(x)$? What point is on the graph of the function?
(c) If $f(x) = -22$, what is x? What point is on the graph of f?
(d) Is $\dfrac{2}{5}$ a zero of f?

2.4 Linear Functions and Models

Objectives

1 Graph Linear Functions

2 Find the Zero of a Linear Function

3 Build Linear Models from Verbal Descriptions

4 Build Linear Models from Data

Are You Prepared for This Section?

Before getting started, complete the following problems. If you get a problem wrong, go back to the section cited and review the material.

P1. Graph: $y = 2x - 3$ [Section 1.6, pp. 114–115]

P2. Graph: $\frac{1}{2}x + y = 2$ [Section 1.6, pp. 106–107]

P3. Graph: $y = -4$ [Section 1.6, pp. 107–108]

P4. Graph: $x = 5$ [Section 1.6, pp. 107–108]

P5. Find and interpret the slope of the line through $(-1, 3)$ and $(3, -4)$.   [Section 1.6, pp. 108–110]

P6. Find the equation of the line through $(1, 3)$ and $(4, 9)$.   [Section 1.6, pp. 115–116]

P7. Solve: $0.5(x - 40) + 100 = 84$ [Section 1.1, pp. 50–55]

P8. Solve: $4x + 20 \geq 32$ [Section 1.4, pp. 86–88]

▶ **1** Graph Linear Functions

Recall from Chapter 1 that a linear equation in two variables has the form $Ax + By = C$, where A, B, and C are real numbers, and A and B are not both zero.

Consider the graphs of the four lines in Figure 17. Notice the lines that rise from left to right have positive slope, and lines that fall from left to right have negative slope. Lines that have zero slope are horizontal lines, and lines that have undefined slope are vertical lines. Remember, just as we read a text from left to right, we also read graphs from left to right.

Figure 17

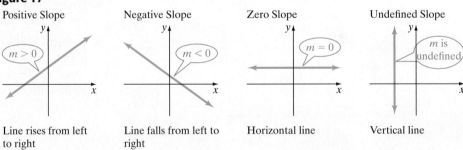

Positive Slope   Negative Slope   Zero Slope   Undefined Slope

$m > 0$ $m < 0$ $m = 0$ m is undefined

Line rises from left to right   Line falls from left to right   Horizontal line   Vertical line

In Figure 17, all the graphs except one pass the vertical line test for identifying graphs of functions. This leads to the conclusion that **all linear equations except those of the form $x = a$, vertical lines, are functions.**

Therefore any linear equation that is in the form $Ax + By = C$ can be written using function notation, provided $B \neq 0$, as follows:

$$Ax + By = C \quad B \neq 0$$

Subtract Ax from both sides: $By = -Ax + C$

Divide both sides by B: $\dfrac{By}{B} = \dfrac{-Ax + C}{B}$

Simplify: $y = -\dfrac{A}{B}x + \dfrac{C}{B}$

$$\qquad\qquad \updownarrow \qquad \updownarrow \qquad \updownarrow$$

$$f(x) = mx + b$$

Prepared?...Answers

P1.   **P2.**

P3.   **P4.**

P5. $-\dfrac{7}{4}$; y decreases by 7 when x increases by 4.   **P6.** $y = 2x + 1$ **P7.** $\{8\}$

P8. $\{x \mid x \geq 3\}$; $[3, \infty)$

This leads to the following definition:

> **Definition**
>
> A **linear function** is a function of the form
> $$f(x) = mx + b$$
> where m is the slope and $(0, b)$ is the y-intercept. The graph of a linear function is a line.

Linear functions can be graphed using the same techniques used to graph linear equations written in slope-intercept form, $y = mx + b$ (See Section 1.6).

EXAMPLE 1   **Graphing a Linear Function**

Graph the linear function: $f(x) = 3x - 5$

Figure 18

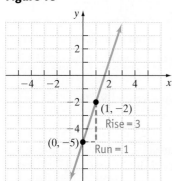

Solution
Since $f(x) = 3x - 5$ is in the form $f(x) = mx + b$, the y-intercept is $(0, -5)$, so plot that point in the coordinate plane. Because the slope $m = 3 = \dfrac{3}{1} = \dfrac{\Delta y}{\Delta x} = \dfrac{\text{rise}}{\text{run}}$, go to the right 1 unit and up 3 units and end up at $(1, -2)$. The line through these two points is the graph of $f(x) = 3x - 5$, shown in Figure 18.   ●

Quick ✓

1. For the graph of a linear function $f(x) = mx + b$, m is the _____ and $(0, b)$ is the _____.

2. The graph of a linear function is called a ___.

3. *True or False* All linear equations are functions.

4. For the linear function $G(x) = -2x + 3$, the slope is ___ and the y-intercept is _____.

In Problems 5–8, graph each linear function.

5. $f(x) = 2x - 3$ 6. $G(x) = -5x + 4$ 7. $h(x) = \dfrac{3}{2}x + 1$ 8. $f(x) = 4$

Work Smart

Because $\dfrac{3}{1}$ is equivalent to $\dfrac{-3}{-1}$, we could go to the left 1 and down 3 and still remain on the line.

▶ ❷ **Find the Zero of a Linear Function**

In Section 2.3 we stated that if r is a zero of a function f, then $f(r) = 0$. To find the zero of any function f, solve the equation $f(x) = 0$.

EXAMPLE 2   **Finding the Zero of a Linear Function**

Find the zero of $f(x) = -4x + 12$.

Solution
Find the zero by solving $f(x) = 0$.

$$f(x) = 0$$
$$-4x + 12 = 0$$

Subtract 12 from both sides of the equation: $-4x = -12$

Divide both sides of the equation by -4: $x = 3$

Check: Since $f(3) = -4(3) + 12 = 0$, the zero of f is 3.   ●

Quick ✓

In Problems 9–11, find the zero of each linear function.

9. $f(x) = 3x - 15$ **10.** $G(x) = \dfrac{1}{2}x + 4$ **11.** $F(p) = -\dfrac{2}{3}p + 8$

▶ Applications of Linear Functions

Linear functions have many applications. For example, the cost of cab fare, sales commissions, and the cost of breakfast as a function of the number of eggs ordered can be modeled by linear functions.

EXAMPLE 3   **Sales Commissions**

Tony's weekly salary at Apple Chevrolet is 0.75% of his weekly sales plus $450, so $S(x) = 0.0075x + 450$ describes Tony's weekly salary S as a linear function of his weekly sales x.

(a) What is the implied domain of the function?

(b) If Tony sells cars worth a total of $50,000 one week, what is his salary?

(c) If Tony earned $600 one week, what was the value of the cars that he sold?

(d) Draw a graph of the function.

(e) For what value of cars sold will Tony's weekly salary exceed $1200?

Solution

(a) The independent variable is weekly sales, x. Because negative weekly sales do not make sense, the function's domain is $\{x \mid x \geq 0\}$ or, using interval notation, $[0, \infty)$.

(b) If Tony's weekly sales are $x = \$50{,}000$, then he earns

$$S(50{,}000) = 0.0075(50{,}000) + 450$$
$$= \$825$$

Tony will earn $825 for the week, if he sells $50,000 worth of cars.

(c) Here, solve the equation $S(x) = 600$.

$$0.0075x + 450 = 600$$

Subtract 450 from both sides: $0.0075x = 150$

Divide both sides by 0.0075: $x = \$20{,}000$

If Tony earned $600, then he sold $20,000 worth of cars in a week.

(d) Plot the independent variable, weekly sales, along the horizontal axis and the dependent variable, salary, along the vertical axis. Graph the equation by plotting points. Part (b) shows that (50,000, 825) is on the graph. Part (c) indicates that (20,000, 600) is on the graph. Why do we also know that (0, 450) is on the graph? Plot these points to get the graph in Figure 19.

Figure 19

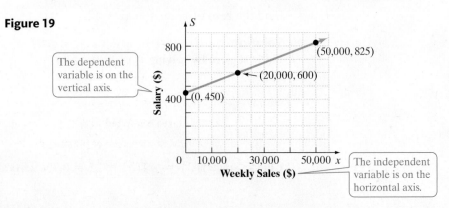

The dependent variable is on the vertical axis.

The independent variable is on the horizontal axis.

(e) Solve the inequality $S(x) > 1200$.

$$0.0075x + 450 > 1200$$

Subtract 450 from both sides: $\qquad 0.0075x > 750$

Divide both sides of the inequality by 0.0075: $\qquad x > 100,000$

If Tony sells more than $100,000 worth of cars for the week, his salary will exceed $1200. ●

Notice that in Figure 19 the function was graphed only over its domain, $[0, \infty)$ — that is, only in quadrant I. Also notice that the horizontal axis is labeled x for the independent variable, weekly sales, and the vertical axis is labeled S for the dependent variable, salary. For this reason, the intercept on the vertical axis is the S-intercept, not the y-intercept. It is also indicated on the axes what x and S represent. Labeling your axes is always a good practice.

Quick ✓

12. The cost, C, of renting a 12-foot moving truck for a day is $40 plus $0.35 times the number of miles driven. The linear function $C(x) = 0.35x + 40$ describes the cost C of driving the truck x miles.

(a) What is the domain of this linear function?

(b) Determine the C-intercept of the graph of the linear function.

(c) What is the rental cost if the truck is driven 80 miles?

(d) How many miles was the truck driven if the rental cost is $85.50?

(e) Graph the linear function.

(f) How many miles can you drive if you can spend up to $127.50?

❸ Build Linear Models from Verbal Descriptions

A linear function has the form $f(x) = mx + b$, where m is the slope of the linear function and $(0, b)$ is its y-intercept. In Section 1.6, we learned that slope can be thought of as an average rate of change. Slope describes how much a dependent variable changes for a given change in the independent variable. For example, in the linear function $f(x) = 4x + 3$, the slope is $4 = \dfrac{4}{1} = \dfrac{\Delta y}{\Delta x}$, so the dependent variable y increases by 4 units for every 1-unit increase in x (the independent variable). When the average rate of change of a function is constant, a linear function is used to model the situation. For example, if your phone company charges $0.05 per minute to talk regardless of the number of minutes on the phone, then a linear function can be used to model the cost of talking, with slope $m = \dfrac{0.05 \text{ dollar}}{1 \text{ minute}}$.

▶ **EXAMPLE 4** **Cost Function**

In the linear cost function $C(x) = ax + b$, b represents the fixed costs of operating a business, and a represents the costs associated with manufacturing one additional item. Suppose that a bicycle manufacturer has daily fixed costs of $2000 and each bicycle costs $80 to manufacture.

(a) Write a linear function showing the cost to manufacture x bicycles in a day.

(b) What is the cost to manufacture 5 bicycles in a day?

(c) How many bicycles can be manufactured for $2800?

(d) Graph the linear function.

(continued)

Solution

(a) Because each bicycle costs \$80 to manufacture, $a = 80$. The fixed costs are \$2000, so $b = 2000$. Therefore, the cost function is

$$C(x) = 80x + 2000$$

(b) Evaluate the function for $x = 5$.

$$C(5) = 80(5) + 2000$$
$$= \$2400$$

It will cost \$2400 to manufacture 5 bicycles.

(c) Solve $C(x) = 2800$.

$$C(x) = 2800$$
$$80x + 2000 = 2800$$

Subtract 2000 from both sides: $\quad 80x = 800$

Divide both sides by 80: $\quad x = 10$

Ten bicycles can be manufactured for \$2800.

(d) Label the horizontal axis x and the vertical axis C. See Figure 20 for the graph of the cost function.

Figure 20

Number of Bicycles

Quick ✓

13. Suppose the business presented in Example 4 must pay a tax of \$1 per bicycle produced.

(a) Write a linear function that expresses the cost C of producing x bicycles in a day.

(b) What is the cost of manufacturing 5 bicycles in a day?

(c) How many bicycles can be manufactured for \$2810?

(d) Graph the linear function.

▶ **EXAMPLE 5**   **Straight-Line Depreciation**

The *book value* of an asset such as a building or piece of machinery is the value that the company uses to create its balance sheet. Some companies use *straight-line depreciation* so that the book value of the asset declines by a constant amount each year. The amount of the decline depends on the useful life that the company places on the asset. Suppose Pearson Publishing Company just purchased a new fleet of cars for its sales force at a cost of \$29,400 per car. The company uses the straight-line depreciation method for 7 years.

(a) Write a linear function that expresses the book value V of each car as a function of its age, x.

(b) What is the domain of this linear function?

(c) What is the book value of each car after 3 years?

(d) When will the book value of each car be \$12,600?

(e) Graph the linear function.

Solution

(a) Let the linear function $V(x) = mx + b$ represent the book value of each car after x years. The original value of each car is \$29,400, so $V(0) = 29,400$. Thus the V-intercept of the function is $(0, 29,400)$. After 7 years, the book value of the car is \$0. Use the ordered pairs $(0, 29,400)$ and $(7, 0)$ to find the slope, or amount of yearly (annual) depreciation, of V:

$$m = \frac{0 - 29,400}{7 - 0} = \frac{-29,400}{7} = -4200$$

So each car depreciates by $4200 per year. The linear function that represents the book value of each car after x years is

$$V(x) = -4200x + 29,400$$

(b) A car cannot have a negative age, so the age x must be greater than or equal to zero. In addition, each car is depreciated over 7 years, at which time $V(7) = 0$. Therefore, the domain of the function is $\{x \mid 0 \leq x \leq 7\}$, or $[0, 7]$ using interval notation.

(c) The book value of each car after $x = 3$ years is given by $V(3)$.

$$V(3) = -4200(3) + 29,400$$
$$= \$16,800$$

(d) To find when the book value is $12,600, solve the equation

$$V(x) = 12,600$$
$$-4200x + 29,400 = 12,600$$

Subtract 29,400 from both sides: $\qquad -4200x = -16,800$

Divide both sides by -4200: $\qquad x = 4$

Each car will have a book value of $12,600 in 4 years.

(e) Label the horizontal axis x and the vertical axis V. Since $V(0) = 29,400$, the point $(0, 29,400)$ is on the graph. Since $V(7) = 0$, the point $(7, 0)$ is on the graph. To graph the function, we use these intercepts, along with the points $(3, 16,800)$ and $(4, 12,600)$ from parts (c) and (d). See Figure 21.

Figure 21

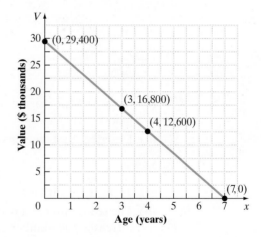

Quick ✓

14. Roberta's monthly payments for her new car are $250 per month. She estimates that maintenance and gas cost her $0.18 per mile.

(a) Write a linear function that expresses the monthly cost C of operating the car as a function of miles driven, x.

(b) What is the domain of this linear function?

(c) What is the monthly cost of driving 320 miles?

(d) How many miles can Roberta drive each month if she can afford the monthly cost to be $282.40?

(e) Graph the linear function.

▶ ❹ Build Linear Models from Data

If we have a set of data with more than two points, how can we tell whether the data (the two variables) are related linearly? There are some methods (beyond the scope of this course) for determining whether two variables are linearly related, but we can also draw a picture of the data, a *scatter diagram*, and learn whether the variables *might* be linearly related.

Scatter Diagrams

The graph consisting of the ordered pairs that make up a relation is called a **scatter diagram**.

EXAMPLE 6 **Drawing a Scatter Diagram**

The on-base percentage for a baseball team is the percent of time that the team safely reaches base. Table 4 shows the number of runs scored and the on-base percentage for various teams in the 2015 season.

Table 4

Team	On-Base Percentage, x	Runs Scored, y	(x, y)
NY Yankees	32.3	764	(32.3, 764)
Los Angeles Angels	30.7	661	(30.7, 661)
Texas Rangers	32.5	751	(32.5, 751)
Toronto Blue Jays	34.0	891	(34.0, 891)
Minnesota Twins	30.5	696	(30.5, 696)
Oakland A's	31.2	694	(31.2, 694)
Kansas City Royals	32.2	724	(32.2, 724)
Baltimore Orioles	30.7	713	(30.7, 713)

SOURCE: *espn.com*

(a) Draw a scatter diagram of the data, treating on-base percentage as the independent variable.

(b) Describe what happens as the on-base percentage increases.

Figure 22

On-Base Percentage

Solution

(a) To draw a scatter diagram, plot the ordered pairs listed in Table 4. See Figure 22.

(b) The scatter diagram reveals that for the most part, as the on-base percentage increases, the number of runs scored also increases. While this relation is not perfectly linear (because the points don't all fall on a straight line), the *pattern* of the data is linear.

Quick ✓

15. The data listed below represent the total cholesterol (in mg/dL) and age of males.

Age, x	Total Cholesterol, y	Age, x	Total Cholesterol, y
25	180	38	239
25	195	48	204
28	186	51	243
32	180	62	228
32	197	65	269

(a) Draw a scatter diagram, treating age as the independent variable.

(b) Describe the relation between age and total cholesterol.

Recognizing the Type of Relation Between Two Variables

Scatter diagrams help us see the type of relation between two variables. This section will focus on distinguishing linear relations from nonlinear relations. See Figure 23.

Figure 23

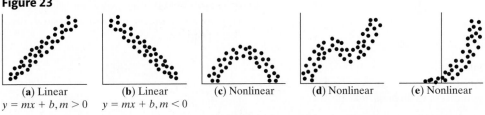

(a) Linear
$y = mx + b, m > 0$

(b) Linear
$y = mx + b, m < 0$

(c) Nonlinear

(d) Nonlinear

(e) Nonlinear

EXAMPLE 7   **Distinguishing Between Linear and Nonlinear Relations**

Determine whether each relation in Figure 24 is linear or nonlinear. If it is linear, state whether the slope is positive or negative.

Figure 24

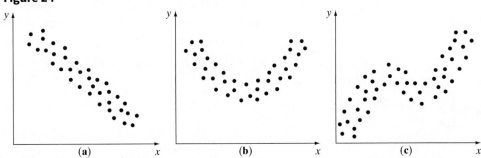

(a)

(b)

(c)

Solution

(a) Linear with negative slope   **(b)** Nonlinear   **(c)** Nonlinear   ●

> **Quick ✓**
>
> *In Problems 16 and 17, determine whether each relation is linear or nonlinear. If it is linear, state whether the slope is positive or negative.*
>
> **16.**
>
>
>
> **17.**
>
>

Fitting a Line to Data

Suppose a scatter diagram indicates a linear relationship, as in Figure 23(a) or (b). To find a linear equation for such data, draw a line through two points on the scatter diagram. Then determine the equation of that line by using the point-slope form of a line, $y - y_1 = m(x - x_1)$. To review using this formula, work Problem P6 in "*Are You Prepared for This Section?*" on page 176, if you haven't already done so.

EXAMPLE 8   **Finding a Model for Linearly Related Data**

Using the data in Table 4 from Example 6,

(a) Select two points and find a linear function for the line connecting the points.

(b) Graph the line on the scatter diagram from Example 6(a).

(c) Use the function found in part (a) to predict the number of runs scored by a team whose on-base percentage is 33.5%.

(*continued*)

(d) Interpret the slope as a rate of change in the context of the problem. Does it make sense to interpret the y-intercept? Explain.

Solution

(a) Select any two points—for example, $(30.7, 661)$ and $(32.5, 751)$. (You should select your own two points and work through the solution.) The slope of the line joining these points is

$$m = \frac{751 - 661}{32.5 - 30.7} = \frac{90}{1.8} = 50$$

Work Smart

If necessary, round the slope and y-intercept to three decimal places.

Use the point-slope form to find the equation of the line that has slope 50 passing through $(30.7, 661)$:

$$\begin{array}{rl} \text{Point-slope form:} & y - y_1 = m(x - x_1) \\ m = 50; x_1 = 30.7, y_1 = 661: & y - 661 = 50(x - 30.7) \\ \text{Distribute:} & y - 661 = 50x - 1535 \\ \text{Add 661 to both sides:} & y = f(x) = 50x - 874 \end{array}$$

(b) Figure 25 shows the scatter diagram with the graph of the line found in part (a). The line through the two points selected in part (a) is graphed.

(c) Evaluate $f(x) = 50x - 874$ at $x = 33.5$.

$$\begin{aligned} f(33.5) &= 50(33.5) - 874 \\ &= 801 \end{aligned}$$

We predict that a team whose on-base percentage is 33.5% will score 801 runs.

(d) The slope of the linear function is 50, so if the on-base percentage increases by 1, then the number of runs scored will increase by 50. The y-intercept, -874, represents the runs scored when the on-base percentage is 0. Since negative runs scored does not make sense, interpreting the y-intercept does not make sense.

Figure 25

On-Base Percentage

Quick ✓

18. Using the data from Quick Check Problem 15 on page 182:

(a) Select two points and find a linear model that describes the relation between the points.

(b) Graph the line on the scatter diagram obtained in Quick Check Problem 15 (page 182).

(c) Predict the total cholesterol of a 39-year-old male. Round to the nearest whole number.

(d) Interpret the slope as a rate of change in the context of the problem. Does it make sense to interpret the y-intercept?

2.4 Exercises MyMathLab® Exercise numbers in green have complete video solutions in MyMathLab or may be accessed using the QR code to the right.

Problems 1–18. are the Quick ✓s that follow the EXAMPLES.

Building Skills

For Problems 19–30, graph each linear function. See Objective 1.

19. $F(x) = 5x - 2$

20. $F(x) = 4x + 1$

21. $G(x) = -3x + 7$

22. $G(x) = -2x + 5$

23. $H(x) = -2$

24. $P(x) = 5$

25. $f(x) = \frac{1}{2}x - 4$

26. $f(x) = \frac{1}{3}x - 3$

27. $F(x) = -\frac{5}{2}x + 5$

28. $P(x) = -\frac{3}{5}x - 1$

29. $G(x) = -\frac{3}{2}x$

30. $f(x) = \frac{4}{5}x$

In Problems 31–38, find the zero of the linear function.
See Objective 2.

31. $f(x) = 2x + 10$ **32.** $f(x) = 3x + 18$

33. $G(x) = -5x + 40$ **34.** $H(x) = -4x + 36$

35. $s(t) = \dfrac{1}{2}t - 3$ **36.** $p(q) = \dfrac{1}{4}q + 2$

37. $P(z) = -\dfrac{4}{3}z + 12$ **38.** $F(t) = -\dfrac{3}{2}t + 6$

In Problems 39–42, determine whether the scatter diagram indicates that a linear relation may exist between the two variables. If a linear relation does exist, indicate whether the slope is positive or negative. See Objective 4.

39.

40.

41.

42.

In Problems 43–46,

(a) *Draw a scatter diagram of the data.*

(b) *Select two points from the scatter diagram and find the equation of the line containing the points selected.*

(c) *Graph the line found in part (b) on the scatter diagram. See Objective 4.*

43.

x	2	4	5	8	9
y	1.4	1.8	2.1	2.3	2.6

44.

x	2	3	5	6	7
y	5.7	5.2	2.8	1.9	1.8

45.

x	1.2	1.8	2.3	3.5	4.1
y	8.4	7.0	7.3	4.5	2.4

46.

x	0	0.5	1.4	2.1	3.9
y	0.8	1.3	1.9	2.5	5.0

Mixed Practice

47. Suppose that $f(x) = 3x + 2$.
   (a) What is the slope?
   (b) What is the y-intercept?
   (c) What is the zero of f?
   (d) Solve $f(x) = 5$. What point is on the graph of f?
   (e) Solve $f(x) \le -1$.
   (f) Graph f.

48. Suppose that $g(x) = 8x + 3$.
   (a) What is the slope?
   (b) What is the y-intercept?
   (c) What is the zero of g?
   (d) Solve $g(x) = 19$. What point is on the graph of g?
   (e) Solve $g(x) > -5$.
   (f) Graph g.

49. Suppose that $f(x) = x - 5$ and $g(x) = -3x + 7$.
   (a) Solve $f(x) = g(x)$. What is the value of f at the solution? What point is on the graph of f? What point is on the graph of g?
   (b) Solve $f(x) > g(x)$.
   (c) Graph f and g in the same Cartesian plane. Label the intersection point.

50. Suppose that $f(x) = \dfrac{4}{3}x + 5$ and $g(x) = \dfrac{1}{3}x + 1$.
   (a) Solve $f(x) = g(x)$. What is the value of f at the solution? What point is on the graph of f? What point is on the graph of g?
   (b) Solve $f(x) \le g(x)$.
   (c) Graph f and g in the same Cartesian plane. Label the intersection point.

51. Find a linear function f such that $f(2) = 6$ and $f(5) = 12$. What is $f(-2)$?

52. Find a linear function g such that $g(1) = 5$ and $g(5) = 17$. What is $g(-3)$?

53. Find a linear function h such that $h(3) = 7$ and $h(-1) = 14$. What is $h\left(\dfrac{1}{2}\right)$?

54. Find a linear function F such that $F(2) = 5$ and $F(-3) = 9$. What is $F\left(-\dfrac{3}{2}\right)$?

55. In parts (a)–(e), use the figure shown below.

   (a) Solve $f(x) = 1$. **(b)** Solve $f(x) = -3$.
   (c) What is $f(4)$?
   (d) What are the intercepts of the function $y = f(x)$?
   (e) Write the equation of the function whose graph is given in the form $f(x) = mx + b$.

56. In parts (a)–(e), use the figure shown below.

(a) Solve $g(x) = 1$.
(b) Solve $g(x) = -1$.
(c) What is $g(-4)$?
(d) What are the intercepts of the function $y = g(x)$?
(e) Write the equation of the function whose graph is given in the form $g(x) = mx + b$.

Applying the Concepts

57. Cab Fare The linear function $C(m) = 1.5m + 2$ describes the cab fare C for a ride of m miles.

(a) What is the domain of this linear function?
(b) What is $C(0)$? Explain what this result means.
(c) What is cab fare for a 5-mile ride?
(d) Graph the linear function.
(e) How many miles can you ride in a cab if you have $13.25?
(f) Over what range of miles can you ride if you can spend no more than $39.50?

58. Sales Commissions Tanya works for Pearson Education as a book representative. The linear function $I(s) = 0.01s + 20,000$ describes the annual income I of Tanya when she has total sales s.

(a) What is the domain of this linear function?
(b) What is $I(0)$? Explain what this result means.
(c) What is Tanya's salary if she sells $500,000 in books for the year?
(d) Graph the linear function.
(e) At what level of sales will Tanya's income be $45,000?

59. Taxes The function $T(x) = 0.15(x - 9226) + 922.50$ represents the federal income tax bill T of a single person whose adjusted gross income in 2015 was x dollars for income between $9226 and $37,450, inclusive. source: *Internal Revenue Service*

(a) What is the domain of this linear function?
(b) What was a single filer's tax bill if adjusted gross income was $20,000?
(c) Which variable is independent and which is dependent?
(d) Graph the linear function over the domain specified in part (a).
(e) What was a single filer's adjusted gross income if his or her tax bill was $2996.25?

60. Luxury Tax In 2002, Major League Baseball signed a labor agreement with the players. In this agreement, any team whose payroll exceeds

$189 million in 2015 will have to pay a luxury tax of 17.5% (for first offenses). The linear function $T(p) = 0.175(p - 189)$ describes the luxury tax T of a team whose payroll was p (in millions).

(a) What is the domain of this linear function?
(b) What was the luxury tax for the Boston Red Sox whose payroll was $200 million in 2015?
(c) Graph the linear function.
(d) What was the payroll of the San Francisco Giants, who paid a luxury tax of $1.3 million in 2015?

61. Health Costs The annual cost of health insurance H as a function of age a is given by the function $H(a) = 22.8a - 117.5$ for $15 \le a \le 90$.
source: *Statistical Abstract*

(a) What are the independent and dependent variables?
(b) What is the domain of this linear function?
(c) What is the health insurance premium of a 30-year-old?
(d) Graph the linear function over its domain.
(e) What is the age of an individual whose health insurance premium is $976.90?

62. Birth Rate A multiple birth is any birth with 2 or more children born. The birth rate is the number of births per 1000 women. The birth rate B of multiple births as a function of age a is given by the function $B(a) = 1.73a - 14.56$ for $15 \le a \le 44$.
source: *Centers for Disease Control*

(a) What are the independent and dependent variables?
(b) What is the domain of this linear function?
(c) What is the multiple birth rate of women who are 22 years of age, according to the model?
(d) Graph the linear function over its domain.
(e) What is the age of women whose multiple birth rate is 49.45?

63. Phone Charges Sprint has a long-distance phone plan that charges a monthly fee of $5.95 plus $0.05 per minute. source: *Sprint.com*

(a) Find a linear function that expresses the monthly bill B as a function of minutes used m.
(b) What are the independent and dependent variables?
(c) What is the domain of this linear function?
(d) What is the monthly bill if 300 minutes are used for long-distance phone calls?
(e) How many minutes were used for long distance if the long-distance phone bill was $17.95?
(f) Graph the linear function.
(g) Over what range of minutes can you talk each month if you don't want to spend more than $18.45?

64. RV Rental The weekly rental cost R of a class C 20-foot recreational vehicle is $129 plus $0.32 per mile, up to a maximum of 500 miles. source: *westernrv.com*

(a) Find a linear function that expresses the cost R as a function of miles driven m.

(b) What are the independent and dependent variables?

(c) What is the domain of this linear function?

(d) What is the rental cost if 360 miles are driven?

(e) How many miles were driven if the rental cost is $275.56?

(f) Graph the linear function.

(g) Over what range of miles can you drive if you have a budget of $273?

65. Depreciation Suppose that a company has just purchased a new computer for $2700. The company chooses to depreciate the computer using the straight-line method over 3 years.

(a) Find a linear function that expresses the book value V of the computer as a function of its age x.

(b) What is the domain of this linear function?

(c) What is the book value of the computer after the first year?

(d) What are the intercepts of the graph of the linear function?

(e) When will the book value of the computer be $900?

(f) Graph the linear function.

66. Depreciation Suppose that a company just purchased a new machine for its manufacturing facility for $1,200,000. The company chooses to depreciate the machine using the straight-line method over 20 years.

(a) Find a linear function that expresses the book value V of the machine as a function of its age x.

(b) What is the domain of this linear function?

(c) What is the book value of the machine after three years?

(d) What are the intercepts of the graph of the linear function?

(e) When will the book value of the machine be $480,000?

(f) Graph the linear function.

67. Diamonds The relation between the cost of a diamond and its weight is linear. In looking at two diamonds, we find that one of the diamonds weighs 0.7 carat and costs $3340, while the other diamond weighs 0.8 carat and costs $4065. source: *diamonds.com*

(a) Find a linear function that relates the price of a diamond, C, to its weight, x, treating weight as the independent variable.

(b) Predict the price of a diamond that weighs 0.77 carat.

(c) Interpret the slope as a rate of change.

(d) If a diamond costs $5300, what do you think it should weigh?

68. Apartments In the North Chicago area, an 820-square-foot apartment rents for $1507 per month. A 970-square-foot apartment rents for $1660. Suppose that the relation between area and rent is linear. source: *apartments.com*

(a) Find a linear function that relates the rent of a North Chicago apartment, R, to its area, x, treating area as the independent variable.

(b) Predict the rent of a 900-square-foot apartment in North Chicago.

(c) Interpret the slope as a rate of change in the context of the problem.

(d) If the rent of a North Chicago apartment is $1300 per month, how big would you expect it to be?

69. The Consumption Function A famous theory in economics developed by John Maynard Keynes states that personal consumption expenditures are a linear function of disposable income. An economist wishes to develop a model that relates income and consumption and obtains the following information from the United States Bureau of Economic Analysis. In 2010, personal disposable income was $11,380 billion and personal consumption expenditures were $10,349 billion. In 2015, personal disposable income was $15,605 billion and personal consumption expenditures were $13,618 billion.

(a) Find a linear function that relates personal consumption expenditures, C, to disposable income, x, treating disposable income as the independent variable.

(b) In 2013, personal disposable income was $14,221 billion. Use this information to find personal consumption expenditures in 2013 to the nearest billion.

(c) Interpret the slope as a rate of change in the context of the problem. In economics, this slope is called the **marginal propensity to consume.**

(d) If personal consumption expenditures were $13,117 billion, what do you think that disposable income was to the nearest billion?

70. Birth Weight According to the National Center for Health Statistics, the average birth weight of babies born to 22-year-old mothers is 3280 grams. The average birth weight of babies born to 32-year-old mothers is 3370 grams. Suppose that the relation between age of mother and birth weight is linear.

(a) Find a linear function that relates age of mother a to birth weight W, treating age of mother as the independent variable.

(b) Predict the birth weight of a baby born to a mother who is 30 years old.

(c) Interpret the slope as a rate of change in the context of the problem.

(d) If a baby weighs 3310 grams, how old do you expect the mother to be?

71. Concrete As concrete cures, it gains strength. The data represents the 7-day and 28-day strength (in pounds per square inch) of a certain type of concrete.

7-day Strength, x	28-day Strength, y
2300	4070
3390	5220
2430	4640
2890	4620
3330	4850
2480	4120
3380	5020
2660	4890
2620	4190
3340	4630

(a) Draw a scatter diagram of the data, treating 7-day strength as the independent variable.
(b) What type of relation appears to exist between 7-day strength and 28-day strength?
(c) Select two points and find an equation of the line containing the points.
(d) Graph the line on the scatter diagram drawn in part (a).
(e) Predict the 28-day strength of a slab of concrete if its 7-day strength is 3000 psi.
(f) Interpret the slope of the line as a rate of change in the context of the problem found in part (c).

72. Candy The following data represent the weight (in grams) of various candy bars and the corresponding number of calories.

Candy Bar	Weight, x	Calories, y
Hershey's Milk Chocolate	44.28	230
Nestle Crunch	44.84	230
Butterfinger	61.30	270
Baby Ruth	66.45	280
Almond Joy	47.33	220
Twix (with Caramel)	58.00	280
Snickers	61.12	280
Heath	39.52	210

SOURCE: *Megan Pocius, student at Joliet Junior College*

(a) Draw a scatter diagram of the data, treating weight as the independent variable.
(b) What type of relation appears to exist between the weight of a candy bar and the number of calories?
(c) Select two points and find an equation of the line containing the points.
(d) Graph the line on the scatter diagram drawn in part (a).

(e) Predict the number of calories in a candy bar that weighs 62.3 grams.
(f) Interpret the slope of the line as a rate of change in the context of the problem found in part (c).

73. Raisins The following data represent the weight (in grams) of a box of raisins and the number of raisins in the box.

Weight, w	Number of Raisins, N
42.3	87
42.7	91
42.8	93
42.4	87
42.6	89
42.4	90
42.3	82
42.5	86
42.7	86
42.5	86

SOURCE: *Jennifer Maxwell, student at Joliet Junior College*

(a) Does the relation defined by the set of ordered pairs (w, N) represent a function?
(b) Draw a scatter diagram of the data, treating weight as the independent variable.
(c) Select two points and find the equation of the line containing the points.
(d) Graph the line on the scatter diagram drawn in part (b).
(e) Express the relationship found in part (c) using function notation.
(f) Predict the number of raisins in a box that weighs 42.5 grams.
(g) Interpret the slope of the line as a rate of change in the context of the problem found in part (c).

74. Height versus Head Circumference The following data represent the height (in inches) and head circumference (in inches) of 9 randomly selected children.

Height, h	Head Circumference, C
25.25	16.4
25.75	16.9
25	16.9
27.75	17.6
26.50	17.3
27.00	17.5
26.75	17.3
26.75	17.5
27.5	17.5

SOURCE: *Denise Slucki, student at Joliet Junior College*

(a) Does the relation defined by the set of ordered pairs (h, C) represent a function?

(b) Draw a scatter diagram of the data, treating height as the independent variable.

(c) Select two points and find the equation of the line containing the points.

(d) Graph the line on the scatter diagram drawn in part (b).

(e) Express the relationship found in part (c) using function notation.

(f) Predict the head circumference of a child who is 26.5 inches tall.

(g) Interpret the slope of the line as a rate of change in the context of the problem found in part (c).

Extending the Concepts

75. Math for the Future: Calculus The **average rate of change** of a function $y = f(x)$ from c to x is defined as

$$\text{Average rate of change} = \frac{\Delta y}{\Delta x} = \frac{f(x) - f(c)}{x - c}, \quad x \neq c$$

provided that c is in the domain of f. The average rate of change of a function is simply the slope of the line joining the points $(c, f(c))$ and $(x, f(x))$. The line joining these points is called a **secant line.** The slope of the secant line is

$$m_{\text{sec}} = \frac{f(x) - f(c)}{x - c}$$

The following figure illustrates the idea.

Below, we show the graph of the function $f(x) = 2x^2 - 4x + 1$.

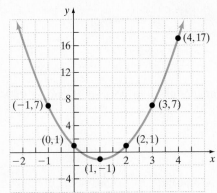

(a) On the graph of $f(x) = 2x^2 - 4x + 1$, draw a line through the points $(1, f(1))$ and $(x, f(x))$, where $x = 4$.

(b) Find the slope of the secant line through $(1, f(1))$ and $(x, f(x))$, where $x = 4$.

(c) Find the equation of the secant line through $(1, f(1))$ and $(x, f(x))$, where $x = 4$.

(d) Repeat parts (a)–(c) for $x = 3$, $x = 2$, $x = 1.5$, and $x = 1.1$.

(e) What happens to the slope of the secant line as x gets closer to 1?

76. A strain of *E. coli* Beu 397-recA441 is placed into a Petri dish at 30° Celsius and allowed to grow. The population is estimated by means of an optical device in which the amount of light that passes through the Petri dish is measured. The data below are collected. Do you think that a linear function could be used to describe the relation between the two variables? Why or why not?

Time, x	Population, y
0	0.09
2.5	0.18
3.5	0.26
4.5	0.35
6	0.50

SOURCE: *Dr. Polly Lavery, Joliet Junior College*

Technology Exercises

The equation of the line obtained in Example 8 depends on the points selected, which will vary from person to person. So the line we found might be different from the line that you found. Although the line that we found in Example 8 fits the data well, there may be a line that "fits better." Do you think that your line fits the data better? Is there a line of *best fit*? As it turns out, there is a method for finding the line that best fits linearly related data (called the *line of best fit*).*

Graphing utilities can be used to draw scatter diagrams and find the line of best fit. Figure 26(a) shows a scatter diagram of the data presented in Table 4 from Example 6, drawn on a graphing calculator. Figure 26(b) shows the line of best fit from a graphing calculator.

The line of best fit is $y = 53.529x - 963.453$.

Figure 26   **(a)**   **(b)**

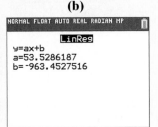

In Problems 77–80,

(a) *Draw a scatter diagram using a graphing calculator.*

(b) *Find the line of best fit for the data in the problem specified using technology.*

77. Problem 71   **78.** Problem 72

79. Problem 73   **80.** Problem 74

---

* This text will not discuss the underlying mathematics of lines of best fit. Books on elementary statistics discuss this topic.

2.5 Compound Inequalities

Objectives

① Determine the Intersection or Union of Two Sets
② Solve Compound Inequalities Involving "and"
③ Solve Compound Inequalities Involving "or"
④ Solve Problems Using Compound Inequalities

Are You Prepared for This Section?

Before getting started, complete the following problems. If you get a problem wrong, go back to the section cited and review the material.

P1. Use set-builder notation and interval notation to name the set of all real numbers x such that $-2 \leq x \leq 5$. [Section 1.4, pp. 82–85]

P2. Graph the inequality $x \geq 4$. [Section 1.4, pp. 82–85]

P3. Use interval notation to express the inequality shown in the graph. [Section 1.4, pp. 82–85]

$$\begin{array}{c} \llap{}\vdash\!\!\!\!+\!\!\!\!+\!\!\!\!+\!\!\!\!+\!\!\!\!+\!\!\!\!+ \\ -2\ -1\ \ 0\ \ 1\ \ 2\ \ 3\ \ 4 \end{array}$$

P4. Solve: $2(x + 3) - 5x = 15$ [Section 1.1, pp. 50–53]

P5. Solve: $2x + 3 > 11$ [Section 1.4, pp. 86–88]

P6. Solve: $x + 8 \geq 4(x - 1) - x$ [Section 1.4, pp. 86–88]

▶ ① Determine the Intersection or Union of Two Sets

Table 5 contains information about students in an Intermediate Algebra course. These people can be classified in various sets. For example, if set A is defined as the set of all students whose age is less than 25. Then

$$A = \{\text{Grace, Sophia, Kevin, Jack, George, Teresa}\}$$

If set B is defined as the set of all students who are female, then

$$B = \{\text{Grace, Sophia, Mary, Nancy, Teresa}\}$$

Now list all the students who are in set A and set B, that is, students who are under 25 years of age and female.

$$A \text{ and } B = \{\text{Grace, Sophia, Teresa}\}$$

Now list all the students who are in set A or in set B or in both sets.

$$A \text{ or } B = \{\text{Grace, Sophia, Kevin, Jack, George, Teresa, Mary, Nancy}\}$$

Figure 27 shows a Venn diagram illustrating the relations among A, B, A and B, and A or B. Notice that Grace, Sophia, and Teresa are in both A and B, while Robert is in neither A nor B.

Table 5

Student	Age	Gender
Grace	19	Female
Sophia	23	Female
Kevin	20	Male
Robert	32	Male
Jack	19	Male
Mary	35	Female
Nancy	40	Female
George	22	Male
Teresa	20	Female

Figure 27

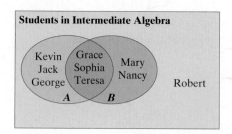

When the word "and" was used, elements common to both set A and set B were listed. When the word "or" was used elements in set A or in set B or in both sets were listed. These results lead to the following definitions.

Prepared?...Answers

P1. $\{x | -2 \leq x \leq 5\}$; $[-2, 5]$

P2.
$$\begin{array}{c} \llap{}+\!\!\!\!+\!\!\!\!+\!\!\!\!+\!\!\!\!+\!\!\!\!\bullet\!\!\!\!\!-\!\!\!\!\!\longrightarrow \\ 0\ \ 1\ \ 2\ \ 3\ \ 4\ \ 5\ \ 6 \end{array}$$

P3. $(-1, 3]$ **P4.** $\{-3\}$

P5. $\{x | x > 4\}$; $(4, \infty)$;
$$\begin{array}{c} \llap{}+\!\!\!\!+\!\!\!\!\circ\!\!\!\!-\!\!\!\!+\!\!\!\!+\!\!\!\!\longrightarrow \\ 2\ \ 3\ \ 4\ \ 5\ \ 6\ \ 7 \end{array}$$

P6. $\{x | x \leq 6\}$; $(-\infty, 6]$;
$$\begin{array}{c} \llap{}\longleftarrow\!\!\!\!+\!\!\!\!+\!\!\!\!+\!\!\!\!\bullet\!\!\!\!+ \\ 2\ \ 3\ \ 4\ \ 5\ \ 6\ \ 7 \end{array}$$

Definitions

• The **intersection** of two sets A and B, denoted $A \cap B$, is the set of all elements in both set A and set B.

• The **union** of two sets A and B, denoted $A \cup B$, is the set of all elements in set A or in set B or in both set A and set B.

• The word **and** implies intersection, while the word **or** implies union.

EXAMPLE 1 **Finding the Intersection and Union of Sets**

Let $A = \{1, 3, 5, 7, 9\}$, $B = \{1, 2, 3, 4, 5\}$, and $C = \{2, 4\}$. Find

(a) $A \cap B$ (b) $A \cup B$ (c) $A \cup C$

Solution

(a) $A \cap B$ is the set of all elements that are in both A and B. So,

$$A \cap B = \{1, 3, 5\}$$

Work Smart

When finding the union of two sets, list each element only once, even if it occurs in both sets.

(b) $A \cup B$ is the set of all elements that are in A or B, or both. So,

$$A \cup B = \{1, 2, 3, 4, 5, 7, 9\}$$

(c) $A \cup C$ is the set of all elements that are in A or C, or both. So, $A \cup C$ is the empty set which can be notated so $\varnothing$ or $\{\ \}$.   ●

Quick ✓

1. The _____ of two sets A and B, denoted $A \cap B$, is the set of all elements that belong to both set A and set B.

2. The word ___ implies intersection. The word ___ implies union.

3. *True or False* The intersection of two sets can be the empty set.

4. *True or False* The symbol for the union of two sets is $\cap$.

In Problems 5–10, let $A = \{1, 2, 3, 4, 5, 6\}$, $B = \{1, 3, 5, 7\}$, and $C = \{2, 4, 6, 8\}$.

5. Find $A \cap B$.   6. Find $A \cap C$.

7. Find $A \cup B$.   8. Find $A \cup C$.

9. Find $B \cap C$.   10. Find $B \cup C$.

▶ **EXAMPLE 2** **Finding the Intersection and Union of Two Sets**

Suppose $A = \{x \mid x \le 5\}$, $B = \{x \mid x \ge 1\}$, and $C = \{x \mid x < -2\}$.

(a) Determine $A \cap B$. Graph the set on a real number line and write it in set-builder notation and interval notation.

(b) Determine $B \cup C$. Graph the set on a real number line and write it in set-builder notation and interval notation.

Solution

(a) $A \cap B$ is the set of all real numbers less than or equal to 5 and greater than or equal to 1. Identify this set by determining where the graphs of the inequalities overlap. See Figure 28.

Figure 28

$A \cap B$ is $\{x \mid 1 \le x \le 5\}$ in set-builder notation, and $[1, 5]$ in interval notation.

Figure 29

(b) $B \cup C$ is the set of all real numbers greater than or equal to 1 or less than -2. See Figure 29. Therefore, $B \cup C$ is $\{x \mid x < -2 \text{ or } x \ge 1\}$ in set-builder notation, and $(-\infty, -2) \cup [1, \infty)$ in interval notation.   ●

Work Smart

Throughout the text when a solution contains more than one interval, and is written in set-builder notation, use the word "or." When a solution is written in interval notation and contains more than one interval, use the union symbol, $\cup$. See the solution to Quick ✓ 12.

Quick ✓

Let $A = \{x \mid x > 2\}, B = \{x \mid x < 7\},$ *and* $C = \{x \mid x \leq -3\}.$

11. Determine $A \cap B$. Graph the set on a real number line and write it in set-builder notation and interval notation.

12. Determine $A \cup C$. Graph the set on a real number line and write it in set-builder notation and interval notation.

▶ ❷ Solve Compound Inequalities Involving "and"

A **compound inequality** is formed by joining two inequalities with the word "and" or "or." For example,

$$3x + 1 > 4 \quad \text{and} \quad 2x - 3 < 7$$
$$5x - 2 \leq 13 \quad \text{or} \quad 2x - 5 > 3$$

are compound inequalities. To **solve a compound inequality** means to find all possible values of the variable such that the compound inequality results in a true statement. For example, the compound inequality

$$3x + 1 > 4 \quad \text{and} \quad 2x - 3 < 7$$

is true for $x = 2$ but false for $x = 0$. It is false for $x = 0$ because the solution is only true for the inequality $2x - 3 < 7$. It is not true for both inequalities.

Look at an example that illustrates how to solve compound inequalities involving the word "and."

EXAMPLE 3 **How to Solve a Compound Inequality Involving "and"**

Solve $3x + 2 > -7$ and $4x + 1 \leq 9$. Express your solution using set-builder and interval notation. Graph the solution set.

Step-by-Step Solution

Step 1: Solve each inequality separately.

	$3x + 2 > -7$		$4x + 1 \leq 9$
Subtract 2 from both sides:	$3x > -9$	Subtract 1 from both sides:	$4x \leq 8$
Divide both sides by 3:	$x > -3$	Divide both sides by 4:	$x \leq 2$

Step 2: Find the intersection of the solution sets, which will represent the solution set to the compound inequality.

To find the intersection of the two solution sets, graph each inequality separately. See Figures 30(a) and (b).

Figure 30

(a)
$x > -3$

(b)
$x \leq 2$

(c)
$-3 < x \leq 2$

The intersection of $x > -3$ and $x \leq 2$ is $-3 < x \leq 2$, so the solution set is $\{x \mid -3 < x \leq 2\}$ or, using interval notation, $(-3, 2]$. The graph of the solution set is shown in Figure 30(c). ●

The steps below summarize the procedure for solving compound inequalities involving "and."

Work Smart

The words "and" and "intersection" suggest overlap. When solving these types of problems, look for the overlap of the graphs.

Solving Compound Inequalities Involving "and"

Step 1: Solve each inequality separately.

Step 2: Find the INTERSECTION of the solution sets of the respective inequalities.

EXAMPLE 4   **Solving a Compound Inequality with "and"**

Solve $-2x + 5 > -1$ and $5x + 6 \leq -4$. Express the solution using set-builder notation and interval notation. Graph the solution set.

Solution
Solve each inequality separately:

$$-2x + 5 > -1 \qquad\qquad\qquad\qquad 5x + 6 \leq -4$$

Subtract 5 from both sides: $\quad -2x > -6 \qquad$ Subtract 6 from both sides: $\quad 5x \leq -10$

Divide both sides by -2 and reverse the direction of the inequality since you are dividing by a negative number: $\qquad x < 3$ $\qquad$ Divide both sides by 5: $\qquad x \leq -2$

The intersection of the solution sets is the solution set to the compound inequality. See Figures 31(a) and (b).

Figure 31   (a)
$x < 3$

(b)
$x \leq -2$

(c)
$x \leq -2$ and $x < 3$

The intersection of $x < 3$ and $x \leq -2$ is $x \leq -2$. The solution set is $\{x \mid x \leq -2\}$ or, using interval notation, $(-\infty, -2]$. The graph of the solution set is shown in Figure 31(c).

Quick ✓
In Problems 13–15, solve each compound inequality. Express the solution using set-builder notation and interval notation. Graph the solution set.

13. $2x + 1 \geq 5$ and $-3x + 2 < 5$

14. $4x - 5 < 7$ and $3x - 1 > -10$

15. $-8x + 3 < -5$ and $\dfrac{2}{3}x + 1 < 3$

EXAMPLE 5   **Solving a Compound Inequality with "and"**

Solve $x - 5 > -1$ and $2x - 3 \leq -5$. Express the solution using set-builder and interval notation. Graph the solution set.

Solution
Solve each inequality separately:

$$x - 5 > -1 \qquad\qquad\qquad\qquad 2x - 3 \leq -5$$

Add 5 to both sides: $\qquad x > 4 \qquad$ Add 3 to both sides: $\qquad 2x \leq -2$

Divide both sides by 2: $\qquad x \leq -1$

Figure 32
$x > 4$

$x \leq -1$

The intersection of the solution sets is the solution set to the compound inequality. See Figure 32. The intersection is the empty set, so the solution set is { } or $\varnothing$.

Quick ✓
In Problems 16 and 17, solve each compound inequality. Express the solution using set-builder notation and interval notation. Graph the solution set.

16. $3x - 5 < -8$ and $2x + 1 > 5$

17. $5x + 1 \leq 6$ and $3x + 2 \geq 5$

Sometimes, "and" inequalities can be combined into a more streamlined notation.

> **Writing Inequalities Involving "and" Compactly**
>
> If $a < b$, then we can write
>
> $$a < x \quad \text{and} \quad x < b$$
>
> more compactly as
>
> $$a < x < b$$

For example,

$$-3 < -4x + 1 \quad \text{and} \quad -4x + 1 < 13$$

can be written as

$$-3 < -4x + 1 < 13$$

Such compound inequalities can be solved by isolating the variable "in the middle" with a coefficient of 1.

▶ **EXAMPLE 6** | **Solving a Compound Inequality**

Solve $-3 < -4x + 1 < 13$. Express the solution using set-builder notation and interval notation. Graph the solution set.

Solution

Isolate the variable "in the middle" with a coefficient of 1:

$$-3 < -4x + 1 < 13$$

Subtract 1 from all three parts (Addition Property): $\quad -3 - 1 < -4x + 1 - 1 < 13 - 1$

$$-4 < -4x < 12$$

Divide all three parts by -4 and
reverse the direction of both inequalities: $\quad \dfrac{-4}{-4} > \dfrac{-4x}{-4} > \dfrac{12}{-4}$

$$1 > x > -3$$

If $b > x > a$, then $a < x < b$: $\quad -3 < x < 1$

The solution set is $\{x \mid -3 < x < 1\}$ or, in interval notation, $(-3, 1)$. Figure 33 shows the graph of the solution set.

Figure 33

<!-- number line from -5 to 3, open circles at -3 and 1 -->
$$-5 \quad -4 \quad -3 \quad -2 \quad -1 \quad 0 \quad 1 \quad 2 \quad 3$$

To visualize the results of Example 6, look at Figure 34, which shows the graph of $f(x) = -3$, $g(x) = -4x + 1$, and $h(x) = 13$. Note that the graph of $g(x) = -4x + 1$ is between the graphs of $f(x) = -3$ and $h(x) = 13$ for $-3 < x < 1$. Thus the solution set of $-3 < -4x + 1 < 13$ is $\{x \mid -3 < x < 1\}$.

Figure 34

> **Quick ✓**
>
> *In Problems 18–20, solve each compound inequality. Express the solution using set-builder notation and interval notation. Graph the solution set.*
>
> **18.** $-2 < 3x + 1 < 10$
>
> **19.** $0 < 4x - 5 \le 3$
>
> **20.** $3 \le -2x - 1 \le 11$

▶ ❸ **Solve Compound Inequalities Involving "or"**

The solution to compound inequalities involving the word "or" is the union of the solutions to each inequality.

EXAMPLE 7 **How to Solve a Compound Inequality Involving "or"**

Solve $3x - 5 < -2$ or $4 - 5x \le -16$. Express the solution using set-builder notation and interval notation. Graph the solution set.

Step-by-Step Solution

Step 1: Solve each inequality separately.

	$3x - 5 < -2$		$4 - 5x \le -16$
Add 5 to each side:	$3x < 3$	Subtract 4 from both sides:	$-5x \le -20$
Divide both sides by 3:	$x < 1$	Divide both sides by −5; Reverse the direction of the inequality:	$x \ge 4$

Step 2: Find the union of the solution sets, which will represent the solution set to the compound inequality.

The union of the two solution sets is $x < 1$ or $x \ge 4$. The solution set using set-builder notation is $\{x \mid x < 1 \text{ or } x \ge 4\}$. The solution set using interval notation is $(-\infty, 1) \cup [4, \infty)$. Figure 35 shows the graph of the solution set.

Figure 35

> **Steps for Solving Compound Inequalities Involving "or"**
>
> **Step 1:** Solve each inequality separately.
>
> **Step 2:** Find the UNION of the solution sets of each inequality.

Work Smart

A common error is to write the solution $x < 1$ or $x \ge 4$ as $1 > x \ge 4$, which is incorrect. There are no real numbers that are less than 1 *and* greater than or equal to 4. Another common error is to "mix" symbols, as in $1 < x \ge 4$. This notation does not make sense!

Quick ✓

In Problems 21–24, solve each compound inequality. Express the solution using set-builder notation and interval notation. Graph the solution set.

21. $x + 3 < 1$ or $x - 2 > 3$

22. $3x + 1 \le 7$ or $2x - 3 > 9$

23. $2x - 3 \ge 1$ or $6x - 5 \ge 1$

24. $\dfrac{3}{4}(x + 4) < 6$ or $\dfrac{3}{2}(x + 1) > 15$

EXAMPLE 8 **Solving Compound Inequalities Involving "or"**

Solve $\dfrac{1}{2}x - 1 < 1$ or $\dfrac{2x - 1}{3} \ge -1$. Express the solution using set-builder notation and interval notation. Graph the solution set.

Solution

Solve each inequality separately:	$\dfrac{1}{2}x - 1 < 1$		$\dfrac{2x - 1}{3} \ge -1$
Add 1 to each side:	$\dfrac{1}{2}x < 2$	Multiply both sides by 3:	$2x - 1 \ge -3$
Multiply both sides by 2:	$x < 4$	Add 1 to both sides:	$2x \ge -2$
		Divide both sides by 2:	$x \ge -1$

Find the union of these solution sets. The graphs in Figure 36 on the following page shows that the union of the two solution sets is the set of all real numbers.

(continued)

Figure 36

The solution set is $\{x \mid x \text{ is any real number}\}$ in set-builder notation, and $(-\infty, \infty)$ in interval notation.

> **Quick ✓**
>
> *In Problems 25 and 26, solve each compound inequality. Express the solution using set-builder notation and interval notation. Graph the solution set.*
>
> **25.** $3x - 2 > -5$ or $2x - 5 \leq 1$
>
> **26.** $-5x - 2 \leq 3$ or $7x - 9 > 5$

▶ **④ Solve Problems Using Compound Inequalities**

We now look at an application involving compound inequalities.

EXAMPLE 9   **Federal Income Taxes**

In 2015, married couples filing a joint federal tax return who were in the 25% tax bracket paid between $10,313 and $29,388 in federal income taxes. These couples' federal income taxes equal $10,313 plus 25% of their taxable income over $74,900. Find the range of taxable incomes for which a married couple would have been in the 25% tax bracket.

SOURCE: *Internal Revenue Service*

Solution

Step 1: Identify We want to find the range of taxable incomes for married couples in the 25% tax bracket. This direct translation problem involves an inequality.

Step 2: Name Let t represent the taxable income.

Step 3: Translate The federal tax bill equals $10,313 plus 25% of the taxable income over $74,900. If the couple had taxable income equal to $75,900, their tax bill was $10,313 plus 25% of $1000 ($1000 is the amount over $74,900). In general, if the couple has taxable income t, then their tax bill will be

$$
\underbrace{\$10,313}_{\$10,313} \quad \underbrace{\text{plus}}_{+} \quad \underbrace{25\%}_{0.25} \quad \cdot \quad \underbrace{\text{of the amount over } \$74,900}_{(t - \$74,900)}
$$

Work Smart

The word "range" tells us that an inequality is to be solved, since it indicates more than one number.

Because the tax bill was between $10,313 and $29,388, we have

$$10{,}313 \leq 10{,}313 + 0.25(t - 74{,}900) \leq 29{,}388 \quad \text{The Model}$$

Step 4: Solve

$$10{,}313 \leq 10{,}313 + 0.25(t - 74{,}900) \leq 29{,}388$$

Distribute 0.25: $\quad 10{,}313 \leq 10{,}313 + 0.25t - 18{,}725 \leq 29{,}388$

Combine like terms: $\quad 10{,}313 \leq -8412 + 0.25t \leq 29{,}388$

Add 8412 to all three parts: $\quad 18{,}725 \leq 0.25t \leq 37{,}800$

Divide all three parts by 0.25: $\quad 74{,}900 \leq t \leq 151{,}200$

Step 5: Check If a married couple had taxable income of $74,900, then their tax bill was $10,313 + 0.25($74,900 − $74,900) = $10,313. If a married couple had taxable income of $151,200, then their tax bill was $10,313 + 0.25($151,200 − $74,900) = $29,388.

Step 6: Answer the Question A married couple who filed a joint federal income tax return with a tax bill between $10,313 and $29,388 had taxable income between $74,900 and $151,200.

Quick ✓

27. In 2015, an individual filing a federal tax return whose income placed him or her in the 25% tax bracket paid federal income taxes between $5156.25 and $18,481.25. The individual had to pay federal income taxes equal to $5156.25 plus 25% of the amount over $37,450. Find the range of taxable income in order for an individual to have been in the 25% tax bracket. (SOURCE: *Internal Revenue Service*)

28. Juan's monthly payments for his new car are $260 per month and he estimates that maintenance and gas cost him $0.20 per mile. During the course of a year, Juan's total monthly cost for his car ranges between $420 and $560. What is the range of miles he drives per month?

 2.5 Exercises MyMathLab® Exercise numbers in green have complete video solutions in MyMathLab or may be accessed using the QR code to the right.

*Problems **1–28** are the Quick ✓s that follow the EXAMPLES.*

Building Skills

In Problems 29–34, use A = {4, 5, 6, 7, 8, 9}, *B* = {1, 5, 7, 9}, *and C* = {2, 3, 4, 6} *to find each set. See Objective 1.*

29. $A \cup B$ **30.** $A \cup C$

31. $A \cap B$ **32.** $A \cap C$

33. $B \cap C$ **34.** $B \cup C$

In Problems 35–38, use the graph of the inequality to find each set. See Objective 1.

35. $A = \{x \mid x \le 5\}; B = \{x \mid x > -2\}$
Find (a) $A \cap B$ and (b) $A \cup B$.

36. $A = \{x \mid x \ge 4\}; B = \{x \mid x < 1\}$
Find (a) $A \cap B$ and (b) $A \cup B$.

37. $E = \{x \mid x > 3\}; F = \{x \mid x < -1\}$
Find (a) $E \cap F$ and (b) $E \cup F$.

38. $E = \{x \mid x \le 2\}; F = \{x \mid x \ge -2\}$
Find (a) $E \cap F$ and (b) $E \cup F$.

In Problems 39–42, use the graph to solve the compound inequality. Graph the solution set. See Objectives 2 and 3.

39. (a) $-5 \le 2x - 1 \le 3$
(b) $2x - 1 < -5$ or $2x - 1 > 3$

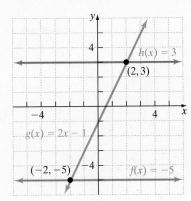

40. (a) $-1 \le \frac{1}{2}x + 1 \le 3$

(b) $\frac{1}{2}x + 1 < -1$ or $\frac{1}{2}x + 1 > 3$

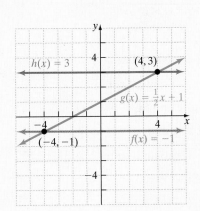

41. (a) $-4 < -\dfrac{5}{3}x + 1 < 6$

   (b) $-\dfrac{5}{3}x + 1 \le -4$ or $-\dfrac{5}{3}x + 1 \ge 6$

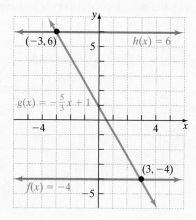

42. (a) $-3 < \dfrac{5}{4}x + 2 < 7$

   (b) $\dfrac{5}{4}x + 2 \le -3$ or $\dfrac{5}{4}x + 2 \ge 7$

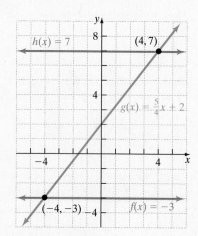

In Problems 43–66, solve each compound inequality. Graph the solution set. See Objective 2.

43. $x < 3$ and $x \ge -2$

44. $x \le 5$ and $x > 0$

45. $4x - 4 < 0$ and $-5x + 1 \le -9$

46. $6x - 2 \le 10$ and $10x > -20$

47. $4x - 3 < 5$ and $-5x + 3 > 13$

48. $x - 3 \le 2$ and $6x + 5 \ge -1$

49. $7x + 2 \ge 9$ and $4x + 3 \le 7$

50. $-4x - 1 < 3$ and $-x - 2 > 3$

51. $-3 \le 5x + 2 < 17$

52. $-10 < 6x + 8 \le -4$

53. $-3 \le 6x + 1 \le 10$

54. $-12 < 7x + 2 \le 6$

55. $3 \le -5x + 7 < 12$

56. $-6 < -3x + 6 \le 4$

57. $-1 \le \dfrac{1}{2}x - 1 \le 3$

58. $0 < \dfrac{3}{2}x - 3 \le 3$

59. $3 \le -2x - 1 \le 11$

60. $-3 < -4x + 1 < 17$

61. $\dfrac{2}{3}x + \dfrac{1}{2} < \dfrac{5}{6}$ and $-\dfrac{1}{5}x + 1 < \dfrac{3}{10}$

62. $x - \dfrac{3}{2} \le \dfrac{5}{4}$ and $-\dfrac{2}{3}x - \dfrac{2}{9} < \dfrac{8}{9}$

63. $-2 < \dfrac{3x + 1}{2} \le 8$

64. $-4 \le \dfrac{4x - 3}{3} < 3$

65. $-8 \le -2(x + 1) < 6$

66. $-6 < -3(x - 2) < 15$

In Problems 67–80, solve each compound inequality. Graph the solution set. See Objective 3.

67. $x < -2$ or $x > 3$

68. $x < 0$ or $x \ge 6$

69. $x - 2 < -4$ or $x + 3 > 8$

70. $x + 3 \le 5$ or $x - 2 \ge 3$

71. $6(x - 2) < 12$ or $4(x + 3) > 12$

72. $4x + 3 > -5$ or $8x - 5 < 3$

73. $-8x + 6x - 2 > 0$ or $5x > 3x + 8$

74. $3x \ge 7x + 8$ or $x < 4x - 9$

75. $2x + 5 \le -1$ or $\dfrac{4}{3}x - 3 > 5$

76. $-\dfrac{4}{5}x - 5 > 3$ or $7x - 3 > 4$

77. $\dfrac{1}{2}x < 3$ or $\dfrac{3x - 1}{2} > 4$

78. $\dfrac{2}{3}x + 2 \le 4$ or $\dfrac{5x - 3}{3} \ge 4$

79. $3(x - 1) + 5 < 2$ or $-2(x - 3) < 1$

80. $2(x + 1) - 5 \le 4$ or $-(x + 3) \le -2$

Mixed Practice

In Problems 81–94, solve each compound inequality. Graph the solution set.

81. $3a + 5 < 5$ and $-2a + 1 \leq 7$

82. $5x - 1 < 9$ and $5x > -20$

83. $5(x + 2) < 20$ or $4(x - 4) > -20$

84. $3(x + 7) < 24$ or $6(x - 4) > -30$

85. $-4 \leq 3x + 2 \leq 10$

86. $-8 \leq 5x - 3 \leq 4$

87. $2x + 7 < -13$ or $5x - 3 > 7$

88. $3x - 8 < -14$ or $4x - 5 > 7$

89. $5 < 3x - 1 < 14$

90. $-5 < 2x + 7 \leq 5$

91. $\dfrac{x}{3} \leq -1$ or $\dfrac{4x - 1}{2} > 7$

92. $\dfrac{x}{2} \leq -4$ or $\dfrac{2x - 1}{3} \geq 2$

93. $-3 \leq -2(x + 1) < 8$

94. $-15 < -3(x + 2) \leq 1$

Applying the Concepts

In Problems 95–100, use the Addition Property and/or Multiplication Properties to find a and b.

95. If $-3 < x < 4$, then $a < x + 4 < b$.

96. If $-2 < x < 3$, then $a < x - 3 < b$.

97. If $4 < x < 10$, then $a < 3x < b$.

98. If $2 < x < 12$, then $a < \dfrac{1}{2}x < b$.

99. If $-2 < x < 6$, then $a < 3x + 5 < b$.

100. If $-4 < x < 3$, then $a < 2x - 7 < b$.

101. Systolic Blood Pressure Blood pressure is measured using two numbers. One of the numbers measures systolic blood pressure. The systolic blood pressure represents the pressure while the heart is beating. In a healthy person, the systolic blood pressure should be greater than 90 and less than 140. If we let the variable x represent a person's systolic blood pressure, express the systolic blood pressure of a healthy person using a compound inequality.

102. Diastolic Blood Pressure Blood pressure is measured using two numbers. One of the numbers measures diastolic blood pressure. The diastolic blood pressure represents the pressure while the

heart is resting between beats. In a healthy person, the diastolic blood pressure should be greater than 60 and less than 90. If we let the variable x represent a person's diastolic blood pressure, express the diastolic blood pressure of a healthy person using a compound inequality.

103. Computing Grades Joanna desperately wants to earn a B in her history class. Her current test scores are 74, 86, 77, and 89. Her final exam is worth two test scores. In order to earn a B, Joanna's average must lie between 80 and 89, inclusive. What range of scores can Joanna receive on the final and earn a B in the course?

104. Computing Grades Jack needs to earn a C in his sociology class. His current test scores are 67, 72, 81, and 75. His final exam is worth three test scores. In order to earn a C, Jack's average must lie between 70 and 79, inclusive. What range of scores can Jack receive on the final exam and earn a C in the course?

105. Federal Tax Withholding The percentage method of withholding for federal income tax (2014) states that a single person whose weekly wages, after subtracting withholding allowances, are over $436, but not over $1506, shall have $34.90 plus 15% of the excess over $436 withheld. Over what range does the amount withheld vary if the weekly wages vary from $800 to $900, inclusive? SOURCE: *Internal Revenue Service*

106. Federal Tax Withholding Rework Problem 105 if the weekly wages vary from $1000 to $1100, inclusive.

107. Commission Gerard had an offer for a medical equipment sales position that pays $2500 per month plus 1% of all sales. What total sales is required to earn between $3000 and $5000 per month?

108. Commission Juanita had a job offer to be an automobile sales position that pays $1500 per month plus 2.5% of all sales. What total sales is required to earn between $4000 and $6000 per month?

109. Electric Bill In North Carolina, Duke Energy charges $42.41 plus $0.092897 for each additional kilowatt hour (kwh) used during the months from July through October for usage in excess of 350 kwh. Suppose one homeowner's electric bill ranged from a low of $88.86 to a high of $137.16 during this time period. Over what range (in kwh) did the usage vary?

110. Electric Bill In North Carolina, Duke Energy charges $42.41 plus $0.084192 for each additional kilowatt hour (kwh) used during the months from November through June for usage in excess of

350 kwh. Suppose one homeowner's electric bill ranged from a low of $55.04 to a high of $89.56 during this time period. Over what range (in kwh) did the usage vary?

Extending the Concepts

111. The Arithmetic Mean If $a < b$, show that $a < \dfrac{a + b}{2} < b$. We call $\dfrac{a + b}{2}$ the **arithmetic mean** of a and b.

△**112. Identifying Triangles** A triangle has the property that the length of the longest side is greater than the difference of the other sides, and the length of the longest side is less than the sum of the other sides. That is, if $a, b,$ and c are sides such that $a \leq b \leq c$, then $b - a < c < b + a$. Determine which of the following could be lengths of the sides of a triangle.

(a) $3, 4, 5$

(b) $4, 7, 12$

(c) $3, 3, 5$

(d) $1, 9, 10$

113. Solve $2x + 1 \leq 5x + 7 \leq x - 5$.

114. Solve $x - 3 \leq 3x + 1 \leq x + 11$.

115. Solve $4x + 1 > 2(2x + 1)$. Provide an explanation that generalizes the result.

116. Solve $4x - 2 > 2(2x - 1)$. Provide an explanation that generalizes the result.

117. Consider the following analysis, assuming that $x < 2$.

$$5 > 2$$
$$5(x - 2) > 2(x - 2)$$
$$5x - 10 > 2x - 4$$
$$3x > 6$$
$$x > 2$$

How can it be that the final line in the analysis states that $x > 2$, when the original assumption stated that $x < 2$?

2.6 Absolute Value Equations and Inequalities

Objectives

1. Solve Absolute Value Equations
2. Solve Absolute Value Inequalities Involving $<$ or $\leq$
3. Solve Absolute Value Inequalities Involving $>$ or $\geq$
4. Solve Applied Problems Involving Absolute Value Inequalities

Figure 37
$|-5| = 5$

Are You Prepared for This Section?

Before getting started, complete the following problems. If you get a problem wrong, go back to the section cited and review the material.

In Problems P1–P4, evaluate each expression.

P1. $|3|$ **P2.** $|-4|$ **P3.** $|-1.6|$ **P4.** $|0|$ [Section R.3, p. 19]

P5. Express the distance between the origin, 0, and 5 as an absolute value.   [Section R.3, p. 19]

P6. Express the distance between the origin, 0, and -8 as an absolute value.   [Section R.3, p. 19]

P7. Solve: $4x + 5 = -9$ [Section 1.1, pp. 50–52]

P8. Solve: $-2x + 1 > 5$ [Section 1.4, pp. 86–88]

Recall that the absolute value of a number is its distance from the origin on the real number line. For example, $|-5| = 5$ because the distance on the real number line from 0 to -5 is 5 units. See Figure 37.

▶ **1** Solve Absolute Value Equations

EXAMPLE 1   **Solving an Absolute Value Equation**

Solve the equation $|x| = 4$.

Solution
Two geometric solutions and one algebraic solution are presented.

Geometric Solution 1: The equation $|x| = 4$ asks, "Which real numbers x are 4 units from 0 on the number line?" Figure 38 on the next page shows the two numbers, -4 and 4. The solution set is $\{-4, 4\}$.

Prepared?...Answers **P1.** 3 **P2.** 4
P3. 1.6 **P4.** 0 **P5.** $|5|$ **P6.** $|-8|$
P7. $\left\{-\dfrac{7}{2}\right\}$ **P8.** $\{x | x < -2\}$; $(-\infty, -2)$

Figure 38

Figure 39

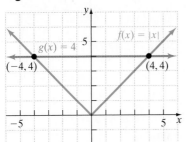

Geometric Solution 2: $|x| = 4$ can be solved by graphing $f(x) = |x|$ and $g(x) = 4$ on the same xy-plane, as in Figure 39. The x-coordinates of the points of intersection are -4 and 4, which are the solutions to the equation $f(x) = g(x)$. Again, the solution set is $\{-4, 4\}$.

Algebraic Solution: Recall that $|a| = a$ if $a \geq 0$ and $|a| = -a$ if $a < 0$. Because it is not known whether x is positive or negative in the equation $|x| = 4$, solve the problem for $x \geq 0$ or $x < 0$.

If $x \geq 0$	If $x < 0$
$|x| = 4$	$|x| = 4$
$|x| = x$ since $x \geq 0$: $x = 4$	$|x| = -x$ since $x < 0$: $-x = 4$
	Multiply both sides by -1: $x = -4$

The solution set is $\{-4, 4\}$.

> **Quick ✓**
> *In Problems 1 and 2, solve the equation geometrically and algebraically.*
>
> **1.** $|x| = 7$ **2.** $|z| = 1$

The results of Example 1 and Quick Checks 1 and 2 lead us to the following result.

Figure 40

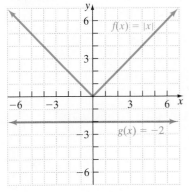

Equations Involving Absolute Value

If a is a positive real number and u is any algebraic expression, then

$$|u| = a \quad \text{is equivalent to} \quad u = a \quad \text{or} \quad u = -a$$

Note: If $a = 0$, the equation $|u| = 0$ is equivalent to $u = 0$. If $a < 0$, the equation $|u| = a$ has no solution, as explained below.

In the equation $|u| = a$, it is required that a be nonnegative (greater than or equal to 0). If a is negative, the equation has no real solution. To see why, consider the equation $|x| = -2$. Figure 40 shows the graph of $f(x) = |x|$ and $g(x) = -2$. Notice that the graphs do not intersect, which shows that the equation $|x| = -2$ has no real solution. The solution set is the empty set, $\varnothing$ or $\{\,\}$.

EXAMPLE 2   **How to Solve an Equation Involving Absolute Value**

Solve the equation: $|2x - 1| + 3 = 12$

Step-by-Step Solution

Step 1: Isolate the expression containing the absolute value.

$$|2x - 1| + 3 = 12$$

Subtract 3 from both sides: $|2x - 1| = 9$

Step 2: Rewrite the equation $|u| = a$ as: $u = a$ or $u = -a$, where u is the algebraic expression in the absolute value symbol.

In the equation $|2x - 1| = 9, u = 2x - 1$ and $a = 9$.

$$2x - 1 = 9 \qquad \text{or} \qquad 2x - 1 = -9$$

Step 3: Solve each equation.

Add 1 to each side: $2x = 10$ Add 1 to each side: $2x = -8$

Divide both sides by 2: $x = 5$ Divide both sides by 2: $x = -4$

(continued)

Step 4: Check: Verify each solution.

$$|2x - 1| + 3 = 12 \qquad\qquad |2x - 1| + 3 = 12$$

Let $x = 5$: $\quad |2(5) - 1| + 3 \overset{?}{=} 12 \qquad$ Let $x = -4$: $\quad |2(-4) - 1| + 3 \overset{?}{=} 12$

$$|10 - 1| + 3 \overset{?}{=} 12 \qquad\qquad |-8 - 1| + 3 \overset{?}{=} 12$$

$$9 + 3 \overset{?}{=} 12 \qquad\qquad 9 + 3 \overset{?}{=} 12$$

$$12 = 12 \quad \text{True} \qquad\qquad 12 = 12 \quad \text{True}$$

Both solutions check, so the solution set is $\{-4, 5\}$.

Solving Absolute Value Equations with One Absolute Value

Step 1: Isolate the expression containing the absolute value.

Step 2: Rewrite the absolute value equation as two equations: $u = a$ and $u = -a$, where u is the expression in the absolute value symbol.

Step 3: Solve each equation.

Step 4: Verify your solution.

Quick ✓

3. $|u| = a$ is equivalent to $u = \underline{}$ or $u = \underline{}$

4. $|2x + 3| = 5$ is equivalent to $2x + 3 = 5$ or _____.

In Problems 5–8, solve each equation.

5. $|2x - 3| = 7$ **6.** $|3x - 2| + 3 = 10$

7. $|-5x + 2| - 2 = 5$ **8.** $3|x + 2| - 4 = 5$

▶ (**EXAMPLE 3**)   **Solving an Equation Involving Absolute Value with No Solution**

Solve the equation: $|x + 5| + 7 = 5$

Solution

$$|x + 5| + 7 = 5$$

Subtract 7 from both sides: $\qquad |x + 5| = -2$

Work Smart

The equation $|u| = a$, where a is a negative real number, has no real solution. Why? See Figure 40 on the previous page.

Since the absolute value of any real number is always nonnegative (greater than or equal to zero), the equation has no real solution. The solution set is $\{\ \}$ or $\varnothing$.

Quick ✓

9. *True or False* $|x| = -4$ has no real solution.

In Problems 10–12, solve each equation.

10. $|5x + 3| = -2$

11. $|2x + 5| + 7 = 3$

12. $|x + 1| + 3 = 3$

What if an absolute value equation has two absolute values, as in $|3x - 1| = |x + 5|$? The signs of the algebraic expressions in the absolute value symbol have four possibilities:

1. both algebraic expressions are nonnegative,

2. both are negative,

3. the left is positive, and the right is negative, or

4. the left is negative and the right is nonnegative.

To see how the solution works, use the fact that $|a| = a$, if $a \geq 0$ and $|a| = -a$, if $a < 0$.

Thus, if $3x - 1 \geq 0$, then $|3x - 1| = 3x - 1$. However, if $3x - 1 < 0$, then $|3x - 1| = -(3x - 1)$. This leads to a method for solving absolute value equations with two absolute values.

Case 1: Both Expressions Are Nonnegative	Case 2: Both Expressions Are Negative	Case 3: The Expression on the Left Is Nonnegative, and That on the Right Is Negative	Case 4: The Expression on the Left Is Negative, and That on the Right Is Nonnegative
$\|3x - 1\| = \|x + 5\|$ $3x - 1 = x + 5$	$\|3x - 1\| = \|x + 5\|$ $-(3x - 1) = -(x + 5)$ $3x - 1 = x + 5$	$\|3x - 1\| = \|x + 5\|$ $3x - 1 = -(x + 5)$	$\|3x - 1\| = \|x + 5\|$ $-(3x - 1) = x + 5$

When both algebraic expressions are nonnegative, or both negative, we end up with equivalent equations. Therefore, Case 1 and Case 2 result in equivalent equations. Also, if one side is nonnegative and the other is negative, we end up with equivalent equations. Therefore, Case 3 and Case 4 result in equivalent equations. The four possibilities reduce to two possibilities.

Equations Involving Two Absolute Values

If u and v are any algebraic expressions, then

$$|u| = |v| \quad \text{is equivalent to} \quad u = v \quad \text{or} \quad u = -v$$

▷ **EXAMPLE 4**   **Solving an Absolute Value Equation Involving Two Absolute Values**

Solve the equation: $|2x - 3| = |x + 6|$

Solution

The equation is in the form $|u| = |v|$, where $u = 2x - 3$ and $v = x + 6$. Rewrite the equation in the form $u = v$ or $u = -v$ and then solve each equation.

$u = v$: $2x - 3 = x + 6$ $u = -v$: $2x - 3 = -(x + 6)$

Add 3 to both sides: $2x = x + 9$ Distribute the -1: $2x - 3 = -x - 6$

Subtract x from both sides: $x = 9$ Add 3 to both sides: $2x = -x - 3$

Add x to both sides: $3x = -3$

Divide both sides by 3: $x = -1$

Check

$x = 9$: $|2(9) - 3| \overset{?}{=} |9 + 6|$ $x = -1$: $|2(-1) - 3| \overset{?}{=} |-1 + 6|$

$|18 - 3| \overset{?}{=} |15|$ $|-2 - 3| \overset{?}{=} |5|$

$|15| \overset{?}{=} 15$ $|-5| \overset{?}{=} 5$

$15 = 15$ True $5 = 5$ True

Both solutions check, so the solution set is $\{-1, 9\}$. ●

Quick ✓

13. $|u| = |v|$ is equivalent to ___ = ___ or ___ = ___.

In Problems 14–17, solve each equation.

14. $|x - 3| = |2x + 5|$

15. $|8z + 11| = |6z + 17|$

16. $|3 - 2y| = |4y + 3|$

17. $|2x - 3| = |5 - 2x|$

▶ ❷ Solve Absolute Value Inequalities Involving < or ≤

The method for solving absolute value equations relies on the geometric interpretation of absolute value. That is, the absolute value of a real number x is its distance from the origin on the real number line. The geometric interpretation of absolute value can also be used to solve absolute value inequalities.

EXAMPLE 5 **Solving an Absolute Value Inequality**

Solve the inequality $|x| < 4$. Graph the solution set.

Solution

The inequality $|x| < 4$ asks for all real numbers x that are less than 4 units from the origin on the real number line. See Figure 41. Notice from the figure that any number between -4 and 4 satisfies the inequality. The solution set is $\{x \mid -4 < x < 4\}$ or, using interval notation, $(-4, 4)$.

These results could also be visualized by graphing $f(x) = |x|$ and $g(x) = 4$. See Figure 42. To solve $f(x) < g(x)$, look for all x-coordinates such that the graph of $f(x)$ is below the graph of $g(x)$. Notice that this is true for all x between -4 and 4, as found above. ●

The results of Example 5 lead to the following.

Figure 41

Figure 42

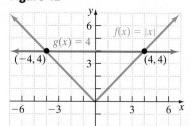

Inequalities of the Form < or ≤ Involving Absolute Value

If a is a positive real number and if u is an algebraic expression, then

$$|u| < a \quad \text{is equivalent to} \quad -a < u < a$$
$$|u| \le a \quad \text{is equivalent to} \quad -a \le u \le a$$

Note: If $a = 0$, $|u| < 0$ has no solution; $|u| \le 0$ is equivalent to $u = 0$. If $a < 0$, the inequality has no solution.

Quick ✓

18. If $a > 0$, then $|u| < a$ is equivalent to _____.

19. To solve $|3x + 4| < 10$, solve ____ $< 3x + 10 <$ __.

In Problems 20 and 21, solve each inequality. Graph the solution set.

20. $|x| \le 5$ 21. $|x| < \dfrac{3}{2}$

EXAMPLE 6 **How to Solve an Absolute Value Inequality Involving ≤**

Solve the inequality $|2x + 3| \le 5$. Express the solution set in set-builder notation and interval notation. Graph the solution set.

Step-by-Step Solution

Step 1: The inequality is in the form $|u| \le a$, where $u = 2x + 3$ and $a = 5$. Rewrite the inequality as a compound inequality that does not involve absolute value.

$$|2x + 3| \le 5$$
$$|u| \le a \text{ means } -a \le u \le a: \quad -5 \le 2x + 3 \le 5$$

Step 2: Solve the resulting compound inequality.

Subtract 3 from all three parts: $\quad -5 - 3 \le 2x + 3 - 3 \le 5 - 3$
$$-8 \le 2x \le 2$$

Divide all three parts by 2: $\qquad \dfrac{-8}{2} \le \dfrac{2x}{2} \le \dfrac{2}{2}$
$$-4 \le x \le 1$$

Figure 43

The solution is $\{x \mid -4 \le x \le 1\}$ or, in interval notation, $[-4, 1]$. Figure 43 shows the graph of the solution set. ●

Work Smart

As a partial check of the solution of Example 6, we can check a number in the interval. Let's try $x = -3$.

$$|2(-3) + 3| \overset{?}{\leq} 5$$
$$|-6 + 3| \overset{?}{\leq} 5$$
$$|-3| \overset{?}{\leq} 5$$
$$3 \leq 5 \quad \text{True}$$

Quick ✓

In Problems 22–24, solve each inequality. Express the solution set in set-builder notation and interval notation. Graph the solution set.

22. $|x + 3| < 5$

23. $|2x - 3| \leq 7$

24. $|7x + 2| < -3$

EXAMPLE 7 **Solving an Absolute Value Inequality Involving** $<$

Solve the inequality $|-3x + 2| + 4 < 14$. Express the solution set in set-builder notation and interval notation. Graph the solution set.

Solution

First, isolate the absolute value.

$$|-3x + 2| + 4 < 14$$

Subtract 4 from both sides: $\qquad\qquad |-3x + 2| < 10$

$|u| < a$ means $-a < u < a$: $\qquad\qquad -10 < -3x + 2 < 10$

Subtract 2 from all three parts: $\qquad -10 - 2 < -3x + 2 - 2 < 10 - 2$

$$-12 < -3x < 8$$

Divide all three parts by -3.
Reverse the direction of the inequalities: $\qquad \dfrac{-12}{-3} > \dfrac{-3x}{-3} > \dfrac{8}{-3}$

$$4 > x > -\dfrac{8}{3}$$

$b > x > a$ is equivalent to $a < x < b$: $\qquad -\dfrac{8}{3} < x < 4$

Figure 44

The solution is $\left\{ x \middle| -\dfrac{8}{3} < x < 4 \right\}$, using set-builder notation, and $\left(-\dfrac{8}{3}, 4 \right)$, using interval notation. Figure 44 shows the graph of the solution set. ●

Quick ✓

25. *True or False* $|x + 1| + 2 < 7$ is equivalent to $-7 < x + 1 + 2 < 7$.

In Problems 26–29, solve each inequality. Express the solution set in set-builder notation and interval notation. Graph the solution set.

26. $|x| + 4 < 6$

27. $|x - 3| + 4 \leq 8$

28. $3|2x + 1| \leq 9$

29. $|-3x + 1| - 5 < 3$

▶ ❸ **Solve Absolute Value Inequalities Involving** $>$ **or** $\geq$

EXAMPLE 8 **Solving an Absolute Value Inequality Involving** $>$

Solve the inequality $|x| > 3$. Graph the solution set.

Solution

The inequality $|x| > 3$ asks for all real numbers x that are more than 3 units from the origin on the real number line. See Figure 45.

Figure 45

More than 3 units from origin, 0.   More than 3 units from origin, 0.

(continued)

Figure 46

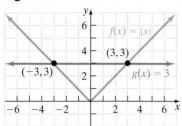

Notice that any number less than -3 or greater than 3 satisfies the inequality. The solution set is $\{x \mid x < -3 \text{ or } x > 3\}$ or, using interval notation, $(-\infty, -3) \cup (3, \infty)$.

These results could also be visualized by graphing $f(x) = |x|$ and $g(x) = 3$. See Figure 46. To solve $f(x) > g(x)$, look for all x-coordinates such that the graph of $f(x)$ is above the graph of $g(x)$. Notice that this is true for all x less than -3 or all x greater than 3, as we found above.

Based on Example 8, we have the following results.

Inequalities of the Form $>$ or $\geq$ Involving Absolute Value

If a is a positive real number and u is an algebraic expression, then

$$|u| > a \quad \text{is equivalent to} \quad u < -a \ \text{ or } \ u > a$$
$$|u| \geq a \quad \text{is equivalent to} \quad u \leq -a \ \text{ or } \ u \geq a$$

Quick ✓

30. $|u| > a$ is equivalent to _____ or _____.

31. $|5x - 2| \geq 7$ is equivalent to $5x - 2 \leq$ ___ or $5x - 2 \geq$ ___.

In Problems 32 and 33, solve each inequality. Graph the solution set.

32. $|x| \geq 6$ **33.** $|x| > \dfrac{5}{2}$

EXAMPLE 9 **How to Solve an Inequality Involving $>$**

Solve the inequality $|2x - 5| > 3$. Express the solution set in set-builder notation and interval notation. Graph the solution set.

Step-by-Step Solution

Step 1: The inequality is in the form $|u| > a$, where $u = 2x - 5$ and $a = 3$. Rewrite the inequality as a compound inequality that does not involve absolute value.

$$|2x - 5| > 3$$
$|u| > a$ means $u < -a$ or $u > a$: $2x - 5 < -3$ or $2x - 5 > 3$

Step 2: Solve each inequality separately.

	$2x - 5 < -3$		$2x - 5 > 3$
Add 5 to both sides:	$2x < 2$	Add 5 to both sides:	$2x > 8$
Divide both sides by 2:	$x < 1$	Divide both sides by 2:	$x > 4$

Step 3: Find the union of the solution sets of each inequality.

The solution set is $\{x \mid x < 1 \text{ or } x > 4\}$ or, using interval notation, $(-\infty, 1) \cup (4, \infty)$. See Figure 47 for the graph of the solution set.

Figure 47

-2 -1 0 1 2 3 4 5 6

Work Smart

$$|u| > a$$
CANNOT be written as
$$-a > u > a$$

Quick ✓

34. $|x - 9| > 6$ is equivalent to $x - 9 > 6$ or $x - 9$ ___ -6.

In Problems 35–40, solve each inequality. Express the solution set in set-builder notation and interval notation. Graph the solution set.

35. $|x + 3| > 4$ **36.** $|4x - 3| \geq 5$

37. $|-3x + 2| > 7$ **38.** $|2x + 5| - 2 > -2$

39. $|6x - 5| \geq 0$ **40.** $|2x + 1| > -3$

Summary   Solving Absolute Value Equations and Inequalities

Absolute Value Form	Equation/Inequality Form	Example	
		Algebraic Solution	**Graphical Solution**
$\|u\| = a$	$u = a$ or $u = -a$	$\|x + 3\| = 5$ $x + 3 = 5$ or $x + 3 = -5$ $x = 2$ or $x = -8$ Solution set: $\{-8, 2\}$	
$\|u\| = \|v\|$	$u = v$ or $u = -v$	$\|x - 2\| = \|2x\|$ $x - 2 = 2x$ or $x - 2 = -2x$ $-2 = x$ or $3x = 2$ $x = -2$ or $x = \dfrac{2}{3}$ Solution set: $\left\{-2, \dfrac{2}{3}\right\}$	
$\|u\| < a$ $\|u\| \leq a$	$-a < u < a$ $-a \leq u \leq a$	$\|x - 1\| \leq 4$ $-4 \leq x - 1 \leq 4$ $-3 \leq x \leq 5$ Solution set: $\{x \mid -3 \leq x \leq 5\}$ or $[-3, 5]$	
$\|u\| > a$ $\|u\| \geq a$	$u < -a$ or $u > a$ $u \leq -a$ or $u \geq a$	$\|x + 1\| > 3$ $x + 1 < -3$ or $x + 1 > 3$ $x < -4$ or $x > 2$ Solution set: $\{x \mid x < -4$ or $x > 2\}$ or $(-\infty, -4) \cup (2, \infty)$	

❹ Solve Applied Problems Involving Absolute Value Inequalities

You may have read phrases such as "margin of error" and "tolerance" in the newspaper. For example, according to a recent Gallup poll, 67% of those polled said that recent increases in gas prices have caused financial hardship for their household. The poll had a margin of error of 4%. The 67% is an estimate of the true percentage of Americans who have experienced financial hardship as a consequence of increases in gas prices. If p represents the true percentage of Americans who have experienced financial hardship due to the increase in gas prices, then the poll's margin of error can be represented mathematically as

$$|p - 67| \leq 4$$

As another example, the tolerance of a belt used in a pulley system whose width is 6 inches is $\dfrac{1}{16}$ inch. If x represents the actual width of the belt, then the acceptable belt widths can be represented as

$$|x - 6| \leq \dfrac{1}{16}$$

EXAMPLE 10   **Analyzing the Margin of Error in a Poll**

The inequality

$$|p - 67| \leq 4$$

(continued)

represents the percentage of Americans who said they experienced financial hardship due to increase in gas prices. Solve the inequality and interpret the results.

Solution

$$|p - 67| \leq 4$$

$|u| \leq a$ means $-a \leq u \leq a$: $-4 \leq p - 67 \leq 4$

Add 67 to all three parts: $63 \leq p \leq 71$

The percentage of Americans who have experienced financial hardship due to increases in gas prices is between 63% and 71%, inclusive. ●

Quick ✓

41. The inequality $|x - 4| \leq \dfrac{1}{32}$ represents the acceptable belt width x (in inches) for a belt that is manufactured for a pulley system. Determine the acceptable belt width.

42. In a poll conducted by ABC News, 9% of Americans stated that they have been shot at. The margin of error in the poll was 1.7%. If we let p represent the true percentage of people who have been shot at, we can represent the margin of error as

$$|p - 9| \leq 1.7$$

Solve the inequality and interpret the results.

2.6 Exercises MyMathLab®   Exercise numbers in green have complete video solutions in MyMathLab or may be accessed using the QR code to the right.

Problems 1–42 are the Quick ✓s that follow the EXAMPLES.

Building Skills

In Problems 43–68, solve each absolute value equation. See Objective 1.

43. $|x| = 10$

44. $|z| = 9$

45. $|p| = -2$

46. $|4| = -1$

47. $|y - 3| = 4$

48. $|x + 3| = 5$

49. $|-3x + 5| = 8$

50. $|-4y + 3| = 9$

51. $|y| - 7 = -2$

52. $|x| + 3 = 5$

53. $|2x + 3| - 5 = 3$

54. $|3y + 1| - 5 = -3$

55. $-2|x - 3| + 10 = -4$

56. $3|y - 4| + 4 = 16$

57. $|-3x| - 5 = -5$

58. $|-2x| + 9 = 9$

59. $\left|\dfrac{3x - 1}{4}\right| = 2$

60. $\left|\dfrac{2x - 3}{5}\right| = 2$

61. $|3x + 2| = |2x - 5|$

62. $|5y - 2| = |4y + 7|$

63. $|8 - 3x| = |2x - 7|$

64. $|5x + 3| = |12 - 4x|$

65. $|4y - 7| = |9 - 4y|$

66. $|5x - 1| = |9 - 5x|$

67. $|2x - 1| = -|x|$

68. $-|x + 1| = |3x - 2|$

In Problems 69–82, solve each absolute value inequality. Graph the solution set on a real number line. See Objective 2.

69. $|x| < 9$

70. $|x| \leq \dfrac{5}{4}$

71. $|x - 4| \leq 7$

72. $|y + 4| < 6$

73. $|3x + 1| < 8$

74. $|4x - 3| \leq 9$

75. $|6x + 5| < -1$ **76.** $|4x + 3| \le 0$

(b) $f(x) \le g(x)$
(c) $f(x) > g(x)$

77. $2|x - 3| + 3 < 9$ **78.** $3|y + 2| - 2 < 7$

79. $|2 - 5x| + 3 < 10$ **80.** $|-3x + 2| - 7 \le -2$

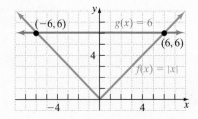

81. $|(2x - 3) - 1| < 0.01$ **82.** $|(3x + 2) - 8| < 0.01$

97. $f(x) = |x + 2|, g(x) = 3$

In Problems 83–94, solve each absolute value inequality. Graph the solution set on the real number line. See Objective 3.

(a) $f(x) = g(x)$
(b) $f(x) < g(x)$
(c) $f(x) \ge g(x)$

83. $|y - 5| > 2$ **84.** $|x + 4| \ge 7$

85. $|-4x - 3| \ge 5$ **86.** $|-5y + 3| > 7$

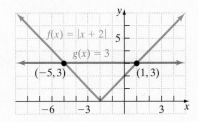

87. $2|y| + 3 > 1$ **88.** $3|z| + 8 > 2$

89. $|-5x - 3| - 7 > 0$ **90.** $|-9x + 2| - 11 \ge 0$

98. $f(x) = |2x|, g(x) = 10$

(a) $f(x) = g(x)$
(b) $f(x) < g(x)$
(c) $f(x) \ge g(x)$

91. $4|-2x + 1| > 4$ **92.** $3|8x + 3| \ge 9$

93. $|1 - 2x| \ge |-5|$ **94.** $|3 - 5x| > |-7|$

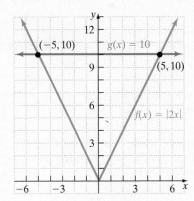

Mixed Practice

In Problems 95–98, use the graphs of the functions given to solve each problem.

95. $f(x) = |x|, g(x) = 5$
(a) $f(x) = g(x)$
(b) $f(x) \le g(x)$
(c) $f(x) > g(x)$

In Problems 99–120, solve each absolute value equation or inequality. For absolute value inequalities, graph the solution set on a real number line.

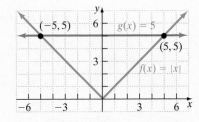

99. $|x| > 5$ **100.** $|x| \ge \dfrac{8}{3}$

96. $f(x) = |x|, g(x) = 6$
(a) $f(x) = g(x)$

101. $|2x + 5| = 3$ **102.** $|4x + 3| = 1$

103. $7|x| = 35$

104. $8|y| = 32$

105. $|2x + 1| - 2 = 5$

106. $|3x - 2| + 1 = 8$

107. $|-2x + 3| + 4 = 0$

108. $|7y - 3| < 11$

109. $|-2x + 3| = -4$

110. $|3x - 4| = -9$

111. $|3x + 2| \geq 5$

112. $|5y + 3| > 2$

113. $|3x - 2| + 7 > 9$

114. $|4y + 3| - 8 \geq -3$

115. $|5x + 3| = |3x + 5|$

116. $|3z - 2| = |z + 6|$

117. $|4x + 7| + 6 < 5$

118. $|4x + 1| > 0$

119. $\left|\dfrac{x - 2}{4}\right| = \left|\dfrac{2x + 1}{6}\right|$

120. $\left|\dfrac{1}{2}x - 3\right| = \left|\dfrac{2}{3}x + 1\right|$

Applying the Concepts

121. Express the fact that x differs from 5 by less than 3 as an inequality involving absolute value. Solve for x.

122. Express the fact that x differs from -4 by less than 2 as an inequality involving absolute value. Solve for x.

123. Express the fact that twice x differs from -6 by more than 3 as an inequality involving absolute value. Solve for x.

124. Express the fact that twice x differs from 7 by more than 3 as an inequality involving absolute value. Solve for x.

125. Tolerance A certain rod in an internal combustion engine is supposed to be 5.7 inches long. The tolerance on the rod is 0.0005 inch. If x represents the length of a rod, the acceptable lengths of a rod can be expressed as $|x - 5.7| \leq 0.0005$. Determine the acceptable lengths of the rod.
SOURCE: *WiseCo Piston*

126. Tolerance A certain rod in an internal combustion engine is supposed to be 6.125 inches long. The tolerance on the rod is 0.0005 inch. If x represents the length of a rod, the acceptable lengths of a rod can be expressed as $|x - 6.125| \leq 0.0005$. Determine the acceptable lengths of the rod.

127. IQ Scores According to the Stanford-Binet IQ test, a normal IQ score is 100. It can be shown that anyone with an IQ x that satisfies the inequality $\left|\dfrac{x - 100}{15}\right| > 1.96$ has an unusual IQ score. Determine the IQ scores that would be considered unusual.

128. Gestation Period The length of human pregnancy is about 266 days. It can be shown that a mother whose gestation period x satisfies the inequality $\left|\dfrac{x - 266}{16}\right| > 1.96$ has an unusual length of pregnancy. Determine the length of pregnancy that would be considered unusual.

Extending the Concepts

In Problems 129–136, solve each equation.

129. $|x| - x = 5$

130. $|y| + y = 3$

131. $z + |-z| = 4$

132. $y - |-y| = 12$

133. $|4x + 1| = x - 2$

134. $|2x + 1| = x - 3$

135. $|x + 5| = -(x + 5)$

136. $|y - 4| = y - 4$

Explaining the Concepts

137. Explain why $|2x - 3| + 1 = 0$ has no solution.

138. Explain why the solution set of $|5x - 3| > -5$ is the set of all real numbers.

139. Explain why $|4x + 3| + 3 < 0$ has the empty set as the solution set.

140. Solve $|x - 5| = |5 - x|$. Explain why the result is reasonable. What do we call this type of equation?

Chapter 2 Activity: Shifting Discovery

Focus: Using graphing skills, discover the possible "rules" for graphing functions.

Time: 30–35 minutes

Group size: 4

Materials Needed: Graph paper (2–3 pieces)

Each member of the group needs to:

1. Draw a coordinate plane and label the *x*-axis and *y*-axis.

2. By plotting points, graph the primary function: $f(x) = x^2$.

3. In the same coordinate plane, each group member graphs *one* of the following functions by plotting points. Be sure to graph the primary function and one of the functions (a)–(d) in the same coordinate plane.

    (a) $f(x) = x^2 + 3$ (b) $f(x) = (x - 3)^2$

    (c) $f(x) = x^2 - 3$ (d) $f(x) = (x + 3)^2$

As a group, discuss the following:

4. What shape are the graphs?

5. Each member of the group should share the difference between the graph of your primary function and the other function you chose.

6. As a group, can you develop rules for these differences?

With this possible rule in mind, each member of the group needs to:

7. Draw a coordinate plane and label the *x*-axis, the *y*-axis, and −10 to 10 on each axis.

8. Graph the primary function by plotting points: $f(x) = |x|$.

9. In the same coordinate plane, each group member graphs *one* of the following functions by plotting points. Be sure to graph the primary function and one of the functions (a)–(d) on the same coordinate plane.

    (a) $f(x) = |x| + 4$ (b) $f(x) = |x - 4|$

    (c) $f(x) = |x| - 4$ (d) $f(x) = |x + 4|$

10. Did the rules you developed in Problem 6 hold true? Discuss.

Chapter 2 Review

Section 2.1   Relations

KEY CONCEPT

- **Relation**
  A correspondence between two variables *x* and *y*, where *y* depends on *x*. Relations can be represented through maps, sets of ordered pairs, equations, or graphs.

KEY TERMS

Relation	Inputs
Corresponds	Outputs
Depends on	Domain
Mapping	Range

You Should Be Able To...	EXAMPLE	Review Exercises
❶ Understand relations (p. 148)	Example 1	1–4
❷ Find the domain and the range of a relation (p. 149)	Examples 2 through 4	1–16
❸ Graph a relation defined by an equation (p. 151)	Example 5	9–14

In Problems 1 and 2, write each relation as a set of ordered pairs. Then identify the domain and range of the relation.

1.

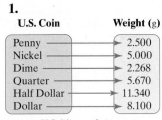

SOURCE: *U.S. Mint website*

2.

SOURCE: *Lego website*

In Problems 3 and 4, write each relation as a map. Then identify the domain and range of the relation.

3. $\{ (2, 7), (-4, 8), (3, 5), (6, -1), (-2, -9) \}$

4. $\{ (3, 1), (3, 7), (5, 1), (-2, 8), (1, 4) \}$

In Problems 5–8, identify the domain and range of the relation from the graph.

5.

6.

7.

8.

In Problems 9–14, graph the relation. Use the graph of the relation to identify the domain and range of the relation.

9. $y = x + 2$

10. $2x + y = 3$

11. $y = -x^2 + 4$

12. $y = |x + 2| - 1$

13. $y = x^3 + 2$

14. $x = y^2 + 1$

15. Cell Phones A cell phone company offers a plan for $40 per month for 3000 minutes with additional minutes costing $0.05 per minute. The graph to the right shows the monthly cost, in dollars, when x minutes are used.

(a) What are the domain and range of the relation?
(b) Explain why the domain obtained in part (a) is reasonable.

16. Vertical Motion The graph below shows the height, in feet, of a ball thrown straight up with an initial speed of 40 feet per second from an initial height of 96 feet after t seconds. What are the domain and the range of the relation?

Section 2.2   An Introduction to Functions

KEY CONCEPTS

- **Functions**
  A special type of relation where any given input, x, corresponds to only one output y. Functions can be represented through maps, sets of ordered pairs, equations, or graphs.

- **Vertical Line Test**
  A set of points in the xy-plane is the graph of a function if and only if every vertical line intersects the graph in at most one point.

- **Domain of a Function**
  When only an equation of a function is given, the domain of the function is the set of real numbers x for which $f(x)$ is a real number. However, in applications, the domain of a function is the largest set of real numbers for which the output of the function is reasonable.

KEY TERMS

Function
Vertical Line Test
Value of f at the number x
Independent variable
Dependent variable
Argument
Domain of f

You Should Be Able To...	EXAMPLE	Review Exercises
❶ Determine whether a relation expressed as a map or ordered pairs represents a function (p. 154)	Examples 1 and 2	17, 18
❷ Determine whether a relation expressed as an equation represents a function (p. 157)	Examples 3 and 4	19–22
❸ Determine whether a relation expressed as a graph represents a function (p. 157)	Example 5	23–26
❹ Find the value of a function (p. 158)	Examples 6 and 7	27–30
❺ Find the domain of a function (p. 160)	Examples 8 and 9	31–34
❻ Work with applications of functions (p. 161)	Example 10	35, 36

In Problems 17 and 18, determine whether the given relation represents a function. State the domain and the range of each relation.

17. (a) $\{\,(-1,-2),(-1,3),(5,0),(7,2),(9,4)\,\}$

(b)

Animal	Typical Lifespan (years)
Camel	50
Macaw	35
Deer	14
Fox	22
Tiger	45
Crocodile	

18. (a) $\{\,(-2,4),(2,3),(-3,-1),(5,7),(4,7)\,\}$

(b)

Color	Type of Car
Red	Camry
Blue	Taurus
Green	Windstar
Black	Durango

In Problems 19–22, determine whether each relation shows y as a function of x.

19. $3x - 5y = 18$

20. $x^2 + y^2 = 81$

21. $y = \pm 10x$

22. $y = x^2 - 14$

In Problems 23–26, determine whether the graph is that of a function.

23.

24.

25.

26.

In Problems 27–30, find the indicated values for the given functions.

27. $f(x) = x^2 + 2x - 5$

(a) $f(-2)$

(b) $f(3)$

28. $g(z) = \dfrac{2z + 1}{z - 3}$

(a) $g(0)$

(b) $g(2)$

29. $F(x) = -2x + 7$

(a) $F(5)$

(b) $F(-x)$

30. $G(x) = 2x + 1$

(a) $G(7)$

(b) $G(x + h)$

For Problems 31–34, find the domain of each function.

31. $f(x) = -\dfrac{3}{2}x + 5$

32. $g(w) = \dfrac{w - 9}{2w + 5}$

33. $h(t) = \dfrac{t + 2}{t - 5}$

34. $G(t) = 3t^2 + 4t - 9$

35. Population Using census data from 1900 to 2014, the function

$$P(t) = 0.136t^2 - 5.043t + 46.927$$

represents the population, P, of Orange County in Florida (in thousands) t years after 1900.

(a) Identify the dependent and independent variables.

(b) Evaluate $P(120)$ and explain what it represents.

(c) Evaluate $P(-70)$ and explain what it represents. Is the result reasonable? Explain.

36. Education The function

$$P(a) = -0.0064a^2 + 0.6826a - 6.82$$

represents the percent of the population a years of age with an advanced degree, where $a \geq 25$. SOURCE: *Current Population Survey*

(a) Identify the dependent and independent variables.

(b) Evaluate $P(30)$ and explain what it represents.

Section 2.3   Functions and Their Graphs

KEY CONCEPTS

- **Graph of a Function**
  The graph of a function, f, is the set of all ordered pairs $(x, f(x))$.

- **Zero of a Function**
  If $f(r) = 0$ for some number r, then r is a zero of f.

KEY TERMS

Zero of a function

You Should Be Able To...	EXAMPLE	Review Exercises
1 Graph a function (p. 165)	Example 1	37–40
2 Obtain information from the graph of a function (p. 167)	Examples 2 through 5	41–45, 47, 48
3 Know properties and graphs of basic functions (p. 170)	p. 170	46
4 Interpret graphs of functions (p. 170)	Example 6	49, 50

In Problems 37–40, graph each function.

37. $f(x) = 2x - 5$ **38.** $g(x) = x^2 - 3x + 2$

39. $h(x) = (x - 1)^3 - 3$ **40.** $f(x) = |x + 1| - 4$

In Problems 41–44, for each function whose graph is shown, find (a) the domain and the range, and (b) the intercepts, if any.

41.

42.

43.

44.

45. The graph of $y = f(x)$ is shown.

   (a) What is $f(-3)$?
   (b) For what value of x does $f(x) = -4$?
   (c) What are the zeros of f?

46. Graph each of the following functions.
   (a) $f(x) = x^2$ **(b)** $f(x) = x$

In Problems 47 and 48, answer the questions about the given function.

47. $h(x) = 2x - 7$
   (a) Is the point $(3, -1)$ on the graph of the function?
   (b) If $x = -2$, what is $h(x)$? What point is on the graph of the function?
   (c) If $h(x) = 4$, what is x? What point is on the graph of h?

48. $g(x) = \dfrac{3}{5}x + 4$
   (a) Is the point $(-5, 2)$ on the graph of the function?
   (b) If $x = 3$, what is $g(x)$? What point is on the graph of the function?
   (c) If $g(x) = -2$, what is x? What point is on the graph of g?

49. Travel by Train A Metrolink train leaves L.A. Union Station and travels 6 miles at a constant speed for 10 minutes, arriving at the Glendale station, where it waits 1 minute for passengers to board and depart. The train continues traveling at the same speed for 5 more minutes to reach downtown Burbank, which is 3 miles from Glendale. Sketch a graph that represents the distance of the train as a function of time until it reaches downtown Burbank.

50. Filling a Tub With the faucet running at a constant rate, it takes Angie 7 minutes to fill her bathtub. She turns off the faucet when the tub is full and realizes the water is too hot. She then opens the drain, letting water out at a constant rate that is half the rate of the faucet. After draining for 2 minutes, she stops the drain and turns on the faucet at the same rate as before (but at a cooler temperature) until the tub is full. Sketch a graph that represents the amount of water in the tub as a function of time.

Section 2.4   Linear Functions and Models

KEY CONCEPT

- **Linear Function**
  A linear function is a function of the form $f(x) = mx + b$, where m is the slope and $(0, b)$ is the y-intercept of the graph.

KEY TERMS

Linear function
Line
Scatter diagram

You Should Be Able To...	EXAMPLE	Review Exercises
1 Graph linear functions (p. 176)	Example 1	51–54
2 Find the zero of a linear function (p. 177)	Example 2	51–54
3 Build linear models from verbal descriptions (p. 179)	Examples 4 and 5	57–60
4 Build linear models from data (p. 181)	Examples 6 through 8	61–64

In Problems 51–54, graph each linear function. Find the zero of each function.

51. $g(x) = 2x - 6$

52. $H(x) = -\dfrac{4}{3}x + 5$

53. $F(x) = -x - 3$

54. $f(x) = \dfrac{3}{4}x - 3$

55. Monthly Car Cost Sarah's monthly payments for her new car are $260 per month. She estimates that maintenance and gas cost her $0.20 per mile. The equation for her monthly cost is $C(m) = 0.20m + 260$, where C is the monthly cost in dollars and m is the number of miles driven that month.

   (a) What is the domain of this function?

   (b) What is $C(0)$? Explain what this result means in the context of the problem.

   (c) What is Sarah's monthly cost if she drives 1000 miles in a month?

   (d) Graph the linear function.

   (e) Over what range of miles can she drive in a month if she wants to keep her monthly cost to no more than $550?

56. Straight-Line Depreciation Using straight-line depreciation, the value V of a particular computer x years after purchase is given by the linear function $V(x) = 1800 - 360x$ for $0 \le x \le 5$.

   (a) What are the independent and dependent variables?

   (b) What is the domain of this linear function?

   (c) What is the initial value of the computer?

   (d) What is the value of the computer 2 years after purchase?

   (e) Graph the linear function over its domain.

   (f) After how long will the value of the computer be $0?

57. Credit Scores Your Fair Isaacs Corporation (FICO) credit score is used to determine your ability to get a loan. The higher your credit score, the better your credit history. Suppose a bank offers a 36-month auto loan at 9% for a FICO score of 675 and at 5% for a FICO score of 750.

   (a) Assuming a linear relation between FICO credit score and auto loan rate, find a linear function that relates the FICO credit score to the auto loan rate (as a percentage), treating the FICO credit score as the independent variable.

   (b) Predict the auto loan rate for a FICO credit score of 710, to the nearest whole percent.

   (c) Interpret the slope.

   (d) For what FICO credit score will a bank offer a 36-month auto loan at 6.5%?

58. Heart Rates According to the American Geriatric Society, the maximum recommended heart rate for a 20-year-old man under stress is 200 beats per minute. The maximum recommended heart rate for a 60-year-old man under stress is 160 beats per minute.

   (a) Find a linear function that relates the maximum recommended heart rate for men to age.

   (b) Predict the maximum recommended heart rate for a 45-year-old man under stress.

   (c) Interpret the slope.

   (d) For what age would the maximum recommended heart rate under stress be 168 beats per minute?

59. Car Rental The daily rental charge for a particular car is $35 plus $0.12 per mile.

   (a) Find a linear function that expresses the rental cost C as a function of the miles driven m.

   (b) What are the independent and dependent variables?

   (c) What is the domain of this linear function?

   (d) For a one-day rental, what is the rental cost if 124 miles are driven?

   (e) For a one-day rental, how many miles were driven if the rental cost was $67.16?

   (f) Graph the linear function.

60. Satellite Television Bill A satellite television company charges $33.99 per month for a 100-channel package, plus $3.50 for each pay-per-view movie watched that month.

(a) Find a linear function that expresses the monthly bill B as a function of x, the number of pay-per-view movies watched that month.

(b) What are the independent and dependent variables?

(c) What is the domain of this linear function?

(d) What is the monthly bill if 5 pay-per-view movies are watched that month?

(e) For one month, how many pay-per-view movies were watched if the bill was $58.49?

(f) Graph the linear function.

In Problems 61 and 62,

(a) *Draw a scatter diagram of the data.*

(b) *Select two points from the scatter diagram and find the equation of the line containing the points selected.*

(c) *Graph the line found in part (b) on the scatter diagram.*

61.

x	2	5	8	11	14
y	13.3	11.6	8.4	7.2	4.6

62.

x	0	0.4	1.5	2.3	4.2
y	0.6	1.1	1.3	1.8	3.0

63. The table below gives the number of calories and the total carbohydrates (in grams) for a one-cup serving of seven name-brand cereals (not including milk).

Cereal	Calories, x	Total Carbohydrates (in grams), y
Rice Krispies®	96	23.2
Life®	160	33.3
Lucky Charms®	120	25.0
Kellogg's Complete®	120	30.7
Wheaties®	110	24.0
Cheerios®	110	22.0
Honey Nut Chex®	160	34.7

SOURCE: *Quaker Oats, General Mills, and Kellogg*

(a) Draw a scatter diagram of the data, treating calories as the independent variable.

(b) What type of relation appears to exist between calories and total carbohydrates in a one-cup serving of cereal?

(c) Select two points and find an equation of the line containing the points.

(d) Graph the line on the scatter diagram drawn in part (a).

(e) Predict the total carbohydrates in a one-cup serving of cereal that has 140 calories.

(f) Interpret the slope of the line found in part (c).

64. Second-Day Delivery Costs The table below lists some selected prices charged by Federal Express for FedEx 2Day delivery, depending on the weight of the package.

Weight (in pounds), x	FedEx 2Day® Delivery Charge, y
1	$11.60
3	$12.20
6	$13.55
8	$16.20
9	$16.90
11	$18.90

SOURCE: *Federal Express Corporation*

(a) Draw a scatter diagram of the data, treating weight as the independent variable.

(b) What type of relation appears to exist between the weight of the package and the FedEx 2Day delivery charge?

(c) Select two points and find an equation of the line containing the points.

(d) Graph the line on the scatter diagram drawn in part (a).

(e) Predict the FedEx 2Day delivery charge for shipping a 5-pound package.

(f) Interpret the slope of the line found in part (c).

Section 2.5   Compound Inequalities

KEY CONCEPT

- If $a < b$, then we can write $a < x$ and $x < b$ as $a < x < b$.

KEY TERMS

Intersection   Compound inequality

Union   Solve a compound inequality

You Should Be Able To...	EXAMPLE	Review Exercises
❶ Determine the intersection or union of two sets (p. 190)	Examples 1 and 2	65–70
❷ Solve compound inequalities involving "and" (p. 192)	Examples 3 through 6	71, 72, 75, 76, 80
❸ Solve compound inequalities involving "or" (p. 194)	Examples 7 and 8	73, 74, 77, 78, 79
❹ Solve problems using compound inequalities (p. 196)	Example 9	81, 82

In Problems 65–68, use $A = \{2, 4, 6, 8\}$,
$B = \{-1, 0, 1, 2, 3, 4\}$, and $C = \{1, 2, 3, 4\}$ to find
each set.

65. $A \cup B$ **66.** $A \cap C$

67. $B \cap C$ **68.** $A \cup C$

In Problems 69 and 70, use the graph of the inequality to
find each set.

69. $A = \{x \mid x \le 4\}; B = \{x \mid x > 2\}$
Find **(a)** $A \cap B$ and **(b)** $A \cup B$.

70. $E = \{x \mid x \ge 3\}; F = \{x \mid x < -2\}$
Find **(a)** $E \cap F$ and **(b)** $E \cup F$.

In Problems 71–80, solve each compound inequality.
Graph the solution set.

71. $x < 4$ and $x + 3 > 2$

72. $3 < 2 - x < 7$

73. $x + 3 < 1$ or $x > 2$

74. $x + 6 \ge 10$ or $x \le 0$

75. $3x + 2 \le 5$ and $-4x + 2 \le -10$

76. $1 \le 2x + 5 < 13$

77. $x - 3 \le -5$ or $2x + 1 > 7$

78. $3x + 4 > -2$ or $4 - 2x \ge -6$

79. $\frac{1}{3}x > 2$ or $\frac{2}{5}x < -4$

80. $x + \frac{3}{2} \ge 0$ and $-2x + \frac{3}{2} > \frac{1}{4}$

81. Heart Rates The normal heart rate for healthy adults between the ages of 21 and 60 should be between 70 and 75 beats per minute (inclusive). If we let x represent the heart rate of an adult between the ages of 21 and 60, express the normal range of values using a compound inequality.

82. Heating Bills For usage above 800 kilowatt hours, the non-space heat energy charge for American Electric Power residential service was $43.56 plus $0.038752 per kilowatt hour. During one winter, a customer's charge ranged from a low of $52.62 to a high of $88.22. Over what range of values did electrical usage vary (in kilowatt hours)? Express answers rounded to the nearest tenth of a kilowatt hour.

Section 2.6   Absolute Value Equations and Inequalities

KEY CONCEPTS

- **Equations Involving Absolute Value**
  If a is a positive real number and if u is any algebraic expression, then $|u| = a$ is equivalent to $u = a$ or $u = -a$.

- **Equations Involving Two Absolute Values**
  If u and v are any algebraic expressions, then $|u| = |v|$ is equivalent to $u = v$ or $u = -v$.

- **Inequalities of the Form $<$ or $\le$ Involving Absolute Value**
  If a is a positive real number and if u is any algebraic expression, then $|u| < a$ is equivalent to $-a < u < a$, and $|u| \le a$ is equivalent to $-a \le u \le a$.

- **Inequalities of the Form $>$ or $\ge$ Involving Absolute Value**
  If a is a positive real number and if u is any algebraic expression, then $|u| > a$ is equivalent to $u < -a$ or $u > a$, and $|u| \ge a$ is equivalent to $u \le -a$ or $u \ge a$.

You Should Be Able To...	EXAMPLE	Review Exercises
1 Solve absolute value equations (p. 200)	Examples 1 through 4	83–88
2 Solve absolute value inequalities involving $<$ or $\leq$ (p. 204)	Examples 5 through 7	89, 91, 94, 95
3 Solve absolute value inequalities involving $>$ or $\geq$ (p. 205)	Examples 8 and 9	90, 92, 93, 96
4 Solve applied problems involving absolute value inequalities (p. 207)	Example 10	97, 98

In Problems 83–88, solve the absolute value equation.

83. $|x| = 4$

84. $|3x - 5| = 4$

85. $|-y + 4| = 9$

86. $-3|x + 2| - 5 = -8$

87. $|2w - 7| = -3$

88. $|x + 3| = |3x - 1|$

In Problems 89–96, solve each absolute value inequality. Graph the solution set on a real number line.

89. $|x| < 2$

90. $|x| \geq \dfrac{7}{2}$

91. $|x + 2| \leq 3$

92. $|4x - 3| \geq 1$

93. $3|x| + 6 \geq 1$

94. $|7x + 5| + 4 < 3$

95. $|(x - 3) - 2| \leq 0.01$ **96.** $\left|\dfrac{2x - 3}{4}\right| > 1$

97. Tolerance The diameter of a certain ball bearing is required to be 0.503 inch. The tolerance on the bearing is 0.001 inch. If x represents the diameter of a bearing, the acceptable diameters of the bearing can be expressed as $|x - 0.503| \leq 0.001$. Determine the acceptable diameters of the bearing.

98. Tensile Strength The tensile strength of paper used to make grocery bags is about 40 lb/in.² A paper grocery bag whose tensile strength satisfies the inequality $\left|\dfrac{x - 40}{2}\right| > 1.96$ has an unusual tensile strength. Determine the tensile strengths that would be considered unusual.

Chapter 2 Test

Step-by-step test solutions are found on the Chapter Test Prep Videos available in MyMathLab®, on YouTube, or may be accessed using the QR code to the right.

1. Write the relation as a map. Then identify the domain and the range of the relation.

$\{(2, 8), (5, -2), (7, 12), (-4, -7), (7, 3), (5, -1)\}$

2. Identify the domain and range of the relation from the graph.

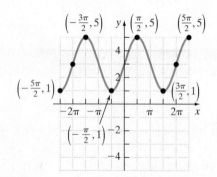

3. Graph the relation $y = x^2 - 3$ by plotting points. Use the graph of the relation to identify the domain and range.

In Problems 4 and 5, determine whether each relation represents a function. Identify the domain and the range of each relation.

4.

5.

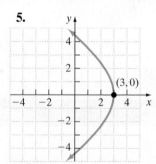

6. Does the equation $y = \pm 5x$ represent a function? Why or why not?

7. For $f(x) = -3x + 11$, find $f(x + h)$.

8. For $g(x) = 2x^2 + x - 1$, find the indicated values.

  (a) $g(-2)$ **(b)** $g(0)$ **(c)** $g(3)$

9. Sketch the graph of $f(x) = x^2 + 3$.

10. Using data from 1996 to 2015, the function $P(x) = 0.22x + 4.44$ approximates the average movie ticket price (in dollars) x years after 1996.

SOURCE: *National Association of Theater Owners*

   (a) Identify the dependent and independent variables.

   (b) Evaluate $P(25)$ and explain what it represents.

   (c) In what year will the average movie ticket be $12.00?

11. Find the domain of $f(x) = \dfrac{-15}{x + 2}$.

12. $h(x) = -5x + 12$

   (a) Is the point $(2, 2)$ on the graph of the function?

   (b) If $x = 3$, what is $h(x)$? What point is on the graph of the function?

   (c) If $h(x) = 27$, what is x? What point is on the graph of h?

   (d) What is the zero of h?

13. The following graph represents the speed of a car as a function of time.

   (a) When does the car stop accelerating?

   (b) For how long does the car maintain a constant speed?

14. Crafts Fair Sales Henry plans to sell small wooden shelves at a crafts fair for $30 each. A booth at the fair costs $100 to rent. Henry estimates his expenses for producing the shelves to be $12 each, so the amount he will make per shelf is $18.

   (a) Write a function that expresses Henry's profit P as a function of the number of shelves sold x.

   (b) What is the implied domain of this linear function?

   (c) What is the profit if Henry sells 34 shelves?

   (d) Graph the linear function.

   (e) If Henry's profit is $764, how many shelves did he sell?

15. Shetland Pony Weights The table below lists the average weight of a Shetland pony, depending on the age of the pony.

Age (months), x	Average Weight (kilograms), y
3	60
6	95
12	140
18	170
24	185

SOURCE: *The Merck Veterinary Manual*

   (a) Draw a scatter diagram of the data, treating age as the independent variable.

   (b) What type of relation appears to exist between the age and the weight of a Shetland pony?

   (c) Select two points and find an equation of the line containing the points.

   (d) Graph the line on the scatter diagram drawn in part (a).

   (e) Predict the weight of a 9-month-old Shetland pony.

   (f) Interpret the slope of the line found in part (c).

16. Solve: $|2x + 5| - 3 = 0$

In Problems 17–20, solve each inequality and graph the solution set on a real number line.

17. $x + 2 < 8$ and $2x + 5 \geq 1$

18. $x > 4$ or $2(x - 1) + 3 < -2$

19. $2|x - 5| + 1 < 7$

20. $|-2x + 1| \geq 5$

CHAPTER

3 Systems of Linear Equations and Inequalities

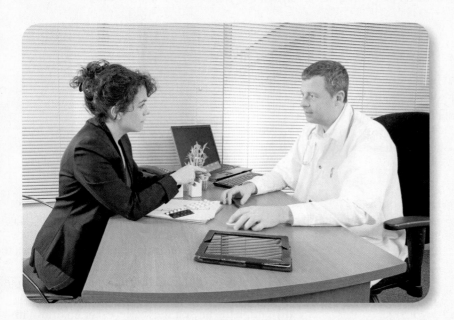

Finally! You've landed an offer for a sales position at your ideal job: pharmaceutical sales. You have been offered two salary options: an annual base salary of $15,000 plus a commission of 1% on all sales, or an annual base salary of $25,000 plus a commission of 0.75% on all sales. What annual sales are required for these options to result in the same annual earnings? See Problem 49 in Section 3.2.

The Big Picture: Putting It Together

Recall from the first part of Chapter 1 that equations in one variable can have no solution, one solution, or infinitely many solutions. In the second part of Chapter 1, linear equations and inequalities in two variables were graphed. Graphs of linear equations in this chapter will help to visualize results.

In this chapter, solving *systems of equations*, two or more linear equations involving two or more variables, will be discussed. A variety of techniques for solving these systems will be covered. In addition, the chapter will demonstrate that these systems can have no solution, one solution, or infinitely many solutions, just like equations in one variable. The chapter will conclude by looking at systems of linear inequalities.

Outline

3.1 Systems of Linear Equations in Two Variables

Objectives

1. Determine Whether an Ordered Pair Is a Solution of a System of Linear Equations
2. Solve a System of Two Linear Equations by Graphing
3. Solve a System of Two Linear Equations by Substitution
4. Solve a System of Two Linear Equations by Elimination
5. Identify Inconsistent Systems
6. Write the Solution of a System with Dependent Equations

Are You Prepared for This Section?

Before getting started, complete the following problems. If you get a problem wrong, go back to the section cited and review the material.

P1. Evaluate $2x - 3y$ for $x = 5, y = 4$.   [Section R.5, pp. 40–41]

P2. Determine whether the point $(4, -1)$ is on the graph of the equation $2x - 3y = 11$.   [Section 1.5, pp. 95–96]

P3. Graph: $y = 3x - 7$ [Section 1.6, pp. 104–105]

P4. Find the equation of the line parallel to $y = -3x + 1$ containing the point $(2, 3)$.   [Section 1.7, p. 123]

P5. Determine the slope and y-intercept of $4x - 3y = 15$.   [Section 1.6, pp. 114–115]

P6. What is the additive inverse of 4?   [Section R.3, p. 21]

P7. Solve: $2x - 3(-3x + 1) = -36$ [Section 1.1, pp. 50–53]

Recall from Section 1.6 that an equation in two variables is linear if it can be written in the form $Ax + By = C$, where $A, B,$ and C are real numbers and A and B are not both zero.

A **system of linear equations** is a grouping of two or more linear equations, each of which contains one or more variables.

EXAMPLE 1

Examples of Systems of Linear Equations

(a) $\begin{cases} 2x + y = 5 \\ x - 5y = -10 \end{cases}$ Two linear equations containing two variables, x and y

(b) $\begin{cases} x + 3y + z = 8 \\ 3x - y + 6z = 12 \\ -4x - y + 2z = -1 \end{cases}$ Three linear equations containing three variables, $x, y,$ and z ●

Use a brace, as shown in the systems in Example 1, to indicate that a *system* of equations is involved. In this section, concentrate on systems of two linear equations containing two variables, such as the system in Example 1(a).

1 Determine Whether an Ordered Pair Is a Solution of a System of Linear Equations

A **solution** of a system of equations consists of values for the variables that are solutions of each equation of the system. Represent the solution of a system of two linear equations containing two unknowns as an ordered pair, (x, y).

EXAMPLE 2

Determining Whether an Ordered Pair Is a Solution of a System of Linear Equations

Determine whether the given ordered pairs are solutions of the system of equations.

$$\begin{cases} 2x + 3y = 9 \\ -5x - 3y = 0 \end{cases}$$

(a) $(6, -1)$ **(b)** $(-3, 5)$

Solution

To make it easier to describe this process, call $2x + 3y = 9$ equation (1) and $-5x - 3y = 0$ equation (2).

$$\begin{cases} 2x + 3y = 9 & (1) \\ -5x - 3y = 0 & (2) \end{cases}$$

Prepared?...Answers   **P1.** -2 **P2.** Yes

P3.

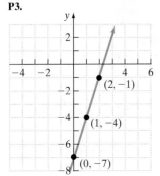

P4. $y = -3x + 9$ **P5.** slope $= \dfrac{4}{3}$;

y-intercept $= -5$ **P6.** -4 **P7.** $\{-3\}$

(a) Let $x = 6$ and $y = -1$ in both equations (1) and (2). If both equations are true, then $(6, -1)$ is a solution.

Equation (1): $\qquad 2x + 3y = 9$ $\qquad$ Equation (2): $\qquad -5x - 3y = 0$

$x = 6, y = -1$: $\quad 2(6) + 3(-1) \overset{?}{=} 9$ $\qquad\qquad$ $x = 6, y = -1$: $\quad -5(6) - 3(-1) \overset{?}{=} 0$

$\qquad\qquad\qquad 12 - 3 \overset{?}{=} 9$ $\qquad\qquad\qquad\qquad\qquad -30 + 3 \overset{?}{=} 0$

$\qquad\qquad\qquad\qquad 9 = 9 \quad$ True $\qquad\qquad\qquad\qquad\qquad -27 = 0 \quad$ False

Work Smart

A solution of a system must satisfy *all* of the equations in the system.

Although $x = 6$, and $y = -1$ satisfy equation (1), they do not satisfy equation (2); therefore, $(6, -1)$ is not a solution of the system of equations.

(b) Let $x = -3$ and $y = 5$ in both equations (1) and (2). If both equations are true, then $(-3, 5)$ is a solution.

Equation (1): $\qquad 2x + 3y = 9$ $\qquad$ Equation (2): $\qquad -5x - 3y = 0$

$x = -3, y = 5$: $\quad 2(-3) + 3(5) \overset{?}{=} 9$ $\qquad\qquad$ $x = -3, y = 5$: $\quad -5(-3) - 3(5) \overset{?}{=} 0$

$\qquad\qquad\qquad -6 + 15 \overset{?}{=} 9$ $\qquad\qquad\qquad\qquad\qquad 15 - 15 \overset{?}{=} 0$

$\qquad\qquad\qquad\qquad 9 = 9 \quad$ True $\qquad\qquad\qquad\qquad\qquad 0 = 0 \quad$ True

Because $x = -3$, and $y = 5$ satisfy both equations (1) and (2), the ordered pair $(-3, 5)$ is a solution of the system of equations. ●

When solving homework problems, number each equation as in Example 1.

Work Smart

It is a good idea to number the equations in a system so that it is easier to keep track of your work.

Quick ✓

1. A _____ __ ____ _____ is a grouping of two or more linear equations, each of which contains one or more variables.

2. A _____ of a system of equations consists of values of the variables that satisfy each equation of the system.

3. Which of the following points is a solution of the system of equations?

$$\begin{cases} 2x + 3y = 7 \\ 3x + y = -7 \end{cases}$$

(a) $(2, 1)$ $\qquad\qquad$ **(b)** $(-4, 5)$ $\qquad\qquad$ **(c)** $(-2, -1)$

Visualizing the Solutions in a System of Two Linear Equations

The graph of each equation in a system of linear equations is a line, so a system of two linear equations containing two variables represents a pair of lines. The graphs of the two lines can appear in one of three ways:

1. **Intersect:** If the lines intersect, then the system of equations has one solution given by the point of intersection. The system is **consistent** and the equations are **independent.** See Figure 1(a).

2. **Parallel:** If the lines are parallel, then the system of equations has no solution because the lines never intersect. The system is **inconsistent** and the equations are **independent.** See Figure 1(b).

3. **Coincident:** If the lines lie on top of each other (are coincident), then the system of equations has infinitely many solutions. The solution set is the set of all points on the line. The system is **consistent** and the equations are **dependent.** See Figure 1(c).

Figure 1

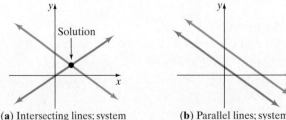

(a) Intersecting lines; system has one solution $\qquad$ **(b)** Parallel lines; system has no solution $\qquad$ **(c)** Coincident lines; system has infinitely many solutions

Quick ✓

4. If a system of equations has no solution, it is said to be _____

5. If a system of equations has infinitely many solutions, the system is said to be _____ and the equations are _____

6. *True or False* A system of two linear equations containing two variables always has at least one solution.

7. *True or False* When the lines in a system of equations are parallel, then the system is inconsistent and has no solution.

8. *True or False* The visual representation of a consistent system consisting of dependent equations is one line.

For now, concentrate on solving systems that have a single solution.

▶ ❷ Solve a System of Two Linear Equations by Graphing

The solution of a system of equations is the point or points of intersection, if any, of the two graphs, because both equations are satisfied at the point(s) of intersection.

EXAMPLE 3 | **Solving a System of Two Linear Equations Using Graphing**

Solve the following system by graphing: $\begin{cases} x + y = -1 \\ -2x + y = -7 \end{cases}$

Solution

First, name $x + y = -1$ equation (1) and name $-2x + y = -7$ equation (2).

$$\begin{cases} x + y = -1 & (1) \\ -2x + y = -7 & (2) \end{cases}$$

Figure 2

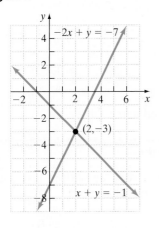

In order to graph each equation, put them in slope-intercept form. Equation (1) in slope-intercept form is $y = -x - 1$, which has slope -1 and y-intercept $(0, -1)$. Equation (2) in slope-intercept form is $y = 2x - 7$, which has slope 2 and y-intercept $(0, -7)$. Figure 2 shows their graphs. Note that the lines could have been graphed using the intercepts. The lines appear to intersect at $(2, -3)$, so the ordered pair $(2, -3)$ is the solution to the system.

Check Let $x = 2$ and $y = -3$ in both equations in the system:

Equation (1): $\quad x + y = -1$ Equation (2): $\quad -2x + y = -7$

$x = 2, y = -3$: $\quad 2 + (-3) \overset{?}{=} -1 \quad$ $x = 2, y = -3$: $\quad -2(2) + (-3) \overset{?}{=} -7$

$\qquad\qquad\quad 2 - 3 \overset{?}{=} -1 \qquad\qquad\qquad\qquad -4 - 3 \overset{?}{=} -7$

$\qquad\qquad\quad -1 = -1 \quad$ True $\qquad\qquad\qquad\qquad -7 = -7 \quad$ True

Both equations are satisfied, so the solution is the ordered pair $(2, -3)$.   ●

Quick ✓

In Problems 9 and 10, solve the system by graphing.

9. $\begin{cases} y = -3x + 10 \\ y = 2x - 5 \end{cases}$

10. $\begin{cases} 2x + y = -1 \\ -2x + 2y = 10 \end{cases}$

▶ ❸ Solve a System of Two Linear Equations by Substitution

Work Smart

Obtaining exact solutions using graphical methods can be difficult. Therefore, algebraic methods are usually preferred.

Finding the exact point of intersection of two lines can be difficult if the x- and y-coordinates of that point are not integers. Therefore, algebraic methods, rather than graphical methods, for solving systems of equations may be preferred. One algebraic method is *substitution*. The goal of the substitution method is to obtain a single linear equation involving a single unknown.

EXAMPLE 4 **How to Solve a System of Two Equations by Substitution**

Solve the following system by substitution: $\begin{cases} 3x + y = -9 & (1) \\ -2x + 3y = 17 & (2) \end{cases}$

Step-by-Step Solution

Step 1: Solve one of the equations for one of the unknowns.

It is easier to solve equation (1) for y since the coefficient of y is 1.

$$3x + y = -9$$

Subtract $3x$ from both sides: $\qquad y = -3x - 9$

Step 2: Substitute $-3x - 9$ for y in equation (2).

Equation (2): $\qquad -2x + 3y = 17$

$$-2x + 3(-3x - 9) = 17$$

Step 3: Solve the equation for x.

Distribute the 3: $\qquad -2x - 9x - 27 = 17$

Combine like terms: $\qquad -11x - 27 = 17$

Add 27 to both sides: $\qquad -11x = 44$

Divide both sides by -11: $\qquad x = -4$

Step 4: Substitute -4 for x in the equation from Step 1.

$$y = -3x + 9$$
$$y = -3(-4) - 9$$
$$y = 12 - 9$$
$$y = 3$$

Step 5: Check Verify that $x = -4$ and $y = 3$ satisfy each equation in the original system.

Equation (1): $\qquad 3x + y = -9$

$$3(-4) + 3 \overset{?}{=} -9$$
$$-12 + 3 \overset{?}{=} -9$$
$$-9 = -9 \quad \text{True}$$

Equation (2): $\qquad -2x + 3y = 17$

$$-2(-4) + 3(3) \overset{?}{=} 17$$
$$8 + 9 \overset{?}{=} 17$$
$$17 = 17 \quad \text{True}$$

Both equations are satisfied, so the solution is the ordered pair $(-4, 3)$. ●

Work Smart

When using substitution, if possible, solve for the variable whose coefficient is 1 or -1 in order to simplify the algebra.

> **Solving a System of Two Linear Equations by Substitution**
>
> **Step 1:** Solve one of the equations for one of the unknowns. Choose the equation that is easier to solve for a variable. Typically, this would be an equation that has a variable whose coefficient is 1 or -1.
>
> **Step 2:** Substitute the expression found in Step 1 into the *other* equation. The result will be a single linear equation in one unknown.
>
> **Step 3:** Solve the linear equation in one unknown found in Step 2.
>
> **Step 4:** Substitute the value of the variable found in Step 3 into one of the original equations to find the value of the other variable.
>
> **Step 5:** Check your answer by substituting the ordered pair into both of the original equations.

EXAMPLE 5 **Solving a System of Two Equations by Substitution**

Solve the following system by substitution: $\begin{cases} 2x - 3y = -6 & (1) \\ -8x + 3y = 3 & (2) \end{cases}$

Solution

Solve equation (1) for x because it seems easiest.

Equation (1): $\quad 2x - 3y = -6$

Add $3y$ to both sides: $\qquad 2x = 3y - 6$

Divide both sides by 2: $\qquad x = \dfrac{3y - 6}{2}$

Divide both terms in the numerator by 2: $\qquad x = \dfrac{3}{2}y - 3$

Now substitute $\dfrac{3}{2}y - 3$ for x in equation (2) and then solve for y.

Work Smart

Notice how substitution is used to reduce a system of two linear equations involving two unknowns down to one linear equation involving one unknown. Again, use algebraic techniques to reduce a problem down to one we already know how to solve!

Equation (2): $\qquad -8x + 3y = 3$

$$-8\left(\dfrac{3}{2}y - 3\right) + 3y = 3$$

Distribute the -8: $\quad -8 \cdot \dfrac{3}{2}y - (-8) \cdot 3 + 3y = 3$

Multiply: $\qquad -12y + 24 + 3y = 3$

Combine like terms: $\qquad -9y + 24 = 3$

Subtract 24 from both sides: $\qquad -9y = -21$

Divide both sides by -9: $\qquad y = \dfrac{-21}{-9}$

$$y = \dfrac{7}{3}$$

Now substitute $\dfrac{7}{3}$ for y in $x = \dfrac{3}{2}y - 3$ to find the value of x.

$$x = \dfrac{3}{2}\left(\dfrac{7}{3}\right) - 3$$

Multiply: $\quad x = \dfrac{7}{2} - 3$

$$x = \dfrac{1}{2}$$

Check Let $x = \dfrac{1}{2}$ and $y = \dfrac{7}{3}$ in both equations in the system:

Equation (1): $\qquad 2x - 3y = -6$ $\qquad\qquad$ Equation (2): $\qquad -8x + 3y = 3$

$$2\left(\dfrac{1}{2}\right) - 3\left(\dfrac{7}{3}\right) \overset{?}{=} -6 \qquad\qquad -8\left(\dfrac{1}{2}\right) + 3\left(\dfrac{7}{3}\right) \overset{?}{=} 3$$

$$1 - 7 \overset{?}{=} -6 \qquad\qquad\qquad -4 + 7 \overset{?}{=} 3$$

$$-6 = -6 \quad \text{True} \qquad\qquad\qquad\qquad 3 = 3 \quad \text{True}$$

Both equations are satisfied, so the solution is the ordered pair $\left(\dfrac{1}{2}, \dfrac{7}{3}\right)$. $\qquad\qquad$ ●

Quick ✓

In Problems 11 and 12, solve the system using substitution.

11. $\begin{cases} y = -3x - 5 \\ 5x + 3y = 1 \end{cases}$ $\qquad\qquad$ **12.** $\begin{cases} 2x + y = -2 \\ -3x - 2y = -2 \end{cases}$

▶ ④ Solve a System of Two Linear Equations by Elimination

Work Smart

Substitution is a good method to use if one of the variables has a coefficient of 1 or if one of the variables is already isolated. Otherwise, elimination may be preferred.

Using substitution in Example 5 led to equations containing fractions. A second algebraic method for solving a system of linear equations is the *elimination method*. This method is usually preferred over the substitution method if substitution leads to fractions.

The basic idea in using elimination is to get the coefficients of one of the variables to be additive inverses, or opposites, such as 5 and -5, so that the equations can be added together to yield a single linear equation involving one unknown.

Let's go over an example that shows how to solve a system of linear equations by elimination.

EXAMPLE 6 **How to Solve a System of Linear Equations by Elimination**

Solve: $\begin{cases} 5x + 2y = -5 & (1) \\ -2x - 4y = -14 & (2) \end{cases}$

Step-by-Step Solution

Step 1: The first goal is to get the coefficients on one of the variables to be additive inverses.

Make the coefficients of y additive inverses by multiplying equation (1) by 2.

$\begin{cases} 5x + 2y = -5 & (1) \\ -2x - 4y = -14 & (2) \end{cases}$

Multiply both sides of (1) by 2: $\begin{cases} 2(5x + 2y) = 2(-5) & (1) \\ -2x - 4y = -14 & (2) \end{cases}$

Use the Distributive Property: $\begin{cases} 10x + 4y = -10 & (1) \\ -2x - 4y = -14 & (2) \end{cases}$

Step 2: Now add equations (1) and (2) to eliminate the variable y and then solve for x.

$\begin{cases} 10x + 4y = -10 & (1) \\ -2x - 4y = -14 & (2) \end{cases}$

Add (1) and (2): $\quad 8x \quad\quad = -24$

Divide both sides by 8: $\quad x = -3$

The x-coordinate of the solution is -3.

Step 3: Substitute for x in either equation (1) or (2) and solve for y. We will use equation (1).

Equation (1): $\quad 5x + 2y = -5$

$x = -3$: $\quad 5(-3) + 2y = -5$

$-15 + 2y = -5$

Add 15 to both sides: $\quad 2y = 10$

Divide both sides by 2: $\quad y = 5$

The y-coordinate of the solution is 5. The solution appears to be $(-3, 5)$.

Step 4: Check Verify that $x = -3$ and $y = 5$ satisfy each equation in the original system.

Equation (1): $\quad 5x + 2y = -5$

$5(-3) + 2(5) \overset{?}{=} -5$

$-15 + 10 \overset{?}{=} -5$

$-5 = -5 \quad$ True

Equation (2): $\quad -2x - 4y = -14$

$-2(-3) - 4(5) \overset{?}{=} -14$

$6 - 20 \overset{?}{=} -14$

$-14 = -14 \quad$ True

The solution is the ordered pair $(-3, 5)$. ●

The following steps can be used to solve a system of linear equations by elimination.

> ### Solving a System of Linear Equations by Elimination
>
> **Step 1:** Write both equations of the system in standard form, $Ax + By = C$. If necessary, multiply both sides of one equation or both equations by a nonzero constant so that the coefficients of one of the variables are additive inverses.
>
> **Step 2:** Add equations (1) and (2) to eliminate the variable whose coefficients are now additive inverses. Solve the resulting equation for the unknown.
>
> **Step 3:** Substitute the value of the variable found in Step 2 into one of the original equations to find the value of the remaining variable.
>
> **Step 4:** Check the answer by substituting the ordered pair in both of the original equations.

What allows us to add two equations and use the result to replace an equation? Remember, an equation is a statement that the left side equals the right side. When equation (2) is added to equation (1), the same quantity is added to both sides of equation (1).

EXAMPLE 7 **Solving a System of Linear Equations by Elimination**

Solve: $\begin{cases} \dfrac{5}{2}x + 2y = 5 & (1) \\[2mm] \dfrac{3}{2}x + \dfrac{3}{2}y = \dfrac{9}{4} & (2) \end{cases}$

Solution

Because both equations have fractions, the first goal is to clear the fractions by multiplying both sides of equation (1) by 2 and both sides of equation (2) by 4.

$\begin{cases} \dfrac{5}{2}x + 2y = 5 & (1) \\[2mm] \dfrac{3}{2}x + \dfrac{3}{2}y = \dfrac{9}{4} & (2) \end{cases}$

Multiply both sides of (1) by 2: $\begin{cases} 2\left(\dfrac{5}{2}x + 2y\right) = 2 \cdot 5 & (1) \end{cases}$

Multiply both sides of (2) by 4: $\begin{cases} 4\left(\dfrac{3}{2}x + \dfrac{3}{2}y\right) = 4 \cdot \dfrac{9}{4} & (2) \end{cases}$

Use the Distributive Property: $\begin{cases} 5x + 4y = 10 & (1) \\ 6x + 6y = 9 & (2) \end{cases}$

Divide both sides of equation (2) by 3: $\begin{cases} 5x + 4y = 10 & (1) \\ 2x + 2y = 3 & (2) \end{cases}$

Multiply both sides of equation (2) by -2: $\begin{cases} 5x + 4y = 10 & (1) \\ -4x - 4y = -6 & (2) \end{cases}$

Add (1) and (2): $\qquad\qquad x \qquad\quad = 4$

Work Smart

If you are not afraid of fractions, you can eliminate x by multiplying both sides of equation (1) by -3 and both sides of equation (2) by 5. Try it!

Work Smart

Although it is not necessary to divide both sides of equation (2) by 3, it makes solving the problem easier. Do you see why?

Now substitute 4 for x into either original equation. In this case, substitute into equation (1).

Equation (1): $\qquad \dfrac{5}{2}x + 2y = 5$

$x = 4$: $\qquad \dfrac{5}{2}(4) + 2y = 5$

$10 + 2y = 5$

Subtract 10 from both sides: $\qquad 2y = -5$

Divide both sides by 2: $\qquad y = -\dfrac{5}{2}$

It appears that $x = 4$ and $y = -\dfrac{5}{2}$.

The check is left to you. The solution is the ordered pair $\left(4, -\dfrac{5}{2}\right)$.

Quick ✓

13. The basic idea in using the elimination method is to get the coefficients of one of the variables to be _____ _____, such as 5 and -5.

In Problems 14–16, solve the system using elimination. **Note:** *Problem 14 is the system that we solved by using substitution in Example 5.*

14. $\begin{cases} 2x - 3y = -6 \\ -8x + 3y = 3 \end{cases}$

15. $\begin{cases} -2x + y = 4 \\ -5x + 3y = 7 \end{cases}$

16. $\begin{cases} -3x + 2y = 3 \\ 4x - 3y = -6 \end{cases}$

Work Smart

The figure below shows a consistent and independent system.

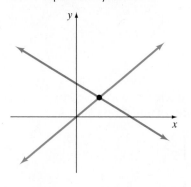

Which Method Should I Use?

When should each of the three methods be used for solving systems of two linear equations containing two unknowns? Suggestions are given below.

Method	Advantages/Disadvantages	When Should I Use It?
Graphical	This method allows the answer to be seen, but if the solutions are not integers, it can be difficult to determine the solution.	When a visual solution is required
Substitution	This method gives exact solutions. The algebra can be easy if one of the variables has a coefficient of 1.	If one of the coefficients of the variables is 1 or one of the variables is already isolated (as in $x =$ or $y =$)
Elimination	This method gives exact solutions. It is easy to use when none of the variables has a coefficient of 1.	If both equations are in standard form ($Ax + By = C$)

▶ ❺ Identify Inconsistent Systems

Examples 2 through 7 dealt only with consistent and independent systems of equations—that is, systems with a single solution. Let's now consider systems that are inconsistent, which means the lines representing their equations are parallel.

EXAMPLE 8   **An Inconsistent System**

Solve: $\begin{cases} 3x + 2y = 2 & (1) \\ -6x - 4y = 8 & (2) \end{cases}$

Solution

Use the elimination method to solve this system because none of the variables has a coefficient of 1. Notice that the coefficients of x can be made additive inverses by multiplying equation (1) by 2.

$$\begin{cases} 3x + 2y = 2 & (1) \\ -6x - 4y = 8 & (2) \end{cases}$$

Multiply (1) by 2: $\begin{cases} 2(3x + 2y) = 2(2) & (1) \\ -6x - 4y = 8 & (2) \end{cases}$

Use the Distributive Property: $\begin{cases} 6x + 4y = 4 & (1) \\ -6x - 4y = 8 & (2) \end{cases}$

Add (1) and (2): $\qquad\qquad\qquad 0 = 12$

The equation $0 = 12$ is false. Conclude that the system has no solution, so the solution set is $\varnothing$ or $\{\,\}$. The system is inconsistent.   ●

Work Smart

When solving a system of equations, if you end up with a statement "0 = some nonzero constant," or a false statement with no variables, the system is inconsistent. Graphically, the lines representing the system are parallel.

Figure 3 shows the graph of the lines from Example 8. Notice that the two lines both have slope $-\dfrac{3}{2}$. Equation (1) has a y-intercept of $(0, 1)$, and equation (2) has a y-intercept of $(0, -2)$. Therefore, the lines are parallel and do not intersect. This geometric statement is equivalent to the algebraic statement that the system has no solution.

Figure 3

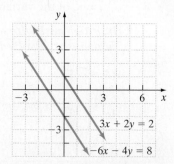

Quick ✓

17. While a system of equations is being solved, a *false* statement, such as $-8 = 0$, results. This means that the solution of the system is __ or __.

In Problem 18, show that the system is inconsistent. Draw a graph to support your result.

18. $\begin{cases} -3x + y = 2 \\ 6x - 2y = 1 \end{cases}$

▶ ❻ **Write the Solution of a System with Dependent Equations**

Let's now consider systems that are consistent but have dependent equations, which means the lines representing their equations are coinciding (in other words, the same line).

EXAMPLE 9   **Solving a System with Dependent Equations**

Solve: $\begin{cases} 3x + y = 1 & (1) \\ -6x - 2y = -2 & (2) \end{cases}$

Solution
Use the substitution method because solving equation (1) for y is straightforward.

Equation (1):	$3x + y = 1$
Subtract $3x$ from both sides:	$y = -3x + 1$

Substitute $-3x + 1$ for y in equation (2).

Equation (2):	$-6x - 2y = -2$
	$-6x - 2(-3x + 1) = -2$
Distribute the -2:	$-6x + 6x - 2 = -2$
	$-2 = -2$

The equation $-2 = -2$ is true. This means that any values of x and y that satisfy $3x + y = 1$ or $-6x - 2y = -2$ are solutions. Some ordered pairs that satisfy both equations are $(0, 1)$, $(1, -2)$, and $(2, -5)$.

The system is consistent and its equations are dependent (the value of y that makes the equation true depends on the value of x), so there are infinitely many solutions. Write the solution in either of two equivalent ways:

$$\{ (x, y) \mid 3x + y = 1 \} \quad \text{or} \quad \{ (x, y) \mid -6x - 2y = -2 \} \qquad ●$$

Figure 4 illustrates the system in Example 9. The graphs of the two equations are lines, and both have slope -3 and y-intercept $(0, 1)$. The lines are coincident.

Look back at the equations in Example 9. Notice that the terms in equation (2) are -2 times the terms in equation (1). This is another way to identify dependent equations when given two equations with two unknowns.

Figure 4

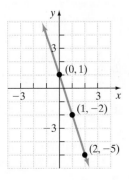

In Other Words
The solution to a consistent system with dependent equations is "the set of all ordered pairs such that one of the equations in the system is true."

Quick ✓

19. When solving a system of equations, a true statement, such as $11 = 11$, results. This means that the equations are _____.

In Problems 20–22, solve the system. Draw a graph to support your result.

20. $\begin{cases} -3x + 2y = 8 \\ 6x - 4y = -16 \end{cases}$ **21.** $\begin{cases} 2x - 3y = -16 \\ -3x + 2y = 19 \end{cases}$ **22.** $\begin{cases} \dfrac{1}{3}x + \dfrac{1}{5}y = 7 \\ \dfrac{1}{6}x - \dfrac{2}{5}y = -4 \end{cases}$

3.1 Exercises   MyMathLab®

Exercise numbers in green have complete video solutions in MyMathLab or may be accessed using the QR code to the right.

*Problems **1–22** are the Quick ✓ s that follow the **EXAMPLES**.*

Building Skills

In Problems 23–26, determine whether the ordered pairs listed are solutions of the system of linear equations. See Objective 1.

23. $\begin{cases} 2x + y = 13 \\ -5x + 3y = 6 \end{cases}$ 24. $\begin{cases} x - 2y = -11 \\ 3x + 2y = -1 \end{cases}$
    (a) $(5, 3)$ (a) $(-5, 3)$
    (b) $(3, 7)$ (b) $(-3, 4)$

25. $\begin{cases} 5x + 2y = 9 \\ -10x - 4y = -18 \end{cases}$ 26. $\begin{cases} -3x + y = 5 \\ 6x - 2y = 6 \end{cases}$
    (a) $(1, 2)$ (a) $(-2, -1)$
    (b) $\left(2, -\dfrac{1}{2}\right)$ (b) $(2, 0)$

In Problems 27–30, use the graph of the system to determine whether the system is consistent or inconsistent. If consistent, indicate whether the equations are independent or dependent. See Objective 1.

27. $\begin{cases} x + y = 1 \\ x + 2y = 0 \end{cases}$ 28. $\begin{cases} -2x + y = 4 \\ 2x + y = 0 \end{cases}$

29. $\begin{cases} x - 2y = -2 \\ x - 2y = 2 \end{cases}$ 30. $\begin{cases} 3x + y = 1 \\ -6x - 2y = -2 \end{cases}$

In Problems 31–34, solve the system of equations by graphing. See Objective 2.

31. $\begin{cases} y = 3x \\ y = -2x + 5 \end{cases}$ 32. $\begin{cases} y = -2x + 4 \\ y = 2x - 4 \end{cases}$

33. $\begin{cases} 2x + y = 2 \\ x + 3y = -9 \end{cases}$ 34. $\begin{cases} -x + 2y = -9 \\ 2x + y = -2 \end{cases}$

In Problems 35–42, solve the system of equations using substitution. See Objective 3.

35. $\begin{cases} y = -\dfrac{1}{2}x + 1 \\ 2x + y = 10 \end{cases}$ 36. $\begin{cases} 3x + y = -4 \\ y = 4x + 17 \end{cases}$

37. $\begin{cases} x = \dfrac{2}{3}y \\ 3x - y = -3 \end{cases}$ 38. $\begin{cases} y = \dfrac{1}{2}x \\ x - 4y = -4 \end{cases}$

39. $\begin{cases} 2x - 4y = 2 \\ x + 2y = 0 \end{cases}$ 40. $\begin{cases} 3x + 2y = 0 \\ 6x + 2y = 5 \end{cases}$

41. $\begin{cases} x + y = 10{,}000 \\ 0.05x + 0.07y = 650 \end{cases}$

42. $\begin{cases} x + y = 5000 \\ 0.04x + 0.08y = 340 \end{cases}$

In Problems 43–50, solve the system of equations using elimination. See Objective 4.

43. $\begin{cases} x + y = -5 \\ -x + 2y = 14 \end{cases}$ 44. $\begin{cases} x + y = -6 \\ -2x - y = 0 \end{cases}$

45. $\begin{cases} x + 2y = -5 \\ 3x + 3y = 9 \end{cases}$ 46. $\begin{cases} -3x + 2y = -5 \\ 2x - y = 10 \end{cases}$

47. $\begin{cases} 2x + 5y = -3 \\ x + \dfrac{5}{4}y = -\dfrac{1}{2} \end{cases}$ 48. $\begin{cases} x + 2y = -\dfrac{8}{3} \\ 3x - 3y = 5 \end{cases}$

49. $\begin{cases} 0.05x + 0.1y = 5.25 \\ 0.08x - 0.02y = 1.2 \end{cases}$

50. $\begin{cases} 0.04x + 0.06y = 2.1 \\ 0.06x - 0.03y = 0.15 \end{cases}$

In Problems 51–54, use either substitution or elimination to show that the system is inconsistent. Draw a graph to support your result. See Objective 5.

51. $\begin{cases} 3x + y = 1 \\ -6x - 2y = -4 \end{cases}$ 52. $\begin{cases} -2x + 4y = 9 \\ x - 2y = -3 \end{cases}$

53. $\begin{cases} 5x - 2y = 2 \\ -10x + 4y = 3 \end{cases}$ 54. $\begin{cases} 6x - 4y = 6 \\ -3x + 2y = 3 \end{cases}$

In Problems 55–60, use either substitution or elimination to show that the system is consistent and its equations are dependent. Solve the system and draw a graph to support your solution. See Objective 6.

55. $\begin{cases} y = \dfrac{1}{2}x + 1 \\ 2x - 4y = -4 \end{cases}$ 56. $\begin{cases} y = -\dfrac{2}{3}x + 3 \\ 2x + 3y = 9 \end{cases}$

57. $\begin{cases} x + 3y = 6 \\ -\dfrac{x}{3} - y = -2 \end{cases}$

58. $\begin{cases} -4x + y = 8 \\ x - \dfrac{y}{4} = -2 \end{cases}$

59. $\begin{cases} \dfrac{1}{3}x - 2y = 6 \\ -\dfrac{1}{2}x + 3y = -9 \end{cases}$

60. $\begin{cases} \dfrac{5}{4}x - \dfrac{1}{2}y = 6 \\ -\dfrac{5}{3}x + \dfrac{2}{3}y = -8 \end{cases}$

Mixed Practice

In Problems 61–68, solve the system of equations using either substitution or elimination.

61. $\begin{cases} x + 3y = 0 \\ -2x + 4y = 30 \end{cases}$

62. $\begin{cases} 2x + y = -1 \\ -3x - 2y = 7 \end{cases}$

63. $\begin{cases} x = 5y - 3 \\ -3x + 15y = 9 \end{cases}$

64. $\begin{cases} y = \dfrac{1}{2}x + 2 \\ x - 2y = -4 \end{cases}$

65. $\begin{cases} 2x - 4y = 18 \\ 3x + 5y = -3 \end{cases}$

66. $\begin{cases} 12x + 45y = 0 \\ 8x + 6y = 24 \end{cases}$

67. $\begin{cases} \dfrac{5}{6}x - \dfrac{1}{3}y = -5 \\ -x + \dfrac{2}{5}y = 1 \end{cases}$

68. $\begin{cases} \dfrac{1}{3}x - \dfrac{1}{2}y = -5 \\ -\dfrac{4}{5}x + \dfrac{6}{5}y = 1 \end{cases}$

Applying the Concepts

In Problems 69–72, write each equation in the system of equations in slope-intercept form. Use the slope-intercept form to determine the number of solutions the system has.

69. $\begin{cases} 2x + y = -5 \\ 5x + 3y = 1 \end{cases}$

70. $\begin{cases} 4x - 2y = 8 \\ -10x + 5y = 5 \end{cases}$

71. $\begin{cases} 3x - 2y = -2 \\ -6x + 4y = 4 \end{cases}$

72. $\begin{cases} 2x - y = -5 \\ -4x + 3y = 9 \end{cases}$

△ **73. Parallelogram** Use the parallelogram shown to answer parts (a) and (b).

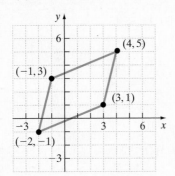

(a) Find the equation of the line for the diagonal through the points $(-1, 3)$ and $(3, 1)$. Find the equation of the line for the diagonal through the points $(-2, -1)$ and $(4, 5)$.

(b) Find the point of intersection of the diagonals.

△ **74. Rhombus** A rhombus is a parallelogram whose adjacent sides are congruent. Use the rhombus to answer parts (a), (b), and (c).

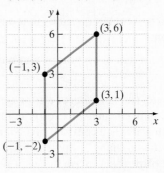

(a) Find the equation of the line for the diagonal through the points $(-1, 3)$ and $(3, 1)$. Find the equation of the line for the diagonal through the points $(-1, -2)$ and $(3, 6)$.

(b) Find the point of intersection of the diagonals.

(c) Compare the slopes of the diagonals. What can be said about the diagonals of a rhombus?

Extending the Concepts

75. Which of the following ordered pairs could be a solution to the system graphed below?

(a) $(2, 4)$ (b) $(-2, 0)$ (c) $(-3, 1)$
(d) $(5, -2)$ (e) $(-1, -3)$ (f) $(-1, 3)$

76. Which of the following systems of equations could have the graph below?

(a) $\begin{cases} 2x + 3y = 12 \\ 2x + y = -2 \end{cases}$

(b) $\begin{cases} 2x + 3y = 3 \\ -2x + y = 2 \end{cases}$

(c) $\begin{cases} 2x - 3y = 12 \\ x + 2y = 2 \end{cases}$

77. For the system $\begin{cases} Ax + 3By = 2 \\ -3Ax + By = -11 \end{cases}$, find A and B such that $x = 3, y = 1$ is a solution.

78. Write a system of equations that has $(3, 5)$ as a solution.

79. Write a system of equations that has $(-1, 4)$ as a solution.

△ **80. Centroid** The medians of a triangle are the line segments from each vertex to the midpoint of the opposite side. The centroid of a triangle is the point where the medians of the triangle intersect. Use the information given in the figure of the triangle to find its centroid.

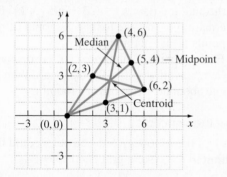

In Problems 81–84, solve each system using either substitution or elimination. Draw a graph to support your solution.

81. $\begin{cases} 3x + y = 5 \\ x + y = 3 \\ x + 3y = 7 \end{cases}$

82. $\begin{cases} x + y = -2 \\ -3x + 2y = 16 \\ 2x - 4y = -16 \end{cases}$

83. $\begin{cases} y = \dfrac{2}{3}x - 5 \\ 4x - 6y = 30 \\ x - 5y = 11 \end{cases}$

84. $\begin{cases} -4x + 3y = 33 \\ 3x - 4y = -37 \\ 2x - 3y = 15 \end{cases}$

Explaining the Concepts

85. In this section, we presented two algebraic methods for solving a system of linear equations. Are there any circumstances where one method is preferable to the other? What are these circumstances?

86. Describe geometrically the three possibilities for a solution of a system of two linear equations containing two variables.

87. The solution of a system of two linear equations in two unknowns is $x = 3, y = -2$. Where do the lines in the system intersect? Why?

88. In the process of solving a system of linear equations, what tips you off that the system is consistent and the equations are dependent? What tips you off that the system is inconsistent?

Technology Exercises

Technology can be used to approximate the point of intersection between two equations. We illustrate this feature by doing Example 3 using Desmos. Start by graphing each equation in the system. Then click on one of the lines. This will bring up the points of interest. Hover your mouse over the intersection point and it will be displayed. See Figure 5. The solution is the ordered pair $(2, -3)$.

Figure 5

In Problems 89–96, use technology to solve each system of equations. If necessary, express your solution rounded to two decimal places.

89. $\begin{cases} y = 3x - 1 \\ y = -2x + 5 \end{cases}$

90. $\begin{cases} y = \dfrac{3}{2}x - 4 \\ y = -\dfrac{1}{4}x + 3 \end{cases}$

91. $\begin{cases} 3x - y = -1 \\ -4x + y = -3 \end{cases}$

92. $\begin{cases} -6x - 2y = 4 \\ 5x + 3y = -2 \end{cases}$

93. $\begin{cases} 4x - 3y = 1 \\ -8x + 6y = -2 \end{cases}$

94. $\begin{cases} -2x + 5y = -2 \\ 4x - 10y = 1 \end{cases}$

95. $\begin{cases} 2x - 3y = 12 \\ 5x + y = -2 \end{cases}$

96. $\begin{cases} x - 3y = 21 \\ x + 6y = -2 \end{cases}$

3.2 Problem Solving: Systems of Two Linear Equations Containing Two Unknowns

Objectives

1. Model and Solve Direct Translation Problems
2. Model and Solve Geometry Problems
3. Model and Solve Mixture Problems
4. Model and Solve Uniform Motion Problems
5. Find the Intersection of Two Linear Functions

Are You Prepared for This Section?

Before getting started, complete the following problems. If you get a problem wrong, go back to the section cited and review the material.

P1. List the problem-solving steps given in Section 1.2. [Section 1.2, p. 61]

P2. A total of $25,000 is invested in stocks and bonds, with *s* representing the amount in stocks. Write an algebraic expression for the amount invested in bonds. [Section 1.2, pp. 65–66]

P3. Suppose that you have a credit card balance of $3500 and the credit card company charges 12.5% annual interest on outstanding balances. How much interest will you have to pay after 1 month? [Section 1.2, pp. 64–65]

P4. Write a linear cost function if the fixed costs are $500 and the cost of producing additional units is $15 per unit. [Section 2.4, pp. 179–180]

In Sections 1.2 and 1.3, problems were modeled and solved using a single variable. In this section, problems will be modeled using two variables. The variables will be related through a system of equations. This approach is different from the problems in Section 1.2 because the Section 1.2 problems had a single unknown whose value was determined from a single equation.

▶ 1 Model and Solve Direct Translation Problems

Let's first look at examples of direct translation.

EXAMPLE 1   **What Are the Numbers?**

The sum of one number and three times another is -12. The difference of twice the first number and the second number is 32. What are the numbers?

Solution

Step 1: Identify Find the value of the two numbers.

Step 2: Name Let x represent the first number. Let y represent the second number.

Step 3: Translate "The sum of one number and three times the other is -12," can be represented by $x + 3y = -12$. Call this equation (1). "The difference of twice the first number and the second number is 32" can be represented by $2x - y = 32$. Call this equation (2). Combine the equations to form the following system:

$$\begin{cases} x + 3y = -12 & (1) \\ 2x - y = 32 & (2) \end{cases} \quad \text{The Model}$$

Step 4: Solve Solve the system using the elimination method.

Eliminate y by multiplying equation (2) by 3:
$$\begin{cases} x + 3y = -12 & (1) \\ 3(2x - y) = 3(32) & (2) \end{cases}$$

Distribute:
$$\begin{cases} x + 3y = -12 \\ 6x - 3y = 96 \end{cases}$$

Add equations (1) and (2): $7x \qquad = 84$

Divide both sides by 7: $x \qquad = 12$

Prepared?...Answers **P1.** See page 61.
P2. $25,000 - s$ **P3.** $36.46
P4. $C(x) = 15x + 500$

Substitute 12 for x in equation (1) and solve for y.

$$x + 3y = -12$$
$$12 + 3y = -12$$

Subtract 12 from both sides: $\qquad 3y = -24$

Divide both sides by 3: $\qquad y = -8$

Step 5: Check If $x = 12$ (the first number) and $y = -8$ (the second number), then "the sum of one number and three times the other" is $12 + 3(-8) = 12 + (-24) = -12$, which we want. "The difference of twice the first number and the second number" is $2(12) - (-8) = 24 + 8 = 32$, which is correct.

Step 6: Answer The two numbers are 12 and -8.

Quick ✓

1. The sum of twice one number and three times another is 9. The difference of the numbers is 22. Find the numbers.

▶ EXAMPLE 2 | **Take Me Out to the Ball Game**

At a baseball game, Adrienne bought her family 4 hot dogs and 3 large Cokes for \$22.25. Dave bought his family 5 hot dogs and 4 large Cokes for \$28.50. How much does each hot dog cost? How much does each large Coke cost?

Solution

Step 1: Identify Find the cost of each hot dog and large Coke.

Step 2: Name Let h represent the cost of a hot dog in dollars. Let c represent the cost of a large Coke in dollars.

Step 3: Translate If one hot dog costs h dollars, then 4 hot dogs cost $4h$ dollars. Similarly 3 large Cokes cost $3c$ dollars. Add these costs together to get \$22.25, Adrienne's purchase.

Adrienne's purchase: $\quad 4h + 3c = 22.25 \quad$ (1)

Using the same notation, describe Dave's purchase:

Dave's purchase: $\quad 5h + 4c = 28.50 \quad$ (2)

Combine equations (1) and (2) to form the following system:

$$\begin{cases} 4h + 3c = 22.25 & (1) \\ 5h + 4c = 28.50 & (2) \end{cases} \quad \text{The Model}$$

Step 4: Solve

Eliminate h by multiplying equation (1) by 5 and equation (2) by -4:
$$\begin{cases} 5(4h + 3c) = 5(22.25) & (1) \\ -4(5h + 4c) = -4(28.50) & (2) \end{cases}$$

Distribute: $\qquad \begin{cases} 20h + 15c = 111.25 & (1) \\ -20h - 16c = -114 & (2) \end{cases}$

Add equations (1) and (2): $\qquad -c = -2.75$

Divide both sides by -1: $\qquad c = 2.75$

Substitute 2.75 for c in equation (1) and solve for h.

$$4h + 3(2.75) = 22.25$$
$$4h + 8.25 = 22.25$$

Subtract 8.25 from both sides: $\qquad 4h = 14$

Divide both sides by 4: $\qquad h = 3.50$

Step 5: Check If a hot dog costs $3.50 and a Coke costs $2.75, and Adrienne buys 4 hot dogs and 3 Cokes, then she'll spend $4($3.50$) + 3($2.75$) = $14.00 + $8.25 = 22.25. If Dave buys 5 hot dogs and 4 Cokes, then he'll spend $5($3.50$) + 4($2.75$) = $17.50 + $11.00 = 28.50. It checks.

Step 6: Answer Each hot dog costs $3.50 and each large Coke costs $2.75. ●

> **Quick ✓**
>
> **2.** At a fast-food establishment, 4 cheeseburgers and 2 medium shakes cost $24.54. At the same fast-food establishment, 3 cheeseburgers and 3 medium shakes cost $22.74. What is the cost of a cheeseburger? What is the cost of a medium shake?

❷ Model and Solve Geometry Problems

Geometry formulas are used to solve many types of problems. The exact formula needed will depend on the problem. Remember that a formula is a model–and in geometry, formulas describe relationships and shapes.

⊙ EXAMPLE 3   Enclosing a Yard with a Fence

The Fitzgeralds just bought a dog, so they need to enclose a large area of their backyard with a fence. The perimeter of the rectangular enclosure is 180 feet. The length is 25 feet longer than the width. What are the dimensions of the enclosure?

Solution

Step 1: Identify Find the length and the width of the enclosure.

Step 2: Name Let l represent the length and w represent the width of the enclosure, both in feet.

Step 3: Translate The perimeter P of a rectangle is $P = 2l + 2w$, where l is the length and w is the width. Thus

$$\text{Equation (1):} \quad 2l + 2w = 180$$

It is given that the length is 25 feet longer than the width, so

$$\text{Equation (2):} \quad l = w + 25$$

Equations (1) and (2) are combined to form the following system:

$$\begin{cases} 2l + 2w = 180 & (1) \\ \quad\quad l = w + 25 & (2) \end{cases} \quad \text{The Model}$$

Step 4: Solve Use the method of substitution because equation (2) is solved for l.

$$\text{Let } l = w + 25 \text{ in equation (1):} \quad 2(w + 25) + 2w = 180$$
$$\text{Distribute:} \quad 2w + 50 + 2w = 180$$
$$\text{Combine like terms:} \quad 4w + 50 = 180$$
$$\text{Subtract 50 from both sides:} \quad 4w = 130$$
$$\text{Divide both sides by 4:} \quad w = 32.5$$

Substitute 32.5 for w in $l = w + 25$.

$$l = 32.5 + 25$$
$$= 57.5$$

Step 5: Check If $l = 57.5$ and $w = 32.5$, then the perimeter is $2(57.5) + 2(32.5) = 180$ feet. The length, 57.5, is 25 feet more than the width, 32.5.

Step 6: Answer The width of the backyard enclosure is 32.5 feet, and the length is 57.5 feet. ●

Quick ✓

3. A rectangular field has a perimeter of 360 yards. The length of the field is twice the width. What is the length of the field? What is the width of the field?

 EXAMPLE 4 **Solving a Geometry Problem**

Two intersecting lines are shown in Figure 6. Suppose the measure of angle a is $(x + 3y)°$, the measure of angle b is $(x + y)°$, and the measure of angle c is $(3x - 2y - 5)°$. Find x and y.

Solution

Step 1: Identify This geometry problem involves supplementary angles. Angle a and b are supplementary, which means that the sum of the measures of angles a and b equals $180°$. In addition, angles b and c are vertical angles. Therefore, the measure of angle b equals the measure of angle c.

Step 2: Name The names of the unknowns, x and y, were given in the problem.

Figure 6

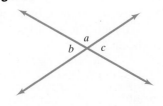

Work Smart

Two angles are supplementary if the sum of the angles is $180°$. When two lines intersect, the nonadjacent angles are vertical.

Step 3: Translate The measure of angle a plus the measure of angle b equals $180°$, so

$$\overbrace{x + 3y}^{\text{Measure of Angle } a} + \overbrace{x + y}^{\text{Measure of Angle } b} = 180$$

This simplifies to

$$\text{Equation (1): } 2x + 4y = 180$$

Also, the measure of angle b equals the measure of angle c, so

$$\overbrace{x + y}^{\text{Measure of Angle } b} = \overbrace{3x - 2y - 5}^{\text{Measure of Angle } c}$$

This simplifies to

$$\text{Equation (2): } -2x + 3y = -5$$

We form a system using equations (1) and (2).

$$\begin{cases} 2x + 4y = 180 & (1) \\ -2x + 3y = -5 & (2) \end{cases} \quad \text{The Model}$$

Step 4: Solve Use the elimination method by adding equations (1) and (2):

$$\begin{cases} 2x + 4y = 180 & (1) \\ -2x + 3y = -5 & (2) \end{cases}$$

Add: $\qquad\qquad\qquad 7y = 175$

Divide both sides by 7: $\qquad y = 25$

Let $y = 25$ in equation (1) and find that $x = 40$.

Step 5: Check If $x = 25$ and $y = 40$, then the measure of angle a is $(x + 3y)° = (40 + 3(25))° = 115°$. The measure of angle b is $(x + y)° = (40 + 25)° = 65°$. The measure of angle c is $(3x - 2y - 5)° = (3(40) - 2(25) - 5)° = 65°$. The sum of the measures of angle a and angle b is $115° + 65° = 180°$, so angle a and angle b are supplementary. In addition, the measures of the vertical angles, b and c, are equal.

Step 6: Answer The value of x is 40 and the value of y is 25. ●

Quick ✓

4. Suppose that lines m and n are parallel. Line p transverses lines m and n. Suppose that the measure of angle a is $(x + 3y)°$, the measure of angle c is $(3x + y)°$, and the measure of angle e is $(5x + y)°$. Find x and y.

❸ Model and Solve Mixture Problems

In Section 1.2, a variety of mixture problems were solved by writing a single equation with a single unknown. An alternative approach to solving these problems is to write a system of equations. This is one of the beauties of modeling: Different models can be developed to solve the exact same problem! Let's revisit a mixture problem and solve it by developing a model that uses a system of equations.

▶ (**EXAMPLE 5**)   **Financial Planning**

Kevin has $15,000 to invest. His goal is to earn 5%, or $750, annually. His financial advisor recommends investing some of the money in corporate bonds that pay 6% and the rest in government-backed Treasury notes paying 3%. How much should Kevin place in each investment in order to achieve his goal?

Solution

Step 1: Identify This mixture problem involves simple interest. Kevin needs to know how much to place in corporate bonds and how much to place in Treasury notes to earn $750 in interest.

Step 2: Name Let c represent the amount invested in corporate bonds and t represent the amount invested in Treasury notes, both in dollars.

Step 3: Translate Organize the given information in Table 1.

Table 1				
	Principal ($)	Rate (%)	Time (Yr)	Interest ($)
Corporate Bonds	c	0.06	1	0.06c
Treasury Notes	t	0.03	1	0.03t
Total	15,000	0.05	1	0.05(15,000) = $750

Kevin wants to earn 5%, or $750, each year on his principal of $15,000. The total interest will be the sum of the interest from the corporate bonds and the Treasury notes.

Interest from corporate bonds + Interest from Treasury notes = $750

Equation (1): $0.06c + 0.03t = 750$

Since the total investment is $15,000, the sum of the amounts invested in corporate bonds and in Treasury notes must equal $15,000:

Equation (2): $c + t = 15{,}000$

Use equations (1) and (2) to form a system of equations.

$$\begin{cases} 0.06c + 0.03t = 750 & (1) \\ c + t = 15{,}000 & (2) \end{cases} \quad \text{The Model}$$

Step 4: Solve

Use the substitution method by first solving equation (2) for t:	$t = 15{,}000 - c$
Substitute $15{,}000 - c$ for t in equation (1):	$0.06c + 0.03(15{,}000 - c) = 750$
Use the Distributive Property:	$0.06c + 450 - 0.03c = 750$
Combine like terms:	$0.03c + 450 = 750$
Subtract 450 from both sides:	$0.03c = 300$
Divide both sides by 0.03:	$c = 10{,}000$

Since $c = 10{,}000$, then $t = 15{,}000 - c = 15{,}000 - 10{,}000 = 5000$.

Step 5: Check The simple interest earned each year on the corporate bonds is $(\$10{,}000)(0.06)(1) = \$600$. The simple interest earned each year on the Treasury notes is $(\$5000)(0.03)(1) = \$150$. The total interest earned is $\$600 + \$150 = \$750$, the amount Kevin wanted.

Step 6: Answer the Question Kevin will invest $10,000 in corporate bonds and $5000 in Treasury notes.

Quick ✓

5. Maria has recently retired and requires an extra $8000 per year in income. She has $120,000 to invest and can invest in an AAA-rated bond that pays 4% per annum or a B-rated bond paying 8% per annum. How much should be placed in each investment in order for Maria to achieve her goal?

▶ EXAMPLE 6   **Blending Coffees**

A coffee shop manager wishes to form a new blend of coffee. She wants to mix strong, distinctive Sumatra beans that sell for $12 per pound with milder Brazilian beans that sell for $8 per pound to get 50 pounds of the new blend. The new blend will sell for $9 per pound. The revenue from selling the new blend will be the same as that from selling the beans separately. How many pounds of the Sumatra and Brazilian beans are required?

Solution

Step 1: Identify This is a mixture problem. Find the number of pounds of Sumatra beans and the number of pounds of Brazilian beans required in the new blend.

Step 2: Name Let s represent the required number of pounds of Sumatra beans and b represent the required number of pounds of Brazilian beans.

Step 3: Translate There is to be no difference in revenue between selling the Sumatra and Brazilian separately and selling the blend. This means that if the blend contains one pound of Sumatra and one pound of Brazilian, the shop should collect $\$12(1) + \$8(1) = \$20$, which is the amount it would collect if it sold the beans separately.

Table 2 summarizes the information, and Figure 7 illustrates the idea.

Table 2

	Price ($/Pound)	Number of Pounds	Revenue
Sumatra	12	s	$12s$
Brazilian	8	b	$8b$
Blend	9	50	$9(50) = 450$

Figure 7

Sumatra
$12 per pound

Brazilian
$8 per pound

Blend
$9 per pound

$$\left(\begin{array}{c}\text{Price per pound} \\ \text{of Sumatra}\end{array}\right)\left(\begin{array}{c}\text{Pounds of} \\ \text{Sumatra}\end{array}\right) + \left(\begin{array}{c}\text{Price per pound} \\ \text{of Brazilian}\end{array}\right)\left(\begin{array}{c}\text{Pounds of} \\ \text{Brazilian}\end{array}\right) = \left(\begin{array}{c}\text{Price per pound} \\ \text{of blend}\end{array}\right)\left(\begin{array}{c}\text{Pounds of} \\ \text{blend}\end{array}\right)$$

$$\$12 \quad \cdot \quad s \quad + \quad \$8 \quad \cdot \quad b \quad = \quad \$9 \quad \cdot \quad 50$$

This gives the equation

$$12s + 8b = 450 \quad (1)$$

The number of pounds of Sumatra beans plus the number of pounds of Brazilian beans should equal 50 pounds, so

$$s + b = 50 \quad (2)$$

Use equations (1) and (2) to form a system of equations.

$$\begin{cases} 12s + 8b = 450 & (1) \\ s + b = 50 & (2) \end{cases} \quad \text{The Model}$$

Step 4: Solve

Use the substitution method by solving equation (2) for b:	$b = 50 - s$
Substitute $50 - s$ for b in equation (1):	$12s + 8(50 - s) = 450$
Use the Distributive Property:	$12s + 400 - 8s = 450$
Combine like terms:	$4s + 400 = 450$
Subtract 400 from both sides:	$4s = 50$
Divide both sides by 4:	$s = 12.5$

Since $s = 12.5$ pounds, then $b = 50 - s = 50 - 12.5 = 37.5$ pounds.

Step 5: Check If the manager mixes 12.5 pounds of Sumatra with 37.5 pounds of Brazilian, the total revenue will be $\$12(12.5) + \$8(37.5) = \$150 + \$300 = \$450$, which equals $\$9(50)$, the revenue from the blend. This checks with the given information.

Step 6: Answer the Question The manager should mix 12.5 pounds of Sumatra beans with 37.5 pounds of Brazilian beans to make the blend.

●

Quick ✓

6. "We're Nuts!" sells cashews for $7.00 per pound and peanuts for $2.50 per pound. The manager has decided to make a cashew and peanut trail mix. She wants the mix to sell for $4.00 per pound, and the revenue from selling the trail mix should be the same as that from selling the nuts alone. How many pounds of cashews and how many pounds of peanuts are required to create 30 pounds of trail mix?

▶ ④ Model and Solve Uniform Motion Problems

Let's now look at uniform motion problems. Remember, these problems use the fact that distance equals rate times time ($d = rt$).

EXAMPLE 7 **Uniform Motion**

The airspeed of a plane is its speed through the air, assuming there is no wind. This speed contrasts with the plane's groundspeed–its speed relative to the ground. The groundspeed of an airplane is affected by the speed of the wind. Suppose that a Boeing 767 flying west from Washington, D.C., to San Francisco, a distance of 2400 miles, takes 6 hours. The trip east from San Francisco to Washington, D.C., takes 4 hours. Find the airspeed of the plane and the effect wind resistance has on the plane.

Solution

Step 1: Identify In this uniform motion problem, find the airspeed of the plane.

Step 2: Name Let a represent the airspeed of the plane and w represent the effect of wind resistance on the plane, both in miles per hour.

Step 3: Translate Going west, the plane is flying into the jet stream, so the plane is slowed down by the wind. Therefore, the groundspeed of the plane is $a - w$. Going east, the jet stream is helping the plane, so the groundspeed of the plane is $a + w$. We set up Table 3.

Table 3

	Distance	= Rate	· Time
Against Wind (West)	2400	$a - w$	6
With Wind (East)	2400	$a + w$	4

Going against the wind gives the equation

$$\text{Equation (1):} \quad 2400 = 6(a - w) \text{ or } 2400 = 6a - 6w$$

Going with the wind gives the equation

$$\text{Equation (2):} \quad 2400 = 4(a + w) \text{ or } 2400 = 4a + 4w$$

Now, form a system of two linear equations containing two unknowns.

$$\begin{cases} 6a - 6w = 2400 & (1) \\ 4a + 4w = 2400 & (2) \end{cases} \quad \text{The Model}$$

Step 4: Solve Use the elimination method by first multiplying equation (1) by 2 and equation (2) by 3 and then adding equations (1) and (2).

$$\begin{cases} 2(6a - 6w) = 2(2400) & (1) \\ 3(4a + 4w) = 3(2400) & (2) \end{cases}$$

Work Smart

In equation (1), we could have divided both sides by 6 rather than distributing. What could we have done with equation (2)?

Work Smart

When checking your work, make sure that the airspeed of the plane is greater than the effect of wind resistance.

Distribute: $\begin{cases} 12a - 12w = 4800 & (1) \\ 12a + 12w = 7200 & (2) \end{cases}$

Add (1) and (2): $\quad 24a \quad\quad = 12{,}000$

Divide both sides by 24: $\quad a = 500$

We let $a = 500$ in equation (1) to find that $w = 100$.

Step 5: Check Flying west, the plane's groundspeed is $500 - 100 = 400$ miles per hour, so the plane will fly $(400 \text{ miles per hour})(6 \text{ hours}) = 2400$ miles. Flying east, the plane's groundspeed is $500 + 100 = 600$ miles per hour, so the plane will fly 2400 miles in 4 hours.

Step 6: Answer The airspeed of the plane is 500 miles per hour. The effect of wind resistance on the plane is 100 miles per hour. ●

Figure 8

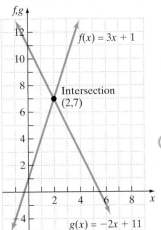

Quick ✓

7. Suppose that a plane flying 1200 miles west requires 4 hours and flying 1200 miles east requires 3 hours. Find the airspeed of the plane and the effect wind resistance has on the plane.

▶ ❺ **Find the Intersection of Two Linear Functions**

It is often useful to know when two linear functions are equal. For example, consider the question "*what value of x will make* $f(x) = 3x + 1$ *equal* $g(x) = -2x + 11$". Answering this question would require solving the equation $f(x) = g(x)$, or $3x + 1 = -2x + 11$. The solution to this equation is represented geometrically as the point of intersection of the graphs of the two functions, as shown in Figure 8. Let's look at an application of intersecting functions from business.

EXAMPLE 8   **Break-Even Analysis**

Business is motivated by profit. A company's profit is the difference between revenue and cost. Companies need to understand how many units of their product they must manufacture and sell in order to be profitable. Suppose a gas grill company sells its entry-level grill for $130. The cost of manufacturing each grill is $80, and the fixed costs per month are $8500.

(a) Write revenue R as a function of the number of grills sold x.

(b) Write cost C as a function of the number of grills manufactured per month x.

(c) Graph the revenue function and cost function in the same Cartesian plane.

(d) The **break-even point** is the point where revenue equals cost. Find the break-even number of grills to be manufactured and sold. What is the revenue when this number of grills is sold? Label the break-even point on the graph drawn in part (c).

Solution

(a) If one grill is sold, the revenue is $130. If two grills are sold, the revenue is $130(2) = 260. If x grills are sold, the revenue is $130x$. The revenue function is $R(x) = 130x$.

(b) The cost function is $C(x) = ax + b$, where a is the cost of manufacturing each unit and b is the fixed cost. With $a = 80$ and $b = 8500$, $C(x) = 80x + 8500$.

(c) Figure 9 on the next page shows the graphs of the revenue and cost functions.

(d) To find the break-even point, solve the equation $R(x) = C(x)$.

$$R(x) = C(x)$$
$$130x = 80x + 8500$$

Subtract 80x from both sides: $\quad 50x = 8500$

Divide both sides by 50: $\quad x = 170$

Figure 9

It appears that the company needs to sell 170 grills each month to break even. We verify this by determining the revenue and cost when $x = 170$.

$$R(170) = 130(170) = \$22,100$$
$$C(170) = 80(170) + 8500 = \$22,100$$

Since the revenue equals the cost when $x = 170$ grills are manufactured and sold, the break-even point is 170 grills per month. Label this point on the graph in Figure 9. Notice that if more than 170 grills are sold, the company will make a profit.

Quick ✓

8. Suppose that a nursery sells 8-foot Austrian Pine trees for $230. The nursery can buy the trees from a tree farm for $160. The fixed costs at the nursery amount to $2100 per month.

  (a) Write revenue R as a function of the number of trees sold x.

  (b) Write cost C as a function of the number of trees purchased from the tree farm x.

  (c) Graph the revenue function and cost function in the same Cartesian plane.

  (d) Find the break-even number of trees to be sold. What is the revenue when this number of trees is sold? Label the break-even point on the graph drawn in part (c).

3.2 Exercises MyMathLab®

Exercise numbers in green have complete video solutions in MyMathLab or may be accessed using the QR code to the right.

*Problems **1–8** are the Quick ✓s that follow the EXAMPLES.*

Building Skills

In Problems 9–14, solve each direct translation problem. See Objective 1.

9. The sum of two numbers is 18. The difference of the two numbers is -2. Find the numbers.

10. The sum of two numbers is 25. The difference of the two numbers is 3. Find the numbers.

11. Janice is thinking of two numbers. She says that two times the first number plus the second number is 47. In addition, the first number plus three times the second number is 81. Find the numbers.

12. Ashad is thinking of two numbers. He says that three times the first number minus the second number is 118. In addition, two times the first number plus the second number is 147. Find the numbers.

13. Rental Costs Gina rented a bounce house for 5 hours at a total cost of $235. Lori rented the same bounce house for 3 hours at a total cost of $165. The cost of renting is based on a flat set-up fee plus a rental rate per hour. How much is the set-up fee? What is the hourly rental fee?

14. Making Change Johnny has $6.75 in dimes and quarters. He has 8 more dimes than quarters. How many quarters does Johnny have? How many dimes does Johnny have?

In Problems 15–18, solve each geometry problem. See Objective 2.

△ **15. Perimeter** The perimeter of a rectangle is 120 meters. If the length of the rectangle is 20 meters more than the width, what are the dimensions of the rectangle?

△ **16. Perimeter** The perimeter of a rectangle is 260 centimeters. If the width of the rectangle is 15 centimeters less than the length, what are the dimensions of the rectangle?

△ **17. Angles** Two lines that are not parallel are shown in the figure. Suppose it is known that the measure of angle 1 is $(10x + 6y)°$, the measure of angle 2 is $(4y)°$, and the measure of angle 3 is $(7x + 2y)°$. Find x and y.

△ **18. Angles** Suppose that lines m and n are parallel. Line p transverses lines m and n. Suppose that it is known that the measure of angle 1 is $(2x + 3y)°$, the measure of angle 2 is $(4x)°$, and the measure of angle 3 is $(5x + y + 1)°$. Find x and y.

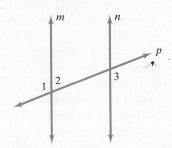

In Problems 19–22, solve each mixture problem. See Objective 3.

19. Investments Suppose that you received an unexpected inheritance of $36,000. You have decided to invest the money by placing some of the money in stocks that pay 4.4% and the rest in bonds that pay 6%. How much should you place in each investment in order to earn 5%, or $1800 annually?

20. Investments Marge and Homer have $80,000 to invest. Their financial advisor has recommended that they diversify by placing their money in two stock plans. On the risky plan they earned 8% annual interest, and on the safer plan they earned 6.5% annual interest. If Marge and Homer earned $5920 last year on their investment, how much did they invest in each of the stock plans?

21. Blending Coffees Suppose that you want to blend two coffees in order to obtain a new blend. The blend will be made with the best Arabica beans that sell for $9.00 per pound and select African Robustas that sell for $11.50 per pound to obtain 100 pounds of the new blend. The new blend will sell for $10.00 per pound and there will be no difference in revenue from selling the new blend versus selling the beans separately. How many pounds of the Arabica and Robusta beans are required?

22. Candy A candy store sells chocolate-covered almonds for $6.50 per pound and chocolate-covered peanuts for $4.00 per pound. The manager decides to make a bridge mix that combines the almonds with the peanuts. She wants the bridge mix to sell for $6.00 per pound, and there should be no loss in revenue from selling the bridge mix versus the almonds and peanuts alone. How many pounds of chocolate-covered almonds and chocolate-covered peanuts are required to create 50 pounds of bridge mix?

In Problems 23–26, solve each uniform motion problem. See Objective 4.

23. Canoeing in the River Jonathon and Samantha paddle their canoe 26 miles downstream in 2 hours. After a picnic, they paddle back upstream. After 3 hours they have traveled only 9 miles back. Assuming that they paddle at a constant rate and the river's current is constant, find the speed at which Jonathon and Samantha can paddle in still water.

24. Against the Wind A Piper Arrow can fly 510 miles in 3 hours with a tailwind. Against this same wind, the plane can fly 390 miles in 3 hours. Find the airspeed of the plane. What is the impact of the wind on the plane?

25. A Car Ride An Infiniti G35 travels at an average speed of 50 miles per hour. A Lincoln Aviator travels at an average speed of 40 miles per hour. In the time it takes the Lincoln to travel a certain distance d, the Infiniti travels 100 miles farther. Find the distance that each car travels and the time of the trip.

26. Runners Enrique leaves his house and starts to run at an average speed of 6 miles per hour. Half an hour later, Enrique's younger (and faster) brother leaves the house to catch up to Enrique running at an average speed of 8 miles per hour. How long will it take for Enrique's brother to run half the distance that Enrique has run?

In Problems 27 and 28, let R represent a company's revenue, let C represent the company's cost, and let x represent the number of units produced and sold each day. See Objective 5.

(a) *Graph the revenue function and cost function in the same Cartesian plane.*

(b) *Find the company's break-even point; that is, find x so that $R(x) = C(x)$. Label this point on the graph drawn in part (a).*

27. $R(x) = 12x$
$C(x) = 5.5x + 9880$

28. $R(x) = 16x$
$C(x) = 7x + 3645$

Applying the Concepts

△ **29. Triangle** An isosceles triangle is one in which two of the sides are the same length (congruent). The perimeter of an isosceles triangle is 35 cm. If the length of each of the congruent sides is 3 times the length of the third side, find the dimensions of the triangle.

△ **30. Trapezoid** A trapezoid is a quadrilateral in which two of the sides are parallel. A trapezoid is an isosceles trapezoid if the two non-parallel sides are congruent (the same length). See the figure. Suppose that the length of each of the congruent sides of an isosceles trapezoid is 12 cm. The perimeter of the trapezoid is 100 cm. Of the remaining two sides, one length is 14 cm shorter than two times the other. Find the length of each of the two remaining sides.

12 cm   12 cm

31. Banking A bank has loaned out $750,000, part of it at 5% per annum and the rest of it at 8% per annum. If the bank receives $52,500 in interest each year, how much was loaned at 5%?

32. One Bad Investment Horace invested $70,000 in two stocks. After 1 year, one of the stocks increased by 13%, while the other stock declined by 5%. His total gain for the year was $2800. How much was invested in the stock that earned 13%? How much was invested in the stock that lost 5%?

33. The sum of twice a first number and three times a second number is 81. If the second number is subtracted from three times the first number, the result is 17. Find the numbers.

34. The sum of four times a first number and a second number is 68. If the first number is decreased by twice the second number, the result is −1. Find the numbers.

35. Moving Walkways Moving walkways are common in airports. Drenell conducts an experiment in which he walks with a 126-foot walkway and then against the walkway. It takes Drenell 18 seconds to walk the 126-foot walkway going with the walkway and it takes him 63 seconds to walk the walkway going against the walkway. What is Drenell's normal walking speed?

36. Biking Suppose that Shannon bikes into the wind for 60 miles and it takes her 6 hours. After a long

rest, she returns (with the wind at her back) in 5 hours. Determine the speed at which Shannon can ride her bike in still air, and determine the effect that the wind had on her speed.

△ **37. Angles** Find the values of x and y based upon the location of the angles in the figure. The measure of angle 1 is $(4x + 2y)°$, the measure of angle 2 is $(10x + 5y)°$, the measure of angle 3 is $(7x + 5y)°$, and the measure of angle 4 is $(9x + 8y)°$.

△ **38. Angles** Find the values of x and y based upon the location of the angles in the parallelogram. The measure of angle 1 is $(5x + 7y)°$, the measure of angle 2 is $(10x + 5y)°$, and the measure of angle 3 is $(15x − 9y)°$.

39. Pharmacy A doctor's prescription calls for a daily intake of liquid containing 40 mg of vitamin C and 30 mg of vitamin D. Your pharmacy stocks two liquids that can be used: Liquid A contains 20% vitamin C and 30% vitamin D; Liquid B contains 50% vitamin C and 25% vitamin D. How many milligrams of each liquid should be mixed to fill the prescription?

40. Pharmacy A doctor's prescription calls for the creation of pills that contain 10 units of vitamin B_{12} and 13 units of vitamin E. Your pharmacy stocks two powders that can be used to make these pills: Powder A contains 20% vitamin B_{12} and 40% vitamin E; Powder B contains 50% vitamin B_{12} and 30% vitamin E. How many units of each powder should be mixed in each pill?

41. Counting Calories Suppose that Kristin ate two orders of McDonald's chicken nuggets and drank one medium Coke, for a total of 770 calories. Kristin's friend Jack ate three orders of nuggets and drank two medium Cokes (Jack takes advantage of free refills), for a total of 1260 calories. How many calories are in an order of chicken nuggets? How many calories are in a medium Coke?

42. Carbs Yvette and José go to McDonald's for breakfast. Yvette ordered two sausage biscuits and one orange juice. The entire meal had 98 grams of carbohydrates. José ordered three sausage biscuits and two orange juices and his meal had 162 grams of carbohydrates. How many grams of carbohydrates are in a sausage biscuit? How many grams of carbohydrates are in an orange juice?

43. Making an Alloy A metallurgist has an alloy with 10% titanium and an alloy with 25% titanium. He needs 100 grams of an alloy with 16% titanium. How much of each alloy should be mixed to obtain the 100 grams of alloy with 16% titanium?

44. Hydrochloric Acid A chemist has a solution that is 20% hydrochloric acid (HCl) and a solution that is 50% HCl. How much of each solution should be mixed to obtain 300 cubic centimeters (cc) of a solution that is 30% HCl?

45. Supply and Demand Suppose that the quantity supplied S and the quantity demanded D of hot dogs at a baseball game are given by these functions:

$$S(p) = -2000 + 3000p$$
$$D(p) = 10,000 - 1000p$$

where p is the price. The **equilibrium price** of a market is defined as the price at which quantity supplied equals quantity demanded $(S = D)$.

(a) Graph each of the two functions in the same Cartesian plane.

(b) Find the equilibrium price for hot dogs at the baseball game. What is the equilibrium quantity? Label the equilibrium point on the graph drawn in part (a).

46. Supply and Demand See Problem 45. Suppose that the quantity supplied S and the quantity demanded D of baseball hats at a baseball game are given by the following functions:

$$S(p) = 9p - 17$$
$$D(p) = -35p + 995$$

where p is the price.

(a) Graph each of the two functions in the same Cartesian plane.

(b) Find the equilibrium price for baseball hats at the baseball game. What is the equilibrium quantity? Label the equilibrium point on the graph drawn in part (a).

47. College Grads The percent of adult males who hold a bachelor's degree M as a function of the year t is $M(t) = 0.25t + 14.4$, where t is the number of years since 1990. The percent of adult females who hold a bachelor's degree F as a function of the year t is $F(t) = 0.29t + 12$, where t is the number of years since 1990. (SOURCE: *Statistical Abstract*)

(a) Graph each of the two functions in the same Cartesian plane.

(b) Assuming that this linear trend continues, find the year in which the percent of male college grads and the percent of female college grads are equal. That is, solve $M(t) = F(t)$. What is the percent of college grads in this year? Label this point on the graph.

48. Choosing Take-Out Pizza The average price in dollars, D, of a 14-inch Domino's Pizza is $D(t) = 1.95t + 13.25$, where t is the number of toppings. The average price in dollars, E, of a 14-inch Extreme Pizza is $E(t) = 1.70t + 14.45$, where t is the number of toppings. (SOURCE: Domino's and Extreme Pizza)

(a) Graph each of the two functions in the same Cartesian plane.

(b) Assuming that this linear trend continues, find the number of toppings in which the price of the Domino's Pizza will equal the price of the Extreme Pizza. That is, solve $D(t) = E(t)$. What is the number of toppings at this price? Label this point on the graph.

49. Salary Suppose that you are offered a sales position for a pharmaceutical company. It offers you two salary options. Option A would pay you an annual base salary of $15,000 plus a commission of 1% on sales. Option B would pay you an annual base salary of $25,000 plus a commission of 0.75% on sales.

(a) Find a function for each salary option.

(b) Graph each function found in part (a) in the same Cartesian plane.

(c) Determine the annual sales required for the options to result in the same annual salary. What would the annual salary be? Label this point on the graph.

50. Salary Suppose you are offered a sales position for a medical equipment company. It offers you two salary options. Option A would pay you an annual base salary of $27,000 plus a commission of 1.5% on sales. Option B would pay you an annual base salary of $45,000 plus a commission of 1% on sales.

(a) Find a function for each salary option.

(b) Graph the functions found in part (a) in the same Cartesian plane.

(c) Determine the annual sales required for the options to result in the same annual salary. What would the annual salary be? Label this point on the graph.

51. Break-Even A wood craftsman makes children's desks. He sells the desks for $60 each. His monthly fixed costs of operating the business are $3500. Each desk costs $35 in material.

(a) Find the revenue function R, treating the number of desks x as the independent variable.

(b) Find the cost function C, treating the number of desks x as the independent variable.

(c) Graph the revenue and cost functions in the same Cartesian plane.

(d) Find the break-even number of desks that must be manufactured and sold. What are the revenue and cost at this number of desks? Label the point on the graph drawn in part (c).

52. Break-Even Audra wants to establish a lemonade stand on her corner. Her father buys some lumber and other materials and makes Audra a lemonade stand for $40. Audra goes to the store and buys ingredients for lemonade. She determines that each cup of lemonade will cost her $0.03 to make, but the actual cup will cost her an additional $0.07. She decides to sell the lemonade for $0.30 a cup.

(a) Find the revenue function R, treating the number of cups of lemonade x as the independent variable.

(b) Find the cost function C, treating the number of cups of lemonade x as the independent variable.

(c) Graph the revenue and cost functions in the same Cartesian plane.

(d) Find the break-even number of cups of lemonade that Audra must sell. What are the revenue and cost at this number of cups of lemonade? Label the point on the graph drawn in part (c).

Year	Men's Time (in seconds)	Women's Time (in seconds)
1968	19.83	22.50
1972	20.00	22.40
1976	20.23	22.37
1980	20.19	22.03
1984	19.80	21.81
1988	19.75	21.34
1992	19.73	21.72
1996	19.32	22.12
2000	20.09	21.84
2004	19.79	22.05
2008	19.30	21.93
2012	19.32	21.69

(c) Draw a line through any two points for the men's winning time. Use these points to find the equation of the line.

(d) Draw a line through any two points for the women's winning time. Use these points to find the equation of the line.

(e) Use the equations found in parts (c) and (d) to find the year in which the winning time for men will equal the winning time for women. What is the winning time?

(f) Do you think that your answer to part (e) is reasonable? Why or why not?

(g) Use the lines from parts (c) and (d) to predict the results from the 2012 Olympic games. Compare your predictions to the actual men's and women's winning times.

Extending the Concepts

53. The Olympics The data in the following table represent the winning times in the 200-meter run in the finals of the Olympics.

(a) Draw a scatter diagram of the men's winning time, treating the year as the independent variable.

(b) On the same graph, draw a scatter diagram of the women's winning time, treating the year as the independent variable. Be sure to use different plotting symbols to label the men's and women's points (such as a □ and a ○).

3.3 Systems of Linear Equations in Three Variables

Objectives

1 Solve Systems of Three Linear Equations

2 Identify Inconsistent Systems

3 Write the Solution of a System with Dependent Equations

4 Model and Solve Problems Involving Three Linear Equations

Are You Prepared for This Section?

Before getting started, complete the following problem. If you get the problem wrong, go back to the section cited and review the material.

P1. Evaluate the expression $3x - 2y + 4z$ for $x = 1, y = -2,$ and $z = 3$. [Section R.5, pp. 40–41]

1 Solve Systems of Three Linear Equations

An example of a linear equation in three variables is $2x - y + z = 8$. An example of a system of three linear equations containing three variables, $x, y,$ and z is

$$\begin{cases} x + 3y + z = 8 \\ 3x - y + 6z = 12 \\ -4x - y + 2z = -1 \end{cases}$$

Systems of three linear equations with three variables have the same possible numbers of solutions as a system of two linear equations containing two variables:

1. **Exactly one solution**—A consistent system with independent equations

2. **No solution**—An inconsistent system

3. **Infinitely many solutions**—A consistent system with dependent equations

Think of a system of three linear equations containing three variables as a geometry problem. The graph of each equation in a system of linear equations containing three variables is a plane in space. A system of three linear equations containing three variables represents three planes in space. Figure 10 illustrates some of the possibilities.

Recall that a **solution** to a system of equations consists of values for the variables that are solutions of each equation of the system. Write the solution to a system of three equations containing three unknowns as an **ordered triple** (x, y, z).

Figure 10

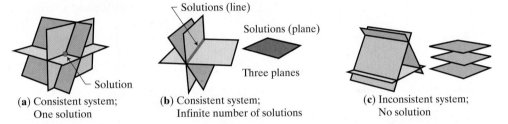

(**a**) Consistent system; One solution Solution

Solutions (line) Solutions (plane) Three planes
(**b**) Consistent system; Infinite number of solutions

(**c**) Inconsistent system; No solution

Quick ✓

1. If a system of equations has no solution, it is said to be _____. If a system of equations has infinitely many solutions, the system is said to be _____ and the equations are _____.

2. A _____ to a system of equations consists of values for the variables that are solutions of each equation of the system.

3. *True or False* A system of three linear equations containing three variables always has at least one solution.

4. *True or False* When the planes in a system of equations are parallel, the system is inconsistent and has no solution.

EXAMPLE 1 **Determining Whether Values Are a Solution of a System**

Determine which of the following ordered triples are solutions of the system.

$$\begin{cases} x + y + z = 0 \\ 2x - y + 3z = 17 \\ -3x + 2y - z = -21 \end{cases}$$

(**a**) $(1, 3, -4)$ (**b**) $(3, -5, 2)$

Solution
First, name the system's equations (1), (2), and (3).

$$\begin{cases} x + y + z = 0 & (1) \\ 2x - y + 3z = 17 & (2) \\ -3x + 2y - z = -21 & (3) \end{cases}$$

(**a**) Let $x = 1, y = 3$, and $z = -4$ in equations (1), (2), and (3). If all three equations are true, then $(1, 3, -4)$ is a solution.

Equation (1): $x + y + z = 0$
$1 + 3 + (-4) \overset{?}{=} 0$
$0 = 0$ True

Equation (2): $2x - y + 3z = 17$
$2(1) - 3 + 3(-4) \overset{?}{=} 17$
$2 - 3 - 12 \overset{?}{=} 17$
$-13 = 17$ False

Equation (3): $-3x + 2y - z = -21$
$-3(1) + 2(3) - (-4) \overset{?}{=} -21$
$-3 + 6 + 4 \overset{?}{=} -21$
$7 = -21$ False

Although this ordered triple satisfies (1), it does not satisfy (2) or (3). Therefore, the ordered triple $(1, 3, -4)$ is not a solution.

(b) Let $x = 3$, $y = -5$, and $z = 2$ in equations (1), (2), and (3).

Equation (1):	Equation (2):	Equation (3):
$x + y + z = 0$	$2x - y + 3z = 17$	$-3x + 2y - z = -21$
$3 + (-5) + (2) \stackrel{?}{=} 0$	$2(3) - (-5) + 3(2) \stackrel{?}{=} 17$	$-3(3) + 2(-5) - (2) \stackrel{?}{=} -21$
$0 = 0$ True	$6 + 5 + 6 \stackrel{?}{=} 17$	$-9 - 10 - 2 \stackrel{?}{=} -21$
	$17 = 17$ True	$-21 = -21$ True

Because the ordered triple $(3, -5, 2)$ satisfies all three equations, it is a solution to the system. ●

Quick ✓

5. Determine which of the following ordered triples are solutions of the system.

$$\begin{cases} x + y + z = 3 \\ 3x + y - 2z = -23 \\ -2x - 3y + 2z = 17 \end{cases}$$

(a) $(3, 2, -2)$ **(b)** $(-4, 1, 6)$

Typically, to solve a system of three linear equations containing three variables, the *elimination method* is used. The first step in solving a system of three linear equations with three unknowns is to reduce the system to one with two linear equations with two unknowns. Then solve the smaller system using the methods of Section 3.1.

To reduce the system of three linear equations to a system of two linear equations, various methods may be used. Eliminate one variable by multiplying equations by nonzero constants to get the coefficients of the variables to be additive inverses. Add these equations to remove that variable. Also interchange any two equations or multiply (or divide) each side of an equation by the same nonzero constant. Another approach is to replace an equation by the sum of that equation and a multiple of a second equation.

⊳ **EXAMPLE 2** **How to Solve a System of Three Linear Equations**

Use the method of elimination to solve the system: $\begin{cases} x + y - z = -1 & (1) \\ 2x - y + 2z = 8 & (2) \\ -3x + 2y + z = -9 & (3) \end{cases}$

Step-by-Step Solution

Step 1: The first step is to eliminate the same variable from two of the equations. Use equation (1) to eliminate the variable x from equations (2) and (3). Do this by multiplying equation (1) by -2 and adding the result to equation (2). The resulting equation is equation (4). Multiply equation (1) by 3 and add the result to equation (3). The resulting equation is equation (5).

$$\begin{aligned} x + y - z &= -1 \quad (1) \\ 2x - y + 2z &= 8 \quad (2) \end{aligned}$$

Multiply (1) by -2:
$$\begin{aligned} -2x - 2y + 2z &= 2 \quad (1) \\ 2x - y + 2z &= 8 \quad (2) \\ \hline \text{Add:} \quad -3y + 4z &= 10 \end{aligned}$$

$$\begin{aligned} x + y - z &= -1 \quad (1) \\ -3x + 2y + z &= -9 \quad (3) \end{aligned}$$

Multiply (1) by 3:
$$\begin{aligned} 3x + 3y - 3z &= -3 \quad (1) \\ -3x + 2y + z &= -9 \quad (3) \\ \hline \text{Add:} \quad 5y - 2z &= -12 \end{aligned}$$

$$\begin{cases} x + y - z = -1 & (1) \\ -3y + 4z = 10 & (4) \\ 5y - 2z = -12 & (5) \end{cases}$$

Step 2: Treat equations (4) and (5) as a system of two equations with two variables. Eliminate z by multiplying equation (5) by 2 and then adding equations (4) and (5). The result is equation (6).

$$\begin{cases} -3y + 4z = 10 & (4) \\ 5y - 2z = -12 & (5) \end{cases}$$

Multiply (5) by 2:
$$\begin{aligned} -3y + 4z &= 10 \quad (4) \\ 10y - 4z &= -24 \quad (5) \\ \hline \text{Add:} \quad 7y &= -14 \quad (6) \end{aligned}$$

Step 3: Solve equation (6) for *y*.

$$7y = -14 \quad (6)$$

Divide both sides by 7:
$$y = -2$$

Step 4: Substitute -2 for *y* in equation (4) and solve for *z*.

$$-3y + 4z = 10 \quad (4)$$
$$-3(-2) + 4z = 10$$
$$6 + 4z = 10$$
$$4z = 4$$
$$z = 1$$

Step 5: Substitute -2 for *y* and for *z* in equation (1) and solve for *x*.

$$x + y - z = -1 \quad (1)$$
$$x + (-2) - 1 = -1$$
$$x - 3 = -1$$
$$x = 2$$

The solution appears to be $(2, -2, 1)$.

Step 6: Verify that $(2, -2, 1)$ is the solution.

Equation (1):
$$x + y - z = -1$$
$$x = 2, y = -2, z = 1: \quad 2 + (-2) - 1 \stackrel{?}{=} -1$$
$$-1 = -1 \quad \text{True}$$

Equation (2):
$$2x - y + 2z = 8$$
$$2(2) - (-2) + 2(1) \stackrel{?}{=} 8$$
$$4 + 2 + 2 \stackrel{?}{=} 8$$
$$8 = 8 \quad \text{True}$$

Equation (3):
$$-3x + 2y + z = -9$$
$$-3(2) + 2(-2) + 1 \stackrel{?}{=} -9$$
$$-6 - 4 + 1 \stackrel{?}{=} -9$$
$$-9 = -9 \quad \text{True}$$

The solution, $(2, -2, 1)$, checks. ●

Work Smart

In Example 2, we eliminated *x* from equations (2) and (3). We could also have eliminated *y* from equation (2) by adding equations (1) and (2). We could then eliminate *y* from equation (3) by multiplying equation (1) by -2 and adding the result to equation (3). Had we done this, we would have ended up with the following system of two equations with two unknowns:

$$\begin{cases} 3x + z = 7 \\ -5x + 3z = -7 \end{cases}$$

The solution of this system is $x = 2$ and $z = 1$. Substituting these values into equation (1) yields $y = -2$, so the solution to the system is $(2, -2, 1)$, which agrees with the solution we found in Example 2. There is more than one way to solve a problem!

Solving a System of Three Linear Equations Containing

Three Unknowns by Elimination

Step 1: Select two of the equations and eliminate one of the variables from one of the equations. Select any two other equations and eliminate the *same variable* from one of the equations.

Step 2: This will leave two equations with two unknowns. Solve this system using the techniques from Section 3.1.

Step 3: Use the two known values of the variables found in Step 2 to find the value of the third variable.

Step 4: Check the answer.

Work Smart

By eliminating a variable in two of the equations in Step 1, we are creating a system of two equations with two unknowns we can solve. Remember, in mathematics it is helpful to relate a problem to one we already know how to solve.

Quick ✓

6. Use the elimination method to solve the system:
$$\begin{cases} x + y + z = -3 \\ 2x - 2y - z = -7 \\ -3x + y + 5z = 5 \end{cases}$$

> **EXAMPLE 3** **Solving a System of Three Linear Equations**

Use the elimination method to solve the system: $\begin{cases} 4x \quad\;\; + \;\; z = 4 & (1) \\ 2x + 3y \quad\quad\;\; = -4 & (2) \\ \quad\quad 2y - 4z = -15 & (3) \end{cases}$

Solution

Eliminate z from equation (3) by multiplying equation (1) by 4 and adding the result to equation (3). The resulting equation is equation (4).

$$\begin{array}{ll} 4x \quad\; + \;\; z = 4 & (1) \\ \quad\; 2y - 4z = -15 & (3) \end{array}$$

Multiply (1) by 4:
$$\begin{array}{ll} 16x \quad\quad + 4z = 16 & (1) \\ \underline{\quad\quad 2y - 4z = -15} & (3) \\ \text{Add:} \quad 16x + 2y \quad\quad = 1 & \end{array}$$

$\longrightarrow \begin{cases} 4x \quad\quad + z = 4 & (1) \\ 2x + 3y \quad\quad = -4 & (2) \\ 16x + 2y \quad\quad = 1 & (4) \end{cases}$

Now focus on the system containing equations (2) and (4). To eliminate the variable x, multiply equation (2) by -8 and then add equations (2) and (4). The result is equation (5).

$$\begin{cases} 2x + 3y = -4 & (2) \\ 16x + 2y = 1 & (4) \end{cases}$$

Multiply (2) by -8:
$$\begin{array}{ll} -16x - 24y = 32 & (2) \\ \underline{\;\;\; 16x + \;\; 2y = \;\; 1} & (4) \\ \text{Add:} \quad\quad\quad -22y = 33 & (5) \end{array}$$

Solve equation (5) for y by dividing both sides of the equation by -22.

$$\text{Equation (5):} \quad -22y = 33$$

$$y = \frac{33}{-22} = -\frac{3}{2}$$

Now substitute $-\dfrac{3}{2}$ for y in equation (2) and solve for x.

$$2x + 3y = -4 \quad (2)$$

Let $y = -\dfrac{3}{2}$ in Equation (2):
$$2x + 3\left(-\frac{3}{2}\right) = -4$$

$$2x - \frac{9}{2} = -4$$

$$2x = \frac{1}{2}$$

$$x = \frac{1}{4}$$

Now substitute $\dfrac{1}{4}$ for x in equation (1) and solve for z.

$$4x + z = 4 \quad (1)$$

$$4\left(\frac{1}{4}\right) + z = 4$$

$$1 + z = 4$$

$$z = 3$$

The check is left to you. The solution is the ordered triple $\left(\dfrac{1}{4}, -\dfrac{3}{2}, 3\right)$.

●

Quick ✓

7. Use the elimination method to solve the system: $\begin{cases} 2x \quad\quad - 4z = -7 \\ x + 6y \quad\quad = 5 \\ \quad\quad 2y - \;\; z = 2 \end{cases}$

▶ ❷ Identify Inconsistent Systems

Examples 2 and 3 were consistent systems with independent equations resulting in a single solution. Now let's look at an inconsistent system.

EXAMPLE 4   **An Inconsistent System of Linear Equations**

Use the elimination method to solve the system: $\begin{cases} x + 2y - z = 4 & (1) \\ -2x + 3y + z = -4 & (2) \\ x + 9y - 2z = 1 & (3) \end{cases}$

Solution

Notice that equation (1) can be used to eliminate x from equations (2) and (3). This can be done by multiplying equation (1) by 2 and adding the result to equation (2). The resulting equation is equation (4).

Work Smart

These figures illustrate two different inconsistent systems geometrically.

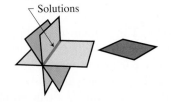

$$\begin{array}{ll} x + 2y - z = 4 & (1) \\ -2x + 3y + z = -4 & (2) \end{array} \quad \text{Multiply by 2:} \quad \begin{array}{ll} 2x + 4y - 2z = 8 & (1) \\ -2x + 3y + z = -4 & (2) \\ \hline \text{Add:} \quad 7y - z = 4 & (4) \end{array}$$

Next, multiply equation (1) by -1 and add the result to equation (3). The resulting equation is equation (5).

$$\begin{array}{ll} x + 2y - z = 4 & (1) \\ x + 9y - 2z = 1 & (3) \end{array} \quad \text{Multiply by} -1: \quad \begin{array}{ll} -x - 2y + z = -4 & (1) \\ x + 9y - 2z = 1 & (3) \\ \hline \text{Add:} \quad 7y - z = -3 & (5) \end{array}$$

Work Smart

Whenever you end up with a false statement such as $0 = -7$, you have an inconsistent system.

Now focus on the system containing equations (4) and (5). Multiply equation (4) by -1 and then add equations (4) and (5).

$$\begin{array}{ll} 7y - z = 4 & (4) \\ 7y - z = -3 & (5) \end{array} \quad \text{Multiply by} -1: \quad \begin{array}{ll} -7y + z = -4 & (4) \\ 7y - z = -3 & (5) \\ \hline \text{Add:} \quad 0 = -7 & (6) \quad \text{False} \end{array}$$

Equation (6) now states that $0 = -7$, which is a false statement. Therefore, the system is inconsistent and has no solution. The solution set is $\varnothing$ or $\{\ \}$.   ●

Quick ✓

8. Use the elimination method to solve the system: $\begin{cases} x - y + 2z = -7 \\ -2x + y - 3z = 5 \\ x - 2y + 3z = 2 \end{cases}$

▶ ❸ Write the Solution of a System with Dependent Equations

Now let's look at a system of dependent equations.

EXAMPLE 5   **Solving a System with Dependent Equations**

Work Smart

These figures illustrate two different dependent systems.

Solutions

Use the elimination method to solve the system: $\begin{cases} x - 3y - z = 4 & (1) \\ x - 2y + 2z = 5 & (2) \\ 2x - 5y + z = 9 & (3) \end{cases}$

Solution

Use equation (1) to eliminate x from equations (2) and (3). Do this by multiplying equation (1) by -1 and adding the result to equation (2). The resulting equation is equation (4).

$$x - 3y - z = 4 \quad (1) \quad \text{Multiply by } -1: \quad -x + 3y + z = -4 \quad (1)$$
$$x - 2y + 2z = 5 \quad (2) \qquad\qquad\qquad \underline{x - 2y + 2z = 5 \quad (2)}$$
$$\text{Add:} \qquad\qquad y + 3z = 1 \quad (4)$$

Next, multiply equation (1) by -2 and add the result to equation (3). The resulting equation is equation (5).

$$x - 3y - z = 4 \quad (1) \quad \text{Multiply by } -2: \quad -2x + 6y + 2z = -8 \quad (1)$$
$$2x - 5y + z = 9 \quad (3) \qquad\qquad\qquad \underline{2x - 5y + z = 9 \quad (3)} \qquad \begin{cases} y + 3z = 1 \quad (4) \\ y + 3z = 1 \quad (5) \end{cases}$$
$$\text{Add:} \qquad\qquad y + 3z = 1 \longrightarrow$$

Now concentrate on the system containing equations (4) and (5).

$$y + 3z = 1 \quad (4) \qquad \text{Multiply by } -1: \quad -y - 3z = -1 \quad (4)$$
$$y + 3z = 1 \quad (5) \qquad\qquad\qquad\qquad \underline{y + 3z = 1 \quad (5)}$$
$$\text{Add:} \qquad\qquad 0 = 0 \quad (6)$$

The statement $0 = 0$ in equation (6) indicates that it may be a consistent system with dependent equations. Show how the values for x and y *depend* on the value of z by letting z represent any real number. Then solve equation (4) for y and obtain y in terms of z.

$$\text{Equation (4):} \quad y + 3z = 1$$
$$\text{Subtract } 3z \text{ from both sides:} \qquad y = -3z + 1$$

Let $y = -3z + 1$ in equation (1) and solve for x in terms of z.

$$\text{Equation (1):} \qquad\qquad x - 3y - z = 4$$
$$\text{Let } y = -3z + 1 \text{ in (1):} \quad x - 3(-3z + 1) - z = 4$$
$$\text{Distribute:} \qquad\qquad x + 9z - 3 - z = 4$$
$$\text{Combine like terms:} \qquad\qquad x + 8z - 3 = 4$$
$$\text{Subtract } 8z \text{ from both sides; add 3 to both sides:} \qquad x = -8z + 7$$

The solution to the system is $\{(x, y, z) \mid x = -8z + 7, y = -3z + 1, z \text{ is any real number}\}$. To find specific solutions to the system, choose any value of z and use the equations $x = -8z + 7$ and $y = -3z + 1$ to determine x and y. Some specific solutions to the system are $(7, 1, 0)$, $(-1, -2, 1)$, and $(15, 4, -1)$.

Quick ✓

9. Use the elimination method to solve the system: $\begin{cases} x - y + 3z = 2 \\ -x + 2y - 5z = -3 \\ 2x - y + 4z = 3 \end{cases}$

Notice that in Examples 2 through 5, we did not always eliminate x first and y second. The order in which variables are eliminated from a system does not matter, and different approaches to solving the problem will lead to the right answer if done correctly. For example, in Example 4 we could have chosen to eliminate z from equations (1) and (3).

▶ ❹ Model and Solve Problems Involving Three Linear Equations

Now let's look at problems involving systems of three equations containing three variables.

EXAMPLE 6) **Production**

A manufacturer makes three different models of swing sets. The Monkey takes 2 hours to cut the wood, 2 hours to stain, and 3 hours to assemble. The Gorilla takes 3 hours to cut the wood, 4 hours to stain, and 4 hours to assemble. The King Kong takes 4 hours to cut the wood, 5 hours to stain, and 5 hours to assemble. The company has 61 hours available to cut the wood, 73 hours available to stain, and 83 hours available to assemble each day. How many of each type of swing set can be manufactured each day?

Solution

Step 1: Identify Determine the number of Monkey, Gorilla, and King Kong swing sets that can be manufactured each day.

Step 2: Name Let m represent the number of Monkey swing sets, g represent the number of Gorilla swing sets, and k represent the number of King Kong swing sets.

Step 3: Translate Organize the given information in Table 4.

Table 4				
	Monkey	Gorilla	King Kong	Total Hours Available
Cut Wood	2	3	4	61
Stain	2	4	5	73
Assemble	3	4	5	83

Making m Monkey swing sets takes $2m$ hours to cut the wood. Similarly, it takes $3g$ and $4k$ hours to cut the wood for g Gorilla and k King Kong sets. Since 61 hours are available,

$$2m + 3g + 4k = 61 \quad \text{Equation (1)}$$

It takes $2m$ hours to stain m Monkey sets, $4g$ hours to stain g Gorilla sets, and $5k$ hours to stain k King Kong sets. Since 73 hours are available,

$$2m + 4g + 5k = 73 \quad \text{Equation (2)}$$

$3m$ hours are needed to assemble m Monkey sets, $4g$ hours to assemble g Gorilla sets, and $5k$ hours to assemble k King Kong sets. Since 83 hours are available,

$$3m + 4g + 5k = 83 \quad \text{Equation (3)}$$

Combine equations (1), (2), and (3) to form the following system:

$$\begin{cases} 2m + 3g + 4k = 61 & (1) \\ 2m + 4g + 5k = 73 & (2) \quad \text{The Model} \\ 3m + 4g + 5k = 83 & (3) \end{cases}$$

Step 4: Solve The solution of this system of equations is $m = 10$, $g = 7$, and $k = 5$.

Step 5: Check Manufacturing 10 Monkeys, 7 Gorillas, and 5 King Kongs requires $2(10) + 3(7) + 4(5) = 61$ hours to cut the wood; $2(10) + 4(7) + 5(5) = 73$ hours to stain; and $3(10) + 4(7) + 5(5) = 83$ hours to assemble.

Step 6: Answer The company can manufacture 10 Monkey swing sets, 7 Gorilla swing sets, and 5 King Kong swing sets each day.

Quick ✓

10. The Mowing 'Em Down lawn mower company makes three styles of lawn mowers. The 21-inch model requires 2 hours to mold, 3 hours for engine manufacturing, and 1 hour to assemble. The 24-inch model requires 3 hours to mold, 3 hours for engine manufacturing, and 1 hour to assemble. The 40-inch model requires 4 hours to mold, 4 hours for engine manufacturing, and 2 hours to assemble. The company has 81 hours available to mold, 95 hours available for engine manufacturing, and 35 hours available to assemble each day. How many of each type of mower can be manufactured each day?

3.3 Exercises MyMathLab®

Exercise numbers in green have complete video solutions in MyMathLab or may be accessed using the QR code to the right.

Problems 1–10 are the Quick ✔s that follow the EXAMPLES.

Building Skills

In Problems 11 and 12, determine whether the ordered triples listed are solutions of the system of linear equations. See Objective 1.

11. $\begin{cases} x + y + 2z = 6 \\ -2x - 3y + 5z = 1 \\ 2x + y + 3z = 5 \end{cases}$

    (a) $(6, 2, -1)$ **(b)** $(-3, 5, 2)$

12. $\begin{cases} 2x + y - 2z = 6 \\ -2x + y + 5z = 1 \\ 2x + 3y + z = 13 \end{cases}$

    (a) $(3, 2, 1)$ **(b)** $(10, -4, 5)$

In Problems 13–20, solve each system of three linear equations containing three unknowns. See Objective 1.

13. $\begin{cases} x + y + z = 5 \\ -2x - 3y + 2z = 8 \\ 3x - y - 2z = 3 \end{cases}$

14. $\begin{cases} x + 2y - z = 4 \\ 2x - y + 3z = 8 \\ -2x + 3y - 2z = 10 \end{cases}$

15. $\begin{cases} x - 3y + z = 13 \\ 3x + y - 4z = 13 \\ -4x - 4y + 2z = 0 \end{cases}$

16. $\begin{cases} x + 2y - 3z = -19 \\ 3x + 2y - z = -9 \\ -2x - y + 3z = 26 \end{cases}$

17. $\begin{cases} x - 4y + z = 5 \\ 4x + 2y + z = 2 \\ -4x + y - 3z = -8 \end{cases}$

18. $\begin{cases} 2x + 2y - z = -7 \\ x + 2y - 3z = -8 \\ 4x - 2y + z = -11 \end{cases}$

19. $\begin{cases} x - 3y = 12 \\ 2y - 3z = -9 \\ 2x + z = 7 \end{cases}$

20. $\begin{cases} 2x + z = -7 \\ 3y - 2z = 17 \\ -4x - y = 7 \end{cases}$

In Problems 21 and 22, show that each system of equations has no solution. See Objective 2.

21. $\begin{cases} x + y - 2z = 6 \\ -2x - 3y + z = 12 \\ -3x - 4y + 3z = 2 \end{cases}$

22. $\begin{cases} -x + 4y - z = 8 \\ 4x - y + 3z = 9 \\ 2x + 7y + z = 0 \end{cases}$

In Problems 23–26, solve each system with dependent equations. See Objective 3.

23. $\begin{cases} x + y + z = 4 \\ -2x - y + 2z = 6 \\ x + 2y + 5z = 18 \end{cases}$

24. $\begin{cases} x + 2y + z = 4 \\ -3x + y + 4z = -2 \\ -x + 5y + 6z = 6 \end{cases}$

25. $\begin{cases} x + 3z = 5 \\ -2x + y = 1 \\ y + 6z = 11 \end{cases}$

26. $\begin{cases} 2x - y = 2 \\ -x + 5z = 3 \\ -y + 10z = 8 \end{cases}$

Mixed Practice

In Problems 27–40, solve each system of equations.

27. $\begin{cases} 2x - y + 2z = 1 \\ -2x + 3y - 2z = 3 \\ 4x - y + 6z = 7 \end{cases}$

28. $\begin{cases} x - y + 3z = 2 \\ -2x + 3y - 8z = -1 \\ 2x - 2y + 4z = 7 \end{cases}$

29. $\begin{cases} x - y + z = 5 \\ -2x + y - z = 2 \\ x - 2y + 2z = 1 \end{cases}$

30. $\begin{cases} x - y + 2z = 3 \\ 2x + y - 2z = 1 \\ 4x - y + 2z = 0 \end{cases}$

31. $\begin{cases} 2y - z = -3 \\ -2x + 3y = 10 \\ 4x + 3z = -11 \end{cases}$

32. $\begin{cases} x \quad\quad - 3z = -3 \\ \quad\quad 3y + 4z = -5 \\ 3x - 2y \quad\quad = 6 \end{cases}$

33. $\begin{cases} x - 2y + z = 5 \\ -2x + y - z = 2 \\ x - 5y - 4z = 8 \end{cases}$

34. $\begin{cases} x + 2y - z = -4 \\ -2x + 4y - z = 6 \\ 2x + 2y + 3z = 1 \end{cases}$

35. $\begin{cases} x + 2y - z = 1 \\ 2x + 7y + 4z = 11 \\ x + 3y + z = 4 \end{cases}$

36. $\begin{cases} x + y - 2z = 3 \\ -2x - 3y + z = -7 \\ x + 2y + z = 4 \end{cases}$

37. $\begin{cases} x + y + z = 5 \\ 3x + 4y + z = 16 \\ -x - 4y + z = -6 \end{cases}$

38. $\begin{cases} x + y + z = 4 \\ 2x + 3y - z = 8 \\ x + y - z = 3 \end{cases}$

39. $\begin{cases} x + y + z = 3 \\ -x + \dfrac{1}{2}y + z = \dfrac{1}{2} \\ -x + 2y + 3z = 4 \end{cases}$

40. $\begin{cases} x + \dfrac{1}{2}y + \dfrac{1}{2}z = \dfrac{3}{2} \\ -x + 2y + 3z = 1 \\ 3x + 4y + 5z = 7 \end{cases}$

Applying the Concepts

41. Role Reversal Write a system of three linear equations containing three unknowns that has the solution $(2, -1, 3)$.

42. Role Reversal Write a system of three linear equations containing three unknowns that has the solution $(-4, 1, -3)$.

43. Curve Fitting The function $f(x) = ax^2 + bx + c$ is a quadratic function, where a, b, and c are constants.

   (a) If $f(1) = 4$, then $4 = a(1)^2 + b(1) + c$ or $a + b + c = 4$. Find two additional linear equations if $f(-1) = -6$ and $f(2) = 3$.

   (b) Use the three linear equations found in part (a) to determine a, b, and c. What is the quadratic function that contains the points $(-1, -6)$, $(1, 4)$, and $(2, 3)$?

44. Curve Fitting The function $f(x) = ax^2 + bx + c$ is a quadratic function, where a, b, and c are constants.

   (a) If $f(-1) = 6$, then $6 = a(-1)^2 + b(-1) + c$ or $a - b + c = 6$. Find two additional linear equations if $f(1) = 2$ and $f(2) = 9$.

   (b) Use the three linear equations found in part (a) to determine a, b, and c. What is the quadratic function that contains the points $(-1, 6)$, $(1, 2)$, and $(2, 9)$?

45. Electricity: Kirchhoff's Rules An application of Kirchhoff's Rule to the circuit shown results in the following system of equations:

$\begin{cases} i_1 + i_3 = i_2 \\ -3 - 3i_1 + 2i_3 = 0 \\ -22 + 4i_2 + 2i_3 = 0 \end{cases}$

In the system circuit, V is the voltage, Ω is resistance, and i is the current. Find the currents i_1, i_2, and i_3.

46. Electricity: Kirchhoff's Rules An application of Kirchhoff's Rule to the circuit shown results in the following system of equations:

$\begin{cases} i_1 + i_3 = i_2 \\ -8 - 5i_1 + 8i_3 = 0 \\ -48 + 6i_2 + 8i_3 = 0 \end{cases}$

In the system circuit, V is the voltage, Ω is resistance, and i is the current. Find the currents i_1, i_2, and i_3.

47. Minor League Baseball In the Joliet Slammers baseball stadium, there are three types of seats available. Club seats are \$14, reserved seats are \$12, and lawn seats are \$7. The stadium capacity is 6000. If all the seats are sold, the total revenue to the club is \$64,125. If all of the club seats are sold, $\dfrac{4}{5}$ of the reserved seats are sold, and $\dfrac{1}{5}$ of the lawn seats are sold, the total revenue is \$45,189. How many are there of each kind of seat?

48. Theater Revenues A theater has 600 seats, divided into orchestra, main floor, and balcony seating. Orchestra seats sell for \$80, main floor seats for \$60, and balcony seats for \$25. If all the seats are sold, the total revenue to the theater is \$33,500. One evening, all the orchestra seats were sold, $\dfrac{3}{5}$ of the main seats were sold, and $\dfrac{4}{5}$ of the balcony seats were sold. The total revenue collected was \$24,640. How many are there of each kind of seat?

49. Nutrition Nancy's dietitian wants her to consume 470 mg of sodium, 89 g of carbohydrates, and 20 g of protein for breakfast. This morning, Nancy wants to have Chex® cereal, 2% milk, and orange juice for breakfast. Each serving of Chex® cereal contains 220 mg of sodium, 26 g of carbohydrates, and 1 g of protein. Each serving of 2% milk contains 125 mg of sodium, 12 g of carbohydrates, and 8 g of protein. Each serving of orange juice contains 0 mg of sodium, 26 g of carbohydrates, and 2 g of protein. How many servings of each does Nancy need?

50. Nutrition Antonio is on a special diet that requires he consume 1325 calories, 172 grams of carbohydrates, and 63 grams of protein for lunch. He wishes to have a Broccoli and Cheese Baked Potato, Chicken BLT Salad, and a medium Coke. Each Broccoli and Cheese Baked Potato has 480 calories, 80 g of carbohydrates, and 9 g of protein. Each Chicken BLT Salad has 310 calories, 10 g of carbohydrates, and 33 g of protein. Each Coke has 140 calories, 37 g of carbohydrates, and 0 g of protein. How many servings of each does Antonio need?

51. Finance Sachi has $25,000 to invest. Her financial planner suggests that she diversify her investment into three investment categories: Treasury bills that yield 1% simple interest annually, municipal bonds that yield 2% simple interest annually, and corporate bonds that yield 6% simple interest annually. Sachi would like to earn $580 per year in income. In addition, Sachi wants her investment in Treasury bills to be $7000 more than her investment in corporate bonds. How much should Sachi invest in each investment category?

52. Finance Delu has $15,000 to invest. She decides to place some of the money into a savings account paying 0.50% annual interest, some in Treasury bonds paying 3% annual interest, and some in a mutual fund paying 10% annual interest. Delu would like to earn $500 per year in income. In addition, Delu wants her investment in the savings account to be twice the amount in the mutual fund. How much should Delu invest in each investment category?

△ **53. Geometry** A circle is inscribed in $\triangle ABC$ as shown in the figure. Suppose that $AB = 6$, $AC = 14$, and $BC = 12$. Find the length of $\overline{AM}$, $\overline{BN}$, and $\overline{CO}$. (*Hint:* $\overline{AM} \cong \overline{AO}$; $\overline{BM} \cong \overline{BN}$; $\overline{NC} \cong \overline{OC}$)

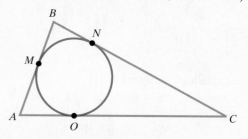

Extending the Concepts

In Problems 54–57, solve each system of equations.

54. $\begin{cases} \dfrac{1}{4}x + \dfrac{1}{4}y + \dfrac{1}{2}z = 6 \\ -\dfrac{1}{8}x + \dfrac{1}{2}y - \dfrac{1}{5}z = -5 \\ \dfrac{1}{2}x + \dfrac{1}{2}y - \dfrac{1}{2}z = -3 \end{cases}$

55. $\begin{cases} \dfrac{2}{5}x + \dfrac{1}{2}y - \dfrac{1}{3}z = 0 \\ \dfrac{3}{5}x - \dfrac{1}{4}y + \dfrac{1}{2}z = 10 \\ -\dfrac{1}{5}x + \dfrac{1}{4}y - \dfrac{1}{6}z = -4 \end{cases}$

56. $\begin{cases} x + y + z + w = 0 \\ 2x - 3y - z + w = -17 \\ 3x + y + 2z - w = 8 \\ -x + 2y - 3z + 2w = -7 \end{cases}$

57. $\begin{cases} x + y + z + w = 3 \\ -2x - y + 3z - w = -1 \\ 2x + 2y - 2z + w = 2 \\ -x + 2y - 3z + 2w = 12 \end{cases}$

Explaining the Concepts

58. Suppose that $(3, 2, 5)$ is the only solution of a system of three linear equations containing three variables. What does this mean geometrically?

59. Why is it necessary to eliminate the same variable in Step 1 of the box "Solving a System of Three Linear Equations Containing Three Unknowns by Elimination" (page 249)?

Putting the Concepts Together (Sections 3.1–3.3)

We designed these problems so that you can review Sections 3.1 through 3.3 and show your mastery of the concepts. Take time to work these problems before proceeding with the next section. The answers are at the back of the text on page AN-21.

1. Use the graph of the system of linear equations to determine the solution.

$$\begin{cases} 5x - 2y = -25 \\ x + y = 2 \end{cases}$$

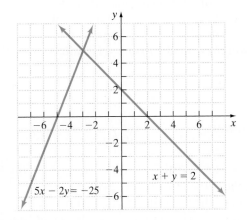

2. Solve the system of linear equations by graphing.

$$\begin{cases} 3x + y = 7 \\ -2x + 3y = -12 \end{cases}$$

In Problems 3–6, solve the system of linear equations using either substitution or elimination.

3. $\begin{cases} x = 2 - 3y \\ 3x + 10y = 5 \end{cases}$ **4.** $\begin{cases} 4x + 3y = -1 \\ 2x - y = 3 \end{cases}$

5. $\begin{cases} x + 3y = 8 \\ \dfrac{1}{5}x + \dfrac{1}{2}y = 1 \end{cases}$ **6.** $\begin{cases} 8x - 4y = 12 \\ -10x + 5y = -15 \end{cases}$

In Problems 7 and 8, solve each system of three linear equations containing three unknowns.

7. $\begin{cases} 2x + y + 3z = 10 \\ x - 2y + z = 10 \\ -4x + 3y + 2z = 5 \end{cases}$ **8.** $\begin{cases} x + 2y - 2z = 3 \\ x + 3y - 4z = 6 \\ 4x + 5y - 2z = 6 \end{cases}$

9. Museum Tickets A museum sells adult tickets for $9.00 and youth tickets for $5.00. On Saturday, the museum collected a total of $5925. It is also known that 825 people passed through the turnstiles. How many adult tickets were sold? How many youth tickets were sold?

10. Theater Seating A theater has 1200 seats that are divided into three sections: orchestra, mezzanine, and balcony. Seats sell for $65 each in the orchestra section, $48 each in the mezzanine section, and $35 each in the balcony section. There are 150 more balcony seats than mezzanine seats. If all the seats are sold, the theater will bring in $55,640 in revenue. Find the number of seats in each section of the theater.

3.4 Using Matrices to Solve Systems

Objectives

❶ Write the Augmented Matrix of a System

❷ Write the System from the Augmented Matrix

❸ Perform Row Operations on a Matrix

❹ Solve Systems Using Matrices

❺ Solve Consistent Systems with Dependent Equations and Solve Inconsistent Systems

Are You Prepared for This Section?

Before getting started, complete the following problems. If you get a problem wrong, go back to the section cited and review the material.

P1. Determine the coefficients of the expression $4x - 2y + z$. [Section R.5, pp. 41–42]

P2. Solve $x - 4y = 3$ for x. [Section 1.3, pp. 74–77]

P3. Evaluate $3x - 2y + z$ when $x = 1, y = -3,$ and $z = 2$. [Section R.5, pp. 40–41]

A way to solve systems of linear equations that streamlines the notation and makes working with systems more manageable is to use a *matrix*.

Definition

A **matrix** is a rectangular array of numbers.

Prepared?...Answers   **P1.** 4, −2, 1
P2. $x = 4y + 3$ **P3.** 11

A matrix has rows and columns. The number of rows and columns determines the size of the matrix. For example, a matrix with 2 rows and 3 columns is called a "2 by 3 matrix," which is denoted "2×3 matrix." Below are some examples of matrices.

2×3 matrix

3 columns

2 rows $\begin{bmatrix} 3 & -1 & 4 \\ 8 & 0 & -5 \end{bmatrix}$

3×3 matrix

3 columns

3 rows $\begin{bmatrix} 2 & -8 & 12 \\ 0 & 7 & -2 \\ 5 & -2 & 1 \end{bmatrix}$

Work Smart

A spreadsheet such as Microsoft Excel is a matrix!

▶ ❶ Write the Augmented Matrix of a System

A system of linear equations may be represented in an *augmented matrix*. To write the **augmented matrix** of a system, write the terms containing the variables of each equation on the left side of the equal sign in the same order (x, y, and z, for example), and write the constants on the right side. A variable that does not appear in an equation has a coefficient of 0. For example, consider the following system of two linear equations containing two unknowns:

coefficients of x coefficients of y

$$\begin{cases} a_1 x + b_1 y = c_1 \\ a_2 x + b_2 y = c_2 \end{cases} \text{ is written as an augmented matrix as } \begin{bmatrix} a_1 & b_1 & c_1 \\ a_2 & b_2 & c_2 \end{bmatrix} \text{ constants}$$

Notice that the first column represents the coefficients of the variable x, the second column represents the coefficients of the variable y, the vertical bar represents the equal signs, and the constants are to the right of the vertical bar. The first row in the matrix is from the first equation, and the second row in the matrix is from the second equation.

EXAMPLE 1 **Writing the Augmented Matrix of a System of Linear Equations**

Write each system of linear equations as an augmented matrix.

(a) $\begin{cases} x - 3y = 2 \\ -2x + 5y = 7 \end{cases}$

(b) $\begin{cases} 2x + 3y - z = 1 \\ x - 2z + 1 = 0 \\ -4x - y + 3z = 5 \end{cases}$

Solution

(a) In the augmented matrix, the first column represents the coefficients of the variable x. The second column represents the coefficients of the variable y. The vertical line signifies the equal signs. The third column represents the constants to the right of the equal sign.

$$\begin{array}{cc} x & y \\ \begin{bmatrix} 1 & -3 & | & 2 \\ -2 & 5 & | & 7 \end{bmatrix} & \begin{array}{l} x - 3y = 2 \\ -2x + 5y = 7 \end{array} \end{array}$$

(b) Write the system with all the variables on the left side of the equal sign and the constants on the right. Write 0 as the coefficient of any missing variable. To this end, rewrite the system

$$\begin{cases} 2x + 3y - z = 1 \\ x - 2z + 1 = 0 \\ -4x - y + 3z = 5 \end{cases} \text{ as } \begin{cases} 2x + 3y - z = 1 \\ x + 0y - 2z = -1 \\ -4x - y + 3z = 5 \end{cases}$$

The augmented matrix is

$$\begin{bmatrix} 2 & 3 & -1 & | & 1 \\ 1 & 0 & -2 & | & -1 \\ -4 & -1 & 3 & | & 5 \end{bmatrix}$$

Quick ✓

1. An m by n rectangular array of numbers is called a(n) _____.

2. The matrix used to represent a system of linear equations is called a(n) _____ matrix.

3. A 4×3 matrix has _ rows and _ columns.

4. *True or False* The augmented matrix of a system of two equations containing two unknowns has 2 rows and 2 columns.

In Problems 5 and 6, write the augmented matrix of each system of equations.

5. $\begin{cases} 3x - y = -10 \\ -5x + 2y = 0 \end{cases}$

6. $\begin{cases} x + 2y - 2z = 11 \\ -x - 2z = 4 \\ 4x - y + z - 3 = 0 \end{cases}$

▶ ❷ Write the System from the Augmented Matrix

EXAMPLE 2   **Writing the System from the Augmented Matrix**

Write the system of linear equations corresponding to each augmented matrix.

(a) $\left[\begin{array}{cc|c} 2 & 1 & -5 \\ -1 & 3 & 2 \end{array}\right]$

(b) $\left[\begin{array}{ccc|c} 1 & -3 & 2 & 5 \\ -2 & 0 & 4 & -3 \\ -1 & 4 & 1 & 0 \end{array}\right]$

Solution

(a) The augmented matrix has two rows, so it represents a system of two equations. There are two columns to the left of the vertical bar, so the system has two variables, which we can call x and y. The system of equations is

$$\begin{cases} 2x + y = -5 \\ -x + 3y = 2 \end{cases}$$

(b) The augmented matrix has three rows, so it represents a system of three equations. The three columns to the left of the vertical bar represent three variables, which we can call x, y, and z. The system of equations is

$$\begin{cases} x - 3y + 2z = 5 \\ -2x + 4z = -3 \\ -x + 4y + z = 0 \end{cases}$$

●

Quick ✓

In Problems 7 and 8, write the system of linear equations corresponding to the given augmented matrix.

7. $\left[\begin{array}{cc|c} 1 & -3 & 7 \\ -2 & 5 & -3 \end{array}\right]$

8. $\left[\begin{array}{ccc|c} 1 & -3 & 2 & 4 \\ 3 & 0 & -1 & -1 \\ -1 & 4 & 0 & 0 \end{array}\right]$

▶ ❸ Perform Row Operations on a Matrix

Use row operations on an augmented matrix to solve the corresponding system of equations.

Row Operations

1. Interchange any two rows.
2. Multiply *all* entries in a row by a nonzero constant.
3. Replace a row by the sum of that row and a nonzero multiple of some other row.

These are the same operations that were performed on systems of equations in Section 3.3. The main reason for using a matrix to solve a system of equations is that its notation is more efficient and helps us organize the mathematics. To see how row operations work, consider the following augmented matrix.

$$\begin{bmatrix} 1 & 3 & | & -1 \\ -2 & 1 & | & 3 \end{bmatrix}$$

Suppose we want to apply a row operation that results in a matrix whose entry in row 2, column 1 is a 0. The current entry here is -2. The row operation to use is

Multiply each entry in row 1 by 2, and then add the result to the corresponding entry in row 2. Have this result replace the current row 2.

To streamline this a little, notation is needed. Use R_2 to represent the new entries in row 2, and use r_1 and r_2 to represent the original entries in rows 1 and 2, respectively, and then express the row operation given above by

The "new" row 2 is obtained by...

Multiplying each entry in row 1 by 2 and adding the result to the corresponding entry in row 2

$$R_2 = 2r_1 + r_2$$

This row operation is demonstrated below.

$$\begin{bmatrix} 1 & 3 & | & -1 \\ -2 & 1 & | & 3 \end{bmatrix} \xrightarrow{R_2 = 2r_1 + r_2} \begin{bmatrix} 1 & 3 & | & -1 \\ 2(1) + (-2) & 2(3) + 1 & | & 2(-1) + 3 \end{bmatrix} = \begin{bmatrix} 1 & 3 & | & -1 \\ 0 & 7 & | & 1 \end{bmatrix}$$

We now have a 0 in row 2, column 1, as desired. Why would this be useful? Notice the second row represents the equation $7y = 1$, which is easy to solve.

EXAMPLE 3 **Applying a Row Operation to an Augmented Matrix**

Apply the row operation $R_2 = -3r_1 + r_2$ to the augmented matrix

$$\begin{bmatrix} 1 & 2 & | & -3 \\ 3 & 4 & | & 5 \end{bmatrix}$$

Solution

The row operation $R_2 = -3r_1 + r_2$ tells us to multiply the entries in row 1 by -3 and then add the result to the entries in row 2. The result should replace the current row 2.

$$\begin{bmatrix} 1 & 2 & | & -3 \\ 3 & 4 & | & 5 \end{bmatrix} \xrightarrow{R_2 = -3r_1 + r_2} \begin{bmatrix} 1 & 2 & | & -3 \\ -3(1) + 3 & -3(2) + 4 & | & -3(-3) + 5 \end{bmatrix} = \begin{bmatrix} 1 & 2 & | & -3 \\ 0 & -2 & | & 14 \end{bmatrix}$$

●

Quick ✓

9. Apply the row operation $R_2 = 4r_1 + r_2$ to the augmented matrix

$$\begin{bmatrix} 1 & -2 & | & 5 \\ -4 & 5 & | & -11 \end{bmatrix}$$

EXAMPLE 4 **Finding a Particular Row Operation**

For the augmented matrix

$$\begin{bmatrix} 1 & -3 & | & -12 \\ 0 & 1 & | & 5 \end{bmatrix}$$

find and perform a row operation that will result in the entry in row 1, column 2 becoming a 0.

Solution

To get a 0 in row 1, column 2, multiply row 2 by 3 and add the result to row 1. That is, apply the row operation $R_1 = 3r_2 + r_1$.

$$\begin{bmatrix} 1 & -3 & | & -12 \\ 0 & 1 & | & 5 \end{bmatrix} \xrightarrow{R_1 = 3r_2 + r_1} \begin{bmatrix} 3(0) + 1 & 3(1) + (-3) & | & 3(5) + (-12) \\ 0 & 1 & | & 5 \end{bmatrix} = \begin{bmatrix} 1 & 0 & | & 3 \\ 0 & 1 & | & 5 \end{bmatrix}$$

●

Work Smart

If you want to change the entries in row 1, then you should multiply the entries in some other row by a "well-chosen" constant and then add this result to the entries in row 1.

Quick ✓

10. For the augmented matrix $\begin{bmatrix} 1 & 5 & | & 13 \\ 0 & 1 & | & 2 \end{bmatrix}$, find and perform a row operation that will result in the entry in row 1, column 2 becoming a 0.

❹ Solve Systems Using Matrices

▶ To solve a system of linear equations using matrices, use row operations on the augmented matrix of the system to obtain a matrix that is in *row echelon form*.

Definition

A matrix is in **row echelon form** when

1. The entry in row 1, column 1 is a 1, and 0s appear below it.
2. The first nonzero entry in each row below the first row is a 1, 0s appear below it, and it appears to the right of the first nonzero entry in any row above.
3. Any rows that contain all 0s to the left of the vertical bar appear at the bottom.

For a system of two equations containing two variables with a single solution, the row echelon form of the augmented matrix is

$$\begin{bmatrix} 1 & a & | & b \\ 0 & 1 & | & c \end{bmatrix}$$

where a, b, and c are real numbers. For a system of three equations containing three variables with a single solution, the row echelon form is

$$\begin{bmatrix} 1 & a & b & | & d \\ 0 & 1 & c & | & e \\ 0 & 0 & 1 & | & f \end{bmatrix}$$

where a, b, c, d, e, and f are real numbers.

The augmented matrix $\begin{bmatrix} 1 & a & | & b \\ 0 & 1 & | & c \end{bmatrix}$ is equivalent to the system $\begin{cases} x + ay = b \ (1) \\ y = c \ (2) \end{cases}$.

Because c is a known number, we know y. Then substitute to find x. For example, the augmented matrix $\begin{bmatrix} 1 & 3 & | & -5 \\ 0 & 1 & | & -3 \end{bmatrix}$ is in row echelon form and is equivalent to the system $\begin{cases} x + 3y = -5 \ (1) \\ y = -3 \ (2) \end{cases}$. Equation (2) states that $y = -3$. Using substitution reveals that $x = 4$, so the solution is $(4, -3)$.

EXAMPLE 5 **How to Solve a System of Two Linear Equations Containing Two Variables Using Matrices**

Solve: $\begin{cases} 3x - 2y = -19 \\ x + 2y = 7 \end{cases}$

Step-by-Step Solution

Step 1: Write the augmented matrix of the system.

$$\begin{bmatrix} 3 & -2 & | & -19 \\ 1 & 2 & | & 7 \end{bmatrix}$$

Step 2: The entry in row 1, column 1 should be 1, so interchange rows 1 and 2.

$$\begin{bmatrix} 1 & 2 & | & 7 \\ 3 & -2 & | & -19 \end{bmatrix}$$

Step 3: The entry in row 2, column 1 should be 0, so use the row operation $R_2 = -3r_1 + r_2$. The entries in row 1 remain unchanged.

$R_2 = -3r_1 + r_2$: $\begin{bmatrix} 1 & 2 & | & 7 \\ 0 & -8 & | & -40 \end{bmatrix}$

Step 4: Now the entry in row 2, column 2 should be 1. This is accomplished by multiplying row 2 by $-\dfrac{1}{8}$. Use the row operation $R_2 = -\dfrac{1}{8} r_2$.

$R_2 = -\dfrac{1}{8} r_2$: $\begin{bmatrix} 1 & 2 & | & 7 \\ 0 & 1 & | & 5 \end{bmatrix}$

Step 5: Row 2 says that $y = 5$. Substitute this value into row 1, which is $x + 2y = 7$. Solve for x.

$$\begin{aligned} x + 2y &= 7 \\ \text{Let } y = 5: \quad x + 2(5) &= 7 \\ x + 10 &= 7 \\ x &= -3 \end{aligned}$$

Step 6: Check: Let $x = -3$ and $y = 5$ in equations (1) and (2) to verify the solution.

Equation (1):
$$\begin{aligned} 3x - 2y &= -19 \\ x = -3, y = 5: \quad 3(-3) - 2(5) &\overset{?}{=} -19 \\ -9 - 10 &\overset{?}{=} -19 \\ -19 &= -19 \quad \text{True} \end{aligned}$$

Equation (2):
$$\begin{aligned} x + 2y &= 7 \\ x = -3, y = 5: \quad -3 + 2(5) &\overset{?}{=} 7 \\ -3 + 10 &\overset{?}{=} 7 \\ 7 &= 7 \quad \text{True} \end{aligned}$$

The solution is the ordered pair $(-3, 5)$. ●

Quick ✓

11. *True or False* The matrix $\begin{bmatrix} 1 & 3 & | & 2 \\ 0 & 1 & | & -5 \end{bmatrix}$ is in row echelon form.

12. Solve the following system using matrices: $\begin{cases} 2x - 4y = 20 \\ 3x + y = 16 \end{cases}$

EXAMPLE 6 **How to Solve a System of Three Linear Equations Containing Three Variables Using Matrices**

Solve: $\begin{cases} x + y + z = 3 \\ -2x - 3y + 2z = 13 \\ 4x + 5y + z = -3 \end{cases}$

Step-by-Step Solution

Step 1: Write the augmented matrix of the system.

$$\begin{bmatrix} 1 & 1 & 1 & 3 \\ -2 & -3 & 2 & 13 \\ 4 & 5 & 1 & -3 \end{bmatrix}$$

Step 2: The entry in row 1, column 1 should be 1.
This is already done.

$$\begin{bmatrix} 1 & 1 & 1 & 3 \\ -2 & -3 & 2 & 13 \\ 4 & 5 & 1 & -3 \end{bmatrix}$$

Step 3: The entry in row 2, column 1 should be 0, so use the row
operation $R_2 = 2r_1 + r_2$. The entry in row 3, column 1
should also be 0, so use the row operation $R_3 = -4r_1 + r_3$.
The entries in row 1 remain unchanged.

$R_2 = 2r_1 + r_2$:
$R_3 = -4r_1 + r_3$:
$$\begin{bmatrix} 1 & 1 & 1 & 3 \\ 0 & -1 & 4 & 19 \\ 0 & 1 & -3 & -15 \end{bmatrix}$$

Step 4: Now the entry in row 2, column 2 should be 1.
This is accomplished by interchanging rows 2 and 3.

$$\begin{bmatrix} 1 & 1 & 1 & 3 \\ 0 & 1 & -3 & -15 \\ 0 & -1 & 4 & 19 \end{bmatrix}$$

Step 5: The entry in row 3, column 2 should be 0.
Use the row operation $R_3 = r_2 + r_3$.

$R_3 = r_2 + r_3$:
$$\begin{bmatrix} 1 & 1 & 1 & 3 \\ 0 & 1 & -3 & -15 \\ 0 & 0 & 1 & 4 \end{bmatrix}$$

Step 6: The entry in row 3, column 3 should be 1.
This is already done.

$$\begin{bmatrix} 1 & 1 & 1 & 3 \\ 0 & 1 & -3 & -15 \\ 0 & 0 & 1 & 4 \end{bmatrix}$$

Step 7: The augmented matrix is in row echelon form. Write the
corresponding system of equations and solve.

$$\begin{cases} x + y + z = 3 & (1) \\ y - 3z = -15 & (2) \\ z = 4 & (3) \end{cases}$$

Equation (3) says $z = 4$. Let $z = 4$
in equation (2) to find y.

Equation (2): $y - 3z = -15$
$z = 4$: $y - 3(4) = -15$
$y - 12 = -15$
$y = -3$

Let $y = -3$ and $z = 4$ in equation (1) to
find that $x = 2$.

Step 8: Check: Let $x = 2$, $y = -3$, and $z = 4$ in each
equation in the original system to verify the solution.

We leave it to you to verify the solution,
$(2, -3, 4)$.

The solution is the ordered triple $(2, -3, 4)$.

Work Smart

Look at Step 4 in Example 6. Other
ways to obtain a 1 in row 2,
column 2 are to use the row operation
$R_2 = 2r_3 + r_2$ or $R_2 = -r_2$.

Quick ✓

In Problems 13 and 14, solve each system of equations using matrices.

13. $\begin{cases} x - y + 2z = 7 \\ 2x - 2y + z = 11 \\ -3x + y - 3z = -14 \end{cases}$

14. $\begin{cases} 2x + 3y - 2z = 1 \\ 2x - 4z = -9 \\ 4x + 6y - z = 14 \end{cases}$

Work Smart

In Step 4, if it is impossible to place a 1 in row 2, column 2, then use operations to try to obtain a 1 in row 2, column 3.

Work Smart

If you get a row with all 0s, then it should be placed as the last row of the matrix.

Solving a System of Linear Equations Using Matrices

Step 1: Write the augmented matrix of the system.

Step 2: Perform row operations so the entry in row 1, column 1 is 1.

Step 3: Perform row operations so all the entries below the 1 in row 1, column 1 are 0s.

Step 4: Perform row operations so the entry in row 2, column 2 is 1. Be sure the entries in column 1 remain unchanged.

Step 5: If solving a system of three linear equations, continue to perform row operations so that the entry below the 1 in row 2, column 2, is 0.

Step 6: Perform row operations so that the entry in row 3, column 3, is 1.

Step 7: With the augmented matrix in row echelon form, write the corresponding system of equations. Substitute to find the values of the variables.

Step 8: Check your answer.

---

⑤ Solve Consistent Systems with Dependent Equations and Solve Inconsistent Systems

The matrix method for solving a system of linear equations also identifies systems that have infinitely many solutions (dependent equations) and systems with no solution (inconsistent systems).

EXAMPLE 7

Solving a Consistent System with Dependent Equations Using Matrices

Solve: $\begin{cases} x + y + 3z = 3 \\ -2x - y - 8z = -5 \\ 3x + 2y + 11z = 8 \end{cases}$

Solution

Write the augmented matrix of the system and perform row operations to get the matrix into row echelon form.

$$\begin{bmatrix} 1 & 1 & 3 & | & 3 \\ -2 & -1 & -8 & | & -5 \\ 3 & 2 & 11 & | & 8 \end{bmatrix} \xrightarrow[\substack{R_2 = 2r_1 + r_2 \\ R_3 = -3r_1 + r_3}]{} \begin{bmatrix} 1 & 1 & 3 & | & 3 \\ 0 & 1 & -2 & | & 1 \\ 0 & -1 & 2 & | & -1 \end{bmatrix} \xrightarrow[R_3 = r_2 + r_3]{} \begin{bmatrix} 1 & 1 & 3 & | & 3 \\ 0 & 1 & -2 & | & 1 \\ 0 & 0 & 0 & | & 0 \end{bmatrix}$$

Notice that the last row is all 0s. The augmented matrix is in row echelon form. The corresponding system of equations is

$$\begin{cases} x + y + 3z = 3 & (1) \\ y - 2z = 1 & (2) \\ 0 = 0 & (3) \end{cases}$$

The statement $0 = 0$ in equation (3) indicates that the equations are dependent. Let z represent any real number, and solve equation (2) for y in terms of z.

$$\text{Equation (2):} \quad y - 2z = 1$$
$$\text{Add } 2z \text{ to both sides:} \quad y = 2z + 1$$

Let $y = 2z + 1$ in equation (1) and solve for x in terms of z.

$$
\begin{aligned}
\text{Equation (1):} && x + y + 3z &= 3 \\
\text{Let } y = 2z + 1: && x + (2z + 1) + 3z &= 3 \\
\text{Combine like terms:} && x + 5z + 1 &= 3 \\
\text{Subtract } 5z + 1 \text{ from both sides:} && x &= -5z + 2
\end{aligned}
$$

The solution to the system is $\{ (x, y, z) \,|\, x = -5z + 2, y = 2z + 1, z \text{ is any real} $ number$\}$. To find specific solutions to the system, choose any value of z and use the equations $x = -5z + 2$ and $y = 2z + 1$ to determine x and y. Some specific solutions to the system are $(2, 1, 0)$, $(-3, 3, 1)$, and $(7, -1, -1)$.

Evaluate the system at some of the specific solutions to see that all of them satisfy the original system. This indicates that our solution is correct. ●

Quick ✓

In Problems 15 and 16, solve each system of equations using matrices.

15. $\begin{cases} 2x + 5y = -6 \\ -6x - 15y = 18 \end{cases}$

16. $\begin{cases} x + y - 3z = 8 \\ 2x + 3y - 10z = 19 \\ -x - 2y + 7z = -11 \end{cases}$

EXAMPLE 8   **Solving an Inconsistent System Using Matrices**

Solve: $\begin{cases} 2x + y - z = 5 \\ -x + 2y + z = 1 \\ 3x + 4y - z = -2 \end{cases}$

Solution
Write the augmented matrix of the system and perform row operations to get the matrix into row echelon form.

$$
\begin{bmatrix} 2 & 1 & -1 & | & 5 \\ -1 & 2 & 1 & | & 1 \\ 3 & 4 & -1 & | & -2 \end{bmatrix}
\xrightarrow{R_1 = r_2 + r_1}
\begin{bmatrix} 1 & 3 & 0 & | & 6 \\ -1 & 2 & 1 & | & 1 \\ 3 & 4 & -1 & | & -2 \end{bmatrix}
\xrightarrow[R_3 = -3r_1 + r_3]{R_2 = r_1 + r_2}
\begin{bmatrix} 1 & 3 & 0 & | & 6 \\ 0 & 5 & 1 & | & 7 \\ 0 & -5 & -1 & | & -20 \end{bmatrix}
$$

$$
\xrightarrow{R_2 = \frac{1}{5} r_2}
\begin{bmatrix} 1 & 3 & 0 & | & 6 \\ 0 & 1 & \frac{1}{5} & | & \frac{7}{5} \\ 0 & -5 & -1 & | & -20 \end{bmatrix}
\xrightarrow{R_3 = 5r_2 + r_3}
\begin{bmatrix} 1 & 3 & 0 & | & 6 \\ 0 & 1 & \frac{1}{5} & | & \frac{7}{5} \\ 0 & 0 & 0 & | & -13 \end{bmatrix}
$$

The entry in row 3, column 3 needs to be 1. This cannot be done (without changing the entries in row 3, column 1 or 2). The corresponding system of equations is

$$
\begin{cases}
x + 3y \phantom{ + \tfrac{1}{5}z} = 6 & (1) \\
 y + \dfrac{1}{5}z = \dfrac{7}{5} & (2) \\
\phantom{x + 3y + \tfrac{1}{5}z} 0 = -13 & (3)
\end{cases}
$$

The statement $0 = -13$ in equation (3) indicates that the system is inconsistent and therefore has no solution. The solution is $\varnothing$ or $\{ \, \}$. ●

Quick ✓

In Problems 17 and 18, solve each system of equations using matrices.

17. $\begin{cases} -2x + 3y = 4 \\ 10x - 15y = 2 \end{cases}$

18. $\begin{cases} -x + 2y - z = 5 \\ 2x + y + 4z = 3 \\ 3x - y + 5z = 0 \end{cases}$

3.4 Exercises MyMathLab® Exercise numbers in green have complete video solutions in MyMathLab or may be accessed using the QR code to the right.

Problems **1–18** *are the Quick* ✓*s that follow the* **EXAMPLES**.

Building Skills

In Problems 19–26, write the augmented matrix of the given system of equations. See Objective 1.

19. $\begin{cases} x - 3y = 2 \\ 2x + 5y = 1 \end{cases}$ **20.** $\begin{cases} -x + y = 6 \\ 5x - y = -3 \end{cases}$

21. $\begin{cases} x + y + z = 3 \\ 2x - y + 3z = 1 \\ -4x + 2y - 5z = -3 \end{cases}$

22. $\begin{cases} x + y - z = 2 \\ -2x + y - 4z = 13 \\ 3x - y - 2z = -4 \end{cases}$

23. $\begin{cases} -x + y - 2 = 0 \\ 5x + y + 5 = 0 \end{cases}$

24. $\begin{cases} 6x + 4y + 2 = 0 \\ -x - y + 1 = 0 \end{cases}$

25. $\begin{cases} x + z = 2 \\ 2x + y = 13 \\ x - y + 4z + 4 = 0 \end{cases}$

26. $\begin{cases} 2x + 7y = 1 \\ -x - 6z = 5 \\ 5x + 2y - 4z + 1 = 0 \end{cases}$

In Problems 27–32, write the system of linear equations corresponding to each augmented matrix. Use x, y; or x, y, z as variables. See Objective 2.

27. $\begin{bmatrix} 2 & 5 & | & 3 \\ -4 & 1 & | & 10 \end{bmatrix}$

28. $\begin{bmatrix} -1 & 4 & | & 0 \\ 2 & -7 & | & -9 \end{bmatrix}$

29. $\begin{bmatrix} 1 & 5 & -3 & | & 2 \\ 0 & 3 & -1 & | & -5 \\ 4 & 0 & 8 & | & 6 \end{bmatrix}$

30. $\begin{bmatrix} -2 & 3 & 0 & | & 3 \\ 4 & 1 & -8 & | & -6 \\ 7 & 2 & -5 & | & 10 \end{bmatrix}$

31. $\begin{bmatrix} 1 & -2 & 9 & | & 2 \\ 0 & 1 & -5 & | & 8 \\ 0 & 0 & 1 & | & \frac{4}{3} \end{bmatrix}$

32. $\begin{bmatrix} 1 & 4 & -1 & | & \frac{2}{3} \\ 0 & 1 & 3 & | & 0 \\ 0 & 0 & 1 & | & 10 \end{bmatrix}$

In Problems 33–38, perform each row operation on the given augmented matrix. See Objective 3.

33. $\begin{bmatrix} 1 & -3 & | & 2 \\ -2 & 5 & | & 1 \end{bmatrix}$

(a) $R_2 = 2r_1 + r_2$ followed by
(b) $R_2 = -r_2$

34. $\begin{bmatrix} 1 & 5 & | & 7 \\ 3 & 11 & | & 13 \end{bmatrix}$

(a) $R_2 = -3r_1 + r_2$ followed by
(b) $R_2 = -\frac{1}{4}r_2$

35. $\begin{bmatrix} 1 & 1 & -1 & | & 4 \\ 2 & 5 & 3 & | & -3 \\ -1 & -3 & 2 & | & 1 \end{bmatrix}$

(a) $R_2 = -2r_1 + r_2$ followed by
(b) $R_3 = r_1 + r_3$

36. $\begin{bmatrix} 1 & -1 & 1 & | & 6 \\ -2 & 1 & -3 & | & 3 \\ 3 & 2 & -2 & | & -5 \end{bmatrix}$

(a) $R_2 = 2r_1 + r_2$ followed by
(b) $R_3 = -3r_1 + r_3$

37. $\begin{bmatrix} 1 & 1 & 1 & | & 4 \\ 0 & 5 & 3 & | & -3 \\ 0 & -4 & 2 & | & 8 \end{bmatrix}$

(a) $R_2 = r_3 + r_2$ followed by
(b) $R_3 = \frac{1}{2}r_3$

38. $\begin{bmatrix} 1 & -3 & 4 & | & 11 \\ 0 & 3 & 6 & | & -12 \\ 0 & -2 & -3 & | & 8 \end{bmatrix}$

(a) $R_2 = \frac{1}{3}r_2$ followed by
(b) $R_3 = 2r_2 + r_3$

In Problems 39–44, solve each system of equations using matrices. See Objective 4.

39. $\begin{cases} 2x + 3y = 1 \\ -x + 4y = -28 \end{cases}$

40. $\begin{cases} 3x + 5y = -1 \\ -2x - 3y = 2 \end{cases}$

41. $\begin{cases} x + 5y - 2z = 23 \\ -2x - 3y + 5z = -11 \\ 3x + 2y + z = 4 \end{cases}$

42. $\begin{cases} x - 4y + 2z = -2 \\ -3x + y - 3z = 12 \\ 2x + 3y - 4z = -20 \end{cases}$

43. $\begin{cases} 4x + 5y - 2z = 0 \\ -8x + 10y - 2z = -13 \\ 15y + 4z = -7 \end{cases}$

44. $\begin{cases} 2x + y - 3z = 4 \\ 2y + 6z = 18 \\ -4x - y = -14 \end{cases}$

In Problems 45–50, solve each system of equations using matrices. See Objective 5.

45. $\begin{cases} x - 3y = 3 \\ -2x + 6y = -6 \end{cases}$

46. $\begin{cases} -4x + 2y = 4 \\ 6x - 3y = -6 \end{cases}$

47. $\begin{cases} x - 2y + z = 3 \\ -2x + 4y - z = -5 \\ -8x + 16y + z = -21 \end{cases}$

48. $\begin{cases} x - 3y + 2z = 4 \\ 3x + y + 4z = -5 \\ -x - 7y = 5 \end{cases}$

49. $\begin{cases} x + y - 2z = 4 \\ -4x + 3z = -4 \\ -2x + 2y - z = 4 \end{cases}$

50. $\begin{cases} -x + 3y + 3z = 0 \\ x + 5y = -2 \\ 8y + 3z = -2 \end{cases}$

Mixed Practice

In Problems 51–56, an augmented matrix of a system of linear equations is given. Write the system of equations corresponding to the given matrix. Use x, y; or x, y, z as variables. Determine whether the system is consistent with independent equations, consistent with dependent equations, or inconsistent. If it is consistent with independent equations, give the solution.

51. $\begin{bmatrix} 1 & 4 & | & -5 \\ 0 & 1 & | & -2 \end{bmatrix}$

52. $\begin{bmatrix} 1 & -2 & | & 3 \\ 0 & 1 & | & -5 \end{bmatrix}$

53. $\begin{bmatrix} 1 & 3 & -2 & | & 6 \\ 0 & 1 & 5 & | & -2 \\ 0 & 0 & 0 & | & 4 \end{bmatrix}$

54. $\begin{bmatrix} 1 & -2 & -4 & | & 6 \\ 0 & 1 & -5 & | & -3 \\ 0 & 0 & 0 & | & 0 \end{bmatrix}$

55. $\begin{bmatrix} 1 & -2 & -1 & | & 3 \\ 0 & 1 & -2 & | & -8 \\ 0 & 0 & 1 & | & 5 \end{bmatrix}$

56. $\begin{bmatrix} 1 & 2 & -1 & | & -7 \\ 0 & 1 & 2 & | & -4 \\ 0 & 0 & 1 & | & -3 \end{bmatrix}$

In Problems 57–78, solve each system of equations using matrices.

57. $\begin{cases} x - 3y = 18 \\ 2x + y = 1 \end{cases}$

58. $\begin{cases} x + 5y = 2 \\ -2x + 3y = 9 \end{cases}$

59. $\begin{cases} 2x + 4y = 10 \\ x + 2y = 3 \end{cases}$

60. $\begin{cases} 5x - 2y = 3 \\ -15x + 6y = -9 \end{cases}$

61. $\begin{cases} x - 6y = 8 \\ 2x + 8y = -9 \end{cases}$

62. $\begin{cases} 3x + 3y = -1 \\ 2x + y = 1 \end{cases}$

63. $\begin{cases} 4x - y = 8 \\ 2x - \dfrac{1}{2}y = 4 \end{cases}$

64. $\begin{cases} 5x - 2y = 10 \\ 2x - \dfrac{4}{5}y = 4 \end{cases}$

65. $\begin{cases} x + y + z = 0 \\ 2x - 3y + z = 19 \\ -3x + y - 2z = -15 \end{cases}$

66. $\begin{cases} x + y + z = 5 \\ 2x - y + 3z = -3 \\ -x + 2y - z = 10 \end{cases}$

67. $\begin{cases} 2x + y - z = 13 \\ -x - 3y + 2z = -14 \\ -3x + 2y - 3z = 3 \end{cases}$

68. $\begin{cases} 2x - y + 2z = 13 \\ -x + 2y - z = -14 \\ 3x + y - 2z = -13 \end{cases}$

69. $\begin{cases} 2x - y + 3z = 1 \\ -x + 3y + z = -4 \\ 3x + y + 7z = -2 \end{cases}$

70. $\begin{cases} -x + 2y + z = 1 \\ 2x - y + 3z = -3 \\ -x + 5y + 6z = 2 \end{cases}$

71. $\begin{cases} 3x + y - 4z = 0 \\ -2x - 3y + z = 5 \\ -x - 5y - 2z = 3 \end{cases}$

72. $\begin{cases} -x + 4y - 3z = 1 \\ 3x + y - z = -3 \\ x + 9y - 7z = -1 \end{cases}$

73. $\begin{cases} 2x - y + 3z = -1 \\ 3x + y - 4z = 3 \\ x + 7y - 2z = 2 \end{cases}$

74. $\begin{cases} x + y + z = 8 \\ 2x + 3y + z = 19 \\ 2x + 2y + 4z = 21 \end{cases}$

75. $\begin{cases} 3x + 5y + 2z = 6 \\ 10y - 2z = 5 \\ 6x + 4z = 8 \end{cases}$

76. $\begin{cases} 2x + y + 3z = 3 \\ 2x - 3y = 7 \\ 4y + 6z = -2 \end{cases}$

77. $\begin{cases} x - z = 3 \\ 2x + y = -3 \\ 2y - z = 7 \end{cases}$

78. $\begin{cases} x + 2y - z = 3 \\ y + z = 1 \\ x - 3z = 2 \end{cases}$

Applying the Concepts

79. Curve Fitting The function $f(x) = ax^2 + bx + c$ is a quadratic function, where a, b, and c are constants.

(a) If $f(-1) = 6$, then $6 = a(-1)^2 + b(-1) + c$ or $a - b + c = 6$. Find two additional linear equations if $f(1) = 0$ and $f(2) = 3$.

(b) Use the three linear equations found in part (a) to determine a, b, and c. What is the quadratic function that contains the points $(-1, 6)$, $(1, 0)$, and $(2, 3)$?

80. Curve Fitting The function $f(x) = ax^2 + bx + c$ is a quadratic function, where a, b, and c are constants.

(a) If $f(-1) = -6$, then $-6 = a(-1)^2 + b(-1) + c$ or $a - b + c = -6$. Find two additional linear equations if $f(1) = 0$ and $f(2) = -3$.

(b) Use the three linear equations found in part (a) to determine a, b, and c. What is the quadratic function that contains the points $(-1, -6)$, $(1, 0)$, and $(2, -3)$?

81. Finance Carissa has $20,000 to invest. Her financial planner suggests that she diversify her investment into three investment categories: Treasury bills that yield 3% simple interest, municipal bonds that yield 4% simple interest, and corporate bonds that yield 6% simple interest. Carissa would like to earn $820 per year in income. In addition, she wants her investment in Treasury bills to be $3000 more than her investment in corporate bonds. How much should Carissa invest in each investment category?

82. Finance Marlon has $12,000 to invest. He decides to place some of the money into a savings account paying 0.50% interest, some in Treasury bonds paying 3% interest, and some in a mutual fund paying 10% interest. Marlon would like to earn $325 per year in income. In addition, Marlon wants his investment in the savings account to be $4000 more than the amount in Treasury bonds. How much should Marlon invest in each investment category?

Extending the Concepts

*Sometimes it is advantageous to write a matrix in **reduced row echelon form**. In this form, row operations are used to obtain entries that are 0 above and below the leading 1 in a row. Augmented matrices for systems in reduced row echelon form with two equations containing two variables and three equations containing three variables, with a single solution, are shown below.*

$$\begin{bmatrix} 1 & 0 & | & a \\ 0 & 1 & | & b \end{bmatrix} \qquad \begin{bmatrix} 1 & 0 & 0 & | & a \\ 0 & 1 & 0 & | & b \\ 0 & 0 & 1 & | & c \end{bmatrix}$$

The obvious advantage to writing an augmented matrix in reduced row echelon form is that the solution to the system is readily seen. In the system with two equations containing two variables, the solution is $x = a$ and $y = b$. In the system with three equations containing three variables, the solution is $x = a, y = b$, and $z = c$.

In Problems 83–86, solve the system of equations by writing the augmented matrix in reduced row echelon form.

83. $\begin{cases} 2x + y = 1 \\ -3x - 2y = -5 \end{cases}$

84. $\begin{cases} 2x + y = -1 \\ -3x - 2y = -3 \end{cases}$

85. $\begin{cases} x + y + z = 3 \\ 2x + y - 4z = 25 \\ -3x + 2y + z = 0 \end{cases}$

86. $\begin{cases} x + y + z = 3 \\ 3x - 2y + 2z = 38 \\ -2x - 3z = -19 \end{cases}$

Explaining the Concepts

87. Write a paragraph that outlines the strategy for putting an augmented matrix in row echelon form.

88. When solving a system of linear equations using matrices, how do you know when the system is inconsistent?

89. What would be the next row operation on the given augmented matrix?

$$\begin{bmatrix} 1 & 3 & | & 8 \\ 0 & 5 & | & 10 \end{bmatrix}$$

90. What would you recommend as the next row operation on the given augmented matrix? Why?

$$\begin{bmatrix} 1 & 3 & -2 & | & 4 \\ 0 & 5 & 3 & | & 2 \\ 0 & -4 & 6 & | & -5 \end{bmatrix}$$

Technology Exercises

Graphing calculators have the ability to solve systems of linear equations by writing the augmented matrix in row echelon form. We enter the augmented matrix into the graphing calculator and name it A. See Figure 11(a). Figure 11(b) shows the results of Example 6 using a graphing calculator.

Figure 11

(a)   (b)

Notice that the row echelon form of the augmented matrix found using the graphing calculator differs from the row echelon form in the algebraic solution presented in Example 6, yet both matrices provide the same solution! This is because the two solutions use different row operations to obtain the row echelon form.

In Problems 91–94, solve each system of equations using technology.

91. $\begin{cases} 2x + 3y = 1 \\ -3x - 4y = -3 \end{cases}$

92. $\begin{cases} 3x + 2y = 4 \\ -5x - 3y = -4 \end{cases}$

93. $\begin{cases} 2x + 3y - 2z = -12 \\ -3x + y + 2z = 0 \\ 4x + 3y - z = 3 \end{cases}$

94. $\begin{cases} 2x - 3y - 4z = 16 \\ -3x + y + 2z = -23 \\ 4x + 3y - z = 13 \end{cases}$

3.5 Determinants and Cramer's Rule

Objectives

1 Evaluate the Determinant of a 2×2 Matrix

2 Use Cramer's Rule to Solve a System of Two Equations

3 Evaluate the Determinant of a 3×3 Matrix

4 Use Cramer's Rule to Solve a System of Three Equations

Are You Prepared for This Section?

Before getting started, complete the following problems. If you get a problem wrong, go back to the section cited and review the material.

P1. Evaluate: $4 \cdot 2 - 3(-3)$ [Section R.4, pp. 34–37]

P2. Simplify: $\dfrac{18}{6}$ [Section R.3, p. 25]

At this point, it is known how to solve systems of linear equations using substitution, elimination, and row operations on augmented matrices. In this section, another method is introduced. This method, which requires that the number of equations equal the number of variables, is called *Cramer's Rule*. It is based on the concept of a *determinant*.

▶ **1 Evaluate the Determinant of a 2×2 Matrix**

Prepared?...Answers   **P1.** 17
P2. 3

A matrix is **square** if the number of rows equals the number of columns. It is possible to computer the *determinant* of any square matrix.

Work Smart

The methods for solving systems presented in this section work *only* when the number of equations equals the number of variables.

Definition

Suppose a, b, c, and d are four real numbers. The **determinant** of a 2 × 2 matrix $\begin{bmatrix} a & b \\ c & d \end{bmatrix}$, denoted $\begin{vmatrix} a & b \\ c & d \end{vmatrix}$, is

$$\begin{vmatrix} a & b \\ c & d \end{vmatrix} = ad - bc$$

Notice that the determinant of a matrix is a number, not a matrix.

EXAMPLE 1 | **Evaluating a 2 × 2 Determinant**

Evaluate each determinant:

(a) $\begin{vmatrix} 5 & 2 \\ 3 & 4 \end{vmatrix}$

(b) $\begin{vmatrix} -1 & 3 \\ -4 & 5 \end{vmatrix}$

Solution

(a) $\begin{vmatrix} 5 & 2 \\ 3 & 4 \end{vmatrix} = 5(4) - 3(2)$

$= 20 - 6$

$= 14$

(b) $\begin{vmatrix} -1 & 3 \\ -4 & 5 \end{vmatrix} = -1(5) - (-4)(3)$

$= -5 - (-12)$

$= -5 + 12$

$= 7$

Quick ✓

1. $D = \begin{vmatrix} a & b \\ c & d \end{vmatrix} = $ _____.

2. A matrix is _____ if the number of rows and the number of columns are equal.

In Problems 3 and 4, evaluate each determinant.

3. $\begin{vmatrix} 5 & 3 \\ 4 & 6 \end{vmatrix}$

4. $\begin{vmatrix} -2 & -5 \\ 1 & 7 \end{vmatrix}$

② Use Cramer's Rule to Solve a System of Two Equations

Use 2 × 2 determinants to solve a system of two equations containing two variables.

Cramer's Rule for Two Equations Containing Two Variables

The solution to the system of equations

$$\begin{cases} ax + by = s & (1) \\ cx + dy = t & (2) \end{cases}$$

is given by

$$x = \frac{\begin{vmatrix} s & b \\ t & d \end{vmatrix}}{\begin{vmatrix} a & b \\ c & d \end{vmatrix}} = \frac{D_x}{D} \qquad y = \frac{\begin{vmatrix} a & s \\ c & t \end{vmatrix}}{\begin{vmatrix} a & b \\ c & d \end{vmatrix}} = \frac{D_y}{D}$$

where

$$D = \begin{vmatrix} a & b \\ c & d \end{vmatrix} = ad - bc \neq 0$$

Look very carefully at the pattern in Cramer's Rule. The denominator in the solution is the determinant of the coefficients of the variables.

$$\begin{cases} ax + by = s \\ cx + dy = t \end{cases} \qquad D = \begin{vmatrix} a & b \\ c & d \end{vmatrix}$$

In the solution for x, the numerator is the determinant formed by replacing the entries in the first column (the coefficients of x) in D by the system's constants. That is,

$$D_x = \begin{vmatrix} s & b \\ t & d \end{vmatrix}$$

In the solution for y, the numerator is the determinant formed by replacing the entries in the second column (the coefficients of y) in D by the constants. That is,

$$D_y = \begin{vmatrix} a & s \\ c & t \end{vmatrix}$$

EXAMPLE 2 **How to Solve a System of Two Linear Equations Containing Two Variables Using Cramer's Rule**

Use Cramer's Rule to solve the system: $\begin{cases} x + 2y = 0 \\ -2x - 8y = -9 \end{cases}$

Step-by-Step Solution

Step 1: Find the determinant of the coefficients of the variables, D.

$$D = \begin{vmatrix} 1 & 2 \\ -2 & -8 \end{vmatrix} = 1(-8) - (-2)(2) = -8 - (-4) = -8 + 4 = -4$$

Step 2: Since $D \neq 0$, find D_x by replacing the first column in D with the system's constants. Find D_y by replacing the second column in D with the constants.

$$D_x = \begin{vmatrix} 0 & 2 \\ -9 & -8 \end{vmatrix} = 0(-8) - (-9)(2) = 0 - (-18) = 18$$

$$D_y = \begin{vmatrix} 1 & 0 \\ -2 & -9 \end{vmatrix} = 1(-9) - (-2)(0) = -9 - 0 = -9$$

Step 3: Find $x = \dfrac{D_x}{D}$ and $y = \dfrac{D_y}{D}$.

$$x = \frac{D_x}{D} = \frac{18}{-4} = -\frac{9}{2} \qquad y = \frac{D_y}{D} = \frac{-9}{-4} = \frac{9}{4}$$

Step 4: Check: $x = -\dfrac{9}{2}$ and $y = \dfrac{9}{4}$

The check of the solution $\left(-\dfrac{9}{2}, \dfrac{9}{4}\right)$ is left to you. ●

If the determinant of the coefficients of the variables, D, is found to be 0, then the system of equations is either consistent with dependent equations, or inconsistent. To determine if the system has no solution or infinitely many solutions, solve the system using the methods of Section 3.1, 3.3, or 3.4.

Quick ✓

In Problems 5 and 6, use Cramer's Rule to solve the system, if possible.

5. $\begin{cases} 3x + 2y = 1 \\ -2x - y = 1 \end{cases}$ **6.** $\begin{cases} 4x - 2y = 8 \\ -6x + 3y = 3 \end{cases}$

▶ ❸ **Evaluate the Determinant of a 3 × 3 Matrix**

To use Cramer's Rule to solve a system of three equations containing three variables, define a 3 × 3 determinant.

The **determinant of a 3 × 3 matrix** is symbolized by

$$\begin{vmatrix} a_{1,1} & a_{1,2} & a_{1,3} \\ a_{2,1} & a_{2,2} & a_{2,3} \\ a_{3,1} & a_{3,2} & a_{3,3} \end{vmatrix}$$

where $a_{1,1}, a_{1,2}, \ldots, a_{3,3}$ are real numbers.

The subscript is used to identify the row and column of an entry. For example, $a_{2,3}$ is the entry in row 2, column 3.

Definition

The **value of the determinant of a 3 × 3 matrix** may be defined in terms of 2 × 2 determinants as follows:

2 × 2 determinant left after removing the row and column containing $a_{1,1}$.	2 × 2 determinant left after removing the row and column containing $a_{1,2}$	2 × 2 determinant left after removing the row and column containing $a_{1,3}$.

The 2 × 2 determinants shown in the definition above are called **minors** of the 3 × 3 determinant. Notice that once again a problem has been reduced to something we already know how to do—here the 3 × 3 determinant was reduced to three different 2 × 2 determinants.

The formula given in the definition above is easiest to remember by noting that each entry in row 1 is multiplied by the 2 × 2 determinant that remains after the row and column containing the entry have been removed.

EXAMPLE 3

Evaluating a 3 × 3 Determinant

Evaluate: $\begin{vmatrix} 3 & 2 & -4 \\ 1 & 7 & -3 \\ 0 & 2 & -5 \end{vmatrix}$

Solution

$$\begin{vmatrix} 3 & 2 & -4 \\ 1 & 7 & -3 \\ 0 & 2 & -5 \end{vmatrix} = 3\begin{vmatrix} 7 & -3 \\ 2 & -5 \end{vmatrix} - 2\begin{vmatrix} 1 & -3 \\ 0 & -5 \end{vmatrix} + (-4)\begin{vmatrix} 1 & 7 \\ 0 & 2 \end{vmatrix}$$

$$= 3(-35 - (-6)) - 2(-5 - 0) - 4(2 - 0)$$
$$= 3(-29) - 2(-5) - 4(2)$$
$$= -87 + 10 - 8$$
$$= -85$$

The definition shows one way to evaluate a 3 × 3 determinant—by *expanding* across row 1. In fact, the expansion can take place across any row or down any column. The terms to be added or subtracted consist of the row (or column) entry times the value of the 2 × 2 determinant that remains after removing the row and column containing the entry. There is only one glitch—the signs of the terms in the expansion change depending on the row or column that is expanded on. The signs of the terms obey the following scheme:

$$\begin{array}{ccc} + & - & + \\ - & + & - \\ + & - & + \end{array}$$

For example, if we choose to expand down column 2, we obtain

$$\begin{vmatrix} a_{1,1} & a_{1,2} & a_{1,3} \\ a_{2,1} & a_{2,2} & a_{2,3} \\ a_{3,1} & a_{3,2} & a_{3,3} \end{vmatrix} = \overset{\text{Minus}}{-a_{1,2}} \begin{vmatrix} a_{2,1} & a_{2,3} \\ a_{3,1} & a_{3,3} \end{vmatrix} + \overset{\text{Plus}}{a_{2,2}} \begin{vmatrix} a_{1,1} & a_{1,3} \\ a_{3,1} & a_{3,3} \end{vmatrix} - \overset{\text{Minus}}{a_{3,2}} \begin{vmatrix} a_{1,1} & a_{1,3} \\ a_{2,1} & a_{2,3} \end{vmatrix}$$

EXAMPLE 4   **Evaluating a 3 × 3 Determinant**

Evaluate: $\begin{vmatrix} 3 & 2 & -4 \\ 1 & 7 & -3 \\ 0 & 2 & -5 \end{vmatrix}$ (from Example 3), by expanding down column 1.

Solution

Expand down column 1 because it has a 0 in it.

$$\begin{vmatrix} 3 & 2 & -4 \\ 1 & 7 & -3 \\ 0 & 2 & -5 \end{vmatrix} = 3\begin{vmatrix} 7 & -3 \\ 2 & -5 \end{vmatrix} - 1\begin{vmatrix} 2 & -4 \\ 2 & -5 \end{vmatrix} + 0\begin{vmatrix} 2 & -4 \\ 7 & -3 \end{vmatrix}$$

$$= 3(-35 - (-6)) - 1(-10 - (-8)) + 0(-6 - (-28))$$
$$= 3(-29) - 1(-2) + 0$$
$$= -87 + 2$$
$$= -85$$

Notice that the results of Example 3 and 4 are the same! However, the computation in Example 4 was a little easier because we expanded down the column that contains a 0.

To make the computation easier, expand across the row that contains the most 0s or down the column that contains the most 0s.

Work Smart

Expand across the row, or down the column, with the most 0s.

Quick ✓

7. Evaluate: $\begin{vmatrix} 2 & -3 & 5 \\ 0 & 4 & -1 \\ 3 & 8 & -7 \end{vmatrix}$

8. Evaluate: $\begin{vmatrix} 3 & 2 & 1 \\ 1 & 1 & -3 \\ -5 & -1 & -5 \end{vmatrix}$

▶ ❹   Use Cramer's Rule to Solve a System of Three Equations

Cramer's Rule can also be applied to a system of three linear equations.

Cramer's Rule for Three Equations Containing Three Variables

For the system of three equations containing three variables

$$\begin{cases} a_1 x + b_1 y + c_1 z = d_1 \\ a_2 x + b_2 y + c_2 z = d_2 \\ a_3 x + b_3 y + c_3 z = d_3 \end{cases}$$

with

$$D = \begin{vmatrix} a_1 & b_1 & c_1 \\ a_2 & b_2 & c_2 \\ a_3 & b_3 & c_3 \end{vmatrix} \neq 0 \quad D_x = \begin{vmatrix} d_1 & b_1 & c_1 \\ d_2 & b_2 & c_2 \\ d_3 & b_3 & c_3 \end{vmatrix} \quad D_y = \begin{vmatrix} a_1 & d_1 & c_1 \\ a_2 & d_2 & c_2 \\ a_3 & d_3 & c_3 \end{vmatrix} \quad D_z = \begin{vmatrix} a_1 & b_1 & d_1 \\ a_2 & b_2 & d_2 \\ a_3 & b_3 & d_3 \end{vmatrix}$$

then

$$x = \frac{D_x}{D} \qquad y = \frac{D_y}{D} \qquad z = \frac{D_z}{D}$$

EXAMPLE 5 **How to Use Cramer's Rule**

Use Cramer's Rule to solve the following system:

$$\begin{cases} x - 2y + z = -9 \\ -3x + y - 2z = 5 \\ 4x + 3z = 1 \end{cases}$$

Step-by-Step Solution

Step 1: Find the determinant of the coefficients of the variables, D. Choose to expand down column 2. Do you see why?

$$D = \begin{vmatrix} 1 & -2 & 1 \\ -3 & 1 & -2 \\ 4 & 0 & 3 \end{vmatrix} = -(-2)\begin{vmatrix} -3 & -2 \\ 4 & 3 \end{vmatrix} + 1\begin{vmatrix} 1 & 1 \\ 4 & 3 \end{vmatrix} - 0\begin{vmatrix} 1 & 1 \\ -3 & -2 \end{vmatrix}$$

$$= 2[-3(3) - 4(-2)] + 1[1(3) - 4(1)] - 0$$

$$= 2(-1) + 1(-1) - 0$$

$$= -2 - 1$$

$$= -3$$

Step 2: Since $D \neq 0$, continue by determining D_x by replacing column 1 in D with the system's constants.

$$D_x = \begin{vmatrix} -9 & -2 & 1 \\ 5 & 1 & -2 \\ 1 & 0 & 3 \end{vmatrix} \overset{\text{Expand down column 2}}{\underset{\downarrow}{=}} -(-2)\begin{vmatrix} 5 & -2 \\ 1 & 3 \end{vmatrix} + 1\begin{vmatrix} -9 & 1 \\ 1 & 3 \end{vmatrix} - 0\begin{vmatrix} -9 & 1 \\ 5 & -2 \end{vmatrix}$$

$$= 2[15 - (-2)] + 1[-27 - 1] - 0$$

$$= 2(17) - 28$$

$$= 6$$

Find D_y by replacing column 2 in D with the system's constants.

$$D_y = \begin{vmatrix} 1 & -9 & 1 \\ -3 & 5 & -2 \\ 4 & 1 & 3 \end{vmatrix} \overset{\text{Expand down column 1}}{\underset{\downarrow}{=}} 1\begin{vmatrix} 5 & -2 \\ 1 & 3 \end{vmatrix} - (-3)\begin{vmatrix} -9 & 1 \\ 1 & 3 \end{vmatrix} + 4\begin{vmatrix} -9 & 1 \\ 5 & -2 \end{vmatrix}$$

$$= -15$$

Find D_z by replacing column 3 in D with the system's constants.

$$D_z = \begin{vmatrix} 1 & -2 & -9 \\ -3 & 1 & 5 \\ 4 & 0 & 1 \end{vmatrix} \overset{\text{Expand down column 2}}{\underset{\downarrow}{=}} -(-2)\begin{vmatrix} -3 & 5 \\ 4 & 1 \end{vmatrix} + 1\begin{vmatrix} 1 & -9 \\ 4 & 1 \end{vmatrix} - 0\begin{vmatrix} 1 & -9 \\ -3 & 5 \end{vmatrix}$$

$$= -9$$

Step 3: Find $x = \dfrac{D_x}{D}, y = \dfrac{D_y}{D}$, and $z = \dfrac{D_z}{D}$.

$$x = \frac{D_x}{D} = \frac{6}{-3} \qquad y = \frac{D_y}{D} = \frac{-15}{-3} \qquad z = \frac{D_z}{D} = \frac{-9}{-3}$$

$$= -2 \qquad\qquad = 5 \qquad\qquad = 3$$

Step 4: Check.

$$\begin{array}{ccc} x - 2y + z = -9 & -3x + y - 2z = 5 & 4x + 3z = 1 \\ -2 - 2(5) + 3 \overset{?}{=} -9 & -3(-2) + 5 - 2(3) \overset{?}{=} 5 & 4(-2) + 3(3) \overset{?}{=} 1 \\ -2 - 10 + 3 \overset{?}{=} -9 & 6 + 5 - 6 \overset{?}{=} 5 & -8 + 9 \overset{?}{=} 1 \\ -9 = -9 \quad \text{True} & 5 = 5 \quad \text{True} & 1 = 1 \quad \text{True} \end{array}$$

The solution is the ordered triple $(-2, 5, 3)$.

Work Smart: Study Skills

At the end of Section 3.1, we summarized the methods that can be used to solve a system of linear equations in two variables. Make up a summary table of your own for solving systems of linear equations in three variables. When is elimination appropriate? When would you use matrices? When is Cramer's Rule best?

Quick ✓

9. Suppose that you wish to solve a system of equations using Cramer's Rule and find that $D = 4, D_x = 2, D_y = -8,$ and $D_z = -4.$ What is the solution of the system?

10. Use Cramer's Rule to solve the system

$$\begin{cases} x - y + 3z = -2 \\ 4x + 3y + z = 9 \\ -2x \quad\quad + 5z = 7 \end{cases}$$

In the case where $D = 0,$ Cramer's Rule does not apply. To determine whether the system has no solution or infinitely many solutions, it is necessary to solve the system using the methods of Section 3.2, 3.3, or 3.4.

3.5 Exercises MyMathLab®

Exercise numbers in green have complete video solutions in MyMathLab or may be accessed using the QR code to the right.

Problems 1–10 are the Quick ✓s that follow the EXAMPLES.

Building Skills

In Problems 11–14, find the value of each determinant. See Objective 1.

11. $\begin{vmatrix} 4 & 2 \\ 1 & 3 \end{vmatrix}$

12. $\begin{vmatrix} 5 & 3 \\ 2 & 4 \end{vmatrix}$

13. $\begin{vmatrix} -2 & -4 \\ 1 & 3 \end{vmatrix}$

14. $\begin{vmatrix} -8 & 5 \\ -4 & 3 \end{vmatrix}$

In Problems 15–22, solve each system of equations using Cramer's Rule, if applicable. See Objective 2.

15. $\begin{cases} x + y = -4 \\ x - y = -12 \end{cases}$

16. $\begin{cases} x + y = 6 \\ x - y = 4 \end{cases}$

17. $\begin{cases} 2x + 3y = 3 \\ -3x + y = -10 \end{cases}$

18. $\begin{cases} 2x + 4y = -6 \\ 3x + 2y = 7 \end{cases}$

19. $\begin{cases} 3x + 4y = 1 \\ -6x + 8y = 4 \end{cases}$

20. $\begin{cases} 4x - 2y = 9 \\ -8x - 2y = -11 \end{cases}$

21. $\begin{cases} 2x - 6y - 12 = 0 \\ 3x - 5y - 11 = 0 \end{cases}$

22. $\begin{cases} 3x - 6y - 2 = 0 \\ x + 2y - 4 = 0 \end{cases}$

In Problems 23–28, find the value of each determinant. See Objective 3.

23. $\begin{vmatrix} 2 & 0 & -1 \\ 3 & 8 & -3 \\ 1 & 5 & -2 \end{vmatrix}$

24. $\begin{vmatrix} -2 & 1 & 6 \\ -3 & 2 & 5 \\ 1 & 0 & -2 \end{vmatrix}$

25. $\begin{vmatrix} -3 & 2 & 3 \\ 0 & 5 & -2 \\ 1 & 4 & 8 \end{vmatrix}$

26. $\begin{vmatrix} 8 & 4 & -1 \\ 2 & -7 & 1 \\ 0 & 5 & -3 \end{vmatrix}$

27. $\begin{vmatrix} 0 & 2 & 1 \\ 1 & -6 & -4 \\ -3 & 4 & 5 \end{vmatrix}$

28. $\begin{vmatrix} -3 & 4 & -2 \\ 1 & -2 & 0 \\ 0 & 6 & 6 \end{vmatrix}$

In Problems 29–42, solve each system of equations using Cramer's Rule, if applicable. See Objective 4.

29. $\begin{cases} x - y + z = -4 \\ x + 2y - z = 1 \\ 2x + y + 2z = -5 \end{cases}$

30. $\begin{cases} x + y - z = 6 \\ x + 2y + z = 6 \\ -x - y + 2z = -7 \end{cases}$

31. $\begin{cases} x + y + z = 4 \\ 5x + 2y - 3z = 7 \\ 2x - y - z = 5 \end{cases}$

32. $\begin{cases} x + y + z = -3 \\ -2x - 3y - z = 1 \\ 2x - y - 3z = -5 \end{cases}$

33. $\begin{cases} 2x + y - z = 4 \\ -x + 2y + 2z = -6 \\ 5x + 5y - z = 6 \end{cases}$

34. $\begin{cases} -x + 2y - z = 2 \\ 2x + y + 2z = -6 \\ -x + 7y - z = 0 \end{cases}$

35. $\begin{cases} 3x + y + z = 5 \\ x + y - 3z = 9 \\ -5x - y - 5z = -10 \end{cases}$

36. $\begin{cases} x - 2y - z = 1 \\ 2x + 2y + z = 3 \\ 6x + 6y + 3z = 6 \end{cases}$

37. $\begin{cases} 2x + z = 27 \\ -x - 3y = 6 \\ x - 2y + z = 27 \end{cases}$

38. $\begin{cases} x + 2y + z = 0 \\ -x - 3y - 2z = 3 \\ 2x - 3z = 7 \end{cases}$

39. $\begin{cases} 5x + 3y = 2 \\ -10x + 3z = -3 \\ y - 2z = -9 \end{cases}$

40. $\begin{cases} 2x + 4y = 0 \\ -2x + z = -5 \\ -4y - 3z = 1 \end{cases}$

41. $\begin{cases} x + y + z = 3 \\ -y + 2z = 1 \\ -x + z = 0 \end{cases}$

42. $\begin{cases} 3y - z = -1 \\ x + 5y - z = -4 \\ -3x + 6y + 2z = 11 \end{cases}$

Mixed Practice

In Problems 43–46, solve for x.

43. $\begin{vmatrix} x & 3 \\ 1 & 2 \end{vmatrix} = 7$

44. $\begin{vmatrix} -2 & x \\ 3 & 4 \end{vmatrix} = 1$

45. $\begin{vmatrix} x & -1 & -2 \\ 1 & 0 & 4 \\ 3 & 2 & 5 \end{vmatrix} = 5$

46. $\begin{vmatrix} 2 & x & -1 \\ 3 & 5 & 0 \\ -4 & 1 & 2 \end{vmatrix} = 0$

Applying the Concepts

Problems 47–50, use the following result. Determinants can be used to find the area of a triangle. If (a_1, b_1), (a_2, b_2), and (a_3, b_3) are the vertices of a triangle, the area of the triangle is $|D|$, where

$$D = \frac{1}{2} \begin{vmatrix} a_1 & a_2 & a_3 \\ b_1 & b_2 & b_3 \\ 1 & 1 & 1 \end{vmatrix}$$

△ **47. Geometry: Area of a Triangle** Given the points $A = (1, 1)$, $B = (5, 1)$, and $C = (5, 6)$,

    (a) Plot the points in the Cartesian plane and form triangle ABC.

    (b) Find the area of the triangle ABC.

△ **48. Geometry: Area of a Triangle** Given the points $A = (-1, -1)$, $B = (3, 2)$, and $C = (0, 6)$,

    (a) Plot the points in the Cartesian plane and form triangle ABC.

    (b) Find the area of the triangle ABC.

△ **49. Geometry: Area of a Parallelogram** Find the area of a parallelogram by doing the following:

    (a) Plot the points $A = (2, 1)$, $B = (7, 2)$, $C = (8, 4)$, and $D = (3, 3)$ in the Cartesian plane.

    (b) Form triangle ABC and find the area of triangle ABC.

    (c) Find the area of the triangle ADC.

    (d) Conclude that the diagonal of the parallelogram forms two triangles of equal area. Use this result to find the area of the parallelogram.

△ **50. Geometry: Area of a Parallelogram** Find the area of a parallelogram by doing the following:

(a) Plot the points $A = (-3, -2)$, $B = (3, 1)$, $C = (4, 4)$, and $D = (-2, 1)$ in the Cartesian plane.

(b) Form triangle ABC and find the area of triangle ABC.

(c) Find the area of the triangle ADC.

(d) Conclude that the diagonal of the parallelogram forms two triangles of equal area. Use this result to find the area of the parallelogram.

△ **51. Geometry: Equation of a Line** An equation of the line containing the two points (x_1, y_1) and (x_2, y_2) may be expressed as the determinant

$$\begin{vmatrix} x & y & 1 \\ x_1 & y_1 & 1 \\ x_2 & y_2 & 1 \end{vmatrix} = 0$$

(a) Find the equation of the line containing $(3, 2)$ and $(5, 1)$ using the determinant.

(b) Verify your result by using the slope formula and the point-slope formula.

△ **52. Geometry: Collinear Points** The distinct points (x_1, y_1), (x_2, y_2), and (x_3, y_3) are collinear if and only if

$$\begin{vmatrix} x_1 & y_1 & 1 \\ x_2 & y_2 & 1 \\ x_3 & y_3 & 1 \end{vmatrix} = 0$$

(a) Plot the points $(-3, -2)$, $(1, 2)$, and $(7, 8)$ in the Cartesian plane.

(b) Show that the points are collinear using the determinant.

(c) Show that the points are collinear using the idea of slopes.

Extending the Concepts

53. Evaluate the determinant $\begin{vmatrix} 3 & -2 \\ 1 & 4 \end{vmatrix}$. Now interchange rows 1 and 2 and recompute the determinant. What do you notice? Do you think that this result is true in general?

54. Evaluate the determinant $\begin{vmatrix} -3 & 1 \\ 6 & 5 \end{vmatrix}$. Multiply the entries in column 2 by 3 and recompute the determinant. What do you notice? Do you think that this result is true in general?

Technology Exercises

Graphing calculators have the ability to evaluate 2×2 determinants. First, enter the matrix into the graphing calculator and name it A. Then compute the determinant of matrix A. Figure 12 shows the results of Example 1(a) using a graphing calculator.

Figure 12

Because graphing calculators have the ability to compute determinants of matrices, they can be used to solve systems of equations using Cramer's Rule. Figure 13 shows the results of solving Example 2 using Cramer's Rule, where $A = D$, $B = D_x$, and $C = D_y$.

Figure 13

NORMAL FLOAT AUTO REAL RADIAN MP

det([B])/det([A])▶Frac
$-\frac{9}{2}$
det([C])/det([A])▶Frac
$\frac{9}{4}$

In Problems 55–60, use technology to solve the system of equations.

55. $\begin{cases} x + y = -4 \\ x - y = -12 \end{cases}$

56. $\begin{cases} x + y = 6 \\ x - y = 4 \end{cases}$

57. $\begin{cases} 2x + 3y = 3 \\ -3x + y = -10 \end{cases}$

58. $\begin{cases} 2x + 4y = -6 \\ 3x + 2y = 7 \end{cases}$

59. $\begin{cases} x - y + z = -4 \\ x + 2y - z = 1 \\ 2x + y + 2z = -5 \end{cases}$

60. $\begin{cases} x + y - z = 6 \\ x + 2y + z = 6 \\ -x - y + 2z = -7 \end{cases}$

3.6 Systems of Linear Inequalities

Objectives

1. Determine Whether an Ordered Pair Is a Solution of a System of Linear Inequalities
2. Graph a System of Linear Inequalities
3. Solve Problems Involving Systems of Linear Inequalities

Are You Prepared for This Section?

Before getting started, complete the following problems. If you get a problem wrong, go back to the section cited and review the material.

P1. Determine whether $x = -2$ satisfies the inequality $-3x + 2 \geq 7$. [Section 1.4, pp. 86–88]

P2. Solve the inequality: $-2x + 1 > 7$ [Section 1.4, pp. 86–88]

P3. Graph the inequality: $3x + 2y > -6$ [Section 1.8, pp. 130–131]

In Section 1.8, a single linear inequality in two variables was graphed. In this section, how to graph a system of linear inequalities in two variables will be discussed.

▶ ❶ Determine Whether an Ordered Pair Is a Solution of a System of Linear Inequalities

An ordered pair **satisfies** a system of linear inequalities if it makes each inequality in the system a true statement.

EXAMPLE 1 Determining Whether an Ordered Pair Is a Solution of a System of Linear Inequalities

Which of the following points, if any, is a solution of the system of linear inequalities?

$$\begin{cases} 3x + y \leq 7 \\ 4x - 2y \leq 8 \end{cases}$$

(a) $(2, 4)$ **(b)** $(-3, 1)$

Solution

(a) Let $x = 2$ and $y = 4$ in each inequality in the system. If each inequality is true, then $(2, 4)$ is a solution of the system of inequalities.

$$3x + y \leq 7 \qquad\qquad 4x - 2y \leq 8$$
$$3(2) + 4 \overset{?}{\leq} 7 \qquad\qquad 4(2) - 2(4) \overset{?}{\leq} 8$$
$$6 + 4 \overset{?}{\leq} 7 \qquad\qquad 8 - 8 \overset{?}{\leq} 8$$
$$10 \leq 7 \quad \text{False} \qquad\qquad 0 \leq 8 \quad \text{True}$$

The inequality $3x + y \leq 7$ is not true when $x = 2$ and $y = 4$, so $(2, 4)$ is not a solution of the system.

(b) Let $x = -3$ and $y = 1$ in each inequality in the system. If each inequality is true, then $(-3, 1)$ is a solution of the system of inequalities.

$$3x + y \leq 7 \qquad\qquad 4x - 2y \leq 8$$
$$3(-3) + 1 \overset{?}{\leq} 7 \qquad\qquad 4(-3) - 2(1) \overset{?}{\leq} 8$$
$$-9 + 1 \overset{?}{\leq} 7 \qquad\qquad -12 - 2 \overset{?}{\leq} 8$$
$$-8 \leq 7 \quad \text{True} \qquad\qquad -14 \leq 8 \quad \text{True}$$

Both inequalities are true, so $(-3, 1)$ is a solution of the system.   ●

Prepared?...Answers P1. Yes
P2. $\{x \mid x < -3\}$ or $(-\infty, -3)$
P3.

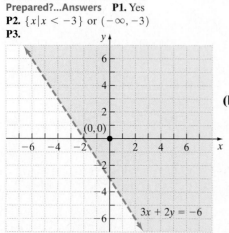

Quick ✓

1. An ordered pair _____ a system of linear inequalities if it makes each inequality in the system a true statement.

2. Determine which of the following ordered pairs is a solution of the system of linear inequalities.

$$\begin{cases} -4x + y < -5 \\ 2x - 5y < 10 \end{cases}$$

    (a) $(1, 2)$ **(b)** $(3, 1)$ **(c)** $(1, -2)$

❷ Graph a System of Linear Inequalities

The graph of a system of inequalities in two variables x and y is the set of all points (x, y) that simultaneously satisfy *each* of the inequalities in the system. To graph a system of linear inequalities, graph each linear inequality individually and then determine where, if at all, they intersect. A graph is the *only* way we show the solution of a system of linear inequalities.

EXAMPLE 2   **Graphing a System of Linear Inequalities**

Graph the system: $\begin{cases} 3x + y \le 7 \\ 4x - 2y \le 8 \end{cases}$

Solution

First, graph the inequality $3x + y \le 7$. To graph $3x + y \le 7$, first graph the solid boundary line $y = -3x + 7$. A test point $(0, 0)$ satisfies the inequality $3x + y \le 7$, so shade below the boundary line. See Figure 14(a).

Then graph the inequality $4x - 2y \le 8$ by first graphing the solid boundary line $4x - 2y = 8$ (or $y = 2x - 4$). A test point $(0, 0)$ satisfies the inequality $4x - 2y \le 8$, so shade above the boundary line. See Figure 14(b).

Now combine the graphs in Figures 14(a) and (b). The intersection of the shaded regions is the solution of the system of linear inequalities. See Figure 14(c). Notice that the point $(-3, 1)$, which was verified to be a solution in Example 1(b), and the point $(0, 0)$ are in the shaded region.

Work Smart

To obtain the boundary line $y = 2x - 4$, solve the equation $4x - 2y = 8$ for y. Or, use intercepts.

Figure 14

Work Smart

Do not forget that the boundary line is graphed as a solid line when the inequality is nonstrict ($\le$ or $\ge$) and as a dashed line when the inequality is strict ($<$ or $>$). In addition, do not forget to reverse the inequality if you multiply or divide each side by a negative number. See Section 1.8.

(a)

(b)

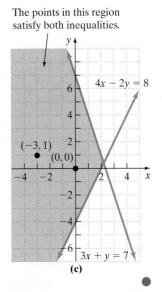
The points in this region satisfy both inequalities.
(c)

Have you noticed that if the inequality is written in the form $y >$ or $y \ge$, the shading is above the line? If the inequality is of the form $y <$ or $y \le$, shade below the line. This will always be true, so instead of using a test point to determine shading, use this method.

Work Smart

Be neat in your shading to make sure that the overlapping region is clear.

Rather than using multiple graphs to obtain the solution set to the system of linear inequalities, use a single graph to determine the overlapping region.

EXAMPLE 3 **Graphing a System of Linear Inequalities**

Graph the system: $\begin{cases} -2x + y > 5 \\ 3x + 2y \geq -4 \end{cases}$

Solution

Graph the inequality $-2x + y > 5$ by adding $2x$ to both sides and obtaining $y > 2x + 5$. Graph $y = 2x + 5$ as a dashed line and shade above (since the inequality is of the form $y >$). To graph the inequality $3x + 2y \geq -4$, solve for y and get $y \geq -\dfrac{3}{2} x - 2$. Graph $y = -\dfrac{3}{2} x - 2$ as a solid line and shade above (since the inequality is of the form $y \geq$).

The overlapping region, shaded purple in Figure 15, represents the solution. The points on the line $3x + 2y \geq -4$ are also solutions of the system because the inequality is not strict.

Figure 15

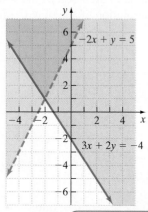

Quick ✓

In Problems 3 and 4, graph the system of linear inequalities.

3. $\begin{cases} 2x + y \leq 5 \\ -x + y \geq -4 \end{cases}$ **4.** $\begin{cases} 3x + y > -2 \\ 2x + 3y < 3 \end{cases}$

▶ **EXAMPLE 4** **Graphing a System of Linear Inequalities with No Solution**

Graph the system: $\begin{cases} 2x + y > 2 \\ 2x + y < -2 \end{cases}$

Solution

Graph the inequality $2x + y > 2$ (or $y > -2x + 2$) using a dashed line because the inequality is strict. In the same Cartesian plane, graph the inequality $2x + y < -2$ (or $y < -2x - 2$) using a dashed line. See Figure 16.

Figure 16

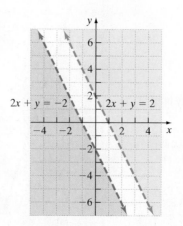

Because no overlapping region results, there are no points in the Cartesian plane that satisfy both inequalities. The system has no solution, so the solution set is ∅ or { }.

Quick ✓

5. *True or False* A system of linear inequalities must always have a nonempty solution set.

Graph the system of linear inequalities.

6. $\begin{cases} -2x + 3y \geq 9 \\ 6x - 9y \geq 9 \end{cases}$

Often, we encounter systems with more than two linear inequalities.

▶ **EXAMPLE 5**) **Graphing a System of Four Linear Inequalities**

Graph the system: $\begin{cases} x + y \leq 5 \\ 2x + y \leq 7 \\ x \qquad \geq 0 \\ \qquad y \geq 0 \end{cases}$

Solution

The two inequalities $x \geq 0$ and $y \geq 5$ require that the graph be in quadrant I. The inequality $x + y \leq 5$, or $y \leq -x + 5$, requires shading below the line $x + y = 5$. The inequality $2x + y \leq 7$, or $y \leq -2x + 7$, requires shading below the line $2x + y = 7$. Figure 17 shows the graph. ●

Notice that in Figure 17 specific points on the graph have been labeled. These are points of intersection between two boundary lines in the system and are called **corner points.** The corner points in Figure 17 are $(0, 0)$, $(0, 5)$, $(2, 3)$, and $(3.5, 0)$. They are found by solving the following systems of equations formed by two boundary lines:

Figure 17

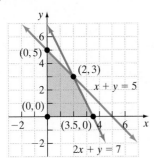

$\begin{cases} x = 0 \\ y = 0 \end{cases}$ $\begin{cases} x + y = 5 \\ x \quad\;\; = 0 \end{cases}$ $\begin{cases} x + y = 5 \\ 2x + y = 7 \end{cases}$ $\begin{cases} 2x + y = 7 \\ y = 0 \end{cases}$

Corner point: $(0, 0)$ Corner point: $(0, 5)$ Corner point: $(2, 3)$ Corner point: $(3.5, 0)$

The graph of the system of linear inequalities in Example 5 is **bounded** because it can be contained within a circle. Look at the systems graphed in Figure 14 or Figure 15. They are **unbounded** because the solution set cannot be contained within a circle. Those graphs extend indefinitely.

Quick ✓

7. The points of intersection between two boundary lines in a system of linear inequalities are called _____ _____.

8. If the solution set of a system of linear inequalities cannot be contained within a circle, then the system is _____.

Graph the system of linear inequalities. Tell whether the graph is bounded or unbounded, and label the corner points.

9. $\begin{cases} x + y \leq 6 \\ 2x + y \leq 10 \\ x \qquad \geq 0 \\ \qquad y \geq 0 \end{cases}$

▶ ❸ **Solve Problems Involving Systems of Linear Inequalities**

In problems that lead to systems of linear inequalities, we use the problem-solving strategy presented in Section 1.2.

EXAMPLE 6 **Breakfast**

Aman likes French toast sticks and orange juice for breakfast. He wants to eat no more than 500 calories and to consume no more than 425 mg of sodium. Each French toast stick (with syrup) has 90 calories and 100 mg of sodium. Each drink orange juice has 140 calories and 25 mg of sodium. Write a system of linear inequalities that represents the possible combinations of French toast sticks and orange juice that Aman can consume. Graph the system.

Solution

Step 1: Identify Find how many French toast sticks and orange juice drinks Aman can consume while staying within his diet restrictions.

Step 2: Name Let x represent the number of French toast sticks and y represent the number of orange juice drinks that Aman consumes.

Step 3: Translate One French toast stick has 90 calories. Two sticks have 180 calories, so x French toast sticks have $90x$ calories. Similarly, y orange juice drinks contain $140y$ calories. The words "no more than" imply a $\leq$ inequality. A linear inequality that describes Aman's options while staying within his calorie restriction is

$$90x + 140y \leq 500$$

A linear inequality that describes Aman's options while staying within his sodium restriction is

$$100x + 25y \leq 425$$

Finally, Aman cannot consume negative quantities of French toast sticks or orange juice, so we have the **nonnegativity constraints**

$$x \geq 0 \quad \text{and} \quad y \geq 0$$

> **In Other Words**
> A nonnegativity constraint means that the value of the variable has to be greater than or equal to zero.

Combining these four linear inequalities, we obtain the system

$$\begin{cases} 90x + 140y \leq 500 \\ 100x + 25y \leq 425 \\ x \geq 0 \\ y \geq 0 \end{cases} \quad \text{The Model}$$

Step 4: Solve Figure 18 shows the graph of the system.

Figure 18

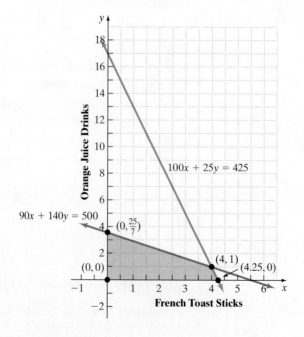

Step 5: Check Although it is not a check, evidence can be gathered that our graph is correct by choosing a point in the shaded region and determining whether all the inequalities in the system are satisfied. For example, the combination of $x = 3$ French toast sticks and $y = 1$ orange juice drink satisfies all four inequalities. ●

Quick ✓

10. Jack and Mary recently retired, and they have \$25,000 to invest. Their financial advisor has recommended that they place at least \$10,000 in Treasury notes yielding 3% and no more than \$15,000 in corporate bonds yielding 6%. Write a system of linear inequalities that represents the possible combinations of investments. Graph the system.

3.6 Exercises   MyMathLab®

Exercise numbers in green have complete video solutions in MyMathLab or may be accessed using the QR code to the right.

Problems **1–10** *are the Quick ✓ s that follow the EXAMPLES.*

Building Skills

In Problems 11–16, determine which of the points, if any, satisfies the system. See Objective 1.

11. $\begin{cases} x + y \le 4 \\ 3x + y \ge 10 \end{cases}$

    (a) $(3, 2)$

    (b) $(5, -1)$

12. $\begin{cases} x + y \ge 2 \\ -3x + y \le 10 \end{cases}$

    (a) $(-3, 6)$

    (b) $(4, 1)$

13. $\begin{cases} 3x + 2y < 12 \\ -2x + 3y > 12 \end{cases}$

    (a) $(3, 2)$

    (b) $(1, 4)$

14. $\begin{cases} 5x + 2y < 10 \\ 4x - 3y < 24 \end{cases}$

    (a) $(1, 3)$

    (b) $(1, 1)$

15. $\begin{cases} x + y \le 8 \\ 3x + y \le 12 \\ x \ge 0 \\ y \ge 0 \end{cases}$

    (a) $(4, 2)$

    (b) $(2, 5)$

16. $\begin{cases} x + y \ge 6 \\ 2x + y \ge 10 \\ x \ge 0 \\ y \ge 0 \end{cases}$

    (a) $(4, 2)$

    (b) $(2, 5)$

In Problems 17–28, graph each system of linear inequalities. See Objective 2.

17. $\begin{cases} x + y \le 4 \\ 3x + y \ge 10 \end{cases}$

18. $\begin{cases} x + y \ge 2 \\ -3x + y \le 10 \end{cases}$

19. $\begin{cases} 3x + 2y < 12 \\ -2x + 3y > 12 \end{cases}$

20. $\begin{cases} 5x + 2y < 10 \\ 4x - 3y < 24 \end{cases}$

21. $\begin{cases} x - \dfrac{1}{2}y \le 3 \\ \dfrac{3}{2}x + y > 3 \end{cases}$

22. $\begin{cases} -x + \dfrac{1}{3}y < 3 \\ \dfrac{4}{3}x + y \ge 4 \end{cases}$

23. $\begin{cases} -2x + y < -8 \\ 2x - y > 8 \end{cases}$

24. $\begin{cases} 3x - 2y < -6 \\ -6x + 4y > 12 \end{cases}$

25. $\begin{cases} -5x + 3y > 12 \\ 5x - 3y > 9 \end{cases}$

26. $\begin{cases} 4x + 3y < -9 \\ -8x - 6y < 12 \end{cases}$

27. $\begin{cases} y \le 8 \\ x \ge 3 \end{cases}$

28. $\begin{cases} y \le 4 \\ x \ge -1 \end{cases}$

In Problems 29–36, graph each system of linear inequalities. Tell whether the graph is bounded or unbounded, and label the corner points. See Objective 2.

29. $\begin{cases} x + y \le 6 \\ 3x + y \le 12 \\ x \ge 0 \\ y \ge 0 \end{cases}$

30. $\begin{cases} x + y \le 9 \\ 3x + 2y \le 24 \\ x \ge 0 \\ y \ge 0 \end{cases}$

31. $\begin{cases} x + y \ge 8 \\ 4x + 2y \ge 28 \\ x \ge 0 \\ y \ge 0 \end{cases}$

32. $\begin{cases} x + y \ge 8 \\ x + 3y \ge 12 \\ x \ge 0 \\ y \ge 0 \end{cases}$

33. $\begin{cases} 2x + 3y \le 30 \\ 3x + 2y \le 25 \\ 5x + 2y \le 35 \\ x \ge 0 \\ y \ge 0 \end{cases}$

34. $\begin{cases} 2x + 3y \le 36 \\ x + y \le 14 \\ 3x + y \le 30 \\ x \ge 0 \\ y \ge 0 \end{cases}$

35. $\begin{cases} 2x + y \ge 13 \\ x + 2y \ge 11 \\ x \ge 4 \\ y \ge 0 \end{cases}$

36. $\begin{cases} 7x + 3y \ge 45 \\ 5x + 3y \ge 39 \\ x \ge 0 \\ y \ge 3 \end{cases}$

Applying the Concepts

37. Breakfast Daria is on a special diet that requires she consume at least 500 mg of potassium and 14 grams of protein for breakfast. One morning, Daria decides to have orange juice, which contains 450 mg of potassium and 2 g of protein per serving, and cereal (with 2% milk), which contains 50 mg of potassium and 6 g of protein per serving.

    (a) Let x denote the number of servings of orange juice and y denote the number of servings of cereal. Write a system of linear inequalities that represents the possible combinations of foods.

    (b) Graph the system and label the corner points.

38. Manufacturing Printers Bob's Printer Emporium manufactures two printers: an ink jet printer and a laser printer. Each ink-jet printer requires 2 hours for molding and 3 hours for assembly; each laser printer requires 3 hours for molding and 4 hours for assembly. There are a total of 80 hours per week available in the work schedule for molding and 120 hours per week available for assembly.

   (a) Using x to denote the number of ink-jet printers and y to denote the number of laser printers, write a system of linear inequalities that describes the possible numbers of each model printer that can be manufactured in a week (it is possible to manufacture a portion of a printer).

   (b) Graph the system and label the corner points.

39. Financial Planning Jack and Mary recently retired, and they have $25,000 to invest. Their financial advisor has recommended that they place at least $5000 in Treasury notes yielding 3% and no more than $15,000 in corporate bonds yielding 6%. In addition, they want to earn at least $900 each year.

   (a) Let x denote the amount invested in Treasury notes and y denote the amount invested in corporate bonds. Write a system of linear inequalities that represents the possible combinations of investments.

   (b) Graph the system and label the corner points.

40. Mixing Nuts You've Got to Be Nuts is a store that specializes in selling nuts. The owner finds that she has excess inventory of 100 pounds (1600 ounces) of cashews and 120 pounds (1920 ounces) of peanuts. She decides to make two types of 1-pound nut mixes from the excess inventory. A premium mix will contain 12 ounces of cashews and 4 ounces of peanuts and the standard mix will contain 6 ounces of cashews and 6 ounces of peanuts.

   (a) Use x to denote the number of premium mixes and y to denote the number of standard mixes. Write a system of linear inequalities that describes the possible numbers of each kind of mix.

   (b) Graph the system and label the corner points.

Extending the Concepts

In Problems 41–44, write a system of linear inequalities that has the given graph.

41.

42.

43.

44.

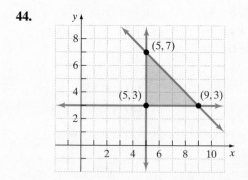

45. Math for the Future: Linear Programming A **linear programming problem** in two variables x and y consists of maximizing (or minimizing) a **linear objective function** $z = Ax + By$, where A and B are real numbers, not both 0, subject to certain constraints. The constraints form a system of linear inequalities. A typical linear programming problem is

$$\text{Maximize:} \quad z = 2x + 3y$$
$$\text{subject to} \quad x + y \le 8$$
$$2x + y \le 10$$
$$x \ge 0, y \ge 0$$

(a) The constraints form the system of linear

inequalities $\begin{cases} x + y \le 8 \\ 2x + y \le 10 \\ x \ge 0 \\ y \ge 0 \end{cases}$. Graph this system.

The shaded region is called the **feasible region.**

(b) The equation $z = 2x + 3y$ is the *objective function.* The goal is to make z as large as possible while (x, y) remains in the feasible region. Graph the objective function for $z = 0, z = 12, z = 18,$ and $z = 24$.

(c) What is the largest value of z that is possible while still remaining within the feasible region? What values of x and y correspond to this value?

Explaining the Concepts

46. Explain how to graph a system of linear inequalities in two variables.

47. If one inequality in a system of two linear inequalities contains a strict inequality and the other inequality contains a nonstrict inequality, is the corner point a solution to the system? Explain your answer.

48. Is it possible for a system of linear inequalities to have the entire Cartesian plane as the solution set? Explain your answer, and if your answer is yes, then provide an example.

49. Is it possible for a system of linear inequalities to have a straight line as the solution set? Explain your answer, and if your answer is yes, then provide an example.

50. What does it mean for a system of linear inequalities to be bounded? What does it mean for a system of linear inequalities to be unbounded?

Technology Exercises

Graphing calculators have the ability to graph systems of linear inequalities. Figure 19 shows the results of Example 3 using a graphing calculator.

Figure 19

In Problems 51–56, graph each system of linear inequalities using technology.

51. $\begin{cases} x + y \le 4 \\ 3x + y \ge 10 \end{cases}$

52. $\begin{cases} x + y \ge 2 \\ -3x + y \le 10 \end{cases}$

53. $\begin{cases} 3x + 2y < 12 \\ -2x + 3y > 12 \end{cases}$

54. $\begin{cases} -5x + 2y < 10 \\ 4x - 3y < 24 \end{cases}$

55. $\begin{cases} x + y \le 6 \\ 3x + y \le 12 \\ x \ge 0 \\ y \ge 0 \end{cases}$

56. $\begin{cases} x + y \le 9 \\ 3x + 2y \le 24 \\ x \ge 0 \\ y \ge 0 \end{cases}$

Chapter 3 Activity: Find the Numbers

Focus: Solving systems of equations

Time: 15 minutes

Group size: 2

Consider the following dialogue between two students.

Michele: Think of two numbers between 1 and 10, and don't tell me what they are.

Rafael: OK, I've thought of two numbers.

Michele: Now tell me the sum of the two numbers and the difference of the two numbers, and I'll tell you what your two numbers are.

Rafael: Their sum is 14 and their difference is 6.

Michele: Your numbers are 10 and 4.

Rafael: That's right! How did you do that?

Michele: I set up a system of equations using the sum and difference that you gave me, along with the variables x and y.

1. Each group member should set up and solve the system of equations described by Michele.

Discuss your results and be sure that you both arrive at the solutions 10 and 4.

2. Now each of you will think of two new numbers, and the other will try to find the numbers by solving a system of equations. But this time the system will be a bit trickier! Give each other the following information about your two numbers, and then figure out each other's numbers:

five more than twice the sum of the numbers
four less than three times the difference of the numbers

3. Would the systems in this activity work if negative numbers were used? Try it and see!

Chapter 3 Review

Section 3.1 Systems of Linear Equations in Two Variables

KEY CONCEPTS

- **Recognizing Solutions to Systems of Two Linear Equations with Two Unknowns**
  - If the lines in a system of two linear equations containing two unknowns intersect, then the point of intersection is the solution, and the system is consistent and its equations are independent.
  - If the lines in a system of two linear equations containing two unknowns are parallel, then the system has no solution, and the system is inconsistent.
  - If the lines in a system of two linear equations containing two unknowns lie on top of each other, then the system has infinitely many solutions. The solution set is the set of all points on the line, the system is consistent, and its equations are dependent.

KEY TERMS

System of linear equations
Solution
Consistent system and independent equations
Inconsistent
Consistent system and dependent equations

You Should Be Able To...	EXAMPLE	Review Exercises
1 Determine whether an ordered pair is a solution of a system of linear equations (p. 221)	Example 2	1–4
2 Solve a system of two linear equations by graphing (p. 223)	Example 3	5–10
3 Solve a system of two linear equations by substitution (p. 223)	Examples 4 and 5	11–14, 19–24
4 Solve a system of two linear equations by elimination (p. 225)	Examples 6 and 7	15–24
5 Identify inconsistent systems (p. 228)	Example 8	9, 23
6 Write the solution of a system with dependent equations (p. 229)	Example 9	12, 17

In Problems 1–4, determine whether the given values of the variables listed are solutions of the system of equations.

1. $\begin{cases} x + 3y = -2 \\ 2x - y = 10 \end{cases}$

    (a) $x = 3, y = -1$
    (b) $x = 4, y = -2$

2. $\begin{cases} -2x + 5y = 4 \\ 4x - 5y = -3 \end{cases}$

    (a) $x = \dfrac{1}{2}, y = 1$
    (b) $x = -1, y = \dfrac{1}{3}$

3. $\begin{cases} 6x - 5y = -12 \\ x - y = -3 \end{cases}$

    (a) $x = 3, y = 6$
    (b) $x = 1, y = 4$

4. $\begin{cases} 3x - y = 9 \\ 8x + 3y = 7 \end{cases}$

    (a) $x = 3, y = 0$
    (b) $x = 2, y = -3$

In Problems 5 and 6, use the graph of the system in the next column to determine the solution.

5. $\begin{cases} x + 4y = 12 \\ 2x - y = 6 \end{cases}$

6. $\begin{cases} 3x + 2y = 4 \\ x - 2y = 4 \end{cases}$

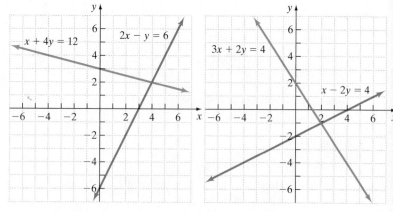

In Problems 7–10, solve each system of linear equations by graphing.

7. $\begin{cases} y = -3x + 1 \\ y = \dfrac{1}{2}x - 6 \end{cases}$

8. $\begin{cases} -2x + 3y = -9 \\ 3x + y = 8 \end{cases}$

9. $\begin{cases} y = -\dfrac{1}{3}x + 2 \\ x + 3y = -9 \end{cases}$

10. $\begin{cases} 2x - 3y = 0 \\ 2x - y = -4 \end{cases}$

In Problems 11–14, solve each system of linear equations using substitution.

11. $\begin{cases} y = -\dfrac{1}{4}x + 2 \\ y = 4x - 32 \end{cases}$ **12.** $\begin{cases} y = -\dfrac{3}{4}x + 2 \\ 3x + 4y = 8 \end{cases}$

13. $\begin{cases} y = 3x - 9 \\ 4x + 3y = -1 \end{cases}$ **14.** $\begin{cases} x - 2y = 7 \\ 3x - y = -4 \end{cases}$

In Problems 15–18, solve each system of linear equations using elimination.

15. $\begin{cases} 2x - y = 9 \\ 3x + y = 11 \end{cases}$ **16.** $\begin{cases} -x + 3y = 4 \\ 3x - 4y = -2 \end{cases}$

17. $\begin{cases} 2x - 4y = 8 \\ -3x + 6y = -12 \end{cases}$ **18.** $\begin{cases} 3x - 4y = -11 \\ 2x - 3y = -7 \end{cases}$

In Problems 19–24, solve each system of linear equations using either substitution or elimination.

19. $\begin{cases} x + y = -4 \\ 2x - 3y = 12 \end{cases}$ **20.** $\begin{cases} 5x - 3y = 2 \\ x + 2y = -10 \end{cases}$

21. $\begin{cases} 3x - 2y = 5 \\ 4x - 5y = 9 \end{cases}$ **22.** $\begin{cases} 12x + 20y = 21 \\ 3x - 2y = 0 \end{cases}$

23. $\begin{cases} 6x + 9y = -3 \\ 8x + 12y = 7 \end{cases}$ **24.** $\begin{cases} 6x + 11y = 2 \\ 5x + 8y = -3 \end{cases}$

Section 3.2 Problem Solving: Systems of Two Linear Equations Containing Two Unknowns

KEY TERM

Break-even point

You Should Be Able To...	EXAMPLE	Review Exercises
1 Model and solve direct translation problems (p. 233)	Examples 1 and 2	25–27
2 Model and solve geometry problems (p. 235)	Examples 3 and 4	28–30
3 Model and solve mixture problems (p. 237)	Examples 5 and 6	31–33
4 Model and solve uniform motion problems (p. 240)	Example 7	34–36
5 Find the intersection of two linear functions (p. 241)	Example 8	37–38

25. Numbers The sum of two numbers is 56. The difference of the two numbers is 14. Find the two numbers.

26. Phi Theta Kappa The local chapter of the Phi Theta Kappa International Honor Society for Two-Year Colleges will induct 73 new members this semester. There are 11 more female inductees than male inductees. How many males and how many females will be inducted?

27. Counting Calories Shawn and Randy ate lunch at a pizza buffet. Shawn ate 3 slices of pepperoni pizza and 5 slices of Italian sausage pizza for a total of 2600 calories. Randy ate 4 slices of pepperoni pizza and 2 slices of Italian sausage pizza for a total of 1880 calories. Assuming the pizza slices are exactly the same size, how many calories are in a slice of pepperoni pizza? How many calories are in a slice of Italian sausage pizza?

△ **28. Rectangle** The length of a rectangle is 5 inches less than twice the width. The perimeter of the rectangle is 68 inches. Find the dimensions of the rectangle.

△ **29. Angles** In the right triangle shown in the figure the measure of angle y is twice the measure of angle x. Find the measures of angles x and y.

30. Angles Two intersecting lines are shown in the figure to the right. Suppose that the measure of angle 1 is $(x + 4y)°$, the measure of angle 2 is $(8x + 5y)°$, and the measure of angle 3 is $(2x + y)°$. Find x and y.

31. Coins Jerome's piggy bank contains a total of 40 nickels and quarters. The value of the coins in the bank is $4.40. How many nickels and how many quarters are in the bank?

32. Hydrochloric Acid Solution For an experiment, a chemist needs 12 liters of a solution that is 30% hydrochloric acid. However, she only has two solutions that are 25% hydrochloric acid and 40% hydrochloric acid. How many liters of each should she mix in order to obtain the needed solution?

33. Investments Verna invested part of a $10,000 inheritance in stocks paying 6.5% simple interest annually. She invested the rest in bonds paying 4.25% simple interest annually. If Verna will receive $582.50 in interest after one year, how much did she invest in stocks? How much did she invest in bonds?

34. Airplane Speed An airplane averaged 160 miles per hour with the wind and 112 miles per hour against the wind. Determine the speed of the plane and the effect of the wind resistance.

35. Catching Up A Pontiac Grand Am enters the Will Rogers Turnpike traveling at 50 miles per hour. One-half hour later, a Ford Mustang enters the turnpike at the same location and travels in the same direction at 80 miles per hour. How long will it take for the Mustang to catch up to the Grand Am? When it catches up, how far will the two cars have traveled on the turnpike?

36. Boat Speed A boat traveled 30 miles upstream in 1.5 hours. The return trip downstream took only 1 hour. Assuming that the boat travels at a constant rate and that the river's current is constant, find the speed of the boat in still water and the speed of the current.

37. High School Sports Participation The number of male students (in millions) who participate in high school sports can be estimated by the function $M(t) = 0.0795t + 3.406$, where t is the number of years since 1990. The number of female students (in millions) who participate in high school sports can be estimated by $F(t) = 0.0997t + 1.892$. (SOURCE: *National Federation of State High School Associations*)

(a) Graph each of the two functions in the same Cartesian plane.

(b) Assuming these linear trends continue, find the year in which participation by male and by female students will be the same.

38. Break-Even Karen bakes pies for a living. She sells pies for $15 each. Her monthly fixed costs are $1200. Each pie costs $2.50 in ingredients.

(a) Find the revenue function R, treating the number of pies x as the independent variable.

(b) Find the cost function C, treating the number of pies x as the independent variable.

(c) Graph the revenue and cost functions in the same Cartesian plane.

(d) Find the break-even number of pies that Karen must make and sell each month. What are the revenue and cost at this number of pies? Label the point on the graph in part (c).

Section 3.3   Systems of Linear Equations in Three Variables

KEY TERMS

Ordered triple	Exactly one solution	No solution	Infinitely many solutions

You Should Be Able To...	EXAMPLE	Review Exercises
1 Solve systems of three linear equations (p. 246)	Examples 1 through 3	39–46
2 Identify inconsistent systems (p. 251)	Example 4	42
3 Write the solution of a system with dependent equations (p. 251)	Example 5	43
4 Model and solve problems involving three linear equations (p. 252)	Example 6	47, 48

In Problems 39–46, solve each system of three linear equations containing three unknowns.

39. $\begin{cases} x + y - z = 1 \\ x - y + z = 7 \\ x + 2y + z = 1 \end{cases}$

40. $\begin{cases} 2x - 2y + z = -10 \\ 3x + y - 2z = 4 \\ 5x + 2y - 3z = 7 \end{cases}$

41. $\begin{cases} x + 2y = -1 \\ 3y + 4z = 7 \\ 2x - z = 6 \end{cases}$

42. $\begin{cases} x + 2y - 3z = -4 \\ x + y + 3z = 5 \\ 3x + 4y + 3z = 7 \end{cases}$

43. $\begin{cases} 3x + y - 2z = 6 \\ x + y - z = -2 \\ -x - 3y + 2z = 14 \end{cases}$

44. $\begin{cases} 9x - y + 2z = -5 \\ -3x - 4y + 4z = 3 \\ 15x + 3y - 2z = -10 \end{cases}$

45. $\begin{cases} 4x - 5y + 2z = -8 \\ 3x + 7y - 3z = 21 \\ 7x - 4y + 2z = -5 \end{cases}$

46. $\begin{cases} 3x - 2y + 5z = -7 \\ 4x + y + 3z = -2 \\ 2x - 3y + 7z = -4 \end{cases}$

△ **47. Angles** In the triangle shown, the measure of angle z is twice the measure of angle y. The measure of angle y is 10° less than three times the measure of angle x. Find the measures of angles x, y, and z.

48. Counting Calories Mike, Clint, and Charlie decided to stop for a snack. Mike consumed 1410 calories by eating a cheeseburger, two medium order of fries, and a medium Coke. Clint consumed 1590 calories by eating two cheeseburgers, a medium order of fries, and two medium Cokes. Charlie consumed 1480 calories by eating three cheeseburgers, no fries, and two medium Cokes. Find the number of calories each in a cheeseburger, in a medium order of fries, and in a medium Coke.

Section 3.4   Using Matrices to Solve Systems

KEY CONCEPTS

- **Augmented Matrix**

$\begin{cases} a_1 x + b_1 y = c_1 \\ a_2 x + b_2 y = c_2 \end{cases}$ is written as an augmented matrix as $\begin{bmatrix} a_1 & b_1 & | & c_1 \\ a_2 & b_2 & | & c_2 \end{bmatrix}$

- **Row Operations**
  1. Interchange any two rows.
  2. Multiply all entries in a row by a nonzero constant.
  3. Replace a row by the sum of that row and a nonzero multiple of some other row.

KEY TERMS

Matrix
Augmented matrix
Row echelon form

You Should Be Able To...	EXAMPLE	Review Exercises
❶ Write the augmented matrix of a system (p. 258)	Example 1	49–52
❷ Write the system from the augmented matrix (p. 259)	Example 4	53–56
❸ Perform row operations on a matrix (p. 259)	Examples 3 and 4	57–60
❹ Solve systems using matrices (p. 261)	Examples 5 and 6	61–70
❺ Solve consistent systems with dependent equations and solve inconsistent systems (p. 264)	Examples 7 and 8	64, 65, 68, 69

In Problems 49–52, write the augmented matrix of the given system of linear equations.

49. $\begin{cases} 3x + y = 7 \\ 2x + 5y = 9 \end{cases}$

50. $\begin{cases} x - 5y = 14 \\ -x + y = -3 \end{cases}$

51. $\begin{cases} 5x - y + 4z = 6 \\ -3x - 3z = -1 \\ x - 2y = 0 \end{cases}$

52. $\begin{cases} 8x - y + 3z = 14 \\ -3x + 5y - 6z = -18 \\ 7x - 4y + 5z = 21 \end{cases}$

In Problems 53–56, write the system of linear equations corresponding to the given augmented matrix.

53. $\begin{bmatrix} 1 & 2 & | & 12 \\ 0 & 3 & | & 15 \end{bmatrix}$

54. $\begin{bmatrix} 3 & -4 & | & -5 \\ -1 & 2 & | & 7 \end{bmatrix}$

55. $\begin{bmatrix} 1 & 3 & 4 & | & 20 \\ 0 & 1 & -2 & | & -16 \\ 0 & 0 & 1 & | & 7 \end{bmatrix}$

56. $\begin{bmatrix} -3 & 7 & 9 & | & 1 \\ 4 & 10 & 7 & | & 5 \\ 2 & -5 & -6 & | & -8 \end{bmatrix}$

In Problems 57–60, perform each row operation on the given augmented matrix.

57. $\begin{bmatrix} 1 & -5 & | & 22 \\ -2 & 9 & | & -40 \end{bmatrix}$

(a) $R_2 = 2r_1 + r_2$ followed by
(b) $R_2 = -r_2$

58. $\begin{bmatrix} 1 & -4 & | & 7 \\ 3 & -7 & | & 6 \end{bmatrix}$

(a) $R_2 = -3r_1 + r_2$ followed by

(b) $R_2 = \dfrac{1}{5} r_2$

59. $\begin{bmatrix} -1 & 2 & 1 & | & 1 \\ 2 & -1 & 3 & | & -3 \\ -1 & 5 & 6 & | & 2 \end{bmatrix}$

(a) $R_2 = 2r_1 + r_2$ followed by

(b) $R_3 = -r_1 + r_3$

60. $\begin{bmatrix} 1 & 3 & 4 & | & 4 \\ 0 & 5 & 10 & | & -15 \\ 0 & -4 & -7 & | & 7 \end{bmatrix}$

(a) $R_2 = \dfrac{1}{5} r_2$ followed by

(b) $R_3 = 4r_2 + r_3$

In Problems 61–70, solve each system of linear equations using matrices.

61. $\begin{cases} x + 2y = 1 \\ x + 3y = -2 \end{cases}$

62. $\begin{cases} 6x - 2y = -7 \\ 4x + 3y = 17 \end{cases}$

63. $\begin{cases} 3x + 2y = -10 \\ 2x - y = -9 \end{cases}$

64. $\begin{cases} -3x + 9y = 15 \\ 5x - 15y = 11 \end{cases}$

65. $\begin{cases} 4x - 2y = 6 \\ 6x - 3y = 9 \end{cases}$

66. $\begin{cases} x + 4y = 4 \\ 4x - 8y = 7 \end{cases}$

67. $\begin{cases} x + 2y - 2z = -11 \\ 2x \quad\quad + z = -6 \\ \quad\quad 5y - 3z = -7 \end{cases}$

68. $\begin{cases} 2x - 5y + 2z = 9 \\ -x + y - 2z = -2 \\ -x - 2y - 4z = 8 \end{cases}$

69. $\begin{cases} 2x - 7y + 11z = -5 \\ 4x - 2y + 6z = 2 \\ -2x + 19y - 27z = 17 \end{cases}$

70. $\begin{cases} 5x + 3y + 7z = 9 \\ 3x + 5y + 4z = 8 \\ x + 3y + 3z = 9 \end{cases}$

Section 3.5   Determinants and Cramer's Rule

KEY CONCEPTS

- **Evaluate a 2 × 2 Determinant**

$$\begin{vmatrix} a & b \\ c & d \end{vmatrix} = ad - bc$$

- **Evaluate a 3 × 3 Determinant**

$$\begin{vmatrix} a_{1,1} & a_{1,2} & a_{1,3} \\ a_{2,1} & a_{2,2} & a_{2,3} \\ a_{3,1} & a_{3,2} & a_{3,3} \end{vmatrix} = a_{1,1} \begin{vmatrix} a_{2,2} & a_{2,3} \\ a_{3,2} & a_{3,3} \end{vmatrix} - a_{1,2} \begin{vmatrix} a_{2,1} & a_{2,3} \\ a_{3,1} & a_{3,3} \end{vmatrix} + a_{1,3} \begin{vmatrix} a_{2,1} & a_{2,2} \\ a_{3,1} & a_{3,2} \end{vmatrix}$$

- **Cramer's Rule:**

The solution of the system of equations $\begin{cases} ax + by = s \\ cx + dy = t \end{cases}$ with $D = \begin{vmatrix} a & b \\ c & d \end{vmatrix} \neq 0$

is given by $x = \dfrac{\begin{vmatrix} s & b \\ t & d \end{vmatrix}}{\begin{vmatrix} a & b \\ c & d \end{vmatrix}} = \dfrac{D_x}{D}$ $y = \dfrac{\begin{vmatrix} a & s \\ c & t \end{vmatrix}}{\begin{vmatrix} a & b \\ c & d \end{vmatrix}} = \dfrac{D_y}{D}$

The solution of the system of equations $\begin{cases} a_1 x + b_1 y + c_1 z = d_1 \\ a_2 x + b_2 y + c_2 z = d_2 \\ a_3 x + b_3 y + c_3 z = d_3 \end{cases}$

with $D = \begin{vmatrix} a_1 & b_1 & c_1 \\ a_2 & b_2 & c_2 \\ a_3 & b_3 & c_3 \end{vmatrix} \neq 0$

KEY TERMS

Square matrix
2 × 2 determinant
3 × 3 determinant
Minor

$$D_x = \begin{vmatrix} d_1 & b_1 & c_1 \\ d_2 & b_2 & c_2 \\ d_3 & b_3 & c_3 \end{vmatrix} \qquad D_y = \begin{vmatrix} a_1 & d_1 & c_1 \\ a_2 & d_2 & c_2 \\ a_3 & d_3 & c_3 \end{vmatrix} \qquad D_z = \begin{vmatrix} a_1 & b_1 & d_1 \\ a_2 & b_2 & d_2 \\ a_3 & b_3 & d_3 \end{vmatrix}$$

is given by

$$x = \frac{D_x}{D} \qquad y = \frac{D_y}{D} \qquad z = \frac{D_z}{D}$$

You Should Be Able To...	EXAMPLE	Review Exercises
❶ Evaluate the determinant of a 2 × 2 matrix (p. 270)	Example 1	71–74
❷ Use Cramer's Rule to solve a system of two equations (p. 271)	Example 2	79–84
❸ Evaluate the determinant of a 3 × 3 matrix (p. 272)	Examples 3 and 4	75–78
❹ Use Cramer's Rule to solve a system of three equations (p. 274)	Example 5	85–90

In Problems 71–78, find the value of each determinant.

71. $\begin{vmatrix} 3 & 4 \\ -1 & 2 \end{vmatrix}$

72. $\begin{vmatrix} 2 & -3 \\ -6 & 9 \end{vmatrix}$

73. $\begin{vmatrix} -5 & 7 \\ -4 & 6 \end{vmatrix}$

74. $\begin{vmatrix} -7 & 2 \\ 6 & -3 \end{vmatrix}$

75. $\begin{vmatrix} 5 & 0 & 1 \\ 2 & -3 & -1 \\ 3 & 6 & -4 \end{vmatrix}$

76. $\begin{vmatrix} 1 & -4 & 5 \\ 0 & 1 & -3 \\ 2 & -6 & 4 \end{vmatrix}$

77. $\begin{vmatrix} 3 & 0 & -1 \\ 2 & 6 & 7 \\ 2 & 5 & 4 \end{vmatrix}$

78. $\begin{vmatrix} 2 & 3 & 1 \\ 1 & -3 & -7 \\ -5 & 4 & 8 \end{vmatrix}$

In Problems 79–90, solve each system of linear equations using Cramer's Rule, if applicable.

79. $\begin{cases} x + 2y = -1 \\ 2x + 3y = 1 \end{cases}$

80. $\begin{cases} 4x + 9y = -13 \\ -5x + 6y = 22 \end{cases}$

81. $\begin{cases} 4x - y = -6 \\ 3x + 5y = 7 \end{cases}$

82. $\begin{cases} x - y = 2 \\ 2x + y = 5 \end{cases}$

83. $\begin{cases} 6x + 2y = 5 \\ 15x + 5y = 8 \end{cases}$

84. $\begin{cases} 12x + y = 6 \\ -6x + 7y = 15 \end{cases}$

85. $\begin{cases} x - y + 2z = 9 \\ 3x + 2y - 4z = 7 \\ 3y + 5z = -1 \end{cases}$

86. $\begin{cases} x - 3y - 3z = -5 \\ 7x + y - 2z = 24 \\ 6x - 5y - 4z = -9 \end{cases}$

87. $\begin{cases} x + y + z = 1 \\ 3x + 2y + 4z = -1 \\ 2x + 2y + 3z = 0 \end{cases}$

88. $\begin{cases} 4x - 3y + z = -6 \\ 4x - 2y + 3z = -3 \\ 8x - 5y - 2z = -12 \end{cases}$

89. $\begin{cases} x - y - 4z = 7 \\ 4x - 3y - 3z = 4 \\ 3x - 2y + z = -3 \end{cases}$

90. $\begin{cases} x + y = -3 \\ 2y - z = -1 \\ 5x + z = 1 \end{cases}$

Section 3.6 Systems of Linear Inequalities

KEY TERMS

Satisfies   Bounded

Corner points   Unbounded

You Should Be Able To...	EXAMPLE	Review Exercises
❶ Determine whether an ordered pair is a solution of a system of linear inequalities (p. 278)	Example 1	91–94
❷ Graph a system of linear inequalities (p. 279)	Examples 2 through 5	95–102
❸ Solve problems involving systems of linear inequalities (p. 281)	Example 6	103, 104

In Problems 91–94, determine which of the ordered pairs, if any, satisfies the system of linear inequalities.

91. $\begin{cases} x - y > 2 \\ 2x + 3y > 8 \end{cases}$

    (a) $(5, 2)$
    (b) $(-3, 5)$

92. $\begin{cases} 7x + 3y \le 21 \\ -2x + y \ge 5 \end{cases}$

    (a) $(-1, 2)$
    (b) $(-3, 4)$

93. $\begin{cases} x + 2y \le 6 \\ 3x - y \ge 2 \\ x \ge 0 \\ y \ge 0 \end{cases}$

    (a) $(2, 1)$
    (b) $(1, 2)$

94. $\begin{cases} 3x - 2y \le 12 \\ 2x + y \le 15 \\ x \ge 0 \\ y \ge 0 \end{cases}$

    (a) $(-2, 3)$
    (b) $(4, 1)$

In Problems 95–100, graph each system of linear inequalities.

95. $\begin{cases} x + y \ge 7 \\ 2x - y \ge 5 \end{cases}$

96. $\begin{cases} 2x + 3y > 9 \\ x - 3y > -18 \end{cases}$

97. $\begin{cases} x - 4y > -4 \\ x + 2y \le 8 \end{cases}$

98. $\begin{cases} x - y < -2 \\ 2x - 2y > 6 \end{cases}$

99. $\begin{cases} 2x - y \ge -2 \\ 2x - 3y \le 6 \end{cases}$

100. $\begin{cases} x \ge 2 \\ y < -3 \end{cases}$

In Problems 101 and 102, graph each system of linear inequalities. Tell whether the graph is bounded or unbounded, and label the corner points.

101. $\begin{cases} 3x + 5y \le 30 \\ 4x - 5y \ge 5 \\ x \le 8 \\ y \ge 0 \end{cases}$

102. $\begin{cases} 3x + 2y \ge 10 \\ x + 2y \ge 6 \\ x \ge 0 \\ y \ge 0 \end{cases}$

103. Investment Options Anna has a maximum of $4000 to invest. One investment option pays 6% simple interest annually, while a second investment option pays 8% simple interest annually. Anna's financial advisor recommends that she invest at least $500 at 6% and no more than $2500 at 8%. Anna wants to earn at least $275 annually in interest.

    (a) Let x denote the amount invested at 6% and y denote the amount invested at 8%. Write a system of linear inequalities that represents Anna's possible combinations of investments.
    (b) Graph the system and label the corner points.

104. Purchasing Tulip Bulbs Jordan has a maximum of $144 to purchase tulip bulbs for her yard. Red tulip bulbs cost $6 per dozen, and yellow tulip bulbs cost $4 per dozen. Jordan wants at least 4 dozen more red tulip bulbs than yellow tulip bulbs.

    (a) Let x denote the number of dozens of red tulip bulbs and y denote the number of dozens of yellow tulip bulbs. Write a system of linear inequalities that represents Jordan's possible combinations of purchases.
    (b) Graph the system and label the corner points.

Chapter 3 Test

Step-by-step test solutions are found on the Chapter Test Prep Videos available in MyMathLab®, or on YouTube®, or may be accessed using the QR code to the right.

1. Solve the system of linear equations by graphing.

$$\begin{cases} 2x - y = 0 \\ 4x - 5y = 12 \end{cases}$$

In Problems 2–5, solve the system of linear equations using either substitution or elimination.

2. $\begin{cases} 5x + 2y = -3 \\ y = 2x - 6 \end{cases}$

3. $\begin{cases} 9x + 3y = 1 \\ x - 2y = 4 \end{cases}$

4. $\begin{cases} 6x - 9y = 5 \\ 8x - 12y = 7 \end{cases}$

5. $\begin{cases} 2x + y = -4 \\ \dfrac{1}{3}x + \dfrac{1}{2}y = 2 \end{cases}$

In Problems 6 and 7, solve each system of three linear equations containing three unknowns.

6. $\begin{cases} x - 2y + 3z = 1 \\ x + y - 3z = 7 \\ 3x - 4y + 5z = 7 \end{cases}$

7. $\begin{cases} 2x + 4y + 3z = 5 \\ 3x - y + 2z = 8 \\ x + y + 2z = 0 \end{cases}$

In Problems 8 and 9, perform each row operation on the given augmented matrix.

8. $\begin{bmatrix} 1 & -3 & | & -2 \\ 2 & -4 & | & 8 \end{bmatrix}$

    (a) $R_2 = -2r_1 + r_2$
    followed by
    (b) $R_2 = \dfrac{1}{2}r_2$

9. $\begin{bmatrix} 1 & -2 & 1 & | & -2 \\ 3 & -5 & 2 & | & 1 \\ 0 & -4 & 5 & | & -32 \end{bmatrix}$

    (a) $R_2 = -3r_1 + r_2$
    followed by
    (b) $R_3 = 4r_2 + r_3$

In Problems 10 and 11, write the augmented matrix for each system of linear equations. Then use the matrix to solve the system.

10. $\begin{cases} x - 5y = 2 \\ 2x + y = 4 \end{cases}$

11. $\begin{cases} x + 2y + z = 3 \\ 4y + 3z = 5 \\ 2x + 3y = 1 \end{cases}$

In Problems 12 and 13, find the value of each determinant.

12. $\begin{vmatrix} 3 & -5 \\ 4 & -8 \end{vmatrix}$

13. $\begin{vmatrix} 0 & 1 & 2 \\ 3 & 3 & -1 \\ -2 & 1 & 2 \end{vmatrix}$

In Problems 14 and 15, solve each system of linear equations using Cramer's Rule, if applicable.

14. $\begin{cases} x - y = -2 \\ 5x + 3y = -8 \end{cases}$

15. $\begin{cases} x + y + z = -2 \\ x + y - 2z = 1 \\ 4x + 2y + 3z = -15 \end{cases}$

In Problems 16 and 17, graph each system of linear inequalities. Label the corner points.

16. $\begin{cases} 2x - y > -2 \\ x - 3y < 9 \end{cases}$

17. $\begin{cases} 3x + 2y \le 12 \\ x - 2y \ge -4 \\ x \quad\quad \ge 0 \\ \quad\quad y \ge 0 \end{cases}$

18. Grain Storage A grain-storage warehouse has a total of 50 bins. Some of the bins hold 25 tons of grain each, and the rest hold 20 tons each. How many of each type of bin are there if the capacity of the warehouse is 1160 tons?

△ **19. Triangle** In the triangle shown below, the measure of angle z is 10° larger than the measure of angle y. Five times the measure of angle x equals the sum of the measures of angles y and z. Find the measures of angles x, y, and z.

20. Shopping It's Margaret's lucky day. While shopping at the outlet mall, she has happened onto a sale where designer blouses are being sold for $12 each and designer sweaters are being sold for $18 each. Unfortunately, Margaret has no more than $180 to spend. She also decides that she can buy at most 13 items.

(a) Let x denote the number of blouses and y denote the number of sweaters that Margaret will buy. Write a system of linear inequalities that represents the possible combinations of purchases.

(b) Graph the system and label the corner points.

Cumulative Review Chapters R–3

1. Consider the following set of numbers:

$\left\{ -13, -\dfrac{7}{8}, 0, \dfrac{\pi}{2}, 2.7, 4\sqrt{2}, 11 \right\}$. List the numbers that are:

(a) Natural numbers
(b) Integers
(c) Rational numbers
(d) Irrational numbers
(e) Real numbers

2. Translate the following statement into a mathematical statement: Twice the sum of a number and four equals three times the number decreased by six.

3. Evaluate: $\dfrac{3 - 7(5)}{4}$

4. Evaluate $\dfrac{x^2 - 5x + 4}{x - 1}$ for $x = 7$.

5. Simplify: $x(x - 4) - 7(x + 5) + 2(x^2 + 4)$

6. Solve: $3|x + 7| - 4 = 20$

7. Solve: $\dfrac{x + 1}{5} \ge \dfrac{5x + 29}{10}$

8. Solve: $|2x - 7| > 3$

9. For $A = \{0, 2, 4, 6, 8, 10, 12\}$ and $B = \{0, 3, 6, 9, 12, 15\}$, find each of the following.

(a) $A \cap B$
(b) $A \cup B$

10. Graph: $y = 2|x| - 5$

11. Find the zero of $f(x) = \dfrac{2}{3}x - 4$.

12. Determine which, if any, of the following graphs are graphs of functions.

(a)

(b)

(continued)

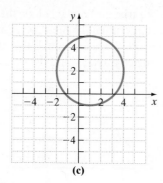

(c)

13. Determine the domain of $f(x) = \dfrac{2x + 7}{2x - 1}$.

14. For $g(x) = 3x^2 + 4x - 7$, find each of the following.

    (a) $g(-4)$

    (b) $g\left(\dfrac{2}{3}\right)$

    (c) $g(-x)$

15. Graph the linear equation $2x - 5y = 20$.

16. Determine whether the graphs of the following linear equations are parallel, perpendicular, or neither.
$$\begin{cases} 3x - 9y = 5 \\ 5x - 15y = -27 \end{cases}$$

17. Find the equation of the line passing through the points $(-6, 14)$ and $(9, -11)$. Write your answer in either slope-intercept or standard form, whichever you prefer.

18. The weight of a collection of quarters varies directly with the number of quarters in the collection. If 50 quarters weigh 281 grams, determine the weight of 35 quarters.

19. Using data from the U.S. Census Bureau for the years 1930 through 2010, the population (in thousands) of New Orleans, Louisiana, x years after 1930 is given by $P(x) = -0.123x^2 + 8.543x + 447.873$.

    (a) Identify the dependent and independent variables.

    (b) What is the domain of this function?

    (c) Find $P(60)$ and explain what this value represents.

    (d) Use the function to find the population of New Orleans in 2010.

20. Cell Phone Verizon offers an Anytime Talk and Text plan that charges $79.99 per month plus $0.40 per minute after 900 minutes.

    (a) Write the monthly cost C as a function of minutes x.

    (b) What is the implied domain of this linear function?

    (c) What is the cost when a person uses 1000 minutes during one month?

    (d) During one month, how many minutes can be enjoyed for $349.99?

In Problems 21 and 22, solve each system of equations.

21. $\begin{cases} y = \dfrac{1}{2}x + 9 \\ 5x + 4y = -20 \end{cases}$
    22. $\begin{cases} 6x - 4y + 3z = 3 \\ 3x + 10y + z = -4 \\ 9x + 2y + 4z = 1 \end{cases}$

In Problems 23 and 24, find the value of each determinant.

23. $\begin{vmatrix} -4 & 2 \\ -5 & 3 \end{vmatrix}$
    24. $\begin{vmatrix} 2 & 3 & 0 \\ -1 & 4 & 2 \\ 2 & -2 & -3 \end{vmatrix}$

25. Graph the following system of linear inequalities.
$$\begin{cases} x + 3y < 12 \\ x \geq 3 \end{cases}$$

Getting Ready for Chapter 4
Laws of Exponents and Scientific Notation

Objectives

1. Simplify Exponential Expressions Using the Product Rule
2. Simplify Exponential Expressions Using the Quotient Rule
3. Evaluate Exponential Expressions with a Zero or Negative Exponent
4. Simplify Exponential Expressions Using the Power Rule
5. Simplify Exponential Expressions Containing Products or Quotients
6. Simplify Exponential Expressions Using the Laws of Exponents
7. Convert Between Scientific Notation and Decimal Notation
8. Use Scientific Notation to Multiply and Divide

Work Smart

Read a^n as "a raised to the power of n" or "a raised to the nth power." We usually read a^2 as "a squared" and a^3 as "a cubed."

In Other Words

When multiplying two exponential expressions with the same base, add the exponents. Then write the common base to the power of this sum.

A quick review of positive integer exponents is in order. Recall that if a is a real number and n is a positive integer, then the symbol a^n means that a should be used as a factor n times. That is,

$$a^n = \underbrace{a \cdot a \cdot \cdots \cdot a}_{n \text{ factors}}$$

For example,

$$4^3 = \underbrace{4 \cdot 4 \cdot 4}_{3 \text{ factors}}$$

In the notation a^n, a is called the **base** and n the **power** or **exponent.**

1 Simplify Exponential Expressions Using the Product Rule

Several general rules can be discovered for simplifying expressions involving *positive* integer exponents. The first rule that we introduce is used when multiplying two exponential expressions that have the same base. Consider the following:

$$x^2 \cdot x^4 = \underbrace{(x \cdot x)}_{2 \text{ factors}} \underbrace{(x \cdot x \cdot x \cdot x)}_{4 \text{ factors}} = \underbrace{x \cdot x \cdot x \cdot x \cdot x \cdot x}_{6 \text{ factors}} = x^6$$

Same base   Same base   Sum of powers 2 and 4

Based on the above, the following results:

Product Rule for Exponents (Positive Integer Exponents)

If a is a real number and m and n are positive integers, then

$$a^m a^n = a^{m+n}$$

EXAMPLE 1   **Using the Product Rule to Simplify Expressions Involving Exponents**

Simplify each expression. All answers should contain only positive integer exponents.

(a) $2^2 \cdot 2^3$

(b) $3z^2 \cdot 4z^4$

Solution

(a) $2^2 \cdot 2^3 = 2^{2+3}$
$= 2^5$
$= 32$

(b) $3z^2 \cdot 4z^4 = 3 \cdot 4 \cdot z^2 \cdot z^4$
$= 12z^{2+4}$
$= 12z^6$

Quick ✓

1. In the notation a^n we call a the ____ and n the _____ or _____.

2. If a is a real number, and m and n are positive integers, then $a^m a^n = $ ____.

In Problems 3–7, simplify each expression. All answers should contain only positive integer exponents.

3. $5^2 \cdot 5$

4. $(-3)^2(-3)^3$

5. $y^4 y^3$

6. $(5x^2)(-2x^5)$

7. $6y^3(-y^2)$

❷ Simplify Exponential Expressions Using the Quotient Rule

To find a general rule for the quotient of two exponential expressions with *positive* integer exponents, use the Reduction Property to divide out common factors. Consider the following:

$$\frac{y^6}{y^2} = \frac{\overbrace{y \cdot y \cdot y \cdot y \cdot y \cdot y}^{6 \text{ factors}}}{\underbrace{y \cdot y}_{2 \text{ factors}}} = \underbrace{y \cdot y \cdot y \cdot y}_{4 \text{ factors}} = y^4$$

Difference of powers 6 and 2

Conclude from this result that

$$\frac{y^6}{y^2} = y^{6-2} = y^4$$

This result is true in general.

In Other Words

When dividing two exponential expressions with a common base, subtract the exponent in the denominator from the exponent in the numerator. Then write the common base to the power of this difference.

Quotient Rule for Exponents (Positive Integer Exponents)

If a is a real number and m and n are positive integers, then

$$\frac{a^m}{a^n} = a^{m-n} \qquad if\ a \neq 0$$

EXAMPLE 2 **Using the Quotient Rule to Simplify Expressions Involving Exponents**

Simplify each expression. Answers should contain only positive integer exponents.

(a) $\dfrac{8^5}{8^3}$ (b) $\dfrac{27z^9}{12z^4}$

Solution

(a) $\dfrac{8^5}{8^3} = 8^{5-3}$ (b) $\dfrac{27z^9}{12z^4} = \dfrac{9 \cdot 3}{4 \cdot 3} z^{9-4}$

$\qquad = 8^2$ $\qquad = \dfrac{9}{4} z^5$

$\qquad = 64$ ●

Quick ✓

8. *True or False* To divide two exponential expressions having the same base, keep the base and subtract the exponents.

In Problems 9–12, simplify each expression. All answers should contain only positive integer exponents.

9. $\dfrac{5^6}{5^4}$ 10. $\dfrac{y^8}{y^6}$ 11. $\dfrac{16a^6}{10a^5}$ 12. $\dfrac{-24b^5}{16b^3}$

❸ Evaluate Exponential Expressions with a Zero or Negative Exponent

Now extend the definition of exponential expressions to *all* integer exponents. That is, evaluate exponential expressions where the exponent can be a positive integer, zero, or a negative integer. Begin with raising a real number to the 0 power.

Zero-Exponent Rule

If a is a nonzero real number, then

$$a^0 = 1 \quad \text{if } a \neq 0$$

This rule is based on the Product Rule and the Identity Property of Multiplication. From the Product Rule for Exponents, it is known that

$$a^0 a^n = a^{0+n}$$
$$= a^n$$
$$= 1 \cdot a^n$$

From the Identity Property of Multiplication, this means that $a^0 = 1$.

Suppose we want to simplify $\dfrac{z^3}{z^5}$. If the Quotient Rule for Exponents is used,

$$\frac{z^3}{z^5} = z^{3-5} = z^{-2}$$

This expression could also be simplified using the Reduction Property.

$$\frac{z^3}{z^5} = \frac{\cancel{z} \cdot \cancel{z} \cdot \cancel{z}}{\cancel{z} \cdot \cancel{z} \cdot \cancel{z} \cdot z \cdot z} = \frac{1}{z^2}$$

This implies that $z^{-2} = \dfrac{1}{z^2}$, which suggests that a raised to a negative power is defined as follows:

Negative-Exponent Rule

If n is a positive integer and if a is a nonzero real number, then

$$a^{-n} = \frac{1}{a^n} \quad \text{or} \quad \frac{1}{a^{-n}} = a^n \quad \text{if } a \neq 0$$

EXAMPLE 3 **Evaluating Exponential Expressions Containing Integer Exponents**

Simplify each expression. All exponents should be positive integers.

(a) 3^{-4} (b) $\dfrac{1}{3^{-2}}$ (c) $5x^0$ (d) $4x^{-5}$

Solution

(a) $3^{-4} = \dfrac{1}{3^4}$ (b) $\dfrac{1}{3^{-2}} = 3^2$ (c) $5x^0 = 5 \cdot 1$ (d) $4x^{-5} = \dfrac{4}{x^5}$

$\quad\quad = \dfrac{1}{81}$ $= 9$ $= 5$

Work Smart

Remember, $3^{-4} = \dfrac{1}{3^4}$ (not -3^4). The negative applies only to the exponent. Similarly, $4x^{-5} = \dfrac{4}{x^5}\left(\text{not } \dfrac{1}{4x^5}\right)$. The negative exponent applies only to the base "x" not to the coefficient.

Quick ✓

13. $a^0 = $ __, provided $a \neq$ __.   **14.** $a^{-n} = $ __, provided $a \neq$ __.

In Problems 15–20, simplify each expression. All exponents should be positive integers.

15. 5^{-3} **16.** $5z^{-7}$ **17.** $\dfrac{1}{x^{-4}}$ **18.** $\dfrac{5}{y^{-3}}$ **19.** -4^0 **20.** $(-10)^0$

⊙ **EXAMPLE 4** **Evaluating Exponential Expressions Containing Integer Exponents**

Simplify each expression. All exponents should be positive integers.

(a) $\left(\dfrac{2}{3}\right)^{-3}$

(b) $\left(\dfrac{1}{7}\right)^{-2}$

Solution

(a) $\left(\dfrac{2}{3}\right)^{-3} = \dfrac{1}{\left(\dfrac{2}{3}\right)^3}$

$= \dfrac{1}{\dfrac{2}{3}\cdot\dfrac{2}{3}\cdot\dfrac{2}{3}}$

$= \dfrac{1}{\dfrac{8}{27}}$

$= \dfrac{27}{8}$

(b) $\left(\dfrac{1}{7}\right)^{-2} = \dfrac{1}{\left(\dfrac{1}{7}\right)^2}$

$= \dfrac{1}{\dfrac{1}{7}\cdot\dfrac{1}{7}}$

$= \dfrac{1}{\dfrac{1}{49}}$

$= 49$

The following shortcut is based on the results of Example 4:

In Other Words

To evaluate $\left(\dfrac{a}{b}\right)^{-n}$, determine the reciprocal of the base and then raise it to the nth power.

If a and b are real numbers and n is an integer, then

$$\left(\frac{a}{b}\right)^{-n} = \left(\frac{b}{a}\right)^{n} \qquad \text{if } a \neq 0, b \neq 0$$

Quick ✓

In Problems 21–24, simplify each expression. All exponents should be positive integers.

21. $\left(\dfrac{4}{3}\right)^{-2}$

22. $\left(-\dfrac{1}{4}\right)^{-3}$

23. $\left(\dfrac{3}{x}\right)^{-2}$

24. $\dfrac{5}{2^{-2}}$

Now that we have definitions for 0 as an exponent and negative exponents, the Product Rule and Quotient Rule for Exponents can be restated, assuming that the exponent is any integer (positive, negative, or zero).

Product Rule for Exponents

If a is a real number and m and n are integers, then

$$a^m a^n = a^{m+n}$$

If m, n, or $m + n$ is 0 or negative, then a cannot be 0.

Quotient Rule for Exponents

If a is a real number and if m and n are integers, then

$$\frac{a^m}{a^n} = a^{m-n} \quad \text{if } a \neq 0$$

Notice that allowing the exponents to be any integer (not just any positive integer) requires restrictions on the value of the base.

EXAMPLE 5 **Using the Product Rule to Simplify Expressions Containing Exponents**

Simplify each expression. All exponents should be positive integers.

(a) $(-3)^2(-3)^{-4}$

(b) $\dfrac{3}{4}y^5 \cdot \dfrac{20}{9}y^{-2}$

Solution

(a) $(-3)^2(-3)^{-4} = (-3)^{2+(-4)}$
$$= (-3)^{-2}$$
$$= \dfrac{1}{(-3)^2}$$
$$= \dfrac{1}{9}$$

(b) $\dfrac{3}{4}y^5 \cdot \dfrac{20}{9}y^{-2} = \dfrac{3}{4} \cdot \dfrac{20}{9}y^{5+(-2)}$
$$= \dfrac{5}{3}y^3$$

EXAMPLE 6 **Using the Quotient Rule to Simplify Expressions Containing Exponents**

Simplify each expression. All exponents should be positive integers.

(a) $\dfrac{w^{-2}}{w^{-5}}$

(b) $\dfrac{20a^3b}{4ab^4}$

Solution

(a) $\dfrac{w^{-2}}{w^{-5}} = w^{-2-(-5)}$
$$= w^{-2+5}$$
$$= w^3$$

(b) $\dfrac{20a^3b}{4ab^4} = 5a^{3-1}b^{1-4}$
$$= 5a^2b^{-3}$$
$$a^{-n} = \dfrac{1}{a^n}; \quad = \dfrac{5a^2}{b^3}$$

Quick ✓

In Problems 25–30, simplify each expression. All exponents should be positive integers.

25. $6^3 \cdot 6^{-5}$

26. $\dfrac{10^{-3}}{10^{-5}}$

27. $(4x^2y^3)(5xy^{-4})$

28. $\left(\dfrac{3}{4}a^3b\right)\left(\dfrac{8}{9}a^{-2}b^3\right)$

29. $\dfrac{-24b^5}{16b^{-3}}$

30. $\dfrac{50s^2t}{15s^5t^{-4}}$

▶ **④ Simplify Exponential Expressions Using the Power Rule**

Another law of exponents applies when an exponential expression containing a power is itself raised to a power.

$$(3^2)^4 = 3^2 \cdot 3^2 \cdot 3^2 \cdot 3^2 = \underbrace{(3 \cdot 3)}_{} \underbrace{(3 \cdot 3)}_{} \underbrace{(3 \cdot 3)}_{} \underbrace{(3 \cdot 3)}_{} = 3^8$$

4 factors   2 factors 2 factors 2 factors 2 factors

$2 \cdot 4 = 8$ factors

The following results:

In Other Words

If an exponential expression contains a power raised to a power, keep the base and multiply the powers.

Power Rule for Exponential Expressions

If a is a real number and m and n are integers, then

$$(a^m)^n = a^{mn}$$

If m or n is 0 or negative, then a must not be 0.

EXAMPLE 7 **Using the Power Rule to Simplify Exponential Expressions**

Simplify each expression. All exponents should be positive integers.

(a) $(y^3)^5$ (b) $[(-3)^3]^2$ (c) $(6^3)^0$

Solution

(a) $(y^3)^5 = y^{3\cdot5}$
$= y^{15}$

(b) $[(-3)^3]^2 = (-3)^{3\cdot2}$
$= (-3)^6$
$= 729$

(c) $(6^3)^0 = 6^{3\cdot0}$
$= 6^0$
$= 1$

Quick ✓

In Problems 31–36, simplify each expression. All exponents should be positive integers.

31. $(2^2)^3$ **32.** $(5^8)^0$ **33.** $[(-4)^3]^2$

34. $(a^3)^5$ **35.** $(z^3)^{-6}$ **36.** $(s^{-3})^{-7}$

❺ Simplify Exponential Expressions Containing Products or Quotients

There are two additional laws of exponents. The first deals with raising a product to a power, and the second deals with raising a quotient to a power. Consider the following product to a power:

$$(xy)^3 = (xy)(xy)(xy)$$
$$= (x\cdot x\cdot x)(y\cdot y\cdot y)$$
$$= x^3y^3$$

The following results:

Product-to-a-Power Rule

If a and b are real numbers and n is an integer, then

$$(ab)^n = a^nb^n$$

If n is 0 or negative, neither a nor b can be 0.

Work Smart

Do not use this rule to try to simplify $(a + b)^2$ as $a^2 + b^2$ or $(a + b)^3$ as $a^3 + b^3$. For this rule to apply, the base must be the *product* of two numbers—not a sum.

EXAMPLE 8 **Using the Product-to-a-Power Rule to Simplify Exponential Expressions**

Simplify each expression. All exponents should be positive integers.

(a) $(3z)^4$ (b) $(-5y^{-2})^{-3}$ (c) $(-4a^2)^{-2}$

Solution

(a) $(3z)^4 = 3^4z^4$
$= 81z^4$

(b) $(-5y^{-2})^{-3} = (-5)^{-3}(y^{-2})^{-3}$
$= \dfrac{y^{-2(-3)}}{(-5)^3}$
$= \dfrac{y^6}{-125}$
$= -\dfrac{y^6}{125}$

(c) $(-4a^2)^{-2} = \dfrac{1}{(-4a^2)^2}$
$= \dfrac{1}{(-4)^2(a^2)^2}$
$= \dfrac{1}{16a^4}$

Quick ✓

In Problems 37–40, simplify each expression. All exponents should be positive integers.

37. $(5y)^3$ **38.** $(6y)^0$ **39.** $(3x^2)^4$ **40.** $(4a^3)^{-2}$

Now look at a quotient raised to a power:

$$\left(\frac{2}{3}\right)^4 = \left(\frac{2}{3}\right)\left(\frac{2}{3}\right)\left(\frac{2}{3}\right)\left(\frac{2}{3}\right) = \frac{2^4}{3^4}$$

The following results:

▶ **Quotient-to-a-Power Rule**

If a and b are real numbers and n is an integer, then

$$\left(\frac{a}{b}\right)^n = \frac{a^n}{b^n} \qquad \text{if } b \neq 0$$

If n is negative or 0, then a cannot be 0.

EXAMPLE 9   **Using the Quotient-to-a-Power Rule to Simplify Exponential Expressions**

Simplify each expression. All exponents should be positive integers.

(a) $\left(\dfrac{w}{4}\right)^3$

(b) $\left(\dfrac{2x^2}{y^3}\right)^4$

Solution

(a) $\left(\dfrac{w}{4}\right)^3 = \dfrac{w^3}{4^3}$

$\qquad = \dfrac{w^3}{64}$

(b) $\left(\dfrac{2x^2}{y^3}\right)^4 = \dfrac{(2x^2)^4}{(y^3)^4}$

$\qquad = \dfrac{2^4(x^2)^4}{(y^3)^4}$

$\qquad = \dfrac{16x^{2\cdot4}}{y^{3\cdot4}}$

$\qquad = \dfrac{16x^8}{y^{12}}$

Quick ✓

In Problems 41–44, simplify each expression. All exponents should be positive integers.

41. $\left(\dfrac{z}{3}\right)^4$

42. $\left(\dfrac{x}{2}\right)^{-5}$

43. $\left(\dfrac{x^2}{y^3}\right)^4$

44. $\left(\dfrac{3a^{-2}}{b^4}\right)^3$

❻ **Simplify Exponential Expressions Using the Laws of Exponents**

The Laws of Exponents are now summarized.

Work Smart: Study Skills

When learning new rules, it is helpful to write each rule on a notecard that you can study with regularly. For example, you could put a^0, $a \neq 0$ on one side of a notecard and its value of 1 on the other side. This will allow you to quiz yourself. There is also a free app called Quizlet that may be used for electronic flash cards.

The Laws of Exponents

If a and b are real numbers and if m and n are integers, then assuming the expression is defined,

Zero-Exponent Rule:	$a^0 = 1$	if $a \neq 0$
Negative-Exponent Rule:	$a^{-n} = \dfrac{1}{a^n}$	if $a \neq 0$
Product Rule:	$a^m a^n = a^{m+n}$	
Quotient Rule:	$\dfrac{a^m}{a^n} = a^{m-n}$	if $a \neq 0$

(continued)

The Laws of Exponents (*continued*)

Power Rule:	$(a^m)^n = a^{mn}$
Product-to-a-Power Rule:	$(ab)^n = a^n b^n$
Quotient-to-a-Power Rule:	$\left(\dfrac{a}{b}\right)^n = \dfrac{a^n}{b^n}$ if $b \neq 0$
Quotient-to-a-Negative-Power Rule:	$\left(\dfrac{a}{b}\right)^{-n} = \left(\dfrac{b}{a}\right)^n$ if $a \neq 0, b \neq 0$

Now do some examples where one or more of the preceding rules are used.

EXAMPLE 10   **Using the Laws of Exponents**

Simplify each expression. All exponents should be positive integers. None of the variables is zero.

▶ **(a)** $\dfrac{a^3 b^{-1}}{(a^2 b)^3}$ ▶ **(b)** $\left(\dfrac{3xy}{x^2 y^{-2}}\right)^2 \left(\dfrac{9x^2 y^{-3}}{x^3 y^2}\right)^{-1}$

Solution

(a)

$$(ab)^n = a^n b^n \quad \downarrow$$

$$\frac{a^3 b^{-1}}{(a^2 b)^3} = \frac{a^3 b^{-1}}{(a^2)^3 b^3}$$

$(a^m)^n = a^{mn}:\quad = \dfrac{a^3 b^{-1}}{a^6 b^3}$

$\dfrac{a^m}{a^n} = a^{m-n}:\quad = a^{3-6} b^{-1-3}$

$\qquad\qquad = a^{-3} b^{-4}$

$a^{-n} = \dfrac{1}{a^n}:\quad = \dfrac{1}{a^3 b^4}$

Work Smart: Study Skills

Many different approaches may be taken to simplify exponential expressions. In Example 10(b), the Quotient-to-a-Power Rule could be used first and then simplified instead of the method shown. Try working a problem one way and then working it again a second way to see if you obtain the same answer and to see which method you prefer.

(b)

$$\frac{a^m}{a^n} = a^{m-n} \quad \downarrow$$

$\left(\dfrac{3xy}{x^2 y^{-2}}\right)^2 \left(\dfrac{9x^2 y^{-3}}{x^3 y^2}\right)^{-1} = (3x^{1-2} y^{1-(-2)})^2 (9x^{2-3} y^{-3-2})^{-1}$

$\qquad\qquad\qquad\qquad = (3x^{-1} y^3)^2 (9x^{-1} y^{-5})^{-1}$

$(a \cdot b)^n = a^n \cdot b^n:\quad = 3^2 (x^{-1})^2 (y^3)^2 (9^{-1})(x^{-1})^{-1}(y^{-5})^{-1}$

$(a^m)^n = a^{mn}:\quad = 9x^{-2} y^6 \left(\dfrac{1}{9}\right) x y^5$

$a^m a^n = a^{m+n}:\quad = x^{-2+1} y^{6+5}$

$a^{-n} = \dfrac{1}{a^n}:\quad = x^{-1} y^{11}$

$\qquad\qquad\qquad = \dfrac{y^{11}}{x}$

Quick ✓

In Problems 45–47, simplify each expression. All exponents should be positive integers.

45. $\dfrac{(3x^2 y)^2}{12xy^{-2}}$ **46.** $(3ab^3)^3 (6a^2 b^2)^{-2}$ **47.** $\left(\dfrac{2x^2 y^{-1}}{x^{-2} y^2}\right)^2 \left(\dfrac{4x^3 y^2}{xy^{-2}}\right)^{-1}$

⑦ Convert Between Scientific Notation and Decimal Notation

Figure 1

Measurements of physical quantities can range from very small to very large. For example, the mass of an electron (Figure 1) is approximately 0.00000000000000000000000000911 gram, and the mass of Earth (Figure 2) is about 5,980,000,000,000,000,000,000,000 kilograms. These numbers are difficult to write and difficult to read, so we use exponents to write them.

> **Definition**
>
> A number written as the product of a number a, where $1 \le |a| < 10$, and a power of 10 is written in **scientific notation.** That is, a number is written in scientific notation when it is of the form
>
> $$a \times 10^N$$
>
> where
>
> $$1 \le |a| < 10 \text{ and } N \text{ is an integer}$$

Figure 2

For example, in scientific notation,

$$\text{Mass of an electron} = 9.11 \times 10^{-28} \text{ gram}$$
$$\text{Mass of Earth} = 5.98 \times 10^{24} \text{ kilograms}$$

⊳ EXAMPLE 11   How to Convert from Decimal Notation to Scientific Notation

Write each number in scientific notation.

(a) 94,873   **(b)** 0.042

Step-by-Step Solution

For a number to be in scientific notation, the decimal must be moved so there is a single nonzero digit to the left of the decimal point. All remaining digits must appear to the right of the decimal point.

(a)

Step 1: The decimal point in 94,873 follows the 3. Therefore, move the decimal point to the left $N = 4$ places until it is between the 9 and the 4. Do you see why?   9 4 8 7 3.

Step 2: The original number is greater than 1, so write 94,873 in scientific notation as 9.4873×10^4

(b)

Step 1: Because 0.042 is less than 1, move the decimal point to the right $N = 2$ places until it is between the 4 and the 2.   0.042

Step 2: The original number is between 0 and 1, so write 0.042 in scientific notation as 4.2×10^{-2} ●

Work Smart

In Example 11(a), the decimal point was moved four places to the left because, when the number is written in scientific notation, we are multiplying 9.4873 by $10^4 = 10,000$. If simplified, this would move the decimal point four places to the right.

Converting a Decimal to Scientific Notation

To change a number from decimal to scientific notation,

Step 1: Count the number N of decimal places that the decimal point must be moved in order to arrive at a number a, where $1 \le |a| < 10$.

Step 2: If the absolute value of the original number is greater than or equal to 1, the scientific notation is $a \times 10^N$. If the absolute value of the original number is between 0 and 1, the scientific notation is $a \times 10^{-N}$.

▶ (EXAMPLE 12) **Converting a Negative Number from Decimal Notation to Scientific Notation**

Write $-520,000,000$ in scientific notation.

Solution

Because the absolute value of the number is greater than 1, move the decimal point to the left $N = 8$ places. Therefore,

$$-520,000,000 = -5.2 \times 10^8$$

●

Quick ✓

In Problems 48–51, write each number in scientific notation.

48. 532 **49.** $-1,230,000$ **50.** 0.034 **51.** -0.0000845

Numbers may also be converted from scientific notation to decimal notation.

Converting a Number from Scientific Notation to Decimal Notation

Determine the exponent on the number 10. If the exponent is negative, then move the decimal $|N|$ decimal places to the left. If the exponent is positive, then move the decimal N decimal places to the right.

(EXAMPLE 13) **Converting from Scientific Notation to Decimal Notation**

Write each number in decimal notation.

(a) 3.2×10^3 **(b)** 7.54×10^{-5}

Solution

(a) The exponent on 10 is 3 so move the decimal point three places to the right.

$$3.2 \times 10^3 = 3.200 \times 10^3 = 3200$$

(b) The exponent on 10 is -5, so move the decimal point five places to the left.

$$7.54 \times 10^{-5} = 000007.54 \times 10^{-5} = 0.0000754$$

●

Work Smart

The decimal point was moved three places to the right in Example 13(a) because $10^3 = 1000$. Thus we are really multiplying 3.2 by 1000. Move the decimal point five places to the left in Example 13(b) because

$$10^{-5} = \frac{1}{10^5} = \frac{1}{100,000} = 0.00001$$

Quick ✓

In Problems 52–55, write each number in decimal notation.

52. 5×10^2 **53.** 9.1×10^5 **54.** 1.8×10^{-4} **55.** 1×10^{-6}

▶ ⑧ Use Scientific Notation to Multiply and Divide

To multiply or divide two numbers written in scientific notation, use two Laws of Exponents:

$$a^m a^n = a^{m+n} \quad \text{and} \quad \frac{a^m}{a^n} = a^{m-n}$$

These laws are used where the base is 10.

EXAMPLE 14 **Multiplying Using Scientific Notation**

Perform the indicated operation. Express the answer in scientific notation.

(a) $(5 \times 10^2)(3 \times 10^8)$
(b) $(2.5 \times 10^{-3})(4.3 \times 10^{-4})$

Solution

(a)
$$(5 \times 10^2)(3 \times 10^8) = (5 \cdot 3) \times (10^2 \cdot 10^8)$$
$$5 \cdot 3 = 15; \text{ use } a^m a^n = a^{m+n}: = 15 \times 10^{10}$$
$$15 = 1.5 \times 10^1: = (1.5 \times 10^1) \times 10^{10}$$
$$= 1.5 \times 10^{11}$$

(b)
$$(2.5 \times 10^{-3})(4.3 \times 10^{-4}) = (2.5 \cdot 4.3) \times (10^{-3} \cdot 10^{-4})$$
$$2.5 \cdot 4.3 = 10.75; \text{ use } a^m a^n = a^{m+n}: = 10.75 \times 10^{-7}$$
$$10.75 = 1.075 \times 10^1: = (1.075 \times 10^1) \times 10^{-7}$$
$$= 1.075 \times 10^{-6}$$ ●

Quick ✓

In Problems 56–58, perform the indicated operation. Express the solution in scientific notation.

56. $(3 \times 10^3)(2 \times 10^5)$ **57.** $(2 \times 10^{-4})(4 \times 10^{-7})$

58. $(6 \times 10^{-5})(4 \times 10^8)$

EXAMPLE 15 **Dividing Using Scientific Notation**

Perform the indicated operation. Express the answer in scientific notation.

(a) $\dfrac{8 \times 10^5}{2 \times 10^3}$ (b) $\dfrac{2.5 \times 10^5}{5 \times 10^{-3}}$

Solution

(a)
$$\frac{8 \times 10^5}{2 \times 10^3} = \frac{8}{2} \times \frac{10^5}{10^3}$$
$$\frac{a^m}{a^n} = a^{m-n}: = 4 \times 10^2$$

(b)
$$\frac{2.5 \times 10^5}{5 \times 10^{-3}} = \frac{2.5}{5} \times \frac{10^5}{10^{-3}}$$
$$\frac{a^m}{a^n} = a^{m-n}: = 0.5 \times 10^{5-(-3)}$$
$$= 0.5 \times 10^8$$
$$0.5 = 5 \times 10^{-1}: = (5 \times 10^{-1}) \times 10^8$$
$$= 5 \times 10^7$$ ●

Quick ✓

In Problems 59–62, perform the indicated operation. Express the solution in scientific notation.

59. $\dfrac{6 \times 10^8}{3 \times 10^6}$ **60.** $\dfrac{6.8 \times 10^{-8}}{3.4 \times 10^{-5}}$ **61.** $\dfrac{4.8 \times 10^7}{9.6 \times 10^3}$ **62.** $\dfrac{3 \times 10^{-5}}{8 \times 10^7}$

EXAMPLE 16　**Using Scientific Notation to Multiply and Divide**

Perform the indicated operation. Express the answer in decimal notation.

(a) $(3,000,000)(90,000)$

(b) $\dfrac{0.00000075}{0.00015}$

Solution

In each of these problems, first write the numbers in scientific notation and then perform the indicated operation.

(a) $(3,000,000)(90,000) = (3 \times 10^6)(9 \times 10^4)$

$a^m a^n = a^{m+n}$:　　$= 27 \times 10^{10}$

$= (2.7 \times 10^1) \times 10^{10}$

$= 2.7 \times 10^{11}$

$= 270,000,000,000$

(b) $\dfrac{0.00000075}{0.00015} = \dfrac{7.5 \times 10^{-7}}{1.5 \times 10^{-4}}$

$= \dfrac{7.5}{1.5} \times \dfrac{10^{-7}}{10^{-4}}$

$\dfrac{a^m}{a^n} = a^{m-n}$:　　$= 5 \times 10^{-7-(-4)}$

$= 5 \times 10^{-3}$

$= 0.005$

Quick ✓

In Problems 63–66, perform the indicated operation. Express the solution in decimal notation.

63. $(8,000,000)(30,000)$

64. $\dfrac{0.000000012}{0.000004}$

65. $(25,000,000)(0.00003)$

66. $\dfrac{0.000039}{13,000,000}$

Getting Ready for Chapter 4 Exercises　MyMathLab®　Exercise numbers in green have complete video solutions in MyMathLab or may be accessed using the QR code to the right.

Problems 1–66 are the Quick ✓s that follow the EXAMPLES.

Mixed Practice

In Problems 67–112, simplify each expression. All exponents should be positive integers.

67. -5^2

68. 5^{-2}

69. -5^{-2}

70. -5^0

71. $-8^2 \cdot 8^{-2}$

72. $\dfrac{8^7}{8^5} \cdot 8^{-2}$

73. $\left(\dfrac{4}{9}\right)^{-2}$

74. $\left(\dfrac{3}{4}\right)^{-3}$

75. $(-3)^2(-3)^{-5}$

76. $(-4)^{-5}(-4)^3$

77. $\dfrac{(-4)^2}{(-4)^{-1}}$

78. $\dfrac{(-3)^3}{(-3)^{-2}}$

79. $\dfrac{2^3 \cdot 3^{-2}}{2^{-2} \cdot 3^{-4}}$

80. $\dfrac{3^{-2} \cdot 5^3}{3^2 \cdot 5}$

81. $(6x)^3(6x)^{-3}$

82. $(5a^2)^5(5a^2)^{-5}$

83. $(2s^{-2}t^4)(-5s^2t)$

84. $(6ab)(3a^3b^{-4})$

85. $\left(\dfrac{1}{4}xy\right) \cdot (20xy^{-2})$

86. $(3xy^3)\left(\dfrac{1}{9}x^2y\right)$

87. $\dfrac{36x^7y^3}{9x^5y^2}$

88. $\dfrac{25a^2b^3}{5ab^6}$

89. $\dfrac{21a^2b}{14a^3b^{-2}}$

90. $\dfrac{25x^{-2}y}{10xy^3}$

91. $(x^{-2})^4$

92. $(z^2)^{-6}$

93. $(3x^2y)^3$

94. $(5a^2b^{-1})^2$

95. $\left(\dfrac{z}{4}\right)^{-3}$

96. $\left(\dfrac{x}{y}\right)^{-8}$

97. $(3a^{-3})^{-2}$

98. $(2y^{-2})^{-4}$

99. $(-2a^2b^3)^{-4}$

100. $(-4a^{-2}b^2)^{-2}$

101. $\dfrac{2^3 \cdot xy^{-2}}{12(x^2)^{-2}y}$

102. $\dfrac{3^2 \cdot x^{-3}(y^2)^3}{15x^2y^8}$

103. $\left(\dfrac{15a^2b^3}{3a^{-4}b^5}\right)^{-2}$

104. $\left(\dfrac{15x^4y^7}{18x^{-3}y}\right)^{-1}$

105. $(4x^4y^{-2})^{-1} \cdot (2x^2y^{-1})^2$

106. $(9a^2b^{-4})^{-1} \cdot (3ab^{-2})^2$

107. $\dfrac{(-2)^2x^3(yz)^2}{-4xy^{-2}z}$

108. $\dfrac{(-3)^3a^3(ab)^{-2}}{9ab^4}$

109. $\dfrac{(3x^{-1}yz^2)^2}{(xy^{-2}z)^3}$

110. $\dfrac{(2ab^2c)^{-1}}{(a^{-1}b^3c^2)^{-2}}$

111. $\dfrac{(6a^3b^{-2})^{-1}}{(2a^{-2}b)^{-2}}\left(\dfrac{3ab^3}{2a^2b^{-3}}\right)^2$

112. $\left(\dfrac{a^{-3}b^{-1}}{2a^4b^{-2}}\right)^2\dfrac{(4a^2b)^2}{(2a^{-2}b)^3}$

In Problems 113–122, perform the indicated operation. Express the solution in scientific notation.

113. $(-5.3 \times 10^{-4})(2.8 \times 10^{-3})$

114. $(6.2 \times 10^3)(-3.8 \times 10^5)$

115. $(4 \times 10^6)^3$

116. $(5 \times 10^8)^2$

117. $\dfrac{5 \times 10^{-6}}{8 \times 10^{-4}}$

118. $\dfrac{1 \times 10^7}{5 \times 10^{-4}}$

119. $\dfrac{(4 \times 10^3)(6 \times 10^7)}{3 \times 10^4}$

120. $\dfrac{(3 \times 10^9)(8 \times 10^{-6})}{4 \times 10^4}$

121. $\dfrac{6.2 \times 10^{-3}}{(3.1 \times 10^4)(2 \times 10^{-7})}$

122. $\dfrac{1.5 \times 10^{13}}{(3 \times 10^4)(5 \times 10^8)}$

In Problems 123–128, perform the indicated operation by changing to scientific notation first. Express the solution in decimal notation.

123. $(4{,}000{,}000)(3{,}000{,}000)$

124. $(15{,}000)(3{,}000{,}000)$

125. $\dfrac{0.00008}{0.002}$

126. $\dfrac{0.00012}{0.0000002}$

127. $\dfrac{(0.000004)1{,}600{,}000}{(0.0008)(0.002)}$

128. $\dfrac{(0.0001)(3{,}500{,}000)}{(0.0005)(1{,}400{,}000)}$

Applying the Concepts

△ **129. Cubes** Suppose the length of a side of a cube is x^2. Find the volume of the cube in terms of x.

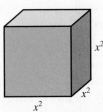

△ **130. Circles** The radius r of a circle is $\dfrac{d}{2}$, where d is the diameter. The area of a circle is given by the formula $A = \pi r^2$. Find the area of a circle in terms of its diameter d.

131. Diameter of a Plant Cell The diameter of a plant cell is 0.00001276 meter. Express this number in scientific notation.

132. Diameter of an Atom The diameter of an atom is about 0.0000000001 meter. Express this diameter as a number in scientific notation.

133. Diameter of Earth The diameter of Earth is 12,760,000 meters. Express this number in scientific notation.

134. National Debt As of February 2016, the debt of the United States federal government was $19,045,000,000,000. Express this number in scientific notation.

135. A Nanometer One nanometer is 1×10^{-9} meter. Express this number in decimal notation.

136. A Light-Year One light-year is 1×10^{16} meters. Express this number in decimal notation.

137. Rubik's Cube The Rubik's Cube has 4×10^{19} possible states. Express this number in decimal notation.

138. Hair Growth Human hair grows at a rate of 1×10^{-8} mile per hour. Express this number in decimal notation. What is the growth rate in miles per year?

139. Per Capita GDP The per capita gross domestic product (GDP) is the GDP divided by the United States population. It represents the average output of each resident of the United States and is measured in dollars per person. In 2015, the gross domestic product of the United States was $\$1.68 \times 10^{13}$. In 2015, the United States population was 3.2×10^8 people. Determine the per capita GDP of the United States in 2015. Express your answer in decimal notation rounded to the nearest dollar.

140. Astronomy The speed of light is 186,000 miles per second. How long does it take a beam of light to reach Earth from the Sun when the Sun is 93,000,000 miles from Earth? Express your answer in seconds, using scientific notation.

141. Visits to Facebook In January 2016, Facebook reported that it had 1.01×10^9 visitors each day. How many visitors to Facebook were there in January 2016? Express the answer in scientific and decimal notation. (SOURCE: *Facebook*)

142. Google Hits Reports suggest that Google.com has 3.5×10^{10} hits per day. How many hits does Google.com get in a 30-day month? Express the answer in scientific and decimal notation.

143. Population Density Population density is the number of people living per some unit of land. In other words, Population Density $= \dfrac{\text{Population}}{\text{Area of Land}}$. The United States is 3.54×10^6 square miles. In 2000, the United States population was 2.81×10^8 people. In 2015, the United States population was 3.19×10^8 people.

  (a) Determine the population density of the United States in 2000. Interpret the result.

  (b) Determine the population density of the United States in 2015. Interpret the result.

  (c) By how much did the population density increase? What does this result mean?

144. Fossil Fuels In 2015, the United States imported 3.3×10^9 barrels of oil. The United States population was 3.19×10^8 people in 2015.

  (a) Determine the per capita number of barrels of oil imported into the United States in 2015.

  (b) One barrel of oil contains 42 gallons. Determine the number of gallons of oil imported into the United States in 2015.

  (c) Use the result from part (b) to compute the per capita number of gallons of oil imported into the United States in 2015.

In Problems 145–148, simplify each algebraic expression by rewriting each factor with a common base. (Hint: Consider that $8 = 2^3$.)

145. $\dfrac{2^{x+3}}{4}$ **146.** $\dfrac{3^{2x}}{27}$

147. $3^x \cdot 27^{3x+1}$ **148.** $9^{-x} \cdot 3^{x+1}$

149. If $3^x = 5$, what does 3^{4x} equal?

150. If $4^x = 6$, what does 4^{5x} equal?

151. If $2^x = 7$, what does 2^{-4x} equal?

152. If $5^x = 3$, what does 5^{-3x} equal?

Explaining the Concepts

153. A friend of yours has a homework problem in which he must simplify $\left(x^4\right)^3$. He tells you that he thinks the answer is x^7. Is he right? If not, explain where he went wrong.

154. A friend of yours is convinced that x^0 must equal 0. Write an explanation that details why $x^0 = 1$. Include any restrictions that must be placed on x.

155. Explain why a cannot be 0 when m, n, or $m + n$ is negative or 0 in the expression a^{m+n}. Use examples to support your explanation.

156. Explain why a cannot be 0 when n is negative or 0 in the expression $\left(a^m\right)^n$. Use examples to support your explanation.

157. Explain why neither a nor b can be 0 when n is 0 or negative in the expression $(ab)^n = a^n b^n$.

158. Provide a justification for the Product Rule for Exponents.

159. Provide a justification for the Quotient Rule for Exponents.

160. Provide a justification for the Power Rule for Exponents.

161. Provide a justification for the Product-to-a-Power Rule for Exponents.

162. Explain how to convert a decimal number to scientific notation. Explain how to convert a number in scientific notation to decimal notation.

163. Explain the benefits of using scientific notation to multiply or divide very large or very small numbers.

4 Polynomials and Polynomial Functions

Did you know that your late 40s and early 50s are your alleged "peak earning years"? In fact, economists develop models that can be used to describe your earnings over your lifetime. Not surprisingly, your earnings will rise from your early 20s to your mid-40s. Then, from your mid-50s and beyond, your earnings start to decrease. See Problem 104 in Section 4.1.

The Big Picture: Putting It Together

In Chapter 2, functions, and linear functions in particular, were discussed. That chapter presented the properties of linear functions and explained how to graph them.

The goal of this chapter is to generalize our discussion of functions a little more. Linear functions belong to a class of functions called *polynomial functions*. This chapter will explain how to add, subtract, multiply, and divide polynomials and polynomial functions. Then it will illustrate the factoring techniques that enable us to "undo" polynomial multiplication. Finally, this chapter introduces the use of factoring to solve equations that have polynomial expressions.

4.1 Adding and Subtracting Polynomials

Objectives

1. Define Monomial and Determine the Coefficient and Degree of a Monomial
2. Define Polynomial and Determine the Degree of a Polynomial
3. Simplify Polynomials by Combining Like Terms
4. Evaluate Polynomial Functions
5. Add and Subtract Polynomial Functions

Are You Prepared for This Section?

Before getting started, complete the following problems. If you get a problem wrong, go back to the section cited and review the material.

P1. What is the coefficient of $-4x^5$? [Section R.5, p. 41]

P2. Combine like terms: $5x^2 - 3x + 1 - 2x^2 - 6x + 3$ [Section R.5, pp. 41–43]

P3. Use the Distributive Property to remove the parentheses: $-4(x - 3)$ [Section R.3, pp. 29–30]

P4. Given $f(x) = -4x + 3$, find $f(3)$. [Section 2.2, pp. 158–160]

▶ ① Define Monomial and Determine the Coefficient and Degree of a Monomial

Recall from Section R.5 that a *term* is a number or the product of a number and one or more variables raised to a power. The numerical factor of a term is the *coefficient*. For example, consider Table 1, where some algebraic expressions are given and their terms are identified.

Table 1

Algebraic Expression	Terms
$5x + 4$	$5x, 4$
$7x^2 - 8x + 3 = 7x^2 + (-8x) + 3$	$7x^2, -8x, 3$
$3x^2 + 7y^{-1}$	$3x^2, 7y^{-1}$

Work Smart

The nonnegative integers are $0, 1, 2, 3, \ldots$.

In this chapter, *polynomials* are studied. Polynomials have terms that are *monomials*.

Definition

A **monomial in one variable** is the product of a constant and a variable raised to a nonnegative integer power. A monomial in one variable is of the form

$$ax^k$$

where a is a constant, x is a variable, and $k \geq 0$ is an integer. The constant a is the **coefficient** of the monomial. If $a \neq 0$, then k is the **degree** of the monomial.

What if $a = 0$? Since $0x = 0x^2 = 0x^3 = \ldots, = 0$, a degree cannot be assigned to 0. Therefore, 0 has no degree.

EXAMPLE 1 Identifying the Coefficient and Degree of Monomials

Monomial	Coefficient	Degree
(a) $5x^3$	5	3
(b) $-\dfrac{2}{3}x^6$	$-\dfrac{2}{3}$	6
(c) $8 = 8x^0$	8	0
(d) $x^2 = 1x^2$	1	2
(e) $-x = -1 \cdot x$	-1	1
(f) 0	0	no degree

Now look at some expressions that are not monomials.

EXAMPLE 2 Expressions That Are Not Monomials

(a) $4x^{\frac{1}{2}}$ is not a monomial because the exponent of the variable x is $\dfrac{1}{2}$, and $\dfrac{1}{2}$ is not a nonnegative integer.

(b) $5x^{-3}$ is not a monomial because the exponent of the variable x is -3, and -3 is not a nonnegative integer.

Quick ✓

1. A _____ in one variable is the product of a number and a variable raised to a nonnegative integer power.

In Problems 2–5, determine whether the expression is a monomial. For those that are monomials, name the coefficient and give the degree.

2. $8x^5$

3. $5x^{-2}$

4. 12

5. $x^{\frac{1}{3}}$

A monomial may contain more than one variable factor. For example, the monomial ax^my^n has two variables, x and y, where m and n are nonnegative integers. The **degree of the monomial** ax^my^n is the sum of the exponents, $m + n$.

EXAMPLE 3   **Monomials in More than One Variable**

(a) $-4x^3y^4$ is a monomial in x and y of degree $3 + 4 = 7$. The coefficient is -4.

(b) $10ab^5$ is a monomial in a and b of degree $1 + 5 = 6$. The coefficient is 10.   ●

Quick ✓

6. The degree of a monomial in the form ax^my^n is _____.

In Problems 7–10, determine whether the expression is a monomial. For those that are monomials, determine the coefficient and degree.

7. $3x^5y^2$

8. $-2m^3n$

9. $4ab^{\frac{1}{2}}$

10. $-xy$

▶ **②** **Define Polynomial and Determine the Degree of a Polynomial**

We begin with a definition.

Definition

A **polynomial** is a monomial or the sum of monomials.

A polynomial is in **standard form** if it is written with the terms in descending order according to degree. The **degree of a polynomial** is the highest degree of all the terms of the polynomial. Remember, the degree of a nonzero constant is 0, and the number 0 has no degree.

EXAMPLE 4   **Examples of Polynomials**

Polynomial	Degree
(a) $7x^3 - 2x^2 + 6x + 4$	3
(b) $3 - 8x + x^2 = x^2 - 8x + 3$	2
(c) $-7x^4 + 24$	4
(d) $x^3y^4 - 3x^3y^2 + 2x^3y$	7
(e) $p^2q - 8p^3q^2 + 3 = -8p^3q^2 + p^2q + 3$	5
(f) 6	0
(g) 0	No Degree

●

EXAMPLE 5   **Is the Algebraic Expression a Polynomial?**

(a) $4x^{-2} - 5x + 1$ is not a polynomial because the exponent on the first term, -2, is negative.

(b) $\dfrac{4}{x^3}$ is not a polynomial because it can be written as $4x^{-3}$, and -3 is less than 0. Remember, the exponents on polynomials must be integers greater than or equal to 0.

(continued)

(c) $\dfrac{8x^2 + 16}{2}$ is a polynomial of degree 2 because it can be written as $4x^2 + 8$ after dividing 2 into each term in the numerator.

(d) $\dfrac{3xy + 1}{xy - 2}$ is not a polynomial because it is the quotient of two polynomials, and the expression cannot be simplified to a polynomial.

Quick ✓

11. *True or False* The degree of the polynomial $2m^2n + 5mn^3 - \dfrac{2}{3}m^2n^3$ is 5.

In Problems 12–16, determine whether the algebraic expression is a polynomial. For those that are polynomials, determine the degree.

12. $-3x^3 + 7x^2 - x + 5$

13. $5z^{-1} + 3$

14. $\dfrac{x - 1}{x + 1}$

15. $\dfrac{3x^2 - 9x + 27}{3}$

16. $5p^3q - 8pq^2 + pq$

Work Smart

The prefix "bi" means "two," as in bicycle. The prefix "tri" means "three," as in tricycle. The prefix "poly" means "many."

Certain polynomials have special names. A polynomial with exactly one term is a monomial; a polynomial that contains two monomials that are not like terms is called a **binomial;** and a polynomial that contains three monomials that are not like terms is called a **trinomial.** So

$-14x$ is a polynomial	but more specifically	$-14x$ is a monomial
$2x^3 - 5x$ is a polynomial	but more specifically	$2x^3 - 5x$ is a binomial
$-x^3 - 4x + 11$ is a polynomial	but more specifically	$-x^3 - 4x + 11$ is a trinomial
$3x^2 + 6xy - 2y^2$ is a polynomial	but more specifically	$3x^2 + 6xy - 2y^2$ is a trinomial

Work Smart

Remember, like terms have the same variable and the same exponent on the variable.

▶ ❸ **Simplify Polynomials by Combining Like Terms**

To simplify a polynomial means to perform all indicated operations such as addition, and combine like terms. **To add polynomials, combine like terms.**

EXAMPLE 6 **Simplifying Polynomials: Addition**

Simplify: $(-4x^3 + 9x^2 + x - 3) + (2x^3 + 6x + 5)$

Solution

The sum can be found using horizontal or vertical addition.

Horizontal Addition: The idea here is to combine like terms. The first step is to remove the parentheses.

$$(-4x^3 + 9x^2 + x - 3) + (2x^3 + 6x + 5) = -4x^3 + 9x^2 + x - 3 + 2x^3 + 6x + 5$$

$$\text{Rearrange terms:} \quad = -4x^3 + 2x^3 + 9x^2 + x + 6x - 3 + 5$$

$$\text{Distributive Property:} \quad = (-4 + 2)x^3 + 9x^2 + (1 + 6)x + (-3 + 5)$$

$$\text{Simplify:} \quad = -2x^3 + 9x^2 + 7x + 2$$

Vertical Addition: Line up like terms in each polynomial vertically and then add the coefficients.

$$
\begin{array}{ccccc}
x^3 & & x^2 & x^1 & x^0 \\
-4x^3 & + & 9x^2 + & x & - 3 \\
2x^3 & & & + 6x & + 5 \\
\hline
-2x^3 & + & 9x^2 + & 7x & + 2 \\
\end{array}
$$

Quick ✓

17. *True or False* $5y^3 + 7y^3 = 12y^6$

In Problems 18–20, simplify by adding the polynomials.

18. $(2x^2 - 3x + 1) + (4x^2 + 5x - 3)$

19. $(5w^4 - 3w^3 + w - 8) + (-2w^4 + w^3 - 7w^2 + 3)$

20. $\left(\dfrac{1}{2}x^2 - \dfrac{4}{3}x + 1\right) + \left(\dfrac{1}{4}x^2 + \dfrac{2}{3}x - 8\right)$

EXAMPLE 7 **Simplifying Polynomials in Two Variables: Addition**

Simplify: $(5a^2b - 3ab + 2ab^2) + (a^2b + 5ab - 4ab^2)$

Solution

Although you may add horizontally or vertically, only the horizontal format is presented. The first step is to remove the parentheses.

$$(5a^2b - 3ab + 2ab^2) + (a^2b + 5ab - 4ab^2) = 5a^2b - 3ab + 2ab^2 + a^2b + 5ab - 4ab^2$$

$$\text{Rearrange terms:} \quad = 5a^2b + a^2b - 3ab + 5ab + 2ab^2 - 4ab^2$$

$$\text{Distributive Property:} \quad = (5 + 1)a^2b + (-3 + 5)ab + (2 - 4)ab^2$$

$$\text{Simplify:} \quad = 6a^2b + 2ab - 2ab^2 \quad \bullet$$

Quick ✓

In Problem 21, simplify by adding the polynomials.

21. $(8x^2y + 2x^2y^2 - 7xy^2) + (-3x^2y + 5x^2y^2 + 3xy^2)$

Polynomials can be subtracted horizontally or vertically as well. Remember,

$$a - b = a + (-b)$$

Thus, to subtract one polynomial from another, add the opposite of *each term* in the polynomial following the subtraction sign and then combine like terms.

EXAMPLE 8 **Simplifying Polynomials: Subtraction**

Simplify: $(5z^3 + 3z^2 - 3) - (-2z^3 + 7z^2 - z + 2)$

Solution

Horizontal Subtraction: Recall that $a - b = a + (-1)b$.

$$(5z^3 + 3z^2 - 3) - (-2z^3 + 7z^2 - z + 2) = 5z^3 + 3z^2 - 3 + (-1)(-2z^3 + 7z^2 - z + 2)$$

$$\text{Distribute the } -1: \quad = 5z^3 + 3z^2 - 3 + 2z^3 - 7z^2 + z - 2$$

$$\text{Rearrange terms:} \quad = 5z^3 + 2z^3 + 3z^2 - 7z^2 + z - 3 - 2$$

$$\text{Combine like terms:} \quad = 7z^3 - 4z^2 + z - 5$$

Vertical Subtraction: Line up like terms, change the sign of each coefficient of the second polynomial, and add.

$$
\begin{array}{cccc}
z^3 & z^2 & z^1 & z^0 \\
5z^3 + 3z^2 & & & - 3 \\
-(-2z^3 + 7z^2 & - z & + 2) \\
\end{array}
\longrightarrow
\begin{array}{cccc}
z^3 & z^2 & z^1 & z^0 \\
5z^3 + 3z^2 & & & - 3 \\
+ 2z^3 - 7z^2 & + z & - 2 \\
\hline
7z^3 - 4z^2 & + z & - 5 \\
\end{array}
\quad \bullet
$$

Work Smart

Be careful with subtraction: The sign of *every term* of the second polynomial must be changed. Vertical subtraction will be used when we divide polynomials.

Quick ✓

In Problems 22–24, simplify by subtracting the polynomials.

22. $(5x^3 - 6x^2 + x + 9) - (4x^3 + 10x^2 - 6x + 7)$

23. $(8y^3 - 5y^2 + 3y + 1) - (-3y^3 + 6y + 8)$

24. $(8x^2y + 2x^2y^2 - 7xy^2) - (-3x^2y + 5x^2y^2 + 3xy^2)$

❹ Evaluate Polynomial Functions

Up to now, only polynomial expressions have been discussed, such as $4x^3 + x^2 - 7x + 1$. If this expression is written in the form $f(x) = 4x^3 + x^2 - 7x + 1$, then it is a *polynomial function*.

Definition

A **polynomial function** is a function whose rule is a polynomial. The domain of every polynomial function is the set of all real numbers. The **degree** of a polynomial function is the highest degree of all the terms.

Recall from Chapter 2 that a linear function is a function of the form $f(x) = mx + b$. Table 2 illustrates some specific types of polynomial functions, their functional forms, specific examples, and their degree.

Table 2

Polynomial	Functional Form	Examples	Degree
Linear	$f(x) = a_1x + a_0,$ where a_1 is the slope and a_0 is the y-intercept	$f(x) = 4x + 5$ $f(x) = -6x$ $f(x) = 3$	1 1 0
Quadratic	$f(x) = a_2x^2 + a_1x + a_0, a_2 \neq 0$	$f(x) = 3x^2 - 5x + 1$	2
Cubic	$f(x) = a_3x^3 + a_2x^2 + a_1x + a_0, a_3 \neq 0$	$f(x) = -2x^3 + 4x - 1$	3

To **evaluate** a polynomial function, substitute the value of the variable and simplify, just as shown in Section 2.2.

EXAMPLE 9 **Evaluating a Polynomial Function**

For the polynomial function $P(x) = 2x^3 - 5x^2 + x - 3$, find

(a) $P(3)$ **(b)** $P(-1)$

Solution

(a) Substitute 3 for x in the expression $2x^3 - 5x^2 + x - 3$ to get

$$P(3) = 2(3)^3 - 5(3)^2 + 3 - 3$$
$$= 2 \cdot 27 - 5 \cdot 9 + 3 - 3$$
$$= 54 - 45 + 3 - 3$$
$$= 9$$

(b) Substitute -1 for x: $\quad P(-1) = 2(-1)^3 - 5(-1)^2 + (-1) - 3$
$$= 2 \cdot (-1) - 5 \cdot 1 + (-1) - 3$$
$$= -2 - 5 - 1 - 3$$
$$= -11$$

Quick ✓

25. Find the following values of the polynomial function $g(x) = -2x^3 + 7x + 1$.

    (a) $g(0)$ **(b)** $g(2)$ **(c)** $g(-3)$

Polynomial functions can be used to model a variety of situations.

EXAMPLE 10   **A Polynomial Model Relating Divorce and Age**

The percentage of Americans who are divorced is a function of age. The polynomial function $D(a) = -0.014a^2 + 1.586a - 28.1889$ models the percent D of Americans of age a who are divorced.

    (a) Use the function to estimate the percent of 35-year-olds who are divorced.

    (b) Use the function to estimate the percent of 65-year-olds who are divorced.

Solution

    (a) The independent variable a represents the age, so $a = 35$. Evaluate $D(35)$.

$$D(a) = -0.014a^2 + 1.586a - 28.1889$$
$$D(35) = -0.014(35)^2 + 1.586(35) - 28.1889$$
$$= -17.15 + 55.51 - 28.1889$$
$$= 10.1711$$

    The percent of 35-year-olds who are divorced is estimated to be 10.17%.

    (b) Evaluate $D(65)$.

$$D(65) = -0.014(65)^2 + 1.586(65) - 28.1889$$
$$= -59.15 + 103.09 - 28.1889$$
$$= 15.7511$$

    The percent of 65-year-olds who are divorced is estimated to be 15.75%.   ●

Quick ✓

26. The polynomial function $B(a) = -1399a^2 + 70{,}573a - 495{,}702$ represents the number of first-time births to women who were a years of age in the United States in 2014.

    (a) Use the function to predict the number of first-time births to 25-year-old women.

    (b) Use the function to predict the number of first-time births to 40-year-old women.

▶ ❺ Add and Subtract Polynomial Functions

Functions, just like numbers, can be added, subtracted, multiplied, and divided. This section will focus on adding and subtracting polynomial functions.

Definition

If f and g are two functions,

    The **sum** $f + g$ is the function defined by

$$(f + g)(x) = f(x) + g(x)$$

    The **difference** $f - g$ is the function defined by

$$(f - g)(x) = f(x) - g(x)$$

(EXAMPLE 11) **Finding the Sum and Difference of Two Polynomial Functions**

Let f and g be two functions defined as

$$f(x) = 3x^2 - x + 1 \quad \text{and} \quad g(x) = -x^2 + 5x - 6$$

Find the following:

(a) $(f + g)(x)$ **(b)** $(f - g)(x)$ **(c)** $(f + g)(2)$ **(d)** $(f - g)(-1)$

Solution

(a) $(f + g)(x) = f(x) + g(x)$
$= (3x^2 - x + 1) + (-x^2 + 5x - 6)$
$= 3x^2 - x + 1 - x^2 + 5x - 6$
$= 3x^2 - x^2 - x + 5x + 1 - 6$
$= 2x^2 + 4x - 5$

(b) $(f - g)(x) = f(x) - g(x)$
$= (3x^2 - x + 1) - (-x^2 + 5x - 6)$
$= 3x^2 - x + 1 + x^2 - 5x + 6$
$= 3x^2 + x^2 - x - 5x + 1 + 6$
$= 4x^2 - 6x + 7$

Work Smart

If you have found $(f + g)(x)$, then take advantage of it when finding its value specific values of x.

(c) Because $(f + g)(x) = 2x^2 + 4x - 5$,

$$(f + g)(2) = 2(2)^2 + 4(2) - 5$$
$$= 8 + 8 - 5$$
$$= 11$$

$(f + g)(2)$ could also have been evaluated as follows:

$$(f + g)(2) = f(2) + g(2)$$
$$= 11 + 0$$
$$= 11$$

(d) Because $(f - g)(x) = 4x^2 - 6x + 7$,

$$(f - g)(-1) = 4(-1)^2 - 6(-1) + 7$$
$$= 4 + 6 + 7$$
$$= 17$$

$(f - g)(-1)$ could also have been evaluated as follows:

$$(f - g)(-1) = f(-1) - g(-1)$$
$$= 5 - (-12)$$
$$= 5 + 12$$
$$= 17$$

Quick ✓

27. Let f and g be two functions defined as

$$f(x) = 3x^2 - x + 1 \quad \text{and} \quad g(x) = -x^2 + 5x - 6$$

Find the following:

(a) $(f + g)(x)$ **(c)** $(f + g)(1)$
(b) $(f - g)(x)$ **(d)** $(f - g)(-2)$

Now let's look at an application of the difference of two functions.

EXAMPLE 12   **The Profit Function**

Profit is defined as total revenue minus total cost. The profit function of a company that manufactures and sells x units of a product is given by

$$P(x) = R(x) - C(x)$$

where P represents the company's profits
R represents the company's revenue
C represents the company's cost

(a) If a company sells a scientific calculator for \$12, its revenue function is $R(x) = 12x$. If the cost of each calculator manufactured is \$7 and fixed costs are \$1200 per week, its cost function is given by $C(x) = 7x + 1200$. Find the company's profit function, $P(x)$.

(b) Determine and interpret $P(800)$.

Solution

(a) The profit function is

$$\begin{aligned}
P(x) &= R(x) - C(x) \\
&= 12x - (7x + 1200) \\
&= 12x - 7x - 1200 \\
&= 5x - 1200
\end{aligned}$$

(b) $\begin{aligned}
P(800) &= 5(800) - 1200 \\
&= 4000 - 1200 \\
&= 2800
\end{aligned}$

If the company manufactures and sells 800 calculators in a week, its profit will be \$2800.   ●

Quick ✓

28. The calculator company presented in Example 12 just gave its employees a raise that increases the cost of each calculator manufactured to \$8. In addition, it renewed its lease on the plant, so its weekly fixed costs increased to \$1250.

(a) Determine the new profit function.

(b) Determine and interpret $P(800)$.

4.1 Exercises   MyMathLab®   Exercise numbers in green have complete video solutions in MyMathLab or may be accessed using the QR code to the right.

Problems 1–28 are the Quick ✓s that follow the EXAMPLES.

Building Skills

In Problems 29–36, determine the coefficient and degree of each monomial. See Objective 1.

29. $3x^2$

30. $5x^4$

31. $-8x^2y^3$

32. $-12xy$

33. $\dfrac{4}{3}x^6$

34. $-\dfrac{5}{3}z^5$

35. 2

36. -7

In Problems 37–40, state why each of the following is not a polynomial. See Objective 2.

37. $2x^{-1} + 3x$

38. $6p^{-3} - p^{-2} + 3p^{-1}$

39. $\dfrac{4}{z - 1}$

40. $\dfrac{x^2 + 2}{x}$

In Problems 41–56, determine whether the algebraic expression is a polynomial (Yes or No). If it is a polynomial, write the polynomial in standard form, determine the degree, and state whether it is a monomial, a binomial, or a trinomial. If it is a polynomial with more than three terms, say the expression is a polynomial. See Objective 2.

41. $5x^2 - 9x + 1$

42. $-3y^2 + 8y + 1$

43. $\dfrac{-20}{n}$

44. $\dfrac{1}{x}$

45. $3y^{\frac{1}{3}} + 2$

46. $8m - 4m^{\frac{1}{2}}$

47. $\dfrac{5}{8}$

48. -12

49. $5 - 8y + 2y^2$

50. $7 - 5p + 2p^2 - p^3$

51. $7x^{-1} + 4$

52. $4y^{-2} + 6y - 1$

53. $3x^2y^2 + 2xy^4 + 4$

54. $4mn^3 - 2m^2n^3 + mn^8$

55. $4pqr + 2p^2q + 3pq^{\frac{1}{4}}$

56. $-2xyz^2 + 7x^3z - 8y^{\frac{1}{2}}z$

In Problems 57–80, simplify each polynomial by adding or subtracting, as indicated. Express your answer as a single polynomial in standard form. See Objective 3.

57. $5z^3 + 8z^3$

58. $10y^4 - 6y^4$

59. $(x^2 + 5x + 1) + (3x^2 - 2x - 3)$

60. $(x^2 - 4x + 1) + (5x^2 + 2x + 7)$

61. $(6p^3 - p^2 + 3p - 4) + (2p^3 - 7p + 3)$

62. $(2w^3 - w^2 + 6w - 5) + (-3w^3 + 5w^2 + 9)$

63. $\begin{aligned} 8x^3 + 4x^2 - 3x + 1 \\ + \ -x^3 - 2x^2 - 3x + 7 \end{aligned}$ **64.** $\begin{aligned} -7n^3 + 2n^2 - 5n - 3 \\ + \ 2n^3 - 5n^2 + n + 1 \end{aligned}$

65. $(5x^2 + 9x + 4) - (3x^2 + 5x + 1)$

66. $(7y^2 + 9y + 12) - (4y^2 + 8y - 3)$

67. $(7s^2t^3 + st^2 - 5t - 8) - (4s^2t^3 + 5st^2 - 7)$

68. $(-2x^3y^3 + 7xy - 3) - (x^3y^3 + 5y^2 + xy - 3)$

69. $(3 - 5x + x^2) + (-2 + 3x - 5x^2)$

70. $(-3 - 5z + 3z^2) + (1 + 2z + z^2)$

71. $(6 - 2y + y^3) - (-2 + y^2 - 2y^3)$

72. $(8 - t^3) - (1 + 3t + 3t^2 + t^3)$

73. $\left(\dfrac{1}{4}x^2 + \dfrac{3}{2}x + 3\right) + \left(\dfrac{1}{2}x^2 - \dfrac{1}{4}x - 2\right)$

74. $\left(\dfrac{3}{4}y^3 - \dfrac{1}{8}y + \dfrac{2}{3}\right) + \left(\dfrac{1}{2}y^3 + \dfrac{5}{12}y - \dfrac{5}{6}\right)$

75. $(5x^2y^2 - 8x^2y + xy^2) + (3x^2y^2 + x^2y - 4xy^2)$

76. $(7a^3b + 9ab^2 - 4a^2b) + (-4a^3b + 3a^2b - 8ab^2)$

77. $(3x^2y + 7xy^2 + xy) - (2x^2y - 4xy^2 - xy)$

78. $(-5xy^2 + 3xy - 9y^2) - (5xy^2 + 7xy - 8y^2)$

79. $\begin{aligned} 9a^3 + 2a^2 - 5a - 8 \\ -(5a^3 - 2a^2 + \ a - 6) \end{aligned}$ **80.** $\begin{aligned} -11p^3 + 8p^2 - p - 7 \\ -(-p^3 + 5p^2 + p + 1) \end{aligned}$

Applying the Concepts

81. Add $5x^3 - 5x + 3$ to $-4x^3 + x^2 - 2x + 1$.

82. Add $2x^3 - 3x^2 - 5x + 7$ to $x^3 + 3x^2 - 6x - 4$.

83. Subtract $4b^3 - b^2 + 3b - 1$ from $2b^3 + 5b^2 - b + 3$.

84. Subtract $2q^3 - 3q^2 + 7q - 2$ from $-5q^3 + q^2 + 2q - 1$.

In Problems 85–88, simplify each of the following.

85. $(2x^2 - 3x + 1) + (x^2 + 9) - (4x^2 - 2x - 5)$

86. $(y^2 + 6y - 2) + (4y^2 - 9) - (2y^2 - 7y - 10)$

87. $(4n - 3) + (n^3 - 9) - (2n^2 - 7n + 3)$

88. $(y^3 - 1) + (y^2 - 9) - (y^3 + 2y^2 - 7y + 3)$

In Problems 89–94, find the following values for each polynomial function. See Objective 4.

   (a) $f(0)$ **(b)** $f(2)$ **(c)** $f(-3)$

89. $f(x) = x^2 - 4x + 1$ **90.** $f(x) = x^2 + 5x - 3$

91. $f(x) = 2x^3 - 7x + 3$ **92.** $f(x) = -2x^3 + 3x - 1$

93. $f(x) = -x^3 + 3x^2 - 2x + 3$

94. $f(x) = -2x^3 + x^2 + 5x - 3$

In Problems 95–100, for the given functions f and g, find the following. See Objective 5.

   (a) $(f + g)(x)$ **(b)** $(f - g)(x)$

   (c) $(f + g)(2)$ **(d)** $(f - g)(1)$

95. $f(x) = 2x + 5; g(x) = -5x + 1$

96. $f(x) = 4x + 3; g(x) = 2x - 3$

97. $f(x) = x^2 - 5x + 3; g(x) = 2x^2 + 3$

98. $f(x) = 3x^2 + x + 2; g(x) = x^2 - 3x - 1$

99. $f(x) = x^3 + 6x^2 + 12x + 2; g(x) = x^3 - 8$

100. $f(x) = 8x^3 + 1; g(x) = x^3 + 3x^2 + 3x + 1$

△ **101. Area** A rectangle has one corner on the graph of $y = 6 - 2x$, another at the origin, a third on the positive y-axis, and the fourth on the positive x-axis (see the figure).

   The polynomial function $A(x) = -2x^2 + 6x$ can be used to find the area A of the rectangle whose vertex is at (x, y).

(a) Find the area of the rectangle whose vertex is at $(2, 2)$.

(b) Find the area of the rectangle whose vertex is at $(1, 4)$.

(c) How can the coordinates of the vertex be used to find the area? What is the area?

△ **102. Area** A rectangle has one corner on the graph of $y = 10 - 2x$, another at the origin, a third on the positive y-axis, and the fourth on the positive x-axis (see the figure).

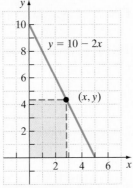

The polynomial function $A(x) = -2x^2 + 10x$ can be used to find the area A of the rectangle whose vertex is at (x, y).

(a) Find the area of the rectangle whose vertex is at $(3, 4)$.

(b) Find the area of the rectangle whose vertex is at $(4, 2)$.

103. Price per Square Foot of a New Home The polynomial function $P(t) = 0.004t^4 - 0.162t^3 + 1.879t^2 - 4.758t + 65.995$ can be used to approximate the average price per square foot P of a new home, where t is the number of years since 1995.

(a) Use the function to estimate the average price per square foot in 1995.

(b) Use the function to predict the average price per square foot in 2021.

104. Income The polynomial function $I(a) = -47.31a^2 + 4646.9a - 57{,}834$ represents the average income for residents of the United States, by age, in 2014. It can be used to approximate the average income I, where a is the age of the individual.

(a) Use the function to estimate the average income of a 20-year-old in 2014.

(b) Use the function to estimate the average income of a 55-year-old in 2014.

105. Profit Function Suppose that the revenue R from selling x cell phones is $R(x) = -1.2x^2 + 220x$. The cost C of selling x cell phones is $C(x) = 0.05x^3 - 2x^2 + 65x + 500$.

(a) Find the profit function, $P(x)$.

(b) Find the profit if $x = 15$ cell phones are sold.

(c) Determine and interpret $P(100)$.

106. Profit Function Suppose that the revenue R from selling x clocks is $R(x) = -0.3x^2 + 30x$. The cost C of selling x clocks is $C(x) = 0.1x^2 + 7x + 400$.

(a) Find the profit function, $P(x)$.

(b) Find the profit if $x = 15$ clocks are sold.

(c) Determine and interpret $P(40)$.

Extending the Concepts

107. If $f(x) = 3x + 2$ and $g(x) = ax - 5$, find a such that $(f + g)(3) = 12$.

108. If $f(x) = -5x + 1$ and $g(x) = ax - 3$, find a such that $(f - g)(2) = 10$.

109. The graph of two functions, f and g, is shown to the right. Use the graph to answer parts (a)–(d).

(a) $(f + g)(2)$

(b) $(f + g)(4)$

(c) $(f - g)(6)$

(d) $(g - f)(6)$

110. Work-Related Disability Suppose that $M(a)$ represents the number of American males with a work-related disability who are a years of age. Suppose that $F(a)$ represents the number of American females with a work-related disability who are a years of age. Determine a function T that represents the total number of Americans with work-related disabilities who are a years of age.

111. Taxes Let $T(x)$ represent the total taxes paid by everyone who worked in the United States in year x. Let $F(x)$ represent the taxes paid to the federal government in year x. Write a function S that represents the total taxes paid other than federal (state, property, sales, and so on) in year x.

Explaining the Concepts

112. Explain the difference between a term and a monomial.

113. What is the degree of a polynomial in one variable that is linear?

114. How many terms are in a binomial? How many terms are in a trinomial?

115. Give a definition of *polynomial* using your own words. Provide examples of polynomials that are monomials, binomials, and trinomials. In addition, give examples of polynomials that are linear.

116. What is the difference between a polynomial and a polynomial function?

117. Explain why the degree of the sum of two polynomials is at most the degree of the polynomial of highest degree.

Technology Exercises

Technology can be used to evaluate any function. Figure 1 shows the result obtained in Example 9(a) and (b) on a graphing calculator with the function to be evaluated, $P(x) = 2x^3 - 5x^2 + x - 3$, in Y_1.

Figure 1

In Problems 118–121, use technology to find the following values for each polynomial function.

    (a) $f(4)$ **(b)** $f(-2)$ **(c)** $f(6)$

118. $f(x) = -2x^2 + 5x + 3$

119. $f(x) = 4x^2 - 7x + 1$

120. $f(x) = -3x^3 + 5x^2 - 8x + 1$

121. $f(x) = 2x^3 - 5x^2 + x + 5$

4.2 Multiplying Polynomials

Objectives

1 Multiply a Monomial and a Polynomial
2 Multiply Two Binomials
3 Multiply Two Polynomials
4 Multiply Special Products
5 Multiply Polynomial Functions

Are You Prepared for This Section?

Before getting started, complete the following problems. If you get a problem wrong, go back to the section cited and review the material.

P1. Simplify: $4x^2 \cdot 3x^3$ [Getting Ready, p. 296]
P2. Simplify: $(-3x)^2$ [Getting Ready, pp. 296–297]
P3. Use the Distributive Property to remove the parentheses: $4(x - 5)$ [Section R.3, pp. 29–30]

Multiplying polynomials is based on the Product Rule for Exponents (Getting Ready, p. 298) and the Distributive Property (Section R.3, p. 29). The following example will help review these concepts.

> **EXAMPLE 1** **Review the Product Rule for Exponents and the Distributive Property**
>
>   **(a)** Simplify: $(2a^3b)(-6a^2b^4)$
>   **(b)** Remove the parentheses: $2(z + 3)$

Solution

    (a) $\qquad (2a^3b)(-6a^2b^4) = (2 \cdot (-6))(a^3 \cdot a^2)(b \cdot b^4)$

          Product Rule: $a^m \cdot a^n = a^{m+n}$: $\qquad = -12a^{3+2}b^{1+4}$

                                    $= -12a^5b^5$

    (b) The Distributive Property is used to remove parentheses:

$$2(z + 3) = 2 \cdot z + 2 \cdot 3$$
$$= 2z + 6$$

Quick ✓

In Problems 1–4, simplify each expression completely. All exponents should be positive.

1. $(3x^5)(2x^2)$
2. $(-7a^3b^2)(3ab^4)$

3. $\left(\frac{2}{3}x^4\right)\left(\frac{15}{8}x\right)$
4. $-3(x + 2)$

▶ ❶ Multiply a Monomial and a Polynomial

Example 1(b) shows that in multiplying a binomial by a monomial, the Distributive Property is used to remove the parentheses. In general, when multiplying a polynomial by a monomial, use the following property:

In Other Words

The Extended Form of the Distributive Property says to multiply each term in parentheses by *a*.

Extended Form of the Distributive Property

$$a(b_1 + b_2 + \cdots + b_n) = a \cdot b_1 + a \cdot b_2 + \cdots + a \cdot b_n$$

where $a, b_1, b_2, \ldots, b_n$ are real numbers.

EXAMPLE 2   **Using the Extended Form of the Distributive Property**

Multiply and simplify each of the following expressions:

(a) $3x^2(x^2 + 4x + 2)$

(b) $\frac{1}{2}xy^3\left(\frac{2}{3}xy^2 + \frac{6}{5}y + \frac{3}{4}\right)$

Solution

(a) $3x^2(x^2 + 4x + 2) = 3x^2 \cdot x^2 + 3x^2 \cdot 4x + 3x^2 \cdot 2$

$$= 3x^4 + 12x^3 + 6x^2$$

(b) $\frac{1}{2}xy^3\left(\frac{2}{3}xy^2 + \frac{6}{5}y + \frac{3}{4}\right) = \frac{1}{2}xy^3 \cdot \frac{2}{3}xy^2 + \frac{1}{2}xy^3 \cdot \frac{6}{5}y + \frac{1}{2}xy^3 \cdot \frac{3}{4}$

$$= \frac{1}{3}x^2y^5 + \frac{3}{5}xy^4 + \frac{3}{8}xy^3$$

Quick ✓

In Problems 5–7, multiply and simplify each expression.

5. $5x(x^2 + 3x + 2)$
6. $2xy(3x^2 - 5xy + 2y^2)$

7. $\frac{3}{4}y^2\left(\frac{4}{3}y^2 + \frac{2}{9}y + \frac{16}{3}\right)$

▶ ❷ Multiply Two Binomials

To find the product of two binomials, distribute the first binomial to each term in the second binomial. Just as with addition and subtraction of polynomials, use either the horizontal or the vertical format.

EXAMPLE 3   **Multiplying Two Binomials**

Find the product: $(3x + 4)(2x - 5)$

Solution

VERTICAL MULTIPLICATION	HORIZONTAL MULTIPLICATION

VERTICAL MULTIPLICATION

$$3x + 4$$
$$\times\ 2x - 5$$
$$\overline{-15x - 20} \leftarrow -5(3x + 4)$$
$$\underline{6x^2 + 8x} \longleftarrow 2x(3x + 4)$$
$$6x^2 - 7x - 20$$

HORIZONTAL MULTIPLICATION

Distribute $3x + 4$ to each term in $2x - 5$.

$$(3x + 4)(2x - 5) = (3x + 4)(2x) + (3x + 4)(-5)$$

Distribute: $= 3x \cdot 2x + 4 \cdot 2x + 3x \cdot (-5) + 4 \cdot (-5)$

$= 6x^2 + 8x - 15x - 20$

Combine like terms: $= 6x^2 - 7x - 20$

In either case, $(3x + 4)(2x - 5) = 6x^2 - 7x - 20$. ●

The FOIL Method

To multiply two binomials, a method known as the **FOIL method** can also be used. The acronym FOIL stands for **F**irst, **O**uter, **I**nner, **L**ast and is used to help remember the order in which terms should be multiplied, as illustrated below.

Work Smart

The FOIL method is a form of the Distributive Property and can be used only when two binomials are being multiplied.

$$\underset{\text{Outer}}{\underset{\text{Inner}}{(ax + b)(cx + d)}} = \overset{F}{ax \cdot cx} + \overset{O}{ax \cdot d} + \overset{I}{b \cdot cx} + \overset{L}{b \cdot d}$$

EXAMPLE 4   **Using the FOIL Method to Multiply Two Binomials**

Find the product:

(a) $(4y - 3)(2y + 5)$

(b) $(2m + n)(m - 3n)$

Solution

(a) $(4y - 3)(2y + 5) = \overset{F}{4y \cdot 2y} + \overset{O}{4y \cdot 5} - \overset{I}{3 \cdot 2y} - \overset{L}{3 \cdot 5}$

$= 8y^2 + 20y - 6y - 15$

$= 8y^2 + 14y - 15$

(b) $(2m + n)(m - 3n) = \overset{F}{2m \cdot m} + \overset{O}{2m \cdot (-3n)} + \overset{I}{n \cdot m} + \overset{L}{n \cdot (-3n)}$

$= 2m^2 - 6mn + mn - 3n^2$

$= 2m^2 - 5mn - 3n^2$ ●

Quick ✓

8. The acronym FOIL stands for ____, ____, ____, ____.

In Problems 9–11, find the product.

9. $(x + 4)(x + 1)$ **10.** $(3v + 5)(2v - 3)$

11. $(2a - b)(a + 5b)$

▶ ❸ **Multiply Two Polynomials**

When multiplying a trinomial by a binomial or a trinomial by a trinomial, make repeated use of the Extended Form of the Distributive Property. The approach is similar to the one

Work Smart

When multiplying polynomials, it is a good idea to write the polynomials in descending order of degree.

used in Example 3. It is a good idea to write each polynomial in standard form. Although either the horizontal or the vertical format may be used, only the horizontal format is presented.

EXAMPLE 5 | **Multiplying Two Polynomials**

Find the product:

(a) $(3x + 1)(x^2 + 4x - 3)$

(b) $(x^2 + 5x + 2)(2x^2 - x + 3)$

Solution

(a) Begin by distributing the binomial $3x + 1$ to each term in the trinomial.

$$(3x + 1)(x^2 + 4x - 3) = (3x + 1) \cdot x^2 + (3x + 1) \cdot 4x + (3x + 1) \cdot (-3)$$

Distributive Property: $= 3x^3 + x^2 + 12x^2 + 4x - 9x - 3$

Combine like terms: $= 3x^3 + 13x^2 - 5x - 3$

(b) Distribute $x^2 + 5x + 2$ to each term in the second trinomial.

$$(x^2 + 5x + 2)(2x^2 - x + 3) = (x^2 + 5x + 2) \cdot 2x^2 + (x^2 + 5x + 2) \cdot (-x)$$
$$+ (x^2 + 5x + 2) \cdot 3$$

Distributive Property: $= x^2 \cdot 2x^2 + 5x \cdot 2x^2 + 2 \cdot 2x^2 + x^2 \cdot (-x)$
$$+ 5x \cdot (-x) + 2 \cdot (-x) + x^2 \cdot 3 + 5x \cdot 3 + 2 \cdot 3$$
$$= 2x^4 + 10x^3 + 4x^2 - x^3 - 5x^2 - 2x + 3x^2$$
$$+ 15x + 6$$

Combine like terms: $= 2x^4 + 9x^3 + 2x^2 + 13x + 6$ ●

Quick ✓

In Problems 12 and 13, find the product.

12. $(2y - 3)(y^2 + 4y + 5)$ **13.** $(z^2 - 3z + 2)(2z^2 + z + 6)$

▶ ❹ **Multiply Special Products**

Certain binomials have products that result in patterns. These products are called special products.

EXAMPLE 6 | **Products of the Form $(A - B)(A + B)$**

Find the product: $(x - 7)(x + 7)$

Solution
Use FOIL and obtain

F   O   I   L

$$(x - 7)(x + 7) = x \cdot x + x \cdot 7 - 7 \cdot x - 7 \cdot 7$$
$$= x^2 + 7x - 7x - 49$$
$$= x^2 - 49$$ ●

The following rule is based on the results of Example 6.

Work Smart

When products of the form $(A - B)(A + B)$ are multiplied, the "middle terms" will always be "opposites" and therefore sum to 0. Also, do not forget multiplication is commutative. Therefore, $(A + B)(A - B) = (A - B)(A + B) = A^2 - B^2$

Difference of Two Squares

$$(A - B)(A + B) = A^2 - B^2$$

EXAMPLE 7 **Using the Difference of Two Squares Formula**

(a)
$$(A + B)(A - B) = A^2 - B^2$$
$$(3x + 2)(3x - 2) = (3x)^2 - 2^2$$
$$= 9x^2 - 4$$

(b)
$$A = 2m \text{ and } B = 5n^2$$
$$\downarrow$$
$$(2m - 5n^2)(2m + 5n^2) = (2m)^2 - (5n^2)^2$$
$$= 4m^2 - 25n^4$$

●

Quick ✓

14. *True or False*   The product of a binomial and a binomial is always a trinomial.

15. $(A - B)(A + B) = $ _____.

In Problems 16 and 17, find each product.

16. $(5y + 2)(5y - 2)$ 17. $(7y + 2z^3)(7y - 2z^3)$

EXAMPLE 8 **Products of the Form $(A + B)^2$ or $(A - B)^2$**

Find each product:

Work Smart

$(A + B)^2 \neq A^2 + B^2$
$(A - B)^2 \neq A^2 - B^2$

Whenever you feel the urge to perform an operation that you're not quite sure about, try it with actual numbers. For example, does

$(3 + 2)^2 = 3^2 + 2^2$? NO!

So $(A + B)^2 \neq A^2 + B^2$

(a) $(4x + 3)^2$ (b) $(3z - 5)^2$

Solution

(a) $(4x + 3)^2 = (4x + 3)(4x + 3)$
$$= (4x)^2 + \underbrace{(4x)(3) + (4x)(3)}_{= 2(4x)(3)} + 3^2$$
$$= 16x^2 + 24x + 9$$

(b) $(3z - 5)^2 = (3z - 5)(3z - 5)$
$$= (3z)^2 \underbrace{- (3z)(5) - (3z)(5)}_{= -2(3z)(5)} + 5^2$$
$$= 9z^2 - 30z + 25$$

●

Example 8 leads to some general results.

Squares of Binomials, or Perfect Square Trinomials

$$(A + B)^2 = A^2 + 2AB + B^2$$
$$(A - B)^2 = A^2 - 2AB + B^2$$

EXAMPLE 9 **Using the Perfect Square Trinomial Formulas**

$$(A + B)^2 = A^2 + 2AB \quad + B^2$$
$$\downarrow$$

(a) $(w + 5)^2 = w^2 + 2 \cdot w \cdot 5 + 5^2$
$$= w^2 + 10w + 25$$

$$(A - B)^2 = A^2 \quad - 2AB \quad + B^2$$

(b) $\quad (6p - 5)^2 = (6p)^2 - 2 \cdot 6p \cdot 5 + 5^2$
$$= 36p^2 - 60p + 25$$

(c) $(3x + 5y^2)^2 = (3x)^2 + 2 \cdot 3x \cdot 5y^2 + (5y^2)^2$
$$= 9x^2 + 30xy^2 + 25y^4$$

Work Smart

If you can't remember the formulas for a perfect square, don't panic! Use the fact that

$$(A + B)^2 = (A + B)(A + B)$$

and then FOIL. Use the same reasoning with perfect squares of the form $(A - B)^2$.

Quick ✓

18. $(A - B)^2 = $ _____ ; $(A + B)^2 = $ _____ .

19. $x^2 + 2xy + y^2$ is referred to as a _____ _____ trinomial.

20. *True or False* $(x + a)^2 = x^2 + a^2$

In Problems 21–23, find each product.

21. $(z - 8)^2$ **22.** $(6p + 5)^2$ **23.** $(4a - 3b)^2$

▶ ❺ Multiply Polynomial Functions

In Section 4.1, functions were added and subtracted. Now functions will be multiplied.

> **Definition**
>
> Let f and g be two functions. The **product $f \cdot g$** is the function defined by
> $$(f \cdot g)(x) = f(x) \cdot g(x)$$

EXAMPLE 10 **Finding the Product of Two Functions**

Suppose that $f(x) = 2x + 5$ and $g(x) = x^2 - 7x + 5$.

(a) Find $f(4) \cdot g(4)$.

(b) Find $(f \cdot g)(x)$.

(c) Use the result from part (b) to determine $(f \cdot g)(4)$.

Solution

(a) Evaluate the functions at $x = 4$ separately.

$$f(4) = 2(4) + 5 \qquad\qquad g(4) = (4)^2 - 7(4) + 5$$
$$= 13 \qquad\qquad\qquad\qquad = 16 - 28 + 5$$
$$\qquad\qquad\qquad\qquad\qquad = -7$$
$$f(4) \cdot g(4) = (13)(-7) = -91$$

(b) $\qquad\qquad (f \cdot g)(x) = f(x) \cdot g(x)$
$$= (2x + 5)(x^2 - 7x + 5)$$

Distribute: $\qquad\qquad = (2x + 5)x^2 + (2x + 5)(-7x) + (2x + 5) \cdot 5$

Distribute: $\qquad\qquad = 2x \cdot x^2 + 5x^2 + (2x)(-7x) + 5(-7x) + 2x \cdot 5 + 5 \cdot 5$

Simplify: $\qquad\qquad = 2x^3 + 5x^2 - 14x^2 - 35x + 10x + 25$

Combine like terms: $\qquad = 2x^3 - 9x^2 - 25x + 25$

(c) $(f \cdot g)(4) = 2(4)^3 - 9(4)^2 - 25(4) + 25$
$$= 2(64) - 9(16) - 100 + 25$$
$$= 128 - 144 - 100 + 25$$
$$= -91$$

Notice that $(f \cdot g)(4) = f(4) \cdot g(4)$, as expected.

Quick ✓

24. $(f \cdot g)(x) = \underline{\quad} \cdot \underline{\quad}.$

25. Suppose that $f(x) = 5x - 3$ and $g(x) = x^2 + 3x + 1$.

    (a) Find $f(2) \cdot g(2)$.

    (b) Find $(f \cdot g)(x)$.

    (c) Use the result from part (b) to determine $(f \cdot g)(2)$.

Section 2.2 showed how to evaluate functions when the argument (the value at which we are evaluating the function) was an algebraic expression.

EXAMPLE 11   **Evaluating Functions**

For the function $f(x) = x^2 + 5x$, find

    (a) $f(x + 3)$

    (b) $f(x + h) - f(x)$

Solution

$$f(\text{input}) = (\text{input})^2 + 5(\text{input})$$

(a) $f(x + 3) = (x + 3)^2 + 5(x + 3)$

$$= x^2 + 6x + 9 + 5x + 15$$

Combine like terms: $= x^2 + 11x + 24$

(b) To evaluate $f(x + h) - f(x)$, first evaluate $f(x + h)$ by replacing x in $f(x)$ with $x + h$. Then subtract $f(x)$ from this result.

$$f(x + h) - f(x) = \overbrace{(x + h)^2 + 5(x + h)}^{f(x+h)} - \overbrace{(x^2 + 5x)}^{f(x)}$$

$$= x^2 + 2xh + h^2 + 5x + 5h - x^2 - 5x$$

Combine like terms: $= 2xh + h^2 + 5h$

Quick ✓

26. For the function $f(x) = x^2 - 2x$, find

    (a) $f(x - 3)$ **(b)** $f(x + h) - f(x)$

4.2 Exercises   MyMathLab®

Exercise numbers in green have complete video solutions in MyMathLab or may be accessed using the QR code to the right.

*Problems **1–26** are the Quick ✓s that follow the EXAMPLES.*

Building Skills

In Problems 27–38, find the product. See Objective 1.

27. $(5xy^2)(-3x^2y^3)$ **28.** $(9a^3b^2)(-3a^2b^5)$

29. $\left(\dfrac{3}{4}yz^3\right)\left(\dfrac{20}{9}y^3z^2\right)$ **30.** $\left(\dfrac{12}{5}x^2y\right)\left(\dfrac{15}{4}x^4y^3\right)$

31. $5x(x^2 + 4x + 2)$ **32.** $6y(y^2 - 4y + 3)$

33. $-4a^2b(3a^2 + 2ab - b^2)$

34. $-3mn^3(4m^2 - mn + 5n^2)$

35. $\dfrac{2}{3}ab\left(\dfrac{3}{4}a^2b - \dfrac{9}{8}ab^3 + 6ab\right)$

36. $\dfrac{5}{2}xy\left(\dfrac{4}{15}x^2y - \dfrac{6}{5}xy + \dfrac{3}{10}xy^2\right)$

37. $0.4x^2(1.2x^2 - 0.8x + 1.5)$

38. $0.8y(0.4y^2 + 1.1y - 2.5)$

In Problems 39–52, find the product of the two binomials. See Objective 2.

39. $(x + 3)(x + 5)$ **40.** $(y - 2)(y - 6)$

41. $(a + 5)(a - 3)$ **42.** $(z - 8)(z + 3)$

43. $(4a + 3)(3a - 1)$ **44.** $(5x - 3)(x + 4)$

45. $(4 - 5x)(3 + 2x)$ **46.** $(2 - 7y)(5 + 2y)$

47. $\left(\dfrac{2}{3}x + 2\right)\left(\dfrac{1}{2}x - 4\right)$ **48.** $\left(\dfrac{3}{2}y + 4\right)\left(\dfrac{4}{3}y - 1\right)$

49. $(4a + 3b)(a - 5b)$ **50.** $(3m - 5n)(m + 2n)$

51. $(2x^2 + 1)(x^2 - 3)$ **52.** $(3x^2 - 5)(x^2 + 2)$

In Problems 53–68, find the product of the polynomials.
See Objective 3.

53. $(x + 1)(x^2 + 4x + 2)$ **54.** $(y - 2)(y^2 + 5y - 3)$

55. $(3a - 2)(2a^2 + a - 5)$

56. $(2b + 3)(3b^2 - 2b + 1)$

57. $(5z^2 + 3z + 2)(4z + 3)$

58. $(3p^2 - 5p + 3)(7p - 2)$

59. $-\dfrac{1}{2}x(2x + 6)(x - 3)$

60. $-\dfrac{4}{3}k(k + 7)(3k - 9)$

61. $(4 + y)(2y^2 - 3 + 5y)$

62. $(3 + 2z)(z^2 + 5 - 3z)$

63. $(w^2 + 2w + 1)(2w^2 - 3w + 1)$

64. $(a^2 + 4a + 4)(3a^2 - a - 2)$

65. $(b + 1)(b - 2)(b + 3)$

66. $(2a - 1)(a + 4)(a + 1)$

67. $(2ab + 5)(4a^2 - 2ab + b^2)$

68. $(xy - 2)(x^2 + 2xy + 4y^2)$

In Problems 69–82, find the special product. See Objective 4.

69. $(x - 6)(x + 6)$ **70.** $(y + 9)(y - 9)$

71. $(a + 8)^2$ **72.** $(b + 3)^2$

73. $(3y - 1)^2$ **74.** $(4z - 5)^2$

75. $(5a + 3b)(5a - 3b)$ **76.** $(8y + 3z)(8y - 3z)$

77. $(8z + y)^2$ **78.** $(4a + 7b)^2$

79. $(10x^2 - y)^2$ **80.** $(7p - 3q^2)^2$

81. $(a^3 + 2b)(a^3 - 2b)$ **82.** $(m^2 - 2n^3)(m^2 + 2n^3)$

In Problems 83–88, for the given functions find
     (a) $(f \cdot g)(x)$ **(b)** $(f \cdot g)(3)$
See Objective 5.

83. $f(x) = x + 4; g(x) = x - 1$

84. $f(x) = x + 5; g(x) = 2x + 1$

85. $f(x) = 4x - 3; g(x) = 2x + 5$

86. $f(x) = 5x - 1; g(x) = 4x + 5$

87. $f(x) = x - 2; g(x) = x^2 + 5x - 3$

88. $f(x) = x + 5; g(x) = x^2 - 2x + 3$

In Problems 89–94, for the given functions, find
     (a) $f(x + 2)$ **(b)** $f(x + h) - f(x)$
See Objective 5.

89. $f(x) = x^2 + 1$

90. $f(x) = x^2 - 4$

91. $f(x) = x^2 + 5x - 2$

92. $f(x) = x^2 - 2x + 3$

93. $f(x) = 3x^2 - x + 1$

94. $f(x) = -2x^2 + x - 5$

Mixed Practice

In Problems 95–116, simplify the expression.

95. $5ab(a - b)^2$ **96.** $-3x(x - 3)^2$

97. $(x^2 + 3)(x^2 - 3)$

98. $(b^3 - 10)(b^3 + 10)$

99. $(z^2 + 9)(z - 3)(z + 3)$

100. $(y^2 + 4)(y + 2)(y - 2)$

101. $(2x^3 + 3)^2$

102. $(3a^4 - 2)^2$

103. $(2m - 3n)(4m + n) - (m - 2n)^2$

104. $(6p + q)(5p - 2q) - (p + 3q)^2$

105. $(x + 3)(x^2 - 3x + 9)$

106. $(2y + 3)(4y^2 - 6y + 9)$

107. $\left(2x - \dfrac{1}{2}\right)^2$

108. $\left(3x + \dfrac{1}{3}\right)^2$

109. $(p + 2)^3$

110. $(z - 3)^3$

111. $(7x - 5y + 2)(3x - 2y + 1)$

112. $(2a + b - 5)(4a - 2b + 1)$

113. $(2p - 1)(p + 3) + (p - 3)(p + 3)$

114. $(3z + 2)(z - 2) + (z + 2)(z - 2)$

115. $(x + 3)(x - 3)(x^2 - 9) - (x + 1)(x^2 - 3)$

116. $(a + 2)(a - 2)(a^2 - 4) - (a + 3)(a^2 - 3)$

Applying the Concepts

△ *The product of polynomials can be visualized by using the area of rectangles. In Problems 117–120, find a polynomial expression for the total area of each of the following figures.*

117.

118.

119.

120.

△ *In Problems 121 and 122, write a polynomial expression for the area of the shaded region of the figure.*

121.

122.

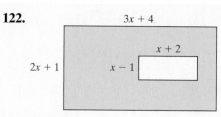

△ **123. Perfect Square** Why is the expression $(a + b)^2$ called a perfect square? Consider the figure below.

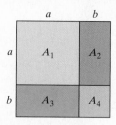

(a) Find the area of each of the four quadrilaterals.

(b) Use the result from part (a) to find the area of the entire region.

(c) Find the length and width of the entire region in terms of a and b. Use this result to find the area of the entire region. What do you notice?

△ **124. Area** Express as a polynomial in standard form the area of the shaded region in the figure shown.

Extending the Concepts

In Problems 125–132, find the product.

125. $[3x - (y + 1)][3x + (y + 1)]$

126. $[5 - (a + b)][5 + (a + b)]$

127. $[2a + (b - 3)]^2$

128. $[(m + 4) - n]^2$

129. $(2^x + 3)(2^x - 4)$

130. $(3^x - 1)(3^x - 9)$

131. $(5^y - 1)^2$

132. $(2^z - 4)^2$

4.3 Dividing Polynomials; Synthetic Division

Objectives

1. Divide a Polynomial by a Monomial
2. Divide Polynomials Using Long Division
3. Divide Polynomials Using Synthetic Division
4. Divide Polynomial Functions
5. Use the Remainder and Factor Theorems

Are You Prepared for This Section?

Before getting started, complete the following problems. If you get a problem wrong, go back to the section cited and review the material.

P1. Simplify: $\dfrac{15x^5}{12x^3}$ [Getting Ready, p. 296]

P2. Add: $\dfrac{2}{7} + \dfrac{5}{7}$ [Section R.3, p. 27]

Previous sections explained how to add, subtract and multiply polynomials. This section will cover how to divide them. We begin with dividing a polynomial by a monomial.

When polynomials are added, subtracted, and multiplied, the result is also a polynomial. A polynomial divided by a polynomial, however, may not be a polynomial.

▶ 1 Divide a Polynomial by a Monomial

Dividing a polynomial by a monomial is based on the Quotient Rule for Exponents (see Getting Ready, p. 298). For example,

$$\frac{24z^5}{18z^2} = \frac{6 \cdot 4}{6 \cdot 3} z^{5-2}$$

$$= \frac{4}{3} z^3$$

Recall that

$$\frac{A}{C} + \frac{B}{C} = \frac{A + B}{C}$$

This process is reversed when dividing a polynomial by a monomial.

For example, if A, B, and C are monomials, then $\dfrac{A + B}{C}$ can be written as shown below.

$$\frac{A + B}{C} = \frac{A}{C} + \frac{B}{C}$$

This result can be extended to polynomials with three or more terms.

EXAMPLE 1

Dividing a Polynomial by a Monomial

Divide and simplify:

(a) $\dfrac{12p^4 + 15p^3 + 3p^2}{3p^2}$ **(b)** $\dfrac{6a^2b^2 - 4a^2b + 3ab^2}{2a^2b^2}$

Solution

Divide the monomial in the denominator into each term in the numerator.

(a) $\dfrac{12p^4 + 15p^3 + 3p^2}{3p^2} = \dfrac{12p^4}{3p^2} + \dfrac{15p^3}{3p^2} + \dfrac{3p^2}{3p^2}$

$\dfrac{a^m}{a^n} = a^{m-n}: \quad = \dfrac{12}{3} p^{4-2} + \dfrac{15}{3} p^{3-2} + \dfrac{3}{3} p^{2-2}$

Simplify: $\quad = 4p^2 + 5p + 1$

Prepared?...Answers

P1. $\dfrac{5}{4} x^2$ **P2.** 1

(continued)

(b)

$$\frac{6a^2b^2 - 4a^2b + 3ab^2}{2a^2b^2} = \frac{6a^2b^2}{2a^2b^2} - \frac{4a^2b}{2a^2b^2} + \frac{3ab^2}{2a^2b^2}$$

$$\frac{a^m}{a^n} = a^{m-n}: \quad = \frac{6}{2}a^{2-2}b^{2-2} - \frac{4}{2}a^{2-2}b^{1-2} + \frac{3}{2}a^{1-2}b^{2-2}$$

Simplify: $\quad = 3a^0b^0 - 2a^0b^{-1} + \frac{3}{2}a^{-1}b^0$

$$a^0 = 1; b^0 = 1; a^{-1} = \frac{1}{a}; b^{-1} = \frac{1}{b}: \quad = 3 - \frac{2}{b} + \frac{3}{2a}$$

Quick ✓

1. The first step in simplifying $\dfrac{4x^4 + 8x^2}{2x}$ is to rewrite $\dfrac{4x^4 + 8x^2}{2x}$ as $\dfrac{}{2x} + \dfrac{}{2x}$.

In Problems 2–4, find the quotient.

2. $\dfrac{9p^4 - 12p^3 + 3p^2}{3p}$

3. $\dfrac{20a^5 - 10a^2 + 1}{5a^3}$

4. $\dfrac{x^4y^4 + 8x^2y^2 - 4xy}{4x^3y}$

▶ ❷ Divide Polynomials Using Long Division

Dividing two polynomials is similar to dividing two integers. Although this procedure should be familiar to you, it is reviewed below.

EXAMPLE 2 **Dividing an Integer by an Integer**

Divide 645 by 14.

Solution

$$
\begin{array}{r}
46 \quad \leftarrow \text{Quotient} \\
\text{Divisor} \rightarrow 14\overline{)645} \quad \leftarrow \text{Dividend} \\
-56 \quad \leftarrow 4 \cdot 14 \text{ (Subtract)} \\
\hline
85 \\
-84 \quad \leftarrow 6 \cdot 14 \text{ (Subtract)} \\
\hline
1 \quad \leftarrow \text{Remainder}
\end{array}
$$

Thus 645 divided by 14 is 46 with a remainder of 1. This can be written as

$$\frac{645}{14} = 46 + \frac{1}{14} = 46\frac{1}{14}$$

In Example 2, the number 14 is the **divisor,** 645 is the **dividend,** 46 is the **quotient,** and 1 is the **remainder.**

A division problem can always be checked by multiplying the quotient by the divisor and adding this product to the remainder. The result should be the dividend:

$$(\text{Quotient})\,(\text{Divisor}) + \text{Remainder} = \text{Dividend}$$

To check Example 2:

$$(46)\,(14) + 1 = 644 + 1 = 645$$

To divide two polynomials, first write each polynomial in standard form (descending order of degree). Use the pattern in Example 2 to guide your work.

EXAMPLE 3 **How to Divide Two Polynomials**

Find the quotient and remainder when

$$4x^2 + 7x + 3 \text{ is divided by } x + 3$$

Step-by-Step Solution

Each polynomial is in standard form. The dividend is $4x^2 + 7x + 3$ and the divisor is $x + 3$.

Step 1: Divide the leading term of the dividend, $4x^2$, by the leading term of the divisor, x. Enter the result over the term $4x^2$.

$$4x \quad \overset{\frac{4x^2}{x} = 4x}{\searrow}$$

$$x + 3\overline{)4x^2 + 7x + 3}$$

Step 2: Multiply $4x$ by $x + 3$. Be sure to vertically align like terms.

$$
\begin{array}{r}
4x \\
x + 3 \overline{)4x^2 + 7x + 3} \\
\underline{4x^2 + 12x} \quad \leftarrow 4x(x + 3) = 4x^2 + 12x
\end{array}
$$

Step 3: Subtract $4x^2 + 12x$ from $4x^2 + 7x + 3$.

$$
\begin{array}{r}
4x \\
x + 3 \overline{)4x^2 + 7x + 3} \\
\underline{-(4x^2 + 12x)} \\
-5x + 3 \quad \leftarrow (4x^2 + 7x + 3) - (4x^2 + 12x) = -5x + 3
\end{array}
$$

Step 4: Repeat Steps 1-3, treating $-5x + 3$ as the dividend by dividing x into $-5x$ to obtain -5.

$$
\begin{array}{r}
4x - 5 \quad \longleftarrow \dfrac{-5x}{x} = -5 \\
x + 3 \overline{)4x^2 + 7x + 3} \\
\underline{-(4x^2 + 12x)} \\
-5x + 3 \\
\underline{-(-5x - 15)} \quad \leftarrow -5(x + 3) = -5x - 15 \\
18 \quad \leftarrow -5x + 3 - (-5x - 15) = 18
\end{array}
$$

Work Smart

When the degree of the remainder is less than the degree of the divisor, you are finished dividing.

Because the degree of 18 is less than the degree of the divisor, $x + 3$, the process ends. The quotient is $4x - 5$ and the remainder is 18.

Step 5: **Check** Verify that $(Quotient)(Divisor) + Remainder = Dividend.$

$$(4x - 5)(x + 3) + 18 = 4x^2 + 12x - 5x - 15 + 18$$

Combine like terms: $= 4x^2 + 7x + 3$

The answer checks, so

$$\frac{4x^2 + 7x + 3}{x + 3} = 4x - 5 + \frac{18}{x + 3}$$

Always write the results of polynomial division as follows:

$$\frac{\text{Dividend}}{\text{Divisor}} = \text{Quotient} + \frac{\text{Remainder}}{\text{Divisor}}$$

EXAMPLE 4   **Dividing Two Polynomials**

Simplify by performing long division: $\dfrac{6x^3 - 11x^2 - 7x + 2}{3x + 2}$

Solution

$$
\begin{array}{r}
2x^2 - 5x + 1 \\
3x + 2 \overline{)6x^3 - 11x^2 - 7x + 2} \\
\underline{-(6x^3 + 4x^2)} \quad \leftarrow 2x^2(3x + 2) \\
-15x^2 - 7x + 2 \\
\underline{-(-15x^2 - 10x)} \quad \leftarrow -5x(3x + 2) \\
3x + 2 \\
\underline{-(3x + 2)} \quad \leftarrow 1(3x + 2) \\
0 \quad \leftarrow \text{Remainder}
\end{array}
$$

$\dfrac{6x^3}{3x} = 2x^2; \dfrac{-15x^2}{3x} = -5x; \dfrac{3x}{3x} = 1$

The quotient is $2x^2 - 5x + 1$ and the remainder is 0.

Check $(\text{Quotient})(\text{Divisor}) + \text{Remainder} = \text{Dividend}$

$$(2x^2 - 5x + 1)(3x + 2) + 0 = 6x^3 - 11x^2 - 7x + 2$$

(*continued*)

The answer checks, so

$$\frac{6x^3 - 11x^2 - 7x + 2}{3x + 2} = 2x^2 - 5x + 1$$

In Example 4, the remainder was 0. Therefore,

$$6x^3 - 11x^2 - 7x + 2 = (2x^2 - 5x + 1)(3x + 2)$$

which means that $2x^2 - 5x + 1$ and $3x + 2$ are *factors* of $6x^3 - 11x^2 - 7x + 2$. This result is true in general: If the remainder is zero, then the divisor and quotient are factors of the dividend.

EXAMPLE 5   **Dividing Two Polynomials with Missing Terms**

Simplify by performing long division: $\dfrac{8 - 15x + x^2 + 4x^3 + 3x^5}{3 + x^2}$

Solution

As usual, the dividend and divisor need to be in standard form. Write the missing x^4 term in the dividend as $0x^4$.

$$
\begin{array}{r}
3x^3 \qquad\quad -\ 5x\ +\ 1 \\
x^2 + 3\overline{)3x^5 + 0x^4 + 4x^3 + x^2 - 15x + 8} \\
-\ (3x^5 \qquad\quad + 9x^3) \qquad\qquad\quad \leftarrow 3x^3(x^2+3)\\
\hline
-5x^3 + x^2 - 15x + 8 \\
-\ (-5x^3 \qquad\quad - 15x) \qquad\quad \leftarrow -5x(x^2+3)\\
\hline
x^2 \qquad\qquad + 8 \\
-\ (x^2 \qquad\qquad + 3) \quad \leftarrow 1(x^2+3)\\
\hline
5 \quad \leftarrow \text{Remainder}
\end{array}
$$

Work Smart

Do not forget to include any missing powers with a coefficient of zero.

The quotient is $3x^3 - 5x + 1$ and the remainder is 5.

Check (Quotient)(Divisor) + Remainder = Dividend

$$(3x^3 - 5x + 1)(x^2 + 3) + 5 = 3x^5 + 4x^3 + x^2 - 15x + 8$$

The answer checks, so

$$\frac{8 - 15x + x^2 + 4x^3 + 3x^5}{3 + x^2} = 3x^3 - 5x + 1 + \frac{5}{3 + x^2}$$

Quick ✓

5. *True or False* To begin polynomial long division, write the divisor and dividend in standard form.

6. Because $\dfrac{3x^2 + 2x - 1}{x + 1} = 3x - 1$, the remainder when dividing $3x^2 + 2x - 1$ by $x + 1$ is ___, and $3x^2 + 2x - 1 = $ _____ · _____. We call $x + 1$ and $3x - 1$ _____ of $3x^2 + 2x - 1$.

7. Given that $\dfrac{6x^3 - x^2 - 9x + 8}{x + 1} = 6x^2 - 7x - 2 + \dfrac{10}{x + 1}$, we call $6x^3 - x^2 - 9x + 8$ the _____, $x + 1$ the _____, and 10 the _____.

8. To check a division problem, verify _____ · _____ + _____ = _____.

In Problems 9–11, simplify by performing long division.

9. $\dfrac{x^3 + 3x^2 - 31x + 21}{x - 4}$ 10. $\dfrac{2x^3 + 7x^2 - 7x - 12}{2x - 3}$ 11. $\dfrac{2 + 12x^2 - 2x^3 - 5x^4 + x^5}{x^2 - 2}$

▶ ❸ Divide Polynomials Using Synthetic Division

To find the quotient and remainder when dividing a polynomial of degree 1 or higher by $x - c$, a version of long division called **synthetic division** can be used.

To see how synthetic division works, use long division to divide the polynomial $2x^3 - 5x^2 - 7x + 20$ by $x - 3$. Synthetic division is just long division written in a more compact form. For example, in the long division below, the terms in red ink are not really necessary because they are identical to the terms directly above them. The subtraction signs are not necessary, because subtraction is understood. With these items removed, the division shown on the right results.

Work Smart

Synthetic division can be used only when the divisor is of the form $x - c$ or $x + c$. Remember that $x + b$ can be rewritten in the form $x - (-b)$.

$$
\begin{array}{r}
2x^2 + \ x \ - 4 \qquad \leftarrow \text{Quotient}\\
x - 3\overline{)2x^3 - 5x^2 - 7x + 20}\\
-(2x^3 - 6x^2)\\
\hline
x^2 - 7x\\
-(x^2 - 3x)\\
\hline
-4x + 20\\
-(-4x + 12)\\
\hline
8 \quad \leftarrow \text{Remainder}
\end{array}
$$

$$
\begin{array}{r}
2x^2 + x - 4\\
x - 3\overline{)2x^3 - 5x^2 - 7x + 20}\\
-6x^2\\
\hline
x^2\\
-3x\\
\hline
-4x\\
12\\
\hline
8
\end{array}
$$

The x's that appear in the division on the right are not necessary if the positioning of each coefficient is done carefully. As long as the rightmost number under the division symbol is the constant, the number to its left is the coefficient of x, and so on, the x's can be removed.

$$
\begin{array}{r}
2x^2 + x - 4x\\
x - 3\overline{)2 -5 - 7 \ \ 20}\\
-6\\
\hline
\boxed{1}\\
-3\\
\hline
\boxed{-4}\\
12\\
\hline
\boxed{8}
\end{array}
$$

This display can be made more compact by moving the rows up so that the boxed numbers align horizontally.

$$
\begin{array}{r}
2x^2 + x - 4\\
x - 3\overline{)2 \ \ -5 \ \ -7 \ \ 20}\\
-6 \ \ -3 \ \ 12\\
\hline
\square \ \ \ 1 \ \ -4 \ \ \ 8
\end{array}
$$

Because the leading coefficient of the divisor is always 1, the leading coefficient of the dividend will always be the leading coefficient of the quotient. Thus the leading coefficient of the quotient, 2, is placed in the boxed position.

$$
\begin{array}{r}
2x^2 + x - 4\\
x - 3\overline{)2 \ \ -5 \ \ -7 \ \ 20}\\
-6 \ \ -3 \ \ 12\\
\hline
2 \ \ \ 1 \ \ -4 \ \ \ 8
\end{array}
$$

Notice that the first three numbers in Row 3 are the coefficients of the quotient. The last number in Row 3 is the remainder. Now, the top row above is not needed.

$$
\begin{array}{rl}
x - 3\overline{)2 \ \ -5 \ \ -7 \ \ 20} & \text{Row 1}\\
-6 \ \ -3 \ \ 12 & \text{Row 2 (subtract)}\\
\hline
2 \ \ \ 1 \ \ -4 \ \ \ 8 & \text{Row 3}
\end{array}
$$

Remember, the entries in Row 3 are obtained by subtracting the entries in Row 2 from the entries in Row 1. Rather than subtracting the entries in Row 2, change the sign of each entry and then add. With this modification, the display becomes

$$
\begin{array}{r}
x - 3\overline{)2 \quad -5 \quad -7 \quad \;\;\;20} \;\; \text{Row 1}\\
\underline{6 \quad\;\; 3 \quad -12} \;\; \text{Row 2 (add)}\\
2 \quad\;\; 1 \quad -4 \quad\;\;\;\; 8 \;\; \text{Row 3}
\end{array}
$$

Notice that each entry in Row 2 is 3 times the entry that is one column to the left in Row 3 (for example, the 6 in Row 2 is 3 times 2; the 3 in Row 2 is 3 times 1, and so on). Remove the $x - 3$ and replace it with 3. The entries in Row 3 give us the quotient and remainder.

Work Smart

If there are any missing powers of x in the dividend, insert a coefficient of 0 for the missing term when doing synthetic division.

$$
\begin{array}{r}
3\overline{)2 \qquad\quad -5 \qquad\quad -7 \qquad\quad 20} \;\; \text{Row 1}\\
6 \qquad\quad\; 3 \qquad\quad -12 \;\; \text{Row 2 (add)}\\
\underline{2^{3(2)\nearrow} \qquad 1^{3(1)\nearrow} \qquad -4^{3(-4)\nearrow} \qquad 8} \;\; \text{Row 3}\\
\underbrace{2x^2 \quad + x \quad - \quad 4}_{\text{Quotient}} \qquad 8 \;\; \boxed{\text{Remainder}}
\end{array}
$$

EXAMPLE 6   **How to Use Synthetic Division to Divide Polynomials**

Use synthetic division to find the quotient and remainder when $3x^3 + 11x^2 + 14$ is divided by $x + 4$.

Step-by-Step Solution

Step 1: Write the dividend in standard form. Then copy the coefficients of the dividend. Remember to insert a 0 for any missing power of x.

$$3x^3 + 11x^2 + 14 = 3x^3 + 11x^2 + 0x + 14$$
$$3 \qquad 11 \qquad 0 \qquad 14 \;\; \text{Row 1}$$

Step 2: Insert the division symbol. Rewrite the divisor in the form $x - c$ and insert the value of c to the left of the division symbol.

$$x + 4 = x - (-4)$$
$$-4\overline{)3 \qquad 11 \qquad 0 \qquad 14} \;\; \text{Row 1}$$

Step 3: Bring the 3 down two rows and enter it in Row 3.

$$
\begin{array}{r}
-4\overline{)3 \qquad 11 \qquad 0 \qquad 14} \;\; \text{Row 1}\\
\downarrow \qquad\qquad\qquad\qquad\qquad \text{Row 2}\\
\underline{}\\
3 \qquad\qquad\qquad\qquad\qquad\; \text{Row 3}
\end{array}
$$

Step 4: Multiply the latest entry in Row 3 by -4 and place the result in Row 2, one column over to the right.

$$
\begin{array}{r}
-4\overline{)3 \qquad\quad 11 \qquad 0 \qquad 14} \;\; \text{Row 1}\\
\downarrow \quad -12 \qquad\qquad\qquad\quad \text{Row 2}\\
\underline{}\\
3^{-4(3)\nearrow} \qquad\qquad\qquad\qquad \text{Row 3}
\end{array}
$$

Step 5: Add the entry in Row 2 to the entry above it in Row 1. Enter the sum in Row 3.

$$
\begin{array}{r}
-4\overline{)3 \qquad\quad 11 \qquad 0 \qquad 14} \;\; \text{Row 1}\\
\downarrow \quad -12 \qquad\qquad\qquad\quad \text{Row 2}\\
\underline{}\\
3^{-4(3)\nearrow} \quad -1 \qquad\qquad\qquad \text{Row 3}
\end{array}
$$

Step 6: Repeat Steps 4 and 5 until no more entries are available in Row 1.

$$
\begin{array}{r}
-4\overline{)3 \qquad\quad 11 \qquad\quad 0 \qquad\quad 14} \;\; \text{Row 1}\\
\downarrow \quad -12 \qquad 4 \quad -16 \;\; \text{Row 2}\\
\underline{}\\
3^{-4(3)\nearrow} \quad -1^{-4(-1)\nearrow} \quad 4^{-4(4)\nearrow} \quad -2 \;\; \text{Row 3}
\end{array}
$$

Step 7: The final entry in Row 3, -2, is the remainder; the other entries in Row 3 (3, -1 and 4) are the coefficients of the quotient, in descending order of degree. The quotient is a polynomial whose degree is one less than the degree of the dividend.

Quotient: $3x^2 - x + 4$
Remainder: -2

Step 8: Check
(Quotient)(Divisor) + Remainder = Dividend

$$(3x^2 - x + 4)(x + 4) + (-2)$$
$$= 3x^3 + 12x^2 - x^2 - 4x + 4x + 16 - 2$$
$$= 3x^3 + 11x^2 + 14$$

Therefore, $\dfrac{3x^3 + 11x^2 + 14}{x + 4} = 3x^2 - x + 4 - \dfrac{2}{x + 4}$.

Let's do one more example, where we consolidate all the steps given in Example 6.

EXAMPLE 7 **Dividing Two Polynomials Using Synthetic Division**

Use synthetic division to find the quotient and remainder when $x^4 - 5x^3 - 6x^2 + 33x - 15$ is divided by $x - 5$.

Solution
The divisor is $x - 5$, so $c = 5$.

$$
\begin{array}{r|rrrrr}
5) & 1 & -5 & -6 & 33 & -15 \\
   & \downarrow & 5 & 0 & -30 & 15 \\
\hline
   & 1 & 0 & -6 & 3 & 0
\end{array}
$$

The dividend is a fourth-degree polynomial, so the quotient is a third-degree polynomial, $x^3 + 0x^2 - 6x + 3 = x^3 - 6x + 3$, and the remainder is 0. Thus

$$\frac{x^4 - 5x^3 - 6x^2 + 33x - 15}{x - 5} = x^3 - 6x + 3$$

In Example 7, because $\dfrac{x^4 - 5x^3 - 6x^2 + 33x - 15}{x - 5} = x^3 - 6x + 3$, we know that $x - 5$ and $x^3 - 6x + 3$ are factors of $x^4 - 5x^3 - 6x^2 + 33x - 15$. Therefore, we can write

$$x^4 - 5x^3 - 6x^2 + 33x - 15 = (x - 5)(x^3 - 6x + 3)$$

Work Smart: Study Skills

Knowing when a method **does not** apply is as essential as knowing when the method **does** apply. When can synthetic division be used to divide polynomials? When can't synthetic division be used?

Quick ✓

12. *True or False* We can divide $-4x^3 + 5x^2 + 10x - 3$ by $x^2 - 2$ using synthetic division.

13. *True or False* We can divide $-4x^3 + 5x^2 + 10x - 3$ by $2x + 1$ using synthetic division.

14. To divide a polynomial by $x + 7$ using synthetic division, use $c = $ ___

In Problems 15–17, use synthetic division to find the quotient.

15. $\dfrac{2x^3 + x^2 - 7x - 13}{x - 2}$ **16.** $\dfrac{x + 1 - 3x^2 + 4x^3}{x + 2}$ **17.** $\dfrac{x^4 + 8x^3 + 15x^2 - 2x - 6}{x + 3}$

▶ ❹ **Divide Polynomial Functions**

So far, adding, subtracting, and multiplying polynomial functions have been discussed. Next, we will look at dividing polynomial functions.

Definition

If f and g are functions, then the **quotient** $\dfrac{f}{g}$ is the function defined by

$$\left(\frac{f}{g}\right)(x) = \frac{f(x)}{g(x)} \quad g(x) \neq 0$$

EXAMPLE 8 · **Dividing Two Polynomial Functions**

If $f(x) = 2x^4 - x^3 - 3x^2 + 13x - 4$ and $g(x) = x^2 - 2$, find

(a) $\left(\dfrac{f}{g}\right)(x)$ **(b)** $\left(\dfrac{f}{g}\right)(2)$

Solution

(a) $\left(\dfrac{f}{g}\right)(x) = \dfrac{f(x)}{g(x)} = \dfrac{2x^4 - x^3 - 3x^2 + 13x - 4}{x^2 - 2}$

Use long division to find the quotient because the divisor is not of the form $x - c$.

$$
\begin{array}{r}
2x^2 - x + 1 \\
x^2 - 2 \overline{)2x^4 - x^3 - 3x^2 + 13x - 4} \\
-(2x^4 \qquad - 4x^2) \\
\hline
-x^3 + x^2 + 13x - 4 \\
-(-x^3 \qquad + 2x) \\
\hline
x^2 + 11x - 4 \\
-(x^2 \qquad - 2) \\
\hline
11x - 2
\end{array}
$$

Thus $\left(\dfrac{f}{g}\right)(x) = \dfrac{f(x)}{g(x)} = 2x^2 - x + 1 + \dfrac{11x - 2}{x^2 - 2}$.

(b) $\left(\dfrac{f}{g}\right)(2) = 2(2)^2 - 2 + 1 + \dfrac{11(2) - 2}{(2)^2 - 2}$

$= 8 - 2 + 1 + \dfrac{20}{2}$

$= 17$ ●

Quick ✓

In Problem 18, find **(a)** $\left(\dfrac{f}{g}\right)(x)$ *and* **(b)** $\left(\dfrac{f}{g}\right)(3)$.

18. $f(x) = 3x^4 - 4x^3 - 3x^2 + 10x - 5; g(x) = x^2 - 2$

❺ Use the Remainder and Factor Theorems

▶ In Example 6, synthetic division was used to find the quotient and remainder when $3x^3 + 11x^2 + 14$ is divided by $x + 4$. Letting $f(x) = 3x^3 + 11x^2 + 14$ yields $f(-4) = -2$, which is equal to the remainder when $3x^3 + 11x^2 + 14$ is divided by $x + 4$. So the value of the function f at $x = -4$ is the same as the remainder when f is divided by $x + 4 = x - (-4)$. This result is true in general! It is called the *Remainder Theorem*.

In Other Words

A theorem is a big idea that can be shown to be true in general. The word comes from a Greek verb meaning "to view."

The Remainder Theorem

Let f be a polynomial function. When $f(x)$ is divided by $x - c$, the remainder is $f(c)$.

EXAMPLE 9   **Using the Remainder Theorem**

Use the Remainder Theorem to find the remainder when $f(x) = 2x^3 - 3x + 8$ is divided by $x + 3$.

Solution

The divisor is $x + 3 = x - (-3)$, so the Remainder Theorem says that the remainder is $f(-3)$:

$$f(x) = 2x^3 - 3x + 8$$
$$f(-3) = 2(-3)^3 - 3(-3) + 8$$
$$= 2(-27) + 9 + 8$$
$$= -54 + 9 + 8$$
$$= -37$$

When $f(x) = 2x^3 - 3x + 8$ is divided by $x + 3$, the remainder is -37.

Check Using synthetic division, we find that the remainder is, in fact, -37.

$$
\begin{array}{r|rrrr}
-3 & 2 & 0 & -3 & 8 \\
   &   & -6 & 18 & -45 \\
\hline
   & 2 & -6 & 15 & -37 \\
\end{array}
$$
$\leftarrow$ Remainder

●

Quick ✓

19. Use the Remainder Theorem to find the remainder when $f(x) = 3x^3 + 10x^2 - 9x - 4$ is divided by

    (a) $x - 2$ **(b)** $x + 4$

Example 7 showed that when the remainder is 0, the quotient and divisor are factors of the dividend. The Remainder Theorem can be used to determine whether an expression of the form $x - c$ is a factor of the dividend. This result is called the *Factor Theorem*.

Work Smart

"If and only if" statements are used to compress two statements into one. The Factor Theorem is two statements:

1. If $f(c) = 0$, then $x - c$ is a factor of f

2. If $x - c$ is a factor of f, then $f(c) = 0$.

The Factor Theorem

Let f be a polynomial function. Then $x - c$ is a factor of $f(x)$ if and only if $f(c) = 0$.

The Factor Theorem can be used to determine whether a polynomial has a particular factor.

EXAMPLE 10   **Using the Factor Theorem**

Use the Factor Theorem to determine whether the function $f(x) = 2x^3 - 3x^2 - 18x - 8$ has the factor

    (a) $x - 3$ **(b)** $x + 2$

Solution

The Factor Theorem states that if $f(c) = 0$, then $x - c$ is a factor of f.

    (a) Because $x - 3$ is of the form $x - c$ with $c = 3$, find the value of $f(3)$.

$$f(3) = 2(3)^3 - 3(3)^2 - 18(3) - 8 = -35 \neq 0$$

    Since $f(3) \neq 0$, $x - 3$ is not a factor of f.

    (b) Because $x + 2 = x - (-2)$ is of the form $x - c$ with $c = -2$, find the value of $f(-2)$. Rather than evaluating the function using substitution, use synthetic division.

$$
\begin{array}{r|rrrr}
-2 & 2 & -3 & -18 & -8 \\
   &   & -4 & 14 & 8 \\
\hline
   & 2 & -7 & -4 & 0 \\
\end{array}
$$
$\leftarrow$ Remainder

(continued)

In Other Words

If $f(c) = 0$, then $f(x)$ can be written in factored form as $f(x) = (x - c)$ (quotient)

The remainder is 0, so $f(-2) = 0$, and therefore $x + 2$ is a factor of f.

This means the dividend can be written as the product of the quotient and divisor. The quotient is $2x^2 - 7x - 4$ and the divisor is $x + 2$, so

$$2x^3 - 3x^2 - 18x - 8 = (x + 2)(2x^2 - 7x - 4)$$

Quick ✓

20. Use the Factor Theorem to determine whether $x - c$ is a factor of $f(x) = 2x^3 - 9x^2 - 6x + 5$ for the given values of c. If $x - c$ is a factor, then write f in factored form. That is, write $f(x) = (x - c)$ (quotient).

   (a) $c = -2$ **(b)** $c = 5$

4.3 Exercises   MyMathLab®

Exercise numbers in green have complete video solutions in MyMathLab or may be accessed using the QR code to the right.

Problems **1–20** *are the* **Quick** ✓ s *that follow the* **EXAMPLES.**

Building Skills

In Problems 21–28, divide and simplify. See Objective 1.

21. $\dfrac{8x^2 + 12x}{4x}$ **22.** $\dfrac{6z^3 + 9z^2}{3z^2}$

23. $\dfrac{2a^3 - 15a^2 + 10a}{5a}$ **24.** $\dfrac{4b^3 + 12b^2 + 24b}{6b}$

25. $\dfrac{2y^3 + 6y}{4y^2}$ **26.** $\dfrac{3z^4 + 12z^2}{6z^3}$

27. $\dfrac{4m^2n^2 + 6m^2n - 18mn^2}{4m^2n^2}$

28. $\dfrac{2x^2y^3 - 9xy^3 + 16x^2y}{2x^2y^2}$

In Problems 29–46, divide using long division. See Objective 2.

29. $\dfrac{x^2 + 5x + 6}{x + 2}$ **30.** $\dfrac{x^2 - 4x - 21}{x + 3}$

31. $\dfrac{2x^2 + x - 4}{x - 2}$ **32.** $\dfrac{3z^2 - 7z - 28}{z - 4}$

33. $\dfrac{2w^2 + 5w - 49}{2w - 7}$ **34.** $\dfrac{4x^2 - 17x - 33}{4x + 7}$

35. $\dfrac{x^3 + 8x^2 + x - 42}{x + 3}$ **36.** $\dfrac{x^3 + x^2 - 22x - 40}{x + 2}$

37. $\dfrac{w^3 - 21w - 20}{w + 4}$ **38.** $\dfrac{a^3 - 49a + 120}{a + 8}$

39. $\dfrac{6x^3 - 13x^2 - 80x - 25}{2x + 5}$ **40.** $\dfrac{4p^3 + 20p^2 + 19p - 15}{2p + 5}$

41. $\dfrac{x^3 - 7x^2 + 3x - 21}{x^2 + 3}$ **42.** $\dfrac{x^3 - 5x^2 - 2x + 10}{x^2 - 2}$

43. $\dfrac{3z^3 + 21z^2 + 5z + 2}{3z^2 + 1}$

44. $\dfrac{2k^3 + 10k^2 - 6k - 8}{2k^2 - 3}$

45. $\dfrac{2x^4 - 11x^3 + 8x^2 - 22x + 144}{x^2 + 2x + 5}$

46. $\dfrac{2x^4 - 7x^3 - 50x^2 - 10x + 96}{x^2 + x - 3}$

In Problems 47–60, divide using synthetic division. See Objective 3.

47. $\dfrac{x^2 - 3x - 10}{x - 5}$ **48.** $\dfrac{x^2 + 4x - 12}{x - 2}$

49. $\dfrac{2x^2 + 11x + 12}{x + 4}$ **50.** $\dfrac{3x^2 + 19x - 40}{x + 8}$

51. $\dfrac{x^2 - 3x - 14}{x - 6}$ **52.** $\dfrac{x^2 + 2x - 17}{x - 4}$

53. $\dfrac{x^3 - 19x - 15}{x - 5}$ **54.** $\dfrac{x^3 - 13x - 17}{x + 3}$

55. $\dfrac{3x^4 - 5x^3 - 21x^2 + 17x + 25}{x - 3}$

56. $\dfrac{2x^4 - x^3 - 38x^2 + 16x + 103}{x + 4}$

57. $\dfrac{x^4 - 40x^2 + 109}{x + 6}$ **58.** $\dfrac{a^4 - 65a^2 + 55}{a - 8}$

59. $\dfrac{2x^3 + 3x^2 - 14x - 15}{x - \dfrac{5}{2}}$ **60.** $\dfrac{3x^3 + 13x^2 + 8x - 12}{x - \dfrac{2}{3}}$

In Problems 61–70, find (a) $\left(\dfrac{f}{g}\right)(x)$, *(b)* $\left(\dfrac{f}{g}\right)(2)$. *See Objective 4.*

61. $f(x) = 4x^3 - 8x^2 + 12x; g(x) = 4x$

62. $f(x) = 3x^3 - 9x^2 + 12x; g(x) = 3x$

63. $f(x) = x^2 - x - 12; g(x) = x - 4$

64. $f(x) = x^2 + 3x - 4; g(x) = x + 4$

65. $f(x) = 2x^2 + 5x - 1; g(x) = x + 3$

66. $f(x) = 3x^2 - 6x + 5; g(x) = 2x + 1$

67. $f(x) = 2x^3 + 9x^2 + x - 12; g(x) = 2x + 3$

68. $f(x) = 3x^3 - 2x^2 - 19x - 6; g(x) = 3x + 1$

69. $f(x) = x^3 - 13x - 12; g(x) = x^2 - 9$

70. $f(x) = x^3 - 19x + 30; g(x) = x^2 - x - 6$

In Problems 71–78, use the Remainder Theorem to find the remainder. See Objective 5.

71. $f(x) = x^2 - 5x + 1$ is divided by $x - 2$

72. $f(x) = x^2 + 4x - 5$ is divided by $x + 2$

73. $f(x) = x^3 - 2x^2 + 5x - 3$ is divided by $x + 4$

74. $f(x) = x^3 + 3x^2 - x + 1$ is divided by $x - 3$

75. $f(x) = 2x^3 - 4x + 1$ is divided by $x - 5$

76. $f(x) = 3x^3 + 2x^2 - 5$ is divided by $x + 3$

77. $f(x) = x^4 + 1$ is divided by $x - 1$

78. $f(x) = x^4 - 1$ is divided by $x - 1$

In Problems 79–86, use the Factor Theorem to determine whether $x - c$ is a factor of the given function for the given value of c. If $x - c$ is a factor, then write f in factored form. That is, write $f(x) = (x - c)(quotient)$. See Objective 5.

79. $f(x) = x^2 - 3x + 2; c = 2$

80. $f(x) = x^2 + 5x + 6; c = 3$

81. $f(x) = 2x^2 + 5x + 2; c = -2$

82. $f(x) = 3x^2 + x - 2; c = -1$

83. $f(x) = 4x^3 - 9x^2 - 49x - 30; c = 3$

84. $f(x) = 2x^3 - 9x^2 - 2x + 24; c = 1$

85. $f(x) = 4x^3 - 7x^2 - 5x + 6; c = -1$

86. $f(x) = 5x^3 + 8x^2 - 7x - 6; c = -2$

Mixed Practice

In Problems 87–100, divide using any appropriate method.

87. $\dfrac{3a^3b^2 - 9a^2b + 18ab}{3ab}$

88. $\dfrac{5s^4t^3 - 15s^3t^2 + 50s^2t}{5s^2t}$

89. $\dfrac{3y^2 + 11y + 6}{3y + 2}$

90. $\dfrac{4a^2 + 23a + 15}{4a + 3}$

91. $\dfrac{x^3 + 6x^2 + 8x + 32}{x^2 + 5}$

92. $\dfrac{x^3 - 3x^2 + 5x - 12}{x^2 + 3}$

93. $\dfrac{8x^3 + 6x}{12x^2}$

94. $\dfrac{3x^4 + 6x^2}{9x^3}$

95. $\dfrac{x^3 + 7x^2 + 2x - 46}{x + 4}$

96. $\dfrac{x^3 + 5x^2 - 29x - 97}{x - 5}$

97. $\dfrac{3x^3 + 5x^2 - 24x - 40}{3x + 5}$

98. $\dfrac{4b^3 + 11b^2 - 28b - 17}{4b + 3}$

99. $\dfrac{8x^3 - 27}{2x - 3}$

100. $\dfrac{8a^3 + 125}{2a + 5}$

101. If $\dfrac{f(x)}{x - 5} = 3x + 5$, find $f(x)$.

102. If $\dfrac{f(x)}{x + 3} = 2x + 7$, find $f(x)$.

103. If $\dfrac{f(x)}{x - 3} = x + 8 + \dfrac{4}{x - 3}$, find $f(x)$.

104. If $\dfrac{f(x)}{x - 3} = x^2 + 2 + \dfrac{7}{x - 3}$, find $f(x)$.

Applying the Concepts

△ **105. Area** The area of a rectangle is $15x^2 + x - 2$ square feet. If the width of the rectangle is $3x - 1$ feet, find the length.

△ **106. Area** The area of a rectangle is $10x^2 + 9x - 9$ square centimeters. If the width of the rectangle is $2x + 3$ centimeters, find the length.

△ **107. Volume** The volume of the box shown is $2x^3 + 9x^2 - 20x - 75$ cubic centimeters. Find the length if the width is $x + 5$ centimeters and the height is $x - 3$ centimeters.

$x - 3$

$x + 5$

△**108. Volume** The volume of the box shown is $4x^3 + 35x^2 + 52x + 21$ cubic feet. Find the height if the width is $4x + 3$ feet and the length is $x + 1$ feet.

109. Average Cost The average cost function is defined in economics as $\overline{C}(x) = \dfrac{C(x)}{x}$, where C is the cost of manufacturing x units of a good. Suppose that $C(x) = 0.01x^3 - 0.4x^2 + 13x + 400$, where x is the number of wristwatches manufactured in a day.

(a) What is $\overline{C}(x) = \dfrac{C(x)}{x}$?

(b) What is the average cost of manufacturing $x = 50$ wristwatches in a day?

110. Average Cost The average cost function is defined in economics as $\overline{C}(x) = \dfrac{C(x)}{x}$, where C is the cost of manufacturing x units of a good. Suppose that $C(x) = 0.01x^3 - 0.45x^2 + 16.5x + 600$, where x is the number of wagons manufactured in a day.

(a) What is $\overline{C}(x) = \dfrac{C(x)}{x}$?

(b) What is the average cost of manufacturing $x = 50$ wagons in a day?

Extending the Concepts

111. Find the sum of a, b, c, and d if
$$\frac{2x^3 - 3x^2 - 26x - 37}{x + 2} = ax^2 + bx + c + \frac{d}{x + 2}.$$

112. What is the remainder when $f(x) = 2x^{30} - 3x^{20} + 4x^{10} - 2$ is divided by $x - 1$?

Explaining the Concepts

113. If f is a polynomial of degree n and it is divided by $x + 4$, the quotient will be a polynomial of degree $n - 1$. Explain why.

114. Explain the Remainder Theorem in your own words. Explain the Factor Theorem in your own words.

115. Given that $\dfrac{6x^3 - x^2 - 9x + 4}{3x + 4} = 2x^2 - 3x + 1$, is $3x + 4$ a factor of $6x^3 - x^2 - 9x + 4$? If so, write $6x^3 - x^2 - 9x + 4$ in factored form.

116. Suppose that you were asked to divide $8x^3 - 3x + 1$ by $x + 3$. Would you use long division or synthetic division? Why?

Putting the Concepts Together (Sections 4.1–4.3)

We designed these problems so that you can review Sections 4.1 through 4.3 and show your mastery of the concepts. Take time to work these problems before proceeding with the next section. The answers are located at the back of the text on page AN-28.

1. Write the polynomial in standard form and determine its degree:
$$3m + 5m^4 - 2m^3 + 8$$

2. Add: $(7a^2 - 4a^3 + 7a - 1) + (2a^2 - 6a - 7)$

3. Subtract: $\left(\dfrac{1}{5}y^2 + 2y - 6\right) - (4y^2 - y + 2)$

4. For $f(x) = 2x^3 - x^2 + 4x + 9$, find $f(2)$.

5. For $f(x) = 6x + 5$ and $g(x) = -x^2 + 2x + 3$, find $(f + g)(-3)$.

6. For $f(x) = 2x^2 + 7$ and $g(x) = x^2 - 4x - 3$, find $(f - g)(x)$.

7. Multiply: $2mn^3(m^2n - 4mn + 6)$

8. Multiply: $(3a - 5b)^2$

9. Multiply: $(7n^2 + 3)(7n^2 - 3)$

10. Multiply: $(3a + 2b)(6a^2 - 2ab + b^2)$

11. For $f(x) = x + 2$ and $g(x) = x^2 - 4x + 11$, find $(f \cdot g)(x)$.

12. Divide using long division: $\dfrac{10z^3 + 41z^2 + 7z - 49}{2z + 7}$

13. Divide using synthetic division:
$$\frac{2x^3 + 25x^2 + 62x - 6}{x + 9}$$

14. For $f(x) = x^3 + 2x^2 - 4x + 5$ and $g(x) = x - 1$, find $\left(\dfrac{f}{g}\right)(x)$.

15. Use the Factor Theorem to determine whether $x + 5$ is a factor of $f(x) = 3x^3 + 8x^2 - 23x + 60$. If so, write the function in factored form—that is, in the form $f(x) = (x + 5)(\text{quotient})$.

4.4 Greatest Common Factor; Factoring by Grouping

Objectives

❶ Factor Out the Greatest Common Factor

❷ Factor by Grouping

Are You Prepared for This Section?

Before getting started, complete the following problems. If you get a problem wrong, go back to the section cited and review the material.

P1. Write 24 as the product of prime factors. [Section R.3, p. 28]

P2. Distribute: $4(3x - 5)$ [Section R.3, pp. 29–30]

Consider the following products:

$$5y(y^2 - 2y + 5) = 5y^3 - 10y^2 + 25y$$
$$(3x + 1)(x - 5) = 3x^2 - 14x - 5$$

The polynomials on the left side are called **factors** of the polynomial on the right side. Expressing a polynomial with integer coefficients as the product of two or more other polynomials with integer coefficients is called **factoring over the integers.**

A polynomial with integer coefficients is **prime** if it cannot be written as the product of two other polynomials with integer coefficients (excluding 1 and −1). A polynomial that has been written as a product of prime factors is said to be **factored completely.** The word *prime* has the same meaning in this context as it does for integers. For example, 3, 7, and 13 are prime numbers, while 4 $(= 2 \cdot 2)$, 12 $(= 2 \cdot 6)$, and 35 $(= 5 \cdot 7)$ are not prime. If 4 is written as $2 \cdot 2$, it is factored completely. If 12 is written as $2 \cdot 6$, it is not factored completely, because 6 can be further factored as $2 \cdot 3$. Thus, 12 factored completely would be written $2 \cdot 2 \cdot 3$.

⊙ ❶ Factor Out the Greatest Common Factor

The first step in factoring any polynomial is to look for its **greatest common factor (GCF),** the largest polynomial that is a factor of all the terms in the polynomial.

EXAMPLE 1 **Finding the Greatest Common Factor**

Find the greatest common factor (GCF) of the terms.

(a) $4x, 12$ **(b)** $6x^3, 12x^2, 15x$ **(c)** $4x^3y^4, 8x^2y^3, 12xy^2$

Solution

In each case, look for the largest polynomial that is a factor of each polynomial.

(a) Look at the coefficients first. The largest number that divides into 4 and 12 evenly is 4, so 4, is part of the GCF. Because 12 does not have a variable factor, the GCF is 4.

(b) The largest number that divides evenly into 6, 12, and 15 is 3, so 3 is part of the GCF. Now, look at the variable expressions, x^3, x^2, and x. Since $x^3 = x \cdot x \cdot x$, $x^2 = x \cdot x$, and $x = x$, choose the variable expression with the smallest exponent, $x(= x^1)$, as part of the GCF. The GCF is $3x$.

(c) The largest number that divides into 4, 8, and 12 is 4, so 4 is part of the GCF. Look at the expression involving x. The smallest exponent involving x is 1, so x is part of the GCF. The smallest exponent involving y is y^2, so y^2 is part of the GCF. The GCF is $4xy^2$. ●

Work Smart

When factoring out a variable as a GCF, include the lowest exponent that appears on that variable.

Quick ✓

1. In $(3x + 1)(x - 5) = 3x^2 - 14x - 5$, the polynomials on the left side are called _____ of the polynomial on the right side.

2. If a polynomial cannot be written as the product of two other polynomials (excluding 1 and −1), then the polynomial is said to be _____.

Prepared?...Answers
P1. $24 = 2 \cdot 2 \cdot 2 \cdot 3$ **P2.** $12x - 20$

3. The _____ _____ _____ of a polynomial is the largest polynomial that is a factor of all the terms in the polynomial.

4. To _____ a polynomial means to write the polynomial as the product of two or more polynomials.

5. *True or False* If we write 8 as $2 \cdot 4$, it is factored completely.

In Problems 6–8, find the greatest common factor (GCF) of the terms.

6. $5y, 15$ 7. $4z^3, 10z^2, 12z$ 8. $6x^3y^5, 9x^2y^3, 12xy^4$

Once the GCF is identified, use the Distributive Property to factor out the GCF.

EXAMPLE 2   **How to Factor Out the Greatest Common Factor in a Polynomial**

Factor out the greatest common factor: $4a^2b^2 - 10ab^3 + 18a^3b^4$

Step-by-Step Solution

Step 1: Find the GCF.	$\text{GCF} = 2ab^2$
Step 2: Rewrite each term as the product of the GCF and the remaining factor.	$4a^2b^2 - 10ab^3 + 18a^3b^4 = 2ab^2 \cdot 2a - 2ab^2 \cdot 5b + 2ab^2 \cdot 9a^2b^2$
Step 3: Factor out the GCF.	$= 2ab^2(2a - 5b + 9a^2b^2)$
Step 4: Check	$2ab^2(2a - 5b + 9a^2b^2) = 2ab^2 \cdot 2a - 2ab^2 \cdot 5b + 2ab^2 \cdot 9a^2b^2$ $= 4a^2b^2 - 10ab^3 + 18a^3b^4$

Thus $4a^2b^2 - 10ab^3 + 18a^3b^4 = 2ab^2(2a - 5b + 9a^2b^2)$.

Factoring a Polynomial Using the Greatest Common Factor

Step 1: Identify the greatest common factor (GCF) of the terms.

Step 2: Rewrite each term as the product of the GCF and the remaining factor.

Step 3: Use the Distributive Property to factor out the GCF.

Step 4: Use the Distributive Property to verify that the factorization is correct.

Quick ✓

In Problems 9–11, factor out the greatest common factor.

9. $7z^2 - 14z$ 10. $6y^3 - 14y^2 + 10y$ 11. $2m^4n^2 + 8m^3n^4 - 6m^2n^5$

When the coefficient of the term of highest degree is negative, factor a negative number out of the polynomial as part of the GCF.

EXAMPLE 3   **Factoring Out a Negative GCF**

Factor out the greatest common factor.

(a) $-8a + 16$ **(b)** $-2b^3 + 10b^2 + 8b$

Solution

(a) $-8a + 16 = \overset{\text{GCF} = -8}{-8 \cdot a + (-8) \cdot (-2)}$

$= -8(a - 2)$

Check $-8(a - 2) = -8 \cdot a + (-8) \cdot (-2)$
$= -8a + 16$

$$\text{GCF} = -2b$$

(b) $-2b^3 + 10b^2 + 8b = -2b \cdot b^2 + (-2b) \cdot (-5b) + (-2b) \cdot (-4)$
$= -2b(b^2 - 5b - 4)$

Check $-2b(b^2 - 5b - 4) = -2b \cdot b^2 + (-2b) \cdot (-5b) + (-2b) \cdot (-4)$
$= -2b^3 + 10b^2 + 8b$ ●

Quick ✓

In Problems 12 and 13, factor out the greatest common factor.

12. $-5y^2 + 10y$ **13.** $-3a^3 + 6a^2 - 12a$

Sometimes the greatest common factor is a binomial.

EXAMPLE 4   **Factoring Out a Binomial as the Greatest Common Factor**

Factor out the greatest common factor.

(a) $4x(x - 3) + 5(x - 3)$
(b) $3y(2y + 1) - 5(2y + 1)^2$
(c) $(c + 4)(c - 1) + (5c - 2)(c - 1)$

Solution

$$\text{GCF} = (x - 3)$$

(a) $4x(x - 3) + 5(x - 3) = (x - 3) \cdot 4x + (x - 3) \cdot 5$
$= (x - 3)(4x + 5)$

(b) $3y(2y + 1) - 5(2y + 1)^2 = (2y + 1) \cdot 3y - (2y + 1) \cdot 5(2y + 1)$
$= (2y + 1)(3y - 5(2y + 1))$
$= (2y + 1)(3y - 10y - 5)$
$= (2y + 1)(-7y - 5)$
Factor out -1: $= -(2y + 1)(7y + 5)$

(c) $(c + 4)(c - 1) + (5c - 2)(c - 1) = (c - 1) \cdot (c + 4) + (c - 1) \cdot (5c - 2)$
$= (c - 1)(c + 4 + 5c - 2)$
$= (c - 1)(6c + 2)$
GCF = 2 in $(6c + 2)$: $= 2(c - 1)(3c + 1)$ ●

Quick ✓

In Problems 14 and 15, factor out the greatest common factor.

14. $4a(a - 3) + 3(a - 3)$ **15.** $(w + 2)(w - 5) + (2w + 1)(w - 5)$

▶ ❷ Factor by Grouping

Sometimes a common factor does not occur in every term of a polynomial, but a common factor does occur in some of its terms, and a second common factor occurs in its remaining terms. When this happens, the common factor can be factored out of each group using the Distributive Property. This technique is called **factoring by grouping** and is used often when a polynomial contains four terms.

EXAMPLE 5 How to Factor by Grouping

Factor by grouping: $4x - 4y + ax - ay$

Step-by-Step Solution

Step 1: Group terms with common factors. In this problem the first two terms have a common factor of 4, and the last two terms have a common factor of a.

$$4x - 4y + ax - ay = (4x - 4y) + (ax - ay)$$

Step 2: In each grouping, factor out the common factor.

$$= 4(x - y) + a(x - y)$$

Step 3: Factor out the common factor that remains.

$$= (x - y)(4 + a)$$

Step 4: Check

$$\overset{\text{F} \quad \text{O} \quad \text{I} \quad \text{L}}{(x - y)(4 + a) = 4x + ax - 4y - ay}$$

Rearrange terms: $= 4x - 4y + ax - ay$

Thus $4x - 4y + ax - ay = (x - y)(4 + a)$.

Factoring a Polynomial by Grouping

Step 1: Group terms with common factors. Sometimes it will be necessary to rearrange the terms.

Step 2: In each grouping, factor out the common factor.

Step 3: Factor out the common factor that remains.

Step 4: Check your work.

EXAMPLE 6 Factoring by Grouping

Factor by grouping:

(a) $x^3 + 3x^2 + 2x + 6$

(b) $6x^2 + 9x - 10x - 15$

Solution

(a) First, group terms with common factors. Note that the first two terms, x^3 and $3x^2$, have a common factor of x^2; the last two terms have a common factor of 2.

$$x^3 + 3x^2 + 2x + 6 = (x^3 + 3x^2) + (2x + 6)$$

Factor out common factor: $= x^2(x + 3) + 2(x + 3)$

$$= (x + 3)(x^2 + 2)$$

Check $(x + 3)(x^2 + 2) = x^3 + 2x + 3x^2 + 6$

Rearrange terms: $= x^3 + 3x^2 + 2x + 6$

So $x^3 + 3x^2 + 2x + 6 = (x + 3)(x^2 + 2)$.

(b) $\qquad 6x^2 + 9x - 10x - 15 = (6x^2 + 9x) + (-10x - 15)$

Factor out -5 in the second grouping: $= 3x(2x + 3) + (-5)(2x + 3)$

$$= (2x + 3)(3x - 5)$$

The check is left for you.

Work Smart

In Example 6(a), grouping the first three terms would result in a common factor of x, but factoring out an x would not result in a common factor, such as $x + 3$.

Work Smart

In Example 6(b), notice how we changed $-10x - 15$ to $+ (-10x - 15)$ before we grouped so we would not change the original problem.

Quick ✔

In Problems 16–18, factor by grouping.

16. $5x + 5y + bx + by$ **17.** $w^3 - 3w^2 + 4w - 12$ **18.** $2x^2 + x - 10x - 5$

4.4 Exercises MyMathLab® Exercise numbers in green have complete video solutions in MyMathLab or may be accessed using the QR code to the right.

Problems 1–18 are the Quick ✔ s that follow the **EXAMPLES.**

Building Skills

In Problems 19–36, factor out the greatest common factor. See Objective 1.

19. $5a + 35$

20. $8z + 48$

21. $-3y + 21$

22. $-4b + 32$

23. $14x^2 - 21x$

24. $12a^2 + 45a$

25. $3z^3 - 6z^2 + 18z$

26. $2w^3 + 10w^2 - 14$

27. $-5p^4 + 10p^3 - 25p^2$

28. $-6q^3 + 36q^2 - 48q$

29. $49m^3n + 84mn^3 - 35m^4n^2$

30. $64x^4y^2 - 40x^3y^4 + 96xy^5$

31. $-18z^3 + 14z^2 + 4z$

32. $-18b^3 + 10b^2 + 6b$

33. $5c(3c - 2) - 3(3c - 2)$

34. $6z(5z + 3) + 5(5z + 3)$

35. $(4a + 3)(a - 3) + (2a - 7)(a - 3)$

36. $(5b + 3)(b + 4) + (3b + 1)(b + 4)$

In Problems 37–48, factor by grouping. See Objective 2.

37. $5x + 5y + ax + ay$

38. $8x - 8y + bx - by$

39. $2z^3 + 10z^2 - 5z - 25$

40. $3y^3 + 9y^2 - 5y - 15$

41. $w^2 - 5w + 3w - 15$

42. $p^2 - 3p + 8p - 24$

43. $2x^2 - 8x - 4x + 16$

44. $3a^2 - 15a - 9a + 45$

45. $3x^3 + 15x^2 - 12x^2 - 60x$

46. $2y^3 + 14y^2 - 4y^2 - 28y$

47. $2ax - 2ay - bx + by$

48. $15x^2 - 5xy + 18xy - 6y^2$

Mixed Practice

In Problems 49–58, factor each polynomial completely.

49. $(w + 3)(w - 3) - (w - 2)(w - 3)$

50. $(x + 5)(x - 3) - (x - 1)(x - 3)$

51. $2y^2 + 5y - 4y - 10$

52. $3q^2 + 5q - 12q - 20$

53. $6x^3y^3 + 9x^2y - 21x^3y^2$

54. $8a^4b^2 + 12a^3b^3 - 36ab^4$

55. $x^3 + x^2 + 3x + 3$

56. $c^3 - c^2 + 5c - 5$

57. $x(x - 2) + 3(x - 2)^2$

58. $2y(y + 4) + 3(y + 4)^2$

Math for the Future

In Problems 59–66, expressions that occur in calculus are given. Factor and simplify each expression.

59. $3x^2(4x + 1)^2 + 8x^3(4x + 1)$

60. $2x(3x + 5)^2 + 6x^2(3x + 5)$

61. $3(x + 9)^2(2x + 5) + 2(x + 9)^3$

62. $4(x - 3)^3(3x + 1) + 3(x - 3)^4$

63. $4(2x - 1)(x - 5)^3 + 3(x - 5)^2(2x - 1)^2$

64. $2(x + 3)(6x + 5)^3 + 18(6x + 5)^2(x + 3)^2$

65. $2(x^2 + 1) \cdot 2x(4x - 3)^3 + 3(4x - 3)^2 \cdot 4(x^2 + 1)^2$

66. $3(x^2 + 3)^2 \cdot 2x(2x + 1)^2 + 2(2x + 1) \cdot 2(x^2 + 3)^3$

Applying the Concepts

△ **67.** **Area** Write the area of the shaded region in factored form.

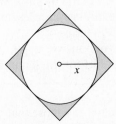

△ **68.** **Area** Write the area of the shaded region in factored form.

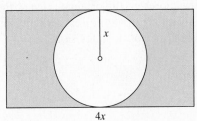

△ **69. Surface Area** The surface area of a cylindrical can whose radius is r inches and height is 4 inches is given by $S = 2\pi r^2 + 8\pi r$. Express the surface area in factored form.

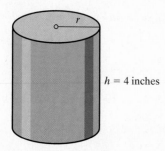

$h = 4$ inches

△ **70. Volume** The volume of a right circular cylinder of height h and radius r inscribed in a sphere of fixed radius R is given by $V = \pi h R^2 - \dfrac{\pi h^3}{4}$. Express the volume of the cylinder in factored form.

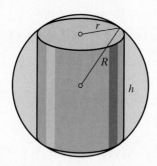

71. Markups and Discounts Suppose that a clothing store marks up its clothes 40% when it buys them from the supplier.

(a) Let x represent the cost of designer shirts purchased from the supplier. Write an algebraic expression representing the selling price of the shirt.

(b) After 1 month, the manager of the clothing store discounts the shirts by 40%. Write a simplified algebraic expression representing the sale price of the shirt in terms of x.

(c) Based on your answer to part (c), is the store selling the shirt for the same price that it paid for it?

72. Summer Clearance Suppose that an electronics store decides to sell last year's model televisions for a 20% discount.

(a) Let x represent the original price of the television. Write an algebraic expression representing the selling price of the television.

(b) After 1 month, the manager of the store discounts the TVs by another 15%. Write a simplified algebraic expression representing the sale price of the televisions in terms of x, the original selling price.

(c) If the original price of the television was $650, what is the sale price after the second discount?

73. Stock Prices Suppose that Christina purchased a stock for x dollars. During the first year, the stock's price rose 15%.

(a) Write an algebraic expression for the price of the stock after the first year in terms of x.

(b) During the second year, the stock's price rose 10%. Write a simplified algebraic expression for the price of the stock after the second year in terms of x.

(c) If the stock was originally purchased for $x = 20, what is the value of the stock after 2 years?

Extending the Concepts

In this section, we factored polynomials over the integers. However, we could also factor polynomials with rational coefficients, such as $\dfrac{1}{3}x + \dfrac{2}{3}$. This polynomial can be factored over the rational numbers by factoring out the greatest common factor $\dfrac{1}{3}$:
$\dfrac{1}{3}x + \dfrac{2}{3} = \dfrac{1}{3}(x + 2)$. *Use this idea to factor out the greatest common factor in Problems 74–77.*

74. $\dfrac{1}{2}x + \dfrac{3}{2}$ **75.** $\dfrac{1}{4}x - \dfrac{7}{4}$

76. $\dfrac{2}{3}x^2 + \dfrac{4}{9}x$ **77.** $\dfrac{1}{5}b^3 + \dfrac{8}{25}b$

78. The Better Deal Which is the better deal: (a) receiving a 30% discount or (b) receiving a 15% discount and then another 15% discount after the first 15% discount was applied? Prove it!

In Problems 79–81, factor out the greater common factor of each expression. Assume n is a positive integer.

79. $x^n + 3x^{n+1} + 6x^{2n}$

80. $2x^{3n} - 8x^{4n} + 16x^{2n}$

81. $4y^{n+3} - 8y^{n+2} + 6y^{n+5}$

4.5 Factoring Trinomials

Objectives

1 Factor Trinomials of the Form $x^2 + bx + c$

2 Factor Trinomials of the Form $ax^2 + bx + c, a \neq 1$

3 Factor Trinomials Using Substitution

Are You Prepared for This Section?

Before getting started, complete the following problems. If you get a problem wrong, go back to the section cited and review the material.

P1. Determine the coefficients of $4x^2 - 9x + 2$. [Section R.5, pp. 41–42]

P2. Find two integers whose sum is 5 and product is 6.

P3. Find two integers whose sum is -4 and product is -32.

P4. Find two integers whose sum is 6 and product is -40.

P5. Find two integers whose sum is -12 and product is 32.

P6. List the factors of 18 whose sum is 11.

P7. List the factors of -24 whose sum is -2.

In this section, trinomials where a, b, and c are integers will be factored. Because the word **quadratic** means "relating to a square," these trinomials are called *quadratic trinomials*.

> ### Definition
> A **quadratic trinomial** is a polynomial of the form $ax^2 + bx + c, a \neq 0$, where a represents the coefficient of the squared (second-degree) term, b represents the coefficient of the linear (first-degree) term, and c represents the constant.

In the trinomial $ax^2 + bx + c$, a is called the leading coefficient. This section begins by looking at quadratic trinomials whose leading coefficient, a, is 1.

1 Factor Trinomials of the Form $x^2 + bx + c$

The goal in factoring a second-degree polynomial of the form $x^2 + bx + c$ is to write it as the product of two first-degree polynomials.

For example,

$$\text{Factored Form} \rightarrow (x - 5)(x + 2) \xrightarrow{\text{Multiplication}} = x^2 - 3x - 10 \leftarrow \text{Product}$$

$$\xleftarrow{\text{Factoring}}$$

The factors of $x^2 - 3x - 10$ are $x - 5$ and $x + 2$. Notice the following:

$$x^2 - 3x - 10 = (x - 5)(x + 2)$$

The sum of -5 and 2 is -3

The product of -5 and 2 is -10

In general, if $x^2 + bx + c = (x + m)(x + n)$, then $mn = c$ and $m + n = b$.

Prepared?...Answers **P1.** 4, -9, 2
P2. 2, 3 **P3.** -8, 4 **P4.** -4, 10
P5. -8, -4 **P6.** 9 and 2 **P7.** -6 and 4

EXAMPLE 1 **How to Factor a Trinomial of the Form $x^2 + bx + c$**

Factor: $x^2 + 8x + 12$

Step-by-Step Solution

Step 1: Look for factors of $c = 12$ whose sum is $b = 8$. Begin by listing all factors of 12 and computing the sum of these factors.

Factors of 12	1, 12	2, 6	3, 4	$-1, -12$	$-2, -6$	$-3, -4$
Sum	13	8	7	-13	-8	-7

Because $2 \cdot 6 = 12$ and $2 + 6 = 8$, let $m = 2$ and $n = 6$.

(continued)

Step 2: Write the trinomial in the form $(x + m)(x + n)$.

$$x^2 + 8x + 12 = (x + 2)(x + 6)$$

Step 3: Check Use FOIL to verify our solution.

$$\begin{array}{cccc} F & O & I & L \end{array}$$
$$(x + 2)(x + 6) = x^2 + 6x + 2x + 2(6)$$
$$= x^2 + 8x + 12$$

Thus $x^2 + 8x + 12 = (x + 2)(x + 6)$. Because multiplication is commutative, the factorization can also be written as $(x + 6)(x + 2)$.

Below is a summary of the steps used in Example 1.

Factoring a Trinomial of the Form $x^2 + bx + c$

Step 1: Find the pair of integers whose product is c and whose sum is b. That is, determine m and n such that $mn = c$ and $m + n = b$.

Step 2: Write $x^2 + bx + c = (x + m)(x + n)$.

Step 3: Check your work by multiplying the binomials.

Work Smart

In a trinomial $x^2 + bx + c$, if both b and c are positive, then m and n must both be positive.

Example 1 shows that if the coefficient of the middle term and the constant are both positive, then both m and n must be positive.

Quick ✓

1. A _____ _____ is a polynomial of the form $ax^2 + bx + c$, where a, b, and c are integers.

2. When factoring a trinomial of the form $x^2 + bx + c$, find pairs of integers m and n such that _____ $= c$ and _____ $= b$.

3. *True or False* $x^2 + 12x + 11 = (x + 11)(x + 1)$

In Problems 4 and 5, factor each trinomial.

4. $y^2 + 9y + 18$ 5. $p^2 + 14p + 24$

EXAMPLE 2

Factoring a Trinomial of the Form $x^2 + bx + c$

Factor: $p^2 - 10p + 24$

Solution

Find factors of $c = 24$ whose sum is $b = -10$. Begin by listing all factors of 24 and computing the sum of these factors.

Factors of 24	1, 24	2, 12	3, 8	4, 6	−1, −24	−2, −12	−3, −8	−4, −6
Sum	25	14	11	10	−25	−14	−11	−10

Because $-4(-6) = 24$ and $-4 + (-6) = -10$, let $m = -4$ and $n = -6$. Write the trinomial in the form $(p + m)(p + n)$.

$$p^2 - 10p + 24 = (p + (-4))(p + (-6))$$
$$= (p - 4)(p - 6)$$

Check

$$\begin{array}{cccc} F & O & I & L \end{array}$$
$$(p - 4)(p - 6) = p^2 - 6p - 4p + (-4)(-6)$$
$$= p^2 - 10p + 24$$

Work Smart

In a trinomial $x^2 + bx + c$, if b is negative and c is positive, then both m and n must be negative.

As illustrated in Example 2, if the coefficient of the middle term is negative and the constant is positive, then m and n must both be negative.

In Problems 6 and 7, factor each trinomial.

6. $q^2 - 6q + 8$ **7.** $x^2 - 8x + 12$

EXAMPLE 3   **Factoring a Trinomial of the Form $x^2 + bx + c, c < 0$**

Factor: $z^2 - 3z - 28$

Solution

Find factors of $c = -28$ whose sum is $b = -3$. Proceed as usual.

Factors of -28	$-1, 28$	$-2, 14$	$-4, 7$	$1, -28$	$2, -14$
Sum	27	12	3	-27	-12

Because $4(-7) = -28$ and $4 + (-7) = -3$, let $m = 4$ and $n = -7$. Write the trinomial in the form $(z + m)(z + n)$.

$$z^2 - 3z - 28 = (z + 4)(z + (-7))$$
$$= (z + 4)(z - 7)$$

Work Smart

In a trinomial $x^2 + bx + c$, if c is negative, then m and n must have opposite signs.

Check

$$(z + 4)(z - 7) = z^2 - 7z + 4z + 4(-7) = z^2 - 3z - 28 \qquad ●$$

Example 3 reveals that if the constant is negative, then m and n must be opposite in sign. In addition, if the coefficient of the middle term is also negative, then the factor with the larger absolute value must be negative.

In Problems 8 and 9, factor each trinomial.

8. $w^2 - 4w - 21$ **9.** $q^2 - 9q - 36$

Remember, a polynomial that cannot be written as the product of two other polynomials is said to be prime.

EXAMPLE 4   **Identifying a Prime Trinomial**

Factor: $y^2 + 4y + 12$

Solution

Look for factors of $c = 12$ whose sum is $b = 4$. Because both b and c are positive, both m and n must be positive, so list only positive factors of 12.

Factors of 12	1, 12	2, 6	3, 4
Sum	13	8	7

There are no factors of 12 whose sum is 4. Therefore, $y^2 + 4y + 12$ is prime.   ●

10. A polynomial that cannot be written as the product of two other polynomials (other than 1 or -1) is a _____ polynomial.

In Problems 11 and 12, factor the trinomial. If the trinomial cannot be factored, say it is prime.

11. $z^2 - 5z + 8$ **12.** $q^2 + 4q - 45$

If a trinomial has more than one variable, take the same approach used for trinomials with one variable. Thus trinomials of the form

$$x^2 + bxy + cy^2$$

factor as

$$(x + my)(x + ny)$$

where $mn = c$ and $m + n = b$.

▶ **EXAMPLE 5** **Factoring Trinomials with Two Variables**

Factor: $p^2 + 6pq - 16q^2$

Solution
The trinomial $p^2 + 6pq - 16q^2$ will factor as $(p + mq)(p + nq)$, where $mn = -16$ and $m + n = 6$. Find factors of $c = -16$ whose sum is $b = 6$. List factors of -16, but because $c = -16$ and $b = 6$, the larger factor of 16 will be positive.

Factors of -16	$-1, 16$	$-2, 8$	$-4, 4$
Sum	15	6	0

Because $-2 \cdot 8 = -16$ and $-2 + 8 = 6$, let $m = -2$ and $n = 8$. Write the trinomial in the form $(p + mq)(p + nq)$.

$$p^2 + 6pq - 16q^2 = (p + (-2)q)(p + 8q)$$
$$= (p - 2q)(p + 8q)$$

Check $(p - 2q)(p + 8q) = p^2 + 8pq - 2pq - 16q^2$
$$= p^2 + 6pq - 16q^2$$ ●

Quick ✓

In Problems 13 and 14, factor each trinomial completely.

13. $x^2 + 8xy + 15y^2$ **14.** $m^2 + mn - 20n^2$

A polynomial is said to be factored completely if each factor in the final factorization is prime. For example, $3x^2 - 6x - 45 = (x - 5)(3x + 9)$ is not factored completely because $(3x + 9)$ has a common factor of 3. However, $3x^2 - 6x - 45 = 3(x - 5)(x + 3)$ is factored completely. The first step in any factoring problem is to look for a greatest common factor.

Work Smart

What's the first thing to look for when factoring any polynomial? A greatest common factor!

▶ **EXAMPLE 6** **Factoring Trinomials with a Common Factor**

Factor: $3u^3 - 9u^2 - 120u$

Solution
Start by noticing the common factor of $3u$ in the trinomial. Factor the $3u$ out:

$$3u^3 - 9u^2 - 120u = 3u(u^2 - 3u - 40)$$

Now factor the trinomial in parentheses, $u^2 - 3u - 40$. Find factors of $c = -40$ whose sum is $b = -3$. List all factors of -40, but since $c = -40$ and $b = -3$, the larger factor (in absolute value) will be negative.

Factors of -40	$1, -40$	$2, -20$	$4, -10$	$5, -8$
Sum	-39	-18	-6	-3

Since $5(-8) = -40$ and $5 + (-8) = -3$, let $m = 5$ and $n = -8$. Write the trinomial in the form $(u + m)(u + n)$.

$$u^2 - 3u - 40 = (u + 5)(u + (-8))$$
$$= (u + 5)(u - 8)$$

Remember, a common factor has already been factored out, so

$$3u^3 - 9u^2 - 120u = 3u(u^2 - 3u - 40)$$
$$= 3u(u + 5)(u - 8)$$

Check

$$3u(u + 5)(u - 8) = 3u(u^2 - 8u + 5u - 40)$$
$$= 3u(u^2 - 3u - 40)$$
$$= 3u^3 - 9u^2 - 120u$$

•

Quick ✓

15. *True or False* The polynomial $(x + 5)(3x - 15)$ is factored completely.

In Problems 16 and 17, factor each trinomial completely.

16. $2x^3 - 12x^2 - 54x$ **17.** $-3z^2 - 21z - 30$

② Factor Trinomials of the Form $ax^2 + bx + c, a \neq 1$

There are two methods for factoring trinomials of the form $ax^2 + bx + c$, where a is not 1:

1. Factoring by grouping

2. Trial and error

Both methods have pros and cons, which will be pointed out on the next page.

▶ Factoring $ax^2 + bx + c, a \neq 1$, by Grouping

Example 7 illustrates how to factor $ax^2 + bx + c, a \neq 1$, by grouping.

EXAMPLE 7   **How to Factor $ax^2 + bx + c, a \neq 1$, by Grouping**

Factor: $3x^2 + 14x + 8$

Step-by-Step Solution

First, notice that $3x^2 + 14x + 8$ has no common factors and that $a = 3, b = 14$, and $c = 8$.

Step 1: Find the value of ac.   The value of $a \cdot c = 3 \cdot 8 = 24$.

Step 2: Find the integers m and n, whose product is 24 and whose sum is 14. Because both 24 and 14 are positive, list only the positive factors of 24.

Factors of 24	1, 24	2, 12	3, 8	4, 6
Sum	25	14	11	14

The integers whose product is 24 and whose sum is 14 are 2 and 12.

Step 3: Write

$$ax^2 + bx + c = ax^2 + mx + nx + c.$$

Write $3x^2 + 14x + 8$ as $3x^2 + 2x + 12x + 8$

$14x = 2x + 12x$

Step 4: Factor the expression in Step 3 by grouping.

$$3x^2 + 2x + 12x + 8 = (3x^2 + 2x) + (12x + 8)$$
$$= x(3x + 2) + 4(3x + 2)$$

Factor out $(3x + 2)$: $= (3x + 2)(x + 4)$

Step 5: Check

$$(3x + 2)(x + 4) = 3x^2 + 12x + 2x + 8$$
$$= 3x^2 + 14x + 8$$

Thus $3x^2 + 14x + 8 = (3x + 2)(x + 4)$.

•

> **Factoring $ax^2 + bx + c, a \neq 1$, by Grouping, Where a, b, and c Have No Common Factors**
>
> **Step 1:** Find the value of ac.
>
> **Step 2:** Find the pair of integers whose product is ac and whose sum is b. That is, find m and n so that $mn = ac$ and $m + n = b$.
>
> **Step 3:** Write $ax^2 + bx + c = ax^2 + mx + nx + c$.
>
> **Step 4:** Factor the expression in Step 3 by grouping.
>
> **Step 5:** Check by multiplying the factors.

EXAMPLE 8 Factoring $ax^2 + bx + c, a \neq 1$, by Grouping

Factor: $12x^2 - x - 6$

Solution

First, notice that $12x^2 - x - 6$ has no common factors and that $a = 12$, $b = -1$, and $c = -6$. The value of $a \cdot c = 12 \cdot (-6) = -72$.

Find the factors of -72 whose sum is -1. Because -72 is negative, one integer will be positive and the other negative. Because -1 is negative, the factor of -72 with the larger absolute value will be negative.

Factors of -72	1, -72	2, -36	3, -24	4, -18	6, -12	8, -9
Sum	-71	-34	-21	-14	-6	-1

The integers whose product is -72 and whose sum is -1 are 8 and -9.

Work Smart

Be careful with the negative sign on $-9x$.

$$-x = 8x - 9x$$

$$12x^2 - x - 6 = 12x^2 + 8x - 9x - 6$$
$$= (12x^2 + 8x) + (-9x - 6)$$
$$= 4x(3x + 2) - 3(3x + 2)$$

Factor out $(3x + 2)$: $= (3x + 2)(4x - 3)$

Check $(3x + 2)(4x - 3) = 12x^2 - 9x + 8x - 6$
$$= 12x^2 - x - 6$$

Thus $12x^2 - x - 6 = (3x + 2)(4x - 3)$.

Quick ✓

18. To factor $2x^2 - 13x + 6$ using grouping, begin by finding factors whose product is ___, and whose sum is ___.

In Problems 19 and 20, factor each trinomial completely.

19. $2b^2 + 7b - 15$ **20.** $10x^2 + 27x + 18$

The advantage of factoring by grouping when factoring trinomials of the form $ax^2 + bx + c, a \neq 1$, is that it is algorithmic (that is, step by step). However, if the product $a \cdot c$ is large, you must find the sum of many pairs of factors. This can be overwhelming. If this is the case, it may be better to try the second method, trial and error.

Using Trial and Error to Factor $ax^2 + bx + c, a \neq 1$

In the trial and error method, list various binomials and use FOIL to find their product until you find the pair of binomials that results in the original trinomial. This method may sound haphazard, but experience and logic can minimize the number of possibilities you must try.

(EXAMPLE 9) **How to Factor $ax^2 + bx + c$, $a \neq 1$, Using Trial and Error**

Factor: $10x^2 + 19x + 6$

Step-by-Step Solution

Always check for common factors first. There are no common factors in $10x^2 + 19x + 6$.

Step 1: List the possibilities for the first terms of each binomial whose product is ax^2.

List all possible ways of representing the first term, $10x^2$.

$$(10x + \_\,)\,(x + \_\,)$$
$$(5x + \_\,)\,(2x + \_\,)$$

Step 2: List the possibilities for the last terms of each binomial whose product is c.

The last term, 6, has the following factors:

$$1 \cdot 6,\, 2 \cdot 3,\, -1 \cdot (-6),\, \text{or} -2 \cdot (-3).$$

Step 3: Write out all the combinations of factors from Steps 1 and 2. Multiply the pairs of binomials until a product that equals the trinomial is found.

Possible Factorization	Product	Possible Factorization	Product
$(10x + 1)\,(x + 6)$	$10x^2 + 61x + 6$	$(5x + 1)\,(2x + 6)$	$10x^2 + 32x + 6$
$(10x + 6)\,(x + 1)$	$10x^2 + 16x + 6$	$(5x + 6)\,(2x + 1)$	$10x^2 + 17x + 6$
$(10x + 2)\,(x + 3)$	$10x^2 + 32x + 6$	$(5x + 2)\,(2x + 3)$	$10x^2 + 19x + 6$
$(10x + 3)\,(x + 2)$	$10x^2 + 23x + 6$	$(5x + 3)\,(2x + 2)$	$10x^2 + 16x + 6$
$(10x - 1)\,(x - 6)$	$10x^2 - 61x + 6$	$(5x - 1)\,(2x - 6)$	$10x^2 - 32x + 6$
$(10x - 6)\,(x - 1)$	$10x^2 - 16x + 6$	$(5x - 6)\,(2x - 1)$	$10x^2 - 17x + 6$
$(10x - 2)\,(x - 3)$	$10x^2 - 32x + 6$	$(5x - 2)\,(2x - 3)$	$10x^2 - 19x + 6$
$(10x - 3)\,(x - 2)$	$10x^2 - 23x + 6$	$(5x - 3)\,(2x - 2)$	$10x^2 - 16x + 6$

The highlighted row is the factorization that works, so $10x^2 + 19x + 6 = (5x + 2)(2x + 3)$.

The solution to Example 9 may make you feel a little overwhelmed—so many possibilities! However, many of the possibilities could have been eliminated with a little thought. For example, both the middle and the last terms are positive, so the factors of c must be positive—this alone eliminates half the possibilities listed in Example 9. Further, because the original polynomial has no common factors, the binomials in the factored form cannot have common factors either. This eliminates four additional possibilities.

When these hints are used, the list in Example 9 becomes

Possible Factorization	Product
$(10x + 1)\,(x + 6)$	$10x^2 + 61x + 6$
$(10x + 3)\,(x + 2)$	$10x^2 + 23x + 6$
$(5x + 6)\,(2x + 1)$	$10x^2 + 17x + 6$
$(5x + 2)\,(2x + 3)$	$10x^2 + 19x + 6$

Not too bad! There are only four possibilities. The steps to factor using trial and error and some helpful hints are summarized on the following page.

> **Factoring $ax^2 + bx + c, a \neq 1$, Using Trial and Error: a, b, and c Have No Common Factors**
>
> **Step 1:** List the possibilities for the first terms of each binomial whose product is ax^2.
> $$(\square x + \ )(\square x + \ ) = ax^2 + bx + c$$
> **Step 2:** List the possibilities for the last terms of each binomial whose product is c:
> $$(\underline{}x + \square)(\underline{}x + \square) = ax^2 + bx + c$$
> **Step 3:** Write all the combinations of factors found in Steps 1 and 2. Multiply the pairs of binomials until you find a product that equals the trinomial.

> **Hints for Using Trial and Error to Factor $ax^2 + bx + c, a \neq 1$**
>
> - Look first for a common factor. If there is a GCF, factor it out. If the leading coefficient is negative, factor it out as part of the GCF.
> - If a and c are positive, then the factors of c must be the same sign as b.
> - The binomials in the factored form cannot have common factors.
> - If the value of b is small, then choose factors of ac that are close to each other. If the value of b is large, then choose factors of ac that are far from each other.
> - If the value of b is correct except for the sign, then interchange the signs in the binomial factors.

▶ (EXAMPLE 10)   **Factoring $ax^2 + bx + c, a \neq 1$, Using Trial and Error**

Factor: $18x^2 + 3x - 10$

Solution
There are no common factors in $18x^2 + 3x - 10$. List all possible ways of representing the first term, $18x^2$.

$$(18x + \underline{})(x + \underline{})$$
$$(9x + \underline{})(2x + \underline{})$$
$$(6x + \underline{})(3x + \underline{})$$

The last term, -10, has the factor pairs $1 \cdot (-10), 2 \cdot (-5), -1 \cdot 10$, and $-2 \cdot 5$.

Before listing some combinations of factors from Steps 1 and 2, notice that the middle term is $3x$. Because this term is positive and small, the binomial factors listed should have outer and inner products that sum to a positive, small number. Therefore, start with $(6x + \underline{})(3x + \underline{})$ and the factors $2 \cdot (-5)$ and $-2 \cdot 5$. Do not use $6x + 2$ or $6x - 2$ as a possible factor, because there is a common factor, 2, in these binomials.

Try $(6x - 5)(3x + 2)$.

$$(6x - 5)(3x + 2) = 18x^2 + 12x - 15x - 10$$
$$= 18x^2 - 3x - 10$$

Close! The only problem is that the middle term has the opposite sign from what is wanted. Therefore, interchange the signs on -5 and 2 in the binomial factors:

$$(6x + 5)(3x - 2) = 18x^2 - 12x + 15x - 10$$
$$= 18x^2 + 3x - 10$$

Thus $18x^2 + 3x - 10 = (6x + 5)(3x - 2)$. ●

The moral of the story in Example 10 is that the name *trial and error* is a bit misleading. You won't have to choose binomial factors haphazardly "until the cows come home," provided that you use the helpful hints and some careful thought.

Quick ✓
In Problems 21 and 22, factor each trinomial completely.

21. $8x^2 + 14x + 5$ **22.** $12y^2 + 32y - 35$

(EXAMPLE 11) **Factoring Trinomials with Two Variables**

Factor: $24x^2 + 13xy - 2y^2$

Solution

There are no common factors in $24x^2 + 13xy - 2y^2$. The trinomial will factor in the form $24x^2 + 13xy - 2y^2 = (\_x + \_y)(\_x + \_y)$.

List all possible ways of representing the first term, $24x^2$.

$$(24x + \_y)(x + \_y)$$
$$(12x + \_y)(2x + \_y)$$
$$(8x + \_y)(3x + \_y)$$
$$(6x + \_y)(4x + \_y)$$

The last term, -2, has the factor pairs $1(-2)$ or $2(-1)$.

Notice that the middle term, $13xy$, is positive and neither large nor small, so the binomial factors listed should have outer and inner products that sum to a positive, midsize number. The only coefficients on y in the factored form are 1 and 2 (ignoring their sign for a moment). But 2 cannot be a factor in the forms $(12x + \_y)(2x + \_y)$ or $(6x + \_y)(4x + \_y)$ because 2 would create a common factor. Therefore, we will start with $(8x + \_y)(3x + \_y)$ and the factors $1(-2)$ and $2(-1)$.

Try $(8x + 1y)(3x - 2y)$.

$$(8x + 1y)(3x - 2y) = 24x^2 - 16xy + 3xy - 2y^2$$
$$= 24x^2 - 13xy - 2y^2$$

Close! The only problem is that the middle term has the opposite sign from what is wanted. Therefore, interchange the signs in the binomial factors:

$$(8x - 1y)(3x + 2y) = 24x^2 + 16xy - 3xy - 2y^2$$
$$= 24x^2 + 13xy - 2y^2$$

Thus $24x^2 + 13xy - 2y^2 = (8x - y)(3x + 2y)$. ●

Quick ✓
In Problems 23–24, factor each trinomial completely.

23. $30x^2 + 7xy - 2y^2$ **24.** $8x^2 - 10xy - 42y^2$

(EXAMPLE 12) **Factoring Trinomials with a Negative Leading Coefficient**

Factor: $-14x^2 + 29x + 15$

Solution

There are no common factors in $-14x^2 + 29x + 15$, but the coefficient of the square term is negative. Factor -1 out of the trinomial to obtain

$$-14x^2 + 29x + 15 = -1(14x^2 - 29x - 15)$$

(continued)

Now factor the expression in parentheses, using either the grouping technique or trial and error, and obtain

$$-14x^2 + 29x + 15 = -1(14x^2 - 29x - 15)$$
$$= -(7x + 3)(2x - 5)$$

Quick ✓

In Problems 25 and 26, factor each trinomial completely.

25. $-6y^2 + 23y + 4$

26. $-9x^2 - 21xy - 10y^2$

▶ ❸ Factor Trinomials Using Substitution

Sometimes a complicated-looking polynomial can be factored by substituting one variable expression for another. This approach is called **factoring by substitution.**

EXAMPLE 13 **Factoring by Substitution**

Factor: $2n^6 - n^3 - 15$

Solution

Notice that $2n^6 - n^3 - 15$ can be written as $2(n^3)^2 - n^3 - 15$ so that the trinomial is in the form $au^2 + bu + c$, where $u = n^3$. Then substitute u for n^3, as shown in the second line below.

Work Smart

To determine whether a trinomial can be factored using substitution, check to see whether the trinomial can be written in the form

$$a(☺)^2 + b(☺) + c$$

where ☺ is some algebraic expression.

$$2n^6 - n^3 - 15 = 2(n^3)^2 - n^3 - 15$$
$$\text{Let } n^3 = u\text{:} \quad = 2u^2 - u - 15$$
$$\text{Factor:} \quad = (2u + 5)(u - 3)$$
$$\text{Let } u = n^3\text{:} \quad = (2n^3 + 5)(n^3 - 3)$$

Thus $2n^6 - n^3 - 15 = (2n^3 + 5)(n^3 - 3)$.

EXAMPLE 14 **Factoring by Substitution**

Factor: $3(x - 3)^2 + 11(x - 3) - 4$

Solution

Notice that $3(x - 3)^2 + 11(x - 3) - 4$ can be written in the form $au^2 + bu + c$, where $u = x - 3$. Substitute u for $x - 3$ to obtain

$$3(x - 3)^2 + 11(x - 3) - 4 = 3u^2 + 11u - 4$$
$$\text{Factor:} \quad = (3u - 1)(u + 4)$$

However, the goal is not to factor an expression in u but to factor an expression in x. Therefore, substitute $x - 3$ for u, as shown in the second line below.

$$3u^2 + 11u - 4 = (3u - 1)(u + 4)$$
$$\text{Let } u = x - 3\text{:} \quad = [3(x - 3) - 1][(x - 3) + 4]$$
$$= (3x - 9 - 1)(x + 1)$$
$$= (3x - 10)(x + 1)$$

Thus $3(x - 3)^2 + 11(x - 3) - 4 = (3x - 10)(x + 1)$.

Quick ✓

27. When factoring $6x^4 - x^2 - 2$ by substitution, let $u = $ __.

28. When factoring $3(2x - 3)^2 + 5(2x - 3) + 2$ by substitution, let $u = $ _____.

In Problems 29 and 30, factor each trinomial by substitution.

29. $y^4 - 2y^2 - 24$

30. $4(x - 3)^2 + 5(x - 3) - 6$

4.5 Exercises MyMathLab®

Exercise numbers in green have complete video solutions in MyMathLab or may be accessed using the QR code to the right.

*Problems **1–30** are the **Quick ✓**s that follow the **EXAMPLES**.*

Building Skills

In Problems 31–46, factor each trinomial completely. See Objective 1.

31. $x^2 + 8x + 15$

32. $x^2 + 8x + 12$

33. $p^2 + 3p - 18$

34. $z^2 + 3z - 28$

35. $r^2 + 10r + 25$

36. $y^2 - 12y + 36$

37. $s^2 + 7s - 60$

38. $q^2 + 2q - 80$

39. $2x^2 - 30x + 112$

40. $3z^2 - 48z + 144$

41. $-w^2 - 2w + 24$

42. $-p^2 + 3p + 54$

43. $x^2 + 7xy + 12y^2$

44. $m^2 + 7mn + 10n^2$

45. $p^3 + 2p^2q - 24pq^2$

46. $x^2y - 4xy^2 - 21y^3$

In Problems 47–62, factor each trinomial completely. See Objective 2.

47. $2p^2 - 15p - 8$

48. $3z^2 - 13z - 10$

49. $4y^2 - 11y + 6$

50. $6x^2 - 37x + 6$

51. $8s^2 + 2s - 3$

52. $12r^2 + 11r - 15$

53. $16z^2 + 8z - 15$

54. $18y^2 + 43y - 5$

55. $-18y^2 - 17y - 4$

56. $-20r^2 - 23r - 6$

57. $2x^2 + 11xy - 21y^2$

58. $3m^2 + 7mn - 6n^2$

59. $12r^2 - 69rs + 45s^2$

60. $12r^2 - 50rs + 8s^2$

61. $24r^2 + 23rs - 12s^2$

62. $18x^2 + 37xy - 20y^2$

In Problems 63–72, factor each trinomial completely. See Objective 3.

63. $x^4 + 3x^2 + 2$

64. $y^4 + 5y^2 + 6$

65. $m^2n^2 + 5mn - 14$

66. $r^2s^2 + 8rs - 48$

67. $(x + 1)^2 - 6(x + 1) - 16$

68. $(y - 3)^2 + 3(y - 3) + 2$

69. $(3r - 1)^2 - 9(3r - 1) + 20$

70. $(5z - 3)^2 - 12(5z - 3) + 32$

71. $2(y - 3)^2 + 13(y - 3) + 15$

72. $3(z + 3)^2 + 14(z + 3) + 8$

Mixed Practice

In Problems 73–104, factor each polynomial completely. If the polynomial cannot be factored, say it is prime. Be sure to look for a greatest common factor.

73. $10w^2 + 41w + 21$

74. $8q^2 + 26q + 11$

75. $4(2y + 1)^2 - 3(2y + 1) - 1$

76. $2(3z - 1)^2 + 3(3z - 1) + 1$

77. $12x^2 + 23xy + 10y^2$

78. $24m^2 + 58mn + 9n^2$

79. $y^2 + 2y - 27$

80. $t^2 - 5t + 8$

81. $x^4 + 6x^2 + 8$

82. $a^6 + 7a^3 + 12$

83. $z^6 + 9z^3 + 20$

84. $r^6 - 6r^3 + 8$

85. $r^2 - 12rs + 32s^2$

86. $p^2 - 14pq + 45q^2$

87. $8(z + 1)^2 + 2(z + 1) - 1$

88. $9(a + 2)^2 - 10(a + 2) + 1$

89. $3x^2 - 7x - 12$

90. $5w^2 - 10w + 12$

91. $2x^2 + 12x - 54$

92. $3y^2 - 6y - 189$

93. $-3r^2 + 39r - 120$

94. $-4s^2 - 32s - 48$

95. $-16m^2 + 12m + 70$

96. $-24y^2 - 39y + 18$

97. $48z^2 + 124z + 28$

98. $48w^2 + 20w - 42$

99. $3x^3 - 6x^2 - 240x$

100. $4x^3 - 52x^2 + 144x$

101. $8x^3y^2 - 76x^2y^2 + 140xy^2$

102. $-24m^3n - 18m^2n + 27mn$

103. $70r^4s - 36r^3s - 16r^2s$

104. $54x^3y + 33x^2y - 72xy$

Applying the Concepts

105. How to Make an Open Box An open box with a rectangular base is to be made from a piece of cardboard that is 24 by 30 inches by cutting a square piece of cardboard from each corner and turning up the sides. See the illustration. The volume V of the box, as a function of the length x of the side of the square cut from each corner, is

$$V(x) = 4x^3 - 108x^2 + 720x$$

(a) Find and interpret $V(3)$.

(b) Completely factor the function
$V(x) = 4x^3 - 108x^2 + 720x$.

(c) Find $V(3)$ using the factored form found in part (b).

(d) Did you find it easier to evaluate $V(3)$ in part (a) or in part (c)?

106. Projectile Motion A boy is standing on a cliff that is 240 feet high and throws a rock out toward the ocean. The height of the rock above the sea can be described by the function in which t is time in seconds and s is the height in feet:

$$s(t) = -16t^2 + 32t + 240$$

(a) Find and interpret $s(3)$.

(b) Completely factor the function
$s(t) = -16t^2 + 32t + 240$.

(c) Find $s(3)$ using the factored form found in part (b).

(d) Did you find it easier to evaluate $s(3)$ in part (a) or in part (c)?

107. Suppose that we know one factor of $6x^2 - 11x - 10$ is $3x + 2$. What is the other factor?

108. Suppose that we know one factor of $8x^2 + 22x - 21$ is $2x + 7$. What is the other factor?

Extending the Concepts

In this section, we factored polynomials with integer coefficients. We can also factor polynomials with rational coefficients. One technique for doing this is to first factor out the coefficient of the highest-degree term as a greatest common factor as follows:

$$\frac{1}{3}x^2 + \frac{4}{3}x + 1 = \frac{1}{3}(x^2 + 4x + 3)$$

$$= \frac{1}{3}(x + 3)(x + 1)$$

Use this technique in Problems 109–114.

109. $\frac{1}{2}x^2 + 3x + 4$

110. $\frac{1}{4}x^2 + 2x + 3$

111. $\frac{1}{3}p^2 - \frac{2}{3}p - 1$

112. $\frac{1}{4}z^2 + \frac{1}{2}z - \frac{15}{4}$

113. $\frac{4}{3}a^2 - \frac{8}{3}a - 32$

114. $\frac{3}{8}b^2 - \frac{15}{8}b - 9$

Math for the Future

In College Algebra, you will be required to factor expressions such as those found in Problems 115–118. Completely factor the following expressions. Hint: Factor by substitution.

115. $2^{2n} - 4 \cdot 2^n - 5$

116. $3^{2x} - 11 \cdot 3^x + 10$

117. $2^{2x} - 2^x - 12$

118. $4^x - 6 \cdot 2^x + 8$

Explaining the Concepts

119. State the circumstances for which trial and error is a better approach than grouping when factoring $ax^2 + bx + c, a \neq 1$.

120. When factoring any polynomial, what is always the first step?

121. A student factored the trinomial $3p^2 - 9p - 30$ as $(3p + 6)(p - 5)$ on an exam but received only partial credit. Can you explain why?

122. How can you tell whether a trinomial is prime? Make up an example to demonstrate your reasoning.

4.6 Factoring Special Products

Objectives

1. Factor Perfect Square Trinomials
2. Factor the Difference of Two Squares
3. Factor the Sum or Difference of Two Cubes

Are You Prepared for This Section?

Before getting started, complete the following problems. If you get a problem wrong, go back to the section cited and review the material.

P1. What is 1^2? 2^2? 3^2? 4^2? 5^2?   [Section R.4, pp. 33–34]

P2. What is $\left(-\dfrac{3}{2}\right)^2$?   [Section R.4, pp. 33–34]

This section will look at polynomials that can be categorized as having special formulas for factoring.

1 Factor Perfect Square Trinomials

Recall the perfect square trinomials from Section 4.2:

$$(A + B)^2 = A^2 + 2AB + B^2$$
$$(A - B)^2 = A^2 - 2AB + B^2$$

Reverse these formulas to obtain a method for factoring perfect square trinomials.

> **Perfect Square Trinomials**
>
> $$A^2 + 2AB + B^2 = (A + B)^2$$
> $$A^2 - 2AB + B^2 = (A - B)^2$$

Work Smart

The perfect squares are $1^2 = 1$, $2^2 = 4$, $3^2 = 9$, and so on. Any variable raised to an even exponent is a perfect square. So $x^2, x^4 = (x^2)^2, x^6 = (x^3)^2$ are all perfect squares.

A polynomial must satisfy two conditions to be a perfect square trinomial:

1. The first and last terms must be perfect squares. Examples of perfect squares are
$$81 = 9^2, \, 4x^2 = (2x)^2, \text{ and } 25a^4 = (5a^2)^2$$

2. The middle term must equal 2 or -2 times the product of the expressions being squared in the first and last terms.

Perfect square trinomials can be factored using the methods introduced in the last section; however, they can be factored more quickly using the formulas above.

EXAMPLE 1 | **Factoring Perfect Square Trinomials**

Factor:

    (a) $z^2 - 10z + 25$ **(b)** $9x^2 + 48xy + 64y^2$ **(c)** $32m^4 - 48m^2 + 18$

Solution

    (a) The first term, z^2, and the third term, $25 = 5^2$, are perfect squares. Because the middle term, $-10z$, equals $-2(z \cdot 5)$, this is a perfect square trinomial. So

$$z^2 - 10z + 25 = z^2 - 2 \cdot z \cdot 5 + 5^2$$

$$A = z, B = 5; A^2 - 2AB + B^2 = (A - B)^2: = (z - 5)^2$$

(continued)

(b) The first term, $9x^2 = (3x)^2$, and the third term, $64y^2 = (8y)^2$, are perfect squares. Because the middle term, $48xy$, equals $2(3x \cdot 8y)$, this is a perfect square trinomial. So

$$9x^2 + 48xy + 64y^2 = (3x)^2 + 2(3x \cdot 8y) + (8y)^2$$
$$A = 3x, B = 8y; A^2 + 2AB + B^2 = (A + B)^2: = (3x + 8y)^2$$

Work Smart

Notice that the sign in the binomial matches the sign of the middle term.

(c) Remember, in any factoring problem first look for a common factor.

$$32m^4 - 48m^2 + 18 = 2(16m^4 - 24m^2 + 9)$$

The first term in parentheses, $16m^4 = (4m^2)^2$, and the third term, $9 = 3^2$, are perfect squares. Because the middle term equals $-2(4m^2 \cdot 3)$, this is a perfect square trinomial. So

$$16m^4 - 24m^2 + 9 = (4m^2)^2 - 2 \cdot (4m^2 \cdot 3) + 3^2$$
$$= (4m^2 - 3)^2$$

Therefore,

$$32m^4 - 48m^2 + 18 = 2(16m^4 - 24m^2 + 9)$$
$$= 2(4m^2 - 3)^2$$

●

Quick ✓

1. A trinomial of the form $A^2 + 2AB + B^2$ or $A^2 - 2AB + B^2$ is called a _____ _____ _____.

2. $A^2 - 2AB + B^2 = $ _____.

3. *True or False* $4x^2 + 6x + 9$ is a perfect square trinomial.

In Problems 4–6, factor each trinomial completely.

4. $x^2 - 18x + 81$ 5. $4x^2 + 20xy + 25y^2$ 6. $18p^4 - 84p^2 + 98$

▶ ❷ **Factor the Difference of Two Squares**

Another special product introduced in Section 4.2 was the difference of two squares:

$$(A - B)(A + B) = A^2 - B^2$$

Difference of Two Squares

$$A^2 - B^2 = (A - B)(A + B)$$

EXAMPLE 2   **Factoring the Difference of Two Squares**

Factor:

(a) $y^2 - 100$ **(b)** $9x^2 - 16y^4$

Solution

(a) Notice that $y^2 - 100$ is the difference of two squares, y^2 and $100 = 10^2$. So

$$y^2 - 100 = y^2 - 10^2$$
$$= (y - 10)(y + 10)$$

(b) Notice that $9x^2 - 16y^4$ is the difference of two squares, $9x^2 = (3x)^2$ and $16y^4 = (4y^2)^2$.

Work Smart

Do not forget that you can always check your answer by multiplying out the factored form.

$$9x^2 - 16y^4 = (3x)^2 - (4y^2)^2$$
$$= (3x - 4y^2)(3x + 4y^2)$$

●

Quick ✓

7. $16z^2 - 25$ is called a _____ of ___ _____ and factors into two binomials.

8. $P^2 - Q^2 =$ _____.

In Problems 9–11, factor completely.

9. $z^2 - 16$ **10.** $16m^2 - 81n^2$

11. $4a^2 - 9b^4$

EXAMPLE 3 **Factoring the Difference of Two Squares**

Factor:

(a) $32x^4 - 2$ **(b)** $x^2 + 10x + 25 - y^2$

Solution

(a) Remember to look for a common factor first. This polynomial has a common factor of 2.

$$32x^4 - 2 = 2(16x^4 - 1)$$

Difference of Two Squares, $16x^4 = (4x^2)^2, 1 = 1^2$: $= 2[\,(4x^2)^2 - 1^2\,]$

$A = 4x^2, B = 1; A^2 - B^2 = (A - B)(A + B)$: $= 2(4x^2 - 1)(4x^2 + 1)$

We are not quite finished. Notice that $4x^2 - 1$ is the difference of two squares, $4x^2 = (2x)^2$ and $1 = 1^2$: $= 2((2x)^2 - 1^2)(4x^2 + 1)$

$A = 2x, B = 1; A^2 - B^2 = (A - B)(A + B)$: $= 2(2x - 1)(2x + 1)(4x^2 + 1)$

Check $2(2x - 1)(2x + 1)(4x^2 + 1) = 2(4x^2 - 1)(4x^2 + 1)$

Use FOIL: $= 2(16x^4 + 4x^2 - 4x^2 - 1)$

$= 2(16x^4 - 1)$

$= 32x^4 - 2$

Thus $32x^4 - 2 = 2(2x - 1)(2x + 1)(4x^2 + 1)$.

Work Smart

Perfect square trinomials are trinomials of the form
$A^2 + 2AB + B^2 = (A + B)^2$
$A^2 - 2AB + B^2 = (A - B)^2$

(b) Remember, when there are four or more terms, try to factor by grouping. However, notice that the first three terms, $x^2 + 10x + 25$, form a perfect square trinomial (which factors into a perfect square) and the last term, y^2, is a perfect square.

$$x^2 + 10x + 25 - y^2 = (x^2 + 10x + 25) - y^2$$

$$= (x^2 + 2(5)(x) + 5^2) - y^2$$

$x^2 + 2(5)x + 5^2 = (x + 5)^2$: $= (x + 5)^2 - y^2$

$(x + 5)^2 - y^2$ is the difference of two squares, where $A = (x + 5)$ and $B = y$: $= (x + 5 - y)(x + 5 + y)$

Check

$$(x + 5 - y)(x + 5 + y) = (x + 5 - y) \cdot x + (x + 5 - y) \cdot 5 + (x + 5 - y) \cdot y$$

$$= x^2 + 5x - xy + 5x + 25 - 5y + xy + 5y - y^2$$

$$= x^2 + 10x + 25 - y^2$$

Thus $x^2 + 10x + 25 - y^2 = (x + 5 - y)(x + 5 + y)$.   ●

You may wonder whether the sum of two squares factors. **The sum of two squares, $a^2 + b^2$, is prime and does not factor over the integers.** Therefore, binomials such as $x^2 + 4$ or $4y^2 + 81$ are prime.

Quick ✓

12. *True or False* The sum of two squares, such as $z^2 + 25$, is prime.

In Problems 13 and 14, factor completely.

13. $3b^4 - 48$

14. $p^2 - 8p + 16 - q^2$

▶ ❸ Factor the Sum or Difference of Two Cubes

Work Smart

The perfect cubes are $1^3 = 1$, $2^3 = 8, 3^3 = 27$, and so on. Any variable raised to a multiple of 3 is a perfect cube. Thus $x^3, x^6 = (x^2)^3, x^9 = (x^3)^3$ are all perfect cubes.

Consider the following products:

$$(A + B)(A^2 - AB + B^2) = A^3 - A^2B + AB^2 + A^2B - AB^2 + B^3$$
$$= A^3 + B^3$$
$$(A - B)(A^2 + AB + B^2) = A^3 + A^2B + AB^2 - A^2B - AB^2 - B^3$$
$$= A^3 - B^3$$

These products show that the sum or difference of two cubes factor as follows:

The Sum of Two Cubes

$$A^3 + B^3 = (A + B)(A^2 - AB + B^2)$$

The Difference of Two Cubes

$$A^3 - B^3 = (A - B)(A^2 + AB + B^2)$$

EXAMPLE 4 **Factoring the Sum or Difference of Two Cubes**

Factor:

(a) $x^3 - 27$ **(b)** $8m^3 + 125n^6$

Solution

(a) $x^3 - 27$ is the difference of two cubes, x^3, and $27 = 3^3$. Let $A = x$ and $B = 3$ in the factoring formula for the difference of two cubes.

$$A^3 - B^3 = (A - B)(A^2 + AB + B^2)$$

$$x^3 - 27 = x^3 - 3^3 = (x - 3)(x^2 + x(3) + 3^2)$$
$$= (x - 3)(x^2 + 3x + 9)$$

(b) $8m^3 + 125n^6$ is the sum of two cubes, $8m^3 = (2m)^3$ and $125n^6 = (5n^2)^3$. Let $A = 2m$ and $B = 5n^2$ in the factoring formula for the sum of two cubes.

$$8m^3 + 125n^6 = (2m)^3 + (5n^2)^3$$
$$= (2m + 5n^2)(2m)^2 - (2m)(5n^2) + (5n^2)^2$$
$$= (2m + 5n^2)(4m^2 - 10mn^2 + 25n^4)$$

Quick ✓

15. $A^3 + B^3 = ($_____$)($_____$)$.

16. $A^3 - B^3 = ($_____$)($_____$)$.

In Problems 17 and 18, factor completely.

17. $z^3 + 64$

18. $125p^3 - 216q^6$

EXAMPLE 5   **Factoring the Sum or Difference of Two Cubes**

Factor:

(a) $27x^3 - 216y^6$ (b) $(x - 4)^3 + 8x^3$

Solution

(a) $27x^3 - 216y^6$ is the difference of two cubes, $27x^3 = (3x)^3$ and $216y^6 = (6y^2)^3$. However, there is a common factor of 27 in the binomial $27x^3 - 216y^6$. How was this recognized as the common factor? Notice that the two cubes, $(3x)^3$ and $(6y^2)^3$, contain a 3 and a 6. Because the common factor, 3, is being cubed, $3^3 = 27$ is a common factor. Therefore, factor out the 27.

$$27x^3 - 216y^6 = 27(x^3 - 8y^6)$$
$$= 27(x^3 - (2y^2)^3)$$
$$A = x, B = 2y^2: = 27(x - 2y^2)(x^2 + x \cdot 2y^2 + (2y^2)^2)$$
$$= 27(x - 2y^2)(x^2 + 2xy^2 + 4y^4)$$

(b) Notice that this is the sum of two cubes, $(x - 4)^3$ and $8x^3 = (2x)^3$. Let $A = x - 4$ and $B = 2x$ in the factoring formula for the sum of two cubes.

$$(x - 4)^3 + 8x^3 = (x - 4)^3 + (2x)^3$$
$$= (x - 4 + 2x)[(x - 4)^2 - (x - 4) \cdot 2x + (2x)^2]$$
Combine like terms; multiply: $= (3x - 4)(x^2 - 8x + 16 - 2x^2 + 8x + 4x^2)$
Combine like terms: $= (3x - 4)(3x^2 + 16)$ ●

Quick ✓

In Problems 19 and 20, factor completely.

19. $32m^3 + 500n^6$

20. $(x + 1)^3 - 27x^3$

4.6 Exercises MyMathLab® Exercise numbers in green have complete video solutions in MyMathLab or may be accessed using the QR code to the right.

Problems 1–20 are the Quick ✓s that follow the EXAMPLES.

Building Skills

In Problems 21–40, factor each perfect square trinomial completely. See Objective 1.

21. $x^2 + 4x + 4$

22. $y^2 + 6y + 9$

23. $36 + 12w + w^2$

24. $49 - 14d + d^2$

25. $4x^2 + 4x + 1$

26. $9z^2 - 6z + 1$

27. $9p^2 - 30p + 25$

28. $16y^2 - 24y + 9$

29. $25a^2 + 90a + 81$

30. $36b^2 + 84b + 49$

31. $9x^2 + 24xy + 16y^2$

32. $4a^2 + 20ab + 25b^2$

33. $3w^2 - 30w + 75$

34. $4c^2 - 24c + 36$

35. $-5t^2 - 70t - 245$

36. $-2a^2 - 32a - 128$

37. $32a^2 - 80ab + 50b^2$

38. $12x^2 - 84xy + 147y^2$

39. $z^4 - 6z^2 + 9$

40. $b^4 + 8b^2 + 16$

In Problems 41–56, factor the difference of two squares completely. See Objective 2.

41. $x^2 - 9$

42. $z^2 - 64$

43. $4 - y^2$

44. $81 - a^2$

45. $4z^2 - 9$

46. $16y^2 - 81$

47. $100m^2 - 81n^2$

48. $x^4 - 9y^2$

49. $m^4 - 36n^6$

50. $x^8 - 100y^4$

51. $8p^2 - 18q^2$

52. $12m^2 - 75n^2$

53. $80p^2r - 245b^2r$

54. $36x^2z - 64y^2z$

55. $(x + y)^2 - 9$

56. $16 - (x - y)^2$

In Problems 57–72, factor the sum or difference of two cubes completely. See Objective 3.

57. $x^3 - 8$

58. $z^3 + 64$

59. $125 + m^3$

60. $216 - n^3$

61. $x^6 - 64y^3$

62. $m^6 - 27n^3$

63. $24x^3 - 375y^3$

64. $16m^3 + 54n^3$

65. $(p + 1)^3 - 27$

66. $(y - 2)^3 - 8$

67. $(3y + 1)^3 + 8y^3$

68. $(2z + 3)^3 + 27z^3$

69. $y^6 + z^9$

70. $m^9 + n^{12}$

71. $y^9 - 1$

72. $x^{12} - 1$

Mixed Practice

In Problems 73–98, factor each polynomial completely.

73. $25x^2 - y^2$

74. $9a^2 - b^2$

75. $8x^3 + 27$

76. $64x^3 - 125$

77. $z^2 - 8z + 16$

78. $p^2 - 20p + 100$

79. $5x^4 - 40xy^3$

80. $3m^4 - 81mn^3$

81. $49m^2 - 42mn + 9n^2$

82. $81p^2 - 72pq + 16q^2$

83. $4x^2 + 16$

84. $-3y^2 - 27$

85. $y^4 - 8y^2 + 16$

86. $p^4 - 18p^2 + 81$

87. $4a^2b^2 + 12ab + 9$

88. $9m^2n^2 - 30mn + 25$

89. $p^2 + 2p + 4$

90. $y^2 - 3y + 9$

91. $n^2 + 13n + 36$

92. $t^2 + 10t + 16$

93. $2n^2 - 2m^2 + 40m - 200$

94. $2a^2 - 2b^2 - 24b - 72$

95. $3y^3 - 24$

96. $-5a^3 - 40$

97. $16x^2 + 24xy + 9y^2 - 100$

98. $36m^2 + 12mn + n^2 - 81$

Applying the Concepts

△ *In Problems 99–102, find an expression in factored form for the area of the shaded region.*

99.

100.

101.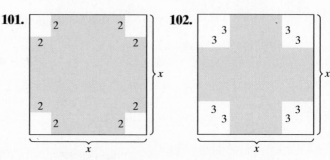

102.

△ *In Problems 103–106, find an expression in factored form for the area or volume of the shaded region.*

103. Circle: $Area = \pi r^2$

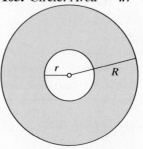

104. Cylinder: $Volume = \pi r^2 h$

105. Rectangular solid: $Volume = lwh$

106. Sphere: $Volume = \frac{4}{3}\pi r^3$

△ **107. A Perfect Square** What is a perfect square? The figure below shows a square whose dimensions are $(a + b)$ by $(a + b)$. The area of the square is therefore $(a + b)^2$. Write the area of the square as the sum of the areas of the four quadrilaterals in the figure. Then show that this sum is equal to $(a + b)^2$.

△ **108. Difference of Two Squares** What is the difference of two squares? The figure below shows two squares. The length of the sides on the "outer" square is a and the length of the sides on the "inner" square (in orange) is b. The area of the shaded region (in blue) is $a^2 - b^2$. Express the area of the blue shaded region as the sum of the two shaded regions in terms of a and b. Conclude that $a^2 - b^2 = (a + b)(a - b)$.

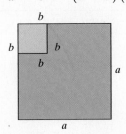

Extending the Concepts

109. Determine two values of b that will make $4x^2 + bx + 81$ a perfect square trinomial. How did you find these values?

110. Determine the value of c that will make $16y^2 + 24y + c$ a perfect square trinomial. How did you find this value?

111. What has to be added to $x^2 + 18x$ to make it a perfect square trinomial?

112. What has to be added to $4x^2 + 36x$ to make it a perfect square trinomial?

In this section, we factored polynomials with integer coefficients. We can also factor polynomials with rational number coefficients. For example,

$$\frac{x^2}{25} - \frac{1}{4} = \left(\frac{x}{5}\right)^2 - \left(\frac{1}{2}\right)^2$$
$$= \left(\frac{x}{5} - \frac{1}{2}\right)\left(\frac{x}{5} + \frac{1}{2}\right)$$

Use this technique to factor Problems 113–120.

113. $b^2 - 0.4b + 0.04$ **114.** $x^2 + 0.6x + 0.09$

115. $9b^2 - \dfrac{1}{25}$ **116.** $100x^2 - \dfrac{1}{81}$

117. $\dfrac{x^2}{9} - \dfrac{y^2}{25}$ **118.** $\dfrac{a^2}{36} - \dfrac{b^2}{49}$

119. $\dfrac{x^3}{8} - \dfrac{y^3}{27}$ **120.** $\dfrac{a^3}{27} + \dfrac{b^3}{64}$

4.7 Factoring: A General Strategy

Objectives

1. Factor Polynomials Completely
2. Write Polynomial Functions in Factored Form

❶ Factor Polynomials Completely

This section begins by putting together all the factoring techniques we have discussed in Sections 4.4–4.6. The following steps should be followed for any factoring problem.

> **Steps for Factoring**
>
> **Step 1:** Factor out the greatest common factor (GCF), if any exists.
>
> **Step 2:** Count the number of terms.
>
> **Step 3:** **(a)** Two terms
>
> - Is it the difference of two squares? If so,
>
> $$A^2 - B^2 = (A - B)(A + B)$$
>
> - Is it the sum of two squares? If so, stop! The expression is prime.

(continued)

- Is it the difference of two cubes? If so,
$$A^3 - B^3 = (A - B)(A^2 + AB + B^2)$$
- Is it the sum of two cubes? If so,
$$A^3 + B^3 = (A + B)(A^2 - AB + B^2)$$

(b) Three terms

- Is it a perfect square trinomial? If so,
$$A^2 + 2AB + B^2 = (A + B)^2 \quad \text{or} \quad A^2 - 2AB + B^2 = (A - B)^2$$
- Is the coefficient of the square term 1? If so,
$$x^2 + bx + c = (x + m)(x + n), \text{where } mn = c \text{ and } m + n = b$$
- Is the coefficient of the square term different from 1? If so, factor using grouping or trial and error.

(c) Four terms

- Use factoring by grouping

Step 4: Check your work by multiplying the factors.

Work Smart: Study Skills

There are a lot of formulas you need to know for factoring. When working problems, write down the formulas you're using each time. This will help get them to your long-term memory.

▶ **EXAMPLE 1** **How to Factor Completely**

Factor: $8x^2 - 16x - 42$

Step-by-Step Solution

Step 1: Factor out the greatest common factor (GCF), if any exists.

The GCF is 2, so factor out 2: $8x^2 - 16x - 42 = 2(4x^2 - 8x - 21)$

Step 2: Count the number of terms of the polynomial in parentheses.

The polynomial in parentheses has three terms.

Step 3: The trinomial in parentheses, $4x^2 - 8x - 21$, is not a perfect square trinomial. Because the leading coefficient, 4, and the constant, -21, aren't that big, factor by grouping.

$$ac = 4(-21) = -84$$

Because $6(-14) = -84$ and $6 + (-14) = -8$,

$$2(4x^2 - 8x - 21) = 2(4x^2 + 6x - 14x - 21)$$
$$= 2[(4x^2 + 6x) + (-14x - 21)]$$

GCF in 1st grouping is 2x:
GCF in 2nd grouping is –7: $= 2[2x(2x + 3) - 7(2x + 3)]$

Factor out 2x + 3: $= 2(2x + 3)(2x - 7)$

Step 4: Check

$$2(2x + 3)(2x - 7) = 2(4x^2 - 14x + 6x - 21)$$
$$= 2(4x^2 - 8x - 21)$$
Distribute: $= 8x^2 - 16x - 42$

The answer checks, so $8x^2 - 16x - 42 = 2(2x + 3)(2x - 7)$. ●

Quick ✓

In Problems 1 and 2, factor each polynomial completely.

1. $2p^2q - 8pq^2 - 90q^3$

2. $-45x^2y + 66xy + 27y$

EXAMPLE 2   **How to Factor Completely**

Factor: $9p^2 - 25q^2$

Step-by-Step Solution

Step 1: Factor out the greatest common factor (GCF), if any exists.

There is no GCF.

Step 2: Count the number of terms.

There are two terms.

Step 3: Because both the first term, $9p^2 = (3p)^2$, and the second term, $25q^2 = (5q)^2$, are perfect squares, this is a difference of two squares.

$$9p^2 - 25q^2 = \overbrace{(3p)}^{A}{}^2 - \overbrace{(5q)}^{B}{}^2$$
$$A^2 - B^2 = (A - B)(A + B): \quad = (3p - 5q)(3p + 5q)$$

Step 4: Check

$$(3p - 5q)(3p + 5q) = 9p^2 + 15pq - 15pq - 25q^2$$
$$\text{Combine like terms:} \quad = 9p^2 - 25q^2$$

Thus $9p^2 - 25q^2 = (3p - 5q)(3p + 5q)$. ●

Quick ✓

In Problems 3 and 4, factor each polynomial completely.

3. $81x^2 - 100y^2$

4. $-3m^2n + 147n$

EXAMPLE 3   **Factoring Completely**

Factor: $16x^2 + 112xy + 196y^2$

Solution

First look for, and find, the greatest common factor, 4.

$$16x^2 + 112xy + 196y^2 = 4(4x^2 + 28xy + 49y^2)$$

The polynomial in parentheses has three terms. The first term is a perfect square, $4x^2 = (2x)^2$. The third term is also a perfect square, $49y^2 = (7y)^2$. The middle term equals $2 \cdot 2x \cdot 7y$. This polynomial is a perfect square trinomial.

$$A = 2x, B = 7y: \quad 4(4x^2 + 28xy + 49y^2) = 4[(2x)^2 + 2(2x)(7y) + (7y)^2]$$
$$A^2 + 2AB + B^2 = (A + B)^2: \quad = 4(2x + 7y)^2$$

Check $4(2x + 7y)^2 = 4(2x + 7y)(2x + 7y)$
$$\text{Use FOIL:} \quad = 4(4x^2 + 14xy + 14xy + 49y^2)$$
$$\text{Combine like terms:} \quad = 4(4x^2 + 28xy + 49y^2)$$
$$\text{Distribute:} \quad = 16x^2 + 112xy + 196y^2$$

Thus $16x^2 + 112xy + 196y^2 = 4(2x + 7y)^2$. ●

Quick ✓

In Problems 5 and 6, factor each polynomial completely.

5. $p^2 - 16pq + 64q^2$

6. $20x^2 + 60x + 45$

EXAMPLE 4 **Factoring Completely**

Factor: $8m^3 + 27n^6$

Solution

The polynomial does not have a common factor and it has two terms. Because the first term is a perfect cube, $8m^3 = (2m)^3$, and the second term is a perfect cube, $27n^6 = (3n^2)^3$, $8m^3 + 27n^6$, is the sum of two cubes.

$$A = 2m, B = 3n^2: \quad 8m^3 + 27n^6 = (2m)^3 + (3n^2)^3$$
$$A^3 + B^3 = (A + B)(A^2 - AB + B^2): \quad = (2m + 3n^2)[(2m)^2 - (2m)(3n^2) + (3n^2)^2]$$
$$= (2m + 3n^2)(4m^2 - 6mn^2 + 9n^4)$$

Check $(2m + 3n^2)(4m^2 - 6mn^2 + 9n^4) = 8m^3 - 12m^2n^2 + 18mn^4$
$$+ 12m^2n^2 - 18mn^4 + 27n^6$$

Combine like terms: $= 8m^3 + 27n^6$

Thus $8m^3 + 27n^6 = (2m + 3n^2)(4m^2 - 6mn^2 + 9n^4)$. ●

Quick ✓

In Problems 7 and 8, factor each polynomial completely.

7. $64y^3 - 125$ **8.** $-16m^3 - 2n^3$

EXAMPLE 5 **Factoring Completely**

Factor: $-4xy^2 + 12xy + 132x$

Solution

Factor out the GCF, $-4x$.

$$-4xy^2 + 12xy + 132x = -4x(y^2 - 3y - 33)$$

The polynomial in parentheses has three terms. Because 33 is not a perfect square, $y^2 - 3y - 33$ is not a perfect square trinomial. Find two factors of -33 whose sum is -3. There are no such factors. Therefore, $y^2 - 3y - 33$ is prime, so

$$-4xy^2 + 12xy + 132x = -4x(y^2 - 3y - 33)$$ ●

Quick ✓

In Problems 9 and 10, factor each polynomial completely.

9. $10z^2 - 15z + 35$ **10.** $6xy^2 + 81x^3$

EXAMPLE 6 **Factoring Completely**

Factor: $6x^3 - 4x^2 - 24x + 16$

Solution

Factor out the GCF, 2.

$$6x^3 - 4x^2 - 24x + 16 = 2(3x^3 - 2x^2 - 12x + 8)$$

Because there are four terms in the parentheses, try to factor by grouping. Group the first two terms and the last two terms. Watch out for the subtraction sign in front of the third term!

Factor by grouping: $2(3x^3 - 2x^2 - 12x + 8) = 2[(3x^3 - 2x^2) + (-12x + 8)]$
Factor out x^2; factor out -4: $= 2[x^2(3x - 2) - 4(3x - 2)]$
Factor out $(3x - 2)$: $= 2(3x - 2)(x^2 - 4)$
$x^2 - 4$ is the difference of two squares: $= 2(3x - 2)(x - 2)(x + 2)$

Check　　$2(3x - 2)(x - 2)(x + 2) = 2(3x - 2)(x^2 - 4)$

FOIL:　　　　　$= 2(3x^3 - 12x - 2x^2 + 8)$

Distribute; rearrange terms:　　$= 6x^3 - 4x^2 - 24x + 16$　　●

Quick ✓

In Problems 11 and 12, factor each polynomial completely.

11. $2x^3 + 5x^2 + 4x + 10$　　　　**12.** $9x^3 + 3x^2 - 9x - 3$

EXAMPLE 7　　**Factoring Completely**

Factor: $x^2 - 6xy + 9y^2 - 25$

Solution

There are no common factors and there are four terms, so try factoring by grouping. Attempts to form two groups of two terms fail. Before thinking that the polynomial is prime, notice that the first three terms form a perfect square trinomial, and the last term is a perfect square.

Group the first three terms:　$x^2 - 6xy + 9y^2 - 25 = (x^2 - 6xy + 9y^2) - 25$

$$= [x^2 - 2x(3y) + (3y)^2] - 25$$

$A = x, B = 3y; A^2 - 2AB + B^2 = (A - B)^2$:　$= (x - 3y)^2 - 5^2$

$A = x - 3y, B = 5; A^2 - B^2 = (A - B)(A + B)$:　$= (x - 3y - 5)(x - 3y + 5)$

Check　　$(x - 3y - 5)(x - 3y + 5) = x^2 - 3xy + 5x - 3xy + 9y^2$

$$-15y - 5x + 15y - 25$$

$$= x^2 - 6xy + 9y^2 - 25$$

Thus $x^2 - 6xy + 9y^2 - 25 = (x - 3y - 5)(x - 3y + 5)$.　　●

Quick ✓

In Problems 13 and 14, factor each polynomial completely.

13. $4x^2 + 4xy + y^2 - 81$　　　　**14.** $16 - m^2 - 8mn - 16n^2$

▶ **②** **Write Polynomial Functions in Factored Form**

Later in this course and in other mathematics courses, you will be required to write polynomial functions in factored form.

EXAMPLE 8　　**Writing a Polynomial Function in Factored Form**

Write each of the following polynomial functions in factored form.

(a) $f(x) = 8x^2 + 14x - 15$　　　　**(b)** $H(z) = -3z^3 - 15z^2 - 27z - 135$

Solution

(a) To write $f(x) = 8x^2 + 14x - 15$ in factored form, factor $8x^2 + 14x - 15$. First, note that there are three terms in the polynomial, and the polynomial has no common factor. Factor the polynomial using trial and error. The first term, $8x^2$, can be represented as

$$(x + \_)(8x + \_) \quad \text{or} \quad (2x + \_)(4x + \_)$$

(*continued*)

The last term, -15, has the factor pairs $1 \cdot (-15), 3 \cdot (-5), -1 \cdot 15,$ and $-5 \cdot 3.$ The middle term's coefficient needs to be 14, which is positive, but not too big. When looking for the correct combination of factors to obtain this result, start with $(2x + \_\_)(4x + \_\_).$ Since the middle term's coefficient is positive, try $+5$ and $-3.$

$$(2x + 5)(4x - 3) = 8x^2 - 6x + 20x - 15$$
$$= 8x^2 + 14x - 15$$

It works! Therefore,

$$f(x) = 8x^2 + 14x - 15$$
$$= (2x + 5)(4x - 3)$$

(b) To write $H(z) = -3z^3 - 15z^2 - 27z - 135$ in factored form, factor $-3z^3 - 15z^2 - 27z - 135.$

$$H(z) = -3z^3 - 15z^2 - 27z - 135$$

Factor out common factor, -3: $= -3(z^3 + 5z^2 + 9z + 45)$

Four terms; factor by grouping: $= -3[(z^3 + 5z^2) + (9z + 45)]$

Factor out z^2; factor out 9: $= -3[z^2(z + 5) + 9(z + 5)]$

Factor out $z + 5$ as a common factor: $= -3(z + 5)(z^2 + 9)$

The check is left for you.

Quick ✓

In Problems 15 and 16, write each polynomial function in factored form.

15. $G(x) = 2x^2 + 3x - 14$ **16.** $F(p) = 27p^2 - 12$

4.7 Exercises   MyMathLab®

*Problems **1–16** are the Quick ✓ s that follow the EXAMPLES.*

Building Skills

In Problems 17–64, factor each polynomial completely.

17. $2x^2 - 12x - 144$

18. $3x^2 + 6x - 105$

19. $-3y^2 + 27$

20. $-5a^2 + 80$

21. $4b^2 + 20b + 25$

22. $8m^2 - 42m + 49$

23. $16w^3 + 2y^6$

24. $54p^6 - 2q^3$

25. $-3z^2 + 12z - 18$

26. $-4c^3 + 16c^2 - 28c$

27. $20y^2 - 9y - 18$

28. $18t^2 - 9t - 20$

29. $x^3 - 4x^2 + 5x - 20$

30. $p^3 + 7p^2 - 3p - 21$

31. $200x^2 + 18y^2$

32. $12p^2 + 50q^2$

33. $x^4 - 81$

34. $16w^4 - 1$

35. $3x^2 - 7x - 16$

36. $4w^2 - 3w - 6$

37. $36q^3 + 24q^2 + 4q$

38. $20k^3 - 60k^2 + 45k$

39. $24m^3n - 66m^2n - 63mn$

40. $20p^3q - 2p^2q - 4pq$

41. $3r^5 - 24r^2s^3$

42. $54p^5 + 16p^2q^3$

43. $2x^3 + 8x^2 - 18x - 72$

44. $3x^3 - 6x^2 - 48x + 96$

45. $9x^4 - 1$

46. $4z^4 - 25$

47. $3w^4 + 4w^2 - 15$

48. $4b^4 + 4b^2 - 15$

49. $(2y + 3)^2 - 5(2y + 3) + 6$

50. $(3x + 5)^2 + 4(3x + 5) - 21$

51. $p^2 - 10p + 25 - 36q^2$

52. $a^2 + 12a + 36 - 4b^2$

53. $y^6 + 6y^3 - 16$

54. $w^6 + 4w^3 - 5$

55. $p^6 - 1$

56. $q^6 + 1$

57. $-3x^3 - 15x^2 + 27x + 135$

58. $-2y^3 - 4y^2 + 32y + 64$

59. $3a - 27a^3$

60. $-5z - 20z^3$

61. $8t^5 + 14t^3 - 72t$

62. $18h^5 + 154h^3 - 72h$

63. $2x^4y + 10x^3y - 18x^2y - 90xy$

64. $2p^4q + 14p^3q - 32p^2q - 224pq$

In Problems 65–74, factor each polynomial function. See Objective 2.

65. $f(x) = x^2 - 2x - 63$

66. $F(x) = x^2 + 3x - 40$

67. $P(m) = 7m^2 + 31m + 12$

68. $H(p) = 5p^2 + 28p - 12$

69. $G(x) = -12x^2 + 147$

70. $g(x) = -100x^2 + 36$

71. $s(t) = -16t^2 + 96t + 256$

72. $S(y) = 2y^3 - 8y^2 - 42y$

73. $H(a) = 2a^3 + 5a^2 - 32a - 80$

74. $G(x) = 2x^3 - x^2 - 18x + 9$

Applying the Concepts

△ *In Problems 75–78, write an algebraic expression in completely factored form for the area that is shaded.*

75.

76.

77.

78.

△ *In Problems 79 and 80, find an algebraic expression in factored form for the difference in the volumes of the two cubes shown.*

79.

80.

Extending the Concepts

81. Show that $x^2 + 9$ is prime.

82. Show that $x^2 - 4x - 8$ is prime.

In Problems 83–86, factor out the indicated common factor of the expression. Then completely factor the expression.

83. $x^{\frac{5}{2}} - 9x^{\frac{1}{2}}; x^{\frac{1}{2}}$

84. $x^{\frac{3}{2}} - 4x^{-\frac{1}{2}}; x^{-\frac{1}{2}}$

85. $1 + 6x^{-1} + 8x^{-2}; x^{-2}$ 　　**86.** $3 - 2x^{-1} - x^{-2}; x^{-2}$

Explaining the Concepts

87. Write out the steps for factoring any polynomial.

88. What does factored completely mean?

4.8 Polynomial Equations

Objectives

1 Solve Polynomial Equations Using the Zero-Product Property

2 Solve Equations Involving Polynomial Functions

3 Model and Solve Problems Involving Polynomials

Are You Prepared for This Section?

Before getting started, complete the following problems. If you get a problem wrong, go back to the section cited and review the material.

P1. Solve: $x + 4 = 0$ 　　　　　　　　　　　[Section 1.1, pp. 50–52]

P2. Solve: $3(x - 2) - 12 = 0$ 　　　　　　　[Section 1.1, pp. 50–53]

P3. Evaluate $2x^2 + 3x + 1$
when (a) $x = 2$ (b) $x = -1$. 　　　　　　[Section R.5, pp. 40–41]

P4. If $f(x) = 4x + 3$, solve $f(x) = 11$.
What point is on the graph of f? 　　　　　[Section 2.3, pp. 167–169]

P5. If $f(x) = -2x + 8$, find $f(4)$.
What point is on the graph of f? 　　　　　[Section 2.3, pp. 167–169]

P6. Find the zero of $f(x) = \dfrac{2}{3}x - 6$. 　　　　[Section 2.4, pp. 177–178]

▶ Sections 4.4–4.7 explained and illustrated how to factor polynomial expressions. Factoring is an important skill, as you will see repeatedly in this course. This section will present one important use of factoring. It turns out that factoring is really handy for solving *polynomial equations.* A **polynomial equation** is an equation that can be written in the form "polynomial expression equals zero." The **degree of a polynomial equation** is the degree of the polynomial expression in the equation. Some examples of polynomial equations are

$4x + 5 = 17$	$2x^2 - 5x - 3 = 0$	$y^3 + 4y^2 = 3y + 18$
Polynomial equation of degree 1	Polynomial equation of degree 2	Polynomial equation of degree 3

Work Smart

Remember, the degree of a polynomial is the value of the largest exponent on a variable. For example, the degree of $4x^3 - 9x^2 + 1$ is 3.

Notice that the polynomial equation $4x + 5 = 17$ is a linear, or first-degree, equation. Linear equations were discussed back in Section 1.1. This section teaches methods for solving polynomial equations when the polynomial expression is of degree 2 or higher and factorable over the integers.

1 Solve Polynomial Equations Using the Zero-Product Property

In this section, use the following property to solve equations.

The Zero-Product Property

If the product of two factors is zero, then at least one of the factors is 0. That is, if $ab = 0$, then $a = 0$ or $b = 0$ or both a and b are 0.

⊙ **EXAMPLE 1** **Using the Zero-Product Property**

Solve: $(x + 5)(2x - 3) = 0$

Solution

The product of two numbers, $x + 5$ and $2x - 3$, is equal to 0. By the Zero-Product Property, at least one of the numbers must equal 0. Therefore, set each factor equal to 0 and solve each equation separately:

$$x + 5 = 0 \quad \text{or} \quad 2x - 3 = 0$$

Subtract 5 from both sides: $\quad x = -5 \qquad$ Add 3 to both sides: $\quad 2x = 3$

Divide both sides by 2: $\quad x = \dfrac{3}{2}$

Check

$$x = -5: \quad (x + 5)(2x - 3) = 0$$
$$(-5 + 5)(2(-5) - 3) \overset{?}{=} 0$$
$$0(-13) \overset{?}{=} 0$$
$$0 = 0 \quad \text{True}$$

$$x = \frac{3}{2}: \quad (x + 5)(2x - 3) = 0$$
$$\left(\frac{3}{2} + 5\right)\left(2\left(\frac{3}{2}\right) - 3\right) \overset{?}{=} 0$$
$$\left(\frac{13}{2}\right)(3 - 3) \overset{?}{=} 0$$
$$\frac{13}{2} \cdot 0 \overset{?}{=} 0$$
$$0 = 0 \quad \text{True}$$

The solution set is $\left\{-5, \dfrac{3}{2}\right\}$. ●

Quick ✓

1. A _____ _____ is an equation that can be written in the form "polynomial expression equals zero."

2. The Zero-Product Property states that if $a \cdot b = 0$, then ____ or ____.

In Problems 3 and 4, use the Zero-Product Property to solve the equation.

3. $x(x + 7) = 0$

4. $(x - 3)(4x + 3) = 0$

Using the Zero-Product Property to Solve Quadratic Equations

The Zero-Product Property can be used to solve *quadratic equations*.

Work Smart

Why can't *a* equal 0 in the definition? If *a* were equal to zero, the equation would be $bx + c = 0$, a linear equation.

Definition

A **quadratic equation** is a polynomial equation that can be written in the form.

$$ax^2 + bx + c = 0$$

where a, b, and c are real numbers and $a \neq 0$.

Some quadratic equations are

$$3x^2 + 5x + 2 = 0 \qquad -7z^2 + 14z = 0 \qquad y^2 - 16 = 0 \qquad p^2 + 8p = 16$$

A quadratic equation written in the form $ax^2 + bx + c = 0$ is said to be in **standard form.** The first three equations listed above are in standard form; the equation $p^2 + 8p = 16$ is not in standard form.

Sometimes, a quadratic equation is called a **second-degree equation** because the equation contains a polynomial of degree 2.

When a quadratic equation is written in standard form, $ax^2 + bx + c = 0$, it may be possible to factor the expression $ax^2 + bx + c$ as the product of two first-degree polynomials. In Sections 7.2 and 7.3, other methods will be presented for solving $ax^2 + bx + c = 0$ when the expression $ax^2 + bx + c$ cannot be factored.

▶ **EXAMPLE 2** **How to Solve a Quadratic Equation by Factoring**

Solve: $2x^2 - 5x = 3$

Step-by-Step Solution

Step 1: Write the quadratic equation in standard form.

$$2x^2 - 5x = 3$$
Subtract 3 from both sides: $2x^2 - 5x - 3 = 0$

Step 2: Factor the expression on the left side of the equation.

$$(2x + 1)(x - 3) = 0$$

Step 3: Set each factor equal to 0.

Use the Zero-Product Property: $2x + 1 = 0$ or $x - 3 = 0$

Step 4: Solve each first-degree equation.

$$2x = -1 \quad \text{or} \quad x = 3$$
$$x = -\frac{1}{2}$$

Step 5: Check: Substitute $-\dfrac{1}{2}$ and 3 into the original equation.

$x = -\dfrac{1}{2}$: $2x^2 - 5x = 3$

$$2\left(-\frac{1}{2}\right)^2 - 5\left(-\frac{1}{2}\right) \stackrel{?}{=} 3$$
$$2\left(\frac{1}{4}\right) + \frac{5}{2} \stackrel{?}{=} 3$$
$$\frac{1}{2} + \frac{5}{2} \stackrel{?}{=} 3$$
$$3 = 3 \quad \text{True}$$

$x = 3$: $2x^2 - 5x = 3$

$$2(3)^2 - 5(3) \stackrel{?}{=} 3$$
$$18 - 15 \stackrel{?}{=} 3$$
$$3 = 3 \quad \text{True}$$

The solution set is $\left\{-\dfrac{1}{2}, 3\right\}$. ●

Solving a Quadratic Equation by Factoring

Step 1: Write the quadratic equation in standard form, $ax^2 + bx + c = 0$.

Step 2: Factor the expression on the left side of the equation.

Step 3: Set each factor found in Step 2 equal to zero using the Zero-Product Property.

Step 4: Solve each first-degree equation for the variable.

Step 5: Check your answers by substituting into the *original* equation.

Quick ✓

5. A _____ _____ is an equation equivalent to one of the form $ax^2 + bx + c = 0$, where $a, b,$ and c are real numbers and $a \neq 0$.

6. Quadratic equations are also known as _____-degree equations.

7. *True or False* The equation $7x - x^2 = 4$ is written in standard form.

In Problems 8–10, solve each quadratic equation by factoring.

8. $p^2 - 5p + 6 = 0$

9. $3t^2 - 14t = 5$

10. $4y^2 + 8y + 3 = y^2 - 1$

EXAMPLE 3 **Solving a Quadratic Equation by Factoring**

Solve: $(m - 1)(3m + 5) = 16m$

Solution
First, write the equation in standard form.

$$(m - 1)(3m + 5) = 16m$$

FOIL: $\qquad 3m^2 + 2m - 5 = 16m$

Subtract 16m from both sides: $\qquad 3m^2 - 14m - 5 = 0$

Factor: $\qquad (3m + 1)(m - 5) = 0$

Set each factor equal to 0: $\qquad 3m + 1 = 0 \quad$ or $\quad m - 5 = 0$

$$3m = -1 \quad \text{or} \quad m = 5$$

$$m = -\frac{1}{3}$$

Work Smart

Do not attempt to solve the equation $(m - 1)(3m + 5) = 16m$ by setting each factor equal to 16m. The Zero-Product Property can be applied only when the product is equal to zero.

Check Substitute $m = -\frac{1}{3}$ and $m = 5$ into the original equation.

$m = -\frac{1}{3}:$ $(m - 1)(3m + 5) = 16m$ $\qquad$ $m = 5:$ $(m - 1)(3m + 5) = 16m$

$$\left(-\frac{1}{3} - 1\right)\left(3\left(-\frac{1}{3}\right) + 5\right) \stackrel{?}{=} 16\left(-\frac{1}{3}\right)$$ $\qquad (5 - 1)(3(5) + 5) \stackrel{?}{=} 16(5)$

$$\left(-\frac{4}{3}\right)(-1 + 5) \stackrel{?}{=} -\frac{16}{3}$$ $\qquad\qquad 4(20) \stackrel{?}{=} 80$

$$-\frac{16}{3} = -\frac{16}{3} \quad \text{True}$$ $\qquad\qquad\qquad 80 = 80 \quad \text{True}$

The solution set is $\left\{-\frac{1}{3}, 5\right\}$.

Quick ✓

11. *True or False* The equation $(x - 2)(x + 3) = 12$ means that $x - 2 = 12$ or $x + 3 = 12$.

In Problems 12 and 13, solve each quadratic equation by factoring.

12. $x(x + 3) = -2$ $\qquad\qquad$ **13.** $(x + 3)(2x - 1) = 7x - 3x^2$

Using the Zero-Product Property to Solve Equations of Degree 3 or Higher

An extended form of the Zero-Product Property can be used to solve polynomial equations of degree 3 or higher. The idea is to write the equation in standard form, factor the expression that equals zero, set each factor equal to zero, and solve.

EXAMPLE 4 **How to Solve a Third-Degree Equation Using the Zero-Product Property**

Solve: $w^3 + 5w^2 - 4w = 20$

Step-by-Step Solution

Step 1: Write the equation in standard form by subtracting 20 from both sides.

$$w^3 + 5w^2 - 4w = 20$$
$$w^3 + 5w^2 - 4w - 20 = 0$$

(continued)

Step 2: Factor the expression on the left side. Because there are four terms, factor by grouping.

Group the 1st two terms; group the last two terms:

$$(w^3 + 5w^2) + (-4w - 20) = 0$$

Factor out the common factor in each group:

$$w^2(w + 5) - 4(w + 5) = 0$$

Factor out $w + 5$:

$$(w + 5)(w^2 - 4) = 0$$

Factor $w^2 - 4$:

$$(w + 5)(w + 2)(w - 2) = 0$$

Step 3: Set each factor equal to 0.

$$w + 5 = 0 \quad \text{or} \quad w + 2 = 0 \quad \text{or} \quad w - 2 = 0$$

Step 4: Solve each first-degree equation.

$$w = -5 \quad \text{or} \quad w = -2 \quad \text{or} \quad w = 2$$

Step 5: Check:
Substitute $w = -5$, $w = -2$, and $w = 2$ into the original equation.

$w = -5$:
$$w^3 + 5w^2 - 4w = 20$$
$$(-5)^3 + 5(-5)^2 - 4(-5) \overset{?}{=} 20$$
$$-125 + 125 + 20 \overset{?}{=} 20$$
$$20 = 20$$
True

$w = -2$:
$$w^3 + 5w^2 - 4w = 20$$
$$(-2)^3 + 5(-2)^2 - 4(-2) \overset{?}{=} 20$$
$$-8 + 20 + 8 \overset{?}{=} 20$$
$$20 = 20$$
True

$w = 2$:
$$w^3 + 5w^2 - 4w = 20$$
$$(2)^3 + 5(2)^2 - 4(2) \overset{?}{=} 20$$
$$8 + 20 - 8 \overset{?}{=} 20$$
$$20 = 20$$
True

The solution set is $\{-5, -2, 2\}$. ●

Quick ✓

14. *True or False* The equation $x^3 - 2x^2 - 8x = 0$ can be solved using the Zero-Product Property.

In Problems 15 and 16, solve the polynomial equation.

15. $y^3 - y^2 + 9 = 9y$

16. $2n^3 - 3n^2 - 2n = 0$

▶ ❷ **Solve Equations Involving Polynomial Functions**

Suppose $f(x) = x^2 + 6x - 3$ and we want to know the values of x such that $f(x) = 4$. This requires solving the equation

$$\underbrace{x^2 + 6x - 3}_{f(x)} = 4$$

EXAMPLE 5 **Solving an Equation Involving a Polynomial Function**

If $f(x) = x^2 + 6x - 3$, find the values of x such that $f(x) = 4$. What points on the graph of f do these values correspond to?

Solution

Solve $f(x) = 4$. That is, solve $x^2 + 6x - 3 = 4$. Start by putting the equation in standard form.

$$x^2 + 6x - 3 = 4$$

Subtract 4 from both sides:
$$x^2 + 6x - 7 = 0$$

Factor:
$$(x + 7)(x - 1) = 0$$

Set each factor equal to 0:
$$x + 7 = 0 \quad \text{or} \quad x - 1 = 0$$
$$x = -7 \quad \text{or} \quad x = 1$$

Work Smart

Remember that you can use the Zero-Product Property only when the expression is in factored form and the equation is equal to 0.

Check

$$x = -7: \quad x^2 + 6x - 3 = 4 \qquad\qquad x = 1: \quad x^2 + 6x - 3 = 4$$

$$(-7)^2 + 6(-7) - 3 \overset{?}{=} 4 \qquad\qquad (1)^2 + 6(1) - 3 \overset{?}{=} 4$$

$$49 - 42 - 3 \overset{?}{=} 4 \qquad\qquad 1 + 6 - 3 \overset{?}{=} 4$$

$$4 = 4 \quad \text{True} \qquad\qquad 4 = 4 \quad \text{True}$$

The values of x such that $f(x) = 4$ are -7 and 1. The points $(-7, 4)$ and $(1, 4)$ are on the graph of f. ●

Figure 2

Quick ✓

17. If $g(x) = x^2 - 8x + 3$, find the values of x such that

(a) $g(x) = 12$ **(b)** $g(x) = -4$

What points do these values correspond to on the graph of g?

Section 2.4 discussed how to find the zero of a linear function. Recall that a zero of a function f is any number r such that $f(r) = 0$. Also, if r is a zero of a function, then $(r, 0)$ is an x-intercept of the graph of the function; that is, the point $(r, 0)$ is on the graph of the function. Figure 2, which is the graph of $y = f(x)$, shows that $(-3, 0)$ and $(3, 0)$ are x-intercepts. Therefore, the zeros of f are -3 and 3.

EXAMPLE 6

Finding the Zeros of a Quadratic Function

Find the zeros of $f(x) = 4x^2 - 5x - 6$. What are the x-intercepts of the graph of the function?

Solution

The zeros are found by solving the equation $f(x) = 0$, or $4x^2 - 5x - 6 = 0$. The solutions to this equation are $-\dfrac{3}{4}$ and 2, so the zeros of $f(x) = 4x^2 - 5x - 6$ are $-\dfrac{3}{4}$ and 2. Because the zeros are $-\dfrac{3}{4}$ and 2, the x-intercepts of the graph of the function are $\left(-\dfrac{3}{4}, 0\right)$ and $(2, 0)$. ●

Quick ✓

For Problems 18 and 19, find the zeros of the function, and state the x-intercepts of the graph of the function.

18. $h(x) = 2x^2 + 3x - 20$ **19.** $F(x) = 5x^2 - 3x - 2$

▶ ❸ **Model and Solve Problems Involving Polynomials**

Many applied problems require solving polynomial equations by factoring. For example, the height of a projectile over time can be described by a polynomial equation. The equation can be used to determine the time at which the projectile is a certain height. As always, use the problem-solving strategy from Section 1.2.

EXAMPLE 7

Geometry: Area of a Rectangle

The length of a rectangle is 8 feet more than its width. If the area of the rectangle is 84 square feet, what are the dimensions of the rectangle? See Figure 3 on the following page.

Solution

Step 1: Identify This is a geometry problem involving the area of a rectangle.

Step 2: Name Let w represent the width of the rectangle and l its length, both in feet.

(continued)

Figure 3

$$\ell$$

Area = 84 square feet w

Step 3: Translate Because the length is 8 feet more than the width, $l = w + 8$. Also, the area of the rectangle is 84 square feet.

$$\text{Area} = (\text{length})(\text{width})$$
$$\text{Area} = lw$$
$$84 = (w + 8)w \quad \text{The Model}$$

Step 4: Solve Do you see that this equation is a quadratic equation? The first step is to write it in standard form, $ax^2 + bx + c = 0$.

$$w(w + 8) = 84$$

Distribute:	$w^2 + 8w = 84$
Subtract 84 from both sides:	$w^2 + 8w - 84 = 0$
Factor:	$(w + 14)(w - 6) = 0$
Set each factor equal to 0:	$w + 14 = 0 \quad$ or $\quad w - 6 = 0$
Solve:	$w = -14 \quad$ or $\quad w = 6$

Work Smart

Always include units in your answer for these types of application problems.

Step 5: Check Since w represents the width of the rectangle, discard the solution $w = -14$. If the width is 6 feet, then the length would be $6 + 8 = 14$ feet. The area of a rectangle that is 6 feet by 14 feet would be $6(14) = 84$ square feet. This is the right answer!

Step 6: Answer The dimensions of the rectangle are 6 feet by 14 feet.

Quick ✓

20. The width of a rectangular plot of land is 6 miles less than its length. If the area of the land is 135 square miles, what are the dimensions of the land?

EXAMPLE 8 **Pricing a Charter**

Chicago Tours offers boat charters along the Chicago coastline on Lake Michigan. John Alfirivich wants to thank his employees by taking his company on a tour. Normally, a ticket costs \$20 per person, but for each person John brings in excess of 30 people, Chicago Tours will lower the ticket price for everyone by \$0.10. Assuming that more than 30 employees will go on the trip and that the capacity of the boat is 120 passengers, how many employees can attend if John is willing to spend \$900 for the tour?

Solution

Step 1: Identify This is a direct translation problem.

Step 2: Name Let x represent the number of employees in excess of 30 that attend.

Step 3: Translate John's cost is equal to the price per person times the number of people. If John brings 30 employees, his cost will be \$20(30). If John brings 31 employees, his cost will be \$19.90(31). If John brings 32 employees, his cost will be \$19.80(32). In general, if John brings x employees in excess of 30, his cost will be $(20 - 0.1x)(x + 30)$. Because John wants to spend \$900,

$$(20 - 0.1x)(x + 30) = 900 \quad \text{The Model}$$

Step 4: Solve

$$(20 - 0.1x)(x + 30) = 900$$

Use FOIL: $20x + 600 - 0.1x^2 - 3x = 900$

Combine like terms; rearrange terms: $-0.1x^2 + 17x + 600 = 900$

Subtract 900 from both sides: $-0.1x^2 + 17x - 300 = 0$

Multiply both sides by -10 to make the coefficient of x^2 equal to 1: $x^2 - 170x + 3000 = 0$

Factor: $(x - 150)(x - 20) = 0$

Set each factor equal to 0: $x - 150 = 0$ or $x - 20 = 0$

Solve: $x = 150$ or $x = 20$

Step 5: Check Remember that x represents the number of passengers in excess of 30. We discard the solution $x = 150$ because the boat holds only 120 passengers. Therefore, $30 + 20 = 50$ passengers can go on the trip. The cost per ticket is $\$20 - 0.1(20) = \$20 - \$2 = \$18$. The cost per ticket times the number of passengers is $\$18(50) = \$900$. The answer checks.

Step 6: Answer A total of 50 employees can attend.

Quick ✓

21. A flash drive manufacturer charges $100 for each box of flash drives ordered. However, for orders in excess of 30 boxes, but less than 65 boxes, it reduces the price for all boxes by $1 per box. If a customer placed an order that qualified for the discount pricing and the bill was $4200, how many boxes of flash drives were ordered?

EXAMPLE 9 **Projectile Motion**

Figure 4

A ball is thrown off a cliff by a child from a height of 240 feet above sea level, as pictured in Figure 4. The height s of the ball above the water (in feet) as a function of time (in seconds) can be modeled by the function

$$s(t) = -16t^2 + 32t + 240$$

(a) When will the ball be 240 feet above sea level?

(b) When will the ball strike the water?

Solution

(a) To determine when the height of the ball will be 240 feet, solve the equation $s(t) = 240$.

$$s(t) = 240$$
$$-16t^2 + 32t + 240 = 240$$

Subtract 240 from both sides: $-16t^2 + 32t = 0$

Factor out $-16t$: $-16t(t - 2) = 0$

Set each factor equal to 0: $-16t = 0$ or $t - 2 = 0$

Solve each equation: $t = 0$ or $t = 2$

The ball will be at a height of 240 feet the instant it leaves the child's hand and, again, after 2 seconds of flight.

(continued)

(b) The ball strikes the water when its height is 0, so solve the equation
$s(t) = 0$.

$$s(t) = 0$$
$$-16t^2 + 32t + 240 = 0$$

Factor out -16: $-16(t^2 - 2t - 15) = 0$

Factor: $-16(t - 5)(t + 3) = 0$

Set each factor equal to 0: $-16 = 0$ or $t - 5 = 0$ or $t + 3 = 0$

Solve each equation: $t = 5$ or $t = -3$

The equation $-16 = 0$ is false, and $t = -3$ makes no sense (since t is time).
Therefore, the ball will strike the water 5 seconds after it is thrown.

Quick ✓

22. A model rocket is fired straight up from the ground. The height s of the rocket
(in feet) as a function of time (in seconds) can be modeled by the function
$s(t) = -16t^2 + 160t$.

  (a) When will the rocket be 384 feet from the ground?

  (b) When will the rocket strike the ground?

4.8 Exercises   MyMathLab®   Exercise numbers in green have
complete video solutions in MyMathLab or
may be accessed using the QR code to the right.

*Problems **1–22** are the Quick ✓s that follow the EXAMPLES.*

Building Skills

In Problems 23–62, solve each equation. See Objective 1.

23. $(x - 3)(x + 1) = 0$ **24.** $(x + 3)(x - 8) = 0$

25. $2x(3x + 4) = 0$ **26.** $4x(2x - 3) = 0$

27. $y(y - 5)(y + 3) = 0$ **28.** $3a(a - 9)(a + 11) = 0$

29. $3p^2 - 12p = 0$ **30.** $5c^2 + 15c = 0$

31. $2w^2 = 16w$ **32.** $4t^2 = -20t$

33. $m^2 + 2m - 15 = 0$ **34.** $x^2 + 3x - 40 = 0$

35. $w^2 - 13w = -36$ **36.** $y^2 + 13y = -42$

37. $p^2 - 6p + 9 = 0$ **38.** $a^2 + 12a + 36 = 0$

39. $5x^2 = 2x + 3$ **40.** $4c^2 + 6 = 25c$

41. $m^2 + 7m + 9 = m$ **42.** $n^2 - 8n = 2n - 25$

43. $3p^2 + 9p - 120 = 0$ **44.** $4y^2 - 20y - 56 = 0$

45. $-4b^2 - 14b + 60 = 0$ **46.** $-6n^2 - 9n + 60 = 0$

47. $\frac{1}{2}x^2 + 2x - 6 = 0$ **48.** $\frac{1}{2}t^2 - 3t - 8 = 0$

49. $\frac{2}{3}x^2 + x = \frac{14}{3}$ **50.** $\frac{2}{3}x^2 + \frac{7}{3}x = 5$

51. $x(x + 8) = 33$ **52.** $y(y + 4) = 45$

53. $(x + 6)(x - 1) = 9 + 3x$

54. $(x + 2)(x + 5) = 3x + 10$

55. $2z^3 - 5z^2 = 3z$

56. $7q^3 + 31q^2 = -12q$

57. $2p^3 + 5p^2 - 8p - 20 = 0$

58. $w^3 + 5w^2 - 16w - 80 = 0$

59. $-30b^3 - 38b^2 = 12b$

60. $-24b^3 + 27b = 18b^2$

61. $(x - 2)^3 = x^3 - 2x$

62. $(x + 2)^3 = x^3 - 2x$

In Problems 63–68, see Objective 2.

63. Suppose that $f(x) = x^2 + 7x + 12$. Find the values of x such that

    (a) $f(x) = 2$ **(b)** $f(x) = 20$

    What points on the graph of f do these values correspond to?

64. Suppose that $f(x) = x^2 + 5x + 3$. Find the values of x such that

    (a) $f(x) = 3$ **(b)** $f(x) = 17$

    What points on the graph of f do these values correspond to?

65. Suppose that $g(x) = 2x^2 - 6x - 5$. Find the values of x such that

    (a) $g(x) = 3$ **(b)** $g(x) = 15$

    What points on the graph of g do these values correspond to?

66. Suppose that $h(x) = 3x^2 - 9x - 8$. Find the values of x such that

    (a) $h(x) = -8$ **(b)** $h(x) = 22$

    What points on the graph of h do these values correspond to?

67. Suppose that $F(x) = -3x^2 + 12x + 5$. Find the values of x such that

    (a) $F(x) = 5$ **(b)** $F(x) = -10$

    What points on the graph of F do these values correspond to?

68. Suppose that $G(x) = -x^2 + 4x + 6$. Find the values of x such that

    (a) $G(x) = 1$ **(b)** $G(x) = 9$

    What points on the graph of G do these values correspond to?

In Problems 69–74, find the zeros of the function. What are the x-intercepts of the graph of the function? See Objective 2.

69. $f(x) = x^2 + 9x + 14$ **70.** $f(x) = x^2 - 13x + 42$

71. $g(x) = 6x^2 - 25x - 9$

72. $h(x) = 8x^2 - 18x - 35$

73. $s(x) = 2x^3 + 2x^2 - 40x$

74. $f(x) = 3x^3 - 15x^2 - 42x$

Mixed Practice

In Problems 75–92, solve each equation.

75. $(x + 3)(x - 5) = 9$ **76.** $(x + 7)(x - 3) = 11$

77. $2q^2 + 3q - 14 = 0$ **78.** $3t^2 + 7t - 20 = 0$

79. $-3b^2 + 21b = 0$ **80.** $-7z^2 + 42z = 0$

81. $(x + 2)(x + 3) = x(x - 2)$

82. $(x + 7)(x - 6) = x(x + 3)$

83. $x^3 + 5x^2 - 4x - 20 = 0$

84. $2c^3 + 3c^2 - 8c - 12 = 0$

85. $(2x + 1)(x - 3) - x^2 = (x - 2)(x - 3)$

86. $(3x - 2)(x + 4) - x(x + 1) = (2x + 1)(x + 4) - 12$

87. $x^3 + x^2 + x + 6 = 3x^2 + 6x$

88. $2x^3 + 16x^2 = 5x^2 - 15x$

89. $4x^4 - 17x^2 + 4 = 0$

90. $9z^4 - 13z^2 + 4 = 0$

91. $(a + 3)^2 - 5(a + 3) = -6$

92. $(2b + 1)^2 + 7(2b + 1) = -12$

In Problems 93–96, find the domain of each function.

93. $f(x) = \dfrac{5}{x^2 - 4}$

94. $f(x) = \dfrac{-9}{x^2 + 6x + 5}$

95. $g(x) = \dfrac{4x + 3}{2x^2 - 3x + 1}$

96. $h(x) = \dfrac{x + 4}{3x^2 - 7x - 6}$

Applying the Concepts

△ **97. Area** The length of a rectangle is 8 centimeters less than its width. What are the dimensions of the rectangle if its area is 128 square centimeters?

△ **98. Area** The length of a rectangle is twice the sum of its width and 3. What are the dimensions of the rectangle if its area is 216 square inches?

△ **99. Area** The height of a triangle is 12 feet more than its base. What are the height and base of the triangle if its area is 110 square feet?

△ **100. Area** The base of a triangle is 4 meters shorter than its height. What are the height and base of the triangle if its area is 48 square meters?

△ **101. Convex Polygons** A **convex polygon** is a polygon whose interior angles are between 0° and 180°. The number of diagonals D in a convex polygon with n sides is given by the formula $D = \dfrac{n(n-3)}{2}$. Determine the number of sides n in a convex polygon that has 20 diagonals.

102. Consecutive Integers The sum S of the consecutive integers $1, 2, 3, \ldots, n$ is given by the formula $S = \dfrac{n(n+1)}{2}$. That is, $1 + 2 + 3 + \cdots + n = \dfrac{n(n+1)}{2}$. How many consecutive integers must be added together to obtain a sum of 36?

103. Enclosing an Area with a Fence A farmer has 100 meters of fencing and wants to enclose a rectangular plot that borders a river. If the farmer does not fence the side along the river, what are the dimensions of the land enclosed if the area enclosed is 800 square meters?

104. Enclosing an Area with a Fence A farmer has 300 feet of fencing and wants to enclose a rectangular corral that borders his barn on one side and then divide it into two plots with a fence parallel to one of the sides (see the figure). Assuming that the farmer will not fence the side along the barn, what are the lengths of the parts of the fence if the total area enclosed is 4800 square feet?

105. Landscape Design Chris Lambton just designed a cloister (a rectangular garden surrounded by a covered walkway on all four sides). The outside dimensions of the garden are 12 feet by 8 feet, and the area of the garden and the walkway together is 252 square feet. What is the width of the walkway?

106. Picture Frame The outside dimensions of a picture frame are 40 inches by 32 inches. The area of the picture within the frame is 1008 square inches. Find the width of the frame.

107. Making a Box A box is to be made from a rectangular piece of corrugated cardboard, where the length is 5 more inches than the width, by cutting a square piece 2 inches on each side from each corner and folding up the sides to form a box. The volume of the box is to be 168 cubic inches. Find the dimensions of the rectangular piece of cardboard.

108. Making a Box A box is to be made from a rectangular piece of corrugated cardboard, where the length is 8 inches more than the width, by cutting a square piece 3 inches on each side from each corner and folding up the sides to form a box. The volume of the box is to be 315 cubic inches. Find the dimensions of the rectangular piece of cardboard.

109. Marginal Cost Marginal cost can be thought of as the cost of producing one additional unit of output. For example, if the marginal cost of producing the 30th unit of output is $9.30, then it cost $9.30 to increase production from 29 to 30 units. The marginal cost C (in dollars) to produce x bicycles is given by $C(x) = x^2 - 40x + 600$.

(a) Find the marginal cost of producing 30 bicycles.
(b) How many bicycles can be manufactured so that the marginal cost equals $200? That is, solve $C(x) = 200$.
(c) Economic theory states that, to maximize profit, production should continue until marginal revenue equals marginal cost. Marginal revenue is the additional revenue received for each bicycle sold. In certain situations, the marginal revenue is the price of the product. Assuming that marginal revenue equals $225, how many bicycles should be manufactured?

110. Marginal Cost (See Problem 109.) Suppose that the marginal cost of manufacturing x cellphones is given by

$$C(x) = \frac{1}{2}x^2 - 30x + 475$$

(a) Find the marginal cost of producing 30 cell phones.
(b) How many cell phones can be manufactured so that marginal cost equals $75? That is, solve $C(x) = 75$.
(c) Economic theory states that, to maximize profit, production should continue until marginal revenue equals marginal cost. Assuming that marginal revenue equals $97, how many cell phones should be manufactured?

111. Projectile Motion Bubba Watson hits a golf ball with an initial speed of 240 feet per second. The height s of the ball (in feet) as a function of time (in seconds) can be modeled by the function

$$s(t) = -16t^2 + 120t$$

(a) When will the height of the ball be 200 feet?
(b) When will the ball hit the ground?

112. Projectile Motion A cannonball is fired from a cliff that is 260 feet high with an initial speed of 128 feet per second. The height s of the cannonball (in feet) as a function of time (in seconds) can be modeled by the function

$$s(t) = -16t^2 + 64t + 260$$

(a) When will the height of the cannonball be 320 feet?
(b) When will the cannonball hit the ground?

Technology Exercises

Technology can be used to find solutions to equations. Solving an equation for x when one side of the equation is 0 is equivalent to finding where the graph of the corresponding function crosses or touches the x-axis. For example, to solve the equation in Example 2, $2x^2 - 5x = 3$, write the quadratic equation in standard form. In Desmos, once you've graphed this equation, if you click on the curve, you'll see gray dots appear at "interesting points" including intercepts. See Figure 5.

Figure 5

In Problems 113–118, solve each equation using technology. Round your answers to two decimal places, if necessary.

113. $2x^2 - 7x - 5 = 0$

114. $2x^2 - x - 10 = 0$

115. $0.2x^2 - 5.1x + 3 = 0$

116. $0.4x^2 - 2.7x + 1 = 0$

117. $-x^2 + 0.6x = -2$

118. $-3.1x^2 - 0.4x = -3$

Chapter 4 Activity: What is the Question?

Focus: Performing operations and solving equations with polynomials

Time: 15–20 minutes

Group size: 2 or 4

In this activity you will work as a team to solve eight multiple-choice questions. However, these questions are different from most multiple-choice questions. You are given the answer to a problem and must determine which of the multiple-choice options has the correct question for the given answer.

Before beginning the activity, decide how you will approach this task as a team. For example:

- If there are two members on your team... one member will always examine choices (a) and (b) and the other will always examine choices (c) and (d).

- If there are four members on your team... one member will always examine choice (a), another member will always examine choice (b), and so on.

1. The answer is $-3x^2 - 10x$. What is the question?

   (a) Simplify: $2x - 3x(x^2 + 4)$

   (b) Find the quotient: $(6x^3 - 20x^2) \div (-2x)$

   (c) Simplify: $-3x(x^2 + 3) + 1$

   (d) Find the quotient: $(-6x^4 - 20x^3) \div (2x^2)$

2. Find $(f + g)(-1)$. The answer is 5. What are f and g?

   (a) $f(x) = 2x + 2, g(x) = -3x - 1$

   (b) $f(x) = 2x + 3, g(x) = -3x + 1$

   (c) $f(x) = -2x + 5, g(x) = -3x + 4$

   (d) $f(x) = -2x - 2, g(x) = 3x - 4$

3. The answer is $(6x + 1)(2x - 3)$. What is the question?

   (a) Factor: $12x^2 - 20x + 3$

   (b) Factor: $12x^2 + 16x - 3$

   (c) Factor: $12x^2 - 16x - 3$

   (d) Factor: $12x^2 + 20x + 3$

4. The answer is $x^2 + 5x + 6$. What is the question?

   (a) Find the product: $(x + 6)(x - 1)$

   (b) Simplify: $2x^2 + 7x + 9 - (x^2 - 2x - 3)$

   (c) Find the product: $(x + 2)(x + 3)$

   (d) Simplify: $(x + 6)^2$

5. The answer is 3. What is the question?

   (a) What is the name of the variable in $16z^2 + 3z - 5$?

   (b) What is the degree of the polynomial $2mn + 6m - 3$?

   (c) How many terms are in the polynomial $2mn + 6m - 3$?

   (d) What is the coefficient of b in the polynomial $3a^2b - 9a + 5b$?

6. The answer is $x = -5$ or $x = 3$. What is the question?

   (a) Solve: $x(x + 2) = 15$

   (b) Find the values of x such that $f(x) = 8$ if $f(x) = x^2 + 4x + 3$.

   (c) Solve: $\dfrac{2}{3}x + 5 = \dfrac{1}{3}x^2$

   (d) Find the zeros of the function $f(x) = x^2 - 2x - 15$.

7. The answer is $8x^2$. What is the question?

   (a) Simplify: $9x^2(x + 1) - 3x^2(3x + 5)$

   (b) Find the greatest common factor: $16x^2y^2 - 8x^3y - 24x^2$

   (c) Find the quotient: $(-8x^3 - 8x^2) \div (x + 1)$

   (d) Factor by grouping: $8x^3 + 8x^2 - x - 1$

8. The answer is $x + 2$. What is the question?

   (a) Find the quotient:
   $(x^3 + x^2 - 7x - 2) \div (x^2 + 3x - 1)$

   (b) Find the binomial factor: $x^3 - 8$

   (c) Find the quotient: $(x^2 + 2x - 3) \div (x - 1)$

   (d) Find the binomial factor: $x^3 + 8$

Chapter 4 Review

Section 4.1   Adding and Subtracting Polynomials

KEY CONCEPTS

- **Monomial**
   A monomial in one variable is the product of a constant and a variable raised to a nonnegative integer power. A monomial in one variable is of the form ax^k, where a is a constant, x is a variable, and $k \geq 0$ is an integer. The constant a is called the **coefficient** of the monomial. If $a \neq 0$, then k is called the **degree** of the monomial.

- **Polynomial**
   A polynomial is a monomial or the sum of monomials.

KEY TERMS

Monomial
Coefficient
Degree
Polynomial
Standard form
Binomial

- **Polynomial Function**
  A polynomial function is a function whose rule is a polynomial. The domain of all polynomial functions is the set of all real numbers.
- **Sum or Difference of Two Functions**
  If f and g are two functions:
  The sum $f + g$ is the function defined by $(f + g)(x) = f(x) + g(x)$.
  The difference $f - g$ is the function defined by
  $(f - g)(x) = f(x) - g(x)$.

Trinomial
Polynomial function
Evaluate
Sum of two functions
Difference of two functions

You Should Be Able To...	EXAMPLE	Review Exercises
1 Define monomial and determine the coefficient and degree of a monomial (p. 310)	Examples 1 through 3	1–2
2 Define polynomial and determine the degree of a polynomial (p. 311)	Examples 4 and 5	3–4
3 Simplify polynomials by combining like terms (p. 312)	Examples 6 through 8	5–10
4 Evaluate polynomial functions (p. 314)	Examples 9 and 10	11–12, 13(b), 14(b), 15(b), 16
5 Add and subtract polynomial functions (p. 315)	Examples 11 and 12	13(a), 14(a), 15(a)

In Problems 1 and 2, determine the coefficient and degree of each monomial.

1. $-7x^4$

2. $\dfrac{1}{9}w^3$

In Problems 3 and 4, write each polynomial in standard form. Then determine the degree of each polynomial.

3. $x + 7x^3 - 8 - 2x^2$ **4.** $3 + 2y - 3y^2 + y^4$

In Problems 5–10, add or subtract as indicated. Express your answer as a single polynomial in standard form.

5. $(x^2 + 2x - 7) + (3x^2 - x - 4)$

6. $(4x^3 - 3x^2 + x - 5) - (x^4 + 2x^2 - 7x + 1)$

7. $\left(\dfrac{1}{4}x^2 - \dfrac{1}{2}x\right) - \left(4x - \dfrac{1}{6}\right)$

8. $\left(\dfrac{1}{2}x^2 - x + \dfrac{1}{4}\right) + \left(\dfrac{1}{3}x^2 + \dfrac{2}{5}\right)$

9. $(x^3y^2 + 6x^2y^2 - xy) + (-x^3y^2 + 4x^2y^2 + xy)$

10. $(a^2b - 4ab^2 + 3) - (2a^2b + 2ab^2 + 7)$

In Problems 11–14, find the indicated function or function value.

11. $f(x) = -3x^2 + 2x - 8$
  (a) $f(-2)$ **(b)** $f(0)$ **(c)** $f(3)$

12. $f(x) = x^3 - 5x^2 + 3x - 1$
  (a) $f(-3)$ **(b)** $f(0)$ **(c)** $f(2)$

13. $f(x) = 4x - 3; g(x) = x^2 + 3x + 2$
  (a) $(f + g)(x)$ **(b)** $(f + g)(3)$

14. $f(x) = 2x^3 + x^2 - 7; g(x) = 3x^2 - x + 5$
  (a) $(f - g)(x)$ **(b)** $(f - g)(2)$

15. Profit Suppose that the revenue function R from selling x graphing calculators is $R(x) = -1.5x^2 + 180x$. The cost C of selling x graphing calculators is $C(x) = x^2 - 100x + 3290$.
  (a) Find the profit function.
  (b) Find the profit if $x = 25$ calculators are sold.

16. Area The area A of the region shown in quadrant I is given by $A(x) = -x^2 + 5x$, where (x, y) is a point in quadrant I on the graph of the line $y = -2x + 5$.
  (a) Find the area of the region when the given point is $(2, 1)$.
  (b) Find the area of the region when the given point is $(1, 3)$.

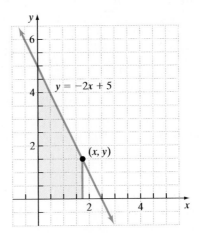

Section 4.2   Multiplying Polynomials

KEY CONCEPTS

- **Product Rule for Exponents**
  If a is a real number and m and n are integers, then $a^m a^n = a^{m+n}$. If m, n, or $m + n$ is 0 or negative, then a cannot be 0.

- **Distributive Property**
  If a, b, and c are real numbers, then $a(b + c) = ab + ac$ and $(a + b) \cdot c = ac + bc$.

- **Extended Form of the Distributive Property**
  $a(b_1 + b_2 + \ldots + b_n) = ab_1 + ab_2 + \ldots + ab_n$, where a, $b_1, b_2, \ldots, b_n$ are real numbers.

- **Difference of Two Squares**
  $$(A - B)(A + B) = A^2 - B^2$$

- **Squares of Binomials or Perfect Square Trinomials**
  $$(A + B)^2 = A^2 + 2AB + B^2$$
  $$(A - B)^2 = A^2 - 2AB + B^2$$

- **Product of Two Functions**
  Let f and g be two functions. The product $f \cdot g$ is the function defined by $(f \cdot g)(x) = f(x) \cdot g(x)$.

KEY TERMS

FOIL
Special products
Product of two functions

You Should Be Able To...	EXAMPLE	Review Exercises
❶ Multiply a monomial and a polynomial (p. 321)	Example 2	17–20
❷ Multiply two binomials (p. 321)	Examples 3 and 4	21–24
❸ Multiply two polynomials (p. 322)	Example 5	25–28
❹ Multiply special products (p. 323)	Examples 6 through 9	29–34
❺ Multiply polynomial functions (p. 325)	Example 10	35–36

In Problems 17–20, find the product.

17. $(-3x^3 y)(4xy^2)$ **18.** $\left(\dfrac{1}{3}mn^4\right)(18m^3 n^3)$

19. $5ab(-2a^2 b + ab^2 - 3ab)$

20. $0.5c(1.7c^2 + 4.3c + 8.9)$

In Problems 21–24, find the product of the binomials.

21. $(x + 2)(x - 9)$ **22.** $(-3x + 1)(2x - 8)$

23. $(m - 4n)(2m + n)$ **24.** $(2a + 15)(-a + 3)$

In Problems 25–28, find the product of the polynomials.

25. $(x + 2)(3x^2 - 5x + 1)$

26. $(w - 4)(w^2 + w - 8)$

27. $(m^2 - 2m + 3)(2m^2 + 5m - 7)$

28. $(2p - 3q)(p^2 + 7pq - 4q^2)$

In Problems 29–34, find the special products.

29. $(3w + 1)(3w - 1)$ **30.** $(2x - 5y)(2x + 5y)$

31. $(6k - 5)^2$ **32.** $(3a + 2b)^2$

33. $(x + 2)(x^2 - 2x + 4)$ **34.** $(2x - 3)(4x^2 + 6x + 9)$

In Problems 35–38, find the indicated function or value.

35. $f(x) = 3x - 7; g(x) = 6x + 5$
  (a) $(f \cdot g)(x)$ **(b)** $(f \cdot g)(-2)$

36. $f(x) = x + 2; g(x) = 3x^2 - x + 1$
  (a) $(f \cdot g)(x)$ **(b)** $(f \cdot g)(4)$

37. $f(x - 3)$ when $f(x) = 5x^2 + 8$

38. $f(x + h) - f(x)$ when $f(x) = -x^2 + 3x - 5$

Section 4.3 Dividing Polynomials; Synthetic Division

KEY CONCEPTS

- **Quotient Rule for Exponents**

  If a is a real number and m and n are integers, then $\dfrac{a^m}{a^n} = a^{m-n}, a \neq 0$.

- **Quotient of Two Functions**

  Let f and g be two functions. The quotient $\dfrac{f}{g}$ is the function defined

  by $\left(\dfrac{f}{g}\right)(x) = \dfrac{f(x)}{g(x)}, g(x) \neq 0$.

- **The Remainder Theorem**

  Let f be a polynomial function. If $f(x)$ is divided by $x - c$, then the remainder is $f(c)$.

- **The Factor Theorem**

  Let f be a polynomial function. Then $x - c$ is a factor of $f(x)$ if and only if $f(c) = 0$.

KEY TERMS

Divisor
Dividend
Quotient
Remainder
Synthetic division

You Should Be Able To...	EXAMPLE	Review Exercises
❶ Divide a polynomial by a monomial (p. 329)	Example 1	39–42
❷ Divide polynomials using long division (p. 330)	Examples 3 through 5	43–48, 63
❸ Divide polynomials using synthetic division (p. 333)	Examples 6 and 7	49–54, 64
❹ Divide polynomial functions (p. 335)	Example 8	55–58
❺ Use the Remainder and Factor Theorems (p. 336)	Examples 9 and 10	59–62

In Problems 39–42, divide and simplify.

39. $\dfrac{12x^3 - 6x^2}{3x}$

40. $\dfrac{15w^5 - 5w^3 + 25w^2 + 10w}{5w}$

41. $\dfrac{7y^3 + 12y^2 - 6y}{2y}$

42. $\dfrac{2m^3n^2 + 8m^2n^2 - 14mn^3}{4m^2n^3}$

In Problems 43–48, divide using long division.

43. $\dfrac{3x^2 - 2x - 8}{x - 2}$

44. $\dfrac{-2x^2 - 3x + 40}{x + 5}$

45. $\dfrac{6z^3 + 9z^2 + 4z - 6}{2z + 3}$

46. $\dfrac{12k^3 - 29k^2 - 14k + 16}{3k - 8}$

47. $\dfrac{16x^4 - 81}{2x - 3}$

48. $\dfrac{2x^4 - 11x^3 + 35x^2 - 54x + 55}{x^2 - 3x + 4}$

In Problems 49–54, divide using synthetic division.

49. $\dfrac{5x^2 + 11x + 8}{x + 2}$

50. $\dfrac{9a^2 - 14a - 8}{a - 2}$

51. $\dfrac{3m^3 + 11m^2 - 5m - 33}{m + 3}$

52. $\dfrac{n^3 + 2n^2 - 39n + 67}{n - 4}$

53. $\dfrac{x^4 + 6x^2 - 7}{x + 1}$

54. $\dfrac{2x^3 + 5x - 8}{x + 2}$

In Problems 55–58, find the indicated function or value.

55. $f(x) = 5x^3 + 25x^2 - 15x$; $g(x) = 5x$

    (a) $\left(\dfrac{f}{g}\right)(x)$ **(b)** $\left(\dfrac{f}{g}\right)(2)$

56. $f(x) = 9x^2 + 54x - 31$; $g(x) = 3x - 2$

    (a) $\left(\dfrac{f}{g}\right)(x)$ **(b)** $\left(\dfrac{f}{g}\right)(-3)$

57. $f(x) = 2x^3 + 12x^2 + 9x - 28$; $g(x) = x + 4$

    (a) $\left(\dfrac{f}{g}\right)(x)$ **(b)** $\left(\dfrac{f}{g}\right)(-2)$

58. $f(x) = 3x^4 - 14x^3 + 31x^2 - 58x + 22$;
$g(x) = x^2 - x + 5$

    (a) $\left(\dfrac{f}{g}\right)(x)$ **(b)** $\left(\dfrac{f}{g}\right)(4)$

In Problems 59 and 60, use the Remainder Theorem to find the remainder.

59. $f(x) = 4x^2 - 7x + 23$ is divided by $x - 4$.

60. $f(x) = x^3 - 2x^2 + 12x - 5$ is divided by $x + 2$.

In Problems 61 and 62, use the Factor Theorem to determine whether $x - c$ is a factor of the given function for the given value of c. If $x - c$ is a factor, then write the function in factored form.

61. $f(x) = 3x^2 + x - 14$; $c = 2$

62. $f(x) = 2x^2 + 13x + 22$; $c = -4$

63. The area of a rectangle is $20x^2 - 11x - 3$ square meters. If the width of the rectangle is $4x - 3$ meters, find an expression for the length.

64. The volume of a rectangular box is $2x^3 + x^2 - 7x - 6$ cubic centimeters. If the height of the box is $x - 2$ centimeters, find an expression for the area of the top of the box.

Section 4.4   Greatest Common Factor; Factoring by Grouping

KEY CONCEPTS

- **Factoring out the greatest common factor**
  Identify the greatest common factor (GCF) of each term. Rewrite each term as the product of the GCF and the remaining factor. Use the Distributive Property to factor out the GCF. Check your work using the Distributive Property.

- **Factoring by grouping**
  Group the terms with common factors. Sometimes it will be necessary to rearrange the terms. In each grouping, factor out the common factor. Factor out the common factor that remains. Check your work.

KEY TERMS

Factors
Factoring a polynomial
Prime
Factored completely
Greatest common factor (GCF)
Factoring by grouping

You Should Be Able To...	EXAMPLE	Review Exercises
❶ Factor out the greatest common factor (p. 341)	Examples 1 through 4	65–72, 79, 80
❷ Factor by grouping (p. 343)	Examples 5 and 6	73–78

In Problems 65–72, factor out the greatest common factor.

65. $4z + 24$

66. $-7y^2 + 91y$

67. $14x^3y^2 + 2xy^2 - 8x^2y$

68. $30a^4b^3 + 15a^3b - 25a^2b^2$

69. $3x(x + 5) - 4(x + 5)$

70. $-4c(2c + 9) + 3(2c + 9)$

71. $(5x + 3)(x - 5y) + (x + 2)(x - 5y)$

72. $(3a - b)(a + 7) - (a + 1)(a + 7)$

In Problems 73–78, factor by grouping.

73. $x^2 + 6x - 3x - 18$

74. $c^2 + 2c - 5c - 10$

75. $14z^2 + 16z - 21z - 24$

76. $21w^2 - 28w + 6w - 8$

77. $2x^3 + 2x^2 - 18x^2 - 18x$

78. $10a^4 + 15a^3 + 70a^3 + 105a^2$

79. Integers The sum of the first n positive integers is given by $\frac{1}{2}n^2 + \frac{1}{2}n$.

(a) Write this expression in factored form.

(b) Use the factored form to determine the sum of the first 32 positive integers.

80. Revenue A computer manufacturer estimates that its revenue for selling x laptop computers can be approximated by the function $R(x) = 5200x - 2x^3$. Express the revenue function in factored form.

Section 4.5 Factoring Trinomials

KEY CONCEPTS

- **Factoring $x^2 + bx + c$**
  $x^2 + bx + c = (x + m)(x + n)$, where $mn = c$ and $m + n = b$
- **Factoring $ax^2 + bx + c$ by grouping**
  See page 352.
- **Factoring $ax^2 + bx + c$ by trial and error**
  See page 354.

KEY TERMS

Factoring by substitution

You Should Be Able To...	EXAMPLE	Review Exercises
❶ Factor trinomials of the form $x^2 + bx + c$ (p. 347)	Examples 1 through 6	81–86
❷ Factor trinomials of the form $ax^2 + bx + c, a \neq 1$ (p. 351)	Examples 7 through 12	87–94
❸ Factor trinomials using substitution (p. 356)	Examples 13 and 14	95–98

In Problems 81–98, factor each trinomial completely. If the polynomial cannot be factored, say it is prime.

81. $w^2 - 11w - 26$

82. $x^2 - 9x + 15$

83. $-t^2 + 6t + 72$

84. $m^2 + 10m + 21$

85. $x^2 + 4xy - 320y^2$

86. $r^2 - 5rs + 6s^2$

87. $5x^2 + 13x - 6$

88. $6m^2 + 41m + 44$

89. $4y^2 - 5y + 7$

90. $8t^2 + 22t - 6$

91. $6x^2 - 13x + 5$

92. $21r^2 - rs - 2s^2$

93. $20x^2 - 57xy + 27y^2$

94. $-2s^2 + 12s + 14$

95. $x^4 - 10x^2 - 11$

96. $10x^2y^2 + 41xy + 4$

97. $(a + 4)^2 - 9(a + 4) - 36$

98. $2(w - 1)^2 + 11(w - 1) + 9$

Section 4.6 Factoring Special Products

KEY CONCEPTS

- **Perfect Square Trinomials**
  $A^2 + 2AB + B^2 = (A + B)^2$
  $A^2 - 2AB + B^2 = (A - B)^2$
- **Difference of Two Squares**
  $A^2 - B^2 = (A - B)(A + B)$
- **Sum or Difference of Two Cubes**
  $A^3 + B^3 = (A + B)(A^2 - AB + B^2)$
  $A^3 - B^3 = (A - B)(A^2 + AB + B^2)$

KEY TERMS

Perfect square trinomial
Difference of two squares
Sum of two squares
Sum of two cubes
Difference of two cubes

You Should Be Able To...	EXAMPLE	Review Exercises
❶ Factor perfect square trinomials (p. 359)	Example 1	99–104
❷ Factor the difference of two squares (p. 360)	Examples 2 and 3	105–110
❸ Factor the sum or difference of two cubes (p. 362)	Examples 4 and 5	111–116

In Problems 99–116, factor completely.

99. $x^2 + 22x + 121$

100. $w^2 - 34w + 289$

101. $144 - 24c + c^2$

102. $x^2 - 8x + 16$

103. $64y^2 + 80y + 25$

104. $12z^2 + 48z + 48$

105. $x^2 - 196$

106. $49 - y^2$

107. $t^2 - 225$

108. $4w^2 - 81$

109. $36x^4 - 25y^2$

110. $80mn^2 - 20m$

111. $x^3 - 343$

112. $729 - y^3$

113. $27x^3 - 125y^3$

114. $8m^6 + 27n^3$

115. $2a^6 - 2b^6$

116. $(y - 1)^3 + 64$

Section 4.7 Factoring: A General Strategy

Steps for Factoring

Step 1: Factor out the greatest common factor (GCF), if any exists.

Step 2: Count the number of terms.

Step 3: **(a)** Two terms

- Is it the difference of two squares? If so,
$$A^2 - B^2 = (A - B)(A + B)$$

- Is it the sum of two squares? If so, stop! The expression is prime.

- Is it the difference of two cubes? If so,
$$A^3 - B^3 = (A - B)(A^2 + AB + B^2)$$

- Is it the sum of two cubes? If so,
$$A^3 + B^3 = (A + B)(A^2 - AB + B^2)$$

(b) Three terms

- Is it a perfect square trinomial? If so,
$$A^2 + 2AB + B^2 = (A + B)^2 \quad \text{or} \quad A^2 - 2AB + B^2 = (A - B)^2$$

- Is the coefficient of the square term 1? If so,
$$x^2 + bx + c = (x + m)(x + n), \text{where } mn = c \text{ and } m + n = b$$

- Is the coefficient of the square term different from 1? If so, factor using grouping or trial and error.

(c) Four terms

- Use factoring by grouping.

Step 4: Check your work by multiplying out the factors.

You Should Be Able To...	EXAMPLE	Review Exercises
❶ Factor polynomials completely (p. 365)	Examples 1 through 7	117–131
❷ Write polynomial functions in factored form (p. 369)	Example 8	132–136

In Problems 117–131, factor each polynomial completely.

117. $x^2 + 7x + 6$

118. $-8x^2y^3 + 12xy^3$

119. $7x^3 - 35x^2 + 28x$

120. $3x^2 - 3x - 18$

121. $4z^2 - 60z + 225$

122. $12x^2 + 7x - 49$

123. $10n^2 - 33n - 7$

124. $8 - 2y - y^2$

125. $2x^3 - 10x^2 + 6x - 30$

126. $(3h + 2)^3 + 64$

127. $5p^3q^2 - 80p$

128. $m^4 - 5m^2 + 4$

129. $686 - 16m^6$

130. $h^3 + 2h^2 - h - 2$

131. $108x^3 + 4y^3$

In Problems 132–136, factor each polynomial function.

132. $F(c) = c^2 - 24c + 144$

133. $f(x) = -3x^2 + 6x + 45$

134. $g(x) = 16x^2 - 100$

135. $G(y) = 16y^3 + 250$

136. $f(x) = -4x^2 - 16$

In Problems 137 and 138, write an expression for the shaded area in factored form.

137.

138.

Section 4.8 Polynomial Equations

KEY CONCEPTS

- **The Zero-Product Property**
  If the product of two numbers is zero, then at least one of the numbers is 0. That is, if $ab = 0$, then $a = 0$ or $b = 0$ or both a and b are 0.

- **Zero of a Function**
  A zero of a function f is any number r such that $f(r) = 0$. If r is a zero of a function, then $(r, 0)$ is also an x-intercept of the graph of the function.

KEY TERMS

Polynomial equation
Degree of a polynomial
  equation
Zero-product property
Quadratic equation
Standard form
Second-degree equation
Zero

You Should Be Able To...	EXAMPLE	Review Exercises
1 Solve polynomial equations using the Zero-Product Property (p. 372)	Examples 1 through 4	139–148
2 Solve equations involving polynomial functions (p. 376)	Examples 5 and 6	149–152
3 Model and solve problems involving polynomials (p. 377)	Examples 7 through 9	153–154

In Problems 139–148, solve each equation.

139. $(w + 5)(w - 13) = 0$ **140.** $x^2 + 21x + 54 = 0$

141. $y^2 + 2y = 15$ **142.** $5a^2 = -20a$

143. $x(x + 1) = 110$ **144.** $15x^2 + 29x - 14 = 0$

145. $\frac{1}{2}x^2 + 5x + 12 = 0$ **146.** $(b + 1)(b - 3) = 5$

147. $2x^3 + 5x^2 - 8x = 20$ **148.** $5x^3 + x^2 - 45x - 9 = 0$

149. Suppose that $f(x) = x^2 + 5x - 18$. Find values of x such that
(a) $f(x) = 6$ **(b)** $f(x) = -4$
What points on the graph of f do these points correspond to?

150. Suppose that $f(x) = 5x^2 - 4x + 3$. Find values of x such that
(a) $f(x) = 3$ **(b)** $f(x) = 4$
What points on the graph of f do these values correspond to?

In Problems 151 and 152, find the zeros of the function. What are the x-intercepts of the graph of the function?

151. $f(x) = 3x^3 + 18x^2 + 24x$

152. $f(x) = -4x^2 + 22x + 42$

153. **Falling Object** At one point, a 4-foot flagpole on top of the KXJB-TV mast in Galesburg, North Dakota, made it the world's tallest structure, standing 2064 feet tall. If an object is dropped from the top of this mast, the height s of the object (in feet) as a function of time (in seconds) can be modeled by the function $s(t) = -16t^2 + 2064$. When will the object be 1280 feet above the ground?

154. **Reliability** A simple parallel system with two identical components has a reliability given by $R = 1 - (1 - r)^2$, where r is the reliability of the individual components.
(a) What is the reliability of the individual components if the system reliability is $R = 0.96$?
(b) What is the reliability of the individual components if the system reliability is $R = 0.99$?

Chapter 4 Test

Step-by-step test solutions are found on the Chapter Test Prep Videos available in MyMathLab®, on You Tube®, or may be accessed using the QR code to the right.

1. Write the polynomial in standard form and determine its degree.
$$7x^2 + x^4 - 5x^7 + 1 - x$$

2. Add:
$$\left(-2a^3b^2 + 5a^2b + ab + 1\right) + \left(\frac{1}{3}a^3b^2 + 4a^2b - 6ab - 5\right)$$

3. For $f(x) = x^3 + 3x^2 - x + 1$, find $f(-2)$.

4. For $f(x) = 7x^3 - 1$ and $g(x) = 4x^2 + 3x - 2$, find $(f - g)(x)$.

In Problems 5–7, find each product.

5. $\dfrac{1}{2}a^2b\left(4ab^2 - 6ab + 8\right)$

6. $(3x - 1)(4x + 17)$

7. $(2m - n)^2$

8. Divide using long division: $\dfrac{6z^3 - 14z^2 + z + 4}{2z^2 + 1}$

9. Divide using synthetic division:
$$\dfrac{5x^2 - 27x - 18}{x - 6}$$

10. For $f(x) = 6x^2 + x - 12$ and $g(x) = 2x + 3$, find $\left(\dfrac{f}{g}\right)(2)$.

11. Use the Remainder Theorem to determine the remainder when $f(x) = 2x^3 - 3x^2 - 4x + 7$ is divided by $x - 3$.

12. Factor out the greatest common factor:
$12a^3b^2 + 8a^2b^2 - 16ab^3$

In Problems 13–18, factor completely.

13. $6c^2 + 21c - 4c - 14$

14. $x^2 - 13x - 48$

15. $-14p^2 - 17p + 6$

16. $5(z - 1)^2 + 17(z - 1) - 12$

17. $-98x^2 + 112x - 32$

18. $16x^2 - 196$

19. Solve: $3m^2 - 5m = 5m - 7$

20. One side of a rectangular patio is 3 meters longer than the other. If the area of the patio is 108 square meters, what are the dimensions of the patio?

Getting Ready for Chapter 5: A Review of Operations on Rational Numbers

Objectives

1 Write Rational Numbers in Lowest Terms

2 Multiply and Divide Rational Numbers

3 Add and Subtract Rational Numbers

The purpose of this Getting Ready section is to review the operations on rational numbers. These methods will be used in Chapter 5 when operations on rational expressions are discussed.

1 Write Rational Numbers in Lowest Terms

It is preferred to write rational numbers in **lowest terms**—that is, without any common factors in the numerator and the denominator. The *Reduction Property* is used to obtain rational numbers in lowest terms.

> **Reduction Property**
>
> If a, b, and c are real numbers, then
>
> $$\frac{ac}{bc} = \frac{a}{b} \qquad \text{if } b \neq 0, c \neq 0$$

EXAMPLE 1   **Writing Rational Numbers in Lowest Terms**

Divide out the common factor
$$\downarrow$$
$$\frac{45}{18} = \frac{9 \cdot 5}{9 \cdot 2} = \frac{5}{2}$$

Quick ✓

1. When a rational number is written so there are no common factors in the numerator and the denominator, the rational number is in _____ _____.

In Problems 2 and 3, write each rational number in lowest terms.

2. $\dfrac{13 \cdot 5}{13 \cdot 6}$ **3.** $\dfrac{80}{12}$

2 Multiply and Divide Rational Numbers

We now review the methods for multiplying and dividing rational numbers.

> **Multiplying Rational Numbers**
>
> **Step 1:** Completely factor each integer in the numerator and denominator.
>
> **Step 2:** Use the fact that if $\dfrac{a}{b}$ and $\dfrac{c}{d}$, $b \neq 0, d \neq 0$, are two rational numbers, then $\dfrac{a}{b} \cdot \dfrac{c}{d} = \dfrac{ac}{bd}$.
>
> **Step 3:** Divide out common factors in the numerator and denominator.

In Other Words

To multiply two rational numbers, multiply the numerators and then multiply the denominators. To divide two rational numbers, multiply the rational number in the numerator by the reciprocal of the rational number in the denominator.

> **Dividing Rational Numbers**
>
> To divide rational numbers, use the fact that if $\dfrac{a}{b}$ and $\dfrac{c}{d}$, $b \neq 0, c \neq 0, d \neq 0$, are two rational numbers, then $\dfrac{\dfrac{a}{b}}{\dfrac{c}{d}} = \dfrac{a}{b} \div \dfrac{c}{d} = \dfrac{a}{b} \cdot \dfrac{d}{c} = \dfrac{ad}{bc}$.

EXAMPLE 2 **Multiplying and Dividing Rational Numbers**

Perform the indicated operation. Be sure to express the result in lowest terms.

▷ **(a)** $\dfrac{10}{3} \cdot \dfrac{18}{25}$ ▷ **(b)** $\dfrac{\frac{14}{5}}{\frac{21}{10}}$

Solution

(a)

$$\dfrac{10}{3} \cdot \dfrac{18}{25} = \dfrac{2 \cdot 5}{3} \cdot \dfrac{9 \cdot 2}{5 \cdot 5}$$

$$= \dfrac{2 \cdot 5}{3} \cdot \dfrac{3 \cdot 3 \cdot 2}{5 \cdot 5}$$

Multiply; use the Reduction Property to divide out common factors: $= \dfrac{2 \cdot 5 \cdot 3 \cdot 3 \cdot 2}{3 \cdot 5 \cdot 5}$

$$= \dfrac{2 \cdot 3 \cdot 2}{5}$$

Multiply: $= \dfrac{12}{5}$

(b) Rewrite the division problem as a multiplication problem by multiplying the numerator, $\dfrac{14}{5}$, by the reciprocal of the denominator, $\dfrac{21}{10}$. The reciprocal of $\dfrac{21}{10}$ is $\dfrac{10}{21}$.

$$\dfrac{\frac{14}{5}}{\frac{21}{10}} = \dfrac{14}{5} \cdot \dfrac{10}{21}$$

Factor: $= \dfrac{7 \cdot 2}{5} \cdot \dfrac{5 \cdot 2}{7 \cdot 3}$

Multiply; divide out common factors: $= \dfrac{7 \cdot 2 \cdot 5 \cdot 2}{5 \cdot 7 \cdot 3}$

$$= \dfrac{2 \cdot 2}{3} = \dfrac{4}{3}$$

Work Smart

In Example 2(a), rather than factoring each numerator and denominator individually and then dividing out like factors, you could proceed as follows:

$$\dfrac{10}{3} \cdot \dfrac{18}{25} = \dfrac{\overset{2}{10} \cdot \overset{6}{18}}{\underset{1}{3} \cdot \underset{5}{25}}$$

Multiply numerators; multiply denominators: $= \dfrac{2 \cdot 6}{1 \cdot 5}$

$$= \dfrac{12}{5}$$

Quick ✔

In Problems 4–7, perform the indicated operation. Express your answer in lowest terms.

4. $\dfrac{5}{7}\left(-\dfrac{21}{10}\right)$ **5.** $\dfrac{35}{15} \cdot \dfrac{3}{14}$ **6.** $\dfrac{\frac{4}{5}}{\frac{12}{25}}$ **7.** $\dfrac{24}{35} \div \left(-\dfrac{8}{7}\right)$

❸ Add and Subtract Rational Numbers

▷ We now review addition and subtraction of rational numbers.

In Other Words

To add two rational numbers with a common denominator, add the numerators and write the sum over the common denominator.

Adding or Subtracting Rational Numbers

Step 1: If $\dfrac{a}{c}$ and $\dfrac{b}{c}$, $c \neq 0$, are two rational numbers, then $\dfrac{a}{c} + \dfrac{b}{c} = \dfrac{a+b}{c}$ and $\dfrac{a}{c} - \dfrac{b}{c} = \dfrac{a-b}{c}$.

Step 2: Write the result in lowest terms.

EXAMPLE 3   **Adding or Subtracting Rational Numbers with Common Denominators**

Perform the indicated operation. Be sure to express the result in lowest terms.

(a) $\dfrac{7}{24} + \dfrac{11}{24}$ **(b)** $\dfrac{2}{15} - \dfrac{7}{15}$

Solution

(a)

$$\dfrac{7}{24} + \dfrac{11}{24} = \dfrac{7 + 11}{24}$$

$$= \dfrac{18}{24}$$

Factor numerator and denominator: $= \dfrac{6 \cdot 3}{6 \cdot 4}$

Divide out common factors: $= \dfrac{3}{4}$

(b)

$$\dfrac{2}{15} - \dfrac{7}{15} = \dfrac{2 - 7}{15}$$

$$= \dfrac{-5}{15}$$

Factor numerator and denominator: $= \dfrac{-1 \cdot 5}{3 \cdot 5}$

Divide out common factors: $= \dfrac{-1}{3} = -\dfrac{1}{3}$ ●

Quick ✓

In Problems 8 and 9, perform the indicated operation. Express your answer in lowest terms.

8. $\dfrac{11}{12} + \dfrac{5}{12}$ **9.** $\dfrac{3}{18} - \dfrac{13}{18}$

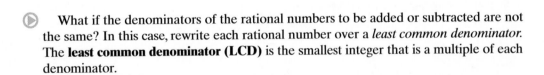   What if the denominators of the rational numbers to be added or subtracted are not the same? In this case, rewrite each rational number over a *least common denominator*. The **least common denominator (LCD)** is the smallest integer that is a multiple of each denominator.

EXAMPLE 4   **How to Find the Least Common Denominator**

Find the least common denominator of the rational numbers $\dfrac{7}{30}$ and $\dfrac{5}{12}$. Then rewrite each rational number with the least common denominator.

Step-by-Step Solution

Step 1: Factor each denominator as a product of prime factors, arranging like factors vertically.	$30 = 5 \cdot 6 = 5 \cdot 3 \cdot 2$ $12 = 3 \cdot 4 = \begin{vmatrix} \;\;\;3 \cdot 2 \cdot 2 \\ \downarrow \;\downarrow \;\downarrow \;\downarrow \end{vmatrix}$	**Work Smart** Line up the factors vertically to find the LCD.
Step 2: Find the product of each of the prime factors the greatest number of times they appear in any factorization.	$\begin{aligned} \text{LCD} &= 5 \cdot 3 \cdot 2 \cdot 2 \\ &= 60 \end{aligned}$	

(continued)

The Multiplicative Identity Property, $1 \cdot a = a$, is used to write $\dfrac{7}{30}$ and $\dfrac{5}{12}$ with the denominator 60. Multiply $\dfrac{7}{30}$ by $1 = \dfrac{2}{2}$ and multiply $\dfrac{5}{12}$ by $1 = \dfrac{5}{5}$:

$$\dfrac{7}{30} = \dfrac{7}{30} \cdot \dfrac{2}{2} \qquad \dfrac{5}{12} = \dfrac{5}{12} \cdot \dfrac{5}{5}$$

$$= \dfrac{7 \cdot 2}{30 \cdot 2} \qquad\qquad = \dfrac{5 \cdot 5}{12 \cdot 5}$$

$$= \dfrac{14}{60} \qquad\qquad\quad = \dfrac{25}{60}$$

Quick ✓

In Problems 10 and 11, find the least common denominator (LCD) of each pair of rational numbers. Then rewrite each rational number with the LCD.

10. $\dfrac{3}{25}$ and $\dfrac{2}{15}$

11. $\dfrac{5}{18}$ and $-\dfrac{1}{63}$

EXAMPLE 5   **How to Add or Subtract Fractions Using the Least Common Denominator**

Perform the indicated operation:

(a) $\dfrac{5}{2} + \dfrac{4}{3}$

(b) $\dfrac{5}{28} - \dfrac{5}{12}$

Step-by-Step Solution

(a) $\dfrac{5}{2} + \dfrac{4}{3}$

Step 1: Find the least common denominator.

Each denominator is prime, so the LCD $= 2 \cdot 3 = 6$.

Step 2: Rewrite each rational number with the common denominator.

$$\dfrac{5}{2} + \dfrac{4}{3} = \dfrac{5}{2} \cdot \dfrac{3}{3} + \dfrac{4}{3} \cdot \dfrac{2}{2}$$

$$= \dfrac{15}{6} + \dfrac{8}{6}$$

Step 3: Add the numerators and write the result over the common denominator.

$$= \dfrac{15 + 8}{6}$$

$$= \dfrac{23}{6}$$

Step 4: Write the result in lowest terms.

The rational number is already in lowest terms.

So, $\dfrac{5}{2} + \dfrac{4}{3} = \dfrac{23}{6}$

(b) $\dfrac{5}{28} - \dfrac{5}{12}$

Step 1: Find the least common denominator.

$$28 = 7 \cdot 4 \quad = 7 \cdot 2 \cdot 2$$
$$12 = \quad 4 \cdot 3 = \; | \; 2 \cdot 2 \cdot 3$$
$$\downarrow \; \downarrow \; \downarrow \; \downarrow$$
$$\text{LCD} = 7 \cdot 2 \cdot 2 \cdot 3$$
$$= 84$$

Step 2: Rewrite each rational number with the common denominator.

$$\dfrac{5}{28} - \dfrac{5}{12} = \dfrac{5}{28} \cdot \dfrac{3}{3} - \dfrac{5}{12} \cdot \dfrac{7}{7}$$
$$= \dfrac{15}{84} - \dfrac{35}{84}$$

Step 3: Subtract the numerators and write the result over the common denominator.

$$= \dfrac{15 - 35}{84}$$
$$= \dfrac{-20}{84}$$

Step 4: Write the result in lowest terms.

$$= \dfrac{-5 \cdot 4}{21 \cdot 4}$$
$$= \dfrac{-5}{21}$$
$$= -\dfrac{5}{21}$$

The steps used in Example 5 are summarized below.

> **Adding or Subtracting Rational Numbers with Unlike Denominators**
>
> **Step 1:** Find the least common denominator.
> **Step 2:** Rewrite each rational number with the common denominator.
> **Step 3:** Add or subtract the numerators, and write the result over the common denominator.
> **Step 4:** Write the result in lowest terms.

Quick ✓

In Problems 12–14, perform the indicated operation. Express your answer in lowest terms.

12. $\dfrac{3}{4} + \dfrac{1}{5}$ 　　　　　 **13.** $\dfrac{3}{20} + \dfrac{2}{15}$ 　　　　　 **14.** $\dfrac{5}{14} - \dfrac{11}{21}$

Getting Ready for Chapter 5 Exercises

MyMathLab®

Exercise numbers in green have complete video solutions in MyMathLab or may be accessed using the QR code to the right.

*Problems **1–14** are the Quick ✓ s that follow the EXAMPLES.*

Building Skills

In Problems 15–20, write each rational number in lowest terms. See Objective 1.

15. $\dfrac{4}{12}$

16. $\dfrac{6}{18}$

17. $-\dfrac{15}{35}$

18. $-\dfrac{12}{28}$

19. $\dfrac{-50}{-10}$

20. $\dfrac{81}{-27}$

In Problems 21–32, multiply or divide the rational numbers. Express each product or quotient as a rational number in lowest terms. See Objective 2.

21. $\dfrac{3}{4} \cdot \dfrac{20}{9}$

22. $\dfrac{2}{3} \cdot \dfrac{15}{6}$

23. $-\dfrac{5}{6} \cdot \dfrac{18}{5}$

24. $-\dfrac{9}{8} \cdot \dfrac{16}{3}$

25. $\dfrac{5}{8} \cdot \dfrac{2}{15}$

26. $\dfrac{3}{14} \cdot \dfrac{7}{12}$

27. $\dfrac{5}{2} \div \dfrac{25}{4}$

28. $\dfrac{2}{3} \div \dfrac{8}{9}$

29. $\dfrac{-\dfrac{6}{5}}{\dfrac{8}{15}}$

30. $\dfrac{\dfrac{12}{7}}{\dfrac{18}{21}}$

31. $\dfrac{-\dfrac{9}{2}}{-\dfrac{3}{4}}$

32. $\dfrac{-\dfrac{10}{7}}{-\dfrac{5}{14}}$

In Problems 33–51, add or subtract the rational numbers. Express each sum or difference as a rational number in lowest terms. See Objective 3.

33. $\dfrac{5}{3} + \dfrac{1}{3}$

34. $\dfrac{3}{4} + \dfrac{9}{4}$

35. $\dfrac{11}{6} - \dfrac{1}{6}$

36. $\dfrac{19}{6} - \dfrac{5}{6}$

37. $-\dfrac{3}{4} + \dfrac{1}{3}$

38. $\dfrac{-1}{7} + \dfrac{4}{9}$

39. $\dfrac{1}{4} + \dfrac{5}{6}$

40. $\dfrac{5}{8} + \dfrac{5}{12}$

41. $-\dfrac{3}{10} - \dfrac{7}{15}$

42. $\dfrac{7}{16} - \dfrac{9}{20}$

43. $\dfrac{5}{18} - \dfrac{11}{15}$

44. $\dfrac{5}{24} + \dfrac{7}{32}$

45. $-\dfrac{7}{8} + \dfrac{3}{10}$

46. $-\dfrac{7}{12} + \dfrac{2}{15}$

47. $\dfrac{7}{24} - \dfrac{3}{20}$

48. $\dfrac{3}{28} + \dfrac{5}{12}$

49. $-\dfrac{7}{9} - \dfrac{2}{15}$

50. $-\dfrac{5}{18} - \dfrac{1}{45}$

51. $\dfrac{-4}{25} - \dfrac{7}{30}$

Explaining the Concepts

52. Explain how to write a rational number in lowest terms.

53. Explain how to find the least common denominator of two rational numbers.

54. Explain how to add two rational numbers that do not have a common denominator.

5 Rational Expressions and Rational Functions

Can a mathematical formula be used to predict the number of runs a major league baseball team will score in a season? Yes! In fact, his knowledge of this mathematical formula is one of the reasons why Billy Beane, former general manager of the Oakland A's, sought out baseball players who got a lot of walks. (Billy's baseball philosophy was the basis for the 2011 movie *Moneyball*, starring Brad Pitt.) See Problem 71 in Section 5.4.

The Big Picture: Putting It Together

In Chapter 4, we learned how to add, subtract, multiply, and divide polynomials. We then learned how to factor polynomial expressions and use the result (along with the Zero-Product Property) to solve a polynomial equation.

We will do the same thing with *rational expressions*. A rational expression is a polynomial divided by another polynomial. Factoring comes in handy when working with rational expressions. The methods for reducing, adding, subtracting, multiplying, and dividing rational expressions are identical to the methods used to do these operations on rational numbers. The point is this—the algebra of rational expressions can be thought of as generic arithmetic of rational numbers.

In this chapter, a new category of problem is introduced called Synthesis Review. These problems will help you see the "big picture" of algebra. For example, you may be asked to add various objects (polynomials, rational expressions, and so on) and then to discuss similarities and differences in performing the operation on the objects.

Outline

5.1 Multiplying and Dividing Rational Expressions

Objectives

❶ Determine the Domain of a Rational Expression

❷ Simplify Rational Expressions

❸ Multiply Rational Expressions

❹ Divide Rational Expressions

❺ Work with Rational Functions

Are You Prepared for This Section?

Before getting started, complete the following problems. If you get a problem wrong, go back to the section cited and review the material.

P1. Factor: $2x^2 - 11x - 21$ [Section 4.5, pp. 351–356]

P2. Solve: $q^2 - 16 = 0$ [Section 4.8, pp. 372–376]

P3. Determine the reciprocal of $\dfrac{5}{2}$. [Section R.3, pp. 23–24]

P4. Explain what *domain* means. [Section 2.1, pp. 149–151]

The quotient of two polynomials is a **rational expression.** Some examples of rational expressions are

(a) $\dfrac{x - 5}{2x + 1}$ (b) $\dfrac{x^2 - 7x - 18}{x^2 - 4}$ (c) $\dfrac{1}{x - 3}$ (d) $\dfrac{2a^2 + 5ab + 2b^2}{a^2 - 6ab + 8b^2}$

Expressions (a), (b), and (c) are rational expressions in one variable, x. Expression (d) is a rational expression in two variables, a and b.

Rational expressions are described the same way as rational numbers. In the expression $\dfrac{x - 5}{2x + 1}$, the expression $x - 5$ is the **numerator** and $2x + 1$ is the **denominator.** When the numerator and denominator have no common factors (except 1 and -1), the rational expression is written in **lowest terms,** or **simplified.**

▶ ❶ Determine the Domain of a Rational Expression

To determine the domain of a rational expression, find all values of the variable that cause the denominator to equal 0 and exclude these values from the domain because division by 0 is not defined. Knowing the domain of a rational expression will help when solving rational equations later in this chapter.

EXAMPLE 1 **Determining the Domain of a Rational Expression**

Determine the domain of each of the following rational expressions.

(a) $\dfrac{2x}{x + 3}$ (b) $\dfrac{p^2 + 5p + 6}{p^2 - 4}$

Solution

(a) Find all values of x that cause $x + 3$ to equal 0.

$$x + 3 = 0$$

Subtract 3 from both sides: $x = -3$

Because -3 causes the denominator, $x + 3$, to equal 0, the domain of $\dfrac{2x}{x + 3}$ is $\{x \mid x \neq -3\}$.

(b) Find all values of p that cause $p^2 - 4$ to equal 0.

$$p^2 - 4 = 0$$

Factor the difference of two squares: $(p - 2)(p + 2) = 0$

Zero-Product Property: $p - 2 = 0$ or $p + 2 = 0$

$p = 2$ or $p = -2$

The domain of $\dfrac{p^2 + 5p + 6}{p^2 - 4}$ is $\{p \mid p \neq -2, p \neq 2\}$. ●

Work Smart

Because we are working with the set of real numbers, the domain of a variable is understood to be all real numbers except those listed. For example, the notation $\{x \mid x \neq -3\}$ means x is any real number except -3.

Prepared?...Answers

P1. $(2x + 3)(x - 7)$

P2. $\{-4, 4\}$ **P3.** $\dfrac{2}{5}$

P4. The set of all inputs for which an algebraic expression is defined.

Quick ✓

1. The quotient of two polynomials is called a _____ _____.

2. In the expression $\dfrac{x + 3}{3x - 5}$, $x + 3$ is called the _____, and $3x - 5$ is called the

   _____.

3. *True or False* The domain of all rational functions is the set of all real numbers.

In Problems 4 and 5, determine the domain of the rational expression.

4. $\dfrac{x - 4}{x + 6}$

5. $\dfrac{z^2 - 9}{z^2 + 3z - 28}$

▶ ❷ Simplify Rational Expressions

A rational expression is simplified by completely factoring the numerator and the denominator and dividing out any common factors using the Reduction Property.

$$\frac{ac}{bc} = \frac{a\cancel{c}}{b\cancel{c}} = \frac{a}{b} \qquad \text{if } b \neq 0, c \neq 0$$

Simplify rational expressions in the same way that rational numbers are written in lowest terms. For example, write $\dfrac{12}{20}$ in lowest terms as follows:

$$\frac{12}{20} = \frac{\cancel{4} \cdot 3}{\cancel{4} \cdot 5} = \frac{3}{5}$$

EXAMPLE 2 | **Simplifying a Rational Expression**

Simplify each rational expression:

(a) $\dfrac{x^2 + 2x - 15}{2x^2 - 3x - 9}$

(b) $\dfrac{q^3 - 8}{3q^2 - 6q}$

Solution

(a) Factor the numerator and denominator. Then divide out common factors using the Reduction Property.

$$\frac{x^2 + 2x - 15}{2x^2 - 3x - 9} = \frac{(x + 5)\,\cancel{(x - 3)}}{(2x + 3)\,\cancel{(x - 3)}}$$

$$= \frac{x + 5}{2x + 3} \qquad x \neq -\frac{3}{2}, x \neq 3$$

(b) Factor the numerator and denominator. Then divide out common factors.

$$\frac{q^3 - 8}{3q^2 - 6q} = \frac{\cancel{(q - 2)}\,(q^2 + 2q + 4)}{3q\,\cancel{(q - 2)}}$$

$$= \frac{q^2 + 2q + 4}{3q} \qquad q \neq 0, q \neq 2 \qquad \bullet$$

Work Smart

In dividing out like factors, the quotient is 1. For example,

$$\frac{\cancel{(x + 3)}}{2x\,\cancel{(x + 3)}} = \frac{1}{2x}$$

To keep the original rational expression and the simplified rational expression equivalent, restrict from the domain all values of the variable that are not in the domain of the *original* rational expression.

Consider Example 2(a). The restriction $x \neq -\dfrac{3}{2}, x \neq 3$ is included for two reasons:

1. To remind us of the restrictions on the variable x.

2. To keep the rational expressions equal. Without the restriction $x \neq -\dfrac{3}{2}, x \neq 3$,

   the expression $\dfrac{x^2 + 2x - 15}{2x^2 - 3x - 9}$ is not equal to $\dfrac{x + 5}{2x + 3}$ because in $\dfrac{x^2 + 2x - 15}{2x^2 - 3x - 9}$,

   x *cannot* take on the value 3, while in $\dfrac{x + 5}{2x + 3}$, x *can* take on the value 3. Not

   allowing x to equal 3 in either instance keeps the expressions equal.

The same logic justifies the restrictions on q in Example 2(b). For the remainder of the text, the restrictions on the variable will not be listed, but you should be aware that the restrictions are necessary to maintain equality.

Sometimes common factors in the numerator and denominator of a rational expression can be found by factoring -1 out of one of the factors. Consider the rational expression $\dfrac{3 - 4x}{4x - 3}$. If -1 is factored out of $3 - 4x$, then $-1(-3 + 4x)$ or $-1(4x - 3)$ so that the common factor, $4x - 3$, can be divided out.

EXAMPLE 3 **Simplifying a Rational Expression**

Simplify: $\dfrac{3x^2 + 11x - 4}{1 - 3x}$

Solution

$$\frac{3x^2 + 11x - 4}{1 - 3x} = \frac{(3x - 1)(x + 4)}{1 - 3x}$$

Factor -1 from $1 - 3x$; divide out common factors:
$$= \frac{\cancel{(3x - 1)}(x + 4)}{-1\cancel{(3x - 1)}}$$

$$= \frac{(x + 4)}{-1}$$

$$\frac{a}{-1} = -a: \quad = -(x + 4)$$

Work Smart

When simplifying, divide out only common factors, not common terms!

WRONG! $\dfrac{x + 1}{x} = \dfrac{\cancel{x} + 1}{\cancel{x}} = 1$ **WRONG!** $\dfrac{x^2 + x + 2}{x + 2} = \dfrac{x^2 + \cancel{x} + \cancel{2}}{\cancel{x} + \cancel{2}} = x^2$

If you aren't sure what you can divide out, try the computation with numbers to see whether it works. For example, does $\dfrac{4}{3} = \dfrac{3 + 1}{3} = \dfrac{\cancel{3} + 1}{\cancel{3}} = 1$? NO! So, $\dfrac{x + 1}{x} \neq 1$.

Quick ✓

6. *True or False* $\dfrac{x^2 + 3x + 5}{x^2 + 7} = \dfrac{3x + 5}{7}$

In Problems 7–9, simplify each rational expression.

7. $\dfrac{x^2 - 7x + 12}{x^2 + 4x - 21}$ **8.** $\dfrac{z^3 - 64}{2z^2 - 3z - 20}$ **9.** $\dfrac{3w^2 + 13w - 10}{2 - 3w}$

▶ ❸ **Multiply Rational Expressions**

We show how to multiply rational expressions in the next example.

EXAMPLE 4 **How to Multiply Rational Expressions**

Multiply and simplify: $\dfrac{x^2 + 2x - 15}{x + 1} \cdot \dfrac{x^2 + 7x}{x^2 + 4x - 21}$

Step-by-Step Solution

Step 1: Completely factor each polynomial in the numerator and denominator.

$$\frac{x^2 + 2x - 15}{x + 1} \cdot \frac{x^2 + 7x}{x^2 + 4x - 21} = \frac{(x + 5)(x - 3)}{x + 1} \cdot \frac{x(x + 7)}{(x + 7)(x - 3)}$$

Step 2: Multiply.

$$= \frac{(x + 5)(x - 3) \cdot x(x + 7)}{(x + 1) \cdot (x + 7)(x - 3)}$$

Step 3: Divide out common factors in the numerator and denominator.

Work Smart: Study Skills

It is a good idea to use a different slash mark for each pair of factors that divides out. Or, use colored pencils to highlight the like factors that divide out.

$$= \frac{(x + 5)(x - 3)\,x(x + 7)}{(x + 1)(x + 7)(x - 3)}$$

$$= \frac{x(x + 5)}{x + 1}$$

The steps for multiplying rational expressions are the same as the steps for multiplying rational numbers.

Multiplying Rational Expressions

Step 1: Completely factor each polynomial in the numerator and denominator.

Step 2: Use the fact that if $\dfrac{a}{b}$ and $\dfrac{c}{d}$, where $b \neq 0$ and $d \neq 0$, are two rational

expressions, then $\dfrac{a}{b} \cdot \dfrac{c}{d} = \dfrac{ac}{bd}$.

Step 3: Divide out common factors in the numerator and denominator.

Always leave the product in factored form. The factored form will be needed to solve rational equations (Section 5.4) and rational inequalities (Section 5.5).

Quick ✓

In Problem 10, multiply and simplify the rational expressions.

10. $\dfrac{p^2 - 9}{p^2 + 5p + 6} \cdot \dfrac{3p^2 - p - 2}{2p - 6}$

EXAMPLE 5   **Multiplying Rational Expressions**

Multiply and simplify each of the following rational expressions.

(a) $\dfrac{y^2 + 2y + 1}{3y^2 + y - 2} \cdot \dfrac{2 - 3y}{y^2 + 5y + 4}$

(b) $\dfrac{p^2 + 5pq + 6q^2}{2p^2 + 7pq + 3q^2} \cdot \dfrac{2p + q}{3p + 6q}$

Solution

Factor the numerator and denominator

(a) $\dfrac{y^2 + 2y + 1}{3y^2 + y - 2} \cdot \dfrac{2 - 3y}{y^2 + 5y + 4} = \dfrac{(y + 1)(y + 1)}{(3y - 2)(y + 1)} \cdot \dfrac{-1(3y - 2)}{(y + 4)(y + 1)}$

Multiply: $= \dfrac{(y + 1)(y + 1)(-1)(3y - 2)}{(3y - 2)(y + 1)(y + 4)(y + 1)}$

Divide out common factors: $= \dfrac{(y + 1)(y + 1)(-1)(3y - 2)}{(3y - 2)(y + 1)(y + 4)(y + 1)}$

$= -\dfrac{1}{y + 4}$

Factor the numerator and denominator

(b) $\dfrac{p^2 + 5pq + 6q^2}{2p^2 + 7pq + 3q^2} \cdot \dfrac{2p + q}{3p + 6q} = \dfrac{(p + 2q)(p + 3q)}{(2p + q)(p + 3q)} \cdot \dfrac{2p + q}{3(p + 2q)}$

Multiply: $= \dfrac{(p + 2q)(p + 3q)(2p + q)}{(2p + q)(p + 3q)(3)(p + 2q)}$

Divide out common factors: $= \dfrac{(p + 2q)(p + 3q)(2p + q)}{(2p + q)(p + 3q)(3)(p + 2q)}$

$= \dfrac{1}{3}$

Work Smart

Did you notice in Example 5(b) that all the factors in the numerator divide out? This means that a factor of 1 remains in the numerator, not 0.

Quick ✓

In Problems 11 and 12, multiply and simplify the rational expressions.

11. $\dfrac{2x + 8}{2x^2 + 11x + 12} \cdot \dfrac{2x^2 - 3x - 9}{6 - 2x}$

12. $\dfrac{m^2 + 2mn + n^2}{2m^2 + 3mn + n^2} \cdot \dfrac{2m^2 - 5mn - 3n^2}{3n - m}$

▶ ❹ Divide Rational Expressions

The rule for dividing rational expressions is the same as the rule for dividing rational numbers.

> **Dividing Rational Expressions**
>
> To divide rational expressions use the fact that if $\dfrac{a}{b}$ and $\dfrac{c}{d}$, where $b \neq 0, c \neq 0$,
>
> and $d \neq 0$, are two rational expressions, then $\dfrac{\dfrac{a}{b}}{\dfrac{c}{d}} = \dfrac{a}{b} \div \dfrac{c}{d} = \dfrac{a}{b} \cdot \dfrac{d}{c}$. Then follow the
>
> steps for multiplying two rational expressions.

In Other Words

To divide two rational expressions, multiply the rational expression in the numerator by the reciprocal of the rational expression in the denominator.

EXAMPLE 6 | **Dividing Rational Expressions**

Divide each of the following rational expressions. Simplify the quotient, if possible.

(a) $\dfrac{\dfrac{20x^5}{3y}}{\dfrac{4x^2}{15y^5}}$

(b) $\dfrac{\dfrac{x^2 - 4x - 12}{4x^3 - 6x^2}}{\dfrac{x^3 + 8}{2x^3 - 4x^2 + 8x}}$

Solution

Rewrite the division problem as a multiplication problem by multiplying the numerator by the reciprocal of the denominator.

Work Smart

When you see $\dfrac{\dfrac{20x^5}{3y}}{\dfrac{4x^2}{15y^5}}$, it may be helpful

to think

$\dfrac{20x^5}{3y} \div \dfrac{4x^2}{15y^5}$

(a)

$$\dfrac{\dfrac{20x^5}{3y}}{\dfrac{4x^2}{15y^5}} = \dfrac{20x^5}{3y} \cdot \dfrac{15y^5}{4x^2}$$

Multiply: $= \dfrac{20 \cdot 15 \cdot x^5 \cdot y^5}{3 \cdot 4 \cdot y \cdot x^2}$

Divide out common factors: $= \dfrac{\overset{5}{\cancel{20}} \cdot \overset{5}{\cancel{15}}}{\underset{1}{\cancel{3}} \cdot \underset{1}{\cancel{4}}} x^{5-2} y^{5-1}$

Simplify: $= 25x^3 y^4$

(b)

$$\dfrac{\dfrac{x^2 - 4x - 12}{4x^3 - 6x^2}}{\dfrac{x^3 + 8}{2x^3 - 4x^2 + 8x}} = \dfrac{x^2 - 4x - 12}{4x^3 - 6x^2} \cdot \dfrac{2x^3 - 4x^2 + 8x}{x^3 + 8}$$

Factor the numerator and denominator: $= \dfrac{(x - 6)(x + 2)}{2x^2(2x - 3)} \cdot \dfrac{2x(x^2 - 2x + 4)}{(x + 2)(x^2 - 2x + 4)}$

Multiply; divide out common factors: $= \dfrac{(x - 6)\cancel{(x + 2)} \cdot \cancel{2x}\cancel{(x^2 - 2x + 4)}}{2x^{\cancel{2}}(2x - 3) \cdot \cancel{(x + 2)}\cancel{(x^2 - 2x + 4)}}$

Simplify: $= \dfrac{x - 6}{x(2x - 3)}$

Quick ✓

In Problems 13 and 14, divide the rational expressions. Simplify the quotient, if possible.

13. $\dfrac{\dfrac{12a^4}{5b^2}}{\dfrac{4a^2}{15b^5}}$

14. $\dfrac{\dfrac{m^2 - 5m}{m - 7}}{\dfrac{2m}{m^2 - 6m - 7}}$

❺ Work with Rational Functions

Ratios of integers, such as $\dfrac{3}{5}$, are called *rational numbers*. Ratios of polynomial expressions are called *rational expressions*. Similarly, ratios of polynomial functions are called *rational functions*.

> **Definition**
>
> A **rational function** is a function of the form
>
> $$R(x) = \frac{p(x)}{q(x)}$$
>
> where p and q are polynomial functions and q is not the zero polynomial. The domain consists of all real numbers except those for which the denominator q is 0.

We find the domain of a rational function because the behavior of the graph of a rational function is interesting near values excluded from its domain. See Problem 97.

EXAMPLE 7 **Finding the Domain of a Rational Function**

Find the domain of $R(x) = \dfrac{x + 2}{x^2 - 5x - 14}$.

Solution

The domain is the set of all real numbers such that the denominator, $x^2 - 5x - 14$, does not equal zero. Find the values of x such that $x^2 - 5x - 14 = 0$ and exclude these values from the domain.

$$x^2 - 5x - 14 = 0$$

Factor: $(x - 7)(x + 2) = 0$

Set each factor equal to 0: $\quad x - 7 = 0 \quad$ or $\quad x + 2 = 0$

Solve: $\quad x = 7 \quad$ or $\quad x = -2$

The values $x = 7$ and $x = -2$ cause the denominator to equal 0, so the domain of R is $\{x \mid x \neq 7, x \neq -2\}$. ●

Quick ✓

In Problem 15, find the domain of the rational function.

15. $R(x) = \dfrac{2x}{x^2 + x - 30}$

Rational functions may be multiplied and divided just as polynomial functions are multiplied and divided. The domain of the product or quotient of two rational functions is the set of all numbers that are in the domains of the functions being multiplied or divided. In addition, for $R(x) = \dfrac{f(x)}{g(x)}$, exclude values of x such that $g(x) = 0$.

EXAMPLE 8   **Multiplying and Dividing Rational Functions**

Given that $f(x) = \dfrac{x^2 - 4}{3x^2 + 9x}$, $g(x) = \dfrac{x + 3}{x^2 - 2x - 8}$, and $h(x) = \dfrac{2x^2 + 7x + 6}{x^2 + 5x}$, find

▶ **(a)** $R(x) = f(x) \cdot g(x)$ ▶ **(b)** $P(x) = \dfrac{f(x)}{h(x)}$

and state the domain of each function.

Solution

(a)
$$R(x) = f(x) \cdot g(x) = \frac{x^2 - 4}{3x^2 + 9x} \cdot \frac{x + 3}{x^2 - 2x - 8}$$

Factor the numerator and denominator:
$$= \frac{(x + 2)(x - 2)}{3x(x + 3)} \cdot \frac{x + 3}{(x - 4)(x + 2)}$$

Multiply; divide out common factors:
$$= \frac{\cancel{(x + 2)}\,(x - 2)\,\cancel{(x + 3)}}{3x\,\cancel{(x + 3)}\,(x - 4)\,\cancel{(x + 2)}}$$

Simplify:
$$= \frac{x - 2}{3x(x - 4)}$$

The domain of $f(x)$ is $\{x | x \neq -3, x \neq 0\}$. The domain of $g(x)$ is $\{x | x \neq -2, x \neq 4\}$. Therefore, the domain of $R(x)$ is $\{x | x \neq -3, x \neq -2, x \neq 0, x \neq 4\}$.

(b)
$$P(x) = \frac{f(x)}{h(x)} = \frac{\dfrac{x^2 - 4}{3x^2 + 9x}}{\dfrac{2x^2 + 7x + 6}{x^2 + 5x}}$$

$$= \frac{x^2 - 4}{3x^2 + 9x} \cdot \frac{x^2 + 5x}{2x^2 + 7x + 6}$$

Factor the numerator and denominator:
$$= \frac{(x - 2)(x + 2)}{3x(x + 3)} \cdot \frac{x(x + 5)}{(2x + 3)(x + 2)}$$

Multiply; divide out common factors:
$$= \frac{(x - 2)\,\cancel{(x + 2)}\,\cancel{x}\,(x + 5)}{3\cancel{x}\,(x + 3)(2x + 3)\,\cancel{(x + 2)}}$$

Simplify:
$$= \frac{(x - 2)(x + 5)}{3(x + 3)(2x + 3)}$$

Work Smart

In Example 8(b), $3x^2 + 9x$ cannot equal 0. Why? In addition, $2x^2 + 7x + 6$ cannot equal 0. Why? And $x^2 + 5x$ cannot equal 0. Why?

The domain of $f(x)$ is $\{x | x \neq -3, x \neq 0\}$. The domain of $h(x)$ is $\{x | x \neq -5, x \neq 0\}$. Because the denominator of $P(x)$ cannot equal 0, also exclude those values of x such that $2x^2 + 7x + 6 = 0$. These values are $x = -2$ and $x = -\dfrac{3}{2}$. Therefore, the domain of $P(x)$ is

$$\left\{ x \,\middle|\, x \neq -5, x \neq -3, x \neq -2, x \neq -\frac{3}{2}, x \neq 0 \right\}.$$

●

Quick ✓

16. Given that $f(x) = \dfrac{x^2 - 4x - 5}{3x - 5}$, $g(x) = \dfrac{3x^2 + 4x - 15}{x^2 - 2x - 15}$, and $h(x) = \dfrac{4x^2 + 7x + 3}{9x^2 - 15x}$, find

(a) $R(x) = f(x) \cdot g(x)$ **(b)** $H(x) = \dfrac{f(x)}{h(x)}$

and state the domain of each function.

5.1 Exercises MyMathLab® Exercise numbers in green have complete video solutions in MyMathLab or may be accessed using the QR code to the right.

*Problems **1–16** are the Quick ✔s that follow the **EXAMPLES**.*

Building Skills

In Problems 17–26, state the domain of each rational expression. See Objective 1.

17. $\dfrac{3}{x + 5}$

18. $\dfrac{4}{x - 7}$

19. $\dfrac{x - 1}{x^2 - 6x - 16}$

20. $\dfrac{2x + 1}{x^2 + 4x - 45}$

21. $\dfrac{p^2 - 4}{2p^2 + p - 10}$

22. $\dfrac{m^2 + 5m + 6}{3m^2 + 4m - 4}$

23. $\dfrac{x + 1}{x^2 + 1}$

24. $\dfrac{x - 2}{x^2 + 4}$

25. $\dfrac{3x - 2}{(x - 1)^2}$

26. $\dfrac{x + 5}{x^2 + 8x + 16}$

In Problems 27–46, simplify each rational expression. See Objective 2.

27. $\dfrac{2x + 8}{x^2 - 16}$

28. $\dfrac{x^2 - 3x}{x^2 - 9}$

29. $\dfrac{p^2 + 4p + 3}{p + 1}$

30. $\dfrac{a^2 - 2a - 24}{a + 4}$

31. $\dfrac{5x + 25}{x^3 + 5x^2}$

32. $\dfrac{6x - 42}{x^3 - 7x^2}$

33. $\dfrac{q^2 - 3q - 18}{q^2 - 8q + 12}$

34. $\dfrac{w^2 + 5w - 14}{w^2 + 6w - 16}$

35. $\dfrac{2y^2 - 3y - 20}{2y^2 + 15y + 25}$

36. $\dfrac{3n^2 + n - 2}{3n^2 - 20n + 12}$

37. $\dfrac{9 - x^2}{x^2 + 2x - 15}$

38. $\dfrac{25 - k^2}{k^2 + 2k - 35}$

39. $\dfrac{x^3 + 2x^2 - 8x}{2x^4 - 32x^2}$

40. $\dfrac{2z^2 - 10z - 28}{4z^3 - 32z^2 + 28z}$

41. $\dfrac{x^2 - xy - 6y^2}{x^2 - 4y^2}$

42. $\dfrac{a^2 + 5ab + 4b^2}{a^2 + 8ab + 16b^2}$

43. $\dfrac{x^3 - 5x^2 + 3x - 15}{x^2 - 10x + 25}$

44. $\dfrac{v^3 + 3v^2 - 5v - 15}{v^2 + 6v + 9}$

45. $\dfrac{x^3 + 8}{x^2 - 5x - 14}$

46. $\dfrac{27q^3 + 1}{6q^2 - 7q - 3}$

In Problems 47–58, multiply and simplify each rational expression. See Objective 3.

47. $\dfrac{3x}{x^2 - x - 12} \cdot \dfrac{x - 4}{12x^2}$

48. $\dfrac{5x^2}{x + 3} \cdot \dfrac{x^2 + 7x + 12}{20x}$

49. $\dfrac{2x^2 - x - 6}{x^2 + 3x - 4} \cdot \dfrac{x^2 - x - 20}{2x^2 - 7x - 15}$

50. $\dfrac{3x^2 + 14x - 5}{x^2 + x - 30} \cdot \dfrac{x^2 - 2x - 15}{3x^2 + 8x - 3}$

51. $\dfrac{x^2 - 9}{x^2 - 25} \cdot \dfrac{x^2 - 2x - 15}{x^2 + 4x - 21}$

52. $\dfrac{p^2 - 16}{p^2 - 25} \cdot \dfrac{p^2 + 2p - 24}{p^2 + 3p - 4}$

53. $\dfrac{2q^2 - 5q - 3}{3q^2 + 19q + 6} \cdot \dfrac{3q^2 + 7q + 2}{3 - q}$

54. $\dfrac{2y^2 - 5y - 12}{2y^2 - y - 6} \cdot \dfrac{4y^2 - 5y - 6}{4 - y}$

55. $\dfrac{x^2 - 5x + 6}{x^2 + 2x - 8} \cdot (x + 4)$

56. $\dfrac{p^2 - 4p - 5}{p^2 - 5p - 6} \cdot (p - 6)$

57. $\dfrac{m^2 - n^2}{5m - 5n} \cdot \dfrac{10m + 5n}{2m^2 + 3mn + n^2}$

58. $\dfrac{a^2 + 2ab + b^2}{3a + 3b} \cdot \dfrac{b - a}{a^2 - b^2}$

In Problems 59–66, divide each rational expression. Simplify the quotient, if possible. See Objective 4.

59. $\dfrac{\dfrac{x + 3}{2x - 8}}{\dfrac{4x}{9}}$

60. $\dfrac{\dfrac{x - 2}{3x}}{\dfrac{5x - 10}{x}}$

61. $\dfrac{\dfrac{4a}{b^2}}{\dfrac{2a^2}{b}}$

62. $\dfrac{\dfrac{9m^3}{2n^2}}{\dfrac{3m}{8n^4}}$

63. $\dfrac{\dfrac{p^2 - 4p - 5}{2p^2 - 3p - 2}}{\dfrac{p^2 + p}{p^2 + p - 6}}$

64. $\dfrac{\dfrac{y^2 - 9}{2y^2 - y - 15}}{\dfrac{3y^2 + 10y + 3}{2y^2 + y - 10}}$

65. $\dfrac{\dfrac{x^3 - 1}{x^2 - 1}}{\dfrac{3x^2 + 3x + 3}{x^2 + 3x + 1}}$

66. $\dfrac{\dfrac{8x^3 + 1}{2x}}{\dfrac{x^3 + 2x^2 - 15x}{2x^2 - 5x - 3}}$

In Problems 67–76, determine the domain of each rational function.
See Objective 5.

67. $R(x) = \dfrac{2}{x - 1}$

68. $R(x) = \dfrac{5}{x + 3}$

69. $R(x) = \dfrac{x - 2}{(2x + 1)(x - 4)}$

70. $R(x) = \dfrac{3x + 2}{(4x - 1)(x + 5)}$

71. $R(x) = \dfrac{x + 9}{x^2 + 6x + 5}$

72. $R(x) = \dfrac{5x - 2}{x^2 - 6x - 16}$

73. $R(x) = \dfrac{x - 2}{2x^2 - 9x + 10}$

74. $R(x) = \dfrac{x + 3}{3x^2 + 7x - 6}$

75. $R(x) = \dfrac{x - 1}{x^2 + 1}$

76. $R(x) = \dfrac{4x}{4x^2 + 1}$

For Problems 77–80, see Objective 5.

77. If $f(x) = \dfrac{x^2 - 2x - 15}{x + 6}$, $g(x) = \dfrac{x^2 + 5x - 6}{2x^2 - 7x - 15}$, and
$h(x) = \dfrac{x + 3}{3x^2 + 17x - 6}$, find **(a)** $R(x) = f(x) \cdot g(x)$
and state its domain **(b)** $R(x) = \dfrac{f(x)}{h(x)}$ and state its
domain.

78. If $f(x) = \dfrac{x^2 - 7x - 8}{2x - 5}$, $g(x) = \dfrac{2x^2 + 3x - 20}{x^2 - 10x + 16}$, and
$h(x) = \dfrac{x^2 - 3x - 40}{x + 9}$, find **(a)** $R(x) = f(x) \cdot g(x)$
and state its domain **(b)** $R(x) = \dfrac{f(x)}{h(x)}$ and state its
domain.

79. If $f(x) = \dfrac{3x^2 - x - 10}{x^3 - 1}$, $g(x) = \dfrac{x^2 + 5x - 6}{2x^2 + 3x - 14}$, and
$h(x) = \dfrac{3x^2 + 8x + 5}{x^2 - 1}$, find **(a)** $R(x) = (f \cdot g)(x)$
and state its domain **(b)** $R(x) = \left(\dfrac{f}{h}\right)(x)$ and state its
domain.

80. If $f(x) = \dfrac{4x^2 - 9x - 9}{x^3 - 8}$, $g(x) = \dfrac{x^2 + 7x - 18}{5x^2 - 14x - 3}$,
and $h(x) = \dfrac{x^2 - 6x + 9}{x^2 - 4}$, **(a)** find $R(x) = (f \cdot g)(x)$
and state its domain **(b)** find $R(x) = \left(\dfrac{f}{h}\right)(x)$ and
state its domain.

Mixed Practice

In Problems 81–88, multiply or divide each rational
expression, as indicated. Simplify the product or quotient,
if possible.

81. $\dfrac{z^3 + 8}{z^2 - 3z - 10} \cdot \dfrac{z^2 - 2z - 15}{2z^2 - 4z + 8}$

82. $\dfrac{x^3 - 27}{2x^2 + 5x - 25} \cdot \dfrac{x^2 + 2x - 15}{x^3 + 3x^2 + 9x}$

83. $\dfrac{\dfrac{m^2 - 4n^2}{m^3 - n^3}}{\dfrac{2m + 4n}{m^2 + mn + n^2}}$

84. $\dfrac{\dfrac{x^2 + 2xy + y^2}{x^2 + 3xy + 2y^2}}{\dfrac{x^2 - y^2}{x + 2y}}$

85. $\dfrac{4w + 8}{w^2 - 4w} \cdot \dfrac{w^2 - 3w - 4}{w^2 + 3w + 2}$

86. $\dfrac{5m - 5}{m^2 + 6m} \cdot \dfrac{m^2 + 2m - 24}{m^2 + 3m - 4}$

87. $\dfrac{\dfrac{2x - 6}{x^2 + x}}{\dfrac{x^2 - 4x + 3}{x^2}} \cdot \dfrac{x^2 + 3x + 2}{x^2 + x}$

88. $\dfrac{\dfrac{3x + 15}{2x + 4}}{\dfrac{x + 5}{x^2 - 4}} \cdot \dfrac{4x + 8}{3x^2 - 12}$

Applying the Concepts

89. Make up a rational expression that is undefined at $x = 3$.

90. Make up a rational expression that is undefined at $x = -2$.

91. Make up a rational expression that is undefined at $x = -4$ and at $x = 5$.

92. Make up a rational expression that is undefined at $x = -6$ and at $x = 0$.

93. Develop a rational function R that is undefined at $x = -2$ and at $x = 1$ such that $R(-3) = 1$.

94. Develop a rational function R that is undefined at $x = -4$ and at $x = 3$ such that $R(4) = 1$.

95. Gravity In physics, it is established that the acceleration due to gravity g (in meters per second²) at a height h meters above sea level is given by

$$g(h) = \frac{3.99 \times 10^{14}}{(6.374 \times 10^6 + h)^2}$$

where 6.374×10^6 is the radius of Earth in meters.

(a) What is the acceleration due to gravity at sea level?

(b) What is the acceleration due to gravity in Denver, Colorado, elevation 1600 meters?

(c) What is the acceleration due to gravity on the peak of Mount Everest, elevation 8848 meters?

96. Economics The gross domestic product (GDP) is the total value of all goods and services manufactured within the United States. A model for determining the change in GDP of a government spending plan is given by the function

$$G(s) = \frac{s}{1 - b}$$

where G is the change in gross domestic product, s is the amount spent by the government, and b is the marginal propensity to consume with $0 < b < 1$. The marginal propensity to consume can be thought of as the amount an individual would spend for each additional dollar of income earned. For example, if $b = 0.9$, then individuals will spend $0.90 for each additional dollar earned. If $b = 0.95$, then individuals will spend $0.95 for each additional dollar earned.

(a) Suppose the government spends $100 million on highway infrastructure and the marginal propensity to consume in the United States is $b = 0.9$. What will be the change in GDP?

(b) Suppose the government spends $100 million on highway infrastructure and the marginal propensity to consume in the United States is $b = 0.95$. What will be the change in GDP?

Extending the Concepts

97. Math for the Future Consider the function

$$f(x) = \frac{1}{x - 2}$$

(a) Determine the domain of f.

(b) Fill in the following table. What happens to the values of f as x approaches 2 but remains greater than 2?

x	3	2.5	2.1	2.01	2.001	2.0001
$f(x)$						

(c) Fill in the following table. What happens to the values of f as x approaches 2 but remains less than 2?

x	1	1.5	1.9	1.99	1.999	1.9999
$f(x)$						

(d) Below is the graph of $f(x) = \dfrac{1}{x - 2}$. What happens to the graph of the function as x approaches 2 for values of x larger than 2? What happens to the graph of the function as x approaches 2 for values of x smaller than 2? Compare your results to the results obtained in parts (b) and (c).

98. Math for the Future Consider the function

$$R(x) = \frac{2x + 1}{x - 2}$$

(a) Fill in the following table. What happens to the values of R as x gets larger in the positive direction?

x	5	10	50	100	1000
$R(x)$					

(b) Fill in the following table. What happens to the values of R as x gets larger in the negative direction?

x	−5	−10	−50	−100	−1000
R(x)					

(c) What is the term of highest degree in the numerator? What is the term of highest degree in the denominator? What is the ratio of the coefficients on the terms of highest degree in the numerator and denominator? Compare this result to your results in parts (a) and (b).

(d) Below is the graph of $R(x) = \dfrac{2x + 1}{x - 2}$. What happens to the graph of the function as x gets larger? That is, what happens to the graph of the function as x approaches ∞? What happens to the graph of the function as x gets smaller? That is, what happens to the graph of the function as x approaches $-\infty$? Compare your results to the results obtained in parts (b) and (c).

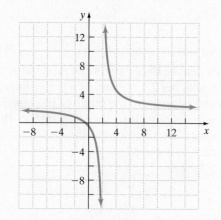

99. Write $\dfrac{x^{-1}}{x + 1}$ as a rational expression.

Explaining the Concepts

100. Define rational expression. Explain how a rational expression is related to a rational number.

101. What does it mean when we say that a rational expression is simplified?

102. Explain why we can divide out only common factors, not common terms.

103. Why is $\dfrac{\sqrt{x}}{x + 1}$ not a rational expression?

104. What is the difference between $f(x) = \dfrac{x^2 - 3x + 2}{x - 1}$ and $g(x) = x - 2$?

Synthesis Review

In Problems 105–110, graph each function.

105. $f(x) = 2x - 3$ **106.** $g(x) = 3x - 6$

107. $F(x) = -4x + 8$ **108.** $G(x) = -5x + 10$

109. $h(x) = x^2 - 4$ **110.** $H(x) = -x^2 + 4$

111. Discuss the methods that you used to graph each function. What methods were the same? What was different?

5.2 Adding and Subtracting Rational Expressions

Objectives

1 Add or Subtract Rational Expressions with a Common Denominator

2 Find the Least Common Denominator of Two or More Rational Expressions

3 Add or Subtract Rational Expressions with Different Denominators

Are You Prepared for This Section?

Before getting started, complete the following problems. If you get a problem wrong, go back to the section cited and review the material.

P1. Evaluate: $\dfrac{1}{6} - \dfrac{5}{8}$ [Getting Ready, pp. 396–397]

P2. Determine the additive inverse of **(a)** 5 **(b)** $x - 2$ [Section R.3, p. 21]

Prepared?...Answers   **P1.** $-\dfrac{11}{24}$

P2.   **(a)** -5 **(b)** $-(x - 2)$ or $2 - x$

▷ ❶ Add or Subtract Rational Expressions with a Common Denominator

---

EXAMPLE 1 **How to Add or Subtract Rational Expressions with a Common Denominator**

Perform the indicated operation.

(a) $\dfrac{x^2 - 3x + 6}{x + 3} + \dfrac{7x - 3}{x + 3}$

(b) $\dfrac{3x - 5}{x + 1} - \dfrac{x + 3}{x + 1}$

Step-by-Step Solution

(a) The rational expressions in the sum have a common denominator, $x + 3$.

Step 1: Add the numerators and write the result over the common denominator.

$$\frac{a}{c} + \frac{b}{c} = \frac{a + b}{c} : \frac{x^2 - 3x + 6}{x + 3} + \frac{7x - 3}{x + 3} = \frac{x^2 - 3x + 6 + (7x - 3)}{x + 3}$$

Combine like terms in the numerator: $= \dfrac{x^2 + 4x + 3}{x + 3}$

Step 2: Simplify the rational expression.

Factor the numerator: $= \dfrac{(x + 3)(x + 1)}{x + 3}$

Divide out common factors: $= x + 1$

(b) The rational expressions in the difference have a common denominator, $x + 1$.

Step 1: Subtract the numerators and write the result over the common denominator.

$$\frac{a}{c} - \frac{b}{c} = \frac{a - b}{c} : \frac{3x - 5}{x + 1} - \frac{x + 3}{x + 1} = \frac{3x - 5 - (x + 3)}{x + 1}$$

Distribute the -1: $= \dfrac{3x - 5 - x - 3}{x + 1}$

Combine like terms in the numerator: $= \dfrac{2x - 8}{x + 1}$

Step 2: Simplify the rational expression.

Factor the numerator: $= \dfrac{2(x - 4)}{x + 1}$ ●

The rules for adding and subtracting rational expressions are the same as the rules for adding and subtracting rational numbers.

In Other Words

To add or subtract rational expressions with the same denominator, add or subtract the numerators and write the result over the common denominator.

Adding or Subtracting Rational Expressions

Step 1: If $\dfrac{a}{c}$ and $\dfrac{b}{c}$, $c \neq 0$, are two rational expressions, then $\dfrac{a}{c} + \dfrac{b}{c} = \dfrac{a + b}{c}$ and $\dfrac{a}{c} - \dfrac{b}{c} = \dfrac{a - b}{c}$.

Step 2: Simplify the result.

Quick ✓

In Problems 1 and 2, perform the indicated operation. Be sure to simplify the result.

1. $\dfrac{x^2 - 3x - 1}{x - 2} + \dfrac{x^2 - 2x + 3}{x - 2}$

2. $\dfrac{4x + 3}{x + 5} - \dfrac{x - 6}{x + 5}$

EXAMPLE 2   **Adding Rational Expressions with Denominators That Are Additive Inverses**

Perform the indicated operation and simplify the result.

$$\frac{3x}{x-2} + \frac{2}{2-x}$$

Solution

Notice that the denominators are additive inverses (opposites) of each other.

$$2 - x = -x + 2 = -1(x - 2)$$

$$\frac{3x}{x-2} + \frac{2}{2-x} = \frac{3x}{x-2} + \frac{2}{-1(x-2)}$$

$$\text{Use } \frac{a}{-b} = \frac{-a}{b}: \quad = \frac{3x}{x-2} + \frac{-2}{x-2}$$

$$\text{Use } \frac{a}{c} + \frac{b}{c} = \frac{a+b}{c}: \quad = \frac{3x-2}{x-2}$$

Quick ✓

In Problem 3, perform the indicated operation. Be sure to simplify the result.

3. $\dfrac{4x}{x-5} + \dfrac{3}{5-x}$

⏵ ❷ **Find the Least Common Denominator of Two or More Rational Expressions**

If the denominators of the rational expressions to be added or subtracted are not the same, then rewrite each rational expression with a *least common denominator*. The **least common denominator (LCD)** is the polynomial of least degree that is a multiple of each denominator in the rational expressions to be added or subtracted. The idea is the same one used to add rational numbers without common denominators. In fact, it would be a great idea to review Example 4 on pages 395–396 in Getting Ready for Chapter 5 before reading Example 3.

EXAMPLE 3   **How to Find the Least Common Denominator of Rational Expressions**

Find the least common denominator of the rational expressions.

(a) $\dfrac{4}{3x^2y^2}$ and $\dfrac{5}{6xy^3}$ **(b)** $\dfrac{3x-5}{x^3+2x^2+x}$ and $\dfrac{x-1}{x^2+4x+3}$

Step-by-Step Solution

(a) $\dfrac{4}{3x^2y^2}$ and $\dfrac{5}{6xy^3}$

Step 1: Write each denominator as the product of prime factors.

$$3x^2y^2 = 3 \cdot x^2 \cdot y^2$$
$$6xy^3 = 2 \cdot 3 \cdot x \cdot y^3$$

Step 2: Find the product of each prime factor the greatest number of times it appears in any factorization.

$$\text{LCD} = 2 \cdot 3 \cdot x^2 \cdot y^3$$
$$= 6x^2y^3$$

(b) $\dfrac{3x-5}{x^3+2x^2+x}$ and $\dfrac{x-1}{x^2+4x+3}$

Step 1: Write each denominator as the product of prime factors.

$$x^3+2x^2+x = x(x^2+2x+1) = x(x+1)^2$$

$$x^2+4x+3 = \qquad\qquad (x+1)(x+3)$$

Step 2: Find the product of each prime factor the greatest number of times it appears in any factorization.

$$\text{LCD} = x(x+1)^2(x+3) \quad \bullet$$

Finding the Least Common Denominator

Step 1: Factor each denominator completely. When factoring, write the factored form using powers. For example, write x^2+4x+4 as $(x+2)^2$.

Step 2: The LCD is the product of each prime factor the greatest number of times it appears in any factorization.

Quick ✓

4. The _____ is the polynomial of least degree that is a multiple of each denominator in the rational expressions to be added or subtracted.

In Problems 5 and 6, find the least common denominator of the rational expressions.

5. $\dfrac{5}{8x^2y}$ and $\dfrac{1}{12xy^3}$

6. $\dfrac{4x-3}{x^2-5x-14}$ and $\dfrac{x+1}{x^2+4x+4}$

▶ ❸ **Add or Subtract Rational Expressions with Different Denominators**

EXAMPLE 4 **How to Add Rational Expressions with Unlike Denominators**

Add $\dfrac{3}{8x^2}+\dfrac{1}{12x}$. Simplify the result.

Step-by-Step Solution

Step 1: Find the least common denominator.

$$8x^2 = 2\cdot2\cdot2\cdot x^2$$
$$12x = 2\cdot2\cdot3\cdot x$$
The LCD is $2\cdot2\cdot2\cdot3\cdot x^2 = 24x^2$.

Step 2: Rewrite each rational expression with the common denominator.

The denominator needs to be $24x^2$, so multiply $\dfrac{3}{8x^2}$ by $1=\dfrac{3}{3}$. Multiply $\dfrac{1}{12x}$ by $1=\dfrac{2x}{2x}$. Do you see why?

$$\dfrac{3}{8x^2} = \dfrac{3}{8x^2}\cdot\dfrac{3}{3} = \dfrac{9}{24x^2}$$

$$\dfrac{1}{12x} = \dfrac{1}{12x}\cdot\dfrac{2x}{2x} = \dfrac{2x}{24x^2}$$

Step 3: Add the rational expressions found in Step 2.

$$\dfrac{3}{8x^2}+\dfrac{1}{12x} = \dfrac{9}{24x^2}+\dfrac{2x}{24x^2}$$

$$\dfrac{a}{c}+\dfrac{b}{c}=\dfrac{a+b}{c}: \quad = \dfrac{9+2x}{24x^2}$$

$$= \dfrac{2x+9}{24x^2}$$

(continued)

Step 4: Simplify the result.

The rational expression is simplified.

Thus $\dfrac{3}{8x^2} + \dfrac{1}{12x} = \dfrac{2x + 9}{24x^2}$.

Adding or Subtracting Rational Expressions with Unlike Denominators

Step 1: Find the least common denominator.

Step 2: Rewrite each rational expression as an equivalent rational expression with the common denominator. Multiply out the numerator, but leave the denominator in factored form.

Step 3: Add or subtract the rational expressions found in Step 2.

Step 4: Simplify the result.

Quick ✓

In Problems 7 and 8, perform the indicated operation and simplify the result.

7. $\dfrac{3}{10a} + \dfrac{4}{15a^2}$ **8.** $\dfrac{3}{8y} - \dfrac{13}{24y}$

EXAMPLE 5   **Adding Rational Expressions with Unlike Denominators**

Perform the indicated operation and simplify the result.

(a) $\dfrac{x - 1}{x + 3} + \dfrac{x}{x + 2}$ **(b)** $\dfrac{x - 1}{x^2 + 2x - 8} + \dfrac{x - 1}{x^2 - 16}$

Solution

(a) $\dfrac{x - 1}{x + 3} + \dfrac{x}{x + 2}$

The LCD is $(x + 3)(x + 2)$. To rewrite each rational expression as an equivalent rational expression with the LCD, multiply $\dfrac{x - 1}{x + 3}$ by $1 = \dfrac{x + 2}{x + 2}$ and multiply $\dfrac{x}{x + 2}$ by $\dfrac{x + 3}{x + 3}$.

$$\dfrac{x - 1}{x + 3} + \dfrac{x}{x + 2} = \dfrac{x - 1}{x + 3} \cdot \dfrac{x + 2}{x + 2} + \dfrac{x}{x + 2} \cdot \dfrac{x + 3}{x + 3}$$

Multiply the numerators: $\quad = \dfrac{x^2 + x - 2}{(x + 3)(x + 2)} + \dfrac{x^2 + 3x}{(x + 3)(x + 2)}$

Use $\dfrac{a}{c} + \dfrac{b}{c} = \dfrac{a + b}{c}$: $\quad = \dfrac{x^2 + x - 2 + (x^2 + 3x)}{(x + 3)(x + 2)}$

Combine like terms: $\quad = \dfrac{2x^2 + 4x - 2}{(x + 3)(x + 2)}$

Factor the numerator: $\quad = \dfrac{2(x^2 + 2x - 1)}{(x + 3)(x + 2)}$

(b) $\dfrac{x - 1}{x^2 + 2x - 8} + \dfrac{x - 1}{x^2 - 16}$

First, factor the denominators to find the LCD.

$$x^2 + 2x - 8 = (x + 4)(x - 2)$$
$$x^2 - 16 = (x + 4)(x - 4)$$
$$\text{LCD} = (x + 4)(x - 2)(x - 4)$$

$$\text{Multiply } \frac{x-1}{(x+4)(x-2)} \text{ by } \frac{x-4}{x-4}; \text{ Multiply } \frac{x-1}{(x+4)(x-4)} \text{ by } \frac{x-2}{x-2}$$

$$\frac{x-1}{x^2+2x-8} + \frac{x-1}{x^2-16} = \frac{x-1}{(x+4)(x-2)} \cdot \frac{x-4}{x-4} + \frac{x-1}{(x+4)(x-4)} \cdot \frac{x-2}{x-2}$$

$$\text{Multiply the numerators: } = \frac{x^2-5x+4}{(x+4)(x-2)(x-4)} + \frac{x^2-3x+2}{(x+4)(x-2)(x-4)}$$

$$\text{Use } \frac{a}{c}+\frac{b}{c}=\frac{a+b}{c}: = \frac{x^2-5x+4+(x^2-3x+2)}{(x+4)(x-2)(x-4)}$$

$$\text{Combine like terms: } = \frac{2x^2-8x+6}{(x+4)(x-2)(x-4)}$$

$$\text{Factor the numerator: } = \frac{2(x-3)(x-1)}{(x+4)(x-2)(x-4)}$$

Quick ✓

In Problems 9 and 10, perform the indicated operation and simplify the result.

9. $\dfrac{3x}{x-1} + \dfrac{x+5}{x+2}$

10. $\dfrac{y+4}{3y^2-y-2} - \dfrac{1}{3y^2+14y+8}$

EXAMPLE 6 **How to Subtract Rational Expressions with Unlike Denominators**

Perform the indicated operation and simplify the result.

$$\frac{2x-1}{2x^2-7x-4} - \frac{x-1}{2x^2+3x+1}$$

Step-by-Step Solution

Step 1: Find the least common denominator.

$$2x^2-7x-4 = (2x+1)(x-4)$$
$$2x^2+3x+1 = (2x+1)(x+1)$$
$$\text{LCD} = (2x+1)(x-4)(x+1)$$

Step 2: Rewrite each rational expression as an equivalent rational expression with the common denominator. Multiply the numerators, but leave the denominators in factored form.

$$\frac{2x-1}{2x^2-7x-4} - \frac{x-1}{2x^2+3x+1} = \frac{2x-1}{(2x+1)(x-4)} \cdot \frac{x+1}{x+1} - \frac{x-1}{(2x+1)(x+1)} \cdot \frac{x-4}{x-4}$$

$$\text{Multiply the numerators: } = \frac{2x^2+x-1}{(2x+1)(x-4)(x+1)} - \frac{x^2-5x+4}{(2x+1)(x-4)(x+1)}$$

Step 3: Subtract the rational expressions found in Step 2.

$$\text{Use } \frac{a}{c}-\frac{b}{c}=\frac{a-b}{c}: = \frac{2x^2+x-1-(x^2-5x+4)}{(2x+1)(x-4)(x+1)}$$

$$\text{Distribute the } -1: = \frac{2x^2+x-1-x^2+5x-4}{(2x+1)(x-4)(x+1)}$$

$$\text{Combine like terms: } = \frac{x^2+6x-5}{(2x+1)(x-4)(x+1)}$$

Step 4: Simplify the rational expression.

The rational expression in Step 3 above is simplified.

Quick ✓

In Problem 11, perform the indicated operation and simplify the result.

11. $\dfrac{3x + 4}{2x^2 + x - 6} - \dfrac{x + 4}{x^2 + 4x + 4}$

EXAMPLE 7 **Adding and Subtracting Three Rational Expressions**

Perform the indicated operations and simplify the result.

$$\frac{6}{x^2 - 9} + \frac{x + 1}{x + 3} - \frac{x - 2}{x - 3}$$

Solution
Factor each denominator.

$$x^2 - 9 = (x + 3)(x - 3)$$
$$x + 3 = x + 3$$
$$x - 3 = x - 3$$
$$\text{LCD} = (x + 3)(x - 3)$$

Work Smart

Remember when subtracting rational expressions to subtract the entire numerator.

$$\frac{6}{x^2 - 9} + \frac{x + 1}{x + 3} - \frac{x - 2}{x - 3} = \frac{6}{(x + 3)(x - 3)} + \frac{x + 1}{x + 3} \cdot \frac{x - 3}{x - 3} - \frac{x - 2}{x - 3} \cdot \frac{x + 3}{x + 3}$$

Multiply the numerators: $= \dfrac{6}{(x + 3)(x - 3)} + \dfrac{x^2 - 2x - 3}{(x + 3)(x - 3)} - \dfrac{x^2 + x - 6}{(x + 3)(x - 3)}$

Use $\dfrac{a}{c} + \dfrac{b}{c} = \dfrac{a + b}{c}; \dfrac{a}{c} - \dfrac{b}{c} = \dfrac{a - b}{c}$: $= \dfrac{6 + x^2 - 2x - 3 - (x^2 + x - 6)}{(x + 3)(x - 3)}$

Distribute the -1: $= \dfrac{6 + x^2 - 2x - 3 - x^2 - x + 6}{(x + 3)(x - 3)}$

Combine like terms: $= \dfrac{-3x + 9}{(x + 3)(x - 3)}$

Factor: $= \dfrac{-3(x - 3)}{(x + 3)(x - 3)}$

Divide out common factors: $= \dfrac{-3}{x + 3}$

$= -\dfrac{3}{x + 3}$

Quick ✓

In Problem 12, perform the indicated operations and simplify the result.

12. $\dfrac{4}{x^2 - 4} - \dfrac{x + 3}{x - 2} + \dfrac{x + 3}{x + 2}$

5.2 Exercises MyMathLab® Exercise numbers in green have complete video solutions in MyMathLab or may be accessed using the QR code to the right.

Problems 1–12 are the Quick ✓s that follow the EXAMPLES.

Building Skills

In Problems 13–24, perform the indicated operation and simplify the result. See Objective 1.

13. $\dfrac{3x}{x + 1} + \dfrac{5}{x + 1}$

14. $\dfrac{5x}{x - 3} + \dfrac{2}{x - 3}$

15. $\dfrac{2x}{2x + 5} - \dfrac{1}{2x + 5}$

16. $\dfrac{9x}{6x - 5} - \dfrac{2}{6x - 5}$

17. $\dfrac{2x}{2x^2 - 7x - 15} + \dfrac{3}{2x^2 - 7x - 15}$

18. $\dfrac{x}{3x^2 + 8x - 3} + \dfrac{3}{3x^2 + 8x - 3}$

19. $\dfrac{2x^2 - 5x + 7}{x^2 - 2x - 15} - \dfrac{x^2 + 3x - 8}{x^2 - 2x - 15}$

20. $\dfrac{3x^2 + 8x - 1}{x^2 - 3x - 28} - \dfrac{2x^2 + 2x - 9}{x^2 - 3x - 28}$

21. $\dfrac{3x}{x - 5} + \dfrac{1}{5 - x}$ **22.** $\dfrac{3x}{x - 6} + \dfrac{2}{6 - x}$

23. $\dfrac{2x^2 - 4x - 1}{x - 3} - \dfrac{x^2 - 6x + 4}{3 - x}$

24. $\dfrac{x^2 + 2x - 5}{x - 4} - \dfrac{x^2 - 5x - 15}{4 - x}$

In Problems 25–34, find the least common denominator. See Objective 2.

25. $\dfrac{3}{4x^3}$ and $\dfrac{9}{8x}$ **26.** $\dfrac{5}{3a^3}$ and $\dfrac{2}{9a^2}$

27. $\dfrac{1}{15xy^2}$ and $\dfrac{7}{18x^3y}$ **28.** $\dfrac{1}{8a^3b}$ and $\dfrac{5}{12ab^2}$

29. $\dfrac{5x}{x - 4}$ and $\dfrac{3}{x + 2}$ **30.** $\dfrac{x - 3}{x + 2}$ and $\dfrac{x + 7}{x - 5}$

31. $\dfrac{x - 4}{x^2 - x - 12}$ and $\dfrac{2x + 1}{x^2 - 9x + 20}$

32. $\dfrac{2m - 7}{m^2 + 3m - 18}$ and $\dfrac{5m + 1}{m^2 - 7m + 12}$

33. $\dfrac{p + 1}{2p^2 + 3p - 2}$ and $\dfrac{4p - 1}{p^3 + 2p^2}$

34. $\dfrac{x - 6}{x^2 - 9}$ and $\dfrac{3x}{x^3 - 3x^2}$

In Problems 35–60, add or subtract, as indicated, and simplify the result. See Objective 3.

35. $\dfrac{3}{4x^2} + \dfrac{5}{8x}$ **36.** $\dfrac{2}{9x} + \dfrac{5}{3x^2}$

37. $\dfrac{5}{12a^2 b} - \dfrac{4}{15ab^2}$ **38.** $\dfrac{3}{14mn^3} - \dfrac{2}{21m^2 n}$

39. $\dfrac{y + 2}{y - 5} - \dfrac{y - 4}{y + 3}$ **40.** $\dfrac{x + 2}{x - 3} - \dfrac{x + 2}{x + 1}$

41. $\dfrac{a + 5}{a - 2} - \dfrac{5a + 18}{a^2 - 4}$ **42.** $\dfrac{z + 1}{z + 3} - \dfrac{z + 17}{z^2 - z - 12}$

43. $\dfrac{3}{(x - 2)(x + 3)} - \dfrac{5}{(x + 3)(x + 4)}$

44. $\dfrac{1}{(x - 1)(x + 3)} + \dfrac{5}{(x + 1)(x - 1)}$

45. $\dfrac{x - 3}{x^2 + 3x + 2} + \dfrac{x - 1}{x^2 - 4}$ **46.** $\dfrac{x - 5}{x^2 + 4x + 3} + \dfrac{x - 2}{x^2 - 1}$

47. $\dfrac{w - 4}{2w^2 + 3w + 1} - \dfrac{w + 3}{2w^2 - 5w - 3}$

48. $\dfrac{x - 1}{2x^2 + 7x + 6} + \dfrac{x - 1}{x^2 + 6x + 8}$

49. $\dfrac{x + y}{x^2 - 6xy + 9y^2} + \dfrac{x + 2y}{x^2 - 2xy - 3y^2}$

50. $\dfrac{m - 2n}{m^2 + 4mn + 4n^2} + \dfrac{m - n}{m^2 - mn - 6n^2}$

51. $\dfrac{3}{x^2 - 4x - 5} - \dfrac{2}{x^2 - 6x + 5}$

52. $\dfrac{3}{x^2 + 7x + 10} - \dfrac{4}{x^2 + 6x + 5}$

53. $\dfrac{p^2 - 3p - 10}{p^2 - 16} + \dfrac{p^2 - 3p - 10}{16 - p^2}$

54. $\dfrac{y^2 + 4y + 4}{y^2 - 9} + \dfrac{y^2 + 4y + 4}{9 - y^2}$

55. $\dfrac{2}{w + 2} - \dfrac{3}{w} + \dfrac{w + 10}{w^2 - 4}$

56. $\dfrac{7}{m - 3} - \dfrac{5}{m} - \dfrac{2m + 6}{m^2 - 9}$

57. $\dfrac{p - 2}{p^2 + 6p + 9} + \dfrac{1}{p + 3} - \dfrac{2p + 1}{2p^2 + p - 15}$

58. $\dfrac{x - 1}{x^2 - 16} + \dfrac{1}{x + 4} - \dfrac{4x + 1}{3x^2 - 7x - 20}$

59. $\dfrac{2}{x} - \dfrac{2}{x - 1} + \dfrac{3}{(x - 1)^2}$

60. $\dfrac{2}{x} - \dfrac{2}{x + 2} + \dfrac{2}{(x + 2)^2}$

Mixed Practice

In Problems 61–72, perform the indicated operation and simplify the result.

61. $\dfrac{1}{x - 3} - \dfrac{x^2 + 18}{x^3 - 27}$ **62.** $\dfrac{1}{x + 2} + \dfrac{x - 10}{x^3 + 8}$

63. $\dfrac{2x^2 - x}{x + 3} + \dfrac{3x}{x + 3} - \dfrac{x^2 + 3}{x + 3}$

64. $\dfrac{2x^2}{x - 1} - \dfrac{x^2 - 2x}{x - 1} + \dfrac{x - 4}{x - 1}$

65. $6 + \dfrac{x - 3}{x + 3}$ **66.** $3 + \dfrac{x + 4}{x - 4}$

67. $\dfrac{b + 3}{b^2 + 2b - 8} - \dfrac{b + 2}{b^2 - 4}$ **68.** $\dfrac{a + 3}{a^2 - 8a + 15} + \dfrac{a + 3}{a^2 - 9}$

69. $\dfrac{y - 1}{y + 4} + \dfrac{y - 2}{y + 3} - \dfrac{y^2 + 3y + 1}{y^2 + 7y + 12}$

70. $\dfrac{z+3}{z-6} + \dfrac{z-1}{z-2} - \dfrac{6z}{z^2 - 8z + 12}$

71. $\dfrac{x+4}{x^2 - 5x + 6} + \dfrac{x-1}{x^2 - 2x - 3} - \dfrac{2x+1}{x^2 - x - 2}$

72. $\dfrac{x-1}{x^2 + 4x - 5} + \dfrac{3x-1}{x^2 + 3x - 10} - \dfrac{4x+1}{x^2 - 3x + 2}$

73. Given that $f(x) = \dfrac{3}{x-2}$ and $g(x) = \dfrac{2}{x+1}$,

    (a) find $R(x) = f(x) + g(x)$,
    (b) state the domain of $R(x)$.

74. Given that $f(x) = \dfrac{5}{x+2}$ and $g(x) = \dfrac{3}{x-1}$,

    (a) find $R(x) = f(x) + g(x)$,
    (b) state the domain of $R(x)$.

75. Given that $f(x) = \dfrac{x+1}{x^2 - 3x - 4}$

    and $g(x) = \dfrac{x+4}{x^2 - x - 12}$,

    (a) find $R(x) = f(x) + g(x)$,
    (b) state the domain of $R(x)$,
    (c) find $H(x) = f(x) - g(x)$,
    (d) state the domain of $H(x)$.

76. Given that $f(x) = \dfrac{x+5}{x^2 - 5x + 6}$ and

    $g(x) = \dfrac{x+1}{x^2 - 4x - 12}$,

    (a) find $R(x) = f(x) + g(x)$,
    (b) state the domain of $R(x)$,
    (c) find $H(x) = f(x) - g(x)$,
    (d) state the domain of $H(x)$.

Applying the Concepts

△ **77. Surface Area of a Box** The volume of a closed box with a square base is 2000 cubic inches. Its surface area S, as a function of the length of the base x, is given by

$$S(x) = 2x^2 + \dfrac{8000}{x}$$

    (a) Write S over a common denominator. That is, write S so that the rule is a single rational expression.
    (b) Find and interpret $S(10)$.

△ **78. Surface Area of a Can** The volume of a cylindrical can is 200 cubic centimeters. Its surface area S, as a function of the radius r of the can, is given by

$$S(r) = 2\pi r^2 + \dfrac{400}{r}$$

    (a) Write S over a common denominator. That is, write S so that the rule is a single rational expression.
    (b) Find and interpret $S(4)$. Round your answer to two decimal places.

79. Road Trip Suppose you and a group of your friends decide to go on a road trip to a neighboring university that is 200 miles away. Your average speed for the first 50 miles of the trip is 10 miles an hour slower than the average speed for the remaining 150 miles of the trip. If we let s represent your average speed for the first 50 miles of the trip, the function

$$T(s) = \dfrac{50}{s} + \dfrac{150}{s+10}$$

represents the amount of time T it will take to get to the neighboring university.

    (a) Write T over a common denominator. That is, write T so that the rule is a single rational expression.
    (b) Find and interpret $T(50)$.

80. Vacation The distance from Chicago, Illinois, to Naples, Florida, is approximately 1200 miles. Atlanta, Georgia, is approximately the midpoint between Chicago and Naples. On a recent trip to Naples, the Sullivan family averaged s miles per hour between Chicago and Atlanta, and they averaged 5 miles per hour faster between Atlanta and Naples. The time T of their trip, as a function of their average speed s between Chicago and Atlanta, is given by the function

$$T(s) = \dfrac{600}{s} + \dfrac{600}{s+5}$$

    (a) Write T over a common denominator. That is, write T so that the rule is a single rational expression.
    (b) Find and interpret $T(50)$. Round your answer to two decimal places.
    (c) Using the result from part (b), compute the average speed of the Sullivans for the entire trip. Are you surprised by the result?

Extending the Concepts

81. Write $x^{-1} + y^{-1}$ as a single rational expression with no negative exponents.

82. Write $\left(\dfrac{a}{b}\right)^{-1} + \left(\dfrac{b}{a}\right)^{-1}$ as a single rational expression with no negative exponents.

Explaining the Concepts

83. Explain how to find the least common denominator when adding or subtracting two rational expressions with unlike denominators.

84. Explain how to add or subtract rational expressions with unlike denominators.

Synthesis Review

In Problems 85–91, multiply the expressions.

85. $4a(a - 3)$

86. $-5z(z + 4)$

87. $(p - 3)(p + 3)$

88. $(3q + 1)(3q - 1)$

89. $(w - 2)(w^2 + 2w + 4)$

90. $(2v + 1)(4v^2 - 2v + 1)$

91. $\dfrac{x^3 - 8}{2x^2 + 5x + 2} \cdot \dfrac{2x^2 - x - 1}{x^2 - 4} \cdot \dfrac{x^2 + 4x + 4}{x - 1}$

5.3 Complex Rational Expressions

Objectives

① Simplify a Complex Rational Expression by Simplifying the Numerator and Denominator Separately (Method I)

② Simplify a Complex Rational Expression Using the Least Common Denominator (Method II)

> **In Other Words**
>
> A complex rational expression is simplified when it is of the form *polynomial over polynomial* and the polynomials have no common factors.

Prepared?...Answers

P1. $(3y + 2)(2y - 3)$ **P2.** $\dfrac{4b^6}{9a^4}$

Are You Prepared for This Section?

Before getting started, complete the following problems. If you get a problem wrong, go back to the section cited and review the material.

P1. Factor: $6y^2 - 5y - 6$ [Section 4.5, pp. 351–356]

P2. Simplify: $\left(\dfrac{3ab^2}{2a^{-1}b^5}\right)^{-2}$ [Getting Ready, pp. 301–302]

A **complex rational expression** is a quotient whose numerator and/or denominator contains a rational expression. The following are examples of complex rational expressions.

$$\frac{3}{1 + \dfrac{x}{7}} \quad \text{and} \quad \frac{3 - \dfrac{1}{x}}{1 + \dfrac{1}{x}} \quad \text{and} \quad \frac{\dfrac{x + 1}{x - 2} - \dfrac{3}{x + 2}}{\dfrac{2x + 3}{x - 2} + 1}$$

To **simplify** a complex rational expression means to write the rational expression in the form $\dfrac{p}{q}$, where p and q are polynomials that have no common factors. This can be done using one of two methods. In Method I, simplify the numerator and the denominator separately. In Method II, simplify by using the least common denominator. We'll begin with Method I.

▶ ① Simplify a Complex Rational Expression by Simplifying the Numerator and Denominator Separately (Method I)

EXAMPLE 1   **How to Simplify a Complex Rational Expression Using Method I**

Simplify: $\dfrac{\dfrac{1}{3} + \dfrac{1}{x}}{\dfrac{x + 3}{2}}$

Step-by-Step Solution

Notice that x cannot equal -3 because that would make the denominator, $\dfrac{x + 3}{2}$, equal zero; x cannot equal 0 because it would make the expression $\dfrac{1}{x}$ in the numerator undefined.

(continued)

Step 1: Write the numerator of the complex rational expression as a single rational expression.

The least common denominator of $\frac{1}{3}$ and $\frac{1}{x}$ is $3x$.

$$\frac{1}{3} + \frac{1}{x} = \frac{1}{3} \cdot \frac{x}{x} + \frac{1}{x} \cdot \frac{3}{3}$$

$$= \frac{x}{3x} + \frac{3}{3x}$$

$$\frac{a}{c} + \frac{b}{c} = \frac{a+b}{c}: \quad = \frac{x+3}{3x}$$

Step 2: Write the denominator of the complex rational expression as a single rational expression.

This is already done.

Step 3: Rewrite the complex rational expression using the rational expressions found in Steps 1 and 2.

$$\frac{\frac{1}{3} + \frac{1}{x}}{\frac{x+3}{2}} = \frac{\frac{x+3}{3x}}{\frac{x+3}{2}}$$

Step 4: Simplify the rational expression using the techniques for dividing rational expressions from Section 5.1.

Multiply the numerator by the reciprocal of the denominator:
$$= \frac{x+3}{3x} \cdot \frac{2}{x+3}$$

Multiply; divide out common factors:
$$= \frac{(x+3) \cdot 2}{3x(x+3)}$$

$$= \frac{2}{3x}$$

The steps to simplify a complex rational expression using Method I are summarized below.

> **Simplifying a Complex Rational Expression by Simplifying the Numerator and Denominator Separately (Method I)**
>
> **Step 1:** Write the numerator of the complex rational expression as a single rational expression.
>
> **Step 2:** Write the denominator of the complex rational expression as a single rational expression.
>
> **Step 3:** Rewrite the complex rational expression using the rational expressions found in Steps 1 and 2.
>
> **Step 4:** Simplify the rational expression using the techniques for dividing rational expressions from Section 5.1.

Quick ✓

1. An expression such as $\dfrac{\frac{x}{5} - \frac{y}{7}}{\frac{xy}{35}}$ is called a _____ _____ _____.

2. *True or False* To simplify a complex rational expression means to write it in the form $\dfrac{p}{q}$ where p and q are polynomials with no common factors.

In Problems 3 and 4, simplify each expression using Method I.

3. $\dfrac{\frac{3}{2} - \frac{1}{3}}{\frac{5}{6}}$

4. $\dfrac{\frac{z}{4} - \frac{4}{z}}{\frac{z+4}{16}}, z \neq 0, -4$

The domain restrictions for the remaining examples will not be stated, but you should be aware that the restrictions are needed to maintain equality.

EXAMPLE 2   **Simplifying a Complex Rational Expression Using Method I**

Simplify: $\dfrac{\dfrac{2x}{x+4} - \dfrac{x-7}{x^2-16}}{x - \dfrac{x^2+4}{x+4}}$

Solution

Write the numerator and denominator of the complex rational expression as a single rational expression.

NUMERATOR:

$$\text{LCD} = (x-4)(x+4)$$

$$\frac{2x}{x+4} - \frac{x-7}{x^2-16} \overset{\downarrow}{=} \frac{2x}{x+4} \cdot \frac{x-4}{x-4} - \frac{x-7}{x^2-16}$$

$$= \frac{2x^2-8x}{x^2-16} - \frac{x-7}{x^2-16}$$

Use $\dfrac{a}{c} - \dfrac{b}{c} = \dfrac{a-b}{c}$: $\quad = \dfrac{2x^2-9x+7}{x^2-16}$

DENOMINATOR:

$$\text{LCD} = x+4$$

$$x - \frac{x^2+4}{x+4} \overset{\downarrow}{=} \frac{x}{1} \cdot \frac{x+4}{x+4} - \frac{x^2+4}{x+4}$$

$$= \frac{x^2+4x}{x+4} - \frac{x^2+4}{x+4}$$

Use $\dfrac{a}{c} - \dfrac{b}{c} = \dfrac{a-b}{c}$: $\quad = \dfrac{4x-4}{x+4}$

Rewrite the complex rational expression using the numerator and denominator just found, and then simplify.

$$\frac{\dfrac{2x}{x+4} - \dfrac{x-7}{x^2-16}}{x - \dfrac{x^2+4}{x+4}} = \frac{\dfrac{2x^2-9x+7}{x^2-16}}{\dfrac{4x-4}{x+4}}$$

Rewrite the division problem as a multiplication problem: $= \dfrac{2x^2-9x+7}{x^2-16} \cdot \dfrac{x+4}{4x-4}$

Factor; Multiply; Divide out common factors: $= \dfrac{(2x-7)\,\cancel{(x-1)}\,\cancel{(x+4)}}{(x-4)\,\cancel{(x+4)}\cdot 4\,\cancel{(x-1)}}$

$$= \dfrac{2x-7}{4(x-4)}$$ ●

Quick ✓

5. *True or False* $\dfrac{x-y}{\dfrac{1}{x}+\dfrac{1}{y}} = \dfrac{x-y}{1} \cdot \left(\dfrac{x}{1}+\dfrac{y}{1}\right)$

6. Simplify using Method I: $\dfrac{\dfrac{2x}{x+1} - \dfrac{x^2-3}{x^2+3x+2}}{4 + \dfrac{4}{x+2}}$

▶ ❷ Simplify a Complex Rational Expression Using the Least Common Denominator (Method II)

We now simplify complex rational expressions using the least common denominator. Example 1 is redone using this second method so that you can compare the two methods.

EXAMPLE 3 **How to Simplify a Complex Rational Expression Using Method II**

Simplify: $\dfrac{\dfrac{1}{3} + \dfrac{1}{x}}{\dfrac{x+3}{2}}$

Step-by-Step Solution

Step 1: Find the least common denominator of all the denominators in the complex rational expression.

The denominators of the complex rational expression are 3, x, and 2. The least common denominator is $2 \cdot 3 \cdot x = 6x$.

Step 2: Multiply both the numerator and the denominator of the complex rational expression by the LCD found in Step 1.

$$\dfrac{\dfrac{1}{3} + \dfrac{1}{x}}{\dfrac{x+3}{2}} \cdot \dfrac{6x}{6x} = \dfrac{\left(\dfrac{1}{3} + \dfrac{1}{x}\right) \cdot 6x}{\left(\dfrac{x+3}{2}\right) \cdot 6x}$$

Distribute $6x$ to each term: $\quad = \dfrac{\dfrac{1}{3} \cdot \overset{2}{6x} + \dfrac{1}{x_1} \cdot 6x^1}{\dfrac{x+3}{2} \cdot \overset{3}{6x}}$

Step 3: Simplify the rational expression.

Simplify: $\quad = \dfrac{2x+6}{3x(x+3)}$

Factor and divide out common factors: $\quad = \dfrac{2(x+3)}{3x(x+3)}$

$\quad = \dfrac{2}{3x}$

Example 3 has the same result as Example 1!

> **Simplifying a Complex Rational Expression Using the Least Common Denominator (Method II)**
>
> **Step 1:** Find the least common denominator of all the denominators in the complex rational expression.
>
> **Step 2:** Multiply both the numerator and the denominator of the complex rational expression by the least common denominator found in Step 1.
>
> **Step 3:** Simplify the rational expression.

Quick ✓

In Problems 7 and 8, simplify each expression using Method II.

7. $\dfrac{\dfrac{3}{2} - \dfrac{1}{3}}{\dfrac{5}{6}}$

8. $\dfrac{\dfrac{z}{4} - \dfrac{4}{z}}{\dfrac{z+4}{16}}$

EXAMPLE 4 **Simplifying Complex Rational Expressions Using Method II**

Simplify: $\dfrac{\dfrac{1}{x} + \dfrac{1}{x-2}}{\dfrac{x}{x^2-4} + \dfrac{1}{x-2}}$

Solution

Because $x^2 - 4 = (x - 2)(x + 2)$, the least common denominator among all denominators is $x(x - 2)(x + 2)$. So multiply the numerator and denominator by $x(x - 2)(x + 2)$.

$$\frac{\dfrac{1}{x} + \dfrac{1}{x - 2}}{\dfrac{x}{x^2 - 4} + \dfrac{1}{x - 2}} = \frac{\dfrac{1}{x} + \dfrac{1}{x - 2}}{\dfrac{x}{x^2 - 4} + \dfrac{1}{x - 2}} \cdot \frac{x(x - 2)(x + 2)}{x(x - 2)(x + 2)}$$

Distribute the LCD to each term:
$$= \frac{\dfrac{1}{x} \cdot x(x - 2)(x + 2) + \dfrac{1}{x - 2} \cdot x(x - 2)(x + 2)}{\dfrac{x}{x^2 - 4} \cdot x(x - 2)(x + 2) + \dfrac{1}{x - 2} \cdot x(x - 2)(x + 2)}$$

Factor and divide out like factors:
$$= \frac{\dfrac{1}{\cancel{x}} \cdot \cancel{x}(x - 2)(x + 2) + \dfrac{1}{\cancel{x - 2}} \cdot x \cancel{(x - 2)}(x + 2)}{\dfrac{x}{\cancel{(x - 2)}\cancel{(x + 2)}} \cdot x \cancel{(x - 2)}\cancel{(x + 2)} + \dfrac{1}{\cancel{x - 2}} \cdot x \cancel{(x - 2)}(x + 2)}$$

$$= \frac{(x - 2)(x + 2) + x(x + 2)}{x^2 + x(x + 2)}$$

Factor out $(x + 2)$:
$$= \frac{(x + 2)(x - 2 + x)}{x^2 + x^2 + 2x}$$

$$= \frac{(x + 2)(2x - 2)}{2x^2 + 2x}$$

Factor and divide our common factors:
$$= \frac{2(x + 2)(x - 1)}{2x(x + 1)}$$

$$= \frac{(x + 2)(x - 1)}{x(x + 1)}$$

Quick ✓

9. Simplify using Method II: $\dfrac{\dfrac{x + 2}{x + 5} - \dfrac{x + 2}{x + 1}}{\dfrac{2x + 1}{x + 1} - 1}$

▶ Comparing Methods

Both methods will be used in the next example. As you work through the exercise set, develop a sense about when you prefer Method I over Method II, and vice versa.

EXAMPLE 5 **Comparing Methods I and II**

Simplify $\dfrac{x^{-1} + y^{-1}}{x^{-3} + y^{-3}}$ as a rational expression that contains no negative exponents.

Solution

First, rewrite the expression without negative exponents.

Work Smart

Remember, $x^{-1} = \dfrac{1}{x}$ and $y^{-1} = \dfrac{1}{y}$,

but $x^{-1} + y^{-1} \neq \dfrac{1}{x + y}$.

Rather, $x^{-1} + y^{-1} = \dfrac{1}{x} + \dfrac{1}{y}$.

$$\frac{x^{-1} + y^{-1}}{x^{-3} + y^{-3}} = \frac{\dfrac{1}{x} + \dfrac{1}{y}}{\dfrac{1}{x^3} + \dfrac{1}{y^3}}$$

Method I

Write the numerator of the complex rational expression as a single quotient. The LCD is xy.

$$\frac{1}{x} + \frac{1}{y} = \frac{1}{x} \cdot \frac{y}{y} + \frac{1}{y} \cdot \frac{x}{x}$$

$$= \frac{y}{xy} + \frac{x}{xy}$$

$$= \frac{y + x}{xy}$$

$$= \frac{x + y}{xy}$$

Write the denominator of the complex rational expression as a single quotient. The LCD is $x^3 y^3$.

$$\frac{1}{x^3} + \frac{1}{y^3} = \frac{1}{x^3} \cdot \frac{y^3}{y^3} + \frac{1}{y^3} \cdot \frac{x^3}{x^3}$$

$$= \frac{y^3}{x^3 y^3} + \frac{x^3}{x^3 y^3}$$

$$= \frac{y^3 + x^3}{x^3 y^3}$$

$$= \frac{x^3 + y^3}{x^3 y^3}$$

Now rewrite the complex rational expression using the numerator and denominator just found and then simplify.

$$\frac{x^{-1} + y^{-1}}{x^{-3} + y^{-3}} = \frac{\dfrac{1}{x} + \dfrac{1}{y}}{\dfrac{1}{x^3} + \dfrac{1}{y^3}}$$

$$= \frac{\dfrac{x + y}{xy}}{\dfrac{x^3 + y^3}{x^3 y^3}}$$

Multiply the numerator by the reciprocal of the denominator:

$$= \frac{x + y}{xy} \cdot \frac{x^3 y^3}{x^3 + y^3}$$

Factor:

$$= \frac{x + y}{xy} \cdot \frac{x^3 y^3}{(x + y)(x^2 - xy + y^2)}$$

Divide out common factors:

$$= \frac{\cancel{x + y}}{\cancel{x}\,\cancel{y}} \cdot \frac{x^{3^2} y^{3^2}}{\cancel{(x + y)}\,(x^2 - xy + y^2)}$$

$$= \frac{x^2 y^2}{x^2 - xy + y^2}$$

Work Smart

See Section 4.6 for a review of factoring special products.

Method II

The LCD of all the denominators is $x^3 y^3$. Multiply the numerator and denominator of the complex rational expression by $x^3 y^3$.

$$\frac{\dfrac{1}{x} + \dfrac{1}{y}}{\dfrac{1}{x^3} + \dfrac{1}{y^3}} = \frac{\dfrac{1}{x} + \dfrac{1}{y}}{\dfrac{1}{x^3} + \dfrac{1}{y^3}} \cdot \frac{x^3 y^3}{x^3 y^3}$$

Distribute the LCD to each term:

$$= \frac{\dfrac{1}{x} \cdot x^3 y^3 + \dfrac{1}{y} \cdot x^3 y^3}{\dfrac{1}{x^3} \cdot x^3 y^3 + \dfrac{1}{y^3} \cdot x^3 y^3}$$

$$= \frac{x^2 y^3 + x^3 y^2}{y^3 + x^3}$$

Factor out the GCF in the numerator:
Factor the sum of two cubes in the denominator: $= \dfrac{x^2 y^2 (x + y)}{(x + y)(x^2 - xy + y^2)}$

Divide out common factors: $= \dfrac{x^2 y^2}{x^2 - xy + y^2}$

In this particular problem, Method II seems more efficient.

Quick ✓

Simplify the expression so that it does not contain any negative exponents. Use both methods and decide which method you prefer for this problem.

10. $\dfrac{3a^{-1} + b^{-1}}{9a^{-2} - b^{-2}}$

5.3 Exercises   MyMathLab®

Exercise numbers in green have complete video solutions in MyMathLab or may be accessed using the QR code to the right.

*Problems **1–10** are the Quick ✓s that follow the **EXAMPLES**.*

Building Skills

In Problems 11–18, simplify the expression using Method I. See Objective 1.

11. $\dfrac{\frac{5}{6} + \frac{1}{9}}{\frac{5}{2} - \frac{3}{8}}$

12. $\dfrac{\frac{3}{10} - \frac{3}{4}}{\frac{2}{5} + \frac{7}{10}}$

13. $\dfrac{1 + \frac{1}{x}}{1 - \frac{1}{x}}$

14. $\dfrac{1 + \frac{1}{x^2}}{1 - \frac{1}{x^2}}$

15. $\dfrac{1 - \frac{a}{a - 2}}{2 - \frac{a + 2}{a}}$

16. $\dfrac{\frac{a}{a + 1} - 1}{\frac{a + 3}{a} - 2}$

17. $\dfrac{\frac{x + 2}{x - 1} - \frac{x + 5}{x + 3}}{x + 11}$

18. $\dfrac{\frac{x + 5}{x - 2} - \frac{x + 3}{x - 1}}{3x + 1}$

In Problems 19–26, simplify the expression using Method II. See Objective 2.

19. $\dfrac{\frac{5}{6} + \frac{7}{10}}{\frac{1}{5} - 1}$

20. $\dfrac{2 + \frac{7}{4}}{\frac{3}{8} - 2}$

21. $\dfrac{w - \frac{1}{w}}{w + \frac{1}{w}}$

22. $\dfrac{\frac{7}{w} + \frac{9}{x}}{\frac{9}{w} - \frac{7}{x}}$

23. $\dfrac{\frac{x - 1}{x - 4} - \frac{x}{x - 2}}{1 - \frac{3}{x - 4}}$

24. $\dfrac{\frac{x - 4}{x - 1} - \frac{x}{x - 3}}{3 + \frac{12}{x - 3}}$

25. $\dfrac{\frac{z + 2}{z - 2} + \frac{z - 2}{z + 2}}{\frac{z + 2}{z - 2} - \frac{z - 2}{z + 2}}$

26. $\dfrac{\frac{m + 3}{m - 3} - \frac{m - 3}{m + 3}}{\frac{m + 3}{m - 3} + \frac{m - 3}{m + 3}}$

Mixed Practice

In Problems 27–52, simplify the complex rational expression using either Method I or Method II.

27. $\dfrac{\frac{3}{y} - 1}{\frac{9}{y} - y}$

28. $\dfrac{1 - \frac{4}{z}}{z - \frac{16}{z}}$

29. $\dfrac{\frac{4n}{m} - \frac{4m}{n}}{\frac{2}{m} + \frac{2}{n}}$

30. $\dfrac{\frac{n^2}{m} - \frac{m^2}{n}}{\frac{1}{m} - \frac{1}{n}}$

31. $\dfrac{2 + \frac{3}{x}}{\frac{2x^2}{x + 3} - 3}$

32. $\dfrac{1 + \frac{5}{x}}{1 + \frac{1}{x + 4}}$

33. $\dfrac{\frac{x}{3} - \frac{3}{x}}{\frac{3}{x^2} - \frac{1}{3}}$

34. $\dfrac{\frac{5}{x} - \frac{x}{5}}{\frac{1}{5} - \frac{5}{x^2}}$

35. $\dfrac{\frac{2x + 1}{x - 1} - \frac{x - 1}{x - 3}}{x^2 - 3x - 4}$

36. $\dfrac{\frac{x + 5}{x - 3} - \frac{x}{x + 4}}{3x^2 - 4x - 15}$

37. $\dfrac{\dfrac{x-4}{x+4}+\dfrac{x-4}{x-2}}{1-\dfrac{2}{x-2}}$

38. $\dfrac{\dfrac{x-3}{x+3}+\dfrac{x-3}{x-4}}{1+\dfrac{x+3}{x-4}}$

39. $\dfrac{\dfrac{b^2}{b^2-25}-\dfrac{b}{b+5}}{\dfrac{b}{b^2-25}-\dfrac{1}{b-5}}$

40. $\dfrac{\dfrac{-6}{x^2+5x+6}}{\dfrac{2}{x+3}-\dfrac{3}{x+2}}$

41. $\dfrac{3}{\dfrac{2}{x}-1}$

42. $\dfrac{4}{\dfrac{5}{x}+3}$

43. $\dfrac{1}{\dfrac{4}{x-1}+2}$

44. $\dfrac{1}{\dfrac{1}{x+2}+2}$

45. $\dfrac{\dfrac{4}{x-1}-1}{\dfrac{4}{x-1}+3}$

46. $\dfrac{\dfrac{7}{x-2}+4}{\dfrac{3}{x-2}-1}$

47. $\dfrac{3x^{-1}+3y^{-1}}{x^{-2}-y^{-2}}$

48. $\dfrac{2x^{-1}+2y^{-1}}{xy^{-1}-x^{-1}y}$

49. $\dfrac{(m+n)^{-1}}{m^{-1}+n^{-1}}$

50. $\dfrac{(x-y)^{-1}}{x^{-1}-y^{-1}}$

51. $\dfrac{a^{-2}b^{-1}-a^{-1}b^{-2}}{4a^{-2}-4b^{-2}}$

52. $\dfrac{a^{-3}+8b^{-3}}{a^{-2}-4b^{-2}}$

Applying the Concepts

53. Electric Circuits An electric circuit contains two resistors connected in parallel, as shown in the figure. If the resistances of these resistors are R_1 and R_2 ohms, respectively, then their combined resistance R is given by the formula

$$R = \dfrac{1}{\dfrac{1}{R_1}+\dfrac{1}{R_2}}$$

    (a) Express R as a simplified rational expression.
    (b) Evaluate the rational expression if $R_1 = 4$ ohms and $R_2 = 10$ ohms.

54. Future Value of Money The value of an account V in which P dollars is deposited every year for the next 5 years paying an interest rate i (expressed as a decimal) is given by the formula

$$V = P \cdot \dfrac{1-\dfrac{1}{(1+i)^5}}{\dfrac{i}{(1+i)^5}}$$

    (a) Express V as a simplified rational expression.
    (b) Determine the value of an account paying 5% when the annual deposit is $1000. Express your answer to the nearest penny.

55. The Lensmaker's Equation The focal length f of a convex lens with index of refraction n is given by

$$f = \dfrac{1}{(n-1)\left[\dfrac{1}{R_1}+\dfrac{1}{R_2}\right]}$$

where R_1 and R_2 are the radii of curvature of the front and back surfaces of the lens. See the figure.

    (a) Express f as a simplified rational expression.
    (b) Determine the focal length of a lens for $n = 1.5$, $R_1 = 0.5$ meter, and $R_2 = 0.3$ meter.

56. Harmonic Mean The harmonic mean is used to determine an average value when data are measured as a rate of change, such as 50 miles per hour. The harmonic mean is found using the formula

$$H = \dfrac{n}{\dfrac{1}{x_1}+\dfrac{1}{x_2}+\cdots+\dfrac{1}{x_n}}$$

where $x_1, x_2, \ldots, x_n$ are the n rates of change.

    (a) Suppose that a family drove from Chicago, Illinois, to Naples, Florida. The distance each way is about 1200 miles. The trip from Chicago to Naples resulted in an average speed of 48 miles per hour, while the return trip resulted in an average speed of 52 miles per hour. Compute the average speed of the entire trip.
    (b) A 300-KB file is downloaded four separate times, resulting in download speeds of 4.3 KB/s, 4.1 KB/s, 3.8 KB/s, and 4.3 KB/s. Compute the average download time.

Extending the Concepts

57. Fibonacci Strikes Again! Write each of the following expressions in the form $\dfrac{ax+b}{bx+c}$. (*Hint:* Use Method I.)

    (a) $1+\dfrac{1}{1+\dfrac{1}{x}}$

(b) $1 + \dfrac{1}{1 + \dfrac{1}{1 + \dfrac{1}{x}}}$

(c) $1 + \dfrac{1}{1 + \dfrac{1}{1 + \dfrac{1}{1 + \dfrac{1}{x}}}}$

(d) $1 + \dfrac{1}{1 + \dfrac{1}{1 + \dfrac{1}{1 + \dfrac{1}{1 + \dfrac{1}{x}}}}}$

(e) Write down the values of a, b, and c from part (a). Now write down the values of a, b, and c from part (b), followed by the values of a, b, and c from part (c), followed by the values of a, b, and c from part (d). What is the pattern? Write the sequence of numbers in increasing order. This sequence of numbers forms the first six numbers in the **Fibonacci sequence.**

Math for the Future *In Calculus, you will be asked to simplify the expression* $\dfrac{f(x + h) - f(x)}{h}$, *which is called the* **difference quotient.** *In Problems 58–61, simplify the difference quotient for the given functions.*

58. $f(x) = \dfrac{1}{x}$

59. $f(x) = \dfrac{1}{x^2}$

60. $f(x) = \dfrac{3}{x^2}$

61. $f(x) = \dfrac{1}{x - 1}$

Explaining the Concepts

62. In your own words, provide a definition for a complex rational expression.

63. Which of the two methods for simplifying complex rational expressions do you prefer? Write a paragraph supporting your opinion.

Synthesis Review

In Problems 64–69, solve each equation.

64. $4x + 3 = 15$

65. $-5a + 2 = 22$

66. $\dfrac{1}{2}w + 3 = 5$

67. $\dfrac{2}{3}x - \dfrac{5}{7}(x + 21) = \dfrac{11}{21}x + \dfrac{4}{7}$

68. $y^2 - 5y = 50$

69. $3p^2 + 19p = 14$

Putting the Concepts Together (Sections 5.1–5.3)

We designed these problems so that you can review Sections 5.1 through 5.3 and show your mastery of the concepts. Take time to work these problems before proceeding with the next section. The answers are located at the back of the text on page AN-31.

1. Determine the domain of the rational function

$$g(x) = \dfrac{3x + 1}{3x^2 - 17x - 6}$$

In Problems 2 and 3, write each rational expression in lowest terms.

2. $\dfrac{24n - 4n^2}{2n^2 - 9n - 18}$

3. $\dfrac{2p^2 - pq - 10q^2}{3p^2 + 2pq - 8q^2}$

In Problems 4–8, perform the indicated operations.

4. $\dfrac{a^2 - 16}{12a^2 + 48a} \cdot \dfrac{6a^3 - 30a^2}{a^2 + 2a - 24}$

5. $\dfrac{\dfrac{x^2 + x - 2}{3x^2 - 5x - 2}}{\dfrac{3x^2 - 2x - 1}{x^2 - 9x + 14}}$

6. $\dfrac{x^2 - 10}{x^2 - 4} - \dfrac{3x}{x^2 - 4}$

7. $\dfrac{3n}{n^2 - 7n + 10} - \dfrac{2n}{n^2 - 8n + 15}$

8. $\dfrac{3y + 2}{y^2 + 5y - 24} + \dfrac{7}{y^2 + 4y - 32}$

In Problems 9 and 10, use the functions $f(x) = \dfrac{2x + 1}{x^2 - 11x + 28}$, $g(x) = \dfrac{3x - 12}{4x^2 + 4x + 1}$, *and* $h(x) = \dfrac{3x}{x - 7}$ *to find each difference or product.*

9. $P(x) = f(x) \cdot g(x)$

10. $D(x) = h(x) - f(x)$

In Problems 11 and 12, simplify each complex rational expression using either method.

11. $\dfrac{\dfrac{1}{m^2} - \dfrac{1}{n^2}}{\dfrac{1}{m} - \dfrac{1}{n}}$

12. $\dfrac{\dfrac{z^2 - 2}{z^2 - 4} + \dfrac{7}{z - 2}}{\dfrac{z^2 + z - 24}{z^2 - 4} - \dfrac{2}{z + 2}}$

5.4 Rational Equations

Objectives

1. Solve Equations Containing Rational Expressions
2. Solve Equations Involving Rational Functions

Are You Prepared for This Section?

Before getting started, complete the following problems. If you get a problem wrong, go back to the section cited and review the material.

P1. Solve: $\dfrac{2}{3}x + \dfrac{1}{2} = \dfrac{3}{4}$ [Section 1.1, pp. 50–55]

P2. Factor: $3z^2 + 11z - 4$ [Section 4.5, pp. 351–356]

P3. Solve: $6y^2 - y - 12 = 0$ [Section 4.8, pp. 372–375]

P4. Determine which of the following is in the domain of $\dfrac{x + 4}{x^2 - 5x - 24}$.

   (a) $x = -4$ **(b)** $x = 8$ [Section R.5, p. 44]

P5. If $f(x) = x^2 - 3x - 15$, solve $f(x) = 3$. [Section 4.8, pp. 376–377]

P6. If $g(4) = 3$, what point is on the graph of g? [Section 2.3, pp. 165–166]

Prepared?...Answers **P1.** $\left\{ \dfrac{3}{8} \right\}$

P2. $(3z - 1)(z + 4)$ **P3.** $\left\{ -\dfrac{4}{3}, \dfrac{3}{2} \right\}$

P4. (a) Yes **(b)** No **P5.** $\{-3, 6\}$
P6. $(4, 3)$

1 Solve Equations Containing Rational Expressions

So far, linear equations (Section 1.1), quadratic equations (Section 4.8), and equations that contain polynomial expressions that can be factored (Section 4.8) have been solved. We now look at another type of equation, the *rational equation*. A **rational equation** is an equation that contains a rational expression. Examples of rational equations are

$$\frac{3}{x + 4} = \frac{5}{x - 1} + \frac{1}{x^2 + 3x - 4} \quad \text{and} \quad \frac{x - 5}{x^2 + 4x - 12} = 3$$

As usual, exclude from the domain all values of the variable that result in division by zero.

EXAMPLE 1 **How to Solve a Rational Equation**

Solve: $\dfrac{2x - 1}{x - 3} = \dfrac{2(x + 1)}{x - 2}$

Step-by-Step Solution

Step 1: Determine the domain of the variable in the rational equation.

Because $x = 2$ and $x = 3$ result in division by zero, the domain of x is

$$\{x \mid x \neq 2, x \neq 3\}$$

Step 2: Determine the least common denominator (LCD) of all the denominators.

The LCD is $(x - 3)(x - 2)$.

Step 3: Multiply both sides of the equation by the LCD, and simplify the expressions on each side.

$$\frac{2x - 1}{x - 3} = \frac{2(x + 1)}{x - 2}$$

Multiply both sides by $(x - 3)(x - 2)$: $(x - 3)(x - 2) \cdot \dfrac{2x - 1}{x - 3} = (x - 3)(x - 2) \cdot \dfrac{2(x + 1)}{x - 2}$

Divide out common factors:	$(x - 2)(2x - 1) = 2(x - 3)(x + 1)$
Multiply:	$2x^2 - 5x + 2 = 2(x^2 - 2x - 3)$
Distribute the 2:	$2x^2 - 5x + 2 = 2x^2 - 4x - 6$

Step 4: Solve the resulting equation.

Subtract $2x^2$ from both sides:	$-5x + 2 = -4x - 6$
Add $4x$ to both sides;	
Subtract 2 from both sides:	$-x = -8$
Multiply both sides by -1:	$x = 8$

Step 5: Check Verify your solution using the original equation.

Let $x = 8$:
$$\frac{2(8) - 1}{8 - 3} \overset{?}{=} \frac{2(8 + 1)}{8 - 2}$$

$$\frac{16 - 1}{5} \overset{?}{=} \frac{2(9)}{6}$$

$$\frac{15}{5} \overset{?}{=} \frac{18}{6}$$

$$3 = 3 \text{ True}$$

The solution checks, so the solution set is $\{8\}$.

The steps that can be used to solve a rational equation are summarized below.

Solving a Rational Equation

Step 1: Determine the domain of the variable in the rational equation.

Step 2: Determine the least common denominator (LCD) of all the denominators.

Step 3: Multiply both sides of the equation by the LCD, and simplify the expressions on each side of the equation.

Step 4: Solve the resulting equation.

Step 5: Verify your solution using the original equation.

Quick ✓

1. A _____ _____ is an equation that contains a rational expression.

In Problems 2 and 3, solve each equation. Be sure to verify your results.

2. $\dfrac{x - 4}{x^2 + 4} = \dfrac{3}{3x + 2}$

3. $\dfrac{4}{x + 2} = \dfrac{7}{x + 4}$

$\boxed{\text{EXAMPLE 2}}$ **Solving a Rational Equation**

Solve: $\dfrac{2}{x} - \dfrac{1}{6} = \dfrac{5}{2x} - \dfrac{1}{3}$

Solution

$\{-6\}$

The domain of the variable is $\{x \mid x \neq 0\}$. The LCD of all denominators is $6x$.

$$\frac{2}{x} - \frac{1}{6} = \frac{5}{2x} - \frac{1}{3}$$

Multiply both sides by the LCD, $6x$: $6x\left(\dfrac{2}{x} - \dfrac{1}{6}\right) = 6x\left(\dfrac{5}{2x} - \dfrac{1}{3}\right)$

Distribute the 6x: $\quad 6x \cdot \dfrac{2}{x} - 6x \cdot \dfrac{1}{6} = 6x \cdot \dfrac{5}{2x} - 6x \cdot \dfrac{1}{3}$

Simplify: $\quad 12 - x = 15 - 2x$

Add 2x to both sides: $\quad 12 + x = 15$

Subtract 12 from both sides: $\quad x = 3$

Check   Let $x = 3$ in the original equation: $\quad \dfrac{2}{3} - \dfrac{1}{6} \overset{?}{=} \dfrac{5}{2 \cdot 3} - \dfrac{1}{3}$

$$\dfrac{4}{6} - \dfrac{1}{6} \overset{?}{=} \dfrac{5}{6} - \dfrac{2}{6}$$

$$\dfrac{3}{6} = \dfrac{3}{6} \quad \text{True}$$

The solution checks, so the solution set is $\{3\}$.

> **Quick ✓**
>
> *In Problems 4 and 5, solve each equation. Be sure to verify your results.*
>
> **4.** $\dfrac{5}{x} + \dfrac{1}{4} = \dfrac{3}{2x} - \dfrac{3}{2}$ **5.** $\dfrac{5}{x} + 2 = \dfrac{10}{3x} + 1$

EXAMPLE 3   **Solving a Rational Equation**

Solve: $\dfrac{3}{p^2 - 4p + 3} + \dfrac{6}{p^2 - 2p - 3} = \dfrac{5}{p^2 - 1}$

Solution

First, factor the denominator of each term to find the domain of the variable, p.

$p^2 - 4p + 3 = (p - 3)(p - 1)$, so $p \neq 3, p \neq 1$ in the first term.
$p^2 - 2p - 3 = (p - 3)(p + 1)$, so $p \neq 3, p \neq -1$ in the second term.
$p^2 - 1 = (p - 1)(p + 1)$, so $p \neq 1, p \neq -1$ in the third term.

The domain of the variable p is $\{p \mid p \neq -1, p \neq 1, p \neq 3\}$. It is known from the factored form of each denominator above that the LCD of all the denominators is $(p - 1)(p + 1)(p - 3)$, so multiply both sides of the equation by $(p - 1)(p + 1)(p - 3)$.

$$\dfrac{3}{p^2 - 4p + 3} + \dfrac{6}{p^2 - 2p - 3} = \dfrac{5}{p^2 - 1}$$

$$(p - 1)(p + 1)(p - 3)\left(\dfrac{3}{(p - 3)(p - 1)} + \dfrac{6}{(p - 3)(p + 1)} \right) = (p - 1)(p + 1)(p - 3) \cdot \dfrac{5}{(p - 1)(p + 1)}$$

Distribute the LCD and simplify: $\quad 3(p + 1) + 6(p - 1) = 5(p - 3)$

Distribute: $\quad 3p + 3 + 6p - 6 = 5p - 15$

Combine like terms: $\quad 9p - 3 = 5p - 15$

Subtract 5p from both sides; add 3 to both sides: $\quad 4p = -12$

Divide both sides by 4: $\quad p = -3$

The check is left to you. The solution set is $\{-3\}$.

> **Quick ✓**
>
> **6.** Solve: $\dfrac{3}{x^2 + 5x + 4} + \dfrac{2}{x^2 - 3x - 4} = \dfrac{4}{x^2 - 16}$

EXAMPLE 4 **Solving a Rational Equation with No Solution**

Solve: $\dfrac{3}{y^2 - 5y + 4} + \dfrac{2}{y^2 - 10y + 24} = \dfrac{2}{y^2 - 7y + 6}$

Solution

First, factor the denominator of each term to find the domain of the variable, y.

$y^2 - 5y + 4 = (y - 4)(y - 1)$, so $y \neq 4, y \neq 1$ in the first term.

$y^2 - 10y + 24 = (y - 6)(y - 4)$, so $y \neq 6, y \neq 4$ in the second term.

$y^2 - 7y + 6 = (y - 6)(y - 1)$, so $y \neq 6, y \neq 1$ in the third term.

The domain of the variable y is $\{y \mid y \neq 1, y \neq 4, y \neq 6\}$. The LCD of all the denominators is $(y - 1)(y - 4)(y - 6)$. Multiply both sides of the equation by the LCD.

$$\frac{3}{y^2 - 5y + 4} + \frac{2}{y^2 - 10y + 24} = \frac{2}{y^2 - 7y + 6}$$

$$(y - 1)(y - 4)(y - 6)\left(\frac{3}{(y - 4)(y - 1)} + \frac{2}{(y - 6)(y - 4)}\right) = (y - 1)(y - 4)(y - 6)\frac{2}{(y - 6)(y - 1)}$$

Distribute the LCD and simplify: $3(y - 6) + 2(y - 1) = 2(y - 4)$

Distribute: $3y - 18 + 2y - 2 = 2y - 8$

Combine like terms: $5y - 20 = 2y - 8$

Subtract $2y$ from both sides; add 20 to both sides: $3y = 12$

Divide both sides by 3: $y = 4$

Notice that $y = 4$ is not in the domain of the variable y, so the equation has no solution. The solution set is $\varnothing$ or $\{\ \}$.

The answer $y = 4$ is an *extraneous solution*. **Extraneous solutions** are results that develop through the solution process but do not satisfy the original equation.

In Other Words

The word *extraneous* means "not constituting a vital part."

Quick ✓

7. _____ _____ are results that develop through the solution process but do not satisfy the original equation.

8. *True or False* Some rational equations have no solution.

In Problems 9 and 10, solve each equation. Be sure to verify your results.

9. $\dfrac{5}{z^2 + 2z - 3} - \dfrac{3}{z^2 + z - 2} = \dfrac{1}{z^2 + 5z + 6}$

10. $\dfrac{5}{x - 4} + \dfrac{3}{x - 2} = \dfrac{11}{x - 4}$

EXAMPLE 5 **Solving a Rational Equation That Leads to a Quadratic Equation**

Solve: $\dfrac{w + 3}{w - 1} + \dfrac{w + 5}{w} = \dfrac{3w + 1}{w - 1}$

Solution

The domain of the variable w is $\{w \mid w \neq 0, w \neq 1\}$. The LCD of all the denominators is $w(w - 1)$, so multiply both sides of the equation by $w(w - 1)$.

$$\frac{w + 3}{w - 1} + \frac{w + 5}{w} = \frac{3w + 1}{w - 1}$$

$$w(w - 1)\left(\frac{w + 3}{w - 1} + \frac{w + 5}{w}\right) = w(w - 1)\frac{3w + 1}{w - 1}$$

(continued)

Distribute the LCD:	$w(w + 3) + (w - 1)(w + 5) = w(3w + 1)$
	$w^2 + 3w + w^2 + 4w - 5 = 3w^2 + w$
Combine like terms:	$2w^2 + 7w - 5 = 3w^2 + w$
Put equation in standard form:	$0 = w^2 - 6w + 5$
If $a = b$, then $b = a$:	$w^2 - 6w + 5 = 0$
Factor:	$(w - 5)(w - 1) = 0$
Use the Zero-Product Property:	$w = 5$ or $w = 1$

Because $w = 1$ is not in the domain of the variable, it is an extraneous solution. The only potential solution is 5.

Check Let $w = 5$ in the original equation.
$$\frac{5 + 3}{5 - 1} + \frac{5 + 5}{5} \overset{?}{=} \frac{3(5) + 1}{5 - 1}$$
$$\frac{8}{4} + \frac{10}{5} \overset{?}{=} \frac{16}{4}$$
$$2 + 2 = 4 \text{ True}$$

The solution set is $\{5\}$.

Quick ✓

In Problems 11 and 12, solve each equation. Be sure to verify your results.

11. $2 - \dfrac{3}{p + 2} = \dfrac{6}{p}$

12. $\dfrac{z + 1}{z + 4} + \dfrac{z + 1}{z - 3} = \dfrac{z^2 + z + 16}{z^2 + z - 12}$

▶ ❷ **Solve Equations Involving Rational Functions**

Now look at a problem involving a rational function that leads to a rational equation.

EXAMPLE 6 **Working with Rational Functions**

For the function $f(x) = x + \dfrac{4}{x}$, solve $f(x) = 5$. What point(s) on the graph of f do these values correspond to?

Solution

Solve the equation $x + \dfrac{4}{x} = 5$. The domain of the variable is $\{x | x \neq 0\}$.

The LCD of all denominators is x, so multiply both sides of the equation by x.

	$x\left(x + \dfrac{4}{x}\right) = 5 \cdot x$
Distribute:	$x^2 + 4 = 5x$
Subtract 5x from both sides:	$x^2 - 5x + 4 = 0$
Factor:	$(x - 4)(x - 1) = 0$
Use the Zero-Product Property:	$x - 4 = 0$ or $x - 1 = 0$
	$x = 4$ or $x = 1$

The check is left to you. If $f(1) = 5$, then the point $(1, 5)$ is on the graph of f. If $f(4) = 5$, then the point $(4, 5)$ is on the graph of f.

Quick ✓

13. *True or False* To solve $f(x) = 4$ when $f(x) = x + \dfrac{3}{x}$, substitute 4 for x in the function and evaluate.

14. For the function $g(x) = x - \dfrac{5}{x}$, solve $g(x) = 4$. What point(s) on the graph of g do these values correspond to?

15. For the function $f(x) = 2x - \dfrac{3}{x}$, solve $f(x) = 1$. What point(s) on the graph of f do these values correspond to?

EXAMPLE 7 **An Application of Rational Functions: Drug Concentration**

The concentration C of a drug in a patient's bloodstream in milligrams per liter t hours after ingestion is modeled by

$$C(t) = \frac{40t}{t^2 + 9}$$

When will the concentration of the drug be 4 milligrams per liter?

Solution
To know when the concentration of the drug is 4, solve the equation $C(t) = 4$.

$$\frac{40t}{t^2 + 9} = 4$$

Multiply both sides by $t^2 + 9$: $40t = 4(t^2 + 9)$

Divide both sides by 4: $10t = t^2 + 9$

Subtract $10t$ from both sides: $0 = t^2 - 10t + 9$

Factor: $0 = (t - 1)(t - 9)$

Use the Zero-Product Property: $t = 1 \quad \text{or} \quad t = 9$

The concentration of the drug will be 4 milligrams per liter 1 hour and 9 hours after ingestion.

Quick ✓

16. The concentration C of a drug in a patient's bloodstream in milligrams per liter t hours after ingestion is modeled by $C(t) = \dfrac{50t}{t^2 + 6}$. When will the concentration of the drug be 4 milligrams per liter?

5.4 Exercises MyMathLab® Exercise numbers in green have complete video solutions in MyMathLab or may be accessed using the QR code to the right.

*Problems **1–16** are the* **Quick ✓**s *that follow the* **EXAMPLES.**

Building Skills

In Problems 17–42, solve each equation. Be sure to verify your results. See Objective 1.

17. $\dfrac{3}{z} - \dfrac{1}{2z} = -\dfrac{5}{8}$

18. $\dfrac{8}{p} + \dfrac{1}{4p} = \dfrac{11}{8}$

19. $\dfrac{y + 2}{y - 5} = \dfrac{y + 6}{y + 1}$

20. $\dfrac{w - 4}{w + 1} = \dfrac{w - 3}{w + 3}$

21. $\dfrac{x + 8}{x + 4} = \dfrac{x + 2}{x - 2}$

22. $\dfrac{2x + 1}{x + 3} = \dfrac{4(x - 1)}{2x + 3}$

23. $a - \dfrac{5}{a} = 4$

24. $m + \dfrac{8}{m} = 6$

25. $6p - \dfrac{3}{p} = 7$

26. $8b - \dfrac{3}{b} = 2$

27. $\dfrac{5 - p}{p - 5} + 2 = \dfrac{1}{p}$

28. $\dfrac{3 - y}{y - 3} + 2 = \dfrac{2}{y}$

29. $\dfrac{3}{2} + \dfrac{5}{x - 3} = \dfrac{x + 9}{2x - 6}$

30. $\dfrac{4}{3} + \dfrac{7}{x - 4} = \dfrac{x - 1}{3x - 12}$

31. $1 + \dfrac{3}{x + 3} = \dfrac{4}{x - 3}$

32. $\dfrac{5}{x + 2} = 1 - \dfrac{3}{x - 2}$

33. $\dfrac{4}{x - 5} + \dfrac{3}{x - 2} = \dfrac{x + 1}{x^2 - 7x + 10}$

34. $\dfrac{4}{x + 3} + \dfrac{5}{x - 6} = \dfrac{4x + 1}{x^2 - 3x - 18}$

35. $\dfrac{1}{x + 5} = \dfrac{2x}{x^2 - 25} - \dfrac{3}{x - 5}$

36. $\dfrac{3}{x - 4} = \dfrac{5x + 4}{x^2 - 16} - \dfrac{4}{x + 4}$

37. $\dfrac{3}{x - 1} - \dfrac{2}{x + 4} = \dfrac{x^2 + 8x + 6}{x^2 + 3x - 4}$

38. $\dfrac{5}{z - 4} + \dfrac{3}{z - 2} = \dfrac{z^2 - z - 2}{z^2 - 6z + 8}$

39. $\dfrac{7}{y^2 + y - 12} - \dfrac{4y}{y^2 + 7y + 12} = \dfrac{6}{y^2 - 9}$

40. $\dfrac{3}{x^2 - 5x - 6} + \dfrac{3}{x^2 - 7x + 6} = \dfrac{6}{x^2 - 1}$

41. $\dfrac{2x + 3}{x - 3} + \dfrac{x + 6}{x - 4} = \dfrac{x + 6}{x - 3}$

42. $\dfrac{x + 5}{x + 1} + 1 = \dfrac{x - 5}{x - 2}$

In Problems 43–50, answer each question regarding the rational function given. See Objective 2.

43. For the function $f(x) = x + \dfrac{9}{x}$, solve $f(x) = 10$.

What point(s) on the graph of f do these values correspond to?

44. For the function $f(x) = x + \dfrac{7}{x}$, solve $f(x) = 8$.

What point(s) on the graph of f do these values correspond to?

45. For the function $f(x) = 2x + \dfrac{4}{x}$, solve $f(x) = -9$.

What point(s) on the graph of f do these values correspond to?

46. For the function $f(x) = 2x + \dfrac{8}{x}$, solve $f(x) = -10$.

What point(s) on the graph of f do these values correspond to?

47. For the function $f(x) = \dfrac{x + 3}{x - 4}$, solve $f(x) = \dfrac{9}{2}$.

What point(s) on the graph of f do these values correspond to?

48. For the function $f(x) = \dfrac{x + 5}{x - 3}$, solve $f(x) = \dfrac{1}{5}$.

What point(s) on the graph of f do these values correspond to?

49. Let $f(x) = \dfrac{x + 2}{2x + 9}$ and $g(x) = \dfrac{x - 1}{x + 3}$. For what value(s) of x does $f(x) = g(x)$? What are the point(s) of intersection of the graphs of f and g?

50. Let $f(x) = \dfrac{4x + 1}{8x + 5}$ and $g(x) = \dfrac{x - 4}{2x - 7}$. For what value(s) of x does $f(x) = g(x)$? What are the point(s) of intersection of the graphs of f and g?

Mixed Practice

In Problems 51–60, solve each equation. Be sure to verify your results.

51. $\dfrac{4}{z + 4} - \dfrac{3}{4} = \dfrac{5z + 2}{4z + 16}$

52. $\dfrac{2b - 1}{b + 5} - \dfrac{2}{3} = \dfrac{1}{3b + 15}$

53. $x + \dfrac{9}{x} = 6$

54. $p + \dfrac{25}{p} = 10$

55. $\dfrac{2}{z^2 + 2z - 3} + \dfrac{3}{z^2 + 4z + 3} = \dfrac{6}{z^2 - 1}$

56. $\dfrac{3}{a^2 + 3a - 10} + \dfrac{2}{a^2 + 7a + 10} = \dfrac{4}{a^2 - 4}$

57. $\dfrac{4}{x} - \dfrac{5}{2x} = \dfrac{3}{4}$

58. $\dfrac{9}{b} + \dfrac{4}{5b} = \dfrac{7}{10}$

59. $\dfrac{3y + 1}{y - 1} + 3 = \dfrac{y + 2}{y + 1}$

60. $\dfrac{x + 3}{x - 2} + 4 = \dfrac{x + 2}{x + 1}$

Recall that to find the zero of a function f, we solve the equation $f(x) = 0$. The zeros of the function also represent the x-intercepts of the graph of the function. To find the zeros of a rational function R, the function must first be simplified. In Problems 61–66, find the zeros of each rational function and list the x-intercept(s) of the graph of the function.

61. $R(x) = \dfrac{3x + 1}{x^2 - 4}$

62. $R(x) = \dfrac{2x + 5}{x^2 - 9}$

63. $R(x) = \dfrac{2x^2 + 5x - 12}{3x^2 + 5x + 2}$

64. $R(x) = \dfrac{4x^2 + 7x - 2}{6x^2 + 5x - 4}$

65. $R(x) = \dfrac{x^3 + 3x^2 - 4x - 12}{4x^3 + 12x^2 + x + 3}$

66. $R(x) = \dfrac{x^3 - 2x^2 - 9x + 18}{x^3 - 2x^2 + 4x - 8}$

Applying the Concepts

67. **Average Cost** Suppose that the average daily cost $\overline{C}$ of manufacturing x bicycles is given by the function
$$\overline{C}(x) = \frac{x^2 + 75x + 5000}{x}$$
Determine the level of production for which the average daily cost will be \$225.

68. **Population** When loggers began cutting in a region in the Amazon rain forest, a rare insect species was discovered. To protect the species, government scientists declared the insects endangered and moved them into a protected area. The population P of the insect t months after being transplanted is modeled by
$$P(t) = \frac{200(1 + 0.4t)}{2(1 + 0.01t)}$$
Predict when the population will be 1350 insects. Round your answer to two decimal places.

69. **Cost-Benefit Model** Environmental scientists often use cost-benefit models to estimate the cost of removing a pollutant from the environment as a function of the percentage of pollutant removed. Suppose a cost-benefit function for the cost C (in millions of dollars) of removing x percent of the pollutants from Maple Lake is given by
$$C(x) = \frac{25x}{100 - x}$$

   (a) If the federal government budgets \$100 million to clean up the lake, what percent of the pollutants can be removed?

   (b) If the federal government budgets \$225 million to clean up the lake, what percent of the pollutants can be removed?

70. **The Learning Curve** Suppose that a student is given 500 vocabulary words to learn. The function
$$P(x) = \frac{0.8x - 0.8}{0.8x + 0.1}$$
models the proportion P of words learned after x hours of studying.

   (a) How long would a student need to study to learn 70% (0.7) of the words?

   (b) How long would a student need to study to learn 400 words?

71. **Runs in Baseball** In his book *Moneyball*, author Michael Lewis cites a formula for predicting the number of runs a team will score in a season. According to the formula, the number of runs R a team will score is given by
$$R = \frac{(h + w)t}{b + w}$$
where h is the number of hits, w is the number of walks, t is the total number of bases, and b is the number of official at-bats. Suppose that the Oakland Athletics scored 750 runs in a season and had 1400 hits, 2250 total bases, and 5500 total at-bats. Use the formula to predict the number of walks that the Athletics had.

△ 72. **Regular Polygons** A regular polygon is a polygon that is both equilateral and equiangular. The measure I of each interior angle of a regular polygon with n sides is $I = \dfrac{180°(n - 2)}{n}$. Find the number of sides of a regular polygon whose interior angles measure 135°.

Extending the Concepts

73. Make up a rational equation that has one real solution.

74. Make up a rational equation that has no real solution.

75. Solve: $\left(\dfrac{4}{x + 3}\right)^2 - 5\left(\dfrac{4}{x + 3}\right) + 6 = 0$

76. Solve: $2 + 11a^{-1} = -12a^{-2}$

Explaining the Concepts

77. Explain the role that domain plays in solving a rational equation.

78. Is the solution set to the equation $\dfrac{x - 6}{x - 6} = 1$ the set of all real numbers? Explain.

Synthesis Review

In Problems 79–83, simplify or solve.

79. $\dfrac{2a^6}{(a^3)^2} - \dfrac{5a^2}{a^3} = \dfrac{3a}{a^3}$

80. $\left(\dfrac{z^{-2}}{2z^{-3}}\right)^{-1} + 3(z - 1)^{-1}$

81. $\dfrac{3}{x - 2} - \dfrac{2x + 1}{x + 1}$

82. $\dfrac{5}{x - 6} + \dfrac{2}{x + 2} = \dfrac{1}{x^2 - 4x - 12}$

83. $\dfrac{x + 1}{2x + 3} - \dfrac{3}{x - 4} = \dfrac{-3}{2x^2 - 5x - 12}$

84. Write a sentence or two explaining the difference between "simplify" and "solve."

Figure 1

Technology Exercises

Technology can be used to approximate solutions. Specifically, with a graphing calculator, the INTERSECT or ZERO (or ROOT) feature can be used. Use the ZERO or ROOT feature of the graphing calculator when one side of the equation is 0; use the INTERSECT feature when neither side of the equation is 0. In Desmos the intersection points and zeros can be found by simply clicking on the curve(s) once graphed to display the "interesting points." Hover your mouse over whatever point you are looking for to display its coordinates. For example, to solve $\dfrac{x-4}{x+1} = 4$, graph $Y_1 = \dfrac{x-4}{x+1}$ and $Y_2 = 4$ in Desmos. The x-coordinates of the point(s) of intersection represent the solution set, as shown in Figure 1.

When approximating solutions to equations, express the solution as a decimal rounded to two decimal places, if necessary. For example, the solution to the equation $\dfrac{x-4}{x+1} = 4$ is $x = -2.67$ rounded to two decimal places.

In Problems 85–90, use technology to solve the equation by graphing each side of the equation and finding the point(s) of intersection.

85. $\dfrac{x-4}{x+4} = 4$ **86.** $\dfrac{x-6}{x+1} = \dfrac{1}{4}$

87. $\dfrac{3}{5} + \dfrac{4}{x+6} = \dfrac{x+12}{5x+30}$ **88.** $\dfrac{4}{3} + \dfrac{7}{x-4} = \dfrac{-7}{3x-12}$

89. $\dfrac{2x^2 + 11x + 12}{x+4} = -5$ **90.** $\dfrac{3x^2 + 10x + 3}{x+3} = -8$

5.5 Rational Inequalities

Objective

1 Solve a Rational Inequality

Are You Prepared for This Section?

Before getting started, complete the following problems. If you get a problem wrong, go back to the section cited and review the material.

P1. Write $-1 < x \le 8$ in interval notation.   [Section 1.4, pp. 82–85]
P2. Solve: $2x + 3 > 4x - 9$ [Section 1.4, pp. 86–88]

In Section 1.4, linear inequalities in one variable, such as $2x - 3 > 4x + 5$, were solved. These inequalities were solved using methods similar to those used in solving linear equations. Set-builder notation and interval notation were used to represent the solution sets for these inequalities.

Although the approach to solving inequalities involving rational expressions is not a simple extension of solving rational equations, the skills developed in solving rational equations are used to solve rational inequalities.

1 Solve a Rational Inequality

A **rational inequality** is an inequality that contains a rational expression. Examples of rational inequalities include

$$\frac{1}{x} > 1 \qquad \frac{x-1}{x+5} \le 0 \qquad \frac{x^2+3x+2}{x-5} > 0 \qquad \frac{3}{x-5} < \frac{4x}{2x-1} + \frac{1}{x}$$

To solve rational inequalities, remember the following:

1. The quotient of two positive numbers is positive; the quotient of a positive and a negative number is negative; and the quotient of two negative numbers is positive.

2. The value of a rational expression may be positive, negative, zero, or undefined, depending on the value of its variable.

The value of a rational expression may change signs on either side of a value of the variable that makes the rational expression equal to 0. For example, if $x = 4$, then the value of the rational expression $\dfrac{x-4}{x+5}$ is $\dfrac{4-4}{4+5} = \dfrac{0}{9} = 0$. If $x = 3$, then the value of

$\dfrac{x-4}{x+5}$ is $\dfrac{3-4}{3+5} = \dfrac{-1}{8} = -\dfrac{1}{8}$, a negative number. If $x = 5$, then the value of $\dfrac{x-4}{x+5}$ is

$\dfrac{5-4}{5+5} = \dfrac{1}{10}$, a positive number.

> **In Other Words**
> When $x = 4$, the expression
> $\dfrac{x-4}{x+5}$ is equal to 0. When $x = 3$,
> the expression is negative, and when
> $x = 5$, the expression is positive.

   The value of a rational expression may also change signs on either side of a value of the variable that makes the rational expression undefined. The rational expression $\dfrac{x-4}{x+5}$ is undefined at $x = -5$. If $x = -6$, then the value of $\dfrac{x-4}{x+5}$ is $\dfrac{-6-4}{-6+5} = \dfrac{-10}{-1} = 10$, and if $x = -4$, then $\dfrac{-4-4}{-4+5} = \dfrac{-8}{1} = -8$.

EXAMPLE 1   How to Solve a Rational Inequality

Solve $\dfrac{x+3}{x-4} \geq 0$. Graph the solution set.

Step-by-Step Solution

Step 1: Write the inequality with 0 on one side of the inequality and with a rational expression, written as a single quotient, on the other side.

The inequality is in the form needed.
$$\dfrac{x+3}{x-4} \geq 0$$

Step 2: Determine the numbers for which the rational expression equals 0 or is undefined.

The rational expression equals 0 when $x = -3$. The rational expression is undefined when $x = 4$.

Step 3: Use the numbers found in Step 2 to separate the real number line into intervals.

Separate the real number line into the following intervals:

Because the rational expression is undefined at $x = 4$, plot an open circle at 4.

Step 4: Choose a test point within each interval formed in Step 3 to determine the sign of the numerator, $x + 3$, and the sign of the denominator, $x - 4$. Then determine the sign of the quotient.

When a test point is chosen in each interval, if one number in an interval satisfies the inequality, then all the numbers in that interval will satisfy the inequality.

- In the interval $(-\infty, -3)$, choose a test point of -4. When $x = -4$, the numerator, $x + 3$, equals -1, and the denominator, $x - 4$, equals -8. Since the quotient of two negatives is positive, the expression $\dfrac{x+3}{x-4} = \dfrac{-1}{-8} = \dfrac{1}{8}$ is positive when $x = -4$. Furthermore, the expression $\dfrac{x+3}{x-4}$ is positive for all x in the interval $(-\infty, -3)$.

- In the interval $(-3, 4)$, let 0 be the test point. When $x = 0$, $x + 3$ is positive, and $x - 4$ is negative. A positive divided by a negative is negative, so $\dfrac{x+3}{x-4} = \dfrac{0+3}{0-4} = \dfrac{3}{-4} = -\dfrac{3}{4}$ is negative when $x = 0$. The expression $\dfrac{x+3}{x-4}$ is negative for all x in the interval $(-3, 4)$.

- In the interval $(4, \infty)$, let 5 be the test point. When $x = 5$, both $x + 3$ and $x - 4$ are positive. A positive divided by a positive is positive, so $\dfrac{x+3}{x-4} = \dfrac{5+3}{5-4} = \dfrac{8}{1} = 8$ is positive. Furthermore, $\dfrac{x+3}{x-4}$ is positive for all x in $(4, \infty)$.

Table 1 on the next page shows these results and the sign of $\dfrac{x+3}{x-4}$ in each interval.

Table 1

Interval					
	$(-\infty, -3)$		$(-3, 4)$		$(4, \infty)$
	$x < -3$	$x = -3$	$-3 < x < 4$	$x = 4$	$x > 4$
Test Point	-4	-3	0	4	5
Sign of $x + 3$	Negative	0	Positive	Positive	Positive
Sign of $x - 4$	Negative	Negative	Negative	0	Positive
Sign of $\dfrac{x + 3}{x - 4}$	Positive	0	Negative	Undefined	Positve
Conclusion about $\dfrac{x + 3}{x - 4}$	$\dfrac{x + 3}{x - 4}$ is positive, so $(-\infty, -3)$ is part of the solution set.	Because the inequality is nonstrict, -3 is part of the solution set.	$\dfrac{x + 3}{x - 4}$ is negative, so $(-3, 4)$ is not part of the solution set.	4 cannot be part of the solution set because it causes division by 0.	$\dfrac{x + 3}{x - 4}$ is positive, so $(4, \infty)$ is part of the solution set.

Find where $\dfrac{x + 3}{x - 4}$ is greater than or equal to zero. The solution set is $\{x \mid x \leq -3 \text{ or } x > 4\}$ using set-builder notation, and $(-\infty, -3] \cup (4, \infty)$ using interval notation. Notice that -3 is part of the solution set because $\dfrac{-3 + 3}{-3 - 4} = \dfrac{0}{-7} = 0$, but 4 is not part of the solution set because it is not in the domain of $\dfrac{x + 3}{x - 4}$. Figure 2 shows the graph of the solution set.

Figure 2

Solving Rational Inequalities

Step 1: Write the inequality so that a rational expression is on one side of the inequality and 0 is on the other. Be sure to write the rational expression as a single quotient in factored form.

Step 2: Determine the values for which the rational expression equals 0 or is undefined.

Step 3: Use the values found in Step 2 to separate the real number line into intervals.

Step 4: Choose a test point within each interval formed in Step 3 to determine the sign of each factor in the numerator and denominator. Then determine the sign of the quotient.

- If the quotient is positive, then the rational expression is positive for all values x in the interval.
- If the quotient is negative, then the rational expression is negative for all values x in the interval.

Also determine the value of the rational expression at each value found in Step 2. If the inequality is not strict ($\leq$ or $\geq$), include the values of the variable for which the rational expression equals 0 in the solution set, but do not include the values for which the rational expression is undefined!

Quick ✓

1. The inequality $\dfrac{2x - 3}{x + 6} > 1$ is an example of a(n) _____ inequality.

2. Solve $\dfrac{x - 7}{x + 3} \geq 0$. Graph the solution set.

3. Solve $\dfrac{1 - x}{x + 5} > 0$. Graph the solution set.

⊚ **EXAMPLE 2** **Solving a Rational Inequality**

Solve $\dfrac{x+3}{x-1} > 2$. Graph the solution set.

Solution

First, write the inequality so that a rational expression is on one side of the inequality and 0 is on the other.

$$\frac{x+3}{x-1} > 2$$

Subtract 2 from both sides: $\qquad \dfrac{x+3}{x-1} - 2 > 0$

LCD $= x - 1$; multiply -2 by $\dfrac{x-1}{x-1}$: $\qquad \dfrac{x+3}{x-1} - 2 \cdot \dfrac{x-1}{x-1} > 0$

Write the rational expression with the LCD: $\qquad \dfrac{x+3-2(x-1)}{x-1} > 0$

Distribute -2: $\qquad \dfrac{x+3-2x+2}{x-1} > 0$

Combine like terms in the numerator: $\qquad \dfrac{-x+5}{x-1} > 0$

The rational expression equals 0 when $x = 5$. The rational expression is undefined when $x = 1$. Separate the real number line into the intervals as shown in Figure 3.

Figure 3

Table 2 shows the sign of $-x + 5$, $x - 1$, and $\dfrac{-x+5}{x-1}$ in each interval. It also shows the value of $\dfrac{-x+5}{x-1}$ at $x = 1$ and $x = 5$.

Table 2

	$(-\infty, 1)$		$(1, 5)$		$(5, \infty)$
Interval		1		5	
Test Point	0	1	3	5	6
Sign of $-x + 5$	Positive	Positive	Positive	0	Negative
Sign of $x - 1$	Negative	0	Positive	Positive	Positive
Sign of $\dfrac{-x+5}{x-1}$	Negative	Undefined	Positive	0	Negative
Conclusion about $\dfrac{-x+5}{x-1}$	$\dfrac{-x+5}{x-1}$ is negative, so $(-\infty, 1)$ is not part of the solution set.	Because $\dfrac{-x+5}{x-1}$ is undefined at $x = 1$, $x = 1$ is not part of the solution set.	$\dfrac{-x+5}{x-1}$ is positive, so $(1, 5)$ is part of the solution set.	Because the inequality is strict, 5 is not part of the solution set.	$\dfrac{-x+5}{x-1}$ is negative, so $(5, \infty)$ is not part of the solution set.

Figure 4

Find the values of x such that $\dfrac{x + 3}{x - 1} > 2$. This is equivalent to determining where $\dfrac{-x + 5}{x - 1} > 0$. Thus the solution set is $\{x \mid 1 < x < 5\}$, or $(1, 5)$ using interval notation. The endpoints of the interval are not in the solution set because the inequality in the original problem is strict. Figure 4 shows the graph of the solution set. ●

Quick ✓

4. Solve $\dfrac{4x + 5}{x + 2} < 3$. Graph the solution set.

5.5 Exercises MyMathLab® Exercise numbers in green have complete video solutions in MyMathLab or may be accessed using the QR code to the right.

Problems 1–4 are the Quick ✓s that follow the EXAMPLES.

Building Skills

In Problems 5–20, solve each rational inequality. Graph the solution set. See Objective 1.

5. $\dfrac{x - 4}{x + 1} > 0$

6. $\dfrac{x + 5}{x - 2} > 0$

7. $\dfrac{x + 9}{x - 3} < 0$

8. $\dfrac{x + 8}{x + 2} < 0$

9. $\dfrac{x + 10}{x - 4} \geq 0$

10. $\dfrac{x + 12}{x - 2} \geq 0$

11. $\dfrac{(3x + 5)(x + 8)}{x - 2} \leq 0$

12. $\dfrac{(3x - 2)(x - 6)}{x + 1} \geq 0$

13. $\dfrac{x - 5}{x + 1} < 1$

14. $\dfrac{x + 3}{x - 4} > 1$

15. $\dfrac{2x - 9}{x - 3} > 4$

16. $\dfrac{3x + 20}{x + 6} < 5$

17. $\dfrac{3}{x - 4} + \dfrac{1}{x} \geq 0$

18. $\dfrac{2}{x + 3} + \dfrac{2}{x} \leq 0$

19. $\dfrac{3}{x - 2} \leq \dfrac{4}{x + 5}$

20. $\dfrac{1}{x - 4} \geq \dfrac{3}{2x + 1}$

Mixed Practice

In Problems 21–30, solve each inequality. Graph the solution set.

21. $\dfrac{(2x - 1)(x + 3)}{x - 5} > 0$ **22.** $\dfrac{(5x - 2)(x + 4)}{x - 5} < 0$

23. $3 - 4(x + 1) < 11$ **24.** $2x + 3(x - 2) \geq x + 2$

25. $\dfrac{x + 7}{x - 8} \leq 0$ **26.** $\dfrac{x - 10}{x + 5} \leq 0$

27. $(x - 2)(2x + 1) \geq 2(x - 1)^2$

28. $(x + 2)^2 < 3x^2 - 2(x + 1)(x - 2)$

29. $\dfrac{3x - 1}{x + 4} \geq 2$

30. $\dfrac{3x - 7}{x + 2} \leq 2$

In Problems 31–34, for each function, find the values of x that satisfy the given condition.

31. Solve $R(x) \leq 0$ if $R(x) = \dfrac{x - 6}{x + 1}$.

32. Solve $R(x) \geq 0$ if $R(x) = \dfrac{x + 3}{x - 8}$.

33. Solve $R(x) < 0$ if $R(x) = \dfrac{2x - 5}{x + 2}$.

34. Solve $R(x) < 0$ if $R(x) = \dfrac{3x + 2}{x - 4}$.

Applying the Concepts

35. Average Cost Suppose that the daily cost C of manufacturing x bicycles is given by $C(x) = 80x + 5000$. Then the average daily cost $\overline{C}$ is given by $\overline{C}(x) = \dfrac{80x + 5000}{x}$. How many bicycles must be produced each day in order for the average cost to be no more than $130?

36. Average Cost See Problem 35. Suppose that the government imposes a $10 tax on each bicycle manufactured so that the daily cost C of manufacturing x bicycles is now given by $C(x) = 90x + 5000$. Now the average daily cost $\overline{C}$ is given by $\overline{C}(x) = \dfrac{90x + 5000}{x}$. How many bicycles must be produced each day in order for the average cost to be no more than $130?

Extending the Concepts

37. Write a rational inequality that has $(2, \infty)$ as the solution set.

38. Write a rational inequality that has $(-2, 5]$ as the solution set.

Explaining the Concepts

39. In solving the rational inequality $\dfrac{x - 4}{x + 1} \leq 0$, a student determines that the only interval that makes the inequality true is $(-1, 4)$. He states that the solution set is $\{x \mid -1 \leq x \leq 4\}$. What is wrong with this solution?

40. In Step 2 of the steps for solving a rational inequality, the values for which the rational expression equals 0 or is undefined are determined. Then these values are used to form intervals on the real number line. Explain why this guarantees that there is not a change in the sign of the rational expression within any given interval.

Synthesis Review

In Problems 41–46, find the x-intercept(s) of the graph of each function.

41. $F(x) = 6x - 12$

42. $G(x) = 5x + 30$

43. $f(x) = 2x^2 + 3x - 14$

44. $h(x) = -3x^2 - 7x + 20$

45. $R(x) = \dfrac{3x - 2}{x + 4}$

46. $R(x) = \dfrac{x^2 + 5x + 6}{x + 2}$

Technology Exercises

Solutions to rational inequalities can be approximated using the INTERSECT or ZERO (or ROOT) feature of the graphing calculator. Use the ZERO or ROOT feature of the graphing calculator when one side of the inequality is 0. Use the INTERSECT feature when neither side of the inequality is 0. For example, to solve $\dfrac{x - 4}{x + 1} > \dfrac{7}{2}$, graph $Y_1 = \dfrac{x - 4}{x + 1}$ and $Y_2 = \dfrac{7}{2}$. To determine the x-values such that the graph of Y_1 is above that of Y_2, find x-coordinates of the point(s) of intersection. The graph of Y_1 is above that of Y_2 between $x = -3$ and $x = -1$. See Figure 5.

Figure 5

In using a graphing calculator to approximate solutions to inequalities, the solution is typically expressed as a decimal rounded to two decimal places, if exact answers cannot be found. The solution to the inequality $\dfrac{x - 4}{x + 1} > \dfrac{7}{2}$ is $\{x \mid -3 < x < -1\}$, or $(-3, -1)$ using interval notation.

In Problems 47–50, solve each inequality using technology.

47. $\dfrac{x - 5}{x + 1} \leq 3$

48. $\dfrac{x + 2}{x - 5} > -2$

49. $\dfrac{2x + 5}{x - 7} > 3$

50. $\dfrac{2x - 1}{x + 5} \leq 4$

5.6 Models Involving Rational Expressions

Objectives

❶ Solve for a Variable in a Rational Expression

❷ Model and Solve Ratio and Proportion Problems

❸ Model and Solve Work Problems

❹ Model and Solve Uniform Motion Problems

Are You Prepared for This Section?

Before getting started, complete the following problem. If you get the problem wrong, go back to the section cited and review the material.

P1. Solve for $y : 4x - 2y = 10$ [Section 1.3, pp. 74–77]

▶ ❶ **Solve for a Variable in a Rational Expression**

To "solve for the variable" means to isolate the variable on one side of the equation, with all other variables and constants, if any, on the other side. To solve formulas for a certain variable, use the same steps that are used to solve rational equations.

(**EXAMPLE 1**)   **Solving for a Variable in a Lens Construction Formula**

The formula $\dfrac{1}{f} = \dfrac{1}{p} + \dfrac{1}{q}$ is used in telescope and camera construction, where f is the focal length of the lens. In general, the larger the focal length of the lens, f, the more power the telescope has. The variable p is the distance between the object we wish to see and the lens; the variable q is the distance from the lens to the point of focus. See Figure 6.

Figure 6

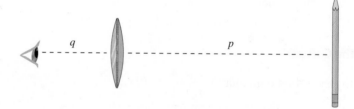

(a) Solve the formula for q.

(b) Suppose a camera has a focal length of 100 mm and the camera is focusing on an object 5000 mm away. What is q, the distance from the lens to the point of focus? Round your answer to the nearest millimeter.

Solution

(a) The goal is to isolate q. Follow the same steps that are used to solve a rational equation. First, note that none of the variables can equal 0. Multiply both sides of the equation by pqf, the LCD of all the denominators.

$$\frac{1}{f} = \frac{1}{p} + \frac{1}{q}$$

$$pqf \cdot \frac{1}{f} = pqf \cdot \left(\frac{1}{p} + \frac{1}{q} \right)$$

Distribute pqf: $pqf \cdot \dfrac{1}{f} = pqf \cdot \dfrac{1}{p} + pqf \cdot \dfrac{1}{q}$

Simplify: $pq = qf + pf$

To get all terms involving q on one side, subtract qf from both sides: $pq - qf = qf + pf - qf$

$$pq - qf = pf$$

Factor out q: $q(p - f) = pf$

Divide both sides by $p - f$: $q = \dfrac{pf}{p - f}$

(b) Substitute $f = 100$ mm and $p = 5000$ mm in $q = \dfrac{pf}{p - f}$.

$$q = \frac{5000 \cdot 100}{5000 - 100}$$

$$\approx 102 \text{ mm}$$

The distance from the lens to the point of focus is approximately 102 mm.

Quick ✓

1. The formula $Y = \dfrac{G}{1 - b}$ is used in economics to determine the impact on gross domestic product (GDP) Y of increasing government spending by G dollars if the proportion of additional income that people spend is b.

(a) Solve the formula for b.

(b) Find b if the government increased spending by \$100 billion and GDP increased by \$1000 billion.

▶ ❷ Model and Solve Ratio and Proportion Problems

The problems in this objective focus on the idea of ratio and proportion. The **ratio** of two numbers a and b can be written as

$$a{:}b \quad \text{or} \quad \frac{a}{b}$$

When solving algebraic problems, write ratios as $\dfrac{a}{b}$. A **proportion** is a statement that two ratios are equal. That is, proportions are equations of the form $\dfrac{a}{b} = \dfrac{c}{d}$. You may be familiar with using proportions to solve problems involving similar figures, such as triangles, from geometry. Two figures are **similar** if their angles have the same measure and their corresponding sides are proportional. Figure 7 shows examples of similar figures.

In Figure 7(a), $\triangle ABC$ is similar to $\triangle DEF$, and in Figure 7(b), quadrilateral $ABCD$ is similar to quadrilateral $EFGH$. Because $\triangle ABC$ is similar to $\triangle DEF$, we know that the ratio of AB to AC equals the ratio of DE to DF. That is,

$$\frac{AB}{AC} = \frac{DE}{DF}$$

Use the principle that the ratios of corresponding sides are equal to find unknown lengths in similar figures.

Figure 7

EXAMPLE 2 **Similar Figures**

Suppose you want to know the height of a building. A 20-foot-tall light post casts a shadow that is 12 feet long. The shadow cast by the building is 660 feet long. Find the height of the building.

Solution
The ratio of the length of the building's shadow to its height equals the ratio of the length of the light post's shadow to its height because the building and its shadow form a triangle that is similar to the triangle formed by the light post and its shadow. See Figure 8 on the next page.

Figure 8

Work Smart

When setting up the proportion, there are multiple ways to do it. For example, we could have set it up this way.

$$\underset{\underset{\underset{\text{shadow}}{\uparrow}}{\text{post}}}{\overset{\text{building}}{\frac{660}{12}}} = \underset{\underset{\text{object height}}{\uparrow}}{\frac{h}{20}}$$

Set up the proportion problem as

$$\frac{\text{shadow length}}{\text{object height}} : \overset{\overset{\text{building}}{\downarrow}}{\frac{660}{h}} = \overset{\overset{\text{post}}{\downarrow}}{\frac{12}{20}}$$

and solve the equation. The domain of h is $\{h \,|\, h > 0\}$ since h represents the height of the building. The LCD of all denominators is $20h$, so multiply both sides of the equation by the LCD, $20h$.

$$20h \cdot \frac{660}{h} = 20h \cdot \frac{12}{20}$$

$$13{,}200 = 12h$$

Divide both sides by 12: $1100 = h$

The height of the building is 1100 feet.

Figure 9

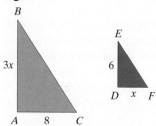

Quick ✓

2. A(n) _____ is a statement that two ratios are equal. Two figures are _____ if their angles have the same measure and their corresponding sides are proportional.

3. Suppose a man who is 6 feet tall casts a shadow that is 2 feet long. At the same time, a building casts a shadow that is 40 feet long. How tall is the building?

4. Suppose that $\triangle ABC$ is similar to $\triangle DEF$, as shown in Figure 9. Find the length of AB and DF.

Proportions come up in direct translation problems as well.

EXAMPLE 3   **A Proportion Problem**

According to the National Vital Statistics Report dated January 15, 2015, the birth rate for unmarried women (15–44 years of age) was 44.3 live births per 1000 population in the United States in 2013. In 2013, there were 1,595,873 births to unmarried women. Determine the population of unmarried women between 15 and 44 years of age in 2013.

Solution

Step 1: Identify This is a direct translation problem involving proportions. Find the population of unmarried women in 2013.

Step 2: Name We let p represent the population of unmarried women in 2013.

Step 3: Translate Since it is known that the rate of live births was 44.3 per 1000 and that there were 1,595,873 births, we have the model

$$\frac{\text{number of births}}{\text{population}} : \quad \frac{44.3}{1000} = \frac{1{,}595{,}873}{p}$$

Step 4: Solve Now proceed to solve the equation.

$$\frac{44.3}{1000} = \frac{1{,}595{,}873}{p}$$

Multiply both sides by the LCD, 1000p: $\quad 1000p \cdot \dfrac{44.3}{1000} = 1000p \cdot \dfrac{1{,}595{,}873}{p}$

Divide out common factors: $\quad 44.3p = 1{,}595{,}873{,}000$

Divide both sides by 44.3: $\quad p \approx 36{,}024{,}221$

Step 5: Check It is always a good idea to make sure your answer is reasonable. According to the U.S. Census Bureau, there were over 62 million women 15–44 years of age in the United States in 2013, so the answer seems reasonable.

Step 6: Answer The population of unmarried women 15–44 years of age in the United States in 2013 was approximately 36,024,221.

Quick ✓

5. According to a Gallup poll, approximately 114 of every 1000 Americans had no health insurance in 2015. There were approximately 36,651,000 uninsured Americans in 2015. Find the population of the United States in 2015.

▶ ❸ Model and Solve Work Problems

Work or "constant-rate jobs" problems assume that jobs, even those performed by people, are performed at a **constant rate**. Although this assumption is reasonable for machines, it is not likely to be true for people simply because of the old phrase "too many chefs spoil the broth." Think of it this way—if you continually add more people to paint a room, the time to complete the job may decrease initially, but eventually the painters get in each other's way and the time to completion actually increases. Models could be built that take this into account, but instead the "constant-rate" assumption will be made for humans as well, in order to keep the mathematics manageable.

The constant-rate assumption states that if it takes t units of time to complete a job, then $\dfrac{1}{t}$ of the job is done in 1 unit of time. For example, if it takes 5 hours to paint a room, then $\dfrac{1}{5}$ of the room should be painted in 1 hour.

Work Smart

Remember, when a model is used, it is common to make simplifying assumptions to make the math easier to deal with.

EXAMPLE 4 **Working Together on a Job**

It's Saturday and Kevin needs to cut and edge the grass. At 9 A.M., Michael asks Kevin to go golfing at 11:00 A.M. Typically, it takes Kevin 3 hours to cut and edge the grass. When Michael cuts and edges the grass, it takes 4 hours. If they work together, will they be able to finish the lawn and still make the golf date?

Solution

Step 1: Identify Find how long it will take Michael and Kevin to finish the lawn, so that it can be determined whether they can make the golf date 2 hours after they start the lawn work.

Step 2: Name Let t represent the time (in hours) that it takes to finish the lawn working together. Then, in 1 hour they will complete $\dfrac{1}{t}$ of the job.

Step 3: Translate Since Kevin can finish the job in 3 hours, Kevin will finish $\frac{1}{3}$ of the job in 1 hour. Michael can finish the job in 4 hours, so Michael will finish $\frac{1}{4}$ of the job in 1 hour. Set up the model using the following logic:

$$\left(\begin{array}{c}\text{Part of job done by Kevin}\\\text{in 1 hour}\end{array}\right) + \left(\begin{array}{c}\text{Part of job done by Michael}\\\text{in 1 hour}\end{array}\right) = \left(\begin{array}{c}\text{Part of job done together}\\\text{in 1 hour}\end{array}\right)$$

$$\frac{1}{3} \qquad + \qquad \frac{1}{4} \qquad = \qquad \frac{1}{t} \quad \text{The Model}$$

Step 4: Solve

$$\frac{1}{3} + \frac{1}{4} = \frac{1}{t}$$

Multiply both sides by the LCD, $12t$: $\quad 12t \cdot \left(\frac{1}{3} + \frac{1}{4}\right) = 12t \cdot \frac{1}{t}$

Distribute: $\quad 12t \cdot \frac{1}{3} + 12t \cdot \frac{1}{4} = 12t \cdot \frac{1}{t}$

Divide common factors: $\quad 4t + 3t = 12$

Combine like terms: $\quad 7t = 12$

Divide both sides by 7: $\quad t = \frac{12}{7} \approx 1.714$

Work Smart

Convert 0.714 hour to minutes by multiplying 0.714 by 60 minutes to obtain approximately 43 minutes.

Step 5: Check Is the answer reasonable? It is expected that the time is greater than 0 but less than 3 (because it takes Kevin 3 hours working by himself). The answer of 1.714 hours, or approximately 1 hour, 43 minutes, seems reasonable.

Step 6: Answer If they start right away, they should finish at 10:43 A.M. If they can get to the course in 17 minutes, they can make the tee time.

Quick ✓

6. Juan and Maria have a pool in their backyard. They can fill their pool in 30 hours using their hose alone. Their neighbor's pool is the same size and can be filled in 24 hours. Suppose their neighbor lets them borrow his hose. How long will it take Juan and Maria to fill their pool with both hoses?

EXAMPLE 5 **The Kitchen Sink**

Suppose the kitchen sink can be filled in 5 minutes. If the sink is full, it takes 8 minutes to drain the sink when the drain is left partially open. If the sink's drain is accidentally left partially open, how long will it take to fill the sink?

Solution

Step 1: Identify Find how long it will take to fill the sink with the drain partially open.

Step 2: Name Let t represent the time (in minutes) that it takes to fill the sink. Then, in 1 minute $\frac{1}{t}$ of the sink will be full.

Step 3: Translate It is known that the sink can be filled in 5 minutes when the drain is closed, so after 1 minute, $\frac{1}{5}$ of the sink is full. It takes 8 minutes to drain, so after 1 minute, $\frac{1}{8}$ of the sink is drained. Set up the model using the following logic:

Work Smart

Note that the portion of sink drained after 1 minute is subtracted, because it is "working against us."

$$\begin{pmatrix} \text{Portion of sink filled} \\ \text{in 1 minute with} \\ \text{closed drain} \end{pmatrix} - \begin{pmatrix} \text{Portion of sink drained} \\ \text{after 1 minute} \end{pmatrix} = \begin{pmatrix} \text{Portion of sink filled} \\ \text{after 1 minute with} \\ \text{open drain} \end{pmatrix}$$

$$\frac{1}{5} \qquad - \qquad \frac{1}{8} \qquad = \qquad \frac{1}{t} \quad \text{The Model}$$

Step 4: Solve

$$\frac{1}{5} - \frac{1}{8} = \frac{1}{t}$$

Multiply both sides by the LCD: 40t: $40t \cdot \left(\frac{1}{5} - \frac{1}{8} \right) = 40t \cdot \frac{1}{t}$

Distribute: $40t \cdot \frac{1}{5} - 40t \cdot \frac{1}{8} = 40t \cdot \frac{1}{t}$

Simplify: $8t - 5t = 40$

Combine like terms: $3t = 40$

Divide both sides by 3: $t = \frac{40}{3} \approx 13.3 \text{ minutes}$

Step 5: Check Is this answer reasonable? It is expected to be greater than 5 since this is the time it takes to fill the sink with the drain closed. The answer of 13.3 minutes, or about 13 minutes, 20 seconds, seems reasonable.

Step 6: Answer It will take 13.3 minutes or 13 minutes, 20 seconds to fill the sink.   ●

Quick ✓

7. A children's inflatable pool takes 20 minutes to fill with an electric air pump. It takes 50 minutes to let the air out of the pool. If the pool's valve is accidentally left open, how long will it take to fill the pool?

▶ ❹ Model and Solve Uniform Motion Problems

Uniform motion problems were introduced in Section 1.2. Recall that uniform motion problems use the fact that distance equals rate times time—that is, $d = rt$. In this section, an alternative form of this model, $t = \frac{d}{r}$, will be used.

EXAMPLE 6   A Round-Trip Flight

A plane flies 990 miles west (into the wind) and makes the return trip following the same flight path. The effect of the wind on the plane is 20 miles per hour. The round trip takes 10 hours. What is the speed of the plane in still air?

Solution

Step 1: Identify In this uniform motion problem, find the speed of the plane in still air.

Step 2: Name Let r represent the speed of the plane, in miles per hour, in still air.

Step 3: Translate Going west, the plane is flying into the wind. The speed of the plane is its rate in still air less the impact of the wind, or $r - 20$. Similar logic tells us that the speed of the plane going east is $r + 20$. We set up Table 3. Remember, since $d = rt$, then $t = \dfrac{d}{r}$ or $\dfrac{d}{r} = t$.

Table 3			
	Distance (miles)	**Rate (miles per hour)**	**Time (hours)**
West	990	$r - 20$	$\dfrac{990}{r - 20}$
East	990	$r + 20$	$\dfrac{990}{r + 20}$

The round trip takes 10 hours, so set up the equation:

$$\text{Time going west} + \text{Time going east} = 10$$

$$\frac{990}{r - 20} + \frac{990}{r + 20} = 10 \quad \text{The Model}$$

Step 4: Solve Solve for r:

$$\frac{990}{r - 20} + \frac{990}{r + 20} = 10$$

Multiply both sides by the LCD, $(r - 20)(r + 20)$: $(r - 20)(r + 20)\left(\dfrac{990}{r - 20} + \dfrac{990}{r + 20}\right) = 10(r - 20)(r + 20)$

Distribute: $(r - 20)(r + 20)\dfrac{990}{r - 20} + (r - 20)(r + 20)\dfrac{990}{r + 20} = 10(r - 20)(r + 20)$

Divide out common factors: $990(r + 20) + 990(r - 20) = 10(r - 20)(r + 20)$

Distribute: $990r + 19{,}800 + 990r - 19{,}800 = 10(r^2 - 400)$

Combine like terms: $1980r = 10(r^2 - 400)$

Divide both sides by 10: $198r = r^2 - 400$

Write the equation in standard form: $0 = r^2 - 198r - 400$

Factor: $0 = (r - 200)(r + 2)$

Use the Zero-Product Property: $r = 200$ or $r = -2$

Step 5: Check The answer -2 is rejected because the rate of the plane must be positive. Thus it appears that the plane will travel at a rate of 200 miles per hour in still air. Flying west, the plane travels at 180 miles per hour, so the trip takes $\dfrac{990 \text{ miles}}{180 \text{ miles per hour}} = 5.5$ hours. Flying east, the plane travels at 220 miles per hour, so the trip takes $\dfrac{990 \text{ miles}}{220 \text{ miles per hour}} = 4.5$ hours. The total trip takes $5.5 + 4.5 = 10$ hours. It checks!

Step 6: Answer The plane travels 200 miles per hour in still air. ●

Quick ✓

8. A canoe travels on a river whose current is running at 2 miles per hour. After traveling 12 miles upstream, the canoe turns around and makes the 12-mile trip back downstream. The trip up and back takes 8 hours. What is the speed of the canoe in still water?

5.6 Exercises MyMathLab® Exercise numbers in green have complete video solutions in MyMathLab or may be accessed using the QR code to the right.

Problems **1–8** *are the Quick* ✓ *s that follow the* **EXAMPLES.**

Building Skills

In Problems 9–18, solve each formula for the indicated variable. See Objective 1.

9. **Chemistry (Gas Laws)** Solve $\dfrac{V_1}{V_2} = \dfrac{P_2}{P_1}$ for P_1.

10. **Chemistry (Gas Laws)** Solve $\dfrac{V_1}{V_2} = \dfrac{P_2}{P_1}$ for V_2.

11. **Finance** Solve $R = \dfrac{r}{1-t}$ for t.

12. **Finance** Solve $P = \dfrac{A}{1+r}$ for r.

13. **Slope** Solve $m = \dfrac{y-y_1}{x-x_1}$ for x.

14. **Slope** Solve $m = \dfrac{y-y_1}{x-x_1}$ for x_1.

15. **Physics** Solve $\omega = \dfrac{rmv}{I+mr^2}$ for v.

16. **Physics** Solve $\omega = \dfrac{rmv}{I+mr^2}$ for I.

17. **Physics** Solve $V = \dfrac{mv}{M+m}$ for m.

18. **Physics** Solve $v_2 = \dfrac{2m_1}{m_1+m_2}\, v_1$ for m_1.

Applying the Concepts

In Problems 19–26, solve the proportion problem. See Objective 2.

△ 19. Suppose that $\triangle ABC$ is similar to $\triangle DEF$, as shown in the figure. Find the length of AB and DF.

△ 20. Suppose that $\triangle ABC$ is similar to $\triangle DEF$, as shown in the figure. Find the length of AB and DF.

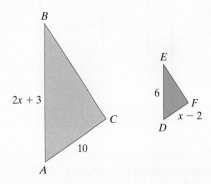

21. **Motor Vehicle Death Rates** According to the Centers for Disease Control, in 2013 the death rate in the United States as a result of a motor vehicle accident was 10.7 per 100,000 population. In 2013, there were 33,866 fatalities in motor vehicle accidents. What was the population of the United States in 2013?

22. **Flight Accidents** According to the *National Transportation Safety Board*, in 2013, with regard to general aviation accidents, there were 5.85 airplane accidents per 100,000 flight hours. Also in 2013, there were a total of 1222 accidents. How many flight hours were flown in 2013?

23. **Road Trip** At current prices, Roberta can drive her car 8.2 miles per dollar of gasoline that she buys. Roberta and three of her friends decide to go on a road trip to a neighboring university that is 105 miles away and agree to split the cost evenly. To the nearest cent, how much money will each have to contribute to get to the university and back?

24. **Car Payments** At current rates, a 60-month term car loan is being offered where the monthly payments are $0.0191 per dollar borrowed. Suppose that Eduardo's car payment is $340 per month. To the nearest dollar, how much did Eduardo borrow?

25. **Pascal's Principle** Pascal's Principle applied to a hydraulic lever states that the ratio of the force exerted on an input piston F_1 to the area displaced A_1 equals the ratio of the force on the output piston F_2 to the area displaced A_2. See the figure on the next page. If a force of 30 pounds is exerted with an area of 12 square feet of water displaced in the right pipe and an area

(continued)

of 5 square feet is displaced in the left pipe, determine the force exerted by the left pipe.

26. **Pascal's Principle** See Problem 25. If a force of 40 pounds is exerted with an area of 15 square feet of water displaced in the right pipe and an area of 8 square feet is displaced in the left pipe, determine the force exerted by the left pipe.

In Problems 27–32, solve the work problem. See Objective 3.

27. **Sharing a Paper Route** Amiri can deliver his newspapers in 80 minutes. It takes Horus 60 minutes to do the same route. How long would it take them to deliver the newspapers if they worked together?

28. **Painting a Room** Latoya can paint five 10-foot-by-14-foot rooms by herself in 14 hours. Lisa can paint five 10-foot-by-14-foot rooms by herself in 10 hours. Working together, how long would it take to paint five 10-foot-by-14-foot rooms?

29. **Cutting the Grass** Avery can cut the grass working by himself in 3 hours. When Avery cuts the grass with his younger brother Connor, it takes 2 hours. How long would it take Connor to cut the grass if he worked by himself?

30. **Assembling a Swing Set** Alexandra and Frank can assemble a King Kong swing set working together in 6 hours. One day, when Frank called in sick, Alexandra was able to assemble a King Kong swing set in 10 hours. How long would it take Frank to assemble a King Kong swing set if he worked by himself?

31. **Emptying a Pool** A swimming pool can be emptied in 6 hours using a 10-horsepower pump along with a 6-horsepower pump. The 6-horsepower pump requires 5 hours more than the 10-horsepower pump to empty the pool when working by itself. How long would it take to empty the pool using just the 10-horsepower pump?

32. **Draining a Pond** A pond can be emptied in $3.75 \left(= \dfrac{15}{4} \right)$ hours using a 10-horsepower pump along with a 4-horsepower pump. The 4-horsepower pump requires 4 hours more than the 10-horsepower pump to empty the pond when working by itself. How long would it take to empty the pond using just the 10-horsepower pump?

In Problems 33–42, solve the uniform motion problem. See Objective 4.

33. **Tough Commute** You have a 20-mile commute into work. Since you leave very early, the trip going to work is easier than the trip home. You can travel to work in the same time that it takes for you to make it 16 miles on the trip back home. Your average speed coming home is 7 miles per hour slower than your average speed going to work. What is your average speed going to work?

34. **Riding Your Bicycle** Every weekend, you ride your bicycle on a forest preserve path. The path is 20 miles long and ends at a waterfall, at which point you relax and then make the trip back to the starting point. One weekend, you find that in the same time it takes you to travel to the waterfall, you are only able to return 12 miles. Your average speed going to the waterfall is 4 miles per hour faster than your average speed on the return trip. What was your average speed going to the waterfall?

35. **Moving Walkway** In order to access the outer part of Terminal 1 at O'Hare International Airport, you must walk quite some distance in a tunnel that travels under part of the airport. To make the walk less difficult, there is a moving walkway that travels at 2 feet per second. Suppose that Hana can travel 152 feet while walking on the walkway in the same amount of time it takes her to travel 72 feet while walking on the pavement without the aid of the moving sidewalk. How fast does Hana walk?

36. **Escalator** When exiting Terminal 1 at O'Hare International Airport, you can either take an escalator up to the main level or take traditional stairs. Suppose that the escalator travels 1.5 feet per second. Karli can walk up the 50-foot escalator in the same amount of time it takes her to walk 30 feet up the stairs. How fast does Karli walk upstairs?

37. Football Rob Gronkowski of the New England Patriots can run 100 yards in 12 seconds, and Matt Forte of the New York Jets can run 100 yards in 9 seconds. Suppose that Gronkowski catches a pass at his own 20-yard line in stride and starts running away from Forte, who is at the 15-yard line directly behind Gronkowski. See the figure. At what yard line will Forte catch up to Gronkowski?

20 yd 15 yd

38. Running a Race Roger can run one mile in 8 minutes. Jeff can run one mile in 6 minutes. If Jeff gives Roger a 1-minute head start, how long will it take before Jeff catches up to Roger? How far will each have run?

39. Uphill/Downhill A bicyclist rides his bicycle 12 miles up a hill and then comes back down. His speed coming downhill is 8 miles per hour faster than his speed going uphill. The roundtrip takes 2 hours and 15 minutes $\left(= \frac{9}{4} \text{hours} \right)$. What was the speed of the bicyclist going uphill?

40. Round Trip A plane flies 600 miles west (into the wind) and makes the return trip following the same flight path. The effect of the jet stream on the plane is 15 miles per hour. The round trip takes 9 hours. What is the speed of the plane in still air?

41. Scenic Drive Joe and Nancy live in Morro Bay, California, right off of Highway 1. They decide to take a trip north to Monterey. The first 50 miles of the drive is pretty easy, while the last 68 miles of the drive is filled with curves. They drive at an average of 9 miles per hour faster for the first 50 miles of the trip. The entire trip takes 3 hours. How fast did Joe and Nancy drive for the first 50 miles of the trip?

42. A Race Dirk and Garret decide to have a 40-mile bicycle race. During the race, Dirk averages 2 miles per hour faster than Garret and beats Garret by $\frac{2}{3}$ of an hour. What was Dirk's average speed?

Extending the Concepts

43. The Olympics The current world record holder in the 100-meter dash is Usain Bolt, with a time of 9.58 seconds. In 1999, Maurice Green set the world record in the 100-meter dash with a time of 9.79 seconds. If these two athletes ran in the same race repeating their respective times, by how many meters would Bolt beat Greene?

Explaining the Concepts

44. When solving work problems, we assume that each individual works at a constant rate and that there is no gain or loss of efficiency when additional individuals are added to the job. Explain what "no gain or loss of efficiency" means. Do you think this assumption is reasonable? If not, then why do we make the assumption?

Synthesis Review

In Problems 45–50, use the Laws of Exponents to simplify each expression.

45. $\left(a^3 \right)^5$

46. $a \left(a^2 b^{-3} \right)^3$

47. $\left(ab \right)^{-2} \left(\dfrac{a^3}{b^2} \right)^2$

48. $\left(\dfrac{13 a^5 b^2}{a b^{-7}} \right)^0$

49. $\left(\dfrac{3 m^3 n^{-1}}{m n^5} \right)^{-2}$

50. $\left(\dfrac{12 p q^{-3}}{3 p^4 q^{-4}} \right)^2$

5.7 Variation

Objectives

1 Model and Solve Direct Variation Problems

2 Model and Solve Inverse Variation Problems

3 Model and Solve Joint Variation and Combined Variation Problems

Are You Prepared for This Section?

Before getting started, complete the following problems. If you get a problem wrong, go back to the section cited and review the material.

P1. Solve: $30 = 5x$ [Section 1.1, pp. 50–53]

P2. Solve: $4 = \dfrac{k}{3}$ [Section 1.1, pp. 50–53]

P3. Graph: $y = 3x$ [Section 1.6, pp. 114–115]

▶ 1 Model and Solve Direct Variation Problems

Often two variables are related in terms of proportions. Examples include "Revenue is proportional to sales" and "Force is proportional to acceleration." When one variable is proportional to another variable, it is called *variation*. **Variation** indicates how one quantity varies in relation to some other quantity. Quantities may vary *directly*, *inversely*, or *jointly*. Direct variation is discussed first.

In Other Words

If y is directly proportional to x, then y and x are related through a linear equation whose y-intercept is $(0, 0)$ and whose slope is k, the constant of proportionality.

Figure 10
$y = kx, k > 0, x \geq 0$

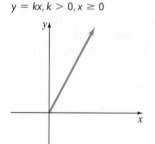

Definition

If x and y represent two quantities, then y **varies directly** with x, or y is **directly proportional to** x, if there is a nonzero number k such that

$$y = kx$$

The number k is called the **constant of proportionality** or the **constant of variation**.

If y varies directly with x, then y is a linear function of x. The graph in Figure 10 shows the relationship between y and x if y varies directly with x and $k > 0, x \geq 0$. The constant of proportionality is the slope of the line, and the y-intercept is $(0, 0)$.

If two quantities vary directly, then knowing the value of each quantity in one instance enables us to write a formula that is true in all cases.

EXAMPLE 1 **Hooke's Law**

Hooke's Law states that the force (or weight) on a spring is directly proportional to the length that the spring stretches from its "at rest" position. That is, $F = kx$, where F is the force exerted, x is the extension of the spring, and k is the proportionality constant that varies from spring to spring.

(a) Suppose a 20-pound weight causes a spring to stretch 10 inches. See Figure 11. Find the constant of proportionality, k.

Prepared?...Answers   **P1.** $\{6\}$
P2. $\{12\}$
P3.

Figure 11

(b) Write the relation between force and stretch length using function notation.

(c) What amount of weight would make the spring stretch 8 inches?

(d) Graph the relation between force and the length the spring stretches.

Solution

(a) It is known that $F = kx$ and that $F = 20$ pounds when $x = 10$ inches. Use the given information to find k, the constant of proportionality.

$$F = kx$$
$$20 = k(10)$$
$$k = 2$$

(b) Because $k = 2$, then $F = 2x$. Using function notation, write

$$F(x) = 2x$$

(c) If the spring stretches $x = 8$ inches, then the force exerted is

$$F(8) = 2(8) = 16 \text{ pounds}$$

(d) Figure 12 shows the relation between force and the length of the spring. ●

Figure 12

Length of Spring (inches)

| EXAMPLE 2 | **Car Payments** |

Suppose Dulce just purchased a used car for $10,000. She decides to put $1000 down on the car and borrows the remaining $9000. The bank lends Dulce $9000 at 4.9% interest for 48 months. Her monthly payment is $206.86. The monthly payment p on a car varies directly with the amount borrowed b.

(a) Find a function that relates the monthly payment p to the amount borrowed b for any car loan with the same terms.

(b) Suppose that Dulce put $2000 down on the car instead. What would her monthly payment be?

(c) Graph the relation between monthly payment and amount borrowed.

Solution

(a) Because p varies directly with b, it is known that for some constant k,

$$p = kb$$

Because $p = \$206.86$ when $b = \$9000$, it follows that

$$206.86 = k(9000)$$

Divide by 9000: $k = 0.022984$

Work Smart

If possible, find the exact value of k. If you cannot avoid approximate values, do not round the value of k to fewer than 5 decimal places.

Rewrite the equation as

$$p = 0.022984b$$

Write this as a linear function:

$$p(b) = 0.022984b$$

(b) When Dulce puts $2000 down, she will need to borrow $8000. When $b = \$8000$,

$$p(8000) = 0.022984(8000)$$
$$\approx \$183.87$$

If Dulce put $2000 down, her monthly payment would be $183.87.

(c) Figure 13 shows the relation between the monthly payment p and the amount borrowed b. ●

Figure 13

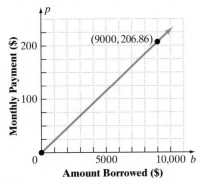

Amount Borrowed ($)

Quick ✓

1. _____ indicates how one quantity varies in relation to some other quantity.

2. If x and y are two quantities, then y is directly proportional to x if there is a nonzero number k such that _____.

3. The cost of gas C varies directly with the number of gallons pumped, g. Suppose that the cost of pumping 8 gallons of gas is $22.00.

   (a) Find a function that relates the cost of gas C to the number of gallons pumped g.

   (b) Suppose that 4.6 gallons are pumped into your car. What will the cost be?

   (c) Graph the relation between cost and number of gallons pumped.

▶ ❷ Model and Solve Inverse Variation Problems

Another kind of variation is now discussed.

Figure 14

$y = \dfrac{k}{x}, k > 0, x > 0$

> **Definition**
>
> If x and y represent two quantities, then y **varies inversely** with x, or y is **inversely proportional to** x, if there is a nonzero number k such that
>
> $$y = \frac{k}{x}$$
>
> where k is the constant of variation (or constant of proportionality).

The graph in Figure 14 illustrates the relationship between y and x if y varies inversely with x, where $k > 0$ and $x > 0$. Notice from the graph that as x increases, the value of y decreases.

EXAMPLE 3 **Weight That Can Be Supported by a Beam**

The weight W that can be safely supported by a 2-inch by 4-inch (2-by-4) piece of lumber varies inversely with its length l. See Figure 15.

Figure 15

   (a) Experiments indicate that the maximum weight a 12-foot pine 2-by-4 can support is 400 pounds. Find a function that relates the maximum weight to the length l for any pine 2-by-4.

   (b) Determine the maximum weight a 15-foot pine 2-by-4 can sustain.

Solution

   (a) Because W varies inversely with l, it is known that for some constant k,

$$W = \frac{k}{l}$$

Because $W = 400$ when $l = 12$,

$$400 = \frac{k}{12}$$

Multiply both sides by 12: $\qquad k = 4800$

Substitute into $W = \dfrac{k}{l}$: $\qquad W = \dfrac{4800}{l}$

Write as a function: $\quad W(l) = \dfrac{4800}{l}$

(b) When $l = 15$ feet,

$$W(15) = \frac{4800}{15}$$

$$= 320 \text{ pounds}$$

The maximum weight that a 15-foot pine 2-by-4 can sustain is 320 pounds. ●

Quick ✔

4. If x and y represent two quantities, then y varies inversely with x, or y is inversely proportional to x, if there is a nonzero number k such that _____.

5. Suppose that y varies inversely with x for $x > 0$.

    (a) Find an equation that relates x and y if $y = 2$ when $x = 3$.

    (b) Use the equation in part (a) to find y when $x = 4$.

6. The rate of vibration (in oscillations per second) V of a string under constant tension varies inversely with the length l.

    (a) If a string is 30 inches long and vibrates 500 times per second, find a function that relates the rate of vibration of a string to its length.

    (b) What is the rate of vibration of a string that is 50 inches long?

▶ ❸ Model and Solve Joint Variation and Combined Variation Problems

When a variable quantity Q is proportional to the product of two or more other variables, Q **varies jointly** with these quantities. For example, $y = kxz$ is read as "y varies jointly with x and z." When direct and inverse variation occur at the same time, it is called **combined variation**. For example, $y = \dfrac{kx}{z}$, is read as "y varies directly with x and inversely with z." Similarly, $y = \dfrac{kmn}{p}$ can be read as "y varies jointly with m and n and inversely with p."

EXAMPLE 4 | **Force of the Wind–Joint Variation**

The force F of the wind on a flat surface positioned at a right angle to the direction of the wind varies jointly with the area A of the surface and the square of the speed v of the wind. A wind of 30 miles per hour blowing on a window measuring 4 feet by 5 feet has a force of 150 pounds. What is the force on a window measuring 2 feet by 6 feet caused by a hurricane-force wind of 100 miles per hour?

Solution

Because F varies jointly with A and the square of v, it is known that for some constant k,

$$F = kAv^2$$

The area A of the window is $(4 \text{ feet})(5 \text{ feet}) = 20$ square feet. Substitute $F = 150$, $A = 20$, and $v = 30$ to get

$$150 = k(20)(30^2)$$

Solve this equation for k.

$$150 = 18,000k$$

(*continued*)

Divide by 18,000: $k = \dfrac{1}{120}$

Substitute into $F = kAv^2$: $F = \dfrac{1}{120}Av^2$

Thus, for a 100-mile-per-hour wind blowing on a window whose area is $A = (2\text{ feet})(6\text{ feet}) = 12$ square feet, the force F is

$$F = \frac{1}{120}(12)(100)^2$$
$$= 1000 \text{ pounds}$$

Quick ✓

7. The equation $t = ksp$ is an example of _____ variation.

8. The kinetic energy K of an object varies jointly with its mass and the square of its velocity. The kinetic energy of a 110-kg linebacker running at 9 meters per second is 4455 joules. Determine the kinetic energy of a 140-kg linebacker running at 5 meters per second.

EXAMPLE 5 **Centripetal Force–Combined Variation**

Figure 16

The force required to keep an object traveling in a circular motion is called centripetal force. The centripetal force F required to keep an object of a fixed mass in circular motion varies directly with the square of the velocity v of the object and inversely with the radius r of the circle. See Figure 16.

The force required to keep a car traveling 30 meters per second on a circular road with radius 50 meters is 21,600 newtons. Determine the force required to keep the same car on a circular road if it is traveling 40 meters per second on a road whose radius is 30 meters.

Solution
The force varies directly with the square of the velocity of the object and inversely with the radius of the circle. Then for some constant k, combined variation is

$$F = \frac{kv^2}{r}$$

Because $F = 21,600$ when $v = 30$ and $r = 50$,

$$21,600 = \frac{k \cdot 30^2}{50}$$

Solving for k yields $k = 1200$. Therefore,

$$F = \frac{1200v^2}{r}$$

For a car traveling 40 meters per second on a circular road with radius 30 meters, the force required to keep the car on the road is

$$F = \frac{1200 \cdot 40^2}{30}$$
$$= 64,000 \text{ newtons}$$

Quick ✓

9. When direct and inverse variation occur at the same time, it is called _____.

10. The electrical resistance R of a wire varies directly with the length l of the wire and inversely with the square of the diameter d of the wire. If a wire 432 feet long and 4 millimeters in diameter has a resistance of 1.24 ohms, find the resistance in a wire that is 282 feet long with a diameter of 3 millimeters. Round your answer to two decimal places.

5.7 Exercises MyMathLab® Exercise numbers in green have complete video solutions in MyMathLab or may be accessed using the QR code to the right.

*Problems **1–10** are the Quick ✓s that follow the EXAMPLES.*

Building Skills

In Problems 11–16, (a) find the constant of proportionality k, (b) write the linear function relating the two variables, and (c) find the quantity indicated. See Objective 1.

11. Suppose that y varies directly with x. When $x = 5$, $y = 30$. Find y when $x = 7$.

12. Suppose that y varies directly with x. When $x = 3$, $y = 15$. Find y when $x = 5$.

13. Suppose that y is directly proportional to x. When $x = 7$, $y = 3$. Find y when $x = 28$.

14. Suppose that y is directly proportional to x. When $x = 20$, $y = 4$. Find y when $x = 35$.

15. Suppose that y is directly proportional to x. When $x = 8$, $y = 4$. Find y when $x = 30$.

16. Suppose that y is directly proportional to x. When $x = 12$, $y = 8$. Find y when $x = 20$.

In Problems 17–20, (a) find the constant of proportionality k, (b) write the function relating the two variables, and (c) find the quantity indicated. See Objective 2.

17. Suppose that y varies inversely with x. When $x = 10$, $y = 2$. Find y if $x = 5$.

18. Suppose that y varies inversely with x. When $x = 3$, $y = 15$. Find y if $x = 5$.

19. Suppose that y is inversely proportional to x. When $x = 7$, $y = 3$. Find y if $x = 28$.

20. Suppose that y is inversely proportional to x. When $x = 20$, $y = 4$. Find y if $x = 35$.

In Problems 21–24 (a) find the constant of proportionality k, (b) write the function relating the variables, and (c) find the quantity indicated. See Objective 3.

21. Suppose that y varies jointly with x and z. When $x = 8$ and $z = 5$, $y = 10$. Find y if $x = 12$ and $z = 9$.

22. Suppose that y varies jointly with x and z. When $x = 6$ and $z = 10$, $y = 20$. Find y if $x = 8$ and $z = 15$.

23. Suppose that Q varies directly with x and inversely with y. When $x = 5$ and $y = 6$, $Q = \dfrac{13}{12}$. Find Q if $x = 9$ and $y = 4$.

24. Suppose that Q varies directly with x and inversely with y. When $x = 4$ and $y = 3$, $Q = \dfrac{14}{5}$. Find Q if $x = 8$ and $y = 3$.

Applying the Concepts

25. Mortgage Payments The monthly payment p on a mortgage varies directly with the amount borrowed b. Suppose that you decide to borrow $190,000 using a 30-year mortgage at 4% interest. You are told that your monthly payment is $907.09.

(a) Write a linear function that relates the monthly payment p to the amount borrowed b for a mortgage with the same terms.

(b) Assume that you have decided to buy a more expensive home that requires you borrow $220,000. What will your approximate monthly payment be?

(c) Graph the relation between monthly payment and amount borrowed.

26. Mortgage Payments The monthly payment p on a mortgage varies directly with the amount borrowed b. Suppose that you decide to borrow $190,000 using a 15-year mortgage at 3% interest. You are told that your payment is $1312.11.

(a) Write a linear function that relates the monthly payment p to the amount borrowed b for a mortgage with the same terms.

(b) Assume that you have decided to buy a more expensive home that requires you borrow $220,000. What will your approximate monthly payment be?

(c) Graph the relation between monthly payment and amount borrowed.

27. Cost Function The cost C of purchasing chocolate-covered almonds varies directly with the weight w in pounds. Suppose that the cost of purchasing 5 pounds of chocolate-covered almonds is $28.

(a) Write a linear function that relates the cost C to the number of pounds of chocolate-covered almonds purchased w.

(b) What would it cost to purchase 3.5 pounds of chocolate-covered almonds?

(c) Graph the relation between cost and weight.

28. Conversion Suppose that you are planning a trip to Europe, so you need to obtain some euros. The amount received in euros varies directly with the amount in U.S. dollars. Your friend just converted $600 into 548 euros.

(a) Write a linear function that relates the number of euros E to the number of U.S. dollars d.

(b) If you convert $700 into euros, how many euros will you receive?

(c) Graph the relation between euros and U.S. dollars.

29. **Falling Objects** The velocity of a falling object (ignoring air resistance) v is directly proportional to the time t of the fall. If, after 2 seconds, the velocity of the object is 64 feet per second, what will its velocity be after 3 seconds?

30. **Circumference of a Circle** The circumference of a circle C is directly proportional to its radius r. If the circumference of a circle whose radius is 5 inches is 10π inches, what is the circumference of a circle whose radius is 8 inches?

31. **Demand** Suppose that the demand D for candy at the movie theater is inversely related to the price p.

(a) When the price of candy is $2.50 per bag, the theater sells 150 bags of candy. Express the demand for candy as a function of its price.

(b) Determine the number of bags of candy that will be sold if the price is raised to $3 a bag.

32. **Driving to School** The time t that it takes you to get to school varies inversely with your average speed s.

(a) Suppose that it takes you 30 minutes to drive to school when your average speed is 35 miles per hour. Express your driving time to school as a function of your average speed.

(b) Suppose that your average speed driving to school is 30 miles per hour. How long will it take you to get to school?

33. **Pressure** The volume of a gas V held at a constant temperature in a closed container varies inversely with its pressure P. If the volume of a gas is 600 cubic centimeters (cc) when the pressure is 150 millimeters of mercury (mm Hg), find the volume when the pressure is 200 mm Hg.

34. **Resistance** The current I in a circuit is inversely proportional to its resistance R measured in ohms. Suppose that when the current in a circuit is 30 amperes, the resistance is 8 ohms. Find the current in the same circuit when the resistance is 10 ohms.

35. **Weight** The weight of an object above the surface of Earth varies inversely with the square of the distance from the center of Earth. Maria weighs 120 pounds when she is on the surface of Earth (3960 miles from the center). Determine Maria's weight if she is at the top of Mount McKinley (3.8 miles from the surface of Earth).

36. **Intensity of Light** The intensity I of light (measured in foot-candles) varies inversely with the square of the distance from the bulb. Suppose the intensity of a 100-watt light bulb at a distance of 2 meters is 0.075 foot-candle. Determine the intensity of the bulb at a distance of 3 meters.

37. **Drag Force** When an object moves through air, a frictionlike drag force tends to slow the object down. The drag force D on a free-falling parachutist varies jointly with the surface area of the parachutist and the square of his velocity. The drag force on a parachutist with surface area 2 square meters falling at 40 meters per second is 1152 newtons. Find the drag force on a parachutist whose surface area is 2.5 square meters falling at 50 meters per second.

38. **Kinetic Energy** The kinetic energy K (measured in joules) of a moving object varies jointly with the mass of the object and the square of its velocity v. The kinetic energy of a linebacker weighing 110 kilograms and running at a speed of 8 meters per second is 3520 joules. Find the kinetic energy of a wide receiver weighing 90 kilograms and running at a speed of 10 meters per second.

39. **Newton's Law of Gravitation** According to Newton's law of universal gravitation, the force F of gravity between any two objects varies jointly with the masses of the objects m_1 and m_2 and inversely with the square of the distance between the objects r. The force of gravity between a 105-kg man and his 80-kg wife when they are separated by a distance of 5 meters is 2.24112×10^{-8} newton. Find the force of gravity between the man and his wife when they are 2 meters apart.

40. **Electrical Resistance** The electrical resistance of a wire varies directly with the length of the wire and inversely with the square of the diameter of the wire. If a wire 50 feet long and 3 millimeters in diameter has a resistance of 0.255 ohm, find the length of a wire of the same material whose resistance is 0.147 ohm and whose diameter is 2.5 millimeters.

41. **Stress of Material** The stress in the material of a pipe subject to internal pressure varies jointly with the internal pressure and internal diameter of the pipe and inversely with the thickness of the pipe. The stress is 100 pounds per square inch when the diameter is 5 inches, the thickness is 0.75 inch, and the internal pressure is 25 pounds per square inch. Find the stress when the internal pressure is 50 pounds per square inch, the diameter is 6 inches, and the thickness is 0.5 inch.

42. **Gas Laws** The volume V of an ideal gas varies directly with the temperature T and inversely with the pressure P. If a cylinder contains oxygen at a temperature of 300 kelvin (K) and a pressure of 15 atmospheres in a volume of 100 liters, what is the constant of proportionality k? If a piston is lowered into the cylinder, decreasing the volume occupied by the gas to 70 liters and raising the temperature to 315 K, what is the pressure?

Extending the Concepts

43. David and Goliath The force F (in newtons) required to maintain an object in a circular path varies jointly with the mass m (in kilograms) of the object and the square of its speed (measured in meters per second) and inversely with the radius r (in meters) of the circular path. Suppose that David has a rope that is 3 meters long. On the end of the rope he has attached a pouch that holds a 0.5-kilogram stone. Suppose that David is able to spin the rope in a circular motion at the rate of 50 revolutions per minute.

(a) The spinning rate of 50 revolutions per minute can be converted into an *angular velocity* ω (lowercase Greek letter omega) using the formula $\omega = 2\pi$ (revolutions per minute). Write the spinning rate as an angular velocity rounded to two decimal places.

(b) Use the fact that $v = \omega r$, where r is the radius of the circle, to find the linear velocity (in meters per minute) of the stone if it were released. Now convert the linear velocity to meters per second.

(c) Suppose that the force on the rope required to keep the rock in a circular motion is 2.3 newtons. Use the result from part (b) to find the constant of proportionality k.

(d) David is fairly certain that he will require more force than this to beat Goliath in a battle, so he increases the circular motion to 80 revolutions per minute and increases the length of the rope to 4 meters. What is the force required to keep the stone in a circular motion?

44. Suppose that y is directly proportional to x^2. If x is doubled, what happens to the value of y?

Chapter 5 Activity: Correct the Quiz

Focus: Performing operations and solving equations and inequalities with rational expressions

Time: 20 minutes

Group size: 2

In this activity you will work as a team to grade the student quiz shown below. One of you will grade the odd questions, and the other will grade the even questions. If an answer is correct, mark it correct. If an answer is wrong, mark it wrong and show the correct answer.

Once all of the quiz questions are graded, explain your results to each other and compute the final score for the quiz. Be prepared to discuss your results with the rest of the class.

Student Quiz

Name: Ima Student   Quiz Score: _____

(1) Multiply: $\dfrac{2x^3 + 54}{5x^2 + 5x - 30} \cdot \dfrac{6x + 12}{3x^2 - 9x + 27}$	Answer: $\dfrac{4(x + 2)}{5(x - 2)}$
(2) Subtract: $\dfrac{xy}{x^2 - y^2} - \dfrac{y}{x + y}$	Answer: $\dfrac{y^2}{x^2 - y^2}$
(3) Solve: $\dfrac{3}{x - 3} + \dfrac{4}{x} = \dfrac{-12}{x^2 - 3x}$	Answer: $\varnothing$
(4) Divide: $\dfrac{x^2 + x - 6}{5x^2 - 7x - 6} \div \dfrac{3x^2 + 13x + 12}{6x^2 + 17x + 12}$	Answer: $\dfrac{2}{5}$
(5) Joe can mow his lawn in 2 hours. Mike can mow the same lawn in 3 hours. If they work together, how long will it take them to mow the lawn?	Answer: 1 hr, 12 min
(6) Simplify: $\dfrac{\dfrac{1}{x + 5} - \dfrac{2}{x - 7}}{\dfrac{4}{x - 7} + \dfrac{1}{x + 5}}$	Answer: $\dfrac{x + 17}{5x + 13}$
(7) Add: $\dfrac{x}{x^2 + 10x + 25} + \dfrac{4}{x^2 + 6x + 5}$	Answer: $\dfrac{x^2 + 5x + 20}{(x + 5)(x + 1)}$
(8) Solve: $\dfrac{4x}{x - 3} \geq 5$	Answer: $(3, 15]$

Chapter 5 Review

Section 5.1   Multiplying and Dividing Rational Expressions

KEY CONCEPTS

KEY TERMS

- **Multiplying Rational Expressions**

  If $\dfrac{a}{b}$ and $\dfrac{c}{d}$, $b \neq 0$, $d \neq 0$, are two rational expressions, then $\dfrac{a}{b} \cdot \dfrac{c}{d} = \dfrac{ac}{bd}$.

- **Dividing Rational Expressions**

  If $\dfrac{a}{b}$ and $\dfrac{c}{d}$, $b \neq 0$, $c \neq 0$, $d \neq 0$, are two rational expressions, then $\dfrac{\dfrac{a}{b}}{\dfrac{c}{d}} = \dfrac{a}{b} \cdot \dfrac{d}{c} = \dfrac{ad}{bc}$.

Rational expression
Numerator
Denominator
Lowest terms
Simplified
Rational function

You Should Be Able To...	EXAMPLE	Review Exercises
❶ Determine the domain of a rational expression (p. 400)	Example 1	1–4
❷ Simplify rational expressions (p. 401)	Examples 2 and 3	5–10
❸ Multiply rational expressions (p. 402)	Examples 4 and 5	11–16
❹ Divide rational expressions (p. 404)	Example 6	17–22
❺ Work with rational functions (p. 405)	Examples 7 through 8	23–26

In Problems 1–4, state the domain of each rational expression.

1. $\dfrac{x - 5}{3x - 2}$

2. $\dfrac{a^2 - 16}{a^2 - 3a - 28}$

3. $\dfrac{m - 3}{m^2 + 9}$

4. $\dfrac{n^2 + 7n + 10}{n^2 - 2n - 8}$

In Problems 5–10, simplify each rational expression.

5. $\dfrac{6x + 30}{x^2 - 25}$

6. $\dfrac{4y^2 - 28y}{2y^5 - 14y^4}$

7. $\dfrac{w^2 - 4w - 21}{w^2 + 7w + 12}$

8. $\dfrac{6a^2 - 7ab - 3b^2}{10a^2 - 11ab - 6b^2}$

9. $\dfrac{7 - m}{3m^2 - 20m - 7}$

10. $\dfrac{n^3 - 4n^2 + 3n - 12}{n^2 - 8n + 16}$

In Problems 11–22, multiply or divide each rational expression, as indicated. Simplify the product or quotient.

11. $\dfrac{4p^2}{p^2 - 3p - 18} \cdot \dfrac{p + 3}{8p}$

12. $\dfrac{q^2 + 6q}{6q + 12} \cdot \dfrac{4q + 8}{q^2 + q - 30}$

13. $\dfrac{x^3 - 4x^2}{x^2 - 4} \cdot \dfrac{x^2 + 4x - 12}{x^3 + 2x^2}$

14. $\dfrac{y^2 - 3y - 28}{y^3 + 4y^2} \cdot \dfrac{2y^2 + 10y}{y^2 - 12y + 35}$

15. $\dfrac{6a^2 + ab - b^2}{3a^2 + 2ab - b^2} \cdot \dfrac{3a^2 + 4ab + b^2}{4a^2 - b^2}$

16. $\dfrac{m^2 + m - 20}{m^3 - 64} \cdot \dfrac{3m^2 + 12m + 48}{m^2 + 3m - 10}$

17. $\dfrac{\dfrac{4c^2}{3d^4}}{\dfrac{8c}{27d}}$

18. $\dfrac{\dfrac{6z - 24}{7z + 21}}{\dfrac{z - 4}{z^2 - 9}}$

19. $\dfrac{\dfrac{x^2 - 11x + 30}{x^2 - 8x + 15}}{\dfrac{x^2 - 5x - 6}{x^2 + 8x + 7}}$

20. $\dfrac{\dfrac{m^2 + mn - 12n^2}{m^3 - 27n^3}}{\dfrac{m + 5n}{m^2 + 3mn + 9n^2}}$

21. $\dfrac{\dfrac{4p^3 - 4pq^2}{p^2 - 5pq - 24q^2}}{\dfrac{2p^3 + 4p^2 q + 2pq^2}{p^2 - 7pq - 8q^2}}$

22. $\dfrac{\dfrac{15a^2 + 11a - 14}{25a^2 - 49}}{\dfrac{27a^3 - 8}{10a^2 + 11a - 35}}$

In Problems 23–26, use the functions

$$f(x) = \frac{2x^2 + 3x - 2}{x - 5}, g(x) = \frac{x^2 - 3x - 10}{2x - 1}, \text{ and}$$

$$h(x) = \frac{2x - 1}{x^2 + 9x + 14} \text{ to find each product or quotient.}$$

State the domain of each product or quotient.

23. $P(x) = f(x) \cdot g(x)$

24. $R(x) = g(x) \cdot h(x)$

25. $Q(x) = \dfrac{g(x)}{f(x)}$

26. $T(x) = \dfrac{f(x)}{h(x)}$

Section 5.2   Adding and Subtracting Rational Expressions

KEY CONCEPTS

- **Adding/Subtracting Rational Expressions**

  If $\dfrac{a}{c}$ and $\dfrac{b}{c}$, $c \neq 0$, are two rational expressions, then $\dfrac{a}{c} + \dfrac{b}{c} = \dfrac{a + b}{c}$

  and $\dfrac{a}{c} - \dfrac{b}{c} = \dfrac{a - b}{c}$.

  Note: If the rational expressions do not have a common denominator, then the least common denominator can be found using the steps on page 413. Then follow the steps listed on page 414 to add or subtract rational expressions with unlike denominators.

KEY TERM

Least common denominator

You Should Be Able To...	EXAMPLE	Review Exercises
❶ Add or subtract rational expressions with a common denominator (p. 411)	Examples 1 and 2	27–32
❷ Find the least common denominator of two or more rational expressions (p. 412)	Example 3	33–36
❸ Add or subtract rational expressions with different denominators (p. 413)	Examples 4 through 7	37–50

In Problems 27–32, perform the indicated operation and simplify the result.

27. $\dfrac{4x}{x - 5} + \dfrac{3}{x - 5}$

28. $\dfrac{4y}{y - 3} - \dfrac{12}{y - 3}$

29. $\dfrac{a^2 - 2a - 4}{a^2 - 6a + 8} + \dfrac{4a - 20}{a^2 - 6a + 8}$

30. $\dfrac{3b^2 + 8b - 5}{2b^2 - 5b - 12} - \dfrac{2b^2 + 7b + 15}{2b^2 - 5b - 12}$

31. $\dfrac{5c^2 - 8c}{c - 8} + \dfrac{2c^2 + 16c}{8 - c}$

32. $\dfrac{2d^2 + d}{d^2 - 1} - \dfrac{d^2 + 1}{d^2 - 1} + \dfrac{d - 2}{d^2 - 1}$

In Problems 33–36, find the least common denominator.

33. $\dfrac{4}{9x^4}$ and $\dfrac{5}{12x^2}$

34. $\dfrac{2y + 1}{y - 9}$ and $\dfrac{3y}{y + 2}$

35. $\dfrac{3p + 4}{2p^2 - 3p - 20}$ and $\dfrac{7p^2}{2p^3 + 5p^2}$

36. $\dfrac{q - 4}{q^2 + 4q - 5}$ and $\dfrac{q - 6}{q^2 + 2q - 15}$

In Problems 37–48, perform the indicated operation and simplify the result.

37. $\dfrac{1}{mn^4} + \dfrac{4}{m^3n^2}$

38. $\dfrac{3}{2xy^3} - \dfrac{7}{6x^2y}$

39. $\dfrac{p}{p - q} - \dfrac{q}{p + q}$

40. $\dfrac{x + 8}{x^2 - 10x + 21} - \dfrac{x - 5}{x^2 - 3x - 28}$

41. $\dfrac{3}{y^2 - 2y + 1} - \dfrac{2}{y^2 + y - 2}$

42. $\dfrac{3a - 5b}{4a^2 - 9b^2} + \dfrac{4}{2a - 3b}$

43. $\dfrac{4x^2 - 10x}{x^2 - 9} + \dfrac{8x - 2x^2}{9 - x^2}$

44. $\dfrac{1}{n + 5} - \dfrac{n^2 - 10n}{n^3 + 125}$

45. $\dfrac{m + n}{m + 3n} - \dfrac{m - 4n}{m - 7n} + \dfrac{7mn + n^2}{m^2 - 4mn - 21n^2}$

46. $\dfrac{z^2 + 10z + 3}{z^2 - 9} - \dfrac{2z}{z - 3} + \dfrac{z}{z + 3}$

47. $\dfrac{y - 1}{y - 2} - \dfrac{y + 1}{y + 2} + \dfrac{y - 6}{y^2 - 4}$

48. $\dfrac{2a}{a^2 - 16} - \dfrac{1}{a - 4} - \dfrac{1}{a^2 + 2a - 8}$

49. $f(x) = \dfrac{5}{x - 4}$ and $g(x) = \dfrac{x}{x + 2}$

   (a) Find $S(x) = f(x) + g(x)$.
   (b) State the domain of $S(x)$.

50. $f(x) = \dfrac{x + 3}{2x^2 + x - 15}$ and $g(x) = \dfrac{x - 7}{4x^2 - 8x - 5}$

   (a) Find $D(x) = f(x) - g(x)$.
   (b) State the domain of $D(x)$.

Section 5.3 Complex Rational Expressions

KEY CONCEPT

- There are two methods that can be used to simplify a complex rational expression. The steps for Method I are presented on page 420, and the steps for Method II are presented on page 422.

KEY TERMS

Complex rational expression
Simplify

You Should Be Able To...	EXAMPLE	Review Exercises
❶ Simplify a complex rational expression by simplifying the numerator and denominator separately (Method I) (p. 419)	Examples 1, 2, and 5	51–54, 59–66
❷ Simplify a complex rational expression using the least common denominator (Method II) (p. 421)	Examples 3 through 5	55–58, 59–66

In Problems 51–54, simplify each complex rational expression by using Method I (that is, by simplifying the numerator and denominator separately).

51. $\dfrac{x - \dfrac{1}{x}}{1 - \dfrac{1}{x}}$

52. $\dfrac{\dfrac{1}{x} - \dfrac{1}{y}}{\dfrac{1}{x^2} - \dfrac{1}{y^2}}$

53. $\dfrac{\dfrac{a}{b} - \dfrac{a - b}{a + b}}{\dfrac{a}{b} + \dfrac{a + b}{a - b}}$

54. $\dfrac{\dfrac{2}{a + 2} - 1}{\dfrac{1}{a + 2} + 1}$

In Problems 55–58, simplify each complex rational expression using Method II (that is, by using the least common denominator).

55. $\dfrac{\dfrac{3}{t} + \dfrac{4}{t^2}}{5 + \dfrac{1}{t^2}}$

56. $\dfrac{\dfrac{1}{a} - \dfrac{1}{b}}{\dfrac{b}{a} - \dfrac{a}{b}}$

57. $\dfrac{\dfrac{1}{z - 1} - \dfrac{1}{z}}{\dfrac{1}{z} - \dfrac{1}{z + 1}}$

58. $\dfrac{1 + \dfrac{x}{x + 1}}{\dfrac{2x + 1}{x - 1}}$

In Problems 59–66, simplify the complex rational expression using either Method I or Method II.

59. $\dfrac{\dfrac{x}{y} + 1}{\dfrac{x}{y} - 1}$

60. $\dfrac{\dfrac{a}{a-b} - \dfrac{b}{a+b}}{\dfrac{b}{a-b} + \dfrac{a}{a+b}}$

61. $\dfrac{\dfrac{1}{x-2} - \dfrac{x}{x^2-4}}{1 - \dfrac{2}{x+2}}$

62. $\dfrac{z - \dfrac{5z}{z+5}}{z + \dfrac{5z}{z-5}}$

63. $\dfrac{\dfrac{m-n}{m+n} + \dfrac{n}{m}}{\dfrac{m}{n} - \dfrac{m-n}{m+n}}$

64. $\dfrac{\dfrac{x+4}{x-2} - \dfrac{x-3}{x+1}}{5x^2 + 4x - 1}$

65. $\dfrac{3x^{-1} - 3y^{-1}}{(x+y)^{-1}}$

66. $\dfrac{2c^{-1} - (3d)^{-1}}{(6d)^{-1}}$

Section 5.4 Rational Equations

KEY CONCEPT

- The steps for solving a rational equation are given on page 429.

KEY TERMS
Rational equation
Extraneous solution

You Should Be Able To...	EXAMPLE	Review Exercises
1 Solve equations containing rational expressions (p. 428)	Examples 1 through 5	67–78
2 Solve equations involving rational functions (p. 432)	Examples 6 and 7	79, 80

In Problems 67–78, solve each equation. Be sure to verify your results.

67. $\dfrac{2}{z} - \dfrac{1}{3z} = \dfrac{1}{6}$

68. $\dfrac{4}{m-4} = \dfrac{-5}{m+2}$

69. $m - \dfrac{14}{m} = 5$

70. $\dfrac{2}{n+3} = \dfrac{1}{n-3}$

71. $\dfrac{s}{s-1} = 1 + \dfrac{2}{s}$

72. $\dfrac{3}{x^2 - 7x + 10} + 2 = \dfrac{x-4}{x-5}$

73. $\dfrac{1}{k-1} + \dfrac{1}{k+2} = \dfrac{3}{k^2+k-2}$

74. $x + \dfrac{3x}{x-3} = \dfrac{9}{x-3}$

75. $\dfrac{2}{a+3} - \dfrac{4}{a^2-4} = \dfrac{a+1}{a^2+5a+6}$

76. $\dfrac{2}{z^2+2z-8} = \dfrac{1}{z^2+9z+20} + \dfrac{4}{z^2+3z-10}$

77. $\dfrac{x-3}{x+4} = \dfrac{14}{x^2+6x+8}$

78. $\dfrac{5}{y-5} + 4 = \dfrac{3y-10}{y-5}$

79. For the function $f(x) = \dfrac{6}{x-2}$, solve $f(x) = 2$. What point(s) are on the graph of f?

80. For the function $g(x) = x - \dfrac{21}{x}$, solve $g(x) = 4$. What point(s) are on the graph of g?

Section 5.5 Rational Inequalities

KEY CONCEPT

- The steps for solving any rational inequality are given on page 438.

KEY TERM
Rational inequality

You Should Be Able To...	EXAMPLE	Review Exercises
1 Solve a rational inequality (p. 436)	Examples 1 and 2	81–90

In Problems 81–88, solve each rational inequality. Graph the solution set.

81. $\dfrac{x-4}{x+2} \geq 0$

82. $\dfrac{y-5}{y+4} < 0$

83. $\dfrac{4}{z^2-9} \leq 0$

84. $\dfrac{w^2+5w-14}{w-4} < 0$

85. $\dfrac{m-5}{m^2+3m-10} \geq 0$

86. $\dfrac{4}{n-2} \leq -2$

87. $\dfrac{a+1}{a-2} > 3$

88. $\dfrac{4}{c-2} - \dfrac{3}{c} < 0$

In Problems 89 and 90, for each function, find the values of x that satisfy the given condition.

89. Solve $Q(x) < 0$ if $Q(x) = \dfrac{2x+3}{x-4}$.

90. Solve $R(x) \geq 0$ if $R(x) = \dfrac{x+5}{x+1}$.

Section 5.6   Models Involving Rational Expressions

KEY TERMS

| Ratio | Proportion | Similar | Constant rate |

You Should Be Able To...	EXAMPLE	Review Exercises
1 Solve for a variable in a rational expression (p. 442)	Example 1	91–94
2 Model and solve ratio and proportion problems (p. 443)	Examples 2 and 3	95–98
3 Model and solve work problems (p. 445)	Examples 4 and 5	99–102
4 Model and solve uniform motion problems (p. 447)	Example 6	103–106

In Problems 91–94, solve each formula for the indicated variable.

91. Electronics (Capacitance) Solve $\dfrac{1}{C_1} + \dfrac{1}{C_2} = \dfrac{1}{C}$ for C.

92. Chemistry (Ideal Gas Law) Solve $\dfrac{P_1V_1}{T_1} = \dfrac{P_2V_2}{T_2}$ for T_2.

93. Physics (Kepler's Third Law) Solve $T = \dfrac{4\pi^2 a^2}{MG}$ for G.

94. Statistics (z-score) Solve $z = \dfrac{x-\mu}{\sigma}$ for x.

In Problems 95–98, solve each proportion problem.

95. Suppose that $\triangle ABC$ is similar to $\triangle DEF$, as shown in the figure. Find the lengths of AB and DF.

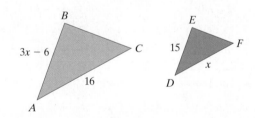

96. Casting Shadows At a particular time of day, a pine tree casts a 30-foot shadow. At the exact same time, a nearby 5-foot post casts an 8-foot shadow. Find the height of the tree.

97. Nutrition According to the nutrition facts on a box of Honey Nut Cheerios® (source: *General Mills*), a $\frac{3}{4}$-cup serving contains 22 grams of total carbohydrates. How many grams of total carbohydrates are in a 3-cup bowl of the cereal?

98. Payroll One day, Jeri earns $48.75 for 5 hours of work. How much will she earn for 8 hours of work?

In Problems 99–102, solve each work problem.

99. Filling a Tank One pipe can fill a tank in 48 minutes. Another pipe can fill the tank in 1 hour and 12 minutes. If both pipes are used, how long will it take to fill the tank?

100. Mowing the Lawn Together, Diane and Craig can mow their lawn in 1 hour and 10 minutes. Working alone, Diane can mow the lawn in 2 hours. How long will it take Craig to mow the lawn when working alone?

101. Carpeting Together, Rick and John can carpet a large room in 12 hours. Alone, Rick can carpet the same size room in 7 hours less time than John. How long would it take Rick to carpet the same size room if working alone? How long would it take John if working alone?

102. Draining a Sink A faucet can fill a sink in 1 minute when the drain is plugged, but when the drain is unplugged, it takes 1 minute and 30 seconds to fill the sink. With the faucet off, how long would it take the drain to empty a full sink?

In Problems 103–106, solve each motion problem.

103. Pleasure Flight In his private plane, Nick can fly 180 miles per hour if the wind is not blowing. One day, Nick took a pleasure flight. He flew 100 miles directly against the wind and then returned (flying with the wind) to his point of origin. If the total time of the flight was 1 hour and 15 minutes, what was the speed of the wind?

104. Boating In his motorboat, Jesse can travel 20 miles downstream in the same amount of time it takes him to travel 10 miles upstream. If the speed of the current is 5 miles per hour, how fast can Jesse's motorboat travel in still water?

105. Running/Walking To stay in shape, Todd first runs 3 miles and then walks 1 mile every morning. Todd's average running speed is 4 times his average walking speed. If Todd spends a total of 35 minutes running and walking each morning, find the average speed at which he walks and the average speed at which he runs.

106. Road Trip Because of heavy traffic, Danielle averaged only 30 miles per hour for the first 20 miles of her trip. If she averaged 50 miles per hour for the entire 100-mile trip, what was her average speed for the last 80 miles?

Section 5.7 Variation

You Should Be Able To...	EXAMPLE	Review Exercises
❶ Model and solve direct variation problems (p. 452)	Examples 1 and 2	107, 108, 113, 114
❷ Model and solve inverse variation problems (p. 454)	Example 3	109, 111, 115, 116
❸ Model and solve joint variation and combined variation problems (p. 455)	Examples 4 and 5	110, 112, 117, 118

In Problems 107–112, (a) find the constant of proportionality k, (b) write the function relating the variables, and (c) find the quantity indicated.

107. Suppose that y varies directly with x. If $y = 30$ when $x = 6$, find y when $x = 10$.

108. Suppose that y varies directly with x. If $y = 18$ when $x = -3$, find y when $x = 8$.

109. Suppose that y varies inversely with x. If $y = 15$ when $x = 4$, find y when $x = 5$.

110. Suppose that y varies jointly with x and z. If $y = 45$ when $x = 6$ and $z = 10$, find y when $x = 8$ and $z = 7$.

111. Suppose that s varies inversely with the square of t. If $s = 18$ when $t = 2$, find s when $t = 3$.

112. Suppose that w varies directly with x and inversely with z. If $w = \frac{4}{3}$ when $x = 10$ and $z = 12$, find w when $x = 9$ and $z = 16$.

113. Snow-Water Equivalent The amount of water in snow is directly proportional to the depth of the snow. Suppose the amount of water in 40 inches of snow is 4.8 inches. How much water is contained in 50 inches of snow?

114. Car Payments Roberta is buying a car. The monthly payment p for the car varies directly with the amount borrowed b. Suppose the dealership tells her that if she borrows \$15,000, her monthly payment will be \$293.49. What will be Roberta's monthly payment if she borrows \$18,000?

115. Radio Signals The frequency of a radio signal varies inversely with the wavelength. A signal of 800 kilohertz has a wavelength of 375 meters. What frequency has a signal of wavelength 250 meters?

116. Ohm's Law The electrical current flowing through a wire varies inversely with the resistance of the wire.

If the current is 8 amperes when the resistance is 15 ohms, for what resistance will the current be 10 amperes?

117. Volume of a Cylinder The volume V of a right circular cylinder varies jointly with the height h and the square of the diameter d. If the volume of a cylinder is 231 cubic centimeters when the diameter is 7 centimeters and the height is 6 centimeters, find the volume when the diameter is 8 centimeters and the height is 14 centimeters.

118. Volume of a Pyramid The volume V of a pyramid varies jointly with the base area B and the height h. If the volume of a pyramid is 270 cubic inches when the base area is 81 square inches and the height is 10 inches, find the volume of a pyramid with base area 125 square inches and height 9 inches.

Chapter 5 Test
Step-by-step test solutions are found on the Chapter Test Prep Videos available in MyMathLab®, on You Tube, or may be accessed using the QR code to the right.

1. Determine the domain of $f(x) = \dfrac{2x + 1}{2x^2 - 13x - 7}$.

In Problems 2 and 3, simplify each rational expression.

2. $\dfrac{2m^2 + 5m - 12}{3m^2 + 11m - 4}$ **3.** $\dfrac{2b - 3a}{3a^2 + 10ab - 8b^2}$

In Problems 4–7, perform the indicated operations.

4. $\dfrac{4x^2 - 12x}{x^2 - 9} \cdot \dfrac{2x^2 + 11x + 15}{8x^3 - 32x^2}$

5. $\dfrac{\dfrac{y^2 + 2y - 8}{4y^2 - 5y - 6}}{\dfrac{3y^2 - 14y - 5}{4y^2 - 17y - 15}}$

6. $\dfrac{3p^2 + 3pq}{p^2 - q^2} - \dfrac{3p - 2q}{p - q}$

7. $\dfrac{9c + 2}{3c^2 - 2c - 8} + \dfrac{7}{3c^2 + c - 4}$

In Problems 8 and 9, use the functions $f(x) = \dfrac{3x}{x^2 - 4}$,
$g(x) = \dfrac{6}{x^2 + 2x}$, *and* $h(x) = \dfrac{9x^2 - 45x}{x^2 - 2x - 8}$ *to find each sum or quotient. State the domain of each.*

8. $Q(x) = \dfrac{f(x)}{h(x)}$ **9.** $S(x) = f(x) + g(x)$

In Problems 10 and 11, simplify each complex rational expression using either Method I or Method II.

10. $\dfrac{1 - \dfrac{1}{a}}{1 - \dfrac{1}{a^2}}$ **11.** $\dfrac{\dfrac{5}{d + 2} - \dfrac{1}{d - 2}}{\dfrac{3}{d + 2} - \dfrac{6}{d - 2}}$

In Problems 12 and 13, solve each rational equation. Be sure to verify the results.

12. $\dfrac{1}{6x} - \dfrac{1}{3} = \dfrac{5}{4x} + \dfrac{3}{4}$

13. $\dfrac{7n}{n + 3} + \dfrac{21}{n - 3} = \dfrac{126}{n^2 - 9}$

14. Solve $\dfrac{x + 5}{x - 2} \geq 3$. Graph the solution set.

15. Electronics (Coulomb's Law) Solve $\dfrac{1}{F} = \dfrac{D^2}{kq_1q_2}$ for k.

16. Printing Documents A particular laser printer can print out a 10-page document in 25 seconds. How long will it take to print out a 48-page document?

17. Cleaning House Linnette can clean the house in 4 hours. Her husband Darrell can do the same job in 6 hours. If the two work together, how long will it take them to clean the house?

18. Kayaking Chuck kayaked 4 miles upstream in the same time it took him to kayak 10 miles downstream. If Chuck can average 7 miles per hour in still water, what was the rate of the current?

19. Using a Lever Using a lever, the force F required to lift a weight is inversely proportional to the length l of the force arm of the lever (assuming all other factors are constant). If a force of 50 pounds is required to lift a granite boulder when the force arm length is 4 feet, how much force will be required to lift the boulder if the force arm length is 10 feet?

20. Lateral Surface Area of a Cylinder The lateral surface area L of a right circular cylinder varies jointly with its radius r and height h. If the lateral surface area of a cylinder with radius 7 centimeters and height 12 centimeters is 528 square centimeters, what would be the lateral surface area if the radius were 9 centimeters and the height were 14 centimeters?

Cumulative Review Chapters R–5

1. Evaluate: $\dfrac{4^3 - 6 \cdot 7 + 14}{4 - 1^2}$

2. Simplify: $2x(x - 3) + 4(x - 2) + 15$

3. Evaluate $\dfrac{3x^2 - 4x - 5}{x - 3}$ for $x = -2$.

4. Solve: $7x + 9 = 3x - 23$

5. Solve: $|3x + 7| \le 8$

6. Determine the domain of $h(x) = \dfrac{x - 5}{2x^2 - 7x - 15}$.

7. For $f(x) = x^2 - 5x$, find each of the following.
   (a) $f(-3)$ **(b)** $f\left(\dfrac{1}{4}\right)$ **(c)** $f(x + 2)$

8. Graph the linear equation $4x + 3y = 15$.

9. Find the equation of the line that passes through the points $(-5, 1)$ and $(10, -8)$. Write your answer in either slope-intercept or standard form, whichever you prefer.

10. Find the equation of the line that passes through the point $(3, 5)$ and is perpendicular to the graph of $x + 4y = 20$. Write your answer in either slope-intercept or standard form, whichever you prefer.

11. The recommended dosage D of a prescription drug varies directly with a person's weight w. If the recommended dosage for a 125-pound person is 1500 milligrams, find the recommended dosage for a 180-pound person.

12. Write the system of linear equations that corresponds to the following augmented matrix.
$$\begin{bmatrix} 2 & -3 & | & 7 \\ 5 & 2 & | & 8 \end{bmatrix}$$

13. Perform each row operation on the given augmented matrix. $\begin{bmatrix} 1 & 1 & 0 & | & -3 \\ 0 & 2 & -1 & | & -1 \\ 5 & 0 & 1 & | & 1 \end{bmatrix}$

   (a) $R_3 = -5r_1 + r_3$ followed by
   (b) $R_2 = \dfrac{1}{2}r_2$

14. Solve the following system of equations
$$\begin{cases} x + 4y = -2 \\ 2x - 12y = -9 \end{cases}$$

15. Evaluate: $\begin{vmatrix} 2 & 0 & 4 \\ 1 & -1 & -2 \\ 2 & -2 & 3 \end{vmatrix}$

16. Graph the following system of linear inequalities:
$$\begin{cases} 3x + 2y < 8 \\ x - 4y \ge 12 \end{cases}$$

In Problems 17–20, perform the indicated operation.

17. $(3x^2 - 4xy + 7y^2) + (5x^2 - 9xy + 2y^2)$

18. $(5x^2 - 3x + 12) - (2x^2 - 4x - 15)$

19. $(2x - 3)(x^2 - 4x + 6)$

20. $\dfrac{4x^3 - 7x + 45}{2x + 5}$

In Problems 21–22, factor completely.

21. $x^4 + 5x^3 - 8x - 40$ **22.** $6x^2 + x - 15$

In Problems 23–24, perform the indicated operations. Be sure to simplify each result.

23. $\dfrac{3x^2 - 2x - 1}{3x^2 - 5x - 2} \cdot \dfrac{x^2 - 9x + 14}{x^2 + x - 2}$

24. $\dfrac{3}{x^2 + x - 6} - \dfrac{2}{x^2 + 2x - 3}$

25. Grading Tests Bill can grade a set of test papers in 80 minutes. Karl can grade the same number of test papers in 120 minutes. How long will it take Bill and Karl to grade the test papers if they work together?

Getting Ready for Chapter 6:
Square Roots

Objectives

1. Evaluate Square Roots of Perfect Squares

2. Determine Whether a Square Root Is Rational, Irrational, or Not a Real Number

3. Find Square Roots of Variable Expressions

In Other Words

Taking the square root of a number "undoes" squaring a number.

In Other Words

The notation $b = \sqrt{a}$ means "produce the number b greater than or equal to 0 whose square is a."

In Section R.4, exponents were introduced. Exponents indicate repeated multiplication. For example, 4^2 means $4 \cdot 4$, so $4^2 = 16$; $(-6)^2$ means $(-6) \cdot (-6) = 36$. This section will show how to "undo" the process of raising a number to the second power and ask questions such as, "What number, or numbers, when squared, give me 16?"

▶ 1 Evaluate Square Roots of Perfect Squares

A real number is squared when it is raised to the power 2. The inverse of squaring a number is finding the **square root.** For example, since $5^2 = 25$ and $(-5)^2 = 25$, the square roots of 25 are -5 and 5. The square roots of $\dfrac{16}{49}$ are $-\dfrac{4}{7}$ and $\dfrac{4}{7}$.

If only the positive square root of a number is wanted, use the symbol $\sqrt{}$, called a **radical sign,** to denote the **principal square root,** or nonnegative (zero or positive) square root.

Definition

If a is a nonnegative real number, the nonnegative real number b such that $b^2 = a$ is the **principal square root** of a and is denoted by $b = \sqrt{a}$.

For example, the positive square root of 25 is written $\sqrt{25} = 5$. We read $\sqrt{25} = 5$ as "the principal (or positive) square root of 25 is 5." If we want the negative square root of 25, use the expression $-\sqrt{25} = -5$.

Properties of Square Roots

- Every positive real number has two square roots, one positive and one negative.
- The square root of 0 is 0. That is, $\sqrt{0} = 0$.
- Use the symbol $\sqrt{}$, called a radical sign, to denote the nonnegative square root of a real number. The nonnegative square root is called the principal square root.
- The number under the radical sign is called the **radicand**. For example, the radicand in $\sqrt{25}$ is 25.
- For any real number c, such that $c \geq 0$, $(\sqrt{c})^2 = c$. For example, $(\sqrt{4})^2 = 4$ and $(\sqrt{8.3})^2 = 8.3$.

To **evaluate** a square root, ask, "What is the nonnegative number whose square is equal to the radicand?"

EXAMPLE 1 | **Evaluating Square Roots**

Evaluate each square root.

(a) $\sqrt{36}$ (b) $\sqrt{\dfrac{1}{9}}$ (c) $\sqrt{0.01}$ (d) $(\sqrt{2.3})^2$

Solution

(a) Is there a positive number whose square is 36? Because $6^2 = 36$, $\sqrt{36} = 6$.

(b) $\sqrt{\dfrac{1}{9}} = \dfrac{1}{3}$ because $\left(\dfrac{1}{3}\right)^2 = \dfrac{1}{9}$.

(c) $\sqrt{0.01} = 0.1$ because $0.1^2 = 0.01$.

(d) $(\sqrt{2.3})^2 = 2.3$ because $(\sqrt{c})^2 = c$ when $c \geq 0$.

Figure 1

6 units

Area = 36 square units

6 units

A rational number is a **perfect square** if it is the square of a rational number. Examples 1(a), (b), and (c) are square roots of perfect squares since $6^2 = 36$, $\left(\frac{1}{3}\right)^2 = \frac{1}{9}$, and $0.1^2 = 0.01$. Perfect squares can be thought of geometrically as shown in Figure 1, where there is a square whose area is 36 square units. The square root of the area, $\sqrt{36}$, gives the length of each side of the square, 6 units.

Quick ✓

1. The symbol $\sqrt{}$ is called a _____ ___.
2. If a is a nonnegative real number, the nonnegative number b such that $b^2 = a$ is the _____ _____ ___ of a and is denoted by $b = \sqrt{a}$.
3. The square roots of 16 are ___ and ___.

In Problems 4–8, evaluate each square root.

4. $\sqrt{81}$ **5.** $\sqrt{900}$ **6.** $\sqrt{\dfrac{9}{4}}$ **7.** $\sqrt{0.16}$ **8.** $\left(\sqrt{13}\right)^2$

EXAMPLE 2   **Evaluating an Expression Containing Square Roots**

Evaluate each expression:

    (a) $-4\sqrt{36}$ **(b)** $\sqrt{9} + \sqrt{16}$ **(c)** $\sqrt{9 + 16}$ **(d)** $\sqrt{64 - 4 \cdot 7 \cdot 1}$

Solution

    (a) The expression $-4\sqrt{36}$ represents -4 times the positive square root of 36. To simplify, first find the positive square root of 36 and then multiply this result by -4.

$$-4\sqrt{36} = -4 \cdot 6$$
$$= -24$$

Work Smart

In Examples 2(b) and (c), notice that
$$\sqrt{9} + \sqrt{16} \neq \sqrt{9 + 16}$$

In general,
$$\sqrt{a} + \sqrt{b} \neq \sqrt{a + b}$$

The radical sign acts as a grouping symbol, just like parentheses, so always simplify the radicand before taking the square root.

    (b) $\sqrt{9} + \sqrt{16} = 3 + 4$
$$= 7$$

    (c) $\sqrt{9 + 16} = \sqrt{25}$
$$= 5$$

    (d) $\sqrt{64 - 4 \cdot 7 \cdot 1} = \sqrt{64 - 28}$
$$= \sqrt{36}$$
$$= 6$$

Quick ✓

In Problems 9–12, evaluate each expression.

   9. $5\sqrt{9}$ **10.** $\sqrt{36 + 64}$ **11.** $\sqrt{36} + \sqrt{64}$ **12.** $\sqrt{25 - 4 \cdot 3 \cdot (-2)}$

▶ **②** **Determine Whether a Square Root Is Rational, Irrational, or Not a Real Number**

Work Smart

Recall from Section R.2 that a rational number is a number that can be written as a quotient of two integers.

Work Smart

The square roots of negative real numbers are not real numbers.

Not all radical expressions will simplify to a rational number. For example, because there is no rational number whose square is 5, $\sqrt{5}$ is not a rational number. In fact, $\sqrt{5}$ is an *irrational* number. Remember, an irrational number is a number that cannot be written as the quotient of two integers.

What about evaluating $\sqrt{-16}$? Because any positive real number squared is positive, any negative real number squared is also positive, and 0 squared is 0, there is no real number whose square is -16. **Therefore, negative real numbers do not have square roots that are real numbers!**

These points are summarized below.

> **More Properties of Square Roots**
> - The square root of a perfect square is a rational number.
> - The square root of a positive rational number that is not a perfect square is an irrational number. For example, $\sqrt{20}$ is an irrational number because 20 is not a perfect square.
> - The square root of a negative real number is not a real number. For example, $\sqrt{-2}$ is not a real number.

If a radical has a radicand that is not a perfect square, one of two things can be done:

1. Write a decimal approximation of the radical.
2. Simplify the radical using properties of radicals, if possible (Section 6.3).

Figure 2

NORMAL FLOAT AUTO REAL RADIAN MP
$\sqrt{5}$
2.236067977

EXAMPLE 3 **Writing a Radical as a Decimal Using Technology**

Write $\sqrt{5}$ as a decimal rounded to two decimal places.

Solution
Figure 2 shows the result from a graphing calculator: $\sqrt{5} \approx 2.24$.

EXAMPLE 4 **Determining Whether a Square Root of an Integer Is Rational, Irrational, or Not a Real Number**

Determine whether each square root is rational, irrational, or not a real number. Evaluate each real square root that is rational. For each square root that is irrational, express the square root as a decimal rounded to two decimal places.

(a) $\sqrt{51}$ **(b)** $\sqrt{169}$ **(c)** $\sqrt{-81}$

Solution

Work Smart
$\sqrt{-81}$ is not a real number, but $-\sqrt{81}$ is a real number because $-\sqrt{81} = -9$. Note the placement of the negative sign.

(a) $\sqrt{51}$ is irrational because 51 is not a perfect square. That is, there is no rational number whose square is 51. Using technology, $\sqrt{51} \approx 7.14$.

(b) $\sqrt{169}$ is a rational number because $13^2 = 169$, so $\sqrt{169} = 13$.

(c) $\sqrt{-81}$ is not a real number. There is no real number whose square is -81.

Quick ✓

13. *True or False* Negative numbers do not have square roots that are real numbers.

In Problems 14–17, determine whether each square root is rational, irrational, or not a real number. Evaluate each square root that is rational. For each square root that is irrational, approximate the square root rounded to two decimal places.

14. $\sqrt{400}$ **15.** $\sqrt{40}$ **16.** $\sqrt{-25}$ **17.** $-\sqrt{196}$

▶ ❸ Find Square Roots of Variable Expressions

What is $\sqrt{4^2}$? Because $4^2 = 16$, $\sqrt{4^2} = \sqrt{16} = 4$. This result suggests that $\sqrt{a^2} = a$ for any real number a. But wait. Does $\sqrt{(-4)^2} = -4$? No! $\sqrt{(-4)^2} = \sqrt{16} = 4$, so both $\sqrt{4^2}$ and $\sqrt{(-4)^2}$ equal 4. Regardless of whether the "a" in $\sqrt{a^2}$ is positive or negative, the result is positive. Therefore, $\sqrt{a^2} = a$ is incorrect, because the sign of a is unknown. How can this "formula" be fixed? Section R.3 demonstrated that $|a|$ is a positive number if a is nonzero. From this, the following results:

In Other Words

The square root of a nonzero number squared will always be positive. The absolute value ensures this.

For any **real number** a,

$$\sqrt{a^2} = |a|$$

The bottom line is this—the square root of a variable expression raised to the second power is the absolute value of the variable expression.

EXAMPLE 5 **Evaluating Square Roots**

Evaluate each square root.

(a) $\sqrt{7^2}$ (b) $\sqrt{(-15)^2}$ (c) $\sqrt{x^2}$ (d) $\sqrt{(3x-1)^2}$
(e) $\sqrt{x^2 + 6x + 9}$

Solution

(a) $\sqrt{7^2} = 7$

(b) $\sqrt{(-15)^2} = |-15| = 15$

(c) It is unknown whether the real number x is positive, negative, or zero. To ensure that the result is positive or zero, write $\sqrt{x^2} = |x|$.

(d) $\sqrt{(3x-1)^2} = |3x - 1|$

(e) Notice that the radicand, $x^2 + 6x + 9$, factors to $(x+3)^2$. Therefore,
$$\sqrt{x^2 + 6x + 9} = \sqrt{(x+3)^2}$$
$$= |x + 3|$$

●

Quick ✓

18. $\sqrt{a^2} =$ ___.

In Problems 19–22, evaluate each square root.

19. $\sqrt{(-14)^2}$ 20. $\sqrt{z^2}$ 21. $\sqrt{(2x+3)^2}$ 22. $\sqrt{p^2 - 12p + 36}$

Getting Ready for Chapter 6 Exercises

MyMathLab® Exercise numbers in green have complete video solutions in MyMathLab or may be accessed using the QR code to the right.

Problems 1–22 are the Quick ✓s that follow the EXAMPLES.

Building Skills

In Problems 23–32, evaluate each square root. See Objective 1.

23. $\sqrt{1}$ 24. $\sqrt{9}$

25. $-\sqrt{100}$ 26. $-\sqrt{144}$

27. $\sqrt{\dfrac{1}{4}}$ 28. $\sqrt{\dfrac{4}{81}}$

29. $\sqrt{0.36}$ 30. $\sqrt{0.25}$

31. $\left(\sqrt{1.6}\right)^2$ 32. $\left(\sqrt{3.7}\right)^2$

In Problems 33–44, tell whether the square root is rational, irrational, or not a real number. If the square root is rational, find the exact value; if the square root is irrational, write the approximate value rounded to two decimal places. See Objective 2.

33. $\sqrt{-14}$ 34. $\sqrt{-50}$

35. $\sqrt{64}$ 36. $\sqrt{121}$

37. $\sqrt{\dfrac{1}{16}}$ 38. $\sqrt{\dfrac{49}{100}}$

39. $\sqrt{44}$ 40. $\sqrt{24}$

41. $\sqrt{50}$ 42. $\sqrt{12}$

43. $\sqrt{-16}$ 44. $\sqrt{-64}$

In Problems 45–56, simplify each square root. See Objective 3.

45. $\sqrt{8^2}$

46. $\sqrt{5^2}$

47. $\sqrt{(-19)^2}$

48. $\sqrt{(-13)^2}$

49. $\sqrt{r^2}$

50. $\sqrt{w^2}$

51. $\sqrt{(x+4)^2}$

52. $\sqrt{(x-8)^2}$

53. $\sqrt{(4x-3)^2}$

54. $\sqrt{(5x+2)^2}$

55. $\sqrt{4y^2+12y+9}$

56. $\sqrt{9z^2-24z+16}$

Mixed Practice

In Problems 57–74, simplify each expression.

57. $\sqrt{25+144}$

58. $\sqrt{9+16}$

59. $\sqrt{25}+\sqrt{144}$

60. $\sqrt{9}+\sqrt{16}$

61. $\sqrt{-144}$

62. $\sqrt{-36}$

63. $3\sqrt{25}$

64. $-10\sqrt{16}$

65. $5\sqrt{\dfrac{16}{25}}-\sqrt{144}$

66. $2\sqrt{\dfrac{9}{4}}-\sqrt{4}$

67. $\sqrt{8^2-4\cdot1\cdot7}$

68. $\sqrt{9^2-4\cdot1\cdot20}$

69. $\sqrt{(-5)^2-4\cdot2\cdot5}$

70. $\sqrt{(-3)^2-4\cdot3\cdot2}$

71. $\dfrac{-(-1)+\sqrt{(-1)^2-4\cdot6(-2)}}{2(-1)}$

72. $\dfrac{-7+\sqrt{7^2-4\cdot2\cdot6}}{2\cdot2}$

73. $\sqrt{(6-1)^2+(15-3)^2}$

74. $\sqrt{(2-(-1))^2+(6-2)^2}$

75. What are the square roots of 36? What is $\sqrt{36}$?

76. What are the square roots of 64? What is $\sqrt{64}$?

Math for the Future: Statistics *For Problems 77 and 78, use the formula* $Z=\dfrac{X-\mu}{\dfrac{\sigma}{\sqrt{n}}}$ *from statistics (a formula used to determine the value of one observation relative to that of another) to evaluate the expression for the given values. Write the exact value and then write your answer rounded to two decimal places.*

77. $X=120, \mu=100, \sigma=15, n=13$

78. $X=40, \mu=50, \sigma=10, n=5$

Explaining the Concepts

79. Explain why $\sqrt{a^2}=|a|$. Provide examples to support your explanation.

6

Radicals and Rational Exponents

Wind chill chart (Fahrenheit)   temperature (°F)

calm	40	35	30	25	20	15	10	5	0	−5	−10	−15	−20	−25	−30	−35	−40
5	36	31	25	19	13	7	1	−5	−11	−16	−22	−28	−34	−40	−46	−52	−57
10	34	27	21	15	9	3	−4	−10	−16	−22	−28	−35	−41	−47	−53	−59	−66
15	32	25	19	13	6	0	−7	−13	−19	−26	−32	−39	−45	−51	−58	−64	−71
20	30	24	17	11	4	−2	−9	−15	−22	−29	−35	−42	−48	−55	−61	−68	−74
25	29	23	16	9	3	−4	−11	−17	−24	−31	−37	−44	−51	−58	−64	−71	−78
30	28	22	15	8	1	−5	−12	−19	−26	−33	−39	−46	−53	−60	−67	−73	−80
35	28	21	14	7	0	−7	−14	−21	−27	−34	−41	−48	−55	−62	−69	−76	−82
40	27	20	13	6	−1	−8	−15	−22	−29	−36	−43	−50	−57	−64	−71	−78	−84
45	26	19	12	5	−2	−9	−16	−23	−30	−37	−44	−51	−58	−65	−72	−79	−86
50	26	19	12	4	−3	−10	−17	−24	−31	−38	−45	−52	−60	−67	−74	−81	−88
55	25	18	11	4	−3	−11	−18	−25	−32	−39	−46	−54	−61	−68	−75	−82	−89
60	25	17	10	3	−4	−11	−19	−26	−33	−40	−48	−55	−62	−69	−76	−84	−91

wind speed (mph)

wind chill (°F) = $35.74 + 0.6215T - 35.75(V^{0.16}) + 0.4275T(V^{0.16})$ T = air temperature (°F)

frostbite times   ▨ 30 minutes   ▨ 10 minutes   ▨ 5 minutes V = wind speed (mph)

Source: U.S. National Weather Service; Meteorological Services of Canada

According to the National Weather Service, the wind chill temperature (WCT) index uses science, technology, and computer modeling to provide an accurate, understandable, and useful formula for calculating the dangers from winter winds and freezing temperatures. The wind chill temperature is defined only for temperatures at or below 50 degrees Fahrenheit (50°F) and wind speeds above 3 miles per hour.

What is the wind chill temperature if the temperature is 0°F and the wind speed is 10 miles per hour? See Problem 137 in Section 6.1.

The Big Picture: Putting It Together

In Chapter 4, polynomial expressions were simplified by adding, subtracting, multiplying, and dividing. Polynomials were also factored. Chapter 5 explained how to use the skills learned in Chapter 4 to simplify rational expressions and perform operations on rational expressions.

This chapter will present a similar discussion for radical expressions. The discussion begins with adding, subtracting, multiplying, and dividing radical expressions. Then factoring is used to simplify radical expressions. Throughout this discussion, keep in mind that radicals perform the "inverse" of raising a real number to a positive integer exponent. For example, a square root undoes the squaring operation.

Outline

6.1 *nth* Roots and Rational Exponents

Objectives

❶ Evaluate *n*th Roots

❷ Simplify Expressions of the Form $\sqrt[n]{a^n}$

❸ Evaluate Expressions of the Form $a^{\frac{1}{n}}$

❹ Evaluate Expressions of the Form $a^{\frac{m}{n}}$

Are You Prepared for This Section?

Before getting started, complete the following problems. If you get a problem wrong, go back to the section cited and review the material.

P1. Simplify: $\left(\dfrac{x^2y}{xy^{-2}}\right)^{-3}$ [Getting Ready: Laws of Exponents, pp. 301–302]

P2. Simplify: $(\sqrt{7})^2$ [Getting Ready: Square Roots, pp. 468–469]

P3. Evaluate: $\sqrt{64}$ [Getting Ready: Square Roots, pp. 468–469]

P4. Evaluate: $\sqrt{(x+1)^2}$ [Getting Ready: Square Roots, pp. 470–471]

P5. Simplify: **(a)** 3^{-2} **(b)** x^{-4} [Getting Ready: Laws of Exponents, pp. 296–298]

In the Getting Ready: Square Roots section, skills for evaluating square roots were reviewed. This section will extend this skill to other types of roots.

▶ ❶ Evaluate *n*th Roots

A real number is cubed when it is raised to the power 3. The inverse of cubing a number is finding the *cube root*. For example, because $2^3 = 8$, the cube root of 8 is 2; because $(-2)^3 = -8$, the cube root of -8 is -2. In general, *n*th roots of numbers can be found.

In Other Words

The notation $\sqrt[n]{a} = b$ means "a number b such that raising that number to the *n*th power gives a." Thus $\sqrt[3]{125} = x$ means "a number x such that raising x to the third power gives 125."

Definition

The **principal *n*th root of a number *a*,** symbolized by $\sqrt[n]{a}$, where $n \geq 2$ is an integer, is defined as follows:

$$\sqrt[n]{a} = b \qquad \text{means} \qquad a = b^n$$

- If $n \geq 2$ and even, then a and b must be greater than or equal to 0.
- If $n \geq 3$ and odd, then a and b can be any real number.

Work Smart

If the index is even, then the radicand must be greater than or equal to zero in order for a radical to simplify to a real number. If the index is odd, the radicand can be any real number.

In the notation $\sqrt[n]{a}$, the integer n, $n \geq 2$, is called the **index.** A radical written without the index means the square root, so $\sqrt{a}$ represents the square root of a. If the index is 3, $\sqrt[3]{a}$ is called the **cube root** of a.

If the index is even, then the radicand must be greater than or equal to 0. If the index is odd, then the radicand can be any real number. Do you know why? Since $\sqrt[n]{a} = b$ means $b^n = a$, if the index n is even, then $b^n \geq 0$ so $a \geq 0$. If there is an odd index n, then b^n can be any real number, so a can be any real number.

Before *n*th roots are evaluated, some "perfect" powers of 2, 3, 4, and 5 are listed. This list will help you find roots in the following examples. Notice in the display that follows that perfect cubes and perfect fifths can be negative, but perfect squares and perfect fourths cannot. Can you determine why?

Prepared?...Answers   **P1.** $\dfrac{1}{x^3y^9}$ **P2.** 7

P3. 8   **P4.** $|x+1|$ **P5. (a)** $\dfrac{1}{9}$ **(b)** $\dfrac{1}{x^4}$

Perfect Squares	Perfect Cubes		Perfect Fourths	Perfect Fifths	
$(\pm 1)^2 = 1$	$1^3 = 1$	$(-1)^3 = -1$	$(\pm 1)^4 = 1$	$1^5 = 1$	$(-1)^5 = -1$
$(\pm 2)^2 = 4$	$2^3 = 8$	$(-2)^3 = -8$	$(\pm 2)^4 = 16$	$2^5 = 32$	$(-2)^5 = -32$
$(\pm 3)^2 = 9$	$3^3 = 27$	$(-3)^3 = -27$	$(\pm 3)^4 = 81$	$3^5 = 243$	$(-3)^5 = -243$
$(\pm 4)^2 = 16$	$4^3 = 64$	$(-4)^3 = -64$	$(\pm 4)^4 = 256$	$4^5 = 1024$	$(-4)^5 = -1024$
$(\pm 5)^2 = 25$	$5^3 = 125$	$(-5)^3 = -125$	$(\pm 5)^4 = 625$	$5^5 = 3125$	$(-5)^5 = -3125$
and so on	and so on		and so on	and so on	

EXAMPLE 1   **Evaluating *nth* Roots of Real Numbers**

Evaluate:

(a) $\sqrt[3]{1000}$ (b) $\sqrt[4]{16}$ (c) $\sqrt[3]{-8}$ (d) $\sqrt[4]{-81}$

Solution

(a) Look for a number whose cube is 1000. Since $10^3 = 1000$, $\sqrt[3]{1000} = 10$.

(b) $\sqrt[4]{16} = 2$ beacause $2^4 = 16$.

(c) $\sqrt[3]{-8} = -2$ because $(-2)^3 = -8$.

(d) Because there is no real number b such that $b^4 = -81$, $\sqrt[4]{-81}$ is not a real number.

Quick ✓

1. In the notation $\sqrt[n]{a}$, the integer n, $n \geq 2$, is called the _____.

In Problems 2–6, evaluate each root.

2. $\sqrt[3]{64}$ 3. $\sqrt[4]{81}$ 4. $\sqrt[3]{-216}$

5. $\sqrt[4]{-32}$ 6. $\sqrt[5]{\dfrac{1}{32}}$

The *nth* roots in Examples 1(a)–(c) are all rational numbers, but some *nth* roots are not. Just as square roots can be approximated using technology, *nth* roots can also be approximated.

EXAMPLE 2   **Approximating an *nth* Root Using a Calculator**

(a) Write $\sqrt[3]{25}$ as a decimal rounded to two decimal places.

(b) Write $\sqrt[4]{18}$ as a decimal rounded to two decimal places.

Solution

A graphing calculator can be used to approximate both values.

(a) Because $\sqrt[3]{8} = 2$ and $\sqrt[3]{27} = 3$, $\sqrt[3]{25}$ is between 2 and 3 (closer to 3). Figure 1(a) shows that $\sqrt[3]{25} \approx 2.92$.

(b) Because $\sqrt[4]{16} = 2$ and $\sqrt[4]{81} = 3$, $\sqrt[4]{18}$ is between 2 and 3 (closer to 2). Figure 1(b) shows that $\sqrt[4]{18} \approx 2.06$.

Figure 1

(a)

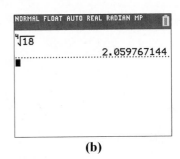
(b)

Quick ✓

In Problems 7 and 8, use technology to write the approximate value of each radical rounded to two decimal places.

7. $\sqrt[3]{50}$ 8. $\sqrt[4]{80}$

▶ ❷ Simplify Expressions of the Form $\sqrt[n]{a^n}$

Remember that $\sqrt{a^2} = |a|$. But what about $\sqrt[3]{a^3}$ or $\sqrt[4]{a^4}$? Recall that the definition of the principal nth root, $\sqrt[n]{a}$, requires that a be nonnegative (greater than or equal to zero) when n is even, but a can be any real number when n is odd.

> **Simplifying $\sqrt[n]{a^n}$**
>
> If $n \geq 2$ is a positive integer and a is a real number, then
>
> $$\sqrt[n]{a^n} = |a| \quad \text{if } n \geq 2 \text{ is even}$$
> $$\sqrt[n]{a^n} = a \quad \text{if } n \geq 3 \text{ is odd}$$

EXAMPLE 3 **Simplifying Radicals**

Simplify:

(a) $\sqrt[3]{x^3}$ **(b)** $\sqrt[4]{(x-7)^4}$ **(c)** $-\sqrt[6]{(-3)^6}$ **(d)** $\sqrt[3]{-\dfrac{8}{125}}$

Solution

(a) Because the index, 3, is odd, $\sqrt[3]{x^3} = x$.

(b) Because the index, 4, is even, $\sqrt[4]{(x-7)^4} = |x-7|$.

(c) $-\sqrt[6]{(-3)^6} = -|-3| = -3$

(d) $\sqrt[3]{-\dfrac{8}{125}} = \sqrt[3]{\dfrac{-8}{125}} = \sqrt[3]{\dfrac{(-2)^3}{5^3}} = \sqrt[3]{\left(\dfrac{-2}{5}\right)^3} = -\dfrac{2}{5}$ ●

Work Smart

Note in Example 3(d) that the negative sign could be applied to 125 instead of 8 and the answer would be the same.

Quick ✓

In Problems 9–13, simplify each radical.

9. $\sqrt[4]{5^4}$ **10.** $\sqrt[6]{z^6}$ **11.** $\sqrt[7]{(3x-2)^7}$

12. $\sqrt[8]{(-2)^8}$ **13.** $\sqrt[5]{-\dfrac{32}{243}}$

▶ ❸ Evaluate Expressions of the Form $a^{\frac{1}{n}}$

In the Getting Ready: Laws of Exponents section in Chapter 4, methods for simplifying algebraic expressions that contain integer exponents were developed. Now the rules that apply to integer exponents will be extended to rational exponents.

Work Smart

Remember, a rational number is a number of the form $\dfrac{p}{q}$, where p and q are integers and $q \neq 0$.

Start by defining "a raised to the power $\dfrac{1}{n}$," where a is a real number and n is a positive integer. This definition needs to obey all the laws of exponents presented earlier. For example, because $a^2 = a \cdot a$,

$$(5^{\frac{1}{2}})^2 = 5^{\frac{1}{2}} \cdot 5^{\frac{1}{2}}$$
$$a^m \cdot a^n = a^{m+n}: \quad = 5^{\frac{1}{2}+\frac{1}{2}}$$
$$= 5^1$$
$$= 5$$

Also, because $(\sqrt{5})^2 = 5$, it is reasonable to conclude that

$$5^{\frac{1}{2}} = \sqrt{5}$$

This suggests the following definition:

> **Definition of $a^{\frac{1}{n}}$**
>
> If a is a real number and n is an integer with $n \geq 2$, then
>
> $$a^{\frac{1}{n}} = \sqrt[n]{a}$$
>
> provided that $\sqrt[n]{a}$ exists.

EXAMPLE 4   **Evaluating Expressions Containing Exponents of the Form $a^{\frac{1}{n}}$**

Write each of the following expressions as a radical and simplify, if possible.

(a) $9^{\frac{1}{2}}$ **(b)** $(-64)^{\frac{1}{3}}$ **(c)** $-100^{\frac{1}{2}}$ **(d)** $(-100)^{\frac{1}{2}}$ **(e)** $z^{\frac{1}{2}}$

Solution

(a) $9^{\frac{1}{2}} = \sqrt{9}$
$= 3$

(b) $(-64)^{\frac{1}{3}} = \sqrt[3]{-64}$
$= -4$

(c) $-100^{\frac{1}{2}} = -1 \cdot 100^{\frac{1}{2}}$
$= -1 \cdot \sqrt{100}$
$= -1 \cdot 10$
$= -10$

(d) $(-100)^{\frac{1}{2}} = \sqrt{-100}$ is not a real number because there is no real number whose square is -100.

(e) $z^{\frac{1}{2}} = \sqrt{z}$

Work Smart

Notice how parentheses are used or not used in Examples 4(c) and (d). In Example 4(c), $-100^{\frac{1}{2}}$ means evaluate $100^{\frac{1}{2}}$ first and then multiply the result by -1. Remember the order of operations: Evaluate exponents before multiplying.

Quick ✓

14. If a is a nonnegative real number and $n \geq 2$ is an integer, then $a^{\frac{1}{n}} =$ ___.

In Problems 15–19, write each of the following expressions as a radical, and simplify, if possible.

15. $25^{\frac{1}{2}}$ **16.** $(-27)^{\frac{1}{3}}$ **17.** $-64^{\frac{1}{2}}$ **18.** $(-64)^{\frac{1}{2}}$ **19.** $b^{\frac{1}{2}}$

EXAMPLE 5   **Writing Radicals with Rational Exponents**

Rewrite each of the following radicals with a rational exponent.

(a) $\sqrt[4]{7a}$

(b) $\sqrt[5]{\dfrac{xy^3}{4}}$

Solution

(a) The index on the radical is 4, so 4 becomes the denominator of the rational exponent. Use parentheses because the radicand is $7a$, and the exponent $\dfrac{1}{4}$ is applied to both 7 and a.

$$\sqrt[4]{7a} = (7a)^{\frac{1}{4}}$$

(b) The index on the radical is 5, so the rational exponent is $\dfrac{1}{5}$.

$$\sqrt[5]{\dfrac{xy^3}{4}} = \left(\dfrac{xy^3}{4}\right)^{\frac{1}{5}}$$

Quick ✓

In Problems 20–22, rewrite each of the following radicals with a rational exponent.

20. $\sqrt[5]{8b}$

21. $\sqrt[8]{\dfrac{mn^5}{3}}$

22. $4\sqrt[3]{y}$

▶ ❹ **Evaluate Expressions of the Form $a^{\frac{m}{n}}$**

Now $a^{\frac{m}{n}}$ is defined, where m and n are integers.

Definition of $a^{\frac{m}{n}}$

If a is a real number and $\dfrac{m}{n}$ is a rational number in lowest terms with $n \geq 2$, then

$$a^{\frac{m}{n}} = \sqrt[n]{a^m} = \left(\sqrt[n]{a}\right)^m$$

provided that $\sqrt[n]{a}$ exists.

In Other Words
The expression $a^{\frac{m}{n}} = \sqrt[n]{a^m}$ means raise a to the mth power first, and then take the nth root of the result. The expression $a^{\frac{m}{n}} = (\sqrt[n]{a})^m$ means take the nth root of a first, and then raise that result to the power of m.

In the expression $a^{\frac{m}{n}}$, $\dfrac{m}{n}$ is expressed in lowest terms, and $n \geq 2$. The definition obeys all the laws of exponents presented earlier. For example,

$$a^{\frac{m}{n}} = a^{m \cdot \frac{1}{n}} = (a^m)^{\frac{1}{n}} = \sqrt[n]{a^m}$$

and

$$a^{\frac{m}{n}} = a^{\frac{1}{n} \cdot m} = (a^{\frac{1}{n}})^m = (\sqrt[n]{a})^m$$

When evaluating or simplifying $a^{\frac{m}{n}}$, use either $\sqrt[n]{a^m}$ or $(\sqrt[n]{a})^m$, whichever makes simplifying the expression easier. Generally, taking the root first, as in $(\sqrt[n]{a})^m$, is easier.

EXAMPLE 6 **Evaluating Expressions of the Form $a^{\frac{m}{n}}$**

Evaluate each of the following expressions, if possible.

    (a) $25^{\frac{3}{2}}$ **(b)** $64^{\frac{2}{3}}$ **(c)** $-9^{\frac{5}{2}}$ **(d)** $(-8)^{\frac{4}{3}}$ **(e)** $(-81)^{\frac{7}{2}}$

Solution

    (a) $25^{\frac{3}{2}} = (\sqrt{25})^3 = 5^3 = 125$

    (b) $64^{\frac{2}{3}} = (\sqrt[3]{64})^2 = 4^2 = 16$

    (c) $-9^{\frac{5}{2}} = -1 \cdot 9^{\frac{5}{2}} = -1(\sqrt{9})^5 = -1 \cdot 3^5 = -1 \cdot 243 = -243$

    (d) $(-8)^{\frac{4}{3}} = (\sqrt[3]{-8})^4 = (-2)^4 = 16$

    (e) $(-81)^{\frac{7}{2}}$ is not a real number because $(-81)^{\frac{7}{2}} = (\sqrt{-81})^7$, and $\sqrt{-81}$ is not a real number.

Quick ✓

23. If a is a real number and $\dfrac{m}{n}$ is a rational number in lowest terms with $n \geq 2$, then $a^{\frac{m}{n}} =$ _____ or _____, provided that $\sqrt[n]{a}$ exists.

In Problems 24–28, evaluate each expression, if possible.

24. $16^{\frac{3}{2}}$ **25.** $27^{\frac{2}{3}}$ **26.** $-16^{\frac{3}{4}}$

27. $(-64)^{\frac{2}{3}}$ **28.** $(-25)^{\frac{5}{2}}$

The expressions in Examples 6(a)–(d) all simplified to rational numbers. Not all expressions involving rational exponents simplify to rational numbers.

EXAMPLE 7 **Approximating Expressions Involving Rational Exponents**

Write $34^{\frac{3}{4}}$ as a decimal rounded to two decimal places.

Figure 2

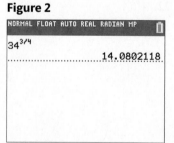

Solution

Figure 2 shows the results obtained from a graphing calculator. So $34^{\frac{3}{4}} \approx 14.08$.

Quick ✓

In Problems 29 and 30, approximate the expression rounded to two decimal places.

29. $50^{\frac{2}{3}}$ **30.** $40^{\frac{3}{20}}$

EXAMPLE 8 **Writing Radicals with Rational Exponents**

Rewrite each of the following radicals with a rational exponent.

(a) $\sqrt[3]{x^2}$ **(b)** $\left(\sqrt[5]{10a^2b}\right)^4$

Solution

(a) The index, 3, is the denominator of the rational exponent, and the power on the radicand, 2, is the numerator of the rational exponent.

$$\sqrt[3]{x^2} = x^{\frac{2}{3}}$$

(b) The index, 5, is the denominator of the rational exponent, and the power, 4, is the numerator of the rational exponent.

$$\left(\sqrt[5]{10a^2b}\right)^4 = (10a^2b)^{\frac{4}{5}}$$

Quick ✓

In Problems 31–33, write each radical with a rational exponent.

31. $\sqrt[8]{a^3}$ **32.** $\sqrt[4]{t^{12}}$ **33.** $\left(\sqrt[4]{12ab^3}\right)^9$

If a rational exponent is negative, then use the rule for negative rational exponents given below.

Negative-Exponent Rule

If $\dfrac{m}{n}$ is a rational number, and if a is a nonzero real number (that is, if $a \neq 0$), then define

$$a^{-\frac{m}{n}} = \frac{1}{a^{\frac{m}{n}}} \quad \text{and} \quad \frac{1}{a^{-\frac{m}{n}}} = a^{\frac{m}{n}} \quad \text{if } a \neq 0, \quad \text{and } \sqrt[n]{a} \text{ is defined.}$$

EXAMPLE 9 **Evaluating Expressions with Negative Rational Exponents**

Rewrite each of the following with positive exponents, and completely simplify, if possible.

(a) $36^{-\frac{1}{2}}$ **(b)** $\dfrac{1}{27^{-\frac{2}{3}}}$ **(c)** $(6a)^{-\frac{5}{4}}$

Solution

(a) $36^{-\frac{1}{2}} = \dfrac{1}{36^{\frac{1}{2}}} = \dfrac{1}{\sqrt{36}} = \dfrac{1}{6}$

(b) Because the negative exponent is in the denominator, use $\dfrac{1}{a^{-\frac{m}{n}}} = a^{\frac{m}{n}}$ to simplify:

$$\frac{1}{27^{-\frac{2}{3}}} = 27^{\frac{2}{3}} = \left(\sqrt[3]{27}\right)^2 = 3^2 = 9$$

(c) $(6a)^{-\frac{5}{4}} = \dfrac{1}{(6a)^{\frac{5}{4}}}$

Quick ✓

In Problems 34–36, rewrite each of the following with positive exponents, and completely simplify, if possible.

34. $81^{-\frac{1}{2}}$ **35.** $\dfrac{1}{8^{-\frac{2}{3}}}$ **36.** $(13x)^{-\frac{3}{2}}$

*Problems **1–36** are the Quick ✓s that follow the **EXAMPLES**.*

Building Skills

In Problems 37–46, evaluate each root. See Objective 1.

37. $\sqrt[3]{125}$ **38.** $\sqrt[3]{216}$

39. $\sqrt[3]{-27}$ **40.** $\sqrt[3]{-64}$

41. $-\sqrt[4]{625}$ **42.** $-\sqrt[4]{256}$

43. $\sqrt[3]{-\dfrac{1}{8}}$ **44.** $\sqrt[3]{\dfrac{8}{125}}$

45. $-\sqrt[5]{-243}$ **46.** $-\sqrt[5]{-1024}$

In Problems 47–50, use technology to write each expression as a decimal rounded to two decimal places. See Objective 1.

47. $\sqrt[3]{25}$ **48.** $\sqrt[3]{85}$

49. $\sqrt[4]{12}$ **50.** $\sqrt[4]{2}$

In Problems 51–58, simplify each radical. See Objective 2.

51. $\sqrt[3]{5^3}$ **52.** $\sqrt[4]{6^4}$

53. $\sqrt[4]{m^4}$ **54.** $\sqrt[5]{n^5}$

55. $\sqrt[9]{(x-3)^9}$ **56.** $\sqrt[6]{(2x-3)^6}$

57. $-\sqrt[4]{(3p+1)^4}$ **58.** $-\sqrt[3]{(6z-5)^3}$

In Problems 59–72, evaluate each expression, if possible. See Objective 3.

59. $4^{\frac{1}{2}}$ **60.** $16^{\frac{1}{2}}$

61. $-36^{\frac{1}{2}}$ **62.** $-25^{\frac{1}{2}}$

63. $8^{\frac{1}{3}}$ **64.** $27^{\frac{1}{3}}$

65. $-16^{\frac{1}{4}}$ **66.** $-81^{\frac{1}{4}}$

67. $\left(\dfrac{4}{25}\right)^{\frac{1}{2}}$ **68.** $\left(\dfrac{8}{27}\right)^{\frac{1}{3}}$

69. $(-125)^{\frac{1}{3}}$ **70.** $(-216)^{\frac{1}{3}}$

71. $(-4)^{\frac{1}{2}}$ **72.** $(-81)^{\frac{1}{2}}$

In Problems 73–76, rewrite each of the following radicals with a rational exponent. See Objective 3.

73. $\sqrt[3]{3x}$ **74.** $\sqrt[5]{2y}$

75. $\sqrt[4]{\dfrac{x}{3}}$ **76.** $\sqrt{\dfrac{w}{2}}$

In Problems 77–92, evaluate each expression. See Objective 4.

77. $4^{\frac{5}{2}}$ **78.** $25^{\frac{3}{2}}$

79. $-16^{\frac{3}{2}}$ **80.** $-100^{\frac{5}{2}}$

81. $8^{\frac{4}{3}}$ **82.** $27^{\frac{4}{3}}$

83. $(-64)^{\frac{2}{3}}$ **84.** $(-125)^{\frac{2}{3}}$

85. $-(-32)^{\frac{3}{5}}$ **86.** $-(-216)^{\frac{2}{3}}$

87. $144^{-\frac{1}{2}}$ **88.** $121^{-\frac{1}{2}}$

89. $\dfrac{1}{25^{-\frac{3}{2}}}$ **90.** $\dfrac{1}{49^{-\frac{3}{2}}}$

91. $\dfrac{1}{8^{-\frac{5}{3}}}$ **92.** $27^{-\frac{4}{3}}$

In Problems 93–100, rewrite each of the following radicals with a rational exponent. See Objective 4.

93. $\sqrt[4]{x^3}$ **94.** $\sqrt[3]{p^5}$

95. $\left(\sqrt[5]{3x}\right)^2$ **96.** $\left(\sqrt[4]{6z}\right)^3$

97. $\sqrt{\left(\dfrac{5x}{y}\right)^3}$ **98.** $\sqrt[6]{\left(\dfrac{2a}{b}\right)^5}$

99. $\sqrt[3]{(9ab)^4}$ **100.** $\sqrt[4]{(3pq)^7}$

In Problems 101–106, use technology to write each expression as a decimal rounded to two decimal places. See Objective 4.

101. $20^{\frac{1}{2}}$ **102.** $5^{\frac{1}{2}}$

103. $4^{\frac{5}{3}}$ **104.** $100^{\frac{3}{4}}$

105. $10^{0.1}$ **106.** $100^{0.25}$

Mixed Practice

In Problems 107–130, simplify each expression, if possible.

107. $\sqrt[3]{x^3} + 4\sqrt[5]{x^5}$ **108.** $\sqrt[3]{(x-1)^3} + \sqrt[5]{32}$

109. $-16^{\frac{3}{4}} - \sqrt[5]{32}$ **110.** $\dfrac{5\sqrt[7]{x^7}}{\sqrt[3]{x^3}}, x \neq 0$

111. $\sqrt[3]{512}$ **112.** $\sqrt[3]{-125}$

113. $9^{2.5}$ **114.** $100^{1.5}$

115. $\sqrt[4]{-16}$ **116.** $\sqrt[4]{-1}$

117. $144^{-\frac{1}{2}} \cdot 3^2$ **118.** $125^{-\frac{1}{3}}$

119. $\sqrt[3]{0.008}$ **120.** $\sqrt[4]{0.0081}$

121. $4^{0.5} + 25^{1.5}$

122. $100^{0.5} - 4^{1.5}$

123. $(-25)^{\frac{5}{2}}$

124. $(-125)^{-\frac{1}{3}}$

125. $\sqrt[3]{(3p - 5)^3}$

126. $\sqrt[5]{(6b - 1)^5}$

127. $-9^{\frac{3}{2}} + \dfrac{1}{27^{-\frac{2}{3}}}$

128. $\dfrac{1}{64^{\frac{1}{2}}} - 4^{-\frac{3}{2}}$

129. $\sqrt[6]{(-2)^6}$

130. $\sqrt[4]{(-10)^4}$

In Problems 131–134, evaluate each function.

131. $f(x) = x^{\frac{3}{2}}$; find $f(4)$

132. $g(x) = x^{-\frac{3}{2}}$; find $g(16)$

133. $F(z) = z^{\frac{4}{3}}$; find $F(-8)$

134. $G(a) = a^{\frac{5}{3}}$; find $G(-8)$

Applying the Concepts

135. What is the cube root of 1000? What is $\sqrt[3]{1000}$?

136. What is the cube root of 729? What is $\sqrt[3]{729}$?

137. Wind Chill According to the National Weather Service, the wind chill temperature is how cold people and animals feel when outside. Wind chill is based on the rate of heat loss from exposed skin caused by wind and cold. The formula for computing wind chill W is

$$W = 35.74 + 0.6215T - 35.75v^{0.16} + 0.4275Tv^{0.16}$$

where T is the air temperature in degrees Fahrenheit and v is the wind speed in miles per hour.

(a) What is the wind chill if it is 30°F and the wind speed is 10 miles per hour?

(b) What is the wind chill if it is 30°F and the wind speed is 20 miles per hour?

(c) What is the wind chill if it is 0°F and the wind speed is 10 miles per hour?

138. Money The annual rate of interest r (expressed as a decimal) required to have A dollars after t years from an initial deposit of P dollars is given by

$$r = \left(\frac{A}{P}\right)^{\frac{1}{t}} - 1$$

(a) If you deposit $100 in a mutual fund today and have $144 in the account in 2 years, what was your annual rate of interest earned?

(b) If you deposit $100 in a mutual fund today and have $337.50 in 3 years, what was your annual rate of interest earned?

(c) The Rule of 72 states that your money will double in $\dfrac{72}{100r}$ years, where r is the rate of interest earned (expressed as a decimal). Suppose that you deposit $1000 in a mutual fund today and have $2000 in 8 years. What rate of interest did you earn? Compute $\dfrac{72}{100r}$ for this rate of interest. Is it close?

139. Terminal Velocity Terminal velocity is the maximum speed that a body falling through air can reach (because of air resistance). Terminal velocity is given by the formula $v_t = \sqrt{\dfrac{2mg}{C\rho A}}$, where m is the mass of the falling object, g is acceleration due to gravity (≈ 9.81 meters per second^2), C is a drag coefficient with $0.5 \le C \le 1.0$, ρ is the density of air (≈ 1.2 kg/m^3), and A is the cross-sectional area of the object. Suppose that a raindrop whose radius is 1.5 mm falls from the sky. The mass of the raindrop is given by $m = \dfrac{4}{3}\pi r^3 \rho_w$, where r is its radius and $\rho_w = 1000$ kg/m^3. The cross-sectional area of the raindrop is $A = \pi r^2$.

(a) Substitute the formulas for the mass and area of a raindrop into the formula for terminal speed, and simplify the expression.

(b) Determine the terminal velocity of a raindrop whose radius is 0.0015 m with $C = 0.6$.

140. Kepler's Law Early in the seventeenth century, Johannes Kepler (1571–1630) discovered that the square of the period T of a planet to rotate around the Sun varies directly with the cube of its mean distance r from the Sun. The period of a planet is the amount of time (in Earth years) for the planet to complete one orbit around the Sun. Kepler's Law can be expressed using rational exponents as $T = kr^{\frac{3}{2}}$, where k is the constant of proportionality.

(a) The period of Mercury is 0.241 year, and its mean distance from the Sun is 5.79×10^{10} meters. Use this information to state Kepler's Law (find the value of k).

(b) The mean distance of Mars from the Sun is 2.28×10^{11}m. Use this information, along with the result of part (a), to find the amount of time it takes Mars to complete one orbit around the Sun.

Explaining the Concepts

141. Explain why $(-9)^{\frac{1}{2}}$ is not a real number, but $-9^{\frac{1}{2}}$ is a real number.

142. In your own words, provide a justification for why $a^{\frac{1}{n}} = \sqrt[n]{a}$.

143. Under what conditions is $a^{\frac{m}{n}}$ a real number?

Synthesis Review

In Problems 144–147, simplify each expression completely.

144. $\left(\dfrac{x^2 y}{y^{-2}}\right)^3$

145. $\dfrac{(x + 2)^2 (x - 1)^4}{(x + 2)(x - 1)}$

146. $\dfrac{(3a^2 + 5a - 3) - (a^2 - 2a - 9)}{4a^2 + 12a + 9}$

147. $\dfrac{(4z^2 - 7z + 3) + (-3z^2 - z + 9)}{(4z^2 - 2z - 7) + (-3z^2 - z + 9)}$

6.2 Simplifying Expressions Using the Laws of Exponents

Objectives

4 Simplify Expressions Involving Rational Exponents

5 Simplify Radical Expressions

6 Factor Expressions Containing Rational Exponents

Are You Prepared for This Section?

Before getting started, complete the following problems. If you get a problem wrong, go back to the section cited and review the material.

P1. Simplify: z^{-3} [Getting Ready: Laws of Exponents, pp. 296–298]

P2. Simplify: $x^{-2} \cdot x^5$ [Getting Ready: Laws of Exponents, pp. 298–299]

P3. Simplify: $\left(\dfrac{2a^2}{b^{-1}}\right)^3$ [Getting Ready: Laws of Exponents, pp. 301–302]

P4. Evaluate: $\sqrt{144}$ [Getting Ready: Square Roots, pp. 468–469]

▶ 1 Simplify Expressions Involving Rational Exponents

The Laws of Exponents presented in the Getting Ready: Laws of Exponents section applied to integer exponents. These laws apply to rational exponents as well.

Work Smart: Study Skills

When learning new rules, it is often helpful to write each rule on a notecard that you can study regularly. For example, you could put a^0 on one side of a notecard and its value of 1 on the other side. This will allow you to quiz yourself. There is also a free app called Quizlet that can be used.

The Laws of Exponents

If a and b are real numbers and if r and s are rational numbers, then, assuming the expression is defined, the following rules apply.

Zero-Exponent Rule: $a^0 = 1$ if $a \neq 0$

Negative-Exponent Rule: $a^{-r} = \dfrac{1}{a^r}$ if $a \neq 0$

Product Rule: $a^r \cdot a^s = a^{r+s}$

Quotient Rule: $\dfrac{a^r}{a^s} = a^{r-s} = \dfrac{1}{a^{s-r}}$ if $a \neq 0$

Power Rule: $(a^r)^s = a^{r \cdot s}$

Product-to-a-Power Rule: $(a \cdot b)^r = a^r \cdot b^r$

Quotient-to-a-Power Rule: $\left(\dfrac{a}{b}\right)^r = \dfrac{a^r}{b^r}$ if $b \neq 0$

Quotient-to-a-Negative-Power Rule: $\left(\dfrac{a}{b}\right)^{-r} = \left(\dfrac{b}{a}\right)^r$ if $a \neq 0, b \neq 0$

Work Smart

We *simplify* expressions (no equal sign) and *solve* equations (these have equal signs).

The direction **simplify** shall mean the following:

- All the exponents are positive. For example, x^{-3} is written as $\frac{1}{x^3}$ when simplified.
- Each base occurs only once. For example, $x^3 y^2 x^2$ is simplified as $x^5 y^2$.
- There are no powers written to powers. For example, $(x^3)^2$ is simplified as x^6.

Prepared?...Answers P1. $\dfrac{1}{z^3}, z \neq 0$

P2. x^3 **P3.** $8a^6 b^3$ **P4.** 12

EXAMPLE 1 | **Simplifying Expressions Involving Rational Exponents**

Simplify each of the following:

(a) $27^{\frac{1}{2}} \cdot 27^{\frac{5}{6}}$

(b) $\dfrac{8^{\frac{1}{3}}}{8^{\frac{5}{3}}}$

Solution

(a)

$$a^r \cdot a^s = a^{r+s}$$
$$\downarrow$$
$$27^{\frac{1}{2}} \cdot 27^{\frac{5}{6}} = 27^{\frac{1}{2}+\frac{5}{6}}$$
$$= 27^{\frac{3}{6}+\frac{5}{6}}$$
$$= 27^{\frac{8}{6}}$$
$$= 27^{\frac{4}{3}}$$
$$a^{\frac{m}{n}} = (\sqrt[n]{a})^m: \quad = (\sqrt[3]{27})^4$$
$$= 3^4$$
$$= 81$$

Work Smart

Example 1(b) could also be simplified as follows:

$$\dfrac{8^{\frac{1}{3}}}{8^{\frac{5}{3}}} = \dfrac{1}{8^{\frac{5}{3}-\frac{1}{3}}}$$
$$= \dfrac{1}{8^{\frac{4}{3}}}$$
$$= \dfrac{1}{(\sqrt[3]{8})^4}$$
$$= \dfrac{1}{16}$$

(b)

$$\dfrac{a^r}{a^s} = a^{r-s}$$
$$\downarrow$$
$$\dfrac{8^{\frac{1}{3}}}{8^{\frac{5}{3}}} = 8^{\frac{1}{3}-\frac{5}{3}}$$
$$= 8^{-\frac{4}{3}}$$
$$= \dfrac{1}{8^{\frac{4}{3}}}$$
$$a^{\frac{m}{n}} = (\sqrt[n]{a})^m: \quad = \dfrac{1}{(\sqrt[3]{8})^4}$$
$$= \dfrac{1}{2^4}$$
$$= \dfrac{1}{16}$$

●

EXAMPLE 2 | **Simplifying Expressions Involving Rational Exponents**

Simplify each of the following. Assume all variables are positive.

(a) $\left(36^{\frac{2}{5}}\right)^{\frac{5}{4}}$

(b) $\left(x^{\frac{1}{2}} \cdot y^{\frac{2}{3}}\right)^{\frac{3}{2}}$

Solution

(a)

$$(a^r)^s = a^{r \cdot s}$$
$$\downarrow$$
$$\left(36^{\frac{2}{5}}\right)^{\frac{5}{4}} = 36^{\frac{2}{5} \cdot \frac{5}{4}}$$
$$= 36^{\frac{10}{20}}$$
$$= 36^{\frac{1}{2}}$$
$$a^{1/2} = \sqrt{a}: \quad = \sqrt{36}$$
$$= 6$$

(b)

$$(ab)^r = a^r \cdot b^r$$
$$\downarrow$$
$$\left(x^{\frac{1}{2}} \cdot y^{\frac{2}{3}}\right)^{\frac{3}{2}} = \left(x^{\frac{1}{2}}\right)^{\frac{3}{2}} \cdot \left(y^{\frac{2}{3}}\right)^{\frac{3}{2}}$$
$$(a^r)^s = a^{r \cdot s}: \quad = x^{\frac{1 \cdot 3}{2 \cdot 2}} \cdot y^{\frac{2 \cdot 3}{3 \cdot 2}}$$
$$= x^{\frac{3}{4}} y$$

●

Quick ✓

1. If a and b are real numbers and if r is a rational number, then assuming the expression is defined, $(ab)^r = $ ___.

2. If a is a real number and if r and s are rational numbers, then assuming the expression is defined, $a^r \cdot a^s = $ ____.

In Problems 3–7, simplify each expression. Assume all variables are positive.

3. $5^{\frac{3}{4}} \cdot 5^{\frac{1}{6}}$

4. $\dfrac{32^{\frac{6}{5}}}{32^{\frac{3}{5}}}$

5. $\left(100^{\frac{3}{8}}\right)^{\frac{4}{3}}$

6. $\left(a^{\frac{3}{2}} \cdot b^{\frac{5}{4}}\right)^{\frac{2}{3}}$

7. $\dfrac{x^{\frac{1}{2}} \cdot x^{\frac{1}{3}}}{\left(x^{\frac{1}{12}}\right)^2}$

EXAMPLE 3 | **Simplifying Expressions Involving Rational Exponents**

Simplify each of the following. Assume all variables are positive.

(a) $\left(x^{\frac{2}{3}} y^{-1}\right)\left(x^{-1} y^{\frac{1}{2}}\right)^{\frac{2}{3}}$

(b) $\left(\dfrac{9xy^{\frac{4}{3}}}{x^{\frac{5}{6}} y^{-\frac{2}{3}}}\right)^{\frac{1}{2}}$

(continued)

Solution

(a)

Product-to-a-Power Rule: $(ab)^r = a^r b^r$

$$\left(x^{\frac{2}{3}}y^{-1}\right)\left(x^{-1}y^{\frac{1}{2}}\right)^{\frac{2}{3}} = x^{\frac{2}{3}}y^{-1}\left(x^{-1}\right)^{\frac{2}{3}}\left(y^{\frac{1}{2}}\right)^{\frac{2}{3}}$$

Power Rule: $(a^r)^s = a^{rs}$: $\quad = x^{\frac{2}{3}}y^{-1}x^{-\frac{2}{3}}y^{\frac{1}{3}}$

Product Rule: $a^r \cdot a^s = a^{r+s}$: $\quad = x^{\frac{2}{3}+\left(-\frac{2}{3}\right)}y^{-1+\frac{1}{3}}$

$$= x^0 y^{-\frac{2}{3}}$$

$a^0 = 1$; Negative-Exponent Rule: $a^{-r} = \dfrac{1}{a^r}$: $\quad = \dfrac{1}{y^{\frac{2}{3}}}$

(b)

Quotient Rule: $\dfrac{a^r}{a^s} = a^{r-s}$

$$\left(\frac{9xy^{\frac{4}{3}}}{x^{\frac{5}{6}}y^{-\frac{2}{3}}}\right)^{\frac{1}{2}} = \left(9x^{1-\frac{5}{6}}y^{\frac{4}{3}-\left(-\frac{2}{3}\right)}\right)^{\frac{1}{2}}$$

$x^{1-\frac{5}{6}} = x^{\frac{6}{6}-\frac{5}{6}} = x^{\frac{1}{6}}$;

$y^{\frac{4}{3}-\left(-\frac{2}{3}\right)} = y^{\frac{4}{3}+\frac{2}{3}} = y^{\frac{6}{3}} = y^2$: $\quad = \left(9x^{\frac{1}{6}}y^2\right)^{\frac{1}{2}}$

Product-to-a-Power Rule: $(a \cdot b)^r = a^r \cdot b^r$: $\quad = 9^{\frac{1}{2}} \cdot \left(x^{\frac{1}{6}}\right)^{\frac{1}{2}}\left(y^2\right)^{\frac{1}{2}}$

$9^{\frac{1}{2}} = \sqrt{9} = 3$; Power Rule: $(a^r)^s = a^{rs}$: $\quad = 3x^{\frac{1}{12}}y$

Quick ✓

In Problems 8–10, simplify each expression. Assume all variables are positive.

8. $\left(8x^{\frac{3}{4}}y^{-1}\right)^{\frac{2}{3}}$ **9.** $\left(\dfrac{25x^{\frac{1}{2}}y^{\frac{3}{4}}}{x^{-\frac{3}{4}}y}\right)^{\frac{1}{2}}$ **10.** $8\left(125a^{\frac{3}{4}}b^{-1}\right)^{\frac{2}{3}}$

▶ ❷ Simplify Radical Expressions

Rational exponents can be used to simplify radicals.

EXAMPLE 4 | **Simplifying Radicals Using Rational Exponents**

Use rational exponents to simplify the radicals. Assume all variables are positive.

(a) $\sqrt[8]{16^4}$ **(b)** $\sqrt[3]{-64x^6y^3}$ **(c)** $\dfrac{\sqrt{x}}{\sqrt[3]{x^2}}$ **(d)** $\sqrt{\sqrt[3]{z}}$

Solution
The idea in all these problems is to rewrite the radical as an expression involving a rational exponent. Then use the Laws of Exponents to simplify the expression. Finally, write the simplified expression as a radical.

(a)

Write radical as a rational exponent using $\sqrt[n]{a^m} = a^{\frac{m}{n}}$.

$$\sqrt[8]{16^4} = 16^{\frac{4}{8}}$$

Simplify the exponent: $\quad = 16^{\frac{1}{2}}$

Write rational exponent as a radical: $\quad = \sqrt{16}$

$$= 4$$

(b)

$$\sqrt[n]{a} = a^{\frac{1}{n}}$$

$$\sqrt[3]{-64x^6y^3} = \left(-64x^6y^3\right)^{\frac{1}{3}}$$

$(ab)^r = a^r \cdot b^r$: $\quad = (-64)^{\frac{1}{3}}\left(x^6\right)^{\frac{1}{3}}\left(y^3\right)^{\frac{1}{3}}$

$(a^r)^s = a^{rs}$: $\quad = (-64)^{\frac{1}{3}} \cdot x^{6\cdot\frac{1}{3}} \cdot y^{3\cdot\frac{1}{3}}$

$(-64)^{\frac{1}{3}} = \sqrt[3]{-64} = -4$: $\quad = -4x^2y$

(c)

$$\sqrt{a} = a^{\frac{1}{2}};\ \sqrt[n]{a^m} = a^{\frac{m}{n}}$$
$$\downarrow$$

$$\frac{\sqrt{x}}{\sqrt[3]{x^2}} = \frac{x^{\frac{1}{2}}}{x^{\frac{2}{3}}}$$

$$\frac{a^r}{a^s} = a^{r-s}: \quad = x^{\frac{1}{2}-\frac{2}{3}}$$

$$\frac{1}{2}-\frac{2}{3} = \frac{3}{6}-\frac{4}{6} = -\frac{1}{6}: \quad = x^{-\frac{1}{6}}$$

$$a^{-r} = \frac{1}{a^r}: \quad = \frac{1}{x^{\frac{1}{6}}}$$

Write rational exponent as a radical: $\quad = \dfrac{1}{\sqrt[6]{x}}$

(d)

Write radicand with a rational exponent.
$$\downarrow$$

$$\sqrt{\sqrt[3]{z}} = \sqrt{z^{\frac{1}{3}}}$$

$$\sqrt{a} = a^{\frac{1}{2}}: \quad = \left(z^{\frac{1}{3}}\right)^{\frac{1}{2}}$$

$$\left(a^r\right)^s = a^{rs}: \quad = z^{\frac{1}{3}\cdot\frac{1}{2}}$$

$$= z^{\frac{1}{6}}$$

Write rational exponent as a radical: $\quad = \sqrt[6]{z}$

Quick ✓

In Problems 11–14, use rational exponents to simplify each radical. Assume all variables are positive.

11. $\sqrt[10]{36^5}$

12. $\sqrt[4]{16a^8b^{12}}$

13. $\dfrac{\sqrt[3]{x^2}}{\sqrt[4]{x}}$

14. $\sqrt[4]{\sqrt[3]{a^2}}$

▶ ❸ **Factor Expressions Containing Rational Exponents**

Often, expressions involving rational exponents contain a common factor. When this occurs, factor out the common factor to write the expression in simplified form. The goal of this type of problem is to write the expression as either a single product or a single quotient. The following examples illustrate this idea.

EXAMPLE 5 **Writing an Expression Containing Rational Exponents as a Single Product**

Simplify $9x^{\frac{4}{3}} + 4x^{\frac{1}{3}}(3x + 5)$ by factoring out the common factor of both terms, $x^{\frac{1}{3}}$.

Solution
$x^{\frac{1}{3}}$ is a factor of the second term, $4x^{\frac{1}{3}}(3x + 5)$. It is also a factor of the first term, $9x^{\frac{4}{3}}$. This can be seen by rewriting $9x^{\frac{4}{3}}$ as $9x^{\frac{3}{3}+\frac{1}{3}} = 9x^{\frac{3}{3}} \cdot x^{\frac{1}{3}} = 9x \cdot x^{\frac{1}{3}}$. Now factor out $x^{\frac{1}{3}}$.

$$9x^{\frac{4}{3}} + 4x^{\frac{1}{3}}(3x + 5) = 9x \cdot x^{\frac{1}{3}} + 4x^{\frac{1}{3}}(3x + 5)$$

Factor out $x^{1/3}$: $= x^{\frac{1}{3}}(9x + 4(3x + 5))$

Distribute the 4: $= x^{\frac{1}{3}}(9x + 12x + 20)$

Combine like terms: $= x^{\frac{1}{3}}(21x + 20)$

Work Smart

When factoring out the greatest common factor, factor out the variable expression raised to the largest exponent that the expressions have in common. For example,

$$3x^5 + 12x^2 = 3x^2(x^3 + 4)$$

Quick ✓

15. Simplify $8x^{\frac{3}{2}} + 3x^{\frac{1}{2}}(4x + 3)$ by factoring out the common factor, $x^{\frac{1}{2}}$.

EXAMPLE 6 **Writing an Expression Containing Rational Exponents as a Single Quotient**

Simplify $4x^{\frac{1}{2}} + x^{-\frac{1}{2}}(2x + 1)$ by factoring out the common factor, $x^{-\frac{1}{2}}$.

Solution

$x^{-\frac{1}{2}}$ is a factor of the second term, $x^{-\frac{1}{2}}(2x + 1)$. It is also a factor of the first term, $4x^{\frac{1}{2}}$. This can be seen by rewriting $4x^{\frac{1}{2}}$ as $4x^{\frac{2}{2}-\frac{1}{2}} = 4x^{\frac{2}{2}} \cdot x^{-\frac{1}{2}} = 4x \cdot x^{-\frac{1}{2}}$. Factor out $x^{-\frac{1}{2}}$ as the greatest common factor (GCF).

$$4x^{\frac{1}{2}} + x^{-\frac{1}{2}}(2x + 1) = 4x \cdot x^{-\frac{1}{2}} + x^{-\frac{1}{2}}(2x + 1)$$

Factor out $x^{-1/2}$: $= x^{-\frac{1}{2}}(4x + (2x + 1))$

Combine like terms: $= x^{-\frac{1}{2}}(6x + 1)$

Rewrite without negative exponents: $= \dfrac{6x + 1}{x^{\frac{1}{2}}}$

Quick ✓

16. Simplify $9x^{\frac{1}{3}} + x^{-\frac{2}{3}}(3x + 1)$ by factoring out the common factor, $x^{-\frac{2}{3}}$.

6.2 Exercises MyMathLab® Exercise numbers in green have complete video solutions in MyMathLab or may be accessed using the QR code to the right.

*Problems **1–16** are the Quick ✓s that follow the EXAMPLES.*

Building Skills

In Problems 17–38, simplify each of the following expressions. Assume all variables are positive. See Objective 1.

17. $5^{\frac{1}{2}} \cdot 5^{\frac{3}{2}}$

18. $3^{\frac{1}{3}} \cdot 3^{\frac{5}{3}}$

19. $\dfrac{8^{\frac{5}{4}}}{8^{\frac{1}{4}}}$

20. $\dfrac{10^{\frac{7}{5}}}{10^{\frac{2}{5}}}$

21. $2^{\frac{1}{3}} \cdot 2^{-\frac{1}{2}}$

22. $9^{-\frac{5}{4}} \cdot 9^{\frac{1}{3}}$

23. $\dfrac{x^{\frac{1}{4}}}{x^{\frac{5}{6}}}$

24. $\dfrac{y^{\frac{1}{5}}}{y^{\frac{9}{10}}}$

25. $\left(4^{\frac{4}{3}}\right)^{\frac{3}{8}}$

26. $\left(9^{\frac{9}{5}}\right)^{\frac{5}{6}}$

27. $\left(25^{\frac{3}{4}} \cdot 4^{-\frac{3}{4}}\right)^2$

28. $\left(36^{-\frac{1}{4}} \cdot 9^{\frac{3}{4}}\right)^{-2}$

29. $\left(x^{\frac{3}{4}} \cdot y^{\frac{1}{3}}\right)^{\frac{2}{3}}$

30. $\left(a^{\frac{5}{4}} \cdot b^{\frac{3}{2}}\right)^{\frac{2}{5}}$

31. $\left(x^{-\frac{1}{3}} \cdot y\right)\left(x^{\frac{1}{2}} \cdot y^{-\frac{4}{3}}\right)$

32. $\left(a^{\frac{4}{3}} \cdot b^{-\frac{1}{2}}\right)\left(a^{-2} \cdot b^{\frac{5}{2}}\right)$

33. $\left(4a^2 b^{-\frac{3}{2}}\right)^{\frac{1}{2}}$

34. $\left(25p^{\frac{2}{5}}q^{-1}\right)^{\frac{1}{2}}$

35. $\left(\dfrac{x^{\frac{2}{3}}y^{-\frac{1}{3}}}{8x^{\frac{1}{2}}y}\right)^{\frac{1}{3}}$

36. $\left(\dfrac{64m^{\frac{1}{2}}n}{m^{-2}n^{\frac{4}{3}}}\right)^{\frac{1}{2}}$

37. $\left(\dfrac{50x^{\frac{3}{4}}y}{2x^{\frac{1}{2}}}\right)^{\frac{1}{2}} + \left(\dfrac{x^{\frac{1}{2}}y^{\frac{1}{2}}}{9x^{\frac{3}{4}}y^{\frac{3}{2}}}\right)^{-\frac{1}{2}}$

38. $\left(\dfrac{27x^{\frac{1}{2}}y^{-1}}{y^{-\frac{2}{3}}x^{-\frac{1}{2}}}\right)^{\frac{1}{3}} - \left(\dfrac{4x^{\frac{1}{3}}y^{\frac{4}{9}}}{x^{-\frac{1}{3}}y^{\frac{2}{3}}}\right)^{\frac{1}{2}}$

In Problems 39–54, use rational exponents to simplify each radical. Assume all variables are positive. See Objective 2.

39. $\sqrt{x^8}$

40. $\sqrt[3]{x^6}$

41. $\sqrt[12]{8^4}$

42. $\sqrt[9]{125^6}$

43. $\sqrt[3]{-8a^3 b^{12}}$

44. $\sqrt{25x^4 y^6}$

45. $\dfrac{\sqrt{x}}{\sqrt[4]{x}}$

46. $\dfrac{\sqrt[3]{y^2}}{\sqrt[4]{y}}$

47. $\sqrt{x} \cdot \sqrt[3]{x}$

48. $\sqrt[6]{p^3} \cdot \sqrt[3]{p}$

49. $\sqrt{\sqrt[4]{x^3}}$

50. $\sqrt[3]{\sqrt{x^3}}$

51. $\sqrt{3} \cdot \sqrt[3]{9}$

52. $\sqrt{5} \cdot \sqrt[3]{25}$

53. $\dfrac{\sqrt{6}}{\sqrt[4]{36}}$

54. $\dfrac{\sqrt[4]{49}}{\sqrt{7}}$

For Problems 55–64, see Objective 3.

55. Simplify $5x^{\frac{3}{2}} + 3x^{\frac{1}{2}}(x + 5)$ by factoring out the common factor, $x^{\frac{1}{2}}$.

56. Simplify $5x^{\frac{4}{3}} + 4x^{\frac{1}{3}}(2x - 3)$ by factoring out the common factor, $x^{\frac{1}{3}}$.

57. Simplify $5(x + 2)^{\frac{2}{3}}(3x - 2) + 9(x + 2)^{\frac{5}{3}}$ by factoring out the common factor, $(x + 2)^{\frac{2}{3}}$.

58. Simplify $3(x - 5)^{\frac{1}{2}}(3x + 1) + 6(x - 5)^{\frac{3}{2}}$ by factoring out the common factor, $(x - 5)^{\frac{1}{2}}$.

59. Simplify $x^{-\frac{1}{2}}(2x + 5) + 4x^{\frac{1}{2}}$ by factoring out the common factor, $x^{-\frac{1}{2}}$.

60. Simplify $x^{-\frac{2}{3}}(3x + 2) + 9x^{\frac{1}{3}}$ by factoring out the common factor, $x^{-\frac{2}{3}}$.

61. Simplify $2(x - 4)^{-\frac{1}{3}}(4x - 3) + 12(x - 4)^{\frac{2}{3}}$ by factoring out the common factor, $2(x - 4)^{-\frac{1}{3}}$.

62. Simplify $4(x + 3)^{\frac{1}{2}} + (x + 3)^{-\frac{1}{2}}(2x + 1)$ by factoring out the common factor, $(x + 3)^{-\frac{1}{2}}$.

63. Simplify $15x(x^2 + 4)^{\frac{1}{2}} + 5(x^2 + 4)^{\frac{3}{2}}$.

64. Simplify $24x(x^2 - 1)^{\frac{1}{3}} + 9(x^2 - 1)^{\frac{4}{3}}$.

Mixed Practice

In Problems 65–76, simplify each expression. Assume all variables are positive.

65. $\sqrt[8]{4^4}$

66. $\sqrt[6]{27^2}$

67. $(-2)^{\frac{1}{2}} \cdot (-2)^{\frac{3}{2}}$

68. $25^{\frac{3}{4}} \cdot 25^{\frac{3}{4}}$

69. $(100^{\frac{1}{5}})^{\frac{3}{2}}$

70. $(8^4)^{\frac{5}{12}}$

71. $(\sqrt[4]{25})^2$

72. $(\sqrt[6]{27})^2$

73. $\sqrt[4]{x^2} - \dfrac{\sqrt[4]{x^6}}{x}$

74. $\sqrt[9]{a^6} - \dfrac{\sqrt[6]{a^5}}{\sqrt[6]{a}}$

75. $(4 \cdot 9^{\frac{1}{4}})^{-2}$

76. $(4^{-1} \cdot 81^{\frac{1}{2}})^{\frac{1}{2}}$

In Problems 77–82, distribute and simplify.

77. $x^{\frac{1}{2}}(x^{\frac{3}{2}} - 2)$

78. $x^{\frac{1}{3}}(x^{\frac{5}{3}} + 4)$

79. $2y^{-\frac{1}{3}}(1 + 3y)$

80. $3a^{-\frac{1}{2}}(2 - a)$

81. $4z^{\frac{3}{2}}(z^{\frac{3}{2}} - 8z^{-\frac{3}{2}})$

82. $8p^{\frac{2}{3}}(p^{\frac{4}{3}} - 4p^{-\frac{2}{3}})$

Applying the Concepts

83. If $3^x = 25$, what does $3^{\frac{x}{2}}$ equal?

84. If $5^x = 64$, what does $5^{\frac{x}{3}}$ equal?

85. If $7^x = 9$, what does $\sqrt{7^x}$ equal?

86. If $5^x = 27$, what does $\sqrt[3]{5^x}$ equal?

Extending the Concepts

In Problems 87 and 88, simplify the expression using rational exponents.

87. $\sqrt[4]{\sqrt[3]{\sqrt{x}}}$

88. $\sqrt[5]{\sqrt[3]{\sqrt{x^2}}}$

89. Without using technology, determine the value of $(6^{\sqrt{2}})^{\sqrt{2}}$.

90. Determine the domain of $g(x) = (x - 3)^{\frac{1}{2}}(x - 1)^{-\frac{1}{2}}$.

91. Determine the domain of $f(x) = (x + 3)^{\frac{1}{2}}(x + 1)^{-\frac{1}{2}}$.

Synthesis Review

In Problems 92–95, simplify each epression.

92. $(2x - 1)(x + 4) - (x + 1)(x - 1)$

93. $3a(a - 3) + (a + 3)(a - 2)$

94. $\dfrac{\sqrt{x^2 + 4x + 4}}{x + 2}, x + 2 > 0$

95. $\dfrac{x^2 - 4}{x + 2}(x + 5) - (x + 4)(x - 1)$

6.3 Simplifying Radical Expressions Using Properties of Radicals

Objectives

1 Use the Product Property to Multiply Radical Expressions

2 Use the Product Property to Simplify Radical Expressions

3 Use the Quotient Property to Simplify Radical Expressions

4 Multiply Radicals with Unlike Indices

Are You Prepared for This Section?

Before getting started, complete the following problems. If you get a problem wrong, go back to the section cited and review the material.

P1. List the perfect squares that are less than 200.

P2. List the perfect cubes that are less than 200.

P3. Simplify: **(a)** $\sqrt{16}$ **(b)** $\sqrt{p^2}$ [Getting Ready: Square Roots, pp. 468–471]

1 Use the Product Property to Multiply Radical Expressions

Perhaps you are noticing a trend: After a new algebraic expression is introduced, how to multiply, divide, add, and subtract the algebraic expression is explained. Well, here we go again! We begin with multiplying radical expressions when they have the same index.

Prepared?...Answers   P1. 1, 4, 9, 16, 25, 36, 49, 64, 81, 100, 121, 144, 169, 196
P2. 1, 8, 27, 64, 125   **P3. (a)** 4 **(b)** $|p|$

Consider that

$$\sqrt{4 \cdot 25} = \sqrt{100} = 10 \quad \text{and} \quad \sqrt{4} \cdot \sqrt{25} = 2 \cdot 5 = 10$$

This suggests the following result:

In Other Words

$\sqrt[n]{a} \cdot \sqrt[n]{b} = \sqrt[n]{ab}$ means the product of the roots equals the root of the product, provided the index is the same.

> **Product Property Of Radicals**
>
> If $\sqrt[n]{a}$ and $\sqrt[n]{b}$ are real numbers and $n \geq 2$ is an integer, then
>
> $$\sqrt[n]{a} \cdot \sqrt[n]{b} = \sqrt[n]{ab}$$

This formula can be justified using rational exponents.

$$\sqrt[n]{a} \cdot \sqrt[n]{b} = a^{\frac{1}{n}} \cdot b^{\frac{1}{n}}$$

Product-to-a-Power Rule: $= (a \cdot b)^{\frac{1}{n}}$

$a^{\frac{1}{n}} = \sqrt[n]{a}: \quad = \sqrt[n]{a \cdot b}$

EXAMPLE 1 **Using the Product Property to Multiply Radicals**

Multiply.

(a) $\sqrt{5} \cdot \sqrt{3}$ **(b)** $\sqrt[3]{2} \cdot \sqrt[3]{13}$ **(c)** $\sqrt{x-3} \cdot \sqrt{x+3}$ **(d)** $\sqrt[5]{6c} \cdot \sqrt[5]{7c^2}$

Solution

(a) $\sqrt{5} \cdot \sqrt{3} = \sqrt{5 \cdot 3} = \sqrt{15}$

(b) $\sqrt[3]{2} \cdot \sqrt[3]{13} = \sqrt[3]{2 \cdot 13} = \sqrt[3]{26}$

(c) $\sqrt{x-3} \cdot \sqrt{x+3} = \sqrt{(x-3)(x+3)} = \sqrt{x^2 - 9}$ if $x \geq 3$

(d) $\sqrt[5]{6c} \cdot \sqrt[5]{7c^2} = \sqrt[5]{6c \cdot 7c^2} = \sqrt[5]{42c^3}$ ●

Work Smart

In Example 1(c), notice that $\sqrt{x^2 - 9}$ does *not* equal $\sqrt{x^2} - \sqrt{9}$.

Quick ✓

1. If $\sqrt[n]{a}$ and $\sqrt[n]{b}$ are real numbers and $n \geq 2$ is an integer, then $\sqrt[n]{a} \cdot \sqrt[n]{b} =$ _____.

In Problems 2–5, multiply each radical expression.

2. $\sqrt{11} \cdot \sqrt{7}$ **3.** $\sqrt[4]{6} \cdot \sqrt[4]{7}$ **4.** $\sqrt{x-5} \cdot \sqrt{x+5}, x \geq 5$ **5.** $\sqrt[7]{5p} \cdot \sqrt[7]{4p^3}$

▶ ❷ **Use the Product Property to Simplify Radical Expressions**

So far, radicals have been simplified only when the radicand was a perfect square, such as $\sqrt{81} = 9$, or a perfect cube, such as $\sqrt[3]{\dfrac{1}{8}} = \dfrac{1}{2}$. When a radical does not simplify to a rational number, one of two things can be done:

1. Write a decimal approximation of the radical.

2. Simplify the radical using properties of radicals, if possible.

Using technology to find the decimal approximation of radicals was illustrated in the Getting Ready: Square Roots section and in Section 6.1. Now how to use properties of radicals to write the radical in simplified form will be presented.

The advantage of simplifying over approximating is that simplifying maintains an *exact* radical rather than an approximate value.

Recall that a number that is the square of a rational number is called a perfect square. Thus $1^2 = 1, 2^2 = 4, 3^2 = 9$, and so on are perfect squares. A number that is the cube of a rational number is called a perfect cube. Thus $1^3 = 1, 2^3 = 8, 3^3 = 27$, $(-1)^3 = -1, (-2)^3 = -8$, and so on are perfect cubes. In general, if n is the index of a radical, then a^n is a perfect power of a, where a is a rational number.

Work Smart

index $\rightarrow \sqrt[n]{a} \leftarrow$ radicand

A radical expression with index n is **simplified** if the radicand does not contain any factors that are perfect nth powers. For example, $\sqrt{50}$ is not simplified because 25 is a factor of 50, and 25 is a perfect square; $\sqrt[3]{16}$ is not simplified because 8 is a factor of 16, and 8 is a perfect cube; $\sqrt[3]{x^5}$ is not simplified because $x^5 = x^3 \cdot x^2$, and x^3 is a perfect cube.

To simplify radicals, use the Product Property of Radicals "in reverse." That is, use $\sqrt[n]{ab} = \sqrt[n]{a} \cdot \sqrt[n]{b}$.

EXAMPLE 2   **How to Use the Product Property to Simplify a Radical**

Simplify: $\sqrt{18}$

Step-by-Step Solution

Step 1: What is the index on the radical? Because the index is 2, write each factor of the radicand as the product of two factors, one of which is a perfect square.

The perfect squares are 1, 4, 9, 16, 25, ... Because 9 is a factor of 18, and 9 is a perfect square, write 18 as $9 \cdot 2$: $\sqrt{18} = \sqrt{9 \cdot 2}$

Step 2: Write the radical as the product of two radicals, one of which contains a perfect square.

$\sqrt{ab} = \sqrt{a} \cdot \sqrt{b}$: $= \sqrt{9} \cdot \sqrt{2}$

Step 3: Take the square root of each perfect square.

$= 3\sqrt{2}$ ●

The steps used in Example 2 are summarized below.

Work Smart

When performing Step 1 with real numbers, look for the *largest* factor of the radicand that is a perfect power of the index.

> **Simplifying a Radical Expression of the Form $\sqrt[n]{a}$**
>
> **Step 1:** Write each factor of the radicand as the product of two factors, one of which is a perfect nth power.
>
> **Step 2:** Using the Product Property of Radicals, write the radicand as the product of two radicals, one of which contains perfect nth powers.
>
> **Step 3:** Take the nth root of each perfect nth power.

EXAMPLE 3   **Using the Product Property to Simplify a Radical**

Simplify each of the following:

(a) $5\sqrt[3]{24}$ **(b)** $\sqrt{128x^2}$ **(c)** $\sqrt[4]{20}$

Solution

(a) First, notice that the index is 3. Therefore, find the largest factor of 24 that is a perfect cube. The positive perfect cubes are 1, 8, 27, ... Because 8 is a factor of 24 and is a perfect cube, write 24 as $8 \cdot 3$.

$$5\sqrt[3]{24} = 5 \cdot \sqrt[3]{8 \cdot 3}$$
$$\sqrt[n]{ab} = \sqrt[n]{a} \cdot \sqrt[n]{b}: \quad = 5 \cdot \sqrt[3]{8} \cdot \sqrt[3]{3}$$
$$= 5 \cdot 2 \cdot \sqrt[3]{3}$$
$$= 10\sqrt[3]{3}$$

(b) The index is 2. Because 64 is a factor of 128, and 64 is a perfect square, write 128 as $64 \cdot 2$; x^2 is a perfect square.

$$\sqrt{128x^2} = \sqrt{64x^2 \cdot 2}$$
$$\sqrt[n]{ab} = \sqrt[n]{a} \cdot \sqrt[n]{b}: \quad = \sqrt{64x^2} \cdot \sqrt{2}$$
$$\sqrt{64x^2} = \sqrt{64} \cdot \sqrt{x^2} = 8|x|: \quad = 8|x|\sqrt{2}$$

(continued)

(c) In $\sqrt[4]{20}$, the index is 4. The fourth powers (or perfect fourths) are 1, 16, 81,.... There are no factors of 20 that are fourth powers, so the radical $\sqrt[4]{20}$ cannot be simplified. ●

Quick ✓

6. List the first six positive integers that are perfect squares.

7. List the first six positive integers that are perfect cubes.

In Problems 8–11, simplify each of the radical expressions.

8. $\sqrt{48}$ 9. $4\sqrt[3]{54}$ 10. $\sqrt{200a^2}$ 11. $\sqrt[4]{40}$

EXAMPLE 4 **Simplifying an Expression Involving a Square Root**

Simplify: $\dfrac{4 - \sqrt{20}}{2}$

Solution

This expression can be simplified using two different approaches.

Method 1:

4 is the largest perfect square factor of 20
$$\frac{4 - \sqrt{20}}{2} = \frac{4 - \sqrt{4 \cdot 5}}{2}$$

Use $\sqrt{a \cdot b} = \sqrt{a} \cdot \sqrt{b}$: $= \dfrac{4 - \sqrt{4} \cdot \sqrt{5}}{2}$

$= \dfrac{4 - 2 \cdot \sqrt{5}}{2}$

Factor out the 2 in the numerator: $= \dfrac{2(2 - \sqrt{5})}{2}$

Divide out common factor: $= 2 - \sqrt{5}$

Method 2:

4 is the largest perfect square factor of 20
$$\frac{4 - \sqrt{20}}{2} = \frac{4 - \sqrt{4 \cdot 5}}{2}$$

Use $\sqrt{a \cdot b} = \sqrt{a} \cdot \sqrt{b}$: $= \dfrac{4 - \sqrt{4} \cdot \sqrt{5}}{2}$

$= \dfrac{4 - 2 \cdot \sqrt{5}}{2}$

Use $\dfrac{A + B}{C} = \dfrac{A}{C} + \dfrac{B}{C}$: $= \dfrac{4}{2} - \dfrac{2 \cdot \sqrt{5}}{2}$

Divide out common factor: $= 2 - \sqrt{5}$ ●

Quick ✓

In Problems 12 and 13, simplify the expression.

12. $\dfrac{6 + \sqrt{45}}{3}$

13. $\dfrac{-2 + \sqrt{32}}{4}$

Recall how to simplify $\sqrt[n]{a^n}$ from Section 6.1.

Simplifying $\sqrt[n]{a^n}$

If $n \geq 2$ is a positive integer and a is a real number, then

$$\sqrt[n]{a^n} = |a| \quad \text{if } n \geq 2 \text{ is even}$$
$$\sqrt[n]{a^n} = a \quad \text{if } n \geq 3 \text{ is odd}$$

This means that

$$\sqrt{a^2} = |a| \quad \sqrt[3]{a^3} = a \quad \sqrt[4]{a^4} = |a| \quad \sqrt[5]{a^5} = a \quad \text{and so on.}$$

In order to make our mathematical lives a little easier, assume throughout this section that all variables in the radicand are greater than or equal to zero (nonnegative) unless stated otherwise. Thus

$$\sqrt{a^2} = a \qquad \sqrt[3]{a^3} = a \qquad \sqrt[4]{a^4} = a \qquad \sqrt[5]{a^5} = a \qquad \text{and so on.}$$

What if the exponent on the radicand is greater than the index, as in $\sqrt{x^3}$ or $\sqrt[3]{x^6}$? The Laws of Exponents along with the rule for simplifying $\sqrt[n]{a^n}$ could be used, or rational exponents could be used.

$$\sqrt[3]{x^6} = \sqrt[3]{(x^2)^3} = x^2 \quad \text{or} \quad \sqrt[3]{x^6} = x^{\frac{6}{3}} = x^2$$
$$\sqrt[3]{x^{12}} = \sqrt[3]{(x^4)^3} = x^4 \quad \text{or} \quad \sqrt[3]{x^{12}} = x^{\frac{12}{3}} = x^4$$

EXAMPLE 5   **Simplifying a Radical with a Variable Radicand**

Simplify $\sqrt{20x^{10}}$.

Solution

Because 4 is the largest perfect square factor of 20, write 20 as $4 \cdot 5$.

$$\sqrt{20x^{10}} = \sqrt{4 \cdot 5 \cdot x^{10}}$$
$$= \sqrt{4x^{10}} \cdot \sqrt{5}$$

$\sqrt{4} = 2, \sqrt{x^{10}} = \sqrt{(x^5)^2} = x^5$

or $\sqrt{x^{10}} = x^{\frac{10}{2}} = x^5$: $= 2x^5\sqrt{5}$ ●

> **Quick ✓**
>
> **14.** Simplify $\sqrt{75a^6}$.

What if the index does not divide evenly into the exponent on the variable in the radicand, as in $\sqrt[3]{x^8}$? In this case, rewrite the variable expression as the product of two variable expressions where one of the factors has an exponent that is a multiple of the index:

3 divides evenly into 6: $\sqrt[3]{x^8} = \sqrt[3]{x^6 \cdot x^2}$

So

$$\sqrt[3]{x^8} = \sqrt[3]{x^6 \cdot x^2}$$
$$= \sqrt[3]{x^6} \cdot \sqrt[3]{x^2}$$
$$= x^2\sqrt[3]{x^2}$$

EXAMPLE 6   **Simplifying Radicals**

Simplify:

(a) $\sqrt{80a^3}$ **(b)** $\sqrt[3]{27m^4n^{14}}$

Solution

(a) $80 = 16 \cdot 5; a^3 = a^2 \cdot a$

$$\sqrt{80a^3} = \sqrt{16 \cdot 5 \cdot a^2 \cdot a}$$
$$= \sqrt{16a^2 \cdot 5a}$$

$\sqrt[n]{ab} = \sqrt[n]{a} \cdot \sqrt[n]{b}$: $= \sqrt{16a^2} \cdot \sqrt{5a}$

$\sqrt{a^2} = a$ assuming $a \geq 0$: $= 4a\sqrt{5a}$

(continued)

(b)

$$m^4 = m^3 \cdot m; \ n^{14} = n^{12} \cdot n^2$$

$$\sqrt[3]{27m^4n^{14}} = \sqrt[3]{27 \cdot m^3 \cdot m \cdot n^{12} \cdot n^2}$$

$$= \sqrt[3]{27 \cdot m^3 \cdot n^{12} \cdot m \cdot n^2}$$

$$\sqrt[n]{ab} = \sqrt[n]{a} \cdot \sqrt[n]{b}: \quad = \sqrt[3]{27m^3n^{12}} \cdot \sqrt[3]{mn^2}$$

$$\sqrt[3]{27} = 3; \ \sqrt[3]{m^3} = m; \ \sqrt[3]{n^{12}} = n^{\frac{12}{3}} = n^4: \quad = 3mn^4\sqrt[3]{mn^2}$$

Quick ✓

In Problems 15–17, simplify each radical.

15. $\sqrt{18a^5}$ **16.** $\sqrt[3]{-128x^6y^{10}}$ **17.** $\sqrt[4]{16a^5b^{11}}$

⊳ (**EXAMPLE 7**) **Multiplying and Simplifying Radicals**

Multiply and simplify:

(a) $\sqrt{3} \cdot \sqrt{15}$ **(b)** $3\sqrt[3]{4x} \cdot \sqrt[3]{2x^4}$ **(c)** $\sqrt[4]{27a^2b^5} \cdot \sqrt[4]{6a^3b^6}$

Solution

Remember, to multiply two radicals, the index must be the same. When radicals have the same index, multiply the radicands and then simplify the product.

(a) Notice that neither radical can be simplified. However, the index on both radicals is 2, so multiply the radicands.

$$\sqrt{3} \cdot \sqrt{15} = \sqrt{3 \cdot 15}$$

$$= \sqrt{45}$$

$$\text{9 is the largest perfect square factor of 45:} \quad = \sqrt{9 \cdot 5}$$

$$\text{Product Property of Radicals:} \quad = \sqrt{9} \cdot \sqrt{5}$$

$$= 3\sqrt{5}$$

(b) The index on both radicals is 3, so multiply the radicands.

$$3\sqrt[3]{4x} \cdot \sqrt[3]{2x^4} = 3\sqrt[3]{4x \cdot 2x^4}$$

$$= 3\sqrt[3]{8x^5}$$

$$x^3 \text{ is a perfect cube; 8 is a perfect cube:} \quad = 3\sqrt[3]{8x^3 \cdot x^2}$$

$$\text{Product Property of Radicals:} \quad = 3\sqrt[3]{8x^3} \cdot \sqrt[3]{x^2}$$

$$\sqrt[3]{8} = 2; \ \sqrt[3]{x^3} = x: \quad = 3 \cdot 2 \cdot x \cdot \sqrt[3]{x^2}$$

$$= 6x\sqrt[3]{x^2}$$

(c) The index on both radicals is 4, so multiply the radicands.

$$\sqrt[4]{27a^2b^5} \cdot \sqrt[4]{6a^3b^6} = \sqrt[4]{3^3 \cdot a^2b^5 \cdot 3 \cdot 2 \cdot a^3b^6}$$

$$= \sqrt[4]{3^4 \cdot 2 \cdot a^5b^{11}}$$

$$3^4, a^4, \text{ and } b^8 \text{ are perfect fourths:} \quad = \sqrt[4]{3^4 \cdot 2 \cdot a^4 \cdot a \cdot b^8 \cdot b^3}$$

$$\text{Product Property of Radicals:} \quad = \sqrt[4]{3^4a^4b^8} \cdot \sqrt[4]{2ab^3}$$

$$= 3ab^2\sqrt[4]{2ab^3}$$

Work Smart

Notice that $27 = 3^3$ and that $6 = 3 \cdot 2$, so that $27 \cdot 6 = 3^3 \cdot 3 \cdot 2 = 3^4 \cdot 2$. This makes finding the perfect fourth powers a lot easier!

Quick ✓

In Problems 18–20, multiply and simplify the radicals.

18. $\sqrt{6} \cdot \sqrt{8}$ **19.** $\sqrt[3]{12a^2} \cdot \sqrt[3]{10a^4}$ **20.** $4\sqrt[3]{8a^2b^5} \cdot \sqrt[3]{6a^2b^4}$

▶ ❸ Use the Quotient Property to Simplify Radical Expressions

Now consider that

$$\sqrt{\frac{64}{4}} = \sqrt{\frac{4 \cdot 16}{4}} = \sqrt{16} = 4 \quad \text{and} \quad \frac{\sqrt{64}}{\sqrt{4}} = \frac{8}{2} = 4$$

This suggests the following result:

In Other Words

$$\sqrt[n]{\frac{a}{b}} = \frac{\sqrt[n]{a}}{\sqrt[n]{b}}$$

means "the root of the quotient equals the quotient of the roots" provided that the radicals have the same index.

> **Quotient Property of Radicals**
>
> If $\sqrt[n]{a}$ and $\sqrt[n]{b}$ are real numbers, $b \neq 0$, and $n \geq 2$ is an integer, then
>
> $$\frac{\sqrt[n]{a}}{\sqrt[n]{b}} = \sqrt[n]{\frac{a}{b}}$$

This formula can be justified using rational exponents.

$$\frac{\sqrt[n]{a}}{\sqrt[n]{b}} = \frac{a^{\frac{1}{n}}}{b^{\frac{1}{n}}} = \left(\frac{a}{b}\right)^{\frac{1}{n}} = \sqrt[n]{\frac{a}{b}}$$

EXAMPLE 8 **Using the Quotient Property to Simplify Radicals**

Simplify:

(a) $\sqrt{\dfrac{18}{25}}$ **(b)** $\sqrt[3]{\dfrac{6z^3}{125}}$ **(c)** $\sqrt[4]{\dfrac{10a^2}{81b^4}}, b \neq 0$

Solution
In each of these problems, the expression in the denominator is a perfect nth power, and the index is n. Therefore, simplify the expression using the Quotient Rule "in reverse": $\sqrt[n]{\dfrac{a}{b}} = \dfrac{\sqrt[n]{a}}{\sqrt[n]{b}}$.

(a) $\sqrt{\dfrac{18}{25}} = \dfrac{\sqrt{18}}{\sqrt{25}}$

$= \dfrac{3\sqrt{2}}{5}$

(b) $\sqrt[3]{\dfrac{6z^3}{125}} = \dfrac{\sqrt[3]{6z^3}}{\sqrt[3]{125}}$

$= \dfrac{z\sqrt[3]{6}}{5}$

(c) $\sqrt[4]{\dfrac{10a^2}{81b^4}} = \dfrac{\sqrt[4]{10a^2}}{\sqrt[4]{81b^4}}$

$= \dfrac{\sqrt[4]{10a^2}}{3b}$ ●

Quick ✓
In Problems 21–23, simplify the radicals.

21. $\sqrt{\dfrac{13}{49}}$ **22.** $\sqrt[3]{\dfrac{27p^3}{8}}$ **23.** $\sqrt[4]{\dfrac{3q^4}{16}}$

▶ EXAMPLE 9 **Using the Quotient Property to Simplify Radicals**

Simplify:

(a) $\dfrac{\sqrt{24a^3}}{\sqrt{6a}}$ **(b)** $\dfrac{-2\sqrt[3]{54a}}{\sqrt[3]{2a^4}}$ **(c)** $\dfrac{\sqrt[3]{-375x^2y}}{\sqrt[3]{3x^{-1}y^7}}$

(continued)

Solution

In these problems, the radical expression in the denominator cannot be simplified. However, the index on the numerator and denominator of each expression is the same, so write each expression as a single radical.

(a) $\dfrac{\sqrt{24a^3}}{\sqrt{6a}} = \sqrt{\dfrac{24a^3}{6a}}$

$\qquad\qquad = \sqrt{4a^2}$

$\qquad\qquad = 2a$

(b) $\dfrac{-2\sqrt[3]{54a}}{\sqrt[3]{2a^4}} = -2 \cdot \sqrt[3]{\dfrac{54a}{2a^4}}$

$\qquad\qquad\qquad = -2 \cdot \sqrt[3]{\dfrac{27}{a^3}}$

$\qquad\qquad\qquad = -2 \cdot \dfrac{3}{a}$

$\qquad\qquad\qquad = -\dfrac{6}{a}$

(c) $\dfrac{\sqrt[3]{-375x^2y}}{\sqrt[3]{3x^{-1}y^7}} = \sqrt[3]{\dfrac{-375x^2y}{3x^{-1}y^7}}$

$\qquad\qquad\qquad = \sqrt[3]{\dfrac{-125x^{2-(-1)}}{y^{7-1}}}$

$\qquad\qquad\qquad = \sqrt[3]{\dfrac{-125x^3}{y^6}}$

$\qquad\qquad\qquad = -\dfrac{5x}{y^2}$

Quick ✓

In Problems 24–26, simplify the radicals.

24. $\dfrac{\sqrt{12a^5}}{\sqrt{3a}}$

25. $\dfrac{\sqrt[3]{-24x^2}}{\sqrt[3]{3x^{-1}}}$

26. $\dfrac{\sqrt[3]{250a^5b^{-2}}}{\sqrt[3]{2ab}}$

▶ ❹ Multiply Radicals with Unlike Indices

The rule $\sqrt[n]{a} \cdot \sqrt[n]{b} = \sqrt[n]{ab}$ only works when the index on each radical is the same. What if the indices on the radicals are different? Can the product still be simplified? The answer is yes! Use the fact that $\sqrt[n]{a} = a^{\frac{1}{n}}$.

EXAMPLE 10 | **Multiplying Radicals with Unlike Indices**

Multiply and simplify: $\sqrt[4]{8} \cdot \sqrt[3]{5}$

Solution

Notice that the indices are not the same, so $\sqrt[n]{a} \cdot \sqrt[n]{b} = \sqrt[n]{ab}$ cannot be used. Use rational exponents along with $\sqrt[n]{a} = a^{\frac{1}{n}}$ instead.

$$\sqrt[4]{8} \cdot \sqrt[3]{5} = 8^{\frac{1}{4}} \cdot 5^{\frac{1}{3}}$$

LCD = 12: $\qquad\qquad = 8^{\frac{3}{12}} \cdot 5^{\frac{4}{12}}$

$a^{\frac{r}{s}} = (a^r)^{\frac{1}{s}}: \qquad = \left[(8^3)^{\frac{1}{12}} \cdot (5^4)^{\frac{1}{12}} \right]$

$a^r \cdot b^r = (ab)^r: \qquad = \left[(8^3)(5^4) \right]^{\frac{1}{12}}$

$\qquad\qquad\qquad = (320{,}000)^{\frac{1}{12}}$

$a^{\frac{1}{n}} = \sqrt[n]{a}: \qquad\quad = \sqrt[12]{320{,}000}$

Quick ✓

In Problems 27 and 28, multiply and simplify.

27. $\sqrt[4]{5} \cdot \sqrt[3]{3}$

28. $\sqrt{10} \cdot \sqrt[3]{12}$

6.3 Exercises MyMathLab®

Exercise numbers in green have complete video solutions in MyMathLab or may be accessed using the QR code to the right.

*Problems **1–28** are the Quick ✓ s that follow the EXAMPLES.*

Building Skills

In Problems 29–36, use the Product Property to multiply. Assume that all variables can be any real number unless stated otherwise. See Objective 1.

29. $\sqrt[3]{6} \cdot \sqrt[3]{10}$

30. $\sqrt[3]{-5} \cdot \sqrt[3]{7}$

31. $\sqrt{3a} \cdot \sqrt{5b}$

32. $\sqrt[4]{6a^2} \cdot \sqrt[4]{7b^2}$

33. $\sqrt{x-7} \cdot \sqrt{x+7}, x \geq 7$

34. $\sqrt{p-5} \cdot \sqrt{p+5}, p \geq 5$

35. $\sqrt{\dfrac{5x}{3}} \cdot \sqrt{\dfrac{3}{x}}, x > 0$

36. $\sqrt[3]{\dfrac{-9x^2}{4}} \cdot \sqrt[3]{\dfrac{4}{3x}}, x \neq 0$

In Problems 37–64, simplify each radical using the Product Property. Assume that all variables can be any real number. See Objective 2.

37. $\sqrt{50}$

38. $\sqrt{32}$

39. $\sqrt[3]{54}$

40. $\sqrt[4]{162}$

41. $\sqrt{48x^2}$

42. $\sqrt{20a^2}$

43. $\sqrt[3]{-27x^3}$

44. $\sqrt[3]{-64p^3}$

45. $\sqrt[4]{32m^4}$

46. $\sqrt[4]{48z^4}$

47. $\sqrt{12p^2q}$

48. $\sqrt{45m^2n}$

49. $\sqrt{162m^4}$

50. $\sqrt{98w^8}$

51. $\sqrt{y^{13}}$

52. $\sqrt{s^9}$

53. $\sqrt[3]{c^8}$

54. $\sqrt[5]{x^{12}}$

55. $\sqrt{125p^3q^4}$

56. $\sqrt{243ab^5}$

57. $\sqrt[3]{-16x^9}$

58. $\sqrt[3]{-54q^{12}}$

59. $\sqrt[5]{-16m^8n^2}$

60. $\sqrt{75x^6y}$

61. $\sqrt[4]{(x-y)^5}, x > y$

62. $\sqrt[3]{(a+b)^5}$

63. $\sqrt[3]{8x^3 - 8y^3}$

64. $\sqrt[3]{8a^3 + 8b^3}$

In Problems 65–70, simplify each expression. See Objective 2.

65. $\dfrac{4 + \sqrt{36}}{2}$

66. $\dfrac{5 - \sqrt{100}}{5}$

67. $\dfrac{9 + \sqrt{18}}{3}$

68. $\dfrac{10 - \sqrt{75}}{5}$

69. $\dfrac{7 - \sqrt{98}}{14}$

70. $\dfrac{-6 + \sqrt{108}}{6}$

In Problems 71–88, multiply and simplify. Assume that all variables are greater than or equal to zero. See Objective 2.

71. $\sqrt{5} \cdot \sqrt{5}$

72. $\sqrt{6} \cdot \sqrt{6}$

73. $\sqrt{2} \cdot \sqrt{8}$

74. $\sqrt{3} \cdot \sqrt{12}$

75. $\sqrt[3]{4} \cdot \sqrt[3]{2}$

76. $\sqrt[3]{9} \cdot \sqrt[3]{3}$

77. $\sqrt{5x} \cdot \sqrt{15x}$

78. $\sqrt{6x} \cdot \sqrt{30x}$

79. $\sqrt[3]{4b^2} \cdot \sqrt[3]{6b^2}$

80. $\sqrt[3]{9a} \cdot \sqrt[3]{6a^2}$

81. $2\sqrt{6ab} \cdot 3\sqrt{15ab^3}$

82. $3\sqrt{14pq^3} \cdot 2\sqrt{7pq}$

83. $\sqrt[4]{27p^3q^2} \cdot \sqrt[4]{12p^2q^2}$

84. $\sqrt[3]{16m^2n} \cdot \sqrt[3]{27m^2n}$

85. $\sqrt[5]{-8a^3b^4} \cdot \sqrt[5]{12a^3b}$

86. $\sqrt[5]{-27x^4y^2} \cdot \sqrt[5]{18x^3y^4}$

87. $\sqrt[4]{8(x-y)^2} \cdot \sqrt[4]{6(x-y)^3}, x > y$

88. $\sqrt[3]{9(a+b)^2} \cdot \sqrt[3]{6(a+b)^5}$

In Problems 89–96, simplify each expression. Assume that all variables are greater than zero. See Objective 3.

89. $\sqrt{\dfrac{3}{16}}$

90. $\sqrt{\dfrac{5}{36}}$

91. $\sqrt[4]{\dfrac{5x^4}{16}}$

92. $\sqrt[4]{\dfrac{2a^8}{81}}$

93. $\sqrt{\dfrac{9y^2}{25x^2}}$

94. $\sqrt{\dfrac{4a^4}{81b^2}}$

95. $\sqrt[3]{\dfrac{-27x^9}{64y^{12}}}$

96. $\sqrt[5]{\dfrac{-32a^{15}}{243b^{10}}}$

In Problems 97–110, divide and simplify. Assume that all variables are greater than zero. See Objective 3.

97. $\dfrac{\sqrt{8}}{\sqrt{2}}$

98. $\dfrac{\sqrt{27}}{\sqrt{3}}$

99. $\dfrac{\sqrt[3]{128}}{\sqrt[3]{2}}$

100. $\dfrac{\sqrt[4]{64}}{\sqrt[4]{4}}$

101. $\dfrac{\sqrt{48a^3}}{\sqrt{6a}}$

102. $\dfrac{\sqrt{54y^5}}{\sqrt{3y}}$

103. $\dfrac{\sqrt{24a^5b}}{\sqrt{3ab^3}}$

104. $\dfrac{\sqrt{360m^7n^3}}{\sqrt{5mn^5}}$

105. $\dfrac{\sqrt{512a^7b}}{3\sqrt{2ab^3}}$

106. $\dfrac{\sqrt{375x^2y^7}}{10\sqrt{3y}}$

107. $\dfrac{\sqrt[3]{104a^5}}{\sqrt[3]{4a^{-1}}}$

108. $\dfrac{\sqrt[3]{-128x^8}}{\sqrt[3]{2x^{-1}}}$

109. $\dfrac{\sqrt{90x^3y^{-1}}}{\sqrt{2x^{-3}y}}$

110. $\dfrac{\sqrt{96a^5b^{-3}}}{\sqrt{3a^{-5}b}}$

In Problems 111–118, multiply and simplify. See Objective 4.

111. $\sqrt{3} \cdot \sqrt[3]{4}$

112. $\sqrt{2} \cdot \sqrt[3]{7}$

113. $\sqrt[3]{2} \cdot \sqrt[6]{3}$

114. $\sqrt[4]{3} \cdot \sqrt[8]{5}$

115. $\sqrt{3} \cdot \sqrt[3]{18}$

116. $\sqrt{6} \cdot \sqrt[3]{9}$

117. $\sqrt[4]{9} \cdot \sqrt[6]{12}$

118. $\sqrt[5]{8} \cdot \sqrt[10]{16}$

Mixed Practice

In Problems 119–134, perform the indicated operation and simplify.
Assume all variables are greater than zero.

119. $\sqrt[3]{\dfrac{5x}{8}}$

120. $\sqrt[3]{\dfrac{7a^2}{64}}$

121. $\sqrt[3]{5a} \cdot \sqrt[3]{9a}$

122. $\sqrt[5]{8b^2} \cdot \sqrt[5]{3b}$

123. $\sqrt{72a^4}$

124. $\sqrt{24b^6}$

125. $\sqrt[3]{6a^2b} \cdot \sqrt[3]{9ab}$

126. $\sqrt[4]{8x^3y^2} \cdot \sqrt[4]{4x^2y^3}$

127. $\dfrac{\sqrt[3]{-32a}}{\sqrt[3]{2a^4}}$

128. $\dfrac{\sqrt[3]{-250p^2}}{\sqrt[3]{2p^5}}$

129. $-5\sqrt[3]{32m^3}$

130. $-7\sqrt[3]{250p^3}$

131. $\sqrt[3]{81a^4b^7}$

132. $\sqrt[5]{32p^7q^{11}}$

133. $\sqrt[3]{12} \cdot \sqrt[3]{18}$

134. $\sqrt[4]{8} \cdot \sqrt[4]{18}$

Applying the Concepts

△ **135. Length of a Line Segment** The length of the line segment joining the points $(2, 5)$ and $(-1, -1)$ is given by

$$\sqrt{(5 - (-1))^2 + (2 - (-1))^2}$$

(a) Plot the points in the Cartesian plane, and draw a line segment connecting the points.

(b) Express the length of the line segment as a radical in simplified form.

△ **136. Length of a Line Segment** The length of the line segment joining the points $(4, 2)$ and $(-2, 4)$ is given by

$$\sqrt{(4 - 2)^2 + (-2 - 4)^2}$$

(a) Plot the points in the Cartesian plane, and draw a line segment connecting the points.

(b) Express the length of the line segment as a radical in simplified form.

137. Revenue Growth Suppose that the annual revenue R (in millions of dollars) of a company after t years of operating is modeled by the function

$$R(t) = \sqrt[3]{\dfrac{t}{2}}$$

(a) Predict the revenue of the company after 8 years of operation.

(b) Predict the revenue of the company after 27 years of operation.

△ **138. Sphere** The radius r of a sphere whose volume is V is given by

$$r = \sqrt[3]{\dfrac{3V}{4\pi}}$$

(a) Write the radius of a sphere whose volume is 9 cubic centimeters as a radical in simplified form.

(b) Write the radius of a sphere whose volume is 32π cubic meters as a radical in simplified form.

Extending the Concepts

139. Suppose that $f(x) = \sqrt{2x}$ and $g(x) = \sqrt{8x^3}$.

(a) Find $(f \cdot g)(x)$.

(b) Evaluate $(f \cdot g)(3)$.

In Problems 140–143, evaluate the formula

$$x = \dfrac{-b \pm \sqrt{b^2 - 4ac}}{2a}$$

for the given values of a, b, and c. Note that the symbol $\pm$ is shorthand notation to indicate that there are two solutions. One solution is obtained when you add the quantity after the $\pm$ symbol, and another is obtained when you subtract. This formula can be used to solve any equation of the form $ax^2 + bx + c = 0, a \neq 0$.

140. $a = 1, b = 4, c = 1$

141. $a = 1, b = 6, c = 3$

142. $a = 2, b = 1, c = -1$

143. $a = 3, b = 4, c = -1$

Explaining the Concepts

144. Explain how you would simplify $\sqrt[3]{16a^5}$.

145. To use the Product Property to multiply radicals, what must be true about the index in each radical?

146. In Example 1(c) the restriction $x \geq 3$ was included.

(a) Find the domain of each factor in $(\sqrt{x + 3})(\sqrt{x - 3})$.

(b) The domain of the product, $\sqrt{x^2 - 9}$, is $x \leq -3$ or $x \geq 3$. Show that the product is a real number if $x = -4$ or $x = 4$ but is not a real number if $x = 0$. What conclusion can you draw about the domain of the product of two square roots?

Synthesis Review

In Problems 147–150, solve the following.

147. $4x + 3 = 13$

148. $2|7x - 1| + 4 = 16$

149. $\dfrac{3}{5}x + 2 \leq 28$

150. $\dfrac{5}{2}|x + 1| + 1 \leq 11$

151. How are the solutions in Problems 147 and 149 similar? How are the solutions in Problems 148 and 150 similar?

Technology Exercises

152. Exploration To understand the circumstances under which absolute value symbols are required when simplifying radicals, use technology to complete the following.

(a) Graph $Y_1 = \sqrt{x^2}$ and $Y_2 = x$. Do you think that $\sqrt{x^2} = x$? Now graph $Y_1 = \sqrt{x^2}$ and $Y_2 = |x|$. Do you think that $\sqrt{x^2} = |x|$?

(b) Graph $Y_1 = \sqrt[3]{x^3}$ and $Y_2 = x$. Do you think that $\sqrt[3]{x^3} = x$? Now graph $Y_1 = \sqrt[3]{x^3}$ and $Y_2 = |x|$. Do you think that $\sqrt[3]{x^3} = |x|$?

(c) Graph $Y_1 = \sqrt[4]{x^4}$ and $Y_2 = x$. Do you think that $\sqrt[4]{x^4} = x$? Now graph $Y_1 = \sqrt[4]{x^4}$ and $Y_2 = |x|$. Do you think that $\sqrt[4]{x^4} = |x|$?

(d) In your own words, make a generalization about $\sqrt[n]{x^n}$.

6.4 Adding, Subtracting, and Multiplying Radical Expressions

Objectives

1 Add or Subtract Radical Expressions

2 Multiply Radical Expressions

Are You Prepared for This Section?

Before getting started, complete the following problems. If you get a problem wrong, go back to the section cited and review the material.

P1. Add: $4y^3 - 2y^2 + 8y - 1 + (-2y^3 + 7y^2 - 3y + 9)$ [Section 4.1, pp. 312–313]

P2. Subtract: $5z^2 + 6 - (3z^2 - 8z - 3)$ [Section 4.1, pp. 313–314]

P3. Multiply: $(4x + 3)(x - 5)$ [Section 4.2, pp. 321–322]

P4. Multiply: $(2y - 3)(2y + 3)$ [Section 4.2, pp. 323–324]

▶ **1** Add or Subtract Radical Expressions

Recall that a radical expression is an algebraic expression that contains a radical. **Like radicals** have the same index and the same radicand. For example, the expressions $3\sqrt{5}$ and $8\sqrt{5}$ contain like radicals because each radical has the same index, 2, and the same radicand, 5; $4\sqrt[3]{x - 4}$ and $10\sqrt[3]{x - 4}$ also contain like radicals because each radical has the same index, 3, and the same radicand, $x - 4$.

Just as like terms are added or subtracted using the Distributive Property, radical expressions can be added or subtracted by combining like radicals using the Distributive Property.

EXAMPLE 1   **Adding and Subtracting Radical Expressions**

Add or subtract, as indicated. Assume all variables are greater than or equal to zero.

(a) $5\sqrt{2x} + 9\sqrt{2x}$

(b) $3\sqrt[3]{10} + 7\sqrt[3]{10} - 5\sqrt[3]{10}$

Work Smart

To add or subtract radicals, *both* the index and the radicand must be the same.

Prepared?...Answers
P1. $2y^3 + 5y^2 + 5y + 8$
P2. $2z^2 + 8z + 9$ **P3.** $4x^2 - 17x - 15$
P4. $4y^2 - 9$

Solution

(a) Both radicals have the same index, 2, and the same radicand, $2x$, so they are like radicals.

$$5\sqrt{2x} + 9\sqrt{2x} = (5 + 9)\sqrt{2x}$$
$$5 + 9 = 14: \quad = 14\sqrt{2x}$$

(continued)

(b) All three radicals have the same index, 3, and the same radicand, 10, so they are like radicals.

$$3\sqrt[3]{10} + 7\sqrt[3]{10} - 2\sqrt[3]{10} = (3 + 7 - 2)\sqrt[3]{10}$$
$$3 + 7 - 2 = 8: \quad = 8\sqrt[3]{10}$$

Quick ✓

1. Two radicals are _____ if each radical has the same index and the same radicand.

2. *True or False* $\sqrt{7} + \sqrt{5} = \sqrt{7 + 5}$

In Problems 3 and 4, add or subtract, as indicated.

3. $9\sqrt{13y} + 4\sqrt{13y}$ 4. $\sqrt[4]{5} + 9\sqrt[4]{5} - 3\sqrt[4]{5}$

Sometimes, before adding or subtracting, the radicals must be simplified so that the radicands are the same.

EXAMPLE 2 **Adding and Subtracting Radical Expressions**

Add or subtract, as indicated. Assume all variables are greater than or equal to zero.

(a) $3\sqrt{12} + 7\sqrt{3}$ **(b)** $3x\sqrt{20x} - 7\sqrt{5x^3}$ **(c)** $3\sqrt{5} + 7\sqrt{13}$

Solution

(a) The index on each radical is the same, but the radicands are different. However, the radicals can be simplified to make the radicands the same.

$$\sqrt{12} = \sqrt{4} \cdot \sqrt{3}$$
$$3\sqrt{12} + 7\sqrt{3} = 3\sqrt{4} \cdot \sqrt{3} + 7\sqrt{3}$$
$$\sqrt{4} = 2: \quad = 3 \cdot 2\sqrt{3} + 7\sqrt{3}$$
$$= 6\sqrt{3} + 7\sqrt{3}$$
$$\text{Factor out } \sqrt{3}: \quad = (6 + 7)\sqrt{3}$$
$$= 13\sqrt{3}$$

(b) The index on each radical is the same, but the radicands are different. However, each square root can be simplified.

$$\sqrt{20x} = \sqrt{4} \cdot \sqrt{5x} \qquad \sqrt{5x^3} = \sqrt{x^2} \cdot \sqrt{5x}$$
$$3x\sqrt{20x} - 7\sqrt{5x^3} = 3x \cdot \sqrt{4} \cdot \sqrt{5x} - 7 \cdot \sqrt{x^2} \cdot \sqrt{5x}$$
$$\text{Simplify radicals:} \quad = 3x \cdot 2 \cdot \sqrt{5x} - 7 \cdot x \cdot \sqrt{5x}$$
$$\text{Multiply:} \quad = 6x\sqrt{5x} - 7x\sqrt{5x}$$
$$\text{Factor out } \sqrt{5x}: \quad = (6x - 7x)\sqrt{5x}$$
$$= -x\sqrt{5x}$$

(c) For $3\sqrt{5} + 7\sqrt{13}$, the index on each radical is the same, but the radicands are different and cannot be simplified to make the radicands the same. Therefore, the terms cannot be added.

Quick ✓

In Problems 5–7, add or subtract, as indicated. Assume all variables are greater than or equal to zero.

5. $4\sqrt{18} - 3\sqrt{8}$ 6. $-5x\sqrt[3]{54x} + 7\sqrt[3]{2x^4}$ 7. $7\sqrt{10} - 6\sqrt{3}$

EXAMPLE 3   **Adding or Subtracting Radical Expressions**

Add or subtract, as indicated.

(a) $\sqrt[3]{16x^4} - 7x\sqrt[3]{-2x} + \sqrt[3]{54x}$ (b) $3\sqrt[4]{m^4n} - 5m\sqrt[8]{n^2}, m \geq 0$

Solution

(a) The index on each radical is the same. The radicands are different but can be simplified to make the radicands the same.

$$\boxed{\sqrt[3]{16x^4} = \sqrt[3]{8x^3} \cdot \sqrt[3]{2x}} \quad \boxed{\sqrt[3]{-2x} = \sqrt[3]{-1} \cdot \sqrt[3]{2x}} \quad \boxed{\sqrt[3]{54x} = \sqrt[3]{27} \cdot \sqrt[3]{2x}}$$

$$\sqrt[3]{16x^4} - 7x\sqrt[3]{-2x} + \sqrt[3]{54x} = \sqrt[3]{8x^3} \cdot \sqrt[3]{2x} - 7x\sqrt[3]{-1} \cdot \sqrt[3]{2x} + \sqrt[3]{27} \cdot \sqrt[3]{2x}$$

$$\sqrt[3]{8x^3} = 2x; \ \sqrt[3]{-1} = -1; \ \sqrt[3]{27} = 3: \quad = 2x\sqrt[3]{2x} - 7x(-1)\sqrt[3]{2x} + 3\sqrt[3]{2x}$$

$$-7x(-1) = 7x: \quad = 2x\sqrt[3]{2x} + 7x\sqrt[3]{2x} + 3\sqrt[3]{2x}$$

$$\text{Factor out } \sqrt[3]{2x}: \quad = (2x + 7x + 3)\sqrt[3]{2x}$$

$$\text{Simplify:} \quad = (9x + 3)\sqrt[3]{2x}$$

$$\text{Factor out } 3: \quad = 3(3x + 1)\sqrt[3]{2x}$$

Work Smart: Study Skills

Contrast adding radicals with multiplying radicals:

Add: $3\sqrt{5} + 8\sqrt{5} = (3 + 8)\sqrt{5}$
$= 11\sqrt{5}$

Multiply:
$3\sqrt{5} \cdot 8\sqrt{5} = 3 \cdot 8 \cdot \sqrt{5} \cdot \sqrt{5}$
$= 24\sqrt{25}$
$= 24 \cdot 5$
$= 120$

Ask yourself these questions:
How must the radicals be "like" to be added?
How must the radicals be "like" to be multiplied?

(b) Here, the index on the radicals *and* the radicands are different. Start by dealing with the index using rational exponents to try to get the same index on both radicals.

$$\sqrt[n]{a^m} = a^{\frac{m}{n}}$$
$$\downarrow$$

$$3\sqrt[4]{m^4n} - 5m\sqrt[8]{n^2} = 3\sqrt[4]{m^4n} - 5m \cdot n^{\frac{2}{8}}$$

$$\text{Simplify the rational exponent:} \quad = 3\sqrt[4]{m^4n} - 5m \cdot n^{\frac{1}{4}}$$

$$\text{Rewrite as radical:} \quad = 3\sqrt[4]{m^4n} - 5m \cdot \sqrt[4]{n}$$

Now the index is the same, but the radicands are still different.

$$\sqrt[4]{m^4n} = \sqrt[4]{m^4} \cdot \sqrt[4]{n}: \quad = 3\sqrt[4]{m^4} \cdot \sqrt[4]{n} - 5m \cdot \sqrt[4]{n}$$

$$\text{Simplify:} \quad = 3m \cdot \sqrt[4]{n} - 5m \cdot \sqrt[4]{n}$$

$$\text{Factor out } \sqrt[4]{n}: \quad = (3m - 5m) \cdot \sqrt[4]{n}$$

$$\text{Simplify:} \quad = -2m\sqrt[4]{n}$$

●

Quick ✓

In Problems 8 and 9, add or subtract, as indicated. Assume all variables are greater than or equal to zero.

8. $\sqrt[3]{8z^4} - 2z\sqrt[3]{-27z} + \sqrt[3]{125z}$ **9.** $\sqrt{25m} - 3\sqrt[4]{m^2}$

❷ Multiply Radical Expressions

Now let's look at multiplying radical expressions involving more than one radical. Multiply these expressions the same way polynomials are multiplied.

▶ EXAMPLE 4   **Multiplying Radical Expressions**

Multiply and simplify:

(a) $\sqrt{5}(3 - 4\sqrt{5})$ (b) $\sqrt[3]{2}(3 + \sqrt[3]{4})$ (c) $(3 + 2\sqrt{7})(2 - 3\sqrt{7})$

(continued)

Solution

(a) Use the Distributive Property and multiply each term in the parentheses by $\sqrt{5}$.

$$\sqrt{5}(3 - 4\sqrt{5}) = \sqrt{5} \cdot 3 - \sqrt{5} \cdot 4\sqrt{5}$$

Multiply radicals: $= 3\sqrt{5} - 4 \cdot \sqrt{25}$

$\sqrt{25} = 5$: $= 3\sqrt{5} - 4 \cdot 5$

Simplify: $= 3\sqrt{5} - 20$

(b) Distribute $\sqrt[3]{2}$ to each term in the parentheses.

$$\sqrt[3]{2}(3 + \sqrt[3]{4}) = \sqrt[3]{2} \cdot 3 + \sqrt[3]{2} \cdot \sqrt[3]{4}$$

Multiply radicals: $= 3\sqrt[3]{2} + \sqrt[3]{8}$

$\sqrt[3]{8} = 2$: $= 3\sqrt[3]{2} + 2$

(c) Treat this just like the product of two binomials and multiply using FOIL.

First   Last

$$(3 + 2\sqrt{7})(2 - 3\sqrt{7}) = 3 \cdot 2 - 3 \cdot 3\sqrt{7} + 2\sqrt{7} \cdot 2 - 2\sqrt{7} \cdot 3\sqrt{7}$$

Inner

Outer

Multiply: $= 6 - 9\sqrt{7} + 4\sqrt{7} - 6\sqrt{49}$

$\sqrt{49} = 7$: $= 6 - 9\sqrt{7} + 4\sqrt{7} - 6 \cdot 7$

$= 6 - 9\sqrt{7} + 4\sqrt{7} - 42$

Simplify: $= -36 - 5\sqrt{7}$

Work Smart

$-36 - 5\sqrt{7} \neq -41\sqrt{7}$

Do you know why?

Quick ✓

In Problems 10–12, multiply and simplify.

10. $\sqrt{6}(3 - 5\sqrt{6})$ **11.** $\sqrt[3]{12}(3 - \sqrt[3]{2})$ **12.** $(2 - 7\sqrt{3})(5 + 4\sqrt{3})$

The formulas for perfect squares, $(A + B)^2 = A^2 + 2AB + B^2$ and $(A - B)^2 = A^2 - 2AB + B^2$, as well as the formula for the difference of two squares, $(A + B)(A - B) = A^2 - B^2$, can be used to multiply radicals.

▶ **EXAMPLE 5**   **Multiplying Radical Expressions Involving Special Products**

Multiply and simplify.

(a) $(2\sqrt{3} + \sqrt{5})^2$ **(b)** $(3 + \sqrt{7})(3 - \sqrt{7})$

Solution

(a) Notice that $(2\sqrt{3} + \sqrt{5})^2$ is in the form $(A + B)^2$, so

$$(A + B)^2 = A^2 + 2 \cdot A \cdot B + B^2$$
$$(2\sqrt{3} + \sqrt{5})^2 = (2\sqrt{3})^2 + 2 \cdot 2\sqrt{3} \cdot \sqrt{5} + (\sqrt{5})^2$$

Multiply: $= 4\sqrt{9} + 4\sqrt{15} + \sqrt{25}$

Work Smart

Notice, in Example 5(a), that
$(2\sqrt{3} + \sqrt{5})^2 \neq (2\sqrt{3})^2 + (\sqrt{5})^2$

Simplify: $= 4 \cdot 3 + 4\sqrt{15} + 5$

Combine like terms: $= 17 + 4\sqrt{15}$

(b) Notice that $(3 + \sqrt{7})(3 - \sqrt{7})$ is in the form $(A + B)(A - B)$, so

$$(A + B)(A - B) = A^2 - B^2$$
$$(3 + \sqrt{7})(3 - \sqrt{7}) = 3^2 - (\sqrt{7})^2$$
$$= 9 - 7$$
$$= 2$$

Notice that the product found in Example 5(b) is an integer. There are no radicals in the product. Radical expressions such as $3 + \sqrt{7}$ and $3 - \sqrt{7}$ are called **conjugates** of each other. When square roots that are conjugates are multiplied, the result never contains a radical. This result plays a huge role in the next section.

Quick ✓

13. *True or False* $(\sqrt{a} + \sqrt{b})^2 = (\sqrt{a})^2 + (\sqrt{b})^2$

14. The radical expressions $4 + \sqrt{5}$ and $4 - \sqrt{5}$ are examples of _____.

15. *True or False* The conjugate of $-5 + \sqrt{2}$ is $5 - \sqrt{2}$.

In Problems 16–18, multiply and simplify.

16. $(2 - \sqrt{5})^2$ **17.** $(\sqrt{7} - 3\sqrt{2})^2$ **18.** $(\sqrt{3} + \sqrt{2})(\sqrt{3} - \sqrt{2})$

6.4 Exercises MyMathLab® Exercise numbers in green have complete video solutions in MyMathLab or may be accessed using the QR code to the right.

*Problems **1–18** are the Quick ✓s that follow the EXAMPLES.*

Building Skills

In Problems 19–26, add or subtract as indicated. Assume all variables are greater than or equal to zero. See Objective 1.

19. $3\sqrt{2} + 7\sqrt{2}$ **20.** $6\sqrt{3} + 8\sqrt{3}$

21. $5\sqrt[3]{x} - 3\sqrt[3]{x}$ **22.** $12\sqrt[4]{z} - 5\sqrt[4]{z}$

23. $8\sqrt{5x} - 3\sqrt{5x} + 9\sqrt{5x}$

24. $4\sqrt[3]{3y} + 8\sqrt[3]{3y} - 10\sqrt[3]{3y}$

25. $4\sqrt[3]{5} - 3\sqrt{5} + 7\sqrt[3]{5} - 8\sqrt{5}$

26. $12\sqrt{7} + 5\sqrt[4]{7} - 5\sqrt{7} + 6\sqrt[4]{7}$

In Problems 27–48, add or subtract as indicated. Assume all variables are greater than or equal to zero. See Objective 1.

27. $\sqrt{8} + 6\sqrt{2}$ **28.** $6\sqrt{3} + \sqrt{12}$

29. $\sqrt[3]{24} - 4\sqrt[3]{3}$ **30.** $\sqrt[3]{32} - 5\sqrt[3]{4}$

31. $\sqrt[3]{54} - 7\sqrt[3]{128}$ **32.** $7\sqrt[4]{48} - 4\sqrt[4]{243}$

33. $5\sqrt{54x} - 3\sqrt{24x}$ **34.** $2\sqrt{48z} - \sqrt{75z}$

35. $2\sqrt{8} + 3\sqrt{10}$ **36.** $4\sqrt{12} + 2\sqrt{20}$

37. $\sqrt{12x^3} + 5x\sqrt{108x}$ **38.** $3\sqrt{63z^3} + 2z\sqrt{28z}$

39. $\sqrt{12x^2} + 3x\sqrt{2} - 2\sqrt{98x^2}$

40. $\sqrt{48y^2} - 4y\sqrt{12} + \sqrt{108y^2}$

41. $\sqrt[3]{-54x^3} + 3x\sqrt[3]{16} - 2\sqrt[3]{128}$

42. $2\sqrt[3]{-5x^3} + 4x\sqrt[3]{40} - \sqrt[3]{135}$

43. $\sqrt{9x - 9} + \sqrt{4x - 4}$

44. $\sqrt{4x + 12} - \sqrt{9x + 27}$

45. $\sqrt{16x} - \sqrt[6]{x^3}$ **46.** $\sqrt{25x} - \sqrt[4]{x^2}$

47. $\sqrt[3]{27x} + 2\sqrt[9]{x^3}$ **48.** $\sqrt[4]{16y} + \sqrt[8]{y^2}$

In Problems 49–82, multiply and simplify. Assume all variables are greater than or equal to zero. See Objective 2.

49. $\sqrt{3}(2 - 3\sqrt{2})$ **50.** $\sqrt{5}(5 + 3\sqrt{3})$

51. $\sqrt{3}(\sqrt{2} + \sqrt{6})$ **52.** $\sqrt{2}(\sqrt{5} - 2\sqrt{10})$

53. $\sqrt[3]{4}(\sqrt[3]{3} - \sqrt[3]{6})$ **54.** $\sqrt[3]{6}(\sqrt[3]{2} + \sqrt[3]{12})$

55. $\sqrt{2x}(3 - \sqrt{10x})$ **56.** $\sqrt{5x}(6 + \sqrt{15x})$

57. $(3 + \sqrt{2})(4 + \sqrt{3})$ **58.** $(5 + \sqrt{5})(3 + \sqrt{6})$

59. $(6 + \sqrt{3})(2 - \sqrt{7})$ **60.** $(7 - \sqrt{3})(6 + \sqrt{5})$

61. $(4 - 2\sqrt{7})(3 + 3\sqrt{7})$

62. $(9 + 5\sqrt{10})(1 - 3\sqrt{10})$

63. $(\sqrt{2} + 3\sqrt{6})(\sqrt{3} - 2\sqrt{2})$

64. $(2\sqrt{3} + \sqrt{10})(\sqrt{5} - 2\sqrt{2})$

65. $(2\sqrt{5} + \sqrt{3})(4\sqrt{5} - 3\sqrt{3})$

66. $(\sqrt{6} - 2\sqrt{2})(2\sqrt{6} + 3\sqrt{2})$

67. $(1 + \sqrt{3})^2$ **68.** $(2 - \sqrt{3})^2$

69. $(\sqrt{2} - \sqrt{5})^2$ **70.** $(\sqrt{7} - \sqrt{3})^2$

71. $(\sqrt{x} - \sqrt{2})^2$ **72.** $(\sqrt{z} + \sqrt{5})^2$

73. $(\sqrt{2} - 1)(\sqrt{2} + 1)$

74. $(\sqrt{3} - 1)(\sqrt{3} + 1)$

75. $(3 - 2\sqrt{5})(3 + 2\sqrt{5})$

76. $(6 + 3\sqrt{2})(6 - 3\sqrt{2})$

77. $(\sqrt{2x} + \sqrt{3y})(\sqrt{2x} - \sqrt{3y})$

78. $(\sqrt{5a} + \sqrt{7b})(\sqrt{5a} - \sqrt{7b})$

79. $(\sqrt[3]{x} + 4)(\sqrt[3]{x} - 3)$

80. $(\sqrt[3]{y} - 6)(\sqrt[3]{y} + 3)$

81. $(\sqrt[3]{2a} - 5)(\sqrt[3]{2a} + 5)$

82. $(\sqrt[3]{4p} - 1)(\sqrt[3]{4p} + 3)$

Mixed Practice

In Problems 83–106, perform the indicated operation and simplify. Assume all variables are greater than or equal to zero.

83. $\sqrt{5}(\sqrt{3} + \sqrt{10})$

84. $\sqrt{7}(\sqrt{14} + \sqrt{3})$

85. $\sqrt{28x^5} - x\sqrt{7x^3} + 5\sqrt{175x^5}$

86. $\sqrt{180a^5} + a^2\sqrt{20} - a\sqrt{80a^3}$

87. $(2\sqrt{3} + 5)(2\sqrt{3} - 5)$

88. $(4\sqrt{2} - 2)(4\sqrt{2} + 2)$

89. $\sqrt[3]{7}(2 + \sqrt[3]{4})$

90. $\sqrt[3]{9}(5 + 2\sqrt[3]{2})$

91. $(2\sqrt{2} + 5)(4\sqrt{2} - 4)$

92. $(5\sqrt{5} - 3)(3\sqrt{5} - 4)$

93. $4\sqrt{18} + 2\sqrt{32}$

94. $5\sqrt{20} + 2\sqrt{80}$

95. $(\sqrt{5} - \sqrt{3})^2 - \sqrt{60}$

96. $(\sqrt{2} - \sqrt{7})^2 - \sqrt{56}$

97. $3\sqrt[3]{5x^3y} + \sqrt[3]{40y}$

98. $5\sqrt[3]{3m^3n} + \sqrt[3]{81n}$

99. $(\sqrt{2x} - \sqrt{7y})(\sqrt{2x} + \sqrt{7y}) - 2\sqrt{x^2}$

100. $(\sqrt{3a} - \sqrt{4b})(\sqrt{3a} + \sqrt{4b}) + 4\sqrt{b^2}$

101. $(2 + \sqrt{x + 1})^2$

102. $(4 + \sqrt{2x + 3})^2$

103. $(5 - \sqrt{x + 2})^2$

104. $(1 - \sqrt{3x - 4})^2$

105. $-\dfrac{3}{5} \cdot \left(-\dfrac{\sqrt{5}}{5}\right) - \dfrac{4}{5} \cdot \left(-\dfrac{2\sqrt{5}}{5}\right)$

106. $\dfrac{4}{5} \cdot \left(-\dfrac{\sqrt{5}}{5}\right) + \left(-\dfrac{3}{5}\right) \cdot \left(-\dfrac{2\sqrt{5}}{5}\right)$

107. Suppose that $f(x) = \sqrt{3x}$ and $g(x) = \sqrt{12x}$; find
  (a) $(f + g)(x)$ **(b)** $(f + g)(4)$
  (c) $(f \cdot g)(x)$

108. Suppose that $f(x) = \sqrt{4x - 4}$ and
  $g(x) = \sqrt{25x - 25}$; find
  (a) $(f + g)(x)$ **(b)** $(f + g)(10)$
  (c) $(f \cdot g)(x)$

109. Show that $-2 + \sqrt{5}$ is a solution to the equation $x^2 + 4x - 1 = 0$. Show that $-2 - \sqrt{5}$ is also a solution.

110. Show that $3 + \sqrt{7}$ is a solution to the equation $x^2 - 6x + 2 = 0$. Show that $3 - \sqrt{7}$ is also a solution.

Applying the Concepts

In Problems 111 and 112, find the perimeter and area of the figures shown. Express your answer as a radical in simplified form.

111. **112.**

Problems 113 and 114, use **Heron's Formula** *for finding the area of a triangle whose sides are known. Heron's Formula states that the area A of a triangle with sides a, b, and c is*

$$A = \sqrt{s(s-a)(s-b)(s-c)}$$

where

$$s = \frac{1}{2}(a+b+c)$$

Find the area of the shaded region by computing the difference in the areas of the triangles. That is, compute "area of larger triangle minus area of smaller triangle." Write your answer as a radical in simplified form.

113.   **114.**

Explaining the Concepts

115. Explain how to add or subtract radicals.

116. Multiply $(\sqrt{a} - \sqrt{b})(\sqrt{a} + \sqrt{b})$ and provide a general result regarding the product of conjugates involving square roots.

Synthesis Review

In Problems 117–122, multiply each of the following.

117. $(3a^3b)(4a^2b^4)$ **118.** $(3p - 1)(2p + 3)$

119. $(3y + 2)(2y - 1)$ **120.** $(m - 4)(m + 4)$

121. $(5w + 2)(5w - 2)$ **122.** $(\sqrt{x} + 2)(\sqrt{x} - 2)$

6.5 Rationalizing Radical Expressions

Objectives

1 Rationalize a Denominator Containing One Term

2 Rationalize a Denominator Containing Two Terms

> **In Other Words**
> The process of "rationalizing the denominator" is called this because the denominator becomes a rational number (no radicals).

Are You Prepared for This Section?

Before getting started, complete the following problems. If you get a problem wrong, go back to the section cited and review the material.

P1. What would we need to multiply 12 by in order to make it the smallest perfect square that is a multiple of 12?

P2. Simplify: $\sqrt{25x^2}, x > 0$ [Section 6.3, pp. 488–492]

When radical expressions appear in the denominator of a quotient, it is customary to rewrite the quotient without radicals in the denominator. This process is called **rationalizing the denominator**. This section discusses how to rationalize denominators that contain one or two terms.

▶ **1** Rationalize a Denominator Containing One Term

To rationalize a denominator containing a single square root, multiply the numerator and denominator of the quotient by a square root so that the radicand in the denominator becomes a perfect square.

For example, if the denominator of a quotient contains $\sqrt{5}$, multiply the numerator and denominator by $\sqrt{5}$ because $5 \cdot 5 = 25$, a perfect square. If the denominator contains $\sqrt{8}$, multiply the numerator and denominator by $\sqrt{2}$ because $8 \cdot 2 = 16$, a perfect square. Because both the numerator and the denominator are being multiplied by the same value, the radical expression is being multiplied by 1, so the value of the radical expression is unchanged.

EXAMPLE 1 **Rationalizing a Denominator Containing a Square Root**

Rationalize the denominator of each expression:

(a) $\dfrac{1}{\sqrt{7}}$ **(b)** $\dfrac{\sqrt{5}}{\sqrt{12}}$ **(c)** $\dfrac{2}{3\sqrt{2x}}, x > 0$

(continued)

Solution

(a) The denominator contains $\sqrt{7}$. Ask, "What can I multiply 7 by to obtain a perfect square?" Because $7 \cdot 7 = 49$, a perfect square, multiply the numerator and denominator by $\sqrt{7}$.

$$\frac{1}{\sqrt{7}} = \frac{1}{\sqrt{7}} \cdot \frac{\sqrt{7}}{\sqrt{7}}$$

$$\text{Multiply numerators;} \atop \text{multiply denominators:} = \frac{\sqrt{7}}{\sqrt{49}}$$

$$\sqrt{49} = 7: \; = \frac{\sqrt{7}}{7}$$

(b) $\sqrt{12}$ is in the denominator. Although multiplying 12 by itself would result in a perfect square, is this the best choice? No, because multiplying 12 by 3 gives 36, which is the smallest integer that makes the radicand a perfect square.

$$\frac{\sqrt{5}}{\sqrt{12}} = \frac{\sqrt{5}}{\sqrt{12}} \cdot \frac{\sqrt{3}}{\sqrt{3}}$$

$$\text{Multiply numerators;} \atop \text{multiply denominators:} = \frac{\sqrt{15}}{\sqrt{36}}$$

$$\sqrt{36} = 6: \; = \frac{\sqrt{15}}{6}$$

(c)

$$\text{Multiply by } 1 = \frac{\sqrt{2x}}{\sqrt{2x}}$$

$$\frac{2}{3\sqrt{2x}} = \frac{2}{3\sqrt{2x}} \cdot \frac{\sqrt{2x}}{\sqrt{2x}}$$

$$\text{Multiply numerators;} \atop \text{multiply denominators:} = \frac{2\sqrt{2x}}{3\sqrt{4x^2}}$$

$$\sqrt{4x^2} = 2x \text{ because } x > 0: \; = \frac{2\sqrt{2x}}{3 \cdot 2x}$$

$$\text{Divide out the common factor, 2:} \; = \frac{\sqrt{2x}}{3x}$$

> **Work Smart**
>
> Remember that $a = 1 \cdot a$ and that 1 can take many forms. In Example 1(a), $1 = \dfrac{\sqrt{7}}{\sqrt{7}}$.

> **Work Smart**
>
> An alternative approach to Example 1(b) would be to simplify $\sqrt{12}$ first, as follows:
>
> $$\frac{\sqrt{5}}{\sqrt{12}} = \frac{\sqrt{5}}{\sqrt{4 \cdot 3}}$$
> $$= \frac{\sqrt{5}}{2\sqrt{3}}$$
> $$= \frac{\sqrt{5}}{2\sqrt{3}} \cdot \frac{\sqrt{3}}{\sqrt{3}}$$
> $$= \frac{\sqrt{15}}{6}$$

Quick ✓

1. Rewriting a quotient to remove radicals from the denominator is called

   _____ ___ _____.

2. To rationalize the denominator of $\dfrac{\sqrt{5}}{\sqrt{11}}$, multiply the numerator and

   denominator by ____.

In Problems 3–5, rationalize each denominator.

3. $\dfrac{1}{\sqrt{3}}$ **4.** $\dfrac{\sqrt{5}}{\sqrt{8}}$ **5.** $\dfrac{5}{\sqrt{10x}}$

In general, to rationalize a denominator containing a single radical with index n, multiply the numerator and denominator of the quotient by a radical so that the product in the denominator has a radicand that is a perfect nth power. So, if the denominator contains a radical whose index is 3, multiply the numerator and denominator by a cube root so that the radicand in the denominator becomes a perfect cube. For example, if the denominator of a quotient contains $\sqrt[3]{4}$, multiply the numerator and denominator of the quotient by $\sqrt[3]{2}$ because $4 \cdot 2 = 8$, a perfect cube.

EXAMPLE 2

Rationalizing a Denominator Containing Cube Roots and Fourth Roots

Rationalize the denominator of each expression:

(a) $\dfrac{1}{\sqrt[3]{6}}$

(b) $\sqrt[3]{\dfrac{5}{18}}$

(c) $\dfrac{6}{\sqrt[4]{4z^3}}, z > 0$

Solution

(a) Notice that $\sqrt[3]{6}$ is in the denominator. To make the radicand in the denominator a perfect cube (because the index is 3), multiply the numerator and denominator by $\sqrt[3]{6^2} = \sqrt[3]{36}$ because $6 \cdot 36 = 6 \cdot 6^2 = 6^3 = 216$, a perfect cube.

$$\frac{1}{\sqrt[3]{6}} = \frac{1}{\sqrt[3]{6}} \cdot \frac{\sqrt[3]{6^2}}{\sqrt[3]{6^2}}$$

Multiply numerators; multiply denominators: $\quad = \dfrac{\sqrt[3]{36}}{\sqrt[3]{6^3}}$

$\sqrt[3]{6^3} = 6: \quad = \dfrac{\sqrt[3]{36}}{6}$

Work Smart

The radicand in $\sqrt[3]{6}$ is 6, or 6^1. To make it a perfect cube, 6^3, multiply by 6^2. Then $6^1 \cdot 6^2 = 6^3$. The cube root of 6^3, or 216, is 6.

(b) First, use the Quotient Property $\left(\sqrt[n]{\dfrac{a}{b}} = \dfrac{\sqrt[n]{a}}{\sqrt[n]{b}} \right)$ to rewrite the radical as the quotient of two radicals.

$$\sqrt[3]{\frac{5}{18}} = \frac{\sqrt[3]{5}}{\sqrt[3]{18}}$$

What does $\sqrt[3]{18}$ need to be multiplied by to make the radicand a perfect cube? First, write 18 as $9 \cdot 2 = 3^2 \cdot 2^1$. If $3^2 \cdot 2^1$ is multiplied by $3^1 \cdot 2^2 = 12$, then the radicand will be a perfect cube in the denominator.

$$\sqrt[3]{\frac{5}{18}} = \frac{\sqrt[3]{5}}{\sqrt[3]{18}} = \frac{\sqrt[3]{5}}{\sqrt[3]{3^2 \cdot 2}} \cdot \frac{\sqrt[3]{3 \cdot 2^2}}{\sqrt[3]{3 \cdot 2^2}}$$

Multiply numerators; multiply denominators: $\quad = \dfrac{\sqrt[3]{60}}{\sqrt[3]{3^3 \cdot 2^3}}$

$\sqrt[3]{3^3 \cdot 2^3} = 3 \cdot 2 = 6: \quad = \dfrac{\sqrt[3]{60}}{6}$

(c) Rewrite the denominator as $\sqrt[4]{2^2 \cdot z^3}$. To make the radicand a perfect fourth power, multiply $\sqrt[4]{2^2 \cdot z^3}$ by $\sqrt[4]{2^2 \cdot z}$ to obtain $\sqrt[4]{2^4 z^4}$ in the denominator.

$$\frac{6}{\sqrt[4]{4z^3}} = \frac{6}{\sqrt[4]{2^2 \cdot z^3}} \cdot \frac{\sqrt[4]{2^2 \cdot z}}{\sqrt[4]{2^2 \cdot z}}$$

Multiply numerators; multiply denominators: $\quad = \dfrac{6\sqrt[4]{4z}}{\sqrt[4]{2^4 \cdot z^4}}$

$\sqrt[4]{2^4 \cdot z^4} = 2z: \quad = \dfrac{6\sqrt[4]{4z}}{2z}$

Simplify: $\quad = \dfrac{3\sqrt[4]{4z}}{z}$

Quick ✔

In Problems 6–8, rationalize each denominator.

6. $\dfrac{4}{\sqrt[3]{3}}$

7. $\sqrt[3]{\dfrac{3}{20}}$

8. $\dfrac{3}{\sqrt[4]{p}}, p > 0$

▶ ❷ Rationalize a Denominator Containing Two Terms

To rationalize a denominator containing two terms involving square roots, use the fact that

$$(A + B)(A - B) = A^2 - B^2$$

and multiply both the numerator and the denominator of the quotient by the conjugate of the denominator. For example, if the quotient is $\dfrac{3}{\sqrt{3} + 2}$, multiply both the numerator and the denominator by the conjugate of $\sqrt{3} + 2$, which is $\sqrt{3} - 2$. Recall from the last section that the product $(\sqrt{3} + 2)(\sqrt{3} - 2)$ will not contain a radical.

EXAMPLE 3 **Rationalizing a Denominator Containing Two Terms**

Rationalize the denominator: $\dfrac{\sqrt{2}}{\sqrt{6} + 2}$

Solution

Because $\sqrt{6} + 2$ is in the denominator of the quotient, multiply the numerator and denominator by the conjugate of $\sqrt{6} + 2$, $\sqrt{6} - 2$.

$$\frac{\sqrt{2}}{\sqrt{6} + 2} = \frac{\sqrt{2}}{\sqrt{6} + 2} \cdot \frac{\sqrt{6} - 2}{\sqrt{6} - 2}$$

Multiply the numerators
and denominators:
$$= \frac{\sqrt{2}(\sqrt{6} - 2)}{(\sqrt{6} + 2)(\sqrt{6} - 2)}$$

Distribute $\sqrt{2}$ in the numerator;
$(A + B)(A - B) = A^2 - B^2$ in the denominator:
$$= \frac{\sqrt{12} - 2\sqrt{2}}{(\sqrt{6})^2 - 2^2}$$

$\sqrt{12} = 2\sqrt{3}$:
$$= \frac{2\sqrt{3} - 2\sqrt{2}}{6 - 4}$$

Factor out the GCF of 2 in numerator:
$$= \frac{2(\sqrt{3} - \sqrt{2})}{2}$$

Divide out common factor, 2:
$$= \sqrt{3} - \sqrt{2}$$

●

Quick ✔

9. To rationalize the denominator of $\dfrac{4 - \sqrt{3}}{-2 + \sqrt{7}}$, multiply the numerator and denominator by _____.

In Problems 10 and 11, rationalize the denominator.

10. $\dfrac{4}{\sqrt{3} + 1}$

11. $\dfrac{\sqrt{2}}{\sqrt{6} - \sqrt{2}}$

EXAMPLE 4   **Rationalizing a Denominator Containing Two Terms**

Rationalize the denominator: $\dfrac{\sqrt{6} - 3}{\sqrt{10} - \sqrt{6}}$

Solution

Multiply the numerator and denominator by the conjugate of the denominator, $\sqrt{10} + \sqrt{6}$.

$$\dfrac{\sqrt{6} - 3}{\sqrt{10} - \sqrt{6}} = \dfrac{\sqrt{6} - 3}{\sqrt{10} - \sqrt{6}} \cdot \dfrac{\sqrt{10} + \sqrt{6}}{\sqrt{10} + \sqrt{6}}$$

Multiply the numerators and denominators: $= \dfrac{(\sqrt{6} - 3)(\sqrt{10} + \sqrt{6})}{(\sqrt{10} - \sqrt{6})(\sqrt{10} + \sqrt{6})}$

Use FOIL in the numerator;
$(A + B)(A - B) = A^2 - B^2$ in the denominator: $= \dfrac{\sqrt{60} + \sqrt{36} - 3\sqrt{10} - 3\sqrt{6}}{(\sqrt{10})^2 - (\sqrt{6})^2}$

Simplify radicals: $= \dfrac{2\sqrt{15} + 6 - 3\sqrt{10} - 3\sqrt{6}}{10 - 6}$

$$= \dfrac{2\sqrt{15} + 6 - 3\sqrt{10} - 3\sqrt{6}}{4}$$

Quick ✓

In Problem 12, rationalize the denominator.

12. $\dfrac{\sqrt{5} + 4}{\sqrt{5} - \sqrt{2}}$

6.5 Exercises   MyMathLab®   Exercise numbers in green have complete video solutions in MyMathLab or may be accessed using the QR code to the right.

Problems 1–12 are the Quick ✓s that follow the EXAMPLES.

Building Skills

In Problems 13–36, rationalize each denominator. Assume all variables are positive. See Objective 1.

13. $\dfrac{1}{\sqrt{2}}$

14. $\dfrac{2}{\sqrt{3}}$

15. $-\dfrac{6}{5\sqrt{3}}$

16. $-\dfrac{3}{2\sqrt{3}}$

17. $\dfrac{3}{\sqrt{12}}$

18. $\dfrac{5}{\sqrt{20}}$

19. $\dfrac{\sqrt{2}}{\sqrt{6}}$

20. $\dfrac{\sqrt{3}}{\sqrt{11}}$

21. $\sqrt{\dfrac{2}{p}}$

22. $\sqrt{\dfrac{5}{z}}$

23. $\dfrac{\sqrt{8}}{\sqrt{y^3}}$

24. $\dfrac{\sqrt{32}}{\sqrt{a^5}}$

25. $\dfrac{2}{\sqrt[3]{2}}$

26. $\dfrac{5}{\sqrt[3]{3}}$

27. $\sqrt[3]{\dfrac{7}{q}}$

28. $\sqrt[3]{\dfrac{-4}{p}}$

29. $\sqrt[3]{\dfrac{-3}{50}}$

30. $\sqrt[3]{\dfrac{-5}{72}}$

31. $\dfrac{2}{\sqrt[3]{20y}}$

32. $\dfrac{8}{\sqrt[3]{36z^2}}$

33. $\dfrac{-4}{\sqrt[4]{3x^3}}$

34. $\dfrac{6}{\sqrt[4]{9b^2}}$

35. $\dfrac{12}{\sqrt[5]{m^3n^2}}$

36. $\dfrac{-3}{\sqrt[5]{ab^3}}$

In Problems 37–56, rationalize each denominator. Assume all variables are positive. See Objective 2.

37. $\dfrac{4}{\sqrt{6} - 2}$

38. $\dfrac{6}{\sqrt{7} - 2}$

39. $\dfrac{5}{\sqrt{5} + 2}$

40. $\dfrac{10}{\sqrt{10} + 3}$

41. $\dfrac{8}{\sqrt{7} - \sqrt{3}}$

42. $\dfrac{12}{\sqrt{11} - \sqrt{7}}$

43. $\dfrac{\sqrt{2}}{\sqrt{10} - \sqrt{6}}$

44. $\dfrac{\sqrt{3}}{\sqrt{15} - \sqrt{6}}$

45. $\dfrac{\sqrt{p}}{\sqrt{p} + \sqrt{q}}$

46. $\dfrac{\sqrt{a}}{\sqrt{a} + \sqrt{b}}$

47. $\dfrac{18}{2\sqrt{3} + 3\sqrt{2}}$

48. $\dfrac{15}{3\sqrt{5} + 4\sqrt{3}}$

49. $\dfrac{\sqrt{7}+3}{\sqrt{7}-3}$ **50.** $\dfrac{\sqrt{5}+3}{\sqrt{5}-3}$ **51.** $\dfrac{\sqrt{3}-4\sqrt{2}}{2\sqrt{3}+5\sqrt{2}}$

52. $\dfrac{3\sqrt{6}+5\sqrt{7}}{2\sqrt{6}-3\sqrt{7}}$ **53.** $\dfrac{\sqrt{p}+2}{\sqrt{p}-2}$ **54.** $\dfrac{\sqrt{x}-4}{\sqrt{x}+4}$

55. $\dfrac{\sqrt{2}-3}{\sqrt{8}-\sqrt{2}}$ **56.** $\dfrac{2\sqrt{3}+3}{\sqrt{12}-\sqrt{3}}$

Mixed Practice

In Problems 57–64, perform the indicated operation and simplify.

57. $\sqrt{3}+\dfrac{1}{\sqrt{3}}$ **58.** $\sqrt{5}-\dfrac{1}{\sqrt{5}}$

59. $\dfrac{\sqrt{10}}{2}-\dfrac{1}{\sqrt{2}}$ **60.** $\dfrac{\sqrt{5}}{2}+\dfrac{3}{\sqrt{5}}$

61. $\sqrt{\dfrac{1}{3}}+\sqrt{12}+\sqrt{75}$ **62.** $\sqrt{\dfrac{2}{5}}+\sqrt{20}-\sqrt{45}$

63. $\dfrac{3}{\sqrt{18}}-\sqrt{\dfrac{1}{2}}$ **64.** $\sqrt{\dfrac{4}{3}}+\dfrac{4}{\sqrt{48}}$

In Problems 65–76, simplify each expression so that the denominator does not contain a radical. Work smart because in some of the problems, it will be easier if you divide the radicands before attempting to rationalize the denominator.

65. $\dfrac{\sqrt{3}}{\sqrt{12}}$ **66.** $\dfrac{\sqrt{2}}{\sqrt{18}}$ **67.** $\dfrac{3}{\sqrt{72}}$

68. $\dfrac{7}{\sqrt{98}}$ **69.** $\sqrt{\dfrac{4}{3}}$ **70.** $\sqrt{\dfrac{9}{5}}$

71. $\dfrac{\sqrt{3}-3}{\sqrt{3}+3}$ **72.** $\dfrac{\sqrt{2}-5}{\sqrt{2}+5}$ **73.** $\dfrac{2}{\sqrt{5}+2}$

74. $\dfrac{5}{\sqrt{6}+4}$ **75.** $\dfrac{\sqrt{8}}{\sqrt{2}}$ **76.** $\dfrac{\sqrt{75}}{\sqrt{3}}$

In Problems 77–82, find the reciprocal of the given number. Be sure to rationalize the denominator.

77. $\sqrt{3}$ **78.** $\sqrt{7}$ **79.** $\sqrt[3]{12}$

80. $\sqrt[3]{18}$ **81.** $\sqrt{3}+5$ **82.** $7-\sqrt{2}$

Applying the Concepts

Math for the Future: Trigonometry *In Problems 83 and 84, simplify each expression.*

83. $\dfrac{1}{\sqrt{2}}\cdot\dfrac{\sqrt{3}}{2}-\dfrac{1}{\sqrt{2}}\cdot\dfrac{1}{2}$

84. $-\sqrt{\dfrac{2}{3}}\cdot\left(-\dfrac{2}{\sqrt{5}}\right)+\dfrac{1}{\sqrt{3}}\cdot\dfrac{1}{\sqrt{5}}$

Sometimes it may be useful to rationalize a numerator. In Problems 85–88, rationalize each expression by multiplying the numerator and denominator by the conjugate of the numerator.

85. $\dfrac{\sqrt{2}+1}{3}$ **86.** $\dfrac{\sqrt{3}+2}{2}$

87. $\dfrac{\sqrt{x}-\sqrt{h}}{\sqrt{x}}$ **88.** $\dfrac{\sqrt{a}-\sqrt{b}}{\sqrt{2}}$

Extending the Concepts

89. When two quantities a and b are positive, you can verify that $a=b$ by showing that $a^2=b^2$. Verify that $\dfrac{\sqrt{6}+\sqrt{2}}{4}=\dfrac{\sqrt{2}+\sqrt{3}}{2}$ by squaring each side.

90. Rationalize the denominator: $\dfrac{2}{\sqrt{2}+\sqrt{3}-\sqrt{9}}$

91. Math for the Future: Calculus The following problem comes up in calculus. Consider the rational expression:

$$\dfrac{\sqrt{x+h}-\sqrt{x}}{h}$$

  (a) Rationalize the numerator by multiplying the numerator and denominator by $\sqrt{x+h}+\sqrt{x}$. Be sure to simplify the expression completely.
  (b) Evaluate the expression found in part (a) at $h=0$.
  (c) The expression found in part (b) represents the formula for the slope of the line tangent to the graph of $f(x)=\sqrt{x}$ at any value of $x\geq 0$. Find the slope of the line tangent to the graph of $f(x)=\sqrt{x}$ at $x=4$.
  (d) If $f(x)=\sqrt{x}$, what is $f(4)$? What point is on the graph of $f(x)=\sqrt{x}$?
  (e) Find the equation of the line tangent to $f(x)=\sqrt{x}$ at $x=4$, using the slope found in part (c) and the point found in part (d).
  (f) Graph the function $f(x)=\sqrt{x}$ and the tangent line in the same Cartesian plane.

Explaining the Concepts

92. Explain why it is necessary to multiply the numerator and denominator by the conjugate of the denominator when rationalizing a denominator containing two terms.

93. Explain why removing irrational numbers from the denominator is called "rationalization."

Synthesis Review

In Problems 94–97, graph each of the following functions using point plotting.

94. $f(x)=5x-3$ **95.** $g(x)=-3x+9$

96. $G(x)=x^2$ **97.** $F(x)=x^3$

Putting the Concepts Together (Sections 6.1–6.5)

We designed these problems so that you can review Sections 6.1 through 6.5 and show your mastery of the concepts. Take time to work these problems before proceeding with the next section. The answers are located at the back of the text on page AN-35.

1. Evaluate: $-25^{\frac{1}{2}}$ 2. Evaluate: $(-64)^{-\frac{2}{3}}$

3. Write the expression $\sqrt[4]{3x^3}$ with a rational exponent.

4. Write the expression $7z^{\frac{4}{5}}$ as a radical expression.

5. Simplify using rational exponents: $\sqrt[3]{\sqrt{64x^3}}$

6. Distribute and simplify: $c^{\frac{1}{2}}(c^{\frac{3}{2}} + c^{\frac{5}{2}})$

In Problems 7–9, use Laws of Exponents to simplify each expression. Assume all variables in the radicand are greater than or equal to zero. Express answers with positive exponents.

7. $(a^{\frac{2}{3}}b^{-\frac{1}{3}})(a^{\frac{4}{3}}b^{-\frac{5}{3}})$ 8. $\dfrac{x^{\frac{3}{4}}}{x^{\frac{1}{8}}}$

9. $(x^{\frac{3}{4}}y^{-\frac{1}{8}})^8$

In Problems 10–19, perform the indicated operation and simplify. Assume all variables in the radicand are greater than or equal to zero.

10. $\sqrt{15a} \cdot \sqrt{2b}$

11. $\sqrt{10m^3n^2} \cdot \sqrt{20mn}$

12. $\sqrt[3]{\dfrac{-32xy^4}{4x^{-2}y}}$

13. $2\sqrt{108} - 3\sqrt{75} + \sqrt{48}$

14. $-5b\sqrt{8b} + 7\sqrt{18b^3}$

15. $\sqrt[3]{16y^4} - y\sqrt[3]{2y}$

16. $(3\sqrt{x})(4\sqrt{x})$

17. $3\sqrt{x} + 4\sqrt{x}$

18. $(2 - 3\sqrt{2})(10 + \sqrt{2})$

19. $(4\sqrt{2} - 3)^2$

In Problems 20 and 21, rationalize the denominator.

20. $\dfrac{3}{2\sqrt{32}}$ 21. $\dfrac{4}{\sqrt{3} - 8}$

6.6 Functions Involving Radicals

Objectives

❶ Evaluate Functions Involving Radicals

❷ Find the Domain of a Function Involving a Radical

❸ Graph Functions Involving Square Roots

❹ Graph Functions Involving Cube Roots

Are You Prepared for This Section?

Before getting started, complete the following problems. If you get a problem wrong, go back to the section cited and review the material.

P1. Evaluate: $\sqrt{121}$ [Getting Ready: Square Roots, pp. 468–469]

P2. Simplify: $\sqrt{p^2}$ [Getting Ready: Square Roots, pp. 470–471]

P3. Given $f(x) = x^2 - 4$, find $f(3)$.   [Section 2.2, pp. 154–162]

P4. Solve: $-2x + 3 \geq 0$ [Section 1.4, pp. 86–88]

P5. Graph $f(x) = x^2 + 1$
using point plotting.   [Section 2.3, pp. 158–160]

▶ ❶ **Evaluate Functions Involving Radicals**

Functions whose rule contains a radical can be evaluated by substituting the value of the independent variable into the rule, just as in Section 2.2.

EXAMPLE 1   **Evaluating Functions for Which the Rule Is a Radical Expression**

For the functions $f(x) = \sqrt{x + 2}$ and $g(x) = \sqrt[3]{3x + 1}$, find

Prepared?...Answers
See page 510.

(a) $f(7)$ **(b)** $f(10)$ **(c)** $g(-3)$

(continued)

Solution

(a) $f(x) = \sqrt{x + 2}$
$f(7) = \sqrt{7 + 2}$
$= \sqrt{9}$
$= 3$

(b) $f(x) = \sqrt{x + 2}$
$f(10) = \sqrt{10 + 2}$
$= \sqrt{12}$
$= 2\sqrt{3}$

(c) $g(x) = \sqrt[3]{3x + 1}$
$g(-3) = \sqrt[3]{3(-3) + 1}$
$= \sqrt[3]{-8}$
$= -2$ ●

Quick ✓

In Problems 1 and 2, find the values for each function.

1. $f(x) = \sqrt{3x + 7}$
   (a) $f(3)$ (b) $f(7)$

2. $g(x) = \sqrt[3]{2x + 7}$
   (a) $g(-4)$ (b) $g(10)$

▶ ❷ Find the Domain of a Function Involving a Radical

Recall $\sqrt[n]{a} = b$ means $a = b^n$. This definition showed that for $n \geq 2$ and even, the radicand, a, must be greater than or equal to 0. For $n \geq 3$ and odd, the radicand, a, can be any real number. This leads to a procedure for finding the domain of a function whose rule contains a radical.

> **Finding the Domain of a Function Involving a Radical**
>
> - If the index on a radical is even, then the radicand must be greater than or equal to zero.
> - If the index on a radical is odd, then the radicand can be any real number.

EXAMPLE 2   **Finding the Domain of a Radical Function**

Find the domain of each of the following functions:

   (a) $f(x) = \sqrt{x - 5}$ (b) $G(x) = \sqrt[3]{2x + 1}$ (c) $h(t) = \sqrt[4]{5 - 2t}$

Solution

(a) The function $f(x) = \sqrt{x - 5}$ tells us to take the square root of $x - 5$. Square roots can be taken only of numbers greater than or equal to zero, so the radicand, $x - 5$, must be greater than or equal to zero. This requires that

$$x - 5 \geq 0$$
Add 5 to both sides: $x \geq 5$

The domain of f is $\{x \mid x \geq 5\}$, or the interval $[5, \infty)$.

(b) The function $G(x) = \sqrt[3]{2x + 1}$ tells us to take the cube root of $2x + 1$. The cube root of any real number can be taken, so the domain of G is $\{x \mid x$ is any real number$\}$, or the interval $(-\infty, \infty)$.

(c) The function $h(t) = \sqrt[4]{5 - 2t}$ tells us to take the fourth root of $5 - 2t$. Fourth roots can be taken only of numbers greater than or equal to zero, so the radicand, $5 - 2t$, must be greater than or equal to zero:

$$5 - 2t \geq 0$$
Subtract 5 from both sides: $-2t \geq -5$
Divide both sides by -2 (Don't forget to change the direction of the inequality!): $t \leq \dfrac{5}{2}$

The domain of h is $\left\{t \mid t \leq \dfrac{5}{2}\right\}$, or the interval $\left(-\infty, \dfrac{5}{2}\right]$. ●

Prepared?...Answers
P1. 11 **P2.** $|p|$ **P3.** 5

P4. $\left\{x \mid x \leq \dfrac{3}{2}\right\}$ or $\left(-\infty, \dfrac{3}{2}\right]$

P5.

Quick ✓

3. If the index on a radical expression is ____, then the radicand must be greater than or equal to zero. If the index on a radical expression is ____, then the radicand can be any real number.

In Problems 4–6, find the domain of each function.

4. $H(x) = \sqrt{x + 6}$ **5.** $g(t) = \sqrt[5]{3t - 1}$ **6.** $F(m) = \sqrt[4]{6 - 3m}$

▶ ❸ Graph Functions Involving Square Roots

The **square root function** is given by $f(x) = \sqrt{x}$. The domain of the square root function is $\{x \mid x \geq 0\}$ or, in interval notation, $[0,\infty)$. Graph $f(x) = \sqrt{x}$ by finding some ordered pairs (x, y) such that $y = \sqrt{x}$. Then plot the ordered pairs in the xy-plane and connect the points. To make life easy, choose values of x that are perfect squares $(0, 1, 4, 9,$ and so on$)$. Table 1 shows some points on the graph of $f(x) = \sqrt{x}$. Figure 3 shows the graph of $f(x) = \sqrt{x}$ and illustrates that the range of $f(x) = \sqrt{x}$ is $[0,\infty)$.

Table 1

x	$f(x) = \sqrt{x}$	(x, y) or $(x, f(x))$
0	$f(0) = \sqrt{0} = 0$	$(0, 0)$
1	$f(1) = 1$	$(1, 1)$
4	$f(4) = 2$	$(4, 2)$
9	$f(9) = 3$	$(9, 3)$
16	$f(16) = 4$	$(16, 4)$

Figure 3

$f(x) = \sqrt{x}$

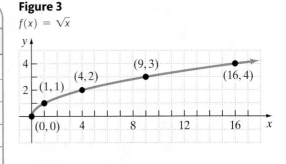

The point-plotting method can be used to graph a variety of functions involving square roots.

EXAMPLE 3   **Graphing a Function Involving a Square Root**

Consider the function $f(x) = \sqrt{x - 2}$.

(a) Find the domain.

(b) Graph the function using point plotting.

(c) Based on the graph, determine the range.

Solution

(a) Because $f(x) = \sqrt{x - 2}$ is a square root function, the radicand, $x - 2$, must be greater than or equal to zero. This requires that

$$x - 2 \geq 0$$

Add 2 to both sides: $x \geq 2$

The domain of f is $\{x \mid x \geq 2\}$, or the interval $[2,\infty)$.

(b) Choose values of x that are greater than or equal to 2 and that will make the radicand a perfect square. See Table 2 on the next page. Figure 4 shows the graph of $f(x) = \sqrt{x - 2}$.

(continued)

Table 2

x	$f(x) = \sqrt{x-2}$	(x, y) or $(x, f(x))$
2	$f(2) = \sqrt{2-z} = 0$	$(2, 0)$
3	$f(3) = 1$	$(3, 1)$
6	$f(6) = 2$	$(6, 2)$
11	$f(11) = 3$	$(11, 3)$
18	$f(18) = 4$	$(18, 4)$

Figure 4

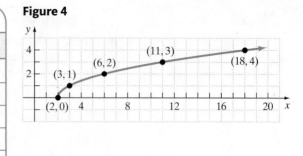

(c) From the graph of $f(x) = \sqrt{x-2}$ given in Figure 4, we can see that the range of $f(x) = \sqrt{x-2}$ is $[0, \infty)$.

Quick ✓

7. Consider the function $f(x) = \sqrt{x+3}$.

    (a) Find the domain.

    (b) Graph the function using point plotting.

    (c) Based on the graph, determine the range.

▶ ❹ Graph Functions Involving Cube Roots

The **cube root function** is given by $f(x) = \sqrt[3]{x}$. The domain of the cube root function is $\{x \mid x \text{ is any real number}\}$ or, in interval notation, $(-\infty, \infty)$. Graph $f(x) = \sqrt[3]{x}$ by finding some ordered pairs (x, y) such that $y = \sqrt[3]{x}$. Then plot the ordered pairs in the xy-plane and connect the points. To make life easy, choose values of x that are perfect cubes ($-8, -1, 0, 1, 8$, and so on). See Table 3. Figure 5 shows the graph of $f(x) = \sqrt[3]{x}$.

Table 3

x	$f(x) = \sqrt[3]{x}$	(x, y) or $(x, f(x))$
-8	$f(-8) = \sqrt[3]{-8} = -2$	$(-8, -2)$
-1	$f(-1) = -1$	$(-1, -1)$
$-\dfrac{1}{8}$	$f\left(-\dfrac{1}{8}\right) = -\dfrac{1}{2}$	$\left(-\dfrac{1}{8}, -\dfrac{1}{2}\right)$
0	$f(0) = 0$	$(0, 0)$
$\dfrac{1}{8}$	$f\left(\dfrac{1}{8}\right) = \dfrac{1}{2}$	$\left(\dfrac{1}{8}, \dfrac{1}{2}\right)$
1	$f(1) = 1$	$(1, 1)$
8	$f(8) = 2$	$(8, 2)$

Figure 5

$f(x) = \sqrt[3]{x}$

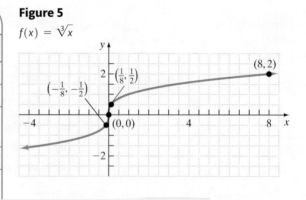

The graph of $f(x) = \sqrt[3]{x}$ in Figure 5 shows that the range of the function is the set of all real numbers or $(-\infty, \infty)$.

The point-plotting method can be used to graph a variety of functions involving cube roots.

EXAMPLE 4 **Graphing a Function Involving a Cube Root**

Consider the function $g(x) = \sqrt[3]{x} + 2$.

(a) Find the domain.

(b) Graph the function using point plotting.

(c) Based on the graph, determine the range.

Solution

(a) The function $g(x) = \sqrt[3]{x} + 2$ tells us to take the cube root of x and then add 2. The cube root is defined for all real numbers, so the domain of g is $\{x \mid x \text{ is any real number}\}$, or the interval $(-\infty, \infty)$.

(b) Choose values of x that make the radicand a perfect cube. See Table 4. Figure 6 shows the graph of $g(x) = \sqrt[3]{x} + 2$.

Table 4

x	$g(x) = \sqrt[3]{x} + 2$	(x, y) or $(x, f(x))$
-8	$g(-8) = \sqrt[3]{-8} + 2 = 0$	$(-8, 0)$
-1	$g(-1) = 1$	$(-1, 1)$
0	$g(0) = 2$	$(0, 2)$
1	$g(1) = 3$	$(1, 3)$
8	$g(8) = 4$	$(8, 4)$

Figure 6

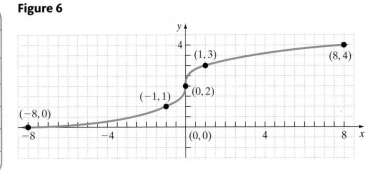

(c) The graph of $g(x) = \sqrt[3]{x} + 2$ shows that the range of $g(x) = \sqrt[3]{x} + 2$ is the set of all real numbers or $(-\infty, \infty)$.

Quick ✓

8. Consider the function $G(x) = \sqrt[3]{x} - 1$.

(a) Find the domain.

(b) Graph the function using point plotting.

(c) Based on the graph, determine the range.

6.6 Exercises MyMathLab® Exercise numbers in green have complete video solutions in MyMathLab or may be accessed using the QR code to the right.

*Problems **1–8** are the Quick ✓s that follow the EXAMPLES.*

Building Skills

In Problems 9–20, evaluate each radical function at the indicated values. See Objective 1.

9. $f(x) = \sqrt{x + 6}$
  (a) $f(3)$
  (b) $f(8)$
  (c) $f(-2)$

10. $f(x) = \sqrt{x + 10}$
  (a) $f(6)$
  (b) $f(2)$
  (c) $f(-6)$

11. $g(x) = -\sqrt{2x + 3}$
  (a) $g(11)$
  (b) $g(-1)$
  (c) $g\left(\dfrac{1}{8}\right)$

12. $g(x) = -\sqrt{4x + 5}$
  (a) $g(1)$
  (b) $g(10)$
  (c) $g\left(\dfrac{1}{8}\right)$

13. $G(m) = 2\sqrt{5m - 1}$
  (a) $G(1)$
  (b) $G(5)$
  (c) $G\left(\dfrac{1}{2}\right)$

14. $G(p) = 3\sqrt{4p + 1}$
  (a) $G(2)$
  (b) $G(11)$
  (c) $G\left(\dfrac{1}{8}\right)$

15. $H(z) = \sqrt[3]{z + 4}$
    (a) $H(4)$
    (b) $H(-12)$
    (c) $H(-20)$

16. $G(t) = \sqrt[3]{t - 6}$
    (a) $G(7)$
    (b) $G(-21)$
    (c) $G(22)$

17. $f(x) = \sqrt{\dfrac{x - 2}{x + 2}}$
    (a) $f(7)$
    (b) $f(6)$
    (c) $f(10)$

18. $f(x) = \sqrt{\dfrac{x - 4}{x + 4}}$
    (a) $f(5)$
    (b) $f(8)$
    (c) $f(12)$

19. $g(z) = \sqrt[3]{\dfrac{2z}{z - 4}}$
    (a) $g(-4)$
    (b) $g(8)$
    (c) $g(12)$

20. $H(z) = \sqrt[3]{\dfrac{3z}{z + 5}}$
    (a) $H(3)$
    (b) $H(4)$
    (c) $H(-1)$

In Problems 21–36, find the domain of the radical function. See Objective 2.

21. $f(x) = \sqrt{x - 7}$

22. $f(x) = \sqrt{x + 4}$

23. $g(x) = \sqrt{2x + 7}$

24. $g(x) = \sqrt{3x + 7}$

25. $F(x) = \sqrt{4 - 3x}$

26. $G(x) = \sqrt{5 - 2x}$

27. $H(z) = \sqrt[3]{2z + 1}$

28. $G(z) = \sqrt[3]{5z - 3}$

29. $W(p) = \sqrt[4]{7p - 2}$

30. $C(y) = \sqrt[4]{3y - 2}$

31. $g(x) = \sqrt[5]{x - 3}$

32. $g(x) = \sqrt[5]{x + 9}$

33. $f(x) = \sqrt{\dfrac{3}{x + 5}}$

34. $f(x) = \sqrt{\dfrac{3}{x - 3}}$

35. $H(x) = \sqrt[5]{\dfrac{x + 3}{x - 3}}$

36. $H(x) = \sqrt[3]{\dfrac{x - 5}{x + 2}}$

In Problems 37–52, (a) determine the domain of the function; (b) graph the function using point plotting; and (c) based on the graph, determine the range of the function. See Objective 3.

37. $f(x) = \sqrt{x - 4}$

38. $f(x) = \sqrt{x - 1}$

39. $g(x) = \sqrt{x + 2}$

40. $g(x) = \sqrt{x + 5}$

41. $G(x) = \sqrt{2 - x}$

42. $F(x) = \sqrt{4 - x}$

43. $f(x) = \sqrt{x} + 3$

44. $f(x) = \sqrt{x} + 1$

45. $g(x) = \sqrt{x} - 4$

46. $g(x) = \sqrt{x} - 2$

47. $H(x) = 2\sqrt{x}$

48. $h(x) = 3\sqrt{x}$

49. $f(x) = \dfrac{1}{2}\sqrt{x}$

50. $g(x) = \dfrac{1}{4}\sqrt{x}$

51. $G(x) = -\sqrt{x}$

52. $F(x) = \sqrt{-x}$

In Problems 53–58, (a) determine the domain of the function; (b) graph the function using point plotting; and (c) based on the graph, determine the range of the function. See Objective 4.

53. $h(x) = \sqrt[3]{x} + 2$

54. $g(x) = \sqrt[3]{x} - 4$

55. $f(x) = \sqrt[3]{x} - 3$

56. $H(x) = \sqrt[3]{x} + 3$

57. $G(x) = 2\sqrt[3]{x}$

58. $F(x) = 3\sqrt[3]{x}$

Applying the Concepts

59. Distance to a Point on a Graph Suppose that $P = (x, y)$ is a point on the graph of $y = x^2 - 4$. The distance from P to $(0, 1)$ is given by the function

$$d(x) = \sqrt{x^4 - 9x^2 + 25}$$

See the figure.

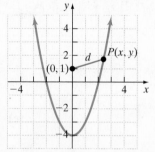

    (a) What is the distance from $P = (0, -4)$ to $(0, 1)$? That is, what is $d(0)$?
    (b) What is the distance from $P = (1, -3)$ to $(0, 1)$? That is, what is $d(1)$?
    (c) What is the distance from $P = (5, 21)$ to $(0, 1)$? That is, what is $d(5)$?

60. Distance to a Point on a Graph Suppose that $P = (x, y)$ is a point on the graph of $y = x^2 - 2$. The distance from P to $(0, 2)$ is given by the function

$$d(x) = \sqrt{x^4 - 7x^2 + 16}$$

See the figure.

    (a) What is the distance from $P = (0, -2)$ to $(0, 2)$? That is, what is $d(0)$?
    (b) What is the distance from $P = (1, -1)$ to $(0, 2)$? That is, what is $d(1)$?
    (c) What is the distance from $P = (4, 14)$ to $(0, 2)$? That is, what is $d(4)$?

△ **61. Area** A rectangle is inscribed in a semicircle of radius 3, as shown in the figure. Let $P = (x, y)$ be the point in quadrant I that is a vertex of the rectangle and is on the circle.

The area A of the rectangle as a function of x is given by

$$A(x) = 2x\sqrt{9 - x^2}$$

(a) What is the area of the rectangle whose vertex is at $(1, 2\sqrt{2})$?

(b) What is the area of the rectangle whose vertex is at $(2, \sqrt{5})$?

(c) What is the area of the rectangle whose vertex is at $(\sqrt{2}, \sqrt{7})$?

△ **62. Area** A rectangle is inscribed in a semicircle of radius 4 as shown in the figure. Let $P = (x, y)$ be the point in quadrant I that is a vertex of the rectangle and is on the circle.

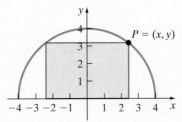

The area A of the rectangle as a function of x is given by

$$A(x) = 2x\sqrt{16 - x^2}$$

(a) What is the area of the rectangle whose vertex is at $(1, \sqrt{15})$?

(b) What is the area of the rectangle whose vertex is at $(2, 2\sqrt{3})$?

(c) What is the area of the rectangle whose vertex is at $(2\sqrt{2}, 2\sqrt{2})$?

Extending the Concepts

63. Use the results of Problems 37–40 to make a generalization about how to obtain the graph of $g(x) = \sqrt{x + c}$ from the graph of $f(x) = \sqrt{x}$.

64. Use the results of Problems 43–46 to make a generalization about how to obtain the graph of $g(x) = \sqrt{x} + c$ from the graph of $f(x) = \sqrt{x}$.

Synthesis Review

In Problems 65–69, add each of the following.

65. $\dfrac{1}{3} + \dfrac{1}{2}$

66. $\dfrac{1}{5} + \dfrac{3}{4}$

67. $\dfrac{1}{x} + \dfrac{3}{x + 1}$

68. $\dfrac{5}{x - 3} + \dfrac{2}{x + 1}$

69. $\dfrac{4}{x - 1} + \dfrac{3}{x + 1}$

70. Explain how adding rational numbers that do not have denominators with any common factors is similar to adding rational expressions that do not have denominators with any common factors.

Technology Exercises

Technology can graph square root and cube root functions. The figure below shows the graph of $f(x) = \sqrt{x - 2}$ using GeoGebra. Note how the graph shown exists only for $x \geq 2$. Graphing functions is useful in verifying the domain that we find algebraically.

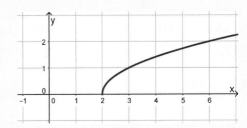

In Problems 71–92, graph the function using technology. Compare the graphs obtained by technology to the hand-drawn graphs in Problems 37–58.

71. $f(x) = \sqrt{x - 4}$

72. $f(x) = \sqrt{x - 1}$

73. $g(x) = \sqrt{x + 2}$

74. $g(x) = \sqrt{x + 5}$

75. $G(x) = \sqrt{2 - x}$

76. $F(x) = \sqrt{4 - x}$

77. $f(x) = \sqrt{x} + 3$

78. $f(x) = \sqrt{x} + 1$

79. $g(x) = \sqrt{x} - 4$

80. $g(x) = \sqrt{x} - 2$

81. $H(x) = 2\sqrt{x}$

82. $h(x) = 3\sqrt{x}$

83. $f(x) = \dfrac{1}{2}\sqrt{x}$

84. $g(x) = \dfrac{1}{4}\sqrt{x}$

85. $G(x) = -\sqrt{x}$

86. $F(x) = \sqrt{-x}$

87. $h(x) = \sqrt[3]{x + 2}$

88. $g(x) = \sqrt[3]{x - 4}$

89. $f(x) = \sqrt[3]{x} - 3$

90. $H(x) = \sqrt[3]{x} + 3$

91. $G(x) = 2\sqrt[3]{x}$

92. $F(x) = 3\sqrt[3]{x}$

6.7 Radical Equations and Their Applications

Objectives

1. Solve Radical Equations Containing One Radical
2. Solve Radical Equations Containing Two Radicals
3. Solve for a Variable in a Radical Equation

Are You Prepared for This Section?

Before getting started, complete the following problems. If you get a problem wrong, go back to the section cited and review the material.

P1. Solve: $3x - 5 = 0$ [Section 1.1, pp. 49–51]

P2. Solve: $2p^2 + 4p - 6 = 0$ [Section 4.8, pp. 373–375]

P3. Simplify: $\sqrt[3]{(x - 5)^3}$ [Section 6.1, p. 476]

When the variable in an equation occurs in a radicand, the equation is called a **radical equation.** Examples of radical equations are

$$\sqrt{3x + 1} = 5 \qquad \sqrt[3]{x - 5} - 5 = 12 \qquad \sqrt{x - 2} - \sqrt{2x + 5} = 2$$

This section discusses how to solve radical equations involving one or two radicals.

▶ ① Solve Radical Equations Containing One Radical

To solve an equation containing a single radical, isolate the radical. Then, as stated in Section 6.3, use the fact that $\left(\sqrt[n]{a}\right)^n = a$, assuming the radicand is greater than or equal to zero.

EXAMPLE 1   **How to Solve a Radical Equation Containing One Radical**

Solve: $\sqrt{2x - 3} - 5 = 0$

Step-by-Step Solution

Step 1: Isolate the radical.

$$\sqrt{2x - 3} - 5 = 0$$

Add 5 to both sides: $\sqrt{2x - 3} = 5$

Step 2: Raise both sides to the power of the index.

The index is 2, so square both sides: $\left(\sqrt{2x - 3}\right)^2 = 5^2$

$$2x - 3 = 25$$

Step 3: Solve the equation that results.

Add 3 to both sides: $2x - 3 + 3 = 25 + 3$

$$2x = 28$$

Divide both sides by 2: $\dfrac{2x}{2} = \dfrac{28}{2}$

$$x = 14$$

Step 4: Check

$$\sqrt{2x - 3} - 5 = 0$$

Let $x = 14$ in the original equation: $\sqrt{2 \cdot 14 - 3} - 5 \overset{?}{=} 0$

$$\sqrt{28 - 3} - 5 \overset{?}{=} 0$$

$$5 - 5 = 0 \quad \text{True}$$

The solution set is $\{14\}$.

  ●

It has always been emphasized to check answers whenever solving an equation. This is particularly important when solving equations involving radicals. Why? When radical equations containing an even index are solved, apparent solutions that are not solutions to the original equation can creep in. As seen when rational equations were solved in Chapter 5, these solutions are extraneous.

Prepared?...Answers P1. $\left\{\dfrac{5}{3}\right\}$

P2. $\{-3, 1\}$ **P3.** $x - 5$

> **Solving a Radical Equation Containing One Radical**
>
> **Step 1:** Isolate the radical. That is, get the radical by itself on one side of the equation.
> **Step 2:** Raise both sides of the equation to the power of the index. This will eliminate the radical from the equation.
> **Step 3:** Solve the equation that results.
> **Step 4:** Check your answer in the original equation.

Quick ✓

1. When the variable in an equation occurs in a radical, the equation is called a
_____ _____.

2. When an apparent solution is not a solution of the original equation, the apparent solution is a(n) _____ solution.

3. *True or False* The first step in solving $x + \sqrt{x - 3} = 5$ is to square both sides of the equation.

4. Solve: $\sqrt{3x + 1} - 4 = 0$

EXAMPLE 2 **Solving a Radical Equation Containing One Radical**

Solve:

(a) $\sqrt{3x + 10} + 2 = 4$ **(b)** $\sqrt{5x - 1} + 7 = 5$

Solution

(a)

$$\sqrt{3x + 10} + 2 = 4$$

Subtract 2 from both sides: $\sqrt{3x + 10} + 2 - 2 = 4 - 2$

$$\sqrt{3x + 10} = 2$$

The index is 2, so square both sides: $(\sqrt{3x + 10})^2 = 2^2$

$$3x + 10 = 4$$

Subtract 10 from both sides: $3x + 10 - 10 = 4 - 10$

$$3x = -6$$

Divide both sides by 3: $\dfrac{3x}{3} = \dfrac{-6}{3}$

$$x = -2$$

Work Smart

Just because the apparent solution is negative does not automatically make it extraneous. Determine whether the apparent solution makes the radicand negative by checking the solution.

Check

$$\sqrt{3x + 10} + 2 = 4$$

Let $x = -2$ in the original equation: $\sqrt{3 \cdot (-2) + 10} + 2 \stackrel{?}{=} 4$

$$\sqrt{-6 + 10} + 2 \stackrel{?}{=} 4$$

$$\sqrt{4} + 2 \stackrel{?}{=} 4$$

$$2 + 2 \stackrel{?}{=} 4$$

$$4 = 4 \quad \text{True}$$

The solution set is $\{-2\}$.

(b)

$$\sqrt{5x - 1} + 7 = 5$$

Subtract 7 from both sides: $\sqrt{5x - 1} + 7 - 7 = 5 - 7$

$$\sqrt{5x - 1} = -2$$

The equation has no real solution because the principal square root of a number cannot be less than 0. Put another way, there is no real number whose principal square root is -2, so the equation has no real solution. The solution set is $\varnothing$ or $\{\ \}$. ●

Quick ✓

5. Solve: $\sqrt{2x + 35} - 2 = 3$ **6.** Solve: $\sqrt{2x + 3} + 8 = 6$

⊙ **EXAMPLE 3** **Solving a Radical Equation Containing One Radical**

Solve: $\sqrt{x + 5} = x - 1$

Solution

Work Smart
$(x - 1)^2 = x^2 - 2x + 1;$
$(x - 1)^2 \neq x^2 + 1$

$$\sqrt{x + 5} = x - 1$$

The index is 2, so square both sides: $(\sqrt{x + 5})^2 = (x - 1)^2$

$$x + 5 = x^2 - 2x + 1$$

Subtract x and 5 from both sides: $x + 5 - x - 5 = x^2 - 2x + 1 - x - 5$

$$0 = x^2 - 3x - 4$$

Factor: $0 = (x - 4)(x + 1)$

Zero-Product Property: $x - 4 = 0$ or $x + 1 = 0$

$$x = 4 \text{ or } \qquad x = -1$$

Check $\sqrt{x + 5} = x - 1$

$x = 4$: $\sqrt{4 + 5} \overset{?}{=} 4 - 1$ $x = -1$: $\sqrt{-1 + 5} \overset{?}{=} -1 - 1$

$\sqrt{9} \overset{?}{=} 3$ $\sqrt{4} \overset{?}{=} -2$

$3 = 3$ True $2 = -2$ False

The apparent solution $x = -1$ does not check, so it is an extraneous solution. The solution set is $\{4\}$.

EXAMPLE 4 **Solving a Radical Equation Containing One Radical**

Solve: $\sqrt[3]{3x - 12} + 4 = 1$

Solution

$$\sqrt[3]{3x - 12} + 4 = 1$$

Subtract 4 from both sides: $\sqrt[3]{3x - 12} + 4 - 4 = 1 - 4$

$$\sqrt[3]{3x - 12} = -3$$

The index is 3, so cube both sides: $(\sqrt[3]{3x - 12})^3 = (-3)^3$

$$3x - 12 = -27$$

Add 12 to both sides: $3x - 12 + 12 = -27 + 12$

$$3x = -15$$

Divide both sides by 3: $x = -5$

Check

$$\sqrt[3]{3x - 12} + 4 = 1$$

Let $x = -5$ in the original equation: $\sqrt[3]{3 \cdot (-5) - 12} + 4 \overset{?}{=} 1$

$$\sqrt[3]{-15 - 12} + 4 \overset{?}{=} 1$$

$$\sqrt[3]{-27} + 4 \overset{?}{=} 1$$

$$-3 + 4 \overset{?}{=} 1$$

$$1 = 1 \quad \text{True}$$

The solution set is $\{-5\}$.

Quick ✓

7. Solve: $\sqrt{2x + 1} = x - 1$ **8.** Solve: $\sqrt[3]{3x + 1} - 4 = -6$

Sometimes, an equation will contain rational exponents rather than radicals. When solving these problems, rewrite the equation with a radical or use the fact that $(a^r)^s = a^{r \cdot s}$.

EXAMPLE 5 Solving an Equation Containing a Rational Exponent

Solve: $(5x - 1)^{\frac{1}{2}} + 3 = 10$

Solution

$$(5x - 1)^{\frac{1}{2}} + 3 = 10$$

Subtract 3 from both sides: $(5x - 1)^{\frac{1}{2}} = 7$

Square both sides: $\left((5x - 1)^{\frac{1}{2}} \right)^2 = 7^2$

Use $(a^r)^s = a^{r \cdot s}$: $(5x - 1)^{\frac{1}{2} \cdot 2} = 49$

$$5x - 1 = 49$$

Add 1 to both sides: $5x = 50$

Divide both sides by 5: $x = 10$

Check

$$(5x - 1)^{\frac{1}{2}} + 3 = 10$$

Let $x = 10$ in the original equation: $(5 \cdot 10 - 1)^{\frac{1}{2}} + 3 \stackrel{?}{=} 10$

$$(49)^{\frac{1}{2}} + 3 \stackrel{?}{=} 10$$

$$7 + 3 = 10$$

$$10 = 10 \quad \text{True}$$

The solution set is $\{10\}$.

Quick ✓

9. Solve: $(2x - 3)^{\frac{1}{3}} - 7 = -4$

▶ ❷ Solve Radical Equations Containing Two Radicals

EXAMPLE 6 How to Solve a Radical Equation Containing Two Radicals

Solve: $\sqrt[3]{p^2 - 4p - 4} = \sqrt[3]{-3p + 2}$

Step-by-Step-Solution

Step 1: Isolate one of the radicals.

The radical on the left side of the equation is isolated:

$$\sqrt[3]{p^2 - 4p - 4} = \sqrt[3]{-3p + 2}$$

Step 2: Raise both sides to the power of the index.

The index is 3, so cube both sides:

$$\left(\sqrt[3]{p^2 - 4p - 4} \right)^3 = \left(\sqrt[3]{-3p + 2} \right)^3$$

$$p^2 - 4p - 4 = -3p + 2$$

Step 3: Because there is no radical, solve the equation that results.

Add $3p$ to both sides; subtract 2 from both sides: $p^2 - 4p - 4 + 3p - 2 = -3p + 2 + 3p - 2$

Combine like terms: $p^2 - p - 6 = 0$

Factor: $(p - 3)(p + 2) = 0$

Use the Zero-Product Property: $p - 3 = 0 \quad \text{or} \quad p + 2 = 0$

$$p = 3 \quad \text{or} \quad p = -2$$

Step 4: Check

$$\sqrt[3]{p^2 - 4p - 4} = \sqrt[3]{-3p + 2}$$

$p = -2$:

$$\sqrt[3]{(-2)^2 - 4(-2) - 4} \stackrel{?}{=} \sqrt[3]{-3(-2) + 2}$$

$$\sqrt[3]{4 + 8 - 4} \stackrel{?}{=} \sqrt[3]{6 + 2}$$

$$\sqrt[3]{8} = \sqrt[3]{8} \quad \text{True}$$

$p = 3$:

$$\sqrt[3]{(3)^2 - 4(3) - 4} \stackrel{?}{=} \sqrt[3]{-3(3) + 2}$$

$$\sqrt[3]{9 - 12 - 4} \stackrel{?}{=} \sqrt[3]{-9 + 2}$$

$$\sqrt[3]{-7} = \sqrt[3]{-7} \quad \text{True}$$

Both apparent solutions check. The solution set is $\{-2, 3\}$.

When a radical equation contains two radicals, use the following steps to solve the equation.

> **Solving a Radical Equation Containing Two Radicals**
>
> **Step 1:** Isolate one of the radicals. That is, get one of the radicals by itself on one side of the equation.
>
> **Step 2:** Raise both sides of the equation to the power of the index. This will eliminate one or both radicals from the equation.
>
> **Step 3:** If a radical remains in the equation, then follow the steps for solving a radical equation containing one radical. Otherwise, solve the equation that results.
>
> **Step 4:** Check your answer. Discard any extraneous solutions.

Quick ✓

10. Solve: $\sqrt[3]{m^2 + 4m + 4} = \sqrt[3]{2m + 7}$

EXAMPLE 7 **Solving a Radical Equation—Squaring Twice**

Solve: $\sqrt{3x + 6} - \sqrt{x + 6} = 2$

Solution

$$\sqrt{3x + 6} - \sqrt{x + 6} = 2$$

Add $\sqrt{x + 6}$ to both sides:
$$\sqrt{3x + 6} = 2 + \sqrt{x + 6}$$

Square both sides:
$$(\sqrt{3x + 6})^2 = (2 + \sqrt{x + 6})^2$$

Use $(A + B)^2 = A^2 + 2AB + B^2$:
$$3x + 6 = 4 + 4\sqrt{x + 6} + (\sqrt{x + 6})^2$$

$(\sqrt{x + 6})^2 = x + 6$:
$$3x + 6 = 4 + 4\sqrt{x + 6} + x + 6$$

Isolate the radical:
$$2x - 4 = 4\sqrt{x + 6}$$

Factor out 2:
$$2(x - 2) = 4\sqrt{x + 6}$$

Divide both sides by 2:
$$x - 2 = 2\sqrt{x + 6}$$

Square both sides:
$$(x - 2)^2 = (2\sqrt{x + 6})^2$$

$$x^2 - 4x + 4 = 4(x + 6)$$

Distribute:
$$x^2 - 4x + 4 = 4x + 24$$

Subtract 4x and 24 from both sides:
$$x^2 - 8x - 20 = 0$$

Factor:
$$(x - 10)(x + 2) = 0$$

Use the Zero-Product Property:
$$x - 10 = 0 \quad \text{or} \quad x + 2 = 0$$
$$x = 10 \quad \text{or} \quad x = -2$$

Work Smart

$(2 + \sqrt{x + 6})^2 \neq 2^2 + (\sqrt{x + 6})^2$

Remember,

$(A + B)^2 = (A + B)(A + B)$
$\qquad\quad = A^2 + 2AB + B^2$

Work Smart

When there is more than one radical, it is best to isolate the radical with the more complicated radicand.

Check $x = -2$: $\sqrt{3 \cdot (-2) + 6} - \sqrt{-2 + 6} \stackrel{?}{=} 2$ $\qquad$ $x = 10$: $\sqrt{3 \cdot 10 + 6} - \sqrt{10 + 6} \stackrel{?}{=} 2$

$\sqrt{0} - \sqrt{4} \stackrel{?}{=} 2$ $\qquad\qquad\qquad\qquad\qquad$ $\sqrt{36} - \sqrt{16} \stackrel{?}{=} 2$

$0 - 2 \stackrel{?}{=} 2$ $\qquad\qquad\qquad\qquad\qquad\qquad\qquad$ $6 - 4 \stackrel{?}{=} 2$

$-2 = 2$ False $\qquad\qquad\qquad\qquad\qquad\qquad\qquad$ $2 = 2$ True

The apparent solution $x = -2$ does not check, so it is an extraneous solution. The solution set is $\{10\}$.

Quick ✓

11. Solve: $\sqrt{2x + 1} - \sqrt{x + 4} = 1$

▶ ❸ Solve for a Variable in a Radical Equation

In many situations, you will need to solve for a variable in a formula. For instance, Example 8 assesses how much error there is in an estimate based on a statistical study. This commonly used formula from statistics contains a radical.

EXAMPLE 8 **Solving for a Variable**

A formula from statistics for finding the margin of error in estimating a population mean is given by

$$E = z \cdot \frac{\sigma}{\sqrt{n}}$$

where E represents the margin of error, σ (sigma) represents a measure of variability, z is a measure of relative position, and n is the sample size.

(a) Solve this equation for n, the sample size.

(b) Find n when $\sigma = 12$, $z = 2$, and $E = 3$.

Solution

(a)
$$E = z \cdot \frac{\sigma}{\sqrt{n}}$$

Multiply both sides by $\sqrt{n}$: $\sqrt{n} \cdot E = z\sigma$

Divide both sides by E: $\sqrt{n} = \dfrac{z\sigma}{E}$

Square both sides: $n = \left(\dfrac{z\sigma}{E}\right)^2$

(b) $n = \left(\dfrac{z\sigma}{E}\right)^2 = \left(\dfrac{2 \cdot 12}{3}\right)^2$

$$= 64$$

The sample size needs to be $n = 64$.

Quick ✓

12. The period of a pendulum is the time it takes to complete one trip back and forth. The period T, in seconds, of a pendulum of length L, in feet, may be approximated using the formula $T = 2\pi\sqrt{\dfrac{L}{32}}$.

(a) Solve the equation for L.

(b) Determine the length of a pendulum whose period is 2π seconds.

6.7 Exercises MyMathLab®

Exercise numbers in green have complete video solutions in MyMathLab or may be accessed using the QR code to the right.

*Problems **1–12** are the Quick ✓s that follow the EXAMPLES.*

Building Skills

In Problems 13–44, solve each equation. See Objective 1.

13. $\sqrt{x} = 4$

14. $\sqrt{p} = 6$

15. $\sqrt{x - 3} = 2$

16. $\sqrt{y - 5} = 3$

17. $\sqrt{2t + 3} = 5$

18. $\sqrt{3w - 2} = 4$

19. $\sqrt{4x + 3} = -2$

20. $\sqrt{6p - 5} = -5$

21. $\sqrt[3]{4t} = 2$

22. $\sqrt[3]{9w} = 3$

23. $\sqrt[3]{5q + 4} = 4$

24. $\sqrt[3]{7m + 20} = 5$

25. $\sqrt{y} + 3 = 8$

26. $\sqrt{q} - 5 = 2$

27. $\sqrt{x + 5} - 3 = 1$

28. $\sqrt{x - 4} + 4 = 7$

29. $\sqrt{2x + 9} + 5 = 6$

30. $\sqrt{4x + 21} + 2 = 5$

31. $3\sqrt{x} + 5 = 8$

32. $4\sqrt{t} - 2 = 10$

33. $\sqrt{4 - x} - 3 = 0$

34. $\sqrt{6 - w} - 3 = 1$

35. $\sqrt{p} = 2p$

36. $\sqrt{q} = 3q$

37. $\sqrt{x + 6} = x$

38. $\sqrt{2p + 8} = p$

39. $\sqrt{w} = 6 - w$

40. $\sqrt{m} = 12 - m$

41. $\sqrt{17 - 2x} + 1 = x$

42. $\sqrt{1 - 4x} - 5 = x$

43. $\sqrt{w^2 - 11} + 5 = w + 4$

44. $\sqrt{z^2 - z - 7} + 3 = z + 2$

In Problems 45–58, solve each equation. See Objective 2.

45. $\sqrt{x + 9} = \sqrt{2x + 5}$

46. $\sqrt{3x + 1} = \sqrt{2x + 7}$

47. $\sqrt[3]{4x - 3} = \sqrt[3]{2x - 9}$

48. $\sqrt[3]{3y - 2} = \sqrt[3]{5y + 8}$

49. $\sqrt{2w^2 - 3w - 4} = \sqrt{w^2 + 6w + 6}$

50. $\sqrt{2x^2 + 7x - 10} = \sqrt{x^2 + 4x + 8}$

51. $\sqrt{3w + 4} = 2 + \sqrt{w}$

52. $\sqrt{3y - 2} = 2 + \sqrt{y}$

53. $\sqrt{x + 1} - \sqrt{x - 2} = 1$

54. $\sqrt{2x - 1} - \sqrt{x - 1} = 1$

55. $\sqrt{2x + 6} - \sqrt{x - 1} = 2$

56. $\sqrt{2x + 6} - \sqrt{x - 6} = 3$

57. $\sqrt{2x + 5} - \sqrt{x - 1} = 2$

58. $\sqrt{4x + 1} - \sqrt{2x + 1} = 2$

In Problems 59–72, solve each equation.

59. $(2x + 3)^{\frac{1}{2}} = 3$

60. $(4x + 1)^{\frac{1}{2}} = 5$

61. $(6x - 1)^{\frac{1}{4}} = (2x + 15)^{\frac{1}{4}}$

62. $(6p + 3)^{\frac{1}{5}} = (4p - 9)^{\frac{1}{5}}$

63. $(x + 3)^{\frac{1}{2}} - (x - 5)^{\frac{1}{2}} = 2$

64. $(3x + 1)^{\frac{1}{2}} - (x - 1)^{\frac{1}{2}} = 2$

65. $(x - 1)^{2/3} = 4$

66. $(2x + 3)^{3/5} = 8$

67. $(x - 2)^{3/2} = 64$

68. $(3x - 2)^{3/4} = 8$

69. $(2x^4 - 81)^{1/4} + 3x = 4x$

70. $(17x^4 - 256)^{1/4} + 5x = 7x$

71. $\dfrac{1}{(x - 4)^{-2/3}} - 3 = 6$

72. $\dfrac{1}{(x + 3)^{-2/3}} + 5 = 9$

For Problems 73–78, see Objective 3.

73. Finance Solve $A = P\sqrt{1 + r}$ for r.

74. Centripetal Acceleration Solve $v = \sqrt{ar}$ for a.

75. Volume of a Sphere Solve $r = \sqrt[3]{\dfrac{3V}{4\pi}}$ for V.

76. Surface Area of a Sphere Solve $r = \sqrt{\dfrac{S}{4\pi}}$ for S.

77. Coulomb's Law Solve $r = \sqrt{\dfrac{4F\pi\epsilon_0}{q_1 q_2}}$ for F.

78. Potential Energy Solve $V = \sqrt{\dfrac{2U}{C}}$ for U.

Mixed Practice

In Problems 79–94, solve each equation.

79. $\sqrt{5p - 3} + 7 = 3$

80. $\sqrt{3b - 2} + 8 = 5$

81. $\sqrt{x + 12} = x$

82. $\sqrt{x + 20} = x$

83. $\sqrt{2p + 12} = 4$

84. $\sqrt{3a - 5} = 2$

85. $\sqrt[4]{x + 7} = 2$

86. $\sqrt[5]{x + 23} = 2$

87. $(3x + 1)^{\frac{1}{3}} + 2 = 0$

88. $(5x - 2)^{\frac{1}{3}} + 3 = 0$

89. $\sqrt{10 - x} - x = 10$

90. $\sqrt{9 - x} - x = 11$

91. $\sqrt{2x + 5} = \sqrt{3x - 4}$

92. $\sqrt{4c - 5} = \sqrt{3c + 1}$

93. $\sqrt{x - 1} + \sqrt{x + 4} = 5$

94. $\sqrt{x - 3} + \sqrt{x + 4} = 7$

95. Solve $\sqrt{x^2} = x + 4$ using the techniques of this section. Now solve $\sqrt{x^2} = x + 4$ using the fact that $\sqrt{x^2} = |x|$. Which approach do you like better?

96. Solve $\sqrt{p^2} = 3p + 4$ using the techniques of this section. Now solve $\sqrt{p^2} = 3p + 4$ using the fact that $\sqrt{p^2} = |p|$. Which approach do you like better?

97. Suppose that $f(x) = \sqrt{x - 2}$.

(a) Solve $f(x) = 0$. What point is on the graph of f?

(b) Solve $f(x) = 1$. What point is on the graph of f?

(c) Solve $f(x) = 2$. What point is on the graph of f?

(d) Use the information obtained in parts (a)–(c) to graph $f(x) = \sqrt{x - 2}$.

(e) Use the graph and the concept of the range of a function to explain why the equation $f(x) = -1$ has no solution.

98. Suppose that $g(x) = \sqrt{x + 3}$.

(a) Solve $g(x) = 0$. What point is on the graph of g?

(b) Solve $g(x) = 1$. What point is on the graph of g?

(c) Solve $g(x) = 2$. What point is on the graph of g?

(d) Use the information obtained in parts (a)–(c) to graph $g(x) = \sqrt{x + 3}$.

(e) Use the graph and the concept of the range of a function to explain why the equation $g(x) = -1$ has no solution.

Applying the Concepts

△ **99. Finding a y-Coordinate** The solutions to the equation

$$\sqrt{4^2 + (y-2)^2} = 5$$

represent the y-coordinates such that the distance from the point $(3,2)$ to $(-1,y)$ in the Cartesian plane is 5 units.

(a) Solve the equation for y.

(b) Plot $(3,2)$ and the points found in part (a) in the Cartesian plane. Draw and label the lengths of the sides of the figure formed.

△ **100. Finding an x-Coordinate** The solutions to the equation

$$\sqrt{x^2 + 4^2} = 5$$

represent the x-coordinates such that the distance from the point $(0,3)$ to $(x,-1)$ in the Cartesian plane is 5 units.

(a) Solve the equation for x.

(b) Plot $(0,3)$ and the points found in part (a) in the Cartesian plane. Draw and label the lengths of the sides of the figure formed.

101. Revenue Growth Suppose that the annual revenue R (in millions of dollars) of a company after t years of operating is modeled by the function

$$R(t) = \sqrt[3]{\frac{t}{2}}$$

(a) After how many years can the company expect to have annual revenue of $1 million?

(b) After how many years can the company expect to have annual revenue of $2 million?

△ **102. Sphere** The radius r of a sphere whose volume is V is given by

$$r = \sqrt[3]{\frac{3V}{4\pi}}$$

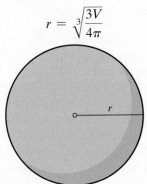

(a) Find the volume of a sphere whose radius is 3 meters.

(b) Find the volume of a sphere whose radius is 2 meters.

103. Birth Rates A plural birth is a live birth to twins, triplets, and so forth. The function $R(t) = 26 \cdot \sqrt[10]{t}$ models the plural birth rate R (live births per 1000 live births), where t is the number of years since 1995.

(a) Use the model to predict the year in which the plural birth rate will be 39. Round to the nearest year.

(b) Use the model to predict the year in which the plural birth rate will be 36. Round to the nearest year.

104. Money The annual rate of interest r (expressed as a decimal) required to have A dollars after t years from an initial deposit of P dollars is given by

$$r = \sqrt[t]{\frac{A}{P}} - 1$$

(a) Suppose that you deposit $1000 in an account that pays 5% annual interest so that $r = 0.05$. How much will you have after $t = 2$ years?

(b) Suppose that you deposit $1000 in an account that pays 5% annual interest so that $r = 0.05$. How much will you have after $t = 3$ years?

Extending the Concepts

105. $\sqrt{\sqrt{x} + 25} = 5$ **106.** $\sqrt{\sqrt{x} + 3} = 2$

107. Solve: $\sqrt{3\sqrt{x} + 1} = \sqrt{2x + 3}$

108. Solve: $\sqrt[3]{2\sqrt{x} - 2} = \sqrt[3]{x - 1}$

Explaining the Concepts

109. Why is it always necessary to check solutions when solving radical equations?

110. How can you tell by inspection that the equation $\sqrt{x - 2} + 5 = 0$ will have no real solution?

111. Using the concept of domain, explain why radical equations with an even index may have extraneous solutions, but radical equations with an odd index do not have extraneous solutions.

112. Which step in the process of solving a radical equation leads to the possibility of extraneous solutions?

Synthesis Review

In Problems 113–116, consider the set

$$\left\{ 0, -4, 12, \frac{2}{3}, 1.\overline{56}, \sqrt{2^3}, \pi, \sqrt{-5}, \sqrt[3]{-4} \right\}$$

Indicate which of the numbers in the set are ...

113. Integers

114. Rational numbers

115. Irrational numbers

116. Real numbers

117. State the difference between a rational number and an irrational number. Why is $\sqrt{-1}$ not real?

Technology Exercises

Technology can be used to verify solutions obtained algebraically. To solve $\sqrt{2x-3}=5$ presented in Example 1, graph $Y_1 = \sqrt{2x-3}$ and $Y_2 = 5$ and determine the x-coordinate of the point of intersection. The figure below shows the solution to this equation using Desmos.

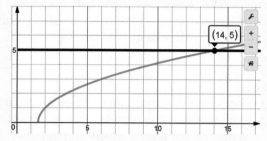

The x-coordinate of the point of intersection is 14, so the solution set is $\{14\}$.

118. Verify your solution to Problem 17 by graphing $Y_1 = \sqrt{2x+3}$ and $Y_2 = 5$ and then finding the x-coordinate of their point of intersection.

119. Verify your solution to Problem 18 by graphing $Y_1 = \sqrt{3x-2}$ and $Y_2 = 4$ and then finding the x-coordinate of their point of intersection.

120. When you solved Problem 19 algebraically, you should have determined that the equation has no real solution. Verify this result by graphing $Y_1 = \sqrt{4x+3}$ and $Y_2 = -2$. Explain what the algebraic solution means graphically.

121. When you solved Problem 20 algebraically, you should have determined that the equation has no real solution. Verify this result by graphing $Y_1 = \sqrt{6x-5}$ and $Y_2 = -5$. Explain what the algebraic solution means graphically.

6.8 The Complex Number System

Objectives

❶ Evaluate the Square Root of Negative Real Numbers

❷ Add or Subtract Complex Numbers

❸ Multiply Complex Numbers

❹ Divide Complex Numbers

❺ Evaluate the Powers of i

Are You Prepared for This Section?

Before getting started, complete the following problems. If you get a problem wrong, go back to the section cited and review the material.

P1. List the numbers in the set $\left\{8, -\dfrac{1}{3}, -23, 0, \sqrt{2}, 1.\overline{26}, -\dfrac{12}{3}, \sqrt{-5}\right\}$ that are

    (a) Natural numbers   **(b)** Whole numbers

    (c) Integers   **(d)** Rational numbers

    (e) Irrational numbers   **(f)** Real numbers   [Section R.2, pp. 11–14]

P2. Multiply: $3x(4x-3)$ [Section 4.2, p. 321]

P3. Multiply: $(z+4)(3z-2)$ [Section 4.2, pp. 321–322]

P4. Multiply: $(2y+5)(2y-5)$ [Section 4.2, pp. 323–324]

▶ If you look back at Section R.2, where the various number systems were introduced, you should notice that each time a situation was encountered where a number system could not handle a problem, the number system was expanded. For example, if only the whole numbers were considered, then we could not describe a negative balance in a checking account, so integers were introduced. If the world had to be described only by integers, then parts of a whole, as in ½ a pizza or ¾ of a dollar, could not be talked about, so rational numbers were introduced. If only rational numbers were considered, then it wouldn't be possible to find a number whose square is 2. So the irrational numbers were introduced, such that $(\sqrt{2})^2 = 2$. By combining the rational numbers with the irrational numbers, the real number system was created.

Now, suppose a number whose square is -1 is wanted. Recall that the square of any real number is never negative. This is the *Nonnegativity Property*.

Nonnegativity Property of Real Numbers

For any real number a, $a^2 \geq 0$.

Because of the Nonnegativity Property, there is no real number solution to the equation

$$x^2 = -1$$

To remedy this situation, a new number is introduced.

Definition

The **imaginary unit,** denoted by *i,* is the number whose square is −1. That is,

$$i^2 = -1$$

Introducing the number *i* establishes a new number system called the **complex number system**.

Definition

Complex numbers are numbers of the form *a* + *bi,* where *a* and *b* are real numbers. The real number *a* is called the **real part** of the number *a* + *bi*; the real number *b* is called the **imaginary part** of *a* + *bi*.

For example, the complex number 6 + 2*i* has the real part 6 and the imaginary part 2. The complex number 4 − 3*i* = 4 + (−3)*i* has the real part 4 and the imaginary part −3.

When a complex number is written in the form *a* + *bi*, where *a* and *b* are real numbers, it is in **standard form**. The complex number *a* + 0*i* is typically written as *a*. This reminds us that the real number system is a subset of the complex number system. The complex number 0 + *bi* is usually written as *bi*. Any number of the form *bi* is called a **pure imaginary number**. Figure 7 shows the relation among the number systems.

Figure 7
The Complex Number System

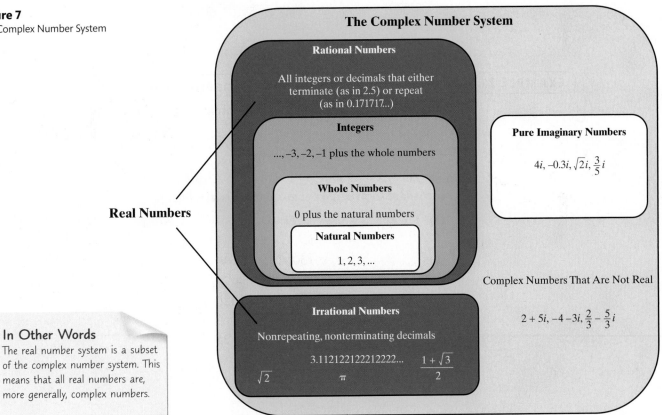

The Complex Number System

Rational Numbers

All integers or decimals that either terminate (as in 2.5) or repeat (as in 0.171717...)

Integers

..., −3, −2, −1 plus the whole numbers

Whole Numbers

0 plus the natural numbers

Natural Numbers

1, 2, 3, ...

Real Numbers

Irrational Numbers

Nonrepeating, nonterminating decimals

3.1121221222212222... $\dfrac{1 + \sqrt{3}}{2}$

$\sqrt{2}$ π

Pure Imaginary Numbers

$4i, -0.3i, \sqrt{2}i, \dfrac{3}{5}i$

Complex Numbers That Are Not Real

$2 + 5i, -4 - 3i, \dfrac{2}{3} - \dfrac{5}{3}i$

In Other Words
The real number system is a subset of the complex number system. This means that all real numbers are, more generally, complex numbers.

▶ **①** **Evaluate the Square Root of Negative Real Numbers**

First, square roots of negative numbers are defined.

Definition

If *N* is a positive real number, define the **principal square root of** −*N*, denoted by $\sqrt{-N}$, as

$$\sqrt{-N} = \sqrt{N}\,i$$

where *i* is the number whose square is −1.

EXAMPLE 1 **Evaluating the Square Root of a Negative Number**

Write each of the following as a pure imaginary number.

(a) $\sqrt{-16}$ (b) $\sqrt{-3}$ (c) $\sqrt{-18}$

Solution

(a) $\sqrt{-16} = \sqrt{16}i$ (b) $\sqrt{-3} = \sqrt{3}i$ (c) $\sqrt{-18} = \sqrt{18}i$

$\qquad\qquad = 4i$ $= 3\sqrt{2}i$ ●

Work Smart

When writing complex numbers whose imaginary part is a radical, as in $\sqrt{3}i$, be sure that the "i" is not written under the radical.

Quick ✓

1. The _____, denoted by i, is the number whose square is -1.
2. Any number of the form bi is called a ___ _____ _____.
3. If N is a positive real number, define the principal square root of $-N$, denoted by $\sqrt{-N}$, as $\sqrt{-N} = $ ___.
4. *True or False* All real numbers are complex numbers.

In Problems 5–7, write each radical as a pure imaginary number.

5. $\sqrt{-36}$ 6. $\sqrt{-5}$ 7. $\sqrt{-12}$

EXAMPLE 2 **Writing Complex Numbers in Standard Form**

Write each of the following in standard form.

(a) $2 - \sqrt{-25}$ (b) $3 + \sqrt{-50}$ (c) $\dfrac{4 - \sqrt{-12}}{2}$

Solution

The standard form of a complex number is $a + bi$.

(a) $2 - \sqrt{-25} = 2 - \sqrt{25}i$

$\qquad\qquad\quad = 2 - 5i$

(b) $\qquad\qquad 3 + \sqrt{-50} = 3 + \sqrt{50}i$

$\sqrt{50} = \sqrt{25} \cdot \sqrt{2} = 5\sqrt{2}:\qquad = 3 + 5\sqrt{2}i$

(c) $\qquad\qquad \dfrac{4 - \sqrt{-12}}{2} = \dfrac{4 - \sqrt{12}i}{2}$

$\sqrt{12} = \sqrt{4} \cdot \sqrt{3} = 2\sqrt{3}:\qquad = \dfrac{4 - 2\sqrt{3}i}{2}$

Factor out 2: $\qquad\qquad = \dfrac{2(2 - \sqrt{3}i)}{2}$

Divide out the 2's: $\qquad = 2 - \sqrt{3}i$ ●

Quick ✓

In Problems 8–10, write each expression in the standard form of a complex number, $a + bi$.

8. $4 + \sqrt{-100}$ 9. $-2 - \sqrt{-8}$

10. $\dfrac{6 - \sqrt{-72}}{3}$

⏵ ❷ Add or Subtract Complex Numbers

You have learned to add, subtract, multiply, and divide using various number systems. Now, we will explore how to perform these operations on complex numbers.

The sum of complex numbers is found by adding the real parts and then adding the imaginary parts.

> **Sum of Complex Numbers**
> $$(a + bi) + (c + di) = (a + c) + (b + d)i$$

In Other Words
To add two complex numbers, add the real parts, and then add the imaginary parts. To subtract two complex numbers, subtract the real parts, and then subtract the imaginary parts.

To subtract two complex numbers, use this rule:

> **Difference of Complex Numbers**
> $$(a + bi) - (c + di) = (a - c) + (b - d)i$$

EXAMPLE 3 **Adding Complex Numbers**

Add:

(a) $(4 - 3i) + (-2 + 5i)$ **(b)** $(4 + \sqrt{-25}) + (6 - \sqrt{-16})$

Solution

(a) $(4 - 3i) + (-2 + 5i) = [4 + (-2)] + (-3 + 5)i$
$$= 2 + 2i$$

$\sqrt{-25} = 5i$ $\sqrt{-16} = 4i$

(b) $(4 + \sqrt{-25}) + (6 - \sqrt{-16}) = (4 + 5i) + (6 - 4i)$
$$= (4 + 6) + (5 - 4)i$$
$$= 10 + 1i$$
$$= 10 + i$$

EXAMPLE 4 **Subtracting Complex Numbers**

Work Smart

Adding or subtracting complex numbers is just like combining like terms. For example,
$(4 - 3x) + (-2 + 5x)$
$= 4 + (-2) - 3x + 5x$
$= 2 + 2x$
so
$(4 - 3i) + (-2 + 5i)$
$= 4 + (-2) - 3i + 5i$
$= 2 + 2i$

Subtract:

(a) $(-3 + 7i) - (5 - 4i)$ **(b)** $(3 + \sqrt{-12}) - (-2 - \sqrt{-27})$

Solution

(a) $(-3 + 7i) - (5 - 4i) = (-3 - 5) + (7 - (-4))i$
$$= -8 + 11i$$

$\sqrt{-12} = 2\sqrt{3}i$ $\sqrt{-27} = 3\sqrt{3}i$

(b) $(3 + \sqrt{-12}) - (-2 - \sqrt{-27}) = (3 + 2\sqrt{3}i) - (-2 - 3\sqrt{3}i)$
$$= (3 + 2) + [2\sqrt{3} - (-3\sqrt{3})]i$$
$$= 5 + 5\sqrt{3}i$$

Quick ✔

In Problems 11–13, add or subtract, as indicated.

11. $(4 + 6i) + (-3 + 5i)$ **12.** $(4 - 2i) - (-2 + 7i)$

13. $(4 - \sqrt{-4}) + (-7 + \sqrt{-9})$

▶ ❸ **Multiply Complex Numbers**

Multiply complex numbers using the same ideas used to multiply polynomials.

EXAMPLE 5 **Multiplying Complex Numbers**

Multiply:

(a) $4i(3 - 6i)$ **(b)** $(-2 + 4i)(3 - i)$

Solution

(a) Distribute the $4i$ to each term in the parentheses.

$$4i(3 - 6i) = 4i \cdot 3 - 4i \cdot 6i$$
$$= 12i - 24i^2$$
$$i^2 = -1: \quad = 12i - 24(-1)$$
$$= 24 + 12i$$

(b)

$$\begin{aligned}
(-2 + 4i)(3 - i) &= -2 \cdot 3 - 2(-i) + 4i \cdot 3 + 4i(-i) \\
&= -6 + 2i + 12i - 4i^2 \\
\text{Combine like terms; } i^2 = -1: \quad &= -6 + 14i - 4(-1) \\
&= -6 + 14i + 4 \\
&= -2 + 14i
\end{aligned}$$

Quick ✓

In Problems 14 and 15, multiply.

14. $3i(5 - 4i)$ **15.** $(-2 + 5i)(4 - 2i)$

Review the Product Property of Radicals. Notice that the property applies only when $\sqrt[n]{a}$ and $\sqrt[n]{b}$ are real numbers. This means that

$$\sqrt{a} \cdot \sqrt{b} \neq \sqrt{ab} \quad \text{if } a < 0 \text{ or } b < 0$$

How, then, do we perform this multiplication? First write the radical as a complex number using the fact that $\sqrt{-N} = \sqrt{N}i$, and then multiply.

EXAMPLE 6 **Multiplying Square Roots of Negative Numbers**

Multiply:

(a) $\sqrt{-25} \cdot \sqrt{-4}$ **(b)** $(2 + \sqrt{-16})(1 - \sqrt{-9})$

Solution

(a) The Product Property of Radicals cannot be used on this product because $\sqrt{-25}$ and $\sqrt{-4}$ are not real numbers. Therefore, express the radicals as pure imaginary numbers and then multiply.

Work Smart

$\sqrt{-25} \cdot \sqrt{-4} \neq \sqrt{(-25)(-4)}$

because the Product Property of Radicals applies only when the radical is a real number.

$$\sqrt{-25} \cdot \sqrt{-4} = 5i \cdot 2i$$
$$= 10i^2$$
$$i^2 = -1: \quad = 10(-1)$$
$$= -10$$

(b) First, rewrite each expression as a complex number in standard form.

$$(2 + \sqrt{-16})(1 - \sqrt{-9}) = (2 + 4i)(1 - 3i)$$

$$\text{FOIL:} \quad = 2 \cdot 1 + 2(-3i) + 4i \cdot 1 + 4i(-3i)$$

$$= 2 - 6i + 4i - 12i^2$$

$$\text{Combine like terms; } i^2 = -1: \quad = 2 - 2i - 12(-1)$$

$$= 2 - 2i + 12$$

$$= 14 - 2i$$

Quick ✓

In Problems 16 and 17, multiply.

16. $\sqrt{-9} \cdot \sqrt{-36}$ **17.** $(2 + \sqrt{-36})(4 - \sqrt{-25})$

▶ **Complex Conjugates**

A special product that involves the *conjugate* of a complex number is now introduced.

> **Complex Conjugate**
>
> If $a + bi$ is a complex number, then its **conjugate** is defined as $a - bi$.

In Other Words

The complex conjugate of $a + bi$ is $a - bi$. The complex conjugate of $a - bi$ is $a + bi$.

For example,

Complex Number	Conjugate
$4 + 7i$	$4 - 7i$
$-10 - 3i$	$-10 + 3i$

EXAMPLE 7   **Multiplying a Complex Number by Its Conjugate**

Find the product of $4 + 3i$ and its conjugate, $4 - 3i$.

Solution

$$(4 + 3i)(4 - 3i) = 4 \cdot 4 + 4(-3i) + 3i \cdot 4 + 3i(-3i)$$

$$= 16 - 12i + 12i - 9i^2$$

$$= 16 - 9(-1)$$

$$= 16 + 9$$

$$= 25$$

Wow! The product of $4 + 3i$ and its conjugate $4 - 3i$ is 25—a real number! In fact, the result of Example 7 is true in general:

> **Product of a Complex Number and its Conjugate**
>
> The product of a complex number and its conjugate is a nonnegative real number. That is,
>
> $$(a + bi)(a - bi) = a^2 + b^2$$

Notice how multiplying a complex number by its conjugate is different from multiplying $(a + b)(a - b)$:

$$(a + b)(a - b) = a^2 - b^2, \quad \text{whereas} \quad (a + bi)(a - bi) = a^2 + b^2.$$

Do you see why?

Quick ✓

18. The complex conjugate of $-3 + 5i$ is _____ .

In Problems 19 and 20, multiply.

19. $(3 - 8i)(3 + 8i)$ **20.** $(-2 + 5i)(-2 - 5i)$

▶ ❹ Divide Complex Numbers

Knowing that the product of a complex number and its conjugate is a nonnegative real number makes it possible to divide complex numbers. In Example 8, notice how the approach is similar to rationalizing a denominator. (See Examples 3 and 4 in Section 6.5.)

EXAMPLE 8   **How to Divide Complex Numbers**

Divide: $\dfrac{-3 + i}{5 + 3i}$

Step-by-Step Solution

Step 1: Write the numerator and denominator in standard form, $a + bi$.

The numerator and denominator are already in standard form.

Step 2: Multiply the numerator and denominator by the complex conjugate of the denominator.

Multiply by $1 = \dfrac{5 - 3i}{5 - 3i}$

$$\frac{-3 + i}{5 + 3i} = \frac{-3 + i}{5 + 3i} \cdot \frac{5 - 3i}{5 - 3i}$$

$$= \frac{(-3 + i)(5 - 3i)}{(5 + 3i)(5 - 3i)}$$

Step 3: Simplify by writing the quotient in standard form, $a + bi$.

Multiply in the numerator and denominator; $(a + bi)(a - bi) = a^2 + b^2$:

$$= \frac{-3 \cdot 5 - 3(-3i) + i \cdot 5 + i(-3i)}{5^2 + 3^2}$$

$$= \frac{-15 + 9i + 5i - 3i^2}{25 + 9}$$

Combine like terms; $i^2 = -1$:

$$= \frac{-15 + 14i + 3}{34}$$

$$= \frac{-12 + 14i}{34}$$

Divide 34 into each term in the numerator to write in standard form:

$$= \frac{-12}{34} + \frac{14}{34}i$$

Write each fraction in lowest terms:

$$= -\frac{6}{17} + \frac{7}{17}i$$

●

Below, the steps used in Example 8 are summarized.

Dividing Complex Numbers

Step 1: Write the numerator and denominator in standard form, $a + bi$.

Step 2: Multiply the numerator and denominator by the complex conjugate of the denominator.

Step 3: Simplify by writing the quotient in standard form, $a + bi$.

EXAMPLE 9 | **Dividing Complex Numbers**

Divide: $\dfrac{3 + 4i}{2i}$

Solution

The denominator, $2i$, can be written as $0 + 2i$, and the conjugate of $0 + 2i$ is $0 - 2i$. Because $0 - 2i = -2i$, multiply the numerator and denominator by $-2i$.

$$\frac{3 + 4i}{2i} = \frac{3 + 4i}{2i} \cdot \frac{-2i}{-2i}$$

Multiply numerator; multiply denominator: $= \dfrac{(3 + 4i)(-2i)}{(2i)(-2i)}$

Distribute $-2i$: $= \dfrac{-6i - 8i^2}{-4i^2}$

$i^2 = -1$: $= \dfrac{-6i + 8}{4}$

Divide 4 into each term in the numerator to write in standard form: $= \dfrac{8}{4} - \dfrac{6}{4}i$

Write each fraction in lowest terms: $= 2 - \dfrac{3}{2}i$ ●

In Example 5(b), multiplication was used to find $(-2 + 4i)(3 - i) = -2 + 14i$. Now divide $-2 + 14i$ by $3 - i$. What result do you expect? Verify that $\dfrac{-2 + 14i}{3 - i} = -2 + 4i$. Remember that $\dfrac{A}{B} = C$ can be checked by verifying that $A = B \cdot C$.

Quick ✓

In Problems 21 and 22, divide.

21. $\dfrac{-4 + i}{3i}$ **22.** $\dfrac{4 + 3i}{1 - 3i}$

▶ **⑤ Evaluate the Powers of *i***

The **powers of *i*** follow a pattern.

$i^1 = i$ $i^2 = -1$ $i^3 = i^2 \cdot i^1 = -1 \cdot i = -i$ $i^4 = i^2 \cdot i^2 = (-1)(-1) = 1$

$i^5 = i^4 \cdot i = 1 \cdot i = i$ $i^6 = i^4 \cdot i^2 = 1(-1) = -1$ $i^7 = i^4 \cdot i^3 = 1(-i) = -i$ $i^8 = i^4 \cdot i^4 = 1 \cdot 1 = 1$

$i^9 = i^8 \cdot i = 1 \cdot i = i$ $i^{10} = i^8 \cdot i^2 = 1(-1) = -1$ $i^{11} = i^8 \cdot i^3 = 1(-i) = -i$ $i^{12} = (i^4)^3 = (1)^3 = 1$

Do you see the pattern? In the first column, all the expressions simplify to i; in the second column, all the expressions simplify to -1; in the third column, all the expressions simplify to $-i$; in the fourth column, all the expressions simplify to 1. That is, the powers of i repeat with every fourth power. Figure 8 shows the pattern.

Figure 8

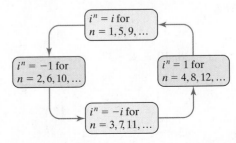

Thus any power of i can be expressed as $1, -1, i,$ or $-i$. Use the steps on the next page to simplify any power of i.

> **Simplifying the Powers of i**
>
> **Step 1:** Divide the exponent of i by 4. Rewrite i^n as $(i^4)^q \cdot i^r$, where q is the quotient and r is the remainder of the division.
>
> **Step 2:** Simplify the product in Step 1 to i^r since $i^4 = 1$.

EXAMPLE 10 **Simplifying Powers of i**

Simplify:

    (a) i^{34} **(b)** i^{101}

Solution

Work Smart

An alternative approach to Example 10:

(a) $i^{34} = (i^2)^{17} = (-1)^{17} = -1$

(b) $i^{101} = i^{100} \cdot i^1 = (i^2)^{50} \cdot i$
$$= (-1)^{50} \cdot i$$
$$= 1 \cdot i$$
$$= i$$

(a) Divide 34 by 4 and obtain a quotient of $q = 8$ and a remainder of $r = 2$, so
$$i^{34} = (i^4)^8 \cdot i^2$$
$$= (1)^8(-1)$$
$$= -1$$

(b) Divide 101 by 4 and obtain a quotient of $q = 25$ and a remainder of $r = 1$, so
$$i^{101} = (i^4)^{25} \cdot i^1$$
$$= (1)^{25} \cdot i$$
$$= i$$

Quick ✓

In Problems 23 and 24, simplify the power of i.

    23. i^{43} **24.** i^{98}

6.8 Exercises   MyMathLab®

Problems 1–24 are the Quick ✓s that follow the EXAMPLES.

Building Skills

In Problems 25–34, write each expression as a pure imaginary number. See Objective 1.

25. $\sqrt{-4}$ **26.** $\sqrt{-25}$ **27.** $-\sqrt{-81}$

28. $-\sqrt{-100}$ **29.** $\sqrt{-45}$ **30.** $\sqrt{-48}$

31. $\sqrt{-300}$ **32.** $\sqrt{-162}$ **33.** $\sqrt{-7}$

34. $\sqrt{-13}$

In Problems 35–42, write each expression as a complex number in standard form. See Objective 1.

35. $5 + \sqrt{-49}$ **36.** $4 - \sqrt{-36}$

37. $-2 - \sqrt{-28}$ **38.** $10 + \sqrt{-32}$

39. $\dfrac{4 + \sqrt{-4}}{2}$ **40.** $\dfrac{10 - \sqrt{-25}}{5}$

41. $\dfrac{4 + \sqrt{-8}}{12}$ **42.** $\dfrac{15 - \sqrt{-50}}{5}$

In Problems 43–50, add or subtract, as indicated. See Objective 2.

43. $(4 + 5i) + (2 - 7i)$ **44.** $(-6 + 2i) + (3 + 12i)$

45. $(4 + i) - (8 - 5i)$ **46.** $(-7 + 3i) - (-3 + 2i)$

47. $(4 - \sqrt{-4}) - (2 + \sqrt{-9})$

48. $(-4 + \sqrt{-25}) + (1 - \sqrt{-16})$

49. $(-2 + \sqrt{-18}) + (5 - \sqrt{-50})$

50. $(-10 + \sqrt{-20}) - (-6 + \sqrt{-45})$

In Problems 51–74, multiply. See Objective 3.

51. $6i(2 - 4i)$ **52.** $3i(-2 - 6i)$

53. $-\dfrac{1}{2}i(4 - 10i)$ **54.** $\dfrac{1}{3}i(12 + 15i)$

55. $(2 + i)(4 + 3i)$ **56.** $(3 - i)(1 + 2i)$

57. $(-3 - 5i)(2 + 4i)$ **58.** $(5 - 2i)(-1 + 2i)$

59. $(2 - 3i)(4 + 6i)$ **60.** $(6 + 8i)(-3 + 4i)$

61. $(3 - \sqrt{2}i)(-2 + \sqrt{2}i)$ **62.** $(1 + \sqrt{3}i)(-4 - \sqrt{3}i)$

63. $\left(\frac{1}{2} - \frac{1}{4}i\right)\left(\frac{2}{3} + \frac{3}{4}i\right)$ **64.** $\left(-\frac{2}{3} + \frac{4}{3}i\right)\left(\frac{1}{2} - \frac{3}{2}i\right)$

65. $(3 + 2i)^2$ **66.** $(2 + 5i)^2$

67. $(-4 - 5i)^2$ **68.** $(2 - 7i)^2$

69. $\sqrt{-9} \cdot \sqrt{-4}$ **70.** $\sqrt{-36} \cdot \sqrt{-4}$

71. $\sqrt{-8} \cdot \sqrt{-10}$ **72.** $\sqrt{-12} \cdot \sqrt{-15}$

73. $(2 + \sqrt{-81})(-3 - \sqrt{-100})$

74. $(1 - \sqrt{-64})(-2 + \sqrt{-49})$

In Problems 75–80, (a) find the conjugate of the complex number, and (b) multiply the complex number by its conjugate. See Objective 3.

75. $3 + 5i$ **76.** $5 + 2i$

77. $2 - 7i$ **78.** $9 - i$

79. $-7 + 2i$ **80.** $-1 - 4i$

In Problems 81–94, divide. See Objective 4.

81. $\dfrac{1 + i}{3i}$ **82.** $\dfrac{2 - i}{2i}$

83. $\dfrac{-5 + 2i}{5i}$ **84.** $\dfrac{-4 + 5i}{6i}$

85. $\dfrac{3}{2 + i}$ **86.** $\dfrac{2}{4 + i}$

87. $\dfrac{-2}{-3 - 7i}$ **88.** $\dfrac{-4}{-5 - 3i}$

89. $\dfrac{2 + 3i}{3 - 2i}$ **90.** $\dfrac{2 + 5i}{5 - 2i}$

91. $\dfrac{4 + 2i}{1 - i}$ **92.** $\dfrac{-6 + 2i}{1 + i}$

93. $\dfrac{4 - 2i}{1 + 3i}$ **94.** $\dfrac{5 - 3i}{2 + 4i}$

In Problems 95–102, simplify. See Objective 5.

95. i^{53} **96.** i^{72}

97. i^{43} **98.** i^{110}

99. i^{153} **100.** i^{-45}

101. i^{131} **102.** i^{-26}

Mixed Practice

In Problems 103–116, perform the indicated operation.

103. $(-4 - i)(4 + i)$ **104.** $(-5 + 2i)(5 - 2i)$

105. $(3 + 2i)^2$ **106.** $(-3 + 2i)^2$

107. $\dfrac{-3 + 2i}{3i}$ **108.** $\dfrac{5 - 3i}{4i}$

109. $\dfrac{-4 + i}{-5 - 3i}$ **110.** $\dfrac{-4 + 6i}{-5 - i}$

111. $(10 - 3i) + (2 + 3i)$ **112.** $(-4 + 5i) + (4 - 2i)$

113. $5i^{37}(-4 + 3i)$ **114.** $2i^{57}(3 - 4i)$

115. $\sqrt{-10} \cdot \sqrt{-15}$ **116.** $\sqrt{-8} \cdot \sqrt{-12}$

In Problems 117–122, find the reciprocal of the complex number. Write each number in standard form.

117. $5i$ **118.** $7i$

119. $2 - i$ **120.** $3 - 5i$

121. $-4 + 5i$ **122.** $-6 + 2i$

123. Suppose that $f(x) = x^2$; find **(a)** $f(i)$ **(b)** $f(1 + i)$.

124. Suppose that $f(x) = x^2 + x$; find **(a)** $f(i)$ **(b)** $f(1 + i)$.

125. Suppose that $f(x) = x^2 + 2x + 2$; find **(a)** $f(3i)$
(b) $f(1 - i)$.

126. Suppose that $f(x) = x^2 + x - 1$; find **(a)** $f(2i)$
(b) $f(2 + i)$.

Applying the Concepts

127. Impedance (Series Circuit) The total impedance, Z, of an ac circuit containing components in series is equivalent to the sum of the individual impedances. Impedance is measured in ohms (Ω) and is expressed as an imaginary number of the form $Z = R + i \cdot X$. Here, R represents resistance and X represents reactance.

  (a) If the impedance in one part of a series circuit is $7 + 3i$ ohms and the impedance of the remainder of the circuit is $3 - 4i$ ohms, find the total impedance of the circuit.

  (b) What is the total resistance of the circuit?

  (c) What is the total reactance of the circuit?

128. Impedance (Parallel Circuit) The total impedance, Z, of an ac circuit consisting of two parallel pathways is given by the formula $\dfrac{1}{Z} = \dfrac{1}{Z_1} + \dfrac{1}{Z_2}$, where Z_1 and Z_2 are the impedances of the pathways. If the impedances of the individual pathways are $Z_1 = 5$ ohms and $Z_2 = 1 - 2i$ ohms, find the total impedance of the circuit.

Extending the Concepts

129. For the function $f(x) = x^2 + 4x + 5$, find
(a) $f(-2 + i)$ **(b)** $f(-2 - i)$.

130. For the function $f(x) = x^2 - 2x + 2$, find
(a) $f(1 + i)$ **(b)** $f(1 - i)$.

131. For the function $f(x) = x^3 + 1$, find **(a)** $f(-1)$
(b) $f\left(\dfrac{1}{2} + \dfrac{\sqrt{3}}{2}i\right)$ **(c)** $f\left(\dfrac{1}{2} - \dfrac{\sqrt{3}}{2}i\right)$

132. For the function $f(x) = x^3 - 1$, find **(a)** $f(1)$ **(b)** $f\left(-\dfrac{1}{2} + \dfrac{\sqrt{3}}{2}i\right)$ **(c)** $f\left(-\dfrac{1}{2} - \dfrac{\sqrt{3}}{2}i\right)$.

133. Any complex number z such that $f(z) = 0$ is called a **complex zero** of f. Look at the complex zeros in Problems 129–132. Conjecture a general result regarding the complex zeros of polynomials that have real coefficients.

134. The Complex Plane We are able to plot real numbers on a real number line. We are able to plot ordered pairs (x, y) in a Cartesian plane. Can we plot complex numbers? Yes! A complex number $x + yi$ can be interpreted geometrically as the point (x, y) in the xy-plane. Each point in the plane corresponds to a complex number, and each complex number corresponds to a point in the plane (just as in ordered pairs). In the **complex plane,** the x-axis is called the **real axis** and the y-axis is called the **imaginary axis.**

(a) Plot the complex number $4 + 2i$ in the complex plane.

(b) Plot the complex number $-2 + 6i$ in the complex plane.

(c) Plot the complex number $-5i$ in the complex plane.

(d) Plot the complex number -4 in the complex plane.

Explaining the Concepts

135. Explain the relations among the natural numbers, whole numbers, integers, rational numbers, real numbers, and complex number system.

136. Explain why the product of a complex number and its conjugate is a nonnegative real number. That is, explain why $(a + bi)(a - bi) = a^2 + b^2$.

137. How is multiplying two complex numbers related to multiplying two binomials?

138. How is the method used to rationalize denominators related to the method used to write the quotient of two complex numbers in standard form?

Synthesis Review

139. Expand: $(x + 2)^3$ **140.** Expand: $(y - 4)^2$

141. Evaluate: $(3 + i)^3$ **142.** Evaluate: $(4 - 3i)^2$

143. How is raising a complex number to a positive integer power related to raising a binomial to a positive integer power?

Technology Exercises

Technology gives us the ability to add, subtract, multiply, and divide complex numbers. Using a graphing calculator, first put the calculator into complex mode, as shown in Figure 9(a). Figure 9(b) shows the results of Examples 3(a) and 3(b). Figure 9(c) shows the results of Examples 5(b) and 8.

Figure 9

(a)   (b)

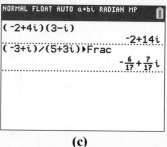

(c)

In Problems 144–151, use technology to simplify each expression. Write your answers in standard form, $a + bi$.

144. $(1.3 - 4.3i) + (-5.3 + 0.7i)$

145. $(-3.4 + 1.9i) - (6.5 - 5.3i)$

146. $(0.3 - 5.2i)(1.2 + 3.9i)$

147. $(-4.3 + 0.2i)(7.2 - 0.5i)$

148. $\dfrac{4}{3 - 8i}$ **149.** $\dfrac{1 - 7i}{-4 + i}$

150. $3i^6 + (5 - 0.3i)(3 + 2i)$

151. $-6i^{14} + 5i^6 - (2 + i)(3 + 11i)$

Chapter 6 Activity: Which One Does Not Belong?

Focus: Simplifying Radicals and Rational Exponents

Time: 15–25 minutes

Group Size: 2–4

5–10 Minutes: Individually, evaluate each row below. Decide which item does not belong and why. Be creative with your reasons!

10–15 Minutes: Discuss each row, and make a list of all members' responses. Is there any example that everyone agrees with?

	A	B	C	D
1	$16^{\frac{3}{2}}$	$-16^{\frac{3}{2}}$	$-16^{\frac{2}{3}}$	$-16^{-\frac{2}{3}}$
2	$\sqrt{(a-b)^3}, a > b$	$\sqrt{a^3 - b^3}$	$\sqrt{(a-b)^3}, a < b$	$\sqrt{a-b}, a > b$
3	The range of $f(x) = x^2$	The range of $f(x) = \sqrt[3]{x}$	The range of $f(x) = |x|$	The range of $f(x) = x^3$
4	i^{212}	i^0	i^{-108}	i

Chapter 6 Review

Section 6.1   *n*th Roots and Rational Exponents

KEY CONCEPTS

- **Simplifying $\sqrt[n]{a^n}$**

  If $n \geq 2$ is a positive integer and a is a real number, then

  $$\sqrt[n]{a^n} = |a| \quad \text{if } n \geq 2 \text{ is even}$$
  $$\sqrt[n]{a^n} = a \quad \text{if } n \geq 3 \text{ is odd}$$

- If a is a real number and $n \geq 2$ is an integer, then $a^{\frac{1}{n}} = \sqrt[n]{a}$ provided that $\sqrt[n]{a}$ exists.

- If a is a real number, $\dfrac{m}{n}$ is a rational number in lowest terms with $n \geq 2$, then $a^{\frac{m}{n}} = \sqrt[n]{a^m} = (\sqrt[n]{a})^m$ provided that $\sqrt[n]{a}$ exists.

- If $\dfrac{m}{n}$ is a rational number and if a is a nonzero real number, then

  $$a^{-\frac{m}{n}} = \frac{1}{a^{\frac{m}{n}}} \text{ or } \frac{1}{a^{-\frac{m}{n}}} = a^{\frac{m}{n}}, \text{ if } a \neq 0 \text{ and the } n\text{th root of } a \text{ is defined.}$$

KEY TERMS

Principal *n*th root of a number *a*

Index

Cube root

You Should Be Able To...	EXAMPLE	Review Exercises
❶ Evaluate *n*th roots (p. 474)	Examples 1 and 2	1–5, 19, 20
❷ Simplify expressions of the form $\sqrt[n]{a^n}$ (p. 476)	Example 3	6–8
❸ Evaluate expressions of the form $a^{\frac{1}{n}}$ (p. 476)	Examples 4 and 5	9–12, 17, 21
❹ Evaluate expressions of the form $a^{\frac{m}{n}}$ (p. 477)	Examples 6 through 9	13–16, 18, 22–24

In Problems 1–8, simplify each radical.

1. $\sqrt[3]{343}$

2. $\sqrt[3]{-125}$

3. $\sqrt[3]{\dfrac{8}{27}}$

4. $\sqrt[4]{81}$

5. $-\sqrt[5]{-243}$

6. $\sqrt[3]{10^3}$

7. $\sqrt[5]{z^5}$

8. $\sqrt[4]{(5p-3)^4}$

In Problems 9–16, evaluate each of the following expressions.

9. $81^{\frac{1}{2}}$

10. $(-256)^{\frac{1}{4}}$

11. $-4^{\frac{1}{2}}$

12. $729^{\frac{1}{3}}$

13. $16^{\frac{7}{4}}$

14. $-(-27)^{\frac{2}{3}}$

15. $-121^{\frac{3}{2}}$

16. $\dfrac{1}{36^{-\frac{1}{2}}}$

In Problems 17–20, use a calculator to write each expression rounded to two decimal places.

17. $(-65)^{\frac{1}{3}}$

18. $4^{\frac{3}{5}}$

19. $\sqrt[3]{100}$

20. $\sqrt[4]{10}$

In Problems 21–24, rewrite each of the following radicals with a rational exponent.

21. $\sqrt[3]{5a}$

22. $\sqrt[5]{p^7}$

23. $\left(\sqrt[4]{10z}\right)^3$

24. $\sqrt[6]{(2ab)^5}$

Section 6.2   Simplifying Expressions Using the Laws of Exponents

KEY CONCEPTS

KEY TERM

Simplify

- **If a and b are real numbers and if r and s are rational numbers, then, assuming the expression is defined, the following rules apply.**

Zero-Exponent Rule:	$a^0 = 1$	if $a \neq 0$
Negative-Exponent Rule:	$a^{-r} = \dfrac{1}{a^r}$	if $a \neq 0$
Product Rule:	$a^r \cdot a^s = a^{r+s}$	
Quotient Rule:	$\dfrac{a^r}{a^s} = a^{r-s} = \dfrac{1}{a^{s-r}}$	if $a \neq 0$
Power Rule:	$(a^r)^s = a^{r \cdot s}$	
Product-to-a-Power Rule:	$(a \cdot b)^r = a^r \cdot b^r$	
Quotient-to-a-Power Rule:	$\left(\dfrac{a}{b}\right)^r = \dfrac{a^r}{b^r}$	if $b \neq 0$
Quotient-to-a-Negative-Power Rule:	$\left(\dfrac{a}{b}\right)^{-r} = \left(\dfrac{b}{a}\right)^r$	if $a \neq 0, b \neq 0$

You Should Be Able To...	EXAMPLE	Review Exercises
❶ Simplify expressions involving rational exponents (p. 482)	Examples 1 through 3	25–30
❷ Simplify radical expressions (p. 484)	Example 4	31–34
❸ Factor expressions containing rational exponents (p. 485)	Examples 5 and 6	35, 36

In Problems 25–34, simplify the expression. Assume all variables are positive.

25. $4^{\frac{2}{3}} \cdot 4^{\frac{7}{3}}$

26. $\dfrac{k^{\frac{1}{2}}}{k^{\frac{3}{4}}}$

27. $\left(p^{\frac{4}{3}} \cdot q^4\right)^{\frac{3}{2}}$

28. $\left(32a^{-\frac{3}{2}} \cdot b^{\frac{1}{4}}\right)^{\frac{1}{5}}$

29. $5m^{-\frac{2}{3}}\left(2m + m^{-\frac{1}{3}}\right)$

30. $\left(\dfrac{16x^{\frac{1}{3}}}{x^{-\frac{1}{3}}}\right)^{-\frac{1}{2}} + \left(\dfrac{x^{-\frac{3}{2}}}{64x^{-\frac{1}{2}}}\right)^{\frac{1}{3}}$

31. $\sqrt[8]{x^6}$

32. $\sqrt{121x^4y^{10}}$

33. $\sqrt[3]{m^2} \cdot \sqrt{m^3}$

34. $\dfrac{\sqrt[3]{c}}{\sqrt[6]{c^4}}$

In Problems 35 and 36, factor the expression.

35. $2(3m - 1)^{\frac{1}{4}} + (m - 7)(3m - 1)^{\frac{5}{4}}$

36. $3(x^2 - 5)^{\frac{1}{3}} - 4x(x^2 - 5)^{-\frac{2}{3}}$

Section 6.3   Simplifying Radical Expressions Using Properties of Radicals

KEY CONCEPTS

- **Product Property of Radicals**
  If $\sqrt[n]{a}$ and $\sqrt[n]{b}$ are real numbers and $n \geq 2$ is an integer, then $\sqrt[n]{a} \cdot \sqrt[n]{b} = \sqrt[n]{ab}$.
- **Quotient Property of Radicals**
  If $\sqrt[n]{a}$ and $\sqrt[n]{b}$ are real numbers, $b \neq 0$, and $n \geq 2$ is an integer, then $\dfrac{\sqrt[n]{a}}{\sqrt[n]{b}} = \sqrt[n]{\dfrac{a}{b}}$.

KEY TERM

Simplify

You Should Be Able To...	EXAMPLE	Review Exercises
❶ Use the Product Property to multiply radical expressions (p. 487)	Example 1	37, 38
❷ Use the Product Property to simplify radical expressions (p. 488)	Examples 2 through 7	39–52
❸ Use the Quotient Property to simplify radical expressions (p. 493)	Examples 8 and 9	53–60
❹ Multiply radicals with unlike indices (p. 494)	Example 10	61, 62

In Problems 37 and 38, use the Product Property to multiply. Assume that all variables can be any real number.

37. $\sqrt{15} \cdot \sqrt{7}$ **38.** $\sqrt[4]{2ab^2} \cdot \sqrt[4]{6a^2b}$

In Problems 39–44, simplify each radical using the Product Property. Assume that all variables can be any real number.

39. $\sqrt{80}$ **40.** $\sqrt[3]{-500}$

41. $\sqrt[3]{162m^6n^4}$ **42.** $\sqrt[4]{50p^8q^4}$

43. $2\sqrt{16x^6y}$ **44.** $\sqrt{(2x+1)^3}$

In Problems 45–48, simplify each radical using the Product Property. Assume that all variables are greater than or equal to zero.

45. $\sqrt{w^3z^2}$ **46.** $\sqrt{45x^4yz^3}$

47. $\sqrt[3]{16a^{12}b^5}$ **48.** $\sqrt{4x^2 + 8x + 4}$

In Problems 49–52, multiply and simplify. Assume that all variables are greater than or equal to zero.

49. $\sqrt{15} \cdot \sqrt{18}$ **50.** $\sqrt[3]{20} \cdot \sqrt[3]{30}$

51. $\sqrt[3]{-3x^4y^7} \cdot \sqrt[3]{24x^3y^2}$ **52.** $3\sqrt{4xy^2} \cdot 5\sqrt{3x^2y}$

In Problems 53–60, simplify. Assume that all variables are greater than zero.

53. $\sqrt{\dfrac{121}{25}}$ **54.** $\sqrt{\dfrac{5a^4}{64b^2}}$

55. $\sqrt[3]{\dfrac{54k^2}{9k^5}}$ **56.** $\sqrt[3]{\dfrac{-160w^{11}}{343w^{-4}}}$

57. $\dfrac{\sqrt{12h^3}}{\sqrt{3h}}$ **58.** $\dfrac{\sqrt{50a^3b^3}}{\sqrt{8a^5b^{-3}}}$

59. $\dfrac{\sqrt[3]{-8x^7y}}{\sqrt[3]{27xy^4}}$ **60.** $\dfrac{\sqrt[4]{48m^2n^7}}{\sqrt[4]{3m^6n}}$

In Problems 61 and 62, multiply and simplify.

61. $\sqrt{5} \cdot \sqrt[3]{2}$ **62.** $\sqrt[4]{8} \cdot \sqrt[6]{4}$

Section 6.4   Adding, Subtracting, and Multiplying Radical Expressions

KEY CONCEPTS

- To add or subtract radicals, the radicals must have the same index and the same radicand.
- To multiply radicals, the index on each radical must be the same. Then multiply the radicands.

KEY TERM

Like radicals

You Should Be Able To...	EXAMPLE	Review Exercises
❶ Add or subtract radical expressions (p. 497)	Examples 1 through 3	63–72
❷ Multiply radical expressions (p. 499)	Examples 4 and 5	73–82

In Problems 63–72, add or subtract as indicated. Assume all variables are greater than or equal to zero.

63. $2\sqrt[4]{x} + 6\sqrt[4]{x}$

64. $7\sqrt[3]{4y} + 2\sqrt[3]{4y} - 3\sqrt[3]{4y}$

65. $5\sqrt{2} - 2\sqrt{12}$

66. $\sqrt{18} + 2\sqrt{50}$

67. $\sqrt[3]{-16z} + \sqrt[3]{54z}$

68. $7\sqrt[3]{8x^2} - \sqrt[3]{-27x^2}$

69. $\sqrt{16a} + \sqrt[6]{729a^3}$

70. $\sqrt{27x^2} - x\sqrt{48} + 2\sqrt{75x^2}$

71. $5\sqrt[3]{4m^5y^2} - \sqrt[6]{16m^{10}y^4}$

72. $\sqrt{y^3 - 4y^2} - 2\sqrt{y - 4} + \sqrt[4]{y^2 - 8y + 16}$

In Problems 73–82, multiply and simplify. Assume all variables are greater than or equal to zero.

73. $\sqrt{3}(\sqrt{5} - \sqrt{15})$

74. $\sqrt[3]{5}(3 + \sqrt[3]{4})$

75. $(3 + \sqrt{5})(4 - \sqrt{5})$

76. $(7 + \sqrt{3})(6 + \sqrt{2})$

77. $(1 - 3\sqrt{5})(1 + 3\sqrt{5})$

78. $(\sqrt[3]{x} + 1)(9\sqrt[3]{x} - 4)$

79. $(\sqrt{x} - \sqrt{5})^2$

80. $(11\sqrt{2} + \sqrt{5})^2$

81. $(\sqrt{2a} - b)(\sqrt{2a} + b)$

82. $(\sqrt[3]{6s} + 2)(\sqrt[3]{6s} - 7)$

Section 6.5   Rationalizing Radical Expressions

KEY CONCEPTS

- To rationalize the denominator containing a single radical with index n, multiply the numerator and denominator by a radical so that the radicand in the denominator is a perfect nth power.
- To rationalize a denominator containing two terms, use the fact that $(A + B)(A - B) = A^2 - B^2$.

KEY TERM

Rationalizing the denominator

You Should Be Able To...	EXAMPLE	Review Exercises
❶ Rationalize a denominator containing one term (p. 503)	Examples 1 and 2	83–90, 99
❷ Rationalize a denominator containing two terms (p. 506)	Examples 3 and 4	91–98, 100

In Problems 83–98, rationalize the denominator. Assume all variables are greater than or equal to zero.

83. $\dfrac{2}{\sqrt{6}}$

84. $\dfrac{6}{\sqrt{3}}$

85. $\dfrac{\sqrt{48}}{\sqrt{p^3}}$

86. $\dfrac{5}{\sqrt{2a}}$

87. $\dfrac{-2}{\sqrt{6y^3}}$

88. $\dfrac{3}{\sqrt[3]{5}}$

89. $\sqrt[3]{\dfrac{-4}{45}}$

90. $\dfrac{27}{\sqrt[5]{8p^3q^4}}$

91. $\dfrac{6}{7 - \sqrt{6}}$

92. $\dfrac{3}{\sqrt{3} - 9}$

93. $\dfrac{\sqrt{3}}{3 + \sqrt{2}}$

94. $\dfrac{\sqrt{k}}{\sqrt{k} - \sqrt{m}}$

95. $\dfrac{\sqrt{10} + 2}{\sqrt{10} - 2}$

96. $\dfrac{3 - \sqrt{y}}{3 + \sqrt{y}}$

97. $\dfrac{4}{2\sqrt{3} + 5\sqrt{2}}$

98. $\dfrac{\sqrt{5} - \sqrt{6}}{\sqrt{10} + \sqrt{3}}$

99. Simplify: $\dfrac{\sqrt{7}}{3} + \dfrac{6}{\sqrt{7}}$

100. Find the reciprocal: $4 - \sqrt{7}$

Section 6.6   Functions Involving Radicals

KEY CONCEPT

- **Finding the Domain of a Function Involving Radicals**
  1. If the index on a radical expression is even, then the radicand must be greater than or equal to zero.
  2. If the index on a radical expression is odd, then the radicand can be any real number.

KEY TERMS

Square root function
Cube root function

You Should Be Able To...	EXAMPLE	Review Exercises
❶ Evaluate functions involving radicals (p. 509)	Example 1	101–104
❷ Find the domain of a function involving a radical (p. 510)	Example 2	105–110, 111(a), 112(a), 113(a), 114(a)
❸ Graph functions involving square roots (p. 511)	Example 3	111(b)–113(b)
❹ Graph functions involving cube roots (p. 512)	Example 4	114(b)

In Problems 101–104, evaluate each radical function at the indicated values.

101. $f(x) = \sqrt{x + 4}$
  (a) $f(-3)$
  (b) $f(0)$
  (c) $f(5)$

102. $g(x) = \sqrt{3x - 2}$
  (a) $g\left(\dfrac{2}{3}\right)$
  (b) $g(2)$
  (c) $g(6)$

103. $H(t) = \sqrt[3]{t + 3}$
  (a) $H(-2)$
  (b) $H(-4)$
  (c) $H(5)$

104. $G(z) = \sqrt{\dfrac{z - 1}{z + 2}}$
  (a) $G(1)$
  (b) $G(-3)$
  (c) $G(2)$

In Problems 105–110, find the domain of the radical function.

105. $f(x) = \sqrt{3x - 5}$

106. $g(x) = \sqrt[3]{2x - 7}$

107. $h(x) = \sqrt[4]{6x + 1}$

108. $F(x) = \sqrt[5]{2x - 9}$

109. $G(x) = \sqrt{\dfrac{4}{x - 2}}$

110. $H(x) = \sqrt{\dfrac{x - 3}{x}}$

In Problems 111–114, (a) determine the domain of the function; (b) graph the function using point plotting; and (c) based on the graph, determine the range of the function.

111. $f(x) = \dfrac{1}{2}\sqrt{1 - x}$

112. $g(x) = \sqrt{x + 1} - 2$

113. $h(x) = -\sqrt{x + 3}$

114. $F(x) = \sqrt[3]{x + 1}$

Section 6.7   Radical Equations and Their Applications

KEY TERMS

Radical equation   Extraneous solution

You Should Be Able To...	EXAMPLE	Review Exercises
❶ Solve radical equations containing one radical (p. 516)	Examples 1 through 5	115–124, 129, 130
❷ Solve radical equations containing two radicals (p. 519)	Examples 6 and 7	125–128
❸ Solve for a variable in a radical equation (p. 521)	Example 8	131, 132

In Problems 115–130, solve each equation.

115. $\sqrt{m} = 13$

116. $\sqrt[3]{3t + 1} = -2$

117. $\sqrt[4]{3x - 8} = 3$

118. $\sqrt{2x + 5} + 4 = 2$

119. $\sqrt{4 - k} - 3 = 0$

120. $3\sqrt{t} - 4 = 11$

121. $2\sqrt[3]{m} + 5 = -11$

122. $\sqrt{q + 2} = q$

123. $\sqrt{w + 11} + 3 = w + 2$

124. $\sqrt{p^2 - 2p + 9} = p + 1$

125. $\sqrt{a + 10} = \sqrt{2a - 1}$

126. $\sqrt{5x + 9} = \sqrt{7x - 3}$

127. $\sqrt{c - 8} + \sqrt{c} = 4$

128. $\sqrt{x + 2} - \sqrt{x + 9} = 7$

129. $(4x - 3)^{1/3} - 3 = 0$

130. $(x^2 - 9)^{1/4} = 2$

131. Height of a Cone Solve $r = \sqrt{\dfrac{3V}{\pi h}}$ for h.

132. Ball Slide Speed Factor Solve $f_s = \sqrt[3]{\dfrac{30}{v}}$ for v.

Section 6.8 The Complex Number System

KEY CONCEPTS

- **Nonnegativity Property of Real Numbers**
  For any real number a, $a^2 \geq 0$.
- **Imaginary Unit**
  The imaginary unit, denoted by i, is the number whose square is -1.
  That is, $i^2 = -1$.
- **Complex numbers**
  Complex numbers are numbers of the form $a + bi$, where a and b are
  real numbers. The real number a is called the real part of the number
  $a + bi$; the real number b is called the imaginary part of $a + bi$.
- **Square Roots of Negative Numbers**
  If N is a positive real number, then the principal square root of $-N$,
  denoted by $\sqrt{-N}$, is $\sqrt{-N} = \sqrt{N}i$, where i is the number whose square
  is -1.
- **Sum of Complex Numbers**
  $(a + bi) + (c + di) = (a + c) + (b + d)i$
- **Difference of Complex Numbers**
  $(a + bi) - (c + di) = (a - c) + (b - d)i$
- **Complex Conjugate**
  If $a + bi$ is a complex number, then its conjugate is $a - bi$.
- **Product of a Complex Number and Its Conjugate**
  $(a + bi)(a - bi) = a^2 + b^2$

KEY TERMS

Imaginary unit
Complex number system
Complex number
Real part
Imaginary part
Standard form
Pure imaginary number
Principal square root
Conjugate
Powers of i

You Should Be Able To...	EXAMPLE	Review Exercises
① Evaluate the square root of negative real numbers (p. 525)	Examples 1 and 2	133–136
② Add or subtract complex numbers (p. 527)	Examples 3 and 4	137–140
③ Multiply complex numbers (p. 528)	Examples 5 through 7	141–146
④ Divide complex numbers (p. 530)	Examples 8 and 9	147–150
⑤ Evaluate the powers of i (p. 531)	Example 10	151–152

*In Problems 133 and 134, write each expression as a pure
imaginary number.*

133. $\sqrt{-29}$ **134.** $\sqrt{-54}$

*In Problems 135 and 136, write each expression as a
complex number in standard form.*

135. $14 - \sqrt{-162}$ **136.** $\dfrac{6 + \sqrt{-45}}{3}$

In Problems 137–150, perform the indicated operation.

137. $(3 - 7i) + (-2 + 5i)$

138. $(4 + 2i) - (9 - 8i)$

139. $(8 - \sqrt{-45}) - (3 + \sqrt{-80})$

140. $(1 + \sqrt{-9}) + (-6 + \sqrt{-16})$

141. $(4 - 5i)(3 + 7i)$

142. $\left(\dfrac{1}{2} + \dfrac{2}{3}i\right)(4 - 9i)$

143. $\sqrt{-3} \cdot \sqrt{-27}$

144. $(1 + \sqrt{-36})(-5 - \sqrt{-144})$

145. $(1 + 12i)(1 - 12i)$

146. $(7 + 2i)(5 + 4i)$

147. $\dfrac{4}{3 + 5i}$

148. $\dfrac{-3}{7 - 2i}$

In Problems 151 and 152, simplify.

151. i^{59}

152. i^{173}

149. $\dfrac{2 - 3i}{5 + 2i}$

150. $\dfrac{4 + 3i}{1 - i}$

Chapter 6 Test   *Step-by-step test solutions are found on the Chapter Test Prep Videos available in MyMathLab®, on YouTube, or may be accessed using the QR code to the right.*

---

1. Evaluate: $49^{-\frac{1}{2}}$

In Problems 2 and 3, simplify using rational exponents.

2. $\sqrt[3]{8x^{\frac{1}{2}}y^3} \cdot \sqrt{9xy^{\frac{1}{2}}}$

3. $\sqrt[5]{\left(2a^4b^3\right)^7}$

In Problems 4–9, perform the indicated operation and simplify. Assume all variables are greater than zero.

4. $\sqrt{3m} \cdot \sqrt{13n}$

5. $\sqrt{32x^7y^4}$

6. $\dfrac{\sqrt{9a^3b^{-3}}}{\sqrt{4ab}}$

7. $\sqrt{5x^3} + 2\sqrt{45x}$

8. $\sqrt{9a^2b} - \sqrt[4]{16a^4b^2}$

9. $\left(11 + 2\sqrt{x}\right)\left(3 - \sqrt{x}\right)$

In Problems 10 and 11, rationalize the denominator.

10. $\dfrac{-2}{3\sqrt{72}}$

11. $\dfrac{\sqrt{5}}{\sqrt{5} + 2}$

12. For $f(x) = \sqrt{-2x + 3}$, find the following:

(a) $f(1)$

(b) $f(-3)$

13. Determine the domain of the function

$$g(x) = \sqrt{-3x + 5}.$$

14. For $f(x) = \sqrt{x} - 3$, do the following:

(a) Determine the domain of the function.

(b) Graph the function using point plotting.

(c) From the graph, determine the range of the function.

In Problems 15–17, solve the given equations.

15. $\sqrt{x + 3} = 4$

16. $\sqrt{x + 13} - 4 = x - 3$

17. $\sqrt{x - 1} + \sqrt{x + 2} = 3$

In Problems 18–20, perform the indicated operation.

18. $(13 + 2i) + (4 - 15i)$

19. $(4 - 7i)(2 + 3i)$

20. $\dfrac{7 - i}{12 + 11i}$

CHAPTER

7 Quadratic Equations and Functions

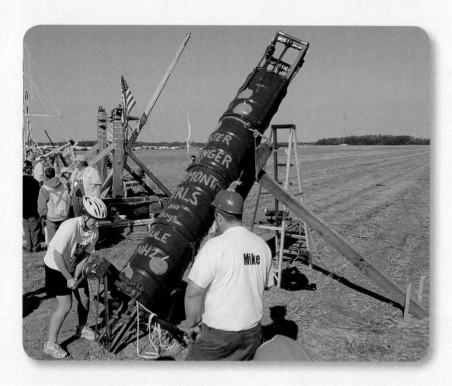

One of the more unusual sports—Punkin Chunkin—holds its World Championship competition every year in Millsboro, Delaware. Participants catapult or fire pumpkins to see who can toss them the farthest (and most accurately). Interestingly, this activity is an application of a quadratic function at work. See Problems 79–82 in Section 7.5.

The Big Picture: Putting It Together

In Part I of Chapter 1, we reviewed solving linear equations and inequalities in one variable. In Part II of Chapter 1, our discussion of "everything linear" concluded by covering linear equations and inequalities in two variables.

This chapter is dedicated to extending the discussion of quadratic equations, $ax^2 + bx + c = 0$, that began in Section 4.8. In Section 4.8, quadratic equations $ax^2 + bx + c = 0$, were solved by factoring and using the Zero-Product Property. Remember, if the expression $ax^2 + bx + c$ was not factorable, there was not a method for solving the equation. The missing piece for solving this type of quadratic equation was the idea of a radical. The familiarity with radicals acquired in Chapter 6 supplies the tools necessary to solve all quadratic equations $ax^2 + bx + c = 0$, regardless of whether the expression $ax^2 + bx + c$ is factorable. In addition to solving quadratic equations, graphing quadratic functions and solving quadratic and polynomial inequalities will be covered.

7.1 Solving Quadratic Equations by Completing the Square

Objectives

1. Solve Quadratic Equations Using the Square Root Property
2. Complete the Square in One Variable
3. Solve Quadratic Equations by Completing the Square
4. Solve Problems Using the Pythagorean Theorem

Are You Prepared for This Section?

Before getting started, complete the following problems. If you get a problem wrong, go back to the section cited and review the material.

P1. Multiply: $(2p + 3)^2$ [Section 4.2, pp. 324–325]

P2. Factor: $y^2 - 8y + 16$ [Section 4.6, pp. 359–360]

P3. Solve: $x^2 + 5x - 14 = 0$ [Section 4.8, pp. 373–375]

P4. Solve: $x^2 - 16 = 0$ [Section 4.8, pp. 373–375]

P5. Simplify: **(a)** $\sqrt{36}$ **(b)** $\sqrt{45}$ **(c)** $\sqrt{-12}$ [Getting Ready, pp. 468–469; Section 6.3, pp. 488–490; Section 6.8, pp. 525–526]

P6. Find the complex conjugate of $-3 + 2i$.   [Section 6.8, p. 529]

P7. Simplify: $\sqrt{x^2}$ [Getting Ready, pp. 470–471]

P8. Define: $|x|$ [Section R.3, p. 19]

▶ ① Solve Quadratic Equations Using the Square Root Property

Suppose we want to solve the quadratic equation

$$x^2 = p$$

where p is any real number. This equation is saying, "Give me all numbers whose square is p." For example, the equation $x^2 = 16$ means "Find all numbers whose square is 16." Because those numbers are -4 and 4, the solution set to the equation $x^2 = 16$ is $\{-4, 4\}$.

In general, to solve an equation of the form $x^2 = p$, take the square root of both sides of the equation.

$$x^2 = p$$

Take the square root of both sides: $\sqrt{x^2} = \sqrt{p}$

$\sqrt{x^2} = |x|$: $\quad |x| = \sqrt{p}$

$|x| = -x$ if $x < 0$; $|x| = x$ if $x \geq 0$: $\quad -x = \sqrt{p} \quad$ or $\quad x = \sqrt{p}$

Solve for x: $\quad x = -\sqrt{p} \quad$ or $\quad x = \sqrt{p}$

This gives the following result.

Work Smart

The Square Root Property is useful for solving equations of the form "some unknown squared equals a real number." To solve this equation, take the square root of both sides of the equation, but do not forget the $\pm$ symbol to obtain the positive and negative square root.

The Square Root Property

If $x^2 = p$, then $x = \sqrt{p}$ or $x = -\sqrt{p}$.

In using the Square Root Property to solve an equation such as $x^2 = p$, the solutions are usually written as $x = \pm\sqrt{p}$, which is read "x equals plus or minus the square root of p." For example, the two solutions of the equation

$$x^2 = 16$$

are

$$x = \pm\sqrt{16}$$
$$= \pm 4$$

Let's use the Square Root Property in an example now.

Prepared?...Answers

P1. $4p^2 + 12p + 9$

P2. $(y - 4)^2$ **P3.** $\{-7, 2\}$ **P4.** $\{-4, 4\}$

P5. (a) 6   **(b)** $3\sqrt{5}$ **(c)** $2\sqrt{3}i$

P6. $-3 - 2i$

P7. $|x|$ **P8.** $|x| = -x$ if $x < 0$
$\qquad\qquad |x| = x$ if $x \geq 0$

EXAMPLE 1   **How to Solve a Quadratic Equation Using the Square Root Property**

Solve: $x^2 - 9 = 0$

Step-by-Step Solution

Step 1: Isolate the expression containing the square term.

$$x^2 - 9 = 0$$

Add 9 to both sides: $x^2 = 9$

Step 2: Use the Square Root Property. Don't forget the $\pm$ symbol.

$$x = \pm\sqrt{9}$$

Simplify the radical: $= \pm 3$

Step 3: Isolate the variable, if necessary.

The variable is already isolated.

Step 4: Check Verify your solution(s).

$x = -3: \quad (-3)^2 - 9 \overset{?}{=} 0 \qquad\qquad x = 3: \quad 3^2 - 9 \overset{?}{=} 0$

$9 - 9 = 0$ True $\qquad\qquad\qquad 9 - 9 = 0$ True

The solution set is $\{-3, 3\}$.

The steps that are used to solve a quadratic equation using the Square Root Property are summarized below.

> **Solving a Quadratic Equation Using the Square Root Property**
>
> **Step 1:** Isolate the expression containing the squared term.
> **Step 2:** Use the Square Root Property: If $x^2 = p$, then $x = \pm\sqrt{p}$.
> **Step 3:** Isolate the variable, if necessary.
> **Step 4:** Verify your solution(s).

The equation in Example 1 could also be solved by factoring the difference of two squares:

$$x^2 - 9 = 0$$
$$(x - 3)(x + 3) = 0$$

Zero-Product Property: $x - 3 = 0$ or $x + 3 = 0$
$$x = 3 \quad \text{or} \quad x = -3$$

Work Smart

As our mathematical knowledge develops, there is often more than one way to solve a problem.

So there is more than one way to find the solution! However, in solving equations of the form $x^2 = p$, factoring works nicely only when p is a perfect square.

EXAMPLE 2   **Solving a Quadratic Equation Using the Square Root Property**

Solve: $3x^2 - 60 = 0$

Solution

$$3x^2 - 60 = 0$$

Add 60 to both sides of the equation: $3x^2 = 60$

Divide both sides by 3: $x^2 = 20$

Use the Square Root Property: $x = \pm\sqrt{20}$

Simplify the radical: $= \pm 2\sqrt{5}$

Check

$$x = -2\sqrt{5}: \quad 3(-2\sqrt{5})^2 - 60 \overset{?}{=} 0 \qquad x = 2\sqrt{5}: \quad 3(2\sqrt{5})^2 - 60 \overset{?}{=} 0$$

$$(ab)^2 = a^2b^2: \quad 3(-2)^2(\sqrt{5})^2 - 60 \overset{?}{=} 0 \qquad 3(2)^2(\sqrt{5})^2 - 60 \overset{?}{=} 0$$

$$3 \cdot 4 \cdot 5 - 60 \overset{?}{=} 0 \qquad 3 \cdot 4 \cdot 5 - 60 \overset{?}{=} 0$$

$$60 - 60 = 0 \quad \text{True} \qquad 60 - 60 = 0 \quad \text{True}$$

The solution set is $\left\{-2\sqrt{5}, 2\sqrt{5}\right\}$. ●

Quick ✓

1. If $x^2 = p$, then $x = $ ___ or $x = $ ___.

In Problems 2–4, solve the quadratic equation using the Square Root Property.

2. $p^2 = 48$ **3.** $3b^2 = 75$ **4.** $s^2 - 81 = 0$

As the next example illustrates, the solution to a quadratic equation does not have to be real.

EXAMPLE 3   **Solving a Quadratic Equation Using the Square Root Property**

Solve: $y^2 + 14 = 2$

Solution

Work Smart

For a review of complex numbers, see Section 6.8.

$$y^2 + 14 = 2$$

Subtract 14 from both sides: $y^2 = -12$

Use the Square Root Property: $y = \pm\sqrt{-12}$

Simplify the radical: $= \pm 2\sqrt{3}i$

Check

$$y = -2\sqrt{3}i: \quad (-2\sqrt{3}i)^2 + 14 \overset{?}{=} 2 \qquad y = 2\sqrt{3}i: \quad (2\sqrt{3}i)^2 + 14 \overset{?}{=} 2$$

$$(-2\sqrt{3})^2 i^2 + 14 \overset{?}{=} 2 \qquad (2\sqrt{3})^2 i^2 + 14 \overset{?}{=} 2$$

$$i^2 = -1: \quad 4 \cdot 3 \cdot (-1) + 14 \overset{?}{=} 2 \qquad 4 \cdot 3 \cdot (-1) + 14 \overset{?}{=} 2$$

$$-12 + 14 \overset{?}{=} 2 \qquad -12 + 14 \overset{?}{=} 2$$

$$2 = 2 \quad \text{True} \qquad 2 = 2 \quad \text{True}$$

The solution set is $\left\{-2\sqrt{3}i, 2\sqrt{3}i\right\}$. ●

Quick ✓

In Problems 5 and 6, solve the quadratic equation using the Square Root Property.

5. $d^2 = -72$ **6.** $3q^2 + 27 = 0$

EXAMPLE 4   **Solving Quadratic Equations Using the Square Root Property**

Solve:

   (a) $(x - 2)^2 = 25$ **(b)** $(y + 5)^2 + 24 = 0$

Solution

(a)

$$(x - 2)^2 = 25$$

Use the Square Root Property: $x - 2 = \pm\sqrt{25}$

Simplify the radical: $x - 2 = \pm 5$

Add 2 to each side: $x = 2 \pm 5$

2 ± 5 means $2 - 5$ or $2 + 5$: $x = 2 - 5$ or $x = 2 + 5$

$$= -3 \qquad\qquad = 7$$

(continued)

Check

$x = -3:$ $(-3 - 2)^2 \overset{?}{=} 25$ $x = 7:$ $(7 - 2)^2 \overset{?}{=} 25$

$(-5)^2 \overset{?}{=} 25$ $5^2 \overset{?}{=} 25$

$25 = 25$ True $25 = 25$ True

The solution set is $\{-3, 7\}$.

(b) $(y + 5)^2 + 24 = 0$

Subtract 24 from both sides: $(y + 5)^2 = -24$

Use the Square Root Property: $y + 5 = \pm\sqrt{-24}$

$\sqrt{-24} = \sqrt{24} \cdot i = \sqrt{4 \cdot 6} i = 2\sqrt{6} i:$ $y + 5 = \pm 2\sqrt{6} i$

Subtract 5 from each side: $y = -5 \pm 2\sqrt{6} i$

$y = -5 - 2\sqrt{6} i$ or $y = -5 + 2\sqrt{6} i$

Check

$y = -5 - 2\sqrt{6} i:$ $y = -5 + 2\sqrt{6} i:$

$(-5 - 2\sqrt{6} i + 5)^2 + 24 \overset{?}{=} 0$ $(-5 + 2\sqrt{6} i + 5)^2 + 24 \overset{?}{=} 0$

$(-2\sqrt{6} i)^2 + 24 \overset{?}{=} 0$ $(2\sqrt{6} i)^2 + 24 \overset{?}{=} 0$

$4 \cdot 6 \cdot i^2 + 24 \overset{?}{=} 0$ $4 \cdot 6 \cdot i^2 + 24 \overset{?}{=} 0$

$-24 + 24 = 0$ True $-24 + 24 = 0$ True

The solution set is $\{-5 - 2\sqrt{6} i, -5 + 2\sqrt{6} i\}$.   ●

Quick ✓

In Problems 7 and 8, solve the quadratic equation using the Square Root Property.

7. $(y + 3)^2 = 100$ **8.** $(q - 5)^2 + 20 = 4$

▶ ❷ **Complete the Square in One Variable**

The idea behind **completing the square** in one variable is to "adjust" the left side of a quadratic equation of the form $x^2 + bx + c = 0$ to make it a perfect square trinomial. Recall that perfect square trinomials have the form

$$A^2 + 2AB + B^2 = (A + B)^2 \text{ or } A^2 - 2AB + B^2 = (A - B)^2$$

For example, $x^2 + 6x + 9$ is a perfect square trinomial because $x^2 + 6x + 9 = x^2 + 2 \cdot x \cdot 3 + 3^2 = (x + 3)^2$.

"Adjust" $x^2 + bx + c$ by adding a number to make it a perfect square trinomial. For example, to make $x^2 + 6x$ a perfect square, add 9. Why choose 9? Dividing the coefficient of the first-degree term, 6, by 2 and then squaring the result yields 9. That is $\left(\dfrac{6}{2}\right)^2 = 3^2 = 9$. This approach works in general.

Obtaining a Perfect Square Trinomial

Work Smart

To complete the square, the coefficient of x^2 must be 1.

Multiply the coefficient of the first-degree term by $\dfrac{1}{2}$. Square the result. That is, for $x^2 + bx$, compute $\left(\dfrac{1}{2}b\right)^2$, and add the result to the expression $x^2 + bx$ to obtain

$$x^2 + bx + \left(\frac{1}{2}b\right)^2.$$

EXAMPLE 5 **Obtaining a Perfect Square Trinomial**

Determine the number that must be added to each expression to make it a perfect square trinomial. Then factor the expression.

Start	Add	Result	Factored Form
$y^2 + 8y$	$\left(\dfrac{1}{2} \cdot 8\right)^2 = 4^2 = 16$	$y^2 + 8y + 16$	$(y + 4)^2$
$x^2 + 12x$	$\left(\dfrac{1}{2} \cdot 12\right)^2 = 6^2 = 36$	$x^2 + 12x + 36$	$(x + 6)^2$
$a^2 - 20a$	$\left(\dfrac{1}{2}(-20)\right)^2 = (-10)^2 = 100$	$a^2 - 20a + 100$	$(a - 10)^2$
$p^2 - 5p$	$\left(\dfrac{1}{2}(-5)\right)^2 = \left(\dfrac{-5}{2}\right)^2 = \dfrac{25}{4}$	$p^2 - 5p + \dfrac{25}{4}$	$\left(p - \dfrac{5}{2}\right)^2$

Work Smart

It is important to notice that the expression at the "Start" is not equivalent to the "Result" because we are adding a quantity to the expression at the "Start."

Work Smart

It is common to write the value $\left(\dfrac{1}{2}b\right)^2$ as a fraction, not a decimal.

Notice in the factored form that the perfect square trinomial always factors so that

$$x^2 + bx + \left(\frac{b}{2}\right)^2 = \left(x + \frac{b}{2}\right)^2 \quad \text{or} \quad x^2 - bx + \left(\frac{b}{2}\right)^2 = \left(x - \frac{b}{2}\right)^2$$

That is, the perfect square trinomial always factors as $\left(x \pm \dfrac{b}{2}\right)^2$, where the $+$ is used if the coefficient of the first-degree term is positive and the $-$ is used if the coefficient of the first-degree term is negative. The $\dfrac{b}{2}$ represents $\dfrac{1}{2}$ the value of the coefficient of the first-degree term.

Quick ✓

In Problems 9 and 10, determine the number that must be added to the expression to make it a perfect square trinomial. Then factor the expression.

9. $p^2 + 14p$ **10.** $w^2 - 3w$

Figure 1

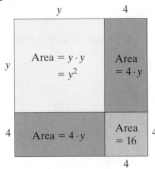

Why is this process called "completing the square"? Consider the expression $y^2 + 8y$ given in Example 5, which is geometrically represented in Figure 1. The yellow area is y^2 and each orange area is $4y$ (for a total area of $8y$). But what is the area of the green region to make the square complete? The dimensions of the green region are 4 by 4, so its area is 16. The area of the entire square region, $(y + 4)^2$, equals the sum of the area of the four regions that make up the square: $y^2 + 4y + 4y + 16 = y^2 + 8y + 16$.

❸ Solve Quadratic Equations by Completing the Square

Until now, we have solved quadratic equations of the form $ax^2 + bx + c = 0$ only when $ax^2 + bx + c$ was factorable. Now we can solve $ax^2 + bx + c = 0$ when $ax^2 + bx + c$ is not factorable. We begin with equations of the form $x^2 + bx + c = 0$. That is, the coefficient of the square term is 1.

▶ **EXAMPLE 6** **How to Solve a Quadratic Equation by Completing the Square**

Solve: $x^2 + 6x + 1 = 0$

Step-by-Step Solution

Step 1: Rewrite $x^2 + bx + c = 0$ as $x^2 + bx = -c$ by subtracting the constant from both sides of the equation.

Subtract 1 from both sides:

$$x^2 + 6x + 1 = 0$$
$$x^2 + 6x = -1$$

Step 2: Complete the square in the expression $x^2 + bx$ by making it a perfect square trinomial. Add the same number to each side of the equation to maintain the equality.

$$\left(\frac{1}{2} \cdot 6\right)^2 = 3^2 = 9;\ \text{add 9 to both sides:}$$

$$x^2 + 6x + 9 = -1 + 9$$
$$x^2 + 6x + 9 = 8$$

Step 3: Factor the perfect square trinomial on the left side of the equation.

$$x^2 + 6x + 9 = (x+3)^2:$$

$$(x+3)^2 = 8$$

Step 4: Solve the equation using the Square Root Property.

$$\sqrt{8} = 2\sqrt{2}:$$

Subtract 3 from both sides:

$a \pm b$ means $a - b$ or $a + b$:

$$x + 3 = \pm\sqrt{8}$$
$$x + 3 = \pm 2\sqrt{2}$$
$$x = -3 \pm 2\sqrt{2}$$
$$x = -3 - 2\sqrt{2} \quad \text{or} \quad x = -3 + 2\sqrt{2}$$

Step 5: Check Verify your solution(s).

$$x^2 + 6x + 1 = 0$$

$x = -3 - 2\sqrt{2}:$

$$(-3 - 2\sqrt{2})^2 + 6(-3 - 2\sqrt{2}) + 1 \overset{?}{=} 0$$
$$9 + 12\sqrt{2} + 8 - 18 - 12\sqrt{2} + 1 \overset{?}{=} 0$$
$$0 = 0 \quad \text{True}$$

$x = -3 + 2\sqrt{2}:$

$$(-3 + 2\sqrt{2})^2 + 6(-3 + 2\sqrt{2}) + 1 \overset{?}{=} 0$$
$$9 - 12\sqrt{2} + 8 - 18 + 12\sqrt{2} + 1 \overset{?}{=} 0$$
$$0 = 0 \quad \text{True}$$

The solution set is $\left\{-3 - 2\sqrt{2}, -3 + 2\sqrt{2}\right\}$. ●

Solving a Quadratic Equation By Completing the Square

Step 1: Rewrite $x^2 + bx + c = 0$ as $x^2 + bx = -c$ by subtracting the constant from both sides of the equation.

Step 2: Complete the square in the expression $x^2 + bx$ by making it a perfect square trinomial. Add the same number to each side of the equation.

Step 3: Factor the perfect square trinomial on the left side of the equation.

Step 4: Solve the equation using the Square Root Property.

Step 5: Verify your solution(s).

Quick ✓

In Problems 11 and 12, solve the equation by completing the square.

11. $b^2 + 2b - 8 = 0$ **12.** $z^2 - 8z + 9 = 0$

Work Smart

If the coefficient of the square term is not 1, then before using the method of completing the square, divide each side of the equation by the coefficient of the square term so that it becomes 1.

When the coefficient of the square term is not 1, multiply or divide both sides of the equation by a nonzero constant so this coefficient becomes 1. The next example demonstrates this method.

EXAMPLE 7 **Solving a Quadratic Equation by Completing the Square When the Coefficient of the Square Term Is Not 1**

Solve: $2x^2 + 4x + 3 = 0$

Solution

To make the coefficient of the square term be 1, divide both sides of the equation by 2.

$$2x^2 + 4x + 3 = 0$$

Divide both sides of the equation by 2: $\dfrac{2x^2 + 4x + 3}{2} = \dfrac{0}{2}$

Simplify: $x^2 + 2x + \dfrac{3}{2} = 0$

Now solve the equation by completing the square.

Subtract $\dfrac{3}{2}$ from both sides: $x^2 + 2x = -\dfrac{3}{2}$

$\left(\dfrac{1}{2} \cdot 2\right)^2 = 1^2 = 1$; add 1 to both sides: $x^2 + 2x + 1 = -\dfrac{3}{2} + 1$

Simplify: $x^2 + 2x + 1 = -\dfrac{1}{2}$

Factor: $(x + 1)^2 = -\dfrac{1}{2}$

Use the Square Root Property: $x + 1 = \pm\sqrt{-\dfrac{1}{2}}$

$\sqrt{-N} = \sqrt{N}\,i$ $\qquad x + 1 = \pm\sqrt{\dfrac{1}{2}}\,i$

$\sqrt{\dfrac{1}{2}} = \dfrac{\sqrt{1}}{\sqrt{2}} = \dfrac{\sqrt{1}}{\sqrt{2}} \cdot \dfrac{\sqrt{2}}{\sqrt{2}} = \dfrac{\sqrt{2}}{2}$: $\qquad x + 1 = \pm\dfrac{\sqrt{2}}{2}\,i$

Subtract 1 from both sides: $x = -1 \pm \dfrac{\sqrt{2}}{2}\,i$

$a \pm b$ means $a - b$ or $a + b$: $\quad x = -1 - \dfrac{\sqrt{2}}{2}\,i \quad$ or $\quad x = -1 + \dfrac{\sqrt{2}}{2}\,i$

Work Smart

Notice the solutions in Example 7 are complex conjugates of each other.

The solution set is $\left\{-1 - \dfrac{\sqrt{2}}{2}\,i, -1 + \dfrac{\sqrt{2}}{2}\,i\right\}$. The check is left to you.

Quick ✓

In Problems 13 and 14, solve the quadratic equation by completing the square.

13. $2q^2 + 6q - 1 = 0$

14. $3m^2 + 2m + 7 = 0$

Figure 2

Hypotenuse c
b
Leg
$90°$
a
Leg

④ Solve Problems Using the Pythagorean Theorem

The Pythagorean Theorem is a statement about *right triangles*. A **right triangle** is one that contains a **right angle**—that is, an angle of 90°. The side of the triangle opposite the 90° angle is the **hypotenuse;** the remaining two sides are the **legs.** In Figure 2, c represents the length of the hypotenuse; a and b represent the lengths of the legs. Notice the use of the symbol $\llcorner$ to show the 90° angle.

The Pythagorean Theorem

In a right triangle, the square of the length of the hypotenuse equals the sum of the squares of the lengths of the legs. That is, in the right triangle shown on the previous page in Figure 2,

$$c^2 = a^2 + b^2$$

EXAMPLE 8 **Finding the Hypotenuse of a Right Triangle**

In a right triangle, one leg is of length 5 inches and the other is of length 12 inches. What is the length of the hypotenuse?

Solution

Because this is a right triangle, use the Pythagorean Theorem with $a = 5$ and $b = 12$ to find the length c of the hypotenuse.

$$c^2 = a^2 + b^2$$
$$c^2 = 5^2 + 12^2$$
$$= 25 + 144$$
$$= 169$$

Now use the Square Root Property to find c, the length of the hypotenuse.

$$c = \sqrt{169} = 13$$

The length of the hypotenuse is 13 inches. Because c represents a length, which must be positive, find only the positive square root of 169.   ●

Quick ✓

15. The side of a right triangle opposite the 90° angle is called the _____; the remaining two sides are called ___.

16. *True or False* The Pythagorean Theorem states that for any triangle, the length of the hypotenuse is equal to the sum of the squares of the lengths of the legs.

In Problem 17, the lengths of the legs of a right triangle are given. Find the length of the hypotenuse.

17. $a = 3, b = 4$

EXAMPLE 9 **How Far Can You See?**

The Currituck Lighthouse is located in Corolla, in the Outer Banks of North Carolina. The lighthouse, completed in 1875, stands 162 feet tall, however, a person standing in the observation deck is located 158 feet above the ground. See Figure 3.

Figure 3

The website for the Currituck Lighthouse states that a person standing on the observation deck could see approximately 18 miles. See Figure 4. Assuming that the radius of the Earth is 3960 miles, verify this claim.

Figure 4

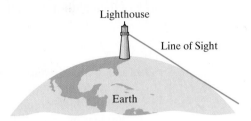

Solution

Step 1: Identify Find how far a person can see from the lighthouse.

Step 2: Name Call this unknown distance d, in miles.

Step 3: Translate To help with the translation, draw a picture. From the center of Earth, draw two lines: one through the lighthouse and the other to the farthest point a person can see from the lighthouse. See Figure 5.

Figure 5

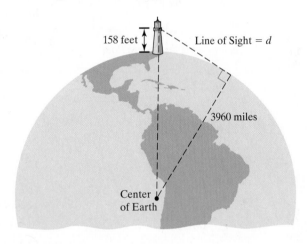

The line of sight and the two lines drawn from the center of Earth form a right triangle, so the angle where the line of sight touches the horizon measures 90°. From the Pythagorean Theorem,

$$\text{Hypotenuse}^2 = \text{Leg}^2 + \text{Leg}^2$$

The length of the hypotenuse is 3960 miles plus 158 feet. Add 3960 miles to 158 feet by converting the height of the tower to miles. Because 158 feet $=$ 158 feet $\cdot \dfrac{1 \text{ mile}}{5280 \text{ feet}} = \dfrac{158}{5280}$ mile, the hypotenuse is $\left(3960 + \dfrac{158}{5280}\right)$ miles. One of the legs is 3960 miles. The length of the other leg is our unknown, d. Thus

$$3960^2 + d^2 = \left(3960 + \frac{158}{5280}\right)^2 \quad \text{The Model}$$

Step 4: Solve

$$3960^2 + d^2 = \left(3960 + \frac{158}{5280}\right)^2$$

Subtract 3960^2 from both sides: $d^2 = \left(3960 + \dfrac{158}{5280}\right)^2 - 3960^2$

Use a calculator: $d^2 \approx 237.000895$

Square Root Property: $d \approx \sqrt{237.000895}$

≈ 15.39 miles

Step 5: Check The solution is less than the distance given on the website.

Step 6: Answer The distance given on the Currituck Lighthouse website appears to overstate the actual distance a person could see. Someone standing on the observation deck of the lighthouse could see about 15.39 miles. ●

Quick ✓

18. The USS *Constitution* (aka *Old Ironsides*) is perhaps the most famous ship from United States Naval history. The mainmast of the Constitution is 220 feet high. Suppose that a sailor climbs the mainmast to a height of 200 feet in order to look for enemy vessels. How far could the sailor see? Assume the radius of the Earth is 3960 miles.

7.1 Exercises MyMathLab® Exercise numbers in green have complete video solutions in MyMathLab or may be accessed using the QR code to the right.

Problems 1–18 are the Quick ✓ s that follow the EXAMPLES.

Building Skills

In Problems 19–44, solve each equation using the Square Root Property. See Objective 1.

19. $y^2 = 100$

20. $x^2 = 81$

21. $p^2 = 50$

22. $z^2 = 48$

23. $m^2 = -25$

24. $n^2 = -49$

25. $w^2 = \dfrac{5}{4}$

26. $z^2 = \dfrac{8}{9}$

27. $x^2 + 5 = 13$

28. $w^2 - 6 = 14$

29. $3z^2 = 48$

30. $4y^2 = 100$

31. $3x^2 = 8$

32. $5y^2 = 32$

33. $2p^2 + 23 = 15$

34. $-3x^2 - 5 = 22$

35. $(d - 1)^2 = -18$

36. $(z + 4)^2 = -24$

37. $3(q + 5)^2 - 1 = 8$

38. $5(x - 3)^2 + 2 = 27$

39. $(3q + 1)^2 = 9$

40. $(2p + 3)^2 = 16$

41. $\left(x - \dfrac{2}{3}\right)^2 = \dfrac{5}{9}$

42. $\left(y + \dfrac{3}{2}\right)^2 = \dfrac{3}{4}$

43. $x^2 + 8x + 16 = 81$

44. $q^2 - 6q + 9 = 16$

In Problems 45–52, complete the square in each expression. Then factor the perfect square trinomial. See Objective 2.

45. $x^2 + 10x$

46. $y^2 + 16y$

47. $z^2 - 18z$

48. $p^2 - 4p$

49. $y^2 + 7y$

50. $x^2 + x$

51. $w^2 + \dfrac{1}{2}w$

52. $z^2 - \dfrac{1}{3}z$

In Problems 53–72, solve each quadratic equation by completing the square. See Objective 3.

53. $x^2 + 4x = 12$

54. $y^2 + 3y = 18$

55. $x^2 - 4x + 1 = 0$

56. $p^2 - 6p + 4 = 0$

57. $a^2 - 4a + 5 = 0$

58. $m^2 - 2m + 5 = 0$

59. $b^2 + 5b - 2 = 0$

60. $q^2 + 7q + 7 = 0$

61. $m^2 = 8m + 3$

62. $n^2 = 10n + 5$

63. $p^2 - p + 3 = 0$

64. $z^2 - 3z + 5 = 0$

65. $2y^2 - 5y - 12 = 0$

66. $3a^2 - 4a - 4 = 0$

67. $3y^2 - 6y + 2 = 0$

68. $2y^2 - 2y - 1 = 0$

69. $2z^2 - 5z + 1 = 0$ **70.** $2x^2 - 7x + 2 = 0$

71. $2x^2 + 4x + 5 = 0$ **72.** $2z^2 + 6z + 5 = 0$

In Problems 73–82, the lengths of the legs of a right triangle are given. Find the hypotenuse. Give exact answers and decimal approximations rounded to two decimal places. See Objective 4.

73. $a = 6, b = 8$ **74.** $a = 7, b = 24$

75. $a = 12, b = 16$ **76.** $a = 15, b = 8$

77. $a = 5, b = 5$ **78.** $a = 3, b = 3$

79. $a = 1, b = \sqrt{3}$ **80.** $a = 2, b = \sqrt{5}$

81. $a = 6, b = 10$ **82.** $a = 8, b = 10$

In Problems 83–86, use the right triangle shown below and find the missing length. Give exact answers and decimal approximations rounded to two decimal places. See Objective 4.

83. $a = 4, c = 8$ **84.** $a = 4, c = 10$

85. $b = 8, c = 12$ **86.** $b = 2, c = 10$

Mixed Practice

87. Given that $f(x) = (x - 3)^2$, find all x such that $f(x) = 36$. What points are on the graph of f?

88. Given that $f(x) = (x - 5)^2$, find all x such that $f(x) = 49$. What points are on the graph of f?

89. Given that $g(x) = (x + 2)^2$, find all x such that $g(x) = 18$. What points are on the graph of g?

90. Given that $h(x) = (x + 1)^2$, find all x such that $h(x) = 32$. What points are on the graph of h?

Applying the Concepts

In Problems 91 and 92, find the exact length of the diagonal in each rectangle.

 91.

 92.

In Problems 93–100, express your answer as a decimal rounded to three decimal places.

93. **Golf** A golfer hits an errant tee shot that lands in the rough. The golfer finds that the ball is exactly 30 yards to the right of the 100-yard marker, which indicates the distance to the center of the green as shown in the figure. How far is the ball from the center of the green?

94. **Baseball** Avisail Garcia plays right field for the Chicago White Sox. He catches a fly ball 40 feet from the right field foul line, as indicated in the figure. How far is it to home plate?

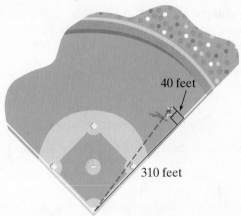

95. **Guy Wire** A guy wire is a wire used to support telephone poles. Suppose that a guy wire is located 30 feet up a telephone pole and is anchored to the ground 10 feet from the base of the pole. How long is the guy wire?

96. Guy Wire A guy wire is used to support an antenna on a rooftop. The wire is located 40 feet up on the antenna and anchored to the roof 8 feet from the base of the antenna. What is the length of the guy wire?

97. Ladder Bob needs to wash the windows on his house. He has a 25-foot ladder and places the base of the ladder 10 feet from the wall on the house.

  (a) How far up the wall will the ladder reach?

  (b) If his windows are 20 feet above the ground, what is the farthest distance the base of the ladder can be from the wall?

98. Fire Truck Ladder A fire truck has a 75-foot ladder. If the truck can safely park 20 feet from a building, how far up the building can the ladder reach, assuming that the base of the ladder is resting on top of the truck and the truck is 10 feet tall?

99. Gravity The distance s that an object falls (in feet) after t seconds, ignoring air resistance, is given by the equation $s = 16t^2$.

  (a) How long does it take an object to fall 16 feet?

  (b) How long does it take an object to fall 48 feet?

  (c) How long does it take an object to fall 64 feet?

△ **100. Equilateral Triangles** An equilateral triangle is one whose sides are all the same length. The area A of an equilateral triangle whose sides are each length x is given by $A = \dfrac{\sqrt{3}}{4}x^2$.

  (a) What is the length of each side of an equilateral triangle whose area is $\dfrac{8\sqrt{3}}{9}$ square feet?

  (b) What is the length of each side of an equilateral triangle whose area is $\dfrac{25\sqrt{3}}{4}$ square meters?

Problems 101 and 102 are based on the following information. If P dollars are invested today at an annual interest rate r compounded once a year, then the value of the account A after 2 years is given by the formula $A = P(1 + r)^2$.

101. Value of Money Find the rate of interest required to turn an investment of $1000 into $1200 after 2 years. Express your answer as a percent rounded to two decimal places.

102. Value of Money Find the rate of interest required to turn an investment of $5000 into $6200 after 2 years. Express your answer as a percent rounded to two decimal places.

Extending the Concepts

*If you look carefully at the Pythagorean Theorem, it states, "If we have a right triangle, then $c^2 = a^2 + b^2$, where c is the length of the hypotenuse." In this theorem "a right triangle" represents the hypothesis, and "$c^2 = a^2 + b^2$" represents the conclusion. The **converse** of a theorem interchanges the hypothesis and conclusion. The converse of the Pythagorean Theorem is true.*

> ### Converse of the Pythagorean Theorem
> In a triangle, if the square of the length of one side equals the sum of the squares of the lengths of the other two sides, then the triangle is a right triangle. The 90° angle is opposite the longest side.

In Problems 103–106, the lengths of the sides of a triangle are given. Determine whether the triangle is a right triangle. If it is, identify the hypotenuse.

103. 8, 15, 17   **104.** 4, 6, 8

105. 14, 18, 20   **106.** 20, 48, 52

107. Pythagorean Triples Suppose that m and n are positive integers with $m > n$. If $a = m^2 - n^2$, $b = 2mn$, and $c = m^2 + n^2$, show that a, b, and c are the lengths of the sides of a right triangle using the Converse of the Pythagorean Theorem. Any numbers a, b, and c found from the above formulas are called **Pythagorean Triples.**

108. Solve $ax^2 + bx + c = 0$ for x by completing the square.

Synthesis Review

In Problems 109–112, solve each equation.

109. $a^2 - 5a - 36 = 0$ **110.** $p^2 + 4p = 32$

111. $|4q + 1| = 3$ **112.** $\left|\dfrac{3}{4}w - \dfrac{2}{3}\right| = \dfrac{5}{2}$

113. In Problems 109–112, you are asked to solve quadratic and absolute value equations. In solving both types of equations, you must reduce the equation to a simpler equation. What type of equation is the simpler equation?

7.2 Solving Quadratic Equations by the Quadratic Formula

Objectives

❶ Solve Quadratic Equations Using the Quadratic Formula

❷ Use the Discriminant to Determine the Nature of Solutions of a Quadratic Equation

❸ Model and Solve Problems Involving Quadratic Equations

Are You Prepared for This Section?

Before getting started, complete the following problems. If you get a problem wrong, go back to the section cited and review the material.

P1. Simplify: **(a)** $\sqrt{121}$ **(b)** $\sqrt{54}$ [Getting Ready, pp. 468–469; Section 6.3, pp. 488–490]

P2. Simplify: **(a)** $\sqrt{-9}$ **(b)** $\sqrt{-72}$ [Section 6.8, pp. 525–526]

P3. Simplify: $\dfrac{3 + \sqrt{18}}{6}$ [Section 6.3, pp. 488–490]

So far, three methods have been used for solving quadratic equations: (1) factoring, (2) the Square Root Property, and (3) completing the square. Why are three methods needed? Because each method provides the quickest route to the solution when used appropriately. For example, if the quadratic expression is easy to factor, the method of factoring will yield a solution most easily. If the equation is in the form $x^2 = p$ or $(ax \pm b)^2 = p$, using the Square Root Property is fastest. If the quadratic expression is not factorable, completing the square is needed. But the method of completing the square is tedious. Is there an alternative to this method? Yes!

❶ Solve Quadratic Equations Using the Quadratic Formula

▶ The method of completing the square leads to a general formula for solving the quadratic equation

$$ax^2 + bx + c = 0 \qquad a \neq 0$$

To complete the square, first get the constant c on the right-hand side of the equation.

$$ax^2 + bx = -c \qquad a \neq 0$$

Divide both sides of the equation by a: $x^2 + \dfrac{b}{a}x = -\dfrac{c}{a}$

Now complete the square on the left side by adding the square of $\dfrac{1}{2}$ of the coefficient of x to both sides of the equation. That is, add

$$\left(\frac{1}{2} \cdot \frac{b}{a}\right)^2 = \frac{b^2}{4a^2}$$

to both sides of the equation. This gives

$$x^2 + \frac{b}{a}x + \frac{b^2}{4a^2} = -\frac{c}{a} + \frac{b^2}{4a^2}$$

$$x^2 + \frac{b}{a}x + \frac{b^2}{4a^2} = \frac{b^2}{4a^2} - \frac{c}{a}$$

Combine like terms on the right-hand side. The least common denominator on the right-hand side is $4a^2$, so multiply $-\dfrac{c}{a}$ by $\dfrac{4a}{4a}$:

$$x^2 + \frac{b}{a}x + \frac{b^2}{4a^2} = \frac{b^2}{4a^2} - \frac{c}{a} \cdot \frac{4a}{4a}$$

$$x^2 + \frac{b}{a}x + \frac{b^2}{4a^2} = \frac{b^2}{4a^2} - \frac{4ac}{4a^2}$$

$$x^2 + \frac{b}{a}x + \frac{b^2}{4a^2} = \frac{b^2 - 4ac}{4a^2}$$

Prepared?...Answers   P1. (a) 11
(b) $3\sqrt{6}$ **P2. (a)** $3i$ **(b)** $6\sqrt{2}i$
P3. $\dfrac{1}{2} + \dfrac{\sqrt{2}}{2}$ or $\dfrac{1 + \sqrt{2}}{2}$

Work Smart

To factor any perfect square trinomial of the form $x^2 + bx + c$, write

$$\left(x + \frac{b}{2}\right)^2.$$

Factor the left-hand side and obtain

$$\left(x + \frac{b}{2a}\right)^2 = \frac{b^2 - 4ac}{4a^2}$$

Assume that $a > 0$ (you'll see why in a little while). This assumption does not affect the results because if $a < 0$, both sides of the equation $ax^2 + bx + c = 0$ could be multiplied by -1 to make it positive. With this assumption, use the Square Root Property and get

$$x + \frac{b}{2a} = \pm\sqrt{\frac{b^2 - 4ac}{4a^2}}$$

$\sqrt{\dfrac{a}{b}} = \dfrac{\sqrt{a}}{\sqrt{b}}:$ $x + \dfrac{b}{2a} = \pm\dfrac{\sqrt{b^2 - 4ac}}{\sqrt{4a^2}}$

$\sqrt{4a^2} = 2a$ since $a > 0$: $x + \dfrac{b}{2a} = \pm\dfrac{\sqrt{b^2 - 4ac}}{2a}$

Subtract $\dfrac{b}{2a}$ from both sides: $x = -\dfrac{b}{2a} \pm \dfrac{\sqrt{b^2 - 4ac}}{2a}$

Write over a common denominator: $x = \dfrac{-b \pm \sqrt{b^2 - 4ac}}{2a}$

Work Smart: Study Skills

When solving homework problems, always write down the quadratic formula before you begin solving the equation. This will help you memorize the formula.

This is called the *quadratic formula*.

In Other Words

The quadratic formula says that the solution(s) to the equation $ax^2 + bx + c = 0$ is (are) "the opposite of b plus or minus the square root of b squared minus $4ac$, all over $2a$."

The Quadratic Formula

The solution(s) to the quadratic equation $ax^2 + bx + c = 0$, $a \neq 0$, is (are) given by the **quadratic formula**

$$x = \frac{-b \pm \sqrt{b^2 - 4ac}}{2a}$$

▶ (**EXAMPLE 1**) **How to Solve a Quadratic Equation Using the Quadratic Formula**

Solve: $12x^2 + 5x - 3 = 0$

Step-by-Step Solution

$\boxed{a = 12}\ \boxed{b = 5}\ \boxed{c = -3}$

Step 1: Write the equation in standard form $ax^2 + bx + c = 0$ and identify the values of a, b, and c.

$12x^2 + 5x - 3 = 0$

Step 2: Substitute the values of a, b, and c into the quadratic formula.

$$x = \frac{-b \pm \sqrt{b^2 - 4ac}}{2a}$$

$$x = \frac{-5 \pm \sqrt{5^2 - 4(12)(-3)}}{2(12)}$$

Step 3: Simplify the expression found in Step 2.

$$= \frac{-5 \pm \sqrt{25 + 144}}{24}$$

$$= \frac{-5 \pm \sqrt{169}}{24}$$

$\sqrt{169} = 13:$ $= \dfrac{-5 \pm 13}{24}$

$$a \pm b \text{ means } a - b \text{ or } a + b: \quad x = \frac{-5 - 13}{24} \quad \text{or} \quad x = \frac{-5 + 13}{24}$$

$$= \frac{-18}{24} \quad \text{or} \quad = \frac{8}{24}$$

$$\text{Simplify:} \quad = -\frac{3}{4} \quad \text{or} \quad = \frac{1}{3}$$

Step 4: Check

$$12x^2 + 5x - 3 = 0$$

$$x = -\frac{3}{4}: \quad 12\left(-\frac{3}{4}\right)^2 + 5\left(-\frac{3}{4}\right) - 3 \overset{?}{=} 0 \qquad x = \frac{1}{3}: \quad 12\left(\frac{1}{3}\right)^2 + 5\left(\frac{1}{3}\right) - 3 \overset{?}{=} 0$$

$$12 \cdot \frac{9}{16} - \frac{15}{4} - 3 \overset{?}{=} 0 \qquad\qquad 12 \cdot \frac{1}{9} + \frac{5}{3} - 3 \overset{?}{=} 0$$

$$\frac{27}{4} - \frac{15}{4} - 3 \overset{?}{=} 0 \qquad\qquad \frac{4}{3} + \frac{5}{3} - 3 \overset{?}{=} 0$$

$$\frac{12}{4} - 3 \overset{?}{=} 0 \qquad\qquad \frac{9}{3} - 3 \overset{?}{=} 0$$

$$0 = 0 \quad \text{True} \qquad\qquad 0 = 0 \quad \text{True}$$

The solution set is $\left\{ -\frac{3}{4}, \frac{1}{3} \right\}$.   ●

Work Smart

If $b^2 - 4ac$ is a perfect square, then the quadratic equation can be solved by factoring. In Example 1, $b^2 - 4ac = 169$, a perfect square.

$$12x^2 + 5x - 3 = 0$$
$$(4x + 3)(3x - 1) = 0$$
$$4x + 3 = 0 \text{ or } 3x - 1 = 0$$
$$x = -\frac{3}{4} \quad \text{or} \quad x = \frac{1}{3}$$

Notice that the solutions to the equation in Example 1 are rational numbers and that the expression $b^2 - 4ac$ under the radical in the quadratic formula, 169, is a perfect square. This leads to a generalization. Whenever the expression $b^2 - 4ac$ is a perfect square, the quadratic equation will have rational solutions, and the quadratic equation can be solved by factoring (provided the coefficients of the quadratic equation are rational numbers).

With that said, *any* quadratic equation can be solved using the quadratic formula.

> **Solving a Quadratic Equation Using the Quadratic Formula**
>
> **Step 1:** Write the equation in standard form $ax^2 + bx + c = 0$ and identify the values of a, b, and c.
>
> **Step 2:** Substitute the values of a, b, and c into the quadratic formula.
>
> **Step 3:** Simplify the expression found in Step 2.
>
> **Step 4:** Verify the solution(s).

Quick ✓

1. The solution(s) to the quadratic equation $ax^2 + bx + c = 0$, $a \neq 0$, are given

by the quadratic formula $x = \underline{\hspace{2cm}}$.

In Problems 2 and 3, solve each equation using the quadratic formula.

2. $2x^2 - 3x - 9 = 0$ **3.** $2x^2 + 7x = 4$

▶ **EXAMPLE 2**   **Solving a Quadratic Equation Using the Quadratic Formula**

Solve: $3p^2 = 6p - 1$

Solution

First, write the equation in standard form to identify a, b, and c.

$$3p^2 = 6p - 1$$

Subtract $6p$ from both sides;
Add 1 to both sides: $\quad 3p^2 - 6p + 1 = 0$

(continued)

Write the quadratic equation using "p =": $\quad p = \dfrac{-b \pm \sqrt{b^2 - 4ac}}{2a}$

$a = 3, b = -6, c = 1:\quad p = \dfrac{-(-6) \pm \sqrt{(-6)^2 - 4(3)(1)}}{2(3)}$

$= \dfrac{6 \pm \sqrt{36 - 12}}{6}$

$= \dfrac{6 \pm \sqrt{24}}{6}$

$\sqrt{24} = \sqrt{4 \cdot 6} = 2\sqrt{6}:\quad = \dfrac{6 \pm 2\sqrt{6}}{6}$

$\dfrac{a + b}{c} = \dfrac{a}{c} + \dfrac{b}{c}:\quad = \dfrac{6}{6} \pm \dfrac{2\sqrt{6}}{6}$

Simplify: $\quad = 1 \pm \dfrac{\sqrt{6}}{3}$

$a \pm b$ means $a - b$ or $a + b$: $\quad p = 1 - \dfrac{\sqrt{6}}{3}$ or $p = 1 + \dfrac{\sqrt{6}}{3}$

Work Smart

$\dfrac{6 \pm 2\sqrt{6}}{6}$ can also be simplified by factoring:

$\dfrac{6 \pm 2\sqrt{6}}{6} = \dfrac{2(3 \pm \sqrt{6})}{6}$

$= \dfrac{3 \pm \sqrt{6}}{3}$

This is equivalent to $1 \pm \dfrac{\sqrt{6}}{3}$. Ask your instructor which form of the solution is preferred, if any.

We leave it to you to verify the solutions. The solution set is $\left\{1 - \dfrac{\sqrt{6}}{3}, 1 + \dfrac{\sqrt{6}}{3}\right\}$. ●

Notice in Example 2 that the value of $b^2 - 4ac$, 24, is positive, but not a perfect square. There are two solutions to the quadratic equation and they are irrational.

Quick ✓

4. *True or False* For $3x^2 - 2x = 9$, $a = 3$, $b = -2$, and $c = 9$.

5. Solve: $4z^2 + 1 = 8z$

EXAMPLE 3   **Solving a Rational Equation That Leads to a Quadratic Equation**

Solve: $9m + \dfrac{4}{m} = 12$

Solution

First note that m cannot equal 0. To clear the equation of rational expressions, multiply both sides of the equation by the LCD, m.

$$m\left(9m + \dfrac{4}{m}\right) = 12 \cdot m$$

Distribute m: $\quad 9m^2 + 4 = 12m$

Subtract $12m$ from both sides: $\quad 9m^2 - 12m + 4 = 0$

$m = \dfrac{-b \pm \sqrt{b^2 - 4ac}}{2a}$

Work Smart

Notice that the entire expression $-b \pm \sqrt{b^2 - 4ac}$ is in the numerator of the quadratic formula.

$a = 9, b = -12, c = 4:\quad m = \dfrac{-(-12) \pm \sqrt{(-12)^2 - 4(9)(4)}}{2(9)}$

$= \dfrac{12 \pm \sqrt{144 - 144}}{18}$

$$= \frac{12 \pm \sqrt{0}}{18}$$

$$= \frac{12}{18} = \frac{2}{3}$$

We leave it to you to verify the solution. The solution set is $\left\{\frac{2}{3}\right\}$. ●

In Example 3, there was one solution rather than two (as in Examples 1 and 2). The solution $x = \frac{2}{3}$ is called a **repeated root** because it occurs twice, once for $\frac{12 + 0}{18} = \frac{12}{18} = \frac{2}{3}$ and once for $\frac{12 - 0}{18} = \frac{12}{18} = \frac{2}{3}$. Notice that the value of $b^2 - 4ac$ equals 0, which is the reason for the repeated root. More will be said about this soon.

Quick ✓

In Problems 6 and 7, solve each equation.

6. $4w + \dfrac{25}{w} = 20$

7. $2x = 8 - \dfrac{3}{x}$

▶ **EXAMPLE 4**   **Solving a Quadratic Equation Using the Quadratic Formula**

Solve: $y^2 - 4y + 13 = 0$

Solution

$\left\{ -\dfrac{1}{2} - \dfrac{\sqrt{7}}{2}i, -\dfrac{1}{2} + \dfrac{\sqrt{7}}{2}i \right\}$

$$y^2 - 4y + 13 = 0$$

$$1y^2 - 4y + 13 = 0$$

$$y = \frac{-b \pm \sqrt{b^2 - 4ac}}{2a}$$

$a = 1, b = -4, c = 13:$ $y = \dfrac{-(-4) \pm \sqrt{(-4)^2 - 4(1)(13)}}{2(1)}$

$$= \frac{4 \pm \sqrt{16 - 52}}{2}$$

$\sqrt{-36} = 6i:$ $= \dfrac{4 \pm \sqrt{-36}}{2} = \dfrac{4 \pm 6i}{2}$

$\dfrac{a + b}{c} = \dfrac{a}{c} + \dfrac{b}{c}:$ $= \dfrac{4}{2} \pm \dfrac{6}{2}i = 2 \pm 3i$

$a \pm b$ means $a - b$ or $a + b:$ $y = 2 - 3i$ or $y = 2 + 3i$

We leave it to you to verify the solution. The solution set is $\{2 - 3i, 2 + 3i\}$. ●

Notice in Example 4 that the value of $b^2 - 4ac$ is negative and that the equation has two complex solutions that are not real.

Quick ✓

8. Solve: $z^2 + 2z + 26 = 0$

---

❷ **Use the Discriminant to Determine the Nature of Solutions of a Quadratic Equation**

Work Smart: Study Skills

Think of the *discriminant* as the expression that helps you *discriminate* what types of solutions are possible.

In the quadratic formula $x = \dfrac{-b \pm \sqrt{b^2 - 4ac}}{2a}$, the quantity $b^2 - 4ac$ is called the **discriminant** because its value tells us the number and the type of solutions to expect from the quadratic equation when $a, b,$ and c are rational numbers.

		The Discriminant and the Nature of the Solutions of a Quadratic Equation $ax^2 + bx + c = 0$, Where a, b, and c Are Rational Numbers		
Example	Value of Discriminant	Description of Discriminant	Number of Solutions	Type of Solution
1	169	Positive and a perfect square	Two	Rational
2	24	Positive and not a perfect square	Two	Irrational
3	0	Zero	One (repeated root)	Rational
4	−36	Negative	Two	Complex, nonreal

Work Smart

The rules in the box to the right apply only if the coefficients of the quadratic equation are rational numbers.

In Example 4, the solutions are complex conjugates. In general, for any quadratic equation of the form $ax^2 + bx + c = 0$, where a, b, and c are real numbers and $b^2 - 4ac < 0$, the equation has two complex solutions that are not real and are complex conjugates.

This result is a consequence of the quadratic formula. Suppose that $b^2 - 4ac = -N < 0$. Then, by the quadratic formula, the solutions are

$$x = \frac{-b \pm \sqrt{b^2 - 4ac}}{2a} = \frac{-b \pm \sqrt{-N}}{2a}$$

$$= \frac{-b \pm \sqrt{N}i}{2a} = \frac{-b}{2a} \pm \frac{\sqrt{N}}{2a}i$$

which are complex conjugates.

⊙ **EXAMPLE 5** **Determining the Nature of the Solutions of a Quadratic Equation**

For each quadratic equation, determine the discriminant. Use the value of the discriminant to determine whether the quadratic equation has two rational solutions, two irrational solutions, one repeated real solution, or two complex solutions that are not real.

(a) $x^2 - 5x + 2 = 0$ **(b)** $9y^2 + 6y + 1 = 0$ **(c)** $3p^2 - p = -5$

Solution

(a) Compare $x^2 - 5x + 2 = 0$ to the standard form $ax^2 + bx + c = 0$.

$$x^2 - 5x + 2 = 0$$

$a = 1$ $b = -5$ $c = 2$

Substitute the values for a, b, and c shown above into the discriminant formula $b^2 - 4ac$:

$$b^2 - 4ac = (-5)^2 - 4(1)(2) = 25 - 8 = 17$$

Because $b^2 - 4ac = 17$ is positive but not a perfect square, the quadratic equation has two irrational solutions.

(b) For the quadratic equation $9y^2 + 6y + 1 = 0$, $a = 9$, $b = 6$, and $c = 1$. Substituting these values into $b^2 - 4ac$ gives

$$b^2 - 4ac = 6^2 - 4(9)(1) = 36 - 36 = 0$$

Because $b^2 - 4ac = 0$, the quadratic equation has one repeated rational solution.

(c) Is the quadratic equation $3p^2 - p = -5$ in standard form? No! Add 5 to both sides of the equation and write the equation as $3p^2 - p + 5 = 0$. For $a = 3$, $b = -1$, and $c = 5$, the value of the discriminant, $b^2 - 4ac$, is

$$b^2 - 4ac = (-1)^2 - 4(3)(5) = 1 - 60 = -59$$

Because $-59 < 0$, the quadratic equation has two complex solutions that are not real and are conjugates.   ●

Quick ✓

9. In the quadratic formula, the quantity $b^2 - 4ac$ is called the _____ of the quadratic equation.

10. *True or False* The discriminant of the quadratic equation $4x^2 - 5x + 1 = 0$ is $\sqrt{25 - 4(4)(1)} = \sqrt{9} = 3$.

11. If the discriminant of a quadratic equation is _____, then the quadratic equation has two complex solutions that are not real.

12. *True or False* If the discriminant of a quadratic equation is zero, then the equation has no solution.

13. *True or False* The solutions to a quadratic equation in which the solutions are complex numbers that are not real are complex conjugates.

In Problems 14–16, use the value of the discriminant to determine whether the quadratic equation has two rational solutions, two irrational solutions, one repeated real solution, or two complex solutions that are not real.

14. $2z^2 + 5z + 4 = 0$ 15. $4y^2 + 12y = -9$ 16. $2x^2 - 4x + 1 = 0$

▶ Which Method Should I Use?

We now have four methods for solving quadratic equations:

1. Factoring

2. Square Root Property

3. Completing the Square

4. The Quadratic Formula

You may be asking yourself, "Which method should I use?" and "Does it matter which method I use?" The answer to the second question is that it does not matter which method you use, but one method may be more efficient than the others. Table 1 on the next page contains guidelines to help you solve any quadratic equation. The method that is most efficient for solving a quadratic equation depends on the equation. The value of the discriminant can be used as a guide in choosing the most efficient method.

Notice in Table 1, on the next page, that completing the square is not one of the methods recommended for use in solving a quadratic equation. This is because the quadratic formula is based on completing the square of $ax^2 + bx + c = 0$. Besides, completing the square may be a cumbersome task, whereas the quadratic formula is fairly straightforward. Learning how to complete the square was worth the time, however, because it is used to derive the quadratic formula. Also, completing the square is a skill that will be needed later in this course and in future math courses.

Table 1

Form of the Quadratic Equation	Most Efficient Method	Example
$x^2 = p$, where p is any real number	Square Root Property	$x^2 = 45$ Square Root Property: $x = \pm\sqrt{45}$ $= \pm 3\sqrt{5}$
$ax^2 + c = 0$	Square Root Property	$3p^2 + 12 = 0$ Subtract 12 from both sides: $3p^2 = -12$ Divide both sides by 3: $p^2 = -4$ Square Root Property: $p = \pm\sqrt{-4}$ $= \pm 2i$
$(ax + b)^2 = p$, where p is any real number	Square Root Property	$(2x + 7)^2 = 18$ Square Root Property: $2x + 7 = \pm 3\sqrt{2}$ Subtract 7 from both sides: $2x = -7 \pm 3\sqrt{2}$ Divide by 2: $x = \dfrac{-7 \pm 3\sqrt{2}}{2}$
$ax^2 + bx + c = 0$, where $b^2 - 4ac$ is a perfect square	Factoring or the quadratic formula. Factor if the quadratic expression is easy to factor. Otherwise, use the quadratic formula.	$a = 2, b = 1, c = -10$: $\quad 2m^2 + m - 10 = 0$ $b^2 - 4ac = 1^2 - 4(2)(-10) = 1 + 80 = 81$ 81 is a perfect square, so use factoring: $2m^2 + m - 10 = 0$ $(2m + 5)(m - 2) = 0$ $2m + 5 = 0 \quad \text{or} \quad m - 2 = 0$ $m = -\dfrac{5}{2} \quad \text{or} \quad m = 2$
$ax^2 + bx + c = 0$, where $b^2 - 4ac$ is not a perfect square	Quadratic formula	$a = 2, b = 4, c = -1$: $\quad 2x^2 + 4x - 1 = 0$ $b^2 - 4ac = 4^2 - 4(2)(-1) = 16 + 8 = 24$ 24 is not a perfect square, so use the quadratic formula (because it's easier than completing the square): $x = \dfrac{-b \pm \sqrt{b^2 - 4ac}}{2a}$ $= \dfrac{-4 \pm \sqrt{24}}{2(2)}$ $= \dfrac{-4 \pm 2\sqrt{6}}{4}$ $= -1 \pm \dfrac{\sqrt{6}}{2}$ $x = -1 - \dfrac{\sqrt{6}}{2} \text{ or } x = -1 + \dfrac{\sqrt{6}}{2}$
$ax^2 + bx + c = 0$, where $b^2 - 4ac$ is negative	Quadratic formula	$a = 3, b = -4, c = 7$: $\quad 3x^2 - 4x + 7 = 0$ $b^2 - 4ac = (-4)^2 - 4(3)(7) = 16 - 84 = -68$ The discriminant is negative, so use the quadratic formula: $x = \dfrac{-b \pm \sqrt{b^2 - 4ac}}{2a}$ $x = \dfrac{-(-4) \pm \sqrt{-68}}{2(3)}$ $= \dfrac{4 \pm \sqrt{-1 \cdot 4 \cdot 17}}{6}$ $= \dfrac{4 \pm 2\sqrt{17}i}{6}$ $= \dfrac{2}{3} \pm \dfrac{1}{3}\sqrt{17}i$ $x = \dfrac{2}{3} - \dfrac{1}{3}\sqrt{17}i \text{ or } x = \dfrac{2}{3} + \dfrac{1}{3}\sqrt{17}i$

Quick ✓

17. *True or False* If the discriminant of a quadratic equation is a perfect square, then the equation can be solved by factoring.

In Problems 18–20, solve each quadratic equation using any appropriate method.

18. $5n^2 - 45 = 0$ **19.** $-2y^2 + 5y - 6 = 0$ **20.** $3w^2 + 2w = 5$

▶ ❸ Model and Solve Problems Involving Quadratic Equations

Many applied problems, such as the one below, require solving quadratic equations. As always, use the problem-solving strategy from Section 1.2.

EXAMPLE 6 **Revenue**

The revenue R received by a company selling x specialty T-shirts per week is given by the function $R(x) = -0.005x^2 + 30x$.

(a) How many T-shirts must be sold in order for revenue to be $25,000 per week?

(b) How many T-shirts must be sold in order for revenue to be $45,000 per week?

Solution

(a) Step 1: Identify Determine the number x of T-shirts required so that $R = \$25{,}000$.

Step 2: Name x represents the number of T-shirts sold.

Step 3: Translate The function $R(x) = 25{,}000$ can be used to solve for the number of T-shirts, x.

$$R(x) = 25{,}000$$
$$-0.005x^2 + 30x = 25{,}000$$
$$-0.005x^2 + 30x - 25{,}000 = 0$$

Step 4: Solve Letting $a = -0.005$, $b = 30$, and $c = -25{,}000$, we have
$$b^2 - 4ac = 30^2 - 4(-0.005)(-25{,}000)$$
$$= 400$$

Because 400 is a perfect square, solve the equation by factoring or using the quadratic formula. It is not obvious how to factor $-0.005x^2 + 30x - 25000$, so use the quadratic formula:

$$\overset{\overset{\textstyle b^2 - 4ac = 400}{\downarrow}}{x = \frac{-30 \pm \sqrt{400}}{2(-0.005)}}$$

$$= \frac{-30 \pm 20}{-0.01}$$

$$x = \frac{-30 - 20}{-0.01} = 5000 \quad \text{or} \quad x = \frac{-30 + 20}{-0.01} = 1000$$

Step 5: Check If 1000 T-shirts are sold, then revenue is $R(1000) = -0.005(1000)^2 + 30(1000) = \$25{,}000$. If 5000 T-shirts are sold, then revenue is $R(5000) = -0.005(5000)^2 + 30(5000) = \$25{,}000$.

Step 6: Answer The company needs to sell either 1000 or 5000 T-shirts each week to earn $25,000 in revenue.

(b) Step 1: Identify Determine the number x of T-shirts required so that $R = \$45{,}000$. That is, solve the equation $R(x) = 45{,}000$.

Step 2: Name x represents the number of T-shirts sold.

Step 3: Translate The function $R(x) = 45{,}000$ can be used to solve for the number of T-shirts, x.

$$R(x) = 45{,}000$$
$$-0.005x^2 + 30x = 45{,}000$$
$$-0.005x^2 + 30x - 45{,}000 = 0$$

Step 4: Solve Letting $a = -0.005$, $b = 30$, and $c = -45{,}000$, we have

$$b^2 - 4ac = 30^2 - 4(-0.005)(-45{,}000) = 0$$

Because the discriminant is 0, the quadratic equation has a single rational solution. Solve the equation by factoring or using the quadratic formula. It is not obvious how to factor $-0.005x^2 + 30x - 45{,}000$, so use the quadratic formula to solve the equation.

$$x = \frac{-30 \pm \sqrt{0}}{2(-0.005)} \qquad b^2 - 4ac = 0$$
$$= \frac{-30 \pm 0}{-0.01}$$
$$= \frac{-30}{-0.01} = 3000$$

Step 5: Check If 3000 T-shirts are sold, then revenue is $R(3000) = -0.005(3000)^2 + 30(3000) = \$45{,}000$.

Step 6: Answer The company needs to sell 3000 T-shirts each week to earn $45,000 in revenue. ●

Quick ✓

21. The revenue R received by a Redbox kiosk renting x DVDs per day is given by the function $R(x) = -0.005x^2 + 4x$.

(a) How many DVDs must be rented in order for revenue to be $600 per day?

(b) How many DVDs must be rented in order for revenue to be $800 per day?

(**EXAMPLE 7**)   **Designing a Window**

A window designer wishes to design a window so that the diagonal is 20 feet. In addition, the width of the window needs to be 4 feet more than the height. What are the dimensions of the window?

Solution

Step 1: Identify Determine the dimensions of the window. That is, find the width and height of the window.

Figure 6

Step 2: Name Let h represent the height of the window in feet. Because the width is 4 feet more than the height, $h + 4$ is the width of the window.

Step 3: Translate Figure 6 illustrates the situation. Notice that the two sides and the diagonal form a right triangle. Express the relationship among the sides using the Pythagorean Theorem.

$$\text{leg}^2 + \text{leg}^2 = \text{hypotenuse}^2: \qquad h^2 + (h+4)^2 = 20^2$$
$$(A+B)^2 = A^2 + 2AB + B^2: \qquad h^2 + h^2 + 8h + 16 = 400$$

Combine like terms: $2h^2 + 8h + 16 = 400$

Subtract 400 from both sides: $2h^2 + 8h - 384 = 0$

Divide both sides by 2: $h^2 + 4h - 192 = 0$

Step 4: Solve In the model, $a = 1$, $b = 4$, and $c = -192$. The discriminant is $b^2 - 4ac = 4^2 - 4(1)(-192) = 784 = 28^2$. Thus solve the equation by factoring.

$$h^2 + 4h - 192 = 0$$
$$(h + 16)(h - 12) = 0$$
$$h + 16 = 0 \quad \text{or} \quad h - 12 = 0$$
$$h = -16 \quad \text{or} \quad h = 12$$

Step 5: Check Disregard the solution $h = -16$ because h represents the height of the window. Determine whether a window whose dimensions are 12 feet by $12 + 4 = 16$ feet has a diagonal that is 20 feet by verifying that $12^2 + 16^2 = 20^2$.

$$12^2 + 16^2 \stackrel{?}{=} 20^2$$
$$144 + 256 \stackrel{?}{=} 400$$
$$400 = 400$$

Step 6: Answer The dimensions of the window are 12 feet by 16 feet.

Quick ✓

22. A rectangular plot of land is designed so that its length is 14 meters more than its width. The diagonal of the land is known to be 34 meters. What are the dimensions of the land?

7.2 Exercises MyMathLab® Exercise numbers in green have complete video solutions in MyMathLab or may be accessed using the QR code to the right.

*Problems **1–22** are the Quick ✓s that follow the EXAMPLES.*

Building Skills

In Problems 23–40, solve each equation using the quadratic formula. See Objective 1.

23. $x^2 - 4x - 12 = 0$

24. $p^2 - 4p - 32 = 0$

25. $6y^2 - y - 15 = 0$

26. $10x^2 + x - 2 = 0$

27. $4m^2 - 8m + 1 = 0$

28. $2q^2 - 4q + 1 = 0$

29. $3w - 6 = \dfrac{1}{w}$

30. $x + \dfrac{1}{x} = 3$

31. $3p^2 = -2p + 4$

32. $5w^2 = -3w + 1$

33. $x^2 - 2x + 7 = 0$

34. $y^2 - 4y + 5 = 0$

35. $2z^2 + 7 = 2z$

36. $2z^2 + 7 = 4z$

37. $4x^2 = 2x + 1$

38. $6p^2 = 4p + 1$

39. $1 = 3q^2 + 4q$

40. $1 = 5w^2 + 6w$

In Problems 41–50, determine the discriminant of each quadratic equation. Use the value of the discriminant to determine whether the quadratic equation has two rational solutions, two irrational solutions, one repeated rational solution, or two complex solutions that are not real. See Objective 2.

41. $x^2 - 5x + 1 = 0$

42. $p^2 + 4p - 2 = 0$

43. $3z^2 + 2z + 5 = 0$

44. $2y^2 - 3y + 5 = 0$

45. $9q^2 - 6q + 1 = 0$

46. $16x^2 + 24x + 9 = 0$

47. $3w^2 = 4w - 2$ **48.** $6x^2 - x = -4$

49. $6x = 2x^2 - 1$ **50.** $10w^2 = 3$

Mixed Practice

In Problems 51–76, solve each equation.

51. $w^2 - 5w + 5 = 0$ **52.** $q^2 - 7q + 7 = 0$

53. $3x^2 + 5x = 8$ **54.** $4p^2 + 5p = 9$

55. $2x^2 = 3x + 35$ **56.** $3x^2 + 5x = 2$

57. $q^2 + 2q + 8 = 0$ **58.** $w^2 + 4w + 9 = 0$

59. $2z^2 = 2(z + 3)^2$ **60.** $3z^2 = 3(z + 1)(z - 2)$

61. $7q - 2 = \dfrac{4}{q}$ **62.** $5m - 4 = \dfrac{5}{m}$

63. $5a^2 - 80 = 0$ **64.** $4p^2 - 100 = 0$

65. $8n^2 + 1 = 4n$ **66.** $4q^2 + 1 = 2q$

67. $27x^2 + 36x + 12 = 0$ **68.** $8p^2 - 40p + 50 = 0$

69. $\dfrac{1}{3}x^2 + \dfrac{2}{9}x - 1 = 0$ **70.** $\dfrac{1}{2}x^2 + \dfrac{3}{4}x - 1 = 0$

71. $(x - 5)(x + 1) = 4$ **72.** $(a - 3)(a + 1) = 2$

73. $\dfrac{x - 2}{x + 2} = x - 3$ **74.** $\dfrac{x - 5}{x + 3} = x - 3$

75. $\dfrac{x - 4}{x^2 + 2} = 2$ **76.** $\dfrac{x - 1}{x^2 + 4} = 1$

77. Suppose that $f(x) = x^2 + 4x - 21$.
 (a) Solve $f(x) = 0$ for x.
 (b) Solve $f(x) = -21$ for x. What points are on the graph of f?

78. Suppose that $f(x) = x^2 + 2x - 8$.
 (a) Solve $f(x) = 0$ for x.
 (b) Solve $f(x) = -8$ for x. What points are on the graph of f?

79. Suppose that $H(x) = -2x^2 - 4x + 1$.
 (a) Solve $H(x) = 0$ for x.
 (b) Solve $H(x) = 2$ for x.

80. Suppose that $g(x) = 3x^2 + x - 1$.
 (a) Solve $g(x) = 0$ for x.
 (b) Solve $g(x) = 4$ for x.

81. What are the zeros of $G(x) = 3x^2 + 2x - 2$?

82. What are the zeros of $F(x) = x^2 + 3x - 3$?

Applying the Concepts

In Problems 83–86, use the Pythagorean Theorem to determine the value of x and the measurements of all sides of the right triangle.

△ **83.**

△ **84.**

△ **85.**

△ **86.**

△ **87. Area** The area of a rectangle is 40 square inches. The width of the rectangle is 4 inches more than the length. What are the dimensions of the rectangle?

△ **88. Area** The area of a rectangle is 60 square inches. The width of the rectangle is 6 inches more than the length. What are the dimensions of the rectangle?

△ **89. Area** The area of a triangle is 25 square inches. The height of the triangle is 3 inches less than the base. What are the base and height of the triangle?

△ **90. Area** The area of a triangle is 35 square inches. The height of the triangle is 2 inches less than the base. What are the base and height of the triangle?

91. Revenue The revenue R received by a company selling x pairs of sunglasses per week is given by the function $R(x) = -0.1x^2 + 70x$.

(a) Find and interpret the values of $R(17)$ and $R(25)$.

(b) How many pairs of sunglasses must be sold in order for revenue to be $10,000 per week?

(c) How many pairs of sunglasses must be sold in order for revenue to be $12,250 per week?

92. Revenue The revenue R received by a company selling x "all-day passes" to a small amusement park per day is given by the function $R(x) = -0.02x^2 + 24x$.

(a) Find and interpret the values of $R(300)$ and $R(800)$.

(b) How many tickets must be sold in order for revenue to be $4000 per day?

(c) How many tickets must be sold in order for revenue to be $7200 per day?

93. Projectile Motion The height s of a ball after t seconds, when thrown straight up with an initial speed of 70 feet per second from an initial height of 5 feet, can be modeled by the function

$$s(t) = -16t^2 + 70t + 5$$

(a) When will the height of the ball be 40 feet? Round your answer to the nearest tenth of a second.

(b) When will the height of the ball be 70 feet? Round your answer to the nearest tenth of a second.

(c) Will the ball ever reach a height of 150 feet? How does the result of the equation tell you this?

94. Projectile Motion The height s of a toy rocket after t seconds, when fired straight up with an initial speed of 150 feet per second from an initial height of 2 feet, can be modeled by the function

$$s(t) = -16t^2 + 150t + 2$$

(a) When will the height of the rocket be 200 feet? Round your answer to the nearest tenth of a second.

(b) When will the height of the rocket be 300 feet? Round your answer to the nearest tenth of a second.

(c) Will the rocket ever reach a height of 500 feet?

△ **95. Similar Triangles** Consult the figure. Suppose that $\triangle ABC$ is similar to $\triangle DEC$. The length of $\overline{BC}$ is 24 inches and the length of $\overline{DE}$ is 6 inches. If the length of $\overline{AB}$ equals the length of $\overline{CE}$, which we call x, find x.

96. Number Sense Three times the square of a number equals the sum of two times the number and 5. Find the number(s).

97. Life Cycle Hypothesis The Life Cycle Hypothesis from economics was presented by Franco Modigliani in 1954. One of its components states that income is a function of age. The function $I(a) = -55a^2 + 5119a - 54,448$ represents the relation between average annual income I and age a.

(a) For what age does average income I equal $40,000? Round your answer to the nearest year.

(b) For what age does average income I equal $50,000? Round your answer to the nearest year.

98. Population The function $P(a) = -0.001a^2 + 0.0573a + 3.6941$ represents the population (in millions) of Americans in 2014, P, that are a years of age. (SOURCE: *United States Census Bureau*)

(a) For what age(s) was the population 4 million in 2014? Round your answer to the nearest year.

(b) For what age(s) was the population 3 million in 2014? Round your answer to the nearest year.

99. Upstream and Back Zene decides to canoe 4 miles upstream on a river to a waterfall and then canoe back. The total trip (excluding the time spent at the waterfall) takes 6 hours. Zene knows she can canoe at an average speed of 5 miles per hour in still water. What is the speed of the current?

100. Round Trip A Cessna aircraft flies 200 miles due west into the jet stream and flies back home on the same route. The total time of the trip (excluding the time on the ground) is 4 hours. The Cessna aircraft can fly 120 miles per hour in still air. What is the net effect of the jet stream on the aircraft?

101. Work Robert and Susan have a newspaper route. When they work the route together, it takes 2 hours to deliver all the newspapers. One morning Robert tells Susan he is too sick to deliver the papers. Susan doesn't remember how long it takes her to deliver the newspapers working alone, but she does remember that Robert can finish the route one hour sooner than she can when working alone. How long will it take Susan to finish the route?

102. Work Demitrius needs to fill his pool. When he rents a water tanker to fill the pool with the help of the hose from his house, it takes 5 hours to fill the pool. This year, money is tight and he can't afford to rent the water tanker to fill the pool. He doesn't remember how long it takes for his house hose to fill the pool, but he does remember that the tanker hose filling the pool alone can finish the job in 8 fewer hours than using his house hose alone. How long will it take Demitrius to fill his pool using only his house hose?

Extending the Concepts

103. Show that the sum of the solutions to a quadratic equation is $-\dfrac{b}{a}$.

104. Show that the product of the solutions to a quadratic equation is $\dfrac{c}{a}$.

105. Show that the real solutions to the equation $ax^2 + bx + c = 0$ are the negatives of the real solutions to the equation $ax^2 - bx + c = 0$. Assume that $b^2 - 4ac \geq 0$.

106. Show that the real solutions to the equation $ax^2 + bx + c = 0$ are the reciprocals of the real solutions to the equation $cx^2 + bx + a = 0$. Assume that $b^2 - 4ac \geq 0$.

Explaining the Concepts

107. Explain the circumstances for which you would use factoring to solve a quadratic equation.

108. Explain the circumstances for which you would use the Square Root Property to solve a quadratic equation.

Synthesis Review

109. (a) Graph $f(x) = x^2 + 3x + 2$ by plotting points.
(b) Solve the equation $x^2 + 3x + 2 = 0$.

(c) Compare the solutions to the equation in part (b) to the x-intercepts of the graph drawn in part (a). What do you notice?

110. (a) Graph $f(x) = x^2 - x - 6$ by plotting points.
(b) Solve the equation $x^2 - x - 6 = 0$.
(c) Compare the solutions to the equation in part (b) to the x-intercepts of the graph drawn in part (a). What do you notice?

111. (a) Graph $g(x) = x^2 - 2x + 1$ by plotting points.
(b) Solve the equation $x^2 - 2x + 1 = 0$.
(c) Compare the solutions to the equation in part (b) to the x-intercepts of the graph drawn in part (a). What do you notice?

112. (a) Graph $g(x) = x^2 + 4x + 4$ by plotting points.
(b) Solve the equation $x^2 + 4x + 4 = 0$.
(c) Compare the solutions to the equation in part (b) to the x-intercepts of the graph drawn in part (a). What do you notice?

Technology Exercises

In Problems 113–116, the graph of the quadratic function f is given. For each function, determine the discriminant of the equation $f(x) = 0$ in order to determine the nature of the solutions of the equation $f(x) = 0$. Compare the nature of solutions based on the discriminant to the graph of the function.

113. $f(x) = x^2 - 7x + 3$ **114.** $f(x) = -x^2 - 3x + 1$

115. $f(x) = -x^2 - 3x - 4$ **116.** $f(x) = x^2 - 6x + 9$

117. (a) Solve the equation $x^2 - 5x - 24 = 0$ algebraically.
(b) Graph $Y_1 = x^2 - 5x - 24$. Compare the x-intercepts of the graph to the solutions found in part (a).

118. (a) Solve the equation $x^2 - 4x - 45 = 0$ algebraically.
(b) Graph $Y_1 = x^2 - 4x - 45$. Compare the x-intercepts of the graph to the solutions found in part (a).

119. (a) Solve the equation $x^2 - 6x + 9 = 0$ algebraically.
  (b) Graph $Y_1 = x^2 - 6x + 9$. Compare the x-intercepts of the graph to the solutions found in part (a).

120. (a) Solve the equation $x^2 + 10x + 25 = 0$ algebraically.
  (b) Graph $Y_1 = x^2 + 10x + 25$. Compare the x-intercepts of the graph to the solutions found in part (a).

121. (a) Solve the equation $x^2 + 5x + 8 = 0$ algebraically.
  (b) Graph $Y_1 = x^2 + 5x + 8$. How is the result of part (a) related to the graph?

122. (a) Solve the equation $x^2 + 2x + 5 = 0$ algebraically.
  (b) Graph $Y_1 = x^2 + 2x + 5$. How is the result of part (a) related to the graph?

7.3 Solving Equations Quadratic in Form

Objective

① Solve Equations That Are Quadratic in Form

Are You Prepared for This Section?

Before getting started, complete the following problems. If you get a problem wrong, go back to the section cited and review the material.

P1. Factor: $x^4 - 5x^2 - 6$ [Section 4.5, p. 356]

P2. Factor: $2(p + 3)^2 + 3(p + 3) - 5$ [Section 4.5, p. 356]

P3. Simplify: **(a)** $(x^2)^2$ **(b)** $(p^{-1})^2$ [Getting Ready, pp. 299–300]

▶ ① Solve Equations That Are Quadratic in Form

Consider the equation $x^4 - 4x^2 - 12 = 0$. Even though this equation is not in the form of a quadratic equation, $ax^2 + bx + c = 0$, the equation can be written as $(x^2)^2 - 4x^2 - 12 = 0$. Let $u = x^2$ in the equation, to get $u^2 - 4u - 12 = 0$, which is of the form $ax^2 + bx + c = 0$. Now solve $u^2 - 4u - 12 = 0$ for u by factoring. Then, using the fact that $u = x^2$, find x, which was the goal in the first place.

In general, if a substitution u transforms an equation into one of the form

$$au^2 + bu + c = 0$$

then the original equation is an **equation quadratic in form.**

It is often hard to tell whether an equation is quadratic in form. Table 2 shows some equations that are quadratic in form and the appropriate substitution.

Prepared? ... Answers
P1. $(x^2 - 6)(x^2 + 1)$
P2. $(2p + 11)(p + 2)$
P3. (a) x^4 **(b)** $p^{-2} = \dfrac{1}{p^2}$

Table 2

Original Equation	Substitution	Equation with Substitution
$2x^4 - 3x^2 + 5 = 0$ $2(x^2)^2 - 3x^2 + 5 = 0$	$u = x^2$	$2u^2 - 3u + 5 = 0$
$3(z - 5)^2 + 4(z - 5) + 1 = 0$	$u = z - 5$	$3u^2 + 4u + 1 = 0$
$-2y + 5\sqrt{y} - 2 = 0$ $-2(\sqrt{y})^2 + 5\sqrt{y} - 2 = 0$	$u = \sqrt{y}$	$-2u^2 + 5u - 2 = 0$

EXAMPLE 1 How to Solve Equations That Are Quadratic in Form

Solve: $x^4 + x^2 - 12 = 0$

Step-by-Step Solution

Step 1: Determine the appropriate substitution and write the equation in the form $au^2 + bu + c = 0$.

$$x^4 + x^2 - 12 = 0$$
$$(x^2)^2 + x^2 - 12 = 0$$
Let $u = x^2$: $\qquad u^2 + u - 12 = 0$

(continued)

Step 2: Solve the equation $au^2 + bu + c = 0$.

$$(u + 4)(u - 3) = 0$$
$$u + 4 = 0 \quad \text{or} \quad u - 3 = 0$$
$$u = -4 \quad \text{or} \quad u = 3$$

Step 3: Rewrite the solution in terms of the original variable.

We want to know x, so replace u with x^2: $\quad x^2 = -4 \quad \text{or} \quad x^2 = 3$

Square Root Property: $\quad x = \pm\sqrt{-4} \quad \text{or} \quad x = \pm\sqrt{3}$
$$= \pm 2i$$

Step 4: Check Verify your solutions.

$x = 2i$:
$$(2i)^4 + (2i)^2 - 12 \overset{?}{=} 0$$
$$2^4 i^4 + 2^2 i^2 - 12 \overset{?}{=} 0$$
$$16(1) + 4(-1) - 12 \overset{?}{=} 0$$
$$16 - 4 - 12 \overset{?}{=} 0$$
$$0 = 0 \quad \text{True}$$

$x = -2i$:
$$(-2i)^4 + (-2i)^2 - 12 \overset{?}{=} 0$$
$$(-2)^4 i^4 + (-2)^2 i^2 - 12 \overset{?}{=} 0$$
$$16(1) + 4(-1) - 12 \overset{?}{=} 0$$
$$16 - 4 - 12 \overset{?}{=} 0$$
$$0 = 0 \quad \text{True}$$

$x = \sqrt{3}$:
$$(\sqrt{3})^4 + (\sqrt{3})^2 - 12 \overset{?}{=} 0$$
$$\sqrt{3^4} + \sqrt{3^2} - 12 \overset{?}{=} 0$$
$$\sqrt{81} + 3 - 12 \overset{?}{=} 0$$
$$9 + 3 - 12 \overset{?}{=} 0$$
$$0 = 0 \quad \text{True}$$

$x = -\sqrt{3}$:
$$(-\sqrt{3})^4 + (-\sqrt{3})^2 - 12 \overset{?}{=} 0$$
$$(-1)^4 \sqrt{3^4} + (-1)^2 \sqrt{3^2} - 12 \overset{?}{=} 0$$
$$\sqrt{81} + 3 - 12 \overset{?}{=} 0$$
$$9 + 3 - 12 \overset{?}{=} 0$$
$$0 = 0 \quad \text{True}$$

The solution set is $\{2i, -2i, \sqrt{3}, -\sqrt{3}\}$.

The steps used to solve an equation that is quadratic in form are summarized below.

> **Solving Equations Quadratic in Form**
>
> **Step 1:** Determine the appropriate substitution and write the equation in the form $au^2 + bu + c = 0$.
> **Step 2:** Solve the equation $au^2 + bu + c = 0$.
> **Step 3:** Rewrite the solution in terms of the original variable.
> **Step 4:** Verify your solution(s).

Quick ✓

1. If a substitution u transforms an equation into one of the form $au^2 + bu + c = 0$, then the original equation is called an equation _____ __ ____.

2. For the equation $2(3x + 1)^2 - 5(3x + 1) + 2 = 0$, an appropriate substitution would be $u = $ _____.

3. *True or False* The equation $3\left(\dfrac{x}{x-2}\right)^2 - \dfrac{5x}{x-2} + 3 = 0$ is quadratic in form.

4. What is the appropriate choice for u when solving the equation $2 \cdot \dfrac{1}{x^2} - 6 \cdot \dfrac{1}{x} + 3 = 0$?

In Problems 5 and 6, solve each equation.

5. $x^4 - 13x^2 + 36 = 0$

6. $p^4 - 7p^2 = 18$

EXAMPLE 2

Solving Equations That Are Quadratic in Form

Solve: $(z^2 - 5)^2 - 3(z^2 - 5) - 4 = 0$

Solution

$$(z^2 - 5)^2 - 3(z^2 - 5) - 4 = 0$$

Let $u = z^2 - 5$: $\qquad u^2 - 3u - 4 = 0$

$$(u - 4)(u + 1) = 0$$

$u - 4 = 0 \qquad$ or $\quad u + 1 = 0$

$u = 4 \qquad$ or $\qquad u = -1$

Replace u with $z^2 - 5$ and solve for z: $\quad z^2 - 5 = 4 \quad$ or $\quad z^2 - 5 = -1$

$z^2 = 9 \qquad$ or $\qquad z^2 = 4$

Square Root Property: $\qquad z = \pm 3 \quad$ or $\qquad z = \pm 2$

Check

$z = -3$:

$((-3)^2 - 5)^2 - 3((-3)^2 - 5) - 4 \overset{?}{=} 0$

$(9 - 5)^2 - 3(9 - 5) - 4 \overset{?}{=} 0$

$4^2 - 3(4) - 4 \overset{?}{=} 0$

$16 - 12 - 4 \overset{?}{=} 0$

$0 = 0 \quad$ True

$z = 3$:

$((3)^2 - 5)^2 - 3((3)^2 - 5) - 4 \overset{?}{=} 0$

$(9 - 5)^2 - 3(9 - 5) - 4 \overset{?}{=} 0$

$4^2 - 3(4) - 4 \overset{?}{=} 0$

$16 - 12 - 4 \overset{?}{=} 0$

$0 = 0 \quad$ True

$z = -2$:

$((-2)^2 - 5)^2 - 3((-2)^2 - 5) - 4 \overset{?}{=} 0$

$(4 - 5)^2 - 3(4 - 5) - 4 \overset{?}{=} 0$

$(-1)^2 - 3(-1) - 4 \overset{?}{=} 0$

$1 + 3 - 4 \overset{?}{=} 0$

$0 = 0 \quad$ True

$z = 2$:

$((2)^2 - 5)^2 - 3((2)^2 - 5) - 4 \overset{?}{=} 0$

$(4 - 5)^2 - 3(4 - 5) - 4 \overset{?}{=} 0$

$(-1)^2 - 3(-1) - 4 \overset{?}{=} 0$

$1 + 3 - 4 \overset{?}{=} 0$

$0 = 0 \quad$ True

The solution set is $\{-3, -2, 2, 3\}$. ●

Quick ✓

In Problems 7 and 8, solve each equation.

7. $(p^2 - 2)^2 - 9(p^2 - 2) + 14 = 0$ **8.** $2(2z^2 - 1)^2 + 5(2z^2 - 1) - 3 = 0$

When both sides of an equation are raised to an even power (such as in squaring both sides of the equation), extraneous solutions may be introduced to the equation. Therefore, it is important that to verify all solutions.

EXAMPLE 3

Solving Equations That Are Quadratic in Form

Solve: $3x - 5\sqrt{x} - 2 = 0$

Solution

$$3x - 5\sqrt{x} - 2 = 0$$

$$3(\sqrt{x})^2 - 5\sqrt{x} - 2 = 0$$

Let $u = \sqrt{x}$: $\qquad 3u^2 - 5u - 2 = 0$

$$(3u + 1)(u - 2) = 0$$

(continued)

$$3u + 1 = 0 \quad \text{or} \quad u - 2 = 0$$
$$3u = -1 \quad \text{or} \quad u = 2$$
$$u = -\frac{1}{3}$$

Replace u with $\sqrt{x}$ and solve for x: $\quad \sqrt{x} = -\frac{1}{3} \quad \text{or} \quad \sqrt{x} = 2$

Square both sides: $\quad x = \frac{1}{9} \quad \text{or} \quad x = 4$

Check

$x = \frac{1}{9}$: $\quad 3 \cdot \frac{1}{9} - 5\sqrt{\frac{1}{9}} - 2 = 0$ $\qquad\qquad$ $x = 4$: $\quad 3 \cdot 4 - 5\sqrt{4} - 2 = 0$

$\qquad\qquad \frac{1}{3} - 5 \cdot \frac{1}{3} - 2 = 0$ $\qquad\qquad\qquad\qquad\qquad\quad 12 - 5 \cdot 2 - 2 = 0$

$\qquad\qquad\quad \frac{1}{3} - \frac{5}{3} - \frac{6}{3} = 0$ $\qquad\qquad\qquad\qquad\qquad\qquad 12 - 10 - 2 = 0$

$\qquad\qquad\qquad\qquad -\frac{10}{3} = 0 \quad \text{False}$ $\qquad\qquad\qquad\qquad\qquad\qquad\qquad 0 = 0 \quad \text{True}$

The solution $x = \frac{1}{9}$ is extraneous. The solution set is $\{4\}$. $\qquad\qquad\qquad$ ●

The equation in Example 3 could also have been solved using the methods introduced in Section 6.7 by isolating the radical and squaring both sides.

Quick ✓

In Problems 9 and 10, solve the equation.

9. $3w - 14\sqrt{w} + 8 = 0$ $\qquad\qquad$ **10.** $2q - 9\sqrt{q} - 5 = 0$

EXAMPLE 4 **Solving Equations That Are Quadratic in Form**

Solve: $4x^{-2} + 13x^{-1} - 12 = 0$

Solution

$$4x^{-2} + 13x^{-1} - 12 = 0$$
$$4(x^{-1})^2 + 13x^{-1} - 12 = 0$$

Let $u = x^{-1}$: $\quad 4u^2 + 13u - 12 = 0$

Factor: $\quad (4u - 3)(u + 4) = 0$

$\qquad\qquad 4u - 3 = 0 \quad \text{or} \quad u + 4 = 0$

$\qquad\qquad\quad 4u = 3 \quad \text{or} \quad u = -4$

$\qquad\qquad\qquad\quad u = \frac{3}{4}$

Replace u with x^{-1} and solve for x: $\quad x^{-1} = \frac{3}{4} \quad \text{or} \quad x^{-1} = -4$

$x^{-1} = \frac{1}{x}$: $\qquad\qquad\qquad\qquad\qquad \frac{1}{x} = \frac{3}{4} \quad \text{or} \quad \frac{1}{x} = -4$

Take the reciprocal of both sides of the equation: $\qquad\qquad x = \frac{4}{3} \quad \text{or} \quad x = \frac{1}{-4} = -\frac{1}{4}$

Check

$$x = \frac{4}{3}:$$

$$4\left(\frac{4}{3}\right)^{-2} + 13\left(\frac{4}{3}\right)^{-1} - 12 \stackrel{?}{=} 0$$

$$4\left(\frac{3}{4}\right)^{2} + 13 \cdot \frac{3}{4} - 12 \stackrel{?}{=} 0$$

$$4 \cdot \frac{9}{16} + \frac{39}{4} - 12 \stackrel{?}{=} 0$$

$$\frac{9}{4} + \frac{39}{4} - \frac{48}{4} \stackrel{?}{=} 0$$

$$0 = 0 \quad \text{True}$$

$$x = -\frac{1}{4}:$$

$$4\left(-\frac{1}{4}\right)^{-2} + 13\left(-\frac{1}{4}\right)^{-1} - 12 \stackrel{?}{=} 0$$

$$4(-4)^{2} + 13(-4) - 12 \stackrel{?}{=} 0$$

$$4 \cdot 16 - 52 - 12 \stackrel{?}{=} 0$$

$$64 - 52 - 12 \stackrel{?}{=} 0$$

$$0 = 0 \quad \text{True}$$

The solution set is $\left\{ -\frac{1}{4}, \frac{4}{3} \right\}$.

●

Quick ✓

In Problem 11, solve the equation.

11. $5x^{-2} + 12x^{-1} + 4 = 0$

▶ **EXAMPLE 5** **Solving Equations That Are Quadratic in Form**

Solve: $a^{\frac{2}{3}} + 3a^{\frac{1}{3}} - 28 = 0$

Solution

$$a^{\frac{2}{3}} + 3a^{\frac{1}{3}} - 28 = 0$$

$$\left(a^{\frac{1}{3}}\right)^{2} + 3a^{\frac{1}{3}} - 28 = 0$$

Let $u = a^{\frac{1}{3}}$: $\qquad u^{2} + 3u - 28 = 0$

$$(u + 7)(u - 4) = 0$$

$$u + 7 = 0 \qquad \text{or} \quad u - 4 = 0$$

$$u = -7 \qquad \text{or} \qquad u = 4$$

Replace u with $a^{\frac{1}{3}}$ and solve for a: $\qquad a^{\frac{1}{3}} = -7 \qquad \text{or} \qquad a^{\frac{1}{3}} = 4$

Cube both sides of the equation: $\qquad \left(a^{\frac{1}{3}}\right)^{3} = (-7)^{3} \quad \text{or} \quad \left(a^{\frac{1}{3}}\right)^{3} = 4^{3}$

$$a = -343 \quad \text{or} \qquad a = 64$$

Check $a = -343$:

$$(-343)^{\frac{2}{3}} + 3(-343)^{\frac{1}{3}} - 28 = 0$$

$$\left(\sqrt[3]{-343}\right)^{2} + 3 \cdot \sqrt[3]{-343} - 28 \stackrel{?}{=} 0$$

$$(-7)^{2} + 3(-7) - 28 \stackrel{?}{=} 0$$

$$49 - 21 - 28 \stackrel{?}{=} 0$$

$$0 = 0 \quad \text{True}$$

$a = 64$:

$$(64)^{\frac{2}{3}} + 3(64)^{\frac{1}{3}} - 28 = 0$$

$$\left(\sqrt[3]{64}\right)^{2} + 3 \cdot \sqrt[3]{64} - 28 \stackrel{?}{=} 0$$

$$4^{2} + 3 \cdot 4 - 28 \stackrel{?}{=} 0$$

$$16 + 12 - 28 \stackrel{?}{=} 0$$

$$0 = 0 \quad \text{True}$$

The solution set is $\{-343, 64\}$.

●

Quick ✓

In Problem 12, solve the equation.

12. $p^{\frac{2}{3}} - 4p^{\frac{1}{3}} - 5 = 0$

7.3 Exercises MyMathLab®

Exercise numbers in green have complete video solutions in MyMathLab or may be accessed using the QR code to the right.

*Problems **1–12** are the Quick ✔ s that follow the **EXAMPLES**.*

Building Skills

In Problems 13–48, solve each equation. See Objective 1.

13. $x^4 - 5x^2 + 4 = 0$

14. $x^4 - 10x^2 + 9 = 0$

15. $q^4 + 13q^2 + 36 = 0$

16. $z^4 + 10z^2 + 9 = 0$

17. $4a^4 - 17a^2 + 4 = 0$

18. $4b^4 - 5b^2 + 1 = 0$

19. $p^4 + 6 = 5p^2$

20. $q^4 + 15 = 8q^2$

21. $(x - 3)^2 - 6(x - 3) - 7 = 0$

22. $(x + 2)^2 - 3(x + 2) - 10 = 0$

23. $(x^2 - 1)^2 - 11(x^2 - 1) + 24 = 0$

24. $(p^2 - 2)^2 - 8(p^2 - 2) + 12 = 0$

25. $(y^2 + 2)^2 + 7(y^2 + 2) + 10 = 0$

26. $(q^2 + 4)^2 + 3(q^2 + 4) - 4 = 0$

27. $x - 3\sqrt{x} - 4 = 0$

28. $x - 5\sqrt{x} - 6 = 0$

29. $w + 5\sqrt{w} + 6 = 0$

30. $z + 7\sqrt{z} + 6 = 0$

31. $2x + 5\sqrt{x} = 3$

32. $3x = 11\sqrt{x} + 4$

33. $x^{-2} + 3x^{-1} = 28$

34. $q^{-2} + 2q^{-1} = 15$

35. $10z^{-2} + 11z^{-1} = 6$

36. $10a^{-2} + 23a^{-1} = 5$

37. $x^{\frac{2}{3}} + 3x^{\frac{1}{3}} - 4 = 0$

38. $y^{\frac{2}{3}} - 2y^{\frac{1}{3}} - 3 = 0$

39. $z^{\frac{2}{3}} - z^{\frac{1}{3}} = 2$

40. $w^{\frac{2}{3}} + 2w^{\frac{1}{3}} = 3$

41. $a + a^{\frac{1}{2}} = 30$

42. $b + 3b^{\frac{1}{2}} = 28$

43. $\dfrac{1}{x^2} - \dfrac{5}{x} + 6 = 0$

44. $\dfrac{1}{x^2} - \dfrac{7}{x} + 12 = 0$

45. $\left(\dfrac{1}{x + 2}\right)^2 + \dfrac{4}{x + 2} = 5$

46. $\left(\dfrac{1}{x + 2}\right)^2 + \dfrac{6}{x + 2} = 7$

47. $p^6 - 28p^3 + 27 = 0$

48. $y^6 - 7y^3 - 8 = 0$

Mixed Practice

In Problems 49–62, solve each equation.

49. $8a^{-2} + 2a^{-1} = 1$

50. $6b^{-2} - b^{-1} = 1$

51. $z^4 = 4z^2 + 32$

52. $x^4 + 3x^2 = 4$

53. $x^{\frac{1}{2}} + x^{\frac{1}{4}} - 6 = 0$

54. $c^{\frac{1}{2}} + c^{\frac{1}{4}} - 12 = 0$

55. $w^4 - 5w^2 - 36 = 0$

56. $p^4 - 15p^2 - 16 = 0$

57. $\left(\dfrac{1}{x + 3}\right)^2 + \dfrac{2}{x + 3} = 3$

58. $\left(\dfrac{1}{x - 1}\right)^2 + \dfrac{7}{x - 1} = 8$

59. $x - 7\sqrt{x} + 12 = 0$

60. $x - 8\sqrt{x} + 12 = 0$

61. $2(x - 1)^2 - 7(x - 1) = 4$

62. $3(y - 2)^2 - 4(y - 2) = 4$

63. Suppose that $f(x) = x^4 + 7x^2 + 12$. Find the values of x such that
(a) $f(x) = 12$
(b) $f(x) = 6$

64. Suppose that $f(x) = x^4 + 5x^2 + 3$. Find the values of x such that
(a) $f(x) = 3$
(b) $f(x) = 17$

65. Suppose that $g(x) = 2x^4 - 6x^2 - 5$. Find the values of x such that
(a) $g(x) = -5$ (b) $g(x) = 15$

66. Suppose that $h(x) = 3x^4 - 9x^2 - 8$. Find the values of x such that
(a) $h(x) = -8$ (b) $h(x) = 22$

67. Suppose that $F(x) = x^{-2} - 5x^{-1}$. Find the values of x such that
(a) $F(x) = 6$ (b) $F(x) = 14$

68. Suppose that $f(x) = x^{-2} - 3x^{-1}$. Find the values of x such that
(a) $f(x) = 4$ (b) $f(x) = 18$

In Problems 69–74, find the zeros of the function. (Hint: Remember, r is a zero if f(r) = 0.)

69. $f(x) = x^4 + 9x^2 + 14$

70. $f(x) = x^4 - 13x^2 + 42$

71. $g(t) = 6t - 25\sqrt{t} - 9$

72. $h(p) = 8p - 18\sqrt{p} - 35$

73. $s(d) = \dfrac{1}{(d+3)^2} - \dfrac{4}{d+3} + 3$

74. $f(a) = \dfrac{1}{(a-2)^2} + \dfrac{3}{a-2} - 4$

Applying the Concepts

75. (a) Solve $x^2 - 5x + 6 = 0$.
(b) Solve $(x-3)^2 - 5(x-3) + 6 = 0$. Compare the solutions to part (a).
(c) Solve $(x+2)^2 - 5(x+2) + 6 = 0$. Compare the solutions to part (a).
(d) Solve $(x-5)^2 - 5(x-5) + 6 = 0$. Compare the solutions to part (a).
(e) Conjecture a generalization for the solution of $(x-a)^2 - 5(x-a) + 6 = 0$.

76. (a) Solve $x^2 + 3x - 18 = 0$.
(b) Solve $(x-1)^2 + 3(x-1) - 18 = 0$. Compare the solutions to part (a).

(c) Solve $(x+5)^2 + 3(x+5) - 18 = 0$. Compare the solutions to part (a).
(d) Solve $(x-3)^2 + 3(x-3) - 18 = 0$. Compare the solutions to part (a).
(e) Conjecture a generalization for the solution of $(x-a)^2 + 3(x-a) - 18 = 0$.

77. Consider the function $f(x) = 2x^2 - 3x + 1$.
(a) Solve $f(x) = 0$.
(b) Solve $f(x-2) = 0$. Compare the solutions to part (a).
(c) Solve $f(x-5) = 0$. Compare the solutions to part (a).
(d) Conjecture a generalization for the zeros of $f(x-a)$.

78. Consider the function $f(x) = 3x^2 - 5x - 2$.
(a) Solve $f(x) = 0$.
(b) Solve $f(x-1) = 0$. Compare the solutions to part (a).
(c) Solve $f(x-4) = 0$. Compare the solutions to part (a).
(d) Conjecture a generalization for the zeros of $f(x-a)$.

79. Revenue The function
$$R(x) = \frac{(x-1990)^2}{2} + \frac{3(x-1990)}{2} + 3000$$
models the revenue R (in thousands of dollars) of a start-up computer consulting firm in year x, where $x \geq 1990$.
(a) Determine and interpret $R(1990)$.
(b) Solve and interpret $R(x) = 3065$.
(c) According to the model, in what year can the firm expect to receive \$3350 thousand in revenue?

80. Revenue The function
$$R(x) = \frac{(x-2000)^2}{3} + \frac{5(x-2000)}{3} + 2000$$
models the revenue R (in thousands of dollars) of a start-up computer software firm in year x, where $x \geq 2000$.
(a) Determine and interpret $R(2000)$.
(b) Solve and interpret $R(x) = 2250$.
(c) According to the model, in what year can the firm expect to receive \$2350 thousand in revenue?

Extending the Concepts

All of the problems given in this section resulted in equations quadratic in form that could be factored after the appropriate substitution. However, this is not a necessary requirement to solving equations quadratic in form. In Problems 81–84, determine the appropriate substitution, and then use the quadratic formula to find the value of u. Finally, determine the value of the variable in the equation.

81. $x^4 + 5x^2 + 2 = 0$

82. $x^4 + 7x^2 + 4 = 0$

83. $2(x - 2)^2 + 8(x - 2) - 1 = 0$

84. $3(x + 1)^2 + 6(x + 1) - 1 = 0$

Explaining the Concepts

85. The equation $x - 5\sqrt{x} - 6 = 0$ can be solved either by using the methods of this section or by isolating the radical and squaring both sides. Solve it both ways and explain which approach you prefer.

86. Explain the steps required to solve an equation quadratic in form. Be sure to include an explanation of how to identify the appropriate substitution.

87. Under what circumstances might extraneous solutions occur when one is solving equations quadratic in form?

Synthesis Review

In Problems 88–91, add or subtract the expressions.

88. $(4x^2 - 3x - 1) + (-3x^2 + x + 5)$

89. $(3p^{-2} - 4p^{-1} + 8) - (2p^{-2} - 8p^{-1} - 1)$

90. $3\sqrt{2x} - \sqrt{8x} + \sqrt{50x}$

91. $\sqrt[3]{16a} + \sqrt[3]{54a} - \sqrt[3]{128a^4}$

92. Write a sentence or two that discusses how to add or subtract algebraic expressions, in general.

Technology Exercises

In Problems 93–98, use technology to find the real solutions to the equations. Round your answers to two decimal places, if necessary.

93. $x^4 + 5x^2 - 14 = 0$ **94.** $x^4 - 4x^2 - 12 = 0$

95. $2(x - 2)^2 = 5(x - 2) + 1$

96. $3(x + 3)^2 = 2(x + 3) + 6$

97. $x - 5\sqrt{x} = -3$ **98.** $x + 4\sqrt{x} = 5$

99. (a) Graph $Y = x^2 - 5x - 6$. Find the x-intercepts of the graph.
  (b) Graph $Y = (x + 2)^2 - 5(x + 2) - 6$. Find the x-intercepts of the graph.
  (c) Graph $Y = (x + 5)^2 - 5(x + 5) - 6$. Find the x-intercepts of the graph.
  (d) Make a generalization based upon the results of parts (a), (b), and (c).

100. (a) Graph $Y = x^2 + 4x + 3$. Find the x-intercepts of the graph.
  (b) Graph $Y = (x - 3)^2 + 4(x - 3) + 3$. Find the x-intercepts of the graph.
  (c) Graph $Y = (x - 6)^2 + 4(x - 6) + 3$. Find the x-intercepts of the graph.
  (d) Make a generalization based upon the results of parts (a), (b), and (c).

Putting the Concepts Together (Sections 7.1–7.3)

We designed these problems so that you can review Sections 7.1–7.3 and show your mastery of the concepts. Take time to work these problems before proceeding with the next section. The answers are located at the back of the text on page AN-40.

In Problems 1–3, complete the square in the given expression. Then factor the perfect square trinomial.

1. $z^2 + 10z$

2. $x^2 + 7x$

3. $n^2 - \frac{1}{4}n$

In Problems 4–6, solve each quadratic equation using the stated method.

4. $(2x - 3)^2 - 5 = -1$; Square Root Property

5. $x^2 + 8x + 4 = 0$; completing the square

6. $x(x - 6) = -7$; quadratic formula

In Problems 7–10, solve each equation using the method you prefer.

7. $49x^2 - 80 = 0$

8. $p^2 - 8p + 6 = 0$

9. $3y^2 + 6y + 4 = 0$

10. $\frac{1}{4}n^2 + n = \frac{1}{6}$

In Problems 11–13, determine the discriminant of each quadratic equation. Use the value of the discriminant to determine whether the equation has two rational solutions, two irrational solutions, one repeated rational solution, or two complex solutions that are not real.

11. $9x^2 + 12x + 4 = 0$ **12.** $3x^2 + 6x - 2 = 0$

13. $2x^2 + 6x + 5 = 0$

14. Find the missing length in the right triangle shown below.

In Problems 15 and 16, solve each equation.

15. $2m + 7\sqrt{m} - 15 = 0$

16. $p^{-2} - 3p^{-1} - 18 = 0$

17. Revenue The revenue R received by a company selling x microwave ovens per day is given by the function $R(x) = -0.4x^2 + 140x$. How many microwave ovens must be sold for revenue to be $12,000 per day?

18. Airplane Ride An airplane flies 300 miles into the wind and then flies home against the wind. The total time of the trip (excluding time on the ground) is 5 hours. If the plane can fly 140 miles per hour in still air, what was the speed of the wind? Round your answer to the nearest tenth.

7.4 Graphing Quadratic Functions Using Transformations

Objectives

1 Graph Quadratic Functions of the Form $f(x) = x^2 + k$

2 Graph Quadratic Functions of the Form $f(x) = (x - h)^2$

3 Graph Quadratic Functions of the Form $f(x) = ax^2$

4 Graph Quadratic Functions of the Form $f(x) = ax^2 + bx + c$

5 Find a Quadratic Function from Its Graph

Are You Prepared for This Section?

Before getting started, complete the following problems. If you get a problem wrong, go back to the section cited and review the material.

P1. Graph $y = x^2$ using point plotting.   [Section 1.5, pp. 97–99]

P2. Use the point-plotting method to graph $y = x^2 - 3$.   [Section 1.5, pp. 97–99]

P3. What is the domain of $f(x) = 2x^2 + 5x + 1$?   [Section 2.2, pp. 160–161]

Definition

A **quadratic function** is a function of the form

$$f(x) = ax^2 + bx + c$$

where a, b, and c are real numbers and $a \neq 0$. The domain of a quadratic function consists of all real numbers.

Many situations can be modeled using quadratic functions. For example, Example 10 of Section 2.2 showed that Franco Modigliani used the quadratic function $I(a) = -55a^2 + 5119a - 54,448$ to model the relation between average annual income, I, and age, a.

Also, if the effect of air resistance on a projectile is ignored, then the height of the projectile as a function of horizontal distance traveled can be modeled using a quadratic function. See Figure 7 on the next page.

In this section and the next, new techniques for graphing quadratic functions will be explored. In Section 1.5, equations were graphed using point plotting, but this method is inefficient and can lead to incomplete graphs. Remember, a complete graph shows all of its "interesting features," such as its intercepts and high and low points. Two methods for graphing quadratic functions that are superior to the point-plotting

Prepared?...Answers

P1.

P2.

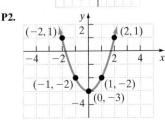

P3. The set of all real numbers

Figure 7

method will be presented here. The first method uses *transformations* and is the subject of this section. The second method, discussed in the next section, uses properties of quadratic functions.

Work Smart

Consider the quadratic function $f(x) = ax^2 + bx + c$, where $a = 1$, $b = 0$, and c is any real number. This is a function of the form $f(x) = x^2 + k$.

▶ ❶ **Graph Quadratic Functions of the Form $f(x) = x^2 + k$**

We begin by learning how to graph any quadratic function of the form $f(x) = x^2 + k$, such as $g(x) = x^2 + 3$ or $h(x) = x^2 - 4$.

Consider the graph of $y = g(x) = x^2$ in Figure 8, which first appeared in Section 1.5. What effect does adding a real number k to the function $g(x) = x^2$ have on its graph? Let's see!

Figure 8

$y = g(x) = x^2$

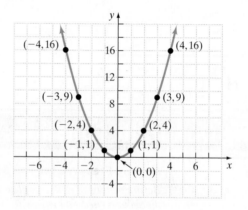

EXAMPLE 1   **Graphing a Quadratic Function of the Form $f(x) = x^2 + k$**

In the same Cartesian plane, graph $g(x) = x^2$ and $f(x) = x^2 + 3$.

Solution

Begin by obtaining some points on the graphs of g and f. For example, when $x = 0$, then $y = g(0) = 0$ and $y = f(0) = 0^2 + 3 = 3$. When $x = 1$, then $y = g(1) = 1$ and $y = f(1) = 1^2 + 3 = 4$. Table 3 lists these points along with a few others. Notice that while the x-coordinates stay the same, the y-coordinates on the graph of $f(x) = x^2 + 3$ are exactly 3 units greater than the corresponding y-coordinates on the graph of $g(x) = x^2$. Figure 9 shows the graphs of f and g.

Table 3

x	$g(x) = x^2$	$(x, g(x))$	$f(x) = x^2 + 3$	$(x, f(x))$
-2	$(-2)^2 = 4$	$(-2, 4)$	$(-2)^2 + 3 = 7$	$(-2, 7)$
-1	$(-1)^2 = 1$	$(-1, 1)$	$(-1)^2 + 3 = 4$	$(-1, 4)$
0	0	$(0, 0)$	3	$(0, 3)$
1	1	$(1, 1)$	4	$(1, 4)$
2	4	$(2, 4)$	7	$(2, 7)$

Figure 9

The graph of f is identical to the graph of g, except that it is shifted up 3 units.

●

Let's look at another example.

EXAMPLE 2 **Graphing a Quadratic Function of the Form** $f(x) = x^2 - k$

In the same Cartesian plane, graph $g(x) = x^2$ and $f(x) = x^2 - 4$.

Solution
Table 4 lists some points on the graphs of g and f. Notice that while the x-coordinates stay the same, the y-coordinates on the graph of $f(x) = x^2 - 4$ are exactly 4 units less than the corresponding y-coordinates on the graph of $g(x) = x^2$. Figure 10 shows the graphs of f and g.

Figure 10

Table 4

x	$g(x) = x^2$	$(x, g(x))$	$f(x) = x^2 - 4$	$(x, f(x))$
-2	4	$(-2, 4)$	$(-2)^2 - 4 = 0$	$(-2, 0)$
-1	1	$(-1, 1)$	$(-1)^2 - 4 = -3$	$(-1, -3)$
0	0	$(0, 0)$	-4	$(0, -4)$
1	1	$(1, 1)$	-3	$(1, -3)$
2	4	$(2, 4)$	0	$(2, 0)$

The graph of f is identical to the graph of g except that f is shifted down 4 units. ●

The results of Examples 1 and 2 lead to the following conclusion.

Graphing a Function of the Form $f(x) = x^2 + k$ **or** $f(x) = x^2 - k$

To obtain the graph of $f(x) = x^2 + k, k > 0$, from the graph of $y = x^2$, shift the graph of $y = x^2$ up k units. To obtain the graph of $f(x) = x^2 - k, k > 0$, from the graph of $y = x^2$, shift the graph of $y = x^2$ down k units.

Quick ✓

1. A _____ _____ is a function of the form $f(x) = ax^2 + bx + c$, where $a, b,$ and c are real numbers and $a \neq 0$.

2. To graph $f(x) = x^2 + k, k > 0$, from the graph of $y = x^2$, shift the graph of $y = x^2$ __ k units. To graph $f(x) = x^2 - k, k > 0$, from the graph of $y = x^2$ shift the graph of $y = x^2$ ____ k units.

In Problems 3 and 4, use the graph of $y = x^2$ to graph the quadratic function. Show at least three points on the graph.

3. $f(x) = x^2 + 5$ 4. $f(x) = x^2 - 2$

▶ ❷ **Graph Quadratic Functions of the Form** $f(x) = (x - h)^2$

Now look at the graph of any quadratic function of the form $f(x) = (x - h)^2$ such as $f(x) = (x + 3)^2$ or $f(x) = (x - 2)^2$.

EXAMPLE 3 **Graphing a Quadratic Function of the Form $f(x) = (x - h)^2$**

In the same Cartesian plane, graph $g(x) = x^2$ and $f(x) = (x - 2)^2$.

Solution

Again, use the point-plotting method. Table 5 lists some points on the graphs of g and f. Note that when $g(x) = 0$, $x = 0$, and when $f(x) = 0$, $x = 2$. Also, when $g(x) = 4$, $x = -2$ or 2, and when $f(x) = 4$, $x = 0$ or 4. The graph of f is identical to that of g, except that f is shifted 2 units to the right of g. See Figure 11.

Table 5

x	$g(x) = x^2$	$(x, g(x))$	$f(x) = (x - 2)^2$	$(x, f(x))$
-2	4	$(-2, 4)$	$(-2 - 2)^2 = 16$	$(-2, 16)$
-1	1	$(-1, 1)$	$(-1 - 2)^2 = 9$	$(-1, 9)$
0	0	$(0, 0)$	4	$(0, 4)$
1	1	$(1, 1)$	1	$(1, 1)$
2	4	$(2, 4)$	0	$(2, 0)$
3	9	$(3, 9)$	1	$(3, 1)$
4	16	$(4, 16)$	4	$(4, 4)$

Figure 11

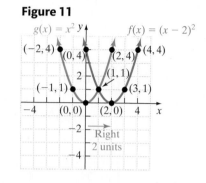

Why does this happen? Because 2 is being subtracted from each x-value in the function $f(x) = (x - 2)^2$, the x-values must be greater by 2 in order to obtain the same y-value that was obtained in the graph of $g(x) = x^2$. •

What if a positive number h is added to x?

EXAMPLE 4 **Graphing a Quadratic Function of the Form $f(x) = (x + h)^2$**

In the same Cartesian plane, graph $g(x) = x^2$ and $f(x) = (x + 3)^2$.

Solution

Table 6 lists some points on the graphs of g and f. Notice that when $g(x) = 0$, then $x = 0$, and when $f(x) = 0$, then $x = -3$. Also, when $g(x) = 4$, then $x = -2$ or 2, and when $f(x) = 4$, then $x = -5$ or -1. The graph of f is identical to that of g, except that f is shifted 3 units to the left of g. See Figure 12.

Table 6

x	$g(x) = x^2$	$(x, g(x))$	$f(x) = (x + 3)^2$	$(x, f(x))$
-5	25	$(-5, 25)$	$(-5 + 3)^2 = 4$	$(-5, 4)$
-4	16	$(-4, 16)$	$(-4 + 3)^2 = 1$	$(-4, 1)$
-3	9	$(-3, 9)$	$(-3 + 3)^2 = 0$	$(-3, 0)$
-2	4	$(-2, 4)$	1	$(-2, 1)$
-1	1	$(-1, 1)$	4	$(-1, 4)$
0	0	$(0, 0)$	9	$(0, 9)$
1	1	$(1, 1)$	16	$(1, 16)$
2	4	$(2, 4)$	25	$(2, 25)$

Figure 12

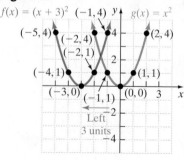

Why does this happen? Because 3 is being added to each x-value in the function $f(x) = (x + 3)^2$, the x-values must be smaller by 3 in order to obtain the same y-value that was obtained in the graph of $g(x) = x^2$. •

Examples 3 and 4 lead to the following conclusion.

Graphing a Function of the Form $f(x) = (x - h)^2$ or $f(x) = (x + h)^2$

To obtain the graph of $f(x) = (x - h)^2, h > 0$, from the graph of $y = x^2$, shift the graph of $y = x^2$ to the right h units. To obtain the graph of $f(x) = (x + h)^2$, $h > 0$, from the graph of $y = x^2$, shift the graph of $y = x^2$ to the left h units.

Quick ✓

5. *True or False* To obtain the graph of $f(x) = (x + 12)^2$ from the graph of $y = x^2$, shift the graph of $y = x^2$ to the right 12 units.

In Problems 6 and 7, use the graph of $y = x^2$ to graph the quadratic function. Show at least three points on the graph.

6. $f(x) = (x + 5)^2$ 7. $f(x) = (x - 1)^2$

Let's do an example where a horizontal shift and a vertical shift are combined.

EXAMPLE 5   **Combining Horizontal and Vertical Shifts**

Graph the function $f(x) = (x + 2)^2 - 3$.

Solution

Graph f in stages. Begin with the graph of $y = x^2$ as shown in Figure 13(a). Shift that graph 2 units to the left to get the graph of $y = (x + 2)^2$ shown in Figure 13(b). Then shift the graph of $y = (x + 2)^2$ down 3 units to get the graph of $y = (x + 2)^2 - 3$ shown in Figure 13(c). At each stage, key points are plotted.

Figure 13

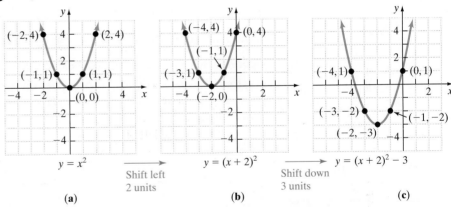

Note: The order in which the stages are performed to obtain the graph does not matter. In Example 5, the graph could just as easily have shifted down 3 units first and then shifted left 2 units.

Quick ✓

In Problems 8 and 9, graph each quadratic function using horizontal and vertical shifts. Show at least three points on the graph.

8. $f(x) = (x - 3)^2 + 2$ 9. $f(x) = (x + 1)^2 - 4$

▶ ③ **Graph Quadratic Functions of the Form $f(x) = ax^2$**

So far, the coefficient of the square term has been equal to 1. What impact does the value of a have on the graph of $f(x) = ax^2 + bx + c$? To answer this question, consider quadratic functions of the form $f(x) = ax^2, a \neq 0$.

Work Smart

In a "taller" graph, points with the same x-values have greater y-values. In a "shorter" graph, points with the same x-values have smaller y-values.

First, consider situations in which the value of a is positive. Table 7 shows values on the graphs of $f(x) = x^2$, $g(x) = \frac{1}{2}x^2$, and $h(x) = 2x^2$. Figure 14 shows the graphs of f, g, and h. First, notice that all three graphs open "up." Next, notice that the values of the y-coordinates on the graph of g are exactly $\frac{1}{2}$ of the values of the y-coordinates on the graph of f. The values of the y-coordinates on the graph of h are exactly 2 times the values of the y-coordinates on the graph of f. Put another way, the larger the value of a, the "taller" the graph is, and the smaller the value of a, the "shorter" the graph is.

Table 7

x	$f(x) = x^2$	$g(x) = \dfrac{1}{2}x^2$	$h(x) = 2x^2$
-2	$(-2)^2 = 4$	$\dfrac{1}{2}(-2)^2 = 2$	$2(-2)^2 = 8$
-1	$(-1)^2 = 1$	$\dfrac{1}{2}(-1)^2 = \dfrac{1}{2}$	$2(-1)^2 = 2$
0	0	0	0
1	1	$\dfrac{1}{2}$	2
2	4	2	8

Figure 14
$f(x) = ax^2$. Since $a > 0$, the graphs open up.

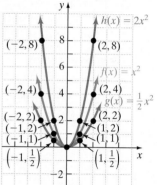

Now consider what happens when a is negative. Table 8 shows values on the graphs of $f(x) = -x^2$, $g(x) = -\frac{1}{2}x^2$, and $h(x) = -2x^2$. Figure 15 shows the graphs of f, g, and h. First, notice that all three graphs open "down." Next, notice that the values of the y-coordinates on the graph of g are exactly $\frac{1}{2}$ times the values of the y-coordinates on the graph of f. The values of the y-coordinates on the graph of h are exactly 2 times the values of the y-coordinates on the graph of f. Put another way, the larger the value of $|a|$, the "taller" the graph is, and the smaller the value of $|a|$, the "shorter" the graph is.

Table 8

x	$f(x) = -x^2$	$g(x) = -\dfrac{1}{2}x^2$	$h(x) = -2x^2$
-2	$-(-2)^2 = -4$	$-\dfrac{1}{2}(-2)^2 = -2$	$-2(-2)^2 = -8$
-1	$-(-1)^2 = -1$	$-\dfrac{1}{2}(-1)^2 = -\dfrac{1}{2}$	$-2(-1)^2 = -2$
0	0	0	0
1	-1	$-\dfrac{1}{2}$	-2
2	-4	-2	-8

Figure 15
$f(x) = ax^2$. Since $a < 0$, the graphs open down.

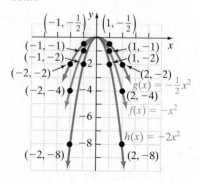

These conclusions are summarized on the following page.

Properties of the Graph of $f(x) = ax^2$

- When $a > 0$, the graph of $f(x) = ax^2$ opens up. If $0 < a < 1$ (a is between 0 and 1), the graph will be "shorter" than that of $y = x^2$. If $a > 1$, the graph will be "taller" than that of $y = x^2$.
- When $a < 0$, the graph of $f(x) = ax^2$ will open down. If $0 < |a| < 1$, the graph will be "shorter" than that of $y = x^2$. If $|a| > 1$, the graph will be "taller" than that of $y = x^2$.
- When $|a| > 1$, we say that the graph is **vertically stretched** by a factor of $|a|$. When $0 < |a| < 1$, we say that the graph is **vertically compressed** by a factor of $|a|$.

Graphing a Function of the Form $f(x) = ax^2$

To obtain the graph of $f(x) = ax^2$ from the graph of $y = x^2$, multiply each y-coordinate on the graph of $y = x^2$ by a.

EXAMPLE 6 **Graphing a Quadratic Function of the Form $f(x) = ax^2$**

Use the graph of $y = x^2$ to obtain the graph of $f(x) = -2x^2$.

Solution
To obtain the graph of $f(x) = -2x^2$ from the graph of $y = x^2$, multiply each y-coordinate on the graph of $y = x^2$ by -2 (the value of a). See Figure 16.

Figure 16

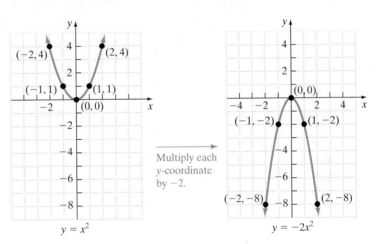

Quick ✓

10. When obtaining the graph of $f(x) = ax^2$ from the graph of $y = x^2$, multiply each ___-coordinate on the graph of $y = x^2$ by __. If $|a| > 1$, the graph is _____ _____ by a factor of $|a|$. If $0 < |a| < 1$, the graph is _____ _____ by a factor of $|a|$.

In Problems 11 and 12, use the graph of $y = x^2$ to graph each quadratic function. Label at least 3 points on the graph.

11. $f(x) = 3x^2$ **12.** $f(x) = -\dfrac{1}{4}x^2$

---

④ Graph Quadratic Functions of the Form $f(x) = ax^2 + bx + c$

The graphs in Examples 1–6 are typical of graphs of all quadratic functions. The graph of a quadratic function is called a **parabola** (pronounced puh-răb-uh-luh). Figure 17 on the next page shows two parabolas.

Figure 17

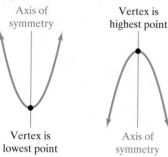

Axis of symmetry

Vertex is highest point

Vertex is lowest point

Axis of symmetry

(a) Opens up
$a > 0$

(b) Opens down
$a < 0$

The parabola in Figure 17(a) **opens up** (since $a > 0$) and has a lowest point; the parabola in Figure 17(b) **opens down** (since $a < 0$) and has a highest point. The lowest or highest point of a parabola is the **vertex**. The vertical line passing through the vertex of a parabola is its **axis of symmetry**. If we take the portion of the parabola to the right of the vertex and fold it over the axis of symmetry, it would lie directly on top of the portion of the parabola to the left of the vertex. Therefore, the parabola is symmetric about its axis of symmetry. Note that the axis of symmetry is not part of the graph of the quadratic function, but it will be useful when quadratic functions are graphed using the methods presented in the next section.

The techniques of shifting horizontally, shifting vertically, stretching, and compressing are collectively referred to as **transformations**. To graph any quadratic function of the form $f(x) = ax^2 + bx + c$ using transformations, take the following steps.

Work Smart

Although there may be multiple orders that work, it is recommended that, when graphing parabolas using transformations, you obtain the graph that results from the vertical compression or stretch first, followed by the horizontal shift, followed by the vertical shift.

Graphing Quadratic Functions Using Transformations

Step 1: Write the function $f(x) = ax^2 + bx + c$ as $f(x) = a(x - h)^2 + k$ by completing the square in x.

Step 2: Graph the function $f(x) = a(x - h)^2 + k$ using transformations.

Notice that the quadratic function $f(x) = ax^2 + bx + c$ must be written as $f(x) = a(x - h)^2 + k$ to determine the horizontal and vertical shifts.

EXAMPLE 7 **How to Graph a Quadratic Function of the Form $f(x) = ax^2 + bx + c$ Using Transformations**

Graph $f(x) = x^2 + 4x + 3$ using transformations. Identify the vertex and axis of symmetry of the parabola. Based on the graph, determine the domain and range of the quadratic function.

Step-by-Step Solution

Step 1: Write the function $f(x) = ax^2 + bx + c$ as $f(x) = a(x - h)^2 + k$ by completing the square in x.

$$f(x) = x^2 + 4x + 3$$

Group the terms involving x: $\qquad = (x^2 + 4x) + 3$

Complete the square in x by taking $\dfrac{1}{2}$ the coefficient

of x and squaring the result: $\left(\dfrac{1}{2} \cdot 4\right)^2 = 4$.

Because 4 was added, it must also be subtracted
so that the value of the function does not change: $\qquad = (x^2 + 4x + 4) + 3 - 4$

$$= (x^2 + 4x + 4) - 1$$

Factor the perfect square trinomial in parentheses: $\qquad = (x + 2)^2 - 1$

Step 2: Graph the function $f(x) = a(x - h)^2 + k$ using transformations.

The graph of $f(x) = (x + 2)^2 - 1$ is the graph of $y = x^2$ shifted 2 units left and 1 unit down. See Figure 18.

Figure 18

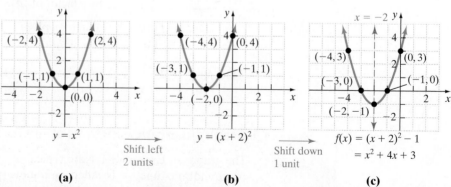

$y = x^2$

$(-2, 4)$ $(2, 4)$ $(-1, 1)$ $(1, 1)$ $(0, 0)$

Shift left 2 units

(a)

$y = (x + 2)^2$

$(-4, 4)$ $(0, 4)$ $(-3, 1)$ $(-1, 1)$ $(-2, 0)$

Shift down 1 unit

(b)

$f(x) = (x + 2)^2 - 1$
$= x^2 + 4x + 3$

$x = -2$

$(-4, 3)$ $(0, 3)$ $(-3, 0)$ $(-1, 0)$ $(-2, -1)$

(c)

From the graph in Figure 18(c), note that the vertex of the parabola is $(-2, -1)$. The axis of symmetry is the line $x = -2$. The domain is the set of all real numbers, or $(-\infty, \infty)$. The range is $\{y \mid y \geq -1\}$, or $[-1, \infty)$.

▶ **EXAMPLE 8**   **Graphing a Quadratic Function of the Form $f(x) = ax^2 + bx + c$ Using Transformations**

Graph $f(x) = -2x^2 + 4x + 1$ using transformations. Identify the vertex and axis of symmetry of the parabola. Based on the graph, determine the domain and range of the quadratic function.

Solution
Write the function $f(x) = ax^2 + bx + c$ as $f(x) = a(x - h)^2 + k$ by completing the square in x.

$$f(x) = -2x^2 + 4x + 1$$

Group the terms involving x: $= (-2x^2 + 4x) + 1$

Factor out the coefficient of x^2, -2, from the parentheses: $= -2(x^2 - 2x) + 1$

Complete the square in x by taking $\dfrac{1}{2}$ the coefficient of x and

squaring the result: $\left(\dfrac{1}{2} \cdot -2\right)^2 = 1$. Add 1 inside the

parentheses. Because 1 is multiplied by -2, we really

subtracted 2, so we must add 2 to offset this: $= -2(x^2 - 2x + 1) + 1 + 2$

Factor the perfect square trinomial in parentheses: $= -2(x - 1)^2 + 3$

Now, graph the function $f(x) = a(x - h)^2 + k$ using transformations. Because $a = -2$, the parabola opens down and is stretched by a factor of 2. Because 1 is subtracted from x, the parabola shifts 1 unit to the right; because $k = 3$, the parabola shifts 3 units up. See Figure 19.

Figure 19

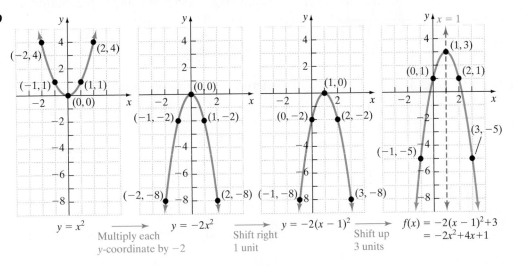

The vertex of the parabola is $(1, 3)$. The axis of symmetry is the line $x = 1$. The domain is the set of all real numbers, or $(-\infty, \infty)$. The range is $\{y \mid y \le 3\}$, or $(-\infty, 3]$. ●

Work Smart: Study Skills

Be sure you know what $y = a(x - h)^2 + k$ represents in reference to the graph of $y = x^2$:

- $|a|$ represents the vertical stretch or compression factor: how tall or short the graph appears.
- The sign of a determines whether the parabola opens up or down.
- h represents the number of units the graph is shifted horizontally.
- k represents the number of units the graph is shifted vertically.
- The vertex of the parabola has coordinates (h, k).

For example, $y = 4(x + 2)^2 - 3$ means that the graph of $y = x^2$ is stretched vertically by a factor of 4, is shifted 2 units to the left, and is shifted down 3 units. The vertex of $y = 4(x + 2)^2 - 3$ is at $(-2, -3)$ and represents the low point of the graph since $a > 0$.

Work Smart

When picking at least three points on the graph, it is best to pick the vertex and one point on each side of the vertex.

Quick ✓

13. *True or False* The graph of $f(x) = -3x^2 + x + 6$ opens down.

In Problems 14 and 15, graph each quadratic function using transformations. Be sure to label at least three points on the graph. Based on the graph, determine the domain and range of each function.

14. $f(x) = -3(x + 2)^2 + 1$ **15.** $f(x) = 2x^2 - 8x + 5$

▶ ❺ Find a Quadratic Function from Its Graph

In Example 7, the quadratic function $f(x) = x^2 + 4x + 3 = (x + 2)^2 - 1$ was graphed. Notice that the vertex of the parabola is $(-2, -1)$. If f is written as $f(x) = (x - (-2))^2 - 1$ and compared to $f(x) = a(x - h)^2 + k$, then $h = -2$ and $k = -1$. Thus, the vertex of a quadratic function in the form $f(x) = a(x - h)^2 + k$ is (h, k). Now look at Example 8, where $f(x) = -2x^2 + 4x + 1 = -2(x - 1)^2 + 3$. The vertex is $(h, k) = (1, 3)$.

Given the vertex, (h, k), and one additional point on the graph of a quadratic function, the quadratic function $f(x) = a(x - h)^2 + k$ that has this graph may be found. For this reason, $f(x) = a(x - h)^2 + k$ is called the **vertex form** of a quadratic function.

(EXAMPLE 9) **Finding the Quadratic Function Given Its Vertex and One Other Point**

Figure 20

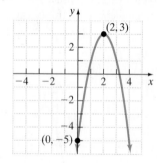

Determine the quadratic function whose graph is given in Figure 20. Write the function in vertex form, $f(x) = a(x - h)^2 + k$.

Solution

The vertex is $(2, 3)$, so $h = 2$ and $k = 3$. Substitute these values into $f(x) = a(x - h)^2 + k$.

$$f(x) = a(x - h)^2 + k$$
$$h = 2, k = 3: \quad = a(x - 2)^2 + 3$$

To determine the value of a, use the fact that $f(0) = -5$ (the y-intercept).

$$f(x) = a(x - 2)^2 + 3$$
$$x = 0, y = f(0) = -5: \quad -5 = a(0 - 2)^2 + 3$$
$$-5 = a(4) + 3$$
$$-5 = 4a + 3$$
$$\text{Subtract 3 from both sides:} \quad -8 = 4a$$
$$a = -2$$

The quadratic function whose graph is shown in Figure 20 is $f(x) = -2(x - 2)^2 + 3$.

●

Quick ✓

In Problem 16, find the quadratic function whose graph is given. Write the function in vertex form, $f(x) = a(x - h)^2 + k$.

16.

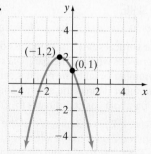

7.4 Exercises MyMathLab® Exercise numbers in green have complete video solutions in MyMathLab or may be accessed using the QR code to the right.

*Problems **1–16** are the Quick ✓s that follow the EXAMPLES.*

Building Skills

17. Match each quadratic function to its graph.

(**I**) $f(x) = x^2 + 3$ (**II**) $f(x) = (x + 3)^2$

(**III**) $f(x) = x^2 - 3$ (**IV**) $f(x) = (x - 3)^2$

(**A**)   (**B**)

(**C**)   (**D**)

18. Match each quadratic function to its graph.

(**I**) $f(x) = (x - 2)^2 - 4$ (**II**) $f(x) = -(x - 2)^2 + 4$

(**III**) $f(x) = -(x + 2)^2 + 4$ (**IV**) $f(x) = 2(x - 2)^2 - 4$

(**A**)   (**B**)

(**C**)   (**D**)

In Problems 19–26, verbally explain how to obtain the graph of the given quadratic function from the graph of $y = x^2$. For example, to obtain the graph of $f(x) = (x - 3)^2 - 6$ from the graph of $y = x^2$, take the graph of $y = x^2$ and shift it 3 units to the right and 6 units down.

19. $f(x) = (x + 10)^2$ **20.** $G(x) = (x - 9)^2$

21. $F(x) = x^2 + 12$ **22.** $g(x) = x^2 - 8$

23. $H(x) = 2(x - 5)^2$ **24.** $h(x) = 4(x + 7)^2$

25. $f(x) = -3(x + 5)^2 + 8$ **26.** $F(x) = -\dfrac{1}{2}(x - 3)^2 - 5$

In Problems 27–30, use the graph of $y = x^2$ to graph the quadratic function. See Objective 1.

27. $f(x) = x^2 + 1$ **28.** $h(x) = x^2 + 6$

29. $f(x) = x^2 - 1$ **30.** $g(x) = x^2 - 7$

In Problems 31–34, use the graph of $y = x^2$ to graph each quadratic function. See Objective 2.

31. $F(x) = (x - 3)^2$ **32.** $F(x) = (x - 2)^2$

33. $h(x) = (x + 2)^2$ **34.** $f(x) = (x + 4)^2$

In Problems 35–40, use the graph of $y = x^2$ to graph each quadratic function. See Objective 3.

35. $g(x) = 4x^2$ **36.** $G(x) = 5x^2$

37. $H(x) = \dfrac{1}{3}x^2$ **38.** $h(x) = \dfrac{3}{2}x^2$

39. $p(x) = -x^2$ **40.** $P(x) = -3x^2$

In Problems 41–54, use the graph of $y = x^2$ to graph each quadratic function. See Objective 4.

41. $f(x) = (x - 1)^2 - 3$ **42.** $g(x) = (x + 2)^2 - 1$

43. $F(x) = (x + 3)^2 + 1$ **44.** $G(x) = (x - 4)^2 + 2$

45. $h(x) = -(x + 3)^2 + 2$ **46.** $H(x) = -(x - 3)^2 + 5$

47. $G(x) = 2(x + 1)^2 - 2$ **48.** $F(x) = 3(x - 2)^2 - 1$

49. $H(x) = -\dfrac{1}{2}(x + 5)^2 + 3$ **50.** $f(x) = -\dfrac{1}{2}(x + 6)^2 + 2$

51. $f(x) = x^2 + 2x - 4$ **52.** $f(x) = x^2 + 4x - 1$

53. $g(x) = x^2 - 4x + 8$ **54.** $G(x) = x^2 - 2x + 7$

In Problems 55–60, determine the quadratic function whose graph is given. Write the function in vertex form, $f(x) = a(x - h)^2 + k$. See Objective 5.

55.   **56.**

57.   **58.**

59.

60.

Mixed Practice

In Problems 61–78, write each function in vertex form,
$f(x) = a(x - h)^2 + k$. *Then graph each quadratic function using transformations. Determine the vertex and axis of symmetry. Based on the graph, determine the domain and range of the quadratic function.*

61. $f(x) = x^2 + 6x - 16$
62. $f(x) = x^2 + 4x + 5$

63. $F(x) = x^2 + x - 12$
64. $h(x) = x^2 - 7x + 10$

65. $H(x) = 2x^2 - 4x - 1$
66. $g(x) = 2x^2 + 4x - 3$

67. $P(x) = 3x^2 + 12x + 13$
68. $f(x) = 3x^2 + 18x + 25$

69. $F(x) = -x^2 - 10x - 21$
70. $g(x) = -x^2 - 8x - 14$

71. $g(x) = -x^2 + 6x - 1$
72. $f(x) = -x^2 + 10x - 17$

73. $H(x) = -2x^2 + 8x - 4$
74. $h(x) = -2x^2 + 12x - 17$

75. $f(x) = \frac{1}{3}x^2 - 2x + 4$
76. $f(x) = \frac{1}{2}x^2 + 2x - 1$

77. $G(x) = -12x^2 - 12x + 1$
78. $h(x) = -4x^2 + 4x$

Applying the Concepts

In Problems 79–88, write a quadratic function in the vertex form,
$f(x) = a(x - h)^2 + k$ *with the properties given.*

79. Opens up; vertex and y-intercept at $(3, 0)$.

80. Opens up; vertex at $(0, 2)$; contains $(1, 3)$.

81. Opens up; vertex at $(-3, 1)$; contains $(-5, 5)$.

82. Opens up; vertex at $(4, -2)$; contains $(2, 2)$.

83. Opens down; vertex at $(5, -1)$; contains $(6, -2)$.

84. Opens down; vertex at $(-4, -7)$; y-intercept $(0, -23)$.

85. Opens up; vertically stretched by a factor of 4; vertex at $(9, -6)$.

86. Opens up; vertically compressed by a factor of $\frac{1}{2}$; vertex at $(-5, 0)$.

87. Opens down; vertically compressed by a factor of $\frac{1}{3}$; vertex at $(0, 6)$.

88. Opens down; vertically stretched by a factor of 5; vertex at $(5, 8)$.

Explaining the Concepts

89. What is the lowest or highest point on a parabola called? How do we know whether this point is a high point or a low point?

90. Why does the graph of a quadratic function open up if $a > 0$ and down if $a < 0$?

91. Can a quadratic function have a range of $(-\infty, \infty)$? Justify your answer.

92. Can the graph of a quadratic function have more than one y-intercept? Justify your answer.

Synthesis Review

In Problems 93–95, divide.

93. $\dfrac{349}{12}$

94. $\dfrac{4x^2 + 19x - 1}{x + 5}$

95. $\dfrac{2x^4 - 11x^3 + 13x^2 - 8x}{2x - 1}$

96. Explain how division of real numbers is related to division of polynomials.

Technology Exercises

In Problems 97–104, graph each quadratic function. Determine the vertex and axis of symmetry. Based on the graph, determine the range of the function.

97. $f(x) = x^2 + 1.3$
98. $f(x) = x^2 - 3.5$

99. $g(x) = (x - 2.5)^2$
100. $G(x) = (x + 4.5)^2$

101. $h(x) = 2.3(x - 1.4)^2 + 0.5$

102. $H(x) = 1.2(x + 0.4)^2 - 1.3$

103. $F(x) = -3.4(x - 2.8)^2 + 5.9$

104. $f(x) = 0.3(x + 3.8)^2 - 8.9$

105. Exploration: Quadratic Functions—a Value *Open the "Quadratic Functions—Vertex Form" Geogebra applet. The applet may be found using the QR code at the beginning of this section or through the Multimedia Library in MyMathLab. If you are using a keyboard, it is easier to move the sliders with the arrow keys instead of a mouse. When you begin the applet, the boxes should be unchecked. If they are not, uncheck them.*

(a) Move the sliders so that $a = 0, h = 0$, and $k = 0$. Now move the a slider to explore the relationship between $y = x^2$ and $f(x) = ax^2$ for $a > 0$ by letting $a = 1, a = 2, a = 3, a = 4$, and $a = 5$. What is the relationship between $y = x^2$ and $f(x) = ax^2$ when $a > 0$?

(b) Move the sliders so that $a = 0, h = 0$, and $k = 0$. Now move the a slider to explore the relationship between $y = x^2$ and $f(x) = ax^2$ for $a < 0$ by letting $a = -1, a = -2, a = -3, a = -4$, and $a = -5$. What is the relationship between $y = x^2$ and $f(x) = ax^2$ when $a < 0$?

(c) Now let $a = 0$. What type of function is $f(x)$?

106. Exploration: Quadratic Functions—Vertex Form *Open the "Quadratic Functions—Vertex Form" Geogebra applet.* The applet may be found using the QR code at the beginning of this section or through the Multimedia Library in MyMathLab. Check the "Vertex Form" and "Vertex" boxes. Explore the relationship between $y = x^2$ and $f(x) = a(x - h)^2 + k$. Move the sliders so that $a = 1$, $h = 0$, and $k = 0$. If you are using a keyboard, it is easier to move the sliders with the arrow keys instead of a mouse.

(a) Explore the relationship between h and the x-coordinate of the vertex of the parabola. Move the slider h to take on the values from -5 to 5. For each value of h, write down the coordinates of the vertex. What relationship exists between h and this point?

(b) Now explore the relationship between k and the y-coordinate of the vertex of the parabola. Reset the slider h back to 0 and let k take on the values from -5 to 5. For each value of k, write down the coordinates of the vertex. What relationship exists between k and this point?

(c) In general, when a quadratic function is written in the form $f(x) = a(x - h)^2 + k$, what are the coordinates of the vertex?

7.5 Graphing Quadratic Functions Using Properties

Objectives

❶ Graph Quadratic Functions of the Form $f(x) = ax^2 + bx + c$

❷ Find the Maximum or Minimum Value of a Quadratic Function

❸ Model and Solve Optimization Problems Involving Quadratic Functions

Are You Prepared for This Section?

Before getting started, complete the following problems. If you get a problem wrong, go back to the section cited and review the material.

P1. Find the intercepts of the graph of $2x + 5y = 20$.   [Section 1.6, pp. 106–107]

P2. Solve: $2x^2 - 3x - 20 = 0$ [Section 4.8, pp. 373–375]

P3. Find the zeros of $f(x) = x^2 - 3x - 4$.   [Section 4.8, p. 377]

In Section 7.4, quadratic functions were graphed using transformations. This section covers graphing a quadratic function by using its intercepts, axis of symmetry, and vertex.

❶ Graph Quadratic Functions of the Form $f(x) = ax^2 + bx + c$

▶ As explained in Section 7.4, a quadratic function $f(x) = ax^2 + bx + c$ can be written in the form $f(x) = a(x - h)^2 + k$ by completing the square in x. The value of a determines whether the graph of the quadratic function (the parabola) opens up or down. In addition, the point with coordinates (h, k) is the vertex of the quadratic function.

A formula that may be used to determine the vertex of a parabola can be found by completing the square in $f(x) = ax^2 + bx + c$, $a \neq 0$, as follows:

$$f(x) = ax^2 + bx + c$$

Group terms involving x: $= (ax^2 + bx) + c$

Factor out a: $= a\left(x^2 + \dfrac{b}{a}x\right) + c$

Complete the square in x by taking $\dfrac{1}{2}$ the coefficient of x and squaring the result: $\left(\dfrac{1}{2} \cdot \dfrac{b}{a}\right)^2 = \dfrac{b^2}{4a^2}$. Because $\dfrac{b^2}{4a^2}$ is added inside the parentheses, subtract $a \cdot \dfrac{b^2}{4a^2} = \dfrac{b^2}{4a}$ outside the parentheses:

Factor the perfect square trinomial; $= a\left(x^2 + \dfrac{b}{a}x + \dfrac{b^2}{4a^2}\right) + c - \dfrac{b^2}{4a}$

multiply c by $\dfrac{4a}{4a}$ to get a common denominator: $= a\left(x + \dfrac{b}{2a}\right)^2 + c \cdot \dfrac{4a}{4a} - \dfrac{b^2}{4a}$

Write expression in the form $f(x) = a(x - h)^2 + k$: $= a\left(x - \left(-\dfrac{b}{2a}\right)\right)^2 + \dfrac{4ac - b^2}{4a}$

Compare $f(x) = a\left(x - \left(-\dfrac{b}{2a}\right)\right)^2 + \dfrac{4ac - b^2}{4a}$ to $f(x) = a(x - h)^2 + k$, and conclude the following:

Work Smart

Another formula for the vertex is

$$\left(-\frac{b}{2a}, \frac{-D}{4a}\right)$$

where $D = b^2 - 4ac$, the discriminant.

The Vertex of a Parabola

Any quadratic function $f(x) = ax^2 + bx + c, a \neq 0$, has vertex

$$\left(-\frac{b}{2a}, \frac{4ac - b^2}{4a}\right)$$

Because the y-coordinate on the graph of any function can be found by evaluating the function at the corresponding x-coordinate, the coordinates of the vertex can be restated as

$$\left(-\frac{b}{2a}, f\left(-\frac{b}{2a}\right)\right)$$

For instance, Example 7 in Section 7.4 showed $f(x) = x^2 + 4x + 3 = (x + 2)^2 - 1$ has a vertex of $(-2, -1)$. In $f(x) = x^2 + 4x + 3, a = 1, b = 4$, and $c = 3$, so the x-coordinate of the vertex is given by $x = -\frac{b}{2a} = -\frac{4}{2(1)} = -2$. The y-coordinate of the vertex is $f(-2) = (-2)^2 + 4(-2) + 3 = -1$.

Because the axis of symmetry intersects the vertex, the axis of symmetry is $x = -\frac{b}{2a}$. In addition, the parabola will open up if $a > 0$ and down if $a < 0$. By using this information along with the intercepts of the graph, we can obtain a complete graph.

The y-intercept is the value of the quadratic function $f(x) = ax^2 + bx + c$ at $f(0) = c$.

Any x-intercepts are found by solving the quadratic equation

$$f(x) = ax^2 + bx + c = 0$$

This equation has two, one, or no real solutions, depending on the value of the discriminant $b^2 - 4ac$. Use the value of the discriminant to determine how many x-intercepts the graph of the quadratic function has.

The x-Intercepts of the Graph of a Quadratic Function

1. If the discriminant $b^2 - 4ac > 0$, the graph of $f(x) = ax^2 + bx + c$ has two x-intercepts. The graph crosses the x-axis at the solutions to the equation $ax^2 + bx + c = 0$.
2. If the discriminant $b^2 - 4ac = 0$, the graph of $f(x) = ax^2 + bx + c$ has one x-intercept. The graph touches the x-axis at the solution to the equation $ax^2 + bx + c = 0$.
3. If the discriminant $b^2 - 4ac < 0$, the graph of $f(x) = ax^2 + bx + c$ has no x-intercepts. The graph does not cross or touch the x-axis.

Figure 21 illustrates these possibilities for parabolas that open up.

Figure 21

$f(x) = ax^2 + bx + c, a > 0$

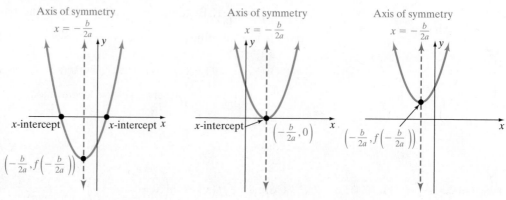

EXAMPLE 1   **How to Graph a Quadratic Function Using Its Properties**

Graph $f(x) = x^2 + 2x - 15$ using its properties.

Step-by-Step Solution

Compare $f(x) = x^2 + 2x - 15$ to $f(x) = ax^2 + bx + c$ and see that $a = 1$, $b = 2$, and $c = -15$.

Step 1: Determine whether the parabola opens up or down.

The parabola opens up because $a = 1 > 0$.

Step 2: Determine the vertex and axis of symmetry.

The x-coordinate of the vertex is

$$x = -\frac{b}{2a} = -\frac{2}{2(1)} = -1$$

The y-coordinate of the vertex is

$$f\left(-\frac{b}{2a}\right) = f(-1)$$
$$= (-1)^2 + 2(-1) - 15$$
$$= 1 - 2 - 15$$
$$= -16$$

The vertex is $(-1, -16)$. The axis of symmetry is the line $x = -\frac{b}{2a} = -1$.

Step 3: Determine the y-intercept, $f(0)$.

$$f(0) = 0^2 + 2(0) - 15$$
$$= -15$$

The y-intercept is $(0, -15)$.

Step 4: Find the discriminant, $b^2 - 4ac$, to determine the number of x-intercepts. Then determine the x-intercepts, if any.

Because $a = 1$, $b = 2$, and $c = -15$, the value of the discriminant is $b^2 - 4ac = (2)^2 - 4(1)(-15) = 64 > 0$, so the parabola has two x-intercepts. Find the x-intercepts by solving

$$f(x) = 0$$
$$x^2 + 2x - 15 = 0$$
$$(x + 5)(x - 3) = 0$$

Factor: $x + 5 = 0$ or $x - 3 = 0$

Zero-Product Property: $x = -5$ or $x = 3$

The x-intercepts are $(-5, 0)$ and $(3, 0)$.

Step 5: Plot the vertex, y-intercept, and x-intercepts. Use the axis of symmetry to find an additional point. Draw the graph of the quadratic function.

Use the axis of symmetry to find the additional point $(-2, -15)$ by recognizing that the y-intercept, $(0, -15)$, is 1 unit to the right of the axis of symmetry, so there must be a point 1 unit to the left of the axis of symmetry. Plot the points and draw the graph. See Figure 22.

Figure 22

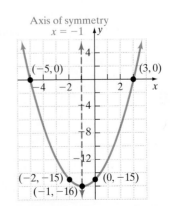

> **Graphing a Quadratic Function Using Its Properties**
>
> To graph any quadratic function of the form $f(x) = ax^2 + bx + c, a \neq 0$, use the following steps:
>
> **Step 1:** Determine whether the parabola opens up or down.
>
> **Step 2:** Determine the vertex and axis of symmetry.
>
> **Step 3:** Determine the y-intercept, $f(0) = c$.
>
> **Step 4:** Determine the discriminant, $b^2 - 4ac$.
> - If $b^2 - 4ac > 0$, then the parabola has two x-intercepts, which are found by solving $f(x) = 0 (ax^2 + bx + c = 0)$.
> - If $b^2 - 4ac = 0$, the vertex is the x-intercept.
> - If $b^2 - 4ac < 0$, there are no x-intercepts.
>
> **Step 5:** Plot the vertex, y-intercept, and any x-intercept(s). Use the axis of symmetry to find an additional point. Draw the graph of the quadratic function.

Quick ✓

1. Any quadratic function $f(x) = ax^2 + bx + c, a \neq 0$, will have a vertex whose x-coordinate is $x =$ ____ .

2. The graph of $f(x) = ax^2 + bx + c$ will have two x-intercepts if $b^2 - 4ac$ ____ 0.

3. How many x-intercepts does the graph of $f(x) = -2x^2 - 3x + 6$ have?

4. What is the vertex of $f(x) = x^2 + 4x - 3$?

5. Graph $f(x) = x^2 - 4x - 12$ using its properties.

In Example 1, the function was factorable, so the x-intercepts were rational numbers. When the quadratic function cannot be factored, use the quadratic formula to find the x-intercepts. For the purpose of graphing the quadratic function, approximate the x-intercepts rounded to two decimal places.

EXAMPLE 2 **Graphing a Quadratic Function Using Its Properties**

Graph $f(x) = -2x^2 + 12x - 5$ using its properties. Based on the graph, determine the domain and range of the quadratic function.

Solution

In $f(x) = -2x^2 + 12x - 5$, $a = -2, b = 12$, and $c = -5$. The parabola opens down because $a = -2 < 0$. The x-coordinate of the vertex is

$$x = -\frac{b}{2a} = -\frac{12}{2(-2)} = 3$$

The y-coordinate of the vertex is

$$y = f\left(-\frac{b}{2a}\right)$$
$$= f(3)$$
$$= -2(3)^2 + 12(3) - 5$$
$$= -18 + 36 - 5$$
$$= 13$$

The vertex is $(3, 13)$. The axis of symmetry is the line $x = -\dfrac{b}{2a} = 3$.

The y-intercept is $f(0) = -2(0)^2 + 12(0) - 5 = -5$. Find the discriminant, $b^2 - 4ac$, to determine the number of x-intercepts. Because $a = -2$, $b = 12$, and $c = -5$, $b^2 - 4ac = (12)^2 - 4(-2)(-5) = 104 > 0$. The parabola will have two x-intercepts. Find the x-intercepts by solving

$$f(x) = 0$$
$$-2x^2 + 12x - 5 = 0$$

The equation cannot be solved by factoring (because $b^2 - 4ac$ is not a perfect square), so use the quadratic formula:

$$x = \frac{-b \pm \sqrt{b^2 - 4ac}}{2a}$$

$a = -2, b = 12, b^2 - 4ac = 104:$
$$= \frac{-12 \pm \sqrt{104}}{2(-2)}$$

$\sqrt{104} = \sqrt{4 \cdot 26} = 2\sqrt{26}:$
$$= \frac{-12 \pm 2\sqrt{26}}{-4}$$

Divide -4 into each term in the numerator and simplify: $\quad x = 3 \pm \dfrac{\sqrt{26}}{2} \quad$ Exact solution

Figure 23

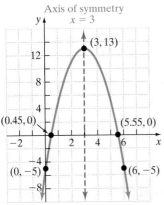

Axis of symmetry
$x = 3$

Evaluate $3 \pm \dfrac{\sqrt{26}}{2}$ and find the x-intercepts are approximately 0.45 and 5.55. Use the axis of symmetry to find the additional point $(6, -5)$. (The y-intercept, $(0, -5)$, is 3 units to the left of the axis of symmetry, therefore, there must be a point 3 units to the right of the axis of symmetry.) Next, plot the vertex, the intercepts, and the additional point. Draw the graph of the quadratic function. See Figure 23. Based on the graph, the domain is $\{x \mid x \text{ is any real number}\}$ or $(-\infty, \infty)$ and the range is $\{y \mid y \le 13\}$ or $(-\infty, 13]$. ●

Work Smart

Notice that the vertex in Example 2 lies in quadrant I (above the x-axis) and the graph opens down. This tells us the graph must have two x-intercepts.

Quick ✓

6. Graph $f(x) = -3x^2 + 12x - 7$ using its properties. Based on the graph, determine the domain and range of the quadratic function.

(**EXAMPLE 3**)   **Graphing a Quadratic Function Using Its Properties**

Graph $g(x) = x^2 - 8x + 16$ using its properties. Based on the graph, determine the domain and range of the quadratic function.

Solution

In $g(x) = x^2 - 8x + 16$, $a = 1$, $b = -8$, and $c = 16$. The parabola opens up because $a = 1 > 0$. The x-coordinate of the vertex is

$$x = -\frac{b}{2a} = -\frac{-8}{2(1)} = 4$$

The y-coordinate of the vertex is

$$y = f\left(-\frac{b}{2a}\right)$$
$$= f(4)$$
$$= 4^2 - 8(4) + 16$$
$$= 16 - 32 + 16$$
$$= 0$$

The vertex is $(4, 0)$. The axis of symmetry is the line $x = -\dfrac{b}{2a} = 4$.

Figure 24

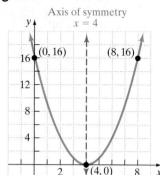

Axis of symmetry
x = 4

The y-intercept is $f(0) = 0^2 - 8(0) + 16 = 16$. Because $a = 1, b = -8,$ and $c = 16, b^2 - 4ac = (-8)^2 - 4(1)(16) = 0$. The vertex $(4, 0)$ is on the x-axis, so the parabola has one x-intercept. Verify this by solving

$$f(x) = 0$$
$$x^2 - 8x + 16 = 0$$
Factor: $(x - 4)^2 = 0$
$$x = 4$$

Use the point $(0, 16)$ and the axis of symmetry to find the additional point $(8, 16)$. The parabola is shown in Figure 24. Based on the graph, the domain is $\{x | x \text{ is any real number}\}$ or $(-\infty, \infty)$ and the range is $\{y | y \geq 0\}$ or $[0, \infty)$. ●

Quick ✓

7. Graph $f(x) = x^2 + 6x + 9$ using its properties. Based on the graph, determine the domain and range of the quadratic function.

▶ (**EXAMPLE 4**)   **Graphing a Quadratic Function Using Its Properties**

Graph $F(x) = 2x^2 + 6x + 5$ using its properties. Based on the graph, determine the domain and range of the quadratic function.

Solution

In $F(x) = 2x^2 + 6x + 5, a = 2, b = 6,$ and $c = 5$. The parabola opens up because $a = 2 > 0$. The x-coordinate of the vertex is

$$x = -\frac{b}{2a} = -\frac{6}{2(2)} = -\frac{3}{2}$$

The y-coordinate of the vertex is

$$y = f\left(-\frac{b}{2a}\right) = f\left(-\frac{3}{2}\right)$$
$$= 2\left(-\frac{3}{2}\right)^2 + 6\left(-\frac{3}{2}\right) + 5$$
$$= 2\left(\frac{9}{4}\right) - 9 + 5$$
$$= \frac{1}{2}$$

Work Smart

Notice that the vertex lies in quadrant II (above the x-axis) and the graph opens up. This tells us the graph has no x-intercepts.

The vertex is $\left(-\frac{3}{2}, \frac{1}{2}\right)$. The axis of symmetry is the line $x = -\frac{b}{2a} = -\frac{3}{2}$.

The y-intercept is $f(0) = 2(0)^2 + 6(0) + 5 = 5$. Because the parabola opens up and the vertex is in the second quadrant, the parabola has no x-intercepts. Verify this by computing the discriminant with $a = 2, b = 6,$ and $c = 5$:

$$b^2 - 4ac = 6^2 - 4(2)(5) = -4 < 0$$

The parabola has no x-intercepts. Use the point $(0, 5)$ and the axis of symmetry to find the additional point $(-3, 5)$. The parabola is shown in Figure 25. Based on the graph, the domain is $\{x | x \text{ is any real number}\}$ or $(-\infty, \infty)$ and the range is $\left\{y | y \geq \frac{1}{2}\right\}$ or $\left[\frac{1}{2}, \infty\right)$. ●

Figure 25

Axis of symmetry
$x = -\frac{3}{2}$

Quick ✓

8. Graph $G(x) = -3x^2 + 9x - 8$ using its properties. Based on the graph, determine the domain and range of the quadratic function.

Figure 26

Vertex is
highest point;
y-coordinate is
maximum value

Vertex is
lowest point;
y-coordinate is
minimum value

(a) Opens up
$a > 0$

(b) Opens down
$a < 0$

▶ ② Find the Maximum or Minimum Value of a Quadratic Function

Recall, the graph of a quadratic function $f(x) = ax^2 + bx + c$ is a parabola with vertex at $\left(-\dfrac{b}{2a}, f\left(-\dfrac{b}{2a}\right)\right)$. The vertex is the highest point on the graph if $a < 0$ and the lowest point on the graph if $a > 0$. If the vertex is the highest point then $f\left(-\dfrac{b}{2a}\right)$, the y-coordinate of the vertex, is the **maximum value** of f. If the vertex is the lowest point, then $f\left(-\dfrac{b}{2a}\right)$ is the **minimum value** of f. See Figure 26.

Questions involving **optimization,** the process of finding the maximum or minimum value(s) of a function, can now be answered. In the case of quadratic functions, the maximum $(a < 0)$ or minimum $(a > 0)$ is found at the vertex.

EXAMPLE 5   **Finding the Maximum or Minimum Value of a Quadratic Function**

Determine whether the quadratic function

$$f(x) = 3x^2 + 12x - 7$$

has a maximum or a minimum value. Then find the maximum or minimum value of the function, and indicate when it occurs.

Solution

In $f(x) = 3x^2 + 12x - 7$, $a = 3$, $b = 12$, and $c = -7$. Because $a = 3 > 0$, the graph of the quadratic function opens up, so the function has a minimum value. The minimum value of the function occurs at

$$x = -\frac{b}{2a} = -\frac{12}{2(3)} = -2$$

Work Smart

The vertex is not the maximum or minimum value—the y-coordinate of the vertex is the maximum or minimum value of the function.

The minimum value of the function is

$$
\begin{aligned}
f\left(-\frac{b}{2a}\right) &= f(-2) \\
&= 3(-2)^2 + 12(-2) - 7 \\
&= 3(4) - 24 - 7 \\
&= -19
\end{aligned}
$$

The minimum value of the function is -19 and occurs at $x = -2$. ●

Quick ✓

9. *True or False* For the quadratic function $f(x) = ax^2 + bx + c$, if $a < 0$ then $f\left(-\dfrac{b}{2a}\right)$ is the maximum value of f.

In Problems 10 and 11, determine whether the quadratic function has a maximum or a minimum value. Then find the maximum or minimum value of the function and where it occurs.

10. $f(x) = 2x^2 - 8x + 1$ **11.** $G(x) = -x^2 + 10x + 8$

Properties of quadratic functions are useful to answer questions regarding applications modeled by quadratic functions.

EXAMPLE 6 **Maximizing Revenue**

Suppose the marketing department of Dell Computer has found that, when a certain model of computer is sold at a price of p dollars, the daily revenue R (in dollars) as a function of the price p is the function

$$R(p) = -\frac{1}{4}p^2 + 400p$$

(a) For what price will the revenue be maximized?

(b) What is the maximum daily revenue?

Solution

(a) Notice that the revenue function is a quadratic function whose graph opens down because $a = -\frac{1}{4} < 0$. Therefore, the function has a maximum at

$$p = -\frac{b}{2a} = -\frac{400}{2\left(-\dfrac{1}{4}\right)} = -\frac{400}{-\dfrac{1}{2}} = 800$$

Revenue will be maximized when the price is $p = \$800$.

(b) The maximum daily revenue is found by letting $p = \$800$ in the revenue function.

$$R(800) = -\frac{1}{4}(800)^2 + 400 \cdot 800$$

$$= \$160,000$$

The maximum daily revenue is $\$160,000$. See Figure 27 for an illustration.

Figure 27

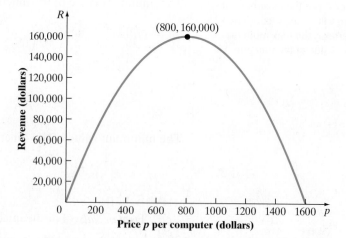

Quick ✓

12. Suppose that the marketing department of Texas Instruments has found that, when a certain model of calculator is sold at a price of p dollars, the daily revenue R (in dollars) as a function of the price p is the function $R(p) = -0.5p^2 + 75p$.

(a) For what price will the daily revenue be maximized?

(b) What is the maximum daily revenue?

▶ ❸ Model and Solve Optimization Problems Involving Quadratic Functions

Models that result in quadratic functions will now be discussed. Use the problem-solving strategy introduced in Section 1.2.

EXAMPLE 7 | **Maximizing the Area Enclosed by a Fence**

A farmer has 3000 feet of fence with which to enclose a rectangular field. What is the maximum area that can be enclosed by the fence? What are the dimensions of the rectangle that encloses the most area?

Solution

Step 1: Identify Find the dimensions of a rectangle that maximize the area.

Step 2: Name Let w represent the width of the rectangle and let l represent its length, both in feet.

Step 3: Translate See Figure 28.

Figure 28

Because 3000 feet of fence is available, the perimeter of the rectangle will be 3000 feet. Use the perimeter formula of a rectangle to get

$$2l + 2w = 3000$$

The area of the rectangle, $A = lw$, uses two variables, l and w. To express A in terms of only one variable, solve the equation $2l + 2w = 3000$ for l and then substitute for l in the area formula $A = lw$.

Work Smart

The equation $2l + 2w = 3000$ could also be solved for w. Ultimately the same solution would be obtained.

$$2l + 2w = 3000$$

Subtract 2w from both sides: $\quad 2l = 3000 - 2w$

Divide both sides by 2: $\quad l = \dfrac{3000 - 2w}{2}$

Simplify: $\quad l = 1500 - w$

Next let $l = 1500 - w$ in the formula $A = lw$.

$$A = (1500 - w)w$$
$$= -w^2 + 1500w$$

Now, A is a quadratic function of w.

$$A(w) = -w^2 + 1500w \quad \text{The Model}$$

Step 4: Solve Find the dimensions that result in a maximum area enclosed by the fence. The model developed in Step 3 is a quadratic function that opens down (because $a = -1 < 0$), so the vertex is a maximum point on the graph of A. The maximum value occurs at

$$w = -\frac{b}{2a} = -\frac{1500}{2(-1)} = 750$$

(continued)

The maximum value of A is

$$A\left(-\frac{b}{2a}\right) = A(750)$$
$$= -750^2 + 1500(750)$$
$$= 562{,}500 \text{ square feet}$$

Because $l = 1500 - w$, then if the width is 750 feet, the length will be $l = 1500 - 750 = 750$ feet.

Step 5: Check With a length and width of 750 feet, the perimeter is $2(750) + 2(750) = 3000$ feet. The area is $(750 \text{ feet})(750 \text{ feet}) = 562{,}500$ square feet. Everything checks!

Step 6: Answer The largest area that can be enclosed by 3000 feet of fence has an area of 562,500 square feet. Its dimensions are 750 feet by 750 feet.

$\bullet$

Quick ✓

13. Roberta has 1000 yards of fence to enclose a rectangular field. What is the maximum area that can be enclosed by the fence? What are the dimensions of the rectangle that encloses the most area?

EXAMPLE 8   **Pricing a Charter**

Chicago Tours offers boat charters along the Chicago coastline on Lake Michigan. Normally, a ticket costs $20 per person, but for any group, Chicago Tours will lower the price of a ticket by $0.10 per person for each person in excess of 30. Determine the group size that will maximize revenue. What is the maximum revenue that can be earned from a group sale?

Solution

Step 1: Identify This is a direct translation problem involving revenue. Remember, revenue is price times quantity.

Step 2: Name Let x represent the number in the group in excess of 30.

Step 3: Translate Revenue is price times quantity. If 30 individuals make up a group, the revenue to Chicago Tours will be $20(30)$. If 31 individuals make up a group, revenue will be $19.90(31)$. The revenue for 32 individuals will be $19.80(32)$. In general, for $30 + x$ individuals, revenue will be $(20 - 0.1x)(30 + x)$. Thus the revenue R for a group that has $30 + x$ people in it is given by

$$R(x) = (20 - 0.1x)(30 + x)$$
$$\text{FOIL:} \quad = 600 + 20x - 3x - 0.1x^2$$
$$= -0.1x^2 + 17x + 600 \qquad \text{The Model}$$

Step 4: Solve Find the revenue-maximizing number of individuals in the group. The function R is a quadratic function with a graph that opens down (because $a = -0.1 < 0$), so the vertex is a maximum point. The value of x that results in a maximum is given by

$$x = -\frac{b}{2a} = -\frac{17}{2(-0.1)}$$
$$= 85$$

The maximum revenue is

$$R(85) = -0.1(85)^2 + 17(85) + 600$$
$$= \$1322.50$$

Step 5: Check Because x represents the number of passengers in excess of 30, $30 + 85 = 115$ tickets should be sold to maximize revenue. The cost per ticket is $\$20 - 0.1(85) = \$20 - \$8.50 = \$11.50$. The cost per ticket times the number of passengers is $\$11.50(115) = \$1322.50$. The answer checks!

Step 6: Answer A group sale of 115 passengers will maximize revenue. The maximum revenue is $1322.50.

Quick ✓

14. A flash drive manufacturer charges $100 for each box of flash drives ordered. However, it reduces the price by $1 per box for each box in excess of 30 boxes but less than 90 boxes. Determine the number of boxes of flash drives that should be sold to maximize revenue. What is the maximum revenue?

7.5 Exercises MyMathLab®

Exercise numbers in green have complete video solutions in MyMathLab or may be accessed using the QR code to the right.

*Problems **1–14** are the Quick ✓s that follow the EXAMPLES.*

Building Skills

In Problems 15–22, (a) find the vertex of each parabola, (b) use the discriminant to determine the number of x-intercepts the graph will have. Then determine the x-intercepts. See Objective 1.

15. $f(x) = x^2 - 6x - 16$

16. $g(x) = x^2 + 4x - 12$

17. $G(x) = -2x^2 + 4x - 5$

18. $H(x) = x^2 - 4x + 5$

19. $h(x) = 4x^2 + 4x + 1$

20. $f(x) = x^2 - 6x + 9$

21. $F(x) = 4x^2 - x - 1$

22. $P(x) = -2x^2 + 3x + 1$

In Problems 23–62, graph each quadratic function using its properties by following Steps 1–5 on page 592. Based on the graph, determine the domain and range of the quadratic function. See Objective 1.

23. $f(x) = x^2 - 4x - 5$

24. $f(x) = x^2 - 2x - 8$

25. $G(x) = x^2 + 12x + 32$

26. $g(x) = x^2 - 12x + 27$

27. $F(x) = -x^2 + 2x + 8$

28. $g(x) = -x^2 + 2x + 15$

29. $H(x) = x^2 - 4x + 4$

30. $h(x) = x^2 + 6x + 9$

31. $g(x) = x^2 + 2x + 5$

32. $f(x) = x^2 - 4x + 7$

33. $h(x) = -x^2 - 10x - 25$

34. $P(x) = -x^2 - 12x - 36$

35. $p(x) = -x^2 + 2x - 5$

36. $f(x) = -x^2 + 4x - 6$

37. $F(x) = 4x^2 - 4x - 3$

38. $f(x) = 4x^2 - 8x - 21$

39. $G(x) = -9x^2 + 18x + 7$

40. $g(x) = -9x^2 - 36x - 20$

41. $H(x) = 4x^2 - 4x + 1$

42. $h(x) = 9x^2 + 12x + 4$

43. $f(x) = -16x^2 - 24x - 9$

44. $F(x) = -4x^2 - 20x - 25$

45. $f(x) = 2x^2 + 8x + 11$

46. $F(x) = 3x^2 + 6x + 7$

47. $P(x) = -4x^2 + 6x - 3$

48. $p(x) = -2x^2 + 6x + 5$

49. $h(x) = x^2 + 5x + 3$

50. $H(x) = x^2 + 3x + 1$

51. $G(x) = -3x^2 + 8x + 2$

52. $F(x) = -2x^2 + 6x + 1$

53. $f(x) = 5x^2 - 5x + 2$

54. $F(x) = 4x^2 + 4x - 1$

55. $H(x) = -3x^2 + 6x$

56. $h(x) = -4x^2 + 8x$

57. $f(x) = x^2 - \dfrac{5}{2}x - \dfrac{3}{2}$

58. $g(x) = x^2 + \dfrac{5}{2}x - 6$

59. $G(x) = \dfrac{1}{2}x^2 + 2x - 6$

60. $H(x) = \dfrac{1}{4}x^2 + x - 8$

61. $F(x) = -\dfrac{1}{4}x^2 + x + 15$

62. $G(x) = -\dfrac{1}{2}x^2 - 8x - 24$

In Problems 63–74, determine whether the quadratic function has a maximum or a minimum value. Then find the maximum or minimum value and where it occurs. See Objective 2.

63. $f(x) = x^2 + 8x + 13$

64. $f(x) = x^2 - 6x + 3$

65. $G(x) = -x^2 - 10x + 3$

66. $g(x) = -x^2 + 4x + 12$

67. $F(x) = -2x^2 + 12x + 5$

68. $H(x) = -3x^2 + 12x - 1$

69. $h(x) = 4x^2 + 16x - 3$

70. $G(x) = 5x^2 + 10x - 1$

71. $f(x) = 2x^2 - 5x + 1$ **72.** $F(x) = 3x^2 + 4x - 3$

73. $H(x) = -3x^2 + 4x + 1$ **74.** $h(x) = -4x^2 - 6x + 1$

Applying the Concepts

75. Revenue Function Suppose that the marketing department of Samsung has found that, when a certain model of DVD player is sold at a price of p dollars, the daily revenue R (in dollars) as a function of the price p is $R(p) = -2.5p^2 + 600p$.

   (a) For what price will the daily revenue be maximized?

   (b) What is the maximum daily revenue?

76. Revenue Function Suppose that the marketing department of Samsung has found that, when a certain model of cell phones is sold at a price of p dollars, the daily revenue R (in dollars) as a function of the price p is $R(p) = -5p^2 + 600p$.

   (a) For what price will the daily revenue be maximized?

   (b) What is the maximum daily revenue?

77. Marginal Cost The marginal cost of a product can be thought of as the cost of producing one additional unit of output. For example, if the marginal cost of producing the fiftieth unit of a product is $6.30, then it costs $6.30 to increase production from 49 to 50 units of output. Suppose that the marginal cost C (in dollars) to produce x digital cameras is given by $C(x) = 0.05x^2 - 6x + 215$. How many digital cameras should be produced to minimize marginal cost? What is the minimum marginal cost?

78. Marginal Cost (See Problem 77.) The marginal cost C (in dollars) of manufacturing x MP3 players is given by $C(x) = 0.05x^2 - 9x + 435$. How many MP3 players should be manufactured to minimize marginal cost? What is the minimum marginal cost?

79. Punkin Chunkin Suppose that an air cannon in the Punkin Chunkin contest whose muzzle is 10 feet above the ground fires a pumpkin at an angle of 45° to the horizontal with a muzzle velocity of 335 feet per second. The model $s(t) = -16t^2 + 240t + 10$ can be used to estimate the height s of a pumpkin after t seconds.

   (a) Determine the time at which the pumpkin is at its maximum height.

   (b) Determine the maximum height of the pumpkin.

   (c) After how long will the pumpkin strike the ground?

80. Punkin Chunkin Suppose that a catapult in the Punkin Chunkin contest releases a pumpkin 8 feet above the ground at an angle of 45° to the horizontal with an initial speed of 220 feet per second. The model $s(t) = -16t^2 + 155t + 8$ can be used to estimate the height s of a pumpkin after t seconds.

   (a) Determine the time at which the pumpkin is at its maximum height.

   (b) Determine the maximum height of the pumpkin.

   (c) After how long will the pumpkin strike the ground?

81. Punkin Chunkin Suppose that an air cannon in the Punkin Chunkin contest whose muzzle is 10 feet above the ground fires a pumpkin at an angle of 45° to the horizontal with a muzzle velocity of 335 feet per second. The model $h(x) = \dfrac{-32}{335^2}x^2 + x + 10$ can be used to estimate the height h of the pumpkin after it has traveled x feet.

   (a) How far from the cannon will the pumpkin reach its maximum height?

   (b) What is the maximum height of the pumpkin?

   (c) How far will the pumpkin travel before it strikes the ground?

   (d) Compare your answer in part (b) of this problem with the answer found in part (b) of Problem 79. Why might the answers differ?

82. Punkin Chunkin Suppose that a catapult in the Punkin Chunkin contest releases a pumpkin 8 feet above the ground at an angle of 45° to the horizontal with an initial speed of 220 feet per second. The model $h(x) = \dfrac{-32}{220^2}x^2 + x + 8$ can be used to estimate the height h of the pumpkin after it has traveled x feet.

   (a) How far from the cannon will the pumpkin reach its maximum height?

   (b) What is the maximum height of the pumpkin?

   (c) How far will the pumpkin travel before it strikes the ground?

   (d) Compare your answer in part (b) of this problem with the answer found in part (b) of Problem 80. Why might the answers differ?

83. Life Cycle Hypothesis The Life Cycle Hypothesis from economics was presented by Franco Modigliani in 1954. One of its components states that income is a function of age. The function $I(a) = -55a^2 + 5119a - 54,448$ represents the relation between average annual income I and age a.

   (a) According to the model, at what age will average income be a maximum?

   (b) According to the model, what is the maximum average income?

84. Age and Divorce The percentage of Americans who divorce is a function of age. The polynomial function $D(a) = -0.014a^2 + 1.586a - 28.1889$ models the percent D of Americans of age a who are divorced.

(a) To the nearest year, what is the age for which the highest percentage of Americans are divorced?

(b) To the nearest percentage, what is the percentage of Americans who are divorced?

85. Fun with Numbers The sum of two numbers is 36. Find the numbers such that their product is a maximum.

86. Fun with Numbers The sum of two numbers is 50. Find the numbers such that their product is a maximum.

87. Fun with Numbers The difference of two numbers is 18. Find the numbers such that their product is a minimum.

88. Fun with Numbers The difference of two numbers is 10. Find the numbers such that their product is a minimum.

89. Enclosing a Rectangular Field Maurice has 500 yards of fencing and wishes to enclose a rectangular area. What is the maximum area that can be enclosed by the fence? What are the dimensions of the area enclosed?

90. Enclosing a Rectangular Field Maude has 800 yards of fencing and wishes to enclose a rectangular area. What is the maximum area that can be enclosed by the fence? What are the dimensions of the area enclosed?

91. Maximizing an Enclosed Area A farmer with 2000 meters of fencing wants to enclose a rectangular plot that borders a river. If the farmer does not fence the side along the river, what is the largest area that can be enclosed? What are the dimensions of the enclosed area? See the figure.

92. Maximizing an Enclosed Area A farmer with 8000 meters of fencing wants to enclose a rectangular plot and then divide it into two plots with a fence parallel to one of the sides. See the figure. What is the largest area that can be enclosed? What are the lengths of the sides of each part of the enclosed area?

93. Constructing Rain Gutters A rain gutter is to be made of aluminum sheets that are 20 inches wide by turning up the edges 90°. What depth will provide maximum cross-sectional area and hence allow the most water to flow?

94. Maximizing the Volume of a Box A box with a rectangular base is to be constructed such that the perimeter of the base of the box is to be 40 inches. The height of the box must be 15 inches. Find the dimensions of the box such that the volume is maximized. What is the maximum volume?

95. Revenue Function Weekly demand for jeans at a department store obeys the demand equation

$$x = -p + 110$$

where x is the quantity demanded and p is the price (in dollars).

(a) Express the revenue R as a function of p. (*Hint: R = xp*)

(b) What price p maximizes revenue? What is the maximum revenue?

(c) How many pairs of jeans will be sold at the revenue-maximizing price?

96. Revenue Function Demand for hot dogs at a baseball game obeys the demand equation

$$x = -800p + 8000$$

where x is the quantity demanded and p is the price (in dollars).

(a) Express the revenue R as a function of p. (*Hint: R = xp*)

(b) What price p maximizes revenue? What is the maximum revenue?

(c) How many hot dogs will be sold at the revenue-maximizing price?

Extending the Concepts

Answer Problems 97 and 98 using the following information: A quadratic function of the form $f(x) = ax^2 + bx + c$ with $b^2 - 4ac > 0$ may also be written in the form $f(x) = a(x - r_1)(x - r_2)$, where r_1 and r_2 are the x-intercepts of the graph of the quadratic function.

97. (a) Find a quadratic function whose x-intercepts are 2 and 6 with $a = 1$, $a = 2$, and $a = -2$.
   (b) How does the value of a affect the intercepts?
   (c) How does the value of a affect the axis of symmetry?
   (d) How does the value of a affect the vertex?

98. (a) Find a quadratic function whose x-intercepts are -1 and 5 with $a = 1$, $a = 2$, and $a = -2$.
   (b) How does the value of a affect the intercepts?
   (c) How does the value of a affect the axis of symmetry?
   (d) How does the value of a affect the vertex?

Explaining the Concepts

99. Explain how the discriminant is used to determine the number of x-intercepts the graph of a quadratic function will have.

100. Provide two methods for finding the vertex of any quadratic function $f(x) = ax^2 + bx + c$.

101. Refer to Example 6 on page 596. Notice that if the price charged for the computer is $0 or $1600, the revenue is $0. It is easy to explain why revenue would be $0 if the price charged were $0, but how can revenue be $0 if the price charged is $1600?

Synthesis Review

In Problems 102–105, graph each function using point plotting.

102. $f(x) = -2x + 12$
103. $G(x) = \frac{1}{4}x - 2$
104. $f(x) = x^2 - 5$
105. $f(x) = (x + 2)^2 + 4$

106. For each function in Problems 102–105, explain an alternative method for graphing the function. Which method do you prefer? Why?

Technology Exercises

Technology can be used to determine the coordinates of the vertex of a parabola. In Desmos, once the function is graphed, click on the curve and gray dots will appear at "interesting points," including the vertex. Hover over the point with your mouse to display the ordered pair. See Figure 29.

Figure 29

In Problems 107–114, use technology to graph each quadratic function and determine the vertex. If necessary, round your answers to two decimal places.

107. $f(x) = x^2 - 7x + 3$ **108.** $f(x) = x^2 + 3x + 8$

109. $G(x) = -2x^2 + 14x + 13$

110. $g(x) = -4x^2 - x + 11$

111. $F(x) = 5x^2 + 3x - 20$ **112.** $F(x) = 3x^2 + 2x - 21$

113. $H(x) = \frac{1}{2}x^2 - \frac{2}{3}x + 5$ **114.** $h(x) = \frac{3}{4}x^2 + \frac{4}{3}x - 1$

115. Exploration: Quadratic Functions—b and c Values
Open the "Quadratic Functions–Standard Form" Geogebra applet. The applet may be found using the QR code at the beginning of this section or through the Multimedia Library in MyMathLab. If you are using a keyboard, it is easier to move the sliders with the arrow keys instead of a mouse. Uncheck the boxes in the applet. Move the sliders so that $a = 1$, $b = 0$, and $c = 0$.
   (a) Move the b slider to take on the values from -5 to 5. What role does b play in the graph of $y = ax^2 + bx + c$?
   (b) Now reset the sliders so that $a = 1$, $b = 0$, and $c = 0$ and check the y-intercept box. Move the slider c to take on the values from -5 to 5. For each value of c write down the corresponding y-intercept of the parabola. What relationship exists between c and this point?

116. Exploration: Quadratic Functions—Discriminant *Open the "Quadratic Functions—Standard Form" Geogebra applet. The applet may be found using the QR code at the beginning of this section or through the Multimedia Library in MyMathLab. If you are using a keyboard, it is easier to move the sliders with the arrow keys instead of a mouse. Uncheck the boxes in the applet. Move the sliders so that $a = 1$, $b = 0$, and $c = 0$ and check the x-intercept and discriminant boxes.*
   (a) Move the a, b, and c sliders to find two different functions each of which has a discriminant that is a positive number. How many x-intercepts does each of these functions have?
   (b) Now move the a, b, and c sliders to find two different functions each of which has a discriminant that is a negative number. How many x-intercepts does each of these functions have?
   (c) Now move the a, b, and c sliders to find two different functions each of which has a discriminant that is zero. How many x-intercepts does each of these functions have?
   (d) In general, what is the relationship between the discriminant of a quadratic function and the number of x-intercepts it has? In other words, when the discriminant is positive, the function has ___ x-intercepts; when the discriminant is negative, the function has ___ x-intercepts; and when the discriminant is zero, the function has ___ x-intercepts.

7.6 Polynomial Inequalities

Objectives

1. Solve Quadratic Inequalities
2. Solve Polynomial Inequalities

Are You Prepared for This Section?

Before getting started, complete the following problems. If you get a problem wrong, go back to the section cited and review the material.

P1. Write $-4 \leq x < 5$ in interval notation.   [Section 1.4, pp. 82–85]

P2. Solve: $3x + 5 > 5x - 3$ [Section 1.4, pp. 86–88]

In Section 1.4, linear inequalities in one variable, such as $2x - 3 > 4x + 5$, were solved using methods similar to solving linear equations. The solution sets of inequalities were represented using either set-builder notation or interval notation.

Solving *polynomial inequalities* is not a simple extension of solving polynomial equations. However, the skills developed in solving polynomial equations are used to solve polynomial inequalities.

> **Definition**
>
> A **polynomial inequality** is an inequality of the form
>
> $$f(x) > 0 \quad f(x) < 0 \quad f(x) \geq 0 \quad f(x) \leq 0$$
>
> where f is a polynomial function.

Some examples of polynomial inequalities are

$$x^2 - 3x - 4 > 0 \quad 2x^2 - 5x + 1 < 0 \quad x^3 - 6x^2 + 7x + 1 \geq 0$$

We begin by looking at quadratic inequalities—a special case of polynomial inequalities.

1 Solve Quadratic Inequalities

> **Definition**
>
> A **quadratic inequality** is an inequality of the form
>
> $$ax^2 + bx + c > 0 \quad \text{or} \quad ax^2 + bx + c < 0 \quad \text{or}$$
> $$ax^2 + bx + c \geq 0 \quad \text{or} \quad ax^2 + bx + c \leq 0$$
>
> where $a \neq 0$.

First a graphical approach to the solution of a quadratic inequality will be presented and then an algebraic method.

To understand the logic behind the graphical approach, consider the following. Suppose we want to solve the inequality $ax^2 + bx + c > 0$. If $f(x) = ax^2 + bx + c$, then we want all x-values such that $f(x) > 0$. Since $f(x)$ represents the y-values of the graph of the function $f(x) = ax^2 + bx + c$, look for all x-values such that the graph of f is *above* the x-axis that is, where y-values are positive. This occurs when the graph lies in either quadrant I or quadrant II of the Cartesian plane. If asked to solve $f(x) < 0$, look for the x-values such that the graph is *below* the x-axis that is, where the y-values are negative. This occurs when the graph lies in either quadrant III or quadrant IV of the Cartesian plane. See Figure 30 on the next page for an illustration of this idea.

Figure 30

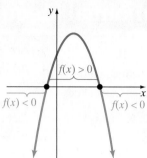

EXAMPLE 1 **How to Solve a Quadratic Inequality Using the Graphical Method**

Solve $x^2 - 4x - 5 \geq 0$ using the graphical method.

Step-by-Step Solution

Step 1: Write the inequality so that $ax^2 + bx + c$ is on one side of the inequality and 0 is on the other.

The inequality is already in the form needed.

$$x^2 - 4x - 5 \geq 0$$

Step 2: Graph the function $f(x) = ax^2 + bx + c$. Label the x-intercepts on the graph.

Graph $f(x) = x^2 - 4x - 5$. Because $a = 1 > 0$, the graph opens up.

x-intercepts: $f(x) = 0$ **y-intercept:** $f(0) = -5$

$x^2 - 4x - 5 = 0$ The y-intercept is $(0, -5)$.

$(x - 5)(x + 1) = 0$

$x = 5$ or $x = -1$

The x-intercepts are $(-1, 0)$ and $(5, 0)$.

$$\textbf{Vertex:} \quad x = -\frac{b}{2a} = -\frac{-4}{2(1)} = 2$$

$$y = f\left(-\frac{b}{2a}\right) = f(2) = -9$$

The vertex is at $(2, -9)$. Figure 31 shows the graph.

Figure 31

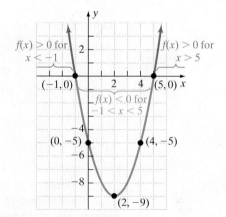

Step 3: From the graph, determine where the function is positive and where it is negative. This information will give the solution set to the inequality.

Figure 31 shows that the graph of $f(x) = x^2 - 4x - 5$ is greater than 0 for $x < -1$ or $x > 5$. Because the inequality is nonstrict ($\geq$), include the x-intercepts in the solution. The solution is $\{x \mid x \leq -1 \text{ or } x \geq 5\}$ using set-builder notation, or $(-\infty, -1] \cup [5, \infty)$ using interval notation. See Figure 32 for a graph of the solution set.

Figure 32

$\xleftarrow{\hspace{1cm}} \quad -4 \ -3 \ -2 \ -1 \quad 0 \quad 1 \quad 2 \quad 3 \quad 4 \quad 5 \quad 6 \quad 7 \xrightarrow{\hspace{1cm}}$

Quick ✓

In Problem 1, solve the quadratic inequality using the graphical method. Graph the solution set.

1. $x^2 + 3x - 10 \geq 0$

The second method for solving inequalities is algebraic. This method uses the Rules of Signs for multiplying or dividing real numbers. Recall that the product of two positive real numbers is positive, the product of a positive real number and a negative real number is negative, and the product of two negative real numbers is positive.

Let's solve the inequality from Example 1 using the algebraic method.

EXAMPLE 2 **How to Solve a Quadratic Inequality Using the Algebraic Method**

Solve the inequality from Example 1, $x^2 - 4x - 5 \geq 0$, using the algebraic method.

Step-by-Step Solution

Step 1: Write the inequality so that $ax^2 + bx + c$ is on one side of the inequality and 0 is on the other.

The inequality is already in the form needed.

$$x^2 - 4x - 5 \geq 0$$

Step 2: Determine the solutions to the equation $ax^2 + bx + c = 0$.

$$x^2 - 4x - 5 = 0$$

Factor: $\quad (x - 5)(x + 1) = 0$

Use the Zero-Product Property: $\qquad x - 5 = 0 \quad$ or $\quad x + 1 = 0$

$$x = 5 \quad \text{or} \qquad x = -1$$

Step 3: Use the solutions to the equation from Step 2 to separate the real number line into intervals.

Separate the real number line into intervals:

$$(-\infty, -1) \qquad (-1, 5) \qquad (5, \infty)$$

Step 4: Write $x^2 - 4x - 5$ in factored form as $(x - 5)(x + 1)$. Choose a test point within each interval formed in Step 3 to determine the sign of each factor. Then determine the sign of the product. Also write the value of $x^2 - 4x - 5$ at each solution found in Step 2.

The sign of a binomial factor will change only on either side of its corresponding zero. For example, the sign of $x + 1$ changes from negative to positive on either side of $x = -1$. Therefore, when a test point is chosen in each interval, if one number in an interval satisfies the inequality, then all the numbers in that interval will satisfy the inequality.

- In the interval $(-\infty, -1)$, choose a test point of -2. When $x = -2$, the factor $x - 5$ equals $-2 - 5 = -7$, and the factor $x + 1$ equals $-2 + 1 = -1$. Because the product of two negatives is positive, the expression $(x - 5)(x + 1)$ is positive for all x in the interval $(-\infty, -1)$.

- In the interval $(-1, 5)$, let 0 be the test point. When $x = 0, x - 5$ is negative and $x + 1$ is positive. Therefore, $(x - 5)(x + 1)$ is negative for all x in the interval $(-1, 5)$.

- In the interval $(5, \infty)$, let 6 be the test point. For $x = 6$, both $x - 5$ and $x + 1$ are positive, so $(x - 5)(x + 1)$ is positive in the interval $(5, \infty)$.

- For $x = -1$ and $x = 5$, the value of $x^2 - 4x - 5 = (x - 5)(x + 1)$ is zero.

(continued)

Table 9 shows the sign of $(x - 5)(x + 1)$ in each interval, as well as the value of $x^2 - 4x - 5$ at $x = -1$ and at $x = 5$.

Table 9

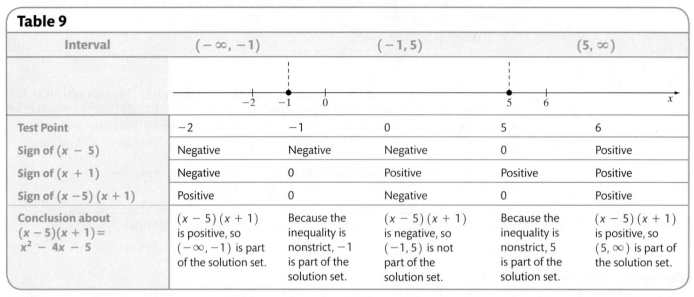

Interval	$(-\infty, -1)$		$(-1, 5)$		$(5, \infty)$
Test Point	-2	-1	0	5	6
Sign of $(x - 5)$	Negative	Negative	Negative	0	Positive
Sign of $(x + 1)$	Negative	0	Positive	Positive	Positive
Sign of $(x - 5)(x + 1)$	Positive	0	Negative	0	Positive
Conclusion about $(x - 5)(x + 1) =$ $x^2 - 4x - 5$	$(x - 5)(x + 1)$ is positive, so $(-\infty, -1)$ is part of the solution set.	Because the inequality is nonstrict, -1 is part of the solution set.	$(x - 5)(x + 1)$ is negative, so $(-1, 5)$ is not part of the solution set.	Because the inequality is nonstrict, 5 is part of the solution set.	$(x - 5)(x + 1)$ is positive, so $(5, \infty)$ is part of the solution set.

Figure 33

To find where $x^2 - 4x - 5$ is greater than or equal to zero, include -1 and 5 in the solution. The solution set is $\{x \mid x \le -1 \text{ or } x \ge 5\}$ in set-builder notation, or $(-\infty, -1] \cup [5, \infty)$ in interval notation. See Figure 33 for a graph of the solution set. ●

Quick ✓

In Problem 2, solve the quadratic inequality using the algebraic method.

2. $x^2 + 3x - 10 \ge 0$

Solving Quadratic Inequalities

Graphical Method	Algebraic Method
Step 1: Write the inequality so that $ax^2 + bx + c$ is on one side of the inequality and 0 is on the other.	**Step 1:** Write the inequality so that $ax^2 + bx + c$ is on one side of the inequality and 0 is on the other.
Step 2: Graph the function $f(x) = ax^2 + bx + c$. Label the x-intercepts of the graph.	**Step 2:** Determine the solutions to the equation $ax^2 + bx + c = 0$.
Step 3: From the graph, determine where the function is positive and where it is negative. This information will give you the solution set to the inequality.	**Step 3:** Use the solutions to the equation from Step 2 to separate the real number line into intervals.
	Step 4: Write $ax^2 + bx + c$ in factored form. Within each interval formed in Step 3, determine the sign of each factor. Then determine the sign of the product. Also write the value of $ax^2 + bx + c$ at each solution found in Step 2.
	(a) If the product of the factors is positive, then $ax^2 + bx + c > 0$ for all numbers x in the interval.
	(b) If the product of the factors is negative, then $ax^2 + bx + c < 0$ for all numbers x in the interval.

If the inequality is not strict ($\le$ or $\ge$), include the solutions of $ax^2 + bx + c = 0$ in the solution set.

Now let's do an example where both methods are used.

▶ (EXAMPLE 3)   **Solving a Quadratic Inequality**

Solve $-x^2 + 10 > 3x$ using both the graphical and the algebraic methods.

Solution

Graphical Method:
To make the computation easier, rearrange the inequality so that the coefficient of x^2 is positive.

$$-x^2 + 10 > 3x$$

Add x^2 to both sides; subtract 10 from both sides: $0 > x^2 + 3x - 10$

$0 > b$ is equivalent to $b < 0$: $x^2 + 3x - 10 < 0$

Graph $f(x) = x^2 + 3x - 10$. Because $a = 1 > 0$, the graph opens up. Find the intercepts and vertex of the function.

x-intercepts:

$$f(x) = 0$$
$$x^2 + 3x - 10 = 0$$
$$(x + 5)(x - 2) = 0$$
$$x + 5 = 0 \quad \text{or} \quad x - 2 = 0$$
$$x = -5 \quad \text{or} \quad x = 2$$

y-intercept: $f(0) = -10$

Figure 34

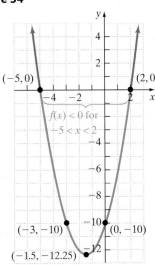

Vertex:

$$x = -\frac{b}{2a} = -\frac{3}{2(1)} = -\frac{3}{2} = -1.5$$

$$y = f\left(-\frac{b}{2a}\right) = f\left(-\frac{3}{2}\right) = -\frac{49}{4} = -12.25$$

The vertex is at $(-1.5, -12.25)$.

Figure 34 shows the graph. The graph of $f(x) = x^2 + 3x - 10$ is below the x-axis (and therefore $x^2 + 3x - 10 < 0$) for $-5 < x < 2$. Because the inequality is strict $(<)$ do not include the x-intercepts in the solution. Thus the solution is $\{x \mid -5 < x < 2\}$ in set-builder notation, or $(-5, 2)$ in interval notation.

Algebraic Method:
Rearrange the inequality so that the coefficient of x^2 is positive.

$$-x^2 + 10 > 3x$$

Add x^2 to both sides; subtract 10 from both sides: $0 > x^2 + 3x - 10$

$0 > b$ is equivalent to $b < 0$: $x^2 + 3x - 10 < 0$

Now solve the equation

$$x^2 + 3x - 10 = 0$$
$$(x + 5)(x - 2) = 0$$
$$x + 5 = 0 \quad \text{or} \quad x - 2 = 0$$
$$x = -5 \quad \text{or} \quad x = 2$$

Use the solutions to separate the real number line into the following intervals:

$$(-\infty, -5) \qquad (-5, 2) \qquad (2, \infty)$$

The factored form of $x^2 + 3x - 10$ is $(x + 5)(x - 2)$. Table 10 on the next page shows the sign of each factor and of $(x + 5)(x - 2)$ in each interval. The value of $(x + 5)(x - 2)$ at $x = -5$ and at $x = 2$ is also listed.

Table 10

Interval	$(-\infty, -5)$		$(-5, 2)$		$(2, \infty)$

Test Point	-6	-5	0	2	3
Sign of $(x + 5)$	Negative	0	Positive	Positive	Positive
Sign of $(x - 2)$	Negative	Negative	Negative	0	Positive
Sign of $(x + 5)(x - 2)$	Positive	0	Negative	0	Positive
Conclusion about $(x + 5)(x - 2) < 0$	$(x + 5)(x - 2)$ is positive in the interval $(-\infty, -5)$, so it is not part of the solution set.	The inequality is strict, so -5 is not part of the solution set.	$(x + 5)(x - 2)$ is negative in the interval $(-5, 2)$, so it is part of the solution set.	The inequality is strict, so 2 is not part of the solution set.	$(x + 5)(x - 2)$ is positive in the interval $(2, \infty)$, so it is not part of the solution set.

The inequality is strict, so -5 and 2 are not included in the solution. The solution set is $\{x \,|\, -5 < x < 2\}$ in set-builder notation, or $(-5, 2)$ in interval notation.

Figure 35 shows the graph of the solution set.

Figure 35

Quick ✓

In Problem 3, solve the quadratic inequality. Graph the solution set.

3. $-x^2 > 2x - 24$

EXAMPLE 4 **Solving a Quadratic Inequality**

Solve $2x^2 > 4x - 1$ using both the graphical and the algebraic methods. Graph the solution set.

Solution

Graphical Method:

$$2x^2 > 4x - 1$$

Subtract 4x from both sides; add 1 to both sides: $2x^2 - 4x + 1 > 0$

Graph $f(x) = 2x^2 - 4x + 1$ to determine where $f(x) > 0$. Because $a = 2 > 0$, the graph opens up.

x-intercepts: $f(x) = 0$ **y-intercept:** $f(0) = 1$

$$2x^2 - 4x + 1 = 0$$

$a = 2; b = -4; c = 1$: $x = \dfrac{-(-4) \pm \sqrt{(-4)^2 - 4(2)(1)}}{2(2)}$

$$= \frac{4 \pm \sqrt{8}}{4}$$

$$= 1 \pm \frac{\sqrt{2}}{2}$$

$$\approx 0.29 \text{ or } 1.71$$

Vertex:
$$x = -\frac{b}{2a} = -\frac{-4}{2(2)} = 1$$

$$f\left(-\frac{b}{2a}\right) = f(1) = -1$$

The vertex is at $(1, -1)$. Figure 36 shows the graph.

Figure 36

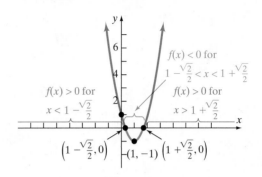

The graph shown in Figure 36 shows that the graph of $f(x) = 2x^2 - 4x + 1$ is greater than 0 for $x < 1 - \dfrac{\sqrt{2}}{2}$ or $x > 1 + \dfrac{\sqrt{2}}{2}$. Because the inequality is strict, do not include the x-intercepts in the solution. Thus the solution is $\left\{ x \,\middle|\, x < 1 - \dfrac{\sqrt{2}}{2} \text{ or } x > 1 + \dfrac{\sqrt{2}}{2} \right\}$ in set-builder notation, or $\left(-\infty, 1 - \dfrac{\sqrt{2}}{2} \right) \cup \left(1 + \dfrac{\sqrt{2}}{2}, \infty \right)$ in interval notation.

Algebraic Method:

First, write the inequality in the form $ax^2 + bx + c > 0$.

$$2x^2 > 4x - 1$$

Subtract 4x from both sides; add 1 to both sides: $2x^2 - 4x + 1 > 0$

Now determine the solutions to the equation $2x^2 - 4x + 1 = 0$.

$$x = \frac{-(-4) \pm \sqrt{(-4)^2 - 4(2)(1)}}{2(2)}$$

$$= \frac{4 \pm \sqrt{8}}{4}$$

$$= 1 \pm \frac{\sqrt{2}}{2}$$

$$\approx 0.29 \text{ or } 1.71$$

Separate the real number line into the following intervals:

$$\left(-\infty, 1 - \frac{\sqrt{2}}{2} \right) \quad \left(1 - \frac{\sqrt{2}}{2}, 1 + \frac{\sqrt{2}}{2} \right) \quad \left(1 + \frac{\sqrt{2}}{2}, \infty \right)$$

Work Smart

The factor $\left(x - \left(1 - \dfrac{\sqrt{2}}{2} \right) \right)$ is close to the factor $x - 0.29$. Similarly, use $x - 1.71$ as an approximation of $\left(x - \left(1 + \dfrac{\sqrt{2}}{2} \right) \right)$.

The solutions to the equation $2x^2 - 4x + 1 = 0$ enable us to factor $2x^2 - 4x + 1$ as $\left(x - \left(1 - \dfrac{\sqrt{2}}{2} \right) \right)\left(x - \left(1 + \dfrac{\sqrt{2}}{2} \right) \right)$. Set up Table 11 on the next page, using 0.29 as an approximation of $1 - \dfrac{\sqrt{2}}{2}$ and 1.71 as an approximation for $1 + \dfrac{\sqrt{2}}{2}$ to help determine the sign of each factor.

Table 11

Interval	$\left(-\infty, 1 - \dfrac{\sqrt{2}}{2}\right)$	$\left(1 - \dfrac{\sqrt{2}}{2}, 1 + \dfrac{\sqrt{2}}{2}\right)$	$\left(1 + \dfrac{\sqrt{2}}{2}, \infty\right)$
		$1 - \frac{\sqrt{2}}{2} \approx 0.29 \qquad\qquad 1 + \frac{\sqrt{2}}{2} \approx 1.71$	
Test Point	$0 \qquad\qquad 1 - \dfrac{\sqrt{2}}{2}$	$1 \qquad\qquad 1 + \dfrac{\sqrt{2}}{2}$	2
Sign of $\left(x - \left(1 - \dfrac{\sqrt{2}}{2}\right)\right)$ $\approx (x - 0.29)$	Negative $\qquad\qquad$ 0	Positive $\qquad\qquad$ Positive	Positive
Sign of $\left(x - \left(1 + \dfrac{\sqrt{2}}{2}\right)\right)$ $\approx (x - 1.71)$	Negative $\qquad\qquad$ Negative	Negative $\qquad\qquad$ 0	Positive
Sign of $\left(x - \left(1 - \dfrac{\sqrt{2}}{2}\right)\right)$ $\left(x - \left(1 + \dfrac{\sqrt{2}}{2}\right)\right)$	Positive $\qquad\qquad$ 0	Negative $\qquad\qquad$ 0	Positive
Conclusion about $2x^2 - 4x + 1 > 0$	$2x^2 - 4x + 1$ is positive in the interval $\left(-\infty, 1 - \dfrac{\sqrt{2}}{2}\right)$, so it is part of the solution set. $\quad$ Since the inequality is strict, $1 - \dfrac{\sqrt{2}}{2}$ is not part of the solution set.	$2x^2 - 4x + 1$ is negative in the interval $\left(1 - \dfrac{\sqrt{2}}{2}, 1 + \dfrac{\sqrt{2}}{2}\right)$, so it is not part of the solution set. $\quad$ Since the inequality is strict, $1 + \dfrac{\sqrt{2}}{2}$ is not part of the solution set.	$2x^2 - 4x + 1$ is positive in the interval $\left(1 + \dfrac{\sqrt{2}}{2}, \infty\right)$, so it is part of the solution set.

Work Smart

Instead of using the signs of the factors, evaluate $2x^2 - 4x + 1$ at each test point. For example, at $x = 0$,

$2x^2 - 4x + 1$
$= 2(0)^2 - 4(0) + 1$
$= 1$
$> 0.$

Therefore, $2x^2 - 4x + 1 > 0$ for all $x < 1 - \dfrac{\sqrt{2}}{2}.$

To find where $2x^2 - 4x + 1$ is positive, do not include $1 - \dfrac{\sqrt{2}}{2}$ or $1 + \dfrac{\sqrt{2}}{2}$ in the solution. The solution set is $\left\{ x \,\middle|\, x < 1 - \dfrac{\sqrt{2}}{2} \text{ or } x > 1 + \dfrac{\sqrt{2}}{2} \right\}$ in set-builder notation, or $\left(-\infty, 1 - \dfrac{\sqrt{2}}{2}\right) \cup \left(1 + \dfrac{\sqrt{2}}{2}, \infty\right)$ in interval notation.

Figure 37 shows the graph of the solution set.

Figure 37

Quick ✓

4. Solve $3x^2 > -x + 5$ using both the graphical and the algebraic methods. Graph the solution set.

▶ ❷ Solve Polynomial Inequalities

To solve polynomial inequalities, focus on the algebraic method, because we do not yet have experience graphing polynomial functions. To solve polynomial inequalities algebraically, use the same steps used to solve quadratic inequalities algebraically.

EXAMPLE 5 **Solving a Polynomial Inequality of Degree 3**

Solve $x^3 + x^2 - 9x - 9 > 0$ algebraically and graph the solution set.

Solution

The inequality is already written so that the polynomial is on one side of the inequality and 0 is on the other side. Therefore, solve the equation $x^3 + x^2 - 9x - 9 = 0$.

$$x^3 + x^2 - 9x - 9 = 0$$

Factor by grouping: $\quad x^2(x+1) - 9(x+1) = 0$

$$(x^2 - 9)(x+1) = 0$$

$$(x+3)(x-3)(x+1) = 0$$

Set each factor equal to 0: $\quad x + 3 = 0 \quad$ or $\quad x - 3 = 0 \quad$ or $\quad x + 1 = 0$

$$x = -3 \qquad\qquad x = 3 \qquad\qquad x = -1$$

Notice that the solutions of the equation $x^3 + x^2 - 9x - 9 = 0$ separate the number line into *four* intervals: $(-\infty, -3), (-3, -1), (-1, 3)$ and $(3, \infty)$. Determine the sign of each of the factors $(x+3)$, $(x-3)$, and $(x+1)$, and then find the sign of the product. Table 12 shows the sign of each factor and the sign of $(x+3)(x-3)(x+1)$ in each interval. The value of $(x+3)(x-3)(x+1)$ at $x = -3, x = -1$, and $x = 3$ is also in the table.

Table 12

Interval	$(-\infty, -3)$		$(-3, -1)$		$(-1, 3)$		$(3, \infty)$
Test Point	-4	-3	-2	-1	1	3	4
Sign of $(x+3)$	Negative	0	Positive	Positive	Positive	Positive	Positive
Sign of $(x-3)$	Negative	Negative	Negative	Negative	Negative	0	Positive
Sign of $(x+1)$	Negative	Negative	Negative	0	Positive	Positive	Positive
Sign of $(x+3)(x-3)(x+1)$	Negative	0	Positive	0	Negative	0	Positive
Conclusion about $(x+3)(x-3)(x+1) > 0$	The product of the factors is negative, so $(-\infty, -3)$ is not part of the solution set.	The inequality is strict, so $x = -3$, is not in the solution set.	The product of the factors is positive, so $(-3, -1)$ is part of the solution set.	The inequality is strict, so $x = -1$, is not in the solution set.	The product is negative, so $(-1, 3)$ is not part of the solution set.	The inequality is strict, so $x = 3$ is not in the solution set.	The product of factors is positive, so $(3, \infty)$ is part of the solution set.

The inequality is strict, so do not include $-3, -1$, or 3 in the solution set. The solution of the inequality $x^3 + x^2 - 9x - 9 > 0$ is $\{x \mid -3 < x < -1 \text{ or } x > 3\}$ in set-builder notation, or $(-3, -1) \cup (3, \infty)$ in interval notation. Figure 38 shows the graph of the solution set.

Figure 38

Quick ✔

Solve Problems 5 and 6 algebraically. Graph the solution set.

5. $(2x - 3)(x + 4)(x - 5) \le 0$

6. $x^3 + 7x^2 - 4x - 28 > 0$

7.6 Exercises MyMathLab® Exercise numbers in green have complete video solutions in MyMathLab or may be accessed using the QR code to the right.

*Problems **1–6** are the Quick ✔s that follow the EXAMPLES.*

Building Skills

In Problems 7–10, use the graphs of the quadratic function f to determine the solution.

7. (a) Solve $f(x) > 0$. **(b)** Solve $f(x) \le 0$.

8. (a) Solve $f(x) > 0$. **(b)** Solve $f(x) \le 0$.

9. (a) $f(x) \ge 0$ **(b)** $f(x) < 0$

10. (a) $f(x) > 0$ **(b)** $f(x) \le 0$

In Problems 11–32, solve each quadratic inequality. Graph the solution set. See Objective 1.

11. $(x - 5)(x + 2) \ge 0$ **12.** $(x - 8)(x + 1) \le 0$

13. $(x + 3)(x + 7) < 0$ **14.** $(x - 4)(x - 10) > 0$

15. $x^2 - 2x - 35 > 0$ **16.** $x^2 + 3x - 18 \ge 0$

17. $n^2 - 6n - 8 \le 0$ **18.** $p^2 + 5p + 4 < 0$

19. $m^2 + 5m \ge 14$ **20.** $z^2 > 7z + 8$

21. $2q^2 \ge q + 15$ **22.** $2b^2 + 5b < 7$

23. $3x + 4 \ge x^2$ **24.** $x + 6 < x^2$

25. $-x^2 + 3x < -10$ **26.** $-x^2 > 4x - 21$

27. $-3x^2 \le -10x - 8$ **28.** $-3m^2 \ge 16m + 5$

29. $x^2 + 4x + 1 < 0$ **30.** $x^2 - 3x - 5 \ge 0$

31. $-2a^2 + 7a \ge -4$ **32.** $-3p^2 < 3p - 5$

In Problems 33–38, solve each quadratic inequality using the graphical method.

33. $z^2 + 2z + 3 > 0$ **34.** $y^2 + y + 5 \ge 0$

35. $2b^2 + 5b \le -6$ **36.** $3w^2 + w < -2$

37. $x^2 - 6x + 9 > 0$ **38.** $p^2 - 8p + 16 \le 0$

In Problems 39–48, solve each polynomial inequality. Graph the solution set. See Objective 2.

39. $(x + 1)(x - 2)(x - 5) > 0$

40. $(x + 3)(x - 1)(x - 3) < 0$

41. $(2x + 1)(x - 4)(x - 9) \le 0$

42. $(3x + 4)(x - 2)(x - 6) \ge 0$

43. $(x + 3)(2x^2 - x - 1) \ge 0$

44. $(x - 2)(3x^2 + x - 2) \le 0$

45. $x^3 + 3x^2 - 4x - 12 \le 0$

46. $x^3 + 4x^2 - 9x - 36 \ge 0$

47. $4x^3 + 16x^2 - 9x - 36 > 0$

48. $3x^3 + 5x^2 - 12x - 20 < 0$

Mixed Practice

In Problems 49–56, for each function find the values of x that satisfy the given condition.

49. Solve $f(x) < 0$ if $f(x) = x^2 - 5x$.

50. Solve $f(x) > 0$ if $f(x) = x^2 + 4x$.

51. Solve $f(x) \ge 0$ if $f(x) = x^2 - 3x - 28$.

52. Solve $f(x) \le 0$ if $f(x) = x^2 + 2x - 48$.

53. Solve $g(x) > 0$ if $g(x) = 2x^2 + x - 10$.

54. Solve $F(x) < 0$ if $F(x) = 2x^2 + 7x - 15$.

55. Solve $f(x) < 0$ if $f(x) = 4x^3 - x^2 - 14x$.

56. Solve $g(x) > 0$ if $g(x) = 2x^3 - 7x^2 - 9x$.

In Problems 57–60, find the domain of the given function.

57. $f(x) = \sqrt{x^2 + 8x}$ **58.** $f(x) = \sqrt{x^2 - 5x}$

59. $g(x) = \sqrt{x^2 - x - 30}$ **60.** $G(x) = \sqrt{x^2 + 2x - 63}$

Applying the Concepts

61. Physics A ball is thrown vertically upward with an initial speed of 80 feet per second from a cliff 500 feet above sea level. The height s (in feet) of the ball from the ground after t seconds is $s(t) = -16t^2 + 80t + 500$. For what time t is the ball more than 596 feet above sea level?

500 feet

62. Physics A water balloon is thrown vertically upward with an initial speed of 64 feet per second from the top of a building 200 feet above the ground. The height s (in feet) of the balloon from the ground after t seconds is $s(t) = -16t^2 + 64t + 200$. For what time t is the balloon more than 248 feet above the ground?

63. Revenue Function Suppose that the marketing department of Panasonic has found that, when a certain model of DVD player is sold at a price of p dollars, the daily revenue R (in dollars) as a function of the price p is $R(p) = -2.5p^2 + 600p$. Determine the prices for which revenue will exceed \$35,750. That is, solve $R(p) > 35{,}750$.

64. Revenue Function Suppose that the marketing department of Samsung has found that, when a certain model of cell phone is sold at a price of p dollars, the daily revenue R (in dollars) as a function of the price p is $R(p) = -5p^2 + 600p$. Determine the prices for which revenue will exceed \$17,500. That is, solve $R(p) > 17{,}500$.

Extending the Concepts

In Problems 65–68, solve each inequality by inspection. Then provide a verbal explanation of the solution.

65. $(x + 3)^2 \le 0$

66. $(x - 4)^2 > 0$

67. $(x + 8)^2 > -2$

68. $(3x + 1)^2 < -2$

69. Write a quadratic inequality that has $[-3, 2]$ as the solution set.

70. Write a quadratic inequality that has $(0, 5)$ as the solution set.

In Problems 71–76, use the techniques presented in this section to solve each inequality algebraically.

71. $(x + 1)(x - 2)(x + 3)(x - 4) > 0$

72. $(x - 4)(x - 1)(x + 5)(x + 6) < 0$

73. $x^4 - 29x^2 + 100 \le 0$

74. $x^4 - 10x^2 + 9 \ge 0$

75. $\dfrac{x^2 + 5x + 6}{x - 2} > 0$

76. $\dfrac{x^2 - 3x - 10}{x + 1} < 0$

Explaining the Concepts

77. The inequalities $(3x + 2)^2 < 2$ and $(3x + 2)^{-2} > \dfrac{1}{2}$ have the same solution set. Why?

78. The inequality $x^2 + 3 < -2$ has no solution. Explain why.

79. The inequality $x^2 - 1 \geq -1$ has the set of all real numbers as the solution. Explain why.

80. Explain when the endpoints of an interval are included in the solution set of a quadratic inequality.

81. Is the inequality $x^2 + 1 > 1$ true for all real numbers? Explain.

82. Explain how to solve the quadratic inequality $f(x) < 0$ from the graph of $y = f(x)$, where f is a quadratic function.

83. Explain how to solve the quadratic inequality $f(x) > 0$ from the graph of $y = f(x)$, where f is a quadratic function.

Synthesis Review

In Problems 84–87, simplify each expression.

84. $\dfrac{3a^4b}{12a^{-3}b^5}$

85. $(4mn^{-3})(-2m^4n)$

86. $\left(\dfrac{3x^4y}{6x^{-2}y^5}\right)^{\frac{1}{2}}$

87. $\left(\dfrac{9a^{\frac{2}{3}}b^{\frac{1}{2}}}{a^{\frac{-1}{9}}b^{\frac{3}{4}}}\right)^{-1}$

88. Do the Laws of Exponents presented in the "Getting Ready: Laws of Exponents" section also apply to the Laws of Exponents for rational exponents presented in Section 6.1

Technology Exercises

Technology may be used to solve the quadratic inequality in Example 1. If using a graphing calculator, graph $Y_1 = x^2 - 4x - 5$. Use the ZERO feature of the calculator to find the x-intercepts of the graph. See Figures 39(a) and (b).

Figure 39

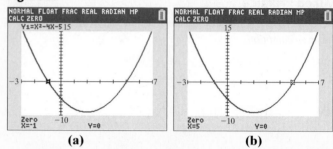

(a)　　　　(b)

From Figures 39(a) and (b), we can see that $x^2 - 4x - 5 \geq 0$ for $x \leq -1$ or $x \geq 5$.

In Problems 89–92, solve each quadratic inequality using technology.

89. $2x^2 + 7x - 49 > 0$

90. $2x^2 + 3x - 27 < 0$

91. $6x^2 + x \leq 40$

92. $8x^2 + 18x \geq 81$

Chapter 7 Activity: Presidential Decision Making

Focus: Developing quadratic equations.

Time: 30–35 minutes

Group size: 2–4

1. Your boss, Huntington Corporation's president, Gerald Cain, is very concerned about his approval rating with his employees. Last year, on January 1, he made some policy changes and saw his approval rating drop. In fact, his approval rating was 48% just before he made the policy changes. One month later, his rating was at 41%, two months later it was at 40%, and at three months it began to climb and was 45%. He discovered that the following function described his approval rating for that year, where x is the month:

$$R(x) = 3x^2 - 10x + 48$$

(a) As a group, use the above function to find when President Cain's approval rating will return to the original rating.

(b) If his approval rating continues to climb, when will he reach a 68% approval rating?

2. On January 1 of this year, President Cain surveyed his employees again and found that 68% of them approved of his leadership skills. At this time, President Cain decided to become very strict with his employees and began a series of new policies. He noticed that his approval rating began steadily to slip and reached an all-time low of 38% on March 30 (3 months later). President Cain is not worried because he knows from last year that this drop in popularity will bottom out and eventually rise. He believes his popularity can be modeled by a quadratic function and needs your help. He has more bad news to deliver but does not want to begin the next round of policy changes until his approval rating is back to approximately 50%.

(a) As a group, write a quadratic function that would model President Cain's approval rating.

(b) As a group, develop different ways to advise President Cain on what date to begin his policy changes. Use graphs and computations to prove your point.

Chapter 7 Review

Section 7.1 Solving Quadratic Equations by Completing the Square

KEY CONCEPTS

- **Square Root Property**
  If $x^2 = p$, then $x = \sqrt{p}$ or $x = -\sqrt{p}$.
- **Pythagorean Theorem**
  In a right triangle, the square of the length of the hypotenuse is equal to the sum of the squares of the lengths of the legs. That is, $\text{leg}^2 + \text{leg}^2 = \text{hypotenuse}^2$.

KEY TERMS

Completing the square
Right triangle
Right angle
Hypotenuse
Legs

You Should Be Able To...	EXAMPLE	Review Exercises
❶ Solve quadratic equations using the Square Root Property (p. 543)	Examples 1 through 4	1–10
❷ Complete the square in one variable (p. 546)	Example 5	11–16
❸ Solve quadratic equations by completing the square (p. 547)	Examples 6 and 7	17–26
❹ Solve problems using the Pythagorean Theorem (p. 549)	Examples 8 and 9	27–36

In Problems 1–10, solve each equation using the Square Root Property.

1. $m^2 = 169$

2. $n^2 = 75$

3. $a^2 = -16$

4. $b^2 = \dfrac{8}{9}$

5. $(x - 8)^2 = 81$

6. $(y - 2)^2 - 62 = 88$

7. $(3z + 5)^2 = 100$

8. $7p^2 = 18$

9. $3q^2 + 251 = 11$

10. $\left(x + \dfrac{3}{4}\right)^2 = \dfrac{13}{16}$

In Problems 11–16, complete the square in each expression. Then factor the perfect square trinomial.

11. $a^2 + 30a$

12. $b^2 - 14b$

13. $c^2 - 11c$

14. $d^2 + 9d$

15. $m^2 - \dfrac{1}{4}m$

16. $n^2 + \dfrac{6}{7}n$

In Problems 17–26, solve each quadratic equation by completing the square.

17. $x^2 - 10x + 16 = 0$

18. $y^2 - 3y - 28 = 0$

19. $z^2 - 6z - 3 = 0$

20. $a^2 - 5a - 7 = 0$

21. $b^2 + b + 7 = 0$

22. $c^2 - 6c + 17 = 0$

23. $2d^2 - 7d + 3 = 0$

24. $2w^2 + 2w + 5 = 0$

25. $3x^2 - 9x + 8 = 0$

26. $3x^2 + 4x - 2 = 0$

In Problems 27–32, the lengths of the legs of a right triangle are given. Find the hypotenuse.

27. $a = 9, b = 12$

28. $a = 8, b = 8$

29. $a = 3, b = 6$

30. $a = 10, b = 24$

31. $a = 5, b = \sqrt{11}$

32. $a = 6, b = \sqrt{13}$

In Problems 33–35, use the right triangle shown below to find the missing length.

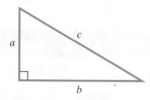

33. $a = 9, c = 12$ **34.** $b = 5, c = 10$

35. $b = 6, c = 17$

36. Baseball Diamond A baseball diamond is really a square that is 90 feet long on each side. (See the figure.) What is the distance between home plate

and second base? Round your answer to the nearest tenth of a foot.

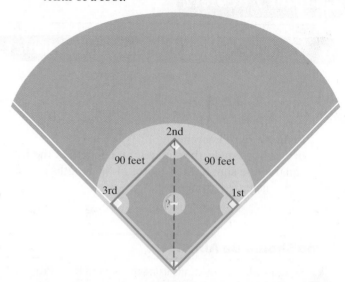

Section 7.2 Solving Quadratic Equations by the Quadratic Formula

KEY CONCEPTS

- **The Quadratic Formula**
  The solutions to the equation $ax^2 + bx + c = 0, a \neq 0$, are given by
  $$x = \frac{-b \pm \sqrt{b^2 - 4ac}}{2a}$$

- **Discriminant**
  For the quadratic equation $ax^2 + bx + c = 0$, where a, b, and c are rational numbers:
  - If $b^2 - 4ac > 0$, the equation has two unequal real solutions.
    - If $b^2 - 4ac$ is a perfect square, the equation has two rational solutions provided a, b, and c are rational numbers.
    - If $b^2 - 4ac$ is not a perfect square, the equation has two irrational solutions provided a, b, and c are rational numbers.
  - If $b^2 - 4ac = 0$, the equation has a repeated rational solution.
  - If $b^2 - 4ac < 0$, the equation has two complex solutions that are not real.

KEY TERM

Discriminant

You Should Be Able To...	EXAMPLE	Review Exercises
❶ Solve quadratic equations using the quadratic formula (p. 555)	Examples 1 through 4	37–46
❷ Use the discriminant to determine the nature of solutions of a quadratic equation (p. 559)	Example 5	47–52
❸ Model and solve problems involving quadratic equations (p. 563)	Examples 6 and 7	63–68

In Problems 37–46, solve each equation using the quadratic formula.

37. $x^2 - x - 20 = 0$ **38.** $4y^2 = 8y + 21$

39. $3p^2 + 8p = -3$ **40.** $2q^2 - 3 = 4q$

41. $3w^2 + w = -3$ **42.** $9z^2 + 16 = 24z$

43. $m^2 - 4m + 2 = 0$ **44.** $5n^2 + 4n + 1 = 0$

45. $5x + 13 = -x^2$ **46.** $-2y^2 = 6y + 7$

In Problems 47–52, determine the discriminant of each quadratic equation. Use the value of the discriminant to determine whether the quadratic equation has two rational solutions, two irrational solutions, one repeated rational solution, or two complex solutions that are not real.

47. $p^2 - 5p - 8 = 0$

48. $m^2 + 8m + 16 = 0$

49. $3n^2 + n = -4$

50. $7w^2 + 3 = 8w$

51. $4x^2 + 49 = 28x$

52. $11z - 12 = 2z^2$

In Problems 53–62, solve each equation using any appropriate method.

53. $x^2 + 8x - 9 = 0$

54. $6p^2 + 13p = 5$

55. $n^2 + 13 = -4n$

56. $5y^2 - 60 = 0$

57. $\frac{1}{4}q^2 - \frac{1}{2}q - \frac{3}{8} = 0$

58. $\frac{1}{8}m^2 + m + \frac{5}{2} = 0$

59. $(w - 8)(w + 6) = -33$

60. $(x - 3)(x + 1) = -2$

61. $9z^2 = 16$

62. $\frac{1 - 2x}{x^2 + 5} = 1$

63. Pythagorean Theorem Use the Pythagorean Theorem to determine the value of x and the measurements of the right triangle shown below.

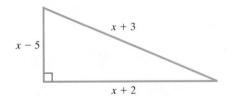

64. Area The area of a rectangle is 108 square centimeters. The width of the rectangle is 3 centimeters less than the length. What are the dimensions of the rectangle?

65. Revenue The revenue R received by a company selling x cell phones per week is given by the function $R(x) = -0.2x^2 + 180x$.

  (a) How many cell phones must be sold in order for revenue to be $36,000 per week?

  (b) How many cell phones must be sold in order for revenue to be $40,500 per week?

66. Projectile Motion The height s of a ball after t seconds, when it is thrown straight up with an initial speed of 50 feet per second from an initial height of 180 feet, can be modeled by the function $s(t) = -16t^2 + 50t + 180$.

  (a) When will the height of the ball be 200 feet? Round your answer to the nearest tenth of a second.

  (b) When will the height of the ball be 100 feet? Round your answer to the nearest tenth of a second.

  (c) Will the ball ever reach a height of 300 feet? How does the result of the equation tell you this?

67. Pleasure Boat Ride A pleasure boat carries passengers 10 miles upstream and then returns to the starting point. The total time of the trip (excluding the time on land) is 2 hours. If the speed of the current is 3 miles per hour, find the speed of the boat in still water. Round your answer to the nearest tenth of a mile per hour.

68. Work Together, Tom and Beth can wash their car in 30 minutes. By himself, Tom can wash the car in 14 minutes less time than Beth can by herself. How long will it take Beth to wash the car by herself? Round your answer to the nearest tenth of a minute.

Section 7.3   Solving Equations Quadratic in Form

KEY TERM

Equation quadratic in form

You Should Be Able To...	EXAMPLE	Review Exercises
❶ Solve equations that are quadratic in form (p. 569)	Examples 1 through 5	69–80

In Problems 69–78, solve each equation.

69. $x^4 + 7x^2 - 144 = 0$ **70.** $4w^4 + 5w^2 - 6 = 0$

71. $3(a + 4)^2 - 11(a + 4) + 6 = 0$

72. $(q^2 - 11)^2 - 2(q^2 - 11) - 15 = 0$

73. $y - 13\sqrt{y} + 36 = 0$ **74.** $5z + 2\sqrt{z} - 3 = 0$

75. $p^{-2} - 4p^{-1} - 21 = 0$ **76.** $2b^{\frac{2}{3}} + 13b^{\frac{1}{3}} - 7 = 0$

77. $m^{\frac{1}{2}} + 2m^{\frac{1}{4}} - 8 = 0$ **78.** $\left(\dfrac{1}{x + 5}\right)^2 + \dfrac{3}{x + 5} = 28$

In Problems 79 and 80, find the zeros of the function.

79. $f(x) = 4x - 20\sqrt{x} + 21$ **80.** $g(x) = x^4 - 17x^2 + 60$

Section 7.4   Graphing Quadratic Functions Using Transformations

KEY CONCEPTS

- **Graphing a Function of the Form $f(x) = x^2 + k$ or $f(x) = x^2 - k$**
  To obtain the graph of $f(x) = x^2 + k, k > 0,$ from the graph of $y = x^2,$ shift the graph of $y = x^2$ vertically up k units. To obtain the graph of $f(x) = x^2 - k, k > 0,$ from the graph of $y = x^2$ shift the graph of $y = x^2$ vertically down k units.

- **Graphing a Function of the Form $f(x) = (x - h)^2$ or $f(x) = (x + h)^2$**
  To obtain the graph of $f(x) = (x - h)^2, h > 0,$ from the graph of $y = x^2,$ shift the graph of $y = x^2$ horizontally to the right h units. To obtain the graph of $f(x) = (x + h)^2, h > 0,$ from the graph of $y = x^2$ shift the graph of $y = x^2$ horizontally left h units.

- **Graphing a Function of the Form $f(x) = ax^2$**
  To obtain the graph of $f(x) = ax^2$ from the graph of $y = x^2,$ multiply each y-coordinate on the graph of $y = x^2$ by a.

KEY TERMS

Quadratic function
Vertically stretched
Vertically compressed
Parabola
Opens up
Opens down
Vertex
Axis of symmetry
Transformations

You Should Be Able To...	EXAMPLE	Review Exercises
❶ Graph quadratic functions of the form $f(x) = x^2 + k$ (p. 578)	Examples 1 and 2	81–82; 87–90
❷ Graph quadratic functions of the form $f(x) = (x - h)^2$ (p. 579)	Examples 3 through 5	83–84; 87–90
❸ Graph quadratic functions of the form $f(x) = ax^2$ (p. 581)	Example 6	85–86; 89–90
❹ Graph quadratic functions of the form $f(x) = ax^2 + bx + c$ (p. 583)	Examples 7 and 8	91–96
❺ Find a quadratic function from its graph (p. 586)	Example 9	97–100

In Problems 81–90, use the graph of $y = x^2$ to graph the quadratic function.

81. $f(x) = x^2 + 4$ **82.** $g(x) = x^2 - 5$

83. $h(x) = (x + 1)^2$ **84.** $F(x) = (x - 4)^2$

85. $G(x) = -4x^2$ **86.** $H(x) = \dfrac{1}{5}x^2$

87. $p(x) = (x - 4)^2 - 3$ **88.** $P(x) = (x + 4)^2 + 2$

89. $f(x) = -(x - 1)^2 + 4$ **90.** $F(x) = \dfrac{1}{2}(x + 2)^2 - 1$

In Problems 91–96, graph each quadratic function using transformations. Determine the vertex and axis of symmetry. Based on the graph, determine the domain and range of each function.

91. $g(x) = x^2 - 6x + 10$ **92.** $G(x) = x^2 + 8x + 11$

93. $h(x) = 2x^2 - 4x - 3$ **94.** $H(x) = -x^2 - 6x - 10$

95. $p(x) = -3x^2 + 12x - 8$ **96.** $P(x) = \dfrac{1}{2}x^2 - 2x + 5$

In Problems 97–100, determine the quadratic function whose graph is given.

97.

98.

99.

100.

Section 7.5   Graphing Quadratic Functions Using Properties

KEY CONCEPTS

- **Vertex of a Parabola**
  Any quadratic function of the form $f(x) = ax^2 + bx + c, a \neq 0$, will have vertex
  $$\left(-\frac{b}{2a}, f\left(-\frac{b}{2a}\right)\right)$$

- **The x-Intercepts of the Graph of a Quadratic Function**
  1. If $b^2 - 4ac > 0$, the graph of $f(x) = ax^2 + bx + c$ has two x-intercepts.
  2. If $b^2 - 4ac = 0$, the graph of $f(x) = ax^2 + bx + c$ has one x-intercept.
  3. If $b^2 - 4ac < 0$, the graph of $f(x) = ax^2 + bx + c$ has no x-intercepts.

KEY TERMS

Maximum value
Minimum value
Optimization

You Should Be Able To...	EXAMPLE	Review Exercises
❶ Graph quadratic functions of the form $f(x) = ax^2 + bx + c$ (p. 589)	Examples 1 through 4	101–108
❷ Find the maximum or minimum value of a quadratic function (p. 595)	Examples 5 and 6	109–112
❸ Model and solve optimization problems involving quadratic functions (p. 597)	Examples 7 and 8	113–118

In Problems 101–108 graph each quadratic function using its properties by following Steps 1–5 on page 592. Based on the graph, determine the domain and range of each function.

101. $f(x) = x^2 + 2x - 8$ **102.** $F(x) = 2x^2 - 5x + 3$

103. $g(x) = -x^2 + 6x - 7$ **104.** $G(x) = -2x^2 + 4x + 3$

105. $h(x) = 4x^2 - 12x + 9$ **106.** $H(x) = \frac{1}{3}x^2 + 2x + 3$

107. $p(x) = \frac{1}{4}x^2 + 3x + 10$ **108.** $P(x) = -x^2 + 4x - 9$

In Problems 109–112, determine whether the quadratic function has a maximum or a minimum value. Then find the maximum or minimum value and where it occurs.

109. $f(x) = -2x^2 + 16x - 10$

110. $g(x) = 6x^2 - 3x - 1$

111. $h(x) = -4x^2 + 8x + 3$

112. $F(x) = -\frac{1}{3}x^2 + 4x - 7$

113. Revenue Suppose that the marketing department of Zenith has found that, when a certain model of television is sold for a price of p dollars, the daily revenue R (in dollars) as a function of the price p is
$$R(p) = -\frac{1}{3}p^2 + 150p.$$

(a) For what price will the daily revenue be maximized?

(b) What is this maximum daily revenue?

114. Electrical Power In a 120-volt electrical circuit having a resistance of 16 ohms, the available power P (in watts) is given by the function $P(I) = -16I^2 + 120I$, where I represents the current (in amperes).

(a) What current will produce the maximum power in the circuit?

(b) What is this maximum power?

115. Fun with Numbers The sum of two numbers is 24. Find the numbers such that their product is a maximum.

116. Maximizing an Enclosed Area Becky has 15 yards of fencing to make a rectangular kennel for her dog. She will build the kennel next to her garage, so she only needs to enclose three sides. (See the figure.)

(a) What dimensions maximize the area of the kennel?

(b) What is this maximum area?

117. Kicking a Football Ted kicks a football at a 45° angle to the horizontal with an initial velocity of 80 feet per second. The model $h(x) = -0.005x^2 + x$ can be used to estimate the height h of the ball after it has traveled x feet.

(a) How far from Ted will the ball reach a maximum height?

(b) What is the maximum height of the ball?

(c) How far will the ball travel before it strikes the ground?

118. Revenue Monthly demand for automobiles at a certain dealership obeys the demand equation $x = -0.002p + 60$, where x is the quantity and p is the price (in dollars).

(a) Express the revenue R as a function of p. (*Hint:* $R = xp$)

(b) What price p maximizes revenue? What is the maximum revenue?

(c) How many automobiles will be sold at the revenue-maximizing price?

Section 7.6　Polynomial Inequalities

KEY CONCEPT

- The steps for solving a quadratic inequality are given on page 606.

KEY TERM

Quadratic inequality
Polynomial inequality

You Should Be Able To	EXAMPLE	Review Exercises
❶ Solve quadratic inequalities (p. 603)	Examples 1 through 4	119–126
❷ Solve polynomial inequalities (p. 610)	Example 5	127, 128

In Problems 119 and 120, use the graphs of the quadratic function f to determine the solution.

119. **(a)** $f(x) > 0$ **(b)** $f(x) < 0$

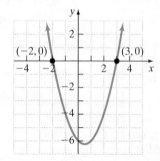

120. **(a)** $f(x) \geq 0$ **(b)** $f(x) \leq 0$

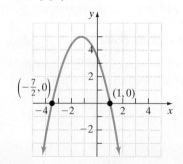

In Problems 121–128, solve each inequality. Graph the solution set.

121. $x^2 - 2x - 24 \leq 0$

122. $y^2 + 7y - 8 \geq 0$

123. $3z^2 - 19z + 20 > 0$

124. $p^2 + 4p - 2 < 0$

125. $4m^2 - 20m + 25 \geq 0$

126. $6w^2 - 19w - 7 \leq 0$

127. $(2x - 3)(x + 1)(x - 2) < 0$

128. $x^3 + 5x^2 - 9x - 45 \geq 0$

Chapter 7 Test *Step-by-step test solutions are found on the Chapter Test Prep Videos available in* MyMathLab®, *on* You Tube™, *or may be accessed using the QR code to the right.*

In Problems 1 and 2, complete the square in the given expression. Then factor the perfect square trinomial.

1. $x^2 - 3x$

2. $m^2 + \dfrac{2}{5}m$

In Problems 3–6, solve each equation using any appropriate method you prefer.

3. $9\left(x + \dfrac{4}{3}\right)^2 = 1$

4. $m^2 - 6m + 4 = 0$

5. $2w^2 - 4w + 3 = 0$

6. $\dfrac{1}{2}z^2 - \dfrac{3}{2}z = -\dfrac{7}{6}$

7. Determine the discriminant of $2x^2 + 5x = 4$. Use the value of the discriminant to determine whether the quadratic equation has two rational solutions, two irrational solutions, one repeated real solution, or two complex solutions that are not real.

8. Find the missing length in the right triangle shown below.

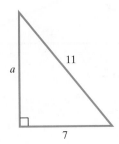

In Problems 9 and 10, solve each equation.

9. $x^4 - 5x^2 - 36 = 0$

10. $6y^{\frac{1}{2}} + 13y^{\frac{1}{4}} - 5 = 0$

In Problems 11 and 12, graph each quadratic function by determining the vertex, intercepts, and axis of symmetry. Based on the graph, determine the domain and range of the quadratic function.

11. $f(x) = (x + 2)^2 - 5$

12. $g(x) = -2x^2 - 8x - 3$

13. Determine the quadratic function whose graph is given.

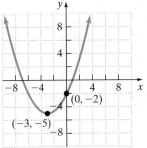

14. Determine whether the quadratic function

$$h(x) = -\frac{1}{4}x^2 + x + 5$$ has a maximum or a

minimum value. Then find the maximum or minimum value.

In Problems 15 and 16, solve each inequality. Graph the solution set.

15. $2m^2 + m - 15 > 0$

16. $x^3 + 5x^2 - 4x - 20 \le 0$

17. Projectile Motion The height s of a rock after t seconds, when propelled straight up with an initial speed of 80 feet per second from an initial height of 20 feet, can be modeled by the function $s(t) = -16t^2 + 80t + 20$. When will the height of the rock be 50 feet? Round your answer to the nearest tenth of a second.

18. Work Together, Lex and Rupert can roof a house in 16 hours. By himself, Rex can roof the house in 4 hours less time than Rupert can by himself. How long will it take Rupert to roof the house by himself? Round your answer to the nearest tenth of an hour.

19. Revenue A small company has found that, when one of its products is sold for a price of p dollars, the weekly revenue (in dollars) as a function of price p is $R(p) = -0.25p^2 + 170p$.
  (a) For what price will the weekly revenue from this product be maximized?
  (b) What is this maximum weekly revenue?

20. Maximizing Volume A box with a rectangular base is to be constructed such that the perimeter of the base of the box is 50 inches. The height of the box must be 12 inches.
  (a) Find the dimensions that maximize the volume of the box.
  (b) What is this maximum volume?

Cumulative Review Chapters R–7

1. Evaluate: $\dfrac{2^4 - 5 \cdot 7 + 3}{1 - 3^2}$

2. Simplify: $2(3c + 1) - (c - 9) - 2c$

3. Solve: $p - 5 = 3(p - 2) - 9$

4. Solve: $5 + 2|x + 1| > 9$

5. Determine the domain of $h(x) = \dfrac{x^2 - 16}{x^2 + x - 12}$.

6. Graph the linear equation: $3x - 4y = 8$

7. Find the equation of the line that passes through the point $(5, -1)$ and is parallel to the graph of $2x + 5y = -15$. Write your answer in either slope-intercept or standard form, whichever you prefer.

8. Solve the following system of equations:
$$\begin{cases} 5x + 2y = 0 \\ 2x - y = -9 \end{cases}$$

9. Evaluate: $\begin{vmatrix} 4 & 1 & 0 \\ 1 & 2 & -1 \\ 3 & -2 & 1 \end{vmatrix}$

10. Graph the following system of linear inequalities:
$$\begin{cases} 2x + 3y < -3 \\ 2x + y > -5 \end{cases}$$

In Problems 11 and 12, add, subtract, multiply, or divide as indicated.

11. $(a^3 - 9a^2 + 11) + (7a^2 - 5a - 8)$

12. $\dfrac{8x^4 + 12x^2 + 17x - 18}{2x^2 - x + 5}$

In Problems 13–14, factor completely.

13. $8m^3 + 27n^3$

14. $7y^2 + 23y - 20$

In Problems 15 and 16, perform the indicated operations. Be sure to express the final answer in lowest terms.

15. $\dfrac{\dfrac{2w^2 - 11w + 12}{w^2 - 16}}{\dfrac{2w^2 - 7w + 6}{w^2 + 9w + 20}}$

16. $\dfrac{3}{k^2 - 5k + 4} - \dfrac{2}{k^2 + 4k - 5}$

17. Solve: $\dfrac{1}{2n} - \dfrac{1}{6} = \dfrac{5}{4n} + \dfrac{1}{3}$

18. Solve the inequality and graph the solution set:
$$\dfrac{x - 4}{x + 5} \geq 0$$

19. **Plane Speeds** Two private planes take off from the same airport at the same time and travel in opposite directions. One plane travels 16 miles per hour faster than the other. After 2 hours, the planes are 536 miles apart. Find the speed of each plane.

20. Add: $\sqrt{48} + \sqrt{75} - 2\sqrt{3}$

21. Rationalize the denominator: $\dfrac{5}{\sqrt[3]{16}}$

22. Solve: $\sqrt{a - 12} + \sqrt{a} = 6$

23. **Pythagorean Theorem** Use the Pythagorean Theorem to determine the value of x for the given measurements of the right triangle shown below.

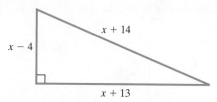

24. Solve: $x^2 - 20 = 8x$

25. Graph: $f(x) = -2x^2 + 4x + 3$

8 Exponential and Logarithmic Functions

We all dream of a long and happy retirement. To make this dream a reality, we must save money during our working years so that our retirement savings can grow through the power of compounding interest. To see how this phenomenon works, see Example 9 in Section 8.2 and Problems 89–92 in Section 8.2.

The Big Picture: Putting It Together

Up to this point, polynomial, rational, and radical expressions and functions have been discussed. We learned how to evaluate, simplify, and solve equations involving these types of algebraic expressions. We also learned how to graph linear functions, quadratic functions, and certain types of radical functions.

This chapter introduces two more types of functions, the *exponential* and *logarithmic functions*. As usual, these functions will be evaluated and graphed, and their properties explored. Equations involving exponential or logarithmic expressions will also be solved.

Outline

8.1 Composite Functions and Inverse Functions

Objectives

1 Form the Composite Function
2 Determine Whether a Function Is One-to-One
3 Find the Inverse of a Function Defined by a Map or Set of Ordered Pairs
4 Obtain the Graph of the Inverse Function from the Graph of a Function
5 Find the Inverse of a Function Defined by an Equation

Are You Prepared for This Section?

Before getting started, complete the following problems. If you get a problem wrong, go back to the section cited and review the material.

P1. Determine the domain: $R(x) = \dfrac{x^2 - 9}{x^2 + 3x - 28}$ [Section 5.1, p. 405]

P2. If $f(x) = 2x^2 - x + 1$, find **(a)** $f(-2)$ **(b)** $f(a + 1)$. [Section 2.2, pp. 158–160]

P3. The graph of a relation is given. Does the relation represent a function? Why or why not? [Section 2.2, pp. 157–158]

▶ 1 Form the Composite Function

Suppose you decide to buy four shirts at $20 per shirt, and the sales tax on your purchase is 5%. First, you determine the before-tax cost of the shirts to be $20(4) = $80. The after-tax cost of the shirts is $80 + $80(0.05) = $80(1 + 0.05) = 1.05($80) = $84. Thus $C(x) = 20x$ represents the cost C of buying x shirts at $20 per shirt. The after-tax cost S of the x shirts is $1.05(20x) = S(C(x))$.

In general, the total cost can be found as a function of the number of shirts by evaluating $S(C(x))$ and obtaining $S(C(x)) = S(20x) = 1.05(20x) = 21x$. The function $S(C(x))$ is a special type of function called a *composite function*.

Here's another example. Consider the function $y = (x - 3)^4$. If $y = f(u) = u^4$ and $u = g(x) = x - 3$, then, by a substitution process, the original function is: $y = f(u) = f(g(x)) = (x - 3)^4$. This process is called **composition.** To evaluate this function, first evaluate $x - 3$ and then raise the result to the fourth power. Technically, two different functions form the function $y = (x - 3)^4$. Figure 1 illustrates the idea for $x = 5$.

In general, suppose that f and g are two functions and that x is a number in the domain of g. Evaluating g at x yields $g(x)$. If $g(x)$ is in the domain of f, then evaluate f at $g(x)$ and obtain the expression $f(g(x))$. The correspondence from x to $f(g(x))$ is called a *composite function $f \circ g$*.

Figure 1

> **In Other Words**
> To find $f(g(x))$, first determine $g(x)$. This is the output of g. Then evaluate f at the output of g. The result is $f(g(x))$.

Definition

Given two functions f and g, the **composite function,** denoted by $f \circ g$ (read as "f composed with g"), is defined by

$$(f \circ g)(x) = f(g(x))$$

The notation $f(g(x))$ is read "f of g of x".

Figure 2 illustrates the definition. Notice that the "inside" function g in $f(g(x))$ is always evaluated first.

Figure 2

EXAMPLE 1 **Evaluating a Composite Function**

Suppose that $f(x) = x^2 - 3$ and $g(x) = 2x + 1$. Find

(a) $(f \circ g)(3)$ (b) $(g \circ f)(3)$ (c) $(f \circ f)(-2)$

Solution

(a) Using the flowchart in Figure 2, evaluate $(f \circ g)(3)$ as follows:

More directly,

$$(f \circ g)(3) = f(g(3)) = f(2(3) + 1) = f(7) = 7^2 - 3 = 49 - 3 = 46$$

(b) $(g \circ f)(3) = g(f(3)) = g(6) = 2(6) + 1 = 13$

$$f(x) = x^2 - 3 \qquad g(x) = 2x + 1$$
$$f(3) = 3^2 - 3$$
$$= 6$$

(c) $(f \circ f)(-2) = f(f(-2)) = f(1) = 1^2 - 3 = -2$

$$f(x) = x^2 - 3 \qquad f(x) = x^2 - 3$$
$$f(-2) = (-2)^2 - 3$$
$$= 1$$

Quick ✓

1. Given two functions f and g, the _____ _____, denoted by $f \circ g$, is defined by $(f \circ g)(x) = f(g(x))$.

2. Suppose that $f(x) = 4x - 3$ and $g(x) = x^2 + 1$. Find

   (a) $(f \circ g)(2)$ (b) $(g \circ f)(2)$ (c) $(f \circ f)(-3)$

Rather than evaluating a composite function at a specific value, a composite function can be written in terms of the independent variable, x.

EXAMPLE 2 **Finding a Composite Function**

Suppose that $f(x) = x^2 + 2x$ and $g(x) = 2x - 1$. Find

(a) $(f \circ g)(x)$ (b) $(g \circ f)(x)$ (c) $(f \circ g)(2)$

Solution

(a) $\qquad\qquad (f \circ g)(x) = f(g(x))$

$g(x) = 2x - 1:\quad = f(2x - 1)$

$f(x) = x^2 + 2x:\quad = (2x - 1)^2 + 2(2x - 1)$

FOIL; distribute: $\quad = 4x^2 - 4x + 1 + 4x - 2$

Combine like terms: $\quad = 4x^2 - 1$

Work Smart

$(f \circ g)(x)$ does not mean $(f \cdot g)(x)$.

(continued)

Work Smart

Notice in Example 2(a) and 2(b) that
$(f \circ g)(x) \neq (g \circ f)(x)$.

(b)

$$(g \circ f)(x) = g(f(x))$$
$$f(x) = x^2 + 2x: \quad = g(x^2 + 2x)$$
$$g(x) = 2x - 1: \quad = 2(x^2 + 2x) - 1$$
$$\text{Distribute:} \quad = 2x^2 + 4x - 1$$

(c) Instead of finding $(f \circ g)(2)$ using the approach presented in Example 1, find $(f \circ g)(2)$ using the results from part (a).

$$(f \circ g)(2) = f(g(2)) = 4(2)^2 - 1 = 4(4) - 1 = 15 \qquad \bullet$$

Quick ✓

3. *True or False* $(f \circ g)(x) = f(x) \cdot g(x)$

4. If $f(g(x)) = (4x - 3)^5$ and $f(x) = x^5$, what is $g(x)$?

5. Suppose that $f(x) = x^2 - 3x + 1$ and $g(x) = 3x + 2$. Find

    (a) $(f \circ g)(x)$ **(b)** $(g \circ f)(x)$ **(c)** $(f \circ g)(-2)$

▶ ❷ **Determine Whether a Function Is One-to-One**

Figures 3 and 4 illustrate two different functions represented as mappings. The function in Figure 3 maps states to their population in millions. The function in Figure 4 maps animals to their life expectancy.

Figure 3

Figure 4

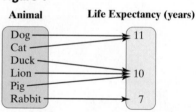

If you were asked to name the state with a population of 0.9 million based on the function in Figure 3, you would say "South Dakota." If you were asked to name the animal whose life expectancy is 11 years based on the function in Figure 4, you might say "dog" or "cat." What is the difference between the functions in Figures 3 and 4? In Figure 3, one (and only one) element in the domain corresponds to each element in the range. In Figure 4, this is not the case – there is more than one element in the domain that corresponds to an element in the range. For example, "dog" corresponds to "11," but "cat" also corresponds to "11." Functions such as the one in Figure 3 have a special name.

In Other Words

A function is NOT one-to-one if two different inputs correspond to the same output.

Definition

A function is **one-to-one** if any two different inputs in the domain correspond to two different outputs in the range. That is, if x_1 and x_2 are two different inputs to a function f, then $f(x_1) \neq f(x_2)$.

Thus a function is not one-to-one if two different elements in the domain correspond to the same element in the range. The function in Figure 3 is one-to-one because no two different elements in the domain correspond to the same element in the range. The function in Figure 4 is not one-to-one because two different elements in the domain, dog and cat, both correspond to 11.

EXAMPLE 3   **Determining Whether a Function Is One-to-One**

Determine which of the following functions is one-to-one.

(a) For the function to the right, the domain represents the age of five males, and the range represents their HDL (good) cholesterol (mg/dL).

(b) $\{(-2, 6), (-1, 3), (0, 2), (1, 4), (2, 8)\}$

Solution

(a) The function is not one-to-one because two different inputs, 55 and 61, correspond to the same output, 38.

(b) The function is one-to-one because no two distinct inputs (the x-coordinates of the ordered pairs) correspond to the same output (the y-coordinates). ●

Quick ✓

6. A function is _____ if any two different inputs in the domain correspond to two different outputs in the range. That is, if x_1 and x_2 are two different inputs to a function f, then $f(x_1) \neq f(x_2)$.

In Problems 7 and 8, determine whether the function is one-to-one.

7.

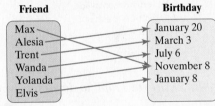

8. $\{(-3, 3), (-2, 2), (-1, 1), (0, 0), (1, -1)\}$

▶ Consider the functions $y = 2x - 5$ and $y = x^2 - 4$ shown in Figure 5.

Figure 5

(a) $y = 2x - 5$

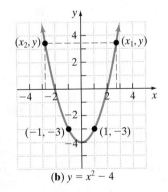

(b) $y = x^2 - 4$

Note, for the function $y = 2x - 5$ shown in Figure 5(a), that any output is the result of only one input. However, for the function $y = x^2 - 4$ shown in Figure 5(b), there are instances where a given output is the result of more than one input. For example, the output -3 is the result of input -1 and of input 1.

If a horizontal line intersects the graph of a function at more than one point, the function cannot be one-to-one because this would mean that two different inputs give the same output. This result is stated formally as the *horizontal line test*.

Horizontal Line Test

If every horizontal line intersects the graph of a function f in at most one point, then f is one-to-one.

EXAMPLE 4   **Using the Horizontal Line Test**

For each function, use the graph to determine whether the function is one-to-one.

(a) $f(x) = -2x^3 + 1$ (b) $g(x) = x^3 - 4x$

Solution

(a) Figure 6(a) illustrates the horizontal line test for $f(x) = -2x^3 + 1$. Because every horizontal line intersects the graph of f exactly once, f must be one-to-one.

(b) Figure 6(b) illustrates the horizontal line test for $g(x) = x^3 - 4x$. The horizontal line $y = 1$ intersects the graph three times, so g is not one-to-one.

Figure 6

(a) Every horizontal line intersects the graph once; f is one-to-one.

(b) A horizontal line intersects the graph three times; g is not one-to-one.

Work Smart

For a graph NOT to be one-to-one you only need to find one horizontal line that intersects it more than once.

Quick ✓

In Problem 9, use the graph to determine whether the given function is one-to-one.

9. (a) $f(x) = x^4 - 4x^2$ (b) $g(x) = x^5 + 4x$

▶ ❸ **Find the Inverse of a Function Defined by a Map or Set of Ordered Pairs**

Now that the concept of a one-to-one function is understood, *inverse functions* can be explored.

> **Definition**
>
> If f is a one-to-one function with ordered pairs of the form (a, b), then the **inverse function**, denoted f^{-1}, is the set of ordered pairs of the form (b, a).

Work Smart

The symbol f^{-1} represents the inverse function of f. The -1 used in f^{-1} is not an exponent. That is,

$$f^{-1}(x) \neq \frac{1}{f(x)}.$$

    In order for the inverse of a function to also be a function, the function must be one-to-one. We begin by finding inverses of functions represented by maps and sets of ordered pairs. To do this, interchange the inputs and outputs. So for each ordered pair (a, b) that is defined in a function f, the ordered pair (b, a) is defined in the inverse function f^{-1}. The inverse undoes what the function does. Suppose a function takes the input 11 and gives the output 3. This can be represented as $(11, 3)$. The inverse would take as input 3 and give the output 11, which can be represented as $(3, 11)$.

EXAMPLE 5 **Finding the Inverse of a Function Defined by a Map**

Find the inverse of the following function. The domain of the function represents the states, and the range represents the states' population in millions. Give the domain and the range of the inverse function.

State	Population (millions)
Indiana	6.6
Washington	7.1
South Dakota	0.9
North Carolina	9.9
Tennessee	6.5

Solution

The elements in the domain represent the inputs to the function. The elements in the range represent the outputs of the function. The function is one-to-one, so the inverse is a function. To find the inverse function, interchange the elements in the domain with the elements in the range. For example, the function receives as input Indiana and outputs 6.6, so the inverse receives as input 6.6 and outputs Indiana. The inverse function is shown below.

Population (millions)	State
6.6	Indiana
7.1	Washington
0.9	South Dakota
9.9	North Carolina
6.5	Tennessee

The domain of the inverse function is $\{6.6, 7.1, 0.9, 9.9, 6.5\}$. The range of the inverse function is $\{$ Indiana, Washington, South Dakota, North Carolina, Tennessee $\}$. ●

Quick ✓

10. Find the inverse of the one-to-one function shown below. The domain of the function represents the lengths of the right humerus (in mm), and the range represents the lengths of the right tibia (in mm) of rats sent to space. Give the domain and the range of the inverse function.

Right Humerus, mm	Right Tibia, mm
24.80	36.05
24.59	35.57
24.29	34.58
23.81	34.20
24.87	34.73

SOURCE: *NASA Life Sciences Data Archive*

EXAMPLE 6 **Finding the Inverse of a Function Defined by a Set of Ordered Pairs**

Find the inverse of the following function. State the domain and the range of the inverse function.

$$\{(-2, 6), (-1, 3), (0, 2), (1, 5), (2, 8)\}$$

Solution

The function is one-to-one, so the inverse is a function. Find the inverse by interchanging the entries in each ordered pair. Thus the inverse function is given by

$$\{(6, -2), (3, -1), (2, 0), (5, 1), (8, 2)\}$$

The domain of the inverse function is $\{6, 3, 2, 5, 8\}$, and the range of the inverse function is $\{-2, -1, 0, 1, 2\}$. ●

Quick ✓

11. Find the inverse of the following function. State the domain and the range of the inverse function.

$$\{(-3, 3), (-2, 2), (-1, 1), (0, 0), (1, -1)\}$$

In Examples 5 and 6, notice that the elements that are in the domain of f are the same elements that are in the range of its inverse. In addition, the elements that are in the range of f are the same elements that are in the domain of its inverse.

> **Relation Between the Domain and Range of a Function and Its Inverse**
>
> All elements in the domain of a function are elements in the range of its inverse.
> All elements in the range of a function are elements in the domain of its inverse.

Figure 7

▶ ❹ Obtain the Graph of the Inverse Function from the Graph of a Function

Recall that if a function is defined by a set of ordered pairs, the inverse can be found by interchanging the entries. Thus if (a, b) is a point on the graph of a function, then (b, a) is a point on the graph of the inverse function. See Figure 7. From the graph, it follows that the point (b, a) on the graph of the inverse function is the reflection about the line $y = x$ of the point (a, b).

The graph of a function f and the graph of its inverse are symmetric with respect to the line $y = x$.

EXAMPLE 7 **Graphing the Inverse Function**

Figure 8(a) shows the graph of a one-to-one function $y = f(x)$. Draw the graph of its inverse.

Solution

Begin by adding the graph of $y = x$ to Figure 8(a). Since the points $(-4, -2)$, $(-3, -1)$, $(-1, 0)$, and $(4, 2)$ are on the graph of f, the points $(-2, -4)$, $(-1, -3)$, $(0, -1)$, and $(2, 4)$ are on the graph of the inverse of f, f^{-1}. Using these points, along with the fact that the graph of f^{-1} is a reflection about the line $y = x$ of the graph of f, draw the graph of f^{-1}. See Figure 8(b).

Figure 8

(a)

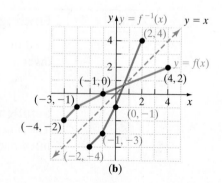

(b)

Quick ✓

12. Below is the graph of a one-to-one function $y = f(x)$. Draw the graph of its inverse.

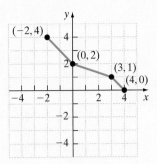

▶ ❺ **Find the Inverse of a Function Defined by an Equation**

We have learned how to find the inverse of a one-to-one function defined by a map, a set of ordered pairs, or a graph. It is now time to discuss how to find the inverse of a function defined by an equation.

Remember that when a function is defined by an equation, the notation $y = f(x)$ is used. The notation $y = f^{-1}(x)$ can be used to denote the equation whose rule is the inverse function of f.

Because the inverse of a function "undoes" what the original function does, the following relation between a function f and its inverse, f^{-1}, exists.

Work Smart

If $f(x) = 2x$, then each input is doubled. The inverse function is $f^{-1}(x) = \frac{x}{2}$, so each input is cut in half. If $x = 10$, then $f(10) = 20$. Now substitute the output of f, 20, into f^{-1} to get $f^{-1}(20) = 10$. Notice we are right back where we started!

$$f^{-1}(f(x)) = x \text{ for every } x \text{ in the domain of } f$$
$$f(f^{-1}(x)) = x \text{ for every } x \text{ in the domain of } f^{-1}$$

Let's now find the inverse of a one-to-one function defined by an equation.

EXAMPLE 8 **How to Find the Inverse of a One-to-One Function**

Find the inverse of $f(x) = 2x - 3$.

Step-by-Step Solution

First, verify that the function is one-to-one. Because the graph of the function f is a line with y-intercept $(0, -3)$ and slope 2, we know by the horizontal line test that the function is one-to-one and therefore has an inverse function.

Step 1: Replace $f(x)$ with y in the equation for $f(x)$.	$f(x) = 2x - 3$ $y = 2x - 3$

Step 2: In $y = f(x)$, interchange the variables x and y to obtain $x = f(y)$.	$x = 2y - 3$

Step 3: Solve the equation found in Step 2 for y in terms of x.

Add 3 to both sides: $x + 3 = 2y$

Divide both sides by 2: $\dfrac{x + 3}{2} = y$

Step 4: Replace y with $f^{-1}(x)$.

$$f^{-1}(x) = \dfrac{x + 3}{2}$$

(continued)

Step 5: Check Verify your result by showing that $f^{-1}(f(x)) = x$ and $f(f^{-1}(x)) = x$.

$$f^{-1}(f(x)) = f^{-1}(2x - 3) \qquad\qquad f(f^{-1}(x)) = f\left(\frac{x+3}{2}\right)$$

$$= \frac{2x - 3 + 3}{2} \qquad\qquad\qquad = 2\left(\frac{x+3}{2}\right) - 3$$

$$= \frac{2x}{2} \qquad\qquad\qquad\qquad = x + 3 - 3$$

$$= x \qquad\qquad\qquad\qquad\qquad = x$$

Everything checks, so $f^{-1}(x) = \dfrac{x+3}{2}$.

The steps used in Example 8 are summarized below.

Finding The Inverse of a One-to-One Function Defined by an Equation

Step 1: Replace $f(x)$ with y in the equation for $f(x)$.

Step 2: In $y = f(x)$, interchange the variables x and y to obtain $x = f(y)$.

Step 3: Solve the equation found in Step 2 for y in terms of x.

Step 4: Replace y with $f^{-1}(x)$.

Step 5: Verify the result by showing that $f^{-1}(f(x)) = x$ and $f(f^{-1}(x)) = x$.

Work Smart

We need to show that *both* $f^{-1}(f(x)) = x$ and $f(f^{-1}(x)) = x$ because function composition is not commutative.

Consider the function $f(x) = 2x - 3$ and its inverse $f^{-1}(x) = \dfrac{x+3}{2}$ from Example 8. Notice that $f(5) = 2(5) - 3 = 7$. What do you think $f^{-1}(7)$ will equal? Notice that $f^{-1}(7) = \dfrac{7+3}{2} = \dfrac{10}{2} = 5$. Thus f^{-1} "undoes" what f did!

Quick ✓

13. *True or False* If f is a one-to-one function so that its inverse is f^{-1}, then $f^{-1}(f(x)) = x$ for every x in the domain of f, and $f(f^{-1}(x)) = x$ for every x in the domain of f^{-1}.

14. *True or False* The notation $f^{-1}(x)$ is equivalent to $\dfrac{1}{f(x)}$.

15. Find the inverse of $g(x) = 5x - 1$.

EXAMPLE 9 **Finding the Inverse of a One-to-One Function**

Find the inverse of $h(x) = x^3 + 4$.

Solution

$$h(x) = x^3 + 4$$

Replace $h(x)$ with y in the equation for $h(x)$: $\qquad y = x^3 + 4$

Interchange the variables x and y: $\qquad x = y^3 + 4$

Solve the equation for y in terms of x: $\qquad x - 4 = y^3$

Take the cube root of both sides: $\qquad \sqrt[3]{x - 4} = y$

Replace y with $h^{-1}(x)$: $\qquad h^{-1}(x) = \sqrt[3]{x - 4}$

Check Verify your result by showing that $h^{-1}(h(x)) = x$ and $h(h^{-1}(x)) = x$.

$$h^{-1}(h(x)) = h^{-1}(x^3 + 4) \qquad\qquad h(h^{-1}(x)) = h(\sqrt[3]{x - 4})$$
$$= \sqrt[3]{x^3 + 4 - 4} \qquad\qquad\qquad = (\sqrt[3]{x - 4})^3 + 4$$
$$= \sqrt[3]{x^3} \qquad\qquad\qquad\qquad = x - 4 + 4$$
$$= x \qquad\qquad\qquad\qquad\qquad = x$$

So $h^{-1}(x) = \sqrt[3]{x - 4}$.

Quick ✔

16. Find the inverse of $f(x) = x^5 + 3$.

8.1 Exercises MyMathLab® Exercise numbers in green have complete video solutions in MyMathLab or may be accessed using the QR code to the right.

*Problems **1–16** are the Quick ✔ s that follow the EXAMPLES.*

Building Skills

In Problems 17–24, for the given functions f and g, find

(a) $(f \circ g)(3)$ (b) $(g \circ f)(-2)$
(c) $(f \circ f)(1)$ (d) $(g \circ g)(-4)$

See Objective 1.

17. $f(x) = 2x + 5; g(x) = x - 4$

18. $f(x) = 4x - 3; g(x) = x + 2$

19. $f(x) = x^2 + 4; g(x) = 2x + 3$

20. $f(x) = x^2 - 3; g(x) = 5x + 1$

21. $f(x) = 2x^3; g(x) = -2x^2 + 5$

22. $f(x) = -2x^3; g(x) = x^2 + 1$

23. $f(x) = |x - 10|; g(x) = \dfrac{12}{x + 3}$

24. $f(x) = \sqrt{x + 8}; g(x) = x^2 - 4$

In Problems 25–36, for the given functions f and g, find

(a) $(f \circ g)(x)$ (b) $(g \circ f)(x)$
(c) $(f \circ f)(x)$ (d) $(g \circ g)(x)$

See Objective 1.

25. $f(x) = x + 1; g(x) = 2x$

26. $f(x) = x - 3; g(x) = 4x$

27. $f(x) = 2x + 7; g(x) = -4x + 5$

28. $f(x) = 3x - 1; g(x) = -2x + 5$

29. $f(x) = x^2; g(x) = x - 3$

30. $f(x) = x^2 + 1; g(x) = x + 1$

31. $f(x) = \sqrt{x}; g(x) = x + 4$

32. $f(x) = \sqrt{x + 2}; g(x) = x - 2$

33. $f(x) = |x + 4|; g(x) = x^2 - 4$

34. $f(x) = |x - 3|; g(x) = x^3 + 3$

35. $f(x) = \dfrac{2}{x + 1}; g(x) = \dfrac{1}{x}$

36. $f(x) = \dfrac{2}{x - 1}; g(x) = \dfrac{4}{x}$

In Problems 37–46, determine which of the following functions is one-to-one. See Objective 2.

37.

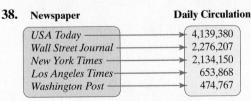

SOURCE: *United States Census Bureau*

38.

SOURCE: *Information Please Almanac*

39.

SOURCE: *Information Please Almanac*

40.

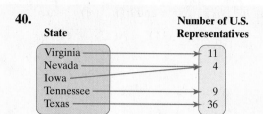

41. $\{(-3,4), (-2,6), (-1,8), (0,10), (1,12)\}$

42. $\{(-2,6), (-1,3), (0,0), (1,-3), (2,6)\}$

43. $\{(-2,4), (-1,2), (0,0), (1,2), (2,4)\}$

44. $\{(-2,-8), (-1,-1), (0,0), (1,1), (2,8)\}$

45. $\{(0,-4), (-1,-1), (-2,0), (1,1), (2,4)\}$

46. $\{(-3,0), (-2,3), (-1,0), (0,-3)\}$

In Problems 47–52, use the horizontal line test to determine whether the function whose graph is given is one-to-one. See Objective 2.

47.

48.

49.

50.

51.

52.

In Problems 53–58, find the inverse of the following one-to-one functions. See Objective 3.

53.

SOURCE: *U.S. Mint website*

54.

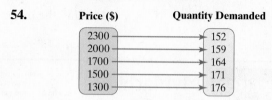

55. $\{(0,3), (1,4), (2,5), (3,6)\}$

56. $\{(-1,4), (0,1), (1,-2), (2,-5)\}$

57. $\{(-2,3), (-1,1), (0,-3), (1,9)\}$

58. $\{(-10,1), (-5,4), (0,3), (5,2)\}$

In Problems 59–64, the graph of a one-to-one function f is given. Draw the graph of the inverse function f^{-1}. See Objective 4.

59.

60.

61.

62.

63.

64.

In Problems 65–72, verify that the functions f and g are inverses of each other. See Objective 5.

65. $f(x) = x + 5; g(x) = x - 5$

66. $f(x) = 10x; g(x) = \dfrac{x}{10}$

67. $f(x) = 5x + 7; g(x) = \dfrac{x - 7}{5}$

68. $f(x) = 3x - 5; g(x) = \dfrac{x + 5}{3}$

69. $f(x) = \dfrac{3}{x - 1}; g(x) = \dfrac{3}{x} + 1$

70. $f(x) = \dfrac{2}{x + 4}; g(x) = \dfrac{2}{x} - 4$

71. $f(x) = \sqrt[3]{x + 4}; g(x) = x^3 - 4$

72. $f(x) = \sqrt[3]{2x + 1}; g(x) = \dfrac{x^3 - 1}{2}$

In Problems 73–92, find the inverse function of the given one-to-one function. See Objective 5.

73. $f(x) = 6x$

74. $f(x) = 12x$

75. $f(x) = x + 4$

76. $g(x) = x + 6$

77. $h(x) = 2x - 7$

78. $H(x) = 3x + 8$

79. $G(x) = 2 - 5x$

80. $F(x) = 1 - 6x$

81. $g(x) = x^3 + 3$

82. $f(x) = x^3 - 2$

83. $p(x) = \dfrac{1}{x + 3}$

84. $P(x) = \dfrac{1}{x + 1}$

85. $F(x) = \dfrac{5}{2 - x}$

86. $G(x) = \dfrac{2}{3 - x}$

87. $f(x) = \sqrt[3]{x - 2}$

88. $f(x) = \sqrt[5]{x + 5}$

89. $R(x) = \dfrac{x}{x + 2}$

90. $R(x) = \dfrac{2x}{x + 4}$

91. $f(x) = \sqrt[3]{x - 1} + 4$

92. $g(x) = \sqrt[3]{x + 2} - 3$

Applying the Concepts

93. Environmental Disaster An oil tanker hits a rock that rips a hole in the hull of the ship. Oil leaking from the ship forms a circular region around the ship. If the radius r of the circle (in feet) as a function of time t (in hours) is $r(t) = 20t$, express the area A of the circular region contaminated with oil as a function of time. What will be the area of the circular region at 3 hours? (*Hint:* $A(r) = \pi r^2$)

94. Volume of a Balloon The volume V of a hot-air balloon (in cubic meters) as a function of its radius r is given by $V(r) = \dfrac{4}{3}\pi r^3$. If the radius r of the balloon is increasing as a function of time t (in minutes) according to $r(t) = 3\sqrt[3]{t}$, for $t \geq 0$, find the volume of the balloon as a function of time t. What will be the volume of the balloon after 30 minutes?

95. Buying Carpet You want to purchase new carpet for your family room. Carpet is sold by the square yard, but you have measured your family room in square feet. The function $A(x) = \dfrac{x}{9}$ converts the area of a room in square feet to an area in square yards. Suppose that the carpet you have selected is \$18 per square yard installed. Then the function $C(A) = 18A$ represents the cost C of installing carpet in a room that measures A square yards.

(a) Find the cost C as a function of the square footage of the room x.

(b) If your family room is 15 feet by 21 feet, what will it cost to install the carpet?

96. Tax Time You have a job that pays \$20 per hour. Your gross salary G as a function of hours worked h is given by $G(h) = 20h$. Federal tax withholding T on your paycheck is equal to 18% of gross earnings G, so federal tax withholding as a function of gross pay is given by the function $T(G) = 0.18G$.

(a) Find federal tax withholding T as a function of hours worked h.

(b) Suppose that you worked 28 hours last week. What will be the federal tax withholding on your paycheck?

97. If $f(4) = 12$ and f is one-to-one, what is $f^{-1}(12)$?

98. If $g(-2) = 7$ and f is one-to-one, what is $g^{-1}(7)$?

99. The domain of a one-to-one function f is $[0, \infty)$, and its range is $[-5, \infty)$. State the domain and the range of f^{-1}.

100. The domain of a one-to-one function f is $[5, \infty)$, and its range is $[0, \infty)$. State the domain and the range of f^{-1}.

101. The domain of a one-to-one function g is $[-4, 10]$, and its range is $(-6, 12)$. State the domain and the range of g^{-1}.

102. The domain of a one-to-one function g is $[0, 15,]$ and its range is $(0, 8)$. State the domain and the range of g^{-1}.

103. Taxes The function $T(x) = 0.15(x - 9275) + 927.50$ represents the tax bill T of a single person whose adjusted gross income is x dollars for income between \$9275 and \$37,650, inclusive. (SOURCE: *Internal Revenue Service*) Find the inverse function that expresses adjusted gross income x as a function of taxes T. That is, find $x(T)$.

104. Health Costs The annual cost of health insurance H as a function of age a is given by the function $H(a) = 22.8a - 117.5$ for $15 \leq a \leq 90$. (SOURCE: *Statistical Abstract*) Find the inverse function that expresses age a as a function of health insurance cost H. That is, find $a(H)$.

Extending the Concepts

105. If $f(x) = 2x^2 - x + 5$ and $g(x) = x + a$, find a so that the y-intercept of the graph of $(f \circ g)(x)$ is $(0, 20)$.

106. If $f(x) = x^2 - 3x + 1$ and $g(x) = x - a$, find a so that the y-intercept of the graph of $(f \circ g)(x)$ is $(0, -1)$.

Explaining the Concepts

107. Explain what it means for a function to be one-to-one. Why must a function be one-to-one in order for its inverse to be a function?

108. Are all linear functions one-to-one? Explain.

109. Explain why domain of f = range of f^{-1} and range of f = domain of f^{-1}.

110. State the horizontal line test. Why does it work?

Technology Exercises

Many types of technology, such as graphing calculators, have the ability to evaluate composite functions. To obtain the results of Example 1, we would let $Y_1 = f(x) = x^2 - 3$ and $Y_2 = g(x) = 2x + 1$.

Figure 9 shows the results of Example 1 using a graphing calculator.

Figure 9

In Problems 111–118, use technology to evaluate the composite functions. Compare your answers with those found in Problems 17–24.

   (a) $(f \circ g)(3)$ **(b)** $(g \circ f)(-2)$
   (c) $(f \circ f)(1)$ **(d)** $(g \circ g)(-4)$

111. $f(x) = 2x + 5; g(x) = x - 4$

112. $f(x) = 4x - 3; g(x) = x + 2$

113. $f(x) = x^2 + 4; g(x) = 2x + 3$

114. $f(x) = x^2 - 3; g(x) = 5x + 1$

115. $f(x) = 2x^3; g(x) = -2x^2 + 5$

116. $f(x) = -2x^3; g(x) = x^2 + 1$

117. $f(x) = |x - 10|; g(x) = \dfrac{12}{x + 3}$

118. $f(x) = \sqrt{x + 8}; g(x) = x^2 - 4$

In Problems 119–122, the functions f and g are inverses. Graph both functions on the same screen, along with the line $y = x$, to see the symmetry of the functions about the line $y = x$.

119. $f(x) = x + 5; g(x) = x - 5$

120. $f(x) = 10x; g(x) = \dfrac{x}{10}$

121. $f(x) = 5x + 7; g(x) = \dfrac{x - 7}{5}$

122. $f(x) = 3x - 5; g(x) = \dfrac{x + 5}{3}$

8.2 Exponential Functions

Objectives

1 Evaluate Exponential Expressions

2 Graph Exponential Functions

3 Define the Number e

4 Solve Exponential Equations

5 Use Exponential Models That Describe Our World

Are You Prepared for This Section?

Before getting started, complete the following problems. If you get a problem wrong, go back to the section cited and review the material.

P1. Evaluate: **(a)** 2^3 **(b)** 2^{-1} **(c)** 3^4 [Getting Ready, pp. 295–298]

P2. Graph: $f(x) = x^2$ [Section 7.4, p. 578]

P3. State the definition of a rational number.   [Section R.2, p. 12]

P4. State the definition of an irrational number.   [Section R.2, p. 13]

P5. Write 3.20349193 as a decimal **(a)** rounded to four decimal places **(b)** truncated to four decimal places.   [Section R.2, pp. 14–15]

P6. Simplify: **(a)** $m^3 \cdot m^5$ **(b)** $\dfrac{a^7}{a^2}$ **(c)** $(z^3)^4$ [Getting Ready, pp. 295–298]

P7. Solve: $x^2 - 5x = 14$ [Section 4.8, pp. 373–375]

Suppose editor Mary Beckwith has just hired you as a proofreader for Pearson Publishing. Mary offers you two options: Option A states that you will be paid $100 for each error you find in the final page proofs of a text. Option B states that you will start with $1 and your payment will double for each error you find in the final page proofs. You know from experience that the final page proofs of a text typically have about 15–20 errors. Which option will you choose?

If there is one error, Option A pays $100, while Option B pays $2. For two errors, Option A pays $2(\$100) = \$200$, while Option B pays $2^2 = \$4$. For three errors, Option A pays $3(\$100) = \$300$, while Option B pays $2^3 = \$8$. Option A seems to be the way to go. To complete the analysis, set up Table 1, which lists the payment amount as a function of the number of errors in the page proofs. Remember, in Option B, the payment amount doubles each time you find an error.

Prepared?...Answers

P1. (a) 8 **(b)** $\dfrac{1}{2}$ **(c)** 81

P2.

P3. A rational number is a number that can be expressed as a quotient $\dfrac{p}{q}$ of two integers. The integer p is called the numerator, and the integer q, which cannot be 0, is called the denominator. The set of rational numbers are the numbers $\mathbb{Q} = \left\{ x \,\middle|\, x = \dfrac{p}{q}, \text{where } p, q \text{ are integers and } q \neq 0 \right\}$.

P4. An irrational number has a decimal representation that neither repeats nor terminates.

P5. (a) 3.2035 **(b)** 3.2034

P6. (a) m^8 **(b)** a^5 **(c)** z^{12}

P7. $\{-2, 7\}$

Table 1					
Number of Errors	Option A Payment	Option B Payment	Number of Errors	Option A Payment	Option B Payment
0	$0	$1	11	$1100	$2048
1	$100	$2	12	$1200	$4096
2	$200	$4	13	$1300	$8192
3	$300	$8	14	$1400	$16,384
4	$400	$16	15	$1500	$32,768
5	$500	$32	16	$1600	$65,536
6	$600	$64	17	$1700	$131,072
7	$700	$128	18	$1800	$262,144
8	$800	$256	19	$1900	$524,288
9	$900	$512	20	$2000	$1,048,576
10	$1000	$1024			

Holy cow! If you find 20 errors, you will get paid over a million dollars! Mary Beckwith better reconsider her offer! If x represents the number of errors, Option A can be expressed as a linear function, $f(x) = 100x$; Option B can be expressed as an *exponential function*, $g(x) = 2^x$.

Definition

An **exponential function** is a function of the form

$$f(x) = a^x$$

where $a \neq 1$ is a positive real number $(a > 0)$. The domain of the exponential function is the set of all real numbers.

The restrictions on the base a will be addressed shortly. The key point is that *the independent variable is in the exponent of the exponential expression.* Contrast this idea with polynomial functions (such as $f(x) = x^2 - 4x$ or $g(x) = 2x^3 + x^2 - 5$), where the independent variable is the base of each exponential expression.

▶ ① Evaluate Exponential Expressions

In Section 6.1, $a^{\frac{m}{n}}$ was defined, where the base a is a positive real number and the exponent $\frac{m}{n}$ is a rational number.

But what if you want to raise the base a to an irrational number? Although the answer to this question requires advanced mathematics, here is an intuitive explanation. Suppose you want to find the value of $3^{\sqrt{2}}$. The calculator says that $\sqrt{2} \approx 1.414213562$, so it should seem reasonable that $3^{\sqrt{2}}$ is approximately $3^{1.4}$, where the 1.4 comes from truncating the decimals to the right of the 4 in the tenths position. A better approximation of $3^{\sqrt{2}}$ would be $3^{1.4142}$, where the digits to the right of the ten-thousandths position have been truncated. The more decimals used in approximating $\sqrt{2}$, the better the approximation of $3^{\sqrt{2}}$.

To evaluate expressions of the form a^x using a scientific calculator, enter the base a, press the $\boxed{x^y}$ key, enter the exponent x, and press $\boxed{=}$. To evaluate expressions of the form a^x using a graphing calculator, enter the base a, press the caret $\boxed{\wedge}$ key, enter the exponent x, and press $\boxed{\text{ENTER}}$.

EXAMPLE 1 **Evaluating Exponential Expressions**

Using a calculator, evaluate each of the following expressions. Write as many places as your calculator allows.

    (a) $3^{1.4}$ **(b)** $3^{1.41}$ **(c)** $3^{1.414}$ **(d)** $3^{1.4142}$ **(e)** $3^{\sqrt{2}}$

Solution

    (a) $3^{1.4} \approx 4.655536722$ **(b)** $3^{1.41} \approx 4.706965002$
    (c) $3^{1.414} \approx 4.727695035$ **(d)** $3^{1.4142} \approx 4.72873393$
    (e) $3^{\sqrt{2}} \approx 4.728804388$ ●

Quick ✓

1. An exponential function is a function of the form $f(x) = a^x$ where a ___ 0 and a ___ 1.

In Problem 2, use a calculator to evaluate each of the following expressions. Write as many places as your calculator allows.

2. (a) $2^{1.7}$ **(b)** $2^{1.73}$ **(c)** $2^{1.732}$ **(d)** $2^{1.7321}$ **(e)** $2^{\sqrt{3}}$

Example 1 illustrates that the value of an exponential expression at any real number can be approximated. This is why the domain of any exponential function is the set of all real numbers.

In the definition of an exponential function, $a = 1$ was ruled out, and it was required that a be positive. The base $a = 1$ is excluded because this function is the constant function $f(x) = 1^x = 1$. Exclude bases that are negative to avoid problems for exponents such as $\frac{1}{2}$ or $\frac{3}{4}$. For example, suppose $f(x) = (-2)^x$. In the real number system, $f\left(\frac{1}{2}\right)$ could not be evaluated because $f\left(\frac{1}{2}\right) = (-2)^{\frac{1}{2}} = \sqrt{-2}$, which is not a real number. Finally, exclude a base of 0, because this function is $f(x) = 0^x$, which equals zero for $x > 0$ and is undefined when $x < 0$. When $x = 0$, $f(x) = 0^x$ is *indeterminate* because its value is not precisely determined.

② Graph Exponential Functions

▶ Let's learn about properties of exponential functions from their graphs.

EXAMPLE 2   **Graphing an Exponential Function**

Graph the exponential function $f(x) = 2^x$ using point plotting. From the graph, state the domain and the range of the function.

Solution
Begin by locating some points on the graph of $f(x) = 2^x$ as shown in Table 2. Plot the points in Table 2 and connect them in a smooth curve. Figure 10 shows the graph of $f(x) = 2^x$.

Figure 10

Table 2

x	$f(x) = 2^x$	$(x, f(x))$
-3	$f(-3) = 2^{-3} = \dfrac{1}{2^3} = \dfrac{1}{8}$	$\left(-3, \dfrac{1}{8}\right)$
-2	$f(-2) = 2^{-2} = \dfrac{1}{2^2} = \dfrac{1}{4}$	$\left(-2, \dfrac{1}{4}\right)$
-1	$f(-1) = 2^{-1} = \dfrac{1}{2^1} = \dfrac{1}{2}$	$\left(-1, \dfrac{1}{2}\right)$
0	$f(0) = 2^0 = 1$	$(0, 1)$
1	$f(1) = 2^1 = 2$	$(1, 2)$
2	$f(2) = 2^2 = 4$	$(2, 4)$
3	$f(3) = 2^3 = 8$	$(3, 8)$

The domain of any exponential function is the set of all real numbers. Notice that no x caused 2^x to be less than or equal to 0. Based on this and the graph, the range of $f(x) = 2^x$ is the set of all positive real numbers $\{y | y > 0\}$, or $(0, \infty)$ in interval notation.   ●

The graph of $f(x) = 2^x$ in Figure 10 is typical of all exponential functions that have a base larger than 1. Figure 11 shows the graphs of two other exponential functions whose bases are larger than 1: $y = 3^x$ and $y = 6^x$. Notice that the larger the base, the steeper the graph is for $x > 0$ and the closer the graph is to the x-axis for $x < 0$.

The information for $f(x) = a^x$, where $a > 1$, is summarized below.

Figure 11

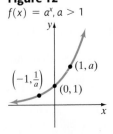

Figure 12
$f(x) = a^x, a > 1$

Properties of the Graph of an Exponential Function $f(x) = a^x, a > 1$

1. The domain is the set of all real numbers. The range is the set of all positive real numbers.

2. There are no x-intercepts; the y-intercept is $(0, 1)$.

3. The graph of f contains the points $\left(-1, \dfrac{1}{a}\right)$, $(0, 1)$, and $(1, a)$.

See Figure 12.

Quick ✓

3. Graph the exponential function $f(x) = 4^x$ using point plotting. From the graph, state the domain and the range of the function.

▶ Now consider $f(x) = a^x, 0 < a < 1$.

EXAMPLE 3 **Graphing an Exponential Function**

Graph the exponential function $f(x) = \left(\dfrac{1}{2}\right)^x$ using point plotting. From the graph, state the domain and the range of the function.

Solution

Begin by locating some points on the graph of $f(x) = \left(\dfrac{1}{2}\right)^x$ as shown in Table 3. Plot the points in Table 3 and connect them in a smooth curve. Figure 13 shows the graph of $f(x) = \left(\dfrac{1}{2}\right)^x$.

Table 3

x	$f(x) = \left(\dfrac{1}{2}\right)^x$	$(x, f(x))$
-3	$f(-3) = \left(\dfrac{1}{2}\right)^{-3} = 2^3 = 8$	$(-3, 8)$
-2	$f(-2) = \left(\dfrac{1}{2}\right)^{-2} = 2^2 = 4$	$(-2, 4)$
-1	$f(-1) = \left(\dfrac{1}{2}\right)^{-1} = 2^1 = 2$	$(-1, 2)$
0	$f(0) = \left(\dfrac{1}{2}\right)^0 = 1$	$(0, 1)$
1	$f(1) = \left(\dfrac{1}{2}\right)^1 = \dfrac{1}{2}$	$\left(1, \dfrac{1}{2}\right)$
2	$f(2) = \left(\dfrac{1}{2}\right)^2 = \dfrac{1}{4}$	$\left(2, \dfrac{1}{4}\right)$
3	$f(3) = \left(\dfrac{1}{2}\right)^3 = \dfrac{1}{8}$	$\left(3, \dfrac{1}{8}\right)$

Figure 13

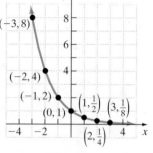

The domain of any exponential function is the set of all real numbers. From the graph, the range of $f(x) = \left(\dfrac{1}{2}\right)^x$ is the set of all positive real numbers $\{y \mid y > 0\}$, or $(0, \infty)$ in interval notation.

The graph of $f(x) = \left(\dfrac{1}{2}\right)^x$ in Figure 13 is typical of all exponential functions that have a base between 0 and 1. Figure 14 shows the graphs of two additional exponential functions whose bases are between 0 and 1: $y = \left(\dfrac{1}{3}\right)^x$ and $y = \left(\dfrac{1}{6}\right)^x$. Notice that the smaller the base, the closer the graph is to the x-axis for $x > 0$ and the steeper the graph is for $x < 0$.

Figure 14

The information about $f(x) = a^x$, where $0 < a < 1$, is now summarized.

Figure 15

$f(x) = a^x, 0 < a < 1$

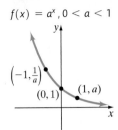

Properties of the Graph of an Exponential Function $f(x) = a^x, 0 < a < 1$

1. The domain is the set of all real numbers. The range is the set of all positive real numbers.

2. There are no x-intercepts; the y-intercept is $(0, 1)$.

3. The graph of f contains the points $\left(-1, \dfrac{1}{a}\right)$, $(0, 1)$, and, $(1, a)$.

   See Figure 15.

Quick ✓

4. The graph of every exponential function $f(x) = a^x$ passes through three points:

   _____ , _____ , and _____ .

5. *True or False* The domain of the exponential function $f(x) = a^x, a > 0, a \neq 1$, is the set of all real numbers.

6. *True or False* The range of the exponential function $f(x) = a^x, a > 0, a \neq 1$, is the set of all real numbers.

7. Graph the exponential function $f(x) = \left(\dfrac{1}{4}\right)^x$ using point plotting. From the graph, state the domain and the range of the function.

EXAMPLE 4 **Graphing an Exponential Function**

Use point plotting to graph $f(x) = 3^{x+1}$. From the graph, state the domain and the range of the function.

Solution
Choose values of x and find the corresponding values of the function. See Table 4. Then plot the ordered pairs and connect them in a smooth curve. See Figure 16.

Table 4

x	$f(x)$	$(x, f(x))$
-3	$f(-3) = 3^{-3+1} = 3^{-2} = \dfrac{1}{3^2} = \dfrac{1}{9}$	$\left(-3, \dfrac{1}{9}\right)$
-2	$f(-2) = 3^{-2+1} = 3^{-1} = \dfrac{1}{3^1} = \dfrac{1}{3}$	$\left(-2, \dfrac{1}{3}\right)$
-1	$f(-1) = 3^{-1+1} = 3^0 = 1$	$(-1, 1)$
0	$f(0) = 3^{0+1} = 3^1 = 3$	$(0, 3)$
1	$f(1) = 3^{1+1} = 3^2 = 9$	$(1, 9)$

Figure 16

The domain is the set of all real numbers. The range is $\{y \mid y > 0\}$, or $(0, \infty)$ in interval notation.

Quick ✓

In Problems 8 and 9, graph each function using point plotting. From the graph, state the domain and the range of each function.

8. $f(x) = 2^{x-1}$

9. $f(x) = 3^x + 1$

▶ ❸ Define the Number e

Many models use an exponential function whose base is an irrational number symbolized by the letter e. The number e can be used to model the growth of a stock's price or to estimate the time of death of a carbon-based life form.

Definition

The **number e** is defined as the number that the expression

$$\left(1 + \frac{1}{n}\right)^n$$

approaches as n increases.

Table 5 shows some values of $\left(1 + \frac{1}{n}\right)^n$ as n increases. The last number in column 4 is the number e correct to nine decimal places.

Table 5

n	$\dfrac{1}{n}$	$1 + \dfrac{1}{n}$	$\left(1 + \dfrac{1}{n}\right)^n$
1	1	2	2
2	0.5	1.5	2.25
5	0.2	1.2	2.48832
10	0.1	1.1	2.59374246
100	0.01	1.01	2.704813829
1,000	0.001	1.001	2.716923932
10,000	0.0001	1.0001	2.718145927
100,000	0.00001	1.00001	2.718268237
1,000,000	0.000001	1.000001	2.718280469
1,000,000,000	10^{-9}	$1 + 10^{-9}$	2.718281827

The exponential function $f(x) = e^x$, whose base is the number e, occurs so often in applications that it is sometimes called *the* exponential function. Most calculators have the key $\boxed{e^x}$ or $\boxed{\exp(x)}$, which you can use to evaluate the exponential function $f(x) = e^x$ for a given value of x.

Use your calculator to approximate the values of $f(x) = e^x$ for $x = -1$, 0, and 1 as was done to create Table 6. The graph of the exponential function is shown in Figure 17(a). Because $2 < e < 3$, the graph of $f(x) = e^x$ lies between the graph of $y = 2^x$ and that of $y = 3^x$. See Figure 17(b).

Table 6

x	$f(x) = e^x$
-2	$e^{-2} \approx 0.135$
-1	$e^{-1} \approx 0.368$
0	$e^0 = 1$
1	$e^1 \approx 2.718$
2	$e^2 \approx 7.389$

Figure 17

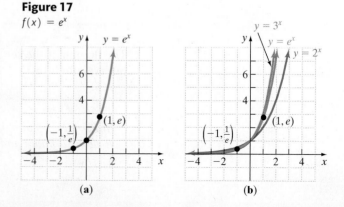

(a)   (b)

Quick ✓

10. What is the value of e rounded to five decimal places?

11. Evaluate each of the following rounded to three decimal places:

   (a) e^4 **(b)** e^{-4}

▶ ❹ **Solve Exponential Equations**

Equations that involve terms of the form $a^x, a > 0, a \neq 1$, are called **exponential equations.** Some exponential equations can be solved using the Laws of Exponents and the following property.

Work Smart

In order to use the Property for Solving Exponential Equations, both sides of the equation must have the *same* base.

> **Property For Solving Exponential Equations**
>
> $$\text{If } a^u = a^v, \text{ then } u = v.$$

This property results from the fact that exponential functions are one-to-one. In exponential functions, any output is the result of one (and only one) input. That is to say, two different inputs cannot yield the same output. Contrast this with $y = x^2$, which is not one-to-one because two different inputs, -2 and 2, correspond to the same output, 4.

(EXAMPLE 5) How to Solve an Exponential Equation

Solve: $2^{x-3} = 32$

Step-by-Step Solution

Step 1: Use the Laws of Exponents to write both sides of the equation with the same base.	$32 = 2^5$:	$2^{x-3} = 32$ $2^{x-3} = 2^5$

Step 2: Set the exponents on each side of the equation equal to each other.	If $a^u = a^v$, then $u = v$:	$x - 3 = 5$

Step 3: Solve the equation resulting from Step 2.	Add 3 to both sides:	$x - 3 + 3 = 5 + 3$ $x = 8$

Step 4: Verify your solution(s).

Work Smart

Did you notice that properties of algebra were used to reduce the exponential equation to a linear equation?

Let $x = 8$:

$$2^{x-3} = 32$$
$$2^{8-3} \stackrel{?}{=} 32$$
$$2^5 \stackrel{?}{=} 32$$
$$32 = 32 \quad \text{True}$$

The solution set is $\{8\}$. ●

> **Solving Exponential Equations of the Form $a^u = a^v$**
>
> **Step 1:** Use the Laws of Exponents to write both sides of the equation with the same base.
>
> **Step 2:** Set the exponents on each side of the equation equal to each other.
>
> **Step 3:** Solve the equation resulting from Step 2.
>
> **Step 4:** Verify the solution(s).

Quick ✓

In Problems 12 and 13, solve each equation.

12. $5^{x-4} = 5^{-1}$ **13.** $3^{x+2} = 81$

EXAMPLE 6   **Solving Exponential Equations**

Solve the following equations:

(a) $4^{x^2} = 32$

(b) $\dfrac{e^{x^2}}{e^{2x}} = e^8$

Solution

(a)

	$4^{x^2} = 32$
Rewrite as exponential expressions with a common base:	$(2^2)^{x^2} = 2^5$
Use $(a^m)^n = a^{mn}$:	$2^{2x^2} = 2^5$
If $a^u = a^v$, then $u = v$:	$2x^2 = 5$
Divide both sides by 2:	$x^2 = \dfrac{5}{2}$
Take the square root of both sides:	$x = \pm\sqrt{\dfrac{5}{2}}$
Rationalize the denominator:	$x = \pm\dfrac{\sqrt{10}}{2}$

Check

$x = -\dfrac{\sqrt{10}}{2}:$ $\qquad 4^{\left(-\frac{\sqrt{10}}{2}\right)^2} \overset{?}{=} 32$

$\qquad\qquad\qquad\qquad 4^{\frac{10}{4}} \overset{?}{=} 32$

$\qquad\qquad\qquad\qquad 4^{\frac{5}{2}} \overset{?}{=} 32$

$a^{\frac{m}{n}} = (\sqrt[n]{a})^m:$ $\qquad (\sqrt{4})^5 \overset{?}{=} 32$

$\qquad\qquad\qquad\qquad 2^5 \overset{?}{=} 32$

$\qquad\qquad\qquad\qquad 32 = 32$ True

$x = \dfrac{\sqrt{10}}{2}:$ $\qquad 4^{\left(\frac{\sqrt{10}}{2}\right)^2} \overset{?}{=} 32$

$\qquad\qquad\qquad\qquad 4^{\frac{10}{4}} \overset{?}{=} 32$

$\qquad\qquad\qquad\qquad 4^{\frac{5}{2}} \overset{?}{=} 32$

$\qquad\qquad\qquad\qquad (\sqrt{4})^5 \overset{?}{=} 32$

$\qquad\qquad\qquad\qquad 2^5 \overset{?}{=} 32$

$\qquad\qquad\qquad\qquad 32 = 32$ True

The solution set is $\left\{-\dfrac{\sqrt{10}}{2}, \dfrac{\sqrt{10}}{2}\right\}$.

(b)

	$\dfrac{e^{x^2}}{e^{2x}} = e^8$
Use $\dfrac{a^m}{a^n} = a^{m-n}$:	$e^{x^2-2x} = e^8$
If $a^u = a^v$, then $u = v$:	$x^2 - 2x = 8$
Subtract 8 from both sides:	$x^2 - 2x - 8 = 0$
Factor:	$(x - 4)(x + 2) = 0$
Zero-Product Property:	$x = 4$ or $x = -2$

Check

$x = 4:$

$\dfrac{e^{4^2}}{e^{2(4)}} \overset{?}{=} e^8$

$\dfrac{e^{16}}{e^8} \overset{?}{=} e^8$

$e^8 = e^8$

$x = -2:$

$\dfrac{e^{(-2)^2}}{e^{2(-2)}} \overset{?}{=} e^8$

$\dfrac{e^4}{e^{-4}} \overset{?}{=} e^8$

$e^8 = e^8$

The solution set is $\{-2, 4\}$.

Quick ✓

In Problems 14 and 15, solve each equation.

14. $e^{x^2} = e^x \cdot e^{4x}$

15. $\dfrac{2^{x^2}}{8} = 2^{2x}$

⑤ Use Exponential Models That Describe Our World

Exponential functions are used in many different disciplines, such as biology (half-life), chemistry (carbon dating), economics (time value of money), and psychology (learning curves). Exponential functions are also used in statistics, as shown in the next example.

EXAMPLE 7   **Exponential Probability**

The manager of a crisis helpline knows that between 3:00 A.M. and 5:00 A.M., 3 calls per hour occur (that's 0.05 call per minute). The following formula from statistics gives the likelihood that a call will occur within t minutes of 3 A.M.

$$F(t) = 1 - e^{-0.05t}$$

Determine the likelihood that a person will call

(a) within 5 minutes of 3:00 A.M.

(b) within 20 minutes of 3:00 A.M.

Solution

(a) The likelihood that a call will occur within 5 minutes of 3:00 A.M. is found by evaluating the function $F(t) = 1 - e^{-0.05t}$ at $t = 5$.

$$F(5) = 1 - e^{-0.05(5)}$$
$$\approx 0.221$$

The likelihood that a call will occur in this time span is $0.221 = 22.1\%$.

(b) The likelihood that a call will occur within 20 minutes of 3:00 A.M. is found by evaluating the function $F(t) = 1 - e^{-0.05t}$ at $t = 20$, or $F(20)$.

$$F(20) = 1 - e^{-0.05(20)}$$
$$\approx 0.632$$

The likelihood that a call will occur in this time span is $0.632 = 63.2\%$. ●

Quick ✓

16. A bank manager knows that between 3:00 P.M. and 5:00 P.M., 15 people arrive per hour (that's 0.25 people per minute). The following formula from statistics gives the likelihood that a person will arrive within t minutes of 3:00 P.M.

$$F(t) = 1 - e^{-0.25t}$$

Find the likelihood that a person will arrive

(a) within 10 minutes of 3 P.M.   (b) within 25 minutes of 3 P.M.

EXAMPLE 8   **Radioactive Decay**

The radioactive **half-life** for a given radioisotope of an element is the time it takes for half the radioactive nuclei in any sample to decay to some other substance. For example, the half-life of plutonium-239 is 24,360 years. Plutonium-239 is particularly dangerous because it emits alpha particles that are absorbed into bone marrow. The maximum amount of plutonium-239 an adult can handle without significant injury is 0.13 microgram ($= 0.000000013$ gram). Suppose a researcher has a 1-gram sample of plutonium-239. The amount A (in grams) of plutonium-239 after t years is given by

$$A(t) = 1\left(\frac{1}{2}\right)^{\frac{t}{24,360}}$$

How much plutonium-239 is left in the sample after

(a) 500 years?   (b) 24,360 years?   (c) 73,080 years?

Solution

(a) The amount of plutonium-239 left in the sample after 500 years is found by evaluating A at $t = 500$. That is, determine $A(500)$.

$$A(500) = 1\left(\frac{1}{2}\right)^{\frac{500}{24,360}}$$

Use a calculator: ≈ 0.986 gram

After 500 years, approximately 0.986 gram of plutonium-239 will be left in the sample.

(b) The amount of plutonium-239 remaining after 24,360 years is found by evaluating A at $t = 24,360$. That is, determine $A(24,360)$.

$$A(24,360) = 1\left(\frac{1}{2}\right)^{\frac{24,360}{24,360}}$$

$$= 1\left(\frac{1}{2}\right)^{1}$$

$$= 0.5 \text{ gram}$$

After 24,360 years, there will be 0.5 gram of plutonium-239 left.

(c) To find the amount left after 73,080 years, evaluate $A(73,080)$.

$$A(73,080) = 1\left(\frac{1}{2}\right)^{\frac{73,080}{24,360}}$$

$$= 1\left(\frac{1}{2}\right)^{3}$$

$$= \frac{1}{8} \text{ gram}$$

After 73,080 years, there will be $\frac{1}{8} = 0.125$ gram of plutonium-239 left in the sample.   ●

Quick ✓

17. The half-life of thorium-227 is 18.72 days. Suppose that a researcher possesses a 10-gram sample of thorium-227. The amount A (in grams) of thorium-227 left after t days is given by

$$A(t) = 10 \cdot \left(\frac{1}{2}\right)^{\frac{t}{18.72}}$$

How much thorium-227 is left in the sample after

(a) 10 days?   (b) 18.72 days?
(c) 74.88 days?   (d) 100 days?

When money is deposited in a bank, the bank pays interest on the balance in the account. When solving interest problems, we use the term **payment period** as shown in Table 7 to indicate how often the bank pays interest.

Table 7

Payment Period	Number of Times Interest Is Paid
Annually	Once per year
Semiannually	Twice per year
Quarterly	4 times per year
Monthly	12 times per year
Daily	360 times per year

When the interest due at the end of a payment period is added to the principal so that the interest computed at the end of the *next* payment period is based on this new principal amount (old principal + interest), the interest has been **compounded.** **Compound interest** is interest paid on the original principal and all previously earned interest.

The following formula can be used to determine the value of an account after a certain period of time.

Work Smart

When using the compound interest formula, be sure to express the interest rate as a decimal.

> **Compound Interest Formula**
>
> The amount A after t years resulting from a principal P invested at an annual interest rate r compounded n times per year is
>
> $$A = P\left(1 + \frac{r}{n}\right)^{nt}$$

For example, if you deposit $500 into an account paying 3% annual interest compounded monthly, then $P = \$500$, $r = 0.03$, and $n = 12$ (twelve compounding periods per year).

EXAMPLE 9 **Future Value of Money**

Suppose you deposit $3000 into a Roth IRA today. If the deposit earns 8% interest compounded quarterly, determine its future value A after

(a) 1 year   **(b)** 10 years   **(c)** 35 years, when you plan on retiring

Solution

Use the compound interest formula with $P = \$3000$, $r = 0.08$, and $n = 4$ (for quarterly compounding), so

$$A = \$3000\left(1 + \frac{0.08}{4}\right)^{4t}$$
$$= \$3000(1 + 0.02)^{4t}$$
$$= \$3000(1.02)^{4t}$$

(a) The value of the account after $t = 1$ year is

$$A = \$3000(1.02)^{4(1)}$$
$$= \$3000(1.02)^{4}$$

Use a calculator; see Figure 18: $= \$3000(1.08243216)$
$$= \$3247.30$$

Figure 18

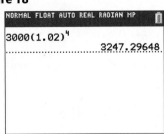

NORMAL FLOAT AUTO REAL RADIAN MP
3000(1.02)⁴
                    3247.29648

(b) The value after $t = 10$ years is

$$A = \$3000(1.02)^{4(10)}$$
$$= \$3000(1.02)^{40}$$

Use a calculator: $= \$3000(2.208039664)$
$$= \$6624.12$$

(c) The value after $t = 35$ years is

$$A = \$3000(1.02)^{4(35)}$$
$$= \$3000(1.02)^{140}$$

Use a calculator: $= \$3000(15.99646598)$
$$= \$47{,}989.40$$

Quick ✓

18. Suppose that you deposit $2000 into a Roth IRA today. Determine the future value A of the deposit if it earns 5% interest compounded monthly (twelve times per year) after

(a) 1 year (b) 15 years (c) 30 years, when you plan on retiring

---

8.2 Exercises MyMathLab® Exercise numbers in green have complete video solutions in MyMathLab or may be accessed using the QR code to the right.

Problems **1–18** *are the Quick ✓s that follow the* **EXAMPLES**.

Building Skills

In Problems 19–22, approximate each number using a calculator. Express your answer rounded to three decimal places. See Objective 1.

19. (a) $3^{2.2}$ **(b)** $3^{2.23}$ **(c)** $3^{2.236}$
(d) $3^{2.2361}$ **(e)** $3^{\sqrt{5}}$

20. (a) $5^{1.4}$ **(b)** $5^{1.41}$ **(c)** $5^{1.414}$
(d) $5^{1.4142}$ **(e)** $5^{\sqrt{2}}$

21. (a) $4^{3.1}$ **(b)** $4^{3.14}$ **(c)** $4^{3.142}$
(d) $4^{3.1416}$ **(e)** 4^{π}

22. (a) $10^{2.7}$ **(b)** $10^{2.72}$ **(c)** $10^{2.718}$
(d) $10^{2.7183}$ **(e)** 10^{e}

In Problems 23–30, the graph of an exponential function is given. Match each graph to one of the following functions. It may prove useful to create a table of values for each function to help you identify the correct graph. See Objective 2.

(a) $f(x) = 2^x$ **(b)** $f(x) = 2^{-x}$ **(c)** $f(x) = 2^{x+1}$
(d) $f(x) = 2^{x-1}$ **(e)** $f(x) = -2^x$ **(f)** $f(x) = 2^x + 1$
(g) $f(x) = 2^x - 1$ **(h)** $f(x) = -2^{-x}$

23.

24.

25.

26.

27.

28.

29.

30.

In Problems 31–42, graph each function. State the domain and the range of the function. See Objective 2.

31. $f(x) = 5^x$ **32.** $f(x) = 7^x$

33. $F(x) = \left(\dfrac{1}{5}\right)^x$ **34.** $F(x) = \left(\dfrac{1}{7}\right)^x$

35. $h(x) = 2^{x+2}$ **36.** $H(x) = 2^{x-2}$

37. $f(x) = 2^x + 3$ **38.** $F(x) = 2^x - 3$

39. $F(x) = \left(\dfrac{1}{2}\right)^x - 1$ **40.** $G(x) = \left(\dfrac{1}{2}\right)^x + 2$

41. $P(x) = \left(\dfrac{1}{3}\right)^{x-2}$ **42.** $p(x) = \left(\dfrac{1}{3}\right)^{x+2}$

In Problems 43–50, approximate each number using a calculator. Express your answer rounded to three decimal places. See Objective 3.

43. (a) $3.1^{2.7}$ **(b)** $3.14^{2.72}$ **(c)** $3.142^{2.718}$
(d) $3.1416^{2.7183}$ **(e)** π^e

44. (a) $2.7^{3.1}$ **(b)** $2.72^{3.14}$ **(c)** $2.718^{3.142}$
(d) $2.7183^{3.1416}$ **(e)** e^{π}

45. e^2 **46.** e^3 **47.** e^{-2}
48. e^{-3} **49.** $e^{2.3}$ **50.** $e^{1.5}$

In Problems 51–54, graph each function. State the domain and the range of the function. See Objective 3.

51. $g(x) = e^{x-1}$ **52.** $f(x) = e^x - 1$

53. $f(x) = -2e^x$ **54.** $F(x) = \dfrac{1}{2}e^x$

In Problems 55–80, solve each equation. See Objective 4.

55. $2^x = 2^5$

56. $3^x = 3^{-2}$

57. $3^{-x} = 81$

58. $4^{-x} = 64$

59. $\left(\dfrac{1}{2}\right)^x = \dfrac{1}{32}$

60. $\left(\dfrac{1}{3}\right)^x = \dfrac{1}{243}$

61. $5^{x-2} = 125$

62. $2^{x+3} = 128$

63. $4^x = 8$

64. $9^x = 27$

65. $2^{-x+5} = 16^x$

66. $3^{-x+4} = 27^x$

67. $3^{x^2-4} = 27^x$

68. $5^{x^2-10} = 125^x$

69. $4^x \cdot 2^{x^2} = 16^2$

70. $9^{2x} \cdot 27^{x^2} = 3^{-1}$

71. $2^x \cdot 8 = 4^{x-3}$

72. $3^x \cdot 9 = 27^x$

73. $\left(\dfrac{1}{5}\right)^x - 25 = 0$

74. $\left(\dfrac{1}{6}\right)^x - 36 = 0$

75. $(2^x)^x = 16$

76. $(3^x)^x = 81$

77. $e^x = e^{3x+4}$

78. $e^{3x} = e^2$

79. $(e^x)^2 = e^{3x-2}$

80. $(e^3)^x = e^2 \cdot e^x$

Mixed Practice

81. Suppose that $f(x) = 2^x$.

   (a) What is $f(3)$? What point is on the graph of f?

   (b) If $f(x) = \dfrac{1}{8}$, what is x? What point is one the graph of f?

82. Suppose that $f(x) = 3^x$.

   (a) What is $f(2)$? What point is on the graph of f?

   (b) If $f(x) = \dfrac{1}{81}$, what is x? What point is on the graph of f?

83. Suppose that $g(x) = 4^x - 1$.

   (a) What is $g(-1)$? What point is on the graph of g?

   (b) If $g(x) = 15$, what is x? What point is on the graph of g?

84. Suppose that $g(x) = 5^x + 1$.

   (a) What is $g(-1)$? What point is on the graph of g?

   (b) If $g(x) = 126$, what is x? What point is on the graph of g?

85. Suppose that $H(x) = 3\left(\dfrac{1}{2}\right)^x$.

   (a) What is $H(-3)$? What point is on the graph of H?

   (b) If $H(x) = \dfrac{3}{4}$, what is x? What point is on the graph of H?

86. Suppose that $F(x) = -2\left(\dfrac{1}{3}\right)^x$.

   (a) What is $F(-1)$? What point is on the graph of F?

   (b) If $F(x) = -18$, what is x? What point is on the graph of F?

Applying the Concepts

87. A Population Model According to the U.S. Census Bureau, the population of the United States in 2016 was 326 million people. In addition, the population of the United States was growing at a rate of 0.99% per year. Assuming that this growth rate continues, the model $P(t) = 326(1.0099)^{t-2016}$ represents the population P (in millions of people) in year t.

   (a) According to this model, what will be the population of the United States in 2021?

   (b) According to this model, what will be the population of the United States in 2050?

   (c) The United States Census Bureau predicts that the United States population will be 439 million in 2050. Compare this estimate to the one obtained in part (b). What might account for any differences?

88. A Population Model According to the U.S. Census Bureau, the population of the world in 2016 was 7563 million people. In addition, the population of the world was growing at a rate of 1.02% per year. Assuming that this growth rate continues, the model $P(t) = 7563(1.0102)^{t-2016}$ represents the population P (in millions of people) in year t.

   (a) According to this model, what will be the population of the world in 2021?

   (b) According to this model, what will be the population of the world in 2025?

   (c) The United States Census Bureau predicts that the world population will be 8000 million (8 billion) in 2025. Compare this estimate to the one obtained in part (b). What might account for any differences?

89. Time Is Money Suppose that you deposit $5000 into a certificate of deposit (CD) today. Determine the future value A of the deposit if it earns 2% interest compounded monthly after

   (a) 1 year.

   (b) 3 years.

   (c) 5 years, when the CD comes due.

90. Time Is Money Suppose that you deposit $8000 into a certificate of deposit (CD) today. Determine the future value A of the deposit if it earns 2% interest compounded quarterly (4 times per year) after

(a) 1 year. **(b)** 3 years.

(c) 5 years, when the CD comes due.

91. Do the Compounding Periods Matter? Suppose that you deposit $2000 into an account that pays 3% annual interest. How much will you have after 5 years if interest is compounded

(a) annually? **(b)** quarterly?

(c) monthly? **(d)** daily?

(e) Based on the results of parts (a)–(d), what impact does increasing the number of compounding periods have on the future value, all other things equal?

92. Do the Compounding Periods Matter? Suppose that you deposit $1000 into an account that pays 6% annual interest. How much will you have after 3 years if interest is compounded

(a) annually? **(b)** quarterly?

(c) monthly? **(d)** daily?

(e) Based on the results of parts (a)–(d), what impact does increasing the number of compounding periods have on the future value, all other things equal?

93. Depreciation Based on data obtained from the *Kelley Blue Book,* the value V of a Ford Focus that is t years old can be modeled by $V(t) = 19{,}841(0.88)^t$.

(a) According to the model, what is the value of a brand-new Focus?

(b) According to the model, what is the value of a 2-year-old Focus?

(c) According to the model, what is the value of a 5-year-old Focus?

94. Depreciation Based on data obtained from the *Kelley Blue Book*, the value V of a Chevy Malibu that is t years old can be modeled by $V(t) = 25{,}258(0.84)^t$.

(a) According to the model, what is the value of a brand-new Chevy Malibu?

(b) According to the model, what is the value of a 2-year-old Chevy Malibu?

(c) According to the model, what is the value of a 5-year-old Chevy Malibu?

95. Radioactive Decay The half-life of beryllium-11 is 13.81 seconds. Suppose that a researcher possesses a 100-gram sample of beryllium-11. The amount A (in grams) of beryllium-11 after t seconds is given by

$$A(t) = 100\left(\frac{1}{2}\right)^{\frac{t}{13.81}}$$

(a) How much beryllium-11 is left in the sample after 1 second?

(b) How much beryllium-11 is left in the sample after 13.81 seconds?

(c) How much beryllium-11 is left in the sample after 27.62 seconds?

(d) How much beryllium-11 is left in the sample after 100 seconds?

96. Radioactive Decay The half-life of carbon-10 is 19.255 seconds. Suppose that a researcher possesses a 100-gram sample of carbon-10. The amount A (in grams) of carbon-10 after t seconds is given by

$$A(t) = 100\left(\frac{1}{2}\right)^{\frac{t}{19.255}}$$

(a) How much carbon-10 is left in the sample after 1 second?

(b) How much carbon-10 is left in the sample after 19.255 seconds?

(c) How much carbon-10 is left in the sample after 38.51 seconds?

(d) How much carbon-10 is left in the sample after 100 seconds?

For Exercises 97 and 98, use Newton's Law of Cooling, which states that the temperature of a heated object decreases exponentially over time toward the temperature of the surrounding medium.

97. Newton's Law of Cooling Suppose a pizza is removed from a 400°F oven and placed in a room whose temperature is 70°F. The temperature u (in °F) of the pizza at time t (in minutes) can be modeled by $u(t) = 70 + 330e^{-0.072t}$.

(a) According to the model, what will be the temperature of the pizza in 5 minutes?

(b) According to the model, what will be the temperature of the pizza in 10 minutes?

(c) If the pizza can be safely consumed when its temperature is 200°F, will it be ready to eat after cooling for 13 minutes?

98. Newton's Law of Cooling Suppose coffee at a temperature of 170°F is poured into a coffee mug and allowed to cool in a room whose temperature is 70°F. The temperature u (in °F) of the coffee at time t (in minutes) can be modeled by $u(t) = 70 + 100e^{-0.045t}$.

(a) According to the model, what will be the temperature of the coffee in 5 minutes?

(b) According to the model, what will be the temperature of the coffee in 10 minutes?

(c) If the coffee doesn't taste good once its temperature reaches 120°F, will it still be tasty after cooling for 20 minutes?

99. Learning Curve Suppose that a student has 200 vocabulary words to learn. If a student learns 20 words in 30 minutes, the function

$$L(t) = 200(1 - e^{-0.0035t})$$

models the number of words L that the student will learn in t minutes.

(a) How many words will the student learn in 45 minutes?

(b) How many words will the student learn in 60 minutes?

100. Learning Curve Suppose that a student has 50 biology terms to learn. If a student learns 10 terms in 30 minutes, the function

$$L(t) = 50(1 - e^{-0.0223t})$$

models the number of terms L that the student will learn in t minutes.

(a) How many words will the student learn in 45 minutes?

(b) How many words will the student learn in 60 minutes?

101. Current in an RL Circuit The equation governing the amount of current I (in amperes) after time t (in seconds) in a single RL circuit consisting of a resistance R (in ohms), an inductance L (in henrys), and an electromotive force E (in volts) is

$$I = \frac{E}{R}\left[1 - e^{-(R/L)t}\right]$$

(a) If $E = 120$ volts, $R = 10$ ohms, and $L = 25$ henrys, how much current I is flowing in 0.05 second?

(b) If $E = 240$ volts, $R = 10$ ohms, and $L = 25$ henrys, how much current I is flowing in 0.05 second?

102. Current in an RC Circuit The equation governing the amount of current I (in amperes) after time t (in microseconds) in a single RC circuit consisting of a resistance R (in ohms), a capacitance C (in microfarads), and an electromotive force E (in volts) is

$$I = \frac{E}{R}e^{-t/(RC)}$$

(a) If $E = 120$ volts, $R = 2500$ ohms, and $C = 100$ microfarads, how much current I is flowing initially $(t = 0)$? In 50 microseconds?

(b) If $E = 240$ volts, $R = 2500$ ohms, and $C = 100$ microfarads, how much current I is flowing initially $(t = 0)$? In 50 microseconds?

Extending the Concepts

In Problems 103 and 104, find the exponential function whose graph is given.

103.

104.

Explaining the Concepts

105. As the base a of an exponential function $f(x) = a^x$ increases (for $a > 1$), what happens to the graph of the exponential function for $x > 0$? What happens to the behavior of the graph for $x < 0$?

106. The graphs of $f(x) = 2^{-x}$ and $g(x) = \left(\dfrac{1}{2}\right)^x$ are identical. Why?

107. Explain the difference between exponential functions and polynomial functions.

108. Can we solve the equation $2^x = 12$ using the fact that if $a^u = a^v$, then $u = v$? Why or why not?

Synthesis Review

In Problems 109–114, evaluate each expression.

109. Evaluate $x^2 - 5x + 1$ at

(a) $x = -2$ **(b)** $x = 3$

110. Evaluate $x^3 - 5x + 2$ at

(a) $x = -4$ **(b)** $x = 2$

111. Evaluate $\dfrac{4}{x + 2}$ at

(a) $x = -2$ **(b)** $x = 3$

112. Evaluate $\dfrac{2x}{x + 1}$ at

(a) $x = -4$ **(b)** $x = 5$

113. Evaluate $\sqrt{2x + 5}$ at

   (a) $x = 2$ **(b)** $x = 11$

114. Evaluate $\sqrt[3]{3x - 1}$ at

   (a) $x = 3$ **(b)** $x = 0$

Technology Exercises

In Problems 115–122, graph each function using technology. State the domain and the range of each function.

115. $f(x) = 1.5^x$

116. $G(x) = 3.1^x$

117. $H(x) = 0.9^x$

118. $F(x) = 0.3^x$

119. $g(x) = 2.5^x + 3$

120. $f(x) = 1.7^x - 2$

121. $F(x) = 1.6^{x-3}$

122. $g(x) = 0.3^{x+2}$

123. Exploration: Graph of Exponential Functions *Open the "Graph of Exponential Functions" Geogebra applet.* The applet may be found using the QR code at the beginning of this section or through the Multimedia Library in MyMathLab. In the upper left corner is one "slider" that represents the value for a in the exponential function $f(x) = a^x$. To start, make sure the box is unchecked. If you are using a keyboard, it may be easier to use the arrow keys, instead of the mouse, to move the slider.

   (a) Move the a slider to 2. This will produce the graph of $f(x) = 2^x$. State the domain, the range, the intercepts, and the values of the three coordinate points that are labeled on the graph.

   (b) Move the a slider to 3. This will produce the graph of $f(x) = 3^x$. State the domain, the range, the intercepts, and the values of the three coordinate points that are labeled on the graph.

   (c) Move the a slider to 4. This will produce the graph of $f(x) = 4^x$. State the domain, the range, the intercepts, and the values of the three coordinate points that are labeled on the graph.

   (d) Use your results from parts (a) through (c) to generalize the following for *all* exponential functions of the form $f(x) = a^x$:
     i. The domain is _____.
     ii. The range is _____.
     iii. How many x-intercepts are there? What is the y-intercept?
     iv. Conclude the graph of $f(x) = a^x$ contains the points $(-1, \_\_)$ $(0, \_)$, and $(1, \_)$.

124. Exploration: Graph of Exponential Functions *Open the "Graph of Exponential Functions" Geogebra applet.* The applet may be found using the QR code at the beginning of this section or through the Multimedia Library in MyMathLab. In the upper left corner is one "slider" that stands for the value of a in the exponential functions $f(x) = a^x$ and $g(x) = \left(\dfrac{1}{a}\right)^x$. To start, make sure the box for $g(x)$ is checked. If you are using a keyboard, it may be easier to use the arrow keys, instead of the mouse, to move the slider.

   (a) Move the a slider to 2. This will produce the graphs of $f(x) = 2^x$ and $g(x) = \left(\dfrac{1}{2}\right)^x$. Notice that the point $(1, 2)$ is on the graph of f and the point $(\_\_, 2)$ is on the graph of g. Also, the point $\left(-1, \dfrac{1}{2}\right)$ is on the graph of f and the point $\left(-, \dfrac{1}{2}\right)$ is on the graph of g. Both graphs contain the point _____.

   (b) Move the a slider to 3. This will produce the graphs of $f(x) = 3^x$ and $g(x) = \left(\dfrac{1}{3}\right)^x$. Notice that the point $(1, 3)$ is on the graph of f and the point $(\_\_, 3)$ is on the graph of g. Also, the point $\left(-1, \dfrac{1}{3}\right)$ is on the graph of f and the point $\left(-, \dfrac{1}{3}\right)$ is on the graph of g. Both graphs contain the point _____.

   (c) If $f(x) = a^x$ and $g(x) = \left(\dfrac{1}{a}\right)^x$, where $a > 1$, in general conclude the following: If the point $(1, a)$ is on the graph of f, the point $(\_\_, a)$ is on the graph of g; if the point $\left(-, \dfrac{1}{a}\right)$ is on the graph of f, the point $\left(-, \dfrac{1}{a}\right)$ is on the graph of g; both graphs contain the point _____.

8.3 Logarithmic Functions

Objectives

1. Change Exponential Equations to Logarithmic Equations
2. Change Logarithmic Equations to Exponential Equations
3. Evaluate Logarithmic Functions
4. Determine the Domain of a Logarithmic Function
5. Graph Logarithmic Functions
6. Work with Natural and Common Logarithms
7. Solve Logarithmic Equations
8. Use Logarithmic Models That Describe Our World

Are You Prepared for This Section?

Before getting started, complete the following problems. If you get a problem wrong, go back to the section cited and review the material.

P1. Solve: $3x + 2 > 0$ [Section 1.4, pp. 86–88]

P2. Solve: $\sqrt{x + 2} = x$ [Section 6.7, pp. 516–519]

P3. Solve: $x^2 = 6x + 7$ [Section 4.8, pp. 372–375]

Consider the "squaring function," $f(x) = x^2$, and the square root function, $g(x) = \sqrt{x}$. How are these two functions related? Well, $f(5) = 25$ and $g(25) = 5$. That is, the input to f is the output of g and the output of f is the input to g, so g "undoes" what f does.

In general, whenever a function is introduced in mathematics, there is a second function that "undoes" it. For example, a square root function undoes the squaring function. A cube root function undoes the cubing function. Is there a function that undoes an exponential function? Yes! This function is the *logarithmic function*.

Work Smart

It is required that a be positive and not equal to 1 for the same reasons that these restrictions existed for the exponential function.

Definition

The **logarithmic function to the base** a, where $a > 0$ and $a \neq 1$, is denoted by $y = \log_a x$ (read as "y is the logarithm to the base a of x"), and

$$y = \log_a x \quad \text{is equivalent to} \quad x = a^y$$

> **In Other Words**
> The logarithm to the base a of x is the number y that we must raise a to in order to obtain x.

To evaluate logarithmic functions, convert them into their equivalent exponential form. Therefore, go from logarithmic form to exponential form, and back. For example,

$$0 = \log_3 1 \quad \text{is equivalent to} \quad 3^0 = 1$$
$$2 = \log_5 25 \quad \text{is equivalent to} \quad 5^2 = 25$$
$$-2 = \log_4 \frac{1}{16} \quad \text{is equivalent to} \quad 4^{-2} = \frac{1}{16}$$

Work Smart: Study Skills

In doing problems similar to Examples 1 and 2, it is a good idea to practice saying, "y equals the logarithm to the base a of x is equivalent to a to the y equals x" so that you memorize the definition of a logarithm.

Notice that the base of the logarithm is the base of the exponential; the argument of the logarithm is what the exponential equals; and the value of the logarithm is the exponent of the exponential expression. A logarithm is just another way of writing an exponential expression.

To see how the logarithmic function undoes the exponential function, consider the function $y = 2^x$. If the input is $x = 3$, then the output is $y = 2^3 = 8$. To undo this function would require that an input of 8 give an output of 3. If $2^3 = 8$, then $3 = \log_2 8$ using the definition of a logarithm. The input of $\log_2 8$ is 8, and its output is 3.

▶ **1 Change Exponential Equations to Logarithmic Equations**

The definition of a logarithm can be used to rewrite exponential equations as logarithmic equations.

EXAMPLE 1   **Changing Exponential Equations to Logarithmic Equations**

Rewrite each exponential equation as an equivalent logarithmic equation.

    (a) $6^2 = 36$ **(b)** $x^2 = 9$ **(c)** $4^3 = y$

Prepared?...Answers

P1. $\left\{x \mid x > -\dfrac{2}{3}\right\}$ or $\left(-\dfrac{2}{3}, \infty\right)$

P2. $\{2\}$ **P3.** $\{-1, 7\}$

Solution

Use the fact that $y = \log_a x$ is equivalent to $a^y = x$ provided that $a > 0$ and $a \neq 1$.

    (a) If $6^2 = 36$, then $2 = \log_6 36$.   **(b)** If $x^2 = 9$, then $2 = \log_x 9$.

    (c) If $4^3 = y$, then $3 = \log_4 y$.

Quick ✓

Quick ✓

1. The logarithm to the base a of x, denoted $y = \log_a x$, can be expressed in exponential form as _____, where a _ 0 and a _ 1.

In Problems 2 and 3, rewrite each exponential equation as an equivalent equation involving a logarithm.

2. $4^3 = 64$　　　　　　　　　**3.** $p^{-2} = 8$

▶ ❷ Change Logarithmic Equations to Exponential Equations

The definition of a logarithm can also be used to rewrite logarithmic equations as exponential equations.

EXAMPLE 2　**Changing Logarithmic Equations to Exponential Equations**

Change each logarithmic equation to an equivalent exponential equation.

　　(a) $4 = \log_3 81$　　　　　**(b)** $-3 = \log_a \dfrac{1}{27}$　　　　　**(c)** $2 = \log_4 x$

Solution
In each of these problems, use the fact that $y = \log_a x$ is equivalent to $a^y = x$ provided that $a > 0$ and $a \neq 1$.

　　(a) If $4 = \log_3 81$, then $3^4 = 81$.

　　(b) If $-3 = \log_a \dfrac{1}{27}$, then $a^{-3} = \dfrac{1}{27}$.

　　(c) If $2 = \log_4 x$, then $4^2 = x$.

Quick ✓

In Problems 4–6, rewrite each logarithmic equation as an equivalent equation involving an exponent.

4. $4 = \log_2 16$　　　　　**5.** $5 = \log_a 20$　　　　　**6.** $-3 = \log_5 z$

▶ ❸ Evaluate Logarithmic Functions

To find the exact value of a logarithm, write the logarithm in exponential notation and use the fact that if $a^u = a^v$, then $u = v$.

EXAMPLE 3　**Finding the Exact Value of a Logarithmic Expression**

Find the exact value of

　　(a) $\log_2 32$　　　　　　　　　**(b)** $\log_4 \dfrac{1}{16}$

Solution

　　(a) Let $y = \log_2 32$ and convert this equation into an exponential equation.

$$y = \log_2 32$$

Write the logarithm as an exponent:　$2^y = 32$

$32 = 2^5$:　$2^y = 2^5$

If $a^u = a^v$, then $u = v$:　$y = 5$

Therefore, $\log_2 32 = 5$.

(b) Let $y = \log_4 \dfrac{1}{16}$ and convert this equation into an exponential equation.

$$y = \log_4 \frac{1}{16}$$

Write the logarithm as an exponent: $4^y = \dfrac{1}{16}$

$$\frac{1}{16} = 4^{-2}: \quad 4^y = 4^{-2}$$

If $a^u = a^v$, then $u = v$: $y = -2$

Therefore, $\log_4 \dfrac{1}{16} = -2$.   ●

Quick ✓

In Problems 7 and 8, find the exact value of each logarithmic expression.

7. $\log_5 25$

8. $\log_2 \dfrac{1}{8}$

The equation $y = \log_a x$ could also be written using function notation as $f(x) = \log_a x$. We use this notation in the next example to evaluate a logarithmic function.

EXAMPLE 4   **Evaluating a Logarithmic Function**

Find the value of each of the following, given that $f(x) = \log_2 x$.

(a) $f(2)$

(b) $f\left(\dfrac{1}{4}\right)$

Solution

(a) Finding $f(2)$ means evaluating $\log_2 x$ at $x = 2$, or evaluating $\log_2 2$. To determine this value, follow the approach of Example 3 by letting $y = \log_2 2$ and converting the equation into an exponential equation.

$$y = \log_2 2$$

Write the logarithm as an exponent: $2^y = 2$

$$2 = 2^1: \quad 2^y = 2^1$$

If $a^u = a^v$, then $u = v$: $y = 1$

Therefore, $f(2) = 1$.

(b) $f\left(\dfrac{1}{4}\right)$ means evaluating $\log_2 x$ at $x = \dfrac{1}{4}$ or $\log_2\left(\dfrac{1}{4}\right)$. Let $y = \log_2\left(\dfrac{1}{4}\right)$.

$$y = \log_2\left(\frac{1}{4}\right)$$

Write the logarithm as an exponent: $2^y = \dfrac{1}{4}$

$$\frac{1}{4} = 2^{-2}: \quad 2^y = 2^{-2}$$

If $a^u = a^v$, then $u = v$: $y = -2$

Therefore, $f\left(\dfrac{1}{4}\right) = -2$.   ●

Quick ✓

In Problems 9 and 10, evaluate the function, given that $g(x) = \log_5 x$.

9. $g(25)$

10. $g\left(\dfrac{1}{5}\right)$

▶ ④ Determine the Domain of a Logarithmic Function

The domain of a function $y = f(x)$ is the set of all x such that the function makes sense, and the range is the set of all outputs of the function. To find the range of the logarithmic function, recognize that $y = f(x) = \log_a x$ is equivalent to $x = a^y$. Because a can be raised to any real number (since $a > 0$ and $a \ne 1$), conclude that y can be any real number in the function $y = f(x) = \log_a x$. In addition, because a^y is positive for any real number, conclude that x must be positive. Since x represents the input of the logarithmic function, the domain of the logarithmic function is the set of all positive real numbers.

Work Smart

Notice that the domain of the logarithmic function is the same as the range of the exponential function. The range of the logarithmic function is the same as the domain of the exponential function.

> **Domain and Range of The Logarithmic Function**
>
> $$\text{Domain of the logarithmic function} = (0, \infty)$$
> $$\text{Range of the logarithmic function} = (-\infty, \infty)$$

Because the domain of the logarithmic function is the set of all positive real numbers, the argument of the logarithmic function must be greater than zero. For example, $f(x) = \log_{10} x$ is defined for $x = 2$, but not for $x = -1$, $x = -8$, or any other $x \le 0$.

EXAMPLE 5 **Finding the Domain of a Logarithmic Function**

Find the domain of each logarithmic function.

(a) $f(x) = \log_6(x - 5)$ (b) $G(x) = \log_3(3x + 1)$

Solution

(a) The argument of the function $f(x) = \log_6(x - 5)$ is $x - 5$. The domain of f is the set of all real numbers x such that $x - 5 > 0$. Solve this inequality:

$$x - 5 > 0$$

Add 5 to both sides: $x > 5$

The domain of f is $\{x | x > 5\}$, or $(5, \infty)$ in interval notation.

(b) The argument of the function $G(x) = \log_3(3x + 1)$ is $3x + 1$. The domain of G is the set of all real numbers x such that $3x + 1 > 0$.

$$3x + 1 > 0$$

Subtract 1 from both sides: $3x > -1$

Divide both sides by 3: $x > -\dfrac{1}{3}$

The domain of G is $\left\{ x \,\middle|\, x > -\dfrac{1}{3} \right\}$, or $\left(-\dfrac{1}{3}, \infty \right)$ in interval notation.   ●

Quick ✓

In Problems 11 and 12, find the domain of each logarithmic function.

11. $g(x) = \log_8(x + 3)$ 12. $F(x) = \log_2(5 - 2x)$

▶ ⑤ Graph Logarithmic Functions

To graph a logarithmic function $y = \log_a x$, it helps to rewrite the function in exponential form: $x = a^y$. Then choose "nice" values of y and use the expression $x = a^y$ to find the corresponding values of x, as shown in Example 6.

EXAMPLE 6 **Graphing a Logarithmic Function**

Graph $f(x) = \log_2 x$ using point plotting. From the graph, state the domain and the range of the function.

Solution
Rewrite $y = f(x) = \log_2 x$ as $x = 2^y$. Table 8 shows various values of y, the corresponding values of x, and points on the graph of $y = f(x) = \log_2 x$. Plot these ordered pairs, connect them in a smooth curve, and obtain the graph of $f(x) = \log_2 x$ in Figure 19. The domain of f is $\{x \mid x > 0\}$, or $(0, \infty)$ in interval notation. The range of f is the set of all real numbers, or $(-\infty, \infty)$ in interval notation.

Table 8

y	$x = 2^y$	(x, y)
-2	$\dfrac{1}{4}$	$\left(\dfrac{1}{4}, -2\right)$
-1	$\dfrac{1}{2}$	$\left(\dfrac{1}{2}, -1\right)$
0	1	$(1, 0)$
1	2	$(2, 1)$
2	4	$(4, 2)$

Figure 19

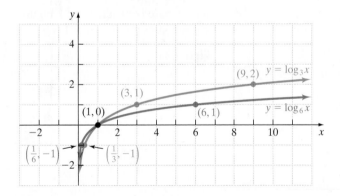

The graph of $f(x) = \log_2 x$ in Figure 19 is typical of all logarithmic functions that have a base larger than 1. Figure 20 shows the graphs of two other logarithmic functions with bases larger than 1, $y = \log_3 x$ and $y = \log_6 x$.

Figure 20

The information about $f(x) = \log_a x$, where $a > 1$, is summarized below.

Figure 21
$f(x) = \log_a x, a > 1$

Properties of the Graph of a Logarithmic Function $f(x) = \log_a x, a > 1$

1. The domain is the set of all positive real numbers. The range is the set of all real numbers.
2. There is no y-intercept; the x-intercept is $(1, 0)$.
3. The graph of f contains the points $\left(\dfrac{1}{a}, -1\right)$, $(1, 0)$, and $(a, 1)$. See Figure 21.

Quick ✓

13. Graph the logarithmic function $f(x) = \log_4 x$ using point plotting. From the graph, state the domain and the range of the function.

Now we consider $f(x) = \log_a x, 0 < a < 1$.

EXAMPLE 7 **Graphing a Logarithmic Function**

Graph $f(x) = \log_{1/2} x$ using point plotting. From the graph, state the domain and the range of the function.

Solution

Rewrite $y = f(x) = \log_{1/2} x$ as $x = \left(\dfrac{1}{2}\right)^y$. Table 9 shows various values of y, the corresponding values of x, and points on the graph of $y = f(x) = \log_{1/2} x$. Plot these points, connect them in a smooth curve, and obtain the graph of $f(x) = \log_{1/2} x$ in Figure 22. The domain of f is $\{x \mid x > 0\}$, or $(0, \infty)$ in interval notation. The range of f is the set of all real numbers, or $(-\infty, \infty)$ in interval notation.

Table 9

y	$x = \left(\dfrac{1}{2}\right)^y$	(x, y)
-2	4	$(4, -2)$
-1	2	$(2, -1)$
0	1	$(1, 0)$
1	$\dfrac{1}{2}$	$\left(\dfrac{1}{2}, 1\right)$
2	$\dfrac{1}{4}$	$\left(\dfrac{1}{4}, 2\right)$

Figure 22

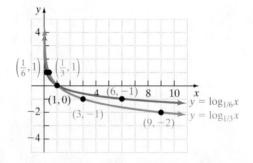

The graph of $f(x) = \log_{1/2} x$ in Figure 22 is typical of all logarithmic functions that have a base between 0 and 1. Figure 23 shows the graph of two more logarithmic functions whose bases are less than 1, $y = \log_{1/3} x$ and $y = \log_{1/6} x$.

Figure 23

The information about $f(x) = \log_a x$, where the base a is between 0 and 1 $(0 < a < 1)$, is summarized below.

Figure 24
$f(x) = \log_a x, 0 < a < 1$

> **Properties of the Graph of a Logarithmic Function $f(x) = \log_a x, 0 < a < 1$**
>
> **1.** The domain is the set of all positive real numbers. The range is the set of all real numbers.
> **2.** There is no y-intercept; the x-intercept is $(1, 0)$.
> **3.** The graph of f contains the points $\left(\dfrac{1}{a}, -1\right)$, $(1, 0)$, and $(a, 1)$. See Figure 24.

Quick ✓

14. Graph the logarithmic function $f(x) = \log_{1/4} x$ using point plotting. From the graph, state the domain and the range of the function.

▶ ⑥ Work with Natural and Common Logarithms

The **natural logarithm function** is a logarithmic function whose base is the number e. This function occurs so frequently in applications that it is given a special symbol, **ln** (from the Latin *logarithmus naturalis*).

Definition

> The natural logarithm: $y = \ln x$ if and only if $x = e^y$

Figure 25
$y = \ln x$

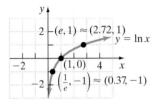

Figure 25 shows the graph of $y = \ln x$.

If the base of a logarithmic function is 10, then we have the **common logarithm function.** If the base a of the logarithmic function is not indicated, it is understood to be 10.

Definition

> The common logarithm: $y = \log x$ if and only if $x = 10^y$

Figure 26 shows the graph of $y = \log x$.

Scientific and graphing calculators have both a natural logarithm button, $\boxed{\ln}$, and a common logarithm button, $\boxed{\log}$. This enables us to approximate the values of logarithms to the base e and the base 10, when the results are not exact.

To evaluate logarithmic expressions using a scientific calculator, enter the argument of the logarithm and then press the $\boxed{\ln}$ or $\boxed{\log}$ button, depending on the base of the logarithm. For example, to evaluate log 80, type in 80 and then press the $\boxed{\log}$ button. The display should show 1.90308999. Try it!

To evaluate logarithmic expressions using a graphing calculator, press the $\boxed{\ln}$ or $\boxed{\log}$ button, depending on the base of the logarithm, and then enter the argument of the logarithm. Finally, press $\boxed{\text{ENTER}}$. For example, to evaluate log 80, press the $\boxed{\log}$ button, type in 80, and then press $\boxed{\text{ENTER}}$. The display should show 1.903089987. Try it!

By the way, it shouldn't be surprising that log 80 is between 1 and 2 since log 10 = 1 (because $10^1 = 10$) and log 100 = 2 (because $10^2 = 100$). It is a good idea to develop number sense for logarithms prior to using your calculator to approximate the value.

Figure 26
$y = \log x$

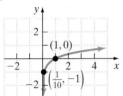

EXAMPLE 8   **Evaluating Natural and Common Logarithms on a Calculator**

Using a calculator, evaluate each of the following. Round your answers to three decimal places.

(a) ln 20   **(b)** log 30   **(c)** ln 0.5

Solution

(a) $\ln 20 \approx 2.996$ **(b)** $\log 30 \approx 1.477$ **(c)** $\ln 0.5 \approx -0.693$ ●

Quick ✓

In Problems 15–17, evaluate each logarithm using a calculator. Round your answers to three decimal places.

15. log 1400   **16.** ln 4.8   **17.** log 0.3

▶ ⑦ Solve Logarithmic Equations

Equations that contain logarithms are called **logarithmic equations.** Take care when solving logarithmic equations, because extraneous solutions (apparent solutions that are not solutions to the original equation) might creep in. To help locate extraneous solutions, remember that in the expression $\log_a M$, both a and M must be positive with $a \neq 1$.

To solve logarithmic equations, rewrite the logarithmic equation as an equivalent exponential equation, using the fact that $y = \log_a x$ is equivalent to $a^y = x$.

EXAMPLE 9 **Solving Logarithmic Equations**

Solve:

(a) $\log_2(3x + 4) = 4$ **(b)** $\log_x 25 = 2$

Solution

(a) Find the solution by writing the logarithmic equation as an exponential equation, using the fact that if $y = \log_a x$, then $a^y = x$.

$$\log_2(3x + 4) = 4$$

If $y = \log_a x$, then $a^y = x$: $2^4 = 3x + 4$

$2^4 = 16$: $16 = 3x + 4$

Subtract 4 from both sides: $12 = 3x$

Divide both sides by 3: $4 = x$

Verify this solution by letting $x = 4$ in the original equation:

$$\log_2(3 \cdot 4 + 4) \overset{?}{=} 4$$
$$\log_2(16) \overset{?}{=} 4$$

Since $2^4 = 16$, $\log_2 16 = 4$. Therefore, the solution set is $\{4\}$.

(b) Change the logarithmic equation to an exponential equation.

$$\log_x 25 = 2$$

If $y = \log_a x$, then $a^y = x$: $x^2 = 25$

Use the Square Root Property: $x = \pm\sqrt{25} = \pm 5$

Since the base of a logarithm must be positive, $x = -5$ is extraneous. We leave it to you to verify that the solution set is $\{5\}$.   ●

Quick ✓

In Problems 18 and 19, solve each equation. Be sure to verify your solution.

18. $\log_3(5x + 1) = 4$ **19.** $\log_x 16 = 2$

EXAMPLE 10 **Solving Logarithmic Equations**

Solve each equation and state the exact solution.

(a) $\ln x = 3$ **(b)** $\log(x + 1) = -2$

Solution

Write each logarithmic equation as an exponential equation.

(a) $$\ln x = 3$$

Write as an exponent; if $y = \ln x$, then $e^y = x$: $e^3 = x$

Verify this solution by letting $x = e^3$ in the original equation:

$$\ln e^3 \overset{?}{=} 3$$

The equation $\ln e^3 = 3$ can be written as $\log_e e^3 = 3$, which is equivalent to $e^3 = e^3$, so this is a true statement. The solution set is $\{e^3\}$.

(b) $$\log(x + 1) = -2$$

Write as an exponent; if $y = \log x$, then $10^y = x$: $10^{-2} = x + 1$

$10^{-2} = 0.01$: $x + 1 = 0.01$

Subtract 1 from both sides: $x = -0.99$

Work Smart

The solution to Example 10(b) illustrates that negative answers may occur in logarithmic equations. Just because an apparent solution is negative does not automatically mean the solution is extraneous.

Verify this solution by letting $x = -0.99$ in the original equation:

$$\log(-0.99 + 1) \overset{?}{=} -2$$
$$\log(0.01) \overset{?}{=} -2$$
$$10^{-2} = 0.01 \quad \text{True}$$

The solution set is $\{-0.99\}$.

Quick ✓

In Problems 20 and 21, solve each equation. Be sure to verify your solution.

20. $\ln x = -2$ **21.** $\log(x - 20) = 4$

⑧ Use Logarithmic Models That Describe Our World

Common logarithms often are used when quantities vary from very large to very small. This is because the common logarithm can "scale down" the measurement. For example, if a certain quantity varied from $0.00000001 = 10^{-8}$ to $100{,}000{,}000 = 10^{8}$, the common logarithm of the same quantity would vary from $\log 10^{-8} = -8$ to $\log 10^{8} = 8$.

Physicists define the **intensity of a sound wave** as the amount of energy the sound wave transmits through a given area. The *loudness L* (measured in **decibels** in honor of Alexander Graham Bell) of a sound of intensity x (measured in watts per square meter) is defined as follows.

> **Definition**
>
> The **loudness** L, measured in decibels, of a sound of intensity x, measured in watts per square meter, is
>
> $$L(x) = 10 \log \frac{x}{10^{-12}}$$

The quantity 10^{-12} watt per square meter in the definition is the least intense sound that a human ear can detect. If $x = 10^{-12}$ watt per square meter, then

$$L(10^{-12}) = 10 \log \frac{10^{-12}}{10^{-12}}$$
$$= 10 \log 1$$
$$= 10(0)$$
$$= 0$$

Thus the loudness of the least intense sound a human ear can detect is 0 decibels.

EXAMPLE 11   **Measuring the Loudness of a Sound**

What is the loudness, in decibels, of normal conversation, which has an intensity level of 10^{-6} watt per square meter?

Solution

Evaluate L at $x = 10^{-6}$.

$$L(10^{-6}) = 10 \log \frac{10^{-6}}{10^{-12}}$$

Laws of exponents; $\dfrac{a^m}{a^n} = a^{m-n}$: $= 10 \log 10^{-6-(-12)}$

Simplify: $= 10 \log 10^{6}$

If $y = \log 10^6$, then $10^y = 10^6$, so $y = 6$: $= 10(6)$

$= 60$ decibels

The loudness of normal conversation is 60 decibels.

Quick ✓

22. An MP3 player has an intensity level of 10^{-2} watt per square meter when set at its maximum level. What is the loudness, in decibels, of the MP3 player on "full blast"?

Problems **1–22** *are the Quick* ✓ *s that follow the EXAMPLES.*

Building Skills

In Problems 23–30, change each exponential equation to an equivalent equation involving a logarithm. See Objective 1.

23. $64 = 4^3$

24. $16 = 2^4$

25. $\dfrac{1}{8} = 2^{-3}$

26. $\dfrac{1}{9} = 3^{-2}$

27. $a^3 = 19$

28. $b^4 = 23$

29. $5^{-6} = c$

30. $10^{-3} = z$

In Problems 31–40, change each logarithmic equation to an equivalent equation involving an exponent. See Objective 2.

31. $\log_2 16 = 4$

32. $\log_3 81 = 4$

33. $\log_3 \dfrac{1}{9} = -2$

34. $\log_2 \dfrac{1}{32} = -5$

35. $\log_5 a = -3$

36. $\log_6 x = -4$

37. $\log_a 4 = 2$

38. $\log_a 16 = 2$

39. $\log_{1/2} 12 = y$

40. $\log_{1/2} 18 = z$

In Problems 41–48, find the exact value of each logarithm without using a calculator. See Objective 3.

41. $\log_3 1$

42. $\log_5 5$

43. $\log_2 8$

44. $\log_4 16$

45. $\log_4\left(\dfrac{1}{16}\right)$

46. $\log_5\left(\dfrac{1}{125}\right)$

47. $\log_{\sqrt{2}} 4$

48. $\log_{\sqrt{3}} 3$

In Problems 49–52, evaluate each function, given that $f(x) = \log_3 x$ and $g(x) = \log_5 x$. See Objective 3.

49. $f(81)$

50. $f(9)$

51. $g\left(\sqrt{5}\right)$

52. $g\left(\sqrt[3]{5}\right)$

In Problems 53–62, find the domain of each function. See Objective 4.

53. $f(x) = \log_2(x - 4)$

54. $f(x) = \log_3(x - 2)$

55. $F(x) = \log_3(2x)$

56. $h(x) = \log_4(5x)$

57. $f(x) = \log_8(3x - 2)$

58. $F(x) = \log_2(4x - 3)$

59. $H(x) = \log_7(2x + 1)$

60. $f(x) = \log_3(5x + 3)$

61. $H(x) = \log_2(1 - 4x)$

62. $G(x) = \log_4(3 - 5x)$

In Problems 63–68, graph each function. From the graph, state the domain and the range of each function. See Objective 5.

63. $f(x) = \log_5 x$

64. $f(x) = \log_7 x$

65. $g(x) = \log_6 x$

66. $G(x) = \log_8 x$

67. $F(x) = \log_{1/5} x$

68. $F(x) = \log_{1/7} x$

In Problems 69–72, change each exponential equation to an equivalent equation involving a logarithm, and change each logarithmic equation to an equivalent equation involving an exponent. See Objective 6.

69. $e^x = 12$

70. $e^4 = M$

71. $\ln x = 4$

72. $\ln(x - 1) = 3$

In Problems 73–76, evaluate each function, given that $H(x) = \log x$ and $P(x) = \ln x$. See Objective 6.

73. $H(0.1)$

74. $H(100,000)$

75. $P(e^3)$

76. $P(e^{-3})$

In Problems 77–88, use a calculator to evaluate each expression. Round your answers to three decimal places. See Objective 6.

77. $\log 67$

78. $\log 106$

79. $\ln 5.4$

80. $\ln 10.4$

81. $\log 0.35$

82. $\log 0.78$

83. $\ln 0.2$

84. $\ln 0.4$

85. $\log \dfrac{5}{4}$

86. $\log \dfrac{10}{7}$

87. $\ln \dfrac{3}{8}$

88. $\ln \dfrac{1}{2}$

In Problems 89–108, solve each logarithmic equation. All answers should be exact. See Objective 7.

89. $\log_3(2x + 1) = 2$

90. $\log_3(5x - 3) = 3$

91. $\log_5(20x - 5) = 3$

92. $\log_4(8x + 10) = 3$

93. $\log_a 36 = 2$

94. $\log_a 81 = 2$

95. $\log_a 18 = 2$

96. $\log_a 28 = 2$

97. $\log_a 1000 = 3$

98. $\log_a 243 = 5$

99. $\ln x = 5$

100. $\ln x = 10$

101. $\log(2x - 1) = -1$

102. $\log(2x + 3) = 1$

103. $\ln e^x = -3$

104. $\ln e^{2x} = 8$

105. $\log_3 81 = x$

106. $\log_4 16 = x + 1$

107. $\log_2(x^2 - 1) = 3$

108. $\log_3(x^2 + 1) = 2$

Mixed Practice

109. Suppose that $f(x) = \log_2 x$.

    (a) What is $f(16)$? What point is on the graph of f?

    (b) If $f(x) = -3$, what is x? What point is on the graph of f?

110. Suppose that $f(x) = \log_5 x$.

    (a) What is $f(5)$? What point is on the graph of f?

    (b) If $f(x) = -2$, what is x? What point is on the graph of f?

111. Suppose that $G(x) = \log_4 (x + 1)$.

    (a) What is $G(7)$? What point is on the graph of G?

    (b) If $G(x) = 2$, what is x? What point is on the graph of G?

112. Suppose that $F(x) = \log_2 x - 3$.

    (a) What is $F(8)$? What point is on the graph of f?

    (b) If $F(x) = -1$, what is x? What point is on the graph of F?

113. Find a so that the graph of $f(x) = \log_a x$ contains the point $(16, 2)$.

114. Find a so that the graph of $f(x) = \log_a x$ contains the point $\left(\dfrac{1}{4}, -2\right)$.

In Problems 115–120, state the domain of each logarithmic function.

115. $f(x) = \log_3(x^2 - 4x - 5)$

116. $f(x) = \log_2(x^2 - 2x - 8)$

117. $f(x) = \log\left(\dfrac{x - 4}{x + 1}\right)$

118. $f(x) = \log\left(\dfrac{x + 3}{x - 2}\right)$

119. $f(x) = \ln|x - 3|$

120. $f(x) = \ln|x + 1|$

Applying the Concepts

121. Loudness of a Whisper A whisper has an intensity level of 10^{-10} watt per square meter. How many decibels is a whisper?

122. Loudness of a Concert If you sit in the front row of a rock concert, you will experience an intensity level of 10^{-1} watt per square meter. How many decibels is a rock concert in the front row? If you move back to the 15th row, you will experience an intensity level of 10^{-2} watt per square meter. How many decibels is a rock concert in the 15th row?

123. Threshold of Pain The threshold of pain has an intensity level of 10^1 watt per square meter. How many decibels is the threshold of pain?

124. Exploding Eardrum Instant perforation of the eardrum occurs at an intensity level of 10^4 watts per square meter. At how many decibels does instant perforation of the eardrum occur?

*Problems 125–128 use the following discussion: The **Richter scale** is one way of converting seismographic readings into numbers that provide an easy reference for measuring the **magnitude M** of an earthquake. All earthquakes are compared to a **zero–level earthquake** whose seismographic reading measures 0.001 millimeter at a distance of 100 kilometers from the epicenter. An earthquake whose seismographic reading measures x millimeters has magnitude M given by*

$$M(x) = \log\left(\frac{x}{10^{-3}}\right)$$

where 10^{-3} is the reading of a zero–level earthquake 100 kilometers from its epicenter.

125. San Francisco, 1906 According to the United States Geological Survey, the San Francisco earthquake of 1906 resulted in a seismographic reading of 63,096 millimeters 100 kilometers from its epicenter. What was the magnitude of this earthquake?

126. Alaska, 1964 According to the United States Geological Survey, an earthquake on March 28, 1964, in Prince William Sound, Alaska, resulted in a seismographic reading of 1,584,893 millimeters 100 kilometers from its epicenter. What was the magnitude of this earthquake? This earthquake was the second largest ever recorded. The largest was the Great Chilean Earthquake of 1960, whose magnitude was 9.5 on the Richter scale.

127. Japan, 2011 According to the United States Geological Survey, an earthquake on March 11, 2011, off the coast of Japan had a magnitude of 8.9. What was the seismographic reading 100 kilometers from its epicenter?

128. Brazil, 2015 According to the United States Geological Survey, an earthquake on November 26, 2015, in Tarauka, Brazil, had a magnitude of 7.6. What was the seismographic reading 100 kilometers from its epicenter?

129. pH The pH of a chemical solution is given by the formula

$$\text{pH} = -\log[H^+]$$

where $[H^+]$ is the concentration of hydrogen ions in moles per liter. Values of pH range from 0 to 14. A solution whose pH is 7 is considered neutral. The pH of pure water at 25 degrees Celsius is 7. A solution whose pH is less than 7 is considered acidic, while a solution whose pH is greater than 7 is considered basic.

(a) What is the pH of household ammonia for which $[H^+]$ is 10^{-12}? Is ammonia basic or acidic?

(b) What is the pH of black coffee for which $[H^+]$ is 10^{-5}? Is black coffee basic or acidic?

(c) What is the pH of lemon juice for which $[H^+]$ is 10^{-2}? Is lemon juice basic or acidic?

(d) What is the concentration of hydrogen ions in human blood $(pH = 7.4)$?

130. Energy of an Earthquake The magnitude and the seismic moment are related to the amount of energy that is given off by an earthquake. The relationship between magnitude and energy is

$$\log E_S = 11.8 + 1.5M$$

where E_S is the energy (in ergs) for an earthquake whose magnitude is M. Note that E_S is the amount of energy given off from the earthquake as seismic waves.

(a) How much energy is given off by an earthquake that measures 5.8 on the Richter scale?

(b) The earthquake on the Rat Islands in Alaska on February 4, 1965, measured 8.7 on the Richter scale. How much energy was given off by this earthquake?

Explaining the Concepts

131. Explain why the base in the logarithmic function $f(x) = \log_a x$ is not allowed to equal 1.

132. Explain why the base in the logarithmic function $f(x) = \log_a x$ must be greater than 0.

133. Explain why the domain of the function $f(x) = \log_a (x^2 + 1)$ is the set of all real numbers.

134. What are the domain and the range of the exponential function $f(x) = a^x$? What are the domain and the range of the logarithmic function $f(x) = \log_a x$? What is the relation between the domain and the range of each function?

Synthesis Review

In Problems 135–140, add or subtract each expression.

135. $(2x^2 - 6x + 1) - (5x^2 + x - 9)$

136. $(x^3 - 2x^2 + 10x + 2) + (3x^3 - 4x - 5)$

137. $\dfrac{3x}{x^2 - 1} + \dfrac{x - 3}{x^2 + 3x + 2}$

138. $\dfrac{x + 5}{x^2 + 5x + 6} - \dfrac{2}{x^2 + 2x - 3}$

139. $\sqrt{8x^3} + x\sqrt{18x}$

140. $\sqrt[3]{27a} + 2\sqrt[3]{a} - \sqrt[3]{8a}$

Technology Exercises

In Problems 141–146, graph each logarithmic function using technology. State the domain and the range of each function.

141. $f(x) = \log(x + 1)$

142. $g(x) = \log(x - 2)$

143. $G(x) = \ln(x) + 1$

144. $F(x) = \ln(x) - 4$

145. $f(x) = 2\log(x - 3) + 1$

146. $G(x) = -\log(x + 1) - 3$

147. Exploration: Graph of Logarithmic Functions *Open the "Graph of Logarithmic Functions" Geogebra applet.* The applet may be found using the QR code at the beginning of this section or through the Multimedia Library in MyMathLab. In the upper left corner is one "slider" that represents the value for a in the logarithmic function $f(x) = \log_a x$. If you are using a keyboard, it may be easier to use the arrow keys, instead of the mouse, to move the slider.

(a) Move the a slider to 2. This will produce the graph of $f(x) = \log_2 x$. State the domain, the range, the intercepts, and the values of the three coordinate points that are labeled on the graph.

(b) Move the a slider to 3. This will produce the graph of $f(x) = \log_3 x$. State the domain, the range, the intercepts, and the values of the three coordinate points that are labeled on the graph.

(c) Move the a slider to 4. This will produce the graph of $f(x) = \log_4 x$. State the domain, the range, the intercepts, and the values of the three coordinate points that are labeled on the graph.

(d) Use your results from parts (a) through (c) to generalize the following for *all* exponential functions of the form $f(x) = \log_a x$:

    i. The domain is _____.

    ii. The range is _____.

    iii. How many y-intercepts are there? What is the x-intercept?

    iv. Conclude the graph of $f(x) = \log_a x$ contains the points $(\_\_, -1)$, $(\_, 0)$, and $(\_, 1)$.

Putting the Concepts Together (Sections 8.1–8.3)

We designed these problems so that you can review Sections 8.1–8.3 and show your mastery of the concepts. Take time to work these problems before proceeding with the next section. The answers are located at the back of the text on page AN-50.

1. Given the functions $f(x) = 2x + 3$ and $g(x) = 2x^2 - 4x$, find

    (a) $(f \circ g)(x)$ **(b)** $(g \circ f)(x)$

    (c) $(f \circ g)(3)$ **(d)** $(g \circ f)(-2)$

    (e) $(f \circ f)(1)$

2. Find the inverse of the given one-to-one functions.

    (a) $f(x) = 3x + 4$ **(b)** $g(x) = x^3 - 4$

3. Sketch the graph of the inverse of the given one-to-one function.

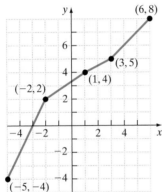

4. Approximate each number using a calculator. Express your answer rounded to three decimal places.

    (a) $2.7^{2.7}$ **(b)** $2.72^{2.72}$ **(c)** $2.718^{2.718}$

    (d) $2.7183^{2.7183}$ **(e)** e^e

5. Change each exponential equation to an equivalent equation involving a logarithm.

    (a) $a^4 = 6.4$ **(b)** $10^x = 278$

6. Change each logarithmic equation to an equivalent equation involving an exponent.

    (a) $\log_2 x = 7$ **(b)** $\ln 16 = M$

7. Find the exact value of each expression without using a calculator.

    (a) $\log_5 625$ **(b)** $\log_{2/3}\left(\dfrac{9}{4}\right)$

8. Determine the domain of the logarithmic function $f(x) = \log_{13}(2x + 12)$.

In Problems 9 and 10, graph each function. State the domain and range of the function.

    9. $f(x) = \left(\dfrac{1}{6}\right)^x$ **10.** $g(x) = \log_{3/2} x$

In Problems 11–14, solve each equation. State the exact solution.

11. $3^{-x+2} = 27$

12. $e^x = e^{2x+5}$

13. $\log_2(2x + 5) = 4$

14. $\ln x = 7$

15. Suppose that a student has 150 anatomy and physiology terms to learn. If the student learns 40 terms in 60 minutes, the function $L(t) = 150(1 - e^{-0.0052t})$ models the number of terms L that the student will learn after t minutes. According to the model, how many terms will the student have learned after 90 minutes?

8.4 Properties of Logarithms

Objectives

1 Understand the Properties of Logarithms

2 Write a Logarithmic Expression as a Sum or Difference of Logarithms

3 Write a Logarithmic Expression as a Single Logarithm

4 Evaluate a Logarithm Whose Base Is Neither 10 Nor e

Are You Prepared for This Section?

Before getting started, complete the following problems. If you get a problem wrong, go back to the section cited and review the material.

P1. Round 3.03468 to three decimal points.   [Section R.2, pp. 14–15]

P2. Evaluate $a^0, a \neq 0$.   [Getting Ready, p. 297]

▶ **1 Understand the Properties of Logarithms**

Logarithms have some very useful properties. These properties will be derived directly from the definition of a logarithm and the laws of exponents.

EXAMPLE 1 **Deriving Properties of Logarithms**

Determine the value of the following logarithmic expressions where a is any positive real number, and $a \neq 1$.

(a) $\log_a 1$ **(b)** $\log_a a$

Solution

(a) Remember, $y = \log_a x$ is equivalent to $x = a^y$, where $a > 0, a \neq 1$.

$$y = \log_a 1$$

Change to an exponent:	$a^y = 1$
Because $a \neq 0, a^0 = 1$:	$a^y = a^0$
If $a^u = a^v$, then $u = v$:	$y = 0$

Thus $\log_a 1 = 0$.

(b) Again, write the logarithm as an exponent in order to evaluate the logarithm. Let $y = \log_a a$, so that

$$y = \log_a a$$

Change to an exponent:	$a^y = a$
$a = a^1$:	$a^y = a^1$
If $a^u = a^v$, then $u = v$:	$y = 1$

Thus $\log_a a = 1$. ●

The results of Example 1 are summarized below.

$$\log_a 1 = 0 \qquad \log_a a = 1$$

Quick ✓

In Problems 1–4, evaluate each logarithm.

1. $\log_5 1$ **2.** $\ln 1$ **3.** $\log_4 4$ **4.** $\log 10$

How can $3^{\log_3 81}$ be evaluated? The exponent, $\log_3 81$, is 4 (because $y = \log_3 81$ means $3^y = 81$, or $3^y = 3^4$, so $y = 4$). Now $3^{\log_3 81} = 3^4 = 81$, so $3^{\log_3 81} = 81$. This result is true in general.

In Other Words

When the number a is raised to the power $\log_a M$, the result is M.

An Inverse Property of Logarithms

If a and M are positive real numbers, with $a \neq 1$, then

$$a^{\log_a M} = M$$

This is an Inverse Property of Logarithms because if the logarithm to the base a of a positive number M is computed and then a is raised to this power, the result is M.

EXAMPLE 2 **Using an Inverse Property of Logarithms**

(a) $5^{\log_5 20} = 20$ **(b)** $0.8^{\log_{0.8} \sqrt{23}} = \sqrt{23}$ ●

Quick ✓

In Problems 5 and 6, evaluate each logarithm.

5. $12^{\log_{12} \sqrt{2}}$ **6.** $10^{\log 0.2}$

How can $\log_5 5^6$ be evaluated? If $y = \log_5 5^6$, then $5^y = 5^6$ (since $y = \log_a x$ means $a^y = x$), so $y = 6$. Thus $\log_5 5^6 = 6$. This result is true in general.

In Other Words

The logarithm to the base a of a raised to the power of r is r.

An Inverse Property of Logarithms

If a is a positive real number, with $a \neq 1$, and r is any real number, then

$$\log_a a^r = r$$

This is another Inverse Property of Logarithms because if a is raised to some power r and then the logarithm to the base a of a^r is computed, the result is r.

EXAMPLE 3 **Using an Inverse Property of Logarithms**

(a) $\log_4 4^3 = 3$ **(b)** $\ln e^{-0.5} = -0.5$ ●

Quick ✓

In Problems 7 and 8, evaluate each logarithm.

7. $\log_8 8^{1.2}$ **8.** $\log 10^{-4}$

▶ ❷ **Write a Logarithmic Expression as a Sum or Difference of Logarithms**

The next two properties deal with logarithms whose arguments are products or quotients.
   Notice that $\log_2 8 = 3$ and that $\log_2 2 + \log_2 4 = 1 + 2 = 3$. Therefore, $\log_2 8 = \log_2 2 + \log_2 4$ and $8 = 4 \cdot 2$. This suggests the following result.

In Other Words

The product rule of logarithms states that the log of a product equals the sum of the logs of the factors.

The Product Rule of Logarithms

If M, N, and a are positive real numbers, with $a \neq 1$, then

$$\log_a (MN) = \log_a M + \log_a N$$

EXAMPLE 4 **Using the Product Rule of Logarithms**

Simplify each of the following logarithms by writing it as the sum of logarithms.

(a) $\log_2 (5 \cdot 3)$ **(b)** $\ln (6z)$

Solution

Do you notice that each argument contains a product? To simplify, use the Product Rule of Logarithms.

(a) $\log_2 (5 \cdot 3) = \log_2 5 + \log_2 3$ **(b)** $\ln (6z) = \ln 6 + \ln z$ ●

Work Smart

$\log_a (M + N)$ does *not* equal $\log_a M + \log_a N$.

Quick ✓

9. *True or False* $\log (x + 4) = \log x + \log 4$

In Problems 10 and 11, write each logarithm as the sum of logarithms.

10. $\log_4 (9 \cdot 5)$ **11.** $\log (5w)$

Notice that $\log_2 8 = 3$ and that $\log_2 16 - \log_2 2 = 4 - 1 = 3$. Therefore, $\log_2 8 = \log_2 16 - \log_2 2$ and $8 = \dfrac{16}{2}$. This suggests the result on the next page.

The Quotient Rule of Logarithms

If M, N, and a are positive real numbers, with $a \neq 1$, then

$$\log_a\left(\frac{M}{N}\right) = \log_a M - \log_a N$$

The relationships in the product and quotient rules for logarithms are not coincidental. After all, when exponential expressions with the same base are multiplied, the exponents are added, and when exponential expressions with the same base are divided, the exponents are subtracted. The same thing is going on here!

EXAMPLE 5 **Using the Quotient Rule of Logarithms**

Simplify each of the following logarithms by writing it as the difference of logarithms.

(a) $\log_2\left(\dfrac{5}{3}\right)$ **(b)** $\log\left(\dfrac{y}{5}\right)$

Solution

Do you notice that each argument contains a quotient? To simplify, use the Quotient Rule of Logarithms.

(a) $\log_2\left(\dfrac{5}{3}\right) = \log_2 5 - \log_2 3$ **(b)** $\log\left(\dfrac{y}{5}\right) = \log y - \log 5$ ●

Work Smart

$\log_a(M - N)$ does *not* equal $\log_a M - \log_a N$.

Quick ✓

In Problems 12 and 13, write each logarithm as the difference of logarithms.

12. $\log_7\left(\dfrac{9}{5}\right)$ **13.** $\ln\left(\dfrac{p}{3}\right)$

A single logarithm can be written as the sum or difference of logs when the argument of the logarithm contains both products and quotients.

EXAMPLE 6 **Writing a Single Logarithm as the Sum or Difference of Logs**

Write $\log_3\left(\dfrac{4x}{y}\right)$ as the sum or difference of logarithms.

Solution

Notice that the argument of the logarithm contains a product and a quotient. Thus the Product Rule and the Quotient Rule of Logarithms can be used. When applying both rules, it is typically easier to use the Quotient Rule first.

The Quotient Rule of Logarithms
↓

$$\log_3\left(\frac{4x}{y}\right) = \log_3(4x) - \log_3 y$$

The Product Rule of Logarithms: $= \log_3 4 + \log_3 x - \log_3 y$ ●

Work Smart

When you need to use both the Product Rule and the Quotient Rule, use the Quotient Rule first.

Quick ✓

In Problems 14 and 15, write each logarithm as the sum or difference of logarithms.

14. $\log_2\left(\dfrac{3m}{n}\right)$ **15.** $\ln\left(\dfrac{q}{3p}\right)$

Another useful property of logarithms allows us to express powers on the argument of a logarithm as factors.

> **The Power Rule of Logarithms**
>
> If M and a are positive real numbers, with $a \neq 1$, and r is any real number, then
>
> $$\log_a M^r = r \log_a M$$

EXAMPLE 7 **Using the Power Rule of Logarithms**

Use the Power Rule of Logarithms to express all powers as factors.

(a) $\log_8 3^5$ (b) $\ln x^{\sqrt{3}}$

Solution

(a) $\log_8 3^5 = 5 \log_8 3$ (b) $\ln x^{\sqrt{3}} = \sqrt{3} \ln x$ ●

Work Smart

Whenever you see the word "factor," you should think product (or multiplication).

Quick ✓

In Problems 16 and 17, write each logarithm so that all powers are factors.

16. $\log_2 5^{1.6}$ **17.** $\log b^5$

Use the direction *expand the logarithm* to mean "Write a logarithm as a sum or difference with all exponents written as factors."

EXAMPLE 8 **Expanding a Logarithm**

Expand the logarithm.

(a) $\log_2(x^2 y^3)$ (b) $\log\left(\dfrac{100x}{\sqrt{y}}\right)$

Solution

(a) The argument of the logarithm contains a product, so use the Product Rule of Logarithms to write the single log as the sum of two logs.

$$\overset{\text{Product Rule}}{\downarrow}$$
$$\log_2(x^2 y^3) = \log_2 x^2 + \log_2 y^3$$

Write exponents as factors: $= 2 \log_2 x + 3 \log_2 y$

(b) The argument of the logarithm contains a quotient and a product.

Work Smart

If the argument of the logarithm contains both a quotient and a product, it is easier to write the quotient as the difference of two logs first.

$$\overset{\text{Quotient Rule}}{\downarrow}$$
$$\log\left(\frac{100x}{\sqrt{y}}\right) = \log(100x) - \log\sqrt{y}$$

$\sqrt{y} = y^{\frac{1}{2}}$: $\qquad = \log(100x) - \log y^{\frac{1}{2}}$

Product Rule: $\qquad = \log 100 + \log x - \log y^{\frac{1}{2}}$

Write exponents as factors: $\qquad = \log 100 + \log x - \dfrac{1}{2}\log y$

$\log 100 = 2$: $\qquad = 2 + \log x - \dfrac{1}{2}\log y$ ●

Quick ✓

In Problems 18 and 19, expand each logarithm.

18. $\log_4(a^2 b)$ **19.** $\log_3\left(\dfrac{9m^4}{\sqrt[3]{n}}\right)$

Work Smart

To write two logs as a single log, the bases on the logs must be the same.

▶ ❸ **Write a Logarithmic Expression as a Single Logarithm**

The Product Rule, Quotient Rule, and Power Rule of Logarithms can be used to write the sums and/or differences of logarithms that have the same base as a single logarithm. This skill will be useful when solving certain logarithmic equations in Section 8.5.

EXAMPLE 9 **Writing Expressions as Single Logarithms**

Write each of the following as a single logarithm.

(a) $\log_6 3 + \log_6 12$ (b) $\log(x - 2) - \log x$

Solution

(a) The base of each logarithm is the same, 6. Use the Product Rule to write the sum of the logs as a single log.

$$\text{Product Rule}$$
$$\downarrow$$
$$\log_6 3 + \log_6 12 = \log_6 (3 \cdot 12)$$
$$3 \cdot 12 = 36: \quad = \log_6 36$$
$$6^2 = 36, \text{ so } \log_6 36 = 2: \quad = 2$$

(b) The base of each logarithm is the same, 10. Use the Quotient Rule to write the difference of the logs as a single log.

$$\log(x - 2) - \log x = \log\left(\frac{x - 2}{x}\right)$$

     ●

Quick ✓

In Problems 20 and 21, write each expression as a single logarithm.

20. $\log_8 4 + \log_8 16$ **21.** $\log_3(x + 4) - \log_3(x - 1)$

Work Smart

If logarithms have coefficients, write them as exponents before using the Product or Quotient Rule.

In order to use the Product Rule or the Quotient Rule to write the sum or difference of logarithms as a single logarithm, the coefficients of the logarithms must be 1. Therefore if logarithms have coefficients, first use the Power Rule to write the coefficient as a power. For example, write $2 \log x$ as $\log x^2$.

EXAMPLE 10 **Writing Expressions as Single Logarithms**

Write each of the following as a single logarithm.

(a) $2 \log_2(x - 1) + \dfrac{1}{2}\log_2 x$ (b) $\log(x - 1) + \log(x + 1) - 3 \log x$

Solution

(a) Each logarithm has the same base. Because the logarithms have coefficients, use the Power Rule to write the coefficients as exponents.

$$2 \log_2(x - 1) + \frac{1}{2}\log_2 x = \log_2(x - 1)^2 + \log_2 x^{\frac{1}{2}}$$
$$x^{\frac{1}{2}} = \sqrt{x}: \quad = \log_2(x - 1)^2 + \log_2 \sqrt{x}$$
$$\text{Product Rule:} \quad = \log_2\left[(x - 1)^2 \sqrt{x}\right]$$

(b) Work from left to right. Since the first two logs are being added, use the Product Rule to write these logs as a single log.

$$\text{Product Rule}$$
$$\downarrow$$

$$\log(x - 1) + \log(x + 1) - 3\log x = \log[(x - 1)(x + 1)] - 3\log x$$

$$(x - 1)(x + 1) = x^2 - 1: \quad = \log(x^2 - 1) - 3\log x$$

Write the coefficient as an exponent: $\quad = \log(x^2 - 1) - \log x^3$

Quotient Rule: $\quad = \log\left(\dfrac{x^2 - 1}{x^3}\right)$ ●

Quick ✓

In Problems 22 and 23, write each expression as a single logarithm.

22. $\log_5 x - 3\log_5 2$

23. $\log_2(x + 1) + \log_2(x + 2) - 2\log_2 x$

▶ ❹ Evaluate a Logarithm Whose Base Is Neither 10 Nor *e*

Section 8.3 discussed how to use a calculator to approximate logarithms whose base was either 10 (the common logarithm) or e (the natural logarithm). But what if the base of the logarithm is neither 10 nor e? To determine how to evaluate these types of logarithms, the following property is needed. In this property, M, N, and a are positive numbers and $a \neq 1$. The property is a result of the fact that the logarithmic function is a one-to-one function.

One-to-One Property of Logarithms

$$\text{If} \quad M = N, \quad \text{then} \quad \log_a M = \log_a N.$$

EXAMPLE 11   **Approximating a Logarithm Whose Base Is Neither 10 Nor *e***

Approximate $\log_2 5$. Round the answer to three decimal places.

Solution

Let $y = \log_2 5$ and then convert the logarithmic equation to an equivalent exponential equation using the fact that if $y = \log_a x$, then $x = a^y$.

$$y = \log_2 5$$
$$2^y = 5$$

Now, use the One-to-One Property and "take the logarithm of both sides." So that a calculator can be used, take either the common log or the natural log of both sides (it doesn't matter which). Let's take the common log of both sides.

> **In Other Words**
>
> "Taking the logarithm" of both sides of an equation is the same type of approach as squaring both sides of an equation.

$$\log 2^y = \log 5$$

Use the Power Rule, $\log a^r = r \log a$: $\quad y \log 2 = \log 5$

Divide both sides by $\log 2$: $\quad y = \dfrac{\log 5}{\log 2}$

Using a calculator, find $\dfrac{\log 5}{\log 2} \approx 2.322$. Thus $\log_2 5 \approx 2.322$. ●

There is another way to approximate a logarithm whose base is neither 10 nor e. It is called the **Change-of-Base Formula.**

Change-of-Base Formula

If $a \neq 1$, $b \neq 1$, and M are positive real numbers, then

$$\log_a M = \frac{\log_b M}{\log_b a}$$

Because many calculators have keys for only the common logarithm, $\boxed{\log}$, and the natural logarithm, $\boxed{\ln}$, we use the Change-of-Base Formula with either $b = 10$ or $b = e$:

$$\log_a M = \frac{\log M}{\log a} \quad \text{or} \quad \log_a M = \frac{\ln M}{\ln a}$$

EXAMPLE 12 **Using the Change-of-Base Formula**

Approximate $\log_4 45$. Round your answer to three decimal places.

Solution
Use the Change-of-Base Formula.

Using common logarithms: $\quad \log_4 45 = \dfrac{\log 45}{\log 4}$

$$\approx 2.746$$

Using natural logarithms: $\quad \log_4 45 = \dfrac{\ln 45}{\ln 4}$

$$\approx 2.746$$

Thus $\log_4 45 \approx 2.746$.

Quick ✓

24. $\log_3 10 = \dfrac{\log \underline{}}{\log \underline{}} = \dfrac{\ln \underline{}}{\ln \underline{}}$

In Problems 25 and 26, approximate each logarithm. Round your answers to three decimal places.

25. $\log_3 32$ **26.** $\log_{\sqrt{2}} \sqrt{7}$

Let's review all the properties of logarithms.

Properties of Logarithms

In the following properties, M, N, a, and b are positive real numbers, $a \neq 1, b \neq 1$, and r is any real number.

- **Inverse Properties of Logarithms**
  $$a^{\log_a M} = M \text{ and } \log_a a^r = r$$
- **The Product Rule of Logarithms**
  $$\log_a(MN) = \log_a M + \log_a N$$
- **The Quotient Rule of Logarithms**
  $$\log_a\left(\frac{M}{N}\right) = \log_a M - \log_a N$$
- $\log_a 1 = 0$

- **The Power Rule of Logarithms**
  $$\log_a M^r = r \log_a M$$
- **Change-of-Base Formula**
  $$\log_a M = \frac{\log_b M}{\log_b a} = \frac{\log M}{\log a} = \frac{\ln M}{\ln a}$$
- **One-to-One Property**
  If $M = N$, then $\log_a M = \log_a N$
- $\log_a a = 1$

8.4 Exercises MyMathLab®

Exercise numbers in green have complete video solutions in MyMathLab or may be accessed using the QR code to the right.

Problems 1–26 are the Quick ✓s that follow the EXAMPLES.

Building Skills

In Problems 27–38, use properties of logarithms to find the exact value of each expression. Do not use a calculator. See Objective 1.

27. $\log_2 2^3$ **28.** $\log_5 5^{-3}$

29. $\ln e^{-7}$ **30.** $\ln e^9$

31. $3^{\log_3 5}$ **32.** $5^{\log_5 \sqrt{2}}$

33. $e^{\ln 2}$ **34.** $e^{\ln 10}$

35. $\log_7 7$ **36.** $\log_5 5$

37. $\log 1$ **38.** $\ln 1$

In Problems 39–46, suppose that $\ln 2 = a$ *and* $\ln 3 = b$. *Use properties of logarithms to write each logarithm in terms of a and b. See Objective 2.*

39. $\ln 6$

40. $\ln \dfrac{3}{2}$

41. $\ln 9$

42. $\ln 4$

43. $\ln 12$

44. $\ln 18$

45. $\ln \sqrt{2}$

46. $\ln \sqrt[4]{3}$

In Problems 47–68, write each expression as a sum or difference of logarithms. Express exponents as factors. See Objective 2.

47. $\log(ab)$

48. $\log_4\left(\dfrac{a}{b}\right)$

49. $\log_5 x^4$

50. $\log_3 z^{-2}$

51. $\log_2(xy^2)$

52. $\log_3(a^3 b)$

53. $\log_5(25x)$

54. $\log_2(8z)$

55. $\log_7\left(\dfrac{49}{y}\right)$

56. $\log_2\left(\dfrac{16}{p}\right)$

57. $\ln(e^2 x)$

58. $\ln\left(\dfrac{x}{e^3}\right)$

59. $\log_3(27\sqrt{x})$

60. $\log_2(32\sqrt[4]{z})$

61. $\log_5\left(x^2 \sqrt{x^2 + 1}\right)$

62. $\log_3\left(x^3 \sqrt{x^2 - 1}\right)$

63. $\log\left(\dfrac{x^4}{\sqrt[3]{x - 1}}\right)$

64. $\ln\left(\dfrac{\sqrt[5]{x}}{(x + 2)^2}\right)$

65. $\log_7\sqrt{\dfrac{x + 1}{x}}$

66. $\log_6\sqrt[3]{\dfrac{x - 2}{x + 1}}$

67. $\log_2\left[\dfrac{x(x - 1)^2}{\sqrt{x + 1}}\right]$

68. $\log_4\left[\dfrac{x^3(x - 3)}{\sqrt[3]{x + 1}}\right]$

In Problems 69–92, write each expression as a single logarithm. See Objective 3.

69. $\log 25 + \log 4$

70. $\log_4 32 + \log_4 2$

71. $\log x + \log 3$

72. $\log_2 6 + \log_2 z$

73. $\log_3 36 - \log_3 4$

74. $\log_2 48 - \log_2 3$

75. $10^{\log 8 - \log 2}$

76. $e^{\ln 24 - \ln 3}$

77. $3 \log_3 x$

78. $8 \log_2 z$

79. $\log_4(x + 1) - \log_4 x$

80. $\log_5(2y - 1) - \log_5 y$

81. $2 \ln x + 3 \ln y$

82. $4 \log_2 a + 2 \log_2 b$

83. $\dfrac{1}{2}\log_3 x + 3 \log_3(x - 1)$

84. $\dfrac{1}{3}\log_4 z + 2 \log_4(2z + 1)$

85. $\log x^5 - 3 \log x$

86. $\log_7 x^4 - 2 \log_7 x$

87. $\dfrac{1}{2}\left[3 \log x + \log y\right]$

88. $\dfrac{1}{3}\left[\ln(x - 1) + \ln(x + 1)\right]$

89. $\log_8(x^2 - 1) - \log_8(x + 1)$

90. $\log_5(x^2 + 3x + 2) - \log_5(x + 2)$

91. $18 \log \sqrt{x} + 9 \log \sqrt[3]{x} - \log 10$

92. $10 \log_4 \sqrt[5]{x} + 4 \log_4 \sqrt{x} - \log_4 16$

In Problems 93–100, use the Change-of-Base Formula and a calculator to evaluate each logarithm. Round your answer to three decimal places. See Objective 4.

93. $\log_2 10$

94. $\log_3 18$

95. $\log_8 3$

96. $\log_7 5$

97. $\log_{1/3} 19$

98. $\log_{1/4} 3$

99. $\log_{\sqrt{2}} 5$

100. $\log_{\sqrt{3}} \sqrt{6}$

Applying the Concepts

101. Find the value of
$\log_2 3 \cdot \log_3 4 \cdot \log_4 5 \cdot \log_5 6 \cdot \log_6 7 \cdot \log_7 8$.

102. Find the value of $\log_2 4 \cdot \log_4 6 \cdot \log_6 8$.

103. Find the value of
$\log_2 3 \cdot \log_3 4 \cdot \; \cdots \; \cdot \log_n(n + 1) \cdot \log_{n+1} 2$.

104. Find the value of
$\log_3 3 \cdot \log_3 9 \cdot \log_3 27 \cdot \; \cdots \; \cdot \log_3 3^n$.

Extending the Concepts

105. Show that
$$\log_a\left(x + \sqrt{x^2 - 1}\right) + \log_a\left(x - \sqrt{x^2 - 1}\right) = 0.$$

106. Show that
$$\log_a\left(\sqrt{x} + \sqrt{x - 1}\right) + \log_a\left(\sqrt{x} - \sqrt{x - 1}\right) = 0.$$

107. If $f(x) = \log_a x$, show that $f(AB) = f(A) + f(B)$.

108. Find the domain of $f(x) = \log_a x^2$ and the domain of $g(x) = 2\log_a x$. Since $\log_a x^2 = 2\log_a x$, how can it be that the domains are not equal? Write a brief explanation.

Explaining the Concepts

109. State the Product Rule for Logarithms in your own words.

110. State the Quotient Rule for Logarithms in your own words.

111. Write an example to illustrate
$$\log_2(x + y) \neq \log_2 x + \log_2 y.$$

112. Write an example to illustrate
$$(\log_a x)^r \neq r\log_a x.$$

Synthesis Review

In Problems 113–118 solve each equation.

113. $4x + 3 = 13$

114. $-3x + 10 = 4$

115. $x^2 + 4x + 2 = 0$

116. $3x^2 = 2x + 1$

117. $\sqrt{x + 2} - 3 = 4$

118. $\sqrt[3]{2x} - 2 = -5$

Technology Exercises

We can use the Change-of-Base Formula to graph any logarithmic function using technology. For example, to graph $f(x) = \log_2 x$ on a graphing calculator, graph $Y_1 = \dfrac{\log x}{\log 2}$ or $Y_1 = \dfrac{\ln x}{\ln 2}$.

In Problems 119–122, graph each logarithmic function using technology. State the domain and the range of each function.

119. $f(x) = \log_3 x$

120. $f(x) = \log_5 x$

121. $F(x) = \log_{1/2} x$

122. $G(x) = \log_{1/3} x$

8.5 Exponential and Logarithmic Equations

Objectives

① Solve Logarithmic Equations Using the Properties of Logarithms

② Solve Exponential Equations

③ Solve Equations Involving Exponential Models

Are You Prepared for This Section?

Before getting started, complete the following problems. If you get a problem wrong, go back to the section cited and review the material.

P1. Solve: $2x + 5 = 13$ [Section 1.1, pp. 50–53]

P2. Solve: $x^2 - 4x = -3$ [Section 4.8, pp. 373–375]

P3. Solve: $3a^2 = a + 5$ [Section 7.2, pp. 555–559]

P4. Solve: $(x + 3)^2 + 2(x + 3) - 8 = 0$ [Section 7.3, pp. 569–573]

▶ ① Solve Logarithmic Equations Using the Properties of Logarithms

Section 8.3 showed how to solve logarithmic equations of the form $\log_a x = y$ by changing the logarithmic equation to an equivalent exponential equation. If a logarithmic equation has more than one logarithm in it, however, properties of logarithms need to be used to solve the equation.

The last section introduced the One-to-One Property that if $M = N$, then $\log_a M = \log_a N$. It turns out that the converse of this property is true as well.

> **One-to-One Property of Logarithms**
>
> In the following property, M, N, and a are positive real numbers, with $a \neq 1$.
>
> $$\text{If } \log_a M = \log_a N, \text{ then } M = N.$$

This property is useful for solving equations that contain logarithms with the same base by setting the arguments equal to each other.

EXAMPLE 1 **Solving a Logarithmic Equation**

Solve: $2 \log_3 x = \log_3 25$

Solution

Both logarithms have the same base, 3, so if the equation is written in the form $\log_a M = \log_a N$, the One-to-One Property can be used.

$$2 \log_3 x = \log_3 25$$

$r \log_a M = \log_a M^r$: $\log_3 x^2 = \log_3 25$

Set the arguments equal to each other: $x^2 = 25$

Square Root Property: $x = -5$ or $x = 5$

The apparent solution $x = -5$ is extraneous because the argument of a logarithm must be greater than zero, and -5 causes the argument to be negative. Now check the other apparent solution.

Check $x = 5$: $2 \log_3 5 \stackrel{?}{=} \log_3 25$

$2 \log_3 5 \stackrel{?}{=} \log_3 5^2$

$\log_a m^r = r \log_a m$: $2 \log_3 5 = 2 \log_3 5$ True

The solution set is $\{5\}$.

Quick ✓

1. If $\log_a M = \log_a N$, then _____.   **2.** Solve: $2 \log_4 x = \log_4 9$

If a logarithmic equation contains more than one logarithm on one side of the equation, then properties of logarithms can be used to rewrite the equation as a single logarithm. Once again, use properties to reduce an equation into a form that is familiar. In this case, use properties of logarithms to express the sum or difference of logarithms as a single logarithm. Then rewrite the equation in exponential form and solve for the unknown.

EXAMPLE 2 **Solving a Logarithmic Equation**

Solve: $\log_2(x - 2) + \log_2 x = 3$

Solution

Begin by rewriting the left side of the equation as a single logarithm.

$$\log_2(x - 2) + \log_2 x = 3$$

$\log_a M + \log_a N = \log_a(MN)$: $\log_2[x(x - 2)] = 3$

If $y = \log_a M$, then $a^y = M$: $2^3 = x(x - 2)$

Distribute: $8 = x^2 - 2x$

Write in standard form: $x^2 - 2x - 8 = 0$

Factor: $(x - 4)(x + 2) = 0$

Zero-Product Property: $x - 4 = 0$ or $x + 2 = 0$

$x = 4$ or $x = -2$

Work Smart

The apparent solution $x = -2$ in Example 2 is extraneous because it results in our attempting to find the log of a negative number, not because -2 is negative.

Check $x = 4$:

$\log_2(4 - 2) + \log_2 4 \stackrel{?}{=} 3$

$\log_2 2 + \log_2 4 \stackrel{?}{=} 3$

$1 + 2 = 3$ True

$x = -2$:

$\log_2(-2 - 2) + \log_2(-2) \stackrel{?}{=} 3$

$x = -2$ is extraneous because it causes the argument to be negative.

The solution set is $\{4\}$.

Quick ✓

3. Solve: $\log_4(x - 6) + \log_4 x = 2$ **4.** Solve: $\log_3(x + 3) + \log_3(x + 5) = 1$

▶ ❷ Solve Exponential Equations

Section 8.2 explained how to solve exponential equations using the fact that if $a^u = a^v$, then $u = v$. However, in many situations both sides of the equation cannot be written with the same base.

EXAMPLE 3 **Using Logarithms to Solve Exponential Equations**

Solve: $3^x = 5$

Solution

Because 5 cannot be written so that it is 3 raised to some integer power, write the equation $3^x = 5$ as a logarithm.

$$3^x = 5$$

If $a^y = x$, then $y = \log_a x$: $x = \log_3 5$ Exact solution

To find a decimal approximation to the solution, use the Change-of-Base Formula.

$$x = \log_3 5 = \frac{\log 5}{\log 3}$$

$$\approx 1.465 \quad \text{Approximate solution}$$

An alternative approach to solving the equation would be to take either the natural logarithm or the common logarithm of both sides of the equation. If the natural logarithm of both sides of the equation is taken, the following results:

$$3^x = 5$$

If $M = N$, then $\ln M = \ln N$: $\ln 3^x = \ln 5$

$\log_a M^r = r \log_a M$: $x \ln 3 = \ln 5$

Divide both sides by $\ln 3$: $x = \dfrac{\ln 5}{\ln 3}$ Exact solution

$$\approx 1.465 \quad \text{Approximate solution}$$

The solution set is $\left\{ \dfrac{\ln 5}{\ln 3} \right\}$. If the common logarithm of both sides had been taken, the solution set would have been $\left\{ \dfrac{\log 5}{\log 3} \right\}$. ●

Quick ✓

In Problems 5 and 6, solve each equation. Express answers in exact form and as a decimal rounded to three decimal places.

5. $2^x = 11$ **6.** $5^{2x} = 3$

EXAMPLE 4 **Using Logarithms to Solve Exponential Equations**

Solve: $4e^{3x} = 10$

Solution

First isolate the exponential expression by dividing both sides of the equation by 4.

$$4e^{3x} = 10$$

$$e^{3x} = \frac{5}{2}$$

Work Smart

You could also take the natural logarithm of both sides:

$$\ln e^{3x} = \ln\left(\frac{5}{2}\right)$$

$$3x = \ln\left(\frac{5}{2}\right)$$

$$x = \frac{\ln\left(\frac{5}{2}\right)}{3}$$

Because $\frac{5}{2}$ cannot be expressed as e raised to an integer power, write the exponential equation as an equivalent logarithmic equation.

$$e^{3x} = \frac{5}{2}$$

If $e^y = x$, then $y = \ln x$: $\ln\left(\frac{5}{2}\right) = 3x$

Divide both sides by 3: $x = \dfrac{\ln\left(\frac{5}{2}\right)}{3}$ Exact solution

$x \approx 0.305$ Approximate solution

The check is left to you. The solution set is $\left\{\dfrac{\ln\left(\frac{5}{2}\right)}{3}\right\}$.

Quick ✓

In Problems 7 and 8, solve each equation. Express answers in exact form and as a decimal rounded to three decimal places.

7. $e^{2x} = 5$

8. $3e^{-4x} = 20$

▶ ❸ Solve Equations Involving Exponential Models

Section 8.2 examined a variety of models from areas such as statistics, biology, and finance. The following examples, rather than evaluating the models at certain values of the independent variable, solve equations that involve the models.

EXAMPLE 5 **Radioactive Decay**

The half-life of plutonium-239 is 24,360 years. The maximum amount of plutonium-239 that an adult can handle without significant injury is 0.13 microgram $(= 0.000000013$ gram$)$. Suppose a researcher has a 1-gram sample of plutonium-239. The amount A (in grams) of plutonium-239 after t years is given by

$$A(t) = 1\left(\frac{1}{2}\right)^{\frac{t}{24,360}}$$

(a) How long will it take until 0.9 gram of plutonium-239 is left in the sample?

(b) How long will it take until the 1-gram sample is safe–that is, until 0.000000013 gram is left?

Solution

(a) To find the time until $A = 0.9$ gram, solve the equation

$$0.9 = 1\left(\frac{1}{2}\right)^{\frac{t}{24,360}}$$

for t. How can t be removed from the exponent? Use the fact that $\log_a M^r = r \log_a M$ to move the variable.

$$\log 0.9 = \log\left(\frac{1}{2}\right)^{\frac{t}{24,360}}$$

$$\log 0.9 = \frac{t}{24,360}\log\left(\frac{1}{2}\right)$$

(continued)

Multiply both sides by 24,360: $\quad 24{,}360 \log 0.9 = t \log\left(\dfrac{1}{2}\right)$

Divide both sides by $\log\left(\dfrac{1}{2}\right)$: $\quad \dfrac{24{,}360 \log 0.9}{\log\left(\dfrac{1}{2}\right)} = t$

Thus $t = \dfrac{24{,}360 \log 0.9}{\log\left(\dfrac{1}{2}\right)} \approx 3702.8$. After approximately 3703 years, there will be 0.9 gram of plutonium-239 left.

(b) To determine the time until $A = 0.000000013$ gram, solve the equation

$$0.000000013 = 1 \cdot \left(\dfrac{1}{2}\right)^{\frac{t}{24{,}360}}$$

Take the logarithm of both sides: $\quad \log(0.000000013) = \log\left(\dfrac{1}{2}\right)^{\frac{t}{24{,}360}}$

$\log_a M^r = r \log_a M$: $\quad \log(0.000000013) = \dfrac{t}{24{,}360} \log\left(\dfrac{1}{2}\right)$

Multiply both sides by 24,360: $\quad 24{,}360 \log(0.000000013) = t \log\left(\dfrac{1}{2}\right)$

Divide both sides by $\log\left(\dfrac{1}{2}\right)$: $\quad \dfrac{24{,}360 \log(0.000000013)}{\log\left(\dfrac{1}{2}\right)} = t$

Thus $t = \dfrac{24{,}360 \log(0.000000013)}{\log\left(\dfrac{1}{2}\right)} \approx 638{,}156.8$. After approximately 638,157 years, the 1-gram sample will be safe! •

Quick ✓

9. The half-life of thorium-227 is 18.72 days. Suppose a researcher has a 10-gram sample of thorium-227. The amount A (in grams) of thorium-227 after t days is given by

$$A(t) = 10\left(\dfrac{1}{2}\right)^{\frac{t}{18.72}}$$

(a) How long will it take until 9 grams of thorium-227 is left in the sample?
(b) How long will it take until 3 grams of thorium-227 is left in the sample?

Now let's look at an example involving compound interest. Remember, the compound interest formula states that the future value of P dollars invested in an account paying an annual interest rate r, compounded n times per year for t years, is given by $A = P\left(1 + \dfrac{r}{n}\right)^{nt}$.

EXAMPLE 6 **Future Value of Money**

Suppose you deposit $5000 into a Roth IRA today. If the deposit earns 8% interest compounded quarterly, when will it be worth

(a) $7500?

(b) $10,000? That is, when will your money have doubled?

Solution

With $P = 5000$, $r = 0.08$, and $n = 4$ (compounded quarterly),

$$A = 5000\left(1 + \frac{0.08}{4}\right)^{4t} \quad \text{or} \quad A = 5000(1.02)^{4t}$$

(a) Find the time t when $A = 7500$. That is, solve

$$7500 = 5000(1.02)^{4t}$$

Divide both sides by 5000: $\qquad 1.5 = (1.02)^{4t}$

Take the logarithm of both sides: $\qquad \log 1.5 = \log(1.02)^{4t}$

$\log_a M^r = r \log_a M$: $\qquad \log 1.5 = 4t \log(1.02)$

Divide both sides by 4 log(1.02): $\qquad \dfrac{\log 1.5}{4 \log(1.02)} = t$

So $t = \dfrac{\log 1.5}{4 \log(1.02)} \approx 5.12$. After approximately 5.12 years (5 years, 1.4 months), the account will be worth \$7500.

(b) Find the time t when $A = 10{,}000$. That is, solve

$$10{,}000 = 5000(1.02)^{4t}$$

Divide both sides by 5000: $\qquad 2 = (1.02)^{4t}$

Take the logarithm of both sides: $\qquad \log 2 = \log(1.02)^{4t}$

$\log_a M^r = r \log_a M$: $\qquad \log 2 = 4t \log(1.02)$

Divide both sides by 4 log(1.02): $\qquad \dfrac{\log 2}{4 \log(1.02)} = t$

Thus $t = \dfrac{\log 2}{4 \log(1.02)} \approx 8.75$. After approximately 8.75 years (8 years, 9 months), the account will be worth \$10,000.

Quick ✓

10. Suppose that you deposit \$2000 into a Roth IRA today. If the deposit earns 6% interest compounded monthly, how long will it be before the account is worth
(a) \$3000?
(b) \$4000? That is, when will your money have doubled?

8.5 Exercises MyMathLab® Exercise numbers in green have complete video solutions in MyMathLab or may be accessed using the QR code to the right.

*Problems **1–10** are the Quick ✓s that follow the EXAMPLES.*

Building Skills

In Problems 11–28, solve each equation. See Objective 1.

11. $\log_2 x = \log_2 7$

12. $\log_5 x = \log_5 13$

13. $2 \log_3 x = \log_3 81$

14. $2 \log_3 x = \log_3 4$

15. $\log_6(3x + 1) = \log_6 10$ **16.** $\log(2x - 3) = \log 11$

17. $\dfrac{1}{2}\ln x = 2 \ln 3$

18. $\dfrac{1}{2}\log_2 x = 2 \log_2 2$

19. $\log_2(x + 3) + \log_2 x = 2$

20. $\log_2(x - 7) + \log_2 x = 3$

21. $\log_2(x + 2) + \log_2(x + 5) = \log_2 4$

22. $\log_2(x + 5) + \log_2(x + 4) = \log_2 2$

23. $\log(x + 3) - \log x = 1$

24. $\log_3(x + 5) - \log_3 x = 2$

25. $\log_4(x + 5) - \log_4(x - 1) = 2$

26. $\log_3(x + 2) - \log_3(x - 2) = 4$

27. $\log_4(x + 8) + \log_4(x + 6) = \log_4 3$

28. $\log_3(x + 8) + \log_3(x + 4) = \log_3 5$

In Problems 29–48, solve each equation. Express irrational solutions in exact form and as a decimal rounded to three decimal places. See Objective 2.

29. $2^x = 10$ **30.** $3^x = 8$

31. $5^x = 20$ **32.** $4^x = 20$

33. $\left(\dfrac{1}{2}\right)^x = 7$ **34.** $\left(\dfrac{1}{2}\right)^x = 10$

35. $e^x = 5$ **36.** $e^x = 3$

37. $10^x = 5$ **38.** $10^x = 0.2$

39. $3^{2x} = 13$ **40.** $2^{2x} = 5$

41. $\left(\dfrac{1}{2}\right)^{4x} = 13$ **42.** $\left(\dfrac{1}{3}\right)^{2x} = 4$

43. $4 \cdot 2^x + 3 = 8$ **44.** $3 \cdot 4^x - 5 = 10$

45. $-3e^x = -18$ **46.** $\dfrac{1}{2}e^x = 4$

47. $0.2^{x+1} = 3^x$ **48.** $0.4^x = 2^{x-3}$

Mixed Practice

In Problems 49–64, solve each equation. Express irrational solutions in exact form and as a decimal rounded to three decimal places.

49. $\log_4 x + \log_4(x - 6) = 2$ **50.** $\log_6 x + \log_6(x + 5) = 2$

51. $5^{3x} = 7$ **52.** $3^{2x} = 4$

53. $3 \log_2 x = \log_2 8$ **54.** $5 \log_4 x = \log_4 32$

55. $\dfrac{1}{3}e^x = 5$ **56.** $-4e^x = -16$

57. $\left(\dfrac{1}{4}\right)^{x+1} = 8^x$ **58.** $9^x = 27^{x-4}$

59. $\log_3 x^2 = \log_3 16$ **60.** $\log_7 x^2 = \log_7 8$

61. $\log_2(x) + \log_2(x - 2) = \log_2(x + 4)$

62. $\log_3 x + \log_3(x - 4) = \log_3(x + 6)$

63. $\log_4 x + \log_4(x - 4) = \log_4 3$

64. $\log_6(x + 1) + \log_6(x - 5) = \log_6 2$

Applying the Concepts

65. A Population Model According to the U.S. Census Bureau, the population of the United States in 2016 was 326 million people. In addition, the population of the United States was growing at a rate of 0.99% per year. Assuming that this growth rate continues, the model $P(t) = 326(1.0099)^{t-2016}$ represents the population P (in millions of people) in year t.

    (a) According to this model, when will the population of the United States be 350 million people?

    (b) According to this model, when will the population of the United States be 471 million people?

66. A Population Model According to the *United States Census Bureau*, the population of the world in 2016 was 7563 million people. In addition, the population of the world was growing at a rate of 1.02% per year. Assuming that this growth rate continues, the model $P(t) = 7563(1.0102)^{t-2016}$ represents the population P (in millions of people) in year t.

(a) According to this model, when will the population of the world be 9.84 billion people?

(b) According to this model, when will the population of the world be 11.58 billion people?

67. Time Is Money Suppose that you deposit $5000 in a certificate of deposit (CD) today. If the deposit earns 2% interest compounded monthly, when will the account be worth

(a) 7000?

(b) 10,000? That is, when will your money have tripled

68. Time Is Money Suppose that you deposit $8000 in a certificate of deposit (CD) today. If the deposit earns 2% interest compounded quarterly, when will it be worth

(a) 10,000?

(b) 24,000? That is, when will your money have tripled?

69. Depreciation Based on data obtained from the *Kelley Blue Book*, the value V of a Ford Focus that is t years old can be modeled by $V(t) = 19,841(0.88)^t$.

(a) According to the model, when will the car be worth $15,000?

(b) According to the model, when will the car be worth $5000?

(c) According to the model, when will the car be worth $1000?

70. Depreciation Based on data obtained from the *Kelley Blue Book*, the value V of a Chevy Malibu that is t years old can be modeled by $V(t) = 25,258(0.84)^t$.

(a) According to the model, when will the car be worth $15,000?

(b) According to the model, when will the car be worth $5000?

(c) According to the model, when will the car be worth $1000?

71. Radioactive Decay The half-life of beryllium-11 is 13.81 seconds. Suppose that a researcher possesses a 100-gram sample of beryllium-11. The amount A (in grams) of beryllium-11 after t seconds is given by

$$A(t) = 100\left(\frac{1}{2}\right)^{\frac{t}{13.81}}$$

(a) When will there be 90 grams of beryllium-11 left in the sample?

(b) When will 25 grams be left?

(c) When will 10 grams of beryllium-11 be left?

72. Radioactive Decay The half-life of carbon-10 is 19.255 seconds. Suppose that a researcher possesses a 100-gram sample of carbon-10. The amount A (in grams) of carbon-10 after t seconds is given by

$$A(t) = 100\left(\frac{1}{2}\right)^{\frac{t}{19.255}}$$

(a) When will there be 90 grams of carbon-10 left in the sample?

(b) When will 25 grams be left?

(c) When will 10 grams be left?

For Exercises 73 and 74, use Newton's Law of Cooling, which states that the temperature of a heated object decreases exponentially over time toward the temperature of the surrounding medium.

73. Newton's Law of Cooling Suppose that a pizza is removed from a 400°F oven and placed in a room where the temperature is 70°F. The temperature u (in °F) of the pizza at time t (in minutes) can be modeled by $u(t) = 70 + 330e^{-0.072t}$.

(a) According to the model, when will the temperature of the pizza be 300°F?

(b) According to the model, when will the temperature of the pizza be 220°F?

74. Newton's Law of Cooling Suppose that coffee that is 170°F is poured into a coffee mug and allowed to cool in a room where the temperature is 70°F. The temperature u (in °F) of the coffee at time t (in minutes) can be modeled by $u(t) = 70 + 100e^{-0.045t}$.

(a) According to the model, when will the temperature of the coffee be 120°F?

(b) According to the model, when will the temperature of the coffee be 100°F?

75. Learning Curve Suppose that a student has 200 vocabulary words to learn. If a student learns 20 words in 30 minutes, the function

$$L(t) = 200(1 - e^{-0.0035t})$$

models the number of words L that the student will learn in t minutes.

(a) How long will it take the student to learn 50 words?

(b) How long will it take the student to learn 150 words?

76. Learning Curve Suppose that a student has 50 biology terms to learn. If a student learns 10 terms in 30 minutes, the function

$$L(t) = 50(1 - e^{-0.0223t})$$

models the number of terms L that the student will learn in t minutes.

(a) How long will it take the student to learn 10 words?

(b) How long will it take the student to learn 40 words?

Extending the Concepts

77. **The Rule of 72** The Rule of 72 states that the time for an investment to double in value is approximately given by 72 divided by the annual interest rate. For example, an investment earning 10% annual interest will double in approximately $\frac{72}{10} = 7.2$ years.

(a) According to the Rule of 72, approximately how long will it take an investment to double if it earns 8% annual interest?

(b) Derive a formula that can be used to find the number of years required for an investment to double. (*Hint*: Let $A = 2P$ in the formula $A = P\left(1 + \dfrac{r}{n}\right)^{nt}$ and solve for t.)

(c) Use the formula derived in part (b) to determine the exact amount of time it takes an investment to double that earns 8% interest compounded monthly. Compare the result to the results given by the Rule of 72.

78. **Critical Thinking** Suppose you need to open a certificate of deposit (CD). Bank A offers 2% interest compounded daily, and Bank B offers 2.1% interest compounded quarterly. Which bank offers the better deal? Why?

79. **Critical Thinking** The bacteria in a 2-liter container double every minute. After 30 minutes the container is full. How long did it take to fill half the container?

Synthesis Review

In Problems 80–85, find the following values for each function.

(a) $f(3)$ (b) $f(-2)$ (c) $f(0)$

80. $f(x) = 5x + 2$

81. $f(x) = -2x + 7$

82. $f(x) = \dfrac{x + 3}{x - 2}$

83. $f(x) = \dfrac{x}{x - 5}$

84. $f(x) = \sqrt{x + 5}$

85. $f(x) = 2^x$

Technology Exercises

The techniques for solving equations introduced in this chapter apply only to certain types of exponential or logarithmic equations. Solutions for other types of equations are usually studied in calculus using numerical methods. However, technology can be used to approximate solutions. Specifically, with a graphing calculator the INTERSECT feature can be used and in Desmos simply by clicking on the curve the "intersting points" including the intersection points will be displayed. For example, to solve $e^x = 3x + 2$, graph $Y_1 = e^x$ and $Y_2 = 3x + 2$ on the same screen in Desmos. Then click on one of the curves to display the "interesting points" and hover over those points with your mouse to see their coordinates. Each x-coordinate represents an approximate solution, as shown in Figure 27.

Figure 27

The solution set is $\{-0.46, 2.13\}$ rounded to two decimal places.

In Problems 86–93, solve each equation using technology. Express your answer rounded to two decimal places.

86. $e^x = -3x + 2$

87. $e^x = -2x + 5$

88. $e^x = x + 2$

89. $e^x = x^2$

90. $e^x + \ln(x) = 2$

91. $e^x - \ln(x) = 4$

92. $\ln x = x^2 + 1$

93. $\ln x = x^2 - 1$

Chapter 8 Activity: Correct the Quiz

Focus: Solving exponential and logarithmic equations

Time: 10–15 minutes

Group size: 2

In this activity you will work as a team to grade the student quiz shown on the next page. One of you will grade the odd questions, and the other will grade the even questions. If an answer is correct, mark it correct. If an answer is wrong, mark it wrong and show the correct answer.

Once all of the quiz questions are graded, explain your results to each other and compute the final score for the quiz. Be prepared to discuss your results with the rest of the class.

Student Quiz

Name: *Ima Student*   Quiz Score: _____

Solve the following equations. Express any irrational answers in exact form.

(1) $\log_2 x = 3$	Answer: $\{9\}$
(2) $\log_{16} x = \dfrac{3}{4}$	Answer: $\{8\}$
(3) $\log(2x) - \log 6 = \log(x - 8)$	Answer: $\{12\}$
(4) $3^x = 27$	Answer: $\{9\}$
(5) $6^{2x} = 18$	Answer: $\left\{ \dfrac{\log 18}{2 \log 6} \right\}$
(6) $4^{x+9} = 7$	Answer: $\left\{ \dfrac{\log 7}{\log 4} - 9 \right\}$
(7) $9^{3x-1} = 27^{4x}$	Answer: $\left\{ -\dfrac{1}{3} \right\}$
(8) $\log_3(2x - 3) = 2$	Answer: $\left\{ \dfrac{11}{2} \right\}$

Chapter 8 Review

Section 8.1   Composite Functions and Inverse Functions

KEY CONCEPTS

- $(f \circ g)(x) = f(g(x))$
- **Horizontal Line Test**
  If every horizontal line intersects the graph of a function f in at most one point, then f is one-to-one.
- For the inverse of a function also to be a function, the function must be one-to-one.
- **Relation Between the Domain and Range of a Function and Its Inverse**
  All elements in the domain of a one-to-one function are also elements in the range of its inverse.
  All elements in the range of a one-to-one function are also elements in the domain of its inverse.
- The graph of a one-to-one function f and the graph of its inverse are symmetric with respect to the line $y = x$.
- $f^{-1}(f(x)) = x$ for every x in the domain of f and
  $f(f^{-1}(x)) = x$ for every x in the domain of f^{-1}

KEY TERMS

Composition
Composite function
One-to-one
Inverse function

You Should be Able To...	EXAMPLE	Review Exercises
❶ Form the composite function (p. 624)	Examples 1 and 2	1–8
❷ Determine whether a function is one-to-one (p. 626)	Examples 3 and 4	9–12
❸ Find the inverse of a function defined by a map or set of ordered pairs (p. 628)	Examples 5 and 6	13–16
❹ Obtain the graph of the inverse function from the graph of a function (p. 630)	Example 7	17, 18
❺ Find the inverse of a function defined by an equation (p. 631)	Examples 8 and 9	19–22

In Problems 1–4, for the given functions f and g, find:

    (a) $(f \circ g)(5)$ **(b)** $(g \circ f)(-3)$

    (c) $(f \circ f)(-2)$ **(d)** $(g \circ g)(4)$

1. $f(x) = 3x + 5; g(x) = 2x - 1$

2. $f(x) = x - 3; g(x) = 5x + 2$

3. $f(x) = 2x^2 + 1; g(x) = x + 5$

4. $f(x) = x - 3; g(x) = x^2 + 1$

In Problems 5–8, for the given functions f and g, find:

    (a) $(f \circ g)(x)$ **(b)** $(g \circ f)(x)$

    (c) $(f \circ f)(x)$ **(d)** $(g \circ g)(x)$

5. $f(x) = x + 1; g(x) = 5x$

6. $f(x) = 2x - 3; g(x) = x + 6$

7. $f(x) = x^2 + 1; g(x) = 2x + 1$

8. $f(x) = \dfrac{2}{x + 1}; g(x) = \dfrac{1}{x}$

In Problems 9–12, determine which of the functions are one-to-one.

9. $\{(-5, 8), (-3, 2), (-1, 8), (0, 12), (1, 15)\}$

10. $\{(-4, 2), (-2, 1), (0, 0), (1, -1), (2, 8)\}$

11.

12.

In Problems 13–16, find the inverse of each one-to-one function.

13.

Age (years)	Height (inches)
24	69
59	71
29	72
81	73
37	74

14.

Price ($)	Quantity Demanded
300	112
200	129
170	144
150	161
130	176

15. $\{(-5, 3), (-3, 1), (1, -3), (2, 9)\}$

16. $\{(-20, 1), (-15, 4), (5, 3), (25, 2)\}$

In Problems 17 and 18, the graph of a one-to-one function f is given. Draw the graph of the inverse function f^{-1}.

17.

18.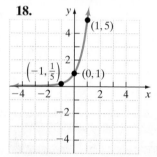

In Problems 19 and 22, the equation of a one-to-one function f is given. Find the equation of the inverse function f^{-1}.

19. $f(x) = 5x$ **20.** $H(x) = 2x + 7$

21. $P(x) = \dfrac{4}{x + 2}$ **22.** $g(x) = 2x^3 - 1$

Section 8.2   Exponential Functions

KEY CONCEPTS

- **Properties of Exponential Functions of the Form** $f(x) = a^x, a > 0, a \neq 1$
  1. The domain is the set of all real numbers. The range is the set of all positive real numbers.
  2. There are no x-intercepts. The y-intercept is $(0, 1)$.
  3. The graph of an exponential function contains the points $\left(-1, \dfrac{1}{a}\right)$, $(0, 1)$, and $(1, a)$.

- **Property for Solving Exponential Equations of the Form** $a^u = a^v$
  If $a^u = a^v$, then $u = v$.

KEY TERMS

Exponential function

The number e

Exponential equation

Half-life

Payment period

Compound interest

You Should Be Able To...	EXAMPLE	Review Exercises
❶ Evaluate exponential expressions (p. 638)	Example 1	25–25
❷ Graph exponential functions (p. 639)	Examples 2 through 4	26–29
❸ Define the number e (p. 642)		30
❹ Solve exponential equations (p. 643)	Examples 5 and 6	31–36
❺ Use exponential models that describe our world (p. 645)	Examples 7 through 9	37–40

In Problems 23–25, approximate each number using a calculator. Express your answer rounded to three decimal places.

23. (a) $7^{1.7}$ (d) $7^{1.7321}$
    (b) $7^{1.73}$ (e) $7^{\sqrt{3}}$
    (c) $7^{1.732}$

24. (a) $10^{3.1}$ (d) $10^{3.1416}$
    (b) $10^{3.14}$ (e) 10^{π}
    (c) $10^{3.142}$

25. (a) $e^{0.5}$ (d) $e^{-0.8}$
    (b) e^{-1} (e) $e^{\sqrt{\pi}}$
    (c) $e^{1.5}$

In Problems 26–29, graph each function. State the domain and the range of the function.

26. $f(x) = 9^x$ **27.** $g(x) = \left(\dfrac{1}{9}\right)^x$

28. $H(x) = 4^{x-2}$ **29.** $h(x) = 4^x - 2$

30. State the definition of the number e.

In Problems 31–36, solve each equation.

31. $2^x = 64$ **32.** $25^{x-2} = 125$

33. $27^x \cdot 3^{x^2} = 9^2$ **34.** $\left(\dfrac{1}{4}\right)^x = 16$

35. $(e^2)^{x-1} = e^x \cdot e^7$ **36.** $(2^x)^x = 512$

37. Future Value of Money Suppose that you deposit $2500 into a traditional IRA that pays 4.5% annual interest. How much money will you have after 25 years if interest is compounded

   (a) annually?   (b) quarterly?
   (c) monthly?   (d) daily?

38. Radioactive Decay The half-life of the radioactive gas radon is 3.5 days. Suppose a researcher possesses a 100-gram sample of radon gas. The amount A (in grams) of radon after t days is given by $A(t) = 100\left(\dfrac{1}{2}\right)^{\frac{t}{3.5}}$. How much radon gas is left in the sample after

   (a) 1 day?
   (b) 3.5 days?
   (c) 7 days?
   (d) 30 days?

39. A Population Model According to the U.S. Census Bureau, the population of Nevada in 2015 was 2.891 million people. In addition, the population of Nevada was growing at a rate of 1.6% per year. Assuming that this growth rate continues, the model $P(t) = 2.891(1.016)^{t-2015}$ represents the population (in millions of people) in year t. According to this model, what will be the population of Nevada

   (a) in 2025?
   (b) in 2030?

40. Newton's Law of Cooling A baker removes a cake from a 350°F oven and places it in a room whose temperature is 72°F. According to Newton's Law of Cooling, the temperature u (in °F) of the cake at time t (in minutes) can be modeled by $u(t) = 72 + 278e^{-0.0835t}$. According to this model, what will be the temperature of the cake

   (a) after 15 minutes?
   (b) after 30 minutes?

Section 8.3   Logarithmic Functions

KEY CONCEPTS

- **The Logarithmic Function to the Base a**
  $y = \log_a x$ is equivalent to $x = a^y$.

- **Properties of the Logarithmic Function of the Form**
  $f(x) = \log_a x, a > 0, a \neq 1$
  1. The domain is the set of all positive real numbers. The range is the set of all real numbers.
  2. There is no y-intercept. The x-intercept is $(0, 1)$.
  3. The graph of the logarithmic function contains the points $\left(\dfrac{1}{a}, -1\right)$, $(1, 0)$, and $(a, 1)$.

KEY TERMS

Logarithmic function
Natural logarithm function
Common logarithm function
Logarithmic equation
Intensity of a sound wave
Decibels
Loudness

You Should Be Able To...	EXAMPLE	Review Exercises
❶ Change exponential equations to logarithmic equations (p. 653)	Example 1	41–44
❷ Change logarithmic equations to exponential equations (p. 654)	Example 2	45–48
❸ Evaluate logarithmic functions (p. 654)	Examples 3 and 4	49–52
❹ Determine the domain of a logarithmic function (p. 656)	Example 5	53–56
❺ Graph logarithmic functions (p. 656)	Examples 6 and 7	57, 58
❻ Work with natural and common logarithms (p. 659)	Example 8	59–62
❼ Solve logarithmic equations (p. 659)	Examples 9 and 10	63–68
❽ Use logarithmic models that describe our world (p. 661)	Example 11	69, 70

In Problems 41–44, change each exponential equation to an equivalent equation involving a logarithm.

41. $3^4 = 81$ **42.** $4^{-3} = \dfrac{1}{64}$

43. $b^3 = 5$ **44.** $10^{3.74} = x$

In Problems 45–48, change each logarithmic equation to an equivalent equation involving an exponent.

45. $\log_8 2 = \dfrac{1}{3}$ **46.** $\log_5 18 = r$

47. $\ln(x + 3) = 2$ **48.** $\log x = -4$

In Problems 49–52, find the exact value of each logarithm without using a calculator.

49. $\log_8 128$ **50.** $\log_6 1$

51. $\log \dfrac{1}{100}$ **52.** $\log_9 27$

In Problems 53–56, find the domain of each function.

53. $f(x) = \log_2(x + 5)$

54. $g(x) = \log_8(7 - 3x)$

55. $h(x) = \ln(3x)$

56. $F(x) = \log_{1/3}(4x + 10)$

In Problems 57 and 58, graph each function.

57. $f(x) = \log_{5/2} x$ **58.** $g(x) = \log_{2/5} x$

In Problems 59–62, use a calculator to evaluate each expression. Round your answers to three decimal places.

59. $\ln 24$ **60.** $\ln \dfrac{5}{6}$

61. $\log 257$ **62.** $\log 0.124$

In Problems 63–68, solve each logarithmic equation.

63. $\log_7(4x - 19) = 2$ **64.** $\log_{1/3}(x^2 + 8x) = -2$

65. $\log_a \dfrac{4}{9} = -2$ **66.** $\ln e^{5x} = 30$

67. $\log(6 - 7x) = 3$ **68.** $\log_b 75 = 2$

69. Loudness of a Vacuum Cleaner A vacuum cleaner has an intensity level of 10^{-4} watt per square meter. What is the loudness, in decibels, of the vacuum cleaner?

70. The Great New Madrid Earthquake According to the United States Geological Survey, an earthquake on December 16, 1811, in New Madrid, Missouri, had a magnitude of approximately 8.0. What would have been the seismographic reading 100 kilometers from its epicenter?

Section 8.4   Properties of Logarithms

KEY CONCEPTS

- **Properties of Logarithms**

  For the following properties, a, M, and N are positive real numbers, with $a \neq 1$, and r is any real number.

  $\log_a a = 1$ $\log_a(MN) = \log_a M + \log_a N$

  $\log_a 1 = 0$ $\log_a M^r = r \log_a M$

  $a^{\log_a M} = M$ $\log_a\left(\dfrac{M}{N}\right) = \log_a M - \log_a N$

  $\log_a a^r = r$ If $M = N$, then $\log_a M = \log_a N$.

- **Change-of-Base Formula**

  If $a \neq 1$, $b \neq 1$, and M are positive real numbers, then $\log_a M = \dfrac{\log_b M}{\log_b a} = \dfrac{\log M}{\log a} = \dfrac{\ln M}{\ln a}$

You Should Be Able To...	EXAMPLE	Review Exercises
❶ Understand the properties of logarithms (p. 665)	Examples 1 through 3	71–76
❷ Write a logarithmic expression as a sum or difference of logarithms (p. 667)	Examples 4 through 8	77–80
❸ Write a logarithmic expression as a single logarithm (p. 670)	Examples 9 and 10	81–84
❹ Evaluate a logarithm whose base is neither 10 nor e (p. 671)	Examples 11 and 12	85–88

In Problems 71–76, use properties of logarithms to find the exact value of each expression. Do not use a calculator.

71. $\log_4 4^{21}$

72. $7^{\log_7 9.34}$

73. $\log_5 5$

74. $\log_9 1$

75. $\log_4 12 - \log_4 3$

76. $12^{\log_{12} 2 + \log_{12} 8}$

In Problems 77–80, write each expression as a sum and/or difference of logarithms. Write exponents as factors.

77. $\log_7\left(\dfrac{xy}{z}\right)$

78. $\log_3\left(\dfrac{81}{x^2}\right)$

79. $\log\left(1000r^4\right)$

80. $\ln\sqrt{\dfrac{x-1}{x}}$

In Problems 81–84, write each expression as a single logarithm.

81. $4\log_3 x + 2\log_3 y$

82. $\dfrac{1}{4}\ln x + \ln 7 - 2\ln 3$

83. $\log_2 3 - \log_2 6$

84. $\log_6(x^2 - 7x + 12) - \log_6(x - 3)$

In Problems 85–88, use the Change-of-Base Formula and a calculator to evaluate each logarithm. Round your answer to three decimal places.

85. $\log_6 50$

86. $\log_\pi 2$

87. $\log_{2/3} 6$

88. $\log_{\sqrt{5}} 20$

Section 8.5 Exponential and Logarithmic Equations

KEY CONCEPT

- In the following property, M, N, and a are positive real numbers, with $a \neq 1$.

  If $\log_a M = \log_a N$, then $M = N$

You Should Be Able To...	EXAMPLE	Review Exercises
❶ Solve logarithmic equations using the properties of logarithms (p. 674)	Examples 1 and 2	89–92
❷ Solve exponential equations (p. 676)	Examples 3 and 4	93–96
❸ Solve equations involving exponential models (p. 677)	Examples 5 and 6	97, 98

In Problems 89–96, solve each equation. Express irrational solutions in exact form and as a decimal rounded to three decimal places.

89. $3\log_4 x = \log_4 1000$

90. $\log_3(x + 7) + \log_3(x + 6) = \log_3 2$

91. $\ln(x + 2) - \ln x = \ln(x + 1)$

92. $\dfrac{1}{3}\log_{12} x = 2\log_{12} 2$

93. $2^x = 15$

94. $10^{3x} = 27$

95. $\dfrac{1}{3}e^{7x} = 13$

96. $3^x = 2^{x+1}$

97. Radioactive Decay The half-life of the radioactive gas radon is 3.5 days. Suppose a researcher possesses a 100-gram sample of radon gas. The amount A (in grams) of radon after t days is given by

$$A(t) = 100\left(\dfrac{1}{2}\right)^{\frac{t}{3.5}}.$$

(a) When will 75 grams of radon gas be left in the sample?

(b) When will 1 gram of radon gas be left in the sample?

98. A Population Model According to the U.S. Census Bureau, the population of Nevada in 2015 was 2.891 million people. In addition, the population of Nevada was growing at a rate of 1.6% per year. Assuming that this growth rate continues, the model

$P(t) = 2.891(1.016)^{t-2015}$ represents the population (in millions of people) in year t. According to this model, when will the population of Nevada be

(a) 3.13 million people?

(b) 5.285 million people?

Chapter 8 Test

Step-by-step test solutions are found on the Chapter Test Prep Videos available in MyMathLab®, on YouTube, or may be accessed using the QR code to the right.

1. Determine whether the following function is one-to-one:

   $\{(1,4),(3,2),(5,8),(-1,4)\}$

2. Find the inverse of $f(x) = 4x - 3$.

3. Approximate each number using a calculator. Express your answer rounded to three decimal places.

   (a) $3.1^{3.1}$

   (b) $3.14^{3.14}$

   (c) $3.142^{3.142}$

   (d) $3.1416^{3.1416}$

   (e) π^{π}

4. Change $4^x = 19$ to an equivalent equation involving a logarithm.

5. Change $\log_b x = y$ to an equivalent equation involving an exponent.

6. Find the exact value of each expression without using a calculator.

   (a) $\log_3\left(\dfrac{1}{27}\right)$

   (b) $\log 10{,}000$

7. Determine the domain of $f(x) = \log_5(7 - 4x)$.

In Problems 8–9, graph each function. State the domain and the range of the function.

8. $f(x) = 6^x$

9. $g(x) = \log_{1/9} x$

10. Use the properties of logarithms to find the exact value of each expression. Do not use a calculator.

    (a) $\log_7 7^{10}$

    (b) $3^{\log_3 15}$

11. Write the expression $\log_4 \dfrac{\sqrt{x}}{y^3}$ as a sum or difference of logarithms. Express exponents as factors.

12. Write the expression $4 \log M + 3 \log N$ as a single logarithm.

13. Use the Change-of-Base Formula and a calculator to evaluate $\log_{3/4} 10$. Round to three decimal places.

In Problems 14–20, solve each equation. Express irrational solutions in exact form and as a decimal rounded to three places.

14. $4^{x+1} = 2^{3x+1}$

15. $5^{x^2} \cdot 125 = 25^{2x}$

16. $\log_a 64 = 3$

17. $\log_2(x^2 - 33) = 8$

18. $2 \log_7(x - 3) = \log_7 3 + \log_7 12$

19. $3^{x-1} = 17$

20. $\log(x - 2) + \log(x + 2) = 2$

21. According to the U.S. Census Bureau, International Data Base, the population of Canada in 2016 was 35.9 million people. In addition, the population of Canada was growing at a rate of 0.8% annually. Assuming that this growth rate continues, the model $P(t) = 35.9(1.008)^{t-2016}$ represents the population (in millions of people) in year t.

    (a) According to the model, what will be the population of Canada in 2020?

    (b) According to the model, in what year will the population of Canada be 50 million?

22. Rustling leaves have an intensity of 10^{-11} watt per square meter. How many decibels are rustling leaves?

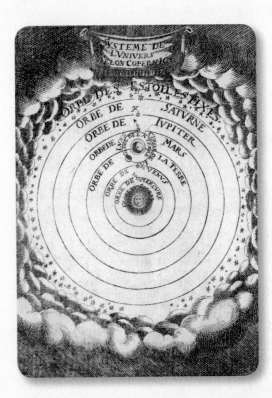

The Big Picture: Putting It Together

In Chapter 1, the Cartesian plane or rectangular coordinate system was introduced. This system lets us make connections between algebra and geometry. In this chapter, that connection will be developed by showing how geometric definitions of certain figures lead to algebraic equations.

This chapter begins by showing how to use algebra to find the distance between any two points in the Cartesian plane. This method is a direct consequence of the Pythagorean Theorem studied in Section 7.1. Knowing the distance formula enables us to present a complete discussion of the *conic sections*.

How could a circle be described to someone? A picture could be drawn to illustrate the verbal description. The Cartesian plane and the distance formula enable us to take the geometric definition of a circle and develop an algebraic equation whose graph represents a circle. This powerful connection between geometry and algebra lets us answer all types of interesting questions. The methods presented in this section form the foundation of an area of mathematics called analytic geometry.

Earth is the center of the universe. Although this statement seems ludicrous to us now, it was commonly believed until the 1500s. A Polish astronomer named Nicolaus Copernicus published a book entitled *De Revolutionibus Orbium Coelestium* (*On the Revolutions of the Celestial Spheres*), which stated that Earth and the other planets orbited the Sun in a circular motion. Copernicus's ideas were not readily accepted by the geocentrists, who held on to the belief that Earth is at the center of the universe. Copernicus's model of planetary motion was later improved upon by the German astronomer Johannes Kepler. In 1609, Kepler published *Astronomia nova* (New Astronomy) in which he proved that the orbit of Mars is an ellipse, with the Sun occupying one of its two foci. See Problems 45–48 in Section 9.4.

9.1 Distance and Midpoint Formulas

Objectives

1 Use the Distance Formula
2 Use the Midpoint Formula

Are You Prepared for This Section?

Before getting started, complete the following problems. If you get a problem wrong, go back to the section cited and review the material.

P1. Simplify: **(a)** $\sqrt{64}$ **(b)** $\sqrt{24}$ **(c)** $\sqrt{(x-2)^2}$ [Section 6.3, pp. 488–491]

P2. Find the length of the hypotenuse in a right triangle whose legs are 6 and 8.   [Section 7.1, pp. 549–550]

▶ **1** Use the Distance Formula

Section 1.5 discussed how to plot points in the Cartesian plane. The distance between any two points plotted in the Cartesian plane can be algebraically computed, so there is a connection between geometry (literally, "measuring the distance") and algebra. This distance is found using the Pythagorean Theorem.

EXAMPLE 1   **Finding the Distance between Two Points**

Find the distance d between the points $(2,4)$ and $(5,8)$.

Solution

First plot the points in the Cartesian plane and connect them with a line, as shown in Figure 1(a). To find the length d, draw a horizontal line through the point $(2,4)$ and a vertical line through the point $(5,8)$ and form a right triangle. The right angle of this triangle is at the point $(5,4)$. The right angle is at this point because in moving horizontally from the point $(2,4)$, there is no "up or down" movement. For this reason, the y-coordinate of the point at the right angle must be 4. Similarly, moving vertically straight down from the point $(5,8)$, reveals that the x-coordinate of the point at the right angle must be 5. See Figure 1(b).

Figure 1

(a)

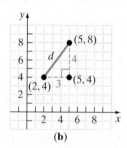

(b)

One leg of the triangle has length 3 (because $5 - 2 = 3$). The other leg has length 4 (because $8 - 4 = 4$). The Pythagorean Theorem says that

$$d^2 = 3^2 + 4^2$$
$$= 9 + 16$$
$$= 25$$

Square Root Property: $d = \pm 5$

Because d is the length of the hypotenuse, we discard the solution $d = -5$ and find that the hypotenuse is 5. Therefore, the distance between the points $(2,4)$ and $(5,8)$ is 5 units. ●

That was quite a bit of work to find the distance between $(2,4)$ and $(5,8)$. The *distance formula* lets us easily find the distance between any two points in the Cartesian plane.

The Distance Formula

The distance between two points $P_1 = (x_1, y_1)$ and $P_2 = (x_2, y_2)$, denoted by $d(P_1, P_2)$, is

$$d(P_1, P_2) = \sqrt{(x_2 - x_1)^2 + (y_2 - y_1)^2}$$

Figure 2

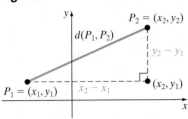

See Figure 2.

Let's justify the formula. Let point P_1 have coordinates (x_1, y_1), and let point P_2 have coordinates (x_2, y_2). If the line joining the points P_1 and P_2 is neither vertical nor horizontal, form a right triangle so that the vertex of the right angle is at the point $P_3 = (x_2, y_1)$, as shown in Figure 3(a).

The vertical distance from P_3 to P_2 is the absolute value of the difference of the y-coordinates, $|y_2 - y_1|$. The horizontal distance from P_1 to P_3 is the absolute value of the difference of the x-coordinates, $|x_2 - x_1|$. See Figure 3(b).

Figure 3

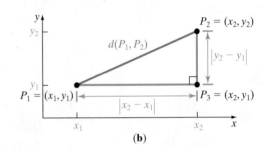

(a)   (b)

The distance $d(P_1, P_2)$ is the length of the hypotenuse of the right triangle, so by the Pythagorean Theorem, it follows that

$$[d(P_1, P_2)]^2 = |x_2 - x_1|^2 + |y_2 - y_1|^2$$
$$= (x_2 - x_1)^2 + (y_2 - y_1)^2$$

Take the square root of both sides: $d(P_1, P_2) = \sqrt{(x_2 - x_1)^2 + (y_2 - y_1)^2}$

Work Smart

Recall that
$$\sqrt{a^2 + b^2} \neq \sqrt{a^2} + \sqrt{b^2}.$$

If the line joining P_1 and P_2 is horizontal, then the y-coordinate of P_1 equals the y-coordinate of P_2; that is, $y_1 = y_2$. See Figure 4(a). In this case, the distance formula still works, because, for $y_1 = y_2$, it becomes

$$d(P_1, P_2) = \sqrt{(x_2 - x_1)^2 + 0^2}$$
$$= \sqrt{(x_2 - x_1)^2} = |x_2 - x_1|$$

A similar argument holds if the line joining P_1 and P_2 is vertical. See Figure 4(b). The distance formula works in all cases.

Figure 4

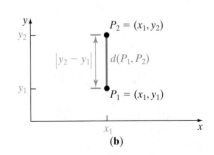

(a)   (b)

The distance between two points $P_1 = (x_1, y_1)$ and $P_2 = (x_2, y_2)$ cannot be negative; it can be 0 only when the two points are identical—that is, when $x_1 = x_2$ and $y_1 = y_2$.

Also, it does not matter whether the distance from P_1 to P_2 or from P_2 to P_1 is computed, because $(x_2 - x_1)^2 = (x_1 - x_2)^2$. This should seem reasonable because the distance from P_1 to P_2 equals the distance from P_2 to P_1.

Figure 5

EXAMPLE 2 | **Finding the Length of a Line Segment**

Find the length of the line segment shown in Figure 5.

Solution

The length of the line segment is the distance between the points $(-2, -1)$ and $(4, 3)$. Use the distance formula with $P_1 = (x_1, y_1) = (-2, -1)$ and $P_2 = (x_2, y_2) = (4, 3)$.

$$d = \sqrt{(x_2 - x_1)^2 + (y_2 - y_1)^2}$$

$x_1 = -2, y_1 = -1, x_2 = 4, y_2 = 3:$
$$= \sqrt{(4 - (-2))^2 + (3 - (-1))^2}$$
$$= \sqrt{6^2 + 4^2}$$
$$= \sqrt{36 + 16}$$
$$= \sqrt{52}$$
$$= 2\sqrt{13} \quad \text{Exact}$$
$$\approx 7.21 \quad \text{Approximate}$$

Quick ✓

1. The distance between two points $P_1 = (x_1, y_1)$ and $P_2 = (x_2, y_2)$, denoted by $d(P_1, P_2)$, is _____.

2. *True or False* The distance between two points can be a negative number.

In Problems 3 and 4, find the distance between the points.

3. $(3, 8)$ and $(0, 4)$ 4. $(-2, -5)$ and $(4, 7)$

The next example shows how algebra (the distance formula) can be used to solve geometry problems.

EXAMPLE 3 | **Using Algebra to Solve Geometry Problems**

Work Smart

When using the distance formula in applications, make sure that both the x-axis and the y-axis have the same unit of measure, such as inches.

Consider the three points $A = (-1, 1)$, $B = (2, -2)$, and $C = (3, 5)$.

  (a) Plot each point in the Cartesian plane and form the triangle ABC.
  (b) Find the length of each side of the triangle.
  (c) Verify that the triangle is a right triangle.
  (d) Find the area of the triangle.

Solution

  (a) Points $A, B,$ and C are plotted in Figure 6.
  (b) Use the distance formula to find the length of each side of the triangle.

$$d(A, B) = \sqrt{(2 - (-1))^2 + (-2 - 1)^2} = \sqrt{3^2 + (-3)^2}$$
$$= \sqrt{9 + 9} = \sqrt{18} = 3\sqrt{2}$$
$$d(A, C) = \sqrt{(3 - (-1))^2 + (5 - 1)^2} = \sqrt{4^2 + 4^2}$$
$$= \sqrt{16 + 16} = \sqrt{32} = 4\sqrt{2}$$
$$d(B, C) = \sqrt{(3 - 2)^2 + (5 - (-2))^2} = \sqrt{1^2 + 7^2} = \sqrt{1 + 49} = \sqrt{50} = 5\sqrt{2}$$

Figure 6

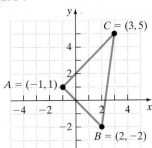

(c) The Pythagorean Theorem says that if you have a right triangle, the sum of squares of the two legs will equal the square of the hypotenuse. The converse of this statement is true as well. That is, if the sum of the squares of two sides of a triangle equals the square of the third side, then the triangle is a right triangle. Thus show that the sum of the squares of two sides of the triangle in Figure 6 equals the square of the third side. In Figure 6, the right angle appears to be at vertex A, which means the side opposite vertex A should be the hypotenuse of the triangle. Show that

$$[d(B, C)]^2 = [d(A, B)]^2 + [d(A, C)]^2$$

The distances from part (b) are known, so

$$(5\sqrt{2})^2 \overset{?}{=} (3\sqrt{2})^2 + (4\sqrt{2})^2$$

$(ab)^n = a^n \cdot b^n: \quad 25 \cdot 2 \overset{?}{=} 9 \cdot 2 + 16 \cdot 2$

$$50 \overset{?}{=} 18 + 32$$

$$50 = 50 \quad \text{True}$$

Because $[d(B, C)]^2 = [d(A, B)]^2 + [d(A, C)]^2$, the triangle ABC is a right triangle.

(d) The area of a triangle is $^1/_2$ times the product of the base and the height. From Figure 6, side AB forms the base and side AC forms the height. The length of side AB is $3\sqrt{2}$ and the length of side AC is $4\sqrt{2}$, so the area of triangle ABC is

$$\text{Area} = \frac{1}{2}(\text{Base})(\text{Height}) = \frac{1}{2}(3\sqrt{2})(4\sqrt{2}) = 12 \text{ square units}$$

Quick ✓

5. Consider the three points $A = (-2, -1)$, $B = (4, 2)$, and $C = (0, 10)$.
   (a) Plot each point in the Cartesian plane and form the triangle ABC.
   (b) Find the length of each side of the triangle.
   (c) Verify that the triangle is a right triangle.
   (d) Find the area of the triangle.

▶ ❷ Use the Midpoint Formula

Consider the two points $P_1 = (x_1, y_1)$ and $P_2 = (x_2, y_2)$ in the Cartesian plane. To find the point $M = (x, y)$ that is on the line segment joining P_1 and P_2 and is the same distance to each of these two points so that $d(P_1, M) = d(M, P_2)$, use the **midpoint formula**.

In Other Words

To find the midpoint of a line segment, average the x-coordinates of the endpoints and average the y-coordinates of the endpoints.

Midpoint Formula

The midpoint $M = (x, y)$ of the line segment from $P_1 = (x_1, y_1)$ to $P_2 = (x_2, y_2)$ is

$$M = \left(\frac{x_1 + x_2}{2}, \frac{y_1 + y_2}{2}\right)$$

EXAMPLE 4 **Finding the Midpoint of a Line Segment**

Find the midpoint of the line segment joining $P_1 = (-2, 3)$ and $P_2 = (4, 7)$. Plot the points P_1, P_2, and their midpoint. Check your answer.

(continued)

Solution

Substitute $x_1 = -2$, $y_1 = 3$, $x_2 = 4$, and $y_2 = 7$ into the midpoint formula. The coordinates (x, y) of the midpoint M are

$$x = \frac{x_1 + x_2}{2} = \frac{-2 + 4}{2} = \frac{2}{2} = 1$$

and

$$y = \frac{y_1 + y_2}{2} = \frac{3 + 7}{2} = \frac{10}{2} = 5$$

Thus the midpoint of the line segment joining $P_1 = (-2, 3)$ and $P_2 = (4, 7)$ is $M = (1, 5)$. See Figure 7.

Figure 7

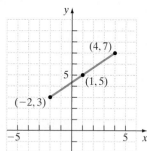

Quick ✓

6. The midpoint $M = (x, y)$ of the line segment from $P_1 = (x_1, y_1)$ to $P_2 = (x_2, y_2)$ is _____.

In Problems 7 and 8, find the midpoint of the line segment joining the points.

7. $(3, 8)$ and $(0, 4)$

8. $(-2, -5)$ and $(4, 10)$

9.1 Exercises MyMathLab®

Exercise numbers in green have complete video solutions in MyMathLab or may be accessed using the QR code to the right.

*Problems **1–8** are the Quick ✓s that follow the **EXAMPLES**.*

Building Skills

In Problems 9–24, find the distance $d(P_1, P_2)$ between the points P_1 and P_2. See Objective 1.

9.

10.

11.

12.

13. $P_1 = (2, 1); P_2 = (6, 4)$

14. $P_1 = (1, 3); P_2 = (4, 7)$

15. $P_1 = (-3, 2); P_2 = (9, -3)$

16. $P_1 = (-10, -3); P_2 = (14, 4)$

17. $P_1 = (-4, 2); P_2 = (2, 2)$

18. $P_1 = (-1, 2); P_2 = (-1, 0)$

19. $P_1 = (0, -3); P_2 = (-3, 3)$

20. $P_1 = (5, 0); P_2 = (-1, -4)$

21. $P_1 = (2\sqrt{2}, \sqrt{5}); P_2 = (5\sqrt{2}, 4\sqrt{5})$

22. $P_1 = (\sqrt{6}, -2\sqrt{2}); P_2 = (3\sqrt{6}, 10\sqrt{2})$

23. $P_1 = (0.3, -3.3); P_2 = (1.3, 0.1)$

24. $P_1 = (-1.7, 1.3); P_2 = (0.3, 2.6)$

In Problems 25–36, find the midpoint of the line segment formed by joining the points P_1 and P_2. See Objective 2.

25. $P_1 = (2, 2); P_2 = (6, 4)$

26. $P_1 = (1, 3); P_2 = (5, 7)$

27. $P_1 = (-3, 2); P_2 = (9, -4)$

28. $P_1 = (-10, -3); P_2 = (14, 7)$

29. $P_1 = (-4, 3); P_2 = (2, 4)$

30. $P_1 = (-1, 2); P_2 = (3, 9)$

31. $P_1 = (0, -3); P_2 = (-3, 3)$

32. $P_1 = (5, 0); P_2 = (-1, -4)$

33. $P_1 = \left(2\sqrt{2}, \sqrt{5}\right); P_2 = \left(5\sqrt{2}, 4\sqrt{5}\right)$

34. $P_1 = \left(\sqrt{6}, -2\sqrt{2}\right); P_2 = \left(3\sqrt{6}, 10\sqrt{2}\right)$

35. $P_1 = (0.3, -3.3); P_2 = (1.3, 0.1)$

36. $P_1 = (-1.7, 1.3); P_2 = (0.3, 2.6)$

Applying the Concepts

37. Consider the three points $A = (0, 3)$, $B = (2, 1)$, and $C = (6, 5)$.

   (a) Plot each point in the Cartesian plane and form the triangle ABC.

   (b) Find the length of each side of the triangle.

   (c) Verify that the triangle is a right triangle.

   (d) Find the area of the triangle.

38. Consider the three points $A = (0, 2)$, $B = (1, 4)$, and $C = (4, 0)$.

   (a) Plot each point in the Cartesian plane and form the triangle ABC.

   (b) Find the length of each side of the triangle.

   (c) Verify that the triangle is a right triangle.

   (d) Find the area of the triangle.

39. Consider the three points $A = (-2, -4)$, $B = (3, 1)$, and $C = (15, -11)$.

   (a) Plot each point in the Cartesian plane and form the triangle ABC.

   (b) Find the length of each side of the triangle.

   (c) Verify that the triangle is a right triangle.

   (d) Find the area of the triangle.

40. Consider the three points $A = (-2, 3)$, $B = (2, 0)$, and $C = (5, 4)$.

   (a) Plot each point in the Cartesian plane and form the triangle ABC.

   (b) Find the length of each side of the triangle.

   (c) Verify that the triangle is a right triangle.

   (d) Find the area of the triangle.

41. Find all points having an x-coordinate of 2 whose distance from the point $(5, 1)$ is 5.

42. Find all points having an x-coordinate of 4 whose distance from the point $(0, 3)$ is 5.

43. Find all points having a y-coordinate of -3 whose distance from the point $(2, 3)$ is 10.

44. Find all points having a y-coordinate of -3 whose distance from the point $(-4, 2)$ is 13.

45. The City of Chicago The city of Chicago's road system is set up like a Cartesian plane, where streets are indicated by the number of blocks they are from Madison Street and State Street. For example, Wrigley Field in Chicago is located at 1060 West Addison, which is 10 blocks west of State Street and 36 blocks north of Madison Street.

City of Chicago, Illinois

Wrigley Field
1060 West Addison

Guaranteed Rate Field
35th and Princeton

   (a) Find the distance "as the crow flies" from Madison and State Street to Wrigley Field. Use city blocks as the unit of measurement.

   (b) Guaranteed Rate Field, home of the White Sox, is located at 35th and Princeton, which is 3 blocks west of State Street and 35 blocks south of Madison. Find the distance "as the crow flies" from Madison and State Street to Guaranteed Rate Field.

   (c) Find the distance "as the crow flies" from Wrigley Field to Guaranteed Rate Field.

46. Baseball A major league baseball "diamond" is actually a square, 90 feet on a side (see the figure). Overlay a Cartesian plane on a major league baseball diamond so that the origin is at home plate, the positive x-axis lies in the direction from home plate to first base, and the positive y-axis lies in the direction from home plate to third base.

(continued)

(a) What are the coordinates of home plate, first base, second base, and third base? Use feet as the unit of measurement.

(b) Suppose the center fielder is located at $(310, 260)$. How far is he from second base?

(c) Suppose the shortstop is located at $(60, 100)$. How far is he from second base?

Extending the Concepts

47. Baseball Refer to Problem 46.

(a) Suppose the right fielder catches a fly ball at $(320, 20)$. How many seconds will it take him to throw the ball to second base if he can throw 130 feet per second? (*Hint:* Time = distance divided by speed.)

(b) Suppose a runner "tagging up" from first base can run 27 feet per second. Would you "send the runner" as the first base coach if the right fielder requires 0.8 second to catch and throw? Why?

48. Let $P = (x, y)$ be a point on the graph of $y = x^2 - 4$.

(a) Express the distance d from P to the point $(1, 2)$ as a function of x using the distance formula. See the figure for an illustration of the problem.

(b) What is d if $x = 0$?

(c) What is d if $x = 3$?

Explaining the Concepts

49. How is the distance formula related to the Pythagorean Theorem?

50. How can the distance formula be used to verify that a point is the midpoint of a line segment? What must be done in addition to using the distance formula to verify a midpoint? Why?

Synthesis Review

51. Evaluate 3^2. What is $\sqrt{9}$?

52. Evaluate 8^2. What is $\sqrt{64}$?

53. Evaluate $(-3)^4$. What is $\sqrt[4]{81}$?

54. Evaluate $(-3)^3$. What is $\sqrt[3]{-27}$?

55. Describe the relationship between raising a number to a positive integer power, n, and the nth root of a number.

9.2 Circles

Objectives

① Write the Standard Form of the Equation of a Circle

② Graph a Circle

③ Find the Center and Radius of a Circle Given an Equation in General Form

Are You Prepared for This Section?

Before getting started, complete the following problem. If you get the problem wrong, go back to the section cited and review the material.

P1. Complete the square in x: $x^2 - 8x$ [Section 7.1, pp. 546–547]

Conics, an abbreviation for conic sections, are curves that result from the intersection of a right circular cone and a plane. The four conics that will be studied are shown in Figure 8. These conics are *circles* (Figure 8(a)), *ellipses* (Figure 8(b)), *parabolas* (Figure 8(c)), and *hyperbolas* (Figure 8(d)).

Figure 8

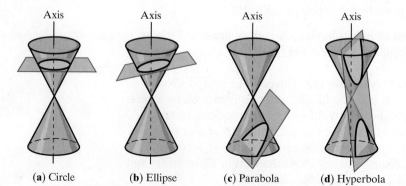

(a) Circle   **(b)** Ellipse   **(c)** Parabola   **(d)** Hyperbola

Prepared?...Answer
P1. $x^2 - 8x + 16$

Circles will be studied in this section, parabolas in Section 9.3, ellipses in Section 9.4, and hyperbolas in Section 9.5.

▶ ❶ Write the Standard Form of the Equation of a Circle

Because of the Cartesian plane, a geometric statement can be translated into an algebraic statement, and vice versa. Consider the following geometric statement that defines a circle.

Figure 9

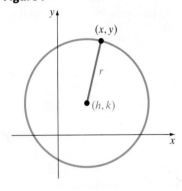

> **Definition**
>
> A **circle** is the set of all points in the Cartesian plane that are a fixed distance r from a fixed point (h, k). The fixed distance r is the **radius,** and the fixed point (h, k) is the **center** of the circle. See Figure 9.

To find an equation that has this graph, let (x, y) represent the coordinates of any point on a circle with radius r and center (h, k). Then the distance between the points (x, y) and (h, k) must always equal r. That is, by the distance formula,

$$\sqrt{(x - h)^2 + (y - k)^2} = r$$

Square both sides of the equation to get

$$(x - h)^2 + (y - k)^2 = r^2$$

and this is the equation of a circle.

Work Smart

Because r represents a distance, and distance is always positive, r is a positive number.

> **Definition**
>
> The **standard form of an equation of a circle** with radius r and center (h, k) is
>
> $$(x - h)^2 + (y - k)^2 = r^2$$

The standard form of a circle of radius r with center at the origin $(0, 0)$ is $x^2 + y^2 = r^2$.

EXAMPLE 1 **Writing the Standard Form of the Equation of a Circle**

Write the standard form of the equation of the circle with radius 4 and center $(2, -3)$.

Solution

Use the equation $(x - h)^2 + (y - k)^2 = r^2$ with $r = 4, h = 2$, and $k = -3$.

$$(x - 2)^2 + (y - (-3))^2 = 4^2$$
$$(x - 2)^2 + (y + 3)^2 = 16$$

●

Quick ✓

1. A _____ is the set of all points in the Cartesian plane that are a fixed distance r from a fixed point (h, k).

2. For a circle, the _____ is the distance from the center to any point on the circle.

In Problems 3 and 4, write the standard form of the equation of each circle whose radius is r and center is (h, k).

3. $r = 5$; $(h, k) = (2, 4)$

4. $r = \sqrt{2}$; $(h, k) = (-2, 0)$

▶ ② Graph a Circle

The graph of any equation of the form $(x - h)^2 + (y - k)^2 = r^2$ is a circle with radius r and center (h, k).

EXAMPLE 2 **Graphing a Circle**

Graph the equation: $(x + 2)^2 + (y - 3)^2 = 9$

Solution

The equation is of the form $(x - h)^2 + (y - k)^2 = r^2$, so its graph is a circle. To graph the equation, first identify the center and radius of the circle by comparing the equation to the standard form of the equation of a circle.

$$(x + 2)^2 + (y - 3)^2 = 9$$
$$(x - (-2))^2 + (y - 3)^2 = 3^2$$
$$(x - h)^2 + (y - k)^2 = r^2$$

Therefore, $h = -2$, $k = 3$, and $r = 3$. The circle has center $(-2, 3)$ and radius 3. To graph this circle, first plot the center $(-2, 3)$. Since the radius is 3 units, go 3 units in any direction from the center and find a point on the circle. It is easiest to find the four points left, right, up, and down from the center. These four points are $(-5, 3)$, $(1, 3)$, $(-2, 6)$, and $(-2, 0)$, respectively. Plot these points in Figure 10(a) and use them to draw the graph of the circle in Figure 10(b).

Figure 10

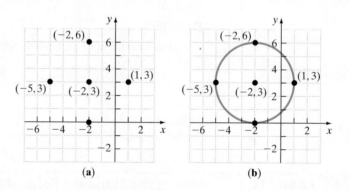

(a)   (b)

Notice that the graph of $(x + 2)^2 + (y - 3)^2 = (x - (-2))^2 + (y - 3)^2 = 9$ is the graph of $x^2 + y^2 = 9$ shifted 2 units left and 3 units up.

> **Quick ✓**
>
> **5.** *True or False* The center of the circle $(x + 1)^2 + (y - 3)^2 = 25$ is $(1, -3)$.
>
> **6.** *True or False* The center of the circle $x^2 + y^2 = 9$ is $(0, 0)$; its radius is 3.
>
> *In Problems 7 and 8, graph the circle.*
>
> **7.** $(x - 3)^2 + (y - 1)^2 = 4$ **8.** $(x + 5)^2 + y^2 = 16$

▶ ③ Find the Center and Radius of a Circle Given an Equation in General Form

Eliminate the parentheses from the standard form of the equation of the circle given in Example 2 to get

$$(x + 2)^2 + (y - 3)^2 = 9$$
$$\text{FOIL:} \quad x^2 + 4x + 4 + y^2 - 6y + 9 = 9$$
$$\text{Subtract 9 from both sides:} \quad x^2 + y^2 + 4x - 6y + 4 = 0$$

Any equation of the form

$$x^2 + y^2 + ax + by + c = 0$$

has a graph that is a circle, a graph that is a point, or no graph at all. For example, the graph of the equation $x^2 + y^2 = 0$ is the single point $(0, 0)$. The equation $x^2 + y^2 + 4 = 0$, or $x^2 + y^2 = -4$, has no graph, because the sum of squares of real numbers is never negative.

In Other Words

The standard form of the equation of a circle is $(x - h)^2 + (y - k)^2 = r^2$. The general form is $x^2 + y^2 + ax + by + c = 0$.

Definition

The **general form of the equation of a circle** is given by the equation

$$x^2 + y^2 + ax + by + c = 0$$

when the graph exists.

If an equation of a circle is in general form, use the method of completing the square to put the equation in standard form, $(x - h)^2 + (y - k)^2 = r^2$, in order to find the center and radius of the circle.

EXAMPLE 3 **Graphing a Circle Whose Equation Is in General Form**

Graph the equation: $x^2 + y^2 + 8x - 2y - 8 = 0$

Solution

To find the center and radius of the circle, put the equation in standard form by completing the square in both x and y (covered in Section 7.1). To do this, group the terms involving x, group the terms involving y, and put the constant on the right side of the equation by adding 8 to both sides.

$$(x^2 + 8x) + (y^2 - 2y) = 8$$

Now complete the square of each expression in parentheses. Remember, any number added to the left side of the equation must also be added to the right.

$$(x^2 + 8x + 16) + (y^2 - 2y + 1) = 8 + 16 + 1$$

$$\left(\frac{1}{2} \cdot 8\right)^2 = 16 \qquad \left(\frac{1}{2} \cdot (-2)\right)^2 = 1$$

Factor: $(x + 4)^2 + (y - 1)^2 = 25$

Figure 11

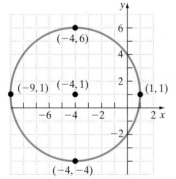

The equation is now in the standard form of the equation of a circle. The center is $(-4, 1)$ and the radius is 5. To graph the equation, plot the center and the four points $r = 5$ units to the left, to the right, above, and below the center. Then draw in the circle. See Figure 11.

Notice that the graph of $(x + 4)^2 + (y - 1)^2 = 25$ is the graph of $x^2 + y^2 = 25$ shifted 4 units left and 1 unit up.

Quick ✓

In Problems 9 and 10, graph each circle.

9. $x^2 + y^2 - 6x - 4y + 4 = 0$

10. $2x^2 + 2y^2 - 16x + 4y - 38 = 0$

9.2 Exercises MyMathLab®

Exercise numbers in green have complete video solutions in MyMathLab or may be accessed using the QR code to the right.

*Problems **1–10** are the **Quick ✓**s that follow the **EXAMPLES**.*

Building Skills

In Problems 11–14, find the center and radius of each circle. Write the standard form of the equation.

11.

12.

13.

14.

In Problems 15–26, write the standard form of the equation of each circle whose radius is r and center is (h, k). See Objective 1.

15. $r = 3$; $(h, k) = (0, 0)$ **16.** $r = 5$; $(h, k) = (0, 0)$

17. $r = 2$; $(h, k) = (1, 4)$ **18.** $r = 4$; $(h, k) = (3, 1)$

19. $r = 6$; $(h, k) = (-2, 4)$ **20.** $r = 3$; $(h, k) = (1, -4)$

21. $r = 4$; $(h, k) = (0, 3)$ **22.** $r = 2$; $(h, k) = (1, 0)$

23. $r = 5$; $(h, k) = (5, -5)$ **24.** $r = 4$; $(h, k) = (-4, 4)$

25. $r = \sqrt{5}$; $(h, k) = (1, 2)$ **26.** $r = \sqrt{7}$; $(h, k) = (5, 2)$

In Problems 27–36, find the center (h, k) and radius r of each circle. Graph each circle. See Objective 2.

27. $x^2 + y^2 = 36$

28. $x^2 + y^2 = 144$

29. $(x - 4)^2 + (y - 1)^2 = 25$

30. $(x - 2)^2 + (y - 3)^2 = 9$

31. $(x + 3)^2 + (y - 2)^2 = 81$

32. $(x - 5)^2 + (y + 2)^2 = 49$

33. $x^2 + (y - 3)^2 = 64$

34. $(x - 6)^2 + y^2 = 36$

35. $(x - 1)^2 + (y + 1)^2 = \dfrac{1}{4}$

36. $(x - 2)^2 + (y + 2)^2 = \dfrac{1}{4}$

In Problems 37–42, find the center (h, k) and radius r of each circle. Graph each circle. See Objective 3.

37. $x^2 + y^2 - 6x + 2y + 1 = 0$

38. $x^2 + y^2 + 2x - 8y + 8 = 0$

39. $x^2 + y^2 + 10x + 4y + 4 = 0$

40. $x^2 + y^2 + 4x - 12y + 36 = 0$

41. $2x^2 + 2y^2 - 12x + 24y - 72 = 0$

42. $2x^2 + 2y^2 - 28x + 20y + 20 = 0$

Mixed Practice

In Problems 43–48, find the standard form of the equation of each circle.

43. Center at the origin and containing the point $(4, -2)$

44. Center at $(0, 3)$ and containing the point $(3, 7)$

45. Center at $(-3, 2)$ and tangent to the y-axis

46. Center at $(2, -3)$ and tangent to the x-axis

47. With endpoints of a diameter at $(2, 3)$ and $(-4, -5)$

48. With endpoints of a diameter at $(-5, -3)$ and $(7, 2)$

Applying the Concepts

△ **49.** Find the area and circumference of the circle $(x - 3)^2 + (y - 8)^2 = 64$.

△ **50.** Find the area and circumference of the circle $(x - 1)^2 + (y - 4)^2 = 49$.

△ **51.** Find the area of the square in the figure.

△ **52.** Find the area of the shaded region in the figure.

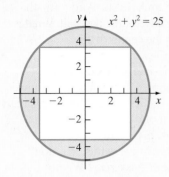

Extending the Concepts

53. Which of the following equation(s) might have the graph shown? (More than one answer may be possible.)

(a) $(x - 2)^2 + y^2 = 1$
(b) $x^2 + (y - 2)^2 = 1$
(c) $(x + 4)^2 + y^2 = 9$
(d) $(x - 5)^2 + y^2 = 25$
(e) $x^2 + y^2 - 8x + 7 = 0$
(f) $x^2 + y^2 + 10x + 18 = 0$

54. Which of the following equation(s) might have the graph shown? (More than one answer may be possible.)

(a) $(x - 2)^2 + (y + 3)^2 = 4$
(b) $(x - 3)^2 + (y - 4)^2 = 4$
(c) $(x + 3)^2 + (y + 4)^2 = 9$
(d) $(x - 5)^2 + (y - 5)^2 = 25$
(e) $x^2 + y^2 + 8x + 10y + 32 = 0$
(f) $x^2 + y^2 - 4x - 6y - 3 = 0$

Explaining the Concepts

55. How is the distance formula related to the definition of a circle?

56. Are circles functions? Why or why not?

57. Is $x^2 = 36 - y^2$ the equation of a circle? Why or why not? If so, what are the center and the radius?

58. Is $3x^2 - 12x + 3y^2 - 15 = 0$ the equation of a circle? Why or why not? If so, what is the center and radius?

Synthesis Review

In Problems 59–63, graph each function using either point plotting or properties of the function. For example, to graph a line, use the slope and y-intercept or the intercepts.

59. $f(x) = 4x - 3$

60. $2x + 5y = 20$

61. $g(x) = x^2 - 4x - 5$

62. $F(x) = -3x^2 + 12x - 12$

63. $G(x) = -2(x + 3)^2 - 5$

64. Present an argument that supports the approach you took to graphing each function in Problems 59–63. For example, in Problem 59, did you use point plotting? Intercepts? Slope? Which method did you use and why?

Technology Exercises

To graph a circle using a graphing calculator, first solve the equation for y. For example, to graph $(x + 2)^2 + (y - 3)^2 = 9$, solve for y as follows:

$$(x + 2)^2 + (y - 3)^2 = 9$$

Subtract $(x + 2)^2$ from both sides: $(y - 3)^2 = 9 - (x + 2)^2$

Take the square root of both sides: $y - 3 = \pm \sqrt{9 - (x + 2)^2}$

Add 3 to both sides: $y = 3 \pm \sqrt{9 - (x + 2)^2}$

Graph the top half $Y_1 = 3 + \sqrt{9 - (x + 2)^2}$ and the bottom half $Y_2 = 3 - \sqrt{9 - (x + 2)^2}$. To get an undistorted view of the graph, be sure to use the ZOOM SQUARE feature on the graphing calculator.

To graph a circle without having to solve the equation for y, you can use Geogebra. Figure 12 shows the graph of $(x + 2)^2 + (y - 3)^2 = 9$.

Figure 12

In Problems 65–74, graph each circle using technology. Compare your graphs to the graphs drawn by hand in Problems 27–36.

65. $x^2 + y^2 = 36$

66. $x^2 + y^2 = 144$

67. $(x - 4)^2 + (y - 1)^2 = 25$

68. $(x - 2)^2 + (y - 3)^2 = 9$

69. $(x + 3)^2 + (y - 2)^2 = 81$

70. $(x - 5)^2 + (y + 2)^2 = 49$

71. $x^2 + (y - 3)^2 = 64$

72. $(x - 6)^2 + y^2 = 36$

73. $(x - 1)^2 + (y + 1)^2 = \dfrac{1}{4}$

74. $(x - 2)^2 + (y + 2)^2 = \dfrac{1}{4}$

75. Exploration: Equation of a Circle *Open the "Equation of a Circle—Standard Form" Geogebra applet. The applet may be found using the QR code at the beginning of this section or through the Multimedia Library in MyMathLab. In the upper left corner are three "sliders." Use your mouse to* change the values of the *h, k,* or *r* sliders by clicking on the appropriate slider and dragging the value left or right. As you are moving the sliders, take note of how the position and equation of the circle change. If you are using a keyboard, it may be easier to use the arrow keys, instead of the mouse, to move the slider. First click on the slider with your mouse, and then use the left and right arrow keys to move the slider. Checking on the *Center* or *Radius* box will help you answer the following questions.

(a) Move the slider so that $h = 0$ and $k = 0$. Where is the center of the circle?

(b) Move the "*r*" slider so that *r* increases from 0 to 10. What happens to the radius of the circle as *r* increases? Change the value of *r* from 2 to 3 to 4. As you do this, what happens to the value on the right side of the equation in green?

(c) Move the "*r*" slider to 3. Move the "*h*" slider so that *h* increases from −10 to 10. What role does *h* play in the graph of the circle? In particular, what do you notice when *h* is negative? Zero? Positive?

(d) Move the "*r*" slider to 3. Move the "*k*" slider so that *k* increases from −10 to 10. What role does *k* play in the graph of the circle? In particular, what do you notice when *k* is negative? Zero? Positive?

(e) Check the "baseball" image box on the right-hand side of the applet. This will place a baseball image on the coordinate plane. Move the sliders to find the formula of the circle, in standard form, that best covers the outside edge of the baseball. What is this equation? What is the center of this circle? What is the radius?

(f) Check the "flower" image box on the right-hand side of the applet. This will place a flower image on the coordinate plane. Move the sliders to find the formula of the circle, in standard form, that best covers the outside edge of the center of the flower. What is this equation? What is the center of this circle? What is the radius?

(g) Check the "half dollar" image box on the right-hand side of the applet. This will place a half dollar image on the coordinate plane. Move the sliders to find the formula of the circle, in standard form, that best covers the outside edge of the half dollar. What is this equation? What is the center of this circle? What is the radius?

(h) Check the "lemon" image box on the right-hand side of the applet. This will place a lemon image on the coordinate plane. Move the sliders to find the formula of the circle, in standard form, that best covers the outside edge of the lemon. What is this equation? What is the center of this circle? What is the radius?

9.3 Parabolas

Objectives

1. Graph Parabolas Whose Vertex Is the Origin
2. Find the Equation of a Parabola
3. Graph a Parabola Whose Vertex Is Not the Origin
4. Solve Applied Problems Involving Parabolas

Are You Prepared for This Section?

Before getting started, complete the following problems. If you get a problem wrong, go back to the section cited and review the material.

P1. Identify the vertex and axis of symmetry of $f(x) = -3(x + 4)^2 - 5$. Does the parabola open up or down? Why?   [Section 7.4, pp. 583–586]

P2. Identify the vertex and axis of symmetry of the quadratic function $f(x) = 2x^2 - 8x + 1$. Does the parabola open up or down? Why?   [Section 7.5, pp. 589–594]

P3. Complete the square of $x^2 - 12x$.   [Section 7.1, pp. 546–547]

P4. Solve: $(x - 3)^2 = 25$ [Section 7.1, pp. 543–546]

A discussion of parabolas began in Section 7.4 when quadratic functions were studied. Refresh your memory with the summary on the next page.

Prepared?...Answers **P1.** Vertex: $(-4, -5)$; axis of symmetry: $x = -4$; opens down since $a = -3 < 0$.
P2. Vertex: $(2, -7)$; axis of symmetry: $x = 2$; opens up since $a = 2 > 0$.
P3. $x^2 - 12x + 36$ **P4.** $\{-2, 8\}$

> **Summary** Parabolas That Open Up or Down
>
> The graph of $y = a(x - h)^2 + k$ or $y = ax^2 + bx + c$ is a parabola that
>
> **1.** opens up if $a > 0$ and opens down if $a < 0$.
> **2.** has vertex (h, k).
> **3.** has a vertex whose x-coordinate is $x = -\dfrac{b}{2a}$. The y-coordinate is found by evaluating the equation at the x-coordinate of the vertex.
>
>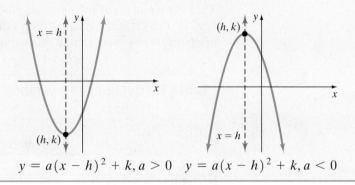
>
> $$y = a(x - h)^2 + k, a > 0 \qquad y = a(x - h)^2 + k, a < 0$$

Figure 13

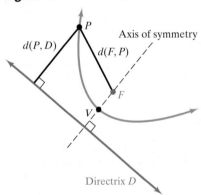

Ideas about parabolas in Sections 7.4 and 7.5 relied on algebra. In this chapter, parabolas (and other conic sections) will be examined from a geometric point of view.

> **Definition**
>
> A **parabola** is the collection of all points P in the plane that are the same distance from a fixed point F as they are from a fixed line D. The point F is called the **focus** of the parabola, and the line D is its **directrix.** In other words, a parabola is the set of points P for which
>
> $$d(F, P) = d(P, D)$$

Figure 13 shows a parabola. The line through the focus F and perpendicular to the directrix D is the **axis of symmetry** of the parabola. The point of intersection of the parabola with its axis of symmetry is the **vertex** V.

Figure 14

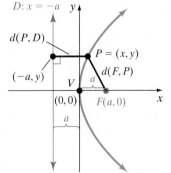

▶ ① Graph Parabolas Whose Vertex Is the Origin

Develop an equation for the parabola based on the definition just given. For example, the distance from the focus to the vertex (a point on the parabola) must equal the distance from the vertex to the directrix. Let a represent the distance from the focus to the vertex. To develop an equation for a parabola, start by looking at parabolas whose vertex is at the origin. Consider four possibilities—parabolas that open left, parabolas that open right, parabolas that open up, and parabolas that open down.

First obtain the equation of a parabola whose vertex is at the origin and opens right. To do this, position a parabola in a Cartesian plane so that its vertex is at the origin, $(0, 0)$, the focus is on the positive x-axis, and the directrix is a vertical line in quadrants II and III. Because a represents the distance from the vertex to the focus, the focus is located at $(a, 0)$. Also, because the distance from the vertex to the directrix is a, the directrix is the line $x = -a$. See Figure 14. If $P = (x, y)$ is any point on the parabola, then the distance from P to the focus F, $(a, 0)$, must equal the distance from P to the directrix, $x = -a$. Use the point $(-a, y)$ on the directrix in the distance formula. That is,

$$d(F, P) = d(P, D)$$

Use the distance formula: $\sqrt{(x - a)^2 + (y - 0)^2} = \sqrt{(x - (-a))^2 + (y - y)^2}$

Square both sides; simplify: $(x - a)^2 + y^2 = (x + a)^2$

Multiply out binomials: $x^2 - 2ax + a^2 + y^2 = x^2 + 2ax + a^2$

Combine like terms: $y^2 = 4ax$

Thus the equation of the parabola whose vertex is at the origin and opens to the right is $y^2 = 4ax$. Obtain the equations of the three other possibilities (opens left, opens up, or opens down) using similar reasoning. These possibilities are summarized in Table 1.

Table 1 Equations of a Parabola: Vertex at (0, 0); Focus on an Axis; $a > 0$

Vertex	Focus	Directrix	Equation	Axis of Symmetry	Opens
$(0, 0)$	$(a, 0)$	$x = -a$	$y^2 = 4ax$	x-axis	Right
$(0, 0)$	$(-a, 0)$	$x = a$	$y^2 = -4ax$	x-axis	Left
$(0, 0)$	$(0, a)$	$y = -a$	$x^2 = 4ay$	y-axis	Up
$(0, 0)$	$(0, -a)$	$y = a$	$x^2 = -4ay$	y-axis	Down

The graphs of the four parabolas are given in Figure 15.

Figure 15

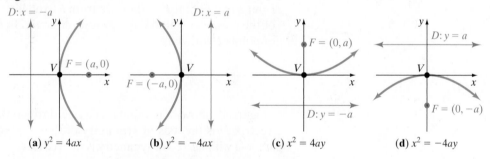

(a) $y^2 = 4ax$ (b) $y^2 = -4ax$ (c) $x^2 = 4ay$ (d) $x^2 = -4ay$

EXAMPLE 1   **Graphing a Parabola That Opens Left**

Graph the equation $y^2 = -12x$.

Solution

The equation $y^2 = -12x$ is of the form $y^2 = -4ax$, where $-4a = -12$, so $a = 3$. The graph of the equation is a parabola with vertex at $(0, 0)$ and focus at $(-a, 0) = (-3, 0)$, so the parabola opens to the left. The directrix is the line $x = 3$. To graph the parabola, it helps to plot the two points on the graph above and below the focus. Because the points are directly above and below the focus, let $x = -3$ in the equation $y^2 = -12x$ and solve for y.

$$y^2 = -12x$$

Let $x = -3$: $= -12(-3)$

$$= 36$$

Take the square root of both sides: $y = \pm 6$

The points on the parabola above and below the focus are $(-3, 6)$ and $(-3, -6)$. These points help in graphing the parabola because they determine the "opening." See Figure 16.

Figure 16

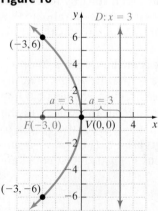

Quick ✓

1. A _____ is the collection of all points P in the plane that are the same distance from a fixed point F as they are from a fixed line D.

2. The point of intersection of the parabola with its axis of symmetry is called the _____.

3. The line through the focus and perpendicular to the directrix is called the ___ __ _____.

In Problems 4 and 5, graph the equation.

4. $y^2 = 8x$ **5.** $y^2 = -20x$

EXAMPLE 2   **Graphing a Parabola That Opens Up**

Graph the equation $x^2 = 8y$.

Solution

The equation $x^2 = 8y$ is of the form $x^2 = 4ay$, where $4a = 8$, so $a = 2$. The graph of the equation is a parabola with vertex at $(0, 0)$ and focus at $(0, a) = (0, 2)$, so the parabola opens up. The directrix is the line $y = -2$. To graph the parabola, plot the two points on the graph to the left and right of the focus. Let $y = 2$ in the equation $x^2 = 8y$ and solve for x.

$$x^2 = 8y = 8(2) = 16$$

Take the square root of both sides: $x = \pm 4$

The points on the parabola to the left and right of the focus are $(-4, 2)$ and $(4, 2)$. See Figure 17.   ●

Figure 17

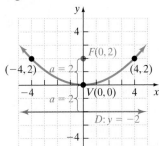

Quick ✓

In Problems 6 and 7, graph the equation.

6. $x^2 = 4y$ **7.** $x^2 = -12y$

▶ ❷ **Find the Equation of a Parabola**

Now use information about a parabola to obtain its equation.

EXAMPLE 3   **Finding an Equation of a Parabola**

Find an equation of the parabola with vertex at $(0, 0)$ and focus at $(5, 0)$. Graph the equation.

Figure 18

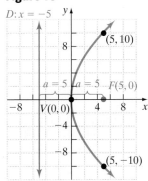

Solution

The distance from the vertex $(0, 0)$ to the focus $(5, 0)$ is $a = 5$. Because the focus lies on the positive x-axis, the parabola opens to the right. This means the equation of the parabola is of the form $y^2 = 4ax$ with $a = 5$:

$$y^2 = 4(5)x$$
$$= 20x$$

Figure 18 shows the graph of $y^2 = 20x$.   ●

EXAMPLE 4 **Finding the Equation of a Parabola**

Find the equation of a parabola with vertex at $(0, 0)$ if its axis of symmetry is the y-axis and its graph contains the point $(4, 3)$. Graph the equation.

Solution

Which of the four equations of parabolas listed in Table 1 should be used? The vertex is at the origin and the axis of symmetry is the y-axis, so the parabola opens either up or down. Because the graph contains the point $(4, 3)$, which is in quadrant I, the parabola must open up. Therefore, the equation of the parabola is of the form $x^2 = 4ay$. Because the point $(4, 3)$ is on the parabola, let $x = 4$ and $y = 3$ in the equation $x^2 = 4ay$ to determine a.

$$x^2 = 4ay$$

$$x = 4, y = 3: \quad 4^2 = 4a(3)$$

$$16 = 12a$$

Divide both sides by 12: $\quad a = \dfrac{16}{12} = \dfrac{4}{3}$

The equation of the parabola is

$$x^2 = 4\left(\dfrac{4}{3}\right)y, \quad \text{or} \quad x^2 = \dfrac{16}{3}y$$

Figure 19

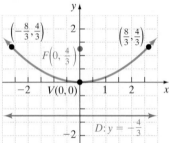

With $a = \dfrac{4}{3}$, the focus is $\left(0, \dfrac{4}{3}\right)$ and the directrix is the line $y = -\dfrac{4}{3}$. Let $y = \dfrac{4}{3}$ to find points left and right of the focus to determine the "opening," so the points $\left(-\dfrac{8}{3}, \dfrac{4}{3}\right)$ and $\left(\dfrac{8}{3}, \dfrac{4}{3}\right)$ are on the graph. Figure 19 shows the graph of the parabola. ●

Quick ✓

In Problems 8 and 9, find the equation of the parabola described. Graph the equation.

8. Vertex at $(0, 0)$; focus at $(0, -8)$

9. Vertex at $(0, 0)$; axis of symmetry the x-axis; contains the point $(3, 2)$

▶ ❸ **Graph a Parabola Whose Vertex Is Not the Origin**

If a parabola with vertex at the origin and axis of symmetry along a coordinate axis is shifted horizontally h units and then vertically k units, the result is a parabola with vertex at (h, k) and axis of symmetry parallel to either the x- or the y-axis. The equations of these parabolas have the same form as those whose vertex is at the origin, except that x is replaced with $x - h$ (the horizontal shift) and y is replaced with $y - k$ (the vertical shift). Table 2 gives the equations of the four parabolas. Figure 20(a)–(d) illustrates the graphs for $h > 0$ and $k > 0$.

Table 2 Parabolas with Vertex at (h, k); Axis of Symmetry Parallel to a Coordinate Axis; $a > 0$

Vertex	Focus	Directrix	Equation	Axis of Symmetry	Opens
(h, k)	$(h + a, k)$	$x = h - a$	$(y - k)^2 = 4a(x - h)$	Parallel to x-axis	Right
(h, k)	$(h - a, k)$	$x = h + a$	$(y - k)^2 = -4a(x - h)$	Parallel to x-axis	Left
(h, k)	$(h, k + a)$	$y = k - a$	$(x - h)^2 = 4a(y - k)$	Parallel to y-axis	Up
(h, k)	$(h, k - a)$	$y = k + a$	$(x - h)^2 = -4a(y - k)$	Parallel to y-axis	Down

Figure 20

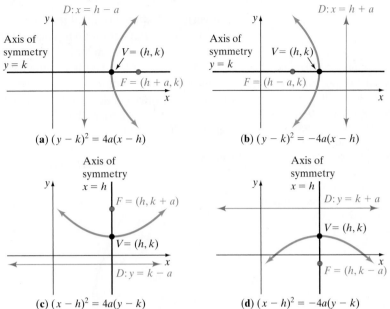

(a) $(y - k)^2 = 4a(x - h)$ **(b)** $(y - k)^2 = -4a(x - h)$

(c) $(x - h)^2 = 4a(y - k)$ **(d)** $(x - h)^2 = -4a(y - k)$

EXAMPLE 5   **Graphing a Parabola Whose Vertex Is Not the Origin**

Graph the parabola $x^2 - 2x + 8y + 25 = 0$.

Solution

Notice the equation is not in any of the forms given in Table 2. Complete the square in x to write the equation in standard form.

$$x^2 - 2x + 8y + 25 = 0$$

Isolate the terms involving x: $x^2 - 2x = -8y - 25$

Complete the square: $x^2 - 2x + 1 = -8y - 25 + 1$

Simplify: $x^2 - 2x + 1 = -8y - 24$

Factor: $(x - 1)^2 = -8(y + 3)$

Work Smart

Notice that the graph of $(x - 1)^2 = -8(y + 3)$ is the graph of $x^2 = -8y$ shifted right 1 unit and down 3 units.

The equation is of the form $(x - h)^2 = -4a(y - k)$. This is a parabola that opens down with vertex $(h, k) = (1, -3)$. Since $-4a = -8$, then $a = 2$. Because the parabola opens down, the focus is $a = 2$ units below the vertex at $(1, -5)$. To find two points on the graph to the left and right of the focus, let $y = -5$ in the equation of the parabola.

Let $y = -5$: $(x - 1)^2 = -8(-5 + 3)$

$(x - 1)^2 = -8(-2)$

$(x - 1)^2 = 16$

Take the square root of both sides: $x - 1 = \pm 4$

Add 1 to both sides: $x = 1 \pm 4$

$x = 1 - 4$ or $x = 1 + 4$

$x = -3$ or $x = 5$

Figure 21

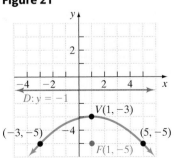

The points $(-3, -5)$ and $(5, -5)$ are on the graph of the parabola. The directrix is $a = 2$ units above the vertex, so $y = -1$ is the directrix. Figure 21 shows the graph.   ●

Quick ✓

10. The vertex of the parabola $(x + 3)^2 = -14(y - 2)$ is _____.

11. Graph the parabola $y^2 - 4y - 12x - 32 = 0$.

▶ ❹ Solve Applied Problems Involving Parabolas

The variety of applications of parabolas is astounding. Suppose that a mirror is shaped like a parabola. If a light bulb is placed at the focus of the parabola, then all the rays from the bulb reflect off the mirror in lines parallel to the axis of symmetry. This concept is used in the design of car headlights, flashlights, and searchlights. See Figure 22.

Suppose that rays of light are received by a parabola. When the rays strike the surface of a parabolic mirror whose axis of symmetry is parallel to these rays, they are all reflected to a single point—the focus. This idea is used in the design of some telescopes and satellite dishes. See Figure 23.

Work Smart

Which is the graph of a parabola and which is the graph of a circle?
$x^2 + y^2 - 2y + 3 = 0$
$2y^2 - 12y - x + 22 = 0$
How can you tell the difference between the equation of a circle and the equation of a parabola?
The equation of a parabola has either an x^2-term or a y^2-term but not both. The equation of a circle has the sum of an x^2-term and a y^2-term.

Figure 22

Rays of light

Light at focus

Figure 23

EXAMPLE 6

A Satellite Dish

A satellite dish is shaped like a parabola. See Figure 24(a). Signals strike the surface of the dish and are reflected to a single point, where the receiver of the dish is located. If a satellite dish is 2 feet across at its opening and 6 inches deep (0.5 foot) at its center, at what position should the receiver be placed?

Solution

To find where to locate the receiver of the satellite dish, locate its focus. Draw a parabola in a Cartesian plane so that the vertex is at the origin and the focus is on the positive y-axis. The width of the dish is 2 feet across and its height is 0.5 feet, so two points on the graph are known, as indicated in Figure 24(b).

Figure 24

(a)

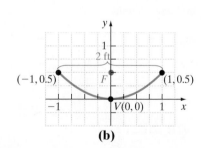

(b)

The parabola is an equation of the form $x^2 = 4ay$. Since $(1, 0.5)$ is a point on the graph, the following can be found:

$$x = 1, y = 0.5: \quad 1^2 = 4a(0.5)$$
$$1 = 2a$$

Divide both sides by 2: $\quad a = \dfrac{1}{2}$

The receiver should be located $\dfrac{1}{2}$ foot from the base of the dish along its axis of symmetry. ●

Quick ✓

12. A satellite dish is shaped like a parabola. Signals strike the surface of the dish and are reflected to a single point, where the receiver of the dish is located. If the dish is 4 feet across at its opening and 6 inches deep (0.5 foot) at its center, at what position should the receiver be placed?

9.3 Exercises MyMathLab® Exercise numbers in green have complete video solutions in MyMathLab or may be accessed using the QR code to the right.

*Problems **1–12** are the Quick ✓ s that follow the EXAMPLES.*

Building Skills

In Problems 13–20, the graph of a parabola is given. Match each graph to its equation.

(a) $y^2 = 8x$ **(b)** $y^2 = -8x$

(c) $x^2 = 8y$ **(d)** $x^2 = -8y$

(e) $(y - 2)^2 = 8(x + 1)$ **(f)** $(y - 2)^2 = -8(x + 1)$

(g) $(x + 1)^2 = 8(y - 2)$ **(h)** $(x + 1)^2 = -8(y - 2)$

13.

14.

15.

16.

17.

18.

19.

20.

In Problems 21–26, find the vertex, focus, and directrix of each parabola. Graph the parabola. See Objective 1.

21. $x^2 = 24y$ **22.** $x^2 = 28y$

23. $y^2 = -6x$ **24.** $y^2 = 10x$

25. $x^2 = -8y$ **26.** $x^2 = -16y$

In Problems 27–36, find the equation of the parabola described. Graph the parabola. See Objective 2.

27. Vertex at $(0, 0)$; focus at $(5, 0)$

28. Vertex at $(0, 0)$; focus at $(0, 5)$

29. Vertex at $(0, 0)$; focus at $(0, -6)$

30. Vertex at $(0, 0)$; focus at $(-8, 0)$

31. Vertex at $(0, 0)$; contains the point $(6, 6)$; axis of symmetry the *y*-axis

32. Vertex at $(0, 0)$; contains the point $(2, 2)$; axis of symmetry the *x*-axis

33. Vertex at $(0, 0)$; directrix the line $y = 3$

34. Vertex at $(0, 0)$; directrix the line $x = -4$

35. Focus at $(-3, 0)$; directrix the line $x = 3$

36. Focus at $(0, -2)$; directrix the line $y = 2$

In Problems 37 and 38, write an equation for each parabola. See Objective 2.

37.

38.
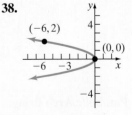

In Problems 39–50, find the vertex, focus, and directrix of each parabola. Graph the parabola. See Objective 3.

39. $(x - 2)^2 = 4(y - 4)$ **40.** $(x + 4)^2 = -4(y - 1)$

41. $(y + 3)^2 = -8(x + 2)$ **42.** $(y - 2)^2 = 12(x + 5)$

43. $(x + 5)^2 = -20(y - 1)$ **44.** $(x - 6)^2 = 2(y - 2)$

45. $x^2 + 4x + 12y + 16 = 0$ **46.** $x^2 + 2x - 8y + 25 = 0$

47. $y^2 - 8y - 4x + 20 = 0$ **48.** $y^2 - 8y + 16x - 16 = 0$

49. $x^2 + 10x + 6y + 13 = 0$ **50.** $x^2 - 4x + 10y + 4 = 0$

Applying the Concepts

51. A Headlight The headlight of a car is in the shape of a parabola. Its diameter is 4 inches and its depth is 1 inch. How far from the vertex should the light bulb be placed so that the rays will be reflected parallel to the axis?

52. A Headlight The headlight of a car is in the shape of a parabola. Suppose the engineers have designed the headlight to be 5 inches wide and want the bulb to be placed at the focus 1 inch from the vertex. What is the depth of the headlight?

---

53. Suspension Bridge The cables of a suspension bridge are in the shape of a parabola, as shown in the figure. The towers supporting the cable are 500 feet apart and 60 feet high. If the cables touch the road surface midway between the towers, what is the height of the cable at a point 150 feet from the center of the bridge?

54. Suspension Bridge The cables of a suspension bridge are in the shape of a parabola. The towers supporting the cable are 400 feet apart and 80 feet high. If the cables touch the road surface midway between the towers, what is the height of the cable at a point 100 feet from the center of the bridge?

55. Parabolic Arch Bridge A bridge is built in the shape of a parabolic arch. The arch has a span of 100 feet and a maximum height of 30 feet. See the illustration. Choose a suitable rectangular coordinate system and find the height of the arch at distances of 10, 30, and 50 feet from the center.

56. Parabolic Arch Bridge A bridge is to be built in the shape of a parabolic arch and is to have a span of 120 feet. The height of the arch at a distance of 30 feet from the center is to be 15 feet. Find the height of the arch at its center.

Extending the Concepts

In Problems 57–60, write an equation for each parabola.

57.

58.

59.

60.

61. For the parabola $x^2 = 8y$, **(a)** verify that $(4, 2)$ is a point on the parabola, and **(b)** show that the distance from the focus to $(4, 2)$ equals the distance from $(4, 2)$ to the directrix.

62. For the parabola $(y - 3)^2 = 12(x + 1)$, **(a)** verify that $(2, 9)$ is a point on the parabola, and **(b)** show that the distance from the focus to $(2, 9)$ equals the distance from $(2, 9)$ to the directrix.

Explaining the Concepts

63. The distance from a point on a parabola to its focus is 8 units. What is the distance from the same point on the parabola to the directrix? Why?

64. Write down the four equations that are parabolas with vertex at (h, k).

65. Draw a parabola and label the vertex, axis of symmetry, focus, and directrix.

66. Explain the difference between the discussion of parabolas presented in this section and the discussion presented in Sections 7.4 and 7.5.

Synthesis Review

67. Graph $y = (x + 3)^2$ using the methods of Section 7.4.

68. Graph $y = (x + 3)^2 = x^2 + 6x + 9$ using the methods of Section 7.5.

69. Graph $y = (x + 3)^2$ using the methods of this section.

70. Graph $4(y + 2) = (x - 2)^2$ using the methods of Section 7.4. (*Hint:* Write the equation in the form $y = a(x - h)^2 + k$.)

71. Graph $4(y + 2) = (x - 2)^2$ using the methods of this section.

72. Compare and contrast the methods of graphing a parabola using the approach in Section 7.4 and the approach in this section. Which do you prefer? Why?

Technology Exercises

To graph a parabola using a graphing calculator, solve the equation for y, just as in graphing circles. This is fairly straightforward if the equation is in the form given in either Table 1 or Table 2. If the equation is not in the form given in Table 1 or Table 2 and it is a parabola that opens up or down, then solving for y is also straightforward. However, if it is a parabola that opens left or right, graph the parabola in two "pieces"—the top half and the bottom half. Use the quadratic formula to accomplish this task. Consider the equation $y^2 + 8y + x - 4 = 0$. This equation is quadratic in y as shown:

$$y^2 + 8y + x - 4 = 0$$

$$\boxed{a = 1} \quad \boxed{b = 8} \quad \boxed{c = x - 4}$$

Use the quadratic formula and obtain

$$y = \frac{-8 \pm \sqrt{8^2 - 4(1)(x - 4)}}{2(1)}$$

so graph

$$Y_1 = \frac{-8 - \sqrt{64 - 4(x - 4)}}{2} \quad \text{and}$$

$$Y_2 = \frac{-8 + \sqrt{64 - 4(x - 4)}}{2}.$$

as shown in Figure 25.

To graph a parabola without having to solve the equation for y, you can use Desmos

Figure 25

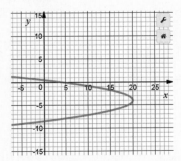

In Problems 73–84, graph each parabola using technology.

73. $x^2 = 24y$ **74.** $x^2 = -8y$

75. $y^2 = -6x$ **76.** $y^2 = 10x$

77. $(x - 2)^2 = 4(y - 4)$ **78.** $(x + 4)^2 = -4(y - 1)$

79. $(y + 3)^2 = -8(x + 2)$ **80.** $(y - 2)^2 = 12(x + 5)$

81. $x^2 + 4x + 12y + 16 = 0$ **82.** $x^2 + 2x - 8y + 25 = 0$

83. $y^2 - 8y - 4x + 20 = 0$ **84.** $y^2 - 8y + 16x - 16 = 0$

9.4 Ellipses

Objectives

1. Graph an Ellipse Whose Center Is the Origin
2. Find the Equation of an Ellipse Whose Center Is the Origin
3. Graph an Ellipse Whose Center Is Not the Origin
4. Solve Applied Problems Involving Ellipses

Are You Prepared for This Section?

Before getting started, complete the following problems. If you get a problem wrong, go back to the section cited and review the material.

P1. Complete the square of $x^2 + 10x$. [Section 7.1, pp. 546–547]

P2. Graph $f(x) = (x + 2)^2 - 1$ using transformations. [Section 7.4, p. 581]

⓵ Graph an Ellipse Whose Center Is the Origin

An ellipse is a conic section that is obtained through the intersection of a plane and a cone. See Figure 8(b) in Section 9.2.

> **Definition**
>
> An **ellipse** is the collection of points in the plane such that the sum of the distances from two fixed points, called the **foci,** is a constant.

Figure 26

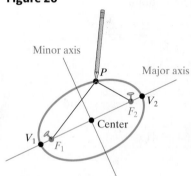

This definition enables us to physically draw an ellipse. To do this, find a piece of string (the length of the string is the constant in the definition). Now stick two thumbtacks on a piece of cardboard so that the distance between them is less than the length of the string. The two thumbtacks represent the foci of the ellipse. Now attach the ends of the string to the thumbtacks and, using the point of a pencil, pull the string taut. Keeping the string taut, rotate the pencil around the two thumbtacks. The pencil traces out an ellipse, as shown in Figure 26.

In Figure 26, the foci are labeled F_1 and F_2. The line containing the foci is called the **major axis.** The midpoint of the line segment joining the foci is called the **center** of the ellipse. The line through the center and perpendicular to the major axis is called the **minor axis.**

The two points of intersection of the ellipse and the major axis are the **vertices,** V_1 and V_2, of the ellipse. The distance from one vertex to the other is called the **length of the major axis.**

In this text, consider only ellipses whose major axis is parallel to (or coincides with) either the x-axis or the y-axis. The equations of an ellipse with center at the origin are given in Table 3.

Table 3 Ellipses with Center at the Origin

Center	Major Axis	Foci	Vertices	Equation
$(0,0)$	x-axis	$(-c, 0)$ and $(c, 0)$	$(-a, 0)$ and $(a, 0)$	$\dfrac{x^2}{a^2} + \dfrac{y^2}{b^2} = 1$
$(0,0)$	y-axis	$(0, -c)$ and $(0, c)$	$(0, -a)$ and $(0, a)$	$\dfrac{x^2}{b^2} + \dfrac{y^2}{a^2} = 1$

The graphs of the two ellipses are given in Figure 27.

Did you notice that in the ellipse $\dfrac{x^2}{a^2} + \dfrac{y^2}{b^2} = 1$ in Figure 27(a) and in the ellipse $\dfrac{x^2}{b^2} + \dfrac{y^2}{a^2} = 1$ in Figure 27(b), the focus, the origin, and the endpoint of the minor axis form a right triangle? In both ellipses, $a > b$ and, from the Pythagorean Theorem, $b^2 + c^2 = a^2$, or $b^2 = a^2 - c^2$.

Use the fact that $a > b$ to determine whether the major axis is the x-axis or y-axis. If the larger denominator is associated with the x^2 term, then the major axis is the

Figure 27

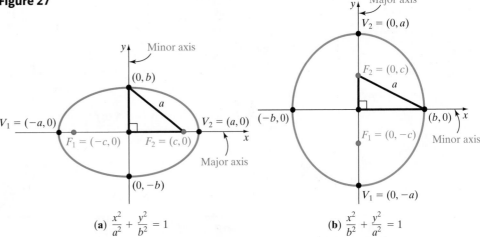

(a) $\dfrac{x^2}{a^2} + \dfrac{y^2}{b^2} = 1$

(b) $\dfrac{x^2}{b^2} + \dfrac{y^2}{a^2} = 1$

Work Smart

The term with the larger denominator tells us which axis is the major axis.

x-axis; if the larger denominator is associated with the y^2 term, then the major axis is the y-axis.

Quick ✓

1. An _____ is the collection of points in the plane such that the sum of the distances from two fixed points, called the ___, is a constant.

2. For an ellipse, the line containing the foci is called the _____ ___.

3. The two points of intersection of the ellipse and the major axis are the _____, V_1 and V_2, of the ellipse.

4. *True or False* If an ellipse has vertex $(a, 0)$ and focus $(c, 0)$, then $c^2 = a^2 + b^2$.

Graphing an ellipse whose center is at the origin is fairly straightforward. Find the intercepts by letting $y = 0$ (x-intercepts) and letting $x = 0$ (y-intercepts). When asked to graph an ellipse, be sure to label the foci.

EXAMPLE 1 **Graphing an Ellipse**

Graph the ellipse: $\dfrac{x^2}{25} + \dfrac{y^2}{9} = 1$

Solution

First, notice that the x^2 term has the larger denominator, 25. This means that the major axis is the x-axis and the equation of the ellipse has the form $\dfrac{x^2}{a^2} + \dfrac{y^2}{b^2} = 1$, so $a^2 = 25$ and $b^2 = 9$. The center of the ellipse is the origin, $(0, 0)$. Because $b^2 = a^2 - c^2$, we know that $c^2 = a^2 - b^2 = 25 - 9 = 16$, so $c = 4$. Since the major axis is the x-axis, the foci are $(-4, 0)$ and $(4, 0)$. Now find the intercepts:

x-intercepts: Let $y = 0$: $\dfrac{x^2}{25} + \dfrac{0^2}{9} = 1$ 　　　**y-intercepts:** Let $x = 0$: $\dfrac{0^2}{25} + \dfrac{y^2}{9} = 1$

$$\dfrac{x^2}{25} = 1 \qquad\qquad\qquad\qquad \dfrac{y^2}{9} = 1$$

$$x^2 = 25 \qquad\qquad\qquad\qquad y^2 = 9$$

$$x = \pm 5 \qquad\qquad\qquad\qquad y = \pm 3$$

(continued)

The intercepts are $(-5, 0)$, $(5, 0)$, $(0, -3)$, and $(0, 3)$. Figure 28 shows the graph of the ellipse.

Figure 28

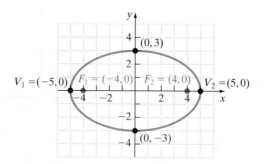

EXAMPLE 2 Graphing an Ellipse

Graph the ellipse: $\dfrac{x^2}{4} + \dfrac{y^2}{16} = 1$

Solution

First, notice that the y^2 term has the larger denominator, 16. This means that the major axis is the y-axis and the equation has the form $\dfrac{x^2}{b^2} + \dfrac{y^2}{a^2} = 1$, so $a^2 = 16$ and $b^2 = 4$. The center of the ellipse is the origin, $(0, 0)$. Because $b^2 = a^2 - c^2$, then $c^2 = a^2 - b^2 = 16 - 4 = 12$, so $c = \sqrt{12} = 2\sqrt{3}$. Since the major axis is the y-axis, the foci are $\left(0, -2\sqrt{3}\right)$ and $\left(0, 2\sqrt{3}\right)$. Now find the intercepts:

Figure 29

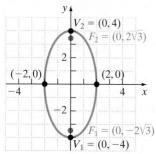

x-intercepts: Let $y = 0$: $\dfrac{x^2}{4} + \dfrac{0^2}{16} = 1$

$$\dfrac{x^2}{4} = 1$$
$$x^2 = 4$$
$$x = \pm 2$$

y-intercepts: Let $x = 0$: $\dfrac{0^2}{4} + \dfrac{y^2}{16} = 1$

$$\dfrac{y^2}{16} = 1$$
$$y^2 = 16$$
$$y = \pm 4$$

The intercepts are $(-2, 0)$, $(2, 0)$, $(0, -4)$, and $(0, 4)$. Figure 29 shows the graph of the ellipse.

Quick ✔

In Problems 5 and 6, graph each ellipse.

5. $\dfrac{x^2}{9} + \dfrac{y^2}{4} = 1$ **6.** $\dfrac{x^2}{16} + \dfrac{y^2}{36} = 1$

⊙ ❷ Find the Equation of an Ellipse Whose Center Is the Origin

Just as in working with parabolas, information about an ellipse is used to find its equation.

EXAMPLE 3 Finding the Equation of an Ellipse

Find the equation of the ellipse whose center is the origin with a focus at $(-2, 0)$ and vertex at $(-5, 0)$. Graph the ellipse.

Solution

The given focus and vertex lie on the x-axis, so the major axis is the x-axis. The equation of the ellipse must be of the form $\dfrac{x^2}{a^2} + \dfrac{y^2}{b^2} = 1$. The distance from the center of the ellipse to the vertex is $a = 5$ units.

The distance from the center of the ellipse to the focus is $c = 2$ units, so $b^2 = a^2 - c^2 = 5^2 - 2^2 = 25 - 4 = 21$. Thus the equation of the ellipse is

$$\frac{x^2}{25} + \frac{y^2}{21} = 1$$

Figure 30 shows the graph. Use $b = \sqrt{21} \approx 4.6$ to graph the y-intercepts.

Figure 30

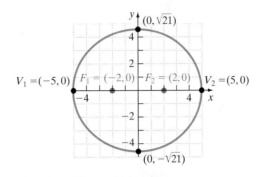

Quick ✓

7. Find the equation of the ellipse whose center is the origin with a focus at $(0, 3)$ and vertex at $(0, 7)$. Graph the ellipse.

⊙ ❸ Graph an Ellipse Whose Center Is Not the Origin

If an ellipse with center at the origin and major axis coinciding with a coordinate axis is shifted horizontally h units and vertically k units, the result is an ellipse with center at (h, k) and major axis parallel to a coordinate axis. The equations of these ellipses have the same forms as those given for ellipses whose center is the origin, except that x is replaced by $x - h$ (the horizontal shift) and y is replaced by $y - k$ (the vertical shift). Table 4 gives the forms of the equations for these ellipses. Figure 31 shows their graphs.

Work Smart

Do not attempt to memorize Table 4. Instead, understand the roles of a and c in an ellipse—a is the distance from the center to each vertex, and c is the distance from the center to each focus.

Table 4 Ellipses with Center at (h, k) and Major Axis Parallel to a Coordinate Axis

Center	Major Axis	Foci	Vertices	Equation
(h, k)	Parallel to x-axis	$(h + c, k)$	$(h + a, k)$	$\dfrac{(x - h)^2}{a^2} + \dfrac{(y - k)^2}{b^2} = 1$
		$(h - c, k)$	$(h - a, k)$	$a > b$ and $b^2 = a^2 - c^2$
(h, k)	Parallel to y-axis	$(h, k + c)$	$(h, k + a)$	$\dfrac{(x - h)^2}{b^2} + \dfrac{(y - k)^2}{a^2} = 1$
		$(h, k - c)$	$(h, k - a)$	$a > b$ and $b^2 = a^2 - c^2$

Figure 31

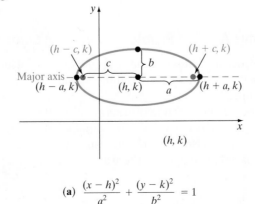

(a) $\dfrac{(x - h)^2}{a^2} + \dfrac{(y - k)^2}{b^2} = 1$

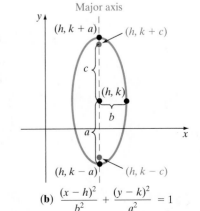

(b) $\dfrac{(x - h)^2}{b^2} + \dfrac{(y - k)^2}{a^2} = 1$

EXAMPLE 4 **Graphing an Ellipse**

Graph the equation: $x^2 + 16y^2 - 4x + 32y + 4 = 0$

Solution

Write the equation in one of the forms given in Table 4. To do this, complete the squares in x and in y.

$$x^2 + 16y^2 - 4x + 32y + 4 = 0$$

Group like variables; place the constant on the right side:
$$(x^2 - 4x) + (16y^2 + 32y) = -4$$

Factor out 16 from $16y^2 + 32y$:
$$(x^2 - 4x) + 16(y^2 + 2y) = -4$$

Complete each square:
$$(x^2 - 4x + 4) + 16(y^2 + 2y + 1) = -4 + 4 + 16$$

Factor:
$$(x - 2)^2 + 16(y + 1)^2 = 16$$

Divide both sides by 16:
$$\frac{(x - 2)^2}{16} + \frac{(y + 1)^2}{1} = 1$$

Work Smart

To complete the square in y, 1 was added, but because it was inside the parentheses with a factor of 16 in front, $16 \cdot 1 = 16$ must be added to the other side.

This is the equation of an ellipse with center at $(2, -1)$. Because the larger value, 16, is the denominator of the x^2 term, the major axis is parallel to the x-axis. Because $a^2 = 16$ and $b^2 = 1$, then $c^2 = a^2 - b^2 = 16 - 1 = 15$. The vertices are $a = 4$ units to the left and right of center at $V_1 = (-2, -1)$ and $V_2 = (6, -1)$. The foci are $c = \sqrt{15}$ units to the left and right of center at $F_1 = (2 - \sqrt{15}, -1)$ and $F_2 = (2 + \sqrt{15}, -1)$. Plot points $b = 1$ unit above and below the center at $(2, 0)$ and $(2, -2)$. Use $2 \pm \sqrt{15} \approx -1.9$ and 5.9 when plotting the foci of the ellipse. See Figure 32.

Figure 32

Work Smart

The graph of $\dfrac{(x - 2)^2}{16} + \dfrac{(y + 1)^2}{1} = 1$

is the graph of $\dfrac{x^2}{16} + \dfrac{y^2}{1} = 1$ shifted

2 units right and 1 unit down.

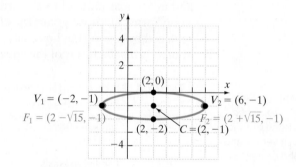

$V_1 = (-2, -1)$ $V_2 = (6, -1)$
$F_1 = (2 - \sqrt{15}, -1)$ $F_2 = (2 + \sqrt{15}, -1)$
$(2, 0)$
$(2, -2)$ $C = (2, -1)$

Quick ✓

8. The center of $\dfrac{(x - 3)^2}{25} + \dfrac{(y + 1)^2}{16} = 1$ is _____.

9. Graph the equation: $9x^2 + y^2 + 54x - 2y + 73 = 0$

▶ ❹ Solve Applied Problems Involving Ellipses

Ellipses are found in many applications in science and engineering. For example, the orbits of the planets around the Sun are elliptical, with the Sun's position at a focus. See Figure 33.

Stone and concrete bridges are often the shape of semielliptical arches. Elliptical gears are used in machinery when a variable rate of motion is required.

Ellipses also have an interesting reflection property. If a source of sound (or light) is placed at one focus, the waves transmitted by the source reflect off the ellipse and

Figure 33

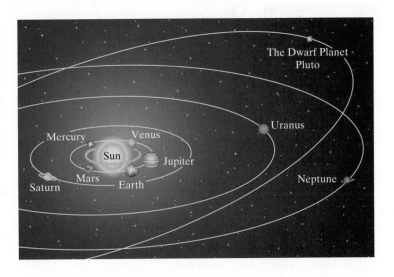

concentrate at the other focus. This is the principle behind whispering galleries, which are rooms with elliptical ceilings. A person standing at one focus of the ellipse can whisper and be heard by a person standing at the other focus, because all the sound waves that reach the ceiling are reflected to the other person. National Statuary Hall in the United States Capital is a whispering gallery.

EXAMPLE 5 **A Whispering Gallery**

The whispering gallery in the Museum of Science and Industry in Chicago is 47.3 feet long. The distance from the center of the room to the foci is 20.3 feet. Find an equation that describes the shape of the room. How high is the room at its center?

Solution
Set up a Cartesian plane so that the center of the ellipse is at the origin and the major axis lies on the x-axis. The equation of the ellipse is

$$\frac{x^2}{a^2} + \frac{y^2}{b^2} = 1$$

Figure 34

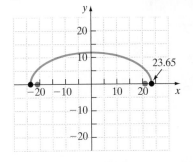

Because the length of the room is 47.3 feet, the distance from the center of the room to each vertex (the end of the room) will be $47.3/2 = 23.65$ feet, so $a = 23.65$ feet. The distance from the center of the room to each focus is $c = 20.3$ feet. See Figure 34. Because $b^2 = a^2 - c^2$, we have $b^2 = 23.65^2 - 20.3^2 = 147.2325$. An equation that describes the shape of the room is given by

$$\frac{x^2}{23.65^2} + \frac{y^2}{147.2325} = 1$$

The height of the room at its center is $b = \sqrt{147.2325} \approx 12.1$ feet. ●

Quick ✓

10. A hall 100 feet in length is to be designed as a whispering gallery. If the foci are to be located 30 feet from the center, determine an equation that describes the room. What is the height of the room at its center?

9.4 Exercises MyMathLab® Exercise numbers in green have complete video solutions in MyMathLab or may be accessed using the QR code to the right.

*Problems **1–10** are the Quick ✔ s that follow the EXAMPLES.*

Building Skills

In Problems 11–14, the graph of an ellipse is given. Match each graph to its equation. See Objective 1.

(a) $x^2 + \dfrac{y^2}{9} = 1$ **(b)** $\dfrac{x^2}{9} + y^2 = 1$

(c) $\dfrac{x^2}{16} + \dfrac{y^2}{9} = 1$ **(d)** $\dfrac{x^2}{9} + \dfrac{y^2}{16} = 1$

11.

12.

13.

14.

In Problems 15–24, find the vertices and foci of each ellipse. Graph each ellipse. See Objective 1.

15. $\dfrac{x^2}{25} + \dfrac{y^2}{16} = 1$ **16.** $\dfrac{x^2}{25} + \dfrac{y^2}{4} = 1$

17. $\dfrac{x^2}{36} + \dfrac{y^2}{100} = 1$ **18.** $\dfrac{x^2}{16} + \dfrac{y^2}{36} = 1$

19. $\dfrac{x^2}{49} + \dfrac{y^2}{4} = 1$ **20.** $\dfrac{x^2}{121} + \dfrac{y^2}{100} = 1$

21. $x^2 + \dfrac{y^2}{49} = 1$ **22.** $\dfrac{x^2}{64} + y^2 = 1$

23. $4x^2 + y^2 = 16$ **24.** $9x^2 + y^2 = 81$

In Problems 25–32, find an equation for each ellipse. Graph each ellipse. See Objective 2.

25. Center at $(0,0)$; focus at $(4,0)$; vertex at $(6,0)$

26. Center at $(0,0)$; focus at $(2,0)$; vertex at $(5,0)$

27. Center at $(0,0)$; focus at $(0,-4)$; vertex at $(0,7)$

28. Center at $(0,0)$; focus at $(0,-1)$; vertex at $(0,5)$

29. Foci at $(\pm 6,0)$; vertices at $(\pm 10,0)$

30. Foci at $(0,\pm 2)$; vertices at $(0,\pm 7)$

31. Foci at $(0,\pm 5)$; length of the major axis is 16

32. Foci at $(\pm 6,0)$; length of the major axis is 20

In Problems 33–42, graph each ellipse. See Objective 3.

33. $\dfrac{(x-3)^2}{9} + \dfrac{(y+2)^2}{25} = 1$

34. $\dfrac{(x-1)^2}{36} + \dfrac{(y+4)^2}{100} = 1$

35. $\dfrac{(x+2)^2}{16} + \dfrac{(y-5)^2}{4} = 1$

36. $\dfrac{(x+5)^2}{64} + \dfrac{(y+1)^2}{16} = 1$

37. $(x-5)^2 + \dfrac{(y+1)^2}{49} = 1$

38. $\dfrac{(x+8)^2}{81} + (y-3)^2 = 1$

39. $4(x+2)^2 + 16(y-1)^2 = 64$

40. $9(x-3)^2 + (y-4)^2 = 81$

41. $4x^2 + y^2 - 24x + 2y - 63 = 0$

42. $16x^2 + 9y^2 - 128x + 54y - 239 = 0$

Applying the Concepts

43. Semielliptical Arch Bridge An arch in the shape of the upper half of an ellipse is used to support a bridge that is to span a river 30 meters wide. The center of the arch is 10 meters above the center of the river. See the figure.

(a) Write the equation for the ellipse in which the x-axis coincides with the water and the y-axis passes through the center of the arch.

(b) Can a rectangular barge that is 18 meters wide and sits 7 meters above the surface of the water fit through the opening of the bridge?

(c) If heavy rains cause the river's level to increase 1.1 meters, will the barge make it through the opening?

44. London Bridge An arch in the shape of the upper half of an ellipse is used to support London Bridge. The main span is 45.6 meters wide. Suppose that the center of the arch is 15 meters above the center of the river.

  (a) Write the equation for the ellipse in which the *x*-axis coincides with the water and the *y*-axis passes through the center of the arch.

  (b) Can a rectangular barge that is 20 meters wide and sits 12 meters above the surface of the water fit through the opening of the bridge?

  (c) If heavy rains cause the river's level to increase 1.5 meters, will the barge make it through the opening?

In Problems 45–48, use the fact that the orbit of a planet about the Sun is an ellipse, with the Sun at one focus. The **aphelion** *of a planet is its greatest distance from the Sun, and the* **perihelion** *is its shortest distance. The* **mean distance** *of a planet from the Sun is the distance from the center to a vertex of the elliptical orbit. See the illustration.*

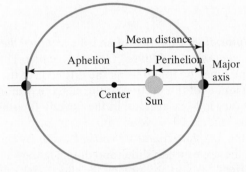

45. Earth The mean distance of Earth from the Sun is 93 million miles. If the aphelion of Earth is 94.5 million miles, what is the perihelion? Write an equation for the orbit of Earth around the Sun.

46. Mars The mean distance of Mars from the Sun is 142 million miles. If the perihelion of Mars is 128.5 million miles, what is the aphelion? Write an equation for the orbit of Mars about the Sun.

47. Jupiter The aphelion of Jupiter is 507 million miles. If the distance from the Sun to the center of its elliptical orbit is 23.2 million miles, what is the perihelion? What is the mean distance? Write an equation for the orbit of Jupiter around the Sun.

48. Dwarf Planet Pluto The perihelion of Pluto is 4551 million miles, and the distance of the Sun from the center of its elliptical orbit is 897.5 million miles. Find the aphelion of Pluto. What is the mean distance of Pluto from the Sun? Write an equation for the orbit of Pluto about the Sun.

Extending the Concepts

In Problems 49–52, write an equation for each ellipse.

49.

50.

51.

52.

53. Show that a circle is a special kind of ellipse by letting $a = b$ in the equation of an ellipse centered at the origin. What is the value of c in a circle? What does this mean regarding the location of the foci?

54. The **eccentricity** e of an ellipse is defined as the number $\dfrac{c}{a}$, where a is the distance from the center of an ellipse to a vertex and c is the distance from the center of an ellipse to a focus. Because $a > c$, it follows that $e < 1$ for an ellipse. Write a paragraph about the general shape of each of the following ellipses. Be sure to justify your conclusion.

  (a) Eccentricity close to 0

  (b) Eccentricity $= 0.5$

  (c) Eccentricity close to 1

Explaining the Concepts

55. In the ellipse drawn below, the center is at the origin and the major axis is the *x*-axis. The point *F* is a focus. In the right triangle (drawn in red), label the lengths of the legs and the hypotenuse using a, b, and c.

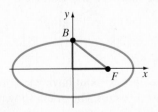

56. How does the ellipse given by the equation $\dfrac{x^2}{25} + \dfrac{y^2}{9} = 1$ differ from the ellipse given by the equation $\dfrac{x^2}{9} + \dfrac{y^2}{25} = 1$? How are they the same?

Synthesis Review

In Problems 57–60, fill in the following table for each function.

x	5	10	100	1000
f(x)				

57. $f(x) = \dfrac{5}{x + 2}$

58. $f(x) = \dfrac{x - 1}{x^2 + 4}$

59. $f(x) = \dfrac{2x + 1}{x - 3}$

60. $f(x) = \dfrac{3x^2 - x + 1}{x^2 + 1}$

In Problems 61 and 62, fill in the following table for each function.

x	5	10	100	1000
f(x)				
g(x)				

61. $f(x) = \dfrac{x^2 + 3x + 1}{x + 1}; g(x) = x + 2$

62. $f(x) = \dfrac{x^2 - 3x + 5}{x + 2}; g(x) = x - 5$

63. For the functions in Problems 57–60, compare the degree of the polynomial in the numerator to the degree of the polynomial in the denominator. Conjecture what happens to a rational function as *x* increases when the degree of the numerator is less than the degree of the denominator. Conjecture what happens to a rational function as *x* increases when the degree of the numerator equals the degree of the denominator.

64. For the functions in Problems 61 and 62, write *f* in the form quotient $+ \dfrac{\text{remainder}}{\text{dividend}}$. That is, perform the division indicated by the rational function. Now compare the values of *f* to those of *g* in the table. What does the function *g* represent?

Technology Exercises

A graphing calculator can be used to graph an ellipse by solving the equation for *y*. Because ellipses are not functions, graph the upper half and the lower half of an ellipse in two pieces. For example, to graph $\dfrac{x^2}{4} + \dfrac{y^2}{9} = 1$, graph

$$Y_1 = 3\sqrt{1 - \dfrac{x^2}{4}} \text{ and } Y_2 = -3\sqrt{1 - \dfrac{x^2}{4}}.$$

To graph an ellipse without having to solve the equation for *y*, you can use Desmos. The graph shown in Figure 35 is obtained.

In Problems 65–72, graph each ellipse using technology.

65. $\dfrac{x^2}{25} + \dfrac{y^2}{16} = 1$

66. $\dfrac{x^2}{25} + \dfrac{y^2}{4} = 1$

67. $4x^2 + y^2 = 16$

68. $9x^2 + y^2 = 81$

Figure 35

69. $\dfrac{(x - 3)^2}{9} + \dfrac{(y + 2)^2}{25} = 1$

70. $\dfrac{(x - 1)^2}{36} + \dfrac{(y + 4)^2}{100} = 1$

71. $\dfrac{(x + 2)^2}{16} + \dfrac{(y - 5)^2}{4} = 1$

72. $\dfrac{(x + 5)^2}{64} + \dfrac{(y + 1)^2}{16} = 1$

73. Exploration: Equation of an Ellipse *Open the "Equation of an Ellipse" Geogebra applet. The applet may be found using the QR code at the beginning of this section or through the Multimedia Library in MyMathLab. In the upper left corner are four "sliders." Use your mouse to change the values of the h, k, horizontal, or vertical sliders by clicking on the appropriate slider and dragging the value left or right. As you are moving the sliders, take note of how the position and equation of the ellipse change. If you are using a keyboard, it may be easier to use the arrow keys, instead of the mouse, to move the slider. First click on the slider with your mouse, and then use the left and right arrow keys to move the slider. Start with all slider values set to the following: h = 5, k = 2, horizontal = 5, vertical = 3.*

(a) Check the *Segment Length* boxes to display the segment lengths from each focus to the ellipse. Click on the blue point on the ellipse to move it. Notice that as you move the blue point on the ellipse, these segment lengths change. What happens to the sum of the lengths of these segments as you move the point? In other words, what happens to the sum of the distances between the foci and the ellipse?

(b) Check the *Center* box to display the center of the ellipse. What is the center of the current ellipse? Move *h* to 3. What did this do to the center of the ellipse? Did this move change the length of the major and minor axes?

(c) Move *k* to 4. What did this do to the center of the ellipse? Did this move change the length of the major and minor axes?

(d) What does the role of *h* and *k* play in the graph of the ellipse?

(e) Move the *horizontal* slider to 4. What did this do to the graph of the ellipse? What did this change in the equation of the ellipse?

(f) Now move the *horizontal* slider to 2. What did this do to the graph of the ellipse? What did this change in the equation of the ellipse?

(g) What role does the number shown on the horizontal slider play in the graph and equation of an ellipse?

(h) Move the *vertical* slider to 4. What did this do to the graph of the ellipse? What did this change in the equation of the ellipse?

(i) Now move the *vertical* slider to 1. What did this do to the graph of the ellipse? What did this change in the equation of the ellipse?

(j) What role does the number shown on the vertical slider play in the graph and equation of an ellipse?

9.5 Hyperbolas

Objectives

1 Graph a Hyperbola Whose Center Is the Origin

2 Find the Equation of a Hyperbola Whose Center Is the Origin

3 Find the Asymptotes of a Hyperbola Whose Center Is the Origin

In Other Words

The distance from F_1 to P minus the distance from F_2 to P is a constant value for any point P on a hyperbola.

Figure 36

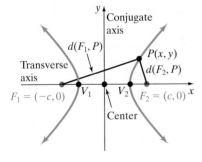

Work Smart

Notice for hyperbolas that $c^2 = a^2 + b^2$, but for ellipses $c^2 = a^2 - b^2$.

Prepared?...Answers P1. $x^2 - 5x + \dfrac{25}{4}$
P2. $\{-8, 8\}$

Are You Prepared for This Section?

Before getting started, complete the following problems. If you get a problem wrong, go back to the section cited and review the material.

P1. Complete the square: $x^2 - 5x$ [Section 7.1, pp. 546–547]
P2. Solve: $y^2 = 64$ [Section 7.1, pp. 543–544]

 1 Graph a Hyperbola Whose Center Is the Origin

Recall from Section 9.2 that a hyperbola is a conic section that is obtained through the intersection of a plane and two cones. See Figure 8(d) in Section 9.2.

A **hyperbola** is the collection of all points in the plane the difference of whose distances from two fixed points, called the **foci,** is a positive constant.

Figure 36 illustrates a hyperbola with foci F_1 and F_2. The line containing the foci is the **transverse axis.** The midpoint of the line segment joining the foci is the **center** of the hyperbola. The line through the center and perpendicular to the transverse axis is the **conjugate axis.** The hyperbola consists of two separate curves called **branches.** The two points of intersection of the hyperbola and the transverse axis are the **vertices,** V_1 and V_2, of the hyperbola.

The equation of a hyperbola can be found using the distance formula. However, this information will not be presented here. Instead, the equations of the hyperbolas whose branches open left and right and the equations of hyperbolas whose branches open up and down, with both hyperbolas centered at the origin will be given. See Table 5.

Table 5 Hyperbolas with Center at the Origin

Center	Transverse Axis	Branches Open	Foci	Vertices	Equation
$(0,0)$	x-axis	Left and right	$(-c, 0)$ and $(c, 0)$	$(-a, 0)$ and $(a, 0)$	$\dfrac{x^2}{a^2} - \dfrac{y^2}{b^2} = 1$, where $b^2 = c^2 - a^2$ or $c^2 = a^2 + b^2$
$(0,0)$	y-axis	Up and down	$(0, -c)$ and $(0, c)$	$(0, -a)$ and $(0, a)$	$\dfrac{y^2}{a^2} - \dfrac{x^2}{b^2} = 1$, where $b^2 = c^2 - a^2$ or $c^2 = a^2 + b^2$

The graphs of the two hyperbolas are given in Figure 37.

Figure 37

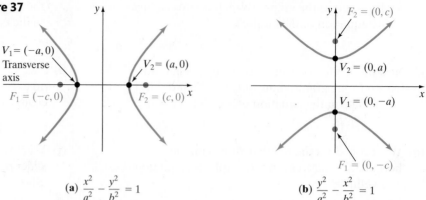

(a) $\dfrac{x^2}{a^2} - \dfrac{y^2}{b^2} = 1$

(b) $\dfrac{y^2}{a^2} - \dfrac{x^2}{b^2} = 1$

Notice the difference between the two equations given in Table 5. When the x^2 term is first, as in Figure 37(a), the transverse axis is the x-axis and the hyperbola opens left and right. When the y^2 term is first, as in Figure 37(b), the transverse axis is the y-axis and the hyperbola opens up and down. In both cases, the value of a^2 is in the denominator of the first term.

Quick ✓

1. A(n) _____ is the collection of points in the plane the difference of whose distances from two fixed points is a positive constant.

2. For a hyperbola, the foci lie on a line called the _____ ___.

3. The line through the center of a hyperbola that is perpendicular to the transverse axis is called the _____ ___.

EXAMPLE 1 **Graphing a Hyperbola**

Graph the equation: $\dfrac{x^2}{16} - \dfrac{y^2}{4} = 1$

Solution

The equation is of the form $\dfrac{x^2}{a^2} - \dfrac{y^2}{b^2} = 1$. The x^2 term is first, so the hyperbola opens left and right, and the transverse axis is the x-axis. The center of the hyperbola is the origin, $(0, 0)$. Because $a^2 = 16$ and $b^2 = 4$, then $c^2 = a^2 + b^2 = 16 + 4 = 20$, so $c = 2\sqrt{5}$. The vertices are at $(\pm a, 0) = (\pm 4, 0)$, and the foci are at $(\pm c, 0) = (\pm 2\sqrt{5}, 0)$.

To obtain the graph, plot the vertices and foci. Then locate points above and below the foci (as in graphing parabolas). Thus let $x = \pm 2\sqrt{5}$ in the equation $\dfrac{x^2}{16} - \dfrac{y^2}{4} = 1$.

$$\frac{(\pm 2\sqrt{5})^2}{16} - \frac{y^2}{4} = 1$$

$$\frac{20}{16} - \frac{y^2}{4} = 1$$

Reduce the fraction: $\dfrac{5}{4} - \dfrac{y^2}{4} = 1$

Subtract $\dfrac{5}{4}$ from both sides: $-\dfrac{y^2}{4} = -\dfrac{1}{4}$

Multiply both sides by -4: $y^2 = 1$

Take the square root of both sides: $y = \pm 1$

The points above and below the foci are $\left(\pm 2\sqrt{5}, -1\right)$ and $\left(\pm 2\sqrt{5}, 1\right)$. Use $\pm 2\sqrt{5} \approx \pm 4.5$ to graph the foci. See Figure 38 for the graph of the hyperbola.

Figure 38

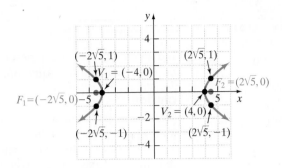

EXAMPLE 2 | **Graphing a Hyperbola**

Graph the equation: $4y^2 - 16x^2 = 16$

Solution

Write the equation in one of the forms given in Table 5. To do this, divide both sides of the equation by 16:

$$\frac{y^2}{4} - \frac{x^2}{1} = 1$$

Because the y^2 term is first, the hyperbola opens up and down and the transverse axis is the y-axis. The center of this hyperbola is the origin, $(0, 0)$. Comparing the equation to $\frac{y^2}{a^2} - \frac{x^2}{b^2} = 1$, we find that $a^2 = 4$ and $b^2 = 1$; therefore, $c^2 = a^2 + b^2 = 4 + 1 = 5$. The vertices are at $(0, \pm a) = (0, \pm 2)$. The foci are at $(0, \pm c) = \left(0, \pm\sqrt{5}\right)$. Locate four more points on the hyperbola to the left and right of each focus by letting $y = \pm\sqrt{5}$ in the equation $\frac{y^2}{4} - \frac{x^2}{1} = 1$ and solving for x.

$$\frac{\left(\pm\sqrt{5}\right)^2}{4} - \frac{x^2}{1} = 1$$

$$\frac{5}{4} - \frac{x^2}{1} = 1$$

Subtract $\frac{5}{4}$ from both sides: $\quad -x^2 = -\frac{1}{4}$

Multiply both sides by -1: $\quad x^2 = \frac{1}{4}$

Take the square root of both sides: $\quad x = \pm\frac{1}{2}$

Figure 39

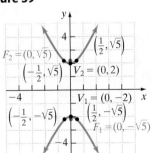

Four additional points on the graph are $\left(-\frac{1}{2}, \pm\sqrt{5}\right)$ and $\left(\frac{1}{2}, \pm\sqrt{5}\right)$. See Figure 39 for the graph of the hyperbola.

Quick ✓

In Problems 4 and 5, graph each hyperbola.

4. $\dfrac{x^2}{36} - \dfrac{y^2}{64} = 1$ **5.** $16y^2 - 9x^2 = 144$

▶ ❷ **Find the Equation of a Hyperbola Whose Center Is the Origin**

Just as in working with parabolas and ellipses, we will use information about a hyperbola to find its equation.

EXAMPLE 3 | **Finding and Graphing the Equation of a Hyperbola**

Find an equation of the hyperbola with one focus at $(0, -3)$ and vertices at $(0, -2)$ and $(0, 2)$. Graph the equation.

Solution
The center of a hyperbola is located at the midpoint of the vertices (or foci). Because the vertices are at $(0, -2)$ and $(0, 2)$, the center must be at the origin, $(0, 0)$. Notice that the given focus and vertices lie on the y-axis. Therefore, the transverse axis is the y-axis, and the hyperbola opens up and down. A vertex is at $(0, 2)$, so $a = 2$. A focus is at $(0, -3)$, so $c = 3$. With $a = 2$ and $c = 3$, $b^2 = c^2 - a^2 = 3^2 - 2^2 = 9 - 4 = 5$.

The equation must be of the form $\dfrac{y^2}{a^2} - \dfrac{x^2}{b^2} = 1$, so let $a^2 = 4$ and $b^2 = 5$ to get

$$\frac{y^2}{4} - \frac{x^2}{5} = 1$$

To find points to the left and right of each focus, let $y = \pm 3$ in the equation $\dfrac{y^2}{4} - \dfrac{x^2}{5} = 1$.

$$\frac{(\pm 3)^2}{4} - \frac{x^2}{5} = 1$$

$$\frac{9}{4} - \frac{x^2}{5} = 1$$

Subtract $\dfrac{9}{4}$ from both sides: $-\dfrac{x^2}{5} = -\dfrac{5}{4}$

Multiply both sides by -5: $x^2 = \dfrac{25}{4}$

Take the square root of both sides: $x = \pm\dfrac{5}{2}$

Figure 40

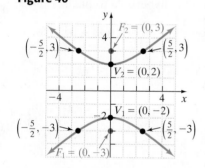

The points to the left and right of the foci are $\left(-\dfrac{5}{2}, \pm 3\right)$ and $\left(\dfrac{5}{2}, \pm 3\right)$. See Figure 40 for the graph of the hyperbola. ●

Look at the equations of the hyperbolas in Examples 1 and 3. For the hyperbola in Example 1, $a^2 = 16$ and $b^2 = 4$, so $a > b$; for the hyperbola in Example 3, $a^2 = 4$ and $b^2 = 5$, so $a < b$. Therefore, for hyperbolas, there are no requirements involving the relative sizes for a and b. For ellipses, however, the relative sizes of a and b determine the major axis.

Quick ✓

6. *True or False* In any hyperbola, it must be the case that $a > b$.

7. Find the equation of a hyperbola whose vertices are $(-4, 0)$ and $(4, 0)$ and has a focus at $(6, 0)$. Graph the hyperbola.

▶ ❸ **Find the Asymptotes of a Hyperbola Whose Center Is the Origin**

As x and y get larger in both the positive and the negative directions, the branches of the hyperbola approach two lines called **asymptotes** of the hyperbola. The asymptotes help us graph hyperbolas. Table 6 gives the asymptotes of the two hyperbolas discussed in this section.

Table 6 Asymptotes of a Hyperbola

Hyperbola	Asymptotes
$\dfrac{x^2}{a^2} - \dfrac{y^2}{b^2} = 1$	$y = -\dfrac{b}{a}x$ and $y = \dfrac{b}{a}x$
$\dfrac{y^2}{a^2} - \dfrac{x^2}{b^2} = 1$	$y = -\dfrac{a}{b}x$ and $y = \dfrac{a}{b}x$

Figure 41 illustrates how the asymptotes can be used to help graph a hyperbola. It is important to remember that the asymptotes are not part of the hyperbola—they only serve as guides in graphing the hyperbola.

Figure 41

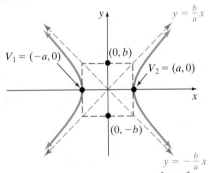

(**a**) Asymptotes on the graph of $\dfrac{x^2}{a^2} - \dfrac{y^2}{b^2} = 1$

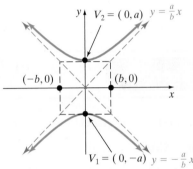

(**b**) Asymptotes on the graph of $\dfrac{y^2}{a^2} - \dfrac{x^2}{b^2} = 1$

For example, to graph the equation $\dfrac{x^2}{a^2} - \dfrac{y^2}{b^2} = 1$, first plot the vertices $(-a, 0)$ and $(a, 0)$. Then plot the points $(0, -b)$ and $(0, b)$. Use these four points to construct a rectangle as shown in Figure 41(a).

The lines through the diagonals of this rectangle have slopes $\dfrac{b}{a}$ and $-\dfrac{b}{a}$. These lines are the asymptotes. The equations of these asymptotes are $y = -\dfrac{b}{a}x$ and $y = \dfrac{b}{a}x$. Using this technique enables us to avoid plotting the four additional points plotted earlier.

EXAMPLE 4   **Graphing a Hyperbola and Finding Its Asymptotes**

Graph the equation $9x^2 - 4y^2 = 36$ using the asymptotes as a guide.

Solution

Divide both sides of the equation by 36 to put the equation in the form $\dfrac{x^2}{a^2} - \dfrac{y^2}{b^2} = 1$.

$$\frac{x^2}{4} - \frac{y^2}{9} = 1$$

The center of the hyperbola is the origin, $(0, 0)$. Because the x^2 term is first, the hyperbola opens to the left and right. The transverse axis is along the x-axis. In addition, notice that $a^2 = 4$ and $b^2 = 9$. Because $c^2 = a^2 + b^2$, then $c^2 = 4 + 9 = 13$, so $c = \pm\sqrt{13}$. The vertices are at $(\pm a, 0) = (\pm 2, 0)$; the foci

(continued)

Figure 42

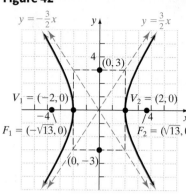

are at $(\pm c, 0) = (\pm\sqrt{13}, 0)$. Since $b^2 = 9$, is then $b = \pm 3$. The equations of the asymptotes are

$$y = \frac{b}{a}x = \frac{3}{2}x \quad \text{and} \quad y = -\frac{b}{a}x = -\frac{3}{2}x$$

To graph the hyperbola, form a rectangle using the points $(\pm a, 0) = (\pm 2, 0)$ and $(0, \pm b) = (0, \pm 3)$. Draw the asymptotes through the vertices of the rectangle. See Figure 42.

●

Quick ✓

8. The asymptotes of the hyperbola $\dfrac{x^2}{a^2} - \dfrac{y^2}{b^2} = 1$ are _____ and _____.

9. Graph the equation $x^2 - 9y^2 = 9$ using the asymptotes as a guide.

10. Graph the equation $\dfrac{y^2}{16} - \dfrac{x^2}{9} = 1$ using the asymptotes as a guide.

9.5 Exercises MyMathLab® Exercise numbers in green have complete video solutions in MyMathLab or may be accessed using the QR code to the right.

Problems 1–10 are the Quick ✓s that follow the EXAMPLES.

Building Skills

In Problems 11–14, the graph of a hyperbola is given. Match each graph to its equation. See Objective 1.

(a) $\dfrac{x^2}{4} - y^2 = 1$

(b) $x^2 - \dfrac{y^2}{4} = 1$

(c) $\dfrac{y^2}{4} - x^2 = 1$

(d) $y^2 - \dfrac{x^2}{4} = 1$

11.

12.

13.

14.

In Problems 15–22, graph each equation. See Objective 1.

15. $\dfrac{x^2}{4} - \dfrac{y^2}{16} = 1$

16. $\dfrac{x^2}{9} - \dfrac{y^2}{16} = 1$

17. $\dfrac{y^2}{25} - \dfrac{x^2}{36} = 1$

18. $\dfrac{y^2}{81} - \dfrac{x^2}{9} = 1$

19. $4x^2 - y^2 = 36$

20. $x^2 - 9y^2 = 36$

21. $25y^2 - x^2 = 100$

22. $4y^2 - 9x^2 = 36$

In Problems 23–28, find the equation for the hyperbola described. Graph the equation. See Objective 2.

23. Center at $(0, 0)$; focus at $(3, 0)$; vertex at $(2, 0)$

24. Center at $(0, 0)$; focus at $(-4, 0)$; vertex at $(-1, 0)$

25. Vertices at $(0, 5)$ and $(0, -5)$; focus at $(0, 7)$

26. Vertices at $(0, 6)$ and $(0, -6)$; focus at $(0, 8)$

27. Foci at $(-10, 0)$ and $(10, 0)$; vertex at $(-7, 0)$

28. Foci at $(-5, 0)$ and $(5, 0)$; vertex at $(-3, 0)$

In Problems 29–34, graph each equation using its asymptotes. See Objective 3.

29. $\dfrac{x^2}{25} - \dfrac{y^2}{9} = 1$

30. $\dfrac{x^2}{16} - \dfrac{y^2}{64} = 1$

31. $\dfrac{y^2}{4} - \dfrac{x^2}{100} = 1$

32. $\dfrac{y^2}{25} - \dfrac{x^2}{100} = 1$

33. $x^2 - y^2 = 4$

34. $y^2 - x^2 = 25$

Mixed Practice

In Problems 35–38, find the equation for the hyperbola described. Graph the equation.

35. Vertices at $(0, -8)$ and $(0, 8)$; asymptote the line $y = 2x$

36. Vertices at $(0, -4)$ and $(0, 4)$; asymptote the line $y = 2x$

37. Foci at $(-3, 0)$ and $(3, 0)$; asymptote the line $y = x$

38. Foci at $(-9, 0)$ and $(9, 0)$; asymptote the line $y = -3x$

Applying the Concepts

In Problems 39–42, write the equation of the hyperbola.

39.

40.

41.

42.

Extending the Concepts

43. Two hyperbolas that have the same set of asymptotes are called **conjugate.** Show that the hyperbolas

$$\frac{x^2}{4} - y^2 = 1 \quad \text{and} \quad y^2 - \frac{x^2}{4} = 1$$

are conjugate. Graph each hyperbola in the same Cartesian plane.

44. The **eccentricity** e of a hyperbola is defined as the number $\frac{c}{a}$. Because $c > a$, it follows that $e > 1$. Describe the general shape of a hyperbola whose eccentricity is close to 1. What is the shape if e is very large?

Explaining the Concepts

45. Explain how the asymptotes of a hyperbola are helpful in obtaining its graph.

46. How can you tell the difference between the equation of a hyperbola and the equation of an ellipse just by looking at the equations?

Synthesis Review

In Problems 47–51, solve each system of equations using either substitution or elimination.

47. $\begin{cases} 2x - 3y = -9 \\ -x + 5y = 8 \end{cases}$ **48.** $\begin{cases} 3x + 4y = 3 \\ -6x + 2y = -\dfrac{7}{2} \end{cases}$

49. $\begin{cases} 2x - 3y = 6 \\ -6x + 9y = -18 \end{cases}$ **50.** $\begin{cases} -2x + y = 8 \\ x - \dfrac{1}{2}y = -4 \end{cases}$

51. $\begin{cases} 6x + 3y = 4 \\ -2x - y = -\dfrac{4}{3} \end{cases}$

52. In Problems 47–51, which method did you use more often? Why? Do you think there are situations where substitution is superior to elimination? Are there situations where elimination is superior to substitution? Describe these circumstances.

Technology Exercises

A graphing calculator can be used to graph hyperbolas by solving the equation for y. Because hyperbolas are not functions, graph the upper half and lower half of the hyperbola in two pieces. For example, to graph $\dfrac{x^2}{4} - \dfrac{y^2}{9} = 1$, graph $Y_1 = 3\sqrt{\dfrac{x^2}{4} - 1}$ and $Y_2 = -3\sqrt{\dfrac{x^2}{4} - 1}$.

To graph a hyperbola without having to solve the equation for y, you can use Desmos. The graph shown in Figure 43 is obtained.

Figure 43

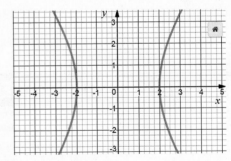

In Problems 53–60, graph each hyperbola using technology.

53. $\dfrac{x^2}{4} - \dfrac{y^2}{16} = 1$ **54.** $\dfrac{x^2}{9} - \dfrac{y^2}{16} = 1$

55. $\dfrac{y^2}{25} - \dfrac{x^2}{36} = 1$ **56.** $\dfrac{y^2}{81} - \dfrac{x^2}{9} = 1$

57. $4x^2 - y^2 = 36$ **58.** $x^2 - 9y^2 = 36$

59. $25y^2 - x^2 = 100$ **60.** $4y^2 - 9x^2 = 36$

Putting the Concepts Together (Sections 9.1–9.5)

We designed these problems so that you can review Sections 9.1–9.5 and show your mastery of the concepts. Take time to work these problems before proceeding with the next section. The answers are located at the back of the text on page AN-56.

1. Find the exact distance $d(P_1, P_2)$ between points $P_1 = (-6, 4)$ and $P_2 = (3, -2)$.

2. Find the midpoint of the line segment formed by joining the points $P_1 = (-3, 1)$ and $P_2 = (5, -7)$.

In Problems 3 and 4, find the center (h, k) and the radius r of each circle. Graph each circle.

3. $(x + 2)^2 + (y - 8)^2 = 36$

4. $x^2 + y^2 + 6x - 4y - 3 = 0$

In Problems 5 and 6, find the standard form of the equation of each circle.

5. Center $(0, 0)$; contains the point $(-5, 12)$

6. With endpoints of a diameter at $(-1, 5)$ and $(5, -3)$.

In Problems 7 and 8, find the vertex, focus, and directrix of each parabola. Graph each parabola.

7. $(x + 2)^2 = -4(y - 4)$

8. $y^2 + 2y - 8x + 25 = 0$

In Problems 9 and 10, find an equation for each parabola described.

9. Vertex at $(-1, -2)$; focus at $(-1, -5)$

10. Vertex at $(-3, 3)$; contains the point $(-1, 7)$; axis of symmetry parallel to the x-axis

In Problems 11 and 12, find the center, vertices, and foci of each ellipse. Graph each ellipse.

11. $x^2 + 9y^2 = 81$

12. $\dfrac{(x + 1)^2}{36} + \dfrac{(y - 2)^2}{49} = 1$

In Problems 13 and 14, find an equation for each ellipse described.

13. Foci at $(0, \pm 6)$; vertices at $(0, \pm 9)$

14. Center at $(3, -4)$; vertex at $(7, -4)$; focus at $(6, -4)$

In Problems 15 and 16, find the vertices, foci, and asymptotes for each hyperbola. Graph each hyperbola using the asymptotes as a guide.

15. $\dfrac{y^2}{81} - \dfrac{x^2}{9} = 1$

16. $25x^2 - y^2 = 25$

17. Find an equation for a hyperbola with center at $(0, 0)$, focus at $(0, -5)$, and vertex at $(0, -2)$.

18. A large floodlight is in the shape of a parabola with diameter 36 inches and depth 12 inches. How far from the vertex should the light bulb be placed so that the rays will be reflected parallel to the axis?

9.6 Systems of Nonlinear Equations

Objectives

1. Solve a System of Nonlinear Equations Using Substitution
2. Solve a System of Nonlinear Equations Using Elimination

Are You Prepared for This Section?

Before getting started, complete the following problems. If you get a problem wrong, go back to the section cited and review the material.

P1. Solve the system using substitution: $\begin{cases} y = 2x - 5 \\ 2x - 3y = 7 \end{cases}$ [Section 3.1, pp. 223–225]

P2. Solve the system using elimination: $\begin{cases} 2x - 4y = -11 \\ -x + 5y = 13 \end{cases}$ [Section 3.1, pp. 225–227]

P3. Solve the system: $\begin{cases} 3x - 5y = 4 \\ -6x + 10y = -8 \end{cases}$ [Section 3.1, pp. 229]

Prepared?...Answers **P1.** $(2, -1)$

P2. $\left(-\dfrac{1}{2}, \dfrac{5}{2}\right)$

P3. $\{(x, y) \mid 3x - 5y = 4\}$

Recall from Section 3.1 that a system of equations is a grouping of two or more equations, each containing one or more variables. Systems of linear equations were solved in Chapter 3. Solving systems of equations where the equations are not linear will now be examined. Systems containing only two equations with two unknowns will be addressed. In a **system of nonlinear equations** in two variables, at least one of the equations is not linear. That is, at least one of the equations cannot be written in the form $Ax + By = C$. Some examples of nonlinear systems of equations containing two unknowns follow.

$$\begin{cases} x + y^2 = 5 & \text{(1) A parabola} \\ 2x + y = 4 & \text{(2) A line} \end{cases} \qquad \begin{cases} x^2 + y^2 = 9 & \text{(1) A circle} \\ -x^2 + y = 9 & \text{(2) A parabola} \end{cases}$$

In Section 3.1, the solution of a system of linear equations were represented geometrically by showing the point of intersection of the equations in the system. The same could be done for nonlinear systems—the point(s) of intersection represent the solution(s) to the system.

As you worked through Chapter 3, you may have developed a sense of when substitution or when elimination was the best approach for solving a system. The same is true of nonlinear systems—sometimes substitution is best, sometimes elimination is best. Experience and a degree of imagination are your friends when solving these problems.

▶ ❶ Solve a System of Nonlinear Equations Using Substitution

The steps for solving a system of nonlinear equations using substitution are identical to the steps for solving a system of linear equations using substitution.

EXAMPLE 1 **How to Solve a System of Nonlinear Equations Using Substitution**

Solve the following system of equations using substitution: $\begin{cases} 3x - y = -2 & \text{(1) A line} \\ 2x^2 - y = 0 & \text{(2) A parabola} \end{cases}$

Step-by-Step Solution

Step 1: Solve equation (1) for y.

Equation (1): $3x - y = -2$

Add y to both sides; add 2 to both sides: $y = 3x + 2$

Step 2: Substitute $3x + 2$ for y in equation (2).

Equation (2): $2x^2 - y = 0$

$2x^2 - (3x + 2) = 0$

Distribute: $2x^2 - 3x - 2 = 0$

Step 3: Solve for x.

Factor: $(2x + 1)(x - 2) = 0$

Zero-Product Property: $2x + 1 = 0$ or $x - 2 = 0$

$x = -\dfrac{1}{2}$ or $x = 2$

Step 4: Let $x = -\dfrac{1}{2}$ and $x = 2$ in equation (1) to determine y.

Equation (1): $3x - y = -2$

$x = -\dfrac{1}{2}$: $3\left(-\dfrac{1}{2}\right) - y = -2$

$-\dfrac{3}{2} - y = -2$

Add $\dfrac{3}{2}$ to both sides: $-y = -\dfrac{1}{2}$

Multiply both sides by -1: $y = \dfrac{1}{2}$

(continued)

$$x = 2: \qquad 3(2) - y = -2$$
$$6 - y = -2$$

Subtract 6 from both sides: $\qquad -y = -8$

Multiply both sides by -1: $\qquad y = 8$

The apparent solutions are $\left(-\dfrac{1}{2}, \dfrac{1}{2}\right)$ and $(2, 8)$.

Step 5: Check

Equation (1)	**Equation (2)**

$$3x - y = -2 \qquad\qquad 2x^2 - y = 0$$

$x = -\dfrac{1}{2}; y = \dfrac{1}{2}:$

$$3\left(-\dfrac{1}{2}\right) - \dfrac{1}{2} \stackrel{?}{=} -2 \qquad\qquad 2\left(-\dfrac{1}{2}\right)^2 - \dfrac{1}{2} \stackrel{?}{=} 0$$

$$-\dfrac{3}{2} - \dfrac{1}{2} \stackrel{?}{=} -2 \qquad\qquad 2\left(\dfrac{1}{4}\right) - \dfrac{1}{2} \stackrel{?}{=} 0$$

$$-\dfrac{4}{2} = -2 \quad \text{True} \qquad\qquad \dfrac{1}{2} - \dfrac{1}{2} = 0 \quad \text{True}$$

$x = 2; y = 8:$

$$3x - y = -2 \qquad\qquad 2x^2 - y = 0$$

$$3(2) - 8 \stackrel{?}{=} -2 \qquad\qquad 2(2)^2 - 8 \stackrel{?}{=} 0$$

$$6 - 8 = -2 \quad \text{True} \qquad\qquad 2(4) - 8 = 0 \quad \text{True}$$

Each solution checks. Figure 44 shows the graphs of each equation in the system with the points of intersection (the solutions of the system) labeled at $\left(-\dfrac{1}{2}, \dfrac{1}{2}\right)$ and $(2, 8)$. ●

Figure 44

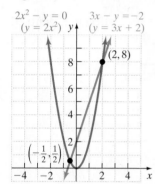

> **Quick ✓**
> 1. Solve the following system of equations using substitution:
> $$\begin{cases} 2x + y = -1 \\ x^2 - y = 4 \end{cases}$$

EXAMPLE 2 **Solving a System of Nonlinear Equations Using Substitution**

Solve the following system of equations using substitution:

$$\begin{cases} x + y = 2 & \text{(1) A line} \\ (x + 2)^2 + (y - 1)^2 = 9 & \text{(2) A circle} \end{cases}$$

Solution
Solve equation (1) for y.

$$x + y = 2$$

Subtract x from both sides: $\qquad y = -x + 2$

Substitute this expression for y in equation (2), $(x + 2)^2 + (y - 1)^2 = 9$, and then solve for x.

Substitute $-x + 2$ for y in equation (2): $\quad (x + 2)^2 + (-x + 2 - 1)^2 = 9$

Combine like terms: $\qquad (x + 2)^2 + (1 - x)^2 = 9$

Use FOIL: $\qquad x^2 + 4x + 4 + 1 - 2x + x^2 = 9$

Combine like terms: $\qquad 2x^2 + 2x + 5 = 9$

Put in standard form: $\qquad 2x^2 + 2x - 4 = 0$

Factor: $\qquad 2(x^2 + x - 2) = 0$

Divide both sides by 2: $\qquad x^2 + x - 2 = 0$

Factor: $\qquad (x + 2)(x - 1) = 0$

Zero-Product Property: $\qquad x + 2 = 0 \quad \text{or} \quad x - 1 = 0$

Solve for x: $\qquad x = -2 \quad \text{or} \quad x = 1$

Let $x = -2$ and $x = 1$ in equation (1) to determine y.

$$x + y = 2$$

$x = -2$: $-2 + y = 2$ $\qquad\qquad$ $x = 1$: $1 + y = 2$

Add 2 to both sides: $y = 4$ $\qquad$ Subtract 1 from both sides: $y = 1$

The apparent solutions are $(-2, 4)$ and $(1, 1)$.

Figure 45

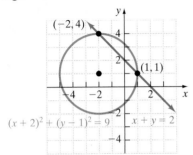

$(x + 2)^2 + (y - 1)^2 = 9 \qquad x + y = 2$

Check	**Equation (1)**	**Equation (2)**

$x = -2; y = 4$: $x + y = 2$ $\qquad$ $(x + 2)^2 + (y - 1)^2 = 9$

$-2 + 4 = 2$ True $\qquad$ $(-2 + 2)^2 + (4 - 1)^2 \overset{?}{=} 9$

$3^2 = 9$ True

$x = 1; y = 1$: $x + y = 2$ $\qquad$ $(x + 2)^2 + (y - 1)^2 = 9$

$1 + 1 = 2$ True $\qquad$ $(1 + 2)^2 + (1 - 1)^2 \overset{?}{=} 9$

$3^2 = 9$ True

Each solution checks. The graphs are shown in Figure 45 with the points of intersection (the solutions of the system) labeled at $(-2, 4)$ and $(1, 1)$.

Quick ✓

2. Solve the following system of equations using substitution:

$$\begin{cases} 2x + y = 0 \\ (x - 4)^2 + (y + 2)^2 = 9 \end{cases}$$

▶ ❷ Solve a System of Nonlinear Equations Using Elimination

Now let's discuss the method of elimination. Use elimination to solve a system of nonlinear equations in the same way it was used to solve systems of linear equations.
    Remember, in using elimination, to get the coefficients of one variable to be additive inverses.

EXAMPLE 3   **How to Solve a System of Nonlinear Equations by Elimination**

Solve the following system of equations using elimination: $\begin{cases} x^2 + y^2 = 13 & \text{(1) A circle} \\ x^2 - y = 7 & \text{(2) A parabola} \end{cases}$

Step-by-Step Solution

Step 1: Multiply equation (2) by -1 so the coefficients of x^2 become additive inverses.

$$\begin{cases} x^2 + y^2 = 13 & (1) \\ -x^2 + y = -7 & (2) \end{cases}$$

Step 2: Add equations (1) and (2) to eliminate x^2. Solve the resulting equation for y.

Add: $\qquad \begin{cases} x^2 + y^2 = 13 & (1) \\ \underline{-x^2 + y = -7} & (2) \end{cases}$

$$y^2 + y = 6$$

Put in standard form: $\qquad y^2 + y - 6 = 0$

Factor: $\qquad (y + 3)(y - 2) = 0$

Zero-Product Property: $y + 3 = 0$ or $y - 2 = 0$

$y = -3$ or $y = 2$

(continued)

Step 3: Solve for x using equation (2).

Work Smart

Use equation (2) in Step 3 because it's easier to work with.

$$x^2 - y = 7 \quad (2)$$

Using $y = -3$: $\quad x^2 - (-3) = 7$

$$x^2 + 3 = 7$$
$$x^2 = 4$$
$$x = \pm 2$$

Using $y = 2$: $\quad x^2 - 2 = 7$

$$x^2 = 9$$
$$x = \pm 3$$

The apparent solutions are $(-2, -3)$, $(2, -3)$, $(-3, 2)$, and $(3, 2)$.

Step 4: Check

The check is left to you to verify that all four of the apparent solutions are solutions to the system. The four points $(-2, -3)$, $(2, -3)$, $(-3, 2)$, and $(3, 2)$ are the points of intersection of the graphs. See Figure 46.

Figure 46

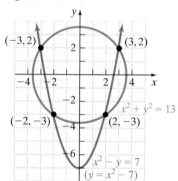

EXAMPLE 4   **Solving a System of Nonlinear Equations by Elimination**

Solve the following system of equations using elimination:

$$\begin{cases} x^2 - y^2 = 4 & \text{(1) A hyperbola} \\ x^2 - y = 0 & \text{(2) A parabola} \end{cases}$$

Solution

Multiply equation (2) by -1 to get the coefficients of x^2 to be additive inverses.

$$\begin{cases} x^2 - y^2 = 4 & (1) \\ -x^2 + y = 0 & (2) \end{cases}$$

Add equations (1) and (2) to eliminate x^2. Solve the resulting equation for y.

Figure 47

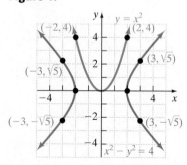

$$\begin{cases} x^2 - y^2 = 4 & (1) \\ \underline{-x^2 + y = 0} & (2) \\ -y^2 + y = 4 \end{cases}$$

Add y^2 to both sides; subtract y from both sides: $\quad y^2 - y + 4 = 0$

This is a quadratic equation whose discriminant is $b^2 - 4ac = (-1)^2 - 4(1)(4) = 1 - 16 = -15$. The equation has no real solution. Therefore, the system of equations is inconsistent. The solution set is $\varnothing$ or $\{\ \}$. Figure 47 shows the graphs of $x^2 - y^2 = 4$ and $x^2 - y = 0$ ($y = x^2$). Figure 47 confirms the algebraic result.

Quick ✓

Solve each system of nonlinear equations using elimination.

3. $\begin{cases} x^2 + y^2 = 16 \\ x^2 - 2y = 8 \end{cases}$

4. $\begin{cases} x^2 - y = -4 \\ x^2 + y^2 = 9 \end{cases}$

9.6 Exercises MyMathLab®

Exercise numbers in green have complete video solutions in MyMathLab or may be accessed using the QR code to the right.

*Problems **1–4** are the Quick ✔s that follow the EXAMPLES.*

Building Skills

In Problems 5–12, solve the system of nonlinear equations using the method of substitution. See Objective 1.

5. $\begin{cases} y = x^2 + 4 \\ y = x + 4 \end{cases}$

6. $\begin{cases} y = x^3 + 2 \\ y = x + 2 \end{cases}$

7. $\begin{cases} y = \sqrt{25 - x^2} \\ x + y = 7 \end{cases}$

8. $\begin{cases} y = \sqrt{100 - x^2} \\ x + y = 14 \end{cases}$

9. $\begin{cases} x^2 + y^2 = 4 \\ y = x^2 - 2 \end{cases}$

10. $\begin{cases} x^2 + y^2 = 16 \\ y = x^2 - 4 \end{cases}$

11. $\begin{cases} xy = 4 \\ x^2 + y^2 = 8 \end{cases}$

12. $\begin{cases} xy = 1 \\ x^2 - y = 0 \end{cases}$

In Problems 13–20, solve the system of nonlinear equations using the method of elimination. See Objective 2.

13. $\begin{cases} x^2 + y^2 = 4 \\ y^2 - x = 4 \end{cases}$

14. $\begin{cases} x^2 + y^2 = 8 \\ x^2 + y^2 + 4y = 0 \end{cases}$

15. $\begin{cases} x^2 + y^2 = 7 \\ x^2 - y^2 = 25 \end{cases}$

16. $\begin{cases} 4x^2 + 16y^2 = 16 \\ 2x^2 - 2y^2 = 8 \end{cases}$

17. $\begin{cases} x^2 + y^2 = 6y \\ x^2 = 3y \end{cases}$

18. $\begin{cases} 2x^2 + y^2 = 18 \\ x^2 - y^2 = 9 \end{cases}$

19. $\begin{cases} x^2 - 2x - y = 8 \\ 6x + 2y = -4 \end{cases}$

20. $\begin{cases} 2x^2 - 5x + y = 12 \\ 14x - 2y = -16 \end{cases}$

Mixed Practice

In Problems 21–36, solve the system of nonlinear equations using any method you wish.

21. $\begin{cases} y = x^2 - 6x + 4 \\ 5x + y = 6 \end{cases}$

22. $\begin{cases} y = x^2 + 4x + 5 \\ x - y = 9 \end{cases}$

23. $\begin{cases} x^2 + y^2 = 16 \\ x^2 - y^2 = 16 \end{cases}$

24. $\begin{cases} x^2 + y^2 = 25 \\ x^2 - y^2 = 25 \end{cases}$

25. $\begin{cases} (x - 4)^2 + y^2 = 25 \\ x - y = -3 \end{cases}$

26. $\begin{cases} (x + 5)^2 + (y - 2)^2 = 100 \\ 8x + y = 18 \end{cases}$

27. $\begin{cases} (x - 1)^2 + (y + 2)^2 = 4 \\ y^2 + 4y - x = -1 \end{cases}$

28. $\begin{cases} (x + 2)^2 + (y - 1)^2 = 4 \\ y^2 - 2y - x = 5 \end{cases}$

29. $\begin{cases} (x + 3)^2 + 4y^2 = 4 \\ x^2 + 6x - y = 13 \end{cases}$

30. $\begin{cases} 9x^2 + 4y^2 = 36 \\ x^2 + (y - 7)^2 = 4 \end{cases}$

31. $\begin{cases} x^2 - y^2 = 21 \\ x + y = 7 \end{cases}$

32. $\begin{cases} y - 2x = 1 \\ 2x^2 + y^2 = 1 \end{cases}$

33. $\begin{cases} x^2 + 2y^2 = 16 \\ 4x^2 - y^2 = 24 \end{cases}$

34. $\begin{cases} 4x^2 + 3y^2 = 4 \\ 6y^2 - 2x^2 = 3 \end{cases}$

35. $\begin{cases} x^2 + y^2 = 25 \\ y = -x^2 + 6x - 5 \end{cases}$

36. $\begin{cases} x^2 + y^2 = 65 \\ y = -x^2 + 9 \end{cases}$

Applying the Concepts

37. Fun with Numbers The difference of two numbers is 2. The sum of their squares is 34. Find the numbers.

38. Fun with Numbers The sum of two numbers is 8. The sum of their squares is 160. Find the numbers.

△ **39. Perimeter and Area of a Rectangle** The perimeter of a rectangle is 48 feet. The area of the rectangle is 140 square feet. Find the dimensions of the rectangle.

△ **40. Perimeter and Area of a Rectangle** The perimeter of a rectangle is 64 meters. The area of the rectangle is 240 square meters. Find the dimensions of the rectangle.

41. Constructing a Box A rectangular piece of cardboard, whose area is 190 square centimeters, is made into an open box by cutting a 2-centimeter square from each corner and turning up the sides. See the figure. If the box is to have a volume of 180 cubic centimeters, what size cardboard should you start with?

42. Fencing A farmer has 132 yards of fencing available to enclose a 900-square-yard region in the shape of adjoining squares with sides of length x and y. See the figure. Find x and y.

Extending the Concepts

In Problems 43–46, solve the system of nonlinear equations. Do not attempt to graph the equations in the system.

43. $\begin{cases} y^2 + y + x^2 - x - 2 = 0 \\ y + 1 + \dfrac{x - 2}{y} = 0 \end{cases}$

44. $\begin{cases} x^3 - 2x^2 + y^2 + 3y - 4 = 0 \\ x - 2 + \dfrac{y^2 - y}{x^2} = 0 \end{cases}$

45. $\begin{cases} \ln x = 4 \ln y \\ \log_3 x = 2 + 2 \log_3 y \end{cases}$

46. $\begin{cases} \ln x = 5 \ln y \\ \log_2 x = 3 + 2 \log_2 y \end{cases}$

47. If r_1 and r_2 are two solutions of a quadratic equation $ax^2 + bx + c = 0$, then it can be shown that

$$r_1 + r_2 = -\frac{b}{a} \quad \text{and} \quad r_1 r_2 = \frac{c}{a}$$

Solve this system of equations for r_1 and r_2.

48. A circle and a line intersect at most twice. A circle and a parabola intersect at most four times. How many times do you think a circle and the graph of a polynomial of degree 3 can intersect? What about a circle and the graph of a polynomial of degree 4? What about a circle and the graph of a polynomial of degree n? Explain your conclusions using an algebraic argument.

Synthesis Review

In Problems 49–53, evaluate the functions given that $f(x) = 3x + 4$ and $g(x) = 2^x$.

49. (a) $f(1)$ **(b)** $g(1)$ **50. (a)** $f(2)$ **(b)** $g(2)$

51. (a) $f(3)$ **(b)** $g(3)$ **52. (a)** $f(4)$ **(b)** $g(4)$

53. (a) $f(5)$ **(b)** $g(5)$

54. Use the results of Problems 49–53 to compute

  (a) $f(2) - f(1)$ **(b)** $f(3) - f(2)$
  (c) $f(4) - f(3)$ **(d)** $f(5) - f(4)$
  (e) $\dfrac{g(2)}{g(1)}$ **(f)** $\dfrac{g(3)}{g(2)}$
  (g) $\dfrac{g(4)}{g(3)}$ **(h)** $\dfrac{g(5)}{g(4)}$
  (i) Make a generalization about $f(n + 1) - f(n)$ for $n \ge 1$ an integer. Make a generalization about $\dfrac{g(n + 1)}{g(n)}$ for $n \ge 1$ an integer.

Technology Exercises

In Problems 55–64, use technology to solve the following systems of nonlinear equations. If necessary, round answers to three decimal places.

55. $\begin{cases} y = x^2 - 6x + 4 \\ 5x + y = 6 \end{cases}$ **56.** $\begin{cases} y = x^2 + 4x + 5 \\ x - y = 9 \end{cases}$

57. $\begin{cases} x^2 + y^2 = 16 \\ x^2 - y^2 = 16 \end{cases}$ **58.** $\begin{cases} x^2 + y^2 = 25 \\ x^2 - y^2 = 25 \end{cases}$

59. $\begin{cases} (x - 4)^2 + y^2 = 25 \\ x - y = -3 \end{cases}$

60. $\begin{cases} (x + 5)^2 + (y - 2)^2 = 100 \\ 8x + y = 18 \end{cases}$

61. $\begin{cases} (x - 1)^2 + (y + 2)^2 = 4 \\ x^2 + 4y - x = -1 \end{cases}$

62. $\begin{cases} (x + 2)^2 + (y - 1)^2 = 4 \\ y^2 - 2y - x = 5 \end{cases}$

63. $\begin{cases} x^2 + 4y^2 = 4 \\ x^2 + 6x - y = -13 \end{cases}$ **64.** $\begin{cases} 9x^2 + 4y^2 = 36 \\ x^2 + (y - 7)^2 = 4 \end{cases}$

Chapter 9 Activity: How Do You Know That...?

Focus: A sharing of ideas to identify topics contained in the study of conics

Time: 30–35 minutes

Group size: 2–4

For each question, every member of the group should spend 1–2 minutes individually listing "How they know ...." At the end of the allotted time, the group should convene and conduct a 2–3-minute discussion of the different responses.

How Do You Know ...

...that a triangle with vertices at $(2, 6)$, $(0, -2)$ and $(5, 1)$ is an isosceles triangle?

...the coordinates for the midpoint between two given ordered pairs?

...that $x^2 + y^2 + 4x - 8y = 16$ is an equation of a circle and not a parabola?

...which way a parabola opens?

...whether an ellipse's center is at the origin or another point?

...that a hyperbola will not intersect its asymptotes?

...that a circle and a parabola can have 0, 1, 2, 3, or 4 points of intersection? (If necessary, use a sketch to support your response.)

Chapter 9 Review

Section 9.1 Distance and Midpoint Formulas

KEY CONCEPTS

- **The Distance Formula**
  The distance between two points $P_1 = (x_1, y_1)$ and $P_2 = (x_2, y_2)$, denoted by $d(P_1, P_2)$, is
  $d(P_1, P_2) = \sqrt{(x_2 - x_1)^2 + (y_2 - y_1)^2}$.
- **The Midpoint Formula**
  The midpoint $M = (x, y)$ of the line segment from $P_1 = (x_1, y_1)$ to $P_2 = (x_2, y_2)$ is $M = \left(\dfrac{x_1 + x_2}{2}, \dfrac{y_1 + y_2}{2} \right)$.

You Should Be Able To...	EXAMPLE	Review Exercises
❶ Use the Distance Formula (p. 690)	Examples 1 through 3	1–6, 12
❷ Use the Midpoint Formula (p. 693)	Example 4	7–11

In Problems 1–6, find the distance $d(P_1, P_2)$ between points P_1 and P_2.

1. $P_1 = (0, 0)$ and $P_2 = (-4, -3)$
2. $P_1 = (-3, 2)$ and $P_2 = (5, -4)$
3. $P_1 = (-1, 1)$ and $P_2 = (5, 3)$
4. $P_1 = (6, -7)$ and $P_2 = (6, -1)$
5. $P_1 = (\sqrt{7}, -\sqrt{3})$ and $P_2 = (4\sqrt{7}, 5\sqrt{3})$
6. $P_1 = (-0.2, 1.7)$ and $P_2 = (1.3, 3.7)$

In Problems 7–11, find the midpoint of the line segment formed by joining the points P_1 and P_2.

7. $P_1 = (-1, 6)$ and $P_2 = (-3, 4)$
8. $P_1 = (7, 0)$ and $P_2 = (5, -4)$

9. $P_1 = (-\sqrt{3}, 2\sqrt{6})$ and $P_2 = (-7\sqrt{3}, -8\sqrt{6})$

10. $P_1 = (5, -2)$ and $P_2 = (0, 3)$

11. $P_1 = \left(\dfrac{1}{4}, \dfrac{2}{3} \right)$ and $P_2 = \left(\dfrac{5}{4}, \dfrac{1}{3} \right)$

12. Consider the three points
    $A = (-2, 2), B = (1, -1), C = (-1, -3)$.

    (a) Plot each point in the Cartesian plane and form the triangle ABC.
    (b) Find the length of each side of the triangle.
    (c) Verify that the triangle is a right triangle.
    (d) Find the area of the triangle.

Section 9.2   Circles

KEY CONCEPTS

- **Standard Form of an Equation of a Circle**

  The standard form of the equation of a circle with radius r and center (h, k) is $(x - h)^2 + (y - k)^2 = r^2$.

- **General Form of the Equation of a Circle**

  The general form of the equation of a circle is $x^2 + y^2 + ax + by + c = 0$ when the graph exists.

KEY TERMS

Conic sections
Circle
Radius
Center

You Should Be Able To...	EXAMPLE	Review Exercises
1 Write the standard form of the equation of a circle (p. 697)	Example 1	13–20
2 Graph a circle (p. 698)	Example 2	15–18; 21–30
3 Find the center and radius of a circle given an equation in general form (p. 698)	Example 3	27–30

In Problems 13 and 14, find the center and radius of each circle. Write the standard form of the equation.

13.

14.
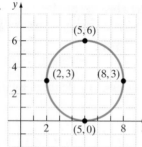

In Problems 15–18, write the standard form of the equation of each circle whose radius is r and center is (h, k). Graph each circle.

15. $r = 4$; $(h, k) = (0, 0)$ **16.** $r = 3$; $(h, k) = (-3, 1)$

17. $r = 1$; $(h, k) = (5, -2)$ **18.** $r = \sqrt{7}$; $(h, k) = (4, 0)$

In Problems 19 and 20, find the standard form of the equation of each circle.

19. Center at $(2, -1)$; contains the point $(5, 3)$.

20. Endpoints of a diameter at $(1, 7)$ and $(-3, -1)$.

In Problems 21–26, find the center (h, k) and the radius r of each circle. Graph each circle.

21. $x^2 + y^2 = 25$

22. $(x - 1)^2 + (y - 2)^2 = 4$

23. $x^2 + (y - 4)^2 = 16$

24. $(x + 1)^2 + (y + 6)^2 = 49$

25. $(x + 2)^2 + \left(y - \dfrac{3}{2}\right)^2 = \dfrac{1}{4}$

26. $(x + 3)^2 + (y + 3)^2 = 4$

In Problems 27–30, find the center (h, k) and the radius r of each circle. Graph each circle.

27. $x^2 + y^2 + 6x + 10y - 2 = 0$

28. $x^2 + y^2 - 8x + 4y + 16 = 0$

29. $x^2 + y^2 + 2x - 4y - 4 = 0$

30. $x^2 + y^2 - 10x - 2y + 17 = 0$

Section 9.3   Parabolas

KEY CONCEPTS

- **Equations of a Parabola: Vertex at $(0, 0)$; $a > 0$**

  See Table 1 on page 704.

- **Equations of a Parabola: Vertex at (h, k); $a > 0$**

  See Table 2 on page 706.

KEY TERMS

Parabola
Focus
Directrix
Axis of symmetry
Vertex

You Should Be Able To...	EXAMPLE	Review Exercises
❶ Graph parabolas whose vertex is the origin (p. 703)	Examples 1 and 2	31–36
❷ Find the equation of a parabola (p. 705)	Examples 3 and 4	31–34
❸ Graph a parabola whose vertex is not the origin (p. 706)	Example 5	37–39
❹ Solve applied problems involving parabolas (p. 708)	Example 6	40

In Problems 31–34, find the equation of the parabola described. Graph each parabola.

31. Vertex at $(0,0)$; focus at $(0,-3)$

32. Focus at $(-4,0)$; directrix the line $x = 4$

33. Vertex at $(0,0)$; contains the point $(8,-2)$; axis of symmetry the x-axis

34. Vertex at $(0,0)$; directrix the line $y = -2$; axis of symmetry the y-axis

In Problems 35–39, find the vertex, focus, and directrix of each parabola. Graph each parabola.

35. $x^2 = 2y$ **36.** $y^2 = 16x$

37. $(x+1)^2 = 8(y-3)$

38. $(y-4)^2 = -2(x+3)$

39. $x^2 - 10x + 3y + 19 = 0$

40. Radio Telescope The U.S. Naval Research Laboratory has a giant radio telescope with a dish that is shaped like a parabola. The signals that are received by the dish strike the surface of the dish and are reflected to a single point, where the receiver is located. If the giant dish is 300 feet across and 44 feet deep at its center, at what position should the receiver be placed?

Section 9.4   Ellipses

KEY CONCEPTS

- **Ellipses with Center at $(0, 0)$**
  See Table 3 on page 712.

- **Ellipses with Center at (h, k)**
  See Table 4 on page 715.

KEY TERMS

Ellipse
Foci
Major axis
Center
Minor axis
Vertices

You Should Be Able To...	EXAMPLE	Review Exercises
❶ Graph an ellipse whose center is the origin (p. 712)	Examples 1 and 2	41, 42
❷ Find the equation of an ellipse whose center is the origin (p. 714)	Example 3	43–45
❸ Graph an ellipse whose center is not the origin (p. 715)	Example 4	46–47
❹ Solve applied problems involving ellipses (p. 716)	Example 5	48

In Problems 41 and 42, find the vertices and foci of each ellipse. Graph each ellipse.

41. $\dfrac{x^2}{9} + y^2 = 1$ **42.** $9x^2 + 4y^2 = 36$

In Problems 43–45, find an equation for each ellipse. Graph each ellipse.

43. Center at $(0,0)$; focus at $(0,3)$; vertex at $(0,5)$

44. Center at $(0,0)$; focus at $(-2,0)$; vertex at $(-6,0)$

45. Foci at $(\pm 8, 0)$; vertices at $(\pm 10, 0)$

In Problems 46 and 47, find the center, vertices, and foci of each ellipse. Graph each ellipse.

46. $\dfrac{(x-1)^2}{49} + \dfrac{(y+2)^2}{25} = 1$

47. $25(x+3)^2 + 9(y-4)^2 = 225$

48. Semielliptical Arch Bridge An arch in the shape of the upper half of an ellipse is used to support a bridge that spans a river 60 feet wide. The center of the arch is 16 feet above the center of the river.

(a) Write the equation for the ellipse in which the x-axis coincides with the water and the y-axis passes through the center of the arch.

(b) Can a rectangular barge that is 25 feet wide and sits 12 feet above the surface of the water fit through the opening of the bridge?

Section 9.5 Hyperbolas

KEY CONCEPTS

- **Hyperbolas with Center at the Origin**
  See Table 5 on page 721.

- **Asymptotes of a Hyperbola**
  See Table 6 on page 725.

KEY TERMS

Hyperbola
Foci
Transverse axis
Center
Conjugate axis
Branches
Vertices

You Should Be Able To...	EXAMPLE	Review Exercises
❶ Graph a hyperbola whose center is the origin (p. 721)	Examples 1 and 2	49–53
❷ Find the equation of a hyperbola whose center is the origin (p. 724)	Example 3	54–56
❸ Find the asymptotes of a hyperbola whose center is the origin (p. 724)	Example 4	52, 53

In Problems 49–51, find the vertices and foci of each hyperbola. Graph each hyperbola.

49. $\dfrac{x^2}{4} - \dfrac{y^2}{9} = 1$

50. $\dfrac{y^2}{25} - \dfrac{x^2}{49} = 1$

51. $16y^2 - 25x^2 = 400$

In Problems 52 and 53, graph each hyperbola using the asymptotes as a guide.

52. $\dfrac{x^2}{36} - \dfrac{y^2}{36} = 1$

53. $\dfrac{y^2}{25} - \dfrac{x^2}{4} = 1$

In Problems 54–56, find an equation for each hyperbola described. Graph each hyperbola.

54. Center at $(0, 0)$; focus at $(-4, 0)$; vertex at $(-3, 0)$

55. Vertices at $(0, -3)$ and $(0, 3)$; focus at $(0, 5)$

56. Vertices at $(0, \pm 4)$; asymptote the line $y = \dfrac{4}{3}x$

Section 9.6 Systems of Nonlinear Equations

KEY TERM

System of nonlinear equations

You Should Be Able To...	EXAMPLE	Review Exercises
❶ Solve a system of nonlinear equations using substitution (p. 729)	Examples 1 and 2	57–60; 65–76
❷ Solve a system of nonlinear equations using elimination (p. 731)	Examples 3 and 4	61–64; 65–76

In Problems 57–60, solve the system of nonlinear equations using the method of substitution.

57. $\begin{cases} 4x^2 + y^2 = 10 \\ y = x \end{cases}$

58. $\begin{cases} y = 2x^2 + 1 \\ y = x + 2 \end{cases}$

59. $\begin{cases} 6x - y = 5 \\ xy = 1 \end{cases}$

60. $\begin{cases} x^2 + y^2 = 26 \\ x^2 - 2y^2 = 23 \end{cases}$

In Problems 61–64, solve the system of nonlinear equations using the method of elimination.

61. $\begin{cases} 4x - y^2 = 0 \\ 2x^2 + y^2 = 16 \end{cases}$

62. $\begin{cases} x^2 - y = -2 \\ x^2 + y = 4 \end{cases}$

63. $\begin{cases} 4x^2 - 2y^2 = 2 \\ -x^2 + y^2 = 2 \end{cases}$

64. $\begin{cases} x^2 + y^2 = 8x \\ y^2 = 3x \end{cases}$

In Problems 65–72, solve the system of nonlinear equations using the method you prefer.

65. $\begin{cases} y = x + 2 \\ y = x^2 \end{cases}$

66. $\begin{cases} x^2 + 2y = 9 \\ 5x - 2y = 5 \end{cases}$

67. $\begin{cases} x^2 + y^2 = 36 \\ x - y = -6 \end{cases}$

68. $\begin{cases} y = 2x - 4 \\ y^2 = 4x \end{cases}$

69. $\begin{cases} x^2 + y^2 = 9 \\ x + y = 7 \end{cases}$

70. $\begin{cases} 2x^2 + 3y^2 = 14 \\ x^2 - y^2 = -3 \end{cases}$

71. $\begin{cases} x^2 + y^2 = 16 \\ x^2 + 4y = 16 \end{cases}$

72. $\begin{cases} x = 4 - y^2 \\ x = 2y + 4 \end{cases}$

73. Fun with Numbers The sum of two numbers is 12. The difference of their squares is 24. Find the two numbers.

△**74. Perimeter and Area of a Rectangle** The perimeter of a rectangle is 34 centimeters. The area of the rectangle is 60 square centimeters. Find the dimensions of the rectangle.

△**75. Dimensions of a Rectangle** The area of a rectangle is 2160 square inches. The diagonal of the rectangle is 78 inches. Find the dimensions of the rectangle.

△**76. Dimensions of a Triangle** A right triangle has a perimeter of 36 feet and a hypotenuse of 15 feet. Find the lengths of the legs of the right triangle.

Chapter 9 Test

Step-by-step test solutions are found on the Chapter Test Prep Videos available in MyMathLab®, *on* YouTube*, or may be accessed using the QR code to the right.*

1. Find the distance $d(P_1, P_2)$ between points $P_1 = (-1, 3)$ and $P_2 = (3, -5)$.

2. Find the midpoint of the line segment formed by joining the points $P_1 = (-7, 6)$ and $P_2 = (5, -2)$.

In Problems 3 and 4, find the center (h, k) and the radius r of each circle. Graph each circle.

3. $(x - 4)^2 + (y + 1)^2 = 9$

4. $x^2 + y^2 + 10x - 4y + 13 = 0$

In Problems 5 and 6, find the standard form of the equation of each circle.

5. Radius $r = 6$ and center $(h, k) = (-3, 7)$

6. Center at $(-5, 8)$; contains the point $(3, 2)$

In Problems 7 and 8, find the vertex, focus, and directrix of each parabola. Graph each parabola.

7. $(y + 2)^2 = 4(x - 1)$

8. $x^2 - 4x + 3y - 8 = 0$

In Problems 9 and 10, find an equation for each parabola described.

9. Vertex at $(0, 0)$; focus at $(0, -4)$

10. Focus at $(3, 4)$; directrix the line $x = -1$

In Problems 11 and 12, find the vertices and foci of each ellipse. Graph each ellipse.

11. $9x^2 + 25y^2 = 225$

12. $\dfrac{(x - 2)^2}{9} + \dfrac{(y + 4)^2}{16} = 1$

In Problems 13 and 14, find an equation for each ellipse described.

13. Center at $(0, 0)$; focus at $(0, -4)$; vertex at $(0, -5)$

14. Vertices at $(-1, 7)$ and $(-1, -3)$; focus at $(-1, -1)$

In Problems 15 and 16, find the vertices, foci, and asymptotes for each hyperbola. Graph each hyperbola using the asymptotes as a guide.

15. $x^2 - \dfrac{y^2}{4} = 1$

16. $16y^2 - 25x^2 = 1600$

17. Find an equation for a hyperbola with foci at $(\pm 8, 0)$ and vertex at $(-3, 0)$.

In Problems 18 and 19, solve the system of nonlinear equations using the method you prefer.

18. $\begin{cases} x^2 + y^2 = 17 \\ x + y = -3 \end{cases}$ **19.** $\begin{cases} x^2 + y^2 = 9 \\ 4x^2 - y^2 = 16 \end{cases}$

20. An arch in the shape of the upper half of an ellipse is used to support a bridge spanning a creek 30 feet wide. The center of the arch is 10 feet above the center of the creek.

   (a) Write the equation for the ellipse in which the x-axis coincides with the creek and the y-axis passes through the center of the arch.

   (b) What is the height of the arch at a distance 12 feet from the center of the creek?

Cumulative Review Chapters R–9

1. Determine the domain of $h(x) = \dfrac{4 - x}{7 - x}$.

2. Graph the linear equation: $5x - 2y = 6$.

3. Find the equation of the line that contains the point $(-3, -2)$ and is parallel to the line $4x - 3y = 15$. Write your answer in either slope-intercept or standard form, whichever you prefer.

4. Solve the following system of linear equations using the method you prefer:

$$\begin{cases} 2x + y = -2 \\ 5x + 2y = -1 \end{cases}$$

5. Evaluate: $\begin{vmatrix} 1 & 2 & 0 \\ 7 & -1 & -2 \\ 4 & -2 & 1 \end{vmatrix}$

In Problems 6 and 7, factor completely.

6. $8m^3 + n^3$ **7.** $9p^2 - 18p + 8$

In Problems 8 and 9, perform the indicated operations. Be sure to express the final answer in lowest terms.

8. $\dfrac{2a^2 - 7a - 4}{3a^2 - 13a + 4} \div \dfrac{a^2 - 2a - 35}{3a^2 + 14a - 5}$

9. $\dfrac{5}{3t^2 - 7t + 2} - \dfrac{2}{3t^2 - 4t + 1}$

10. Solve: $1 + \dfrac{1}{x - 1} = \dfrac{x + 2}{x}$

11. Solve: $\dfrac{x - 5}{x + 4} \geq 0$

12. Simplify: $\sqrt{75} + 7\sqrt{3} - \sqrt{27}$

13. Rationalize the denominator: $\dfrac{9}{2 + \sqrt{7}}$

14. Solve: $5 - \sqrt{x + 4} = 2$

15. Simplify: $(4 - i)(7 + 3i)$

16. Solve: $2x^2 - 4x - 3 = 0$

17. Graph: $f(x) = -x^2 + 2x + 8$

18. Solve: $x^2 - 3x - 4 \leq 0$

In Problems 19 and 20, graph each function. State the domain and the range of the function.

19. $f(x) = 2^{x+3}$ **20.** $g(x) = \log_4 x$

21. Evaluate: $\log_4 32$

In Problems 22 and 23, solve each equation.

22. $6^{x+1} = 216$ **23.** $\log_3(x^2 - 16) = 2$

24. Write the standard form of the equation of a circle with center at $(-3, 1)$ and containing the point $(0, 0)$.

25. Find the vertex, focus, and directrix, and then graph the following parabola: $(y - 3)^2 = 8(x + 3)$.

10 Sequences, Series, and the Binomial Theorem

Population growth has been a topic of debate among scientists for a long time. In fact, over 200 years ago, the English economist and mathematician Thomas Robert Malthus anonymously published a paper predicting that the world's population would overwhelm the Earth's capacity to sustain it—he claimed that food supplies increase arithmetically while population grows geometrically. See Problems 55 and 56 in Section 10.1 and Problem 99 in Section 10.3.

The Big Picture: Putting It Together

In Chapter 2, the domain of a function (inputs) was defined as the set of real numbers so that the outputs of the function make sense. In Sections 10.1–10.3, a special type of function called a *sequence* is introduced. The domain of a sequence is the set of natural numbers (that is, the positive integers). Functions whose domain is the set of all real numbers enable us to model situations in which the value of an independent variable, such as a population of bacteria, changes continuously. In contrast, sequences make it possible to model situations in which the value of a *dependent* variable changes at discrete intervals of time, such as weekly changes in the value of a deposit at a bank.

In Chapter 4, a formula for expanding $(x + a)^2$ was introduced. Does any formula exist for expanding $(x + a)^n$, where n is an integer greater than 2? Yes! This formula is discussed in Section 10.4.

10.1 Sequences

Objectives

1 Write the First Few Terms of a Sequence

2 Find a Formula for the *n*th Term of a Sequence

3 Use Summation Notation

Are You Prepared for This Section?

Before getting started, complete the following problems. If you get a problem wrong, go back to the section cited and review the material.

P1. Evaluate $f(x) = x^2 - 4$ at **(a)** $x = 3$ **(b)** $x = -7$.   [Section 2.2, pp. 158–160]

P2. If $g(x) = 2x - 3$, find $g(1) + g(2) + g(3)$.   [Section 2.2, pp. 158–160]

P3. In the function $f(n) = n^2 - 4$, what is the independent variable?   [Section 2.2, pp. 158–160]

Definition

A **sequence** is a function whose domain is the set of positive integers.

Because a sequence is a function, it has a graph. The function $f(x) = \dfrac{1}{x}$ for $x > 0$ is graphed in Figure 1(a). If all the points on this graph were removed except those whose *x*-coordinates are positive integers—that is, if all points were removed except $(1, 1)$, $\left(2, \dfrac{1}{2}\right)$, $\left(3, \dfrac{1}{3}\right)$, and so on, the remaining points would be the graph of the sequence $f(n) = \dfrac{1}{n}$, as shown in Figure 1(b). Notice that *n* is used to represent the independent variable in a sequence. This serves as a reminder that *n* is a positive integer, or natural number.

Figure 1

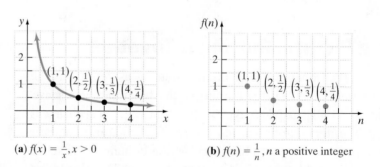

(a) $f(x) = \frac{1}{x}, x > 0$ **(b)** $f(n) = \frac{1}{n}, n$ a positive integer

A sequence may be represented by listing its values in order. For example, the sequence whose graph is given in Figure 1(b) might be represented as

$$f(1), f(2), f(3), f(4), \ldots \quad \text{or} \quad 1, \frac{1}{2}, \frac{1}{3}, \frac{1}{4}, \ldots$$

The numbers in the ordered list are the **terms** of the sequence. Notice that the list never ends, as the **ellipsis** (the three dots) indicates. A sequence that does not end is an **infinite sequence.** In contrast, **finite sequences** have a domain that is the first *n* positive integers. For example,

$$40, 44, 48, 52, 56, 60, 64$$

is a finite sequence because it contains $n = 7$ terms.

▶ 1 Write the First Few Terms of a Sequence

In a sequence, the traditional notation $f(n)$ is not used. This is done to distinguish sequences from functions whose domain is the set of all real numbers. Instead, the notation a_n is used to indicate that the function is a sequence. The value of the subscript represents the value of the independent variable, so a_4 means to evaluate the function defined by the sequence a_n at $n = 4$. For the sequence $f(n) = \dfrac{1}{n}$, write the function as $a_n = \dfrac{1}{n}$ and evaluate the function as follows:

$$a_1 = 1, \quad a_2 = \frac{1}{2}, \quad a_3 = \frac{1}{3}, \quad a_4 = \frac{1}{4}, \quad \ldots, \quad a_n = \frac{1}{n}$$

In addition, just as functions are named f, g, F, G, and so on, sequences can be named. Usually, lowercase letters from the beginning of the alphabet are used to name a sequence, as in a_n, b_n, c_n, and so on although this naming scheme is not necessary.

When a formula for the nth term (sometimes called the **general term**) of a sequence is known, the sequence is represented by placing braces around the formula for the nth term, rather than by writing out the terms of the sequence. For example, the sequence whose nth term is $b_n = \left(\dfrac{1}{3} \right)^n$ can be written as

$$\{ b_n \} = \left\{ \left(\frac{1}{3} \right)^n \right\}$$

or as

$$b_1 = \frac{1}{3}, \quad b_2 = \frac{1}{9}, \quad b_3 = \frac{1}{27}, \quad \ldots, \quad b_n = \left(\frac{1}{3} \right)^n$$

EXAMPLE 1   **Writing the First Five Terms of a Sequence**

Write the first five terms of the sequence $\{ a_n \} = \left\{ \dfrac{n}{n + 1} \right\}$.

Solution
To find the first five terms of the sequence, evaluate the function at $n = 1, 2, 3, 4,$ and 5.

$$a_1 = \frac{1}{1 + 1} = \frac{1}{2}, \quad a_2 = \frac{2}{2 + 1} = \frac{2}{3}, \quad a_3 = \frac{3}{3 + 1} = \frac{3}{4}, \quad a_4 = \frac{4}{4 + 1} = \frac{4}{5}, \quad a_5 = \frac{5}{5 + 1} = \frac{5}{6}$$

The first five terms of the sequence are $\dfrac{1}{2}, \dfrac{2}{3}, \dfrac{3}{4}, \dfrac{4}{5},$ and $\dfrac{5}{6}$.   ●

EXAMPLE 2   **Writing the First Five Terms of a Sequence**

Write the first five terms of the sequence $\{ b_n \} = \{ (-1)^n n^2 \}$.

Solution
The first five terms of the sequence are

$$b_1 = (-1)^1 \cdot 1^2 = -1, \quad b_2 = (-1)^2 \cdot 2^2 = 4, \quad b_3 = (-1)^3 \cdot 3^2 = -9,$$
$$b_4 = (-1)^4 \cdot 4^2 = 16, \quad b_5 = (-1)^5 \cdot 5^2 = -25$$

The first five terms of the sequence are $-1, 4, -9, 16,$ and -25.   ●

Notice in Example 2 that the signs of the terms in the sequence **alternate.** This occurs when there are factors such as $(-1)^{n+1}$, which equals 1 if n is odd and -1 if n is even, or $(-1)^n$, which equals -1 if n is odd and 1 if n is even.

Quick ✓

1. A(n) _____ is a function whose domain is the set of positive integers.

2. A sequence that does not end is said to be a(n) _____ sequence. A(n) ____ sequence has a domain that is the first n positive integers.

3. *True or False* A sequence is a function.

In Problems 4 and 5, write the first five terms of the given sequence.

4. $\{ a_n \} = \{ 2n - 3 \}$ 5. $\{ b_n \} = \{ (-1)^n \cdot 4n \}$

▶ ❷ Find a Formula for the *n*th Term of a Sequence

Sometimes a sequence is indicated by an observed pattern in the first few terms that makes it possible to determine the formula for the *n*th term. In the example that follows, a sufficient number of terms of the sequence are given so that a natural choice for the *n*th term is suggested.

EXAMPLE 3 | **Determining a Sequence from a Pattern**

Find a formula for the *n*th term of each sequence.

- **(a)** $3, 7, 11, 15, \ldots$
- **(b)** $2, 4, 8, 16, \ldots$
- **(c)** $1, -8, 27, -64, 125, \ldots$

Solution
The goal in all these problems is to find a formula in terms of n such that $n = 1$ is the first term, $n = 2$ is the second term, and so on.

- **(a)** When $n = 1$, then $a_1 = 3$; when $n = 2$, then $a_2 = 7$. Notice that each subsequent term increases by 4. A formula for the *n*th term is given by $a_n = 4n - 1$.

- **(b)** The terms of the sequence are all powers of 2 with the first term equaling 2^1, the second term equaling 2^2, and so on. A formula for the *n*th term is given by $b_n = 2^n$.

- **(c)** Notice that the sign of the terms alternate with the first term being positive. Thus $(-1)^{n+1}$ must be part of the formula. Ignoring the sign, notice that the terms are all perfect cubes. A formula for the *n*th term is given by $c_n = (-1)^{n+1}n^3$. ●

Quick ✓

In Problems 6 and 7, find a formula for the nth term of each sequence.

6. $5, 7, 9, 11, \ldots$

7. $\dfrac{1}{2}, -\dfrac{1}{3}, \dfrac{1}{4}, -\dfrac{1}{5}, \ldots$

▶ ❸ Use Summation Notation

In other mathematics courses, such as statistics and calculus, it is important to find the sum of the first n terms of a sequence $\{a_n\}$ — that is,

$$a_1 + a_2 + a_3 + \cdots + a_n$$

Mathematicians express the sum using **summation notation.** With this notation, write $a_1 + a_2 + a_3 + \cdots + a_n$ as

In Other Words
The symbol Σ is read "uppercase sigma," or "sigma" for short.

$$\sum_{i=1}^{n} a_i = a_1 + a_2 + a_3 + \cdots + a_n$$

The symbol Σ is an instruction to sum, or add up, the terms. The integer i is called the **index** of the sum; it tells you where to start the sum and where to end it. The expression

$$\sum_{i=1}^{n} a_i$$

means to add the terms of the sequence $\{a_i\}$ from $i = 1$ through $i = n$. Read the expression $\displaystyle\sum_{i=1}^{n} a_i$ as "the sum of *a* sub *i* from $i = 1$ to $i = n$." When a finite number of terms are to be added, the sum is called a **partial sum.**

EXAMPLE 4 **Finding a Partial Sum**

Write out the sum and determine its value.

(a) $\displaystyle\sum_{i=1}^{4} (2i + 5)$

(b) $\displaystyle\sum_{i=1}^{5} (i^2 - 5)$

Solution

(a) $\displaystyle\sum_{i=1}^{4} (2i + 5) = \underbrace{(2\cdot 1 + 5)}_{i\,=\,1} + \underbrace{(2\cdot 2 + 5)}_{i\,=\,2} + \underbrace{(2\cdot 3 + 5)}_{i\,=\,3} + \underbrace{(2\cdot 4 + 5)}_{i\,=\,4}$

$= 7 + 9 + 11 + 13$

$= 40$

(b) $\displaystyle\sum_{i=1}^{5} (i^2 - 5) = \underbrace{(1^2 - 5)}_{i\,=\,1} + \underbrace{(2^2 - 5)}_{i\,=\,2} + \underbrace{(3^2 - 5)}_{i\,=\,3} + \underbrace{(4^2 - 5)}_{i\,=\,4} + \underbrace{(5^2 - 5)}_{i\,=\,5}$

$= -4 + (-1) + 4 + 11 + 20$

$= 30$

Quick ✓

8. When there is a finite number of terms to be added, the sum is called a

_____ ____.

In Problems 9 and 10, write out the sum and determine its value.

9. $\displaystyle\sum_{i=1}^{3} (4i - 1)$

10. $\displaystyle\sum_{i=1}^{5} (i^3 + 1)$

The index of summation does not have to begin with 1. In addition, the index of summation does not have to be i. For example,

$$\sum_{k=3}^{6} (2k + 1) = \underbrace{(2\cdot 3 + 1)}_{k\,=\,3} + \underbrace{(2\cdot 4 + 1)}_{k\,=\,4} + \underbrace{(2\cdot 5 + 1)}_{k\,=\,5} + \underbrace{(2\cdot 6 + 1)}_{k\,=\,6}$$

$$= 7 + 9 + 11 + 13$$

$$= 40$$

Notice that the terms of this partial sum are identical to those in Example 4(a). What is the moral of the story? The same sum can be made to look entirely different by changing the starting point of the index of summation and changing the variable that represents the index.

Now reverse the process. Rather than writing out a sum, express a sum using summation notation.

EXAMPLE 5 **Writing a Sum in Summation Notation**

Express each sum using summation notation.

(a) $1^2 + 2^2 + 3^2 + \cdots + 10^2$

(b) $2 + 1 + \dfrac{2}{3} + \dfrac{1}{2} + \dfrac{2}{5} + \cdots + \dfrac{1}{6}$

Solution

(a) The sum $1^2 + 2^2 + 3^2 + \cdots + 10^2$ has 10 terms, each of the form i^2, and starts at $i = 1$ and ends at $i = 10$:

$$1^2 + 2^2 + 3^2 + \cdots + 10^2 = \sum_{i=1}^{10} i^2$$

(continued)

(b) First, figure out the pattern. With a little investigation,

$$2 + 1 + \frac{2}{3} + \frac{1}{2} + \frac{2}{5} + \cdots + \frac{1}{6} \text{ can be written as}$$

$$\frac{2}{1} + \frac{2}{2} + \frac{2}{3} + \frac{2}{4} + \frac{2}{5} + \cdots + \frac{2}{12}, \text{ so that the } n\text{th term of the sum is } \frac{2}{n}. \text{ There}$$

are $n = 12$ terms, so

$$2 + 1 + \frac{2}{3} + \frac{1}{2} + \frac{2}{5} + \cdots + \frac{1}{6} = \sum_{n=1}^{12} \left(\frac{2}{n}\right)$$

●

Work Smart

The index of summation can be any variable and can start at any value. Keep this in mind when checking your answers.

Quick ✓

In Problems 11 and 12, write each sum using summation notation.

11. $1 + 4 + 9 + \cdots + 144$

12. $1 + \frac{1}{2} + \frac{1}{4} + \cdots + \frac{1}{32}$

10.1 Exercises **MyMathLab®** Exercise numbers in green have complete video solutions in MyMathLab or may be accessed using the QR code to the right.

*Problems **1–12** are the Quick ✓s that follow the EXAMPLES.*

Building Skills

In Problems 13–24, write the first five terms of each sequence. See Objective 1.

13. $\{3n + 5\}$

14. $\{n - 4\}$

15. $\left\{\dfrac{n}{n + 2}\right\}$

16. $\left\{\dfrac{n + 4}{n}\right\}$

17. $\{(-1)^n n\}$

18. $\{(-1)^{n+1} n\}$

19. $\{2^n + 1\}$

20. $\{3^n - 1\}$

21. $\left\{\dfrac{2n}{2^n}\right\}$

22. $\left\{\dfrac{3n}{3^n}\right\}$

23. $\left\{\dfrac{n}{e^n}\right\}$

24. $\left\{\dfrac{n^2}{2}\right\}$

In Problems 25–32, the given pattern continues. Write the nth term of each sequence suggested by the pattern. See Objective 2.

25. $2, 4, 6, 8, \ldots$

26. $5, 10, 15, 20, \ldots$

27. $\dfrac{1}{2}, \dfrac{2}{3}, \dfrac{3}{4}, \dfrac{4}{5}, \ldots$

28. $\dfrac{1}{2}, 1, \dfrac{3}{2}, 2, \dfrac{5}{2}, \ldots$

29. $3, 6, 11, 18, \ldots$

30. $0, 7, 26, 63, \ldots$

31. $-1, 4, -9, 16, \ldots$

32. $1, -\dfrac{1}{2}, \dfrac{1}{4}, -\dfrac{1}{8}, \ldots$

In Problems 33–44, write out each sum and determine its value. See Objective 3.

33. $\displaystyle\sum_{i=1}^{4} (5i + 1)$

34. $\displaystyle\sum_{i=1}^{5} (3i + 2)$

35. $\displaystyle\sum_{i=1}^{5} \dfrac{i^2}{2}$

36. $\displaystyle\sum_{i=1}^{4} \dfrac{i^3}{2}$

37. $\displaystyle\sum_{k=1}^{3} 2^k$

38. $\displaystyle\sum_{k=1}^{4} 3^k$

39. $\displaystyle\sum_{k=1}^{5} [(-1)^{k+1} \cdot 2k]$

40. $\displaystyle\sum_{k=1}^{8} [(-1)^k \cdot k]$

41. $\displaystyle\sum_{j=1}^{10} 5$

42. $\displaystyle\sum_{j=1}^{8} 2$

43. $\displaystyle\sum_{k=3}^{7} (2k - 1)$

44. $\displaystyle\sum_{j=5}^{10} (k + 4)$

In Problems 45–52, express each sum using summation notation. See Objective 3.

45. $1 + 2 + 3 + \cdots + 15$

46. $1 + 3 + 5 + \cdots + 17$

47. $1 + \dfrac{1}{2} + \dfrac{1}{3} + \cdots + \dfrac{1}{12}$

48. $1 + \dfrac{1}{2} + \dfrac{1}{4} + \cdots + \dfrac{1}{2^{15}}$

49. $1 - \dfrac{1}{3} + \dfrac{1}{9} - \dfrac{1}{27} + \cdots + (-1)^{9+1}\left(\dfrac{1}{3^{9-1}}\right)$

50. $\dfrac{2}{3} - \dfrac{4}{9} + \dfrac{8}{27} + \cdots + (-1)^{15+1}\left(\dfrac{2}{3}\right)^{15}$

51. $5 + (5 + 2 \cdot 1) + (5 + 2 \cdot 2) + (5 + 2 \cdot 3) + \cdots + (5 + 2 \cdot 10)$

52. $3 + 3 \cdot \dfrac{1}{2} + 3 \cdot \dfrac{1}{4} + \cdots + 3 \cdot \left(\dfrac{1}{2}\right)^{11}$

Applying the Concepts

53. The Future Value of Money Suppose that you place $12,000 in your company 401(k) plan, which pays 6% interest compounded quarterly. The balance in the account after n quarters is given by

$$a_n = 12{,}000\left(1 + \frac{0.06}{4}\right)^n$$

(a) Find the value of the account after 1 quarter.
(b) Find the value of the account after 1 year.
(c) Find the value of the account after 10 years.

54. The Future Value of Money Suppose that you place $5000 into a company 401(k) plan that pays 8% interest compounded monthly. The balance in the account after n months is given by

$$a_n = 5000\left(1 + \frac{0.08}{12}\right)^n$$

(a) Find the value of the account after 1 month.
(b) Find the value of the account after 1 year.
(c) Find the value of the account after 10 years.

55. Population Growth According to the U.S. Census Bureau, the population of the United States in 2016 was 326 million people. In addition, the population of the United States was growing at a rate of 0.99% per year. A model for the population of the United States is given by

$$p_n = 326(1.0099)^n$$

where n is the number of years after 2016.

(a) Use this model to predict the United States population in 2020 to the nearest million.
(b) Use this model to predict the United States population in 2050 to the nearest million.

56. Population Growth According to the United States Census Bureau, the population of the world in 2016 was 7563 million people. In addition, the population of the world was growing at a rate of 1.02% per year. A model for the population of the world is given by

$$p_n = 7563(1.0102)^n$$

where n is the number of years after 2016.

(a) Use this model to predict the world population in 2020 to the nearest million.
(b) Use this model to predict the world population in 2050 to the nearest million.

57. Fibonacci Sequence Let

$$u_n = \frac{\left(1 + \sqrt{5}\right)^n - \left(1 - \sqrt{5}\right)^n}{2^n \cdot \sqrt{5}}$$

define the nth term of a sequence. Find the first 10 terms of the sequence. This sequence is called the **Fibonacci sequence.** The terms of the sequence are called **Fibonacci numbers.**

58. Pascal's Triangle The triangular array of numbers shown below is called Pascal's Triangle. Each number in the triangle is found by adding the entries directly above to the left and right of the number. For example, the 4 in the fourth row is found by adding the 1 and 3 above the 4.

Divide the triangular array using diagonal lines (as shown). Now find the sum of the numbers in each of the highlighted diagonal rows. Do you recognize the sequence? (*Hint:* See Problem 57.)

Extending the Concepts

*A second way of defining a sequence is to assign a value to the first (or first few) term(s) and specify the nth term by a formula or equation that involves one or more of the terms preceding it. Sequences defined this way are said to be defined **recursively**, and the rule or formula is called a **recursive formula**. For example, $s_1 = 3$ and $s_n = 2s_{n-1}$ is a recursively defined sequence where $s_1 = 3$, $s_2 = 2s_1 = 2(3) = 6$, $s_3 = 2s_2 = 2(6) = 12$, $s_4 = 2s_3 = 2(12) = 24$, and so on.*

In Problems 59–62, a sequence is defined recursively. Write the first five terms of the sequence.

59. $a_1 = 10,\ a_n = 1.05a_{n-1}$ **60.** $b_1 = 20,\ b_n = 3b_{n-1}$

61. $b_1 = 8,\ b_n = n + b_{n-1}$ **62.** $c_1 = 1000,$
$\qquad\qquad\qquad\qquad\qquad c_n = 1.01c_{n-1} + 100$

63. Fibonacci Sequence Use the result of Problem 57 to do the following problems:

(a) Compute the ratio $\dfrac{u_{n+1}}{u_n}$ for the first 10 terms.

(b) As n gets large, what number does this ratio approach? This number is referred to as the **golden ratio.** Rectangles whose sides are in this ratio were considered pleasing to the eye by the Greeks. For example, the façade of the Parthenon was constructed using the golden ratio.

(c) Compute the ratio $\dfrac{u_n}{u_{n+1}}$ for the first 10 terms.

(d) As n gets large, what number does this ratio approach? This number is also referred to as the **golden ratio.** This ratio is believed to have been used in the construction of the Great Pyramid in Egypt. The ratio equals the sum of the areas of the four face triangles divided by the total surface area of the Great Pyramid.

64. Investigate various applications that lead to a Fibonacci sequence, such as art, architecture, or financial markets. Write an essay on these applications.

Explaining the Concepts

65. Explain how a sequence and a function differ.

66. What does the graph of a sequence look like when compared to the graph of a function? Use the function $f(x) = 3x + 1$ and the sequence $\{a_n\} = \{3n + 1\}$ when doing the comparison.

67. Write a sentence that explains the meaning of the symbol Σ.

68. What does it mean when a sequence alternates?

Synthesis Review

In Problems 69–71 , (a) determine the slope of the linear function, and (b) compute $f(1), f(2), f(3),$ and $f(4)$.

69. $f(x) = 4x - 6$

70. $f(x) = 2x - 10$

71. $f(x) = -5x + 8$

72. Compute $f(2) - f(1), f(3) - f(2),$ and $f(4) - f(3)$ for each of the functions in Problems 69–71. How is the difference in the value of the function for consecutive values of the independent variable related to the slope?

Technology Exercises

Technology can be used to write the terms of a sequence. For example, Figure 2 shows the first five terms of the sequence $\{b_n\} = \{(-1)^n n^2\}$ that we studied in Example 2. Figure 3 shows how the terms of the sequence can be listed in a table using SEQuence mode in a graphing calculator.

Figure 2

Figure 3

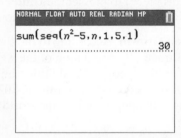

In Problems 73–80, use technology to find the first five terms of the sequence.

73. $\{3n + 5\}$

74. $\{n - 4\}$

75. $\left\{\dfrac{n}{n + 2}\right\}$

76. $\left\{\dfrac{n + 4}{n}\right\}$

77. $\{(-1)^n n\}$

78. $\{(-1)^{n+1} n\}$

79. $\{2^n + 1\}$

80. $\{3^n - 1\}$

Technology can also be used to find the sum of a sequence. For example, Figure 4 shows the result of the sum $\displaystyle\sum_{i=1}^{5} (i^2 - 5)$ that was investigated in Example 4(b) using a graphing calculator.

Figure 4

NORMAL FLOAT AUTO REAL RADIAN MP

sum(seq(n^2-5,n,1,5,1))
                              30

In Problems 81–88, use technology to find the sum.

81. $\displaystyle\sum_{i=1}^{4} (5i + 1)$

82. $\displaystyle\sum_{i=1}^{5} (3i + 2)$

83. $\displaystyle\sum_{i=1}^{5} \dfrac{i^2}{2}$

84. $\displaystyle\sum_{i=1}^{4} \dfrac{i^3}{2}$

85. $\displaystyle\sum_{k=1}^{3} 2^k$

86. $\displaystyle\sum_{k=1}^{4} 3^k$

87. $\displaystyle\sum_{k=1}^{5} [(-1)^{k+1} 2k]$

88. $\displaystyle\sum_{k=1}^{8} [(-1)^k k]$

10.2 Arithmetic Sequences

Objectives

1 Determine Whether a Sequence Is Arithmetic

2 Find a Formula for the *n*th Term of an Arithmetic Sequence

3 Find the Sum of an Arithmetic Sequence

Are You Prepared for This Section?

Before getting started, complete the following problems. If you get a problem wrong, go back to the section cited and review the material.

P1. Determine the slope of $y = -3x + 1$. [Section 1.6 pp. 114–115]

P2. If $g(x) = 5x + 2$, find $g(3)$. [Section 2.2, pp. 158–160]

P3. Solve: $\begin{cases} x - 3y = -17 \\ 2x + y = 1 \end{cases}$ [Section 3.1, pp. 223–228]

In the last section, sequences in general were discussed. Now it's time to consider a specific type of sequence called an *arithmetic sequence.*

1 Determine Whether a Sequence Is Arithmetic

When the difference between successive terms of a sequence is always the same number, the sequence is called an **arithmetic sequence** (sometimes called an **arithmetic progression**). For example, the sequence

$$2, 6, 10, 14, \ldots$$

is arithmetic because the constant difference between consecutive terms is 4. If the first term is called a_1 and the **common difference** between consecutive terms is called d, then the terms of an arithmetic sequence follow the pattern

$$a_1, a_1 + d, a_1 + 2d, a_1 + 3d, \ldots$$

EXAMPLE 1 **Determining Whether a Sequence Is Arithmetic**

Determine whether the sequence $3, 9, 15, 21, \ldots$ is arithmetic. If it is, find the first term a_1 and the common difference d.

Solution

If the difference between consecutive terms is constant, then the sequence is arithmetic. The sequence $3, 9, 15, 21, \ldots$ is arithmetic because this difference is 6 ($= 9 - 3$ or $15 - 9$ or $21 - 15$). The first term is $a_1 = 3$, and the common difference is $d = 6$.   ●

Quick ✓

1. In a(n) _____ sequence, the difference between consecutive terms is constant.

In Problems 2 and 3, determine which of the following sequences is arithmetic. If the sequence is arithmetic, determine the first term a_1 and common difference d.

2. $-3, -1, 1, 3, 5, \ldots$ **3.** $3, 9, 27, 81, \ldots$

EXAMPLE 2 **Determining Whether a Sequence Is Arithmetic**

Show that the sequence $\{s_n\} = \{2n + 7\}$ is arithmetic. Find the first term and the common difference.

Solution

Listing the first few terms and showing that the difference between consecutive terms is the same would be more of a demonstration than a proof. To prove the sequence is arithmetic, show that for *any* consecutive terms, the difference is the same number. This is done by evaluating the sequence at $(n - 1)$ and at n.

Prepared?...Answers
P1. -3 **P2.** 17 **P3.** $(-2, 5)$

(continued)

Then compute the difference between these values. If it is constant, the sequence is arithmetic.

$$s_{n-1} = 2(n-1) + 7 = 2n - 2 + 7 = 2n + 5 \quad \text{and} \quad s_n = 2n + 7$$

Now compute $s_n - s_{n-1}$.

$$
\begin{aligned}
s_n - s_{n-1} &= (2n + 7) - (2n + 5) \\
\text{Distribute:} \quad &= 2n + 7 - 2n - 5 \\
\text{Combine like terms:} \quad &= 2
\end{aligned}
$$

The difference between *any* consecutive terms is 2, so the sequence is arithmetic with common difference $d = 2$. To find the first term, evaluate s_1 and find $s_1 = 2(1) + 7 = 9$, so $a_1 = 9$. ●

EXAMPLE 3 | **Determining Whether a Sequence Defined by a Function Is Arithmetic**

Show that the sequence $\{b_n\} = \{n^2\}$ is not arithmetic.

Solution
As in Example 2, evaluate the sequence at $(n - 1)$ and n. Then show that the difference between consecutive terms is not constant.

$$b_{n-1} = (n-1)^2 = n^2 - 2n + 1 \quad \text{and} \quad b_n = n^2$$

Now compute $b_n - b_{n-1}$.

$$
\begin{aligned}
b_n - b_{n-1} &= n^2 - (n^2 - 2n + 1) \\
\text{Distribute:} \quad &= n^2 - n^2 + 2n - 1 \\
\text{Combine like terms:} \quad &= 2n - 1
\end{aligned}
$$

The difference between consecutive terms is not constant—its value depends on n. Therefore, the sequence is not arithmetic. ●

Quick ✓

In Problems 4–6, determine whether the sequence is arithmetic. If it is, state the first term a_1 and the common difference d.

4. $\{a_n\} = \{3n - 8\}$ **5.** $\{b_n\} = \{n^2 - 1\}$ **6.** $\{c_n\} = \{5 - 2n\}$

▶ **❷ Find a Formula for the *n*th Term of an Arithmetic Sequence**

Suppose a_1 is the first term of an arithmetic sequence whose common difference is d. Find a formula for a_n, the nth term of the sequence. To do this, list the first few terms of the sequence.

$$
\begin{aligned}
a_1 &= a \\
a_2 &= a_1 + d = a_1 + 1 \cdot d \\
a_3 &= a_2 + d = (a_1 + d) + d = a_1 + 2d \\
a_4 &= a_3 + d = (a_1 + 2d) + d = a_1 + 3d \\
a_5 &= a_4 + d = (a_1 + 3d) + d = a_1 + 4d \\
&\vdots \\
a_n &= a_{n-1} + d = [a_1 + (n-2)d] + d = a_1 + (n-1)d
\end{aligned}
$$

The *n*th Term of an Arithmetic Sequence

For an arithmetic sequence $\{a_n\}$ whose first term is a_1 and whose common difference is d, the nth term is determined by the formula

$$a_n = a_1 + (n-1)d$$

EXAMPLE 4 **Finding a Formula for the *n*th Term of an Arithmetic Sequence**

(a) Write a formula for the *n*th term of an arithmetic sequence whose 4th term is 8 and whose common difference is -3.

(b) Find the 14th term of the sequence.

Solution

(a) Find a formula for the *n*th term of an arithmetic sequence. It is known that $d = -3$ and that $a_4 = 8$.

$$\text{nth term:} \quad a_n = a_1 + (n - 1)d$$
$$d = -3; a_4 = 8: \quad a_4 = 8 = a_1 + (4 - 1)(-3)$$

Solve this equation for a_1, the first term of the sequence.

$$8 = a_1 - 9$$
$$\text{Add 9 to both sides:} \quad 17 = a_1$$

The formula for the *n*th term is

$$a_n = a_1 + (n - 1)d: \quad a_n = 17 + (n - 1)(-3)$$
$$\text{Distribute:} \quad = 17 - 3n + 3$$
$$\text{Combine like terms:} \quad = -3n + 20$$

(b) To find the 14th term, let $n = 14$ in $a_n = -3n + 20$.

$$a_{14} = -3(14) + 20$$
$$= -42 + 20$$
$$= -22$$

The 14th term in the sequence is -22.

Quick ✓

7. For an arithmetic sequence $\{a_n\}$ whose first term is a_1 and whose common difference is d, the *n*th term is determined by the formula _____.

8. **(a)** Write a formula for the *n*th term of an arithmetic sequence whose 5th term is 25 and whose common difference is 6.
   (b) Find the 14th term of the sequence.

EXAMPLE 5 **Finding a Formula for the *n*th Term of an Arithmetic Sequence**

The 4th term of an arithmetic sequence is 7, and the 10th term is 31.

(a) Find the first term and the common difference.

(b) Give a formula for the *n*th term of the sequence.

Solution

(a) The *n*th term of an arithmetic sequence is $a_n = a_1 + (n - 1)d$, where a_1 is the first term and d is the common difference. Since $a_4 = 7$ and $a_{10} = 31$:

$$\begin{cases} a_4 = a_1 + (4 - 1)d \\ a_{10} = a_1 + (10 - 1)d \end{cases} \quad \text{or} \quad \begin{cases} 7 = a_1 + 3d \quad (1) \\ 31 = a_1 + 9d \quad (2) \end{cases}$$

This is a system of two linear equations with two variables, a and d. This system can be solved by elimination. Subtracting equation (2) from equation (1) yields

$$-24 = -6d$$
$$\text{Divide both sides by } -6: \quad 4 = d$$

(continued)

Let $d = 4$ in equation (1) to find a_1.

$$7 = a_1 + 3(4)$$
$$7 = a_1 + 12$$

Subtract 12 from both sides: $\quad -5 = a_1$

The first term is $a_1 = -5$ and the common difference is $d = 4$.

(b) A formula for the nth term is

$$a_n = a_1 + (n-1)d$$
$$a_1 = -5; d = 4: \quad = -5 + (n-1)(4)$$
$$\text{Distribute:} \quad = -5 + 4n - 4$$
$$\text{Combine like terms:} \quad = 4n - 9 \qquad \bullet$$

Quick ✓

9. The 5th term of an arithmetic sequence is 7, and the 13th term is 31.
  (a) Find the first term and the common difference.
  (b) Give a formula for the nth term of the sequence.

▷ ❸ Find the Sum of an Arithmetic Sequence

The next result gives a formula for finding the sum of the first n terms of an arithmetic sequence.

Work Smart

There are two formulas for finding the sum of the first n terms of an arithmetic sequence. Use Formula (1) if you know n, the first term a_1, and the common difference d; use Formula (2) if you know n, the first term a_1, and the last term a_n.

Sum of the First n Terms of an Arithmetic Sequence

Let $\{a_n\}$ be an arithmetic sequence with first term a_1 and common difference d. The sum S_n of the first n terms of $\{a_n\}$ is

$$S_n = \frac{n}{2}[2a_1 + (n-1)d] \quad (1) \quad \text{or} \quad S_n = \frac{n}{2}[a_1 + a_n] \quad (2)$$

The following shows where these results come from.

$$S_n = a_1 + a_2 + a_3 + \cdots + a_n$$
$$= \underbrace{a_1}_{a_1} + \underbrace{(a_1 + d)}_{a_2} + \underbrace{(a_1 + 2d)}_{a_3} + \cdots + \underbrace{(a_1 + (n-1)d)}_{a_n}$$

The sequence S_n can also be represented by reversing the order in which the terms are added, so that

$$S_n = a_n + a_{n-1} + \cdots + a_1$$
$$= \underbrace{(a_1 + (n-1)d)}_{a_n} + \underbrace{(a_1 + (n-2)d)}_{a_{n-1}} + \cdots + \underbrace{a_1}_{a_1}$$

Add these two different representations of S_n as follows:

$$
\begin{array}{lllll}
S_n = & a_1 & + (a_1 + d) & + (a_1 + 2d) & + \cdots + (a_1 + (n-2)d) & + (a_1 + (n-1)d) \\
S_n = & (a_1 + (n-1)d) & + (a_1 + (n-2)d) & + (a_1 + (n-3)d) & + \cdots + \quad (a_1 + d) & + \quad a_1 \\
\hline
2S_n = & 2a_1 + (n-1)d & + 2a_1 + (n-1)d & + 2a_1 + (n-1)d & + \cdots + 2a_1 + (n-1)d & + 2a_1 + (n-1)d
\end{array}
$$

Thus $2S_n$ is $2a_1 + (n-1)d$ added to itself n times, or $n[2a_1 + (n-1)d]$. Therefore,

$$2S_n = n[2a_1 + (n-1)d]$$

Divide both sides by 2: $\quad S_n = \dfrac{n}{2}[2a_1 + (n-1)d] \quad$ Formula (1)

This is the first formula in the box. If the expression $2a_1 + (n-1)d$ is rewritten as $a_1 + a_1 + (n-1)d$, notice that $a_1 + (n-1)d$ is a_n.

$$S_n = \frac{n}{2}[a_1 + a_n] \quad \text{Formula (2)}$$

There are two ways to find the sum of the first n terms of an arithmetic sequence. Notice that Formula (1) involves the first term a_1 and the common difference d, whereas Formula (2) involves the first term a_1 and the last term a_n. Use whichever is easier.

EXAMPLE 6   **Finding the Sum of the First n Terms of an Arithmetic Sequence**

Find the sum S_n of the first 50 terms of the arithmetic sequence $2, 5, 8, 11, \ldots$.

Solution
Because the first term is $a_1 = 2$ and the common difference is $d = 5 - 2 = 3$, use the formula $S_n = \dfrac{n}{2}[2a_1 + (n-1)d]$ to find the sum.

$$S_n = \frac{n}{2}[2a_1 + (n-1)d]$$

$$n = 50, a_1 = 2, d = 3: \quad S_{50} = \frac{50}{2}[2(2) + (50-1)(3)]$$

$$= 25[4 + 49(3)]$$

$$= 25[151]$$

$$= 3775$$

The sum of the first 50 terms, S_{50}, of the arithmetic sequence $2, 5, 8, 11, \ldots$ is 3775. ●

> **Quick ✓**
>
> **10.** Find the sum S_n of the first 100 terms of the arithmetic sequence whose first term is 5 and whose common difference is 2.
>
> **11.** Find the sum S_n of the first 70 terms of the arithmetic sequence $1, 5, 9, 13, \ldots$.

EXAMPLE 7   **Finding the Sum of the First n Terms of an Arithmetic Sequence**

Find the sum S_n of the first 40 terms of the arithmetic sequence $\{-2n + 50\}$.

Solution
The first term is $a_1 = -2(1) + 50 = 48$ and the 40th term is $a_{40} = -2(40) + 50 = -30$. Because the first term and the last term of the sequence are known, use the formula $S_n = \dfrac{n}{2}[a_1 + a_n]$ to find the sum.

$$S_n = \frac{n}{2}[a_1 + a_n]$$

$$n = 40, a_1 = 48, a_{40} = -30: \quad S_{40} = \frac{40}{2}[48 + (-30)]$$

$$= 360$$

The sum of the first 40 terms, S_{40}, of the arithmetic sequence $\{-2n + 50\}$ is 360. ●

> **Quick ✓**
>
> **12.** Find the sum S_n of the first 50 terms of the arithmetic sequence whose first term is 4 and 50th term 298.
>
> **13.** Find the sum S_n of the first 75 terms of the arithmetic sequence $\{-3n + 100\}$.

EXAMPLE 8 **Creating a Floor Design**

A ceramic tile floor is designed in the shape of a trapezoid 10 feet wide at the base and 5 feet wide at the top. See Figure 5. The tiles, which measure 6 inches by 6 inches, are to be placed so that each successive row contains one fewer tile than the preceding row. How many tiles are required?

Figure 5

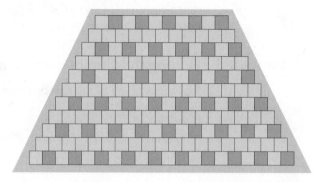

Solution

Each tile is 6 inches (0.5 foot) wide, and the bottom row is 10 feet wide, so the bottom row requires 20 tiles. Similar reasoning tells us the top row requires 10 tiles. Because each successive row has one fewer tile, the total number of tiles required is

$$S = 20 + 19 + 18 + \cdots + 11 + 10$$

This is the sum of an arithmetic sequence with common difference -1. The number of terms to be added is $n = 11$, the first term is $a_1 = 20$, and the last term is $a_{11} = 10$. The sum S is

$$S_n = \frac{n}{2}[a_1 + a_n] \qquad S_{11} = \frac{11}{2}(20 + 10) = 165$$

In all, 165 tiles will be required.

Quick ✓

14. In the corner section of a theater, the first row has 20 seats. Each subsequent row has 2 more seats, and there are a total of 30 rows. How many seats are in this section?

10.2 Exercises MyMathLab®

Exercise numbers in green have complete video solutions in MyMathLab or may be accessed using the QR code to the right.

*Problems **1–14** are the Quick ✓ s that follow the EXAMPLES.*

Building Skills

In Problems 15–22, verify that the given sequence is arithmetic. Then find the first term and common difference. See Objective 1.

15. $\{n + 5\}$ **16.** $\{n - 1\}$ **17.** $\{7n + 2\}$

18. $\{10n + 1\}$ **19.** $\{7 - 3n\}$ **20.** $\{5 - 2n\}$

21. $\left\{\frac{1}{2}n + 5\right\}$ **22.** $\left\{\frac{1}{4}n + \frac{3}{4}\right\}$

In Problems 23–30, find a formula for the nth term of the arithmetic sequence whose first term a_1 and common difference d are given. What is the 5th term? See Objective 2.

23. $a_1 = 4; d = 3$ **24.** $a_1 = 8; d = 3$

25. $a_1 = 10; d = -5$ **26.** $a_1 = 12; d = -3$

27. $a_1 = 2; d = \frac{1}{3}$ **28.** $a_1 = -3; d = \frac{1}{2}$

29. $a_1 = 5; d = -\frac{1}{5}$ **30.** $a_1 = -\frac{4}{3}; d = -\frac{2}{3}$

In Problems 31–36, write a formula for the nth term of each arithmetic sequence. Use the formula to find the 20th term in each arithmetic sequence. See Objective 2.

31. 2, 7, 12, 17, . . .

32. −5, −1, 3, 7, . . .

33. 12, 9, 6, 3, . . .

34. 20, 14, 8, 2, . . .

35. $1, \dfrac{5}{4}, \dfrac{3}{2}, \dfrac{7}{4}, \ldots$

36. $10, \dfrac{19}{2}, 9, \dfrac{17}{2}, \ldots$

In Problems 37–44, find the first term and the common difference of the arithmetic sequence described. Give a formula for the nth term of the sequence. See Objective 2.

37. 3rd term is 17; 7th term is 37

38. 5th term is 7; 9th term is 19

39. 4th term is −2; 8th term is 26

40. 2nd term is −9; 8th term is 15

41. 5th term is −1; 12th term is −22

42. 6th term is −8; 12th term is −38

43. 3rd term is 3; 9th term is 0

44. 5th term is 5; 13th term is 7

For Problems 45–56, see Objective 3.

45. Find the sum of the first 30 terms of the sequence 2, 8, 14, 20, ...

46. Find the sum of the first 40 terms of the sequence 1, 8, 15, 22, ...

47. Find the sum of the first 25 terms of the sequence −8, −5, −2, 1, ...

48. Find the sum of the first 75 terms of the sequence −9, −5, −1, 3, ...

49. Find the sum of the first 40 terms of the sequence 10, 3, −4, −11, ...

50. Find the sum of the first 50 terms of the sequence 12, 4, −4, −12, ...

51. Find the sum of the first 40 terms of the arithmetic sequence $\{4n − 3\}$.

52. Find the sum of the first 80 terms of the arithmetic sequence $\{2n − 13\}$.

53. Find the sum of the first 75 terms of the arithmetic sequence $\{−5n + 70\}$.

54. Find the sum of the first 35 terms of the arithmetic sequence $\{−6n + 25\}$.

55. Find the sum of the first 30 terms of the arithmetic sequence $\left\{5 + \dfrac{2}{3}n\right\}$.

56. Find the sum of the first 28 terms of the arithmetic sequence $\left\{7 − \dfrac{3}{2}n\right\}$.

Applying the Concepts

57. Find x so that $x + 3$, $2x + 1$, and $5x + 2$ are consecutive terms of an arithmetic sequence.

58. Find x so that $2x$, $3x + 2$, and $5x + 3$ are consecutive terms of an arithmetic sequence.

59. A Stack of Cans Suppose that the bottom row in a stack of cans contains 35 cans. Each layer contains one can fewer than the layer below it. The top row has 1 can. How many cans are in the stack?

60. A Pile of Bricks Suppose that the bottom row in a pile of bricks contains 46 bricks. Each layer contains 2 bricks fewer than the layer below it. The top row has 2 bricks. How many bricks are in the stack?

61. The Theater An auditorium has 40 seats in the first row and 25 rows in all. Each successive row contains 2 additional seats. How many seats are in the auditorium?

62. Mosaic A mosaic is designed in the shape of an equilateral triangle, 20 feet on each side. Each tile in the mosaic is in the shape of an equilateral triangle, 12 inches to a side. The tiles are to alternate in color as shown in the illustration. How many tiles of each color will be required?

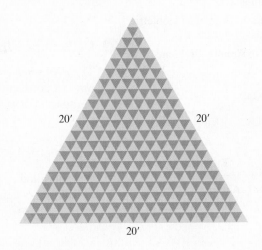

Extending the Concepts

In Problems 63–66, determine the number of terms that are in each arithmetic sequence.

63. $-5, -2, 1, \ldots, 244$

64. $-9, -5, -1, \ldots, 219$

65. $108, 101, 94, \ldots, -326$

66. $99, 93, 87, \ldots, -339$

67. Salary Suppose you have just been hired at the starting salary of $32,000 per year. Your contract guarantees you a $2500 raise each year. How many years will it take before your aggregate salary is $757,500? (*Hint:* Your aggregate salary is $32,000 + ($32,000 + $2500) + \cdots.$)

68. Stadium Construction How many rows are in the corner section of a stadium containing 2030 seats if the first row has 14 seats and each successive row has 4 additional seats?

Explaining the Concepts

69. Explain how you can determine whether a sequence is arithmetic.

70. Provide an explanation that justifies the formula for the sum of the first n terms of an arithmetic sequence.

Synthesis Review

In Problems 71–73, (a) determine the base a of the exponential function, and (b) compute $f(1), f(2), f(3),$ and $f(4)$.

71. $f(x) = 3^x$

72. $f(x) = 4^x$

73. $f(x) = 10\left(\dfrac{1}{2}\right)^x$

74. Compute $\dfrac{f(2)}{f(1)}, \dfrac{f(3)}{f(2)},$ and $\dfrac{f(4)}{f(3)}$ for each of the functions in Problems 71–73. How is the ratio in the value of the function for consecutive values of the independent variable related to the base?

Technology Exercises

In Problems 75–78, use technology to find the sum of each sequence.

75. $\{3.45n + 4.12\}; n = 20$

76. $\{2.67n - 1.23\}; n = 25$

77. $85.9 + 83.5 + 81.1 + \cdots; n = 25$

78. $-11.8 + (-8.2) + (-4.6) + \cdots; n = 30$

10.3 Geometric Sequences and Series

Objectives

❶ Determine Whether a Sequence Is Geometric

❷ Find a Formula for the nth Term of a Geometric Sequence

❸ Find the Sum of a Geometric Sequence

❹ Find the Sum of a Geometric Series

❺ Solve Annuity Problems

Are You Prepared for This Section?

Before getting started, complete the following problems. If you get a problem wrong, go back to the section cited and review the material.

P1. If $g(x) = 4^x$, evaluate $g(1), g(2),$ and $g(3)$.　　[Section 8.2, pp. 638–641]

P2. Simplify: $\dfrac{x^4}{x^3}$　　[Getting Ready, pp. 298–299]

▶ ❶ **Determine Whether a Sequence Is Geometric**

The *difference* between consecutive terms in an arithmetic sequence is constant. If the *ratio* of consecutive terms in a sequence is constant, then the sequence is a **geometric sequence.** For example, the sequence

$$2, 4, 8, 16, \ldots$$

is geometric because the ratio of consecutive terms is $2 \left(= \dfrac{4}{2} = \dfrac{8}{4} = \dfrac{16}{8}\right)$. The first term is called a_1 and the **common ratio** of consecutive terms is called r. The terms of a geometric sequence follow the pattern

$$a_1, \ a_1 r, \ a_1 r^2, \ a_1 r^3, \ldots$$

Prepared?...Answers

P1. $g(1) = 4, g(2) = 16, g(3) = 64$

P2. x

EXAMPLE 1 **Determining Whether a Sequence Is Geometric**

Determine whether the sequence 2, 6, 18, 54, 162, . . . is geometric. If it is, state the first term and the common ratio.

Solution

The sequence is geometric if the ratio of consecutive terms is a constant. Because the ratio of consecutive terms is 3 $\left(= \dfrac{6}{2} = \dfrac{18}{6} = \dfrac{54}{18} = \dfrac{162}{54} \right)$, the sequence is geometric. The first term is $a_1 = 2$ and the common ratio is $r = 3$. ●

Quick ✓

1. In a(n) _____ sequence the ratio of consecutive terms is constant.

In Problems 2–4, determine whether the sequence is geometric. If it is, state the first term and common ratio.

2. 4, 8, 16, 32, 64, . . .   **3.** 5, 10, 16, 23, 31, . . .   **4.** 9, 3, 1, $\dfrac{1}{3}, \dfrac{1}{9}$, . . .

EXAMPLE 2 **Determining Whether a Sequence Is Geometric**

Determine whether the sequence $\{b_n\} = \{3^{-n}\}$ is geometric. If it is, find the common ratio.

Solution

Show that for *any* consecutive terms, the ratio is the same number. Do this by evaluating the sequence at $(n - 1)$ and n. Then compute the ratio of these two values. If the ratio is constant, the sequence is geometric.

$$b_{n-1} = 3^{-(n-1)} = 3^{-n+1} \quad \text{and} \quad b_n = 3^{-n}$$

Work Smart

Notice that the technique used in Example 2 is similar to the approach used in Section 10.2 to show a sequence is arithmetic.

Now compute $\dfrac{b_n}{b_{n-1}}$.

$$\frac{b_n}{b_{n-1}} = \frac{3^{-n}}{3^{-n+1}}$$

$$a^{m+n} = a^m \cdot a^n: \quad = \frac{3^{-n}}{3^{-n} \cdot 3^1}$$

Divide out like factors: $\quad = \dfrac{1}{3}$

The ratio of *any* two consecutive terms is the constant $\dfrac{1}{3}$, so the sequence is geometric with $r = \dfrac{1}{3}$. ●

EXAMPLE 3 **Determining Whether a Sequence Is Geometric**

Determine whether the sequence $\{a_n\} = \{n^2 + 1\}$ is geometric. If it is, find the common ratio.

(continued)

Solution

If the ratio $\dfrac{a_n}{a_{n-1}}$ is constant, then the sequence is geometric.

$$\frac{a_n}{a_{n-1}} = \frac{n^2 + 1}{(n-1)^2 + 1}$$

Use FOIL: $\quad = \dfrac{n^2 + 1}{n^2 - 2n + 1 + 1}$

Combine like terms: $\quad = \dfrac{n^2 + 1}{n^2 - 2n + 2}$

The rational expression cannot be simplified any further. Because the ratio $\dfrac{a_n}{a_{n-1}}$ depends on the value of n, it is not constant. Therefore, the sequence is not geometric. ●

Quick ✓

In Problems 5–7, determine whether the sequence is geometric. If it is geometric, find the common ratio.

5. $\{a_n\} = \{5^n\}$ **6.** $\{b_n\} = \{n^2\}$

7. $\{c_n\} = \left\{5\left(\dfrac{2}{3}\right)^n\right\}$

▶ ❷ **Find a Formula for the *n*th Term of a Geometric Sequence**

If a_1 is the first term of a geometric sequence with common ratio $r \neq 0$, what is a formula for the *n*th term of a_n? Let's list the first few terms to find out.

$$a_1 = 1a_1 = a_1 r^0$$
$$a_2 = ra_1 = a_1 r^1$$
$$a_3 = ra_2 = a_1 r^2$$
$$a_4 = ra_3 = a_1 r^3$$
$$a_5 = ra_4 = a_1 r^4$$
$$\vdots$$
$$a_n = ra_{n-1} = r(a_1 r^{n-2}) = a_1 r^{n-1}$$

This leads to the following result.

The *n*th Term of a Geometric Sequence

If a geometric sequence $\{a_n\}$ has first term a_1 and common ratio r, then the *n*th term is determined by the formula

$$a_n = a_1 r^{n-1} \qquad r \neq 0$$

EXAMPLE 4 **Finding a Particular Term of a Geometric Sequence**

Consider the geometric sequence $8, 6, \dfrac{9}{2}, \dfrac{27}{8}, \ldots$.

(a) Find a formula for the *n*th term.

(b) Find the 11th term of the sequence.

Solution

(a) The sequence is geometric. The first term is $a_1 = 8$. The common ratio, r, is the ratio of any two consecutive terms. So

$$r = \frac{6}{8} = \frac{\frac{9}{2}}{6} = \frac{\frac{27}{8}}{\frac{9}{2}} = \frac{3}{4}. \text{ Now substitute these values in the formula for}$$

the nth term of a geometric sequence.

$$a_n = a_1 r^{n-1}$$

$$a_1 = 8; r = \frac{3}{4}: \quad = 8\left(\frac{3}{4}\right)^{n-1}$$

(b) Find the eleventh term by letting $n = 11$ in the formula found in part (a).

$$a_n = 8\left(\frac{3}{4}\right)^{n-1}$$

$$a_{11} = 8\left(\frac{3}{4}\right)^{11-1}$$

$$= 8\left(\frac{3}{4}\right)^{10}$$

$$\approx 0.45051 \qquad \bullet$$

> **Quick ✓**
>
> *In Problems 8 and 9, find a formula for the nth term of each geometric sequence. Use this result to find the 9th term of the sequence.*
>
> **8.** $a_1 = 5, r = 2$ **9.** $\{50, 25, 12.5, 6.25, \dots\}$

▶ ❸ Find the Sum of a Geometric Sequence

The next result gives us a formula for finding the sum of the first n terms of a geometric sequence.

> **Sum of the First n Terms of a Geometric Sequence**
>
> Let $\{a_n\}$ be a geometric sequence with first term a_1 and common ratio r, where $r \neq 0, r \neq 1$. The sum S_n of the first n terms of $\{a_n\}$ is
>
> $$S_n = a_1 \cdot \frac{1 - r^n}{1 - r} \qquad r \neq 0, r \neq 1$$

Where does this formula come from? The sum S_n of the first n terms of $\{a_n\} = \{a_1 r^{n-1}\}$ is

$$S_n = a_1 + a_1 r + a_1 r^2 + \cdots + a_1 r^{n-1}$$

Multiply both sides by r to obtain

$$rS_n = a_1 r + a_1 r^2 + a_1 r^3 + \cdots + a_1 r^n$$

Now subtract rS_n from S_n:

$$S_n = a_1 + a_1 r + a_1 r^2 + \cdots + a_1 r^{n-1}$$
$$\underline{rS_n = \qquad\quad a_1 r + a_1 r^2 + a_1 r^3 + \cdots + a_1 r^n}$$
$$S_n - rS_n = a_1 - a_1 r^n$$

Factor S_n from the expression on the left and factor a_1 from the expression on the right.

$$S_n(1 - r) = a_1(1 - r^n)$$

Divide both sides by $1 - r$ (since $r \neq 1$) and solve for S_n.

$$S_n = a_1 \cdot \frac{1 - r^n}{1 - r}$$

EXAMPLE 5   **Finding the Sum of the First n Terms of a Geometric Sequence**

Find the sum of the first 10 terms of the sequence $2, 6, 18, 54, \ldots$.

Solution

Find the sum of the first 10 terms of a geometric sequence with $a_1 = 2$ and common ratio $r = 3$.

$$S_n = a_1 \cdot \frac{1 - r^n}{1 - r}$$

$$a_1 = 2, r = 3, n = 10: \quad S_{10} = 2 \cdot \frac{1 - 3^{10}}{1 - 3}$$

$$= 2 \cdot \frac{-59{,}048}{-2}$$

$$= 59{,}048$$

EXAMPLE 6   **Finding the Sum of the First n Terms of a Geometric Sequence in Summation Notation**

Find the sum: $\displaystyle\sum_{n=1}^{8}\left[5\left(\frac{1}{2}\right)^n\right]$

Express your answer to as many decimal places as your calculator allows.

Solution

Expand this sum as follows:

$$\sum_{n=1}^{8}\left[5\left(\frac{1}{2}\right)^n\right] = \frac{5}{2} + \frac{5}{4} + \cdots + \frac{5}{256}$$

Work Smart

$$\sum_{n=1}^{8} 5\left(\frac{1}{2}\right)^n$$

$$= 5\left(\frac{1}{2}\right)^1 + 5\left(\frac{1}{2}\right)^2 + 5\left(\frac{1}{2}\right)^3 + 5\left(\frac{1}{2}\right)^4$$

$$+ \cdots + 5\left(\frac{1}{2}\right)^8$$

Find the sum of the first $n = 8$ terms of a geometric sequence with $a_1 = \frac{5}{2}$ and common ratio $r = \frac{1}{2}$.

$$S_n = a_1 \cdot \frac{1 - r^n}{1 - r}$$

$$a_1 = \frac{5}{2}, r = \frac{1}{2}, n = 8: \quad S_8 = \frac{5}{2} \cdot \frac{1 - \left(\frac{1}{2}\right)^8}{1 - \frac{1}{2}}$$

$$= \frac{5}{2} \cdot \frac{0.99609375}{\frac{1}{2}}$$

$$= 4.98046875$$

Quick ✓

10. For a geometric sequence with first term a_1 and common ratio r, where $r \neq 0, r \neq 1$, the sum of the first n terms is _____.

In Problems 11 and 12, find the sum.

11. $3 + 6 + 12 + 24 + \cdots + 3 \cdot 2^{12}$

12. $\displaystyle\sum_{n=1}^{10} \left[8\left(\frac{1}{2}\right)^n \right]$

▶ ❹ Find the Sum of a Geometric Series

An infinite sum of the form

$$a_1 + a_1 r + a_1 r^2 + \cdots + a_1 r^{n-1} + \cdots$$

whose first term is a_1 and whose common ratio is r, is called an **infinite geometric series** and is denoted by

$$\sum_{n=1}^{\infty} a_1 r^{n-1}$$

The sum of the first n terms of a geometric sequence is given by the formula

$$S_n = a_1 \cdot \frac{1 - r^n}{1 - r}$$

Distribute the a_1:
$$= \frac{a_1 - a_1 r^n}{1 - r}$$

Divide each term by $1 - r$:
$$= \frac{a_1}{1 - r} - \frac{a_1 r^n}{1 - r}$$

As n gets larger and larger, the expression S_n approaches the value $\dfrac{a_1}{1 - r}$ because r^n approaches 0 provided that $-1 < r < 1$.

Sum of a Geometric Series

If $-1 < r < 1$, the sum of the terms of an infinite geometric series with first term a_1 and common ratio r is

$$\sum_{n=1}^{\infty} a_1 r^{n-1} = \frac{a_1}{1 - r}$$

EXAMPLE 7   **Finding the Sum of a Geometric Series**

Find the sum of the geometric series: $4 + 2 + 1 + \dfrac{1}{2} + \cdots$

Solution

This is a geometric series with $a_1 = 4$ and common ratio $r = \dfrac{1}{2}$ $\left(= \dfrac{2}{4} = \dfrac{1}{2} = \dfrac{\frac{1}{2}}{1} \right)$.

Since the common ratio r is between -1 and 1, use the formula for the sum of a geometric series to find that

$$4 + 2 + 1 + \frac{1}{2} + \cdots = \frac{a_1}{1 - r} = \frac{4}{1 - \dfrac{1}{2}} = 8$$

Quick ✓

13. If $-1 < r < 1$, then the sum of the geometric series $\displaystyle\sum_{n=1}^{\infty} a_1 r^{n-1} =$ _____.

In Problems 14 and 15, find the sum of the geometric series.

14. $10 + \dfrac{5}{2} + \dfrac{5}{8} + \dfrac{5}{32} + \cdots$

15. $\displaystyle\sum_{n=1}^{\infty} \left(\dfrac{1}{3}\right)^n$

Work Smart: Study Skills

Let's summarize the formulas for arithmetic and geometric sequences where a_1 is the first term of the sequence, d is the common difference, and r is the common ratio.

	Arithmetic	**Geometric**
nth term	$a_n = a_1 + (n-1)d$	$a_n = a_1 r^{n-1}, r \neq 0$
Sum of the first n terms	$S_n = \dfrac{n}{2}[2a_1 + (n-1)d]$ $= \dfrac{n}{2}(a_1 + a_n)$	$S_n = a_1 \cdot \dfrac{1 - r^n}{1 - r}, r \neq 0, r \neq 1$

If you were asked to find the 5th term of the sequence $5, 20, 80, 320, \ldots$, which formula would you use?

If you were asked to find the sum of the first 5 terms of the sequence $5, 9, 13, 17, \ldots$, which formula would you use?

EXAMPLE 8   **Writing a Repeating Decimal as a Fraction**

Express $0.\overline{1}$ as a fraction in lowest terms.

Solution

The line over the 1 indicates that the 1 repeats indefinitely. That is,

$$0.\overline{1} = 0.11111\ldots$$
$$= 0.1 + 0.01 + 0.001 + \ldots$$

This is a geometric series with $a_1 = 0.1$ and common ratio $r = 0.1 \left(= \dfrac{0.01}{0.1} = \dfrac{0.001}{0.01}\right)$.

Since $-1 < r < 1$, use the formula for the sum of a geometric series to find that

$$0.\overline{1} = 0.1 + 0.01 + 0.001 + \cdots$$

$a_1 = 0.1, r = 0.1$ in $\dfrac{a_1}{1 - r}$:
$$= \dfrac{0.1}{1 - 0.1}$$
$$= \dfrac{0.1}{0.9}$$

Multiply numerator and denominator by 10:
$$= \dfrac{1}{9}$$

Thus $0.\overline{1} = \dfrac{1}{9}$.

Quick ✓

16. Express $0.\overline{2}$ as a fraction in lowest terms.

EXAMPLE 9   **The Multiplier**

Suppose that Americans spend 90% of every additional dollar they earn. Economists would say that an individual's **marginal propensity to consume** is 0.90. For example, if Roberta earns a dollar, she will spend $0.9(\$1) = \$0.90$ of it and save $0.10. Whoever earns $0.90 (from Roberta) will spend 90% of it, or $0.9(\$0.90) = \$0.81$. This process of spending continues and results in a geometric series as follows:

$$\$1 + \$0.90 + \$0.81 + \$0.729 + \cdots$$

The sum of this geometric series is called the **multiplier.** The multiplier is used to determine the total impact of spending on the U.S. economy. Suppose the government gives a child-tax rebate of $500 to Roberta. Determine the impact of the rebate on the U.S. economy if Americans spend 90% of every dollar they earn.

Solution
The total impact of the $500 tax rebate on the U.S. economy is

$$\$500 + \$500(0.9) + \$500(0.9)^2 + \$500(0.9)^3 + \cdots$$

This is a geometric series with first term $a_1 = 500$ and common ratio $r = 0.9$. The sum of this series is

$$\$500 + \$500(0.9) + \$500(0.9)^2 + \$500(0.9)^3 + \cdots = \frac{\$500}{1 - 0.9}$$
$$= \$5000$$

The United States economy will grow by $5000 because of the child-tax credit to Roberta.   ●

Quick ✓
17. Redo Example 9 if the marginal propensity to consume is 95%.

▶ ❺ **Solve Annuity Problems**

Section 8.2 introduced the compound interest formula, which lets us calculate the future value of a lump sum of money that is deposited in an account that pays interest compounded periodically. Often, though, money is invested at periodic intervals of time. An **annuity** is a sequence of equal periodic deposits that may be made annually, quarterly, monthly, or daily.

When deposits are made at the same time the interest is credited, the annuity is called **ordinary.** Only ordinary annuities will be discussed here. The **amount of an annuity** is the sum of all deposits made plus all interest paid.

Suppose the interest an account earns is i percent per payment period (expressed as a decimal). For example, if an account pays 6% compounded monthly (12 times a year), then $i = \dfrac{0.06}{12} = 0.005$. Let's develop a formula for the amount of an annuity.

Suppose $\$P$ is deposited each payment period for n payment periods in an account that earns i% per payment period. When the last deposit is made at the nth payment period, the first deposit has earned interest compounded for $n - 1$ payment periods, the second deposit of $\$P$ has earned interest compounded for $n - 2$ payment periods, and so on. Table 1 shows the value of each deposit after n deposits have been made.

Table 1

Deposit	1	2	3	$n - 1$	n
Future Value of Deposit of $\$P$	$P(1 + i)^{n-1}$	$P(1 + i)^{n-2}$	$P(1 + i)^{n-3}$	$P(1 + i)$	P

Work Smart

The common ratio is $1+i$ because
$$\frac{(1+i)^{n-2}}{(1+i)^{n-3}} = (1+i)^{n-2-(n-3)}$$
$$= (1+i)^{n-2-n+3}$$
$$= (1+i)^1$$
$$= 1+i$$

The amount A of the annuity is the sum of the amounts shown in Table 1, namely,

$$A = P(1+i)^{n-1} + P(1+i)^{n-2} + P(1+i)^{n-3} + \cdots + P(1+i) + P$$
$$= P[(1+i)^{n-1} + (1+i)^{n-2} + (1+i)^{n-3} + \cdots + (1+i) + 1]$$
$$= P[1 + (1+i) + \cdots + (1+i)^{n-3} + (1+i)^{n-2} + (1+i)^{n-1}]$$

The expression is the sum of a geometric sequence with n terms, first term $a_1 = P$, and a common ratio $r = (1+i)$. Using $S_n = a_1 \dfrac{1-r^n}{1-r}$, gives

$$A = P[1 + (1+i) + \cdots + (1+i)^{n-3} + (1+i)^{n-2} + (1+i)^{n-1}]$$
$$= P \cdot \frac{1-(1+i)^n}{1-(1+i)} = P \cdot \frac{1-(1+i)^n}{-i} = P \cdot \frac{(1+i)^n - 1}{i}$$

We have the following result.

> **Amount of an Annuity**
>
> If P represents the deposit in dollars made at each payment period for an annuity at i percent interest per payment period, then the amount A of the annuity after n payment periods is
>
> $$A = P \cdot \frac{(1+i)^n - 1}{i}$$

EXAMPLE 10 **Determining the Amount of an Annuity**

To save for retirement, Alejandro decides to put $100 into a Roth Individual Retirement Account (IRA) every month for the next 30 years. What will be the value of the IRA after 30 years if his account earns 6% interest compounded monthly?

Solution

This is an ordinary annuity with $n = 12 \cdot 30 = 360$ payments and deposits of $P = \$100$. The rate of interest per payment period is $i = \dfrac{0.06}{12} = 0.005$. The amount after 30 years (360 deposits) is

$$A = \$100 \left[\frac{(1+0.005)^{360} - 1}{0.005} \right]$$
$$= \$100 [1004.515042]$$
$$= \$100{,}451.50$$

Quick ✓

18. To save for retirement, Magglio decides to place $500 into a Roth Individual Retirement Account (IRA) every quarter (every three months) for the next 30 years. What will be the value of the IRA after 30 years if his account earns 8% interest compounded quarterly?

10.3 Exercises · MyMathLab®

*Problems **1–18** are the Quick ✓s that follow the EXAMPLES.*

Building Skills

In Problems 19–26, show that the given sequence is a geometric sequence. Then find the first term and common ratio. See Objective 1.

19. $\{4^n\}$

20. $\{(-2)^n\}$

21. $\left\{\left(\dfrac{2}{3}\right)^n\right\}$

22. $\left\{\dfrac{2^n}{3}\right\}$

23. $\{3 \cdot 2^{-n}\}$

24. $\left\{-10\left(\dfrac{1}{2}\right)^n\right\}$

25. $\left\{\dfrac{5^{n-1}}{2^n}\right\}$

26. $\left\{\dfrac{3^{-n}}{2^{n-1}}\right\}$

In Problems 27–34, (a) find a formula for the nth term of the geometric sequence whose first term and common ratio are given, and (b) use the formula to find the 8th term. See Objective 2.

27. $a_1 = 10, r = 2$

28. $a_1 = 2, r = 3$

29. $a_1 = 100, r = \dfrac{1}{2}$

30. $a_1 = 30, r = \dfrac{1}{3}$

31. $a_1 = 1, r = -3$

32. $a_1 = 1, r = -4$

33. $a_1 = 100, r = 1.05$

34. $a_1 = 500, r = 1.04$

In Problems 35–40, find the indicated term of each geometric sequence. See Objective 2.

35. 10th term of $3, 6, 12, 24, \ldots$

36. 12th term of $1, 3, 9, 27, \ldots$

37. 15th term of $4, -2, 1, -\dfrac{1}{2}, \ldots$

38. 8th term of $10, -20, 40, -80, \ldots$

39. 9th term of $0.5, 0.05, 0.005, 0.0005, \ldots$

40. 10th term of $0.4, 0.04, 0.004, 0.0004, \ldots$

In Problems 41–48, find the sum. If necessary, express your answer to as many decimal places as your calculator allows. See Objective 3.

41. $2 + 4 + 8 + \cdots + 2^{12}$

42. $3 + 9 + 27 + \cdots + 3^{10}$

43. $50 + 20 + 8 + \dfrac{16}{5} + \cdots + 50\left(\dfrac{2}{5}\right)^{10-1}$

44. $10 + 5 + \dfrac{5}{2} + \cdots + 10\left(\dfrac{1}{2}\right)^{12-1}$

45. $\displaystyle\sum_{n=1}^{10} [3 \cdot 2^n]$

46. $\displaystyle\sum_{n=1}^{12} [5 \cdot 2^n]$

47. $\displaystyle\sum_{n=1}^{8} \left[\dfrac{4}{2^{n-1}}\right]$

48. $\displaystyle\sum_{n=1}^{14} \left[10\left(\dfrac{1}{2}\right)^{n-1}\right]$

In Problems 49–58, find the sum of each geometric series. If necessary, express your answer to as many decimal places as your calculator allows. See Objective 4.

49. $1 + \dfrac{1}{2} + \dfrac{1}{4} + \cdots$

50. $1 + \dfrac{1}{3} + \dfrac{1}{9} + \cdots$

51. $10 + \dfrac{10}{3} + \dfrac{10}{9} + \cdots$

52. $20 + 5 + \dfrac{5}{4} + \cdots$

53. $6 - 2 + \dfrac{2}{3} - \dfrac{2}{9} + \cdots$

54. $12 - 3 + \dfrac{3}{4} - \dfrac{3}{16} + \cdots$

55. $\displaystyle\sum_{n=1}^{\infty} \left(5\left(\dfrac{1}{5}\right)^n\right)$

56. $\displaystyle\sum_{n=1}^{\infty} \left(10\left(\dfrac{1}{3}\right)^n\right)$

57. $\displaystyle\sum_{n=1}^{\infty} \left(12\left(-\dfrac{1}{3}\right)^{n-1}\right)$

58. $\displaystyle\sum_{n=1}^{\infty} \left(100\left(-\dfrac{1}{2}\right)^{n-1}\right)$

In Problems 59–62, express each repeating decimal as a fraction in lowest terms. See Objective 4.

59. $0.\overline{5}$

60. $0.\overline{3}$

61. $0.\overline{89}$

62. $0.\overline{45}$

Mixed Practice

In Problems 63–74, determine whether the given sequence is arithmetic, geometric, or neither. If the sequence is arithmetic, find the common difference; if it is geometric, find the common ratio.

63. $\{5n + 1\}$

64. $\{8 - 3n\}$

65. $\{2n^2\}$

66. $\{n^2 - 2\}$

67. $\left\{\dfrac{2^{-n}}{5}\right\}$

68. $\left\{\dfrac{2}{3^n}\right\}$

69. $54, 36, 24, 16, \ldots$

70. $100, 20, 4, \dfrac{4}{5}, \ldots$

71. $2, 6, 10, 14, \ldots$

72. $15, 12, 9, 6, \ldots$

73. $1, 2, 3, 5, 8, \ldots$

74. $5, -2, 3, -1, 2, \ldots$

Applying the Concepts

75. Find x so that $x, x + 2$, and $x + 3$ are consecutive terms of a geometric sequence.

76. Find x so that $x - 1, x$, and $x + 2$ are consecutive terms of a geometric sequence.

77. Salary Increases Suppose that you have been hired at an annual salary of $40,000 per year. You have been promised a raise of 5% for each of the next 10 years.

   (a) What will be your salary at the beginning of your 2nd year?

   (b) What will be your salary at the beginning of your 10th year?

   (c) How much will you have earned cumulatively once you have finished your 10th year?

78. Salary Increases Suppose that you have been hired at an annual salary of $45,000 per year. Historically, the typical raise is 4% each year. You expect to be at the company for the next 10 years.

   (a) What will be your salary at the beginning of your 2nd year?

   (b) What will be your salary at the beginning of your 10th year?

   (c) How much will you have earned cumulatively once you have finished your 10th year?

79. Depreciation of a Car Suppose that you have just purchased a Honda Accord for $20,000. Historically, the car depreciates by 8% each year, so that next year the car is worth $20,000(0.92). What will the value of the car be after you have owned it for 5 years?

80. Depreciation of a Car Suppose that you have just purchased a Chevy Impala for $16,000. Historically, the car depreciates by 10% each year, so that next year the car is worth $16,000(0.9). What will the value of the car be after you have owned it for 4 years?

81. Pendulum Swings A pendulum makes its first swing through an arc of 3 feet. On each successive swing, the length of the arc is 0.95 of the previous length.

   (a) What is the length of the arc after 10 swings?

   (b) On which swing is the length of the arc less than 1 foot for the first time?

   (c) After 10 swings, what total length will the pendulum have swung?

   (d) When it stops, what total length will the pendulum have swung?

82. Bouncing Balls A ball is dropped from a height of 30 feet. Each time it strikes the ground, it bounces up to 0.8 of the previous height.

   (a) What height will the ball bounce up to after it strikes the ground for the 4th time?

   (b) How high will it bounce after it strikes the ground for the 5th time?

   (c) How many times does the ball need to strike the ground before its bounce is less than 6 inches?

   (d) What total distance does the ball travel before it stops bouncing?

83. A Job Offer You are interviewing for a job and receive two offers:

A: $30,000 to start, with guaranteed annual increases of 5% for the first 5 years

B: $31,000 to start with guaranteed annual increases of 4% for the first 5 years.

Which offer is best if your goal is to be making as much money as possible after the 5th year? Which is best if your goal is to make as much money as possible over the entire 5 years of the contract?

84. Be an Agent Suppose that you are an agent for a professional baseball player. Management has just offered your client a 5-year contract with a first-year salary of $1,500,000. Beyond that, your client has three choices:

A: A bonus of $75,000 each year (including the first year)

B: An annual increase of 4% per year beginning after the first year

C: An annual increase of $80,000 per year beginning after the first year

Which option provides the most money over the 5-year period? Which the least? Which would you choose? Why?

85. **The Multiplier** Suppose the marginal propensity to consume throughout the U.S. economy is 0.98. What is the multiplier for the U.S. economy?

86. **The Multiplier** Suppose the marginal propensity to consume throughout the U.S. economy is 0.96. What is the multiplier for the U.S. economy?

87. **Stock Price** One method of pricing a stock is based on the stream of future dividends of the stock. Suppose that a stock pays $*P* per year in dividends and, historically, the dividend has been increased by *i*% per year. If you desire an annual rate of return of *r*%, this method of pricing a stock states that the price you should pay is the present value of an infinite stream of payments:

$$\text{Price} = P + P \cdot \frac{1+i}{1+r} + P \cdot \left(\frac{1+i}{1+r}\right)^2 + P \cdot \left(\frac{1+i}{1+r}\right)^3 + \cdots$$

The price of the stock is the sum of a geometric series. Suppose that a stock pays an annual dividend of $2.00 and, historically, the dividend has been increased by 2% per year. You desire an annual rate of return of 9%. What is the most you should pay for the stock?

88. **Stock Price** Refer to Problem 87. Suppose that a stock pays an annual dividend of $3.00 and, historically, the dividend has been increased by 3% per year. You desire an annual rate of return of 10%. What is the most you should pay for the stock?

89. **401(k)** Christine contributes $100 each month into her 401(k) retirement plan. What will be the value of Christine's 401(k) in 30 years if the per annum rate of return is assumed to be 8% compounded monthly?

90. **Saving for a Home** Jolene wants to purchase a new home. Suppose she invests $400 a month into a money market fund. If the per annum interest rate of return on the money market fund is 1.2% compounded monthly, how much will Jolene have for a down payment in 4 years?

91. **Roth IRA** Jackson contributes $500 each quarter into his Roth IRA. What will be the value of Jackson's IRA in 25 years if the per annum rate of return is assumed to be 6% compounded quarterly?

92. **Retirement** Raymont is planning on retiring in 15 years, so he contributes $1500 into his IRA every 6 months (semiannually). What will be the value of the IRA when Raymont retires if the per annum interest rate is 10% compounded semiannually?

93. **What's My Payment?** Suppose that Aaliyah wants to have $1,500,000 in her 401(k) retirement account in 35 years. How much does she need to contribute each month if the account earns 10% interest compounded monthly?

94. **What's My Payment?** Suppose that Sophia wants to have $2,000,000 in her 401(k) retirement account in 25 years. How much does she need to contribute each quarter if the account earns 12% interest compounded quarterly?

Extending the Concepts

95. Express $0.4\overline{9}$ as a fraction in lowest terms.

96. Express $0.85\overline{9}$ as a fraction in lowest terms.

97. Find the sum: $2 + 4 + 8 + \cdots + 1{,}073{,}741{,}824$

98. Can a sequence be both arithmetic and geometric? Give reasons for your answer.

99. Which yields faster growth in the terms of a sequence— an arithmetic sequence or a geometric sequence with $r > 1$? Why? Explain why Thomas Robert Malthus's conjecture that food supplies grow arithmetically while population grows geometrically provides a recipe for disaster unless population growth is curbed.

Explaining the Concepts

100. How do you determine whether a sequence is geometric?

101. How do you determine whether a geometric series has a sum?

Synthesis Review

102. Express $\frac{1}{3}$ as a repeating decimal. Express $\frac{2}{3}$ as a repeating decimal.

103. What is $\frac{1}{3} + \frac{2}{3}$? Now add the repeating decimals found in Problem 102. Conjecture the value of $0.9999999\ldots$.

104. Prove that $0.9999\ldots = 0.\overline{9}$ equals 1.

Technology Exercises

In Problems 105–108, use technology to find the sum of each sequence.

105. $4 + 4.8 + 5.76 + \cdots + 4(1.2)^{15-1}$

106. $3 + 4.8 + 7.68 + \cdots + 3(1.6)^{20-1}$

107. $\displaystyle\sum_{n=1}^{20} [1.2(1.05)^n]$

108. $\displaystyle\sum_{n=1}^{25} [1.3(0.55)^n]$

Putting the Concepts Together (Sections 10.1–10.3)

We designed these problems from Sections 10.1 through 10.3 so that you can review the chapter so far and show your mastery of the concepts. Take time to work these problems before proceeding with the next section. The answers are located at the back of the text on page AN-61.

In Problems 1–6, determine if the sequence is arithmetic, geometric, or neither. If arithmetic or geometric, determine the first term and the common difference or common ratio.

1. $\dfrac{3}{4}, \dfrac{3}{16}, \dfrac{3}{64}, \dfrac{3}{256}, \ldots$

2. $\{2(n + 3)\}$

3. $\left\{\dfrac{7n + 2}{9}\right\}$

4. $1, -4, 9, -16, \ldots$

5. $\{3 \cdot 2^{n+1}\}$

6. $\{n^2 - 5\}$

7. Write out the sum and evaluate: $\displaystyle\sum_{k=1}^{6} [3k + 4]$

8. Express the sum using summation notation:

$$\frac{1}{2(6 + 1)} + \frac{1}{2(6 + 2)} + \frac{1}{2(6 + 3)} + \cdots + \frac{1}{2(6 + 12)}$$

In Problems 9–12, write a formula for the nth term of the indicated sequence. Write the first five terms of each sequence.

9. arithmetic: $a_1 = 25, d = -2$

10. arithmetic: $a_4 = 9, d = 11$

11. geometric: $a_4 = \dfrac{9}{25}, r = \dfrac{1}{5}$

12. geometric: $a_1 = 150, r = 1.04$

In Problems 13–15, find the indicated sum.

13. $2 + 6 + 18 + \cdots + 2 \cdot (3)^{11-1}$

14. $2 + 7 + 12 + 17 + \cdots + [2 + (20 - 1) \cdot 5]$

15. $1000 + 100 + 10 + \cdots$

16. Table Seating A restaurant uses square tables in the main dining area that seat four people. For larger parties, tables can be placed together. Two tables will seat 6 people, three tables will seat 8 people, and so on (see diagram). How many tables are needed to seat a party of 24 people?

10.4 The Binomial Theorem

Objectives

1 Compute Factorials

2 Evaluate a Binomial Coefficient

3 Expand a Binomial

Are You Prepared for This Section?

Before getting started, complete the following problems. If you get a problem wrong, go back to the section cited and review the material.

P1. Multiply: $(x - 5)^2$ [Section 4.2, pp. 324–325]

P2. Multiply: $(2x + 3)^2$ [Section 4.2, pp. 324–325]

▶ **1** **Compute Factorials**

Suppose you want to find the product of the first 12 positive integers, or

$$12 \cdot 11 \cdot 10 \cdot 9 \cdot 8 \cdot 7 \cdot 6 \cdot 5 \cdot 4 \cdot 3 \cdot 2 \cdot 1$$

Prepared?...Answers
P1. $x^2 - 10x + 25$ **P2.** $4x^2 + 12x + 9$

This product equals 479,001,600. Not only is this a big number, but writing all the factors is time-consuming. A shorthand method for writing this product is *factorial notation*.

Definition

If $n \geq 0$ is an integer, the **factorial symbol** $n!$ (read "n factorial") is defined as

$$0! = 1 \qquad 1! = 1$$
$$n! = n(n-1)(n-2) \cdot \cdots \cdot 3 \cdot 2 \cdot 1 \quad \text{if } n \geq 2$$

For example, $2! = 2 \cdot 1 = 2, 3! = 3 \cdot 2 \cdot 1 = 6, 4! = 4 \cdot 3 \cdot 2 \cdot 1 = 24$, and so on. Table 2 lists the values of $n!$ for $0 \leq n \leq 7$.

Table 2

n	0	1	2	3	4	5	6	7
$n!$	1	1	2	6	24	120	720	5040

Because

$$n! = n\underbrace{(n-1)(n-2) \cdot \cdots \cdot 3 \cdot 2 \cdot 1}_{(n-1)!}$$

use the formula

$$n! = n(n-1)!$$

to find successive factorials. For example, because $7! = 5040$,

$$8! = 8 \cdot 7! = 8(5040) = 40{,}320$$

Use your calculator's factorial key to see how fast factorials increase in value. Find the value of 69!. What happens when you try to find 70!? In fact, 70! is larger than 10^{100} (a **googol**), the largest number that most calculators can display.

EXAMPLE 1 **Computing Factorials**

Compute the value of $\dfrac{12!}{9!}$.

Solution

You could directly compute 12! and then 9!, but this would be inefficient. Use the properties of factorials instead.

$$\frac{12!}{9!} = \frac{12 \cdot 11 \cdot 10 \cdot 9!}{9!}$$

Divide out 9!: $\qquad = 12 \cdot 11 \cdot 10$

$$= 1320$$

See Figure 6 for this calculation on a graphing calculator.

Figure 6

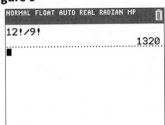

Quick ✔

1. If $n \geq 2$ is an integer, then $n! = $ _____.

2. $0! = $ __; $1! = $ __.

In Problems 3 and 4, find the value of each factorial.

3. $5!$

4. $\dfrac{7!}{3!}$

▶ ❷ Evaluate a Binomial Coefficient

The formula for expanding $(x + a)^n$ for $n = 2$ is known. The *Binomial Theorem* is a formula for expanding $(x + a)^n$ for any positive integer n. If $n = 1, 2, 3$, or 4, the expansion of $(x + a)^n$ is straightforward:

$(x + a)^1 = x + a$	Two terms, beginning with x^1 and ending with a^1
$(x + a)^2 = x^2 + 2ax + a^2$	Three terms, beginning with x^2 and ending with a^2
$(x + a)^3 = x^3 + 3ax^2 + 3a^2x + a^3$	Four terms, beginning with x^3 and ending with a^3
$(x + a)^4 = x^4 + 4ax^3 + 6a^2x^2 + 4a^3x + a^4$	Five terms, beginning with x^4 and ending with a^4

Each expansion of $(x + a)^n$ begins with x^n and ends with a^n. As you read from left to right, the powers of x decrease by 1, while the powers of a increase by 1. The number of terms is $n + 1$. Notice that the degree of each monomial in the expansion equals n. For example, in the expansion of $(x + a)^3$, each monomial $(x^3, 3ax^2, 3a^2x, a^3)$ is of degree 3. As a result, a conjecture might be that the expansion of $(x + a)^n$ would look like this:

$$(x + a)^n = x^n + \_\,ax^{n-1} + \_\,a^2 x^{n-2} + \cdots + \_\,a^{n-1} x + a^n$$

where the blanks are numbers to be found. This is correct, as will be seen shortly.

Before we go any further, we need to introduce the symbol $\binom{n}{j}$, which is read "n taken j at a time" or "n choose j":

Work Smart

Do not write $\binom{n}{j}$ as $\left(\dfrac{n}{j}\right)$.

> ### Definition
>
> If j and n are integers with $0 \le j \le n$, the symbol $\binom{n}{j}$ is defined as
>
> $$\binom{n}{j} = \frac{n!}{j!\,(n - j)!}$$

EXAMPLE 2   Evaluating $\binom{n}{j}$

Find:

(a) $\binom{4}{1}$ (b) $\binom{6}{2}$ (c) $\binom{5}{4}$

Solution

(a) Here, $n = 4$ and $j = 1$, so

$$\binom{4}{1} = \frac{4!}{1!\,(4 - 1)!} = \frac{4!}{1! \cdot 3!} = \frac{4 \cdot 3!}{1! \cdot 3!} = \frac{4 \cdot 3!}{1 \cdot 3!} = 4$$

$$6! = 6 \cdot 5 \cdot 4!$$

(b) $\binom{6}{2} = \dfrac{6!}{2!\,(6 - 2)!} = \dfrac{6!}{2! \cdot 4!} = \dfrac{6 \cdot 5 \cdot 4!}{2 \cdot 1 \cdot 4!} = \dfrac{6 \cdot 5 \cdot 4!}{2 \cdot 1 \cdot 4!} = \dfrac{30}{2} = 15$

(c) $\binom{5}{4} = \dfrac{5!}{4!\,(5 - 4)!} = \dfrac{5!}{4! \cdot 1!} = \dfrac{5 \cdot 4!}{4! \cdot 1} = \dfrac{5 \cdot 4!}{4! \cdot 1} = 5$

> **Quick ✓**
>
> **5.** *True or False* $\binom{n}{j} = \dfrac{j!}{n!\,(n - j)!}$ **6.** *True or False* $\binom{5}{1} = 5$
>
> *In Problems 7 and 8, evaluate each expression.*
>
> **7.** $\binom{7}{1}$ **8.** $\binom{6}{3}$

Four useful formulas involving the symbol $\binom{n}{j}$ are

$$\binom{n}{0} = 1 \qquad \binom{n}{1} = n \qquad \binom{n}{n-1} = n \qquad \binom{n}{n} = 1$$

Suppose that the various values of the symbol $\binom{n}{j}$ are arranged in a triangular display, as shown next and in Figure 7.

$$\binom{0}{0}$$

$$\binom{1}{0} \quad \binom{1}{1}$$

$$\binom{2}{0} \quad \binom{2}{1} \quad \binom{2}{2}$$

$$\binom{3}{0} \quad \binom{3}{1} \quad \binom{3}{2} \quad \binom{3}{3}$$

$$\binom{4}{0} \quad \binom{4}{1} \quad \binom{4}{2} \quad \binom{4}{3} \quad \binom{4}{4}$$

$$\binom{5}{0} \quad \binom{5}{1} \quad \binom{5}{2} \quad \binom{5}{3} \quad \binom{5}{4} \quad \binom{5}{5}$$

Figure 7
Pascal's Triangle

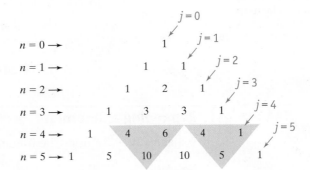

This display is called **Pascal's triangle,** named after Blaise Pascal (1623–1662), a French mathematician.

Pascal's triangle has 1s down the sides. To get any other entry, add the two nearest entries in the row above it. The shaded triangles in Figure 7 illustrate this feature. Using this feature, find the row corresponding to $n = 6$.

$$
\begin{array}{ccccccc}
n = 5 \rightarrow & 1 & 5 & 10 & 10 & 5 & 1 \\
n = 6 \rightarrow & 1 & 6 & 15 & 20 & 15 & 6 & 1
\end{array}
$$

Although Pascal's triangle is an interesting and organized display of the symbol $\binom{n}{j}$, in practice it is not all that helpful. For example, to find the value of $\binom{12}{8}$, you would need to produce 12 rows of the triangle before seeing the answer. It is much faster to use the definition of $\binom{n}{j}$.

▶ ❸ Expand a Binomial

The formula for the **Binomial Theorem** is presented on the next page.

> **The Binomial Theorem**
> Let x and a be real numbers. For any positive integer n,
>
> $$(x + a)^n = \binom{n}{0}x^n + \binom{n}{1}ax^{n-1} + \binom{n}{2}a^2x^{n-2} + \cdots + \binom{n}{j}a^jx^{n-j} + \cdots + \binom{n}{n}a^n$$

You should now see why there is a need to discuss the symbol $\binom{n}{j}$; these symbols are the numerical coefficients that appear in the expansion of $(x + a)^n$. Because of this, the symbol $\binom{n}{j}$ is called a **binomial coefficient.**

EXAMPLE 3 **Expanding a Binomial**

Use the Binomial Theorem to expand $(x + 3)^4$.

Solution
In the Binomial Theorem, let $a = 3$ and $n = 4$. Then

Work Smart
In the expansion, notice that the exponent on x **decreases** by 1 for each term as you move to the right, while the exponent on 3 **increases** by 1.

$$(x + 3)^4 = \binom{4}{0}x^4 + \binom{4}{1}3^1 \cdot x^{4-1} + \binom{4}{2}3^2 \cdot x^{4-2} + \binom{4}{3}3^3 \cdot x^{4-3} + \binom{4}{4}3^4$$

Use Pascal's triangle in Figure 7 or use $\binom{n}{j} = \dfrac{n!}{j!(n-j)!}$ to evaluate the binomial coefficients.

$$= 1 \cdot x^4 + 4 \cdot 3 \cdot x^3 + 6 \cdot 9 \cdot x^2 + 4 \cdot 27 \cdot x + 1 \cdot 81$$
$$= x^4 + 12x^3 + 54x^2 + 108x + 81$$ •

EXAMPLE 4 **Expanding a Binomial**

Expand $(2y - 3)^5$ using the Binomial Theorem.

Solution
First, rewrite the expression $(2y - 3)^5$ as $[2y + (-3)]^5$. Then use the Binomial Theorem with $n = 5$, $x = 2y$, and $a = -3$.

$$[2y + (-3)]^5 = \binom{5}{0}(2y)^5 + \binom{5}{1}(-3)^1(2y)^{5-1} + \binom{5}{2}(-3)^2(2y)^{5-2} + \binom{5}{3}(-3)^3(2y)^{5-3} + \binom{5}{4}(-3)^4(2y)^{5-4} + \binom{5}{5}(-3)^5$$

Use Pascal's triangle or use the formula $\binom{n}{j} = \dfrac{n!}{j!(n-j)!}$ to evaluate binomial coefficients.

$$= 1(2y)^5 + 5(-3)^1(2y)^4 + 10(-3)^2(2y)^3 + 10(-3)^3(2y)^2 + 5(-3)^4(2y)^1 + 1(-3)^5$$
$$= 32y^5 + 5 \cdot (-3) \cdot 16y^4 + 10 \cdot 9 \cdot 8y^3 + 10 \cdot (-27) \cdot 4y^2 + 5 \cdot 81 \cdot 2y - 243$$
$$= 32y^5 - 240y^4 + 720y^3 - 1080y^2 + 810y - 243$$ •

Quick ✓

In Problems 9 and 10, expand each binomial using the Binomial Theorem.

9. $(x + 2)^4$

10. $(2p - 1)^5$

10.4 Exercises   MyMathLab®

Exercise numbers in green have
complete video solutions in MyMathLab or
may be accessed using the QR code to the right.

*Problems **1–10** are the* Quick ✔ s *that follow the* EXAMPLES.

Building Skills

In Problems 11–18, evaluate each factorial expression. See Objective 1.

11. $3!$ **12.** $4!$ **13.** $8!$ **14.** $9!$

15. $\dfrac{10!}{8!}$ **16.** $\dfrac{12!}{10!}$ **17.** $\dfrac{8!}{5!}$ **18.** $\dfrac{9!}{6!}$

In Problems 19–22, evaluate each expression. See Objective 2.

19. $\dbinom{7}{2}$ **20.** $\dbinom{8}{5}$ **21.** $\dbinom{10}{4}$ **22.** $\dbinom{12}{9}$

In Problems 23–38, expand each expression using the Binomial Theorem. See Objective 3.

23. $(x + 1)^5$ **24.** $(x - 1)^4$

25. $(x - 4)^4$ **26.** $(x + 5)^5$

27. $(3p + 2)^4$ **28.** $(2q + 3)^4$

29. $(2z - 3)^5$ **30.** $(3w - 4)^4$

31. $(x^2 + 2)^4$ **32.** $(y^2 - 3)^4$

33. $(2p^3 + 1)^5$ **34.** $(3b^2 + 2)^5$

35. $(x + 2)^6$ **36.** $(p - 3)^6$

37. $(2p^2 - q^2)^4$ **38.** $(3x^2 + y^3)^4$

Applying the Concepts

39. Use the Binomial Theorem to find the numerical value of $(1.001)^4$ correct to five decimal places. (*Hint*: $(1.001)^4 = (1 + 10^{-3})^4$.)

40. Use the Binomial Theorem to find the numerical value of $(1.001)^5$ correct to five decimal places. (*Hint*: $(1.001)^5 = (1 + 10^{-3})^5$.)

41. Use the Binomial Theorem to find the numerical value of $(0.998)^5$ correct to five decimal places.

42. Use the Binomial Theorem to find the numerical value of $(0.997)^5$ correct to five decimal places.

Extending the Concepts

Notice that in the formula for expanding a binomial $(x + a)^n$, the first term is $\dbinom{n}{0}a^0 \cdot x^n$, the second term is $\dbinom{n}{1}a^1 \cdot x^{n-1}$, the third term is $\dbinom{n}{2}a^2 \cdot x^{n-2}$, and so on. In general, the jth term in a binomial expansion of $(x + a)^n$ is $\dbinom{n}{j-1}a^{j-1} x^{n-j+1}$. Use this result to find the indicated term in Problems 43–46.

43. The 3rd term in the expansion of $(x + 2)^7$

44. The 4th term in the expansion of $(x - 1)^{10}$

45. The 6th term in the expansion of $(2p - 3)^8$

46. The 7th term in the expansion of $(3p + 1)^9$

47. Show that $\dbinom{n}{n-1} = n$ and $\dbinom{n}{n} = 1$.

48. Show that if n and j are integers with $0 \le j \le n$, then $\dbinom{n}{j} = \dbinom{n}{n-j}$.

Explaining the Concepts

49. Write the first four rows of Pascal's triangle.

50. Describe the pattern of exponents of x in the expansion of $(x + a)^n$. Describe the pattern of exponents of a in the expansion of $(x + a)^n$.

51. What is true about the degree of each monomial in the expansion of $(x + y)^n$?

52. Explain how you might find a particular term in a binomial expansion. For example, how might you find the 5th term in the expansion of $(x + 3)^8$?

Synthesis Review

53. If $f(x) = x^4$, find $f(a - 2)$ with the aid of the Binomial Formula.

54. If $g(x) = x^5 + 3$, find $g(z + 1)$ with the aid of the Binomial Formula.

55. If $H(x) = x^5 - 4x^4$, find $H(p + 1)$ with the aid of the Binomial Formula.

56. If $h(x) = 2x^5 + 5x^4$, find $h(a + 3)$ with the aid of the Binomial Formula.

Chapter 10 Activity: Discover the Relation

Focus: Review of objectives for arithmetic and geometric sequences and the Binomial Theorem

Time: 30–35 minutes

Group size: 2–4

As a group, discover the relationship between the results found from Column A and Column B (that is, $<$, $>$, or $=$). Be sure to discuss any differences in outcomes.

	Column A	Column B
1	The sum of the first five terms of $$a_n = \frac{3}{5}n + 1 \quad A < B$$	The sum of the first five terms of $$a_n = -\frac{1}{4}(n-1) + 4$$
2	$$\sum_{i=1}^{4}(i^2 + 2i) \quad A = B$$	$$\sum_{i=1}^{4} i^2 + \sum_{i=1}^{4} 2i$$
3	The 8th term of an arithmetic sequence when $a_1 = 3$ and $d = \dfrac{3}{2}$ $\quad A > B$	The 27th term of an arithmetic sequence when $a_1 = \dfrac{5}{3}$ and $d = \dfrac{1}{3}$
4	The 10th term of a geometric sequence with $a_1 = 3$ and $r = \sqrt{2}$ $\quad A < B$	The 9th term of a geometric sequence with $a_1 = 5$ and $r = \sqrt{3}$
5	The coefficient of the 4th term of $(x+y)^{10}$ $\quad A < B$	The coefficient of the 3rd term of $(8x - y)^4$

Chapter 10 Review

Section 10.1   Sequences

KEY TERMS

Sequence
Terms
Ellipsis
Infinite sequence

Finite sequence
General term
Alternate

Summation notation
Index
Partial sum

You Should Be Able To...	EXAMPLE	Review Exercises
❶ Write the first few terms of a sequence (p. 742)	Examples 1 and 2	1–6
❷ Find a formula for the nth term of a sequence (p. 744)	Example 3	7–12
❸ Use summation notation (p. 744)	Examples 4 and 5	13–20

In Problems 1–6, write the first five terms of each sequence.

1. $\{-3n + 2\}$

2. $\left\{\dfrac{n-2}{n+4}\right\}$

3. $\{5^n + 1\}$

4. $\{(-1)^{n-1} \cdot 3n\}$

5. $\left\{\dfrac{n^2}{n+1}\right\}$

6. $\left\{\dfrac{\pi^n}{n}\right\}$

In Problems 7–12, the given pattern continues. Write the nth term of each sequence suggested by the pattern.

7. $-3, -6, -9, -12, -15, \ldots$

8. $\dfrac{1}{3}, \dfrac{2}{3}, 1, \dfrac{4}{3}, \dfrac{5}{3}, \ldots$

9. $5, 10, 20, 40, 80, \ldots$

10. $-\dfrac{1}{2}, 1, -\dfrac{3}{2}, 2, \ldots$

11. $6, 9, 14, 21, 30, \ldots$

12. $0, \dfrac{1}{3}, \dfrac{1}{2}, \dfrac{3}{5}, \ldots$

In Problems 13–16, write out each sum and determine its value.

13. $\displaystyle\sum_{k=1}^{5} (5k - 2)$

14. $\displaystyle\sum_{k=1}^{6} \left(\dfrac{k + 2}{2}\right)$

15. $\displaystyle\sum_{i=1}^{5} (-2i)$

16. $\displaystyle\sum_{i=1}^{4} \dfrac{i^2 - 1}{3}$

In Problems 17–20, express each sum using summation notation.

17. $(4 + 3 \cdot 1) + (4 + 3 \cdot 2) + (4 + 3 \cdot 3) + \cdots + (4 + 3 \cdot 15)$

18. $\dfrac{1}{3^1} + \dfrac{1}{3^2} + \dfrac{1}{3^3} + \cdots + \dfrac{1}{3^8}$

19. $\dfrac{1^3 + 1}{1 + 1} + \dfrac{2^3 + 1}{2 + 1} + \dfrac{3^3 + 1}{3 + 1} + \cdots + \dfrac{10^3 + 1}{10 + 1}$

20. $(-1)^{1-1} \cdot 1^2 + (-1)^{2-1} \cdot 2^2 + (-1)^{3-1} \cdot 3^2 + \cdots + (-1)^{7-1} \cdot 7^2$

Section 10.2   Arithmetic Sequences

KEY CONCEPTS

- **The *n*th Term of an Arithmetic Sequence**
  For an arithmetic sequence whose first term is a_1 and whose common difference is d, the nth term is $a_n = a_1 + (n - 1)d$.

- **Sum of the First *n* Terms of an Arithmetic Sequence**
  For an arithmetic sequence whose first term is a_1 and whose common difference is d, the sum of the first n terms is

  $$S_n = \dfrac{n}{2}[2a_1 + (n - 1)d] \quad \text{or} \quad S_n = \dfrac{n}{2}[a_1 + a_n].$$

KEY TERMS

Arithmetic sequence
Common difference

You Should Be Able To...	EXAMPLE	Review Exercises
❶ Determine whether a sequence is arithmetic (p. 749)	Examples 1 through 3	21–26
❷ Find a formula for the nth term of an arithmetic sequence (p. 750)	Examples 4 and 5	27–32, 37
❸ Find the sum of an arithmetic sequence (p. 752)	Examples 6 through 8	33–36, 38

In Problems 21–26, determine whether the sequence is arithmetic. If so, find the common difference.

21. $4, 10, 16, 22, \ldots$

22. $-1, \dfrac{1}{2}, 2, \dfrac{7}{2}, \ldots$

23. $-2, -5, -9, -14, \ldots$

24. $-1, 3, -5, 7, \ldots$

25. $\{4n + 7\}$

26. $\left\{\dfrac{n + 1}{2n}\right\}$

In Problems 27–32, find a formula for the nth term of the arithmetic sequence. Use the formula to find the 25th term of the sequence.

27. $a_1 = 3; d = 8$

28. $a_1 = -4; d = -3$

29. $7, \dfrac{20}{3}, \dfrac{19}{3}, 6, \ldots$

30. $11, 17, 23, 29, \ldots$

31. The 3rd term is 7 and the 8th term is 25.

32. The 4th term is -20 and the 7th term is -32.

33. Find the sum of the first 30 terms of the sequence $-1, 9, 19, 29, \ldots$.

34. Find the sum of the first 40 terms of the sequence $5, 2, -1, -4, \ldots$.

35. Find the sum of the first 60 terms of the sequence $\{-2n - 7\}$.

36. Find the sum of the first 50 terms of the sequence $\left\{\dfrac{1}{4}n + 3\right\}$.

37. Cicadas Seventeen-year cicadas emerge every 17 years to mate, lay eggs, and start the next 17-year cycle. In 2004, the Brood X cicada (the largest brood of the 17-year cicada) emerged in Maryland, Kentucky, Tennessee, and parts of surrounding states. Determine when the Brood X cicada will first appear in the twenty-second century.

38. Wind Sprints At a certain football practice, players would run wind sprints for exercise. Starting at the goal line, players would sprint to the 10-yard line and back to the goal line. The players would then sprint to the 20-yard line and back to the goal line. This would continue for the 30-yard line, 40-yard line, and 50-yard line. Determine the total distance a player would run during the wind sprints.

Section 10.3 Geometric Sequences and Series

KEY CONCEPTS

- **The _n_th Term of a Geometric Sequence**
  For a geometric sequence whose first term is a_1 and whose common ratio is r, the nth term is $a_n = a_1 r^{n-1}$.

- **Sum of the First _n_ Terms of a Geometric Sequence**
  For a geometric sequence whose first term is a_1 and whose common ratio is r, the sum of the first n terms is $S_n = a_1 \cdot \dfrac{1 - r^n}{1 - r}, r \neq 0, r \neq 1$.

- **Sum of a Geometric Series**
  If $-1 < r < 1$, then the sum of the geometric series $\displaystyle\sum_{n=1}^{\infty} a_1 r^{n-1} = \dfrac{a_1}{1 - r}$.

KEY TERMS

Geometric sequence
Common ratio
Geometric series
Marginal propensity
   to consume
Multiplier

You Should Be Able To...	EXAMPLE	Review Exercises
❶ Determine whether a sequence is geometric (p. 756)	Examples 1 through 3	39–44
❷ Find a formula for the nth term of a geometric sequence (p. 758)	Example 4	45–48, 57
❸ Find the sum of a geometric sequence (p. 759)	Examples 5 and 6	49–52, 58
❹ Find the sum of a geometric series (p. 761)	Examples 7 through 9	53–56
❺ Solve annuity problems (p. 763)	Example 10	59–62

In Problems 39–44, determine whether the given sequence is geometric. If so, determine the common ratio.

39. $\dfrac{1}{3}, 2, 12, 72, \ldots$

40. $-1, 3, -9, 27, \ldots$

41. $1, 1, 2, 6, \ldots$

42. $6, 4, \dfrac{8}{3}, \dfrac{16}{9}, \ldots$

43. $\{5(-2)^n\}$

44. $\{3n - 14\}$

In Problems 45–48, find a formula for the nth term of the geometric sequence. Use the formula to find the 10th term of the sequence.

45. $a_1 = 4, r = 3$

46. $a_1 = 8, r = \dfrac{1}{4}$

47. $a_1 = 5, r = -2$

48. $a_1 = 1000, r = 1.08$

In Problems 49–52, find the sum. If necessary, express your answer to as many decimal places as your calculator allows.

49. $2 + 4 + 8 + \cdots + 2^{15}$

50. $40 + 5 + \dfrac{5}{8} + \cdots + 40\left(\dfrac{1}{8}\right)^{13-1}$

51. $\displaystyle\sum_{n=1}^{12}\left[\dfrac{3}{4}(2)^{n-1}\right]$

52. $\displaystyle\sum_{n=1}^{16}\left[-4(3^{n})\right]$

In Problems 53–56, find the sum of each geometric series. If necessary, express your answer to as many decimal places as your calculator allows.

53. $\displaystyle\sum_{n=1}^{\infty}\left[20\left(\dfrac{1}{4}\right)^{n}\right]$

54. $\displaystyle\sum_{n=1}^{\infty}\left[50\left(-\dfrac{1}{2}\right)^{n-1}\right]$

55. $1 + \dfrac{1}{5} + \dfrac{1}{25} + \cdots$

56. $0.8 + 0.08 + 0.008 + 0.0008 + \cdots$

57. Radioactive Decay The radioactive isotope tritium has a half-life of about 12 years. If there were 200 grams of the isotope initially, use a geometric sequence to determine how much would remain after 72 years.

58. Computer Virus In January 2004, the Mydoom e-mail worm was declared the worst e-mail worm incident in virus history, accounting for roughly 20–30% of worldwide e-mail traffic. Suppose the virus was initially sent to 5 e-mail addresses and that, upon receipt, it sends itself out to 5 e-mail addresses from the address book of the infected computer. If each cycle of e-mails, including the initial sending, takes 1 minute to complete, how many total e-mails will have been sent after 15 minutes?

59. 403(b) Each quarter, Scott contributes $900 to a 403(b) plan at work. His employer agrees to match half of employee contributions up to $600 per quarter. What will be the value of Scott's 403(b) in 25 years if the per annum rate of return is assumed to be 7% compounded quarterly?

60. Lottery Payment The winner of a state lottery has the option of receiving about $2 million per year for 26 years (after taxes) or a lump sum payment of about $28 million (after taxes). Assuming all winnings will be invested at a per annum rate of return of 6.5% compounded annually, which option yields the most money after 26 years?

61. What's My Payment? Sheri starts her career when she is 22 years old and wants to have $2,500,000 in her 401(k) retirement account when she retires in 40 years. How much does she need to contribute each month if the account earns 9% interest compounded monthly?

62. College Savings Plan On Samantha's 8th birthday, her parents open a 529 college savings plan for her. They plan to contribute $400 per month until she turns 18. The per annum rate of return is assumed to be 5.25% compounded monthly, and the cost per credit hour at a private 4-year university is locked in at a rate of $340 per hour. What will be the value of the plan when Samantha turns 18, and how many credit hours will the plan cover?

Section 10.4   The Binomial Theorem

KEY CONCEPTS

- **Factorial symbol $n!$**
  If $n \geq 0$ is an integer, the factorial symbol $n!$ is defined as
  $$0! = 1 \qquad 1! = 1 \qquad n! = n(n-1)(n-2)\cdots\cdot 3\cdot 2\cdot 1 \quad \text{if } n \geq 2.$$

- **The symbol $\dbinom{n}{j}$**
  If j and n are integers with $0 \leq j \leq n$, then $\dbinom{n}{j} = \dfrac{n!}{j!\,(n-j)!}$.

- **The Binomial Theorem**
  Let x and a be real numbers. For any positive integer n,
  $$(x+a)^n = \binom{n}{0}x^n + \binom{n}{1}ax^{n-1} + \binom{n}{2}a^2x^{n-2} + \cdots + \binom{n}{j}a^jx^{n-j} + \cdots + \binom{n}{n}a^n.$$

KEY TERMS

Factorial symbol
Googol
Pascal's triangle
Binomial coefficient

You Should Be Able To...	EXAMPLE	Review Exercises
1 Compute factorials (p. 768)	Example 1	63–66
2 Evaluate a binomial coefficient (p. 770)	Example 2	67–70
3 Expand a binomial (p. 771)	Examples 3 and 4	71–78

In Problems 63–66, evaluate the expression.

63. $5!$

64. $\dfrac{11!}{7!}$

65. $\dfrac{10!}{6!}$

66. $\dfrac{13!}{6!7!}$

In Problems 67–70, evaluate each binomial coefficient.

67. $\dbinom{7}{3}$

68. $\dbinom{10}{5}$

69. $\dbinom{8}{8}$

70. $\dbinom{6}{0}$

In Problems 71–76, expand each expression using the Binomial Theorem.

71. $(z + 1)^4$

72. $(y - 3)^5$

73. $(3y + 4)^6$

74. $(2x^2 - 3)^4$

75. $(3p - 2q)^4$

76. $(a^3 + 3b)^5$

77. Find the 4th term in the expansion of $(x - 2)^8$.

78. Find the 7th term in the expansion of $(2x + 1)^{11}$.

Chapter 10 Test

Step-by-step test solutions are found on the Chapter Test Prep Videos available in MyMathLab®, on You Tube, or may be accessed using the QR code to the right.

In Problems 1–6, determine whether the sequence is arithmetic, geometric, or neither. If arithmetic or geometric, determine the first term and the common difference or common ratio.

1. $-15, -7, 1, 9, \ldots$

2. $\{(-4)^n\}$

3. $\left\{\dfrac{4}{n!}\right\}$

4. $\left\{\dfrac{2n - 3}{5}\right\}$

5. $-3, 2, 0, 5, 3, \ldots$

6. $\{7 \cdot 3^n\}$

7. Write out the sum and evaluate: $\displaystyle\sum_{i=1}^{5}\left[\dfrac{3}{i^2} + 2\right]$

8. Express the sum using summation notation:
$$\dfrac{3}{5} + \dfrac{2}{3} + \dfrac{5}{7} + \dfrac{3}{4} + \cdots + \dfrac{5}{6}$$

In Problems 9–12, write a formula for the nth term of the indicated sequence. Write the first five terms of each sequence.

9. arithmetic: $a_1 = 6, d = 10$

10. arithmetic: $a_1 = 0, d = -4$

11. geometric: $a_1 = 10, r = 2$

12. geometric: $a_3 = 9, r = -3$

In Problems 13–15, find the indicated sum.

13. $-2 + 2 + 6 + \cdots + [4(20 - 1) - 2]$

14. $\dfrac{1}{9} - \dfrac{1}{3} + 1 - 3 + \cdots + \dfrac{1}{9}(-3)^{12-1}$

15. $216 + 72 + 24 + 8 + \cdots$

16. Evaluate $\dfrac{15!}{8!7!}$.

17. Evaluate $\dbinom{12}{5}$.

18. Expand $(5m - 2)^4$ using the Binomial Theorem.

19. A new car sold for $31,000. If the vehicle loses 15% of its value each year, how much will it be worth after 10 years?

20. A weightlifter begins his routine by benching 100 pounds and increases the weight by 30 pounds for each set. If he does 10 repetitions in each set, what is the total weight lifted after 5 sets?

Appendix A: Transformations

Objectives

1 Graph Functions Using Vertical and Horizontal Shifts

2 Graph Functions Using Compressions and Stretches

3 Graph Functions Using Reflections about the x-Axis and the y-Axis

Are You Prepared for This Section?

Before getting started, complete the following problems. If you get a problem wrong, go back to the section cited and review the material.

P1. If $f(3) = 8$, what point is on the graph of $y = f(x)$? [Section 2.3, pp. 164]

P2. Graph the function $f(x) = \sqrt{x}$. [Section 2.3 pp. 168; Section 6.6, pp. 510]

P3. Graph the function $f(x) = \sqrt[3]{x}$. [Section 2.3 pp. 168; Section 6.6, pp. 511]

P4. Graph the function $f(x) = x^2$. [Section 1.5, pp. 96–97; Section 2.3 pp. 168]

P5. Graph the function $f(x) = x^3$. [Section 2.3, pp. 168]

P6. Graph the function $f(x) = |x|$. [Section 2.3, pp. 168]

At this stage, if you were asked to graph any of the functions defined by $y = x$, $y = x^2, y = \sqrt{x}, y = \sqrt[3]{x}$, or $y = |x|$, your response should be "Yes, I recognize these functions and know the general shapes of their graphs." (If this is not your answer, review Section 2.3, Table 3.)

Sometimes we are asked to graph a function that is "almost" like one that we already know how to graph. This section will develop techniques for graphing such functions. Collectively, these techniques are referred to as **transformations**.

▶ **1** Graph Functions Using Vertical and Horizontal Shifts

EXAMPLE 1 **Vertical Shift Up**

Use the graph of $f(x) = |x|$ to obtain the graph of $g(x) = |x| + 4$.

Solution

Begin by obtaining some points on the graphs of f and g. For example, when $x = 0$, then $y = f(0) = 0$ and $y = g(0) = 4$. Table 1 lists these points as well as a few others. Notice that the y-coordinates on the graph of $g(x) = |x| + 4$ are exactly 4 units greater than the corresponding y-coordinates on the graph of $f(x) = |x|$. Figure 1 shows the graphs of f and g.

Table 1

x	$f(x) = \|x\|$	$(x, f(x))$	$g(x) = \|x\| + 4$	$(x, g(x))$
-2	$\|-2\| = 2$	$(-2, 2)$	$\|-2\| + 4 = 6$	$(-2, 6)$
-1	$\|-1\| = 1$	$(-1, 1)$	$\|-1\| + 4 = 5$	$(-1, 5)$
0	0	$(0, 0)$	4	$(0, 4)$
1	1	$(1, 1)$	5	$(1, 5)$
2	2	$(2, 2)$	6	$(2, 6)$

Figure 1

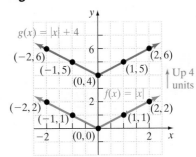

The graph of g is identical to the graph of f, except that it is shifted up 4 units. ●

EXAMPLE 2   **Vertical Shift Down**

Use the graph of $f(x) = |x|$ to obtain the graph of $g(x) = |x| - 5$.

Solution

Table 2 lists some points on the graphs of f and g. Notice that the y-coordinates on the graph of g are 5 units less than the corresponding y-coordinates on the graph of f. Figure 2 shows the graphs of f and g.

Table 2

x	$f(x) = \|x\|$	$(x, f(x))$	$g(x) = \|x\| - 5$	$(x, g(x))$
-2	2	$(-2, 2)$	-3	$(-2, -3)$
-1	1	$(-1, 1)$	-4	$(-1, -4)$
0	0	$(0, 0)$	-5	$(0, -5)$
1	1	$(1, 1)$	-4	$(1, -4)$
2	2	$(2, 2)$	-3	$(2, -3)$

Figure 2

The graph of g is identical to the graph of f, except that it is shifted down 5 units. ●

The results of Examples 1 and 2 lead to the following conclusion.

> If a positive real number k is added to the output of a function, $y = f(x)$, the graph of the new function, $y = f(x) + k$, is the graph of f **shifted vertically up** k units.
>
> If a positive real number k is subtracted from the output of a function, $y = f(x)$, the graph of the new function, $y = f(x) - k$, is the graph of f **shifted vertically down** k units.

Quick ✓

1. If k is a positive real number, to obtain the graph of $y = f(x) + k$ from the graph of $y = f(x)$, shift the graph of $y = f(x)$ _____ k units. To obtain the graph of $y = f(x) - k$ from the graph of $y = f(x)$, shift the graph of $y = f(x)$ _____ k units.

2. Use the graph of $y = |x|$ to graph the function $f(x) = |x| + 5$. Show at least three points.

3. Use the graph of $y = x^2$ to graph the function $f(x) = x^2 - 4$. Show at least three points.

EXAMPLE 3   **Horizontal Shift to the Right**

In the same Cartesian plane, graph $f(x) = x^3$ and $g(x) = (x - 2)^3$.

Solution

Use the point-plotting method. Table 3 on the next page lists some points on the graphs of f and g. Note that when $f(x) = 0$, $x = 0$ and that when $g(x) = 0$, $x = 2$. Also, when $f(x) = -1$, $x = -1$, and when $g(x) = -1$, $x = 1$. The graph of g is identical to that of f, except that g is shifted 2 units to the right. See Figure 3.

Figure 3

Table 3

x	f(x) = x³	(x, f(x))	g(x) = (x − 2)³	(x, g(x))
−2	−8	(−2, −8)	(−2 − 2)³ = −64	(−2, −64)
−1	−1	(−1, 1)	(−1 − 2)³ = −27	(−1, −27)
0	0	(0, 0)	−8	(0, −8)
1	1	(1, 1)	−1	(1, −1)
2	8	(2, 8)	0	(2, 0)

Why does this happen? Because 2 is getting subtracted from each *x*-value in the function $g(x) = (x - 2)^3$, the *x*-values must be greater by 2 in order to obtain the same *y*-value that was obtained in the graph of $f(x) = x^3$. ●

EXAMPLE 4   **Horizontal Shift to the Right**

In the same Cartesian plane, graph $f(x) = x^3$ and $g(x) = (x + 2)^3$.

Solution
Again, use the point-plotting method. Table 4 lists some points on the graphs of *f* and *g*. Notice that when $f(x) = 0$, *x* = 0, and when $g(x) = 0$, *x* = −2. Also, when $f(x) = 1$, *x* = 1, and when $g(x) = 1$, *x* = −1. The graph of *g* is identical to that of *f*, except that *g* is shifted 2 units to the left. See Figure 4.

Table 4

x	f(x) = x³	(x, f(x))	g(x) = (x + 2)³	(x, g(x))
−2	−8	(−2, −8)	0	(−2, 0)
−1	−1	(−1, −1)	1	(−1, 1)
0	0	(0, 0)	8	(0, 8)
1	1	(1, 1)	27	(1, 27)
2	8	(2, 8)	64	(2, 64)

Figure 4

Why does this happen? Because 2 is being added to each *x*-value in the function $g(x) = (x + 2)^3$, the *x*-values must be smaller by 2 in order to obtain the same *y*-value that was obtained in the graph of $f(x) = x^3$. ●

The results of Examples 3 and 4 lead to the following conclusion.

In Other Words
For $y = f(x - h)$, add *h* to each *x*-coordinate on the graph of $y = f(x)$. For $y = f(x + h)$, subtract *h* from each *x*-coordinate on the graph of $y = f(x)$.

If the argument *x* of a function is replaced by $x - h, h > 0$, the graph of the new function, $y = f(x - h)$, is the graph of *f* **shifted horizontally right** *h* units.

If the argument *x* of a function is replaced by $x + h, h > 0$, the graph of the new function, $y = f(x + h)$, is the graph of *f* **shifted horizontally left** *h* units.

Quick ✓

4. *True or False* To obtain the graph of $f(x) = |x - 3|$ from the graph of $y = |x|$, shift the graph of $y = |x|$ to the right 3 units.

5. Use the graph of $y = x^2$ to graph the function $f(x) = (x + 3)^2$. Show at least three points.

6. Use the graph of $y = \sqrt{x}$ to graph the function $f(x) = \sqrt{x} - 1$. Show at least three points.

EXAMPLE 5 **Combining Vertical and Horizontal Shifts**

Use the graph of $f(x) = \sqrt{x}$ to obtain the graphs of $g(x) = \sqrt{x + 2} - 3$.

Solution

Graph g in stages. First, start with the graph of $f(x)$ as shown in Figure 5(a). Next, to get the graph of $y = \sqrt{x + 2}$, shift the graph of f, 2 units to the left. See Figure 5(b). Finally, to get the graph of $y = \sqrt{x + 2} - 3$, shift the graph of $y = \sqrt{x + 2}$ down 3 units. See Figure 5(c). Note the points plotted on each graph. Using key points can be helpful in keeping track of the transformation that has taken place.

Figure 5

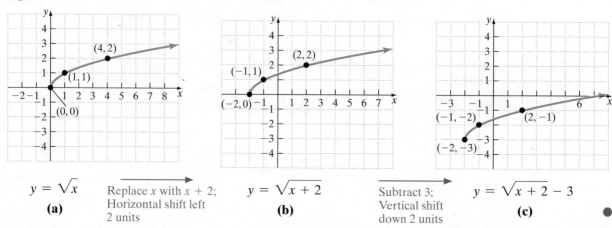

$y = \sqrt{x}$	Replace x with $x + 2$; Horizontal shift left 2 units	$y = \sqrt{x + 2}$	Subtract 3; Vertical shift down 2 units	$y = \sqrt{x + 2} - 3$
(a)		**(b)**		**(c)**

In Example 5, if the vertical shift had been done first, followed by the horizontal shift, the final graph would have been the same. Try it for yourself.

▶ ❷ Graph Functions Using Compressions and Stretches

EXAMPLE 6 **Vertical Stretch**

Use the graph of $f(x) = \sqrt[3]{x}$ to obtain the graph of $g(x) = 3\sqrt[3]{x}$.

Solution

To see the relationship between the graphs of f and g, form Table 5 on the next page, listing points on each graph. For each x-value, the y-coordinate of a point on the graph of g is 3 times as large as the corresponding y-coordinate on the graph of f. To obtain the graph of $g(x) = 3\sqrt[3]{x}$, vertically stretch the graph of $f(x) = \sqrt[3]{x}$ by a factor of 3. For example, $(1, 1)$ is on the graph of f, but $(1, 3)$ is on the graph of g. See Figure 6 on the next page.

Table 5

x	$y = f(x) = \sqrt[3]{x}$	$y = g(x) = 3\sqrt[3]{x}$
−8	$\sqrt[3]{-8} = -2$	$3\sqrt[3]{-8} = 3(-2) = -6$
−1	$\sqrt[3]{-1} = -1$	$3\sqrt[3]{-1} = 3(-1) = -3$
0	0	0
1	1	3
8	2	6

Figure 6

EXAMPLE 7 **Vertical Compression**

Use the graph of $f(x) = x^2$ to obtain the graph of $g(x) = \dfrac{1}{2}x^2$.

Solution

To see the relationship between the graphs of f and g, form Table 6, listing points on each graph. For each x-value, the y-coordinate of a point on the graph of g is $\dfrac{1}{2}$ as large as the corresponding y-coordinate on the graph of f. The graph of $f(x) = x^2$ is vertically compressed by a factor of $\dfrac{1}{2}$ to obtain the graph of $g(x) = \dfrac{1}{2}x^2$. For example, $(2, 4)$ is on the graph of f, but $(2, 2)$ is on the graph of g. See Figure 7.

Table 6

x	$y = f(x) = x^2$	$y = g(x) = \dfrac{1}{2}x^2$
−2	$(-2)^2 = 4$	$\dfrac{1}{2}(-2)^2 = \dfrac{1}{2}(4) = 2$
−1	$(-1)^2 = 1$	$\dfrac{1}{2}(-1)^2 = \dfrac{1}{2}(1) = \dfrac{1}{2}$
0	0	0
1	1	$\dfrac{1}{2}$
2	4	2

Figure 7

Examples 6 and 7 lead to the following conclusion.

In Other Words

For $y = af(x)$, the factor a is "outside" the function, so it affects the y-coordinates. Multiply each y-coordinate on the graph of $y = f(x)$ by a.

When the right side of a function $y = f(x)$ is multiplied by a positive number a, the graph of the new function, $y = af(x)$ is obtained by multiplying each y-coordinate on the graph of $y = f(x)$ by a. The new graph is a **vertically compressed** (if $0 < a < 1$) or a **vertically stretched** (if $a > 0$) version of the graph $y = f(x)$.

Quick ✓

7. To obtain the graph of $y = af(x)$ from the graph of $y = f(x)$, multiply each ___-coordinate on the graph of $y = f(x)$ by ___. If $0 < a < 1$, say that the graph is _____ by a factor of a. If $a > 0$, say that the graph is _____ by a factor of a.

8. Use the graph of $y = |x|$ to graph the function $f(x) = 2|x|$. Show at least three points.

9. Use the graph of $y = \sqrt{x}$ to graph the function $f(x) = \frac{1}{2}\sqrt{x}$. Show at least three points.

What happens if the argument x of a function $y = f(x)$ is multiplied by a positive number a, creating a new function $y = f(ax)$? To find the answer, look at the following exploration.

EXAMPLE 8   **Horizontal Stretch and Compression Technology Exploration**

Use technology to graph the following functions on the same screen:

$$Y_1 = f(x) = x^2 \qquad Y_2 = f(2x) = (2x)^2 \qquad Y_3 = f\left(\frac{1}{2}x\right) = \left(\frac{1}{2}x\right)^2$$

Create a table of values and explore the relation between the x-and y-coordinates of each function.

Solution

You should have obtained the graphs in Figure 8. Look at Table 7. Notice that $(-4, 16), (-2, 4)$ and, $(2, 4)$ are points on the graph of $Y_1 = x^2$ and the points $(-2, 16), (-1, 4)$ and, $(1, 4)$ are points on the graph of $Y_2 = (2x)^2$. For a given y-coordinate, the x-coordinate on the graph of Y_2 is $\frac{1}{2}$ of the x-coordinate on Y_1. It can be concluded that the graph of Y_2 is obtained by multiplying the x-coordinate of each point on the graph of Y_1 by $\frac{1}{2}$. The graph of $Y_2 = (2x)^2$ is the graph of $Y_1 = x^2$ *compressed* horizontally.

Figure 8

$y_2 = (2x)^2$ $y_1 = x^2$

$y_3 = \left(\frac{1}{2}x\right)^2$

Table 7

NORMAL FLOAT AUTO REAL RADIAN MP
PRESS + FOR ΔTbl

X	Y1	Y2	Y3
-5	25	100	6.25
-4	16	64	4
-3	9	36	2.25
-2	4	16	1
-1	1	4	.25
0	0	0	0
1	1	4	.25
2	4	16	1
3	9	36	2.25
4	16	64	4
5	25	100	6.25

X= -5

Now, look at Table 7 again. Notice that $(-1, 1), (1, 1)$, and $(2, 4)$ are points on the graph of $Y_1 = x^2$ and the points $(-2, 1), (2, 1)$, and $(4, 4)$ are points on the graph of $Y_3 = \left(\frac{1}{2}x\right)^2$. For a given y-coordinate, the x-coordinate on the graph of Y_2 is 2 times the x-coordinate on Y_1. It can be concluded that the graph of Y_2 is obtained by multiplying the x-coordinate of each point on the graph of Y_1 by 2. The graph of $Y_3 = \left(\frac{1}{2}x\right)^2$ is the graph of $Y_1 = x^2$ *stretched* horizontally. ●

Based on this exploration, the following results.

> If the argument x of a function $y = f(x)$ is multiplied by a positive number a, the graph of the new function, $y = f(ax)$ is obtained by multiplying each x-coordinate of by $\frac{1}{a}$. A **horizontally compression** results if $a > 0$ and a **horizontal stretch** results if $0 < a < 1$.

Quick ✓

10. To obtain the graph of $y = f(ax)$ from the graph of $y = f(x)$, multiply each ___-coordinate on the graph of $y = f(x)$ by ___. If $0 < a < 1$, say that the graph is _____ _____ by a factor of $\frac{1}{a}$. If $a > 0$, say that the graph is _____ _____ by a factor of $\frac{1}{a}$.

11. Use the graph of $y = x^3$ to graph the function $f(x) = \left(\frac{1}{3}x\right)^3$. Show at least three points.

12. Use the graph of $y = \sqrt[3]{x}$ to graph the function $f(x) = \sqrt[3]{2x}$. Show at least three points.

▶❸ Graph Functions Using Reflections about the x-Axis and the y-Axis

EXAMPLE 9 | **Reflection about the x-Axis**

In the same Cartesian plane, graph $f(x) = |x|$ and $g(x) = -|x|$.

Solution
Use the point-plotting method. Table 8 on the next page lists some points on the graphs of f and g. Notice that when $x = 1$, $f(x) = 1$ and when $x = 1$, $g(x) = -1$. Also, when $x = 2$, $f(x) = 2$, and when $x = 2$, $g(x) = -2$. The graph of g is identical to that of f, except that g is a reflection of f about the x-axis. See Figure 9.

Table 8

x	$f(x) = \lvert x \rvert$	$(x, f(x))$	$g(x) = -\lvert x \rvert$	$(x, g(x))$
-2	2	$(-2, 2)$	-2	$(-2, -2)$
-1	1	$(-1, 1)$	-1	$(-1, -1)$
0	0	$(0, 0)$	0	$(0, 0)$
1	1	$(1, 1)$	-1	$(1, -1)$
2	2	$(2, 2)$	-2	$(2, -2)$

Figure 9

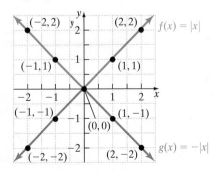

EXAMPLE 10 **Reflection about the y-Axis**

In the same Cartesian plane, graph $f(x) = \sqrt[3]{x}$ and $g(x) = \sqrt[3]{-x}$.

Solution

Again, use the point-plotting method. Table 9 lists some points on the graphs of f and g. Notice when $f(x) = -2, x = -8$ and when $g(x) = -2, x = 8$. Also, when $f(x) = -1, x = -1$, and when $g(x) = -1, x = 1$. The graph of g is identical to that of f, except that g is a reflection of f about the y-axis. See Figure 10.

Table 9

x	$f(x) = \sqrt[3]{x}$	$(x, f(x))$	$g(x) = -|x|$	$(x, g(x))$
-8	$\sqrt[3]{-8} = -2$	$(-8, -2)$	$\sqrt[3]{-(-8)} = \sqrt[3]{8} = 2$	$(-8, 2)$
-1	$\sqrt[3]{-1} = -1$	$(-1, -1)$	$\sqrt[3]{-(-1)} = \sqrt[3]{1} = 1$	$(-1, 1)$
0	0	$(0, 0)$	0	$(0, 0)$
1	1	$(1, 1)$	-1	$(1, -1)$
8	2	$(8, 2)$	-2	$(8, -2)$

Figure 10

When the right side of the function $y = f(x)$ is multiplied by -1, the graph of the new function $y = -f(x)$ is the **reflection about the x-axis** of the graph of the function $y = f(x)$.

When the graph of the function $y = f(x)$ is known, the graph of the new function $y = f(-x)$ is the **reflection about the y-axis** of the graph of the function $y = f(x)$.

Quick ✓

13. The graph of $y = -f(x)$ is a reflection of the graph of $y = f(x)$ about the ___-axis. The graph of $y = f(-x)$ is a reflection of the graph of $y = f(x)$ about the ___-axis.

14. Use the graph of $y = \sqrt{x}$ to graph the function $f(x) = -\sqrt{x}$. Show at least three points.

15. Use the graph of $y = x^3$ to graph the function $f(x) = (-x)^3$. Show at least three points.

EXAMPLE 11 **Combining Graphing Procedures**

Graph the function $f(x) = 2\sqrt{x + 1} - 3$. Find the domain and the range of f.

Solution

Use the following steps to obtain the graph of f:

Step 1: $y = \sqrt{x}$ Square root function

Step 2: $y = 2\sqrt{x}$ Multiply by 2; vertical stretch of the graph of $y = \sqrt{x}$ by a factor of 2.

Step 3: $y = 2\sqrt{x + 1}$ Replace x by $x + 1$; horizontal shift to the left 1 unit.

Step 4: $y = 2\sqrt{x + 1} - 3$ Subtract 3; vertical shift down 3 units.

See Figure 11.

Figure 11

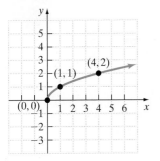

$y = \sqrt{x}$

(a)

Multiply by 2; Vertical stretch of the graph by a factor of 2

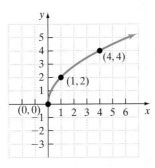

$y = 2\sqrt{x}$

(b)

Replace x by $x + 1$; Horizontal shift left 1 unit

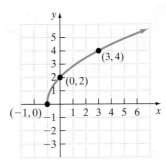

$y = 2\sqrt{x + 1}$

(c)

Subtract 3 Vertical shift down 3 units

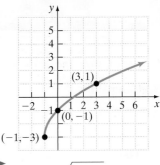

$y = 2\sqrt{x + 1} - 3$

(d)

The domain of f is $[-1, \infty)$ and the range is $[-3, \infty)$

Work Smart

Although the order in which transformations are performed can be altered, you may consider using the following order for consistency:

1. Reflections
2. Compressions and stretches
3. Shifts

Other orderings of the steps shown in Example 11 would also result in the graph of f. For example, try this one:

Step 1: $y = \sqrt{x}$ — Square root function

Step 2: $y = \sqrt{x + 1}$ — Replace x by $x + 1$; horizontal shift to the left 1 unit.

Step 3: $y = 2\sqrt{x + 1}$ — Multiply by 2; vertical stretch of the graph of $y = \sqrt{x}$ by a factor of 2.

Step 4: $y = 2\sqrt{x + 1} - 3$ — Subtract 3; vertical shift down 3 units. ●

Quick ✓

16. Graph $f(x) = -2(x - 1)^2$ using the techniques of shifting, compressing, stretching, and/or reflecting. Start with the graph of the basic function and show at least three points.

Summary of Graphing Techniques

To Graph:	Draw the Graph of f and :	Functional Change to $f(x)$:
Vertical Shifts		
$y = f(x) + k, k > 0$	Raise the graph of f by k units.	Add k to $f(x)$.
$y = f(x) - k, k > 0$	Lower the graph of f by k units.	Subtract k from $f(x)$.
Horizontal Shifts		
$y = f(x + h), h > 0$	Shift the graph of f to the left h units.	Replace x by $x + h$.
$y = f(x - h), h > 0$	Shift the graph of f to the right h units.	Replace x with $x - h$.
Compressing or Stretching		
$y = af(x), a > 0$	Multiply each y-coordinate by a. Stretch the graph of f vertically if $a > 1$. Compress the graph of f vertically if $0 < a < 1$.	Multiply $f(x)$ by a.
$y = f(ax), a > 0$	Multiply each x-coordinate by $\frac{1}{a}$. Stretch the graph of f horizontally if $0 < a < 1$. Compress the graph of f horizontally if $a > 1$.	Replace x by ax.
Reflection about the x-Axis		
$y = -f(x)$	Reflect the graph of f about the x-axis.	Multiply $f(x)$ by -1.
Reflection about the y-Axis		
$y = f(-x)$	Reflect the graph of f about the y-axis.	Replace x by $-x$.

Appendix A: Exercises MyMathLab® Exercise numbers in green have complete video solutions in MyMathLab or may be accessed using the QR code to the right.

*Problems **1–16** are the Quick* ✓ s *that follow the* **EXAMPLES.**

Building Skills

17. Match each function to its graph:

    (I) $f(x) = (x - 1)^3$ **(II)** $f(x) = x^3 + 1$

    (III) $f(x) = -x^3$ **(IV)** $f(x) = (x + 1)^3$

(A)

(B)

(C)

(D)

18. Match each function to its graph:

    (I) $f(x) = -|2x|$ **(II)** $f(x) = |x + 1| - 2$

    (III) $f(x) = |x + 1| + 2$ **(IV)** $f(x) = -|x + 1| + 2$

(A)

(B)

(C)

(D)

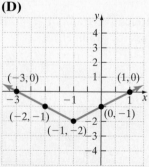

In Problems 19–26, write the function whose graph is the graph of $y = \sqrt{x}$ *but is:*

19. Shifted to the left 3 units

20. Shifted to the right 3 units

21. Shifted down 2 units

22. Shifted up 2 units

23. Reflected about the x-axis

24. Reflected about the y-axis

25. Horizontally stretched by a factor of 2

26. Vertically stretched by a factor of 2

In Problems 27–42, graph each function using the techniques of shifting, compressing, stretching, and/or reflecting. Start with the graph of the basic function (for example, $y = x^2$) and show all stages. Be sure to show at least three key points. Find the domain and the range of each function.

27. $f(x) = x^3 + 1$ **28.** $f(x) = x^3 - 1$

29. $g(x) = (x - 1)^2$ **30.** $g(x) = (x + 1)^2$

31. $h(x) = \sqrt{2x}$ **32.** $h(x) = \sqrt{\dfrac{1}{2}x}$

33. $f(x) = \dfrac{1}{2}|x|$ **34.** $f(x) = 3|x|$

35. $g(x) = \sqrt[3]{x - 1} + 2$ **36.** $g(x) = \sqrt[3]{x + 1} - 2$

37. $h(x) = \sqrt{-x}$ **38.** $h(x) = -\sqrt{x}$

39. $f(x) = 2|x - 1| + 2$ **40.** $f(x) = \dfrac{1}{2}|x - 1| + 3$

41. $f(x) = -\sqrt{x + 2} + 3$ **42.** $f(x) = 2\sqrt{-x} + 1$

43. Use the graph of *f* illustrated below as the first step toward graphing each of the following functions.

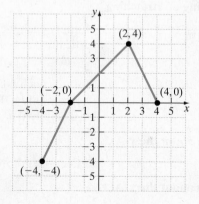

(a) $h(x) = f(x) + 2$

(b) $g(x) = f(x - 1)$

(c) $F(x) = f(2x)$

(d) $k(x) = f\left(\dfrac{1}{2}x\right)$

(e) $G(x) = f(-x) - 2$

(f) $p(x) = f\left(\dfrac{1}{2}x\right) - 1$

(g) $r(x) = 2f(x - 1) + 2$

(h) $j(x) = -f(2x)$

44. Use the graph of f illustrated below as the first step toward graphing each of the following functions.

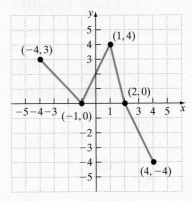

(a) $h(x) = f(x) - 2$

(b) $g(x) = f(x + 1)$

(c) $F(x) = 2f(x)$

(d) $k(x) = \dfrac{1}{2}f(x)$

(e) $G(x) = -f(x) + 2$

(f) $p(x) = \dfrac{1}{2}f(x) + 1$

(g) $r(x) = 2f(x + 1) - 2$

(h) $j(x) = 2f(-x)$

Credits

Photo Credits

Answers to Selected Exercises

Chapter R Review

Section R.1 Success in Mathematics Answers will vary.

Section R.2 Sets and Classification of Numbers **1.** set **2.** elements **3.** set-builder: $\{x \mid x \text{ is a digit less than } 5\}$; roster: $\{0, 1, 2, 3, 4\}$

4. set-builder: $\{x \mid x \text{ is a digit greater than or equal to } 6\}$; roster: $\{6, 7, 8, 9\}$ **5.** True **6.** subset **7.** False **8.** True **9.** False **10.** False

11. True **12.** True **13.** False **14.** True **15.** whole **16.** rational **17.** irrational **18.** False **19.** False **20.** True **21.** True

22. $10, \dfrac{12}{4}$ **23.** $10, \dfrac{0}{3}, \dfrac{12}{4}$ **24.** $-9, 10, \dfrac{0}{3}, \dfrac{12}{4}$ **25.** $\dfrac{7}{3}, -9, 10, 4.\overline{56}, \dfrac{0}{3}, -\dfrac{4}{7}, \dfrac{12}{4}$ **26.** $\pi, 5.7377377737777\ldots, \sqrt{11}$ **27.** All **28. (a)** 5.694

(b) 5.694 **29. (a)** -4.93 **(b)** -4.94 **30.** **31.** $<$ **32.** $<$ **33.** $>$ **34.** $>$ **35.** $=$ **36.** $>$ **37.** $\{0, 1, 2, 3, 4, 5\}$

39. $\{-2, -1, 0, 1, 2, 3, 4\}$ **41.** $\varnothing$ or $\{\ \}$ **43.** True **45.** False **47.** True **49.** True **51.** $\notin$ **53.** $\in$ **55. (a)** 4 **(b)** $-5, 4$ **(c)** $-5, 4, \dfrac{4}{3}, -\dfrac{7}{5}, 5.\overline{1}$

(d) π **(e)** $-5, 4, \dfrac{4}{3}, -\dfrac{7}{5}, 5.\overline{1}, \pi$ **57. (a)** 100 **(b)** $100, -64$ **(c)** $100, -5.423, \dfrac{8}{7}, -64$ **(d)** $\sqrt{2} + 4$ **(e)** $100, -5.423, \dfrac{8}{7}, \sqrt{2} + 4, -64$ **59. (a)** 19.9348

(b) 19.9348 **61. (a)** 0.0 **(b)** 0.1 **63.** **65.** $<$ **67.** $=$ **69.** $>$ **71.** -282 feet **73.** $-\$9.87$ **75.** -6

77. Answers will vary. **79.** Answers will vary. **81.** No, if a number is rational, it cannot be irrational, and vice versa, because a number cannot have a terminating or nonterminating/repeating decimal while simultaneously having a nonterminating/nonrepeating decimal. No; every real number is either a rational or an irrational number. **83.** A set is a well-defined collection of objects. The collection must be well defined so that it is easy to determine whether an object is in the set. **85.** If $A \subseteq B$, then all elements in A are also in B. A could also equal B. If $A \subset B$, then all elements in A are also in B, but $A \ne B$. **87.** If the digit *after* the specified final digit is 4 or less, then truncating and rounding will yield the same decimal approximation.

89. 1.143; 1.142

Section R.3 Operations on Signed Numbers; Properties of Real Numbers **1.** factors **2.** False **3.** 6 **4.** 10 **5.** 12 **6.** -11 **7.** -6.6 **8.** -2.2

9. 0 **10.** additive inverse; opposite; 0 **11.** a **12.** -5 **13.** $-\dfrac{4}{5}$ **14.** 12 **15.** $\dfrac{5}{3}$ **16.** 0 **17.** 4 **18.** -9 **19.** -11 **20.** 9.1 **21.** -5.1 **22.** -16.1

23. True **24.** True **25.** -48 **26.** -60 **27.** 56 **28.** 105 **29.** 5.13 **30.** opposite; reciprocal **31.** $\dfrac{1}{10}$ **32.** $-\dfrac{1}{8}$ **33.** $\dfrac{5}{2}$ **34.** -5 **35.** lowest terms

36. $\dfrac{6}{5}$ **37.** $\dfrac{11}{8}$ **38.** $-\dfrac{6}{5}$ **39.** in lowest terms **40.** $\dfrac{4}{5}$ **41.** $-\dfrac{5}{6}$ **42.** $\dfrac{1}{2}$ **43.** 8 **44.** $\dfrac{5}{11}$ **45.** $-\dfrac{1}{3}$ **46.** -2 **47.** 1 **48.** least common denominator

49. LCD $= 60$; $\dfrac{3}{20} = \dfrac{9}{60}$; $\dfrac{2}{15} = \dfrac{8}{60}$ **50.** LCD $= 90$; $\dfrac{5}{18} = \dfrac{25}{90}$; $-\dfrac{1}{45} = -\dfrac{2}{90}$ **51.** $\dfrac{17}{60}$ **52.** $-\dfrac{1}{6}$ **53.** $-\dfrac{59}{150}$ **54.** $-\dfrac{3}{10}$ **55.** $ab + ac$ **56.** $-a - b$

57. $5x + 15$ **58.** $-6x - 6$ **59.** $-4z + 32$ **60.** $2x + 3$ **61.** $-11p - 8$ **62.** $9 + 4t$ **63.** $\dfrac{2}{3}$ **65.** $\dfrac{8}{3}$ **67.** -9 **69.** 7 **71.** -2.5 **73.** -32 **75.** 84

77. -36.55 **79.** $\dfrac{7}{3}$ **81.** $\dfrac{6}{5}$ **83.** $\dfrac{5}{3}$ **85.** $-\dfrac{3}{7}$ **87.** $\dfrac{2}{3}$ **89.** $\dfrac{5}{4}$ **91.** $\dfrac{1}{6}$ **93.** $\dfrac{11}{12}$ **95.** $2x + 8$ **97.** $-3z + 6$ **99.** $3x - 30$ **101.** $6x - 9$ **103.** $-5z - 17$

105. 3 **107.** $\dfrac{101}{70}$ **109.** $\dfrac{4}{3}$ **111.** 3.3 **113.** -32 **115.** $\dfrac{13}{10}$ **117.** $\dfrac{17}{15}$ **119.** $-\dfrac{139}{32}$ **121.** $\dfrac{1}{9}$ **123.** undefined **125.** 0 **127.** Commutative Property of Multiplication **129.** Multiplicative Inverse Property **131.** Reduction Property **133.** Associative Property of Addition **135.** The difference in age of the oldest and youngest presidents at the time of inauguration is 27 years. **137.** The Bears gained a total of 9 yards in the first 3 plays. Since 9 is less than 10, they did not obtain a first down. **139.** The difference between the highest and lowest elevations is 20,602 feet. **141.** Consider $3 \cdot 6$. This means to add 6 to itself three times. Since the sum of three positive numbers is positive, the product will be positive. Consider $3 \cdot (-6)$. This means to add -6 to itself three times. Since the sum of three negative numbers is negative, the product will be negative.

143. $d(P, Q) = 14$ **145.** $d(P, Q) = 10.4$

147. $d(P, Q) = \dfrac{68}{15}$ **149. (a)** $a \cdot 0 = 0$ **(b)** $a \cdot (b + (-b)) = 0$ **(c)** $ab + a(-b) = 0$ **(d)** $ab < 0$ since the

product of a negative and a positive is negative; $a(-b)$ must be positive so that $ab + a(-b) = 0$. **151.** Zero does not have a multiplicative inverse because division by zero is not defined. **153.** The Reduction Property only applies to dividing out factors and does not apply to sums. **155.** No. For example, $4 - (5 - 3) \ne (4 - 5) - 3$ **157.** No. For example, $16 \div (8 \div 2) \ne (16 \div 8) \div 2$ **159.** -3.8 **161.** -19.2 **163.** $\dfrac{1}{30}$ **165.** 14.4 **167.** $\dfrac{9}{2}$

Section R.4 Order of Operations **1.** base; exponent; power **2.** False **3.** 64 **4.** 49 **5.** -1000 **6.** $\dfrac{8}{27}$ **7.** -64 **8.** 125 **9.** 16 **10.** 36 **11.** 32

12. 48 **13.** 40 **14.** 26 **15.** $\dfrac{10}{13}$ **16.** 18 **17.** 50 **18.** $\dfrac{4}{5}$ **19.** parentheses; exponents; multiplication; division; addition; subtraction **20.** 16 **21.** 48

22. $\frac{6}{7}$ **23.** *14* **24.** *7* **25.** *42* **26.** *−9* **27.** *20* **28.** *−18* **29.** *5* **30.** *−15* **31.** *81* **33.** *−625* **35.** *8* **37.** $-\frac{4}{9}$ **39.** *7* **41.** *15* **43.** *12* **45.** *−28* **47.** *3*

49. *14* **51.** *−2* **53.** *60* **55.** *180* **57.** *9* **59.** *8* **61.** $\frac{5}{2}$ **63.** *2* **65.** $\frac{6}{5}$ **67.** $\frac{2}{3}$ **69.** $-\frac{7}{4}$ **71.** $\frac{3}{2}$ **73.** $\frac{53}{14}$ **75.** $3 \cdot (7 - 2) = 15$ **77.** $3 + 5 \cdot (6 - 3) = 18$

79. *The surface area of the cylinder is about 534.07 square inches.* **81.** *After 3 seconds, the height of the ball is 6 feet.* **83.** *100* **85.** *We cannot use the Reduction Property across addition.* **87.** *Answers will vary. One suggestion:* $5 \cdot 2^3$ *is equivalent to* $5 \cdot 2 \cdot 2 \cdot 2$, *which equals 40. Since* $5 \cdot 2^3 = 5 \cdot 8 = 40$, *we can see that we evaluate exponents before multiplication.* **89.** $\frac{16}{45}$ **91.** $\frac{17}{13}$ **93.** *−212.96* **95.** *−534.53*

Section R.5 Algebraic Expressions
1. *variable* **2.** *constant* **3.** $3 + 11$ **4.** $6 \cdot 7$ **5.** $\frac{y}{4}$ **6.** $3 - z$ **7.** $2(x - 3)$ **8.** $2x - 3$ **9.** *evaluate* **10.** *−7*

11. *41* **12.** *1* **13.** *3* **14.** *11,700 yen; 117,000 yen; 1,170,000 yen* **15.** *0°C; 30°C; 100°C* **16.** *like terms* **17.** *−1* **18.** $-5x$ **19.** $11x^2$ **20.** $-8x + 3$

21. $-4x + 8y$ **22.** $15y - 1$ **23.** $2.3x^2 + 0.9$ **24.** $-6z + 3$ **25.** $4x - 6$ **26.** $-5y + 11$ **27.** $-3z + 10$ **28.** $-6x + 4$ **29.** *1*

30. $\frac{25x + 25}{6} = \frac{25(x + 1)}{6}$ **31.** *domain* **32. (a)** Yes **(b)** Yes **(c)** No **(d)** Yes **33. (a)** Yes **(b)** Yes **(c)** Yes **(d)** No

34. (a) No **(b)** Yes **(c)** Yes **(d)** No **35.** $5 + x$ **37.** $4z$ **39.** $y - 7$ **41.** $2(t + 4)$ **43.** $5x - 3$ **45.** $\frac{y}{3} + 6x$ **47.** *11* **49.** *−9* **51.** *−21* **53.** $\frac{3}{8}$

55. $\frac{6}{7}$ **57.** *29* **59.** *25* **61.** $\frac{3}{2}$ **63.** x **65.** $-6z + 3$ **67.** $-z - 5$ **69.** $\frac{11}{12}x$ **71.** $4x^2 - 3x$ **73.** $1.6x$ **75.** $-3x + 1$ **77.** $-14x + 7$ **79.** $-z + 10$

81. $4x - 3$ **83.** $12v - 11$ **85.** $2x + \frac{9}{10}$ **87.** $\frac{13}{36}x$ **89.** $14.46x - 15.49$ **91.** $-11.08x - 5.44$ **93. (a)** No **(b)** Yes **(c)** Yes **(d)** Yes **95. (a)** Yes

(b) Yes **(c)** Yes **(d)** Yes **97. (a)** No **(b)** Yes **(c)** Yes **(d)** No **99.** 1 in.^{3} ; 8 in.^{3} ; 27 in.^{3} ; 64 in.^{3} **101.(a)** 0 ft; 59 ft; 86 ft; 81 ft; 44 ft **(b)** The ball begins on the ground. When it is hit, it rises in the air (for somewhere around 2 seconds) and then begins to fall back toward the ground.

103. *Let* $x =$ *Bob's age in years; Tony's age* $= x + 5$; *when Bob is 13 years old, Tony is 18 years old.* **105.** *Let* $p =$ *the original price in dollars;* *discounted price* $= \frac{1}{2}p$; *when the original price is $900, the discount price is $450.* **107.** $\frac{4}{3}$ **109.** *Answers will vary. One possible answer is "Twice a number z decreased by 5."* **111.** *Answers will vary. One possible answer is "Twice the difference of a number z and 5."* **113.** *Answers will vary. One possible answer is "One-half the sum of a number z and 3."* **115.** *A variable is a letter used to represent any number from a set of numbers; a constant is either a fixed number or a letter that represents a fixed number.* **117.** *Like terms are terms that have the same variable or variables, along with the same exponents on the variables. We use the Distributive Property "in reverse" to combine like terms, as in* $3x + 4x = (3 + 4)x = 7x$.

119. (a) 3 **(b)** 15 **121. (a)** 63 **(b)** 35 **123. (a)** $-\frac{7}{5}$ **(b)** $\frac{23}{65}$ **125. (a)** 67 **(b)** 32

127. *The calculator displays an error message because* $x = 5$ *makes the denominator equal to 0.*

Chapter 1 Linear Equations and Inequalities

Section 1.1 Linear Equations in One Variable
1. linear; sides **2.** solution **3.** $x = 1$ **4.** $x = -7$ **5.** $z = -1$ **6.** True **7.** isolate **8.** $\{3\}$ **9.** $\{-2\}$

10. $\left\{\frac{1}{5}\right\}$ **11.** $\{2\}$ **12.** $\{-1\}$ **13.** $\left\{\frac{3}{2}\right\}$ **14.** $\{4\}$ **15.** $\{-4\}$ **16.** $\left\{\frac{3}{4}\right\}$ **17.** $\left\{\frac{1}{2}\right\}$ **18.** least common denominator **19.** $\{2\}$ **20.** $\{-5\}$ **21.** $\{-4\}$

22. $\left\{-\frac{3}{5}\right\}$ **23.** $\{10\}$ **24.** $\{32\}$ **25.** conditional equation **26.** contradiction; identity **27.** $\varnothing$ or $\{\,\}$; contradiction **28.** $\{x \mid x$ is any real number$\}$; identity

29. $\{0\}$; conditional **30.** $\{z \mid z$ is any real number$\}$; identity **31.** $x = 2$ **33.** $m = 1$ **35.** $x = 5$ **37.** $\{2\}$ **39.** $\left\{-\frac{1}{4}\right\}$ **41.** $\{9\}$ **43.** $\left\{-\frac{2}{3}\right\}$ **45.** $\{-4\}$

47. $\{2\}$ **49.** $\left\{-\frac{3}{5}\right\}$ **51.** $\{-40\}$ **53.** $\{3\}$ **55.** $\{-5\}$ **57.** $\{\,\}$ or $\varnothing$; contradiction **59.** $\{\,\}$ or $\varnothing$; contradiction **61.** $\{y \mid y$ is any real number$\}$ or $\mathbb{R}$;

identity **63.** $\{-3\}$; conditional **65.** $\{\,\}$ or $\varnothing$; contradiction **67.** $\left\{\frac{5}{2}\right\}$; conditional **69.** $\{z \mid z$ is any real number$\}$ or $\mathbb{R}$; identity **71.** $\left\{-\frac{7}{2}\right\}$; conditional

73. $\left\{\frac{1}{7}\right\}$; conditional **75.** $\left\{\frac{7}{2}\right\}$; conditional **77.** $\{p \mid p$ is any real number$\}$ or $\mathbb{R}$; identity **79.** $\{\,\}$ or $\varnothing$; contradiction **81.** $\left\{-\frac{4}{3}\right\}$; conditional

83. $\{-14\}$; conditional **85.** $\{-3\}$; conditional **87.** $\{-1.6\}$; conditional **89.** $\{2\}$; conditional **91.** $a = -4$ **93.** $a = 3$ **95.** $x = -\frac{1}{2}$ **97.** $x = \frac{3}{4}$ **99.** $x = 1$

101. The card's annual interest rate is 0.15 or 15% **103.** Your adjusted income for 2015 was $32,492. **105.** $4(x + 1) - 2$ is an algebraic expression and $4(x + 1) = 2$ is an equation. An algebraic expression is any combination of variables, grouping symbols, and mathematical operations but does not contain an equal sign. An equation is a statement made up of two algebraic expressions that are equal. **107.** Answers will vary.

Section 1.2 An Introduction to Problem Solving
1. equations **2.** $x + 7 = 12$ **3.** $3y = 21$ **4.** $2(n + 3) = 5$ **5.** $x - 10 = \frac{x}{2}$ **6.** $2n + 3 = 5$
7. False **8.** 18, 20, 22 **9.** 25, 26, 27 **10.** $15 per hour **11.** Connor: $400,000; Olivia: $300,000; Avery: $100,000. **12.** 150 miles **13.** 925 minutes **14.** 100 **15.** 40 **16.** 160 **17.** 75% **18.** $30 **19.** $1.20 **20.** Interest; principal **21.** $32.50 **22.** $10.50; $1410.50 **23.** $67,500 in Aaa-rated bonds; $22,500 in B-rated

bonds **24.** $6250 in CD; $18,750 in corporate bond **25.** 6 pounds of Tea A and 4 pounds of Tea B **26.** 10 pounds of cashews; 20 pounds of peanuts **27.** uniform motion **28.** rate **29.** After 4 hours; 240 miles **30.** After 5 hours; 450 miles **31.** 10 **33.** 40 **35.** 37.5% **37.** $x + 12 = 20$; 8 **39.** $2(y + 3) = 16$; 5 **41.** $w - 22 = 3w$; -11 **43.** $4x = 2x + 14$; 7 **45.** $0.8x = x + 5$; -25 **47.** 13 and 26 **49.** 24, 25, and 26 **51.** Kendra needs an 83 on her final exam to have an average of 80. **53.** Jacob would need to print 2500 pages for the cost to be the same for the two printers. **55.** $12 per hour **57.** The final bill will be $637.04. **59.** The dealer's cost is about $23,755.72. **61.** The flash drives originally cost $23.79. **63.** The Nissan Altima weighs 3193 pounds, the Mazda 6s weighs 3329 pounds, and the Honda Accord EX weighs 3312 pounds. **65.** Adam will get $8500 and Krissy will get $11,500. **67.** You should invest $15,000 in stocks and $9000 in bonds. **69.** The interest charge after one month will be $29.17. **71.** The bank loaned $225,000 at 6% interest. **73.** Pedro should invest $9375 in the 5% bond fund and $15,625 in the 9% stock fund. **75.** 20 pounds of almonds and 30 pounds of peanuts **77.** Bobby has 15 dimes and 32 quarters saved. **79.** Nancy should mix 3/4 cup of cream with 7 1/4 cup of low fat milk. **81.** The race consists of running for 12 miles and biking for 50 miles. **83.** The slow car is traveling at 60 mph, while the faster car travels at 70 mph. **85.** The boats will be 155 miles apart after 2.5 hours. **87.** One person is walking at a rate of 4 miles per hour, and the other is walking at a rate of 6 miles per hour. **89.** Written answers will vary. The average speed of the trip to Florida and back is approximately 54.55 miles per hour. **91.** Answers will vary. **93.** The train is 0.3 mile or 1584 feet long. **95.** Mathematical modeling is the process of developing an equation or inequality to find a solution to a problem. Just as there is more than one way to solve a problem, there is typically more than one way to develop a mathematical model. **97.** Direct translation, mixture, and uniform motion problems. Two types of mixture problems are interest and new blend problems.

Section 1.3 Using Formulas to Solve Problems **1.** formula **2.** $A = \pi r^2$ **3.** $V = \pi r^2 h$ **4.** $C = 175x + 7000$ **5.** $s = \frac{1}{2} g t^2$ **6.** area; volume **7.** False

8. (a) $h = \dfrac{2A}{b}$ **(b)** 5 inches **9. (a)** $b = \dfrac{P - 2a}{2}$ **(b)** 10 cm **10.** $P = \dfrac{I}{rt}$ **11.** $y = \dfrac{C - Ax}{B}$ **12.** $h = \dfrac{4x - 3}{2x - 3}$ **13.** $n = \dfrac{S + d}{a + d}$ **14.** Width: 40 feet; Length:

50 feet **15.** Height: 72 inches; Width: 40 inches **16.** 4.00 inches **17.** $F = ma$ **19.** $V = \dfrac{4}{3}\pi r^3$ **21.** $r = \dfrac{d}{t}$ **23.** $m = \dfrac{y - y_1}{x - x_1}$ **25.** $x = \mu + \sigma Z$ **27.** $m_1 = \dfrac{r^2 F}{Gm_2}$

29. $P = \dfrac{A}{1 + rt}$ **31.** $F = \dfrac{9}{5} C + 32$ **33.** $y = -2x + 13$ **35.** $y = 3x - 5$ **37.** $y = -\dfrac{4}{3}x + \dfrac{13}{3}$ **39.** $y = -3x + 12$ **41. (a)** $h = \dfrac{V}{\pi r^2}$ **(b)** The height of the

cylinder is 8 inches. **43. (a)** $A = \dfrac{206.3 - M}{0.711}$ **(b)** An individual whose maximum heart rate is 160 should be about 65 years old. **45. (a)** $P = \dfrac{A}{(1 + r)^t}$

(b) Approximately $4109.64 should be deposited today to have $5000 in 5 years in an account that pays 4% annual interest. **47.** The smaller angle measures 75° and its supplement measures 105°. **49.** The smaller angle measures 20° and its complement measures 70°. **51.** The window is 5 feet long and 8 feet wide. **53.** The area of the circle is 25π square inches (approximately 78.54 square inches). **55.** The first angle measures 40°, the second measures 55°, and the third measures 85°. **57. (a)** The patio is 17.5 ft wide and 22.5 ft long. **(b)** You would need to purchase 131.25 cubic feet of cement. **59. (a)** The deck has an area of 84π square feet (approximately 264 square feet). **(b)** It would require approximately 97.39 feet of fencing to encircle the pool and deck. **(c)** The fence would cost about $2434.73. **61.** No, the area would increase by a factor of 4. If the length of a side of a cube is doubled, the volume increases by a factor of 8.

Section 1.4 Linear Inequalities in One Variable **1.** closed interval **2.** left endpoint; right endpoint **3.** False

4. $[-3, 2]$ **5.** $[3, 6)$ **6.** $(-\infty, 3]$

7. $\left(\dfrac{1}{2}, \dfrac{7}{2}\right)$ **8.** $0 < x \le 5$

9. $-6 < x < 0$ **10.** $x > 5$ **11.** $x \le \dfrac{8}{3}$

12. $9 < 12$; Addition Property of Inequalities **13.** $x > -9$; Addition Property of Inequalities **14.** $1 < 4$; Multiplication Property of Inequalities **15.** $2 > -3$; Multiplication Property of Inequalities **16.** $x < 6$; Multiplication Property of Inequalities

17. $\{x \mid x > 2\}$; $(2, \infty)$ **18.** $\{x \mid x \le 6\}$; $(-\infty, 6]$

19. $\{x \mid x \ge -6\}$; $[-6, \infty)$ **20.** $\{x \mid x > -3\}$; $(-3, \infty)$

21. $\{x \mid x \ge -2\}$; $[-2, \infty)$ **22.** $\left\{x \mid x > \dfrac{5}{2}\right\}$; $\left(\dfrac{5}{2}, \infty\right)$

23. $\{x \mid x < 4\}$; $(-\infty, 4)$ **24.** $\left\{x \mid x \le -\dfrac{7}{3}\right\}$; $\left(-\infty, -\dfrac{7}{3}\right]$

25. $\{x \mid x \ge 4\}$; $[4, \infty)$ **26.** $\{x \mid x \ge 3\}$; $[3, \infty)$

27. $\left\{x \mid x < \dfrac{1}{2}\right\}$; $\left(-\infty, \dfrac{1}{2}\right)$ **28.** $\{x \mid x > -17\}$; $(-17, \infty)$

29. Any balance over $480 **30.** For any more than 24 boxes, revenue exceeds cost.

31. $[2, 10]$;   **33.** $[-4, 0)$;   **35.** $[6, \infty)$;

37. $\left(-\infty, \dfrac{3}{2}\right)$;   **39.** $1 < x < 8$;

41. $-5 < x \le 1$; **43.** $x < 5$; **45.** $x \ge 3$;

47. $<$; Addition Property of Inequalities **49.** $>$; Multiplication Property of Inequalities **51.** $\le$; Addition Property of Inequalities

53. $\le$; Multiplication Property of Inequalities **55.** $\{x \mid x \le 6\}$; $(-\infty, 6]$

57. $\{x \mid x < 4\}$; $(-\infty, 4)$ **59.** $\{x \mid x > -3\}$; $(-3, \infty)$

61. $\{x \mid x > 6\}$; $(6, \infty)$ **63.** $\{x \mid x > 3\}$; $(3, \infty)$

65. $\{x \mid x < -4\}$; $(-\infty, -4)$ **67.** $\{x \mid x \le -1\}$; $(-\infty, -1]$

69. $\{x \mid x > -2\}$; $(-2, \infty)$ **71.** $\{x \mid x < 17\}$; $(-\infty, 17)$

73. $\{x \mid x \le 4\}$; $(-\infty, 4]$ **75.** $\{x \mid x \le -30\}$; $(-\infty, -30]$

77. $\left\{x \mid x < \dfrac{1}{3}\right\}$; $\left(-\infty, \dfrac{1}{3}\right)$ **79.** $\left\{x \mid x < -\dfrac{11}{4}\right\}$; $\left(-\infty, -\dfrac{11}{4}\right)$

81. $\left\{x \mid x < -\dfrac{16}{15}\right\}$; $\left(-\infty, -\dfrac{16}{15}\right)$ **83.** $\left\{x \mid x \le \dfrac{15}{2}\right\}$; $\left(-\infty, \dfrac{15}{2}\right]$

85. $\left\{x \mid x \ge \dfrac{21}{16}\right\}$; $\left[\dfrac{21}{16}, \infty\right)$ **87.** $\left\{x \mid x > \dfrac{1}{10}\right\}$; $\left(\dfrac{1}{10}, \infty\right)$

89. $\{x \mid x < 3\}$; $(-\infty, 3)$ **91.** $\left\{x \mid x < \dfrac{9}{2}\right\}$; $\left(-\infty, \dfrac{9}{2}\right)$

93. $\{y \mid y > -15\}$; $(-15, \infty)$ **95.** $\{a \mid a \ge 7\}$; $[7, \infty)$

97. $\left\{x \mid x > -\dfrac{1}{3}\right\}$; $\left(-\dfrac{1}{3}, \infty\right)$ **99.** $\{x \mid x \le -14\}$; $(-\infty, -14]$

101. $\left\{x \mid x < \dfrac{3}{2}\right\}$; $\left(-\infty, \dfrac{3}{2}\right)$ **103.** $\{x \mid x \ge -3\}$; $[-3, \infty)$

105. $\left\{x \mid x \ge -\dfrac{3}{4}\right\}$; $\left[-\dfrac{3}{4}, \infty\right)$ **107.** $\{x \mid x \ge 4\}$; $[4, \infty)$ **109.** $\{z \mid z \le 3\}$; $(-\infty, 3]$

111. Jackie must earn at least 182 points on the final exam to earn an A in Mr. Ruffatto's class. **113.** You can order no more than 3 hamburgers to keep the fat content to no more than 69 grams. **115.** The plane can carry up to 18,836 pounds of luggage and cargo. **117.** The monthly benefit did exceed $1000 after the year 2011. **119.** Susan will need to sell at least $5,500,000 in computer systems to earn $100,000. **121.** Supply will exceed demand when the price is greater than $40. **123.** All real numbers are solutions; $\mathbb{R}$ **125.** When we multiply both sides of an inequality by a negative number. Or, if the sides of the inequality are interchanged. **127.** The inequality $4 < x > 7$ means $x > 4$ and $x > 7$, which is equivalent to $x > 7$.

Putting the Concepts Together (Sections 1.1–1.4) 1. (a) $x = -3$ is not a solution. **(b)** $x = 1$ is a solution. **2.** $\{-5\}$ **3.** $\{0\}$ **4.** identity

5. $x - 3 = \dfrac{1}{2}x + 2$ **6.** $\dfrac{x}{2} < x + 5$ **7.** The chemist needs to mix 4 liters of the 20% solution with 12 liters of the 40% solution.

8. After 3.4 hours, the two cars will be 255 miles apart. **9.** $y = \dfrac{3}{2}x - 2$ **10.** $r = \dfrac{A - P}{Pt}$ **11. (a)** $h = \dfrac{V}{\pi r^2}$ **(b)** 6 in.

12. (a) Interval: $(-3, \infty)$ Graph: **(b)** Interval: $(2, 5]$ Graph:

13. (a) Inequality: $x \le -1.5$ Graph: **(b)** Inequality: $-3 < x \le 1$ Graph:

14. $[6, \infty)$ **15.** $(-\infty, 1)$ **16.** $(-\infty, 5]$

17. Logan can invite at most 25 children to the party.

Section 1.5 Rectangular Coordinates and Graphs of Equations **1.** origin **2.** True

3. A: Quadrant I;
B: Quadrant IV;
C: y-axis;
D: Quadrant III

4. A: Quadrant II;
B: x-axis;
C: Quadrant IV;
D: Quadrant I

5. True **6. (a)** No **(b)** Yes **(c)** Yes

7. (a) Yes **(b)** No **(c)** Yes

8. **9.** **10.** **11.** **12.**

13. intercepts **14.** False
15. Intercepts: $(-5, 0)$, $(0, -0.9)$, $(1, 0)$, $(6.7, 0)$; x-intercepts: $(-5, 0)$, $(1, 0)$, $(6.7, 0)$ y-intercept: $(0, -0.9)$

16. (a) $200 thousand **(b)** $350 thousand **(c)** The capacity of the refinery is 700 thousand gallons of gasoline per hour. **(d)** The intercept is $(0, 100)$. The cost of $100 thousand for refining 0 gallons of gasoline can be thought of as fixed costs. **17.** $A: (2, 3)$ I; $B: (-5, 2)$ II; $C: (0, -2)$ y-axis; $D: (-4, -3)$ III; $E: (3, -4)$ IV; $F: (4, 0)$ x-axis

19. A: quadrant I; B: quadrant III; C: x-axis; D: quadrant IV; E: y-axis; F: quadrant II

21. (a) yes **(b)** no **(c)** yes **(d)** yes **23. (a)** yes **(b)** no **(c)** no **(d)** yes
25. (a) no **(b)** yes **(c)** yes **(d)** yes

27. $y = 4x$

29. $y = -\dfrac{1}{2}x$

31. $y = x + 3$

33. $y = -3x + 1$

35. $y = \dfrac{1}{2}x - 4$

37. $2x + y = 7$

39. $y = -x^2$

41. $y = 2x^2 - 8$

43. $y = |x|$

45. $y = |x - 1|$

47. $y = x^3$

49. $y = x^3 + 1$

51. $x^2 - y = 4$

53. $x = y^2 - 1$

55. $(-2, 0)$ and $(0, 3)$
57. $(-2, 0)$, $(1, 0)$, and $(0, -4)$
59. $a = \dfrac{7}{4}$ **61.** $b = 4$ **63. (a)** 400 ft^2 **(b)** 25 feet; 625 ft^2 **(c)** The x-intercepts are $x = 0$ and $x = 50$. These values form the bounds for the width of the opening. The y-intercept is $y = 0$. The area of the opening will be 0 ft^2 when the width is 0 feet. **65. (a)** $39.99; $39.99 **(b)** $3437.49 **(c)** $(0, 39.99)$; The monthly cost will be $39.99 if no minutes are used.

67. Vertical line with an x-intercept of 4.

69. Answers will vary. One possible graph is shown.

71. Answers will vary. One possibility: $y = 0$
73. A complete graph is one that shows enough of the graph so that anyone who is looking at it will visualize the rest of it as an obvious continuation of what is shown. A complete graph should show all the interesting features of the graph, such as intercepts and high/low points.
75. The point-plotting method of graphing an equation requires one to choose certain values of one variable and use the equation to find the corresponding values of the other variable. These points are then plotted and connected in a smooth curve.

77. $y = 3x - 9$

79. $y = -x^2 + 8$

81. $y + 2x^2 = 13$

83. $y = x^3 - 6x + 1$

Section 1.6 Linear Equations in Two Variables 1. linear equation **2.** line

3.

4.

5.

6. True

7.

8.

9.

10. x; y **11.**   **12.** **13.**

14. undefined **15.** $\dfrac{2}{5}$ **16.** False **17.** True **18.** 3; For every 1-unit increase in x, y will increase by 3 units. **19.** $-\dfrac{7}{4}$; For every 4-unit increase in x, y will decrease by 7 units. **20.** 0; For every 1-unit increase in x, there is no change in y; Horizontal line **21.** Undefined; Vertical line **22.** True **23.** 0

24. L_1: $m = \dfrac{1}{5}$; L_2: m is undefined; L_3: $m = -1$; L_4: $m = 0$

25. (a)

(b) $1120 per bicycle. For each bicycle sold, total revenue increased by $1120 when between 0 and 25 bicycles were sold.
(c) $120 per bicycle. For each bicycle sold, total revenue increased by $120 per bicycle when between 102 and 150 bicycles were sold.
(d) No, because the average rate of change (slope) is not constant.

26. (a) **(b)** **(c)**

27. point-slope

28. $y - 5 = 2(x - 3)$

29. $y - 3 = -4(x + 2)$

30. $y + 4 = \dfrac{1}{3}(x - 3)$

31. $y = -2$

32. slope-intercept

33. $m = 3$; y-intercept: $(0, -2)$

34. $m = -3$; y-intercept: $(0, 4)$

35. $m = \dfrac{3}{2}$; y-intercept: $\left(0, -\dfrac{7}{2}\right)$

36. $m = -\dfrac{7}{3}$; y-intercept: $(0, 0)$

37. $y = 2x + 1$

38. $y = -\dfrac{1}{2}x + 3$

39. $y = 6$

40. $x = 3$

41.
43.
45.
47.
49.
51.
53.

55.
57.
59.
61.
63.

65. (a) $\dfrac{4}{3}$ **(b)** For every 3-unit increase in x, y will increase 4 units.

67. (a) $-\dfrac{8}{3}$ **(b)** For every 3-unit increase in x, y will decrease by 8 units. For every 3-unit decrease in x, y will increase by 8 units.

69. $m = 5$
71. $m = -3$
73. $m = \dfrac{4}{5}$
75. $m = 0$

77. m is undefined
79. $m = \dfrac{2}{21}$
81.
83.

85.
87.
89.
91. Answers will vary. One possibility: $(0, 8)$, $(2, 13)$, and $(4, 18)$

93. $y = -\dfrac{4}{9}x - \dfrac{7}{9}$ **95.** $y = 3$ **97.** $y = 2x$ **99.** $y = -3x - 2$

101. $y = \dfrac{4}{3}x - 2$ **103.** $y = -\dfrac{5}{4}x + \dfrac{3}{2}$ **105.** $x = 6$ **107.** $y = \dfrac{7}{5}x$

109. $y = 5x - 13$ **111.** $y = -\dfrac{3}{7}x + \dfrac{1}{7}$ **113.** $x = -1$ **115.** $y = \dfrac{5}{2}x + \dfrac{1}{2}$ **117.** $y = 4$

119. The slope is 2 and the y-intercept is $(0, -1)$.

121. The slope is -4 and the y-intercept is $(0, 0)$.

123. The slope is -2 and the y-intercept is $(0, 3)$.

125. The slope is -2 and the y-intercept is $(0, 4)$.

127. The slope is $\dfrac{1}{4}$ and the y-intercept is $\left(0, -\dfrac{1}{2}\right)$.

129. The slope is undefined and there is no y-intercept.

131. $y = 0$

133. (a)

(b) -0.25 heartbeat per year; Between ages 20 and 30, the maximum number of heartbeats decreases at a rate of 0.25 heartbeat per year. **(c)** -0.25 heartbeat per year; Between ages 50 and 60, the maximum number of heartbeats decreases at a rate of 0.25 heartbeat per year. **(d)** Yes. The maximum number of heartbeats appears to be linearly related to age. The average rate of change (slope) is constant for the data provided.

135. (a)

(b) $2187.80 per year; Between ages 20 and 30, the average individual's income increases at a rate of $2187.80 per year. **(c)** $-$343.50 per year; Between ages 50 and 60, the average individual's income decreases at a rate of $343.50 per year. **(d)** No. The average income is not linearly related to age. The average rate of change (slope) is not constant.

137. $C = \dfrac{5}{9}(°F - 32)$; $60°F$ is equivalent to $15.6°C$. **139.** (c)

141. Horizontal line: $y = b$; Vertical line: $x = a$; Point-slope: $y - y_1 = m(x - x_1)$; Slope-intercept: $y = mx + b$; Standard: $Ax + By = C$

143. Horizontal line of the form $y = b$, where b does not equal 0. **145.** No. All lines must travel through either the x-axis or the y-axis or both.

147. (a) Positive **(b)** Negative **(c)** Zero **(d)** Undefined

Section 1.7 Parallel and Perpendicular Lines
1. slope; y-intercepts **2.** Not parallel **3.** Parallel **4.** Not parallel

5. $y = 3x - 7$

6. $y = -\dfrac{3}{2}x + 1$

7. -1 **8.** $\dfrac{1}{3}$

9. Perpendicular
10. Not perpendicular
11. Perpendicular

12. $y = -\dfrac{1}{2}x$

13. $y = -\dfrac{4}{3}x - 8$

14. $y = -2$

15. (a) $m = 5$ **(b)** $m = -\dfrac{1}{5}$ **17. (a)** $m = -\dfrac{5}{6}$ **(b)** $m = \dfrac{6}{5}$ **19.** Parallel

21. Neither **23.** Perpendicular **25.** Parallel **27.** $y = \dfrac{3}{2}x - 3$ **29.** $y = \dfrac{1}{2}x + 2$ **31.** $y = 2$

33. $y = 2x - 5$

35. $y = \dfrac{1}{2}x + 2$

37. $y = -3$

39. $y = 3$

41. $y = 3x + 2$

43. $y = \dfrac{3}{4}x + 4$

45. $y = -\dfrac{5}{2}x - 8$

47. $m_1 = \dfrac{3}{5}$; $m_2 = -\dfrac{5}{3}$;

Perpendicular

49. $m_1 = -\dfrac{7}{3}$; $m_2 = -\dfrac{7}{3}$;

Parallel

51. $m_1 = -2$; $m_2 = 2$;

Neither

53. (a)

(b) Slope of $\overline{AB} = \dfrac{2}{3}$;

Slope of $\overline{BC} = -\dfrac{3}{2}$

Because the slopes are negative reciprocals, segments $\overline{AB}$ and $\overline{BC}$ are perpendicular. Thus, triangle ABC is a right triangle.

55. (a)

(b) Slope of $\overline{AB} = \dfrac{1}{5}$;

Slope of $\overline{BC} = 3$;

Slope of $\overline{CD} = \dfrac{1}{5}$;

Slope of $\overline{DA} = 3$

Because the slopes of $\overline{AB}$ and $\overline{CD}$ are equal, $\overline{AB}$ and $\overline{CD}$ are parallel. Because the slopes of $\overline{BC}$ and $\overline{DA}$ are equal, $\overline{BC}$ and $\overline{DA}$ are parallel. Thus, quadrilateral $ABCD$ is a parallelogram.

57. $A = -1$ **59.** (c) **61.** No. If the two nonvertical lines have the same x-intercept but different y-intercepts, then their slopes cannot be equal. Therefore, they cannot be parallel lines. **63. (d)** The slopes of the lines are equal; the y-intercepts are different.

Section 1.8 Linear Inequalities in Two Variables
1. half-planes 2. (a) No (b) Yes (c) Yes (d) Yes 3. False 4. True

5.
6.
7.
8. (a) $560x + 300y \leq 900$ (b) Yes (c) No
9. (a) yes (b) no (c) yes
11. (a) yes (b) no (c) yes

13.
15.
17.
19.
21.
23.

25.
27.
29.
31.

33. (a) Let x = the number of Filet-o-Fish orders. Let y = the number of orders of fries. $39x + 29y \leq 125$ (b) yes (c) no 35. (a) Let x = the number of Switch A assemblies. Let y = the number of Switch B assemblies. $2x + 1.5y \leq 80$ (b) no (c) no 37. $y > 2x - 2$ 39. $y \leq -x + 1$

41. Consider the inequality $Ax + By > C$. The equation $Ax + By = C$ separates the Cartesian plane into two half-planes. Any ordered pair (x, y) that makes $Ax + By$ equal to C lies on this line. Any other point will lie in one of the two half-planes. If a point on one side of the line does not satisfy the inequality, then none of the points satisfy the inequality. For this reason, if a point does not satisfy the inequality, then we shade the opposite side.

43. $y > 3$
45. $y < 5x$
47. $y > 2x + 3$

49. $y \leq \dfrac{1}{2}x - 5$
51. $3x + y \leq 4$
$(y \leq -3x + 4)$
53. $2x + 5y \leq -10$
$\left(y \leq -\dfrac{2}{5}x - 2\right)$

Chapter 1 Review
1. $x = 5$ is a solution to the equation. $x = 6$ is not a solution to the equation. 2. $x = -2$ is a solution to the equation. $x = -1$ is a solution to the equation. 3. $y = -2$ is not a solution to the equation. $y = 0$ is a solution to the equation. 4. $w = -14$ is not a solution to the equation. $w = 7$ is not a solution to the equation. 5. conditional; $\{3\}$ 6. conditional; $\{4\}$ 7. conditional; $\{3\}$ 8. conditional; $\{-8\}$ 9. identity; $\{x \mid x$ is any real number$\}$ 10. contradiction; $\{\ \}$ or $\varnothing$ 11. contradiction; $\{\ \}$ or $\varnothing$ 12. conditional; $\{-2\}$ 13. conditional; $\{-47\}$

14. identity; $\{w \mid w$ is any real number$\}$ 15. $x = -\dfrac{3}{2}$ must be excluded from the domain. 16. $x = \dfrac{1}{2}$ must be excluded from the domain.

17. Her Oregon taxable income was $28,900. 18. The regular club price for a DVD is $21.95. 19. $3x + 7 = 22$ 20. $x - 3 = \dfrac{x}{2}$ 21. $0.2x = x - 12$

22. $6x = 2x - 4$ 23. Payton is 5 years old and Shawn is 13 years old. 24. The five odd integers are 21, 23, 25, 27, and 29. 25. Logan needs to get a score of 76.5 on the final exam to have an average of 80. 26. After 1 month, Cherie will accrue about $28.00 in interest. 27. The original price of the sleeping bag was $135.00. 28. The federal minimum wage was $6.55 (per hour). 29. The store should mix 4 pounds of chocolate-covered blueberries with 8 pounds of chocolate-covered strawberries. 30. The store should mix 7.5 pounds of baseball gumballs with 2.5 pounds of soccer gumballs.

31. Angie should invest $4800 at 8% and $3200 at 18%. 32. About 2.14 quarts would need to be drained and replaced with pure antifreeze.

33. Josh drove 90 miles at 60 miles per hour 34. The F14 is traveling at a speed of 1220 miles per hour, and the F15 is traveling at a speed of 1420 miles per hour. 35. $x = \dfrac{k}{y}$ 36. $C = \dfrac{5}{9}(F - 32)$ 37. $W = \dfrac{P - 2L}{2}$ 38. $m_2 = \dfrac{\rho - m_1 v_1}{v_2}$ 39. $T = \dfrac{PV}{nR}$ 40. $W = \dfrac{S - 2LH}{2L + 2H}$ 41. $y = -\dfrac{3}{4}x + \dfrac{1}{2}$

42. $y = \dfrac{5}{4}x + \dfrac{5}{2}$ 43. $y = 4x - 5$ 44. $y = -\dfrac{6}{5}x + 24$ 45. The melting point of platinum is 1772°C. 46. The angles measure 70°, 70°, and 40°.

47. The window measures 15 feet by 23 feet. 48. (a) $x = 2.5C - 750.025$ (b) Debbie can talk for 1000 minutes in one month and not spend more than $100 on her cell phone bill. 49. The patio will be $\dfrac{10}{27}$ of a foot thick (about 4.44 inches). 50. (a) $r = \dfrac{A - \pi Rs}{\pi s}$ (b) The radius of the top of the frustum is 2 feet. 51. (a) $x = \dfrac{C - 7.48}{0.08674}$ (b) Approximately 2289 kwh were used. 52. The angle measures 60°. 53. $(2, 7]$ ———

54. $(-2, \infty)$ ——— 55. $x \leq 4$ ——— 56. $-1 \leq x < 3$ ———

57. $a = 7$ and $b = 15$ **58.** $a = -1$ and $b = 5$ **59.** Solution set: $\{x \mid x \le -4\}$ Interval: $(-\infty, -4]$ Graph:

60. Solution set: $\left\{x \mid x < -\dfrac{1}{3}\right\}$ Interval: $\left(-\infty, -\dfrac{1}{3}\right)$ Graph: **61.** Solution set: $\left\{h \mid h \ge -\dfrac{2}{3}\right\}$ Interval: $\left[-\dfrac{2}{3}, \infty\right)$

Graph: **62.** Solution set: $\{x \mid x > 2\}$ Interval: $(2, \infty)$ Graph:

63. Solution set: $\{p \mid p > 2\}$ Interval: $(2, \infty)$ Graph: **64.** Solution set: $\{x \mid x$ is any real number $\}$ Interval:

$(-\infty, \infty)$ Graph: **65.** $\{\ \}$ or $\varnothing$ **66.** Solution set: $\{x \mid x > 4.2\}$ Interval: $(4.2, \infty)$

Graph: **67.** Solution set: $\{w \mid w > 1\}$ Interval: $(1, \infty)$ Graph:

68. Solution set: $\{y \mid y < -120\}$ Interval: $(-\infty, -120)$ Graph: **69.** To stay within budget, no more than 60 people

can attend the banquet. **70.** To stay within budget, you can drive an average of 216 miles per day. **71.** The band must sell more than 125 candy bars to make a profit. **72.** You can purchase up to 5 DVDs and still be within budget.

73. A: quadrant IV;
B: quadrant III;
C: y-axis; D: quadrant II;
E: x-axis; F: quadrant I

74. A: x-axis;
B: quadrant I;
C: quadrant III; D: quadrant II;
E: quadrant IV;
F: y-axis

75. (a) yes (b) no
(c) no (d) yes
76. (a) no (b) yes
(c) yes (d) no
77. $y = x + 2$

78. $2x + y = 3$

79. $y = -x^2 + 4$

80. $y = |x + 2| - 1$

81. $y = x^3 + 2$

82. $x = y^2 + 1$

83. $(-3, 0), (0, -1), (0, 3)$
x-intercept: $(-3, 0)$;
y-intercepts: $(0, -1), (0, 3)$
84. (a) $599.99 (b) $3499.99

85. **86.**

87.   **88.**   **89.**   **90.**   **91.**   **92.**

93.   **94.**   **95.**

96. (a) $m = \dfrac{5}{4}$ (b) For every 4-unit increase in x, y will increase by 5 units.

97. (a) $m = -\dfrac{1}{4}$ (b) For every 4-unit increase in x, y will decrease by 1 unit. For every 4-unit decrease in x, y will increase by 1 unit.

98. $m = -2$ **99.** $m = \dfrac{3}{2}$ **100.** (a)

(b) 81,493.5 people per year; Between 1940 and 1950, the population of Illinois increased at an average rate of 81,493.5 people per year. (c) 319.3 people per year; Between 1980 and 1990, the population of Illinois increased at an average rate of 319.3 people per year. (d) 41,133.9 people per year; Between 2000 and 2010, the population of Illinois increased at an average rate of 41,133.9 people per year. (e) No. The average rate of change (slope) is not constant.

101.   **102.**   **103.**   **104.**

105. $y = -\dfrac{1}{2}x + 2$ or $x + 2y = 4$

106. $y = 3x$ or $3x - y = 0$

107. $y = -x + 5$ or $x + y = 5$

108. $y = \dfrac{3}{5}x + 2$ or $3x - 5y = -10$ **109.** $y = -\dfrac{1}{3}x + 4$ or $x + 3y = 12$ **110.** $y = 3$ **111.** $y = 2x - 9$ or $2x - y = 9$

112. $y = -\dfrac{1}{3}x + \dfrac{5}{3}$ or $x + 3y = 5$

113. The slope is 4 and the y-intercept is $(0, -6)$. **114.** The slope is $-\dfrac{2}{3}$ and the y-intercept is $(0, 4)$. **115.** $-\dfrac{3}{8}$ **116.** $\dfrac{8}{3}$ **117.** Perpendicular

118. Neither **119.** Parallel
120. Perpendicular

121. $y = -2x + 4$

122. $y = \dfrac{5}{2}x - 7$

123. $x = 1$

124. $y = -\dfrac{1}{3}x + 4$

125. $y = \dfrac{4}{3}x + 2$

126. $y = -4$

127. (a) yes
(b) yes
(c) no
128. (a) no
(b) yes
(c) no

129.

130.

131.

132.

133.

134. $x = -8$

135. (a) Let $x =$ the number of movie tickets. Let $y =$ the number of music downloads. $7.50x + 2y \le 60$ **(b)** no **(c)** yes
136. (a) Let $x =$ the number of candy bars. Let $y =$ the number of candles. $0.50x + 2y \ge 1000$ **(b)** no **(c)** yes

Chapter 1 Test 1. (a) $x = 6$ is a solution to the equation. **(b)** $x = -2$ is not a solution to the equation.
2. (a) Interval: $(-4, \infty)$ **(b)** Interval: $(3, 7]$ **3.** $3x - 8 = x + 4$

4. $\dfrac{2}{3}x + 2(x - 5) > 7$ **5.** $\{2\}$; conditional **6.** $\{\ \}$ or $\varnothing$; contradiction

7. $\{x \mid x \ge 3\}$; $[3, \infty)$ **8.** $\left\{x \mid x > -\dfrac{1}{5}\right\}$; $\left(-\dfrac{1}{5}, \infty\right)$

9. $\left\{x \mid x \ge \dfrac{1}{2}\right\}$; $\left[\dfrac{1}{2}, \infty\right)$ **10.** $y = -\dfrac{7}{4}x + \dfrac{3}{4}$ **11.** Glen's weekly sales must be at least $4375 for him to earn at least $750.

12. There were 14 children at Payton's party. **13.** The sandbox has a width of 4 feet and a length of 6 feet. **14.** The chemist needs to mix 8 liters of the 10% solution with 4 liters of the 40% solution. **15.** It will take Contestant B two hours to catch up to Contestant A.

16. A: quadrant IV;
B: y-axis;
C: x-axis;
D: quadrant I;
E: quadrant III;
F: quadrant II

17. (a) no **(b)** yes **(c)** yes
18. $y = 4x - 1$

19. $y = 4x^2$
20. $(-3, 0), (0, 1), (0, 3)$
x-intercept: $(-3, 0)$;
y-intercepts: $(0, 1), (0, 3)$

21. (a) At 6 seconds the car is traveling 30 miles per hour. **(b)** $(0, 0)$: At the start of the trip, the car is not moving. $(32, 0)$: After 32 seconds, the speed of the car is 0 miles per hour.

22. **23.** **24.** **25.** **26.**

27. $m = -\dfrac{4}{3}$
For every 3-unit increase in x, y will decrease by 4 units. For every 3-unit decrease in x, y will increase by 4 units.

28.

29. Perpendicular **30.** $y = 4x + 13$ or $4x - y = -13$ **31.** $y = -\dfrac{2}{3}x + 5$ or $2x + 3y = 15$
32. $y = \dfrac{1}{5}x - 2$ or $x - 5y = 10$ **33.** $y = -\dfrac{1}{3}x + 4$ or $x + 3y = 12$ **34. (a)** no **(b)** no **(c)** yes

35.   **36.**   **37. (a)**   **(b)** 3.14 square feet per foot; Between diameters of 1 foot and 3 feet, the area of the circle increases at a rate of 3.14 square feet per foot. **(c)** Approximately 18.06 square feet per foot; Between diameters of 10 feet and 13 feet, the area of the circle increases at a rate of approximately 18.06 square feet per foot. **(d)** No. The average rate of change (slope) is not constant.

Cumulative Review Chapters R–1 **1. (a)** (i) 27.235 (ii) 27.236 **(b)** (i) 1.0 (ii) 1.1 **2.**   **3.** -14

4. -6 **5.** 6 **6.** 81 **7.** 74 **8.** $\dfrac{11}{12}$ **9.** 9 **10.** $5a^2 - 4a - 13$ **11. (a)** no **(b)** yes **12.** $-5x + 18$ **13.** $x = 3$ is not a solution to the equation. **14.** $\{4\}$

15. $\left\{\dfrac{13}{2}\right\}$ **16.** $y = \dfrac{2}{5}x - \dfrac{6}{5}$ **17.** $\{x \mid x \geq 7\}$; $[7, \infty)$ **18.** $\{x \mid x \leq 5\}$ or $(-\infty, 5]$

19.   **20.**   **21.**   **22.** $y = -\dfrac{4}{3}x + 2$ or $4x + 3y = 6$ **23.** $y = -3x - 8$ or $3x + y = -8$ **24.**

25. Shawn needs to score at least 91 on the final exam to earn an A (assuming the maximum score on the exam is 100). **26.** A person 62 inches tall would be considered obese if she or he weighed 160 pounds or more. **27.** The angles measure 55° and 125°. **28.** The cylinder should be about 5.96 inches tall. **29.** The three consecutive even integers are 24, 26, and 28.

Chapter 2 Relations, Functions, and More Inequalities

Section 2.1 Relations **1.** corresponds; depends **2.** {(Max, November 8), (Alesia, January 20), (Trent, March 3), (Yolanda, November 8), (Wanda, July 6), (Elvis, January 8)} **3.**   **4.** domain; range **5.** Domain: {Max, Alesia, Trent, Yolanda, Wanda, Elvis}; Range: {January 20, March 3, July 6, November 8, January 8} **6.** Domain: $\{1, 5, 8, 10\}$; Range: $\{3, 4, 13\}$ **7.** Domain: $\{-2, -1, 2, 3, 4\}$; Range: $\{-3, -2, 0, 2, 3\}$ **8.** True **9.** False **10.** Domain: $\{x \mid -2 \leq x \leq 4\}$ or $[-2, 4]$; Range: $\{y \mid -2 \leq y \leq 2\}$ or $[-2, 2]$ **11.** Domain: $\{x \mid x$ is a real number$\}$ or $(-\infty, \infty)$; Range: $\{y \mid y$ is a real number$\}$ or $(-\infty, \infty)$

12. Domain: $\{x \mid x$ is a real number$\}$ or $(-\infty, \infty)$; Range: $\{y \mid y$ is a real number$\}$ or $(-\infty, \infty)$

13. Domain: $\{x \mid x$ is a real number$\}$ or $(-\infty, \infty)$; Range: $\{y \mid y \geq -8\}$ or $[-8, \infty)$

14. Domain: $\{x \mid x \geq 1\}$ or $[1, \infty)$; Range: $\{y \mid y$ is a real number$\}$ or $(-\infty, \infty)$

15. {(*USA Today*, 2.28.), (*Wall Street Journal*, 2.06), (*New York Times*, 1.1), (*Los Angeles Times*, 0.82), (*Washington Post*, 0.70)}; Domain: {*USA Today, Wall Street Journal, New York Times, Los Angeles Times, Washington Post*}; Range: {2.28, 2.06, 1.1, 0.82, 0.70} **17.** {(Less than 9th Grade, $20,791), (9th–12th Grade – no diploma, $23,234), (High School Graduate, $32,456), (Associate's Degree, $42,508), (Bachelor's Degree, $62,240)}; Domain: {Less than 9th Grade, 9th–12th Grade – no diploma, High School Graduate, Associate's Degree, Bachelor's Degree}; Range: {$20,791, $23,234, $32,456, $42,508, $62,240}

19. Domain: $\{-3, -2, -1, 0, 1\}$; Range: $\{4, 6, 8, 10, 12\}$ **21.** Domain: $\{-2, -1, 0, 1, 2\}$; Range: $\{0, 2, 4\}$ **23.** 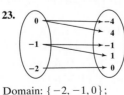 Domain: $\{-2, -1, 0\}$; Range: $\{-4, -1, 0, 1, 4\}$ **25.** Domain: $\{-3, -2, 0, 2, 3\}$; Range: $\{-3, -1, 2, 3\}$ **27.** Domain: $\{x \mid -4 \leq x \leq 4\}$ or $[-4, 4]$; Range: $\{y \mid -2 \leq y \leq 2\}$ or $[-2, 2]$ **29.** Domain: $\{x \mid -1 \leq x \leq 3\}$ or $[-1, 3]$; Range: $\{y \mid 0 \leq y \leq 4\}$ or $[0, 4]$ **31.** Domain: $\{x \mid x$ is a real number$\}$ or $(-\infty, \infty)$; Range: $\{y \mid y \geq -3\}$ or $[-3, \infty)$

33. Domain: $\{x \mid x$ is a real number$\}$ or $(-\infty, \infty)$; Range: $\{y \mid y$ is a real number$\}$ or $(-\infty, \infty)$ **35.** Domain: $\{x \mid x$ is a real number$\}$ or $(-\infty, \infty)$; Range: $\{y \mid y$ is a real number$\}$ or $(-\infty, \infty)$ **37.** Domain: $\{x \mid x$ is a real number$\}$ or $(-\infty, \infty)$; Range: $\{y \mid y$ is a real number$\}$ or $(-\infty, \infty)$ **39.** Domain: $\{x \mid x$ is a real number$\}$ or $(-\infty, \infty)$; Range: $\{y \mid y \leq 0\}$ or $(-\infty, 0]$ **41.** Domain: $\{x \mid x$ is a real number$\}$ or $(-\infty, \infty)$; Range: $\{y \mid y \geq -8\}$ or $[-8, \infty)$ **43.** Domain: $\{x \mid x$ is a real number$\}$ or $(-\infty, \infty)$; Range: $\{y \mid y \geq 0\}$ or $[0, \infty)$ **45.** Domain: $\{x \mid x$ is a real number$\}$ or $(-\infty, \infty)$; Range: $\{y \mid y \geq 0\}$ or $[0, \infty)$ **47.** Domain: $\{x \mid x$ is a real number$\}$ or $(-\infty, \infty)$; Range: $\{y \mid y$ is a real number$\}$ or $(-\infty, \infty)$ **49.** Domain: $\{x \mid x$ is a real number$\}$ or $(-\infty, \infty)$; Range: $\{y \mid y$ is a real number$\}$ or $(-\infty, \infty)$ **51.** Domain: $\{x \mid x$ is a real number$\}$ or $(-\infty, \infty)$; Range: $\{y \mid y \geq -4\}$ or $[-4, \infty)$ **53.** Domain: $\{x \mid x \geq -1\}$ or $[-1, \infty)$; Range: $\{y \mid y$ is a real number$\}$ or $(-\infty, \infty)$

55. (a) Domain: $\{x \mid 0 < x < 50\}$ or $(0, 50)$; Range: $\{y \mid 0 < y \leq 625\}$ or $(0, 625]$ **(b)** Width must be less than $\dfrac{1}{2}$ the perimeter.

57. (a) Domain: $\{m \mid 0 \leq m \leq 15{,}120\}$ or $[0, 15{,}120]$; Range: $\{C \mid 100 \leq C \leq 3380\}$ or $[100, 3380]$ **(b)** $21 \cdot 12 \cdot 60 = 15{,}120$ minutes

59. Actual graphs will vary but all should be horizontal lines. **61.** A relation is a correspondence between two sets called the domain and range. The domain is the set of all inputs, and the range is the set of all outputs.

Section 2.2 An Introduction to Functions
1. function **2.** false **3.** Function; Domain: {Max, Alesia, Trent, Yolanda, Wanda, Elvis}; Range: {January 20, } March 3, July 6, November 8, January 8} **4.** Not a function **5.** Function; Domain: $\{-3, -2, -1, 0, 1\}$; Range: $\{0, 1, 2, 3\}$ **6.** Not a function **7.** Function **8.** Not a function **9.** Function **10.** True **11.** Function **12.** Not a function **13.** 14 **14.** -4 **15.** -18 **16.** 2 **17.** dependent; independent; argument **18.** $2x - 9$ **19.** $2x - 4$ **20.** domain **21.** $(-\infty, \infty)$ **22.** $\{x | x \neq 3\}$ **23.** $\{r | r > 0\}$ or $(0, \infty)$ **24. (a)** Independent variable: t; dependent variable: A **(b)** $A(30) \approx 706.86$ square miles. After 30 days, the area contaminated with oil will be a circle covering about 706.86 square miles. **25.** Function; Domain: { Virginia, Nevada, Arkansas, Tennessee, Texas}; Range: $\{4, 9, 11, 36\}$ **27.** Not a function; Domain: $\{150, 174, 180\}$; Range: $\{118, 130, 140\}$ **29.** Function; Domain: $\{0, 1, 2, 3\}$; Range: $\{3, 4, 5, 6\}$ **31.** Function; Domain: $\{-3, 1, 4, 7\}$; Range: $\{5\}$ **33.** Not a function; Domain: $\{-10, -5, 0\}$; Range: $\{1, 2, 3, 4\}$ **35.** Function **37.** Function **39.** Not a function **41.** Function **43.** Not a function **45.** Function **47.** Not a function **49.** Function **51.** Function **53. (a)** $f(0) = 3$; **(b)** $f(3) = 9$ **(c)** $f(-2) = -1$ **55. (a)** $f(0) = 2$ **(b)** $f(3) = -13$ **(c)** $f(-2) = 12$ **57. (a)** $f(0) = 0$ **(b)** $f(3) = 0$ **(c)** $f(-2) = 10$ **59. (a)** $f(0) = 3$ **(b)** $f(3) = -3$ **(c)** $f(-2) = -3$ **61. (a)** $f(-x) = -2x - 5$ **(b)** $f(x + 2) = 2x - 1$ **(c)** $f(2x) = 4x - 5$ **(d)** $-f(x) = -2x + 5$ **(e)** $f(x + h) = 2x + 2h - 5$ **63. (a)** $f(-x) = 7 + 5x$ **(b)** $f(x + 2) = -3 - 5x$ **(c)** $f(2x) = 7 - 10x$ **(d)** $-f(x) = -7 + 5x$ **(e)** $f(x + h) = 7 - 5x - 5h$ **65.** $f(2) = 7$ **67.** $s(-2) = 16$ **69.** $F(-3) = 5$ **71.** $F(4) = -6$ **73.** $\{x | x$ is any real number $\}$ or $(-\infty, \infty)$

75. $\{z | z \neq 5\}$ **77.** $\{x | x$ is any real number $\}$ or $(-\infty, \infty)$ **79.** $\left\{x \middle| x \neq -\dfrac{1}{3}\right\}$ **81.** $C = -6$ **83.** $A = 5$ **85.** $A(r) = \pi r^2$; 50.27 in.^{2} **87.** $G(h) = 15h$;

$375 **89. (a)** The dependent variable is the population, P, and the independent variable is the age, a. **(b)** $P(20) = 4433.3$ thousand; The population of U.S. residents who were 20 years of age in 2014 was roughly 4,433,000. **(c)** $P(0) = 3694.1$ thousand; $P(0)$ represents residents of the U.S. that are 0 years of age in 2014 was roughly 3,694,000. **91. (a)** The dependent variable is revenue, R, and the independent variable is price, p. **(b)** $R(50) = 7500$; Selling MP3 players for $50 will yield a daily revenue of $7500 for the company. **(c)** $R(120) = 9600$; Selling MP3 players for $120 will yield a daily revenue of $9600 for the company. **93.** $\{r | r > 0\}$ or $(0, \infty)$ **95.** $\{h | 0 \leq h \leq 60\}$ or $[0, 60]$ **97.** $\{p | 0 \leq p \leq 120\}$ or $[0, 120]$ **99. (a) (i)** -5 **(ii)** 1 **(iii)** 1 **(b) (i)** 13 **(ii)** 4 **(iii)** 4 **101.** Answers will vary. **103.** A function is a relation between two sets, the domain and the range. The domain is the set of all inputs to the function, and the range is the set of all outputs. In a function, each input in the domain corresponds to exactly one output in the range. **105.** The four forms of a function are map, ordered pairs, equation, and graph. **107.** $f(2) = 7$ **109.** $F(-3) = 5$ **111.** $H(7) = 5$ **113.** $F(4) = -0.65$ or $-\dfrac{13}{20}$

Section 2.3 Functions and Their Graphs
1. graph; function **2.** 4; -7

6. (a) Domain: $\{x | x$ is a real number $\}$; $(-\infty, \infty)$; Range: $\{y | y \leq 1\}$; $(-\infty, 1]$ **(b)** x-intercepts: $(-2, 0)$ and $(2, 0)$; y-intercept: $(0, 1)$ **7.** $f(3) = 8$; $(-2, 4)$ **8. (a)** $f(-3) = -15$; $f(1) = -3$ **(b)** Domain: $\{x | x$ is a real number $\}$ or $(-\infty, \infty)$ **(c)** Range: $\{y | y$ is a real number $\}$ or $(-\infty, \infty)$ **(d)** x-intercepts: $(-2, 0)$, $(0, 0)$ and $(2, 0)$; y-intercept: $(0, 0)$ **(e)** $\{3\}$

9. (a) No **(b)** $f(3) = -2$; $(3, -2)$ is on the graph **(c)** $x = 5$; $(5, -8)$ is on the graph **10.** yes **11.** no **12.** yes **13.** -2 and 2

14.

15. $f(x) = 4x - 6$ **17.** $h(x) = x^2 - 2$ **19.** $G(x) = |x - 1|$ **21.** $g(x) = x^3$

23. (a) Domain: $\{x | x$ is a real number $\}$ or $(-\infty, \infty)$; Range: $\{y | y$ is a real number $\}$ or $(-\infty, \infty)$ **(b)** $(0, 2)$ and $(1, 0)$ **(c)** 1 **25. (a)** Domain: $\{x | x$ is a real number $\}$ or $(-\infty, \infty)$; Range: $\{y | y \geq -2.25\}$ or $[-2.25, \infty)$ **(b)** $(-2, 0)$, $(4, 0)$, and $(0, -2)$ **(c)** $-2, 4$ **27. (a)** Domain: $\{x | x$ is a real number $\}$ or $(-\infty, \infty)$; Range: $\{y | y$ is a real number $\}$ or $(-\infty, \infty)$ **(b)** $(-3, 0)$, $(-1, 0)$, $(2, 0)$, and $(0, -3)$ **(c)** $-3, -1, 2$ **29. (a)** Domain: $\{x | x$ is a real number $\}$ or $(-\infty, \infty)$; Range: $\{y | y \geq 0\}$ or $[0, \infty)$ **(b)** $(-3, 0)$, $(3, 0)$, and $(0, 9)$ **(c)** $-3, 3$ **31. (a)** Domain: $\{x | x \leq 4\}$ or $(-\infty, 4]$; Range: $\{y | y \leq 3\}$ or $(-\infty, 3]$ **(b)** $(-2, 0)$ and $(0, 2)$ **(c)** -2 **33. (a)** $f(-7) = -2$ **(b)** $f(-3) = 3$ **(c)** $f(6) = 2$ **(d)** negative **(e)** $\{-6, -1, 4\}$ **(f)** $\{x | -7 \leq x \leq 6\}$ or $[-7, 6]$ **(g)** $\{y | -2 \leq y \leq 3\}$ or $[-2, 3]$ **(h)** $(-6, 0)$, $(-1, 0)$ and $(4, 0)$ **(i)** $(0, -1)$ **(j)** $\{-7, 2\}$ **(k)** $x = -3$ **(l)** $-6, -1, 4$ **35. (a)** $F(-2) = 3$ **(b)** $F(3) = -6$ **(c)** $x = -1$ **(d)** $(-4, 0)$ **(e)** $(0, 2)$ **37. (a)** no **(b)** $f(3) = 3$; $(3, 3)$ **(c)** 4; $(4, 7)$ **(d)** no **39. (a)** yes **(b)** $g(6) = 1$; $(6, 1)$ **(c)** -12; $(-12, 10)$ **(d)** yes **41. (c)** **43. (e)** **45. (d)**

47. **49.**

51. (a) III **(b)** I **(c)** IV **(d)** V **(e)** II

53. **55.** Answers will vary.

57. The person's weight increases until age 30, then oscillates back and forth between 158 pounds and 178 pounds, then slowly levels off at about 150 pounds.
59. One possibility: Answers will vary.

61. A function cannot have more than one output for a given input. So, there cannot be two outputs for the input 0.
63. The range of a function is the set of all outputs of the elements in the domain. **65.** The zeros of a function are the input values that make the function equal to zero.

Putting the Concepts Together (Sections 2.1–2.3) **1.** The relation is a function because each element in the domain corresponds to exactly one element in the range. { (−2, −1), (−1, 0), (0, 1), (1, 2), (2, 3) } **2.** (a) Function (b) Not a function **3.** Yes; Domain: {−4, −1, 0, 3, 6}; Range: {−3, −2, 2, 6} **4.** The graph passes the vertical line test. $f(5) = −6$ **5.** 4 **6.** (a) −17 (b) −34 (c) −5x + 20 (d) −5x + 23

7. (a) {h | h is any real number} or (−∞, ∞) (b) $\left\{ w \mid w \neq -\dfrac{1}{3} \right\}$

8. Domain: {x | x is any real number} or (−∞, ∞); Range: {y | y ≥ −2} or [−2, ∞) **9.** (a) h(2.5) = 80; After 2.5 seconds, the height of the ball is 80 feet. (b) {t | 0 ≤ t ≤ 3.8} or [0, 3.8] (c) {h | 0 ≤ h ≤ 105} or [0, 105] (d) 1.25 seconds **10.** (a) no (b) −12; (−2, −12) (c) −4; (−4, −22) (d) yes

Section 2.4 Linear Functions and Models
1. slope; y-intercept **2.** line **3.** False **4.** −2; (0, 3)

5. **6.** **7.** **8.** **9.** 5 **10.** −8 **11.** 12

12. (a) {x | x ≥ 0}; [0, ∞) (b) $40 (c) $68 (d) 130 miles
(e)
(f) You may drive between 0 and 250 miles.

13. (a) C(x) = 81x + 2000 (b) $2405 (c) 10 bicycles (d)

14. (a) C(x) = 0.18x + 250 (b) [0, ∞) (c) $307.60 (d) 180 miles (e)

15. (a) (b) As age increases, total cholesterol also increases.

16. nonlinear **17.** linear, positive slope **18.** (a) Answers will vary. Using (25, 180) and (65, 269): y = f(x) = 2.225x + 124.375 (b) (c) 211 mg/dL (d) Each year, a male's total cholesterol increases by 2.225 mg/dL; No

19. F(x) = 5x − 2 **21.** G(x) = −3x + 7 **23.** H(x) = −2 **25.** $f(x) = \dfrac{1}{2}x − 4$

27. $F(x) = -\dfrac{5}{2}x + 5$ **29.** $G(x) = -\dfrac{3}{2}x$ **31.** −5 **33.** 8 **35.** 6 **37.** 9 **39.** nonlinear **41.** linear; positive slope (b) Answers will vary. Using the points (4, 1.8) and (9, 2.6), the equation is y = 0.16x + 1.16.

43. (a)

45. (a) (b) Answers will vary. Using the points (1.2, 8.4) and (4.1, 2.4), the equation is y = −2.07x + 10.88. (c)

47. (a) 3 (b) (0, 2) (c) $-\dfrac{2}{3}$ (d) 1; (1, 5) (e) {x | x ≤ −1}; (−∞, −1] (f)

49. (a) {3}; −2; (3, −2); (3, −2) (b) {x | x > 3}; (3, ∞) (c)

51. f(x) = 2x + 2; f(−2) = −2 **53.** $h(x) = -\dfrac{7}{4}x + \dfrac{49}{4}; h\left(\dfrac{1}{2}\right) = \dfrac{91}{8}$ **55.** (a) 3 (b) −1 (c) 2 (d) x-intercept: (2, 0); y-intercept: (0, −2) (e) f(x) = x − 2

57. (a) {x | 9926 ≤ x ≤ 37,450}; [9926, 37,450] (b) $2538.60 (c) Independent variable: adjusted gross income; dependent variable: tax bill (d) (e) $23,051

59. (a) $\{m \mid m \geq 0\}$ or $[0, \infty)$

(b) 2; The base fare is $2.00 before any distance is driven.

(c) $9.50

(d)

(e) A person can travel 7.5 miles in a cab for $13.25.

(f) A person can ride between 0 miles and 25 miles.

61. (a) The independent variable is age; the dependent variable is insurance cost.

(b) $\{a \mid 15 \leq a \leq 90\}$ or $[15, 90]$

(c) $566.50

(d)

(e) 48 years

63. (a) $B(m) = 0.05m + 5.95$

(b) The independent variable is minutes; the dependent variable is bill amount.

(c) $\{m \mid m \geq 0\}$ or $[0, \infty)$ **(d)** $20.95

(e) 240 minutes

(f)

(g) You can talk between 0 minutes and 250 minutes.

65. (a) $V(x) = -900x + 2700$

(b) $\{x \mid 0 \leq x \leq 3\}$ or $[0, 3]$

(c) $1800

(d) The V-intercept is $(0, 2700)$, and the x-intercept is $(3, 0)$.

(e) After two years

(f)

67. (a) $C(x) = 7250x - 1735$

(b) $3847.50 **(c)** The cost of diamonds increases at a rate of $7250 per carat.

(d) 0.97 carat

71. (a)

(b) Linear **(c)** Answers will vary. Using the points $(2300, 4070)$ and $(3390, 5220)$, the equation is $y = 1.06x + 1632$.

69. (a) $C(x) = 0.774x + 1543.978$

(b) $12,551 billion

(c) If personal disposable income increase by $1, personal consumption increase by 0.77.

(d) $14.952 billion

(e) 4823 psi

(f) If the 7-day strength is increased by 1 psi, then the 28-day strength will increase by 1.06 psi.

73. (a) No **(b)**

(c) Answers will vary. Using the points $(42.3, 82)$ and $(42.8, 93)$, the equation is $N = 22w - 848.6$.

(d)

(e) $N(w) = 22w - 848.6$ **(f)** approximately 86 raisins

(g) If the weight increases by 1 gram, then the number of raisins increases by 22 raisins.

75. (a)

(b) 6 **(c)** $y = 6x - 7$

(d)

$x = 3$: slope $= 4$; $y = 4x - 5$; $x = 2$: slope $= 2$; $y = 2x - 3$; $x = 1.5$: slope $= 1$; $y = x - 2$; $x = 1.1$: slope $= 0.2$; $y = 0.2x - 1.2$ **(e)** As x approaches 1, the slope decreases and gets closer to 0.

77. (a)

(b) $y = 0.676x + 2675.562$

79. (a)

(b) $y = 11.449x - 399.123$

Section 2.5 Compound Inequalities
1. intersection **2.** and; or **3.** True **4.** False **5.** $\{1, 3, 5\}$ **6.** $\{2, 4, 6\}$ **7.** $\{1, 2, 3, 4, 5, 6, 7\}$

8. $\{1, 2, 3, 4, 5, 6, 8\}$ **9.** $\varnothing$ or $\{\ \}$ **10.** $\{1, 2, 3, 4, 5, 6, 7, 8\}$ **11.** $\{x \mid 2 < x < 7\}$; $(2, 7)$

12. $\{x \mid x \leq -3 \text{ or } x > 2\}$; $(-\infty, -3] \cup (2, \infty)$ **13.** $\{x \mid x \geq 2\}$; $[2, \infty)$

14. $\{x \mid -3 < x < 3\}$; $(-3, 3)$ **15.** $\{x \mid 1 < x < 3\}$; $(1, 3)$ **16.** $\{\ \}$ or $\varnothing$

17. $\{1\}$

18. $\{x|-1 < x < 3\}; (-1, 3)$

19. $\left\{x\left|\frac{5}{4} < x \le 2\right.\right\}; \left(\frac{5}{4}, 2\right]$

20. $\{x|-6 \le x \le -2\}; [-6, -2]$

21. $\{x|x < -2 \text{ or } x > 5\}; (-\infty, -2) \cup (5, \infty)$

22. $\{x|x \le 2 \text{ or } x > 6\}; (-\infty, 2] \cup (6, \infty)$

23. $\{x|x \ge 1\}; [1, \infty)$

24. $\{x|x < 4 \text{ or } x > 9\}; (-\infty, 4) \cup (9, \infty)$

25. $\{x|x \text{ is any real number}\}; (-\infty, \infty)$

26. $\{x|x \ge -1\}; [-1, \infty)$

27. Income between \$37,450 and \$90,750.

28. $[800, 1500]$. **29.** $\{1, 4, 5, 6, 7, 8, 9\}$ **31.** $\{5, 7, 9\}$ **33.** $\varnothing$ or $\{\ \}$

35. (a) $A \cap B = \{x|-2 < x \le 5\}; (-2, 5]$

(b) $A \cup B = \{x|x \text{ is any real number}\}; (-\infty, \infty)$

37. (a) $E \cap F = \varnothing$ or $\{\ \}$ **(b)** $E \cup F = \{x|x < -1 \text{ or } x > 3\}; (-\infty, -1) \cup (3, \infty)$

39. (a) $\{x|-2 \le x \le 2\}; [-2, 2]$ **(b)** $(-\infty, -2) \cup (2, \infty)$

41. (a) $\{x|-3 < x < 3\}; (-3, 3)$ **(b)** $(-\infty, -3] \cup [3, \infty)$

43. $\{x|-2 \le x < 3\}; [-2, 3)$ **45.** $\varnothing$ or $\{\ \}$ **47.** $\{x|x < -2\}; (-\infty, -2)$

49. $\{1\}$ **51.** $\{x|-1 \le x < 3\}; [-1, 3)$

53. $\left\{x\left|-\frac{2}{3} \le x \le \frac{3}{2}\right.\right\}; \left[-\frac{2}{3}, \frac{3}{2}\right]$ **55.** $\left\{x\left|-1 < x \le \frac{4}{5}\right.\right\}; \left(-1, \frac{4}{5}\right]$

57. $\{x|0 \le x \le 8\}; [0, 8]$ **59.** $\{x|-6 \le x \le -2\}; [-6, -2]$

61. $\varnothing$ or $\{\ \}$ **63.** $\left\{x\left|-\frac{5}{3} < x \le 5\right.\right\}; \left(-\frac{5}{3}, 5\right]$ **65.** $\{x|-4 < x \le 3\}; (-4, 3]$

67. $\{x|x < -2 \text{ or } x > 3\}; (-\infty, -2) \cup (3, \infty)$

69. $\{x|x < -2 \text{ or } x > 5\}; (-\infty, -2) \cup (5, \infty)$

71. $\{x|x \text{ is any real number}\}; (-\infty, \infty)$

73. $\{x|x < -1 \text{ or } x > 4\}; (-\infty, -1) \cup (4, \infty)$

75. $\{x|x \le -3 \text{ or } x > 6\}; (-\infty, -3] \cup (6, \infty)$

77. $\{x|x \text{ is any real number}\}; (-\infty, \infty)$

79. $\left\{x\left|x < 0 \text{ or } x > \frac{5}{2}\right.\right\}; (-\infty, 0) \cup \left(\frac{5}{2}, \infty\right)$ **81.** $\{a|-3 \le a < 0\}; [-3, 0)$

83. $\{x|x \text{ is any real number}\}; (-\infty, \infty)$ **85.** $\left\{x\left|-2 \le x \le \frac{8}{3}\right.\right\}; \left[-2, \frac{8}{3}\right]$

87. $\{x|x < -10 \text{ or } x > 2\}; (-\infty, -10) \cup (2, \infty)$ **89.** $\{x|2 < x < 5\}; (2, 5)$

91. $\left\{x\left|x \le -3 \text{ or } x > \frac{15}{4}\right.\right\}; (-\infty, -3] \cup \left(\frac{15}{4}, \infty\right)$

93. $\left\{x\left|-5 < x \le \frac{1}{2}\right.\right\}; \left(-5, \frac{1}{2}\right]$ **95.** $a = 1$ and $b = 8$ **97.** $a = 12$ and $b = 30$ **99.** $a = -1$ and $b = 23$

101. $90 < x < 140$ **103.** Joanna needs to score at least a 77 on the final. That is, $77 \le x \le 100$ (assuming 100 is the max score, otherwise $77 \le x \le 104$).

105. The amount withheld ranges between \$89.50 and \$104.50, inclusive. **107.** Total sales between \$50,000 and \$250,000

109. The electrical usage ranged from 850 kwh and 1370 kwh.

111. Step 1:

$$a < b$$
$$a + a < a + b$$
$$2a < a + b$$
$$\frac{2a}{2} < \frac{a+b}{2}$$
$$a < \frac{a+b}{2}$$

Step 2:

$$a < b$$
$$a + b < b + b$$
$$a + b < 2b$$
$$\frac{a+b}{2} < \frac{2b}{2}$$
$$\frac{a+b}{2} < b$$

Step 3:

Since $a < \dfrac{a+b}{2}$ and $\dfrac{a+b}{2} < b$, it follows that $a < \dfrac{a+b}{2} < b$.

113. { } or ∅ **115.** This is a contradiction. There is no solution. If, during simplification, the variable terms all cancel out and a contradiction results, then there is no solution to the inequality. **117.** If $x < 2$ then $x - 2 < 2 - 2 \Rightarrow x - 2 < 0$. When multiplying both sides of the inequality by $(x - 2)$ in the second step, the direction of the inequality must switch.

Section 2.6 Absolute Value Equations and Inequalities **1.** $\{-7, 7\}$ **2.** $\{-1, 1\}$ **3.** $a; -a$ **4.** $2x + 3 = -5$ **5.** $\{-2, 5\}$ **6.** $\left\{-\frac{5}{3}, 3\right\}$

7. $\left\{-1, \frac{9}{5}\right\}$ **8.** $\{-5, 1\}$ **9.** True **10.** ∅ or { } **11.** ∅ or { } **12.** $\{-1\}$ **13.** $u; v; u; -v$ **14.** $\left\{-8, -\frac{2}{3}\right\}$ **15.** $\{-2, 3\}$ **16.** $\{-3, 0\}$

17. $\{2\}$ **18.** $-a < u < a$ **19.** $-10; 10$ **20.** $\{x | -5 \le x \le 5\}; [-5, 5]$

21. $\left\{x \middle| -\frac{3}{2} < x < \frac{3}{2}\right\}; \left(-\frac{3}{2}, \frac{3}{2}\right)$ **22.** $\{x | -8 < x < 2\}; (-8, 2)$

23. $\{x | -2 \le x \le 5\}; [-2, 5]$ **24.** ∅ or { }

25. False **26.** $\{x | -2 < x < 2\}; (-2, 2)$ **27.** $\{x | -1 \le x \le 7\}; [-1, 7]$

28. $\{x | -2 \le x \le 1\}; [-2, 1]$ **29.** $\left\{x \middle| -\frac{7}{3} < x < 3\right\}; \left(-\frac{7}{3}, 3\right)$

30. $u < -a; u > a$ **31.** $-7; 7$ **32.** $\{x | x \le -6 \text{ or } x \ge 6\}; (-\infty, -6] \cup [6, \infty)$

33. $\left\{x \middle| x < -\frac{5}{2} \text{ or } x > \frac{5}{2}\right\}; \left(-\infty, -\frac{5}{2}\right) \cup \left(\frac{5}{2}, \infty\right)$ **34.** $<$

35. $\{x | x < -7 \text{ or } x > 1\}; (-\infty, -7) \cup (1, \infty)$

36. $\left\{x \middle| x \le -\frac{1}{2} \text{ or } x \ge 2\right\}; \left(-\infty, -\frac{1}{2}\right] \cup [2, \infty)$

37. $\left\{x \middle| x < -\frac{5}{3} \text{ or } x > 3\right\}; \left(-\infty, -\frac{5}{3}\right) \cup (3, \infty)$

38. $\left\{x \middle| x \ne -\frac{5}{2}\right\}; \left(-\infty, -\frac{5}{2}\right) \cup \left(-\frac{5}{2}, \infty\right)$

39. $\{x | x \text{ is any real number}\}; (-\infty, \infty)$

40. $\{x | x \text{ is any real number}\}; (-\infty, \infty)$ **41.** The acceptable belt width is between $\frac{127}{32}$ inches and $\frac{129}{32}$ inches.

42. The percentage of Americans that have been shot at is between 7.3% and 10.7%, inclusive. **43.** $\{-10, 10\}$ **45.** ∅ or { } **47.** $\{-1, 7\}$ **49.** $\left\{-1, \frac{13}{3}\right\}$

51. $\{-5, 5\}$ **53.** $\left\{-\frac{11}{2}, \frac{5}{2}\right\}$ **55.** $\{-4, 10\}$ **57.** $\{0\}$ **59.** $\left\{-\frac{7}{3}, 3\right\}$ **61.** $\left\{-7, \frac{3}{5}\right\}$ **63.** $\{1, 3\}$ **65.** $\{2\}$ **67.** ∅ or { }

69. $\{x | -9 < x < 9\}; (-9, 9)$ **71.** $\{x | -3 \le x \le 11\}; [-3, 11]$

73. $\left\{x \middle| -3 < x < \frac{7}{3}\right\}; \left(-3, \frac{7}{3}\right)$ **75.** ∅ or { }

77. $\{x | 0 < x < 6\}; (0, 6)$ **79.** $\left\{x \middle| -1 < x < \frac{9}{5}\right\}; \left(-1, \frac{9}{5}\right)$

81. $\{x | 1.995 < x < 2.005\}; (1.995, 2.005)$

83. $\{y | y < 3 \text{ or } y > 7\}; (-\infty, 3) \cup (7, \infty)$

85. $\left\{x \middle| x \le -2 \text{ or } x \ge \frac{1}{2}\right\}; (-\infty, -2] \cup \left[\frac{1}{2}, \infty\right)$

87. $\{y | y \text{ is any real number}\}; (-\infty, \infty)$

89. $\left\{x \middle| x < -2 \text{ or } x > \frac{4}{5}\right\}; (-\infty, -2) \cup \left(\frac{4}{5}, \infty\right)$

91. $\{x | x < 0 \text{ or } x > 1\}; (-\infty, 0) \cup (1, \infty)$

93. $\{x | x \le -2 \text{ or } x \ge 3\}; (-\infty, -2] \cup [3, \infty)$

95. **(a)** $\{-5, 5\}$ **(b)** $\{x | -5 \le x \le 5\}; [-5, 5]$ **(c)** $\{x | x < -5 \text{ or } x > 5\}; (-\infty, -5) \cup (5, \infty)$

97. **(a)** $\{-5, 1\}$ **(b)** $\{x | -5 < x < 1\}; (-5, 1)$ **(c)** $\{x | x \le -5 \text{ or } x \ge 1\}; (-\infty, -5] \cup [1, \infty)$

99. $\{x | x < -5 \text{ or } x > 5\}; (-\infty, -5) \cup (5, \infty)$

101. $\{-4, -1\}$ **103.** $\{-5, 5\}$ **105.** $\{-4, 3\}$ **107.** ∅ or { } **109.** ∅ or { }

111. $\left\{x \middle| x \le -\frac{7}{3} \text{ or } x \ge 1\right\}; \left(-\infty, -\frac{7}{3}\right] \cup [1, \infty)$

113. $\left\{x \,\middle|\, x < 0 \text{ or } x > \dfrac{4}{3}\right\}$; $(-\infty, 0) \cup \left(\dfrac{4}{3}, \infty\right)$ **115.** $\{-1, 1\}$ **117.** $\varnothing$ or $\{\ \}$ **119.** $\left\{-8, \dfrac{4}{7}\right\}$

121. $|x - 5| < 3$; $\{x \mid 2 < x < 8\}$; $(2, 8)$ **123.** $|2x - (-6)| > 3$; $\left\{x \,\middle|\, x < -\dfrac{9}{2} \text{ or } x > -\dfrac{3}{2}\right\}$; $\left(-\infty, -\dfrac{9}{2}\right) \cup \left(-\dfrac{3}{2}, \infty\right)$

125. The acceptable rod lengths are between 5.6995 inches and 5.7005 inches, inclusive. **127.** An unusual IQ score would be less than 70.6 or greater

than 129.4. **129.** $\left\{-\dfrac{5}{2}\right\}$ **131.** $\{2\}$ **133.** $\varnothing$ or $\{\ \}$ **135.** $\{x \mid x \le -5\}$; $(-\infty, -5]$ **137.** The absolute value, when isolated, is equal to a negative number,

which is not possible. **139.** The absolute value, when isolated, is less than -3. Since absolute values are always nonnegative, this is not possible.

Chapter 2 Review

1. {(Penny, 2.500), (Nickel, 5.000), (Dime, 2.268), (Quarter, 5.670), (Half Dollar, 11.340), (Dollar, 8.100)};
Domain: { Penny, Nickel, Dime, Quarter, Half Dollar, Dollar }; Range: { 2.268, 2.500, 5.000, 5.670, 8.100, 11.340 }

2. { (16, $12.99), (28, $14.99) (30, $14.99) (59, $24.99) (85, $29.99) }; Domain: {16, 28, 30, 59, 85}; Range: {$12.99, $14.99, $24.99, $29.99}

3. Domain: $\{-4, -2, 2, 3, 6\}$; Range: $\{-9, -1, 5, 7, 8\}$

4. Domain: $\{-2, 1, 3, 5\}$; Range: $\{1, 4, 7, 8\}$

5. Domain: $\{x \mid x \text{ is a real number}\}$ or $(-\infty, \infty)$; Range: $\{y \mid y \text{ is a real number}\}$ or $(-\infty, \infty)$

6. Domain: $\{x \mid -6 \le x \le 4\}$ or $[-6, 4]$; Range: $\{y \mid -4 \le y \le 6\}$ or $[-4, 6]$

7. Domain: $\{2\}$; Range: $\{y \mid y \text{ is a real number}\}$ or $(-\infty, \infty)$

8. Domain: $\{x \mid x \ge -1\}$ or $[-1, \infty)$; Range: $\{y \mid y \ge -2\}$ or $[-2, \infty)$

9. $y = x + 2$
Domain: $\{x \mid x \text{ is a real number}\}$ or $(-\infty, \infty)$; Range: $\{y \mid y \text{ is a real number}\}$ or $(-\infty, \infty)$

10. $2x + y = 3$
Domain: $\{x \mid x \text{ is a real number}\}$ or $(-\infty, \infty)$; Range: $\{y \mid y \text{ is a real number}\}$ or $(-\infty, \infty)$

11. $y = -x^2 + 4$
Domain: $\{x \mid x \text{ is a real number}\}$ or $(-\infty, \infty)$; Range: $\{y \mid y \le 4\}$ or $(-\infty, 4]$

12. $y = |x + 2| - 1$
Domain: $\{x \mid x \text{ is a real number}\}$ or $(-\infty, \infty)$; Range: $\{y \mid y \ge -1\}$ or $[-1, \infty)$

13. $y = x^3 + 2$
Domain: $\{x \mid x \text{ is a real number}\}$ or $(-\infty, \infty)$; Range: $\{y \mid y \text{ is a real number}\}$ or $(-\infty, \infty)$

14. $x = y^2 + 1$
Domain: $\{x \mid x \ge 1\}$ or $[1, \infty)$; Range: $\{y \mid y \text{ is a real number}\}$ or $(-\infty, \infty)$

15. (a) Domain: $\{x \mid 0 \le x \le 44{,}640\}$ or $[0, 44{,}640]$; Range: $\{y \mid 40 \le y \le 2122\}$ or $[40, 2122]$ **(b)** Answers may vary. **16.** Domain: $\{t \mid 0 \le t \le 4\}$ or $[0, 4]$; Range: $\{y \mid 0 \le y \le 121\}$ or $[0, 121]$ **17. (a)** Not a function. Domain: $\{-1, 5, 7, 9\}$; Range: $\{-2, 0, 2, 3, 4\}$ **(b)** Function. Domain: { Camel, Macaw, Deer, Fox, Tiger, Crocodile }; Range: { 14, 22, 35, 45, 50 } **18. (a)** Function; Domain: $\{-3, -2, 2, 4, 5\}$; Range: $\{-1, 3, 4, 7\}$ **(b)** Not a function; Domain: { Red, Blue, Green, Black }; Range: { Camry, Taurus, Windstar, Durango } **19.** Function **20.** Not a function **21.** Not a function **22.** Function **23.** Not a function **24.** Function **25.** Function **26.** Not a function **27. (a)** $f(-2) = -5$ **(b)** $f(3) = 10$

28. (a) $g(0) = -\dfrac{1}{3}$ **(b)** $g(2) = -5$ **29. (a)** $F(5) = -3$ **(b)** $F(-x) = 2x + 7$ **30. (a)** $G(7) = 15$ **(b)** $G(x + h) = 2x + 2h + 1$

31. $\{x \mid x \text{ is a real number}\}$ or $(-\infty, \infty)$ **32.** $\left\{w \,\middle|\, w \ne -\dfrac{5}{2}\right\}$ **33.** $\{t \mid t \ne 5\}$ **34.** $\{t \mid t \text{ is a real number}\}$ or $(-\infty, \infty)$ **35. (a)** The dependent variable is the population, P, and the independent variable is the number of years after 1900, t. **(b)** $P(120) = 1400.167$; The population of Orange County will be roughly 1,400,167 in 2020. **(c)** $P(-70) = 1066.337$; The population of Orange County was roughly 1,066,337 in 1830. This is not reasonable. (The population of the entire Florida territory was roughly 35,000 in 1830.) **36. (a)** The dependent variable is percent of the population with an advanced degree, P, and the independent variable is age, a. **(b)** $P(30) = 7.9$; According to the model, 7.9% of 30-year-olds have an advanced degree.

37. $f(x) = 2x - 5$ **38.** $g(x) = x^2 - 3x + 2$ **39.** $h(x) = (x - 1)^3 - 3$ **40.** $f(x) = |x + 1| - 4$

41. (a) Domain: $\{x \mid x \text{ is a real number}\}$ or $(-\infty, \infty)$; Range: $\{y \mid y \text{ is a real number}\}$ or $(-\infty, \infty)$ **(b)** $(0, 2)$ and $(4, 0)$

42. (a) Domain: $\{x \mid x \text{ is a real number}\}$ or $(-\infty, \infty)$; Range: $\{y \mid y \ge -3\}$ or $[-3, \infty)$ **(b)** $(-2, 0), (2, 0), (0, -3)$

43. (a) Domain: $\{x \mid x \text{ is a real number}\}$ or $(-\infty, \infty)$; Range: $\{y \mid y \text{ is a real number}\}$ or $(-\infty, \infty)$ **(b)** $(0,0)$ and $(2,0)$

44. (a) Domain: $\{x \mid x \geq -3\}$ or $[-3, \infty)$; Range: $\{y \mid y \geq 1\}$ or $[1, \infty)$ **(b)** $(0,3)$ **45. (a)** 4 **(b)** 1 **(c)** -1 and 3

46. (a) **(b)**
47. (a) yes
(b) $h(-2) = -11$; $(-2, -11)$
(c) $x = \dfrac{11}{2}$; $\left(\dfrac{11}{2}, 4\right)$

48. (a) no
(b) $g(3) = \dfrac{29}{5}$; $\left(3, \dfrac{29}{5}\right)$
(c) $x = -10$; $(-10, -2)$

49.

50. **51.** **52.** **53.** **54.**

55. (a) $[0, \infty)$
(b) $260; This is the monthly payment for the car.
(c) $460
(d)
(e) $[0, 1450]$

56. (a) The independent variable is years; the dependent variable is value.
(b) $\{x \mid 0 \leq x \leq 5\}$ or $[0, 5]$
(c) $1800 **(d)** $1080
(e)
(f) After 5 years

57. (a) $L(x) = -\dfrac{4}{75}x + 45$ **(b)** 7% **(c)** The auto loan rate decreases approximately by 0.05% for every 1-unit increase in FICO score.
(d) 722

58. (a) $H(x) = -x + 220$ **(b)** 175 beats per minute **(c)** The maximum recommended heart rate for men under stress decreases at a rate of 1 beat per minute per year.
(d) 52 years

59. (a) $C(m) = 0.12m + 35$
(b) The independent variable is m; the dependent variable is C.
(c) $\{m \mid m \geq 0\}$ or $[0, \infty)$
(d) $49.88 **(e)** 268 miles were driven
(f)

60. (a) $B(x) = 3.50x + 33.99$
(b) The independent variable is x; the dependent variable is B.
(c) $\{x \mid x \geq 0\}$ or $[0, \infty)$
(d) $51.49
(e) 7 pay-per-view movies
(f)

61. (a)
(b) Answers will vary. Using the points $(2, 13.3)$ and $(14, 4.6)$, the equation is $y = -0.725x + 14.75$.
(c)
(d)

62. (a)
(b) Answers will vary. Using the points $(0, 0.6)$ and $(4.2, 3.0)$, the equation is $y = \dfrac{4}{7}x + 0.6$.
(c)
(e) 30.2 grams
(f) In a one-cup serving of cereal, total carbohydrates will increase by 0.158 gram for each 1-calorie increase.

63. (a)
(b) Approximately linear
(c) Answers will vary. Using the points $(96, 23.2)$ and $(160, 33.3)$, the equation is $y = 0.158x + 8.032$.

64. (a)

(b) Linear with positive slope
(c) Answers will vary. Using (3, 12.2) and (9, 16.9), $y = 0.78x + 9.85$

(d)

(e) $13.75
(f) If the weight increases by 1 pound, the shipping charge increases $0.78.

65. $\{-1, 0, 1, 2, 3, 4, 6, 8\}$ **66.** $\{2, 4\}$ **67.** $\{1, 2, 3, 4\}$ **68.** $\{1, 2, 3, 4, 6, 8\}$ **69. (a)** $\{x | 2 < x \le 4\}; (2, 4]$
(b) $\{x | x \text{ is any real number}\}; (-\infty, \infty)$ **70. (a)** $\{\ \}$ or $\varnothing$
(b) $\{x | x < -2 \text{ or } x \ge 3\}; (-\infty, -2) \cup [3, \infty)$

71. $\{x | -1 < x < 4\}; (-1, 4)$ Graph:

72. $\{x | -5 < x < -1\}; (-5, -1)$ Graph:

73. $\{x | x < -2 \text{ or } x > 2\}; (-\infty, -2) \cup (2, \infty)$ Graph:

74. $\{x | x \le 0 \text{ or } x \ge 4\}; (-\infty, 0] \cup [4, \infty)$ Graph:

75. $\{\ \}$ or $\varnothing$ **76.** $\{x | -2 \le x < 4\}; [-2, 4)$ Graph:

77. $\{x | x \le -2 \text{ or } x > 3\}; (-\infty, -2] \cup (3, \infty)$ Graph:

78. $\{x | x \text{ is any real number}\}; (-\infty, \infty)$ Graph:

79. $\{x | x < -10 \text{ or } x > 6\}; (-\infty, -10) \cup (6, \infty)$ Graph:

80. $\left\{x \middle| -\dfrac{3}{2} \le x < \dfrac{5}{8}\right\}; \left[-\dfrac{3}{2}, \dfrac{5}{8}\right)$ Graph: **81.** $70 \le x \le 75$

82. The electric usage varied from roughly 1033.8 kilowatt hours up to roughly 1952.5 kilowatt hours. (recall, x is the number *above* 800).

83. $\{-4, 4\}$ **84.** $\left\{\dfrac{1}{3}, 3\right\}$ **85.** $\{-5, 13\}$ **86.** $\{-3, -1\}$ **87.** $\{\ \}$ or $\varnothing$ **88.** $\left\{-\dfrac{1}{2}, 2\right\}$

89. $\{x | -2 < x < 2\}; (-2, 2)$

90. $\left\{x \middle| x \le -\dfrac{7}{2} \text{ or } x \ge \dfrac{7}{2}\right\}; \left(-\infty, -\dfrac{7}{2}\right] \cup \left[\dfrac{7}{2}, \infty\right)$

91. $\{x | -5 \le x \le 1\}; [-5, 1]$ **92.** $\left\{x \middle| x \le \dfrac{1}{2} \text{ or } x \ge 1\right\}; \left(-\infty, \dfrac{1}{2}\right] \cup [1, \infty)$

93. $\{x | x \text{ is a real number}\}; (-\infty, \infty)$ **94.** $\{\ \}$ or $\varnothing$

95. $\{x | 4.99 \le x \le 5.01\}; [4.99, 5.01]$

96. $\left\{x \middle| x < -\dfrac{1}{2} \text{ or } x > \dfrac{7}{2}\right\}; \left(-\infty, -\dfrac{1}{2}\right) \cup \left(\dfrac{7}{2}, \infty\right)$

97. The acceptable diameters of the bearing are between 0.502 inch and 0.504 inch, inclusive. **98.** Tensile strengths below 36.08 lb/in.^{2} or above 43.92 lb/in.^{2} would be considered unusual.

Chapter 2 Test

1. Domain: $\{-4, 2, 5, 7\}$
Range: $\{-7, -2, -1, 3, 8, 12\}$

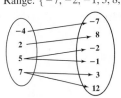

2. Domain:
$\left\{x \middle| -\dfrac{5\pi}{2} \le x \le \dfrac{5\pi}{2}\right\}$ or $\left[-\dfrac{5\pi}{2}, \dfrac{5\pi}{2}\right]$
Range: $\{y | 1 \le y \le 5\}$ or $[1, 5]$

3. $y = x^2 - 3$

Domain: $\{x | x \text{ is a real number}\}$ or $(-\infty, \infty)$; Range: $\{y | y \ge -3\}$ or $[-3, \infty)$

4. Function; Domain: $\{-5, -3, 0, 2\}$; Range: $\{3, 7\}$ **5.** Not a function; Domain: $\{x | x \le 3\}$ or $(-\infty, 3]$; Range: $\{y | y \text{ is a real number}\}$ or $(-\infty, \infty)$
6. No **7.** $f(x + h) = -3x - 3h + 11$ **8. (a)** $g(-2) = 5$ **(b)** $g(0) = -1$ **(c)** $g(3) = 20$

9. $f(x) = x^2 + 3$

10. (a) The dependent variable is the ticket price, P, and the independent variable is the number of years after 1996, x.
(b) $P(25) = 9.94$; According to the model, the average ticket price in 2021 ($x = 25$) will be \$9.94.   **(c)** 2030

11. $\{x \mid x \neq -2\}$ **12.   (a)** yes   **(b)** $h(3) = -3; (3, -3)$ **(c)** $x = -3; (-3, 27)$ **(d)** $\dfrac{12}{5}$

13. (a) The car stops accelerating when the speed stops increasing. Thus, the car stops accelerating after 6 seconds.
(b) The car has a constant speed when the graph is horizontal. Thus, the car maintains a constant speed for 18 seconds.

14. (a) $P(x) = 18x - 100$
(b) $\{x \mid x \geq 0\}$ or $[0, \infty)$
(c) \$512
(d)

Number of Shelves

(e) 48 shelves

15. (a)

(b) Approximately linear.
(c) Answers will vary. Using the points $(6, 95)$

and $(18, 170)$, the equation is $y = 6.25x + 57.5$.
(d)

(e) 113.75 kilograms
(f) A Shetland pony's weight will increase by 6.25 kilograms for each one-month increase in age.

16. $\{-4, -1\}$ **17.** $\{x \mid -2 \leq x < 6\}; [-2, 6)$ ⊢───────⊣

18. $\left\{x \mid x < -\dfrac{3}{2} \text{ or } x > 4\right\}; \left(-\infty, -\dfrac{3}{2}\right) \cup (4, \infty)$ ◄───────►

19. $\{x \mid 2 < x < 8\}; (2, 8)$ ⊢───────⊣

20. $\{x \mid x \leq -2 \text{ or } x \geq 3\}; (-\infty, -2] \cup [3, \infty)$ ◄───────►

Chapter 3 Systems of Linear Equations and Inequalities

Section 3.1 Systems of Linear Equations in Two Variables
1. system of linear equations   **2.** solution   **3. (a)** no   **(b)** yes   **(c)** no
4. inconsistent   **5.** consistent; dependent   **6.** False   **7.** True   **8.** True   **9.** $(3, 1)$ **10.** $(-2, 3)$ **11.** $(-4, 7)$ **12.** $(-6, 10)$ **13.** additive inverses   **14.** $\left(\dfrac{1}{2}, \dfrac{7}{3}\right)$
15. $(-5, -6)$ **16.** $(3, 6)$
17. $\varnothing$ or $\{\ \}$ **18.** $\varnothing$ or $\{\ \}$ **19.** dependent   **20.** $\{(x, y) \mid -3x + 2y = 8\}$ **21.** $(-5, 2)$ **22.** $(12, 15)$

23. (a) no   **(b)** yes   **25. (a)** yes   **(b)** yes   **27.** consistent; independent   **29.** inconsistent

31. $(1, 3)$ **33.** $(3, -4)$ **35.** $(6, -2)$ **37.** $(-2, -3)$ **39.** $\left(\dfrac{1}{2}, -\dfrac{1}{4}\right)$ **41.** $(2500, 7500)$ **51.** $\varnothing$ or $\{\ \}$

43. $(-8, 3)$ **45.** $(11, -8)$ **47.** $\left(\dfrac{1}{2}, -\dfrac{4}{5}\right)$ **49.** $(25, 40)$

53. $\varnothing$ or $\{\ \}$ **55.** $\{(x, y) \mid 2x - 4y = -4\}$ **57.** $\{(x, y) \mid x + 3y = 6\}$ **59.** $\left\{(x, y) \mid \dfrac{1}{3}x - 2y = 6\right\}$ **61.** $(-9, 3)$ **63.** $\{(x, y) \mid x = 5y - 3\}$
65. $\left(\dfrac{39}{11}, -\dfrac{30}{11}\right)$ **67.** $\varnothing$ or $\{\ \}$
69. $y = -2x - 5; y = -\dfrac{5}{3}x + \dfrac{1}{3}$; exactly one solution

71. $y = \dfrac{3}{2}x + 1; y = \dfrac{3}{2}x + 1$; infinite number of solutions

73. (a) $y = -\dfrac{1}{2}x + \dfrac{5}{2}; y = x + 1$ **(b)** $(1, 2)$ **75. (c)** and **(f)**   **77.** $A = \dfrac{7}{6}$ and $B = -\dfrac{1}{2}$ **79.** Answers will vary. One possibility follows: $\begin{cases} x + y = 3 \\ x - y = -5 \end{cases}$

81. $(1, 2)$ **83.** $(6, -1)$ **85.** Yes, typically, we use substitution when one of the equations is solved for one of the variables or when the coefficient on one of the variables is 1 (which makes it easy to solve for that variable). Otherwise, we use elimination.
87. $(3, -2)$

89. $(1.2, 2.6)$
91. $(4, 13)$
93. $\{(x, y) \mid 4x - 3y = 1\}$
95. approximately $(0.35, -3.76)$

Section 3.2 Problem Solving: Systems of Two Linear Equations Containing Two Unknowns

1. 15 and -7 **2.** Cheeseburger: \$4.69: Shake: \$2.89   **3.** Length: 120 yards; Width: 60 yards   **4.** $x = 15$, $y = 30$

5. \$40,000 in Aaa-rated bonds: \$80,000 in B-rated bonds   **6.** 10 pounds of cashews; 20 pounds of peanuts

7. Airspeed: 350 mph; Wind resistance: 50 mph

8. (a) $R(x) = 230x$ **(b)** $C(x) = 160x + 2100$

9. 8 and 10   **11.** The first number is 12 and the second number is 23.

(c)

13. \$60; \$35   **15.** 40 meters by 20 meters   **17.** $x = 4$ and $y = 14$

19. Invest \$22,500 in stocks and \$13,500 in bonds.

21. 60 pounds of Arabica beans and 40 pounds of African Robusta beans

23. Jonathon and Samantha can paddle 8 miles per hour in still water.

25. The Lincoln traveled 400 miles and the Infiniti traveled 500 miles. The time for both trips was 10 hours.

(d) 30 trees; \$6900

27. (a)

(b) $(1520, 18,240)$

29. 15 cm, 15 cm, and 5 cm   **31.** \$250,000

33. 12 and 19

35. Drenell's normal walking speed is 4.5 feet per second.

37. $x = 4$ and $y = 8$

39. Mix 50 mg of Liquid A with 60 mg of Liquid B.

41. An order or chicken nuggets has 280 calories and a medium Coke has 210 calories.

43. Mix 60 g of the 10% alloy with 40 g of the 25% alloy.

45. (a)

(b) \$3; 7000 hot dogs

47. (a)

(b) 2050; 29.47%

49. (a) Let x represent the amount of sales, let A represent the pay from option A, and let B represent the pay from option B.

$$\begin{cases} A(x) = 15,000 + 0.01x \\ B(x) = 25,000 + 0.0075x \end{cases}$$

(b)

(c) Annual sales of \$4,000,000 result in a salary of \$55,000 for both options.

51. (a) $R(x) = 60x$ **(b)** $C(x) = 35x + 3500$

(c)

(d) 140 desks: The cost and revenue will both be \$8400.

53. (a) and **(b)** Let x represent the number of years since 1968.

(c) Answers may vary. Using the points $(4, 20.00)$ and $(36, 19.79)$, the equation is $y = -0.0065625x + 20.02625$. See graph in part (d).

(d) Answers may vary. Using the points $(0, 22.50)$ and $(32, 21.84)$, the equation is $y = -0.020625x + 22.50$.

(e) In the year 2144: approximately 18.87 seconds

(f) Answers will vary.

(g) Answers will vary.

Section 3.3 Systems of Linear Equations in Three Variables

1. inconsistent; consistent; dependent   **2.** solution   **3.** False   **4.** True

5. (a) no   **(b)** yes   **6.** $(-3, 1, -1)$ **7.** $\left(-\dfrac{5}{2}, \dfrac{5}{4}, \dfrac{1}{2}\right)$ **8.** $\varnothing$ or { }; the system is inconsistent   **9.** $\{ (x, y, z) \mid x = -z + 1, y = 2z - 1, z$ is any real number}   **10.** The company can manufacture fourteen 21-inch, eleven 24-inch, and five 40-inch   **11. (a)** no   **(b)** yes   **13.** $(3, -2, 4)$

15. $(3, -4, -2)$ **17.** $\left(\dfrac{1}{3}, -\dfrac{2}{3}, 2\right)$ **19.** $(3, -3, 1)$ **21.** $\varnothing$ or { }   **23.** $\{ (x, y, z) \mid x = 3z - 10, y = -4z + 14, z$ is any real number}

25. $\{ (x, y, z) \mid x = -3z + 5, y = -6z + 11, z$ is any real number}   **27.** $\left(0, 2, \dfrac{3}{2}\right)$ **29.** $\varnothing$ or { }   **31.** $(-5, 0, 3)$ **33.** $\left(-\dfrac{7}{2}, -\dfrac{7}{2}, \dfrac{3}{2}\right)$

35. $\{ (x, y, z) \mid x = 5z - 5, y = -2z + 3, z$ is any real number }   **37.** $\left(\dfrac{11}{2}, 0, -\dfrac{1}{2}\right)$ **39.** $\left\{ (x, y, z) \,\middle|\, x = \dfrac{1}{3}z + \dfrac{2}{3}, y = -\dfrac{4}{3}z + \dfrac{7}{3}, z$ is any real number $\right\}$

41. Answers will vary. One possibility follows. $\begin{cases} x + y + z = 4 \\ x - y + z = 6 \\ x + y - z = -2 \end{cases}$ **43. (a)** $a - b + c = -6$; $4a + 2b + c = 3$ **(b)** $a = -2, b = 5, c = 1$; $f(x) = -2x^2 + 5x + 1$ **45.** $i_1 = 1, i_2 = 4,$ and $i_3 = 3$ **47.** There are 450 club seats, 3795 reserved seats, and 1755 lawn seats in the stadium.

49. Nancy needs 1 serving of Chex® cereal, 2 servings of 2% milk, and 1.5 servings of orange juice.   **51.** \$12,000 in Treasury bills, \$8000 in municipal bonds, and \$5000 in corporate bonds.   **53.** $\overline{AM} = 4, \overline{BN} = 2,$ and $\overline{OC} = 10$ **55.** $(10, -4, 6)$ **57.** $(-2, 1, 0, 4)$

59. To create a system of two equations and two unknowns, something we already know how to solve.

Putting the Concepts Together (Section 3.1–3.3) **1.** $(-3, 5)$

2. $(3, -2)$

3. $(5, -1)$ **4.** $\left(\frac{4}{5}, -\frac{7}{5}\right)$ **5.** $(-10, 6)$ **6.** $\{(x, y) \mid 8x - 4y = 12\}$ **7.** $(-1, -3, 5)$ **8.** $\varnothing$ or $\{\ \}$

9. 450 adult tickets and 375 youth tickets **10.** 290 orchestra seats, 380 mezzanine seats, and 530 balcony seats

Section 3.4 Using Matrices to Solve Systems

1. matrix **2.** augmented **3.** 4; 3 **4.** False **5.** $\begin{bmatrix} 3 & -1 & | & -10 \\ -5 & 2 & | & 0 \end{bmatrix}$

6. $\begin{bmatrix} 1 & 2 & -2 & | & 11 \\ -1 & 0 & -2 & | & 4 \\ 4 & -1 & 1 & | & 3 \end{bmatrix}$ **7.** $\begin{cases} x - 3y = 7 \\ -2x + 5y = -3 \end{cases}$ **8.** $\begin{cases} x - 3y + 2z = 4 \\ 3x - z = -1 \\ -x + 4y = 0 \end{cases}$ **9.** $\begin{bmatrix} 1 & -2 & | & 5 \\ 0 & -3 & | & 9 \end{bmatrix}$ **10.** $R_1 = -5r_2 + r_1;$ $\begin{bmatrix} 1 & 0 & | & 3 \\ 0 & 1 & | & 2 \end{bmatrix}$

11. True **12.** $(6, -2)$ **13.** $(3, -2, 1)$ **14.** $\left(\frac{7}{2}, \frac{2}{3}, 4\right)$ **15.** $\{(x, y) \mid 2x + 5y = -6\}$

16. $\{(x, y, z) \mid x = -z + 5, y = 4z + 3, z \text{ is any real number}\}$ **17.** $\varnothing$ or $\{\ \}$ **18.** $\varnothing$ or $\{\ \}$ **19.** $\begin{bmatrix} 1 & -3 & | & 2 \\ 2 & 5 & | & 1 \end{bmatrix}$ **21.** $\begin{bmatrix} 1 & 1 & 1 & | & 3 \\ 2 & -1 & 3 & | & 1 \\ -4 & 2 & -5 & | & -3 \end{bmatrix}$

23. $\begin{bmatrix} -1 & 1 & | & 2 \\ 5 & 1 & | & -5 \end{bmatrix}$ **25.** $\begin{bmatrix} 1 & 0 & 1 & | & 2 \\ 2 & 1 & 0 & | & 13 \\ 1 & -1 & 4 & | & -4 \end{bmatrix}$ **27.** $\begin{cases} 2x + 5y = 3 \\ -4x + y = 10 \end{cases}$ **29.** $\begin{cases} x + 5y - 3z = 2 \\ 3y - z = -5 \\ 4x + 8z = 6 \end{cases}$ **31.** $\begin{cases} x - 2y + 9z = 2 \\ y - 5z = 8 \\ z = \frac{4}{3} \end{cases}$

33. (a) $\begin{bmatrix} 1 & -3 & | & 2 \\ 0 & -1 & | & 5 \end{bmatrix}$ **(b)** $\begin{bmatrix} 1 & -3 & | & 2 \\ 0 & 1 & | & -5 \end{bmatrix}$ **35. (a)** $\begin{bmatrix} 1 & 1 & -1 & | & 4 \\ 0 & 3 & 5 & | & -11 \\ -1 & -3 & 2 & | & 1 \end{bmatrix}$ **(b)** $\begin{bmatrix} 1 & 1 & -1 & | & 4 \\ 0 & 3 & 5 & | & -11 \\ 0 & -2 & 1 & | & 5 \end{bmatrix}$

37. (a) $\begin{bmatrix} 1 & 1 & 1 & | & 4 \\ 0 & 1 & 5 & | & 5 \\ 0 & -4 & 2 & | & 8 \end{bmatrix}$ **(b)** $\begin{bmatrix} 1 & 1 & 1 & | & 4 \\ 0 & 1 & 5 & | & 5 \\ 0 & -2 & 1 & | & 4 \end{bmatrix}$ **39.** $(8, -5)$ **41.** $(-2, 5, 0)$ **43.** $\left(\frac{29}{34}, -\frac{47}{85}, \frac{11}{34}\right)$ **45.** $\{(x, y) \mid x - 3y = 3\}$ **47.** $\varnothing$ or $\{\ \}$

49. $\left\{(x, y, z) \mid x = \frac{3}{4}z + 1, y = \frac{5}{4}z + 3, z \text{ is any real number}\right\}$ **51.** $\begin{cases} x + 4y = -5 & (1) \\ y = -2 & (2) \end{cases}$ consistent and independent; $(3, -2)$

53. $\begin{cases} x + 3y - 2z = 6 & (1) \\ y + 5z = -2 & (2) \\ 0 = 4 & (3) \end{cases}$ inconsistent; $\varnothing$ or $\{\ \}$ **55.** $\begin{cases} x - 2y - z = 3 & (1) \\ y - 2z = -8 & (2) \\ z = 5 & (3) \end{cases}$ consistent and independent; $(12, 2, 5)$

57. $(3, -5)$ **59.** $\varnothing$ or $\{\ \}$ **61.** $\left(\frac{1}{2}, -\frac{5}{4}\right)$ **63.** $\{(x, y) \mid 4x - y = 8\}$ **65.** $(3, -4, 1)$ **67.** $(4, 0, -5)$

69. $\{(x, y, z) \mid x = -2z - 0.2, y = -z - 1.4, z \text{ is any real number}\}$ **71.** $\varnothing$ or $\{\ \}$ **73.** $\left(\frac{3}{10}, \frac{1}{10}, -\frac{1}{2}\right)$ **75.** $\left(\frac{5}{3}, \frac{2}{5}, -\frac{1}{2}\right)$

77. $(-2, 1, -5)$ **79. (a)** $a + b + c = 0; 4a + 2b + c = 3$ **(b)** $a = 2, b = -3, c = 1; f(x) = 2x^2 - 3x + 1$

81. \$8000 in Treasury bills, \$7000 in municipal bonds, and \$5000 in corporate bonds **83.** $(-3, 7)$ **85.** $(2, 5, -4)$ **87.** Answers will vary.

89. Multiply each entry in row 2 by $\frac{1}{5}$: $R_2 = \frac{1}{5}r_2$. **91.** $(5, -3)$ **93.** $(4, -2, 7)$

Section 3.5 Determinants and Cramer's Rule

1. $ad - bc$ **2.** square **3.** 18 **4.** -9 **5.** $(-3, 5)$ **6.** Cramer's Rule does not apply because $D = 0$.

7. -91 **8.** 20 **9.** $\left(\frac{1}{2}, -2, -1\right)$ **10.** $(-1, 4, 1)$ **11.** 10 **13.** -2 **15.** $(-8, 4)$ **17.** $(3, -1)$ **19.** $\left(-\frac{1}{6}, \frac{3}{8}\right)$ **21.** $\left(\frac{3}{4}, -\frac{7}{4}\right)$ **23.** -9 **25.** -163 **27.** 0

29. $(-2, 1, -1)$ **31.** $(3, -1, 2)$ **33.** Cramer's Rule does not apply. **35.** Cramer's Rule does not apply. **37.** $(12, -6, 3)$

39. $\left(\frac{7}{5}, -\frac{5}{3}, \frac{11}{3}\right)$ **41.** $(1, 1, 1)$ **43.** $x = 5$ **45.** $x = -2$

47. (a)

49. (a)

(b)

51. (a) $x + 2y = 7$ **(b)** $x + 2y = 7$
53. 14: -14: answers may vary.
55. $(-8, 4)$
57. $(3, -1)$
59. $(-2, 1, -1)$

(b) The area of triangle ABC is 10 square units.

The area of triangle ABC is 4.5 square units.

(c) The area of triangle ADC is 4.5 square units.

(d) 9 square units

Section 3.6 Systems of Linear Inequalities   **1.** satisfies   **2. (a)** no   **(b)** yes   **(c)** no

3. $2x + y = 5$

4.

5. False

6.

7. corner points   **8.** unbounded

9.

bounded

10. $\begin{cases} x + y \le 25{,}000 & \text{where } x \text{ is amount in} \\ x \ge 10{,}000 & \text{Treasury notes and } y \\ y \le 15{,}000 & \text{is amount in corporate} \\ y \ge 0 & \text{bonds} \end{cases}$

11. (a) no   **(b)** yes

13. (a) no   **(b)** no

15. (a) no   **(b)** yes

17.

19.

21.

23.

25.

$\varnothing$ or $\{\ \}$

27.

29.

The graph is bounded.

31.

The graph is unbounded.

33.

The graph is bounded.

35.

The graph is unbounded.

37. (a) $\begin{cases} 450x + 50y \ge 500 \\ 2x + 6y \ge 14 \\ x \ge 0 \\ y \ge 0 \end{cases}$

(b)

39. (a) $\begin{cases} x + y \le 25{,}000 \\ 0.03x + 0.06y \ge 900 \\ x \ge 5{,}000 \\ y \le 15{,}0000 \\ y \ge 0 \end{cases}$

(b)

41. $\begin{cases} y \ge x + 4 \\ y \ge -2x + 10 \end{cases}$

43. $\begin{cases} y \le -x + 12 \\ y \le -2x + 16 \\ x \ge 0 \\ y \ge 0 \end{cases}$

45. (a), (b)

(c) $z = 24$; $x = 0$, $y = 8$ **47.** No, because the corner point is not a solution to the strict inequality.   **49.** Yes. Examples may vary. One possibility follows: $\begin{cases} x + y \le 1 \\ x + y \ge 1 \end{cases}$

51.

53.

55. The inequalities $x \ge 0$ and $y \ge 0$ restrict the graph to the first quadrant.

Chapter 3 Review   **1. (a)** no   **(b)** yes   **2. (a)** yes   **(b)** no   **3. (a)** yes   **(b)** no   **4. (a)** no   **(b)** yes   **5.** $(4, 2)$ **6.** $(2, -1)$

7. $(2, -5)$

8. $(3, -1)$

9. $\varnothing$ or $\{\ \}$

10. $(-3, -2)$

11. $(8, 0)$ **12.** $\left\{ (x, y) \mid y = -\dfrac{3}{4}x + 2 \right\}$ **13.** $(2, -3)$ **14.** $(-3, -5)$ **15.** $(4, -1)$ **16.** $(2, 2)$ **17.** $\{ (x, y) \mid 2x - 4y = 8 \}$ **18.** $(-5, -1)$

19. $(0, -4)$ **20.** $(-2, -4)$ **21.** $(1, -1)$ **22.** $\left(\dfrac{1}{2}, \dfrac{3}{4}\right)$ **23.** $\varnothing$ or $\{\,\}$ **24.** $(-7, 4)$ **25.** 35 and 21   **26.** 31 males and 42 females

27. 300 calories in a slice of pepperoni pizza and 340 calories in a slice of Italian sausage pizza   **28.** 21 inches by 13 inches

29. angle $x = 30°$ and angle $y = 60°$ **30.** $x = 15$ and $y = 5$ **31.** 28 nickels and 12 quarters   **32.** 8 liters of the 25%-hydrochloric-acid solution and 4 liters of the 40%-hydrochloric-acid solution   **33.** $7000 in stocks and $3000 in bonds   **34.** The plane's speed is 136 miles per hour, and the effect of the wind's resistance is 24 miles per hour.   **35.** $\dfrac{5}{6}$ hour, or 50 minutes; $66\dfrac{2}{3}$ miles   **36.** The speed of the boat in still water is 25 miles per hour, and the speed of the current is 5 miles per hour.

37. (a)

(b) approximately 2065

38. (a) $R(x) = 15x$ **(b)** $C(x) = 2.50x + 1200$

(c)

(d) 96 pies; $1440

39. $(4, -2, 1)$
40. $(-1, 3, -2)$
41. $(5, -3, 4)$
42. $\{\,\}$ or $\varnothing$

43. $\left\{ (x, y, z) \,\middle|\, x = \dfrac{1}{2}z + 4, y = \dfrac{1}{2}z - 6, z \text{ is any real number} \right\}$ **44.** $\left(-\dfrac{2}{3}, \dfrac{1}{2}, \dfrac{3}{4}\right)$ **45.** $(1, 0, -6)$ **46.** $(-3, 4, 2)$ **47.** angle $x = 21°$, angle $y = 53°$, and angle $z = 106°$ **48.** A cheeseburger contains 300 calories, a medium order of fries contains 410 calories, and a medium Coke contains 290 calories.

49. $\begin{bmatrix} 3 & 1 & | & 7 \\ 2 & 5 & | & 9 \end{bmatrix}$ **50.** $\begin{bmatrix} 1 & -5 & | & 14 \\ -1 & 1 & | & -3 \end{bmatrix}$ **51.** $\begin{bmatrix} 5 & -1 & 4 & | & 6 \\ -3 & 0 & -3 & | & -1 \\ 1 & -2 & 0 & | & 0 \end{bmatrix}$ **52.** $\begin{bmatrix} 8 & -1 & 3 & | & 14 \\ -3 & 5 & -6 & | & -18 \\ 7 & -4 & 5 & | & 21 \end{bmatrix}$ **53.** $\begin{cases} x + 2y = 12 \\ 3y = 15 \end{cases}$

54. $\begin{cases} 3x - 4y = -5 \\ -x + 2y = 7 \end{cases}$ **55.** $\begin{cases} x + 3y + 4z = 20 \\ y - 2z = -16 \\ z = 7 \end{cases}$ **56.** $\begin{cases} -3x + 7y + 9z = 1 \\ 4x + 10y + 7z = 5 \\ 2x - 5y - 6z = -8 \end{cases}$ **57. (a)** $\begin{bmatrix} 1 & -5 & | & 22 \\ 0 & -1 & | & 4 \end{bmatrix}$ **(b)** $\begin{bmatrix} 1 & -5 & | & 22 \\ 0 & 1 & | & -4 \end{bmatrix}$

58. (a) $\begin{bmatrix} 1 & -4 & | & 7 \\ 0 & 5 & | & -15 \end{bmatrix}$ **(b)** $\begin{bmatrix} 1 & -4 & | & 7 \\ 0 & 1 & | & -3 \end{bmatrix}$ **59. (a)** $\begin{bmatrix} -1 & 2 & 1 & | & 1 \\ 0 & 3 & 5 & | & -1 \\ -1 & 5 & 6 & | & 2 \end{bmatrix}$ **(b)** $\begin{bmatrix} -1 & 2 & 1 & | & 1 \\ 0 & 3 & 5 & | & -1 \\ 0 & 3 & 5 & | & 1 \end{bmatrix}$ **60. (a)** $\begin{bmatrix} 1 & 3 & 4 & | & 4 \\ 0 & 1 & 2 & | & -3 \\ 0 & -4 & -7 & | & 7 \end{bmatrix}$

(b) $\begin{bmatrix} 1 & 3 & 4 & | & 4 \\ 0 & 1 & 2 & | & -3 \\ 0 & 0 & 1 & | & -5 \end{bmatrix}$ **61.** $(7, -3)$ **62.** $\left(\dfrac{1}{2}, 5\right)$ **63.** $(-4, 1)$ **64.** $\varnothing$ or $\{\,\}$ **65.** $\{ (x, y) \,|\, 4x - 2y = 6 \}$ **66.** $\left(\dfrac{5}{2}, \dfrac{3}{8}\right)$ **67.** $(-5, 1, 4)$

68. $\varnothing$ or $\{\,\}$ **69.** $\left\{ (x, y, z) \,\middle|\, x = -\dfrac{5}{6}z + 1, y = \dfrac{4}{3}z + 1, z \text{ is any real number} \right\}$ **70.** $(-3, 1, 3)$ **71.** 10   **72.** 0   **73.** -2 **74.** 9

75. 111   **76.** 0   **77.** -31 **78.** 78   **79.** $(5, -3)$ **80.** $\left(-4, \dfrac{1}{3}\right)$ **81.** $(-1, 2)$ **82.** $\left(\dfrac{7}{3}, \dfrac{1}{3}\right)$ **83.** Cramer's Rule does not apply   **84.** $\left(\dfrac{3}{10}, \dfrac{12}{5}\right)$

85. $(5, -2, 1)$ **86.** $(1, 7, -5)$ **87.** $(1, 2, -2)$ **88.** $\left(-\dfrac{1}{8}, 2, \dfrac{1}{2}\right)$ **89.** Cramer's Rule does not apply   **90.** $(2, -5, -9)$ **91. (a)** yes   **(b)** no   **92. (a)** no **(b)** yes   **93. (a)** yes   **(b)** no   **94. (a)** no   **(b)** yes

95.   **96.**   **97.**   **98.**

$\varnothing$ or $\{\,\}$

99.   **100.**   **101.**   **102.**

The graph is bounded.   The graph is unbounded.

103. (a) $\begin{cases} x + y \le 4000 \\ 0.06x + 0.08y \ge 275 \\ x \ge 500 \\ y \le 2500 \\ y \ge 0 \end{cases}$ **(b)**

104. (a) $\begin{cases} 6x + 4y \le 144 \\ x \ge y + 4 \\ x \ge 0 \\ y \ge 0 \end{cases}$ **(b)**

Chapter 3 Test   **1.** $(-2, -4)$ **2.** $(1, -4)$ **3.** $\left(\dfrac{2}{3}, -\dfrac{5}{3}\right)$ **4.** $\varnothing$ or $\{\ \}$ **5.** $(-6, 8)$ **6.** $\{(x, y, z) \mid x = z + 5, y = 2z + 2, z \text{ is any real number}\}$

7. $(5, 1, -3)$ **8. (a)** $\begin{bmatrix} 1 & -3 & | & -2 \\ 0 & 2 & | & 12 \end{bmatrix}$ **(b)** $\begin{bmatrix} 1 & -3 & | & -2 \\ 0 & 1 & | & 6 \end{bmatrix}$ **9. (a)** $\begin{bmatrix} 1 & -2 & 1 & | & -2 \\ 0 & 1 & -1 & | & 7 \\ 0 & -4 & 5 & | & -32 \end{bmatrix}$ **(b)** $\begin{bmatrix} 1 & -2 & 1 & | & -2 \\ 0 & 1 & -1 & | & 7 \\ 0 & 0 & 1 & | & -4 \end{bmatrix}$

10. $\begin{bmatrix} 1 & -5 & | & 2 \\ 2 & 1 & | & 4 \end{bmatrix}$; $(2, 0)$ **11.** $\begin{bmatrix} 1 & 2 & 1 & | & 3 \\ 0 & 4 & 3 & | & 5 \\ 2 & 3 & 0 & | & 1 \end{bmatrix}$; $(2, -1, 3)$ **12.** -4 **13.** 14 **14.** $\left(-\dfrac{7}{4}, \dfrac{1}{4}\right)$ **15.** $(-5, 4, -1)$

16.

17.

18. 32 twenty-five-ton bins and 18 twenty-ton bins

19. angle $x = 30°$, angle $y = 70°$, and angle $z = 80°$

20. (a) $\begin{cases} 12x + 18y \le 180 \\ x + y \le 13 \\ x \ge 0 \\ y \ge 0 \end{cases}$ **(b)**

Cumulative Review Chapter R-3

1. (a) 11 **(b)** $-13, 0, 11$ **(c)** $-13, -\dfrac{7}{8}, 0, 2.7, 11$ **(d)** $\dfrac{\pi}{2}, 4\sqrt{2}$ **(e)** $-13, -\dfrac{7}{8}, 0, \dfrac{\pi}{2}, 2.7, 4\sqrt{2}, 11$ **2.** $2(x + 4) = 3x - 6$ **3.** -8

4. 3 **5.** $3x^2 - 11x - 27$ **6.** $\{-15, 1\}$ **7.** $\{x \mid x \le -9\}$ or $(-\infty, -9]$ **8.** $\{x \mid x < 2 \text{ or } x > 5\}$; $(-\infty, 2) \cup (5, \infty)$ **9. (a)** $\{0, 6, 12\}$

(b) $\{0, 2, 3, 4, 6, 8, 9, 10, 12, 15\}$ **10.**

11. 6 **12. (a)** Function   **(b)** Function   **(c)** Not a function   **13.** $\left\{x \mid x \ne \dfrac{1}{2}\right\}$

14. (a) 25 **(b)** -3 **(c)** $3x^2 - 4x - 7$

15.   **16.** Parallel   **17.** $y = -\dfrac{5}{3}x + 4$; $5x + 3y = 12$ **18.** 196.7 g   **19. (a)** Independent: year; Dependent: population   **(b)** $\{x \mid 0 \le x \le 80\}$

or $[0, 80]$ **(c)** The population in the year 1990 was 517,653   **(d)** $344,113$ **20. (a)** $C(x) = 0.40(x - 900) + 79.99$

(b) $\{x \mid x \ge 0\}$ **(c)** \$79.99   **(d)** 1575 minutes   **21.** $(-8, 5)$ **22.** $\left(\dfrac{2}{3}, -\dfrac{1}{2}, -1\right)$ **23.** -2 **24.** -13

25.

Getting Ready for Chapter 4: Laws of Exponents and Scientific Notation

1. base; power; exponent   **2.** a^{m+n} **3.** 125 **4.** -243 **5.** y^7 **6.** $-10x^7$ **7.** $-6y^5$ **8.** True   **9.** 25 **10.** y^2 **11.** $\dfrac{8}{5}a$ **12.** $-\dfrac{3}{2}b^2$ **13.** $1; 0$ **14.** $\dfrac{1}{a^n}; 0$

15. $\dfrac{1}{125}$ **16.** $\dfrac{5}{z^7}$ **17.** x^4 **18.** $5y^3$ **19.** -1 **20.** 1 **21.** $\dfrac{9}{16}$ **22.** -64 **23.** $\dfrac{x^2}{9}$ **24.** 20 **25.** $\dfrac{1}{36}$ **26.** 100 **27.** $\dfrac{20x^3}{y}$ **28.** $\dfrac{2}{3}ab^4$ **29.** $-\dfrac{3}{2}b^8$ **30.** $\dfrac{10t^5}{3s^3}$

31. 64 **32.** 1 **33.** 4096 **34.** a^{15} **35.** $\dfrac{1}{z^{18}}$ **36.** s^{21} **37.** $125y^3$ **38.** 1 **39.** $81x^8$ **40.** $\dfrac{1}{16a^6}$ **41.** $\dfrac{z^4}{81}$ **42.** $\dfrac{32}{x^5}$ **43.** $\dfrac{x^8}{y^{12}}$ **44.** $\dfrac{27}{a^6b^{12}}$ **45.** $\dfrac{3x^3y^4}{4}$ **46.** $\dfrac{3b^5}{4a}$

47. $\dfrac{x^6}{y^{10}}$ **48.** 5.32×10^2 **49.** -1.23×10^6 **50.** 3.4×10^{-2} **51.** -8.45×10^{-5} **52.** 500 **53.** $910{,}000$ **54.** 0.00018 **55.** 0.000001 **56.** 6×10^8

57. 8×10^{-11} **58.** 2.4×10^4 **59.** 2×10^2 **60.** 2×10^{-3} **61.** 5×10^3 **62.** 3.75×10^{-13} **63.** $240{,}000{,}000{,}000$ **64.** 0.003 **65.** 750 **66.** 0.000000000003

67. -25 **69.** $-\dfrac{1}{25}$ **71.** -1 **73.** $\dfrac{81}{16}$ **75.** $-\dfrac{1}{27}$ **77.** -64 **79.** 288 **81.** 1 (assuming $x \neq 0$) **83.** $-10t^5$ **85.** $\dfrac{5x^2}{y}$ **87.** $4x^2y$ **89.** $\dfrac{3b^3}{2a}$ **91.** $\dfrac{1}{x^8}$ **93.** $27x^6y^3$

95. $\dfrac{64}{z^3}$ **97.** $\dfrac{a^6}{9}$ **99.** $\dfrac{1}{16a^8b^{12}}$ **101.** $\dfrac{2x^5}{3y^3}$ **103.** $\dfrac{b^4}{25a^{12}}$ **105.** 1 **107.** $-x^2y^4z$ **109.** $\dfrac{9y^8z}{x^5}$ **111.** $\dfrac{3b^{16}}{2a^9}$ **113.** -1.484×10^{-6} **115.** 6.4×10^{19} **117.** 6.25×10^{-3}

119. 8×10^6 **121.** 1 **123.** $12{,}000{,}000{,}000{,}000$ **125.** 0.04 **127.** $4{,}000{,}000$ **129.** x^6 cubic units **131.** 1.276×10^{-5} meter **133.** 1.276×10^7 meters

135. 0.000000001 meter **137.** $40{,}000{,}000{,}000{,}000{,}000{,}000$ possible states **139.** $\$52{,}500$ **141.** $3.131 \times 10^{10} = 31{,}310{,}000{,}000$ visitors **143. (a)** 79.4 people per square mile; On average, each square mile of land contained about 79 people. **(b)** 90.1 people per square mile; on average, each square mile of land contained about 90 people. **(c)** about 11 people per square mile; on average, the number of people living on each square mile of land increased by about 11, thereby making living space more crowded. **145.** 2^{x+1} **147.** 3^{10x+3} **149.** 625 **151.** $\dfrac{1}{2401}$ **153.** No, your friend is incorrect. He added exponents instead of multiplying. **155.** Explanations will vary. Essentially, if a were to equal zero and $m, n,$ or $m + n$ were negative, we would end up with division by 0, which is not defined. For example, $0^{-2} = \dfrac{1}{0^2} = \dfrac{1}{0}$, which is undefined. If m, n or $m + n$ were to equal 0, we would end up with 0^0, which is indeterminate.

157. Explanations will vary. If a or b were to equal 0 when n was negative, we would end up with division by 0, which is not defined. If a or b were to equal 0 when n was zero, we end up with 0^0, which is indeterminate. **159.** Explanations will vary. One possibility: $\dfrac{x^5}{x^2} = \dfrac{x \cdot x \cdot x \cdot x \cdot x}{x \cdot x} = x^3 = x^{5-2}$

161. Explanations will vary. One possibility: $(xy)^3 = (xy)(xy)(xy) = x \cdot x \cdot x \cdot y \cdot y \cdot y = x^3y^3$ **163.** Explanations will vary. Basically, using scientific notation enables us to use the Laws of Exponents to perform multiplication or division.

Chapter 4 Polynomials and Polynomial Functions

Section 4.1 Adding and Subtracting Polynomials **1.** monomial **2.** Monomial; 8; 5 **3.** Not a monomial **4.** Monomial; 12; 0 **5.** Not a monomial **6.** $m + n$ **7.** Monomial; coefficient: 3, degree: 7 **8.** Monomial; coefficient: -2, degree: 4 **9.** Not a monomial **10.** Monomial; coefficient: -1, degree: 2 **11.** True **12.** Polynomial; 3 **13.** Not a polynomial **14.** Not a polynomial **15.** Polynomial; 2 **16.** Polynomial; 4 **17.** False **18.** $6x^2 + 2x - 2$

19. $3w^4 - 2w^3 - 7w^2 + w - 5$ **20.** $\dfrac{3}{4}x^2 - \dfrac{2}{3}x - 7$ **21.** $5x^2y + 7x^2y^2 - 4xy^2$ **22.** $x^3 - 16x^2 + 7x + 2$ **23.** $11y^3 - 5y^2 - 3y - 7$

24. $11x^2y - 3x^2y^2 - 10xy^2$ **25. (a)** 1 **(b)** -1 **(c)** 34 **26. (a)** In 2014, there were 394,248 first-time births to women 25 years of age. **(b)** In 2014, there were 88,818 first-time births to women 40 years of age. **27. (a)** $2x^2 + 4x - 5$ **(b)** $4x^2 - 6x + 7$ **(c)** 1 **(d)** 35 **28. (a)** $4x - 1250$ **(b)** $\$1950$; If the company manufactures and sells 800 calculators, its profit will be $1950. **29.** Coefficient: 3; degree: 2 **31.** Coefficient: -8; degree: 5

33. Coefficient: $\dfrac{4}{3}$; degree: 6 **35.** Coefficient: 2; degree: 0 **37.** The exponent in the first term is not an integer that is greater than or equal to 0.

39. There is a variable in the denominator. The expression cannot be written in the standard form for a polynomial. **41.** Yes; $5x^2 - 9x + 1$; 2; trinomial

43. No **45.** No **47.** Yes; $\dfrac{5}{8}$; 0; monomial **49.** Yes; $2y^2 - 8y + 5$; 2; trinomial **51.** No **53.** Yes; $2xy^4 + 3x^2y^2 + 4$; 5; trinomial **55.** No **57.** $13z^3$

59. $4x^2 + 3x - 2$ **61.** $8p^3 - p^2 - 4p - 1$ **63.** $7x^3 + 2x^2 - 6x + 8$ **65.** $2x^2 + 4x + 3$ **67.** $3s^2t^3 - 4st^2 - 5t - 1$ **69.** $-4x^2 - 2x + 1$

71. $3y^3 - y^2 - 2y + 8$ **73.** $\dfrac{3}{4}x^2 + \dfrac{5}{4}x + 1$ **75.** $8x^2y^2 - 7x^2y - 3xy^2$ **77.** $x^2y + 11xy^2 + 2xy$ **79.** $4a^3 + 4a^2 - 6a - 2$ **81.** $x^3 + x^2 - 7x + 4$

83. $-2b^3 + 6b^2 - 4b + 4$ **85.** $-x^2 - x + 15$ **87.** $n^3 - 2n^2 + 11n - 15$ **89. (a)** 1 **(b)** -3 **(c)** 22 **91. (a)** 3 **(b)** 5 **(c)** -30 **93. (a)** 3 **(b)** 3 **(c)** 63 **95. (a)** $-3x + 6$ **(b)** $7x + 4$ **(c)** 0 **(d)** 11 **97. (a)** $3x^2 - 5x + 6$ **(b)** $-x^2 - 5x$ **(c)** 8 **(d)** -6 **99. (a)** $2x^3 + 6x^2 + 12x - 6$ **(b)** $6x^2 + 12x + 10$ **(c)** 58 **(d)** 28 **101. (a)** 4 square units **(b)** 4 square units **(c)** Since the rectangular region is in quadrant I and has one corner at the origin, the coordinates of the vertex give us the length and width of the rectangle. Since the area of a rectangle can be found by multiplying the length and width, we can just multiply the coordinates together to get the area of the region. **103. (a)** The average price per square foot of a home in 1995 was $66.00. **(b)** We predict that the average price per square foot of a home in 2021 will be $193.08. **105. (a)** $P(x) = -0.05x^3 + 0.8x^2 + 155x - 500$ **(b)** $\$1836.25$ **(c)** $P(100) = -27{,}000$; If 100 cell phones are sold, there would be a loss of $27,000. **107.** $a = 2$ **109. (a)** 3 **(b)** -2 **(c)** -1 **(d)** 1 **111.** $S(x) = T(x) - F(x)$ **113.** 1 **115.** A polynomial is the sum or difference of monomials. Monomial: $4x^2$; binomial: $-5y^3 + 8y$; trinomial: $2z^5 - 8z^3 + 10z$. Linear polynomials are of the form $ax + b$, where a and b are constants. **117.** When we add two polynomials, we add the coefficients of the like monomials. Therefore, the highest-degree term of the polynomials being added will be the degree of the polynomial.

119. (a)–(c)

$f(4) = 37$ $f(-2) = 31$ $f(6) = 103$

121. (a)–(c)

$f(4) = 57$ $f(-2) = -33$ $f(6) = 263$

Section 4.2 Multiplying Polynomials 1. $6x^7$ 2. $-21a^4b^6$ 3. $\dfrac{5}{4}x^5$ 4. $-3x - 6$ 5. $5x^3 + 15x^2 + 10x$ 6. $6x^3y - 10x^2y^2 + 4xy^3$

7. $y^4 + \dfrac{1}{6}y^3 + 4y^2$ 8. First, Outer, Inner, Last 9. $x^2 + 5x + 4$ 10. $6v^2 + v - 15$ 11. $2a^2 + 9ab - 5b^2$ 12. $2y^3 + 5y^2 - 2y - 15$

13. $2z^4 - 5z^3 + 7z^2 - 16z + 12$ 14. False 15. $A^2 - B^2$ 16. $25y^2 - 4$ 17. $49y^2 - 4z^6$ 18. $A^2 - 2AB + B^2, A^2 + 2AB + B^2$ 19. perfect square
20. False 21. $z^2 - 16z + 64$ 22. $36p^2 + 60p + 25$ 23. $16a^2 - 24ab + 9b^2$ 24. $f(x) \cdot g(x)$ 25. (a) 77 (b) $5x^3 + 12x^2 - 4x - 3$ (c) 77

26. (a) $x^2 - 8x + 15$ (b) $2xh + h^2 - 2h$ 27. $-15x^3y^5$ 29. $\dfrac{5}{3}y^4z^5$ 31. $5x^3 + 20x^2 + 10x$ 33. $-12a^4b - 8a^3b^2 + 4a^2b^3$ 35. $\dfrac{1}{2}a^3b^2 - \dfrac{3}{4}a^2b^4 + 4a^2b^2$

37. $0.48x^4 - 0.32x^3 + 0.6x^2$ 39. $x^2 + 8x + 15$ 41. $a^2 + 2a - 15$ 43. $12a^2 + 5a - 3$ 45. $-10x^2 - 7x + 12$ 47. $\dfrac{1}{3}x^2 - \dfrac{5}{3}x - 8$ 49. $4a^2 - 17ab - 15b^2$

51. $2x^4 - 5x^2 - 3$ 53. $x^3 + 5x^2 + 6x + 2$ 55. $6a^3 - a^2 - 17a + 10$ 57. $20z^3 + 27z^2 + 17z + 6$ 59. $-x^3 + 9x$ 61. $2y^3 + 13y^2 + 17y - 12$
63. $2w^4 + w^3 - 3w^2 - w + 1$ 65. $b^3 + 2b^2 - 5b - 6$ 67. $8a^3b - 4a^2b^2 + 2ab^3 + 20a^2 - 10ab + 5b^2$ 69. $x^2 - 36$ 71. $a^2 + 16a + 64$
73. $9y^2 - 6y + 1$ 75. $25a^2 - 9b^2$ 77. $64z^2 + 16yz + y^2$ 79. $100x^4 - 20x^2y + y^2$ 81. $a^6 - 4b^2$ 83. (a) $x^2 + 3x - 4$ (b) 14 85. (a) $8x^2 + 14x - 15$
(b) 99 87. (a) $x^3 + 3x^2 - 13x + 6$ (b) 21 89. (a) $x^2 + 4x + 5$ (b) $2xh + h^2$ 91. (a) $x^2 + 9x + 12$ (b) $2xh + 5h + h^2$ 93. (a) $3x^2 + 11x + 11$
(b) $6xh - h + 3h^2$ 95. $5a^3b - 10a^2b^2 + 5ab^3$ 97. $x^4 - 9$ 99. $z^4 - 81$ 101. $4x^6 + 12x^3 + 9$ 103. $7m^2 - 6mn - 7n^2$ 105. $x^3 + 27$

107. $4x^2 - 2x + \dfrac{1}{4}$ 109. $p^3 + 6p^2 + 12p + 8$ 111. $21x^2 - 29xy + 13x + 10y^2 - 9y + 2$ 113. $3p^2 + 5p - 12$ 115. $x^4 - x^3 - 19x^2 + 3x + 84$

117. $x^2 + 6x$ 119. $4x^2 + 29x + 30$ 121. $4x^2 + 11x + 6$ 123. (a) $A_1 = a^2; A_2 = ab; A_3 = ab; A_4 = b^2$ (b) $a^2 + 2ab + b^2$ (c) The length of the
region is $a + b$ and the width is also $a + b$. The area of the region would be $A = (a + b)(a + b) = (a + b)^2$. The result from part (b) is obtained by
multiplying this expression out. 125. $9x^2 - y^2 - 2y - 1$ 127. $4a^2 + 4ab - 12a + b^2 - 6b + 9$ 129. $2^{2x} - 2^x - 12$ 131. $5^{2y} - 2(5^y) + 1$

Section 4.3 Dividing Polynomials; Synthetic Division 1. $4x^4; 8x^2$ 2. $3p^3 - 4p^2 + p$ 3. $4a^2 - \dfrac{2}{a} + \dfrac{1}{5a^3}$ 4. $\dfrac{xy^3}{4} + \dfrac{2y}{x} - \dfrac{1}{x^2}$ 5. True

6. $0; (3x - 1)(x + 1)$; factors 7. dividend; divisor; remainder 8. Quotient; Divisor; Remainder; Dividend 9. $x^2 + 7x - 3 + \dfrac{9}{x - 4}$

10. $x^2 + 5x + 4$ 11. $x^3 - 5x^2 + 2 + \dfrac{6}{x^2 - 2}$ 12. False 13. False 14. -7 15. $2x^2 + 5x + 3 - \dfrac{7}{x - 2}$ 16. $4x^2 - 11x + 23 - \dfrac{45}{x + 2}$ 17. $x^3 + 5x^2 - 2$

18. (a) $3x^2 - 4x + 3 + \dfrac{2x + 1}{x^2 - 2}$ (b) 19 19. (a) 42 (b) 0 20. (a) $f(-2) = -35; x + 2$ is not a factor (b) $f(5) = 0; f(x) = (x - 5)(2x^2 + x - 1)$

21. $2x + 3$ 23. $\dfrac{2}{5}a^2 - 3a + 2$ 25. $\dfrac{1}{2}y + \dfrac{3}{2y}$ 27. $1 + \dfrac{3}{2n} - \dfrac{9}{2m}$ 29. $x + 3$ 31. $2x + 5 + \dfrac{6}{x - 2}$ 33. $w + 6 - \dfrac{7}{2w - 7}$ 35. $x^2 + 5x - 14$

37. $w^2 - 4w - 5$ 39. $3x^2 - 14x - 5$ 41. $x - 7$ 43. $z + 7 + \dfrac{4z - 5}{3z^2 + 1}$ 45. $2x^2 - 15x + 28 + \dfrac{-3x + 4}{x^2 + 2x + 5}$ 47. $x + 2$ 49. $2x + 3$ 51. $x + 3 + \dfrac{4}{x - 6}$

53. $x^2 + 5x + 6 + \dfrac{15}{x - 5}$ 55. $3x^3 + 4x^2 - 9x - 10 - \dfrac{5}{x - 3}$ 57. $x^3 - 6x^2 - 4x + 24 - \dfrac{35}{x + 6}$ 59. $2x^2 + 8x + 6$

61. (a) $\left(\dfrac{f}{g}\right)(x) = x^2 - 2x + 3$ (b) 3 63. (a) $\left(\dfrac{f}{g}\right)(x) = x + 3$ (b) 5 65. (a) $\left(\dfrac{f}{g}\right)(x) = 2x - 1 + \dfrac{2}{x + 3}$ (b) $\dfrac{17}{5}$

67. (a) $\left(\dfrac{f}{g}\right)(x) = x^2 + 3x - 4$ (b) 6 69. (a) $\left(\dfrac{f}{g}\right)(x) = x - \dfrac{4}{x - 3}$ (b) 6 71. -5 73. -119 75. 231 77. 2 79. $x - 2$ is a

factor; $f(x) = (x - 2)(x - 1)$ 81. $x + 2$ is a factor; $f(x) = (x + 2)(2x + 1)$ 83. $x - 3$ is not a factor 85. $x + 1$ is a factor;

$f(x) = (x + 1)(4x^2 - 11x + 6)$ 87. $a^2b - 3a + 6$ 89. $y + 3$ 91. $x + 6 + \dfrac{3x + 2}{x^2 + 5}$ 93. $\dfrac{2}{3}x + \dfrac{1}{2x}$ 95. $x^2 + 3x - 10 - \dfrac{6}{x + 4}$ 97. $x^2 - 8$

99. $4x^2 + 6x + 9$ 101. $f(x) = 3x^2 - 10x - 25$ 103. $f(x) = x^2 + 5x - 20$ 105. $(5x + 2)$ ft 107. $(2x + 5)$ cm

109. (a) $\overline{C}(x) = 0.01x^2 - 0.4x + 13 + \dfrac{400}{x}$ (b) \$26 111. $a = 2, b = -7, c = -12$, and $d = -13$, thus, $a + b + c + d = -30$ 113. The dividend

is the polynomial f and has degree n. Remember, (Divisor)(Quotient) + Remainder = Dividend. Since the remainder must be 0 or a polynomial that
has lower degree than f, the degree n must be obtained from the product of the divisor and the quotient. The divisor is $x + 4$, which is of degree 1, so the
quotient must be of degree $n - 1$. 115. Yes; $(3x + 4)(2x^2 - 3x + 1) = 6x^3 - x^2 - 9x + 4$

Putting the Concepts Together (Sections 4.1–4.3) 1. $5m^4 - 2m^3 + 3m + 8$; Degree: 4 2. $-4a^3 + 9a^2 + a - 8$ 3. $-\dfrac{19}{5}y^2 + 3y - 8$ 4. 29

5. -25 6. $x^2 + 4x + 10$ 7. $2m^3n^4 - 8m^2n^4 + 12mn^3$ 8. $9a^2 - 30ab + 25b^2$ 9. $49n^4 - 9$ 10. $18a^3 + 6a^2b - ab^2 + 2b^3$ 11. $x^3 - 2x^2 + 3x + 22$

12. $5z^2 + 3z - 7$ 13. $2x^2 + 7x - 1 + \dfrac{3}{x + 9}$ 14. $x^2 + 3x - 1 + \dfrac{4}{x - 1}$ 15. yes: $f(x) = (x + 5)(3x^2 - 7x + 12)$

Section 4.4 Greatest Common Factor; Factoring by Grouping 1. factors 2. prime 3. greatest common factor 4. factor 5. False 6. 5
7. $2z$ 8. $3xy^3$ 9. $7z(z - 2)$ 10. $2y(3y^2 - 7y + 5)$ 11. $2m^2n^2(m^2 + 4mn^2 - 3n^3)$ 12. $-5y(y - 2)$ 13. $-3a(a^2 - 2a + 4)$ 14. $(a - 3)(4a + 3)$
15. $3(w - 5)(w + 1)$ 16. $(x + y)(5 + b)$ 17. $(w - 3)(w^2 + 4)$ 18. $(2x + 1)(x - 5)$ 19. $5(a + 7)$ 21. $-3(y - 7)$ 23. $7x(2x - 3)$
25. $3z(z^2 - 2z + 6)$ 27. $-5p^2(p^2 - 2p + 5)$ 29. $7mn(7m^2 + 12n^2 - 5m^3n)$ 31. $-2z(9z^2 - 7z - 2)$ 33. $(3c - 2)(5c - 3)$
35. $2(a - 3)(3a - 2)$ 37. $(x + y)(5 + a)$ 39. $(z + 5)(2z^2 - 5)$ 41. $(w - 5)(w + 3)$ 43. $2(x - 4)(x - 2)$ 45. $3x(x + 5)(x - 4)$
47. $(x - y)(2a - b)$ 49. $5(w - 3)$ 51. $(2y + 5)(y - 2)$ 53. $3x^2y(2xy^2 - 7xy + 3)$ 55. $(x + 1)(x^2 + 3)$ 57. $2(x - 2)(2x - 3)$

59. $x^2(4x + 1)(20x + 3)$ **61.** $(x + 9)^2(8x + 33)$ **63.** $(x - 5)^2(2x - 1)(10x - 23)$ **65.** $4(x^2 + 1)(4x - 3)^2(7x^2 - 3x + 3)$
67. $x^2(4 - \pi)$ square units   **69.** $S = 2\pi r(r + 4)$ square inches   **71. (a)** $1.4x$ **(b)** $0.84x$ **(c)** No   **73.(a)** $x + 0.15x = 1.15x$
(b) $1.265x$ **(c)** \$25.30   **75.** $\frac{1}{4}(x - 7)$ **77.** $\frac{1}{25}b(5b^2 + 8)$ **79.** $x^n(1 + 3x + 6x^n)$ **81.** $2y^{n+2}(2y - 4 + 3y^3)$

Section 4.5 Factoring Trinomials   **1.** quadratic trinomial   **2.** $m \cdot n; m + n$ **3.** True   **4.** $(y + 6)(y + 3)$ **5.** $(p + 2)(p + 12)$ **6.** $(q - 4)(q - 2)$
7. $(x - 2)(x - 6)$ **8.** $(w - 7)(w + 3)$ **9.** $(q - 12)(q + 3)$ **10.** prime   **11.** prime   **12.** $(q + 9)(q - 5)$ **13.** $(x + 3y)(x + 5y)$
14. $(m + 5n)(m - 4n)$ **15.** False   **16.** $2x(x - 9)(x + 3)$ **17.** $-3(z + 2)(z + 5)$ **18.** $12; -13$ **19.** $(2b - 3)(b + 5)$ **20.** $(2x + 3)(5x + 6)$
21. $(2x + 1)(4x + 5)$ **22.** $(6y - 5)(2y + 7)$ **23.** $(6x - y)(5x + 2y)$ **24.** $2(4x + 7y)(x - 3y)$ **25.** $-(6y + 1)(y - 4)$
26. $-(3x + 2y)(3x + 5y)$ **27.** x^2 **28.** $2x - 3$ **29.** $(y^2 + 4)(y^2 - 6)$ **30.** $(x - 1)(4x - 15)$ **31.** $(x + 3)(x + 5)$ **33.** $(p + 6)(p - 3)$
35. $(r + 5)^2$ **37.** $(s + 12)(s - 5)$ **39.** $2(x - 7)(x - 8)$ **41.** $-(w + 6)(w - 4)$ **43.** $(x + 3y)(x + 4y)$ **45.** $p(p - 4q)(p + 6q)$
47. $(2p + 1)(p - 8)$ **49.** $(y - 2)(4y - 3)$ **51.** $(2s - 1)(4s + 3)$ **53.** $(4z - 3)(4z + 5)$ **55.** $-(2y + 1)(9y + 4)$ **57.** $(x + 7y)(2x - 3y)$
59. $3(r - 5s)(4r - 3s)$ **61.** $(3r + 4s)(8r - 3s)$ **63.** $(x^2 + 1)(x^2 + 2)$ **65.** $(mn + 7)(mn - 2)$ **67.** $(x - 7)(x + 3)$ **69.** $3(r - 2)(3r - 5)$
71. $(y + 2)(2y - 3)$ **73.** $(5w + 3)(2w + 7)$ **75.** $2y(8y + 5)$ **77.** $(3x + 2y)(4x + 5y)$ **79.** prime   **81.** $(x^2 + 4)(x^2 + 2)$ **83.** $(z^3 + 5)(z^3 + 4)$
85. $(r - 4s)(r - 8s)$ **87.** $(2z + 3)(4z + 3)$ **89.** prime   **91.** $2(x - 3)(x + 9)$ **93.** $-3(r - 5)(r - 8)$ **95.** $-2(2m - 5)(4m + 7)$
97. $4(4z + 1)(3z + 7)$ **99.** $3x(x - 10)(x + 8)$ **101.** $4xy^2(x - 7)(2x - 5)$ **103.** $2r^2s(5r - 4)(7r + 2)$ **105. (a)** $V(3) = 1296$; when 3-inch
square corners are cut from the piece of cardboard, the resulting box will have a volume of 1296 cubic inches.   **(b)** $V(x) = 4x(x - 15)(x - 12)$
(c) $V(3) = 4(3)(3 - 15)(3 - 12) = 1296$ **(d)** Answers may vary.   **107.** $2x - 5$ **109.** $\frac{1}{2}(x + 4)(x + 2)$ **111.** $\frac{1}{3}(p - 3)(p + 1)$
113. $\frac{4}{3}(a - 6)(a + 4)$ **115.** $(2^n - 5)(2^n + 1)$ **117.** $(2^x - 4)(2^x + 3)$ **119.** The trial-and-error method is better when ac gets large and
there are quite a few factors of the product whose sum must be determined.   **121.** There is a common factor of 3 in the factor $3p + 6$ that should have
been factored out.

Section 4.6 Factoring Special Products   **1.** perfect square trinomial   **2.** $(A - B)^2$ **3.** False   **4.** $(x - 9)^2$ **5.** $(2x + 5y)^2$ **6.** $2(3p^2 - 7)^2$
7. difference; two squares   **8.** $(P - Q)(P + Q)$ **9.** $(z - 4)(z + 4)$ **10.** $(4m - 9n)(4m + 9n)$ **11.** $(2a - 3b^2)(2a + 3b^2)$ **12.** True
13. $3(b + 2)(b - 2)(b^2 + 4)$ **14.** $(p - 4 - q)(p - 4 + q)$ **15.** $(A + B)(A^2 - AB + B^2)$ **16.** $(A - B)(A^2 + AB + B^2)$
17. $(z + 4)(z^2 - 4z + 16)$ **18.** $(5p - 6q^2)(25p^2 + 30pq^2 + 36q^4)$ **19.** $4(2m + 5n^2)(4m^2 - 10mn^2 + 25n^4)$ **20.** $(-2x + 1)(13x^2 + 5x + 1)$
21. $(x + 2)^2$ **23.** $(6 + w)^2$ **25.** $(2x + 1)^2$ **27.** $(3p - 5)^2$ **29.** $(5a + 9)^2$ **31.** $(3x + 4y)^2$ **33.** $3(w - 5)^2$ **35.** $-5(t + 7)^2$ **37.** $2(4a - 5b)^2$
39. $(z^2 - 3)^2$ **41.** $(x - 3)(x + 3)$ **43.** $(2 - y)(2 + y)$ **45.** $(2z - 3)(2z + 3)$ **47.** $(10m - 9n)(10m + 9n)$ **49.** $(m^2 - 6n^3)(m^2 + 6n^3)$
51. $2(2p - 3q)(2p + 3q)$ **53.** $5r(4p - 7b)(4p + 7b)$ **55.** $(x + y - 3)(x + y + 3)$ **57.** $(x - 2)(x^2 + 2x + 4)$ **59.** $(5 + m)(m^2 - 5m + 25)$
61. $(x^2 - 4y)(x^4 + 4x^2y + 16y^2)$ **63.** $3(2x - 5y)(4x^2 + 10xy + 25y^2)$ **65.** $(p - 2)(p^2 + 5p + 13)$ **67.** $(5y + 1)(7y^2 + 4y + 1)$
69. $(y^2 + z^3)(y^4 - y^2z^3 + z^6)$ **71.** $(y - 1)(y^2 + y + 1)(y^6 + y^3 + 1)$ **73.** $(5x - y)(5x + y)$ **75.** $(2x + 3)(4x^2 - 6x + 9)$ **77.** $(z - 4)^2$
79. $5x(x - 2y)(x^2 + 2xy + 4y^2)$ **81.** $(7m - 3n)^2$ **83.** $4(x^2 + 4)$ **85.** $(y - 2)^2(y + 2)^2$ **87.** $(2ab + 3)^2$ **89.** prime   **91.** $(n + 9)(n + 4)$
93. $2(n - m + 10)(n + m - 10)$ **95.** $3(y - 2)(y^2 + 2y + 4)$ **97.** $(4x + 3y - 10)(4x + 3y + 10)$ **99.** $(x - 3)(x + 3)$ square units
101. $(x - 4)(x + 4)$ square units   **103.** $\pi(R - r)(R + r)$ square units   **105.** $10a(a - b)(a + b)$ cubic units
107. $A = a \cdot a + a \cdot b + a \cdot b + b \cdot b = a^2 + 2ab + b^2 = (a + b)^2$ **109.** $b = \pm36$; the middle term must equal $\pm$ twice the product of the quantities
that are squared to get the first and last terms.   **111.** We need to add 81 to make a perfect square trinomial.   **113.** $(b - 0.2)^2$ **115.** $\left(3b - \frac{1}{5}\right)\left(3b + \frac{1}{5}\right)$
117. $\left(\frac{x}{3} - \frac{y}{5}\right)\left(\frac{x}{3} + \frac{y}{5}\right)$ **119.** $\left(\frac{x}{2} - \frac{y}{3}\right)\left(\frac{x^2}{4} + \frac{xy}{6} + \frac{y^2}{9}\right)$

Section 4.7 Factoring: A General Strategy   **1.** $2q(p + 5q)(p - 9q)$ **2.** $-3y(3x + 1)(5x - 9)$ **3.** $(9x - 10y)(9x + 10y)$
4. $-3n(m - 7)(m + 7)$ **5.** $(p - 8q)^2$ **6.** $5(2x + 3)^2$ **7.** $(4y - 5)(16y^2 + 20y + 25)$ **8.** $-2(2m + n)(4m^2 - 2mn + n^2)$ **9.** $5(2z^2 - 3z + 7)$
10. $3x(2y^2 + 27x^2)$ **11.** $(x^2 + 2)(2x + 5)$ **12.** $3(3x + 1)(x - 1)(x + 1)$ **13.** $(2x + y - 9)(2x + y + 9)$ **14.** $(4 - m - 4n)(4 + m + 4n)$
15. $G(x) = (2x + 7)(x - 2)$ **16.** $F(p) = 3(3p + 2)(3p - 2)$ **17.** $2(x - 12)(x + 6)$ **19.** $-3(y - 3)(y + 3)$ **21.** $(2b + 5)^2$
23. $2(2w + y^2)(4w^2 - 2wy^2 + y^4)$ **25.** $-3(z^2 - 4z + 6)$ **27.** $(4y + 3)(5y - 6)$ **29.** $(x - 4)(x^2 + 5)$ **31.** $2(100x^2 + 9y^2)$
33. $(x - 3)(x + 3)(x^2 + 9)$ **35.** prime   **37.** $4q(3q + 1)^2$ **39.** $3mn(4m + 3)(2m - 7)$ **41.** $3r^2(r - 2s)(r^2 + 2rs + 4s^2)$
43. $2(x + 4)(x - 3)(x + 3)$ **45.** $(3x^2 - 1)(3x^2 + 1)$ **47.** $(w^2 + 3)(3w^2 - 5)$ **49.** $2y(2y + 1)$ **51.** $(p - 6q - 5)(p + 6q - 5)$
53. $(y + 2)(y^2 - 2y + 4)(y^3 - 2)$ **55.** $(p - 1)(p + 1)(p^2 + p + 1)(p^2 - p + 1)$ **57.** $-3(x + 5)(x - 3)(x + 3)$ **59.** $3a(1 - 3a)(1 + 3a)$
61. $2t(t^2 + 4)(2t - 3)(2t + 3)$ **63.** $2xy(x + 5)(x - 3)(x + 3)$ **65.** $f(x) = (x - 9)(x + 7)$ **67.** $P(m) = (7m + 3)(m + 4)$
69. $G(x) = -3(2x + 7)(2x - 7)$ **71.** $s(t) = -16(t + 2)(t - 8)$ **73.** $H(a) = (2a + 5)(a - 4)(a + 4)$ **75.** $(x + 3)(3x + 7)$ square units
77. $(x + y)(x - y)$ square units   **79.** $(3x - 4)(9x^2 + 12x + 16)$ cubic units   **81.** The factors of 9 are: 1, 9; 3, 3; $-1, -9; -3, -3$. None of these
factors sum to 0. Therefore, $x^2 + 9$ is prime.   **83.** $x^{\frac{1}{2}}(x - 3)(x + 3)$ **85.** $x^{-2}(x + 4)(x + 2)$ **87.** See pages 365–366.

Section 4.8 Polynomial Equations   **1.** polynomial equation   **2.** $a = 0; b = 0$ **3.** $\{-7, 0\}$ **4.** $\left\{-\frac{3}{4}, 3\right\}$ **5.** quadratic equation   **6.** second
7. False   **8.** $\{2, 3\}$ **9.** $\left\{-\frac{1}{3}, 5\right\}$ **10.** $\left\{-2, -\frac{2}{3}\right\}$ **11.** False   **12.** $\{-2, -1\}$ **13.** $\left\{-\frac{3}{5}, 1\right\}$ **14.** True   **15.** $\{-3, 1, 3\}$ **16.** $\left\{-\frac{1}{2}, 0, 2\right\}$
17. (a) -1 and 9; $(-1, 12)$ and $(9, 12)$ are points on the graph of g.   **(b)** 1 and 7; $(1, -4)$ and $(7, -4)$ are points on the graph of g.

18. Zeros: -4 and $\dfrac{5}{2}$; x-intercepts: $(-4, 0)$ and $\left(\dfrac{5}{2}, 0\right)$ **19.** Zeros: $-\dfrac{2}{5}$ and 1; x-intercepts: $\left(-\dfrac{2}{5}, 0\right)$ and $(1, 0)$ **20.** 9 miles by 15 miles   **21.** 60 boxes

22.(a) 4 seconds and 6 seconds after launch   **(b)** 10 seconds after launch   **23.** $\{-1, 3\}$ **25.** $\left\{-\dfrac{4}{3}, 0\right\}$ **27.** $\{-3, 0, 5\}$ **29.** $\{0, 4\}$ **31.** $\{0, 8\}$

33. $\{-5, 3\}$ **35.** $\{4, 9\}$ **37.** $\{3\}$ **39.** $\left\{-\dfrac{3}{5}, 1\right\}$ **41.** $\{-3\}$ **43.** $\{-8, 5\}$ **45.** $\left\{-6, \dfrac{5}{2}\right\}$ **47.** $\{-6, 2\}$ **49.** $\left\{-\dfrac{7}{2}, 2\right\}$ **51.** $\{-11, 3\}$ **53.** $\{-5, 3\}$

55. $\left\{-\dfrac{1}{2}, 0, 3\right\}$ **57.** $\left\{-\dfrac{5}{2}, -2, 2\right\}$ **59.** $\left\{-\dfrac{3}{5}, -\dfrac{2}{3}, 0\right\}$ **61.** $\left\{\dfrac{4}{3}, 1\right\}$ **63. (a)** $x = -5$ or $x = -2$ **(b)** $x = -8$ or $x = 1$; $(-5, 2)$, $(-2, 2)$, $(-8, 20)$,

and $(1, 20)$ are on the graph of f.   **65. (a)** $x = 4$ or $x = -1$ **(b)** $x = 5$ or $x = -2$; $(-1, 3)$, $(4, 3)$, $(-2, 15)$, and $(5, 15)$ are on the graph of g.

67. (a) $x = 0$ or $x = 4$ **(b)** $x = 5$ or $x = -1$; $(0, 5)$, $(4, 5)$, $(-1, -10)$, and $(5, -10)$ are on the graph of F.   **69.** zeros: -7 and -2; x-intercepts:

$(-7, 0)$ and $(-2, 0)$ **71.** zeros: $-\dfrac{1}{3}$ and $\dfrac{9}{2}$; x-intercepts: $\left(-\dfrac{1}{3}, 0\right)$ and $\left(\dfrac{9}{2}, 0\right)$ **73.** zeros: -5, 0, and 4; x-intercepts: $(-5, 0)$ $(0, 0)$, and $(4, 0)$

75. $\{-4, 6\}$ **77.** $\left\{-\dfrac{7}{2}, 2\right\}$ **79.** $\{0, 7\}$ **81.** $\left\{-\dfrac{6}{7}\right\}$ **83.** $\{-5, -2, 2\}$ **85.** $\varnothing$ or $\{\ \}$ **87.** $\{-2, 1, 3\}$ **89.** $\left\{-2, -\dfrac{1}{2}, \dfrac{1}{2}, 2\right\}$ **91.** $\{-1, 0\}$

93. $\{x \mid x \neq -2, 2\}$ **95.** $\left\{x \mid x \neq \dfrac{1}{2}, 1\right\}$ **97.** The width is 16 cm and the length is 8 cm.   **99.** The base is 10 ft and the height is 22 ft.   **101.** 8 sides

103. 40 meters by 20 meters (along the river) or 10 meters by 80 meters (along the river).   **105.** 3 ft   **107.** The width is 11 in. and the length is 16 in.
109. (a) \$300   **(b)** 20 bicycles   **(c)** 15 bicycles   **111. (a)** after 2.5 seconds (on the way up), and again after 5 seconds (on the way down)
(b) 7.5 seconds after it is hit   **113.** $\{-0.61, 4.11\}$ **115.** $\{0.60, 24.90\}$ **117.** $\{-1.15, 1.75\}$

Chapter 4 Review   **1.** Coefficient: -7; Degree: 4   **2.** Coefficient: $\dfrac{1}{9}$; Degree: 3   **3.** $7x^3 - 2x^2 + x - 8$; degree: 3   **4.** $y^4 - 3y^2 + 2y + 3$; degree: 4

5. $4x^2 + x - 11$ **6.** $-x^4 + 4x^3 - 5x^2 + 8x - 6$ **7.** $\dfrac{1}{4}x^2 - \dfrac{9}{2}x + \dfrac{1}{6}$ **8.** $\dfrac{5}{6}x^2 - x + \dfrac{13}{20}$ **9.** $10x^2y^2$ **10.** $-a^2b - 6ab^2 - 4$ **11. (a)** -24 **(b)** -8 **(c)** -29

12. (a) -82 **(b)** -1 **(c)** -7 **13. (a)** $x^2 + 7x - 1$ **(b)** 29   **14. (a)** $2x^3 - 2x^2 + x - 12$ **(b)** -2 **15. (a)** $P(x) = -2.5x^2 + 280x - 3290$
(b) \$2147.50   **16. (a)** 6 square units   **(b)** 4 square units   **17.** $-12x^4y^3$ **18.** $6m^4n^7$ **19.** $-10a^3b^2 + 5a^2b^3 - 15a^2b^2$ **20.** $0.85c^3 + 2.15c^2 + 4.45c$
21. $x^2 - 7x - 18$ **22.** $-6x^2 + 26x - 8$ **23.** $2m^2 - 7mn - 4n^2$ **24.** $-2a^2 - 9a + 45$ **25.** $3x^3 + x^2 - 9x + 2$ **26.** $w^3 - 3w^2 - 12w + 32$
27. $2m^4 + m^3 - 11m^2 + 29m - 21$ **28.** $2p^3 + 11p^2q - 29pq^2 + 12q^3$ **29.** $9w^2 - 1$ **30.** $4x^2 - 25y^2$ **31.** $36k^2 - 60k + 25$ **32.** $9a^2 + 12ab + 4b^2$
33. $x^3 + 8$ **34.** $8x^3 - 27$ **35. (a)** $18x^2 - 27x - 35$ **(b)** 91   **36. (a)** $3x^3 + 5x^2 - x + 2$ **(b)** 270   **37.** $5x^2 - 30x + 53$ **38.** $-2xh + 3h - h^2$

39. $4x^2 - 2x$ **40.** $3w^4 - w^2 + 5w + 2$ **41.** $\dfrac{7}{2}y^2 + 6y - 3$ **42.** $\dfrac{m}{2n} + \dfrac{2}{n} - \dfrac{7}{2m}$ **43.** $3x + 4$ **44.** $-2x + 7 + \dfrac{5}{x + 5}$ **45.** $3z^2 + 2 - \dfrac{12}{2z + 3}$

46. $4k^2 + k - 2$ **47.** $8x^3 + 12x^2 + 18x + 27$ **48.** $2x^2 - 5x + 12 + \dfrac{2x + 7}{x^2 - 3x + 4}$ **49.** $5x + 1 + \dfrac{6}{x + 2}$ **50.** $9a + 4$ **51.** $3m^2 + 2m - 11$

52. $n^2 + 6n - 15 + \dfrac{7}{n - 4}$ **53.** $x^3 - x^2 + 7x - 7$ **54.** $2x^2 - 4x + 13 - \dfrac{34}{x + 2}$ **55. (a)** $x^2 + 5x - 3$ **(b)** 11   **56. (a)** $3x + 20 + \dfrac{9}{3x - 2}$ **(b)** $\dfrac{112}{11}$

57. (a) $2x^2 + 4x - 7$ **(b)** -7 **58. (a)** $3x^2 - 11x + 5 + \dfrac{2x - 3}{x^2 - x + 5}$ **(b)** $\dfrac{158}{17}$ **59.** 59   **60.** -45 **61.** $x - 2$ is a factor; $f(x) = (x - 2)(3x + 7)$

62. $x + 4$ is not a factor   **63.** $(5x + 1)$ meters   **64.** $(2x^2 + 5x + 3)$ square centimeters   **65.** $4(z + 6)$ **66.** $-7y(y - 13)$ **67.** $2xy(7x^2y - 4x + y)$
68. $5a^2b(6a^2b^2 + 3a - 5b)$ **69.** $(x + 5)(3x - 4)$ **70.** $(2c + 9)(3 - 4c)$ **71.** $(x - 5y)(6x + 5)$ **72.** $(a + 7)(2a - b - 1)$
73. $(x + 6)(x - 3)$ **74.** $(c + 2)(c - 5)$ **75.** $(7z + 8)(2z - 3)$ **76.** $(3w - 4)(7w + 2)$ **77.** $2x(x + 1)(x - 9)$ **78.** $5a^2(2a + 3)(a + 7)$

79. (a) $\dfrac{1}{2}n(n + 1)$ **(b)** 528   **80.** $R(x) = 2x(2600 - x^2)$ **81.** $(w + 2)(w - 13)$ **82.** prime   **83.** $-1(t - 12)(t + 6)$ **84.** $(m + 3)(m + 7)$

85. $(x + 20y)(x - 16y)$ **86.** $(r - 2s)(r - 3s)$ **87.** $(x + 3)(5x - 2)$ **88.** $(2m + 11)(3m + 4)$ **89.** prime   **90.** $2(t + 3)(4t - 1)$
91. $(2x - 1)(3x - 5)$ **92.** $(7r + 2s)(3r - s)$ **93.** $(5x - 3y)(4x - 9y)$ **94.** $-2(s - 7)(s + 1)$ **95.** $(x^2 + 1)(x^2 - 11)$ **96.** $(xy + 4)(10xy + 1)$
97. $(a + 7)(a - 8)$ **98.** $w(2w + 7)$ **99.** $(x + 11)^2$ **100.** $(w - 17)^2$ **101.** $(12 - c)^2$ **102.** $(x - 4)^2$ **103.** $(8y + 5)^2$ **104.** $12(z + 2)^2$
105. $(x - 14)(x + 14)$ **106.** $(7 - y)(7 + y)$ **107.** $(t - 15)(t + 15)$ **108.** $(2w - 9)(2w + 9)$ **109.** $(6x^2 - 5y)(6x^2 + 5y)$
110. $20m(2n - 1)(2n + 1)$ **111.** $(x - 7)(x^2 + 7x + 49)$ **112.** $(9 - y)(y^2 + 9y + 81)$ **113.** $(3x - 5y)(9x^2 + 15xy + 25y^2)$
114. $(2m^2 + 3n)(4m^4 - 6m^2n + 9n^2)$ **115.** $2(a - b)(a + b)(a^2 + ab + b^2)(a^2 - ab + b^2)$ **116.** $(y + 3)(y^2 - 6y + 21)$ **117.** $(x + 1)(x + 6)$
118. $-4xy^3(2x - 3)$ **119.** $7x(x - 1)(x - 4)$ **120.** $3(x - 3)(x + 2)$ **121.** $(2z - 15)^2$ **122.** $(3x + 7)(4x - 7)$ **123.** $(2n - 7)(5n + 1)$
124. $-(y - 2)(y + 4)$ **125.** $2(x - 5)(x^2 + 3)$ **126.** $9(h + 2)(3h^2 + 4)$ **127.** $5p(pq - 4)(pq + 4)$ **128.** $(m - 1)(m + 1)(m - 2)(m + 2)$
129. $-2(2m^2 - 7)(4m^4 + 14m^2 + 49)$ **130.** $(h - 1)(h + 1)(h + 2)$ **131.** $4(3x + y)(9x^2 - 3xy + y^2)$ **132.** $F(c) = (c - 12)^2$
133. $f(x) = -3(x - 5)(x + 3)$ **134.** $g(x) = 4(2x + 5)(2x - 5)$ **135.** $G(y) = 2(2y + 5)(4y^2 - 10y + 25)$ **136.** $f(x) = -4(x^2 + 4)$
137. $(x - 5)(x + 5)$ square units   **138.** $(4 - \pi)x^2$ square units   **139.** $\{-5, 13\}$ **140.** $\{-18, -3\}$ **141.** $\{-5, 3\}$ **142.** $\{-4, 0\}$ **143.** $\{-11, 10\}$

144. $\left\{\dfrac{2}{5}, -\dfrac{7}{3}\right\}$ **145.** $\{-6, -4\}$ **146.** $\{4, -2\}$ **147.** $\left\{-\dfrac{5}{2}, -2, 2\right\}$ **148.** $\left\{-3, -\dfrac{1}{5}, 3\right\}$ **149. (a)** $x = -8$ or $x = 3$ **(b)** $x = -7$ or $x = 2$; $(-8, 6)$,

$(3, 6)$, $(-7, -4)$, and $(2, -4)$ are on the graph.   **150. (a)** $x = 0$ or $x = \dfrac{4}{5}$ **(b)** $x = -\dfrac{1}{5}$ or $x = 1$; $(0, 3)$, $\left(\dfrac{4}{5}, 3\right)$, $\left(-\dfrac{1}{5}, 4\right)$, and $(1, 4)$ are on the graph.

151. zeros: -4, -2, and 0; x-intercepts: $(-4, 0)$, $(-2, 0)$, and $(0, 0)$ **152.** zeros: $-\dfrac{3}{2}$ and 7; x-intercepts: $\left(-\dfrac{3}{2}, 0\right)$ and $(7, 0)$ **153.** after 7 seconds

154. (a) 0.80   **(b)** 0.90

Chapter 4 Test   1. $-5x^7 + x^4 + 7x^2 - x + 1$; Degree: 7   **2.** $-\dfrac{5}{3}a^3b^2 + 9a^2b - 5ab - 4$ **3.** 7   **4.** $7x^3 - 4x^2 - 3x + 1$ **5.** $2a^3b^3 - 3a^3b^2 + 4a^2b$

6. $12x^2 + 47x - 17$ **7.** $4m^2 - 4mn + n^2$ **8.** $3z - 7 + \dfrac{-2z + 11}{2z^2 + 1}$ **9.** $5x + 3$ **10.** 2   **11.** 22   **12.** $4ab^2(3a^2 + 2a - 4b)$ **13.** $(2c + 7)(3c - 2)$

14. $(x - 16)(x + 3)$ **15.** $-(2p + 3)(7p - 2)$ **16.** $(z + 3)(5z - 8)$ **17.** $-2(7x - 4)^2$ **18.** $4(2x - 7)(2x + 7)$ **19.** $\left\{1, \dfrac{7}{3}\right\}$

20. 9 meters by 12 meters

Getting Ready for Chapter 5: A Review of Operations on Rational Numbers

1. lowest terms   **2.** $\dfrac{5}{6}$ **3.** $\dfrac{20}{3}$ **4.** $-\dfrac{3}{2}$ **5.** $\dfrac{1}{2}$ **6.** $\dfrac{5}{3}$ **7.** $-\dfrac{3}{5}$ **8.** $\dfrac{4}{3}$ **9.** $-\dfrac{5}{9}$ **10.** LCD = 75; $\dfrac{3}{25} = \dfrac{9}{75}, \dfrac{2}{15} = \dfrac{10}{75}$ **11.** LCD = 126; $\dfrac{5}{18} = \dfrac{35}{126}, -\dfrac{1}{63} = -\dfrac{2}{126}$

12. $\dfrac{19}{20}$ **13.** $\dfrac{17}{60}$ **14.** $-\dfrac{1}{6}$ **15.** $\dfrac{1}{3}$ **17.** $-\dfrac{3}{7}$ **19.** 5   **21.** $\dfrac{5}{3}$ **23.** -3 **25.** $\dfrac{1}{12}$ **27.** $\dfrac{2}{5}$ **29.** $-\dfrac{9}{4}$ **31.** 6   **33.** 2   **35.** $\dfrac{5}{3}$ **37.** $-\dfrac{5}{12}$ **39.** $\dfrac{13}{12}$ **41.** $-\dfrac{23}{30}$ **43.** $-\dfrac{41}{90}$

45. $-\dfrac{23}{40}$ **47.** $\dfrac{17}{120}$ **49.** $-\dfrac{41}{45}$ **51.** $-\dfrac{59}{150}$ **53.** To find the least common denominator (LCD) of two rational numbers, factor each denominator as the product of prime numbers. Copy first the common factors and then the uncommon factors. The product of these factors is the LCD.

Chapter 5 Rational Expressions and Rational Functions

Section 5.1 Multiplying and Dividing Rational Expressions   1. rational expression   **2.** numerator; denominator   **3.** False   **4.** $\{x | x \ne -6\}$

5. $\{z | z \ne -7, z \ne 4\}$ **6.** False   **7.** $\dfrac{x - 4}{x + 7}$ **8.** $\dfrac{z^2 + 4z + 16}{2z + 5}$ **9.** $-(w + 5)$ **10.** $\dfrac{(3p + 2)(p - 1)}{2(p + 2)}$ **11.** -1 **12.** $-(m + n)$ **13.** $9a^2b^3$

14. $\dfrac{(m - 5)(m + 1)}{2}$ **15.** $\{x | x \ne -6, x \ne 5\}$ **16. (a)** $R(x) = x + 1; \left\{x \middle| x \ne -3, x \ne \dfrac{5}{3}, x \ne 5\right\}$

(b) $H(x) = \dfrac{3x(x - 5)}{4x + 3}; \left\{x \middle| x \ne -1, x \ne -\dfrac{3}{4}, x \ne 0, x \ne \dfrac{5}{3}\right\}$ **17.** $\{x | x \ne -5\}$ **19.** $\{x | x \ne -2, x \ne 8\}$ **21.** $\left\{p \middle| p \ne -\dfrac{5}{2}, p \ne 2\right\}$

23. $\{x | x$ is any real number$\}$ or $(-\infty, \infty)$ **25.** $\{x | x \ne 1\}$ **27.** $\dfrac{2}{x - 4}$ **29.** $p + 3$ **31.** $\dfrac{5}{x^2}$ **33.** $\dfrac{q + 3}{q - 2}$ **35.** $\dfrac{y - 4}{y + 5}$ **37.** $-\dfrac{x + 3}{x + 5}$ **39.** $\dfrac{x - 2}{2x(x - 4)}$

41. $\dfrac{x - 3y}{x - 2y}$ **43.** $\dfrac{x^2 + 3}{x - 5}$ **45.** $\dfrac{x^2 - 2x + 4}{x - 7}$ **47.** $\dfrac{1}{4x(x + 3)}$ **49.** $\dfrac{x - 2}{x - 1}$ **51.** $\dfrac{(x + 3)^2}{(x + 5)(x + 7)}$ **53.** $-\dfrac{(2q + 1)(q + 2)}{q + 6}$ **55.** $x - 3$ **57.** 1

59. $\dfrac{9(x + 3)}{8x(x - 4)}$ **61.** $\dfrac{2}{ab}$ **63.** $\dfrac{(p - 5)(p + 3)}{p(2p + 1)}$ **65.** $\dfrac{x^2 + 3x + 1}{3(x + 1)}$ **67.** $\{x | x \ne 1\}$ **69.** $\left\{x \middle| x \ne -\dfrac{1}{2}, x \ne 4\right\}$ **71.** $\{x | x \ne -5, x \ne -1\}$

73. $\left\{x \middle| x \ne 2, x \ne \dfrac{5}{2}\right\}$ **75.** $\{x | x$ is any real number$\}$ **77. (a)** $\dfrac{(x + 3)(x - 1)}{2x + 3}; \left\{x \middle| x \ne -6, x \ne -\dfrac{3}{2}, x \ne 5\right\}$

(b) $(x - 5)(3x - 1); \left\{x \middle| x \ne -6, x \ne -3, x \ne \dfrac{1}{3}\right\}$ **79. (a)** $\dfrac{(3x + 5)(x + 6)}{(x^2 + x + 1)(2x + 7)}; \left\{x \middle| x \ne -\dfrac{7}{2}, x \ne 1, x \ne 2\right\}$

(b) $\dfrac{x - 2}{x^2 + x + 1}; \left\{x \middle| x \ne -\dfrac{5}{3}, x \ne -1, x \ne 1\right\}$ **81.** $\dfrac{z + 3}{2}$ **83.** $\dfrac{m - 2n}{2(m - n)}$ **85.** $\dfrac{4}{w}$ **87.** $\dfrac{2(x + 2)}{(x + 1)(x - 1)}$ **89.** Answers will vary. One possible rational

expression is $\dfrac{x}{x - 3}$.   **91.** Answers will vary. One possible rational expression is $\dfrac{7}{(x + 4)(x - 5)} = \dfrac{7}{x^2 - x - 20}$.   **93.** $R(x) = \dfrac{4}{x^2 + x - 2}$

95. (a) approximately 9.8208 m/sec^2 **(b)** approximately 9.8159 m/sec^2 **(c)** approximately 9.7936 m/sec^2 **97. (a)** $\{x | x \ne 2\}$

(b)

x	3	2.5	2.1	2.01	2.001	2.0001
$f(x)$	1	2	10	100	1000	10,000

f gets larger in the positive direction.

(c)

x	1	1.5	1.9	1.99	1.999	1.9999
$f(x)$	-1	-2	-10	-100	-1000	$-10,000$

f gets larger in the negative direction.

(d) f gets larger in the positive direction; f gets larger in the negative direction; these results are the same as those in parts (b) and (c).

99. $\dfrac{1}{x(x + 1)}$ **101.** A rational expression is simplified provided there are no common factors between the numerator and denominator.

103. It is not a rational expression because $\sqrt{x} = x^{\frac{1}{2}}$. A rational expression is the ratio of two polynomials, and polynomials must have nonnegative integer exponents (1/2 is not a nonnegative integer).

105.   **107.**   **109.**   **111.** Answers will vary. Ideally, the linear functions would be graphed using the properties of a linear function (i.e., slope and intercepts). The functions in Problems 109 and 110 are likely graphed using point-plotting, which is less efficient than using properties.

Section 5.2 Adding and Subtracting Rational Expressions   1. $2x - 1$ **2.** $\dfrac{3(x + 3)}{x + 5}$ **3.** $\dfrac{4x - 3}{x - 5}$ **4.** least common denominator   **5.** $24x^2y^3$

6. $(x + 2)^2(x - 7)$ **7.** $\dfrac{9a + 8}{30a^2}$ **8.** $-\dfrac{1}{6y}$ **9.** $\dfrac{4x^2 + 10x - 5}{(x - 1)(x + 2)}$ **10.** $\dfrac{y^2 + 7y + 17}{(3y + 2)(y - 1)(y + 4)}$ **11.** $\dfrac{x^2 + 5x + 20}{(2x - 3)(x + 2)^2}$ **12.** $\dfrac{-4}{x - 2}$ **13.** $\dfrac{3x + 5}{x + 1}$

15. $\dfrac{2x - 1}{2x + 5}$ **17.** $\dfrac{1}{x - 5}$ **19.** $\dfrac{x - 3}{x + 3}$ **21.** $\dfrac{3x - 1}{x - 5}$ **23.** $3x - 1$ **25.** LCD = $8x^3$ **27.** LCD = $90x^3y^2$ **29.** LCD = $(x - 4)(x + 2)$

31. LCD $= (x - 4)(x + 3)(x - 5)$ **33.** LCD $= p^2(p + 2)(2p - 1)$ **35.** $\dfrac{5x + 6}{8x^2}$ **37.** $\dfrac{25b - 16a}{60a^2b^2}$ **39.** $\dfrac{14(y - 1)}{(y - 5)(y + 3)}$ **41.** $\dfrac{a + 4}{a + 2}$

43. $\dfrac{-2(x - 11)}{(x + 3)(x - 2)(x + 4)}$ **45.** $\dfrac{2x^2 - 5x + 5}{(x + 2)(x + 1)(x - 2)}$ **47.** $\dfrac{-11w + 9}{(2w + 1)(w + 1)(w - 3)}$ **49.** $\dfrac{2x^2 + xy - 5y^2}{(x - 3y)^2(x + y)}$ **51.** $\dfrac{1}{(x + 1)(x - 1)}$ **53.** 0

55. $\dfrac{6}{w(w - 2)}$ **57.** $\dfrac{(2p + 1)(p - 8)}{(p + 3)^2(2p - 5)}$ **59.** $\dfrac{x + 2}{x(x - 1)^2}$ **61.** $\dfrac{3}{x^2 + 3x + 9}$ **63.** $x - 1$ **65.** $\dfrac{7x + 15}{x + 3}$ **67.** $\dfrac{-1}{(b - 2)(b + 4)}$ **69.** $\dfrac{y - 3}{y + 3}$

71. $\dfrac{7x + 9}{(x - 2)(x - 3)(x + 1)}$ **73. (a)** $\dfrac{5x - 1}{(x - 2)(x + 1)}$ **(b)** $\{x \mid x \neq -1, x \neq 2\}$ **75. (a)** $\dfrac{2x + 7}{(x - 4)(x + 3)}$ **(b)** $\{x \mid x \neq -3, x \neq -1, x \neq 4\}$

(c) $\dfrac{-1}{(x - 4)(x + 3)}$ **(d)** $\{x \mid x \neq -3, x \neq -1, x \neq 4\}$ **77. (a)** $S(x) = \dfrac{2x^3 + 8000}{x}$ **(b)** $S(10) = 1000$; if the length of the base is 10 inches,

then the surface area of the box will be 1000 square inches.   **79. (a)** $\dfrac{200s + 500}{s(s + 10)}$ **(b)** $T(50) = 3.5$; if the average speed for the first 50 miles of the

trip is 50 miles per hour, then it will take a total time of 3.5 hours to get to the neighboring university.   **81.** $\dfrac{x + y}{xy}$ **83.** Completely factor each expression

in the denominator of the rational expressions. Copy all common factors that are numerical. Then copy all common variables and use the largest
exponent for the LCD. Finally, copy all uncommon factors. The product of these factors represents the LCD.   **85.** $4a^2 - 12a$ **87.** $p^2 - 9$
89. $w^3 - 8$ **91.** $x^2 + 2x + 4$

Section 5.3 Complex Rational Expressions   **1.** complex rational expression   **2.** True   **3.** $\dfrac{7}{5}$ **4.** $\dfrac{4(z - 4)}{z}$ **5.** False   **6.** $\dfrac{1}{4}$ **7.** $\dfrac{7}{5}$ **8.** $\dfrac{4(z - 4)}{z}$

9. $\dfrac{-4(x + 2)}{x(x + 5)}$ **10.** $\dfrac{ab}{3b - a}$ **11.** $\dfrac{4}{9}$ **13.** $\dfrac{x + 1}{x - 1}$ **15.** $\dfrac{-2a}{(a - 2)^2}$ **17.** $\dfrac{1}{(x - 1)(x + 3)}$ **19.** $-\dfrac{23}{12}$ **21.** $\dfrac{(w - 1)(w + 1)}{w^2 + 1}$ **23.** $\dfrac{x + 2}{(x - 2)(x - 7)}$

25. $\dfrac{z^2 + 4}{4z}$ **27.** $\dfrac{1}{y + 3}$ **29.** $2(n - m)$ **31.** $\dfrac{x + 3}{x(x - 3)}$ **33.** $-x$ **35.** $\dfrac{1}{(x - 1)(x - 3)}$ **37.** $\dfrac{2(x + 1)}{x + 4}$ **39.** $-b$ **41.** $\dfrac{3x}{2 - x}$ **43.** $\dfrac{x - 1}{2(x + 1)}$

45. $\dfrac{-x + 5}{3x + 1}$ **47.** $\dfrac{3xy}{y - x}$ **49.** $\dfrac{mn}{(m + n)^2}$ **51.** $\dfrac{1}{4(b + a)}$ **53. (a)** $\dfrac{R_1R_2}{R_1 + R_2}$ **(b)** approximately 2.857 ohms   **55. (a)** $\dfrac{R_1R_2}{(n - 1)(R_1 + R_2)}$ **(b)** 0.375 meter

57. (a) $\dfrac{2x + 1}{x + 1}$ **(b)** $\dfrac{3x + 2}{2x + 1}$ **(c)** $\dfrac{5x + 3}{3x + 2}$ **(d)** $\dfrac{8x + 5}{5x + 3}$ **(e)** from part (a): 2, 1, 1; from part (b): 3, 2, 1; from part (c): 5, 3, 2; from part (d): 8, 5, 3; Each

number is the sum of the previous two. The sequence begins 1, 1, 2, 3, 5, 8.   **59.** $-\dfrac{2x + h}{x^2(x + h)^2}$ **61.** $-\dfrac{1}{(x - 1)(x + h - 1)}$ **65.** $\{-4\}$ **67.** $\left\{-\dfrac{109}{4}\right\}$

69. $\left\{-7, \dfrac{2}{3}\right\}$

Putting the Concepts Together (Sections 5.1–5.3)   **1.** $\left\{x \mid x \neq -\dfrac{1}{3}, x \neq 6\right\}$ **2.** $\dfrac{-4n}{2n + 3}$ **3.** $\dfrac{2p - 5q}{3p - 4q}$ **4.** $\dfrac{a(a - 5)}{2(a + 6)}$ **5.** $\dfrac{(x + 2)(x - 7)}{(3x + 1)^2}$

6. $\dfrac{x - 5}{x - 2}$ **7.** $\dfrac{n}{(n - 2)(n - 3)}$ **8.** $\dfrac{3y^2 - 3y - 29}{(y - 3)(y - 4)(y + 8)}$ **9.** $\dfrac{3}{(x - 7)(2x + 1)}$ **10.** $\dfrac{3x^2 - 14x - 1}{(x - 4)(x - 7)}$ **11.** $\dfrac{n + m}{mn}$ **12.** $\dfrac{z + 3}{z - 5}$

Section 5.4 Rational Equations   **1.** rational equation   **2.** $\{-2\}$ **3.** $\left\{\dfrac{2}{3}\right\}$ **4.** $\{-2\}$ **5.** $\left\{-\dfrac{5}{3}\right\}$ **6.** $\{8\}$ **7.** Extraneous solutions   **8.** True

9. $\{\ \}$ or $\varnothing$ **10.** $\{0\}$ **11.** $\left\{-\dfrac{3}{2}, 4\right\}$ **12.** $\{-5\}$ **13.** False   **14.** $\{-1, 5\}$; $(-1, 4), (5, 4)$ **15.** $\left\{-1, \dfrac{3}{2}\right\}$; $(-1, 1)$ and $\left(\dfrac{3}{2}, 1\right)$ **16.** $\dfrac{1}{2}$ hour and

12 hours after ingestion   **17.** $\{-4\}$ **19.** $\{-16\}$ **21.** $\varnothing$ or $\{\ \}$ **23.** $\{-1, 5\}$ **25.** $\left\{-\dfrac{1}{3}, \dfrac{3}{2}\right\}$ **27.** $\{1\}$ **29.** $\{4\}$ **31.** $\{-5, 6\}$ **33.** $\{4\}$

35. $\varnothing$ or $\{\ \}$ **37.** $\{-8\}$ **39.** $\left\{\dfrac{1}{4}\right\}$ **41.** $\{-1\}$ **43.** $x = 9$ or $x = 1$; $(1, 10)$ and $(9, 10)$ **45.** $x = -\dfrac{1}{2}$ or $x = -4$; $\left(-\dfrac{1}{2}, -9\right)$ and $(-4, -9)$

47. $x = 6$; $\left(6, \dfrac{9}{2}\right)$ **49.** $x = -5$ or $x = 3$; $(-5, 3)$ and $\left(3, \dfrac{1}{3}\right)$ **51.** $\left\{\dfrac{1}{4}\right\}$ **53.** $\{3\}$ **55.** $\{-19\}$ **57.** $\{2\}$ **59.** $\left\{0, -\dfrac{3}{5}\right\}$ **61.** $-\dfrac{1}{3}$; x-int: $\left(-\dfrac{1}{3}, 0\right)$

63. $-4, \dfrac{3}{2}$; x-int: $(-4, 0), \left(\dfrac{3}{2}, 0\right)$ **65.** $-2, 2$; x-int: $(-2, 0), (2, 0)$ **67.** either 50 or 100 bicycles   **69. (a)** 80%   **(b)** 90%   **71.** 650 walks

73. Answers will vary. One possibility follows: $\dfrac{2}{x - 1} = \dfrac{3}{x + 1}$ **75.** $\left\{-\dfrac{5}{3}, -1\right\}$ **77.** We find the domain of the rational equation so that we can identify

extraneous solutions.   **79.** $\left\{-\dfrac{1}{2}, 3\right\}$ **81.** $\dfrac{-2x^2 + 6x + 5}{(x - 2)(x + 1)}$ **83.** $\{-1, 10\}$ **85.** $\{-6.67\}$ **87.** $\{-13\}$ **89.** $\{\ \}$ or $\varnothing$

Section 5.5 Rational Inequalities   **1.** rational   **2.** $\{x \mid x < -3 \text{ or } x \geq 7\}$; $(-\infty, -3) \cup [7, \infty)$

3. $\{x \mid -5 < x < 1\}$; $(-5, 1)$

4. $\{x \mid -2 < x < 1\}$; $(-2, 1)$

5. $\{x \mid x < -1 \text{ or } x > 4\}$ or $(-\infty, -1) \cup (4, \infty)$

7. $\{x \mid -9 < x < 3\}$ or $(-9, 3)$

9. $\{x \mid x \le -10 \text{ or } x > 4\}$ or $(-\infty, -10] \cup (4, \infty)$

11. $\left\{x \mid x \le -8 \text{ or } -\dfrac{5}{3} \le x < 2\right\}$ or $(-\infty, -8] \cup \left[-\dfrac{5}{3}, 2\right)$

13. $\{x \mid x > -1\}$ or $(-1, \infty)$ **15.** $\left\{x \mid \dfrac{3}{2} < x < 3\right\}$ or $\left(\dfrac{3}{2}, 3\right)$

17. $\{x \mid 0 < x \le 1 \text{ or } x > 4\}$ or $(0, 1] \cup (4, \infty)$

19. $\{x \mid -5 < x < 2 \text{ or } x \ge 23\}$ or $(-5, 2) \cup [23, \infty)$

21. $\left\{x \mid -3 < x < \dfrac{1}{2} \text{ or } x > 5\right\}$ or $\left(-3, \dfrac{1}{2}\right) \cup (5, \infty)$

23. $\{x \mid x > -3\}$; $(-3, \infty)$ **25.** $\{x \mid -7 \le x < 8\}$ or $[-7, 8)$

27. $\{x \mid x \ge 4\}$; $[4, \infty)$ **29.** $\{x \mid x < -4 \text{ or } x \ge 9\}$ or $(-\infty, -4) \cup [9, \infty)$

31. $\{x \mid -1 < x \le 6\}$ or $(-1, 6]$ **33.** $\left\{x \mid -2 < x < \dfrac{5}{2}\right\}$ or $\left(-2, \dfrac{5}{2}\right)$ **35.** 100 or more bicycles   **37.** Answers will vary. One possibility: $\dfrac{1}{x-2} > 0$

39. Because -1 is not in the domain of the variable x, the solution set is $\{x \mid -1 < x \le 4\}$.   **41.** $(2, 0)$ **43.** $\left(-\dfrac{7}{2}, 0\right), (2, 0)$ **45.** $\left(\dfrac{2}{3}, 0\right)$

47. $\{x \mid x \le -4 \text{ or } x > -1\}$ or $(-\infty, -4] \cup (-1, \infty)$ **49.** $\{x \mid 7 < x < 26\}$ or $(7, 26)$

Section 5.6 Models Involving Rational Expressions   **1. (a)** $b = \dfrac{Y - G}{Y}$ **(b)** 0.9   **2.** proportion; similar   **3.** 120 feet   **4.** $AB = 12$; $DF = 4$

5. approximately 321.5 million people   **6.** $\dfrac{40}{3}$ hours or 13 hours, 20 minutes   **7.** 33.33 minutes or 33 minutes, 20 seconds   **8.** 4 miles per hour

9. $P_1 = \dfrac{V_2 P_2}{V_1}$ **11.** $t = \dfrac{R - r}{R} = 1 - \dfrac{r}{R}$ **13.** $x = \dfrac{y - y_1 + mx_1}{m}$ **15.** $v = \dfrac{\omega(I + mr^2)}{rm}$ **17.** $m = \dfrac{MV}{v - V}$ **19.** $AB = 16$; $DF = 5$ **21.** 316,504,673 people

23. approximately \$6.40   **25.** 12.5 pounds   **27.** approximately 34.3 minutes   **29.** 6 hours   **31.** 10 hours   **33.** 35 miles per hour   **35.** 1.8 feet per second

37. Gronkowski will be on his own 35-yard line when Forte catches up to him.   **39.** 8 miles per hour   **41.** 45 miles per hour   **43.** Approximately 2.15 m

45. a^{15} **47.** $\dfrac{a^4}{b^6}$ **49.** $\dfrac{n^{12}}{9m^4}$

Section 5.7 Variation   **1.** Variation   **2.** $y = kx$

3. (a) $C(g) = 2.75g$ **(b)** \$12.65   **4.** $y = \dfrac{k}{x}$ **5. (a)** $y = \dfrac{6}{x}$ **(b)** 1.5   **6. (a)** $V(l) = \dfrac{15{,}000}{l}$ **(b)** 300 oscillations per second   **7.** joint   **8.** 1750 joules

9. combined variation   **10.** approximately 1.44 ohms   **11. (a)** $k = 6$ **(b)** $y = 6x$ **(c)** $y = 42$ **13. (a)** $k = \dfrac{3}{7}$ **(b)** $y = \dfrac{3}{7}x$

(c) $y = 12$ **15. (a)** $k = \dfrac{1}{2}$ **(b)** $y = \dfrac{1}{2}x$ **(c)** $y = 15$ **17. (a)** $k = 20$ **(b)** $y = \dfrac{20}{x}$ **(c)** $y = 4$ **19. (a)** $k = 21$

(b) $y = \dfrac{21}{x}$ **(c)** $y = \dfrac{3}{4}$ **21. (a)** $k = \dfrac{1}{4}$ **(b)** $y = \dfrac{1}{4}xz$ **(c)** $y = 27$ **23. (a)** $k = \dfrac{13}{10}$ **(b)** $Q = \dfrac{13x}{10y}$ **(c)** $Q = \dfrac{117}{40}$

(c)

Cost (\$) vs Number of Gallons, point (8, 22.00)

25. (a) $p(b) = 0.0047742b$ **(b)** \$1050.32

27. (a) $C(w) = 5.6w$ **(b)** \$19.60

29. 96 feet per second   **31. (a)** $D(p) = \dfrac{375}{p}$ **(b)** 125 bags of candy   **33.** 450 cc

35. approximately 119.8 pounds   **37.** 2250 newtons   **39.** 1.4007×10^{-7} newtons

41. 360 pounds   **43. (a)** 314.16   **(b)** approximately 942.48 meters per minute; approximately 15.708 meters per second   **(c)** $k \approx 0.056$ **(d)** approximately 7.86 newtons

(c)

Monthly Payment (\$) vs Amount Borrowed (\$), points (220,000, 1050.32) and (190,000, 907.09)

(c)

Cost (\$) vs Weight (pounds), points (5, 28), (3.5, 19.60), (0, 0)

Chapter 5 Review   **1.** $\left\{x \mid x \ne \dfrac{2}{3}\right\}$ **2.** $\{a \mid a \ne -4, a \ne 7\}$ **3.** $\{m \mid m \text{ is any real number}\}$ or $(-\infty, \infty)$ **4.** $\{n \mid n \ne -2, n \ne 4\}$ **5.** $\dfrac{6}{x - 5}$

6. $\dfrac{2}{y^3}$ **7.** $\dfrac{w - 7}{w + 4}$ **8.** $\dfrac{3a + b}{5a + 2b}$ **9.** $\dfrac{-1}{3m + 1}$ **10.** $\dfrac{n^2 + 3}{n - 4}$ **11.** $\dfrac{p}{2(p - 6)}$ **12.** $\dfrac{2q}{3(q - 5)}$ **13.** $\dfrac{(x - 4)(x + 6)}{(x + 2)^2}$ **14.** $\dfrac{2(y + 5)}{y(y - 5)}$ **15.** $\dfrac{3a + b}{2a - b}$ **16.** $\dfrac{3}{m - 2}$

17. $\dfrac{9c}{2d^3}$ **18.** $\dfrac{6(z - 3)}{7}$ **19.** $\dfrac{x + 7}{x - 3}$ **20.** $\dfrac{m + 4n}{m + 5n}$ **21.** $\dfrac{2(p - q)}{p + 3q}$ **22.** $\dfrac{2a + 5}{9a^2 + 6a + 4}$ **23.** $(x + 2)^2$; $\left\{x \mid x \ne \dfrac{1}{2}, x \ne 5\right\}$

24. $\dfrac{x-5}{x+7}$; $\left\{x \middle| x \neq -7, x \neq -2, x \neq \dfrac{1}{2}\right\}$ **25.** $\dfrac{(x-5)^2}{(2x-1)^2}$; $\left\{x \middle| x \neq -2, x \neq \dfrac{1}{2}, x \neq 5\right\}$ **26.** $\dfrac{(x+2)^2(x+7)}{x-5}$; $\left\{x \middle| x \neq -7, x \neq -2, x \neq \dfrac{1}{2}, x \neq 5\right\}$

27. $\dfrac{4x+3}{x-5}$ **28.** 4   **29.** $\dfrac{a+6}{a-2}$ **30.** $\dfrac{b+5}{2b+3}$ **31.** 3c   **32.** $\dfrac{d+3}{d+1}$ **33.** LCD $= 36x^4$ **34.** LCD $= (y+2)(y-9)$ **35.** LCD $= p^2(2p+5)(p-4)$

36. LCD $= (q+5)(q-1)(q-3)$ **37.** $\dfrac{m^2+4n^2}{m^3n^4}$ **38.** $\dfrac{9x-7y^2}{6x^2y^3}$ **39.** $\dfrac{p^2+q^2}{(p-q)(p+q)}$ **40.** $\dfrac{20x+17}{(x-7)(x-3)(x+4)}$ **41.** $\dfrac{y+8}{(y-1)^2(y+2)}$

42. $\dfrac{11a+7b}{(2a-3b)(2a+3b)}$ **43.** $\dfrac{6x}{x+3}$ **44.** $\dfrac{5}{n^2-5n+25}$ **45.** $\dfrac{2n}{m-7n}$ **46.** $\dfrac{1}{z-3}$ **47.** $\dfrac{3}{y+2}$ **48.** $\dfrac{a-3}{(a+4)(a-2)}$ **49. (a)** $\dfrac{x^2+x+10}{(x-4)(x+2)}$

(b) $\{x | x \neq -2, x \neq 4\}$ **50. (a)** $\dfrac{x+8}{(2x-5)(2x+1)}$ **(b)** $\left\{x \middle| x \neq -3, x \neq -\dfrac{1}{2}, x \neq \dfrac{5}{2}\right\}$ **51.** $x+1$ **52.** $\dfrac{xy}{x+y}$ **53.** $\dfrac{a-b}{a+b}$ **54.** $\dfrac{-a}{a+3}$

55. $\dfrac{3t+4}{5t^2+1}$ **56.** $\dfrac{1}{a+b}$ **57.** $\dfrac{z+1}{z-1}$ **58.** $\dfrac{x-1}{x+1}$ **59.** $\dfrac{x+y}{x-y}$ **60.** 1   **61.** $\dfrac{2}{x(x-2)}$ **62.** $\dfrac{z-5}{z+5}$ **63.** $\dfrac{n}{m}$ **64.** $\dfrac{2}{(x+1)^2(x-2)}$ **65.** $\dfrac{3y^2-3x^2}{xy}$

66. $\dfrac{12d-2c}{c}$ **67.** $\{10\}$ **68.** $\left\{\dfrac{4}{3}\right\}$ **69.** $\{-2, 7\}$ **70.** $\{9\}$ **71.** $\{2\}$ **72.** $\{3\}$ **73.** $\varnothing$ or $\{\ \}$ **74.** $\{-3\}$ **75.** $\{6\}$ **76.** $\left\{-\dfrac{4}{3}\right\}$ **77.** $\{5\}$

78. $\varnothing$ or $\{\ \}$ **79.** $x = 5$; $(5, 2)$ **80.** $x = -3$ or $x = 7$; $(-3, 4)$ and $(7, 4)$

81. $\{x | x < -2$ or $x \geq 4\}$ or $(-\infty, -2) \cup [4, \infty)$

82. $\{y | -4 < y < 5\}$ or $(-4, 5)$ **83.** $\{z | -3 < z < 3\}$ or $(-3, 3)$

84. $\{w | w < -7$ or $2 < w < 4\}$ or $(-\infty, -7) \cup (2, 4)$

85. $\{m | -5 < m < 2$ or $m \geq 5\}$ or $(-5, 2) \cup [5, \infty)$

86. $\{n | 0 \leq n < 2\}$ or $[0, 2)$ **87.** $\left\{a \middle| 2 < a < \dfrac{7}{2}\right\}$ or $\left(2, \dfrac{7}{2}\right)$

88. $\{c | c < -6$ or $0 < c < 2\}$ or $(-\infty, -6) \cup (0, 2)$ **89.** $\left\{x \middle| -\dfrac{3}{2} < x < 4\right\}$ or $\left(-\dfrac{3}{2}, 4\right)$

90. $\{x | x \leq -5$ or $x > -1\}$ or $(-\infty, -5] \cup (-1, \infty)$ **91.** $C = \dfrac{C_1 C_2}{C_2 + C_1}$ **92.** $T_2 = \dfrac{T_1 P_2 V_2}{P_1 V_1}$ **93.** $G = \dfrac{4\pi^2 a^2}{MT}$ **94.** $x = z \cdot \sigma + \mu$

95. $AB = 24$ and $DF = 10$ **96.** 18.75 feet   **97.** 88 grams   **98.** \$78.00   **99.** 28.8 minutes (or 28 minutes and 48 seconds)   **100.** 2.8 hours (or 2 hours and 48 minutes)   **101.** 21 hours; 28 hours   **102.** 3 minutes   **103.** 60 miles per hour   **104.** 15 miles per hour   **105.** Todd's average walking speed is 3 miles per hour, and his average running speed is 12 miles per hour.   **106.** 60 miles per hour   **107. (a)** $k = 5$ **(b)** $y = 5x$ **(c)** 50

108. (a) $k = -6$ **(b)** $y = -6x$ **(c)** -48 **109. (a)** $k = 60$ **(b)** $y = \dfrac{60}{x}$ **(c)** 12   **110. (a)** $k = \dfrac{3}{4}$ **(b)** $y = \dfrac{3}{4}xz$ **(c)** 42   **111. (a)** $k = 72$ **(b)** $s = \dfrac{72}{t^2}$ **(c)** 8   **112. (a)** $k = \dfrac{8}{5}$ **(b)** $w = \dfrac{8x}{5z}$ **(c)** $\dfrac{9}{10}$ **113.** 6 inches   **114.** \$352.19   **115.** 1200 kilohertz   **116.** 12 ohms   **117.** 704 cubic cm   **118.** 375 cubic inches

Chapter 5 Test   **1.** $\left\{x \middle| x \neq -\dfrac{1}{2}, x \neq 7\right\}$ **2.** $\dfrac{2m-3}{3m-1}$ **3.** $-\dfrac{1}{a+4b}$ **4.** $\dfrac{2x+5}{2x(x-4)}$ **5.** $\dfrac{y+4}{3y+1}$ **6.** $\dfrac{2q}{p-q}$ **7.** $\dfrac{3c-4}{(c-2)(c-1)}$

8. $\dfrac{x-4}{3(x-2)(x-5)}$; $\{x | x \neq -2, x \neq 0, x \neq 2, x \neq 4, x \neq 5\}$ **9.** $\dfrac{3(x^2+2x-4)}{x(x+2)(x-2)}$; $\{x | x \neq -2, x \neq 0, x \neq 2\}$ **10.** $\dfrac{a}{a+1}$ **11.** $\dfrac{4(d-3)}{-3(d+6)}$

12. $\{-1\}$ **13.** $\varnothing$ or $\{\ \}$ **14.** $\left\{x \middle| 2 < x \leq \dfrac{11}{2}\right\}$ or $\left(2, \dfrac{11}{2}\right]$ **15.** $k = \dfrac{FD^2}{q_1 q_2}$ **16.** 120 seconds (or 2 minutes)

17. 2.4 hours (or 2 hours and 24 minutes)   **18.** 3 miles per hour   **19.** 20 pounds   **20.** 792 centimeters^2

Cumulative Review Chapters R–5   **1.** 12   **2.** $2x^2 - 2x + 7$ **3.** -3 **4.** $\{-8\}$ **5.** $\left\{x \middle| -5 \leq x \leq \dfrac{1}{3}\right\}$; $\left[-5, \dfrac{1}{3}\right]$ **6.** $\left\{x \middle| x \neq -\dfrac{3}{2}, x \neq 5\right\}$

7. (a) 24   **(b)** $-\dfrac{19}{16}$ **(c)** $x^2 - x - 6$

8.   **9.** $y = -\dfrac{3}{5}x - 2$; $3x + 5y = -10$ **10.** $y = 4x - 7$; $4x - y = 7$ **11.** 2160 mg   **12.** $\begin{cases} 2x - 3y = 7 \\ 5x + 2y = 8 \end{cases}$

13. (a) $\begin{bmatrix} 1 & 1 & 0 & | & -3 \\ 0 & 2 & -1 & | & -1 \\ 0 & -5 & 1 & | & 16 \end{bmatrix}$ **(b)** $\begin{bmatrix} 1 & 1 & 0 & | & -3 \\ 0 & 1 & -\dfrac{1}{2} & | & -\dfrac{1}{2} \\ 0 & -5 & 1 & | & 16 \end{bmatrix}$ **14.** $\left(-3, \dfrac{1}{4}\right)$ **15.** -14

16.   **17.** $8x^2 - 13xy + 9y^2$ **18.** $3x^2 + x + 27$ **19.** $2x^3 - 11x^2 + 24x - 18$ **20.** $2x^2 - 5x + 9$ **21.** $(x+5)(x-2)(x^2+2x+4)$

22. $(2x-3)(3x+5)$ **23.** $\dfrac{x-7}{x+2}$ **24.** $\dfrac{x+1}{(x-2)(x-1)(x+3)}$ **25.** 48 minutes

Chapter 6 Radicals and Rational Exponents

Getting Ready for Chapter 6: Square Roots **1.** radical sign **2.** principal square root **3.** $-4, 4$ **4.** 9 **5.** 30 **6.** $\dfrac{3}{2}$ **7.** 0.4 **8.** 13 **9.** 15
10. 10 **11.** 14 **12.** 7 **13.** True **14.** rational; 20 **15.** irrational; $= 6.32$ **16.** not a real number **17.** rational; -14 **18.** $|a|$ **19.** 14 **20.** $|z|$
21. $|2x + 3|$ **22.** $|p - 6|$ **23.** 1 **25.** -10 **27.** $\dfrac{1}{2}$ **29.** 0.6 **31.** 1.6 **33.** not a real number **35.** rational; 8 **37.** rational; $\dfrac{1}{4}$ **39.** irrational; 6.63
41. irrational; 7.07 **43.** not a real number **45.** 8 **47.** 19 **49.** $|r|$ **51.** $|x + 4|$ **53.** $|4x - 3|$ **55.** $|2y + 3|$ **57.** 13 **59.** 17 **61.** not a real number
63. 15 **65.** -8 **67.** 6 **69.** not a real number **71.** -4 **73.** 13 **75.** The square roots of 36 are -6 and 6; $\sqrt{36} = 6$ **77.** $\dfrac{4\sqrt{13}}{3}$; 4.81 **79.** Answers will

vary. One reasonable explanation follows. Because $a^2 \ge 0$, the radicand of $\sqrt{a^2}$ is greater than or equal to 0 (nonnegative). The principal square root of
a nonnegative number is nonnegative. The absolute value ensures this result.

Section 6.1 nth Roots and Rational Exponents **1.** index **2.** 4 **3.** 3 **4.** -6 **5.** not a real number **6.** $\dfrac{1}{2}$ **7.** 3.68 **8.** 2.99 **9.** 5 **10.** $|z|$

11. $3x - 2$ **12.** 2 **13.** $-\dfrac{2}{3}$ **14.** $\sqrt[n]{a}$ **15.** 5 **16.** -3 **17.** -8 **18.** not a real number **19.** $\sqrt{b}$ **20.** $(8b)^{\frac{1}{5}}$ **21.** $\left(\dfrac{mn^5}{3}\right)^{\frac{1}{8}}$ **22.** $4y^{\frac{1}{3}}$ **23.** $\sqrt[n]{a^m}; \left(\sqrt[n]{a}\right)^m$

24. 64 **25.** 9 **26.** -8 **27.** 16 **28.** not a real number **29.** 13.57 **30.** 1.74 **31.** $a^{\frac{3}{8}}$ **32.** t^3 **33.** $(12ab^3)^{\frac{9}{4}}$ **34.** $\dfrac{1}{9}$ **35.** 4 **36.** $\dfrac{1}{(13x)^{\frac{3}{2}}}$ **37.** 5 **39.** -3

41. -5 **43.** $-\dfrac{1}{2}$ **45.** 3 **47.** 2.92 **49.** 1.86 **51.** 5 **53.** $|m|$ **55.** $x - 3$ **57.** $-|3p + 1|$ **59.** 2 **61.** -6 **63.** 2 **65.** -2 **67.** $\dfrac{2}{5}$ **69.** -5 **71.** not a real number

73. $(3x)^{\frac{1}{3}}$ **75.** $\left(\dfrac{x}{3}\right)^{\frac{1}{4}}$ **77.** 32 **79.** -64 **81.** 16 **83.** 16 **85.** 8 **87.** $\dfrac{1}{12}$ **89.** 125 **91.** 32 **93.** $x^{\frac{3}{4}}$ **95.** $(3x)^{\frac{2}{3}}$ **97.** $\left(\dfrac{5x}{y}\right)^{\frac{3}{2}}$ **99.** $(9ab)^{\frac{4}{3}}$ **101.** 4.47

103. 10.08 **105.** 1.26 **107.** $5x$ **109.** -10 **111.** 8 **113.** 243 **115.** not a real number **117.** $\dfrac{3}{4}$ **119.** 0.2 **121.** 127 **123.** not a real number **125.** $3p - 5$

127. -18 **129.** 2 **131.** 8 **133.** 16 **135.** 10; 10 **137. (a)** about 21.25°F **(b)** about 17.36°F **(c)** about -15.93°F **139. (a)** $\sqrt{\dfrac{8r\rho_w g}{3C\rho}}$ m/s

(b) about 7.38 m/s **141.** $(-9)^{\frac{1}{2}} = \sqrt{-9}$, but there is no real number whose square is -9. However, $-9^{\frac{1}{2}} = -1 \cdot 9^{\frac{1}{2}} = -1\sqrt{9} = -3$.
143. If $\dfrac{m}{n}$, in lowest terms, is positive, then $a^{\frac{m}{n}}$ is a real number provided $\sqrt[n]{a}$ exists. If $\dfrac{m}{n}$, in lowest terms, is negative, then $a^{\frac{m}{n}}$ is a real number provided
$a \ne 0$ and $\sqrt[n]{a}$ exists. **145.** $(x + 2)(x - 1)^3$ **147.** $\dfrac{z - 6}{z - 1}$

Section 6.2 Simplifying Expressions Using the Laws of Exponents **1.** $a^r b^r$ **2.** a^{r+s} **3.** $5^{\frac{11}{12}}$ **4.** 8 **5.** 10 **6.** $ab^{\frac{5}{6}}$ **7.** $x^{\frac{2}{3}}$ **8.** $\dfrac{4x^{\frac{1}{2}}}{y^{\frac{7}{3}}}$ **9.** $\dfrac{5x^{\frac{5}{8}}}{y^{\frac{1}{8}}}$ **10.** $\dfrac{200a^{\frac{1}{2}}}{b^{\frac{2}{3}}}$

11. 6 **12.** $2a^2 b^3$ **13.** $\sqrt[12]{x^5}$ **14.** $\sqrt[6]{a}$ **15.** $x^{\frac{1}{2}}(20x + 9)$ **16.** $\dfrac{12x + 1}{x^{\frac{2}{3}}}$ **17.** 25 **19.** 8 **21.** $\dfrac{1}{2^{\frac{7}{6}}}$ **23.** $\dfrac{1}{x^{\frac{7}{12}}}$ **25.** 2 **27.** $\dfrac{125}{8}$ **29.** $x^{\frac{1}{3}}y^{\frac{2}{9}}$ **31.** $\dfrac{x^{\frac{1}{6}}}{y^{\frac{1}{3}}}$ **33.** $\dfrac{2a}{b^{\frac{3}{4}}}$

35. $\dfrac{x^{\frac{1}{18}}}{2y^{\frac{4}{9}}}$ **37.** $8x^{\frac{1}{8}}y^{\frac{1}{2}}$ **39.** x^4 **41.** 2 **43.** $-2ab^4$ **45.** $\sqrt[4]{x}$ **47.** $\sqrt[6]{x^5}$ **49.** $\sqrt[8]{x^3}$ **51.** $\sqrt[6]{3^7}$ **53.** 1 **55.** $5x^{\frac{1}{2}}(8x + 15)$ **57.** $8(x + 2)^{\frac{2}{3}}(3x + 1)$ **59.** $\dfrac{6x + 5}{x^{\frac{1}{2}}}$

61. $\dfrac{2(10x - 27)}{(x - 4)^{\frac{1}{3}}}$ **63.** $5(x^2 + 4)^{\frac{1}{2}}(x^2 + 3x + 4)$ **65.** 2 **67.** 4 **69.** 10 **71.** 5 **73.** 0 **75.** $\dfrac{1}{48}$ **77.** $x^2 - 2x^{\frac{1}{2}}$ **79.** $\dfrac{2}{y^{\frac{1}{3}}} + 6y^{\frac{2}{3}}$

81. $4z^3 - 32 = 4(z - 2)(z^2 + 2z + 4)$ **83.** 5 **85.** 3 **87.** $\sqrt[24]{x}$ **89.** 36 **91.** $\{x \mid x > -1\}$ or $(-1, \infty)$ **93.** $4a^2 - 8a - 6$ **95.** -6

Section 6.3 Simplifying Radical Expressions Using Properties of Radicals **1.** $\sqrt[n]{ab}$ **2.** $\sqrt{77}$ **3.** $\sqrt[4]{42}$ **4.** $\sqrt{x^2 - 25}$ **5.** $\sqrt{20p^4}$

6. 1, 4, 9, 16, 25, 36 **7.** 1, 8, 27, 64, 125, 216 **8.** $4\sqrt{3}$ **9.** $12\sqrt[3]{2}$ **10.** $10|a|\sqrt{2}$ **11.** Fully simplified **12.** $2 + \sqrt{5}$ **13.** $\dfrac{-1 + 2\sqrt{2}}{2}$ or $-\dfrac{1}{2} + \sqrt{2}$

14. $5a^3\sqrt{3}$ **15.** $3a^2\sqrt{2a}$ **16.** $-4x^2 y^3\sqrt[3]{2y}$ **17.** $2ab^2\sqrt[4]{ab^3}$ **18.** $4\sqrt{3}$ **19.** $2a^2\sqrt[3]{15}$ **20.** $8ab^3\sqrt[3]{6a}$ **21.** $\dfrac{\sqrt{13}}{7}$ **22.** $\dfrac{3p}{2}$ **23.** $\dfrac{q\sqrt[4]{3}}{2}$ **24.** $2a^2$ **25.** $-2x$

26. $\dfrac{5a\sqrt[3]{a}}{b}$ **27.** $\sqrt[12]{10{,}125}$ **28.** $2\sqrt[6]{2250}$ **29.** $\sqrt[6]{60}$ **30.** $\sqrt{15ab}$ if $a, b \ge 0$ **33.** $\sqrt{x^2 - 49}$ **35.** $\sqrt{5}$ **37.** $5\sqrt{2}$ **39.** $3\sqrt[3]{2}$ **41.** $4|x|\sqrt{3}$ **43.** $-3x$

45. $2|m|\sqrt[4]{2}$ **47.** $2|p|\sqrt{3q}$ **49.** $9m^2\sqrt{2}$ **51.** $y^6\sqrt{y}$ **53.** $c^2\sqrt[3]{c^2}$ **55.** $5pq^2\sqrt{5p}$ **57.** $-2x^3\sqrt[3]{2}$ **59.** $-m\sqrt[5]{16m^3 n^2}$ **61.** $(x - y)\sqrt[4]{x - y}$

63. $2\sqrt[3]{x^3 - y^3}$ **65.** 5 **67.** $3 + \sqrt{2}$ **69.** $\dfrac{1 - \sqrt{2}}{2}$ **71.** 5 **73.** 4 **75.** 2 **77.** $5x\sqrt{3}$ **79.** $2b\sqrt[3]{3b}$ **81.** $18ab^2\sqrt{10}$ **83.** $3pq\sqrt[4]{4p}$ **85.** $-2ab\sqrt[5]{3a}$

87. $2(x - y)\sqrt[4]{3(x - y)}$ **89.** $\dfrac{\sqrt{3}}{4}$ **91.** $\dfrac{x\sqrt[4]{5}}{2}$ **93.** $\dfrac{3y}{5x}$ **95.** $-\dfrac{3x^3}{4y^4}$ **97.** 2 **99.** 4 **101.** $2a\sqrt{2}$ **103.** $\dfrac{2a^2\sqrt{2}}{b}$ **105.** $\dfrac{16a^3}{3b}$ **107.** $a^2\sqrt[3]{26}$ **109.** $\dfrac{3x^3\sqrt{5}}{y}$

111. $\sqrt[6]{432}$ **113.** $\sqrt[6]{12}$ **115.** $3\sqrt[6]{12}$ **117.** $\sqrt[3]{18}$ **119.** $\dfrac{\sqrt[3]{5x}}{2}$ **121.** $\sqrt[3]{45a^2}$ **123.** $6a^2\sqrt{2}$ **125.** $3a\sqrt[3]{2b^2}$ **127.** $\dfrac{-2\sqrt[3]{2}}{a}$ **129.** $-10m\sqrt[3]{4}$ **131.** $3ab^2\sqrt[3]{3ab}$

133. 6 **135. (a)**   **(b)** $3\sqrt{5}$ units **137. (a)** roughly \$1,587,000 **(b)** roughly \$2,381,000 **139. (a)** $4x^2$ **(b)** 36

141. $x = -3 - \sqrt{6}$ or $x = -3 + \sqrt{6}$ **143.** $x = \dfrac{-2 - \sqrt{7}}{3}$ or $x = \dfrac{-2 + \sqrt{7}}{3}$ **145.** The index of each radical must be the same. We can use the Laws

of Exponents to rewrite radicals so that they have common indices. **147.** $\left\{\dfrac{5}{2}\right\}$ **149.** $\left\{x \mid x \le \dfrac{130}{3}\right\}$ **151.** Answers will vary.

Section 6.4 Adding, Subtracting, and Multiplying Radical Expressions **1.** like radicals **2.** False **3.** $13\sqrt{13y}$ **4.** $7\sqrt[4]{5}$ **5.** $6\sqrt{2}$ **6.** $-8x\sqrt[3]{2x}$

7. Fully simplified **8.** $(8z + 5)\sqrt[3]{z}$ **9.** $2\sqrt{m}$ **10.** $3\sqrt{6} - 30$ **11.** $3\sqrt[3]{12} - 2\sqrt[3]{3}$ **12.** $-74 - 27\sqrt{3}$ **13.** False **14.** conjugates **15.** False

16. $9 - 4\sqrt{5}$ **17.** $25 - 6\sqrt{14}$ **18.** 1   **19.** $10\sqrt{2}$ **21.** $2\sqrt[3]{x}$ **23.** $14\sqrt{5x}$ **25.** $11\sqrt[3]{5} - 11\sqrt{5}$ **27.** $8\sqrt{2}$ **29.** $-2\sqrt[3]{3}$ **31.** $-25\sqrt[3]{2}$ **33.** $9\sqrt{6x}$

35. $4\sqrt{2} + 3\sqrt{10}$ **37.** $32x\sqrt{3x}$ **39.** $2x\sqrt{3} - 11x\sqrt{2}$ **41.** $(3x - 8)\sqrt[3]{2}$ **43.** $5\sqrt{x - 1}$ **45.** $3\sqrt{x}$ **47.** $5\sqrt[3]{x}$ **49.** $2\sqrt{3} - 3\sqrt{6}$ **51.** $\sqrt{6} + 3\sqrt{2}$

53. $\sqrt[3]{12} - 2\sqrt[3]{3}$ **55.** $3\sqrt{2x} - 2x\sqrt{5}$ **57.** $12 + 3\sqrt{3} + 4\sqrt{2} + \sqrt{6}$ **59.** $12 - 6\sqrt{7} + 2\sqrt{3} - \sqrt{21}$ **61.** $6\sqrt{7} - 30$ **63.** $\sqrt{6} - 12\sqrt{3} + 9\sqrt{2} - 4$

65. $31 - 2\sqrt{15}$ **67.** $4 + 2\sqrt{3}$ **69.** $7 - 2\sqrt{10}$ **71.** $x - 2\sqrt{2x} + 2$ **73.** 1   **75.** -11 **77.** $2x - 3y$ **79.** $\sqrt[3]{x^2} + \sqrt[3]{x} - 12$ **81.** $\sqrt[3]{4a^2} - 25$

83. $\sqrt{15} + 5\sqrt{2}$ **85.** $26x^2\sqrt{7x}$ **87.** -13 **89.** $2\sqrt[3]{7} + \sqrt[3]{28}$ **91.** $-4 + 12\sqrt{2}$ **93.** $20\sqrt{2}$ **95.** $8 - 4\sqrt{15}$ **97.** $(3x + 2)\sqrt[3]{5y}$ **99.** $-7y$

101. $5 + x + 4\sqrt{x + 1}$ **103.** $27 + x - 10\sqrt{x + 2}$ **105.** $\dfrac{11\sqrt{5}}{25}$ **107. (a)** $3\sqrt{3x}$ **(b)** $6\sqrt{3}$ **(c)** $6x$

109. Check $x = -2 + \sqrt{5}$:

$0 = x^2 + 4x - 1$

$0 \overset{?}{=} (-2 + \sqrt{5})^2 + 4(-2 + \sqrt{5}) - 1$

$0 \overset{?}{=} (-2)^2 + 2(-2)(\sqrt{5}) + (\sqrt{5})^2 - 8 + 4\sqrt{5} - 1$

$0 \overset{?}{=} 4 - 4\sqrt{5} + \sqrt{25} - 8 + 4\sqrt{5} - 1$

$0 \overset{?}{=} 9 - 9$

$0 = 0$ true

The value is a solution.

Check $x = -2 - \sqrt{5}$:

$0 = x^2 + 4x - 1$

$0 \overset{?}{=} (-2 - \sqrt{5})^2 + 4(-2 - \sqrt{5}) - 1$

$0 \overset{?}{=} (-2)^2 - 2(-2)(\sqrt{5}) + (\sqrt{5})^2 - 8 - 4\sqrt{5} - 1$

$0 \overset{?}{=} 4 + 4\sqrt{5} + \sqrt{25} - 8 - 4\sqrt{5} - 1$

$0 \overset{?}{=} 9 - 9$

$0 = 0$ true

The value is a solution.

111. The perimeter is $30\sqrt{2}$ units. The area is 108 square units.   **113.** $12\sqrt{6}$ square units   **115.** To add or subtract radicals, the indices and radicands must be the same. Then use the Distributive Property "in reverse" to add the "coefficients" of the radicals.   **117.** $12a^5b^5$ **119.** $6y^2 + y - 2$ **121.** $25w^2 - 4$

Section 6.5 Rationalizing Radical Expressions   **1.** rationalizing the denominator   **2.** $\sqrt{11}$ **3.** $\dfrac{\sqrt{3}}{3}$ **4.** $\dfrac{\sqrt{10}}{4}$ **5.** $\dfrac{\sqrt{10x}}{2x}$ **6.** $\dfrac{4\sqrt[3]{9}}{3}$ **7.** $\dfrac{\sqrt[3]{150}}{10}$

8. $\dfrac{3\sqrt[4]{p^3}}{p}$ **9.** $-2 - \sqrt{7}$ **10.** $2(\sqrt{3} - 1)$ **11.** $\dfrac{\sqrt{3} + 1}{2}$ **12.** $\dfrac{5 + \sqrt{10} + 4\sqrt{5} + 4\sqrt{2}}{3}$ **13.** $\dfrac{\sqrt{2}}{2}$ **15.** $-\dfrac{2\sqrt{3}}{5}$ **17.** $\dfrac{\sqrt{3}}{2}$ **19.** $\dfrac{\sqrt{3}}{3}$ **21.** $\dfrac{\sqrt{2p}}{p}$ **23.** $\dfrac{2\sqrt{2y}}{y^2}$

25. $\sqrt[3]{4}$ **27.** $\dfrac{\sqrt[3]{7q^2}}{q}$ **29.** $-\dfrac{\sqrt[3]{60}}{10}$ **31.** $\dfrac{\sqrt[3]{50y^2}}{5y}$ **33.** $-\dfrac{4\sqrt[4]{27x}}{3x}$ **35.** $\dfrac{12\sqrt[5]{m^2n^3}}{mn}$ **37.** $2(\sqrt{6} + 2)$ **39.** $5(\sqrt{5} - 2)$ **41.** $2(\sqrt{7} + \sqrt{3})$ **43.** $\dfrac{\sqrt{5} + \sqrt{3}}{2}$

45. $\dfrac{p - \sqrt{pq}}{p - q}$ **47.** $-3(2\sqrt{3} - 3\sqrt{2})$ or $3(3\sqrt{2} - 2\sqrt{3})$ **49.** $-8 - 3\sqrt{7}$ or $-(8 + 3\sqrt{7})$ **51.** $\dfrac{13\sqrt{6} - 46}{38}$ **53.** $\dfrac{p + 4\sqrt{p} + 4}{p - 4}$ **55.** $\dfrac{2 - 3\sqrt{2}}{2}$

57. $\dfrac{4\sqrt{3}}{3}$ **59.** $\dfrac{\sqrt{10} - \sqrt{2}}{2}$ **61.** $\dfrac{22\sqrt{3}}{3}$ **63.** 0   **65.** $\dfrac{1}{2}$ **67.** $\dfrac{\sqrt{2}}{4}$ **69.** $\dfrac{2\sqrt{3}}{3}$ **71.** $\sqrt{3} - 2$ **73.** $2(\sqrt{5} - 2)$ **75.** 2   **77.** $\dfrac{\sqrt{3}}{3}$ **79.** $\dfrac{\sqrt[3]{18}}{6}$ **81.** $\dfrac{5 - \sqrt{3}}{22}$

83. $\dfrac{\sqrt{6} - \sqrt{2}}{4}$ **85.** $\dfrac{1}{3(\sqrt{2} - 1)}$ **87.** $\dfrac{x - h}{x + \sqrt{xh}}$

89. $\dfrac{(\sqrt{6})^2 + 2 \cdot \sqrt{6} \cdot \sqrt{2} + (\sqrt{2})^2}{4^2} \overset{?}{=} \left(\dfrac{\sqrt{2} + \sqrt{3}}{2}\right)^2$

$\dfrac{6 + 2\sqrt{12} + 2}{16} \overset{?}{=} \dfrac{2 + \sqrt{3}}{4}$

$\dfrac{8 + 2 \cdot 2\sqrt{3}}{16} \overset{?}{=} \dfrac{2 + \sqrt{3}}{4}$

$\dfrac{8 + 4\sqrt{3}}{16} \overset{?}{=} \dfrac{2 + \sqrt{3}}{4}$

$\dfrac{2 + \sqrt{3}}{4} = \dfrac{2 + \sqrt{3}}{4}$

91. (a) $\dfrac{1}{\sqrt{x + h} + \sqrt{x}}$ **(b)** $\dfrac{1}{2\sqrt{x}}$ **(c)** $\dfrac{1}{4}$ **(d)** $2; (4, 2)$ **(e)** $y = \dfrac{1}{4}x + 1$

(f)

93. It is called rationalization because we are rewriting the expression so that the denominator is a rational number.

95.

97.

Putting the Concepts Together (Sections 6.1–6.5)   **1.** -5 **2.** $\dfrac{1}{16}$ **3.** $(3x^3)^{\frac{1}{4}}$ **4.** $7\sqrt[5]{z^4}$ **5.** $2x^{\frac{1}{2}}$ or $2\sqrt{x}$ **6.** $c^2 + c^3$ **7.** $\dfrac{a^2}{b^2}$ **8.** $x^{\frac{5}{8}}$ or $\sqrt[8]{x^5}$ **9.** $\dfrac{x^6}{y}$

10. $\sqrt{30ab}$ **11.** $10m^2n\sqrt{2n}$ **12.** $-2xy$ **13.** $\sqrt{3}$ **14.** $11b\sqrt{2b}$ **15.** $y\sqrt[3]{2y}$ **16.** $12x$ **17.** $7\sqrt{x}$ **18.** $14 - 28\sqrt{2} = 14(1 - 2\sqrt{2})$ **19.** $41 - 24\sqrt{2}$

20. $\dfrac{3\sqrt{2}}{16}$ **21.** $-\dfrac{4(\sqrt{3} + 8)}{61}$

Section 6.6 Functions Involving Radicals   **1. (a)** 4   **(b)** $2\sqrt{7}$ **2. (a)** -1 **(b)** 3   **3.** even; odd   **4.** $\{x \mid x \geq -6\}$ or $[-6, \infty)$

5. $\{t \mid t \text{ is any real number}\}$ or $(-\infty, \infty)$ **6.** $\{m \mid m \leq 2\}$ or $(-\infty, 2]$

7. (a) $\{x \mid x \geq -3\}$ or $[-3, \infty)$ **8. (a)** $\{x \mid x \text{ is any real number}\}$ or $(-\infty, \infty)$ **9. (a)** 3   **(b)** $\sqrt{14}$ **(c)** 2   **11. (a)** -5 **(b)** -1 **(c)** $-\dfrac{\sqrt{13}}{2}$

(b)

(b)

13. (a) 4   **(b)** $4\sqrt{6}$ **(c)** $\sqrt{6}$ **15. (a)** 2   **(b)** -2 **(c)** $-2\sqrt[3]{2}$

17. (a) $\dfrac{\sqrt{5}}{3}$ **(b)** $\dfrac{\sqrt{2}}{2}$ **(c)** $\dfrac{\sqrt{6}}{3}$ **19. (a)** 1   **(b)** $\sqrt[3]{4}$ **(c)** $\sqrt[3]{3}$

21. $\{x \mid x \geq 7\}$ or $[7, \infty)$ **23.** $\left\{x \mid x \geq -\dfrac{7}{2}\right\}$ or $\left[-\dfrac{7}{2}, \infty\right)$

(c) $[0, \infty)$

(c) $(-\infty, \infty)$

25. $\left\{ x \middle| x \le \dfrac{4}{3} \right\}$ or $\left(-\infty, \dfrac{4}{3} \right]$ **27.** $\{ z \mid z \text{ is any real number} \}$ or $(-\infty, \infty)$ **29.** $\left\{ p \middle| p \ge \dfrac{2}{7} \right\}$ or $\left[\dfrac{2}{7}, \infty \right)$ **31.** $\{ x \mid x \text{ is any real number} \}$ or $(-\infty, \infty)$

33. $\{ x \mid x > -5 \}$ or $(-5, \infty)$ **35.** $\{ x \mid x \ne 3 \}$ or $(-\infty, 3) \cup (3, \infty)$

37. (a) $\{ x \mid x \ge 4 \}$ or $[4, \infty)$ **39. (a)** $\{ x \mid x \ge -2 \}$ or $[-2, \infty)$ **41. (a)** $\{ x \mid x \le 2 \}$ or $(-\infty, 2]$ **43. (a)** $\{ x \mid x \ge 0 \}$ or $[0, \infty)$

(b)   **(b)**   **(b)**   **(b)**

(c) $[0, \infty)$ **(c)** $[0, \infty)$ **(c)** $[0, \infty)$ **(c)** $[3, \infty)$

45. (a) $\{ x \mid x \ge 0 \}$ or $[0, \infty)$ **47. (a)** $\{ x \mid x \ge 0 \}$ or $[0, \infty)$ **49. (a)** $\{ x \mid x \ge 0 \}$ or $[0, \infty)$ **51. (a)** $\{ x \mid x \ge 0 \}$ or $[0, \infty)$

(b)   **(b)**   **(b)**   **(b)**

(c) $[-4, \infty)$ **(c)** $[0, \infty)$ **(c)** $[0, \infty)$ **(c)** $(-\infty, 0]$

53. (a) $\{ x \mid x \text{ is any real number} \}$ or $(-\infty, \infty)$ **55. (a)** $\{ x \mid x \text{ is any real number} \}$ or $(-\infty, \infty)$ **57. (a)** $\{ x \mid x \text{ is any real number} \}$ or $(-\infty, \infty)$ **59. (a)** 5 units

(b) $\sqrt{17} \approx 4.123$ units

(c) $5\sqrt{17} \approx 20.616$ units

(b)   **(b)**   **(b)**

61. (a) $4\sqrt{2} \approx 5.657$ square units

(b) $4\sqrt{5} \approx 8.944$ square units

(c) $2\sqrt{14} \approx 7.483$ square units

(c) $(-\infty, \infty)$ **(c)** $(-\infty, \infty)$ **(c)** $(-\infty, \infty)$

63. Shift the graph of $f(x)$ $|c|$ units to the right if $c < 0$ or c units to the left if $c > 0$. **65.** $\dfrac{5}{6}$ **67.** $\dfrac{4x + 1}{x(x + 1)}$ **69.** $\dfrac{7x + 1}{(x - 1)(x + 1)}$ or $\dfrac{7x + 1}{x^2 - 1}$

71.

WINDOW
Xmin=0
Xmax=24
Xscl=2
Ymin=-2
Ymax=10
Yscl=1
Xres=1

73.

WINDOW
Xmin=-4
Xmax=20
Xscl=2
Ymin=-2
Ymax=10
Yscl=1
Xres=1

75.

WINDOW
Xmin=-16
Xmax=8
Xscl=2
Ymin=-2
Ymax=10
Yscl=1
Xres=1

77.

WINDOW
Xmin=-4
Xmax=20
Xscl=2
Ymin=-2
Ymax=10
Yscl=1
Xres=1

79.

WINDOW
Xmin=-4
Xmax=20
Xscl=2
Ymin=-8
Ymax=4
Yscl=1
Xres=1

81.

WINDOW
Xmin=-4
Xmax=20
Xscl=2
Ymin=-2
Ymax=10
Yscl=1
Xres=1

83.

WINDOW
Xmin=-4
Xmax=20
Xscl=2
Ymin=-1
Ymax=5
Yscl=1
Xres=1

85.

WINDOW
Xmin=-4
Xmax=20
Xscl=2
Ymin=-6
Ymax=6
Yscl=1
Xres=1

87.

WINDOW
Xmin=-12
Xmax=12
Xscl=2
Ymin=-6
Ymax=6
Yscl=1
Xres=1

89.

WINDOW
Xmin=-12
Xmax=12
Xscl=2
Ymin=-6
Ymax=6
Yscl=1
Xres=1

91.

WINDOW
Xmin=-12
Xmax=12
Xscl=2
Ymin=-6
Ymax=6
Yscl=1
Xres=1

Section 6.7 Radical Equations and Their Applications **1.** radical equation **2.** extraneous **3.** False **4.** $\{5\}$ **5.** $\{-5\}$ **6.** $\varnothing$ or $\{\ \}$ **7.** $\{4\}$

8. $\{-3\}$ **9.** $\{15\}$ **10.** $\{-3,1\}$ **11.** $\{12\}$ **12. (a)** $L = \dfrac{8T^2}{\pi^2}$ **(b)** 32 feet **13.** $\{16\}$ **15.** $\{7\}$ **17.** $\{11\}$ **19.** $\varnothing$ or $\{\ \}$ **21.** $\{2\}$ **23.** $\{12\}$

25. $\{25\}$ **27.** $\{11\}$ **29.** $\{-4\}$ **31.** $\{1\}$ **33.** $\{-5\}$ **35.** $\left\{0, \dfrac{1}{4}\right\}$ **37.** $\{3\}$ **39.** $\{4\}$ **41.** $\{4\}$ **43.** $\{6\}$ **45.** $\{4\}$ **47.** $\{-3\}$ **49.** $\{-1, 10\}$

51. $\{0, 4\}$ **53.** $\{3\}$ **55.** $\{5\}$ **57.** $\{2, 10\}$ **59.** $\{3\}$ **61.** $\{4\}$ **63.** $\{6\}$ **65.** $\{-7, 9\}$ **67.** $\{18\}$ **69.** $\{3\}$ **71.** $\{-23, 31\}$ **73.** $r = \dfrac{A^2 - P^2}{P^2}$

75. $V = \dfrac{4}{3}\pi r^3$ **77.** $F = \dfrac{q_1 q_2 r^2}{4\pi\varepsilon_0}$ **79.** $\varnothing$ or $\{\ \}$ **81.** $\{4\}$ **83.** $\{2\}$ **85.** $\{9\}$ **87.** $\{-3\}$ **89.** $\{-6\}$ **91.** $\{9\}$ **93.** $\{5\}$ **95.** $\{-2\}$

97. (a) $\{2\}$; $(2, 0)$ **(b)** $\{3\}$; $(3, 1)$ **(c)** $\{6\}$; $(6, 2)$ **99. (a)** $y = 5$ or $y = -1$ **101. (a)** after 2 years **(b)** after 16 years

(d) **(e)** The equation $f(x) = -1$ has no solution because the graph of the function does not go below the x-axis.   **(b)**   **103. (a)** in the year 2053 **(b)** in the year 2021

105. $\{0\}$ **107.** $\left\{-\dfrac{3}{4}, 0\right\}$

109. If the index is even, it is important so that you can identify extraneous solutions. Regardless of the index, it is important to make sure that your answer is correct. **111.** In radical expressions with an even index, the radicand must be greater than or equal to zero. After raising both sides of the equation to an even index, it is possible to obtain solutions that result in a negative radicand. We do not have any restrictions on the value of the radicand when the index is odd, so we don't need to be worried about extraneous solutions. **113.** $0, -4, 12$ **115.** $\sqrt{2^3}, \pi, \sqrt[3]{-4}$ **117.** A rational number is a number that can be written as the quotient of two integers, where the denominator is not zero; a rational number is also a number where the decimal either terminates, or does not terminate but repeats. An irrational number has a decimal form that neither terminates nor repeats. The square root of -1 is not real because there is no real number whose square is -1.

119. **121.** [graph] The two graphs do not intersect. Therefore, the equation has no real solution.

Section 6.8 The Complex Number System **1.** imaginary unit **2.** pure imaginary number **3.** $\sqrt{N}i$ **4.** True **5.** $6i$ **6.** $\sqrt{5}i$ **7.** $2\sqrt{3}i$ **8.** $4 + 10i$

9. $-2 - 2\sqrt{2}i$ **10.** $2 - 2\sqrt{2}i$ **11.** $1 + 11i$ **12.** $6 - 9i$ **13.** $-3 + i$ **14.** $12 + 15i$ **15.** $2 + 24i$ **16.** -18 **17.** $38 + 14i$ **18.** $-3 - 5i$ **19.** 73 **20.** 29

21. $\dfrac{1}{3} + \dfrac{4}{3}i$ **22.** $-\dfrac{1}{2} + \dfrac{3}{2}i$ **23.** $-i$ **24.** -1 **25.** $2i$ **27.** $-9i$ **29.** $3\sqrt{5}i$ **31.** $10\sqrt{3}i$ **33.** $\sqrt{7}i$ **35.** $5 + 7i$ **37.** $-2 - 2\sqrt{7}i$ **39.** $2 + i$ **41.** $\dfrac{1}{3} + \dfrac{\sqrt{2}}{6}i$

43. $6 - 2i$ **45.** $-4 + 6i$ **47.** $2 - 5i$ **49.** $3 - 2\sqrt{2}i$ **51.** $24 + 12i$ **53.** $-5 - 2i$ **55.** $5 + 10i$ **57.** $14 - 22i$ **59.** 26 **61.** $-4 + 5\sqrt{2}i$ **63.** $\dfrac{25}{48} + \dfrac{5}{24}i$

65. $5 + 12i$ **67.** $-9 + 40i$ **69.** -6 **71.** $-4\sqrt{5}$ **73.** $84 - 47i$ **75. (a)** $3 - 5i$ **(b)** 34 **77. (a)** $2 + 7i$ **(b)** 53 **79. (a)** $-7 - 2i$ **(b)** 53 **81.** $\dfrac{1}{3} - \dfrac{1}{3}i$

83. $\dfrac{2}{5} + i$ **85.** $\dfrac{6}{5} - \dfrac{3}{5}i$ **87.** $\dfrac{3}{29} - \dfrac{7}{29}i$ **89.** i **91.** $1 + 3i$ **93.** $-\dfrac{1}{5} - \dfrac{7}{5}i$ **95.** i **97.** $-i$ **99.** i **101.** $-i$ **103.** $-15 - 8i$ **105.** $5 + 12i$ **107.** $\dfrac{2}{3} + i$

109. $\dfrac{1}{2} - \dfrac{1}{2}i$ **111.** 12 **113.** $-15 - 20i$ **115.** $-5\sqrt{6}$ **117.** $-\dfrac{1}{5}i$ **119.** $\dfrac{2}{5} + \dfrac{1}{5}i$ **121.** $-\dfrac{4}{41} - \dfrac{5}{41}i$ **123. (a)** -1 **(b)** $2i$

125. (a) $-7 + 6i$ **(b)** $4 - 4i$ **127. (a)** $10 - i$ ohms **(b)** 10 ohms **(c)** -1 ohm **129. (a)** 0 **(b)** 0 **131. (a)** 0 **(b)** 0 **(c)** 0 **133.** For a polynomial with real coefficients, the zeros will be real numbers or will occur in conjugate pairs. If the complex number $a + bi$ is a complex zero of the polynomial, then its conjugate $a - bi$ is also a complex zero. **135.** The set of natural numbers is a subset of the set of whole numbers, which is a subset of the set of integers, which is a subset of the set of rational numbers, which is a subset of the set of real numbers, which is a subset of the set of complex numbers. Thus all real numbers are, more generally, complex numbers. **137.** Both methods can rely on the "FOIL" method or the Distributive Property.

139. $x^3 + 6x^2 + 12x + 8$ **141.** $18 + 26i$ **143.** Both methods rely on special product formulas to obtain the results. **145.** $-9.9 + 7.2i$

147. $-30.86 + 3.59i$ **149.** $-\dfrac{11}{17} + \dfrac{27}{17}i$ **151.** $6 - 25i$

Chapter 6 Review **1.** 7 **2.** -5 **3.** $\dfrac{2}{3}$ **4.** 3 **5.** 3 **6.** 10 **7.** z **8.** $|5p - 3|$ **9.** 9 **10.** not a real number **11.** -2 **12.** 9 **13.** 128 **14.** -9

15. -1331 **16.** 6 **17.** -4.02 **18.** 2.30 **19.** 4.64 **20.** 1.78 **21.** $(5a)^{\frac{1}{3}}$ **22.** $p^{\frac{7}{5}}$ **23.** $(10z)^{\frac{3}{4}}$ **24.** $(2ab)^{\frac{5}{6}}$ **25.** 64 **26.** $\dfrac{1}{k^{\frac{1}{4}}}$ **27.** $p^2 \cdot q^6$ **28.** $\dfrac{2b^{\frac{1}{20}}}{a^{\frac{3}{10}}}$

29. $10m^{\frac{1}{3}} + \dfrac{5}{m}$ **30.** $\dfrac{1}{2x^{\frac{1}{3}}}$ **31.** $\sqrt[4]{x^3}$ **32.** $11x^2 y^5$ **33.** $m^2 \sqrt[6]{m}$ **34.** $\dfrac{1}{\sqrt[3]{c}}$ **35.** $(3m - 1)^{\frac{1}{4}}(3m^2 - 22m + 9)$ **36.** $\dfrac{(3x + 5)(x - 3)}{(x^2 - 5)^{\frac{2}{3}}}$ **37.** $\sqrt{105}$

38. $\sqrt[4]{12a^3b^3}$ **39.** $4\sqrt{5}$ **40.** $-5\sqrt[3]{4}$ **41.** $3m^2n\sqrt[3]{6n}$ **42.** $p^2|q|\sqrt[4]{50}$ **43.** $8|x^3|\sqrt{y}$ as long as $y \geq 0$ **44.** $(2x+1)\sqrt{2x+1}$ as long as $2x+1 \geq 0$

45. $wz\sqrt{w}$ **46.** $3x^2z\sqrt{5yz}$ **47.** $2a^4b\sqrt[3]{2b^2}$ **48.** $2(x+1)$ **49.** $3\sqrt{30}$ **50.** $2\sqrt[3]{75}$ **51.** $-2x^2y^3\sqrt[3]{9x}$ **52.** $30xy\sqrt{3xy}$ **53.** $\dfrac{11}{5}$ **54.** $\dfrac{a^2\sqrt{5}}{8b}$ **55.** $\dfrac{\sqrt[3]{6}}{k}$

56. $\dfrac{-2w^5\sqrt[3]{20}}{7}$ **57.** $2h$ **58.** $\dfrac{5b^3}{2a}$ **59.** $\dfrac{-2x^2}{3y}$ **60.** $\dfrac{2n\sqrt{n}}{m}$ **61.** $\sqrt[6]{500}$ **62.** $2\sqrt[12]{2}$ **63.** $8\sqrt[4]{x}$ **64.** $6\sqrt[3]{4y}$ **65.** $5\sqrt{2}-4\sqrt{3}$ **66.** $13\sqrt{2}$ **67.** $\sqrt[3]{2z}$

68. $17\sqrt[3]{x^2}$ **69.** $7\sqrt{a}$ **70.** $9x\sqrt{3}$ **71.** $4m\sqrt[3]{4m^2y^2}$ **72.** $(y-1)\sqrt{y-4}$ **73.** $\sqrt{15}-3\sqrt{5}$ **74.** $3\sqrt[3]{5}+\sqrt[3]{20}$ **75.** $7+\sqrt{5}$

76. $42+7\sqrt{2}+6\sqrt{3}+\sqrt{6}$ **77.** -44 **78.** $9\sqrt[3]{x^2}+5\sqrt[3]{x}-4$ **79.** $x-2\sqrt{5x}+5$ **80.** $247+22\sqrt{10}$ **81.** $2a-b^2$ **82.** $\sqrt[3]{36s^2}-5\sqrt[3]{6s}-14$

83. $\dfrac{\sqrt{6}}{3}$ **84.** $2\sqrt{3}$ **85.** $\dfrac{4\sqrt{3p}}{p^2}$ **86.** $\dfrac{5\sqrt{2a}}{2a}$ **87.** $-\dfrac{\sqrt{6y}}{3y^2}$ **88.** $\dfrac{3\sqrt[3]{25}}{5}$ **89.** $-\dfrac{\sqrt[3]{300}}{15}$ **90.** $\dfrac{27\sqrt[3]{4p^2q}}{2pq}$ **91.** $\dfrac{42+6\sqrt{6}}{43}$ **92.** $-\dfrac{\sqrt{3}+9}{26}$

93. $\dfrac{\sqrt{3}(3-\sqrt{2})}{7}$ or $\dfrac{3\sqrt{3}-\sqrt{6}}{7}$ **94.** $\dfrac{k+\sqrt{km}}{k-m}$ **95.** $\dfrac{7+2\sqrt{10}}{3}$ **96.** $\dfrac{9-6\sqrt{y}+y}{9-y}$ or $\dfrac{y-6\sqrt{y}+9}{9-y}$ **97.** $\dfrac{10\sqrt{2}-4\sqrt{3}}{19}$ **98.** $\dfrac{8\sqrt{2}-3\sqrt{15}}{7}$

99. $\dfrac{25\sqrt{7}}{21}$ **100.** $\dfrac{4+\sqrt{7}}{9}$ **101. (a)** 1 **(b)** 2 **(c)** 3 **102. (a)** 0 **(b)** 2 **(c)** 4 **103. (a)** 1 **(b)** -1 **(c)** 2 **104. (a)** 0 **(b)** 2 **(c)** $\dfrac{1}{2}$

105. $\left\{x \middle| x \geq \dfrac{5}{3}\right\}$ or $\left[\dfrac{5}{3}, \infty\right)$ **106.** $\{x|x \text{ is any real number}\}$ or $(-\infty, \infty)$ **107.** $\left\{x \middle| x \geq -\dfrac{1}{6}\right\}$ or $\left[-\dfrac{1}{6}, \infty\right)$ **108.** $\{x|x \text{ is any real number}\}$ or $(-\infty, \infty)$

109. $\{x|x > 2\}$ or $(2, \infty)$ **110.** $\{x|x < 0 \text{ or } x \geq 3\}$ or $(-\infty, 0) \cup [3, \infty)$

111. (a) $\{x|x \leq 1\}$ or $(-\infty, 1]$ **112. (a)** $\{x|x \geq -1\}$ or $[-1, \infty)$ **113. (a)** $\{x|x \geq -3\}$ or $[-3, \infty)$ **114. (a)** $\{x|x \text{ is any real number}\}$ or $(-\infty, \infty)$

(b)

(b)

(b)

(b)

(c) $[0, \infty)$ **(c)** $[-2, \infty)$ **(c)** $(-\infty, 0]$ **(c)** $(-\infty, \infty)$

115. $\{169\}$ **116.** $\{-3\}$ **117.** $\left\{\dfrac{89}{3}\right\}$ **118.** $\varnothing$ or $\{\ \}$ **119.** $\{-5\}$ **120.** $\{25\}$ **121.** $\{-512\}$ **122.** $\{2\}$ **123.** $\{5\}$ **124.** $\{2\}$ **125.** $\{11\}$

126. $\{6\}$ **127.** $\{9\}$ **128.** $\varnothing$ or $\{\ \}$ **129.** $\left\{\dfrac{15}{2}\right\}$ **130.** $\{-5, 5\}$ **131.** $h = \dfrac{3V}{\pi r^2}$ **132.** $v = \dfrac{30}{f_s^3}$ **133.** $\sqrt{29}i$ **134.** $3\sqrt{6}i$ **135.** $14 - 9\sqrt{2}i$

136. $2 + \sqrt{5}i$ **137.** $1 - 2i$ **138.** $-5 + 10i$ **139.** $5 - 7\sqrt{5}i$ **140.** $-5 + 7i$ **141.** $47 + 13i$ **142.** $8 - \dfrac{11}{6}i$ **143.** -9 **144.** $67 - 42i$ **145.** 145

146. $27 + 38i$ **147.** $\dfrac{6}{17} - \dfrac{10}{17}i$ **148.** $-\dfrac{21}{53} - \dfrac{6}{53}i$ **149.** $\dfrac{4}{29} - \dfrac{19}{29}i$ **150.** $\dfrac{1}{2} + \dfrac{7}{2}i$ **151.** $-i$ **152.** i

Chapter 6 Test **1.** $\dfrac{1}{7}$ **2.** $6y\sqrt[12]{x^8y^3}$ **3.** $2a^5b^4\sqrt[5]{4a^3b}$ **4.** $\sqrt{39mn}$ **5.** $4x^3y^2\sqrt{2x}$ **6.** $\dfrac{3a}{2b^2}$ **7.** $(x+6)\sqrt{5x}$ **8.** $a\sqrt{b}$ **9.** $33 - 5\sqrt{x} - 2x$ **10.** $-\dfrac{\sqrt{2}}{18}$

11. $5 - 2\sqrt{5}$ **12. (a)** 1 **(b)** 3 **13.** $\left\{x \middle| x \leq \dfrac{5}{3}\right\}$ or $\left(-\infty, \dfrac{5}{3}\right]$ **14. (a)** $\{x|x \geq 0\}$ or $[0, \infty)$ **(b)** **(c)** $[-3, \infty)$ **15.** $\{13\}$

16. $\{3\}$ **17.** $\{2\}$ **18.** $17 - 13i$ **19.** $29 - 2i$ **20.** $\dfrac{73}{265} - \dfrac{89}{265}i$

Chapter 7 Quadratic Equations and Functions

Section 7.1 Solving Quadratic Equations by Completing the Square
1. $\sqrt{p}; -\sqrt{p}$ **2.** $\{-4\sqrt{3}, 4\sqrt{3}\}$ **3.** $\{-5, 5\}$ **4.** $\{-9, 9\}$ **5.** $\{-6\sqrt{2}i, 6\sqrt{2}i\}$

6. $\{-3i, 3i\}$ **7.** $\{-13, 7\}$ **8.** $\{5 - 4i, 5 + 4i\}$ **9.** $49; (p+7)^2$ **10.** $\dfrac{9}{4}; \left(w - \dfrac{3}{2}\right)^2$ **11.** $\{-4, 2\}$ **12.** $\{4 - \sqrt{7}, 4 + \sqrt{7}\}$ **13.** $\left\{\dfrac{-3 - \sqrt{11}}{2}, \dfrac{-3 + \sqrt{11}}{2}\right\}$

14. $\left\{-\dfrac{1}{3} - \dfrac{2\sqrt{5}}{3}i, -\dfrac{1}{3} + \dfrac{2\sqrt{5}}{3}i\right\}$ **15.** hypotenuse; legs **16.** False **17.** $c = 5$ **18.** Approximately 17.32 miles **19.** $\{-10, 10\}$ **21.** $\{-5\sqrt{2}, 5\sqrt{2}\}$

23. $\{-5i, 5i\}$ **25.** $\left\{-\dfrac{\sqrt{5}}{2}, \dfrac{\sqrt{5}}{2}\right\}$ **27.** $\{-2\sqrt{2}, 2\sqrt{2}\}$ **29.** $\{-4, 4\}$ **31.** $\left\{-\dfrac{2\sqrt{6}}{3}, \dfrac{2\sqrt{6}}{3}\right\}$ **33.** $\{-2i, 2i\}$ **35.** $\{1 - 3\sqrt{2}i, 1 + 3\sqrt{2}i\}$

37. $\{-5 - \sqrt{3}, -5 + \sqrt{3}\}$ **39.** $\left\{-\dfrac{4}{3}, \dfrac{2}{3}\right\}$ **41.** $\left\{\dfrac{2}{3} - \dfrac{\sqrt{5}}{3}, \dfrac{2}{3} + \dfrac{\sqrt{5}}{3}\right\}$ **43.** $\{-13, 5\}$ **45.** $x^2 + 10x + 25; (x+5)^2$ **47.** $z^2 - 18z + 81; (z-9)^2$

49. $y^2 + 7y + \dfrac{49}{4}; \left(y + \dfrac{7}{2}\right)^2$ **51.** $w^2 + \dfrac{1}{2}w + \dfrac{1}{16}; \left(w + \dfrac{1}{4}\right)^2$ **53.** $\{-6, 2\}$ **55.** $\{2 - \sqrt{3}, 2 + \sqrt{3}\}$ **57.** $\{2 - i, 2 + i\}$ **59.** $\left\{-\dfrac{5}{2} - \dfrac{\sqrt{33}}{2}, -\dfrac{5}{2} + \dfrac{\sqrt{33}}{2}\right\}$

61. $\{4 - \sqrt{19}, 4 + \sqrt{19}\}$ **63.** $\left\{\dfrac{1}{2} - \dfrac{\sqrt{11}}{2}i, \dfrac{1}{2} + \dfrac{\sqrt{11}}{2}i\right\}$ **65.** $\left\{-\dfrac{3}{2}, 4\right\}$ **67.** $\left\{1 - \dfrac{\sqrt{3}}{3}, 1 + \dfrac{\sqrt{3}}{3}\right\}$ **69.** $\left\{\dfrac{5}{4} - \dfrac{\sqrt{17}}{4}, \dfrac{5}{4} + \dfrac{\sqrt{17}}{4}\right\}$

71. $\left\{-1 - \dfrac{\sqrt{6}}{2}i, -1 + \dfrac{\sqrt{6}}{2}i\right\}$ **73.** 10 **75.** 20 **77.** $5\sqrt{2}; 7.07$ **79.** 2 **81.** $2\sqrt{34}; 11.66$ **83.** $b = 4\sqrt{3} \approx 6.93$ **85.** $a = 4\sqrt{5} \approx 8.94$

87. $\{-3, 9\}$; $(-3, 36), (9, 36)$ **89.** $\{-2 - 3\sqrt{2}, -2 + 3\sqrt{2}\}$; $(-2 - 3\sqrt{2}, 18), (-2 + 3\sqrt{2}, 18)$ **91.** $4\sqrt{5}$ units **93.** approximately 104.403 yards
95. approximately 31.623 feet **97. (a)** approximately 22.913 feet **(b)** 15 feet **99. (a)** 1 second **(b)** approximately 1.732 seconds **(c)** 2 seconds
101. approximately 9.54% **103.** The triangle is a right triangle; the hypotenuse is 17. **105.** The triangle is not a right triangle.
107.
$$c^2 = (m^2 + n^2)^2 = m^4 + 2m^2n^2 + n^4$$
$$a^2 + b^2 = (m^2 - n^2)^2 + (2mn)^2$$
$$= m^4 - 2m^2n^2 + n^4 + 4m^2n^2$$
$$= m^4 + 2m^2n^2 + n^4$$
109. $\{-4, 9\}$ **111.** $\left\{-1, \dfrac{1}{2}\right\}$ **113.** In both cases, the simpler equations are linear.
Because c^2 and $a^2 + b^2$ result in the same
expression, a, b, and c are the lengths of the
sides of a right triangle.

Section 7.2 Solving Quadratic Equations by the Quadratic Formula

1. $\dfrac{-b \pm \sqrt{b^2 - 4ac}}{2a}$ **2.** $\left\{-\dfrac{3}{2}, 3\right\}$ **3.** $\left\{-4, \dfrac{1}{2}\right\}$ **4.** False

5. $\left\{1 - \dfrac{\sqrt{3}}{2}, 1 + \dfrac{\sqrt{3}}{2}\right\}$ **6.** $\left\{\dfrac{5}{2}\right\}$ **7.** $\left\{2 - \dfrac{\sqrt{10}}{2}, 2 + \dfrac{\sqrt{10}}{2}\right\}$ **8.** $\{-1 - 5i, -1 + 5i\}$ **9.** discriminant **10.** False **11.** negative **12.** False **13.** True

14. Two complex solutions that are not real **15.** One repeated rational solution **16.** Two irrational solutions **17.** True **18.** $\{-3, 3\}$

19. $\left\{\dfrac{5}{4} - \dfrac{\sqrt{23}}{4}i, \dfrac{5}{4} + \dfrac{\sqrt{23}}{4}i\right\}$ **20.** $\left\{-\dfrac{5}{3}, 1\right\}$ **21. (a)** 200 or 600 DVDs **(b)** 400 DVDs **22.** 16 meters by 30 meters **23.** $\{-2, 6\}$ **25.** $\left\{-\dfrac{3}{2}, \dfrac{5}{3}\right\}$

27. $\left\{1 - \dfrac{\sqrt{3}}{2}, 1 + \dfrac{\sqrt{3}}{2}\right\}$ **29.** $\left\{1 - \dfrac{2\sqrt{3}}{3}, 1 + \dfrac{2\sqrt{3}}{3}\right\}$ **31.** $\left\{-\dfrac{1}{3} - \dfrac{\sqrt{13}}{3}, -\dfrac{1}{3} + \dfrac{\sqrt{13}}{3}\right\}$ **33.** $\{1 - \sqrt{6}i, 1 + \sqrt{6}i\}$ **35.** $\left\{\dfrac{1}{2} - \dfrac{\sqrt{13}}{2}i, \dfrac{1}{2} + \dfrac{\sqrt{13}}{2}i\right\}$

37. $\left\{\dfrac{1}{4} - \dfrac{\sqrt{5}}{4}, \dfrac{1}{4} + \dfrac{\sqrt{5}}{4}\right\}$ **39.** $\left\{-\dfrac{2}{3} - \dfrac{\sqrt{7}}{3}, -\dfrac{2}{3} + \dfrac{\sqrt{7}}{3}\right\}$ **41.** 21; two irrational solutions **43.** -56; two complex solutions that are not real

45. 0; one repeated rational solution **47.** -8; two complex solutions that are not real **49.** 44; two irrational solutions **51.** $\left\{\dfrac{5}{2} - \dfrac{\sqrt{5}}{2}, \dfrac{5}{2} + \dfrac{\sqrt{5}}{2}\right\}$

53. $\left\{-\dfrac{8}{3}, 1\right\}$ **55.** $\left\{-\dfrac{7}{2}, 5\right\}$ **57.** $\{-1 - \sqrt{7}i, -1 + \sqrt{7}i\}$ **59.** $\left\{-\dfrac{3}{2}\right\}$ **61.** $\left\{\dfrac{1}{7} - \dfrac{\sqrt{29}}{7}, \dfrac{1}{7} + \dfrac{\sqrt{29}}{7}\right\}$ **63.** $\{-4, 4\}$ **65.** $\left\{\dfrac{1}{4} - \dfrac{1}{4}i, \dfrac{1}{4} + \dfrac{1}{4}i\right\}$ **67.** $\left\{-\dfrac{2}{3}\right\}$

69. $\left\{-\dfrac{1}{3} - \dfrac{2\sqrt{7}}{3}, -\dfrac{1}{3} + \dfrac{2\sqrt{7}}{3}\right\}$ **71.** $\{2 - \sqrt{13}, 2 + \sqrt{13}\}$ **73.** $\{1 - \sqrt{5}, 1 + \sqrt{5}\}$ **75.** $\left\{\dfrac{1}{4} - \dfrac{3\sqrt{7}}{4}i, \dfrac{1}{4} + \dfrac{3\sqrt{7}}{4}i\right\}$ **77. (a)** $\{-7, 3\}$

(b) $\{-4, 0\}$; $(-4, -21), (0, -21)$ **79. (a)** $\left\{-1 - \dfrac{\sqrt{6}}{2}, -1 + \dfrac{\sqrt{6}}{2}\right\}$ **(b)** $\left\{-1 - \dfrac{\sqrt{2}}{2}, -1 + \dfrac{\sqrt{2}}{2}\right\}$ **81.** $\dfrac{-1 - \sqrt{7}}{3}, \dfrac{-1 + \sqrt{7}}{3}$ **83.** $x = 3$; the three

sides measure 3, 4, and 5 units **85.** Either $x = 1$ and the three sides measure 3, 4, and 5 units, or $x = 5$ and the three sides measure 7, 24, and 25 units.

87. $-2 + 2\sqrt{11}$ inches by $2 + 2\sqrt{11}$ inches, which is approximately 4.633 inches by 8.633 inches. **89.** The base is $\dfrac{3}{2} + \dfrac{\sqrt{209}}{2}$ inches, which is

approximately 8.728 inches; the height is $-\dfrac{3}{2} + \dfrac{\sqrt{209}}{2}$ inches, which is approximately 5.728 inches. **91. (a)** $R(17) = 1161.1$; if 17 pairs of sunglasses are

sold per week, then the company's revenue will be $1161.10. $R(25) = 1687.5$; if 25 pairs of sunglasses are sold per week, then the company's revenue
will be $1687.50. **(b)** either 200 or 500 pairs of sunglasses **(c)** 350 pairs of sunglasses **93. (a)** after approximately 0.6 second and after approximately
3.8 seconds **(b)** after approximately 1.3 seconds and after approximately 3.0 seconds **(c)** No; the solutions to the equation are
complex solutions that are not real. **95.** 12 inches **97. (a)** ages 25 and 68 **(b)** ages 30 and 63 **99.** approximately 4.3 miles per hour
101. approximately 4.6 hours
103. By the quadratic formula, the solutions of the equation
$ax^2 + bx + c = 0$ are $x = \dfrac{-b - \sqrt{b^2 - 4ac}}{2a}$ and $x = \dfrac{-b + \sqrt{b^2 - 4ac}}{2a}$.
The sum of these two solutions is
$\dfrac{-b - \sqrt{b^2 - 4ac}}{2a} + \dfrac{-b + \sqrt{b^2 - 4ac}}{2a} = \dfrac{-2b}{2a} = -\dfrac{b}{a}$.

105. The solutions of $ax^2 + bx + c = 0$ are $x = \dfrac{-b \pm \sqrt{b^2 - 4ac}}{2a}$.
The solutions of $ax^2 - bx + c = 0$ are
$x = \dfrac{-(-b) \pm \sqrt{(-b)^2 - 4ac}}{2a} = \dfrac{b \pm \sqrt{b^2 - 4ac}}{2a}$.
Now, the negatives of the solutions to $ax^2 - bx + c = 0$
are $-\left(\dfrac{b \pm \sqrt{b^2 - 4ac}}{2a}\right) = \dfrac{-b \mp \sqrt{b^2 - 4ac}}{2a} = \dfrac{-b \pm \sqrt{b^2 - 4ac}}{2a}$,
which are the solutions to $ax^2 + bx + c = 0$.

107. Use factoring if the discriminant is a perfect square.
109. (a) **(b)** $x = -1$ or $x = -2$

(c) The x-intercepts of the function $f(x) = x^2 + 3x + 2$ are -2 and -1,
which are the same as the solutions of the equation $x^2 + 3x + 2 = 0$.

111. (a)   **(b)** $x = 1$

(c) The x-intercept of the function $g(x) = x^2 - 2x + 1$ is 1, which is
the same as the solution of the equation $x^2 - 2x + 1 = 0$.

113. The discriminant is 37; the equation has two irrational solutions. This conclusion based on the discriminant is apparent in the graph because
the graph has two x-intercepts. **115.** The discriminant is -7; the equation has two complex solutions that are not real. This conclusion based on the
discriminant is apparent in the graph because the graph has no x-intercepts.

117. (a) $x = -3$ or $x = 8$

(b) The x-intercepts are -3 and 8.

The x-intercepts of $y = x^2 - 5x - 24$ are the same as the solutions of $x^2 - 5x - 24 = 0$.

119. (a) $x = 3$

(b) The x-intercept is 3.

The x-intercept of $y = x^2 - 6x + 9$ is the same as the solution of $x^2 - 6x + 9 = 0$.

121. (a) $x = -\dfrac{5}{2} \pm \dfrac{\sqrt{7}}{2} i$

(b) The graph has no x-intercepts.

$y = x^2 + 5x + 8$ has no x-intercepts, and the solutions of $x^2 + 5x + 8 = 0$ are not real.

Section 7.3 Solving Equations Quadratic in Form **1.** quadratic in form **2.** $3x + 1$ **3.** True **4.** $u = \dfrac{1}{x}$ **5.** $\{-3, -2, 2, 3\}$ **6.** $\{-3, 3, \sqrt{2}i, -\sqrt{2}i\}$

7. $\{-3, -2, 2, 3\}$ **8.** $\left\{-\dfrac{\sqrt{3}}{2}, \dfrac{\sqrt{3}}{2}, -i, i\right\}$ **9.** $\left\{\dfrac{4}{9}, 16\right\}$ **10.** $\{25\}$ **11.** $\left\{-\dfrac{5}{2}, -\dfrac{1}{2}\right\}$ **12.** $\{-1, 125\}$ **13.** $\{-2, -1, 1, 2\}$ **15.** $\{-3i, 3i, -2i, 2i\}$

17. $\left\{-2, -\dfrac{1}{2}, \dfrac{1}{2}, 2\right\}$ **19.** $\{-\sqrt{3}, -\sqrt{2}, \sqrt{2}, \sqrt{3}\}$ **21.** $\{2, 10\}$ **23.** $\{-3, -2, 2, 3\}$ **25.** $\{-2i, 2i, -\sqrt{7}i, \sqrt{7}i\}$ **27.** $\{16\}$ **29.** $\varnothing$ or $\{\ \}$ **31.** $\left\{\dfrac{1}{4}\right\}$ **33.** $\left\{-\dfrac{1}{7}, \dfrac{1}{4}\right\}$

35. $\left\{-\dfrac{2}{3}, \dfrac{5}{2}\right\}$ **37.** $\{-64, 1\}$ **39.** $\{-1, 8\}$ **41.** $\{25\}$ **43.** $\left\{\dfrac{1}{3}, \dfrac{1}{2}\right\}$ **45.** $\left\{-\dfrac{11}{5}, -1\right\}$ **47.** $\left\{1, 3, -\dfrac{1}{2} - \dfrac{\sqrt{3}}{2}i, -\dfrac{1}{2} + \dfrac{\sqrt{3}}{2}i, -\dfrac{3}{2} - \dfrac{3\sqrt{3}}{2}i, -\dfrac{3}{2} + \dfrac{3\sqrt{3}}{2}i\right\}$

49. $\{-2, 4\}$ **51.** $\{-2i, 2i, -2\sqrt{2}, 2\sqrt{2}\}$ **53.** $\{16\}$ **55.** $\{-3, 3, -2i, 2i\}$ **57.** $\left\{-\dfrac{10}{3}, -2\right\}$ **59.** $\{9, 16\}$ **61.** $\left\{\dfrac{1}{2}, 5\right\}$ **63. (a)** $0, -\sqrt{7}i, \sqrt{7}i$

(b) $-\sqrt{6}i, \sqrt{6}i, -i, i$ **65. (a)** $0, -\sqrt{3}, \sqrt{3}$ **(b)** $-\sqrt{2}i, \sqrt{2}i, -\sqrt{5}, \sqrt{5}$ **67. (a)** $-1, \dfrac{1}{6}$ **(b)** $-\dfrac{1}{2}, \dfrac{1}{7}$ **69.** $-\sqrt{7}i, \sqrt{7}i, -\sqrt{2}i, \sqrt{2}i$ **71.** $\dfrac{81}{4}$ **73.** $-\dfrac{8}{3}, -2$

75. (a) $x = 2$ or $x = 3$ **(b)** $x = 5$ or $x = 6$; comparing these solutions to those in part (a), we note that $5 = 2 + 3$ and $6 = 3 + 3$. **(c)** $x = 0$ or $x = 1$; comparing these solutions to those in part (a), we note that $0 = 2 - 2$ and $1 = 3 - 2$. **(d)** $x = 7$ or $x = 8$; comparing these solutions to those in part (a), we note that $7 = 2 + 5$ and $8 = 3 + 5$. **(e)** The solution set of the equation $(x - a)^2 - 5(x - a) + 6 = 0$ is $\{2 + a, 3 + a\}$.

77. (a) $x = \dfrac{1}{2}$ or $x = 1$ **(b)** $x = \dfrac{5}{2}$ or $x = 3$; comparing these solutions to those in part (a), we note that $\dfrac{5}{2} = \dfrac{1}{2} + 2$ and $3 = 1 + 2$.

(c) $x = \dfrac{11}{2}$ or $x = 6$; comparing these solutions to those in part (a), we note that $\dfrac{11}{2} = \dfrac{1}{2} + 5$ and $6 = 1 + 5$. **(d)** For $f(x) = 2x^2 - 3x + 1$, the zeros

of $f(x - a)$ are $\dfrac{1}{2} + a$ and $1 + a$. **79. (a)** $R(1990) = 3000$; the revenue in 1990 was \$3000 thousand (or \$3,000,000) **(b)** $x = 2000$; in the year 2000,

revenue was \$3065 thousand (or \$3,065,000) **(c)** 2015 **81.** $x = \pm\dfrac{\sqrt{10 + 2\sqrt{17}}}{2}i$ or $x = \pm\dfrac{\sqrt{10 - 2\sqrt{17}}}{2}i$ **83.** $x = \pm\dfrac{3\sqrt{2}}{2}$ **85.** $\{36\}$; Answers will

vary. **87.** Extraneous solutions may result after squaring both sides of the equation. **89.** $p^{-2} + 4p^{-1} + 9$ **91.** $5\sqrt[3]{2a} - 4a\sqrt[3]{2a} = (5 - 4a)\sqrt[3]{2a}$
93. Let $Y_1 = x^4 + 5x^2 - 14$. **95.** Let $Y_1 = 2(x - 2)^2$ and $Y_2 = 5(x - 2) + 1$. **97.** Let $Y_1 = x - 5\sqrt{x}$ and $Y_2 = -3$.

The solution set is approximately $\{-1.41, 1.41\}$.

The solution set is approximately $\{1.81, 4.69\}$.

The solution set is approximately $\{0.49\}$.

99. (a) $Y_1 = x^2 - 5x - 6$

(b) $Y_1 = (x + 2)^2 - 5(x + 2) - 6$

(c) $Y_1 = (x + 5)^2 - 5(x + 5) - 6$

The x-intercepts are -1 and 6.

The x-intercepts are -3 and 4.

The x-intercepts are -6 and 1.

(d) The x-intercepts of the graph of $y = f(x) = x^2 - 5x - 6$ are -1 and 6. The x-intercepts of the graph of
$y = f(x + a) = (x + a)^2 - 5(x + a) - 6$ are $-1 - a$ and $6 - a$.

Putting the Concepts Together (Sections 7.1–7.3) **1.** $z^2 + 10z + 25 = (z + 5)^2$ **2.** $x^2 + 7x + \dfrac{49}{4} = \left(x + \dfrac{7}{2}\right)^2$ **3.** $n^2 - \dfrac{1}{4}n + \dfrac{1}{64} = \left(n - \dfrac{1}{8}\right)^2$

4. $\left\{\dfrac{1}{2}, \dfrac{5}{2}\right\}$ **5.** $\{-4 - 2\sqrt{3}, -4 + 2\sqrt{3}\}$ **6.** $\{3 - \sqrt{2}, 3 + \sqrt{2}\}$ **7.** $\left\{-\dfrac{4\sqrt{5}}{7}, \dfrac{4\sqrt{5}}{7}\right\}$ **8.** $\{4 - \sqrt{10}, 4 + \sqrt{10}\}$ **9.** $\left\{-1 - \dfrac{\sqrt{3}}{3}i, -1 + \dfrac{\sqrt{3}}{3}i\right\}$

10. $\left\{-2 - \dfrac{\sqrt{42}}{3}, -2 + \dfrac{\sqrt{42}}{3}\right\}$ **11.** $b^2 - 4ac = 0$; the quadratic equation has one repeated rational solution. **12.** $b^2 - 4ac = 60$; the quadratic equation has

two irrational solutions. **13.** $b^2 - 4ac = -4$; the quadratic equation has two complex solutions that are not real. **14.** $c = \sqrt{116} = 2\sqrt{29}$

15. $\left\{\dfrac{9}{4}\right\}$ **16.** $\left\{\dfrac{1}{6}, -\dfrac{1}{3}\right\}$ **17.** Revenue will be \$12,000 when either 150 microwaves or 200 microwaves are sold. **18.** The speed of the wind was approximately 52.9 miles per hour.

Section 7.4 Graphing Quadratic Functions Using Transformations   **1.** quadratic function **2.** up; down

3.   **4.**   **5.** False   **6.**   **7.**   **8.**

9.   **10.** y; a; vertically stretched; vertically compressed   **11.**   **12.**   **13.** True
14.   Domain: $\{x \mid x$ is any real number$\}$ or $(-\infty, \infty)$; Range: $\{y \mid y \le 1\}$ or $(-\infty, 1]$

15.   Domain: $\{x \mid x$ is any real number$\}$ or $(-\infty, \infty)$ Range: $\{y \mid y \ge -3\}$ or $[-3, \infty)$ **16.** $f(x) = -(x+1)^2 + 2$ **17.** (I) (D); (II) (A); (III) (C); (IV) (B)   **19.** shift 10 units to the left   **21.** shift 12 units up   **23.** vertically stretch by a factor of 2 (multiply the y-coordinates by 2) and shift 5 units to the right   **25.** multiply the y-coordinates by -3 (which means it opens down and is stretched vertically by a factor of 3), shift 5 units to the left, and shift up 8 units

27.   **29.**   **31.**   **33.**   **35.**   **37.**

39.   **41.**   **43.**   **45.**   **47.**

49.   **51.** $f(x) = (x+1)^2 - 5$ **53.** $g(x) = (x-2)^2 + 4$ **55.** $f(x) = (x+1)^2 - 3$
57. $f(x) = -2(x-3)^2 + 7$
59. $f(x) = (x+4)^2$

61. $f(x) = (x+3)^2 - 25$ vertex is $(-3, -25)$; axis of symmetry is $x = -3$; domain is the set of all real numbers or $(-\infty, \infty)$; range is $\{y \mid y \ge -25\}$ or $[-25, \infty)$ **63.** $F(x) = \left(x + \dfrac{1}{2}\right)^2 - \dfrac{49}{4}$ vertex is $\left(-\dfrac{1}{2}, -\dfrac{49}{4}\right)$; axis of symmetry is $x = -\dfrac{1}{2}$; domain is the set of all real numbers or $(-\infty, \infty)$; range is $\left\{y \mid y \ge -\dfrac{49}{4}\right\}$ or $\left[-\dfrac{49}{4}, \infty\right)$

65. $H(x) = 2(x-1)^2 - 3$ vertex is $(1, -3)$; axis of symmetry is $x = 1$; domain is the set of all real numbers or $(-\infty, \infty)$; range is $\{y \mid y \ge -3\}$ or $[-3, \infty)$ **67.** $P(x) = 3(x+2)^2 + 1$ vertex is $(-2, 1)$; axis of symmetry is $x = -2$; domain is the set of all real numbers or $(-\infty, \infty)$; range is $\{y \mid y \ge 1\}$ or $[1, \infty)$

69. $F(x) = -(x+5)^2 + 4$ vertex is $(-5, 4)$; axis of symmetry is $x = -5$; domain is the set of all real numbers or $(-\infty, \infty)$; range is $\{y \mid y \le 4\}$ or $(-\infty, 4]$ **71.** $g(x) = -(x-3)^2 + 8$ vertex is $(3, 8)$; axis of symmetry is $x = 3$; domain is the set of all real numbers or $(-\infty, \infty)$; range is $\{y \mid y \le 8\}$ or $(-\infty, 8]$

73. $H(x) = -2(x-2)^2 + 4$ vertex is $(2, 4)$; axis of symmetry is $x = 2$; domain is the set of all real numbers or $(-\infty, \infty)$; range is $\{y \mid y \le 4\}$ or $(-\infty, 4]$ **75.** $f(x) = \dfrac{1}{3}(x-3)^2 + 1$ vertex is $(3, 1)$; axis of symmetry is $x = 3$; domain is the set of all real numbers or $(-\infty, \infty)$; range is $\{y \mid y \ge 1\}$ or $[1, \infty)$

77. $G(x) = -12\left(x + \dfrac{1}{2}\right)^2 + 4$ $\left(-\dfrac{1}{2}, 4\right)$ vertex is $\left(-\dfrac{1}{2}, 4\right)$; axis of symmetry is $x = -\dfrac{1}{2}$; domain is the set of all real numbers or $(-\infty, \infty)$; range is $\{y \mid y \le 4\}$ or $(-\infty, 4]$

79. Answers may vary. One possibility: $f(x) = (x - 3)^2$ **81.** Answers may vary. One possibility: $f(x) = (x + 3)^2 + 1$ **83.** Answers may vary.
One possibility: $f(x) = -(x - 5)^2 - 1$ **85.** Answers may vary. One possibility: $f(x) = 4(x - 9)^2 - 6$ **87.** Answers may vary. One possibility:
$f(x) = -\dfrac{1}{3}x^2 + 6$ **89.** The highest or lowest point on a parabola is called the vertex. If $a < 0$, the vertex is the highest point; if $a > 0$, the vertex is the

lowest point. **91.** No. Explanations may vary. **93.** $29 + \dfrac{1}{12}$ **95.** $x^3 - 5x^2 + 4x - 2 + \dfrac{-2}{2x - 1}$

97.
Vertex: $(0, 1.3)$ Axis of
symmetry: $x = 0$ Range:
$\{y \mid y \ge 1.3\}$ or $[1.3, \infty)$

99.
Vertex: $(2.5, 0)$
Axis of symmetry: $x = 2.5$ Range:
$\{y \mid y \ge 0\}$ or $[0, \infty)$

101.
Vertex: $(1.4, 0.5)$
Axis of symmetry: $x = 1.4$ Range:
$\{y \mid y \ge 0.5\}$ or $[0.5, \infty)$

103.
Vertex: $(2.8, 5.9)$
Axis of symmetry: $x = 2.8$
Range: $\{y \mid y \le 5.9\}$ or $(-\infty, 5.9]$

105. (a) The graph opens up, and the larger the value of a, the "taller" the graph is. **(b)** The graph opens down, and the larger the value of $|a|$, the "taller"
the graph is. **(c)** The function is linear.

Section 7.5 Graphing Quadratic Functions Using Properties

1. $-\dfrac{b}{2a}$ **2.** $>$ **3.** 2 **4.** $(-2, -7)$

5. **6.** **7.** **8.**

9. True **10.** minimum; -7 at $x = 2$
11. maximum; 33 at $x = 5$ **12. (a)** The revenue will
be maximized at a price of \$75. **(b)** The maximum
revenue is \$2812.50. **13.** The maximum area that can
be enclosed is 62,500 square feet. The dimensions are
250 feet by 250 feet.

Domain: $\{x \mid x$ is any
real number$\}$ or
$(-\infty, \infty)$; Range:
$\{y \mid y \le 5\}$ or $(-\infty, 5]$

Domain: $\{x \mid x$ is any
real number$\}$ or
$(-\infty, \infty)$; Range:
$\{y \mid y \ge 0\}$ or $[0, \infty)$

Domain: $\{x \mid x$ is any
real number$\}$ or
$(-\infty, \infty)$; Range:
$\left\{y \mid y \le -\dfrac{5}{4}\right\}$ or $\left(-\infty, -\dfrac{5}{4}\right]$

14. The number of boxes that should be sold to maximize revenue is 65, and the maximum revenue is \$4225. **15. (a)** $(3, -25)$ **(b)** The discriminant is
positive; there are two x-intercepts: $(-2, 0)$ and $(8, 0)$. **17. (a)** $(1, -3)$ **(b)** The discriminant is negative; there are no x-intercepts. **19. (a)** $\left(-\dfrac{1}{2}, 0\right)$

(b) The discriminant is zero; there is one x-intercept: $\left(-\dfrac{1}{2}, 0\right)$. **21. (a)** $\left(\dfrac{1}{8}, -\dfrac{17}{16}\right)$ **(b)** The discriminant is positive; there are two x-intercepts: $(-0.39, 0)$
and $(0.64, 0)$.

23.
Domain: $\{x \mid x$ is any real
number$\}$ or $(-\infty, \infty)$
Range: $\{y \mid y \ge -9\}$ or $[-9, \infty)$

25.
Domain: $\{x \mid x$ is any real
number$\}$ or $(-\infty, \infty)$
Range: $\{y \mid y \ge -4\}$ or $[-4, \infty)$

27.
Domain: $\{x \mid x$ is any real
number$\}$ or $(-\infty, \infty)$
Range: $\{y \mid y \le 9\}$ or $(-\infty, 9]$

29.
Domain: $\{x \mid x$ is any real
number$\}$ or $(-\infty, \infty)$
Range: $\{y \mid y \ge 0\}$ or $[0, \infty)$

31.
Domain: $\{x \mid x$ is any real
number$\}$ or $(-\infty, \infty)$
Range: $\{y \mid y \ge 4\}$ or $[4, \infty)$

33.
Domain: $\{x \mid x$ is any real
number$\}$ or $(-\infty, \infty)$
Range: $\{y \mid y \le 0\}$ or $(-\infty, 0]$

35.
Domain: $\{x \mid x$ is any real
number$\}$ or $(-\infty, \infty)$
Range: $\{y \mid y \le -4\}$ or $(-\infty, -4]$

37.
Domain: $\{x \mid x$ is any real
number$\}$ or $(-\infty, \infty)$
Range: $\{y \mid y \ge -4\}$ or $[-4, \infty)$

39.

Domain: $\{x \mid x$ is any real number$\}$ or $(-\infty, \infty)$
Range: $\{y \mid y \le 16\}$ or $(-\infty, 16]$

41.

Domain: $\{x \mid x$ is any real number$\}$ or $(-\infty, \infty)$
Range: $\{y \mid y \ge 0\}$ or $[0, \infty)$

43.

Domain: $\{x \mid x$ is any real number$\}$ or $(-\infty, \infty)$
Range: $\{y \mid y \le 0\}$ or $(-\infty, 0]$

45.

Domain: $\{x \mid x$ is any real number$\}$ or $(-\infty, \infty)$
Range: $\{y \mid y \ge 3\}$ or $[3, \infty)$

47.

Domain: $\{x \mid x$ is any real number$\}$ or $(-\infty, \infty)$
Range: $\left\{y \mid y \le -\dfrac{3}{4}\right\}$ or $\left(-\infty, -\dfrac{3}{4}\right]$

49.

Domain: $\{x \mid x$ is any real number$\}$ or $(-\infty, \infty)$
Range: $\left\{y \mid y \ge -\dfrac{13}{4}\right\}$ or $\left[-\dfrac{13}{4}, \infty\right)$

51.

Domain: $\{x \mid x$ is any real number$\}$ or $(-\infty, \infty)$
Range: $\left\{y \mid y \le \dfrac{22}{3}\right\}$ or $\left(-\infty, \dfrac{22}{3}\right]$

53.

Domain: $\{x \mid x$ is any real number$\}$ or $(-\infty, \infty)$
Range: $\left\{y \mid y \ge \dfrac{3}{4}\right\}$ or $\left[\dfrac{3}{4}, \infty\right)$

55.

Domain: $\{x \mid x$ is any real number$\}$ or $(-\infty, \infty)$
Range: $\{y \mid y \le 3\}$ or $(-\infty, 3]$

57.

Domain: $\{x \mid x$ is any real number$\}$ or $(-\infty, \infty)$
Range: $\left\{y \mid y \ge -\dfrac{49}{16}\right\}$ or $\left[-\dfrac{49}{16}, \infty\right)$

59.

Domain: $\{x \mid x$ is any real number$\}$ or $(-\infty, \infty)$
Range: $\{y \mid y \ge -8\}$ or $[-8, \infty)$

61.

Domain: $\{x \mid x$ is any real number$\}$ or $(-\infty, \infty)$;
Range: $\{y \mid y \le 16\}$ or $(-\infty, 16]$

63. minimum; -3 at $x = -4$ **65.** maximum; 28 at $x = -5$ **67.** maximum; 23 at $x = 3$ **69.** minimum; -19 at $x = -2$ **71.** minimum; $-\dfrac{17}{8}$ at $x = \dfrac{5}{4}$ **73.** maximum; $\dfrac{7}{3}$ at $x = \dfrac{2}{3}$ **75. (a)** \$120 **(b)** \$36,000

77. 60; \$35 **79. (a)** after 7.5 seconds **(b)** 910 feet **(c)** about 15.042 seconds **81. (a)** about 1753.52 feet from the cannon **(b)** about 886.76 feet **(c)** about 3517 feet from the cannon **(d)** The two answers are close. Explanations may vary. **83. (a)** about 46.5 years **(b)** \$64,661.75 **85.** 18 and 18 **87.** -9 and 9

89. 15,625 square yards; 125 yards $\times$ 125 yards **91.** 500,000 square meters; 500 m $\times$ 1000 m and the long side is parallel to the river **93.** 5 inches **95. (a)** $R = -p^2 + 110p$ **(b)** \$55; \$3025 **(c)** 55 pairs

97. (a) $f(x) = x^2 - 8x + 12$; $f(x) = 2x^2 - 16x + 24$; $f(x) = -2x^2 + 16x - 24$ **(b)** The value of a has no effect on the x-intercepts. **(c)** The value of a has no effect on the axis of symmetry.

(d) The x-coordinate of the vertex is 4, which does not depend on a. However, the y-coordinate is $-4a$, which does depend on a. **99.** If the discriminant is positive, the equation $ax^2 + bx + c = 0$ will have two distinct real solutions, which means the graph of $f(x) = ax^2 + bx + c$ will have two x-intercepts. If the discriminant is zero, the equation $ax^2 + bx + c = 0$ will have one real solution, which means the graph of $f(x) = ax^2 + bx + c$ will have one x-intercept. If the discriminant is negative, the equation $ax^2 + bx + c = 0$ will have no real solutions, which means the graph of $f(x) = ax^2 + bx + c$ will have no x-intercepts.

101. If the price is too high, the quantity demanded will be 0.

103.

105.

107. Vertex: $(3.5, -9.25)$

109. Vertex: $(3.5, 37.5)$

111. Vertex: $(-0.3, -20.45)$

113. Vertex: $(0.67, 4.78)$

115. (a) The value of b affects the vertex. **(b)** c is the y-intercept.

Section 7.6 Polynomial Inequalities **1.** $\{x|x \le -5 \text{ or } x \ge 2\}; (-\infty, -5] \cup [2, \infty)$

2. $\{x|x \le -5 \text{ or } x \ge 2\}; (-\infty, -5] \cup [2, \infty)$ **3.** $\{x|-6 < x < 4\}; (-6, 4)$

4. $\left\{x \middle| x < \dfrac{-1 - \sqrt{61}}{6} \text{ or } x > \dfrac{-1 + \sqrt{61}}{6}\right\}; \left(-\infty, \dfrac{-1 - \sqrt{61}}{6}\right) \cup \left(\dfrac{-1 + \sqrt{61}}{6}, \infty\right)$

5. $\left\{x \middle| x \le -4 \text{ or } \dfrac{3}{2} \le x \le 5\right\}$ or $(-\infty, -4] \cup \left[\dfrac{3}{2}, 5\right]$

6. $\{x|-7 < x < -2 \text{ or } x > 2\}$ or $(-7, -2) \cup (2, \infty)$

7. (a) $\{x|x < -6 \text{ or } x > 5\}$ or $(-\infty, -6) \cup (5, \infty)$ **(b)** $\{x|-6 \le x \le 5\}$ or $[-6, 5]$

9. (a) $\left\{x \middle| -6 \le x \le \dfrac{5}{2}\right\}$ or $\left[-6, \dfrac{5}{2}\right]$ **(b)** $\left\{x \middle| x < -6 \text{ or } x > \dfrac{5}{2}\right\}$ or $(-\infty, -6) \cup \left(\dfrac{5}{2}, \infty\right)$

11. $\{x|x \le -2 \text{ or } x \ge 5\}$ or $(-\infty, -2] \cup [5, \infty)$

13. $\{x|-7 < x < -3\}$ or $(-7, -3)$

15. $\{x|x < -5 \text{ or } x > 7\}$ or $(-\infty, -5) \cup (7, \infty)$

17. $\{n|3 - \sqrt{17} \le n \le 3 + \sqrt{17}\}$ or $[3 - \sqrt{17}, 3 + \sqrt{17}]$
$3 - \sqrt{17} \approx -1.12$ $3 + \sqrt{17} \approx 7.12$

19. $\{m|m \le -7 \text{ or } m \ge 2\}$ or $(-\infty, -7] \cup [2, \infty)$

21. $\left\{q \middle| q \le -\dfrac{5}{2} \text{ or } q \ge 3\right\}$ or $\left(-\infty, -\dfrac{5}{2}\right] \cup [3, \infty)$

23. $\{x|-1 \le x \le 4\}$ or $[-1, 4]$

25. $\{x|x < -2 \text{ or } x > 5\}$ or $(-\infty, -2) \cup (5, \infty)$

27. $\left\{x \middle| x \le -\dfrac{2}{3} \text{ or } x \ge 4\right\}$ or $\left(-\infty, -\dfrac{2}{3}\right] \cup [4, \infty)$

29. $\{x|-2 - \sqrt{3} < x < -2 + \sqrt{3}\}$ or $(-2 - \sqrt{3}, -2 + \sqrt{3})$
$-2 - \sqrt{3} \approx -3.73$ $-2 + \sqrt{3} \approx -0.27$

31. $\left\{a \middle| -\dfrac{1}{2} \le a \le 4\right\}$ or $\left[-\dfrac{1}{2}, 4\right]$

33. $\{z|z \text{ is any real number}\}$ or $(-\infty, \infty)$

35. $\varnothing$ or $\{\ \}$

37. $\{x|x \ne 3\}$ or $(-\infty, 3) \cup (3, \infty)$

39. $\{x|-1 < x < 2 \text{ or } x > 5\}$ or $(-1, 2) \cup (5, \infty)$

41. $\left\{x \middle| x \le -\dfrac{1}{2} \text{ or } 4 \le x \le 9\right\}$ or $\left(-\infty, -\dfrac{1}{2}\right] \cup [4, 9]$

43. $\left\{x \middle| -3 \le x \le -\dfrac{1}{2} \text{ or } x \ge 1\right\}$ or $\left[-3, -\dfrac{1}{2}\right] \cup [1, \infty)$

45. $\{x|x \le -3 \text{ or } -2 \le x \le 2\}$ or $(-\infty, -3] \cup [-2, 2]$

47. $\left\{x \middle| -4 < x < -\dfrac{3}{2} \text{ or } x > \dfrac{3}{2}\right\}$ or $\left(-4, -\dfrac{3}{2}\right) \cup \left(\dfrac{3}{2}, \infty\right)$

49. $\{x|0 < x < 5\}$ or $(0, 5)$ **51.** $\{x|x \le -4 \text{ or } x \ge 7\}$ or $(-\infty, -4] \cup [7, \infty)$ **53.** $\left\{x \middle| x < -\dfrac{5}{2} \text{ or } x > 2\right\}$ or $\left(-\infty, -\dfrac{5}{2}\right) \cup (2, \infty)$

55. $\left\{x \middle| x < -\dfrac{7}{4} \text{ or } 0 < x < 2\right\}$ or $\left(-\infty, -\dfrac{7}{4}\right) \cup (0, 2)$ **57.** $\{x|x \le -8 \text{ or } x \ge 0\}$ or $(-\infty, -8] \cup [0, \infty)$ **59.** $\{x|x \le -5 \text{ or } x \ge 6\}$ or
$(-\infty, -5] \cup [6, \infty)$ **61.** between 2 and 3 seconds after the ball is thrown **63.** between \$110 and \$130 **65.** $x = -3$; a perfect square cannot be negative. Therefore, the only solution will be where the perfect square expression equals zero, which is -3. **67.** all real numbers; a perfect square must always be zero or greater. Therefore, it must always be larger than -2. **69.** Answers may vary. One possibility follows: $x^2 + x - 6 \le 0$

71. $(-\infty, -3) \cup (-1, 2) \cup (4, \infty)$ **73.** $[-5, -2] \cup [2, 5]$ **75.** $\{x|-3 < x < -2 \text{ or } x > 2\}$ or $(-3, -2) \cup (2, \infty)$ **77.** Answers will vary. One possibility follows: The inequalities have the same solution set because they are equivalent. **79.** The square of a real number is greater than or equal to zero. Therefore, $x^2 - 1$ is greater than or equal to -1 for all real numbers. **81.** No. The inequality $x^2 + 1 > 1$ is true for all real numbers except $x = 0$. **83.** To solve the quadratic inequality $f(x) > 0$ from the graph of $y = f(x)$, where x is a quadratic function, determine where the graph lies above the x-axis. This information will give you the solution set. **85.** $-\dfrac{8m^5}{n^2}$ **87.** $\dfrac{b^{\frac{1}{4}}}{9a^{\frac{7}{9}}}$

89.

$\{x|x < -7 \text{ or } x > 3.5\}$ or $(-\infty, -7) \cup (3.5, \infty)$

91.

$\{x|-2.67 \le x \le 2.5\}$ or $[-2.67, 2.5]$

Chapter 7 Review **1.** $\{-13, 13\}$ **2.** $\{-5\sqrt{3}, 5\sqrt{3}\}$ **3.** $\{-4i, 4i\}$ **4.** $\left\{-\dfrac{2\sqrt{2}}{3}, \dfrac{2\sqrt{2}}{3}\right\}$ **5.** $\{-1, 17\}$ **6.** $\{2 - 5\sqrt{6}, 2 + 5\sqrt{6}\}$ **7.** $\left\{-5, \dfrac{5}{3}\right\}$

8. $\left\{-\dfrac{3\sqrt{14}}{7}, \dfrac{3\sqrt{14}}{7}\right\}$ **9.** $\{-4\sqrt{5}i, 4\sqrt{5}i\}$ **10.** $\left\{-\dfrac{3}{4} - \dfrac{\sqrt{13}}{4}, -\dfrac{3}{4} + \dfrac{\sqrt{13}}{4}\right\}$ **11.** $a^2 + 30a + 225; (a + 15)^2$ **12.** $b^2 - 14b + 49; (b - 7)^2$

13. $c^2 - 11c + \dfrac{121}{4}; \left(c - \dfrac{11}{2}\right)^2$ **14.** $d^2 + 9d + \dfrac{81}{4}; \left(d + \dfrac{9}{2}\right)^2$ **15.** $m^2 - \dfrac{1}{4}m + \dfrac{1}{64}; \left(m - \dfrac{1}{8}\right)^2$ **16.** $n^2 + \dfrac{6}{7}n + \dfrac{9}{49}; \left(n + \dfrac{3}{7}\right)^2$ **17.** $\{2, 8\}$ **18.** $\{-4, 7\}$

19. $\{3 - 2\sqrt{3}, 3 + 2\sqrt{3}\}$ **20.** $\left\{\dfrac{5}{2} - \dfrac{\sqrt{53}}{2}, \dfrac{5}{2} + \dfrac{\sqrt{53}}{2}\right\}$ **21.** $\left\{-\dfrac{1}{2} - \dfrac{3\sqrt{3}}{2}i, -\dfrac{1}{2} + \dfrac{3\sqrt{3}}{2}i\right\}$ **22.** $\{3 - 2\sqrt{2}i, 3 + 2\sqrt{2}i\}$ **23.** $\left\{\dfrac{1}{2}, 3\right\}$

24. $\left\{-\dfrac{1}{2} - \dfrac{3}{2}i, -\dfrac{1}{2} + \dfrac{3}{2}i\right\}$ **25.** $\left\{\dfrac{3}{2} - \dfrac{\sqrt{15}}{6}i, \dfrac{3}{2} + \dfrac{\sqrt{15}}{6}i\right\}$ **26.** $\left\{-\dfrac{2}{3} - \dfrac{\sqrt{10}}{3}, -\dfrac{2}{3} + \dfrac{\sqrt{10}}{3}\right\}$ **27.** $c = 15$ **28.** $c = 8\sqrt{2}$ **29.** $c = 3\sqrt{5}$ **30.** $c = 26$

31. $c = 6$ **32.** $c = 7$ **33.** $b = 3\sqrt{7}$ **34.** $a = 5\sqrt{3}$ **35.** $a = \sqrt{253}$ **36.** approximately 127.3 feet **37.** $\{-4, 5\}$ **38.** $\left\{-\dfrac{3}{2}, \dfrac{7}{2}\right\}$

39. $\left\{-\dfrac{4}{3} - \dfrac{\sqrt{7}}{3}, -\dfrac{4}{3} + \dfrac{\sqrt{7}}{3}\right\}$ **40.** $\left\{1 - \dfrac{\sqrt{10}}{2}, 1 + \dfrac{\sqrt{10}}{2}\right\}$ **41.** $\left\{-\dfrac{1}{6} - \dfrac{\sqrt{35}}{6}i, -\dfrac{1}{6} + \dfrac{\sqrt{35}}{6}i\right\}$ **42.** $\left\{\dfrac{4}{3}\right\}$ **43.** $\{2 - \sqrt{2}, 2 + \sqrt{2}\}$

44. $\left\{-\dfrac{2}{5} - \dfrac{1}{5}i, -\dfrac{2}{5} + \dfrac{1}{5}i\right\}$ **45.** $\left\{-\dfrac{5}{2} - \dfrac{3\sqrt{3}}{2}i, -\dfrac{5}{2} + \dfrac{3\sqrt{3}}{2}i\right\}$ **46.** $\left\{-\dfrac{3}{2} - \dfrac{\sqrt{5}}{2}i, -\dfrac{3}{2} + \dfrac{\sqrt{5}}{2}i\right\}$ **47.** 57; two irrational solutions **48.** 0; one repeated rational

solution **49.** -47; two complex solutions that are not real **50.** -20; two complex solutions that are not real **51.** 0; one repeated rational solution

52. 25; two rational solutions **53.** $\{-9, 1\}$ **54.** $\left\{-\dfrac{5}{2}, \dfrac{1}{3}\right\}$ **55.** $\{-2 - 3i, -2 + 3i\}$ **56.** $\{-2\sqrt{3}, 2\sqrt{3}\}$ **57.** $\left\{1 - \dfrac{\sqrt{10}}{2}, 1 + \dfrac{\sqrt{10}}{2}\right\}$

58. $\{-4 - 2i, -4 + 2i\}$ **59.** $\{-3, 5\}$ **60.** $\{1 - \sqrt{2}, 1 + \sqrt{2}\}$ **61.** $\left\{-\dfrac{4}{3}, \dfrac{4}{3}\right\}$ **62.** $\{-1 - \sqrt{3}i, -1 + \sqrt{3}i\}$ **63.** $x = 10$; the three sides measure

5, 12, and 13 **64.** 12 centimeters by 9 centimeters **65. (a)** either 300 or 600 cell phones **(b)** 450 cell phones **66. (a)** after approximately 0.5 second and after approximately 2.7 seconds **(b)** after approximately 4.3 seconds **(c)** No; the solutions to the equation are complex solutions that are not real. Another explanation: The vertex is (1.5625, 219.0625). Since $a = -16 < 0$, the parabola opens down, so the maximum value of the function is 219.0625.

67. approximately 10.8 miles per hour **68.** approximately 67.8 minutes **69.** $\{-4i, 4i, -3, 3\}$ **70.** $\left\{-\dfrac{\sqrt{3}}{2}, \dfrac{\sqrt{3}}{2}, -\sqrt{2}i, \sqrt{2}i\right\}$ **71.** $\left\{-\dfrac{10}{3}, -1\right\}$

72. $\{-4, 4, -2\sqrt{2}, 2\sqrt{2}\}$ **73.** $\{16, 81\}$ **74.** $\left\{\dfrac{9}{25}\right\}$ **75.** $\left\{-\dfrac{1}{3}, \dfrac{1}{7}\right\}$ **76.** $\left\{-343, \dfrac{1}{8}\right\}$ **77.** $\{16\}$ **78.** $\left\{-\dfrac{36}{7}, -\dfrac{19}{4}\right\}$ **79.** $\left\{\dfrac{9}{4}, \dfrac{49}{4}\right\}$ **80.** $\{-2\sqrt{3}, -\sqrt{5}, \sqrt{5}, 2\sqrt{3}\}$

81. **82.** **83.** **84.** **85.** **86.**

87. **88.** **89.** **90.** **91.** vertex is $(3, 1)$; axis of symmetry is $x = 3$; Domain: $\{x|x \text{ is any real number}\}$ or $(-\infty, \infty)$; Range: $\{y|y \ge 1\}$ or $[1, \infty)$

92. vertex is $(-4, -5)$; axis of symmetry is $x = -4$ **93.** vertex is $(1, -5)$; axis of symmetry is $x = 1$ **94.** vertex is $(-3, -1)$; axis of symmetry is $x = -3$

Domain: $\{x|x \text{ is any real number}\}$ or $(-\infty, \infty)$ Domain: $\{x|x \text{ is any real number}\}$ or $(-\infty, \infty)$ Domain: $\{x|x \text{ is any real number}\}$ or $(-\infty, \infty)$ Range: $\{y|y \ge -5\}$ or $[-5, \infty)$ Range: $\{y|y \ge -5\}$ or $[-5, \infty)$ Range: $\{y|y \le -1\}$ or $(-\infty, -1]$

95. vertex is $(2, 4)$; axis of symmetry is $x = 2$
Domain: $\{x \mid x \text{ is any real number}\}$ or $(-\infty, \infty)$
Range: $\{y \mid y \leq 4\}$ or $(-\infty, 4]$

96. vertex is $(2, 3)$; axis of symmetry is $x = 2$
Domain: $\{x \mid x \text{ is any real number}\}$ or $(-\infty, \infty)$
Range: $\{y \mid y \geq 3\}$ or $[3, \infty)$

97. $f(x) = 2(x - 2)^2 - 4$ or $f(x) = 2x^2 - 8x + 4$ **98.** $f(x) = -(x - 4)^2 + 3$ or $f(x) = -x^2 + 8x - 13$

99. $f(x) = -\dfrac{1}{2}(x + 2)^2 - 1$ or $f(x) = -\dfrac{1}{2}x^2 - 2x - 3$ **100.** $f(x) = 3(x + 2)^2$ or $f(x) = 3x^2 + 12x + 12$

101.
Domain: $\{x \mid x \text{ is any real number}\}$ or $(-\infty, \infty)$
Range: $\{y \mid y \geq -9\}$ or $[-9, \infty)$

102.
Domain: $\{x \mid x \text{ is any real number}\}$ or $(-\infty, \infty)$
Range: $\left\{y \mid y \geq -\dfrac{1}{8}\right\}$ or $\left[-\dfrac{1}{8}, \infty\right)$

103.
Domain: $\{x \mid x \text{ is any real number}\}$ or $(-\infty, \infty)$
Range: $\{y \mid y \leq 2\}$ or $(-\infty, 2]$

104.
Domain: $\{x \mid x \text{ is any real number}\}$ or $(-\infty, \infty)$
Range: $\{y \mid y \leq 5\}$ or $(-\infty, 5]$

105.
Domain: $\{x \mid x \text{ is any real number}\}$ or $(-\infty, \infty)$
Range: $\{y \mid y \geq 0\}$ or $[0, \infty)$

106.
Domain: $\{x \mid x \text{ is any real number}\}$ or $(-\infty, \infty)$
Range: $\{y \mid y \geq 0\}$ or $[0, \infty)$

107.
Domain: $\{x \mid x \text{ is any real number}\}$ or $(-\infty, \infty)$
Range: $\{y \mid y \geq 1\}$ or $[1, \infty)$

108.
Domain: $\{x \mid x \text{ is any real number}\}$ or $(-\infty, \infty)$
Range: $\{y \mid y \leq -5\}$ or $(-\infty, -5]$

109. maximum; 22 at $x = 4$ **110.** minimum; $-\dfrac{11}{8}$ at $x = \dfrac{1}{4}$ **111.** maximum; 7 at $x = 1$ **112.** maximum; 5 at $x = 6$ **113.** (a) \$225 (b) \$16,875

114. (a) 3.75 amperes (b) 225 watts

115. both numbers are 12 **116.** (a) 3.75 yards by 7.5 yards (b) 28.125 square yards **117.** (a) 100 feet (b) 50 feet (c) 200 feet

118. (a) $R(p) = -0.002p^2 + 60p$ (b) \$15,000; \$450,000 (c) 30 automobiles per month **119.** (a) $\{x \mid x < -2 \text{ or } x > 3\}$ or $(-\infty, -2) \cup (3, \infty)$

(b) $\{x \mid -2 < x < 3\}$ or $(-2, 3)$ **120.** (a) $\left\{x \mid -\dfrac{7}{2} \leq x \leq 1\right\}$ or $\left[-\dfrac{7}{2}, 1\right]$ (b) $\left\{x \mid x \leq -\dfrac{7}{2} \text{ or } x \geq 1\right\}$ or $\left(-\infty, -\dfrac{7}{2}\right] \cup [1, \infty)$

121. $\{x \mid -4 \leq x \leq 6\}$ or $[-4, 6]$

122. $\{y \mid y \leq -8 \text{ or } y \geq 1\}$ or, $(-\infty, -8] \cup [1, \infty)$

123. $\left\{z \mid z < \dfrac{4}{3} \text{ or } z > 5\right\}$ or $\left(-\infty, \dfrac{4}{3}\right) \cup (5, \infty)$

124. $\left\{p \mid -2 - \sqrt{6} < p < -2 + \sqrt{6}\right\}$ or $(-2 - \sqrt{6}, -2 + \sqrt{6})$

125. $\{m \mid m \text{ is any real number}\}$ or $(-\infty, \infty)$

126. $\left\{w \mid -\dfrac{1}{3} \leq w \leq \dfrac{7}{2}\right\}$ or $\left[-\dfrac{1}{3}, \dfrac{7}{2}\right]$

127. $\left\{x \mid x < -1 \text{ or } \dfrac{3}{2} < x < 2\right\}$ or $(-\infty, -1) \cup \left(\dfrac{3}{2}, 2\right)$

128. $\{x \mid -5 \leq x \leq -3 \text{ or } x \geq 3\}$ or $[-5, -3] \cup [3, \infty)$

Chapter 7 Test **1.** $x^2 - 3x + \dfrac{9}{4}; \left(x - \dfrac{3}{2}\right)^2$ **2.** $m^2 + \dfrac{2}{5}m + \dfrac{1}{25}; \left(m + \dfrac{1}{5}\right)^2$ **3.** $\left\{-\dfrac{5}{3}, -1\right\}$ **4.** $\{3 - \sqrt{5}, 3 + \sqrt{5}\}$ **5.** $\left\{1 - \dfrac{\sqrt{2}}{2}i, 1 + \dfrac{\sqrt{2}}{2}i\right\}$

6. $\left\{\dfrac{3}{2} - \dfrac{\sqrt{3}}{6}i, \dfrac{3}{2} + \dfrac{\sqrt{3}}{6}i\right\}$ **7.** 57; two irrational solutions **8.** $a = 6\sqrt{2}$ **9.** $\{-3, 3, -2i, 2i\}$ **10.** $\left\{\dfrac{1}{81}\right\}$

11.
vertex is $(-2, -5)$; axis of symmetry is $x = -2$
x-intercepts: $(-2 - \sqrt{5}, 0), (-2 + \sqrt{5}, 0)$
y-intercept: $(0, -1)$
Domain: $\{x \mid x \text{ is any real number}\}$ or $(-\infty, \infty)$
Range: $\{y \mid y \geq -5\}$ or $[-5, \infty)$

12.
vertex is $(-2, 5)$; axis of symmetry is $x = -2$
x-intercepts: $\left(\dfrac{-4 - \sqrt{10}}{2}, 0\right), \left(\dfrac{-4 + \sqrt{10}}{2}, 0\right)$
y-intercept: $(0, -3)$
Domain: $\{x \mid x \text{ is any real number}\}$ or $(-\infty, \infty)$
Range: $\{y \mid y \leq 5\}$ or $(-\infty, 5]$

13. $f(x) = \frac{1}{3}(x+3)^2 - 5$ or $f(x) = \frac{1}{3}x^2 + 2x - 2$ **14.** maximum; 6

15. $\left\{ m \middle| m < -3 \text{ or } m > \frac{5}{2} \right\}$ or $\left(-\infty, -3 \right) \cup \left(\frac{5}{2}, \infty \right)$

16. $\{x \mid x \le -5 \text{ or } -2 \le x \le 2\}$ or $(-\infty, -5] \cup [-2, 2]$

17. 0.4 second and 4.6 seconds **18.** 34.1 hours **19. (a)** \$340 **(b)** \$28,900 **20. (a)** 12.5 in. by 12.5 in. by 12 in. **(b)** 1875 cubic inches

Cumulative Review Chapters R–7 **1.** 2 **2.** $3c + 11$ **3.** $\{5\}$ **4.** $(-\infty, -3) \cup (1, \infty)$ **5.** $\{x \mid x \ne -4, x \ne 3\}$

6. **7.** $y = -\frac{2}{5}x + 1; 2x + 5y = 5$ **8.** $(-2, 5)$ **9.** -4 **10.** **11.** $a^3 - 2a^2 - 5a + 3$

12. $4x^2 + 2x - 3 + \dfrac{4x - 3}{2x^2 - x + 5}$ **13.** $(2m + 3n)(4m^2 - 6mn + 9n^2)$ **14.** $(7y - 5)(y + 4)$ **15.** $\dfrac{w + 5}{w - 2}$ **16.** $\dfrac{k + 23}{(k - 4)(k - 1)(k + 5)}$

17. $\left\{ -\dfrac{3}{2} \right\}$ **18.** $(-\infty, -5) \cup [4, \infty)$ or $\{x \mid x < -5 \text{ or } x \ge 4\}$

19. One plane is traveling 126 miles per hour and the other plane is traveling 142 miles per hour. **20.** $7\sqrt{3}$ **21.** $\dfrac{5\sqrt[3]{4}}{4}$ **22.** $\{16\}$ **23.** $x = 11$
24. $\{-2, 10\}$ **25.**

Chapter 8 Exponential and Logarithmic Functions

Section 8.1 Composite Functions and Inverse Functions

1. composite function **2. (a)** 17 **(b)** 26 **(c)** -63 **3.** False **4.** $g(x) = 4x - 3$ **5. (a)** $9x^2 + 3x - 1$ **(b)** $3x^2 - 9x + 5$ **(c)** 29
6. one-to-one **7.** not one-to-one **8.** one-to-one **9. (a)** not one-to-one **(b)** one-to-one

10. The domain of the inverse function is $\{36.05, 35.57, 34.58, 34.20, 34.73\}$. The range of the inverse function is $\{24.80, 24.59, 24.29, 23.81, 24.87\}$. **11.** $\{(3, -3), (2, -2), (1, -1), (0, 0), (-1, 1)\}$. The domain of the inverse function is $\{3, 2, 1, 0, -1\}$. The range of the inverse function is $\{-3, -2, -1, 0, 1\}$.

12. 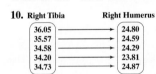 **13.** True **14.** False **15.** $g^{-1}(x) = \dfrac{x + 1}{5}$ **16.** $f^{-1}(x) = \sqrt[5]{x - 3}$ **17. (a)** 3 **(b)** -3 **(c)** 19 **(d)** -12 **19. (a)** 85 **(b)** 19
(c) 29 **(d)** -7 **21. (a)** -4394 **(b)** -507 **(c)** 16 **(d)** -1453 **23. (a)** 8 **(b)** $\dfrac{4}{5}$ **(c)** 1 **(d)** $-\dfrac{4}{3}$ **25. (a)** $(f \circ g)(x) = 2x + 1$
(b) $(g \circ f)(x) = 2x + 2$ **(c)** $(f \circ f)(x) = x + 2$ **(d)** $(g \circ g)(x) = 4x$

27. (a) $(f \circ g)(x) = -8x + 17$ **(b)** $(g \circ f)(x) = -8x - 23$ **(c)** $(f \circ f)(x) = 4x + 21$ **(d)** $(g \circ g)(x) = 16x - 15$ **29. (a)** $(f \circ g)(x) = x^2 - 6x + 9$
(b) $(g \circ f)(x) = x^2 - 3$ **(c)** $(f \circ f)(x) = x^4$ **(d)** $(g \circ g)(x) = x - 6$ **31. (a)** $(f \circ g)(x) = \sqrt{x + 4}$ **(b)** $(g \circ f)(x) = \sqrt{x} + 4$ **(c)** $(f \circ f)(x) = \sqrt[4]{x}$
(d) $(g \circ g)(x) = x + 8$ **33. (a)** $(f \circ g)(x) = x^2$ **(b)** $(g \circ f)(x) = x^2 + 8x + 12$ **(c)** $(f \circ f)(x) = ||x + 4| + 4|$ **(d)** $(g \circ g)(x) = x^4 - 8x^2 + 12$

35. (a) $(f \circ g)(x) = \dfrac{2x}{x + 1}$, where $x \ne -1, 0$ **(b)** $(g \circ f)(x) = \dfrac{x + 1}{2}$, where $x \ne -1$ **(c)** $(f \circ f)(x) = \dfrac{2(x + 1)}{x + 3}$, where $x \ne -1, -3$

(d) $(g \circ g)(x) = x$, where $x \ne 0$ **37.** one-to-one **39.** not one-to-one **41.** one-to-one **43.** not one-to-one **45.** one-to-one **47.** one-to-one
49. not one-to-one **51.** one-to-one

53. **55.** $\{(3, 0), (4, 1), (5, 2), (6, 3)\}$ **57.** $\{(3, -2), (1, -1), (-3, 0), (9, 1)\}$

59. **61.** **63.** **65.** $f(g(x)) = (x - 5) + 5 = x$
$g(f(x)) = (x + 5) - 5 = x$

67. $f(g(x)) = 5\left(\dfrac{x - 7}{5}\right) + 7 = x - 7 + 7 = x$ **69.** $f(g(x)) = \dfrac{3}{\left(\dfrac{3}{x} + 1\right) - 1} = \dfrac{3}{\dfrac{3}{x}} = 3 \cdot \dfrac{x}{3} = x$

$g(f(x)) = \dfrac{(5x + 7) - 7}{5} = \dfrac{5x}{5} = x$

$g(f(x)) = \dfrac{3}{\dfrac{3}{x - 1}} + 1 = 3 \cdot \dfrac{x - 1}{3} + 1 = x - 1 + 1 = x$

71. $f(g(x)) = \sqrt[3]{(x^3 - 4) + 4} = \sqrt[3]{x^3} = x$
$g(f(x)) = (\sqrt[3]{x + 4})^3 - 4 = x + 4 - 4 = x$

73. $f^{-1}(x) = \dfrac{x}{6}$ **75.** $f^{-1}(x) = x - 4$ **77.** $h^{-1}(x) = \dfrac{x + 7}{2}$ **79.** $G^{-1}(x) = \dfrac{2 - x}{5}$

81. $g^{-1}(x) = \sqrt[3]{x - 3}$ **83.** $p^{-1}(x) = \dfrac{1}{x} - 3$ **85.** $F^{-1}(x) = 2 - \dfrac{5}{x}$ **87.** $f^{-1}(x) = x^3 + 2$

89. $R^{-1}(x) = \dfrac{2x}{1 - x}$ **91.** $f^{-1}(x) = (x - 4)^3 + 1$ **93.** $A(t) = 400\pi t^2$; $3600\pi \approx 11{,}309.73$ sq ft **95. (a)** $C(x) = 2x$ **(b)** \$630 **97.** $f^{-1}(12) = 4$

99. Domain of f^{-1}: $[-5, \infty)$; Range of f^{-1}: $[0, \infty)$ **101.** Domain of g^{-1}: $(-6, 12)$; Range of g^{-1}: $[-4, 10]$ **103.** $x(T) = \dfrac{T + 463.75}{0.15}$ for $927.50 \le T \le 5183.75$

105. $\left\{-\dfrac{5}{2}, 3\right\}$ **107.** A function is one-to-one provided no two different inputs correspond to the same output. A function must be one-to-one in order for the inverse to be a function because we interchange the inputs and outputs to obtain the inverse. Thus, if a function is not one-to-one, the inverse will have a single input corresponding to two different outputs. **109.** The domain of f equals the range of f^{-1} because we interchange the roles of the inputs and outputs to obtain the inverse of a function. The same logic explains why the range of f equals the domain of f^{-1}. **111. (a)** 3 **(b)** -3 **(c)** 19 **(d)** -12

113. (a) 85 **(b)** 19 **(c)** 29 **(d)** -7 **115. (a)** -4394 **(b)** -507 **(c)** 16 **(d)** -1453 **117. (a)** 8 **(b)** $\dfrac{4}{5}$ **(c)** 1 **(d)** $-\dfrac{4}{3}$

119. $f(x) = x + 5; g(x) = x - 5$ **121.** $f(x) = 5x + 7; g(x) = \dfrac{x - 7}{5}$

Section 8.2 Exponential Functions

1. $>$; $\ne$ **2. (a)** 3.249009585 **(b)** 3.317278183 **(c)** 3.321880096 **(d)** 3.32211036 **(e)** 3.321997085

3. The domain of f is all real numbers or, in interval notation, $(-\infty, \infty)$. The range of f is $\{y | y > 0\}$ or, in interval notation, $(0, \infty)$.

4. $\left(-1, \dfrac{1}{a}\right); (0, 1); (1, a)$ **5.** True **6.** False

7. The domain of f is all real numbers or, in interval notation, $(-\infty, \infty)$. The range of f is $\{y | y > 0\}$ or, in interval notation, $(0, \infty)$.

8. The domain of f is all real numbers or, in interval notation, $(-\infty, \infty)$. The range of f is $\{y | y > 0\}$ or, in interval notation, $(0, \infty)$.

9. The domain of f is all real numbers or, in interval notation, $(-\infty, \infty)$. The range of f is $\{y | y > 1\}$ or, in interval notation, $(1, \infty)$.

10. 2.71828 **11. (a)** 54.598 **(b)** 0.018 **12.** $\{3\}$ **13.** $\{2\}$ **14.** $\{0, 5\}$ **15.** $\{-1, 3\}$ **16. (a)** 0.918 or 91.8% **(b)** 0.998 or 99.8% **17. (a)** approximately 6.91 grams **(b)** 5 grams **(c)** 0.625 gram **(d)** approximately 0.247 gram **18. (a)** \$2102.32 **(b)** \$4227.41 **(c)** \$8935.49 **19. (a)** 11.212 **(b)** 11.587 **(c)** 11.664 **(d)** 11.665 **(e)** 11.665 **21. (a)** 73.517 **(b)** 77.708 **(c)** 77.924 **(d)** 77.881 **(e)** 77.880 **23.** g **25.** e **27.** f **29.** h

31. Domain: all real numbers or $(-\infty, \infty)$; Range: $\{y | y > 0\}$ or $(0, \infty)$

33. Domain: all real numbers or $(-\infty, \infty)$; Range: $\{y | y > 0\}$ or $(0, \infty)$

35. Domain: all real numbers or $(-\infty, \infty)$; Range: $\{y | y > 0\}$ or $(0, \infty)$

37. Domain: all real numbers or $(-\infty, \infty)$; Range: $\{y | y > 3\}$ or $(3, \infty)$

39. Domain: all real numbers or $(-\infty, \infty)$; Range: $\{y | y > -1\}$ or $(-1, \infty)$

41. Domain: all real numbers or $(-\infty, \infty)$; Range: $\{y | y > 0\}$ or $(0, \infty)$

43. (a) 21.217 **(b)** 22.472 **(c)** 22.460 **(d)** 22.460 **(e)** 22.459 **45.** 7.389 **47.** 0.135 **49.** 9.974

51. Domain: all real numbers or $(-\infty, \infty)$; Range: $\{y | y > 0\}$ or $(0, \infty)$

53. Domain: all real numbers or $(-\infty, \infty)$; Range: $\{y | y < 0\}$ or $(-\infty, 0)$

55. $\{5\}$ **57.** $\{-4\}$ **59.** $\{5\}$ **61.** $\{5\}$ **63.** $\left\{\dfrac{3}{2}\right\}$ **65.** $\{1\}$ **67.** $\{-1, 4\}$ **69.** $\{-4, 2\}$ **71.** $\{9\}$ **73.** $\{-2\}$ **75.** $\{-2, 2\}$ **77.** $\{-2\}$ **79.** $\{2\}$ **81. (a)** $f(3) = 8; (3, 8)$ **(b)** $x = -3; \left(-3, \dfrac{1}{8}\right)$

83. (a) $g(-1) = -\dfrac{3}{4}; \left(-1, -\dfrac{3}{4}\right)$ **(b)** $x = 2; (2, 15)$ **85. (a)** $H(-3) = 24; (-3, 24)$ **(b)** $x = 2; \left(2, \dfrac{3}{4}\right)$

87. (a) approximately 342.5 million people **(b)** approximately 455.7 million people **(c)** Answers may vary. One possibility is that the population is not growing exponentially. **89. (a)** $5100.92 **(b)** $5308.92 **(c)** $5525.29 **91. (a)** $2318.55 **(b)** $2322.37 **(c)** $2323.23 **(d)** $2323.65 **(e)** The future value is higher with more compounding periods. **93. (a)** $19,841 **(b)** $15,365 **(c)** $10,471 **95. (a)** approximately 95.105 grams **(b)** 50 grams **(c)** 25 grams **(d)** approximately 0.661 gram **97. (a)** approximately 300.233°F **(b)** approximately 230.628°F **(c)** yes **99. (a)** approximately 29 words **(b)** approximately 38 words **101. (a)** approximately 0.238 ampere **(b)** approximately 0.475 ampere **103.** $y = 3^x$

105. As x increases, the graph increases very rapidly. As x decreases, the graph approaches the x-axis. **107.** Answers may vary. The big difference is that exponential functions are of the form $f(x) = a^x$ (the variable is in the exponent), while polynomial functions are of the form $f(x) = a_n x^n + a_{n-1} x^{n-1} + \cdots + a_1 x + a_0$ (the variable is a base). **109. (a)** 15 **(b)** -5 **111. (a)** undefined **(b)** $\dfrac{4}{5}$ **113. (a)** 3 **(b)** $3\sqrt{3}$

115. $f(x) = 1.5^x$

Domain: all real numbers or $(-\infty, \infty)$; Range: $\{y \mid y > 0\}$ or $(0, \infty)$

117. $H(x) = 0.9^x$

Domain: all real numbers or $(-\infty, \infty)$; Range: $\{y \mid y > 0\}$ or $(0, \infty)$

119. $g(x) = 2.5^x + 3$

Domain: all real numbers or $(-\infty, \infty)$; Range: $\{y \mid y > 3\}$ or $(3, \infty)$

121. $F(x) = 1.6^{x-3}$

Domain: all real numbers or $(-\infty, \infty)$; Range: $\{y \mid y > 0\}$ or $(0, \infty)$

123. (a) Domain: the set of all real numbers; Range: the set of all positive real numbers; intercepts: $(0, 1)$; $(-1, 1/2)$, $(0, 1)$, and $(1, 2)$ **(b)** Domain: the set of all real numbers; Range: the set of all positive real numbers; intercepts: $(0, 1)$; $(-1, 1/3)$, $(0, 1)$, $(1, 3)$ **(c)** Domain: the set of all real numbers; Range: the set of all positive real numbers; intercepts: $(0, 1)$; $(-1, 1/4)$, $(0, 1)$, and $(1, 4)$ **(d) (i)** the set of all real numbers; **(ii)** the set of all positive real numbers; **(iii)** Zero; $(0, 1)$ **(iv)** $(-1, 1/a)$, $(0, 1)$, and $(1, a)$.

Section 8.3 Logarithmic Functions 1. $x = a^y$; $>$; $\neq$ **2.** $3 = \log_4 64$ **3.** $-2 = \log_p 8$ **4.** $2^4 = 16$ **5.** $a^5 = 20$ **6.** $5^{-3} = z$ **7.** 2 **8.** -3

9. 2 **10.** -1 **11.** $\{x \mid x > -3\}$ or $(-3, \infty)$ **12.** $\left\{ x \mid x < \dfrac{5}{2} \right\}$ or $\left(-\infty, \dfrac{5}{2} \right)$

13. The domain of f is $\{x \mid x > 0\}$ or, in interval notation, $(0, \infty)$. The range of f is all real numbers or, in interval notation, $(-\infty, \infty)$.

14. The domain of f is $\{x \mid x > 0\}$ or, in interval notation, $(0, \infty)$. The range of f is all real numbers or, in interval notation, $(-\infty, \infty)$.

15. 3.146 **16.** 1.569 **17.** -0.523 **18.** $\{16\}$ **19.** $\{4\}$ **20.** $\{e^{-2}\}$ **21.** $\{10,020\}$ **22.** 100 decibels **23.** $3 = \log_4 64$ **25.** $-3 = \log_2 \left(\dfrac{1}{8} \right)$

27. $\log_a 19 = 3$ **29.** $\log_5 c = -6$ **31.** $2^4 = 16$ **33.** $3^{-2} = \dfrac{1}{9}$ **35.** $5^{-3} = a$ **37.** $a^2 = 4$ **39.** $\left(\dfrac{1}{2} \right)^y = 12$ **41.** 0 **43.** 3 **45.** -2 **47.** 4 **49.** 4 **51.** $\dfrac{1}{2}$

53. $\{x \mid x > 4\}$ or $(4, \infty)$ **55.** $\{x \mid x > 0\}$ or $(0, \infty)$ **57.** $\left\{ x \mid x > \dfrac{2}{3} \right\}$ or $\left(\dfrac{2}{3}, \infty \right)$ **59.** $\left\{ x \mid x > -\dfrac{1}{2} \right\}$ or $\left(-\dfrac{1}{2}, \infty \right)$ **61.** $\left\{ x \mid x < \dfrac{1}{4} \right\}$ or $\left(-\infty, \dfrac{1}{4} \right)$

63.

Domain: $\{x \mid x > 0\}$ or $(0, \infty)$ Range: all real numbers or $(-\infty, \infty)$

65.

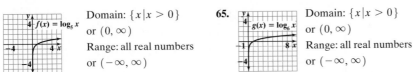

Domain: $\{x \mid x > 0\}$ or $(0, \infty)$ Range: all real numbers or $(-\infty, \infty)$

67.

Domain: $\{x \mid x > 0\}$ or $(0, \infty)$ Range: all real numbers or $(-\infty, \infty)$

69. $\ln 12 = x$ **71.** $e^4 = x$ **73.** -1 **75.** 3 **77.** 1.826 **79.** 1.686 **81.** -0.456 **83.** -1.609 **85.** 0.097 **87.** -0.981 **89.** $\{4\}$ **91.** $\left\{ \dfrac{13}{2} \right\}$ **93.** $\{6\}$

95. $\{3\sqrt{2}\}$ **97.** $\{10\}$ **99.** $\{e^5\}$ **101.** $\left\{ \dfrac{11}{20} \right\}$ **103.** $\{-3\}$ **105.** $\{4\}$ **107.** $\{-3, 3\}$ **109. (a)** $f(16) = 4; (16, 4)$ **(b)** $x = \dfrac{1}{8}; \left(\dfrac{1}{8}, -3 \right)$

111. (a) $G(7) = \dfrac{3}{2}; \left(7, \dfrac{3}{2} \right)$ **(b)** $x = 15; (15, 2)$ **113.** $a = 4$ **115.** $\{x \mid x < -1 \text{ or } x > 5\}$ or $(-\infty, -1) \cup (5, \infty)$ **117.** $\{x \mid x < -1 \text{ or } x > 4\}$ or $(-\infty, -1) \cup (4, \infty)$ **119.** $\{x \mid x \neq 3\}$ or $(-\infty, 3) \cup (3, \infty)$ **121.** 20 decibels **123.** 130 decibels **125.** approximately 7.8 on the Richter scale **127.** approximately 794,328 **129. (a)** 12; basic **(b)** 5; acidic **(c)** 2; acidic **(d)** $10^{-7.4}$ mole per liter **131.** The base of $f(x) = \log_a x$ cannot equal 1 because $y = \log_a x$ is equivalent to $x = a^y$ and a does not equal 1 in the exponential function. In addition, the graph would be a vertical line $(x = 1)$, which is not a function. **133.** The domain of $f(x) = \log_a(x^2 + 1)$ is the set of all real numbers because $x^2 + 1 > 0$ for all x.

135. $-3x^2 - 7x + 10$ **137.** $\dfrac{4x^2 + 2x + 3}{(x + 1)(x - 1)(x + 2)}$ **139.** $5x\sqrt{2x}$

141. $f(x) = \log(x + 1)$

Domain: $\{x \,|\, x > -1\}$ or $(-1, \infty)$; Range: all real number or $(-\infty, \infty)$

143. $G(x) = \ln(x) + 1$

Domain: $\{x \,|\, x > 0\}$ or $(0, \infty)$; Range: all real number or $(-\infty, \infty)$

145. $f(x) = 2\log(x - 3) + 1$

Domain: $\{x \,|\, x > 3\}$ or $(3, \infty)$; Range: all real number or $(-\infty, \infty)$

147. (a) Domain: the set of all positive real numbers; Range: the set of all real numbers; intercepts: $(1, 0)$; $(1/2, -1)$, $(1, 0)$, and $(2, 1)$ **(b)** Domain: the set of all positive real numbers; Range: the set of all real numbers; intercepts: $(1, 0)$; $(1/3, -1)$, $(1, 0)$, and $(3, 1)$ **(c)** Domain: the set of all positive real numbers; Range: the set of all real numbers; intercepts: $(1, 0)$; $(1/4, -1)$, $(1, 0)$, and $(4, 1)$ **(d) (i)** the set of all positive real numbers; **(ii)** the set of all real numbers; **(iii)** Zero; $(1, 0)$; **(iv)** $(1/a, -1)$, $(1, 0)$, and $(a, 1)$.

Putting the Concepts Together (Sections 8.1–8.3) **1. (a)** $(f \circ g)(x) = 4x^2 - 8x + 3$ **(b)** $(g \circ f)(x) = 8x^2 + 16x + 6$ **(c)** $(f \circ g)(3) = 15$
(d) $(g \circ f)(-2) = 6$ **(e)** $(f \circ f)(1) = 13$ **2. (a)** $f^{-1}(x) = \dfrac{x - 4}{3}$ **(b)** $g^{-1}(x) = \sqrt[3]{x + 4}$

3.

4. (a) 14.611 **(b)** 15.206 **(c)** 15.146 **(d)** 15.155 **(e)** 15.154 **5. (a)** $\log_a 6.4 = 4$ **(b)** $\log 278 = x$
6. (a) $2^7 = x$ **(b)** $e^M = 16$ **7. (a)** 4 **(b)** -2 **8.** $\{x \,|\, x > -6\}$ or $(-6, \infty)$

9.

Domain: all real numbers or $(-\infty, \infty)$ Range: $\{y \,|\, y > 0\}$ or $(0, \infty)$

10.

Domain: $\{x \,|\, x > 0\}$ or $(0, \infty)$ Range: all real numbers or $(-\infty, \infty)$

11. $\{-1\}$ **12.** $\{-5\}$ **13.** $\left\{\dfrac{11}{2}\right\}$
14. $\{e^7\}$ **15.** approximately 56 terms

Section 8.4 Properties of Logarithms **1.** 0 **2.** 0 **3.** 1 **4.** 1 **5.** $\sqrt{2}$ **6.** 0.2 **7.** 1.2 **8.** -4 **9.** False **10.** $\log_4 9 + \log_4 5$ **11.** $\log 5 + \log w$
12. $\log_7 9 - \log_7 5$ **13.** $\ln p - \ln 3$ **14.** $\log_2 3 + \log_2 m - \log_2 n$ **15.** $\ln q - \ln 3 - \ln p$ **16.** $1.6 \log_2 5$ **17.** $5 \log b$ **18.** $2 \log_4 a + \log_4 b$
19. $2 + 4 \log_3 m - \dfrac{1}{3}\log_3 n$ **20.** 2 **21.** $\log_3\left(\dfrac{x + 4}{x - 1}\right)$ **22.** $\log_5 \dfrac{x}{8}$ **23.** $\log_2 \dfrac{x^2 + 3x + 2}{x^2}$ **24.** 10; 3; 10; 3 **25.** 3.155 **26.** 2.807 **27.** 3 **29.** -7
31. 5 **33.** 2 **35.** 1 **37.** 0 **39.** $a + b$ **41.** $2b$ **43.** $2a + b$ **45.** $\dfrac{1}{2}a$ **47.** $\log a + \log b$ **49.** $4 \log_5 x$ **51.** $\log_2 x + 2 \log_2 y$ **53.** $2 + \log_5 x$
55. $2 - \log_7 y$ **57.** $2 + \ln x$ **59.** $3 + \dfrac{1}{2}\log_3 x$ **61.** $2 \log_5 x + \dfrac{1}{2}\log_5(x^2 + 1)$ **63.** $4 \log x - \dfrac{1}{3}\log(x - 1)$ **65.** $\dfrac{1}{2}\log_7(x + 1) - \dfrac{1}{2}\log_7 x$
67. $\log_2 x + 2 \log_2(x - 1) - \dfrac{1}{2}\log_2(x + 1)$ **69.** 2 **71.** $\log(3x)$ **73.** 2 **75.** 4 **77.** $\log_3 x^3$ **79.** $\log_4\left(\dfrac{x + 1}{x}\right)$ **81.** $\ln(x^2 y^3)$
83. $\log_3\left[\sqrt{x}(x - 1)^3\right]$ **85.** $\log(x^2)$ **87.** $\log(x\sqrt{xy})$ **89.** $\log_8(x - 1)$ **91.** $\log\left(\dfrac{x^{12}}{10}\right)$ **3.** 3.322 **95.** 0.528 **97.** -2.680 **99.** 4.644 **101.** 3 **103.** 1
105. $\log_a(x + \sqrt{x^2 - 1}) + \log_a(x - \sqrt{x^2 - 1})$ $= \log_a\left[(x + \sqrt{x^2 - 1})(x - \sqrt{x^2 - 1})\right]$ $= \log_a\left[x^2 - x\sqrt{x^2 - 1} + x\sqrt{x^2 - 1} - (x^2 - 1)\right]$ $= \log_a(x^2 - x^2 + 1)$ $= \log_a 1$ $= 0$

107. If $f(x) = \log_a x$, then $f(AB) = \log_a(AB)$ $= \log_a A + \log_a B$ $= f(A) + f(B)$

109. Answers will vary. One possibility: The logarithm of the product of two expressions equals the sum of the logarithms of the two expressions. **111.** Answers will vary. One possibility: $\log_2(2 + 4) \neq \log_2 2 + \log_2 4$

113. $\left\{\dfrac{5}{2}\right\}$ **115.** $\{-2 - \sqrt{2}, -2 + \sqrt{2}\}$ **117.** $\{47\}$

119. $f(x) = \log_3 x = \dfrac{\log x}{\log 3}$

Domain: $\{x \,|\, x > 0\}$ or $(0, \infty)$; Range: all real numbers or $(-\infty, \infty)$

121. $F(x) = \log_{1/2} x = \dfrac{\log x}{\log\left(\frac{1}{2}\right)}$

Domain: $\{x \,|\, x > 0\}$ or $(0, \infty)$; Range: all real numbers or $(-\infty, \infty)$

Section 8.5 Exponential and Logarithmic Equations **1.** $M = N$ **2.** $\{3\}$ **3.** $\{8\}$ **4.** $\{-2\}$ **5.** $\left\{\dfrac{\ln 11}{\ln 2}\right\}$ or $\left\{\dfrac{\log 11}{\log 2}\right\}$; $\{3.459\}$

6. $\left\{\dfrac{\ln 3}{2 \ln 5}\right\}$ or $\left\{\dfrac{\log 3}{2 \log 5}\right\}$; $\{0.341\}$ **7.** $\left\{\dfrac{\ln 5}{2}\right\}$; $\{0.805\}$ **8.** $\left\{-\dfrac{\ln\left(\frac{20}{3}\right)}{4}\right\}$; $\{-0.474\}$ **9. (a)** approximately 2.85 days **(b)** approximately 32.52 days
10. (a) approximately 6.77 years **(b)** approximately 11.58 years **11.** $\{7\}$ **13.** $\{9\}$ **15.** $\{3\}$ **17.** $\{81\}$ **19.** $\{1\}$ **21.** $\{-1\}$
23. $\left\{\dfrac{1}{3}\right\}$ **25.** $\left\{\dfrac{7}{5}\right\}$ **27.** $\{-5\}$ **29.** $\left\{\dfrac{1}{\log 2}\right\} \approx \{3.322\}$ or $\left\{\dfrac{\ln 10}{\ln 2}\right\} \approx \{3.322\}$ **31.** $\left\{\dfrac{\log 20}{\log 5}\right\} \approx \{1.861\}$ or $\left\{\dfrac{\ln 20}{\ln 5}\right\} \approx \{1.861\}$

33. $\left\{\dfrac{\log 7}{\log\left(\frac{1}{2}\right)}\right\} \approx \{-2.807\}$ or $\left\{\dfrac{\ln 7}{\ln\left(\frac{1}{2}\right)}\right\} \approx \{-2.807\}$ **35.** $\{\ln 5\} \approx \{1.609\}$ **37.** $\{\log 5\} \approx \{0.699\}$ **39.** $\left\{\dfrac{\log 13}{2\log 3}\right\} \approx \{1.167\}$ or

$\left\{\dfrac{\ln 13}{2\ln 3}\right\} \approx \{1.167\}$ **41.** $\left\{\dfrac{\log 3}{4\log\left(\frac{1}{2}\right)}\right\} \approx \{-0.396\}$ or $\left\{\dfrac{\ln 3}{4\ln\left(\frac{1}{2}\right)}\right\} \approx \{-0.396\}$ **43.** $\left\{\dfrac{\log\left(\frac{5}{4}\right)}{\log 2}\right\} \approx \{0.322\}$ or $\left\{\dfrac{\ln\left(\frac{5}{4}\right)}{\ln 2}\right\} \approx \{0.322\}$

45. $\{\ln 6\} \approx \{1.792\}$ **47.** $\left\{\dfrac{\log 0.2}{\log 3 - \log 0.2}\right\} \approx \{-0.594\}$ or $\left\{\dfrac{\ln 0.2}{\ln 3 - \ln 0.2}\right\} \approx \{-0.594\}$ **49.** $\{8\}$ **51.** $\left\{\dfrac{\log 7}{3\log 5}\right\} \approx \{0.403\}$ or

$\left\{\dfrac{\ln 7}{3\ln 5}\right\} \approx \{0.403\}$ **53.** $\{2\}$ **55.** $\{\ln 15\} \approx \{2.708\}$ **57.** $\left\{-\dfrac{2}{5}\right\}$ **59.** $\{-4, 4\}$ **61.** $\{4\}$ **63.** $\{2 + \sqrt{7}\} \approx \{4.646\}$

65. (a) 2023 **(b)** 2053   **67. (a)** approximately 16.8 years **(b)** approximately 34.7 years   **69. (a)** approximately 2.188 years **(b)** approximately
10.782 years **(c)** approximately 23.372 years   **71. (a)** approximately 2.099 seconds **(b)** 27.62 seconds **(c)** approximately 45.876 seconds
73. (a) approximately 5 minutes **(b)** approximately 11 minutes   **75. (a)** approximately 82 minutes **(b)** approximately 396 minutes (or 6.6 hours)
77. (a) approximately 9 years **(b)** $t = \dfrac{\log 2}{n\log\left(1 + \dfrac{r}{n}\right)}$ **(c)** approximately 8.693 years, which is about the same as the result from the Rule of 72

79. 29 minutes   **81. (a)** 1 **(b)** 11 **(c)** 7   **83. (a)** $-\dfrac{3}{2}$ **(b)** $\dfrac{2}{7}$ **(c)** 0   **85. (a)** 8 **(b)** $\dfrac{1}{4}$ **(c)** 1   **87.** approximately $\{1.06\}$ **89.** approximately $\{-0.70\}$

91. approximately $\{0.05, 1.48\}$ **93.** approximately $\{0.45, 1\}$

Chapter 8 Review   **1. (a)** 32 **(b)** -9 **(c)** 2 **(d)** 13   **2. (a)** 24 **(b)** -28 **(c)** -8 **(d)** 112   **3. (a)** 201 **(b)** 24 **(c)** 163 **(d)** 14   **4. (a)** 23 **(b)** 37 **(c)** -8
(d) 290   **5. (a)** $(f \circ g)(x) = 5x + 1$ **(b)** $(g \circ f)(x) = 5x + 5$ **(c)** $(f \circ f)(x) = x + 2$ **(d)** $(g \circ g)(x) = 25x$ **6. (a)** $(f \circ g)(x) = 2x + 9$
(b) $(g \circ f)(x) = 2x + 3$ **(c)** $(f \circ f)(x) = 4x - 9$ **(d)** $(g \circ g)(x) = x + 12$ **7. (a)** $(f \circ g)(x) = 4x^2 + 4x + 2$ **(b)** $(g \circ f)(x) = 2x^2 + 3$
(c) $(f \circ f)(x) = x^4 + 2x^2 + 2$ **(d)** $(g \circ g)(x) = 4x + 3$ **8. (a)** $(f \circ g)(x) = \dfrac{2x}{x + 1}$, where $x \neq -1, 0$ **(b)** $(g \circ f)(x) = \dfrac{x + 1}{2}$, where $x \neq -1$
(c) $(f \circ f)(x) = \dfrac{2(x + 1)}{x + 3}$, where $x \neq -1, -3$ **(d)** $(g \circ g)(x) = x$, where $x \neq 0$ **9.** not one-to-one   **10.** one-to-one   **11.** one-to-one
12. not one-to-one

13.   **14.**   **15.** $\{(3, -5), (1, -3), (-3, 1), (9, 2)\}$ **16.** $\{(1, -20), (4, -15), (3, 5), (2, 25)\}$

17.   **18.**   **19.** $f^{-1}(x) = \dfrac{x}{5}$ **20.** $H^{-1}(x) = \dfrac{x - 7}{2}$ **21.** $P^{-1}(x) = \dfrac{4}{x} - 2$ **22.** $g^{-1}(x) = \sqrt[3]{\dfrac{x + 1}{2}}$

23. (a) 27.332 **(b)** 28.975 **(c)** 29.088 **(d)** 29.093 **(e)** 29.091   **24. (a)** 1258.925 **(b)** 1380.384
(c) 1386.756 **(d)** 1385.479 **(e)** 1385.456   **25. (a)** 1.649 **(b)** 0.368 **(c)** 4.482 **(d)** 0.449 **(e)** 5.885

26. Domain: all real numbers or $(-\infty, \infty)$; Range: $\{y | y > 0\}$ or $(0, \infty)$ **27.** Domain: all real numbers or $(-\infty, \infty)$; Range: $\{y | y > 0\}$ or $(0, \infty)$ **28.** Domain: all real numbers or $(-\infty, \infty)$; Range: $\{y | y > 0\}$ or $(0, \infty)$

29. Domain: all real numbers or $(-\infty, \infty)$; Range: $\{y | y > -2\}$ or $(-2, \infty)$ **30.** The number e is defined as the number that the expression $\left(1 + \dfrac{1}{n}\right)^n$ approaches as n increases.   **31.** $\{6\}$ **32.** $\left\{\dfrac{7}{2}\right\}$ **33.** $\{-4, 1\}$ **34.** $\{-2\}$ **35.** $\{9\}$ **36.** $\{-3, 3\}$

37. (a) \$7513.59 **(b)** \$7652.33 **(c)** \$7684.36 **(d)** \$7700.00
38. (a) approximately 82.034 grams **(b)** 50 grams **(c)** 25 grams **(d)** approximately 0.263 gram   **39. (a)** approximately 3.388 million people
(b) approximately 3.668 million people   **40. (a)** approximately 151.449 °F **(b)** approximately 94.706 °F   **41.** $\log_3 81 = 4$ **42.** $\log_4\left(\dfrac{1}{64}\right) = -3$

43. $\log_b 5 = 3$ **44.** $\log x = 3.74$ **45.** $8^{1/3} = 2$ **46.** $5^r = 18$ **47.** $e^2 = x + 3$ **48.** $10^{-4} = x$ **49.** $\dfrac{7}{3}$ **50.** 0   **51.** -2 **52.** $\dfrac{3}{2}$ **53.** $\{x | x > -5\}$ or

$(-5, \infty)$ **54.** $\left\{x \middle| x < \dfrac{7}{3}\right\}$ or $\left(-\infty, \dfrac{7}{3}\right)$ **55.** $\{x | x > 0\}$ or $(0, \infty)$ **56.** $\left\{x \middle| x > -\dfrac{5}{2}\right\}$ or $\left(-\dfrac{5}{2}, \infty\right)$

57.   **58.**   **59.** 3.178   **60.** -0.182 **61.** 2.410   **62.** -0.907 **63.** $\{17\}$ **64.** $\{-9, 1\}$ **65.** $\left\{\dfrac{3}{2}\right\}$ **66.** $\{6\}$ **67.** $\{-142\}$

68. $\{5\sqrt{3}\}$ **69.** 80 decibels   **70.** 100,000 millimeters   **71.** 21   **72.** 9.34   **73.** 1   **74.** 0   **75.** 1   **76.** 16

77. $\log_7 x + \log_7 y - \log_7 z$ **78.** $4 - 2\log_3 x$ **79.** $3 + 4\log r$ **80.** $\dfrac{1}{2}\ln(x - 1) - \dfrac{1}{2}\ln x$

81. $\log_3(x^4y^2)$ **82.** $\ln\left(\dfrac{7\sqrt[4]{x}}{9}\right)$ **83.** -1 **84.** $\log_6(x-4)$ **85.** 2.183 **86.** 0.606 **87.** -4.419 **88.** 3.723 **89.** $\{10\}$ **90.** $\{-5\}$ **91.** $\{\sqrt{2}\} \approx \{1.414\}$

92. $\{64\}$ **93.** $\left\{\dfrac{\log 15}{\log 2}\right\} \approx \{3.907\}$ or $\left\{\dfrac{\ln 15}{\ln 2}\right\} \approx \{3.907\}$ **94.** $\left\{\dfrac{\log 27}{3}\right\} \approx \{0.477\}$ **95.** $\left\{\dfrac{\ln 39}{7}\right\} \approx \{0.523\}$

96. $\left\{\dfrac{\log 2}{\log 3 - \log 2}\right\} \approx \{1.710\}$ or $\left\{\dfrac{\ln 2}{\ln 3 - \ln 2}\right\} \approx \{1.710\}$ **97. (a)** after approximately 1.453 days **(b)** after approximately 23.253 days

98. (a) about 2020 **(b)** about 2053

Chapter 8 Test **1.** not one-to-one **2.** $f^{-1}(x) = \dfrac{x+3}{4}$ **3. (a)** 33.360 **(b)** 36.338 **(c)** 36.494 **(d)** 36.463 **(e)** 36.462 **4.** $\log_4 19 = x$ **5.** $b^y = x$

6. (a) -3 **(b)** 4 **7.** $\left\{x \,\middle|\, x < \dfrac{7}{4}\right\}$ or $\left(-\infty, \dfrac{7}{4}\right)$

8. Domain: all real numbers or $(-\infty, \infty)$; Range: $\{y | y > 0\}$ or $(0, \infty)$

9. Domain: $\{x | x > 0\}$ or $(0, \infty)$; Range: all real numbers or $(-\infty, \infty)$

10. (a) 10 **(b)** 15 **11.** $\dfrac{1}{2}\log_4 x - 3\log_4 y$ **12.** $\log(M^4N^3)$ **13.** -8.004 **14.** $\{1\}$ **15.** $\{1,3\}$ **16.** $\{4\}$ **17.** $\{-17, 17\}$ **18.** $\{9\}$

19. $\left\{\dfrac{\log 17 + \log 3}{\log 3}\right\} \approx \{3.579\}$ or $\left\{\dfrac{\ln 17 + \ln 3}{\ln 3}\right\} \approx \{3.579\}$ **20.** $\{2\sqrt{26}\} \approx \{10.198\}$ **21. (a)** approximately 37.1 million people

(b) about 2058 **22.** 10 decibels

Chapter 9 Conics

Section 9.1 Distance and Midpoint Formulas **1.** $d = \sqrt{(x_2 - x_1)^2 + (y_2 - y_1)^2}$ **2.** False **3.** 5 **4.** $6\sqrt{5} \approx 13.42$

5. (a)   **(b)** $d(A, B) = 3\sqrt{5}; d(A, C) = 5\sqrt{5}; d(B, C) = 4\sqrt{5}$ **(c)** $[d(A, C)]^2 = [d(A, B)]^2 + [d(B, C)]^2$

(d) 30 square units **6.** $M = \left(\dfrac{x_1 + x_2}{2}, \dfrac{y_1 + y_2}{2}\right)$ **7.** $\left(\dfrac{3}{2}, 6\right)$ **8.** $\left(1, \dfrac{5}{2}\right)$ **9.** 5 **11.** $4\sqrt{5} \approx 8.94$

13. 5 **15.** 13 **17.** 6 **19.** $3\sqrt{5} \approx 6.71$ **21.** $3\sqrt{7} \approx 7.94$ **23.** $\sqrt{12.56} \approx 3.54$ **25.** $(4, 3)$ **27.** $(3, -1)$

29. $\left(-1, \dfrac{7}{2}\right)$ **31.** $\left(-\dfrac{3}{2}, 0\right)$ **33.** $\left(\dfrac{7\sqrt{2}}{2}, \dfrac{5\sqrt{5}}{2}\right)$ **35.** $(0.8, -1.6)$

37. (a)

(b) $d(A, B) = 2\sqrt{2} \approx 2.83; d(B, C) = 4\sqrt{2} \approx 5.66;$
$d(A, C) = 2\sqrt{10} \approx 6.32$
(c) $[d(A, B)]^2 + [d(B, C)]^2 \overset{?}{=} [d(A, C)]^2$
$\quad (2\sqrt{2})^2 + (4\sqrt{2})^2 \overset{?}{=} (2\sqrt{10})^2$
$\quad\quad 4 \cdot 2 + 16 \cdot 2 \overset{?}{=} 4 \cdot 10$
$\quad\quad\quad 8 + 32 \overset{?}{=} 40$
$\quad\quad\quad\quad 40 = 40 \leftarrow$ True
Therefore, triangle ABC is a right triangle.
(d) 8 square units

39. (a)

(b) $d(A, B) = 5\sqrt{2} \approx 7.07; d(B, C) = 12\sqrt{2} \approx 16.97;$
$d(A, C) = 13\sqrt{2} \approx 18.38$
(c) $[d(A, B)]^2 + [d(B, C)]^2 \overset{?}{=} [d(A, C)]^2$
$\quad (5\sqrt{2})^2 + (12\sqrt{2})^2 \overset{?}{=} (13\sqrt{2})^2$
$\quad\quad 25 \cdot 2 + 144 \cdot 2 \overset{?}{=} 169 \cdot 2$
$\quad\quad\quad 50 + 288 \overset{?}{=} 338$
$\quad\quad\quad\quad 338 = 338 \leftarrow$ True
Therefore, triangle ABC is a right triangle.
(d) 60 square units

41. $(2, -3), (2, 5)$ **43.** $(-6, -3), (10, -3)$ **45. (a)** approximately 37.36 blocks **(b)** approximately 35.13 blocks **(c)** approximately 71.34 blocks
47. (a) approximately 1.85 seconds **(b)** No. The ball will reach second base (2.65 seconds) before the runner (3.33 seconds). **49.** Answers will vary.
Essentially, the Pythagorean Theorem is a theorem that relates the hypotenuse to the lengths of the legs in a right triangle. The hypotenuse in the
right triangle is the distance between the two points in the Cartesian plane. **51.** 9; 3 **53.** 81; 3 **55.** If n is a positive integer and $a^n = b$, then
$\sqrt[n]{b} = \begin{cases} |a| & \text{if } n \text{ is even} \\ a & \text{if } n \text{ is odd.} \end{cases}$

Section 9.2 Circles **1.** circle **2.** radius **3.** $(x-2)^2 + (y-4)^2 = 25$ **4.** $(x+2)^2 + y^2 = 2$ **5.** False **6.** True

7.
$(x-3)^2 + (y-1)^2 = 4$

8.
$(x+5)^2 + y^2 = 16$

9.
$x^2 + y^2 - 6x - 4y + 4 = 0$

10.
$2x^2 + 2y^2 - 16x + 4y - 38 = 0$

11. $(x-1)^2 + (y-2)^2 = 4$
13. $(x-2)^2 + (y+1)^2 = 16$
15. $x^2 + y^2 = 9$
17. $(x-1)^2 + (y-4)^2 = 4$

19. $(x + 2)^2 + (y - 4)^2 = 36$ **21.** $x^2 + (y - 3)^2 = 16$ **23.** $(x - 5)^2 + (y + 5)^2 = 25$ **25.** $(x - 1)^2 + (y - 2)^2 = 5$

27. $C = (0, 0), r = 6$

29. $C = (4, 1), r = 5$

31. $C = (-3, 2), r = 9$

33. $C = (0, 3), r = 8$

35. $C = (1, -1), r = \dfrac{1}{2}$

37. $C = (3, -1), r = 3$

$x^2 + y^2 - 6x + 2y + 1 = 0$

39. $C = (-5, -2), r = 5$

$x^2 + y^2 + 10x + 4y + 4 = 0$

41. $C = (3, -6), r = 9$

$2x^2 + 2y^2 - 12x + 24y - 72 = 0$

43. $x^2 + y^2 = 20$ **45.** $(x + 3)^2 + (y - 2)^2 = 9$ **47.** $(x + 1)^2 + (y + 1)^2 = 25$ **49.** $A = 64\pi$ square units; $C = 16\pi$ units

51. 32 square units   **53.** (a), (e)   **55.** The distance formula is used along with the definition of a circle to obtain the equation of the circle.

57. Yes, since it can be written as $x^2 + y^2 = 36$. Center: $(0, 0)$; radius $= 6$.

59. $f(x) = 4x - 3$

61. $g(x) = x^2 - 4x - 5$

63. $G(x) = -2(x + 3)^2 - 5$

65.

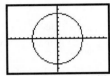

The graph here agrees with that in Problem 27.

67.

The graph here agrees with that in Problem 29.

69.

The graph here agrees with that in Problem 31.

71.

The graph here agrees with that in Problem 33.

73.

The graph here agrees with that in Problem 35.

75. (a) $(0, 0)$ **(e)** $(x - 4.5)^2 + (y - 4.5)^2 = 6.76$; $(4.5, 4.5)$; 2.6   **(f)** $(x + 4)^2 + (y + 4)^2 = 1.44$; $(-4, 4)$; 1.2

(g) $(x + 3.5)^2 + (y + 3.5)^2 = 5.76$; $(-3.5, -3.5)$; 2.4   **(h)** $(x - 3)^2 + (y + 3)^2 = 5.76$; $(3, -3)$; 2.4

Section 9.3 Parabolas   **1.** parabola   **2.** vertex   **3.** axis of symmetry

4. $D: x = -2$

5. $D: x = 5$

6.

7.

8. $x^2 = -32y$

9. $y^2 = \dfrac{4}{3}x$

10. $(-3, 2)$

11. $D: x = -6$

12. The receiver should be located 2 feet from the base of the dish, along its axis of symmetry.

13. c   **15.** a   **17.** b   **19.** e

21. Vertex: $(0, 0)$; focus: $(0, 6)$; directrix: $y = -6$

23. Vertex: $(0, 0)$; focus: $\left(-\dfrac{3}{2}, 0\right)$; directrix: $x = \dfrac{3}{2}$

25. Vertex: $(0, 0)$; focus: $(0, -2)$; directrix: $y = 2$

27. $y^2 = 20x$

$D: x = -5$

29. $x^2 = -24y$

31. $x^2 = 6y$

33. $x^2 = -12y$

35. $y^2 = -12x$

37. $y^2 = x$

39. Vertex: $(2, 4)$; focus: $(2, 5)$; directrix: $y = 3$

41. Vertex: $(-2, -3)$; focus: $(-4, -3)$; directrix: $x = 0$

43. Vertex: $(-5, 1)$; focus: $(-5, -4)$; directrix: $y = 6$

45. Vertex: $(-2, -1)$; focus: $(-2, -4)$; directrix: $y = 2$

47. Vertex: $(1, 4)$; focus: $(2, 4)$; directrix: $x = 0$

49. Vertex: $(-5, 2)$; focus: $\left(-5, \frac{1}{2}\right)$; directrix: $y = \frac{7}{2}$

51. 1 inch from the vertex, along the axis of symmetry **53.** 21.6 feet

55. The height of the bridge is 28.8 feet at a distance of 10 feet from the center, 19.2 feet at a distance of 30 feet from the center, and 0 feet (i.e., ground level) at a distance of 50 feet from the center.

57. $(x - 3)^2 = 4(y + 2)$ **59.** $(y - 3)^2 = -4(x - 2)$

61. (a) Let $x = 4$ and $y = 2$:
$4^2 \stackrel{?}{=} 8 \cdot 2$
$16 = 16 \leftarrow$ True
Thus, $(4, 2)$ is on the parabola.

(b) The focus of the parabola is $F(0, 2)$, and the directrix is $D: y = -2$.
$d(F, P) = \sqrt{(0 - 4)^2 + (2 - 2)^2} = \sqrt{16} = 4$
$d(P, D) = 2 - (-2) = 4$
Thus, $d(F, P) = d(P, D) = 4$.

63. 8 units **65.** See Figure 13.

67. $y = (x + 3)^2$

69. $y = (x + 3)^2$

71. $4(y + 2) = (x - 2)^2$

73.

75.

77.

79.

81.

83.

Section 9.4 Ellipses **1.** ellipse; foci **2.** major axis **3.** vertices **4.** False

5.

6.

7. $\dfrac{x^2}{40} + \dfrac{y^2}{49} = 1$

8. $(3, -1)$

9.

10. $\dfrac{x^2}{2500} + \dfrac{y^2}{1600} = 1$; 40 feet

11. c **13.** d

15. Foci: $(-3, 0)$ and $(3, 0)$; vertices: $(-5, 0)$ and $(5, 0)$

17. Foci: $(0, -8)$ and $(0, 8)$; vertices: $(0, -10)$ and $(0, 10)$

19. Foci: $\left(-3\sqrt{5}, 0\right)$ and $\left(3\sqrt{5}, 0\right)$; vertices: $(-7, 0)$ and $(7, 0)$

21. Foci: $\left(0, -4\sqrt{3}\right)$ and $\left(0, 4\sqrt{3}\right)$; vertices: $(0, -7)$ and $(0, 7)$

23. Foci: $\left(0, -2\sqrt{3}\right)$ and $\left(0, 2\sqrt{3}\right)$; vertices: $(0, -4)$ and $(0, 4)$

25. $\dfrac{x^2}{36} + \dfrac{y^2}{20} = 1$;

27. $\dfrac{x^2}{33} + \dfrac{y^2}{49} = 1$

29. $\dfrac{x^2}{100} + \dfrac{y^2}{64} = 1$

$F_1(-6,0)$ $F_2(6,0)$
$V_1(-10,0)$ $V_2(10,0)$
$(0,8)$ $(0,-8)$

31. $\dfrac{x^2}{39} + \dfrac{y^2}{64} = 1$

$F_2(0,5)$ $V_1(0,8)$
$(-\sqrt{39},0)$ $(\sqrt{39},0)$
$F_1(0,-5)$ $V_2(0,-8)$

33. $\dfrac{(x-3)^2}{9} + \dfrac{(y+2)^2}{25} = 1$

$V_2(3,3)$
$F_2(3,2)$
$(0,-2)$ $(6,-2)$
$C(3,-2)$
$F_1(3,-6)$
$V_1(3,-7)$

35. $\dfrac{(x+2)^2}{16} + \dfrac{(y-5)^2}{4} = 1$

$(-2,7)$ $F_2(-2+2\sqrt{3},5)$
$F_1(-2-2\sqrt{3},5)$
$V_1(-6,5)$ $V_2(2,5)$
$(-2,3)$ $C(-2,5)$

37. $(x-5)^2 + \dfrac{(y+1)^2}{49} = 1$

$V_2(5,6)$ $F_2(5,-1+4\sqrt{3})$
$C(5,-1)$
$(4,-1)$ $(6,-1)$
$F_1(5,-1-4\sqrt{3})$ $V_1(5,-8)$

39. $4(x+2)^2 + 16(y-1)^2 = 64$

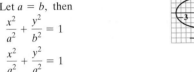

$(-2,3)$
$F_1(-2-2\sqrt{3},1)$ $F_2(-2+2\sqrt{3},1)$
$V_1(-6,1)$ $V_2(2,1)$
$(-2,-1)$ $C(-2,1)$

41. $4x^2 + y^2 - 24x + 2y - 63 = 0$

$V_2(3,9)$
$F_2(3,-1+5\sqrt{3})$
$C(3,-1)$ $(8,-1)$
$(-2,-1)$
$F_1(3,-1-5\sqrt{3})$
$V_1(3,-11)$

43. (a) $\dfrac{x^2}{225} + \dfrac{y^2}{100} = 1$ **(b)** Yes

(c) No

45. Perihelion = 91.5 million miles;

$\dfrac{x^2}{8649} + \dfrac{y^2}{8646.75} = 1$

47. Perihelion = 460.6 million miles; mean distance = 483.8 million miles; $\dfrac{x^2}{234{,}062.44} + \dfrac{y^2}{233{,}524.2} = 1$ **49.** $\dfrac{(x-1)^2}{16} + \dfrac{(y-2)^2}{9} = 1$

51. $\dfrac{(x-2)^2}{4} + \dfrac{y^2}{16} = 1$

53. Let $a = b$, then

$\dfrac{x^2}{a^2} + \dfrac{y^2}{b^2} = 1$

$\dfrac{x^2}{a^2} + \dfrac{y^2}{a^2} = 1$

$a^2\left(\dfrac{x^2}{a^2} + \dfrac{y^2}{a^2}\right) = a^2(1)$

$x^2 + y^2 = a^2$

which is the equation of a circle with center $(0,0)$ and radius a. $c = 0$; The foci are located at the center point.

55.

B a b C F

57.

x	5	10	100	1000
$f(x)$	0.71429	0.41667	0.04902	0.00499

59.

x	5	10	100	1000
$f(x)$	5.5	3	2.07216	2.00702

61.

x	5	10	100	1000
$f(x)$	6.83333	11.90909	101.99010	1001.99900
$g(x)$	7	12	102	1002

63. In Problems 57 and 58, the degree of the numerator is less than the degree of the denominator. In Problems 59 and 60, the degree of the numerator and the degree of the denominator are the same.

Conjecture 1: If the degree of the numerator of a rational function is less than the degree of the denominator, then as x increases, the value of the function will approach zero (0).

Conjecture 2: If the degree of the numerator of a rational function equals the degree of the denominator, then as x increases, the value of the function will approach the ratio of the leading coefficients of the numerator and denominator.

65.

67.

69.

71.

73. (a) The sum stays the same. **(b)** The center is at $(5,2)$ and changes to $(3,2)$. This move did not change the length of the major or minor axis. **(c)** The center changed to $(3,4)$, This move did not change the length of the major or minor axis. **(d)** h and k represent the center (h,k) of the ellipse. **(e)** This changed the length of the horizontal (major) axis and changed the denominator of the first term. **(f)** This changed the length of the horizontal (minor) axis and changed the denominator of the first term. **(g)** The number on the horizontal slider changes the length of the horizontal aixs. **(h)** This changed the length of the vertical (major) axis and changed the denominator of the second term. **(i)** This changed the length of the vertical (minor) axis and changed the denominator of the second term. **(j)** The number on the vertical slider changes the length of the vertical axis.

Section 9.5 Hyperbolas **1.** hyperbola **2.** transverse axis **3.** conjugate axis

4.
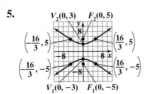

$\left(-10,\dfrac{32}{3}\right)$ $\left(10,\dfrac{32}{3}\right)$
$V_1(-6,0)$ $V_2(6,0)$
$F_1(-10,0)$ $F_2(10,0)$
$\left(-10,-\dfrac{32}{3}\right)$ $\left(10,-\dfrac{32}{3}\right)$

5.
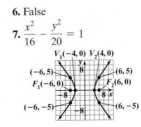

$V_2(0,3)$ $F_2(0,5)$
$\left(-\dfrac{16}{3},5\right)$ $\left(\dfrac{16}{3},5\right)$
$\left(-\dfrac{16}{3},-5\right)$ $\left(\dfrac{16}{3},-5\right)$
$V_1(0,-3)$ $F_1(0,-5)$

6. False

7. $\dfrac{x^2}{16} - \dfrac{y^2}{20} = 1$

$V_1(-4,0)$ $V_2(4,0)$
$(-6,5)$ $(6,5)$
$F_1(-6,0)$ $F_2(6,0)$
$(-6,-5)$ $(6,-5)$

8. $y = -\dfrac{b}{a}x;\ y = \dfrac{b}{a}x$

9.

$y = -\dfrac{1}{3}x$ $(0,1)$ $y = \dfrac{1}{3}x$
$F_1(-\sqrt{10},0)$ $F_2(\sqrt{10},0)$
$V_1(-3,0)$ $V_2(3,0)$
$(-1,0)$

10.

11. (b)   **13.** (a)

15. $\dfrac{x^2}{4} - \dfrac{y^2}{16} = 1$

17. $\dfrac{y^2}{25} - \dfrac{x^2}{36} = 1$

19. $4x^2 - y^2 = 36$

21. $25y^2 - x^2 = 100$

23. $\dfrac{x^2}{4} - \dfrac{y^2}{5} = 1$

25. $\dfrac{y^2}{25} - \dfrac{x^2}{24} = 1$

27. $\dfrac{x^2}{49} - \dfrac{y^2}{51} = 1$

29.

31.

33.

35. $\dfrac{y^2}{64} - \dfrac{x^2}{16} = 1$

37. $\dfrac{x^2}{4.5} - \dfrac{y^2}{4.5} = 1$

39. $x^2 - y^2 = 1$

41. $\dfrac{y^2}{36} - \dfrac{x^2}{9} = 1$

43. The asymptotes of both

hyperbolas are $y = -\dfrac{1}{2}x$ and

$y = \dfrac{1}{2}x$. Thus, they are conjugates.

$$\frac{x^2}{4} - y^2 = 1$$

$$y^2 - \frac{x^2}{4} = 1$$

45. Answers will vary.   **47.** $(-3, 1)$

49. $\{\,(x, y)\,|\,2x - 3y = 6\,\}$

51. $\{\,(x, y)\,|\,6x + 3y = 4\,\}$

53.

55.

57.

59.

Putting the Concepts Together (Sections 9.1–9.5)   **1.** $3\sqrt{13}$ **2.** $(1, -3)$

3. $C = (-2, 8), r = 6$

$(x + 2)^2 + (y - 8)^2 = 36$

4. $C = (-3, 2), r = 4$

$x^2 + y^2 + 6x - 4y - 3 = 0$

5. $x^2 + y^2 = 169$

6. $(x - 2)^2 + (y - 1)^2 = 25$

7. Vertex: $(-2, 4)$; focus: $(-2, 3)$;

directrix: $y = 5$

8. Vertex: $(3, -1)$; focus: $(5, -1)$;

directrix: $x = 1$

9. $(x + 1)^2 = -12(y + 2)$ **10.** $(y - 3)^2 = 8(x + 3)$

11. Center: $(0, 0)$; foci: $\left(-6\sqrt{2}, 0\right)$ and

$\left(6\sqrt{2}, 0\right)$; vertices: $(-9, 0)$ and $(9, 0)$

12. Center: $(-1, 2)$; vertices: $(-1, -5)$

and $(-1, 9)$; foci: $\left(-1, 2 - \sqrt{13}\right)$ and

$\left(-1, 2 + \sqrt{13}\right)$

13. $\dfrac{x^2}{45} + \dfrac{y^2}{81} = 1$

14. $\dfrac{(x - 3)^2}{16} + \dfrac{(y + 4)^2}{7} = 1$

15. Vertices: $(0, -9)$ and $(0, 9)$;
  foci: $\left(0, -3\sqrt{10}\right)$ and $\left(0, 3\sqrt{10}\right)$;
  asymptotes: $y = 3x$ and $y = -3x$

16. Vertices: $(-1, 0)$ and $(1, 0)$;
  foci: $\left(-\sqrt{26}, 0\right)$ and $\left(\sqrt{26}, 0\right)$;
  asymptotes: $y = -5x$ and $y = 5x$

17. $\dfrac{y^2}{4} - \dfrac{x^2}{21} = 1$

18. 6.75 inches from the vertex, along its axis of symmetry

Section 9.6 Systems of Nonlinear Equations

1. $(-3, 5)$ and $(1, -3)$ **2.** $\left(\dfrac{11}{5}, -\dfrac{22}{5}\right)$ and $(1, -2)$ **3.** $(0, -4)$, $(-2\sqrt{3}, 2)$, $(2\sqrt{3}, 2)$

4. $\varnothing$ or $\{\}$ **5.** $(0, 4)$ and $(1, 5)$ **7.** $(3, 4)$ and $(4, 3)$ **9.** $(0, -2)$, $\left(-\sqrt{3}, 1\right)$, and $\left(\sqrt{3}, 1\right)$ **11.** $(-2, -2)$ and $(2, 2)$

13. $(0, -2)$, $(0, 2)$, $\left(-1, -\sqrt{3}\right)$, and $\left(-1, \sqrt{3}\right)$ **15.** $\varnothing$ **17.** $(0, 0)$, $(-3, 3)$, and $(3, 3)$ **19.** $(-3, 7)$ and $(2, -8)$ **21.** $(-1, 11)$ and $(2, -4)$

23. $(-4, 0)$ and $(4, 0)$ **25.** $(0, 3)$ and $(1, 4)$ **27.** $\left(0, -2 - \sqrt{3}\right)$, $\left(0, -2 + \sqrt{3}\right)$, $(1, -4)$, and $(1, 0)$ **29.** $\varnothing$

31. $(5, 2)$ **33.** $\left(-\dfrac{8}{3}, -\dfrac{2\sqrt{10}}{3}\right)$, $\left(-\dfrac{8}{3}, \dfrac{2\sqrt{10}}{3}\right)$, $\left(\dfrac{8}{3}, -\dfrac{2\sqrt{10}}{3}\right)$, and $\left(\dfrac{8}{3}, \dfrac{2\sqrt{10}}{3}\right)$ **35.** $(0, -5)$, $(3, 4)$, $(4, 3)$ and $(5, 0)$

37. Either -5 and -3, or 3 and 5 **39.** 14 feet by 10 feet **41.** 19 cm by 10 cm **43.** $(0, -2)$, $(0, 1)$, and $(2, -1)$ **45.** $(81, 3)$

47. If $r_1 = \dfrac{-b + \sqrt{b^2 - 4ac}}{2a}$, then $r_2 = \dfrac{-b - \sqrt{b^2 - 4ac}}{2a}$; if $r_1 = \dfrac{-b - \sqrt{b^2 - 4ac}}{2a}$, then $r_2 = \dfrac{-b + \sqrt{b^2 - 4ac}}{2a}$.

49. (a) $f(1) = 7$ **(b)** $g(1) = 2$ **51. (a)** $f(3) = 13$ **(b)** $g(3) = 8$ **53. (a)** $f(5) = 19$ **(b)** $g(5) = 32$

55. $(-1, 11)$ and $(2, -4)$ **57.** $(-4, 0)$ and $(4, 0)$ **59.** $(0, 3)$ and $(1, 4)$ **61.** $(0.056, -0.237)$ and $(2.981, -1.727)$ **63.** $\varnothing$

Chapter 9 Review

1. 5 **2.** 10 **3.** $2\sqrt{10} \approx 6.32$ **4.** 6 **5.** $3\sqrt{19} \approx 13.08$ **6.** 2.5 **7.** $(-2, 5)$ **8.** $(6, -2)$ **9.** $\left(-4\sqrt{3}, -3\sqrt{6}\right)$ **10.** $\left(\dfrac{5}{2}, \dfrac{1}{2}\right)$ **11.** $\left(\dfrac{3}{4}, \dfrac{1}{2}\right)$

12. (a)

(b) $d(A, B) = 3\sqrt{2} \approx 4.24$;
  $d(B, C) = 2\sqrt{2} \approx 2.83$;
  $d(A, C) = \sqrt{26} \approx 5.10$

(c) $[d(A, B)]^2 + [d(B, C)]^2 \overset{?}{=} [d(A, C)]^2$
  $\left(3\sqrt{2}\right)^2 + \left(2\sqrt{2}\right)^2 \overset{?}{=} \left(\sqrt{26}\right)^2$
  $9 \cdot 2 + 4 \cdot 2 \overset{?}{=} 26$
  $18 + 8 \overset{?}{=} 26$
  $26 = 26 \leftarrow$ True

Therefore, triangle ABC is a right triangle.

(d) 6 square units

13. $C = (-2, 1)$; $r = 4$;
  $(x + 2)^2 + (y - 1)^2 = 16$
14. $C = (5, 3)$; $r = 3$;
  $(x - 5)^2 + (y - 3)^2 = 9$

15. $x^2 + y^2 = 16$

16. $(x + 3)^2 + (y - 1)^2 = 9$ **17.** $(x - 5)^2 + (y + 2)^2 = 1$

18. $(x - 4)^2 + y^2 = 7$

19. $(x - 2)^2 + (y + 1)^2 = 25$ **21.** $C = (0, 0)$, $r = 5$
20. $(x + 1)^2 + (y - 3)^2 = 20$

22. $C = (1, 2)$, $r = 2$ **23.** $C = (0, 4)$, $r = 4$ **24.** $C = (-1, -6)$, $r = 7$

25. $C = \left(-2, \dfrac{3}{2}\right)$, $r = \dfrac{1}{2}$ **26.** $C = (-3, -3)$, $r = 2$ **27.** $C = (-3, -5)$, $r = 6$ **28.** $C = (4, -2)$, $r = 2$ **29.** $C = (-1, 2)$, $r = 3$

30. $C = (5, 1), r = 3$

31. $x^2 = -12y$

32. $y^2 = -16x$

33. $y^2 = \frac{1}{2}x$

34. $x^2 = 8y$

35. Vertex: $(0, 0)$; focus $\left(0, \frac{1}{2}\right)$; directrix: $y = -\frac{1}{2}$

36. Vertex: $(0, 0)$; focus: $(4, 0)$; directrix: $x = -4$

37. Vertex: $(-1, 3)$; focus: $(-1, 5)$; directrix: $y = 1$

38. Vertex: $(-3, 4)$; focus: $\left(-\frac{7}{2}, 4\right)$; directrix: $x = -\frac{5}{2}$

39. Vertex: $(5, 2)$; focus: $\left(5, \frac{5}{4}\right)$; directrix: $y = \frac{11}{4}$

40. approximately 127.84 feet from the center of the dish, along its axis of symmetry

41. Foci: $\left(-2\sqrt{2}, 0\right)$ and $\left(2\sqrt{2}, 0\right)$; vertices: $(-3, 0)$ and $(3, 0)$

42. Foci: $\left(0, -\sqrt{5}\right)$ and $\left(0, \sqrt{5}\right)$; vertices: $(0, -3)$, and $(0, 3)$

43. $\frac{x^2}{16} + \frac{y^2}{25} = 1$

44. $\frac{x^2}{36} + \frac{y^2}{32} = 1$

45. $\frac{x^2}{100} + \frac{y^2}{36} = 1$

46. Center: $(1, -2)$; vertices: $(-6, -2)$ and $(8, -2)$; foci: $\left(1 - 2\sqrt{6}, -2\right)$ and $\left(1 + 2\sqrt{6}, -2\right)$

47. Center: $(-3, 4)$; vertices: $(-3, -1)$ and $(-3, 9)$; foci: $(-3, 0)$ and $(-3, 8)$

48. (a) $\frac{x^2}{900} + \frac{y}{256} = 1$

(b) Yes

49. Vertices: $(-2, 0)$ and $(2, 0)$; foci: $\left(-\sqrt{13}, 0\right)$ and $\left(\sqrt{13}, 0\right)$

50. Vertices: $(0, -5)$ and $(0, 5)$; foci: $\left(0, -\sqrt{74}\right)$ and $\left(0, \sqrt{74}\right)$

51. Vertices: $(0, -5)$ and $(0, 5)$; foci: $\left(0, -\sqrt{41}\right)$ and $\left(0, \sqrt{41}\right)$

52. Asymptotes: $y = x$ and $y = -x$

53. Asymptotes: $y = \frac{5}{2}x$ and $y = -\frac{5}{2}x$

54. $\dfrac{x^2}{9} - \dfrac{y^2}{7} = 1$

55. $\dfrac{y^2}{9} - \dfrac{x^2}{16} = 1$

56. $\dfrac{y^2}{16} - \dfrac{x^2}{9} = 1$

57. $\left(\sqrt{2}, \sqrt{2}\right)$ and $\left(-\sqrt{2}, -\sqrt{2}\right)$

58. $\left(-\dfrac{1}{2}, \dfrac{3}{2}\right)$ and $(1, 3)$

59. $\left(-\dfrac{1}{6}, -6\right)$ and $(1, 1)$

60. $(-5, -1), (-5, 1), (5, -1),$ and $(5, 1)$

61. $\left(2, -2\sqrt{2}\right)$ and $\left(2, 2\sqrt{2}\right)$

62. $(-1, 3)$ and $(1, 3)$

63. $\left(-\sqrt{3}, -\sqrt{5}\right), \left(-\sqrt{3}, \sqrt{5}\right), \left(\sqrt{3}, -\sqrt{5}\right),$ and $\left(\sqrt{3}, \sqrt{5}\right)$ **64.** $(0,0), \left(5, -\sqrt{15}\right),$ and $\left(5, \sqrt{15}\right)$ **65.** $(2, 4)$ and $(-1, 1)$

66. $(-7, -20)$ and $\left(2, \dfrac{5}{2}\right)$ **67.** $(0, 6)$ and $(-6, 0)$ **68.** $(4, 4)$ and $(1, -2)$ **69.** $\varnothing$ **70.** $(-1, -2), (-1, 2), (1, -2),$ and $(1, 2)$

71. $(-4, 0), (4, 0),$ and $(0, 4)$ **72.** $(4, 0)$ and $(0, -2)$ **73.** 7 and 5   **74.** 12 cm by 5 cm   **75.** 72 inches by 30 inches   **76.** 12 feet and 9 feet

Chapter 9 Test   **1.** $4\sqrt{5}$ **2.** $(-1, 2)$

3. $C = (4, -1), r = 3$

$(x - 4)^2 + (y + 1)^2 = 9$

4. $C = (-5, 2), r = 4$

$x^2 + y^2 + 10x - 4y + 13 = 0$

5. $(x + 3)^2 + (y - 7)^2 = 36$
6. $(x + 5)^2 + (y - 8)^2 = 100$

7. Vertex $(1, -2)$; focus $(2, -2)$; directrix $x = 0$

8. Vertex $(2, 4)$; focus $\left(2, \dfrac{13}{4}\right)$; directrix $y = \dfrac{19}{4}$

9. $x^2 = -16y$
10. $(y - 4)^2 = 8(x - 1)$

11. Foci: $(-4, 0)$ and $(4, 0)$; vertices: $(-5, 0)$ and $(5, 0)$

12. Foci: $\left(2, -4 - \sqrt{7}\right)$ and $\left(2, -4 + \sqrt{7}\right)$ vertices: $(2, -8)$ and $(2, 0)$

13. $\dfrac{x^2}{9} + \dfrac{y^2}{25} = 1$

14. $\dfrac{(x + 1)^2}{16} + \dfrac{(y - 2)^2}{25} = 1$

15. Vertices: $(-1, 0)$ and $(1, 0)$; foci: $\left(-\sqrt{5}, 0\right)$ and $\left(\sqrt{5}, 0\right)$; asymptotes: $y = 2x$ and $y = -2x$

16. Vertices: $(0, -10)$ and $(0, 10)$; foci: $\left(0, -2\sqrt{41}\right)$ and $\left(0, 2\sqrt{41}\right)$; asymptotes: $y = \dfrac{5}{4}x$ and $y = -\dfrac{5}{4}x$

17. $\dfrac{x^2}{9} - \dfrac{y^2}{55} = 1$ **18.** $(-4, 1)$ and $(1, -4)$ **19.** $\left(-\sqrt{5}, -2\right), \left(-\sqrt{5}, 2\right), \left(\sqrt{5}, -2\right),$ and $\left(\sqrt{5}, 2\right)$ **20. (a)** $\dfrac{x^2}{225} + \dfrac{y^2}{100} = 1$ **(b)** 6 feet

Cumulative Review Chapters R-9   **1.** $\{x \mid x \neq 7\}$

2.

3. $y = \dfrac{4}{3}x + 2$ or $4x - 3y = -6$ **4.** $(3, -8)$ **5.** -35 **6.** $(2m + n)(4m^2 - 2mn + n^2)$ **7.** $(3p - 4)(3p - 2)$

8. $\dfrac{2a + 1}{a - 7}$ **9.** $\dfrac{1}{(t - 2)(t - 1)}$ **10.** $\{2\}$ **11.** $\{x \mid x < -4 \text{ or } x \geq 5\}$ or $(-\infty, -4) \cup [5, \infty)$ **12.** $9\sqrt{3}$

13. $-3\left(2 - \sqrt{7}\right)$ **14.** $\{5\}$ **15.** $31 + 5i$ **16.** $\left\{\dfrac{2 - \sqrt{10}}{2}, \dfrac{2 + \sqrt{10}}{2}\right\}$

17.

18. $\{x \mid -1 \leq x \leq 4\}$ or $[-1, 4]$
19. Domain: all real numbers or $(-\infty, \infty)$; Range: $\{y \mid y > 0\}$ or $(0, \infty)$

20. Domain: $\{x \mid x > 0\}$ or $(0, \infty)$; Range: all real numbers or $(-\infty, \infty)$

21. $\dfrac{5}{2}$ **22.** $\{2\}$ **23.** $\{-5, 5\}$

24. $(x + 3)^2 + (y - 1)^2 = 10$

25. Vertex: $(-3, 3)$; focus: $(-1, 3)$; directrix: $x = -5$

Chapter 10 Sequences, Series, and the Binomial Theorem

Section 10.1 Sequences   **1.** sequence   **2.** infinite; finite   **3.** True   **4.** $-1, 1, 3, 5, 7$ **5.** $-4, 8, -12, 16, -20$ **6.** $a_n = 2n + 3$ **7.** $b_n = \dfrac{(-1)^{n+1}}{n+1}$

8. partial sum   **9.** $3 + 7 + 11 = 21$ **10.** $2 + 9 + 28 + 65 + 126 = 230$ **11.** $\displaystyle\sum_{k=1}^{12} k^2$ **12.** $\displaystyle\sum_{n=1}^{6}\left(\dfrac{1}{2}\right)^{n-1}$ **13.** $8, 11, 14, 17, 20$ **15.** $\dfrac{1}{3}, \dfrac{1}{2}, \dfrac{3}{5}, \dfrac{2}{3}, \dfrac{5}{7}$

17. $-1, 2, -3, 4, -5$ **19.** $3, 5, 9, 17, 33$ **21.** $1, 1, \dfrac{3}{4}, \dfrac{1}{2}, \dfrac{5}{16}$ **23.** $\dfrac{1}{e}, \dfrac{2}{e^2}, \dfrac{3}{e^3}, \dfrac{4}{e^4}, \dfrac{5}{e^5}$ **25.** $a_n = 2n$ **27.** $a_n = \dfrac{n}{n+1}$ **29.** $a_n = n^2 + 2$ **31.** $a_n = (-1)^n n^2$

33. 54   **35.** $\dfrac{55}{2}$ **37.** 14   **39.** 6   **41.** 50   **43.** 45   **45.** $\displaystyle\sum_{k=1}^{15} k$ **47.** $\displaystyle\sum_{i=1}^{12} \dfrac{1}{i}$ **49.** $\displaystyle\sum_{i=1}^{9}(-1)^{i+1}\left(\dfrac{1}{3^{i-1}}\right)$ **51.** $\displaystyle\sum_{k=1}^{11}(2k+3)$ **53. (a)** $12,180   **(b)** $12,736.36

(c) $21,768.22   **55. (a)** 339 million   **(b)** 456 million   **57.** $1, 1, 2, 3, 5, 8, 13, 21, 34, 55$ **59.** $10, 10.5, 11.025, 11.57625,$ and 12.1550625 **61.** $8, 10, 13, 17,$ and 22

63. (a) $1; 2; 1.5; 1.\overline{6}; 1.6; 1.625; \dfrac{21}{13} \approx 1.615385; \dfrac{34}{21} \approx 1.619048; \dfrac{55}{34} \approx 1.617647; \dfrac{89}{55} \approx 1.618182$ **(b)** around 1.618   **(c)** $1; 0.5; \dfrac{2}{3} = 0.\overline{6}; 0.6; 0.625;$

$\dfrac{8}{13} \approx 0.615385; \dfrac{13}{21} \approx 0.619048; \dfrac{21}{34} \approx 0.617647; \dfrac{34}{55} \approx 0.618182; \dfrac{55}{89} \approx 0.617978$ **(d)** around 0.618   **65.** The main difference between a function and a sequence is that the domain of a function is based on the real number system, whereas the domain of a sequence is the positive integers. **67.** The symbol $\sum$ means to add up the terms of a sequence.   **69. (a)** $m = 4$ **(b)** $f(1) = -2; f(2) = 2; f(3) = 6; f(4) = 10$ **71. (a)** $m = -5$ **(b)** $f(1) = 3; f(2) = -2; f(3) = -7; f(4) = -12$ **73.** $8, 11, 14, 17,$ and 20 **75.** $\dfrac{1}{3}, \dfrac{1}{2}, \dfrac{3}{5}, \dfrac{2}{3},$ and $\dfrac{5}{7}$ **77.** $-1, 2, -3, 4,$ and -5 **79.** $3, 5, 9, 17,$ and 33

81. 54   **83.** $\dfrac{55}{2}$ **85.** 14   **87.** 6

Section 10.2 Arithmetic Sequences   **1.** arithmetic   **2.** Arithmetic; $a_1 = -3; d = 2$ **3.** Not arithmetic   **4.** Arithmetic $a_1 = -5; d = 3$ **5.** Not arithmetic   **6.** Arithmetic $a_1 = 3; d = -2$ **7.** $a_n = a_1 + (n - 1)d$ **8. (a)** $a_n = 6n - 5$ **(b)** 79   **9. (a)** $a_1 = -5; d = 3$ **(b)** $a_n = 3n - 8$ **10.** 10,400   **11.** 9730   **12.** 7550   **13.** -1050 **14.** 1470 seats   **15.** $a_n - a_{n-1} = d = 1; a_1 = 6$ **17.** $a_n - a_{n-1} = d = 7; a_1 = 9$

19. $a_n - a_{n-1} = d = -3; a_1 = 4$ **21.** $a_n - a_{n-1} = d = \dfrac{1}{2}; a_1 = \dfrac{11}{2}$ **23.** $a_n = 3n + 1; a_5 = 16$ **25.** $a_n = -5n + 15; a_5 = -10$ **27.** $a_n = \dfrac{1}{3}n + \dfrac{5}{3};$ $a_5 = \dfrac{10}{3}$ **29.** $a_n = -\dfrac{1}{5}n + \dfrac{26}{5}; a_5 = \dfrac{21}{5}$ **31.** $a_n = 5n - 3; a_{20} = 97$ **33.** $a_n = -3n + 15; a_{20} = -45$ **35.** $a_n = \dfrac{1}{4}n + \dfrac{3}{4}; a_{20} = \dfrac{23}{4}$

37. $a_1 = 7; d = 5; a_n = 5n + 2$ **39.** $a_1 = -23; d = 7; a_n = 7n - 30$ **41.** $a_1 = 11; d = -3; a_n = -3n + 14$ **43.** $a_1 = 4; d = -\dfrac{1}{2}; a_n = -\dfrac{1}{2}n + \dfrac{9}{2}$

45. $S_{30} = 2670$ **47.** $S_{25} = 700$ **49.** $S_{40} = -5060$ **51.** $S_{40} = 3160$ **53.** $S_{75} = -9000$ **55.** $S_{30} = 460$ **57.** $x = -\dfrac{3}{2}$ **59.** There are 630 cans in the stack.   **61.** There are 1600 seats in the auditorium.   **63.** There are 84 terms in the sequence.   **65.** There are 63 terms in the sequence.   **67.** It will take about 15.22 years.   **69.** A sequence is arithmetic if the difference in consecutive terms is constant. Thus, if the actual terms are given, compute the difference between each pair of consecutive terms and determine whether the differences are constant. If a formula is given for the sequence, compute $a_n - a_{n-1}$ and determine whether the result is a constant–if it is, then the sequence is arithmetic.   **71. (a)** 3   **(b)** $3; 9; 27; 81$ **73. (a)** $\dfrac{1}{2}$ **(b)** $5; \dfrac{5}{2}, \dfrac{5}{4}, \dfrac{5}{8}$ **75.** 806.9   **77.** 1427.5

Section 10.3 Geometric Sequences and Series   **1.** geometric   **2.** Geometric; $a_1 = 4, r = 2$ **3.** Not geometric   **4.** Geometric; $a_1 = 9, r = \dfrac{1}{3}$

5. Geometric; $r = 5$ **6.** Not geometric   **7.** Geometric; $r = \dfrac{2}{3}$ **8.** $a_n = 5 \cdot 2^{n-1}; a_9 = 1280$ **9.** $a_n = 50\left(\dfrac{1}{2}\right)^{n-1}; a_9 = 0.1953125$ or $a_9 = \dfrac{25}{128}$

10. $S_n = a_1 \cdot \dfrac{1 - r^n}{1 - r}$ **11.** 24,573   **12.** 7.9921875   **13.** $\dfrac{a_1}{1 - r}$ **14.** $\dfrac{40}{3}$ **15.** $\dfrac{1}{2}$ **16.** $\dfrac{2}{9}$ **17.** The U.S. economy will grow by $10,000.   **18.** $244,129.08

19. $\dfrac{a_n}{a_{n-1}} = r = 4; a_1 = 4$ **21.** $\dfrac{a_n}{a_{n-1}} = r = \dfrac{2}{3}; a_1 = \dfrac{2}{3}$ **23.** $\dfrac{a_n}{a_{n-1}} = r = \dfrac{1}{2}; a_1 = \dfrac{3}{2}$ **25.** $\dfrac{a_n}{a_{n-1}} = r = \dfrac{5}{2}; a_1 = \dfrac{1}{2}$ **27. (a)** $a_n = 10 \cdot 2^{n-1}$ **(b)** $a_8 = 1280$

29. (a) $a_n = 100 \cdot \left(\dfrac{1}{2}\right)^{n-1}$ **(b)** $a_8 = \dfrac{25}{32}$ **31. (a)** $a_n = (-3)^{n-1}$ **(b)** $a_8 = -2187$ **33. (a)** $a_n = 100 \cdot (1.05)^{n-1}$ **(b)** $a_8 = 100 \cdot (1.05)^7 \approx 140.71$

35. $a_{10} = 1536$ **37.** $a_{15} = \dfrac{1}{4096}$ **39.** $a_9 = 0.000000005$ **41.** 8190   **43.** 83.3245952   **45.** 6138   **47.** 7.96875   **49.** 2   **51.** 15   **53.** $\dfrac{9}{2}$ **55.** $\dfrac{5}{4}$ **57.** 9

59. $\dfrac{5}{9}$ **61.** $\dfrac{89}{99}$ **63.** arithmetic; $d = 5$ **65.** Neither   **67.** geometric; $r = \dfrac{1}{2}$ **69.** geometric; $r = \dfrac{2}{3}$ **71.** arithmetic; $d = 4$ **73.** Neither   **75.** $x = -4$ **77. (a)** $42,000   **(b)** $62,053   **(c)** $503,116   **79.** $13,182   **81. (a)** About 1.891 feet   **(b)** On the 23rd swing   **(c)** About 24.08 feet   **(d)** The pendulum will have swung a total of 60 feet.   **83.** Option A will yield the larger annual salary in the final year of the contract, and option B will yield the larger cumulative salary over the life of the contract.   **85.** The multiplier is 50.   **87.** $31.14 per share   **89.** $149,035.94   **91.** $114,401.52   **93.** $395.09, or about $395

95. $0.4\overline{9} = \dfrac{1}{2}$ **97.** 2,147,483,646   **99.** A geometric sequence with $r > 1$ yields faster growth because of the effect of compounding. Answers will vary. However, a reasonable answer is that because geometric growth is faster than arithmetic growth, the population will grow beyond the ability of food supplies to sustain the population, and hunger will ensue.   **101.** A geometric series has a sum provided that the common ratio r is between -1 and 1. **103.** 1   **105.** approximately 288.1404315   **107.** approximately 41.66310217

Putting the Concepts Together (Sections 10.1–10.3) **1.** Geometric with $a_1 = \dfrac{3}{4}$ and common ratio $r = \dfrac{1}{4}$ **2.** Arithmetic with $a_1 = 8$ and

$d = 2$ **3.** Arithmetic with $a_1 = 1$ and $d = \dfrac{7}{9}$ **4.** Neither arithmetic nor geometric **5.** Geometric with $a_1 = 12$ and $r = 2$ **6.** Neither arithmetic nor

geometric **7.** 87 **8.** $\displaystyle\sum_{i=1}^{12} \dfrac{1}{2(6+i)}$ **9.** $a_n = 27 - 2n$; 25, 23, 21, 19, and 17 **10.** $a_n = 11n - 35$; $-24, -13, -2, 9,$ and 20 **11.** $a_n = 45 \cdot \left(\dfrac{1}{5}\right)^{n-1}$; 45, 9,

$\dfrac{9}{5}, \dfrac{9}{25},$ and $\dfrac{9}{125}$ **12.** $a_n = 150 \cdot (1.04)^{n-1}$; 150, 156, 162.24, 168.7296, and 175.478784 **13.** $S_{11} = 177{,}146$ **14.** $S_{20} = 990$ **15.** $\dfrac{10{,}000}{9}$ **16.** A party of

24 people would require 11 tables.

Section 10.4 The Binomial Theorem **1.** $n(n-1)(n-2)\cdot \cdots \cdot 3\cdot 2\cdot 1$ **2.** $1; 1$ **3.** 120 **4.** 840 **5.** False **6.** True **7.** 7 **8.** 20
9. $x^4 + 8x^3 + 24x^2 + 32x + 16$ **10.** $32p^5 - 80p^4 + 80p^3 - 40p^2 + 10p - 1$ **11.** 6 **13.** 40,320 **15.** 90 **17.** 336 **19.** 21 **21.** 210
23. $x^5 + 5x^4 + 10x^3 + 10x^2 + 5x + 1$ **25.** $x^4 - 16x^3 + 96x^2 - 256x + 256$ **27.** $81p^4 + 216p^3 + 216p^2 + 96p + 16$
29. $32z^5 - 240z^4 + 720z^3 - 1080z^2 + 810z - 243$ **31.** $x^8 + 8x^6 + 24x^4 + 32x^2 + 16$ **33.** $32p^{15} + 80p^{12} + 80p^9 + 40p^6 + 10p^3 + 1$
35. $x^6 + 12x^5 + 60x^4 + 160x^3 + 240x^2 + 192x + 64$ **37.** $16p^8 - 32p^6q^2 + 24p^4q^4 - 8p^2q^6 + q^8$ **39.** 1.00401 **41.** 0.99004 **43.** $84x^5$
45. $-108{,}864p^3$ **47.** $\dbinom{n}{n-1} = \dfrac{n!}{(n-1)!(n-(n-1))!} = \dfrac{n!}{(n-1)!1!} = \dfrac{n\cdot(n-1)!}{(n-1)!} = n; \dbinom{n}{n} = \dfrac{n!}{n!(n-n)!} = \dfrac{n!}{n!0!} = \dfrac{n!}{n!} = 1$

49.
$$\begin{array}{ccccc} & & 1 & & \\ & 1 & & 1 & \\ 1 & & 2 & & 1 \\ 1 & 3 & & 3 & 1 \end{array}$$
51. The degree of each monomial equals n. **53.** $a^4 - 8a^3 + 24a^2 - 32a + 16$ **55.** $p^5 + p^4 - 6p^3 - 14p^2 - 11p - 3$

Chapter 10 Review **1.** $-1, -4, -7, -10,$ and -13 **2.** $-\dfrac{1}{5}, 0, \dfrac{1}{7}, \dfrac{1}{4},$ and $\dfrac{1}{3}$ **3.** 6, 26, 126, 626, and 3126 **4.** 3, -6, 9, -12, and 15 **5.** $\dfrac{1}{2}, \dfrac{4}{3}, \dfrac{9}{4}, \dfrac{16}{5},$

and $\dfrac{25}{6}$ **6.** $\pi, \dfrac{\pi^2}{2}, \dfrac{\pi^3}{3}, \dfrac{\pi^4}{4},$ and $\dfrac{\pi^5}{5}$ **7.** $a_n = -3n$ **8.** $a_n = \dfrac{n}{3}$ **9.** $a_n = 5\cdot 2^{n-1}$ **10.** $a_n = (-1)^n\cdot \dfrac{n}{2}$ **11.** $a_n = n^2 + 5$ **12.** $a_n = \dfrac{n-1}{n+1}$

13. 65 **14.** $\dfrac{33}{2}$ **15.** -30 **16.** $\dfrac{26}{3}$ **17.** $\displaystyle\sum_{i=1}^{15}(4+3i)$ **18.** $\displaystyle\sum_{i=1}^{8}\dfrac{1}{3^i}$ **19.** $\displaystyle\sum_{i=1}^{10}\dfrac{i^3+1}{i+1}$ **20.** $\displaystyle\sum_{i=1}^{7}[(-1)^{i-1}\cdot i^2]$ **21.** arithmetic with $d = 6$ **22.** arithmetic

with $d = \dfrac{3}{2}$ **23.** not arithmetic **24.** not arithmetic **25.** arithmetic with $d = 4$ **26.** not arithmetic **27.** $a_n = 8n - 5$; $a_{25} = 195$

28. $a_n = -3n - 1$; $a_{25} = -76$ **29.** $a_n = -\dfrac{1}{3}n + \dfrac{22}{3}$; $a_{25} = -1$ **30.** $a_n = 6n + 5$; $a_{25} = 155$ **31.** $a_n = \dfrac{18}{5}n - \dfrac{19}{5}$; $a_{25} = \dfrac{431}{5}$

32. $a_n = -4n - 4$; $a_{25} = -104$ **33.** 4320 **34.** -2140 **35.** -4080 **36.** $\dfrac{1875}{4}$, or 468.75 **37.** 2106 **38.** 300 yards **39.** geometric with $r = 6$

40. geometric with $r = -3$ **41.** not geometric **42.** geometric with $r = \dfrac{2}{3}$ **43.** geometric with $r = -2$ **44.** not geometric **45.** $a_n = 4\cdot 3^{n-1}$;

$a_{10} = 78{,}732$ **46.** $a_n = 8\cdot\left(\dfrac{1}{4}\right)^{n-1}$; $a_{10} = \dfrac{1}{32{,}768}$ **47.** $a_n = 5\cdot(-2)^{n-1}$; $a_{10} = -2560$ **48.** $a_n = 1000\cdot(1.08)^{n-1}$; $a_{10} \approx 1999.005$

49. 65,534 **50.** ≈ 45.71428571 **51.** $\dfrac{12{,}285}{4}$ or 3071.25 **52.** $-258{,}280{,}320$ **53.** $\dfrac{20}{3}$ **54.** $\dfrac{100}{3}$ **55.** $\dfrac{5}{4}$ **56.** $\dfrac{8}{9}$ **57.** After 72 years, there will be

3.125 grams of the tritium remaining. **58.** After 15 minutes, about 38.15 billion e-mails will have been sent. **59.** After 25 years, Scott's 403(b) will be
worth $360,114.89. **60.** The lump sum option would yield more money after 26 years. **61.** Sheri would need to contribute $534.04, or about $534,
each month to reach her goal. **62.** When Samantha turns 18, the plan will be worth $62,950.79 and will cover about 185 credit hours. **63.** 120
64. 7920 **65.** 5040 **66.** 1716 **67.** 35 **68.** 252 **69.** 1 **70.** 1 **71.** $z^4 + 4z^3 + 6z^2 + 4z + 1$ **72.** $y^5 - 15y^4 + 90y^3 - 270y^2 + 405y - 243$
73. $729y^6 + 5832y^5 + 19{,}440y^4 + 34{,}560y^3 + 34{,}560y^2 + 18{,}432y + 4096$ **74.** $16x^8 - 96x^6 + 216x^4 - 216x^2 + 81$
75. $81p^4 - 216p^3q + 216p^2q^2 - 96pq^3 + 16q^4$ **76.** $a^{15} + 15a^{12}b + 90a^9b^2 + 270a^6b^3 + 405a^3b^4 + 243b^5$ **77.** $-448x^5$ **78.** $14{,}784x^5$

Chapter 10 Test **1.** arithmetic with $a_1 = -15$ and $d = 8$ **2.** geometric with $a_1 = -4$ and $r = -4$ **3.** neither arithmetic nor geometric

4. arithmetic with $a_1 = -\dfrac{1}{5}$ and $d = \dfrac{2}{5}$ **5.** neither arithmetic nor geometric **6.** geometric with $a_1 = 21$ and $r = 3$ **7.** $\dfrac{17{,}269}{1200}$ **8.** $\displaystyle\sum_{i=1}^{8}\dfrac{i+2}{i+4}$
9. $a_n = 10n - 4$; 6, 16, 26, 36, and 46 **10.** $a_n = 4 - 4n$; 0, -4, -8, -12, and -16 **11.** $a_n = 10\cdot 2^{n-1}$; 10, 20, 40, 80, and 160

12. $a_n = (-3)^{n-1}$; 1, -3, 9, -27, and 81 **13.** 720 **14.** $-\dfrac{132{,}860}{9}$ **15.** 324 **16.** 6435 **17.** 792 **18.** $625m^4 - 1000m^3 + 600m^2 - 160m + 16$

19. $6103.11 **20.** 8000 lb

Appendix A

1. up; down

2.

3.

4. True

5.

6.

7. y; a; vertically compressed; vertically stretched

8.

9.

10. x; $\dfrac{1}{a}$; horizontally stretched; horizontally compressed

11.

12.

13. x; y

14.

15.

16.

17. (I) (C); (II) (B); (III) (A); (IV) (D)

19. $y = \sqrt{x + 3}$

21. $y = \sqrt{x} - 2$

23. $y = -\sqrt{x}$

25. $y = \sqrt{\dfrac{1}{2}x}$

27.

Domain: All real numbers;
Range: All real numbers

29.

Domain: All real numbers;
Range: $[0, \infty)$

31.

Domain: $[0, \infty)$;
Range: $[0, \infty)$

33.

Domain: All real numbers;
Range: $[0, \infty)$

35.

Range: All real numbers;
Domain: All real numbers

37.

Domain: $(-\infty, 0]$;
Range: $[0, \infty)$

39.

Domain: All real numbers;
Range: $[2, \infty)$

41.

Domain: $[-2, \infty)$;
Range: $(-\infty, 3]$

43. (a)

(b)

(c)

(d)

(e)

(f)

(g)

(h)

Applications Index

Subject Index

Properties of the Real Number System (Chapter R)

	Addition	Multiplication
Identity Properties	$a + 0 = 0 + a = a$	$1 \cdot a = a \cdot 1 = a$
Inverse Properties	$a + (-a) = (-a) + a = 0$	$a \cdot \dfrac{1}{a} = \dfrac{1}{a} \cdot a = 1 (a \neq 0)$
Commutative Property	$a + b = b + a$	$a \cdot b = b \cdot a$
Associative Property	$(a + b) + c = a + (b + c)$	$(ab)c = a(bc)$
Distributive Property	$a(b + c) = ab + ac$	
Division Properties	$\dfrac{0}{a} = 0, a \neq 0; \dfrac{a}{a} = 1, a \neq 0; \dfrac{a}{0}$ is undefined, $a \neq 0$	
Reduction Property	$\dfrac{ac}{bc} = \dfrac{a}{b}, b \neq 0, c \neq 0$	

Order of Operations (Chapter R)

How to Evaluate Expressions

1. Do parentheses first. If multiple parentheses, begin with the innermost parentheses and work outward.

2. Then exponents, working from left to right.

3. Next multiplication and division, working from left to right.

4. Perform addition and subtraction, working from left to right.

Geometry Formulas (Chapter 1)

Figure	Formulas	Figure	Formulas
Square	**Area:** $A = s^2$ **Perimeter:** $P = 4s$	Cube	**Volume:** $V = s^3$ **Surface Area:** $S = 6s^2$
Rectangle	**Area:** $A = lw$ **Perimeter:** $P = 2l + 2w$	Rectangular Solid	**Volume:** $V = lwh$ **Surface Area:** $S = 2lw + 2lh + 2wh$
Circle	**Area:** $A = \pi r^2$ **Circumference:** $C = 2\pi r = \pi d$	Sphere	**Volume:** $V = \dfrac{4}{3}\pi r^3$ **Surface Area:** $S = 4\pi r^2$
Triangle	**Area:** $A = \dfrac{1}{2}bh$ **Perimeter:** $P = a + b + c$	Right Circular Cylinder	**Volume:** $V = \pi r^2 h$ **Surface Area:** $S = 2\pi r^2 + 2\pi rh$
Trapezoid	**Area:** $A = \dfrac{1}{2}h(B + b)$ **Perimeter:** $P = a + b + c + B$	Cone	**Volume:** $V = \dfrac{1}{3}\pi r^2 h$
Parallelogram	**Area:** $A = ah$ **Perimeter:** $P = 2a + 2b$		

Formulas for Lines and Slope (Chapter 1)

Standard form of a line	$Ax + By = C$
Equation of a vertical line	$x = a$ where a is the x-intercept
Equation of a horizontal line	$y = b$ where b is the y-intercept
Slope of a line	$m = \dfrac{y_2 - y_1}{x_2 - x_1}, x_1 \neq x_2$ Slope undefined if $x_1 = x_2$
Point-slope form of a line	$y - y_1 = m(x - x_1)$
Slope-intercept form of a line	$y = mx + b$

Functions (Chapter 2)

- A **function** is a special type of relation where any given input, x, corresponds to only one output y. Functions can be represented through maps, sets of ordered pairs, equations, or graphs.
- **Vertical Line Test:** A set of points in the xy-plane is the graph of a function if and only if every vertical line intersects the graph in at most one point.
- The graph of a function, f, is the set of all ordered pairs $(x, f(x))$.
- When only an equation of a function is given, the **domain** of the function is the largest set of real numbers for which $f(x)$ is a real number.
- The **range** of a function is the set of all outputs of the function.

Steps for Factoring (Chapter 4)

Step 1: Factor out the Greatest Common Factor (GCF), if any exists.

Step 2: Count the number of terms.

Step 3: (a) 2 terms
- Is it the difference of two squares? If so, $A^2 - B^2 = (A - B)(A + B)$
- Is it the difference of two cubes? If so, $A^3 - B^3 = (A - B)(A^2 + AB + B^2)$
- Is it the sum of two cubes? If so, $A^3 + B^3 = (A + B)(A^2 - AB + B^2)$

(b) 3 terms
- Is it a perfect square trinomial? If so, $A^2 + 2AB + B^2 = (A + B)^2$ or $A^2 - 2AB + B^2 = (A - B)^2$
- Is the coefficient of the square term 1? If so, $x^2 + bx + c = (x + m)(x + n)$ where $mn = c$ and $m + n = b$
- Is the coefficient of the square term different from 1? If so,
  - **a.** Use factoring by grouping
  - **b.** Use trial and error

(c) 4 terms
- Use factoring by grouping

Step 4: Check your work by multiplying out the factored form.

The Rules of Exponents (Getting Ready for Chapter 4, Chapter 6)

If a and b are real numbers and if r and s are rational numbers, then assuming the expression is defined,

Zero Exponent Rule:	$a^0 = 1$,	if $a \neq 0$
Negative Exponent Rule:	$a^{-r} = \dfrac{1}{a^r}$	if $a \neq 0$
Product Rule:	$a^r \cdot a^s = a^{r+s}$	
Quotient Rule:	$\dfrac{a^r}{a^s} = a^{r-s} = \dfrac{1}{a^{s-r}}$	if $a \neq 0$
Power Rule:	$(a^r)^s = a^{r \cdot s}$	
Product to Power Rule:	$(a \cdot b)^r = a^r \cdot b^r$	
Quotient to Power Rule:	$\left(\dfrac{a}{b}\right)^r = \dfrac{a^r}{b^r}$	if $b \neq 0$
Quotient to a Negative Power Rule:	$\left(\dfrac{a}{b}\right)^{-r} = \left(\dfrac{b}{a}\right)^r$	if $a \neq 0, b \neq 0$

Working with Rational Expressions (Chapter 5)

Multiplying Rational Expressions	$\dfrac{a}{b} \cdot \dfrac{c}{d} = \dfrac{ac}{bd}$	$b \neq 0, d \neq 0$
Adding Rational Expressions	$\dfrac{a}{c} + \dfrac{b}{c} = \dfrac{a + b}{c}$	$c \neq 0$
Subtracting Rational Expressions	$\dfrac{a}{c} - \dfrac{b}{c} = \dfrac{a - b}{c}$	$c \neq 0$
Dividing Rational Expressions	$\dfrac{a}{b} \div \dfrac{c}{d} = \dfrac{\frac{a}{b}}{\frac{c}{d}} = \dfrac{a}{b} \cdot \dfrac{d}{c} = \dfrac{ad}{bc}$	$b \neq 0, c \neq 0,$ $d \neq 0$